ONE WEE!

REFERENCE

Materials
Handbook

Materials Handbook

**An Encyclopedia for Managers,
Technical Professionals,
Purchasing and Production Managers,
Technicians, and Supervisors**

George S. Brady
(Deceased)

Henry R. Clauser
Materials Consultant
Former Editor, Materials Engineering

John A. Vaccari
Materials Consultant
Former Editor, Materials Engineering

Fourteenth Edition

McGraw-Hill

New York   St. Louis   San Francisco   Auckland   Bogotá
Caracas   Hamburg   Lisbon   London   Madrid   Mexico   Milan
Montreal   New Delhi   Paris   San Juan   São Paulo   Singapore
Sydney   Tokyo   Toronto

McGraw-Hill

A Division of The McGraw·Hill Companies

Copyright © 1997, 1991, 1986, 1977, 1971, 1963, 1956 by The McGraw-Hill Companies, Inc. All rights reserved.

Copyright renewed 1979, 1975, 1972, 1968, 1965, 1959, 1957 by George S. Brady. All rights reserved.

Copyright 1951, 1947, 1944, 1940, 1937, 1931, 1929 by McGraw-Hill, Inc. Except as permitted under the United States Copyright Act of 1976, no part of this publication may be reproduced or distributed in any form or by any means, or stored in a data base or retrieval system, without the prior written permission of the publisher.

2 3 4 5 6 7 8 9 0   DOC/DOC   9 0 1 0 9 8 7

ISBN 0-07-007084-9

The sponsoring editor for this book was Harold B. Crawford, the editing supervisor was Caroline Levine, and the production supervisor was Donald F. Schmidt. It was set in New Century Schoolbook by Terry Leaden of McGraw-Hill's Professional Book Group composition unit.

Printed and bound by R. R. Donnelley & Sons Company.

McGraw-Hill books are available at special quantity discounts to use as premiums and sales promotions, or for use in corporate training programs. For more information, please write to the Director of Special Sales, McGraw-Hill, 11 West 19th Street, New York, NY 10011. Or contact your local bookstore.

This book is printed on acid-free paper.

Contents

Preface

The fourteenth edition of the *Materials Handbook* is the cumulative development of an idea that originated with the late Colonel George S. Brady more than sixty years ago. Aware then of the steadily increasing number of materials being developed, Colonel Brady set out to prepare a one-volume encyclopedia that would intelligently describe the important characteristics of commercially available materials without involvement in details—an encyclopedia that would meet the job needs of managers and executives, purchasing and manufacturing managers, supervisors, engineers, students, and others. Attesting to both the validity of his idea and the success of its execution is the fact that over the years the *Materials Handbook* has come to be recognized as the leading reference work of its kind in the world.

The philosophy behind the *Materials Handbook*, as well as its purpose and scope, is best expressed in this excerpt from the foreword to the tenth edition:

> All materials, infinite in possible numbers, derive from only 92 natural elements. It is as basically simple as that, but the varying forms and usages are so intertwined in all the industries that a person can have no real comprehension of the characteristics and economics of any one of the materials which he procures for his own use unless he has an intelligent overall grasp of its varying forms and usages. The selection of data and of examples in the *Materials Handbook* has been made with a view toward giving the reader an intelligent overall insight. It is not the purpose of the book to provide an exhaustive treatise on any materials, as it is assumed that the reader will consult producers of the materials for detailed specifications.
>
> General information, with the most commonly accepted comparative figures, is given on materials in their group classifications in order to give a general picture; selected processed materials and patented and tradenamed materials are then described to give a more specific understanding of commercial applications. The relative position and the length

of description of proprietary materials are for purposes of illustration and bear no relation to the relative merits of the products of any one producer.

Since the first edition of the *Materials Handbook* was published, virtual revolution in materials has been taking place, resulting in new types and grades of materials—plastics, metal alloys, rubbers, textiles, finishes, foodstuffs, chemicals, and animal products—that are being developed at an exponential rate. Each new edition of the *Materials Handbook* reflects this phenomenal proliferation in number and variety. Whereas the first edition covered only a few thousand materials, this latest edition describes some 13,000 different materials. Despite this manyfold increase in coverage, it is virtually impossible to include all commercially available materials in a one-volume work of this kind. Nevertheless, descriptions of the most important and most widely used of the thousands of materials introduced every year are added to each edition of this handbook.

After publication of the tenth edition, Colonel Brady retired and transferred the task of preparing the eleventh and subsequent editions of the *Materials Handbook* to Henry R. Clauser. Mr. Clauser then transferred preparation of this and subsequent editions to me. In serving as co-author I have endeavored to hold to his standards of accuracy and readability. His help and counsel were invaluable to me in maintaining the original purpose and character of the handbook.

John A. Vaccari

Materials, Their Properties and Uses

ABLATIVES. Materials used for the outward dissipation of extremely high heats by mass removal. Their most common use is as an external heat shield to protect supersonic aerospace vehicles from an excessive buildup of heat caused by air friction at the surface. The ablative material must have a low thermal conductivity in order that the heat may remain concentrated in the thin surface layer. As the surface of the ablator melts or sublimes, it is wiped away by the frictional forces that simultaneously heat newly exposed surfaces. The heat is carried off with the material removed. The less material that is lost, the more efficient is the ablative material. The ablative material also should have a high thermal capacity in the solid, liquid, and gaseous states; a high heat of fusion and evaporation; and a high heat of dissociation of its vapors. The ablative agent, or **ablator,** is usually a carbonaceous organic compound, such as a phenolic plastic. As the dissociation products are lost as liquid or vapor, the char is held in place by the refractory reinforcing fibers, still giving a measure of heat resistance. The effective life of an ablative is short, calculated in seconds per millimeter of thickness for the distance traveled in the atmosphere.

Single ablative materials seldom have all the desirable factors, and thus composites are used. Phenolic or epoxy resins are reinforced with asbestos fabric, carbonized cloth, or refractory fibers, such as asbestos, fused silica, and glass. The refractory fibers not only are incorporated for mechanical strength, but also have a function in the ablative process, and surface-active agents may be added to speed the rate of evaporation. **Ablative paint,** for protecting woodwork, may be organic silicones which convert to silica at temperatures above 2000°F (1093°C).

Metals can resist temperatures higher than their melting point by **convection cooling,** or **thermal cooling,** which is heat protection by heat exchange with a coolant. Thus, tungsten can be arc-melted in a copper kettle which is cooled by circulating water. The container metal must have high thermal conductivity, and the heat must be quickly carried away and stored or dissipated. When convection cooling is difficult or not possible, cooling may be accomplished by a heat sink. **Heat-sink cooling** depends on the heat absorption capability of the structural material itself or backed up by another material of higher heat absorption. Copper, beryllium, graphite, and beryllium oxide have been used. A heat-sink material should have high thermal conductivity, high specific heat and melting point, and for aerodynamic applications, a low specific gravity.

ABRASIVES. Materials used for surfacing and finishing metals, stone, wood, glass, and other materials by abrasive action. The natural abrasives include the diamond, emery, corundum, sand, crushed garnet and quartz, tripoli, and pumice. **Artificial abrasives,** or **manufactured abrasives,** are generally superior in uniformity to natural abrasives, and are mostly silicon carbide, aluminum oxide, boron carbide, or boron nitride, marketed under trade names. Artificial diamonds are also now being produced. The massive natural abrasives, such as sandstone, are cut into grinding wheels from the natural block, but most abrasive material is used as grains or built into artificial shapes.

For industrial grinding, artificial abrasives are preferred to natural abrasives because of their greater uniformity. Grading is important because uniform grinding requires grains of the same size. The abrasive grains are used as a grinding powder; are made into wheels, blocks, or stones; or are bonded to paper or cloth. **Abrasive cloth** is made of cotton jean or drills to close tolerances of yarns and weaves, and the grains are attached with glue or resin. But the **Fabricut cloth** of 3M is an open-weave fabric with alumina or silicon-carbide grains of 100 to 400 mesh. The open weave permits easy cleaning of the cloth in an air blast. **Abrasive paper** has the grains, usually aluminum oxide or silicon carbide, glued to one side of 40- to 130-lb kraft paper. The usual grain sizes are No. 16 to No. 500.

Abrasive powder is usually graded in sizes from 8 to 240 mesh. Coarse grain is to 24 mesh; fine grain is 150 to 240. **Blasting abrasive** for blast cleaning of metal castings is usually coarse grain. **Arrowblast,** of Norton Co., is aluminum oxide with grain sizes from 16 to 80 mesh. **Grinding flour** consists of extremely fine grains separated by flotation, usually in grain sizes from 280 to 600 mesh, used for grinding glass and fine polishing. **Levigated abrasives** are fine powders for final burnishing of metals or for metallographic polishing, usually processed to make them chemically neutral. **Green rouge** is levigated chromic oxide, and **mild polish** may be levigated tin oxide; both are used for burnishing soft metals. **Polishing powder** may be aluminum oxide or metal oxide powders of ultrafine particle size down to 600 mesh. **Micria AD,** of Monsanto Co., is alumina; **Micria ZR** is zirconia; and **Micria TIS** is titania. **Gamal,** of Fisher Scientific Co., is a fine aluminum oxide powder, the smaller cubes being 59 μin (1.5 μm), with smaller particles 20 μin (0.5 μm). **Cerox** is cerium oxide used to polish optical lenses and automobile windshields. It cuts fast and gives a smooth surface. **Grinding compounds** for valve grinding are usually aluminum oxide in oil.

Mild abrasives, used in silver polishes and window-cleaning compounds, such as chalk and talc, have a Mohs hardness of 1 to 2. The milder abrasives for dental pastes and powders may be precipitated calcium carbonate, tricalcium phosphate, or combinations of sodium metaphosphate and tricalcium phosphate. Abrasives for metal polishes may also be pumice, diatomite, silica flour, tripoli, whiting, kaolin, tin oxide, or fuller's earth. This type of fine abrasive must be of very uniform grain in order to prevent scratching. **Cuttle bone, or cuttlefish bone,** is a calcareous powder made from the internal shell of a Mediterranean marine mollusk of the genus *Sepia,* and it is used as a fine polishing material for jewelry and in tooth powders. **Ground glass** is regularly marketed as an abrasive for use in scouring compounds and in match-head compositions. **Lapping abrasives,** for finish grinding of hard materials, are diamond dust or boron carbide powder.

Aluminum oxide wheels are used for grinding materials of high tensile strength. **Silicon carbide** is harder but is not as strong as aluminum oxide. It is used for grinding metals that have dense grain structure and for stone. **Vitrified wheels** are made by molding under heat and pressure. They are used for general and precision grinding where the wheel does not exceed a speed of 6,500 surface ft/min (33 m/s). The rigidity gives high precision, and the porosity and strength of bond permit high stock removal. **Silicate wheels** have a silicate binder and are baked. The silicate bond releases the grains more easily than the vitrified, and is used for grinding edge tools to reduce burning of the tool. Synthetic resins are used for bonding where greater strength is required than is obtained with the silicate, but less openness than with the vitrified. Resinoid bonds are used up to 16,000 surface ft/min (81 m/s), and are used especially for thread grinding and cutoff wheels. Shellac binder is used for light work and for high finishing. Rubber is used for precision grinding and for centerless-feed machines.

Grading of abrasive wheels is by grit size number from No. 10 to No. 600, which is 600 mesh; by grade of wheel, or strength of the bond, which is by letter designation, increasing in hardness from A to Z; and by grain spacing or structure number. The ideal condition is with a bond strong enough to hold the grains to accomplish the desired result and then to release them before they become too dull. Essential qualities in the abrasive grain are penetration hardness, body strength sufficient to resist fracture until the points dull and then break to present a new edge, and an attrition resistance suitable to the work. Some wheels are made with a porous honeycombed structure to give free cutting and cooler operation on some types of metal grinding. Some diamond wheels are made with aluminum powder mixed with a thermosetting resin, and the diamond abrasive mix is hot-pressed around this core wheel. Norton diamond wheels are of three types: metal bonded by powder metallurgy, resinoid bonded, and vitrified bonded.

ABRASIVE SAND. Any sand used for abrasive and grinding purposes, but the term does not include the sharp grains obtained by crushing quartz and used for sandpaper. The chief types of abrasive sand include **sandblast sand, glass-grinding sand,** and **stone-cutting sand.** Sand for stone sawing and for marble and glass grinding is usually ungraded, with no preparation other than screening, but it must have tough, uniform grains. **Chats** are sand tailings from the Missouri lead ores, used for sawing stone. **Banding sand** is used for the band grinding of tool handles and for the grinding of plate glass, but is often replaced by artificial abrasives. Banding-sand grains are fine, 95% being retained on a 150-mesh screen. **Burnishing sand,** for metal polishing, is a fine-grain silica sand with rounded grains. It should pass a 65-mesh screen, and be retained on a 100-mesh screen.

ABS PLASTICS. The letters *ABS* identify the family of **acrylonitrile-butadiene-styrene.** Common trade names for these materials are **Cycolac, Magnum,** and **Lustran.** They generally are opaque and distinguished by a good balance of properties, including high impact strength, rigidity, and hardness over a temperature range of -40 to $230°F$ (-40 to $110°C$). Compared to other structural or engineering plastics, they are generally considered to fall at the lower end of the scale. Medium impact grades are hard, rigid, and tough and are used for appearance parts that require high strength, good fatigue resistance, and surface hardness and gloss. High impact grades are formulated for similar products where additional impact strength is gained at some sacrifice in rigidity and hardness. Low-temperature impact grades have high impact strength down to $-40°F$ ($-40°C$). Again, some sacrifice is made in strength, rigidity, and heat resistance. Heat-resistant, high-strength grades provide the best heat resistance—continuous use up to about $200°F$ ($93°C$), and a 264 lb/in² (1.8 MPa) heat deflection temperature of around $215°F$ ($102°C$). Impact strength is about comparable to that of medium impact grades, but strength, modulus of elasticity, and hardness are higher. At stresses above their tensile strength, ABS plastics usually yield plastically instead of rupturing, and impact failures are ductile. Because of relatively low creep, they have good long-term load-carrying ability. This low creep plus low water absorption and relatively high heat resistance provide ABS plastics with good dimensional stability. Transparent grades are available. ABS plastics are readily processed by extrusion, injection molding, blow molding, calendering, and vacuum forming. Resins have been developed especially for cold forming or stamping from extruded sheet. Typical applications are helmets, refrigerator liners, luggage tote trays, housings, grills for hot air systems, and pump impellers. Extruded shapes include tubing and pipe. ABS plated parts are now in wide use, replacing metal parts in the automotive and appliance field.

ACAROID RESIN. A gum resin from the base of the tufted trunk leaves of various species of *Xanthorrhoea* trees of Australia and Tasmania. It is also called **gum accroides** and **yacca gum. Yellow acaroid** from the *X. tateana* is relatively scarce, but a gum of the yellow class comes from the tree *X. preissii* of western Australia, and is in small hollow pieces yellow to reddish. It is known as **black boy resin,** the name coming from the appearance of the tree. **Red acaroid,** known also as **red gum** and **grass tree gum,** comes in small dusty pieces of reddish brown. This variety is from the *X. australis* and about 15 other species of the tree of southeastern Australia. The resins contain 80 to 85% resinotannol with **coumaric acid,** which is a hydroxycinnamic acid, and they also contain free cinnamic acid. They are thus closely related chemically to the balsams. Acaroid resin has the property unique among natural resins of capacity for thermosetting to a hard, insoluble, chemical-resistant film. By treatment with nitric acid it yields picric acid; by treatment with sulfuric acid it yields fast brown to black dyes. The resins are soluble in alcohols and in aniline, only slightly soluble in chlorinated compounds, and insoluble in coal-tar hydrocarbons. Acaroid has some of the physical characteristics of shellac, but is difficult to bleach. It is used for spirit varnishes and metal lacquers, in coatings, in paper sizing, in inks and sealing waxes, in binders, for blending with shellac, in production of picric acid, and in medicine.

ACETAL RESINS. Highly crystalline resins that have the repeating group $(OCH_2)_x$. The resins are **polyformaldehyde.** The natural acetal resin is translucent white and can be readily colored. There are two basic types: a homopolymer, such a Du Pont's **Delrin,** and a copolymer, such as Hoechst Celanese's **Celcon.** In general, the homopolymers are harder and more rigid and have higher tensile, flexural, and fatigue strengths, but lower elongation. The copolymers are more stable in long-term high-temperature service and have better resistance to hot water. Special types of acetals are glass-filled, providing higher strengths and stiffness, and tetrafluoroethylene (TFE) filled, providing exceptional frictional and wear properties.

Acetals are among the strongest and stiffest of the thermoplastics. Their tensile strength ranges from 8,000 to about 13,000 lb/in^2 (55 to 89 MPa), the tensile modulus of elasticity is about 500,000 lb/in^2 (3,445 MPa), and fatigue strength at room temperature is about 5,000 lb/in^2 (34 MPa). Their excellent creep resistance and low moisture absorption (less than 0.4%) give them excellent dimensional stability.

They are useful for continuous service up to about 220°F (104°C). Acetals' low friction and high abrasion resistance, though not as good as nylon's, rates them high among thermoplastics. Their impact resistance is good and remains almost constant over a wide temperature range. Acetals are attacked by some acids and bases, but have excellent resistance to all common solvents. They are processed mainly by molding or extruding. Some parts are also made by blow and rotational molding. Typical parts and products made of acetal include pump impellers, conveyor links, drive sprockets, automobile instrument clusters, spinning reel housings, gear valve components, bearings, and other machine parts. **Acetal homopolymers** are used for mechanical and electrical parts. They have a specific gravity of 1.425, a tensile strength of 10,000 lb/in² (69 MPa), 15% elongation, a dielectric strength of 500 V/mil (19.6 × 10⁶ V/m), and Rockwell hardness M94. They retain their mechanical strength close to the melting point of 347°F (175°C). **Acetal copolymers** are thermoplastic linear acetal resins produced from **trioxane,** which is a cyclic form of formaldehyde. The specific gravity is 1.410, flexural strength 12,000 lb/in² (83 MPa), Rockwell hardness M76, and dielectric strength 1,200 V/mil (47 × 10⁶ V/m). It comes in translucent white pellets for molding.

Tenac SH, a homopolymer from Asahi Chemical, uses a Nylon 3 heat stabilizer to enhance processibility but retains the thermal stability and mechanical properties of the standard homopolymer.

ACETIC ACID. Also known as **ethanoic acid.** A colorless, corrosive liquid of pungent odor and composition $CH_3 \cdot COOH$, having a wide variety of industrial uses as a reagent, solvent, and esterifier. A carboxylic acid, it is employed as a weak acid for etching and for soldering; in stain removers and bleaches; as a preservative; in photographic chemicals; for the manufacture of cellulose acetate and vinyl acetate; as a solvent for essential oils, resins, and gums; as a precipitant for latex; in tanning leather; and in making artificial flavors. Acetic acid is found in the juices of many fruits and in combination in the stems or woody parts of plants. It is the active principle in **vinegar,** giving it the characteristic sour taste, acid flavor, and pungent odor. It is made commercially by oxidation of acetaldehyde (in the presence of manganese, cobalt, or copper acetate), butane, or naphtha. Its specific gravity is 1.049, its boiling point is 244°F (118°C), and it becomes a colorless solid below 61.9°F (16.6°C). The pure 99.9% solid is known as **glacial acetic acid.** Standard and laundry special grades contain 99.5% acid, with water the chief impurity. Standard strengths of water solution are 28, 56, 70, 80, 85, and 90%.

Acetic anhydride, $CH_3COOCOCH_3$, a colorless liquid with boiling point 283°F (139.5°C), is a powerful acetylating agent and is used in making cellulose acetate. It forms acetic acid when water is added. **Hydroxyacetic acid,** $HOCH_2COOH$, or **glycolic acid,** is produced by oxidizing glycol with dilute nitric acid and is intermediate in strength between acetic and formic acids. It is soluble in water, is nontoxic, and is used in foodstuffs, dyeing, tanning, electropolishing, and resins. Its esters are solvents for resins. **Diglycolic acid,** $O(CH_2CO_2H)_2$, is a white solid melting at 298°F (148°C). It is stronger than tartaric and formic acids and is used for making resins and plasticizers. **Thioacetic acid** has the formula of acetamide but with HS replacing the NH_2. It is a pungent liquid used for making esters for synthetic resins.

Chloroacetic acid, $CH_2ClCOOH$, is a white crystalline powder melting at 143°F (61.6°C) and boiling at 372°F (189°C). It is used for producing carboxymethylcellulose, dyes, and drugs. **Sequestrene,** used as a clarifying agent and water softener in soaps and detergents, and to prevent rancidity in foods and sulfonated oils, is ethylene bisaminodiacetic acid, $(HOOCCH_2)_2\text{-}NCH_2CH_2N(CH_2COOH)_2$. It is a liquid, but in the form of its sodium salt is a water-soluble white powder. **Trifluoroacetic acid,** CF_3COOH, is one of the strongest organic acids. It is a colorless, corrosive liquid, boiling at 160°F (71.1°C) and freezing at 4.5°F (-15.3°C). It is used in the manufacture of plastics, dyes, pharmaceuticals, and flame-resistant compounds.

Paracetic acid, $CH_3 \cdot O \cdot COOH$, is a colorless liquid of strong odor with the same solubility as acetic acid. It has 8.6% available oxygen and is used as a bleaching agent, a polymerization catalyst, for making epoxy resins, and as a bactericide. **Acetin** is an ester of acetic acid made from glycerin and acetic acid, used as a solvent for basic dyes and tannins. It is a neutral straw-colored liquid of specific gravity 1.20 and boiling point 271 to 307°F (133 to 153°C). It is also used in low-freezing dynamites and smokeless powder. The triacetic ester, **triacetin,** is a water-white liquid of specific gravity 1.16 and flash point 271°F (133°C), soluble in aromatic hydrocarbons. It is used as a plasticizer and in propellants.

Phenylacetic acid, $C_6H_5CH_2COOH$, is a white flaky solid melting at 166°F (74.5°C). The reactive methylene group makes it useful for the manufacture of fine chemicals. **Cyanoacetic acid,** $CN \cdot CH_2 \cdot COOH$, has an active methylene group and an easily oxidized cyano group, and is used for producing caffeine, while the derivative **ethyl cyanoacetate,** $NC \cdot CH_2COO \cdot CH_2 \cdot CH_3$, a liquid boiling at 405°F (207°C), is used for making many drugs. **Malonic acid,** $CH_2(COOH)_2$, or **propanedioic acid,** is a very reactive acid sometimes used instead of acetic acid for making plastics, drugs, and perfumes. It decomposes at 320°F (160°C), yielding acetic acid and carbon dioxide. **Methyl acetic acid,** CH_3CH_2COOH, is **propionic acid** or **propanoic acid,**

a by-product in the extraction of potash from kelp. Modifications of this acid are used for cross-linking plastics.

ACETONE. An important industrial solvent, used in the manufacture of lacquers, plastics, smokeless powder; for dewaxing lubricating oils; for dissolving acetylene for storage; for dyeing cotton with aniline black; and as a raw material in the manufacture of other chemicals. It is a colorless, flammable liquid with a mintlike odor and is soluble in water and in ether. The composition is $CH_3 \cdot CO \cdot CH_3$, specific gravity 0.790, boiling point 133°F (56°C), and solidification point −137°F (−94°C). Acetone is mainly produced as a by-product in the cleavage of cumene hydroperoxide into phenol. A secondary route is by catalytic dehydrogenation of isopropyl alcohol.

Diacetone, or **diacetone alcohol,** is a colorless liquid of composition $CH_3 \cdot CO \cdot CH_2 \cdot COH(CH_3)_2$ with a pleasant odor. It is used as a solvent for nitrocellulose and cellulose acetate, for gums and resins, in lacquers and thinners, and in ink, paint, and varnish removers. Because of its low freezing point and miscibility with castor oil it is used in hydraulic brake fluids. The specific gravity is 0.938, boiling point 331°F (166°C), and freezing point −65°F (−54°C). Synthetic **methyl acetone** is a mixture of about 50% acetone, 30 methyl acetate, and 20 methanol, used in lacquers, paint removers, and for coagulating latex. **Dihydroxyacetone,** a colorless crystalline solid produced from glycerine by sorbose bacteria reaction, is used in cosmetics, and in preparing foodstuff emulsions, plasticizers, and alkyd resins. It is soluble in water and in alcohol.

ACETYLENE. A colorless gas of the composition $HC\!:\!CH$, used for welding and flame cutting of metals and for producing other chemicals. It contains 92.3% carbon and is therefore nearly gaseous carbon. When pure, it has a sweet odor, but when it contains hydrogen sulfide as an impurity, it has a disagreeable odor. Acetylene burns brightly in air, and was widely used for theater stage lighting before the advent of electric light. When mixed with oxygen as **oxyacetylene** for flame cutting and welding, it gives a temperature of 6332°F (3500°C). In air it is an explosive gas. The maximum explosive effect is with a mixture of 7.7% gas and 92.3% air. Acetylene has a specific gravity of 0.92. It is nontoxic and is soluble in water, alcohol, or acetone. It liquefies under a pressure of 700 lb/in^2 (4.8 MPa) at 70°F (21°C). It is easily generated by the action of water on calcium carbide, but the newest methods involve pyrolysis, or cracking, of hydrocarbons, principally methane. It is also recovered from ethylene feedstock prior to polymerization. About 80% of acetylene usage is for synthesis of industrial chemicals, such as vinyl chloride, vinyl acetate, acrylonitrile, polyvinylpyrrolidone, trichloroethylene, and acetic acid. **Acetylenic alcohols** and **diols** include **propargyl alcohol, butynediol, butenediol, butanediol,** and **butyrolacetone**; these are used in metal pickling and plating and for making agricultural chemicals, polyesters, and vinyl esters. It is marketed compressed in cylinders, dissolved in acetone to make it nonexplosive. One volume of acetone will dissolve 25 volumes of acetylene at atmospheric pressure, or 250 volumes at 10 atm (10.3 kg/cm^2). **Prest-O-Lite** is a trade name of Union Carbide Corp. for acetylene dissolved in acetone. **Acetylene snow,** or solid acetylene, is produced by cooling acetylene below the melting point and compressing. It is insensible to shock and flame and is thus easier to transport. A replacement for acetylene for producing plastics is **methyl acetylene propadiene,** which contains 70% methyl acetylene and 30 of the isomer **propadiene.** It has the reactions of both acetylene and its isomer. **Mapp,** of Dow Chemical Co., for metal cutting, is **methyl acetylene,** $CH{:}C \cdot CH_3$. It is safer to handle and gives about the same flame temperature.

ACRYLIC RESINS. Colorless, highly transparent, thermoplastic, synthetic resins made by the polymerization of acrylic derivatives, chiefly from the esters of **acrylic acid,** $CH_2{:}CH \cdot COOH$, and **methacrylic acid,** $CH_2{:}C(CH_3) \cdot COOH$, **ethyl acrylate** and **methyl acrylate.** **Glacial acrylic acid** is the anhydrous monomer with less than 2% moisture. It can be esterified directly with an alcohol. **Vinyl acrylic acid,** $CH_2{:}CHCH{:}CHCOOH$, with a melting point of 80°C, is made from acrolein and malonic acid. It polymerizes on heating.

The resins vary from soft, sticky semisolids to hard, brittle solids, depending upon the constitution of the monomers and upon the polymerization. They are used for adhesives, protective coatings, finishes, laminated glass, transparent structural sheet, and molded products. Acrylic resins, or **acrylate resins,** are stable and resistant to chemicals. They do not cloud or fade in light when used as laminating material in glass and are used as air-curing adhesives to seal glass to metals or wood. Water-based acrylics are used for the formulation of caulks and sealants. They have better adhesion and weather resistance than butyl rubbers and dry more quickly. The sealants usually contain about 80% solids.

Most **acrylic plastics** are based on polymers of **methyl methacrylate,** which may be modified by copolymerizing or blending with other monomers. Noted for excellent optical properties, they have a light transmission of about 92%. Besides the transparent grades, they can be obtained in translucent or opaque colors as well as the natural color of water white. Moldings have a deep luster and high surface gloss, and for this reason are widely used for decorative parts.

Acrylics have excellent weathering characteristics. Because they are little affected by sunlight, rain, and corrosive atmospheres, they are well suited for outdoor applications. In general, the majority of grades can be used up to about 212°F (100°C). Thermal expansion is relatively high.

Acrylics are hard and stiff. They are also a relatively strong plastic; their tensile strength ranges from 5,000 to about 11,000 lb/in^2 (34 to about 76 MPa). However, regular grades are somewhat brittle. High impact grades are produced by blending with rubber stock. The high strength is useful only for short-term loading. For long-term service, to avoid crazing or surface cracking, tensile stresses must be limited to about 1,500 lb/in^2 (10 MPa).

Acrylic plastics are available as cast sheets, rods, tubes, and blocks. They are also processed by injection or compression molding. Sheets are produced in thicknesses from 0.125 to 0.375 in (0.32 to 0.95 cm) and in sizes up to 10 by 12 ft (3 by 4 m). A special process that produces molecular orientation in the cast product is used to make crack-resistant aircraft cabin windows and fighter plane canopies. Acrylic moldings as large as 1 yd^2 (1 m^2) have been produced. Typical moldings include knobs, handles, escutcheons, parts for vending machines, and a wide variety of lenses for light control, signal lamps, and the like.

Tough molding resins are made by copolymerizing methyl methacrylate with styrene. These molding resins have a flexural strength of 17,600 lb/in^2 (121 MPa). **Acrylate rubbers,** having a tensile strength of 2,500 lb/in^2 (17 MPa) and an elongation of 350%, are used for gaskets, wire insulation, and hose.

Allyl methacrylate is a liquid of the empirical formula $C_7H_{10}O_2$, boiling at 145°F (63°C) and insoluble in water. It can be polymerized to form liquid or hard solid resins, but is used chiefly as a cross-linking agent for other resins to raise the softening point and increase hardness. **Polymethyl alphachloroacrylate,** $(CH_2:CCl \cdot COOCH_3)_x$, is a transparent and craze-resistant resin used for aircraft windows. The heat distortion point is 260°F (127°C), and it has higher tensile and flexural strength than other acrylics. **Cyclohexyl methacrylate** has optical properties similar to those of crown glass and is used for cast lenses, where its softness and low softening point, 160°F (71°C), are not objectionable.

Lucite is methyl methacrylate of Du Pont, marketed as molding powder and in rods, tubes, and cast and molded sheets. **Lucitone,** of Dentsply International, is this material molded in dentures in pink and translucent. **Lucite HM-140** is this material compounded for high-temperature injection molding. **Acrylic syrup** is a liquid Lucite for use as a low-pressure laminating resin. It produces strong, stiff, tough laminates adaptable to translucent or bright colors. Reinforced with glass fibers, a panel with contact cure has a flexural strength of 25,000 lb/in^2 (172 MPa), elongation 1.5%, distortion point at 233°F (112°C), Rockwell hardness R121, and light transmission up to 65%. **Crystalite,** of Rohm & Haas Co., is an acrylic molding powder. **Plexiglas,** of this company, is transparent methyl methacrylate in sheets and rods. All these plastics are used for aircraft windows. **Plexiglas V** is for injection molding, while **Plexiglas VM** is a molding powder to resist heat distortion to 174°F (79°C). **Vernonite,** of Rohm & Haas Co., is an acrylic denture resin. **Quarite** and **Quarite Plus,** of Aristech Chemical, are mineral-filled sheets with a pebbly surface texture for spas, plumbing ware, and architectural applications. **Acrystone,** of the same firm, is a mineral-filled, solid-surfacing, cast acrylic sheet.

The **Acryloid resins,** of the same company, are **acrylic copolymer** solid resins, and the **Acrysol resins** are solutions for coatings. **Plexene M,** of the same company, is a **styrene-acrylic** resin for injection molding. The specific gravity of the molded resin is 1.08, the dielectric strength 350 V/mil (14 × 10^6 V/m), and tensile strength 15,000 lb/in^2 (103 MPa). **Rhoplex resins,** of the same company, are acrylic resin emulsions for paints, textile finishes, and adhesives. Water-soluble acrylic copolymer is used for thickening natural or synthetic rubber latex for paper and textile coatings. Coatings made with acrylics have good adhesion and gloss, are resistant to oils and chemicals, and have good dielectric strength. **Carboset 511** is a water solution of acrylic resin for protecting polished metal surfaces and precision parts against scratching. It is resistant to water, but can be washed off with soap and water. **Cavalon,** of Du Pont, is a **polyacrylic** resin for coatings that has high hardness and resistance to abrasion.

Volan, of the same company, is a **methacrylate-chromic oxychloride,** $CH_2{:}C(CH_3)C(OH)(OCrCl_2)_2$, in which methacrylic acid is joined with two $CrCl_2$ groups to form **resonant bonds.** It is a dark-green liquid with a specific gravity of 1.02, boiling point at 180°F (82°C). When applied to negatively charged surfaces such as cellulose, polyamides, or silica materials, the chromium complex is strongly held while the chlorine is lost. In attaching to glass, the CrO forms a chemical bond to the silica of the glass, $Cr \cdot O \cdot Si$. With polyamides, the CrO attaches to a carbon atom, $Cr \cdot O \cdot C$, and thus provides strong bonds in plastic laminates. **Korad films,** from Polymer Extruded Products, are weatherable, wood-grain, acrylic laminating films for outdoor window and door profiles and for adhesive-free bonding to polypropylene sheet for thermoformed products. **Acumer 3000,** a water-treatment acrylate polymer from Rohm and Haas Co., controls silica and prevents formation of magnesium-silicate scale.

Treated with the polymer, recirculated water can tolerate as much as 300 parts per million of silica without scale formation. **Acumer 5000,** an acrylate terpolymer of the same company, controls iron and inhibits scale formation in boiler water.

ACRYLONITRILE. Also called **vinyl cyanide** and **propene nitrile.** A liquid of composition $CH_2:CHCN$, boiling at 172°F (78°C), used in insecticides and for producing plastics and other chemicals. It is made by the addition of hydrocyanic acid to acetylene, by using propylene as the starter and reacting with ammonia, or from petroleum. **Acrylonitrile fiber,** originally developed in Germany as a textile staple fiber and as a monofilament for screens and weaving, and known as **Redon,** has good dimensional stability and high dielectric strength and is resistant to water and to solvents. The polymerized acrylonitrile has a molecular structure that can be oriented by drawing to give fibers of high strength. **Orlon,** of Du Pont, is a polymerized acrylonitrile fiber. It is nearly as strong as nylon and has a softer feel. It can be crimped to facilitate spinning with wool. It is used for clothing textiles and for filter fabrics. **Dynel,** of Union Carbide Corp., is an **acrylonitrile-vinyl chloride** copolymer staple fiber. It produces textiles with a warmth and feel like those of wool. It has good strength, is resilient, dyes easily, and is mothproof. **Verel,** of Eastman Chemical Products, Inc., is a similar acrylic fiber produced from acrylonitrile and vinylidene chloride, and **Creslan,** of American Cyanamid Co., called **Exlan** in Japan, is an acrylic fiber. **Acrilan,** of Monsanto, is a similar textile fiber and is an **acrylonitrile-vinyl acetate** copolymer. **Acrylonitrile-styrene** is a copolymer for injection molding and extruding that produces rigid thermoplastic parts of higher tensile strength than those of the methacrylates, and has good dimensional stability and scratch resistance. **Saran F-120,** of Dow Chemical Co., is a similar material.

Acrylonitrile also is polymerized with vinyl pyrrolidone or other dye-receptive monomer. The fiber has a molecular structure called a **nitrile alloy,** with a continuous polyacrylonitrile backbond with close-packed hydrophilic groups which hold the dye molecules. It resists heat to 490°F (254°C). Crystal-clear **styrene-acrylonitrile** copolymer is used for molding such articles as dinnerware and food containers. Acrylonitrile-styrene copolymers are also combined with alkyl-substituted phenolic resins to produce hard, glossy, flexible coatings. **Itaconic acid,** or **methylene succinic acid,** $CH_2:CCH(COOH)_2$, is also polymerized with acrylonitrile to produce fibers. When this acid is polymerized with styrene, it produces transparent plastics of good optical properties. Ultrapure **succinonitrile (SCN)** is a transparent organic material that melts at 134°F (58°C) and forms crystals much like common metals. Produced at Rensselaer Polytechnic Institute, it was used by the National Aeronautics and Space Administration on the Space Shuttle to observe dendritic crystal formation in gravity-free space to validate or modify crystal-growth theories for metals on earth. For jet aircraft tires, an extremely wear-resistant rubber is made of **acrylonitrile-butadiene** with an organometallic catalyst that has alternating groups in the copolymer.

Acrylonitrile reacts with cellulose to form a wide range of resins from soluble ethers useful for textile finishes to tough, resistant materials useful for fibers. It can be reacted directly with cotton to improve the fiber. Sodium salts of acrylonitrile are used as soil conditioners. They are more efficient than peat moss.

ACTIVATED CHARCOAL. A nearly chemically pure amorphous carbon made by carbonizing and treating dense material such as coconut shells, peach pits, or hardwood. When made from coal, or in the chemical industry, it is more usually called **activated carbon,** or **filter carbon.** It may be made by dry distillation or by leaching the charcoal with steam or by treatment with zinc chloride or potassium thiocyanate. It is used as an adsorbent material for gas masks, for cigarette filters, and for purifying acids, recovering solvents, and decolorizing liquids. Activated carbon woven into garments protects members of the armed forces from chemical warfare. Garments with superactivated carbon are lighter in weight and much more absorbent. **Coconut charcoal,** valued for gas masks, is an activated charcoal usually made by heating coconut shells in a closed retort, crushing, and steam treating. An activated charcoal made from coconut shells will adsorb 68% of its weight of carbon tetrachloride. A requirement of activated charcoals, besides high adsorbing power, is that they possess strength to retain a porous structure to pass the air or liquid. **Activated carbon CXC4-6,** of Union Carbide Corp., produced from petroleum and used as a catalyst support, is in $\frac{3}{16}$-in (0.48-cm) pellets of high hardness and strength. Powdered activated charcoal is usually ground to 300 mesh. For water purification it should be fine enough to wet easily, but not so light that it will float on the top. For decolorizing or deodorizing oils and chemicals, it is mixed in the liquid and settles out in a few hours. A single drop of water will hold 10,000 particles of powdered charcoal. In sugar and oil refining, it removes color but does not bleach like chemicals. Color removal is measured by the **molasses number,** which is the index of color removed per gram of carbon when tested on a standard molasses solution.

Hydrodarco is a powdered activated carbon. **Norit,** of American Norit Co., Inc., is a group of highly adsorptive activated carbons.

Kelpchar is activated carbon made from seaweed. **Tec-Char,** of Tennessee Eastman Co., is a by-product charcoal obtained in wood distillation and in graded grains for various uses. **Nuchar** is an activated carbon. The activated carbon of the Masonite Corp. is made by subjecting wood chips to high steam pressure and disintegrating by sudden release of the pressure. The doughy mass is briquetted and carbonized. Activated carbon of Calgon Carbon Corp. serves as an adsorbent to stabilize organic chemicals such as halogenated aromatic compounds in contaminated soil and sludge. After stabilization, the treated soil and sludge are mixed with pozzolanic materials such as portland cement, fly ash, and kiln dust for disposal as landfill. Granular activated carbon is used in a system from Envirex Inc. for the removal of benzene, toluene, ethylbenzene, and xylenes from groundwater. It is especially effective in cleaning sites having low levels of these contaminants.

Activated carbon derived from coal is harder than organic carbons and does not crumble easily, permitting a higher flow of liquid to be filtered. It has a high density and high activity. **SGL carbon** has an iodine number of 1,000 compared with 650 for ordinary carbons. Its color-removal index is about 40% higher than that of organic carbons. **Filt-o-cite,** of Shamokin Filler Co., is finely ground anthracite used to replace sand as a filtering agent for industrial wastes.

ADHESIVES. Materials employed for sticking, or adhering, one surface to another. Forms are liquid, paste, powder, and dry film. The commercial adhesives include pastes; glues; pyroxylin cements; rubber cements; latex cement; special cements of chlorinated rubber, synthetic rubbers, or synthetic resins; and the natural mucilages. Adhesives are characterized by degree of tack, or stickiness, by strength of bond after setting or drying, by rapidity of bonding, and by durability. The strength of bond is inherent in the character of the adhesive itself, particularly in its ability to adhere intimately to the surface to be bonded. Adhesives prepared from organic products are in general subject to disintegration on exposure. The life of an adhesive usually depends upon the stability of the ingredient that gives the holding power, although otherwise good cements of synthetic materials may disintegrate by the oxidation of fillers or materials used to increase tack. Plasticizers usually reduce adhesion. Some fillers such as mineral fibers or walnut-shell flour increase the thixotropy and the strength, while some such as starch increase the tack but also increase the tendency to disintegrate.

Adhesives can be grouped into five classifications based on chemical composition. **Natural adhesives** include vegetable- and animal-base adhesives and natural gums. They are inexpensive and easy to apply and have a long shelf life. They develop tack quickly, but provide only low-strength joints. Most are water-soluble. They are supplied as liquids or as dry powders to be mixed with water. **Casein-latex adhesive** is an exception. It consists of combinations of casein with either natural or synthetic rubber latex. It is used to bond metal to wood for panel construction and to join laminated plastics and linoleum to wood and metal. Except for this type, most natural adhesives are used for bonding paper, cardboard, foil, and light wood.

Thermoplastic adhesives can be softened or melted by heating and hardened by cooling. They are based on thermoplastic resins (including asphalt and oleoresin adhesives) dissolved in solvent or emulsified in water. Most become brittle at subzero temperatures and may not be used under stress at temperatures much above 150°F (65°C). Being relatively soft materials, thermoplastic adhesives have poor creep strength. Although lower in strength than all but natural adhesives and suitable only for noncritical service, they are also cheaper than most adhesives. They are also odorless and tasteless and can be made fungus-resistant.

Hot melts, based on polyamides, polyolefins, or polyesters, are compounds modified with waxes and processing aids. They are used in book binding, packaging, carpeting, and making laminates. **Koraprop 210,** an olefin-based hot melt from Kommerling Chemische Fabrik of Germany, bonds dissimilar substrates in the transportation industry.

Elastomeric adhesives, based on natural and synthetic rubbers, are available as solvent dispersions, latexes, or water dispersions. They are primarily used as compounds which have been modified with resins to form some of the adhesive "alloys" discussed below. They are similar to thermoplastics in that they soften with heat, but never melt completely. They generally provide high flexibility and low strength and, without resin modifiers, are used to bond paper and similar materials.

Thermosetting adhesives soften with heat only long enough for the cure to initiate. Once cured, they become relatively infusible up to their decomposition temperature. Although most such adhesives do not decompose at temperatures below 500°F (260°C), some are useful only to about 150°F (65°C). Different chemical types have different curing requirements. Some are supplied as two-part adhesives and mixed before use at room temperature; others require heat and/or pressure to bond. As a group, these adhesives provide stronger bonds than the other three groups. Creep strength is good and peel strength is fair. Generally, bonds are brittle and have little resilience and low impact strength. An adhesive developed at the University of Illinois (Urbana) is based on a cross-linked polyester thermoset made by reacting aromatic carboxylic acids with aromatic acetates. The **polyester adhesive** is stable in air to 662°F (350°C), withstands heat to 392°F (200°C), and is more resistant to moisture than epoxy adhesives. Potential uses include bonding carbon-fiber-reinforced laminates and alumina or silicon-carbide particles in grinding wheels.

Jet-Weld adhesives, of 3M, are one-part, moisture-curing, thermosetting urethane formulations which eliminate the need for clamping. Designed to be applied with a handheld applicator or by bulk dispensing equipment, they are intended for bonding aluminum, glass, plastic, and wood. Two-part **urethane adhesives** from Ashland Chemical permit primerless bonding of auto panels made of sheet molding compound. A 100% reactive two-part urethane is intended for high-pressure lamination of auto, appliance, and building panels. **Terokal 806,** a fast-reacting two-part urethane from Teroson of Germany, primerless-bonds auto seat shells.

Alloy adhesives are adhesives compounded from resins of two or more different chemical families, e.g., thermosetting and thermoplastic, or thermosetting and elastomeric. In such adhesives the performance benefits of two or more types of resins can be combined. For example, thermosetting resins are plasticized by a second resin, resulting in improved toughness, flexibility, and impact resistance.

Paste adhesives are usually water solutions of starches or dextrins, sometimes mixed with gums, resins, or glue to add strength, and containing antioxidants. They are the cheapest of the adhesives, but deteriorate on exposure unless made with chemically altered starches. They are widely employed for the adhesion of paper and paperboard. Much of the so-called **vegetable glue** is **tapioca paste.** It is used for the cheaper plywoods, postage stamps, envelopes, and labeling. It has a quick tack and is valued for pastes for automatic box-making machines. **Latex pastes** of the rub-off type are used for such purposes as photographic mounting, as they do not shrink the paper as do the starch pastes. **Glues** are usually water solutions of animal gelatin, and the only difference between animal glues and edible gelatin is the degree of purity. Hide and bone glues are marketed as dry flake, but fish glue is liquid. **Mucilages** are light vegetable glues, generally from water-soluble gums.

Rubber cements for paper bonding are simple solutions of rubber in a chemical solvent. They are like the latex pastes in that the excess can be rubbed off the paper. Stronger rubber cements are usually compounded with resins, gums, or synthetics. An infinite variety of these cements are possible, and they are all waterproof with good initial bond, but they are subject to deterioration on exposure, as the rubber is uncured. This type of cement is also made from synthetic rubbers which are self-curing. **Curing cements** are rubber compounds to be cured by heat and pressure or by chemical curing agents. When cured, they are stronger, give better adhesion to metal surfaces, and have longer life. **Latex cements** are solvent solutions of rubber latex. They provide excellent tack and give strong bonds to paper, leather, and fabric, but they are subject to rapid disintegration unless cured.

In general, natural rubber has the highest cohesive strength of the rubbers, with rapid initial tack and high bond strength. It also is odorless. **Neoprene** has the highest cohesive strength of the synthetic rubbers, but it requires tackifiers. **Gr-S rubber** (styrene-butadiene) is high in specific adhesion for quick bonding, but has low strength. Reclaimed rubber may be used in cements, but it has low initial tack and needs tackifiers.

Pyroxylin cements may be merely solutions of nitrocellulose in chemical solvents, or they may be compounded with resins, or plasticized with gums or synthetics. They dry by the evaporation of the solvent and have little initial tack, but because of their ability to adhere to almost any type of surface they are called **household cements.** Cellulose acetate may also be used. These cements are used for bonding the soles of women's shoes. The bonding strength is about 10 lb/in^2 (0.07 MPa), or equivalent to the adhesive strength of the outer fibers of the leather to be bonded. For hot-press lamination of wood, the plastic cement is sometimes marketed in the form of thin sheet.

Polyvinyl acetate-crotonic acid copolymer resin is used as a **hot-dip adhesive** for book and magazine binding. It is soluble in alkali solutions, and thus the trim is reusable. **Polyvinyl alcohol,** with fillers of clay and starch, is used for paperboard containers. **Vinyl emulsions** are much used as adhesives for laminates.

Epoxy resin cements give good adhesion to almost any material and are heat-resistant to about 400°F (204°C). An epoxy resin will give a steel-to-steel bond of 3,100 lb/in^2 (21 MPa) and an aluminum-to-aluminum bond to 3,800 lb/in^2 (26 MPa).

Some **pressure-sensitive adhesives** are mixtures of a phenolic resin and an nitrile rubber in a solvent, but adhesive tapes are made with a wide variety of rubber or resin compounds. **Furan cements,** usually made with **furfural-alcohol resins,** are strong and highly resistant to chemicals. They are valued for bonding acid-resistant brick and tile.

Structural adhesives have come to mean those adhesives used to bond metals to other metals, to wood, or to rigid plastics, where bond strength is a critical requirement. They are generally of the alloy or thermosetting type. Three of the most commonly used are the modified epoxies, neoprene-phenolics, and vinyl formal-phenolics. **Modified epoxy adhesives** are thermosetting and may be of either the room-temperature curing type, which cure by addition of a chemical activator, or the heat-curing type. They have high strength and resist temperatures up to nearly 500°F (260°C). **Neoprene-phenolic adhesives** are alloys characterized by excellent peel strength, but lower shear strength than modified epoxies. They are moderately priced and offer good flexibility and vibration absorption. **Vinyl formal-phenolic adhesives** are alloys whose properties fall between those of modified epoxies and the thermoset-elastomer types. They are supplied as solvent dispersions in solution or in film form.

Conventional two-part liquid **epoxy adhesives** are brittle but can be toughened by incorporating a rubber phase. **Scotch-Weld DP-420 and DP-460,** of 3M, are examples, and they are said to provide 10 times greater peel strength. Three others—DP-105, DP-125, and DP-190—provide shear and peel strengths approaching those of the toughest epoxies and 120% elongation versus 40% for the softest conventional epoxies. Thus they are more vibration-resistant and suitable for joining substrates differing more widely in coefficients of thermal expansion. **FM 350 epoxy film adhesive,** of American Cyanamid, cures at 250°F (121°C) whereas prior grades cured at 350°F (177°C). It also increases use temperature, from 300°F (149°C) to 350°F (177°C). **SynSkin HC9872,** of Dexter Aerospace Materials, is a surfacing adhesive film used to reduce labor-intensive finishing in fabricating composite components. **Liquid polymer shim,** also from Dexter, is room-temperature-curing epoxy paste suitable for 300°F (149°C) service.

Acrylic adhesives are solutions of rubber-base polymers in methacrylate monomers. They are two-component systems and have characteristics similar to those of epoxy and urethane adhesives. They bond rapidly at room temperature, and adhesion is not greatly affected by oily or poorly prepared surfaces. Other advantages are low shrinkage during cure, high peel and shear strengths, excellent impact resistance, and good elevated-temperature properties. They can be used to bond a great variety of materials, such as wood, glass, aluminum, brass, copper, steel, most plastics, and dissimilar metals.

Silicone adhesive sealants are room-temperature-vulcanizing elastomers, also called **RTV silicones.** They excel in resistance to ozone, ultraviolet, water, and heat—to 500°F (260°C) or greater—while retaining flexibility at subzero temperatures. They also possess excellent electrical insulative properties, having superior resistance to high-voltage ionization and corona discharge. However, they are relatively weak, having tensile strengths up to about 1,200 lb/in (8.3 MPa) and moderate tear resistance. There are both one-part and two-part systems, with grades suitable for bonding metals, plastics, glass, stoneware, marble, concrete, and wood. Dow Corning and General

Electric are the major producers. Dow's **3146 silicone,** a recent one-part, self-priming grade for electronics packaging, provides excellent adhesion to most metals, glass, select plastics, and FR-boards. It also eliminates emission of volatile organic compounds, cures without exotherm or corrosive by-products, does not emit the vinegar odor associated with other one-part RTVs, has good tear resistance, resists long-term exposure to temperatures up to about 400°F (204°C), and stays flexible at temperatures as low as −85°F (−65°C).

Ultraviolet cure adhesives, of Loctite Corp., are anaerobic structural adhesives formulated specifically for glass bonding applications. The adhesive remains liquid after application until ultraviolet light triggers the curing mechanism.

A **ceramic adhesive** developed by the Air Force for bonding stainless steel to resist heat to 1500°F (816°C) is made with a porcelain enamel frit, iron oxide, and stainless-steel powder. It is applied to both parts and fired at 1750°F (954°C), giving a shear strength of 1,500 lb/in^2 (10 MPa) in the bond. But ceramic cements that require firing are generally classified with ordinary adhesives. **Wash-away adhesives** are used for holding lenses, electronic crystal wafers, or other small parts for grinding and polishing operations. They are based on acrylic or other low-melting thermoplastic resins. They can be removed with a solvent or by heating.

Electrically conductive adhesives are made by adding metallic fillers, such as gold, silver, nickel, copper, or carbon powder. Most conductive adhesives are epoxy-based systems, because of their excellent adhesion to metallic and nonmetallic surfaces. Silicones and polyimides are also frequently the base in adhesives used in bonding conductive gaskets to housings for electromagnetic and radio-frequency interference applications. **ZX adhesives,** of Zymet Inc., are called **anisotropically conductive adhesives.** A monolayer of uniformly sized conductive particles provide Z axis, or through-the-thickness, conductivity.

ADIPIC ACID. Also called **butane dicarboxylic acid** or **hexane-dioic acid.** A white crystalline solid of composition $HOOC(CH_2)_4COOH$, used as a plasticizer in synthetic resins and coatings, and in the production of nylon. It is made by a two-step oxidation of cyclohexane to cyclohexanol and cyclohexanone, followed by treatment with nitric acid. The cyclohexanol can also be derived by phenol dehydrogenation. The melting point is 306°F (152°C). It is soluble in alcohol and slightly soluble in water. Many other dibasic acids useful for making synthetic resins are produced readily from fatty oils. **Suberic acid,** $HOOC(CH_2)_6COOH$, is made by the oxidation of castor oil. It is the same as the **octane-dioic acid** made from butadiene. **Sebacic acid,** $HOOC(CH_2)_8COOH$, called

decane-dioic acid, is produced by heating castor oil with sodium hydroxide. **Azelaic acid,** $HOOC(CH_2)_7COOH$, or **nonane-dioic acid,** is a strong dibasic acid with melting point at 223°F (106°C), made by the oxidation of oleic acid, and is used as a substitute for phthalic anhydride to react with glycerin to form alkyd resins less hard and brittle than those made with phthalic anhydride. It is also used instead of sebacic acid for producing the high-temperature lubricant **ethyl-hexyl sebacate.** Another substitute for this acid is pelagonic acid. **Petroselic acid,** which is an isomer of oleic acid with the double bond in a different position, is made by the hydrogenation of the ricinoleic acid of castor oil and then is oxidized to produce adipic acid. Adipic acid can be used as a substitute for citric acid for the acidulation of beverages, but is less water-soluble. It is also used in protein foods to control the gelling action. The acid also has been used as a substitute for rosin flux in soldering printed wiring boards. Unlike rosin, it leaves no residues that require subsequent cleaning.

ADMIRALTY METAL. These are copper alloys noted for good resistance to low-velocity freshwater and seawater and thus are used for condenser, distiller, and heat-exchanger tubing and related equipment in these environments. Also known as **inhibited admiralty metal** and **admiralty brass,** they comprise nominally 71.5% copper, 28 zinc, plus small amounts of lead (0.07 maximum) and iron (0.06 maximum) and either 0.02 to 0.06 arsenic [**arsenical admiralty metal** (C44300)], 0.02 to 0.10 antimony [**antimonial admiralty metal** (C44400)], or 0.02 to 0.10 phosphorus [**phorphorized admiralty metal** (C44500)]. Though available in the annealed and cold-worked tempers, the annealed condition is often preferred because of the alloys' susceptibility to stress corrosion. Typical tensile properties in the annealed condition are 45,000 to 53,000 lb/in^2 (310 to 365 MPa) ultimate strength, 13,000 to 22,000 lb/in^2 (89 to 152 MPa) yield strength, and 65 to 70% elongation. Cold working appreciably increases strength.

ADSORBENT. A material used to remove odor, taste, haze, and color from oils, foods, pharmaceuticals, or chemicals by selective adsorption of the impurities. Such materials are also called **adsorbates.** Adsorption is the process of adhesion of the molecules of the substance to the surface of the adsorbent; in contrast, absorption entails a physical or chemical reaction. The common adsorbents are **activated carbon,** or **activated clays, alumina, magnesium silicate,** or **silica gel.** The noncarbonaceous adsorbents are used for decolorizing vegetable, animal, or mineral oils, but activated carbon may also be used in conjunction with clays to adsorb color bodies not removed by the clay. Granular adsorbents are employed as filter beds, but powdered adsorbents are stirred into the liquid and are usually more effective. Adsorption from a gas is usually done with activated carbon. Silica gel is usually employed for removing trace quantities of water from water-insoluble liquids, while activated carbon is used for removing trace quantities of oils or chemicals from water. Adsorbents are normally recovered and are regenerated for reuse by heating, steaming, or burning off the adsorbed material.

Adsorbents called **molecular sieves** are used to separate chemicals of different molecular diameters without regard to their boiling points. A double hydrated aluminum calcium silicate marketed by Union Carbide Corp. as a **zeolite** will pass chemicals with molecular diameters less than 5 Å and retain larger ones by selective adsorption. Carbon molecular sieves are used to concentrate nitrogen from air. However, a material used in the separation of liquid mixtures whose components boil too close together for simple fractional distillation is called an **azeotrope.** It is a solvent added to the mixture to increase the relative volatility of one of the components so that it can be separated. The solvent may be alcohols, glycols, or nitrobenzene.

The adsorbents used in vacuum tubes to adsorb or combine with residual gases are called **getters. Flash getters** are pellets or strips of barium or barium alloy used to shorten the exhaust period. The getter is evaporated by induction heating during tube exhaust, and it condenses on the tube walls, adsorbing the gas residues. Later, at operating temperatures of 300 to 400°F (150 to 204°C), the getter formed on the tube wall traps gases liberated during tube life. **Bulk getters** are sheets or wires of zirconium, tantalum, or columbium mounted on the hot electrode to trap gases at temperatures of 900 to 2200°F (482 to 1204°C). **Thorium** or **thorium-misch metal** may be used as getters for high temperatures by a coating sintered on the tube anode.

AEROSOLS. A dispersion of particles in air, particularly the chemical dispensing of a liquid or a finely divided powder substance by a gas propellant under pressure. The common aerosol can system was developed during World War II for dispensing insecticides. Substances commonly dispensed by the aerosol process include resins, paints, waxes, and cosmetics. **Chlorofluorocarbons** (CFCs), **hydrocarbons,** and **carbon dioxide** are used as propellants. The use of CFCs is being phased out because the possibility exists that when they reach the stratosphere, they are subject to attack by ultraviolet radiation, which frees their chlorine atoms, which in turn react with the ozone and thus deplete the protective ozone layer. Their use in aerosols was banned in the United States in 1978, and **CFC** alternatives are being sought in refrigeration, foam blowing, and degreasing. The main propellants now are liquefied hydrocarbons, carbon dioxide, and nitrogen. **Aerothene MM,** of Dow Chemical, is a **methylene chloride** that has properties needed for the efficient functioning of carbon dioxide and hydrocarbon propellants in cosmetic aerosol applications. In noncosmetic aerosol applications, such as paints and insecticides, hydrocarbons claim 45% of the market. The principal objection to their use for cosmetic aerosols has been their flammability.

AGAR-AGAR. The dried bleached gelatinous extract from various species of seaweed, *Algae,* mostly species from the Pacific and Indian Oceans. It is the only one of the seaweed products classified as a strategic material because of its use in medicine, but its use is small compared with the use of the products from other seaweeds. The word *agar* means seaweed. Translations of double words from the primordial languages, such as Malay, Carib, or Gaelic, must be made by taking the first word as a superlative adjective or the second word as a cognate verb. Thus, *agar-agar* means best-quality agar.

When dissolved in hot water, agar forms a transparent jelly and is used for fixing bacteria for counts, as a stabilizer in toilet lotions, and in medicines. It has high thickening power, but, unlike most other seaweed extracts, it is indigestible and is not used in foodstuffs. **Kantan** is a variety of agar from the **tengusa seaweed,** *Gelidium corneum,* of Japan. **Australian agar** is from the abundant seaweed *Gracilaria confervoides.* Commercial agar is colorless, yellowish, or pink to black. It is marketed in strips, blocks, or shredded, and is obtained by boiling the dry seaweed and straining out the insoluble matter.

Most U.S. production of agar, as distinct from the algins of the Atlantic, is from the giant **kelp,** *Macrocystia pyrifera* and *Gelidium cartilagineum* of the coast of California and Mexico, but it is not valued as highly for bacteriologic use as the Asiatic. The kelp grows straight up in water 60 to 100 ft (18.3 to 30.5 m) deep and then spreads out on the water another 60 to 80 ft (18.3 to 24.4 m). It is cut about 3 ft (0.9 m) below the surface, and three crops are harvested annually. The plant is 90% water. The dried kelp is washed with boiling water, cooked with soda ash, filtered, sterilized, and treated with muriatic acid to extract the agar.

AGATE. A natural mixture of crystalline and colloidal silica, but consisting mainly of the mineral chalcedony. It usually occurs in irregular banded layers of various colors derived from mineral salts and, when polished, it has a waxy luster. The specific gravity is about 2.6, and the mineral is sometimes harder than quartz. Agate is used for knife edges and bearings of instruments, for pestles and mortars, for textile rollers, and for ornamental articles; and the finer specimens are employed as gemstones. The finest of the massive agates come from Uruguay and Brazil. Much agate encloses dendritic, or fernlike, patterns of manganese oxide or iron oxide, suggestive of moss. The **moss agates** of Montana and the yellow-green moss agate of California known as **amberine** are used as gemstones. Agate is a water-deposited stone and often occurs in the form of stalactites and in petrified wood. **Agatized wood** of Wyoming and Arizona has a green fluorescence. It is cut into ornaments. Clear translucent yellow agates are called **sard,** while the clear reddish ones are **carnelian.** Both are cut as gemstones. **Sardoine** is a brownish carnelian. **Iris agate,** with rainbow colors, from Montana and Oregon, is highly prized. **Moss opal** of Nevada and California is moss agate intergrown with opal. **Blue moonstone** of California is not a true moonstone but is a blue agate of opalescent luster. Commercial agates may be artificially stained with mineral salts, dyed, or treated with acids to bring out color differences. **White agate** is a cream-colored chalcedony with a waxier appearance than agate.

AJOWAN OIL.   Also called **ptychotis oil.** A yellow essential oil distilled from **ajwan seed** of the herbaceous plant *Carum copticum,* or *Ptychotis ajowan,* of India. The seed yields 3 to 4% oil containing up to 50% **thymol** and some **cymene,** most of the thymol separating out on distillation. Thymol is known as **ajwan ka phul,** meaning flowers of ajwan in Hindustani, and the latter part of the name is Anglicized to *thymol.* Ajowan oil has a specific gravity of 0.900 to 0.930. It is used in pharmaceuticals. Thymol, $(CH_3)_2CHC_6H_3(CH_3)OH$, is a white crystalline solid with a strong thyme odor, soluble in alcohol, and melting at 122°F (50°C). It is used in antiseptics and as a deodorant for leather. Thymol is also obtained from horsemint oil and from eucalyptus oil, or it can be made synthetically from metacresol. It was originally distilled from the **thyme plant,** *Thymus vulgaris,* of the Mediterranean countries, the dried leaves of which are used as a condiment. Cymene, $(CH_3)_2CHC_6H_4CH_3$, is used as a scent in soaps and has high solvent properties. It is also obtained from spruce turpentine. It is a liquid of specific gravity 0.861, boiling at 351°F (177°C).

ALBUMIN.   The water-soluble and alcohol-soluble protein obtained from blood, eggs, or milk and used in adhesives, textile and paper finishes, leather coatings, varnishes, as a clarifying agent for tannins, and in oil emulsions. Crude **blood albumin** is a brown amorphous lumpy material obtained by clotting slaughterhouse blood and dissolving out the albumin. The remaining dark-red material is made into **ground blood** and marketed as a fertilizer. Blood albumin is sold as clear, pale, amber, and colored powders. Blood albumin from human blood is a stable, dry, white powder. It is used in water solution for treatment of shock. The material of **egg white** is sometimes spelled **albumen.** Egg white is a complex mixture of at least eight proteins, with sugar and inorganic salts. More than half of the total is the protein **ovalbumin,** a strong coagulating agent, and another large percentage consists of **conalbumin** which forms metal complexes and unites with iron in the human system. Two of the proteins not so desirable in the human body are **ovomucoid,** which inhibits the action of the digestive enzyme trypsin, and **avidin,** which combines with and destroys the action of the growth vitamin biotin. **Egg albumin** is prepared from the dried egg white and is marketed in yellowish amorphous lumps or powdered. The complexity of proteins is illustrated by the fact that the formula for egg albumin is $C_{1428}H_{2244}N_{462}S_{14}$. The heat of pasteurization damages the proteins of the egg white. A small amount of lactic acid and an aluminum salt will stabilize it and allow pasteurizing at 143°F (62°C). **Milk albumin** is made by coagulating casein. **Soybean albumin** is used to replace egg albumin in confectionery. **Synthetic egg white,** or albumin, was made in Germany from fish by extracting the soluble proteins with acetic acid, removing the fat with trichloroethylene, and hydrolyzing with sodium hydroxide. After neutralization, it is obtained as a white powder. **Fish albumin** is a good emulsifier and can be whipped into a stiff foam for bakery products.

ALCOHOL. The common name for **ethyl alcohol,** but the term properly applies to a large group of organic compounds that have important uses in industry, especially as solvents and in the preparation of other materials. A characteristic of all alcohols is the monovalent —OH group. In the primary alcohols there is always a · CH_2OH group in the molecule. The secondary alcohols have a :CHOH group, and the tertiary alcohols have a distinctive :COH group. Alcohols with one OH group are called **monohydroxy alcohols**; those with more than one OH group are known as **polyhydroxy alcohols** or **polyhydric alcohols.** Another method of classification is by the terms *saturated* and *unsaturated.* The common alcohols used in industry are ethyl, methyl, amyl, butyl, isopropyl, and octyl. The alcohols vary in consistency. **Methyl alcohol** is like water, **amyl alcohol** is oily, and **melissyl alcohol** is a solid. Many of the alcohols are most easily made by fermentation; others are produced from natural gas or from petroleum hydrocarbons. Much of the production of ethyl alcohol is from blackstrap molasses. Alcohols, generally colorless, are similar to water in some ways and are neither alkaline nor acid in reaction.

Methyl alcohol, commonly known as **wood alcohol,** has the chemical name **methanol.** It is also referred to as **carbinol.** A colorless, poisonous liquid of composition CH_3OH, it was originally made by the distillation of hardwoods. It is now produced chiefly by catalytic reduction of carbon monoxide and dioxide by hydrogen. Mobil Corp.'s zeolite-based process has been employed in New Zealand to convert methanol to automobile gasoline. Fuel use, either directly or for making **methyl-tert-butyl ether (MTBE),** is growing. MTBE makes up as much as 15% of **reformulated gasoline.** Methanol is used as a solvent in lacquers, varnishes, and shellac. On oxidation it yields formaldehyde, and it is used in making the latter product for synthetic molding materials. The specific gravity of methyl alcohol is 0.795, the solidifying point is $-144°F$ ($-98°C$), and the boiling point is $149°F$ ($65°C$).

Solidified alcohol, marketed in tins and used as a fuel in small stoves, is a jellylike solution of nitrocellulose in methyl alcohol. It burns with a hot flame. **Sterno** is this material, while **Trioxane,** employed for the same purpose, is an anhydrous formaldehyde trimer, but has the disadvantage of being water soluble.

Butyl alcohol is a colorless liquid used as a solvent for paints and for varnishes and in the manufacture of dyes, plastics, and many chemicals. There are four forms of this alcohol, but the normal or primary butyl alcohol is the most important. **Normal butyl alcohol,** $CH_3(CH_2)_2CH_2 \cdot OH$, has a specific gravity of 0.814 and boiling point of $243°F$ ($117°C$). This form, known as **butanol,** has strong solvent power and is valued where a low evaporation rate is desired, such as in latexes and nitrocellulose lacquers. It is also used for organic synthesis. Several fatty alcohols can be used for production of **Guerbet alcohols,** which have been known since the 1880s, when Marcel Guerbet first synthesized them. They are unsaturated, have high molecular weight, and are liquid at very low temperatures. These properties and their hydrophobic, oily nature suit them for use in synthetic lubricants.

Fluoro alcohols are alcohols in which fluorine is substituted for hydrogen in the nonalcohol branch. They have the general composition $H(CF_2CF_2)_xCH_2OH$, with high specific gravities, 1.48 to 1.66, and high reactivity. As solvents, they dissolve some synthetic resins that resist common solvents. Some of the esters are used as lubricants for temperatures to $500°F$ ($260°C$). **Acetylenic alcohols** are **methyl butynol,** $CH:C \cdot C(CH_3)_2OH$, with a specific gravity of 0.8672, boiling at $219°F$ ($104°C$), and used as a solvent, in metal pickling and plating, and in vitamin manufacture; and **methyl pentynol,** $CH:C \cdot C(CH_3)_2CH_2OH$, boiling at $250°F$ ($121°C$). It is a powerful solvent. It has hypnotic qualities and is also used for tranquilizing fish in transport.

Fatty acid alcohols, made from fatty acids or synthetically, have the general formula $CH_3(CH_2)_xOH$, ranging from the C_8 of **octyl alcohol** to the C_{18} of **stearyl alcohol.** They are easily esterified, oxidized, or ethoxilated and are used for making cosmetics, detergents, emulsifiers, and other chemicals. **Lorol 25,** of Du Pont, is **cetyl alcohol. Elaidyl alcohol,** made from methyl oleate, is an 18-carbon alcohol. It is solid, melting at 75°F (24°C). The fatty acid alcohols vary from water-white liquids to waxy solids. The **Dytols,** of Rohm & Haas, are fatty alcohols, and the **Alfols,** of the Continental Oil Co., are straight-chain primary fatty alcohols made from ethylene and containing even numbers of carbon atoms from 6 to 18. **Polyols** are alcohols containing many hydroxyl, —OH, radicals. They react easily with isocyanates to form urethane.

ALDEHYDE. A group name for substances made by the dehydrogenation or oxidation of alcohols, such as formaldehyde from methyl alcohol. By further oxidation, the aldehydes form corresponding acids, as formic acid. The aldehydes have the radical group —CHO in the molecule, and because of their ease of oxidation they are important reducing agents. They are also used in the manufacture of synthetic resins and many other chemicals. Aldehydes occur in animal tissues and in the odorous parts of plants.

Acetaldehyde is a water-white flammable liquid with an aromatic penetrating odor, used as a reducing agent, preservative, and for silvering mirrors, and in the manufacture of synthetic resins, dyestuffs, and explosives. Also called **ethanal,** it has the composition of $CH_3 \cdot CHO$ and is made by the direct liquid-phase oxidation of ethylene. Less commonly, it is made by oxidation or dehydrogenation of ethanol, acetylene hydration, or partial oxidation of hydrocarbons. Eastman Kodak Co. has a rhodium-catalyzed process that gasifies coal into synthesis gas and then into acetaldehyde. The specific gravity is 0.801, and the boiling point is 69°F (20.8°C). It is soluble in water, alcohol, and naphtha. **Paraldehyde,** $(CH_3 \cdot CHO)_3$, may be used instead of acetaldehyde in resin manufacture; has a higher boiling point, 255°F (124°C), and a higher flash point; but is not as reactive and will not reduce silver solutions to form a mirror. It is used for fulling leather. **Propion aldehyde,** or **propanal,** CH_3CH_2CHO, is made in the same way by oxidation of propyl alcohol. It has a boiling point of 120°F (48.8°C) and has reactions similar to acetaldehyde.

When acetaldehyde is condensed by reaction with a dilute alkali, it forms **acetaldol,** also called **aldol,** a viscous pale-yellow liquid of composition $CH_3 \cdot CH(OH):CH_2CHO$, with a specific gravity of about 1.10, soluble in hot water and in alcohol. It is used to replace formaldehyde for synthetic resins, and for cadmium plating baths and dye baths, and for making butadiene rubber. **Paraldol,** the double molecule of aldol, is a white crystalline material melting at 180°F (82°C). When crude aldol is slightly acidified with acetic acid and heated, it yields **croton aldehyde,** also called **crotonic aldehyde,** and **propylene aldehyde,** $CH_3 \cdot CH:CH \cdot CHO$, with a specific gravity of about 0.855 and a boiling point of 210 to 219°F (99 to 104°C). It is soluble in water, alcohol, and hydrocarbons and is used as a solvent for resins, gums, and rubber and in tanning leather. It has a pungent, suffocating odor and is used in tear gases. Small quantities are sometimes used in city gas mains as a warning agent on the escape of poisonous fuel gas, as even tiny quantities will awaken a sleeping person.

Acrolein is **acrylic aldehyde,** $CH_2:CH \cdot CHO$, a colorless volatile liquid of specific gravity 0.8389, boiling at 127°F (52.7°C). The vapor is irritating to the eyes and nose, and the unpleasant effect of scorching fat is due to the acrolein formed. Acrolein is made by oxidation of propylene with a catalyst. It polymerizes easily and can be copolymerized with ethylene, styrene, epoxies, and other resins to form various types of plastics. Its reactive double bond and carbonyl group make it a useful material for chemical synthesis. It is used as an antimicrobial agent for controlling algae, microbes, mollusks, and aquatic weeds. **Acrolein cyanohydrin,** $CH_2:CHCH(OH)CN$, a water-soluble liquid, boiling at 329°F (165°C), is also used to modify synthetic resins by introducing a nitrile group and a free hydroxyl into the molecular chain. It will copolymerize with ethylene and with acrylonitrile.

ALDER. The wood of several species of tree of the genus *Alnus* of the same family as the birch and beech. The **red alder** is from *A. rubra,* or *A. oregona,* growing in the northwestern United States. The wood is reddish-brown, has a fine even grain, it tough and resilient, can be worked easily, and takes a good polish. It has been much used for furniture, cabinetwork, and interior finish, as it rivals mahogany and walnut in appearance. **Black alder** is from the tree *A. glutinosa,* widely distributed in the northern hemisphere. It is reddish-white and has a smooth, fine grain, with a density of about 35 lb/ft³ (560 kg/m³). It is used for plywood, cabinetwork, and toys. The wood of the alder is also used to produce smoke for curing kippered fish. The smoke is cooled to remove creosote and is filtered. **Formosan alder** is from *A. maritima* of Asia. The wood is light yellow streaked with reddish lines and has a fine texture.

ALKALI. A caustic hydroxide characterized by its ability to neutralize acids and form soluble soaps with fatty acids. Fundamentally, alkalies are inorganic alcohols, with the monovalent hydroxyl group —OH in the molecule, but in the alkalies this group is in combination with a metal or an ammonia group; and alkalies have none of the characteristics of alcohols. All alkalies are basic and have a pH value from 7 to 14. They neutralize acids to form a salt and water. The common alkalies are sodium hydroxide and potassium hydroxide, which are used in making soaps, soluble oils, and cutting compounds, in cleaning solutions, and for etching aluminum. All the alkalies have a brackish taste and a soapy feel; most corrode animal and vegetable tissues.

ALKALI METALS.   A name given to francium, lithium, sodium, potassium, rubidium, strontium, cesium, calcium, and barium because of the basic reaction of their oxides, hydroxides, and carbonates. Carbonates of these metals are called **fixed alkalies.** The metals show a gradation in properties and increase in chemical activity with increase in atomic weight. All are silvery white and very soft. They tarnish rapidly in air and decompose water at ordinary temperatures. In the alkali metals the electron bonding is so weak that even the impact of light rays knocks electrons free. All have remarkable affinity for oxygen. Rubidium and cesium ignite spontaneously in dry oxygen. Calcium, strontium, and barium are also called **earth metals.** Thin films of the alkali metals are transparent to ultraviolet but opaque to visible light.

ALKYD RESINS.   A group of thermosetting synthetic resins known chemically as **hydroxycarboxylic resins,** of which the one produced from phthalic anhydride and glycerol is representative. They are made by the esterification of a polybasic acid with a polyhydric alcohol, and have the characteristics of homogeneity and solubility that make them especially suitable for coatings and finishes, plastic molding compounds, caulking compounds, adhesives, and plasticizers for other resins. The resins have high adhesion to metals; are transparent, easily colored, tough, flexible, heat- and chemical-resistant; and have good dielectric strength. They vary greatly with the raw materials used and with varying percentage compositions, from soft rubbery gums to hard, brittle solids. Phthalic anhydride imparts hardness and stability. Maleic acid makes a higher-melting-point resin. Azelaic acid gives a softer and less brittle resin. The long-chain dibasic acids, such as adipic acid, give resins of great toughness and flexibility. In place of glycerol the glycols yield soft resins, and sometimes the glycerol is modified with a proportion of glycol. The resins are reacted with oils, fatty acids, or other resins, such as urea or melamine, to make them compatible with drying oils and to impart special characteristics.

　　Since alkyd resins are basically esterification products of innumerable polybasic acids and polyhydric alcohols, and can be modified with many types of oils and resins, the actual number of different alkyd resins is unlimited, and the users' specification is normally by service requirements rather than composition. **Short-oil alkyds,** with 30 to 45% nonoxidizing oils, are used in baking enamels, while the **long-oil alkyds,** with 56 to 70% oxidizing oils, are soluble in mineral spirits and are used for brushing enamels.

Alkyd coatings are used for such diverse applications as air-drying water emulsion wall paints and baked enamels for automobiles and appliances. The properties of **oil-modified alkyd coatings** depend upon the specific oil used as well as the percentage of oil in the composition. In general, they are comparatively low-cost and have excellent color retention, durability, and flexibility, but only fair drying speed, chemical resistance, heat resistance, and salt spray resistance. The oil-modified alkyds can be further modified with other resins to produce **resin-modified alkyds.** The resin is sometimes added during manufacture of the alkyd and becomes an integral part of the alkyd, or the modifying resin is blended with the alkyd when the paint is formulated. When mixed with urea formaldehyde or melamine resin, harder and more resistant baked enamels are produced. Alkyds blended with ethyl cellulose are used as tough flexible coatings for electric cable. Other resins blended with alkyds to produce special or improved properties include phenolic, rosin, vinyl, and silicone.

Alkyd plastic molding compounds are composed of a polyester resin and usually a diallyl phthalate monomer plus various inorganic fillers, depending on the desired properties. The raw material is produced in three forms—granular, putty, and glass-fiber-reinforced. As a class, the alkyds have excellent heat resistance up to about 300°F (149°C), high stiffness, and moderate tensile and impact strength. Their low moisture absorption combined with good dielectric strength makes them particularly suitable for electronic and electrical hardware, such as switch gears, insulators, and parts for motor controllers and automotive ignition systems. They are easily molded at low pressures and cure rapidly.

ALLIGATOR LEATHER. A light, tough leather with platelike scales on the surface. It is made from the skins of large saurians, or lizards, of the order *Crocodilia,* abounding in muddy tropical streams. The species *Alligator mississippiensis* inhabits the swamps of the southeastern United States. Alligator leather is valued for luggage, pocketbooks, and shoes. It is much imitated with embossed split sheepskins, but sheepskin is soft and easily scuffed. **Lizard leather,** from the **Java ring lizard,** is another reptile leather valued for women's shoes.

ALLOY. A very old term for the admixture of a precious metal with a metal of lesser value, or with a base metal. This term refers broadly to the combination of at least two atoms or elements. More generally, however, the principal element is a metal, and all are intended for a specific purpose as opposed to the minute quantities of various elements retained from the ore or introduced during the refining process and called **impurities.** In most cases, the two or more elements are metals, **metal alloys** being the most common kind. There are notable exceptions, however, as in the case of steel, which is fundamentally iron alloyed with small amounts of carbon, a nonmetal. In recent years, the term **plastic alloy** also has been applied to plastics.

Metal alloys are more specifically described with reference to the major element by weight, which is also called the **base metal** or **parent metal.** Thus the terms *aluminum alloy, copper alloy,* etc. Elements present in lesser quantities are called **alloying elements.** When one or more alloying elements are present in substantial quantity or, regardless of their amount, have a pronounced effect on the alloy, then they, too, may be reflected in generic designations.

Such designations usually, but not always, list the base metal first. Thus *nickel-copper alloy* refers to an alloy predominantly nickel but also containing copper as a principal alloying element. Alloys so designated may also contain other alloying elements, however. On the other hand, high-carbon steels and nickel steels are largely steel, and beryllium copper is largely copper. Sometimes the base metal and alloying elements are prefixed with numerals indicating their specific amounts by weight or volume in the particular composition.

Metal alloys are also often designated by trade names or by trade association or society designations. Among the more common of the latter are the three-digit designations for the major families of stainless steels and the four-digit ones for aluminum alloys.

Structurally there are two kinds of metal alloys—single-phase and multiphase. **Single-phase alloys** are composed of crystals with the same type of structure. They are formed by "dissolving" together different elements to produce a **solid solution.** The crystalline structure of a solid solution is normally that of the base metal. The atoms of the alloying element (solute) join the base metal (solvent) either as substitution atoms or as interstitial atoms. In the former, alloying atoms occupy some of the lattice sites normally occupied by the host atoms. In the latter, alloying atoms place themselves in between the host atoms. In substitutional solid solutions, the solute and solvent atoms are of approximately similar size; whereas in the interstitial type, the solute atoms must be small enough to fit between the atoms. The nature of solid solutions has an important effect on many alloy properties. For example, strength and hardness increase with the amount of solute present, but ductility usually decreases. Electrical conductivity also is generally lowered by the presence of the solute element.

In contrast to single-phase alloys, **multiphase alloys** are mixtures rather than solid solutions. They are composed of aggregates of two or more different phases. The individual phases making up the alloy are different from one another in their composition or structure. **Solder,** in which the metals lead and tin are present as a mechanical mixture of two separate phases, is an example of the simplest kind of multiphase alloy. In contrast, steel is a complex alloy composed of different phases, some of which are solid solutions. Multiphase alloys far outnumber single-phase alloys in the industrial materials field, chiefly because they provide greater property flexibility. Thus, properties of multiphase alloys are dependent upon many factors, including the composition of the individual phases, the relative amounts of the different phases, and the positions of the various phases relative to one another.

When two different thermoplastic resins are blended, a **plastic alloy** is obtained. Alloying permits resin polymers to be blended that cannot be polymerized. Not all plastics are amenable to alloying. Only resins that are compatible with each other—those that have similar melt traits—can be successfully blended.

ALLOY STEELS. In general, the term applies to all steels exceeding the limits of manganese, silicon, and copper of carbon steels or which contain other alloying ingredients. Alloy steels often take the name of the alloying element or elements having the greatest influence on their performance characteristics or the name of a key characteristic, processing mechanism, or application. Thus, the prevalence of such terms as nickel steels; stainless, or corrosion-resistant, steels; maraging steels; precipitation-hardening steels; tool steels; valve steels, etc. Usually, however, the term excludes high-alloy steels and refers instead to the standard **alloy steels** of the American Iron and Steel Institute (AISI) and SAE International, which contain low to moderate amounts of alloying elements, usually less than 5% total. The AISI or SAE designations of these steels are usually noted by four numerals—13XX to 91XX. The first two numerals pertain to the specific alloying element or elements, and the last two numerals indicate carbon content in hundredths of 1%. Sometimes three numerals are used to denote carbon content, and a letter, such as B for boron and L for lead, follows the first two numerals to indicate an alloying element not indicative of the first two numerals. A letter prefix is used occasionally to designate special furnace practice used to make the steel, and the suffix H is used to designate steels made to specific hardenability requirements. A three-numeral system, 9XX, is commonly used to designate **high-strength, low-alloy (HSLA) steels,** some of which are also called **microalloyed steels** because of the small amount of alloying elements, with the last two numerals indicating minimum tensile yield strength in 1,000 lb/in^2 (6.895 MPa). Although most alloy steels are heat-treated by users and extremely high levels of strength and toughness can be achieved, **HSLA steels** are typically supplied to specific strength levels and are not heat-treated by users.

ALLYL PLASTICS. A group of water-white casting plastics, related to the alkyds, produced by the polymerization of the ester of allyl alcohol or from allyl chloride, both produced from propylene. **Allyl alcohol,** CH_2:$CH \cdot CH_2OH$, is a colorless liquid also known as **propanol,** which can be made by heating glycerol with formic or oxalic acid. The specific gravity is 0.849, boiling point 205°F (96°C), and freezing point -200°F (-129°C). **Allyl chloride,** CH_2:$CH \cdot CH_2Cl$, is a liquid of specific gravity 0.937, boiling point 113°F (45°C), and flash point -25°F (-30°C). **Allyl ester** is a clear, syrupy liquid of specific gravity 1.26 which polymerizes with a peroxide catalyst to form allyl plastics. The liquid monomer can be poured into molds and hardened by polymerization. The polymerized castings are hard and crystal-clear. Allyl plastics have a specific gravity of 1.34 to 1.40, dielectric strength of 1,275 V/mil (50.2 $\times$ 10^6 V/m), refractive index of 1.57, Rockwell M hardness of 116, and compressive strength of 19,600 lb/in^2 (134 MPa). As the plastic has high clarity and less light dispersion than most optical glass, it is used for lenses, prisms, and reflectors. It can be colored easily with dyes and is also used for mechanical and electrical parts. **CR-39,** of PPG Industries, is an optically clear, hard, thermosetting casting resin for clock and instrument faces, windows, and lenses. It is made from diallyl diglycol carbonate and triallyl cyanurate. The sheet material transmits 92% of ordinary light, has a heat deflection temperature of 266°F (130°C), a tensile strength of 5,500 lb/in^2 (38 MPa), a compressive strength of 22,800 lb/in^2 (157 MPa), and a Rockwell hardness of M100. It can be machined with carbide tools. It is craze-resistant and has only half the weight of glass. **Diallyl phenyl phosphate** is a monomer which polymerizes with a catalyst to form a transparent, hard, strong, flame-resistant thermosetting resin. It can also be copolymerized with the thermoplastic vinyl acetate or methyl methacrylate to produce thermosetting resins. **Diallyl phthalate** is a thermosetting resin cured by polymerization without water formation. The molded material, depending on the filler, has a tensile strength from 4,500 to 7,000 lb/in^2 (31 to 48 MPa), a compressive strength up to 30,000 lb/in^2 (207 MPa), a Rockwell hardness to M108, dielectric strength to 430 V/mil (16.9×10^6 V/m), and heat resistance to 450°F (232°C).

Methallyl alcohol, $CH_2{:}CH \cdot CH_3 \cdot CH_2OH$, also forms esters which can be used for the production of plastics. It is a liquid boiling at 238°F (114.5°C). The allyl radical will combine with starch or sugar to form shellaclike resins. **Allyl sucrose,** made by combining sugar and allyl chloride, is a resin that produces varnishes which will withstand temperatures to 400°F (204°C) and are chemical-resistant. **Allyl starch** is resinous material made by treating sweet-potato or grain starches with allyl chloride or allyl bromide. It is soluble in varnish solvents and is a substitute for shellac as a varnishing agent, forming an adherent, resistant film on paper, fabric, wood, or metal. It will withstand temperatures to 400°F (204°C).

ALMOND OIL. An essential oil distilled from the ground, macerated kernels of bitter almond, *Prunus amygdalus* or *communis,* of the Mediterranean countries, and from the kernels of the **apricot,** *P. armeniaca.* The two oils are identical and contain the glucoside **amygdalin.** The U.S. production is mostly from the by-product pits of the apricot canning industry of California. The almond is a small tree closely resembling the peach. The fruit is inedible, but the seeds inside the pits are marketed as roasted and salted nuts and are made into a paste for confections. Oil of bitter almonds is used in perfumery and as a flavor. For flavoring use, the poisonous hydrocyanic acid is extracted. Synthetic almond oil is **benzaldehyde,** C_6H_5CHO, a colorless volatile oil with an almond flavor, produced from benzol or from toluol and used for producing **triphenylmethane dyes** and many chemicals.

ALPACA. A fabric made from the fine, woollike hair of the alpaca, an animal of the llama family in the mountains of Bolivia, Peru, Chile, and Argentina. **Alpaca fiber** is long and fine, with a downy feel, but it does not have the strength or elasticity of fine wool and is more closely allied to hair than to wool. There are two breeds of alpaca animal, Huacaya and Suri, the latter having the longer and finer wool, reaching a fiber length to 30 in (0.762 m). From 3 to 10 lb (1.4 to 4.5 kg) of fiber is obtained per animal. **Llama hair** (pronounced lyahmah) from the llama of Bolivia is marketed as coarse alpaca. In Incan times, llama wool was used by the common people for clothing, while the finer wool from the alpaca and vicuña was reserved for the upper classes. The llama is sheared every 2 years when the wool reaches a length of about 12 in (0.30 m). A considerable amount of stiff guard hairs occur in the fiber.

Vicuña, another animal of the llama family, is almost extinct, and the commercial vicuña cloth is made of alpaca or fine wool, or mixtures. There is a limited production of true vicuña from domestic herds in Bolivia and Peru, raised at an altitude of about 14,000 ft (4,267 m). It is the softest of all animal weaving fibers. Alpaca and vicuña cloths are used for shawls, jackets, and fine goods. Imitation **alpaca fabric** for clothing linings is a lustrous, smooth, and wiry fabric plain-woven with a cotton warp and a worsted mohair filling. When made with a rayon filling, it is called **rayon alpaca.**

ALUM. A colorless to white crystalline **potassium aluminum sulfate,** $KAl(SO_4)_2 \cdot 12H_2O$, or $[KAl(H_2O)_6]SO_4 \cdot 6H_2O$, occurring naturally as the mineral **kalunite,** or **kalinite,** and in combination as the mineral **alunite.** It is also called **potash alum** to distinguish it from other forms. It has a sweetish taste and is very astringent. It is used as an additive in the leather and textile industries, in sizing paper, as a mordant in dyeing, in medicines as an astringent, and in baking powder. It is made commercially by reacting bauxite with sulfuric acid and then potassium sulfate. It is an important water-purifying agent. From a water solution it crystallizes out, forming positively charged particles which attract the negatively charged organic impurities, thus purifying the water as they settle out. Alum has a specific gravity of 1.757, melts in its water of crystallization at 198°F (92°C), and when heated to redness, is converted to **burnt alum,** a porous, friable material which dissolves slowly in water. Alum is marketed as USP, lump, pea, nut, ground, and powdered. The **rice crystal alum** of General Chemical Corp. is 10 to 30 mesh, and the granular is 30 to 60 mesh.

Alumstone is a gray or pinkish, massive form of alunite found in volcanic rocks. A pure variety from Italy is called **Roman alum,** or **roche alum.** The alunite of Australia is used to produce **potassium sulfate,** K_2SO_4, for fertilizer; and the residue, containing 50% alumina, is used for aluminum production. **Soda alum,** in which the potassium is replaced by sodium, occurs in the South American Andes as the mineral **mendozite.** It is more soluble than alum, but is more difficult to purify. The largest use is as a leavening agent in bakeries. **Filter alum,** also called **patent alum** and **aluminous cake,** used for waterworks filtration, is **aluminum sulfate,** $Al_2(SO_4)_3$, plus a varying amount of water of crystallization. The anhydrous form is used as a dehydrating agent for gases. It is a white, crystalline solid readily soluble in water. When filter alum contains a slight excess of alumina, it is called *basic.* The **hexahydrate aluminum sulfate** of Allied-Signal Inc. has composition $Al_2(SO_4)_3 \cdot 6H_2O$ and is used for chemical processing.

Commercial aluminum sulfate is also called **concentrated alum,** and it replaces potash alum for many uses because of its cheapness. It comes in colorless crystals having a strong astringent taste and is used as a mordant in dyeing, in water purification, sizing papers, tanning, printing inks, and dry colors. For use in pickling and tanning leathers, it contains not more than 0.01% iron oxide. **Ammonia alum,** used in tanning sheepskins and fur skins, and in fireproofing and dyeing textiles, is **ammonium-aluminum sulfate,** $NH_4Al(SO_4)_2 \cdot 12H_2O$, a colorless or white crystalline powder soluble in water and, like other alums, insoluble in alcohol. It is valued for water purification because it forms chloramine, and is also used in vegetable glues and porcelain cements. For uses where an alkaline instead of an acid solution is required, **sodium phosphoaluminate** is employed. It is a white, water-soluble powder and is a double salt containing about 70% sodium aluminate and 20% sodium orthophosphate, with the balance water. **Alkophos** of Monsanto Co. is **aluminum phosphate,** $AlPO_4$, a white powder used as a bonding agent for high-temperature cements and as an alkaline flux for ceramics.

ALUMINA. The oxide of aluminum, Al_2O_3. The natural crystalline mineral is called **corundum,** but the synthetic crystals used for abrasives are designated usually as **aluminum oxide** or marketed under trade names. For other uses and as a powder, it is generally called *alumina*. It is widely distributed in nature in combination with silica and other minerals and is an important constituent of the clays for making porcelain, bricks, pottery, and refractories. The **alumina clay** from the large deposits in western Idaho contains an average of 28% Al_2O_3, 5.6% iron oxide, and a high percentage of titanium oxide. Such clays are used for ceramics, but the oxide, alumina, is obtained commercially chiefly by high-temperature fusing of bauxite. It is also produced from alunite with a by-product fertilizer and is obtained from the oil shales of Colorado.

The crushed and graded crystals of alumina, when pure, are nearly colorless, but the fine powder is white. Off colors are due to impurities. The U.S. aluminum oxide used for abrasives is at least 99.5% pure, in nearly colorless crystals melting at 3722°F (2050°C). The chief uses for alumina are for the production of aluminum metal and for abrasives, but it is also used for ceramics, refractories, pigments, and catalyst carriers, and in chemicals.

Aluminum oxide crystals are normally hexagonal and are minute. For abrasives, the grain sizes are usually from 100 to 600 mesh. The larger grain sizes are made up of many crystals, unlike the single-crystal large grains of silicon carbide. The specific gravity is about 3.95, and the hardness is up to Knoop 2,000. The ultrafine alumina abrasive powder of Union Carbide Corp. is of two kinds. Type A is

alpha alumina with hexagonal crystals of particle size 11.8 μin (0.3 μm), specific gravity 4.0, and Mohs hardness 9. Type B is **gamma alumina** with cubic crystals of particle size less than 3.94 μin (0.1 μm), specific gravity of 3.6, and hardness 8. Type A cuts faster, but Type B gives a finer finish. At high temperatures, gamma alumina transforms to the alpha crystal. The aluminum oxide most frequently used for refractories is **beta alumina** in hexagonal crystals heat-stabilized with sodium. It has composition $Na_2O \cdot 11Al_2O_3$, but stabilized alumina is also produced with oxides of calcium, potassium, or magnesium. The alpha and gamma alumina powders of J. T. Baker Chemical Co. for lapping, grinding, and are in dentifrices are of 99.96% purity in controlled particle sizes from 0.8 to 1.6 μin (0.02 to 0.4 μm).

The aluminum oxide abrasives in all forms are sold under trade names. **Alundum,** of Norton Co., **Aloxite,** of Carborundum Co., and **Lionite,** of General Abrasives, are aluminum oxides. **Alundum 32,** of Norton Co., is aluminum oxide with each abrasive grain in a single crystal and, unlike the crushed and ground oxide, having the natural edges of the crystal as the cutting edges. It is about 96% Al_2O_3 and is brown. The white **Alundum 38,** of 99.5% purity, is crushed from large crystals. Another aluminum oxide has high-purity crystals agglomerated into grain sizes from 16 to 60 mesh. The 16 grit is used for metal cleaning, the 40 to 60 grit for granite cutting, and a finer powder, from 100 to 240 mesh, is used for glass and metal polishing.

Activated alumina is partly dehydrated alumina trihydrate, which has a strong affinity for moisture or gases and is used for dehydrating organic solvents. **Hydrated alumina** is **alumina trihydrate,** $Al_2O_3 \cdot 3H_2O$, used as a catalyst carrier. At elevated temperatures it becomes active, and the fine powder is an auxiliary catalyst. It is also used as a filler in plastics and in cosmetics, and it comes in various particle sizes for use in glass and vitreous enamels to increase the strength and luster. **Hydrated alumina C-741** is this material with particles coated with stearic acid for use as a reinforcing pigment in rubber.

Activated alumina F-1, of Aluminum Co. of America, is a porous form of alumina, Al_2O_3, used for drying gases or liquids. It will remove moisture up to 15% of the dry weight of the alumina. **Activated alumina F-6** is this material impregnated with cobaltous chloride which will change color from blue through pale pink to white with progression of the adsorption of moisture. Activated alumina is also used as a catalyst for many chemical processes.

Aluminum hydroxide is a white, bulky, water-insoluble, fine powder of specific gravity 2.42 and refractive index 1.535, used as a base for lake pigments, for making gloss white, as a water repellent in tex-

tile and paper coatings, and as an antacid in medicine. **Boehmite,** of
Du Pont, used in cosmetics, is called **colloidal alumina,** but it is an
aluminum monohydrate, AlOOH, made by polymerizing aluminum
cations in a water solution. **Baymal** of this company, used for coatings, adhesives, paints and for making dense ceramics, is similar. The
powder is in the form of tiny fibrils, 197 nin (5 nm) in diameter and
5,906 nin (150 nm) long. In coatings the fibrils interlock the molecules
into a tight adherent film. In paints the powder acts as a **thixotropic
material** to prevent sagging or running on vertical walls. The powder
is both a **hydrophilic material,** that is, water-soluble, and an
organophilic material soluble in oils and organic solvents. An aluminum monohydrate has been used as a filler in coatings and will
give a hydrophilic coating on normally hydrophobic surfaces. It is in
crystals 0.2 μin (0.005 μm) in size, agglomerated into particles of
about 1,732 μin (44 μm). It has very high bulking properties and is
used also in cosmetics.

Alumina ceramics are the most widely used oxide-type ceramic,
chiefly because alumina is plentiful, relatively low-cost, and equal to
or better than most oxides in mechanical properties. Density can be
varied over a wide range, as can purity—down to about 90% alumina—to meet specific application requirements. Alumina ceramics are
the hardest, strongest, and stiffest of the oxides. They are also outstanding in electrical resistivity and dielectric strength, are resistant
to a wide variety of chemicals, and are unaffected by air, water vapor,
and sulfurous atmospheres. However, with a melting point of only
3700°F (2037°C), they are relatively low in refractoriness, and at
2500°F (1371°C), retain only about 10% of room-temperature
strength. In addition to alumina's wide use as electrical insulators
and in chemical and aerospace applications, its high hardness and
close dimensional tolerance capability make this ceramic suitable for
such abrasion-resistant parts as textile guides, pump plungers, chute
linings, discharge orifices, dies, and bearings. **Calcined alumina** is
made by calcining aluminum hydroxide, and the commercial product
is 99.1% Al_2O_3, with no more than 0.5% adsorbed water and not over
0.4% silicon, iron, sodium, and titanium oxides. The powder, in particle size as small as 394 μin (10 μm), is used for abrasives, glassmaking, and refractory ceramics. The molded material has little porosity
or shrinkage and retains its strength and electrical resistance at high
temperatures. **Alumina A-14,** of Aluminum Co. of America, is a calcined alumina for electrical insulators with a particle size of 79 to 118
μin (2 to 3 μm) and containing not more than 0.08% soda. For electronic and nuclear ceramic parts, a 99.95% pure alumina is produced
by this company. It has less than 0.01% sodium oxide. The alumina
powder of Reynolds Metals Co. has an average particle size down to

39.4 μin (1 μm) and has low soda content. It is made by calcining the powder with silica sand at high temperature. Sintered parts have a service use to 3500°F (1927°C) with tensile strength to 60,000 lb/in^2 (414 MPa). **Alumina Al-200,** of Coors Porcelain Co., for high-frequency insulators, gives a molded product with a tensile strength of 25,000 lb/in^2 (172 MPa), compressive strength of 290,000 lb/in^2 (2,000 MPa), and specific gravity of 3.36. The coefficient of thermal expansion is half that of steel, and the hardness about that of sapphire. **Alumina AD-995,** of this company, is a dense vacuumtight ceramic for high-temperature electronic use. It is 99.5% alumina with no silica. The hardness is Rockwell N80, and the dielectric constant is 9.27. The maximum working temperature is 3200°F (1760°C), and at 2000°F (1093°C) it has a flexural strength of 29,000 lb/in^2 (200 MPa). **Cotronics 902,** of Cotronics Corp., is a high-purity, readily machinable alumina-silicate in bar and plate forms for service temperatures up to 2,100°F (1,149°C).

Koralox, a product of General Electric, is an alumina ceramic material formed by a selective densification process. It is used in the casting of hollow jet engine cores. These cores are then incorporated in molds into which eutectic superalloys are poured to form the turbine blades.

Tabular alumina is alumina converted to the corundum form by calcining at temperatures below the fusing point, and the tabular crystals are larger than those of calcined alumina. It comes as a fine powder or as coarse granules for making refractory ceramics and electrical insulators. **Alumina balls** are marketed in sizes from 0.25 to 0.75 in (0.6 to 1.9 cm) for reactor and catalytic beds. They are usually 99% alumina, having high resistance to heat and chemicals. **Alumina fibers** in the form of short, linear crystals, called **sapphire whiskers,** have high strength up to 200,000 lb/in^2 (1,375 MPa) for use as a filler in plastics to increase heat resistance and dielectric properties. Continuous single-crystal **sapphires (alumina filaments)** have unusual physical properties; high tensile strength [over 300,000 lb/in^2 (2,069 MPa)] and modulus of elasticity of 65 to 70 $\times$ 10^6 lb/in^2 (448 $\times$ 10^3 to 483 $\times$ 10^3 MPa). The filaments are especially needed for use in metal composites at elevated temperatures and in highly corrosive environments. An unusual method for producing single-crystal fibers in lieu of a crystal-growing machine is the floating-zone fiber-drawing process. The fibers are produced directly from a molten ceramic without using a crucible.

FP, a **polycrystalline alumina** (Al_2O_3) fiber, has been developed. The material has greater than 99% purity and a melting point of 3713°F (2045°C), which makes it attractive for use with high-temperature **metal matrix composite** (MMC) processing techniques.

Thanks to a mechanism, currently not explainable by the developer of **FP fibers** (Du Pont), a silica coating results in an increase in tensile strength of the filaments to 275,000 lb/in^2 (1,896 MPa) even though the coating is approximately 10 μin (0.25 μm) thick and the modulus does not change. Fiber FP has been demonstrated as a reinforcement in magnesium, aluminum, lead, copper, and zinc, with emphasis to date on aluminum and magnesium materials. **Alumina particulates** are also used to reinforce aluminum-matrix composites.

Fumed alumina powder of submicrometer size is made by flame reduction of aluminum chloride. It is used in coatings and for plastics reinforcement and in the production of ferrite ceramic magnets.

Alfrax is the trade name of Carborundum Co. for alumina as a catalyst carrier and as a refractory. **Stupalox,** of the same company, for ring and plug gages, is alumina hot-pressed to a specific gravity of 3.95. It is tough and wear-resistant, with a transverse rupture strength of 100,000 lb/in^2 (690 MPa) and compressive strength of 450,000 lb/in^2 (3,100 MPa). Another vitreous ceramic, in tube form and composed of 96% alumina, has been used for protecting thermocouples. It is gastight at 2642°F (1450°C). **Lucalox,** of General Electric Co., is a translucent ceramic pressed from high-purity fine alumina powder and fired at high temperature. It transmits 90% of visible light rays and will withstand temperatures to 3600°F (1982°C). It is polycrystalline with pores removed in the firing. The transverse strength is 50,000 lb/in^2 (345 MPa), and dielectric strength is 1,700 V/mil (67 $\times$ 10^6 V/m). It is used for high-intensity lamps, missile nose cones, and instrument parts. **Alite,** of U.S. Stoneware Corp., used for bearings, valves, nozzles, and extrusion dies, is a white ceramic made by molding or extruding alpha alumina and sintering at 3400°F (1871°C). The bonding is by crystal growth, producing nearly a single crystal. The tensile strength is up to 32,500 lb/in^2 (224 MPa), compressive strength up to 290,000 lb/in^2 (2,000 MPa), and dielectric strength 250 V/mil (9.8 $\times$ 10^6 V/m). It is used for temperatures to 1600°F (871°C). **Aluminum oxide film,** or **alumina film,** used as a supporting material in ionizing tubes, is a strong, transparent sheet made by oxidizing aluminum foil, rubbing off the oxide on one side, and dissolving the foil in an acid solution to leave the oxide film from the other side. It is transparent to electrons. **Alumina bubble brick** is a lightweight refractory brick for kiln lining, made by passing molten alumina in front of an air jet, producing small, hollow bubbles which are then pressed into bricks and shapes. **Alumina foam** bricks and blocks for high-temperature insulation are made by PPG Industries. The foam has a density of 28 lb/ft^3 (448.5 kg/m^3) and porosity of 85%.

ALUMINIDES. Emerging alloys of the intermetallics—**nickel aluminide** (Ni_3Al) and **titanium aluminide** (TiAl and Ti_3Al)—they are envisioned as the next generation of superalloys, or high-temperature structural alloys, because of their high strength and oxidation resistance at elevated temperatures and lower density than traditional superalloys and titanium alloys. The compounds themselves are extremely brittle at room temperature, but ductility can be improved by alloying and controlled thermomechanical processing. Columbium, for example, has proved effective in this regard for Ti_3Al, and boron, in combination with other ingredients and controlled thermomechanical processing, for Ni_3Al. Ni_3Al alloys, the more advanced, have the unique characteristic of providing greater strength at elevated temperatures than at room temperature up to a certain point. One group, for example, containing 18 to 20% aluminum, 10 to 12 iron, 3 to 7 chromium, and 0.22 to 0.25 zirconium or hafnium, has tensile yield strengths of 62,000 to 80,000 lb/in^2 (427 to 552 MPa) and tensile elongations of about 40% at room temperature, and 87,000 to 109,000 lb/in^2 (600 to 752 MPa) and 10 to 35%, respectively, in the 1112 to 1472°F (600 to 800°C) range, according to Oak Ridge National Laboratory, which has been active in developing the aluminides.

A **gamma titanium aluminide,** developed at General Electric Research and Development, contains by weight 49% titanium, 47 aluminum, 2 chromium, and 2 columbium. Weighing about half as much as nickel superalloys, it is just as strong at temperatures up to 1400°F (760°C) and about 50% more rigid than titanium alloys. Blades for a low-pressure aircraft turbine wheel, successfully tested, were 43% more lightweight than those made of a nickel superalloy and demonstrated that the aluminide's much lower ductility could be accommodated.

Sermaloy J, of Sermatech International, is a silicon-enriched aluminide coating diffused into nickel and cobalt superalloys for protection against hot corrosion (sulfidation), oxidation, and particulate erosion at temperatures up to 2200°F (1200°C). The coating comprises aluminum, silicon, chromium, cobalt, and nickel. **Platinum-modified aluminide coatings,** assessed by the Naval Surface Warfare Center, improve the hot-corrosion resistance of Rene 80 nickel superalloy. A continuous overlay of $PtAl_2$ gave the best resistance to 500-h exposure at 1650°F (900°C).

ALUMINIZED STEEL. Standard Type 1 aluminized steel consists of a hot-dipped coating of 92% aluminum and 8% silicon on cold-rolled sheet steel for oxidation and corrosion resistance. It is available with coating weights (total both sides) of 0.40 oz/ft^2 (120 g/m^2) or 0.25 oz/ft^2 (75 g/m^2), provides good oxidation resistance in continuous service to

1250°F (675°C), and is used for auto exhaust-system components and other applications. The thicker coating provides the better environmental resistance, but the thinner one is preferred for parts requiring severe forming because of its better adhesion. Type 2 aluminized steel has a coating weight of 0.60 oz/ft^2 (182 g/m^2) and thus provides still better environmental resistance, but is limited to parts requiring still less severe forming.

Galvalume, a cold-rolled steel having a thin hot-dipped coating of 55% aluminum, 43.4 zinc, and 1.6 silicon on both sides, was developed by Bethlehem Steel as an alternative to Type 2 aluminized steel. It is substantially stronger and correspondingly less ductile than the aluminized steel, scale-resistant to 1250°F (675°C), and superior to conventional galvanized steel in corrosion resistance in rural, industrial, and marine atmospheres. Applications include roofing for preengineered buildings, auto exhaust-system parts, agricultural equipment, and appliance parts.

The advantage of two other aluminized steels stem from modifications to the base steel, the coating being the standard hot-dipped aluminum-silicon alloy. Both were developed as low-cost alternatives to 409 stainless steel for auto exhaust systems. The base steel for Inland Steel's **Aluma-Ti** contains a bit of titanium, stabilizing the carbon and nitrogen contents of the steel and thus increasing the spalling resistance of the coating above 1250°F (675°C). Armco's **Alumi-Therm** uses the company's columbium-bearing interstitial-free steel as the base steel, thus providing better formability than standard aluminized steels.

ALUMINUM. Called **aluminium** in England. A white metal with a bluish tinge, symbol Al, atomic weight 26.97, obtained chiefly from bauxite. It is the most widely distributed of the elements next to oxygen and silicon, occurring in all common clays. Aluminum metal is produced by first extracting alumina (aluminum oxide) from the bauxite by a chemical process. The alumina is then dissolved in a molten electrolyte, and an electric current is passed through it, causing the metallic aluminum to be deposited on the cathode. The metal was discovered in 1727, but was obtained only in small amounts until it was reduced electrolytically in 1885.

Pure (99.99%) aluminum has a specific gravity of 2.70 or a density of 0.097 lb/in^3 (2,685 kg/m^3), a melting point of 1220°F (660°C), electrical and thermal conductivities about two-thirds that of copper, and a tensile modulus of elasticity of 9×10^6 lb/in^2 (62,000 MPa). The metal is nonmagnetic and highly reflective and has a face-centered-cubic crystal structure. Soft and ductile in the annealed condition, it is readily cold-worked to moderate strength. It resists corrosion in many environments due to the presence of a thin aluminum oxide film.

Iron, silicon, and copper are the principal impurities in commercially pure aluminum, wrought products containing at least 99% aluminum and a foil product containing at least 99.99%. Such unalloyed aluminum is available in a wide variety of mill forms and constitutes the **aluminum 1XXX** series in the designation system for wrought aluminum and aluminum alloys. Annealed sheet is quite ductile—35 to 45% tensile elongation—but weak, having a tensile yield strength of 4,000 to 5,000 lb/in^2 (28 to 35 MPa). Cold reduction of 75% increases yield strength to 18,000 to 24,000 lb/in^2 (124 to 166 MPa). Thus, the unalloyed metal is used far more for its electrical, thermal, corrosion-resistant, and cosmetic characteristics than for its mechanical properties. Applications include electrical and thermal conductors, capacitor electrodes, heat exchangers, chemical equipment, packaging foil, heat and light reflectors, and decorative trim.

Aluminum flake enhances the reflectance and durability of paints. **Aluminum powder** is used for powder-metal parts. **Aluminum powder** and **aluminum paste** are used in catalysts, soaps, explosives, fuels, and thermite welding. **Aluminum shot** is used to deoxidize steel, and **aluminum foam,** made by foaming the metal with zirconium hydride or other hydrides, is an effective core material for lightweight structures. Anodize conversion coatings are often applied to aluminum and aluminum alloys for corrosion protection.

ALUMINUM ALLOYS. Alloying aluminum with various elements markedly improves mechanical properties, strength primarily, at only a slight sacrifice in density, thus increasing specific strength, or strength-to-weight ratio. Traditionally, wrought alloys have been produced by thermomechanically processing cast ingot into mill products such as billet, bar, plate, sheet extrusions, and wire. For some alloys, however, such mill products are now made by similarly processing "ingot" consolidated from powder. Such alloys are called **PM (powder metal) wrought alloys** or simply **PM alloys.** To distinguish the traditional type from these, they are now sometimes referred to as **ingot-metallurgy (IM) alloys** or **ingot-cast alloys.** Another class of PM alloys is those used to make PM parts by pressing and sintering the powder to near-net shape. There are also many cast alloys. All told, there are about 100 commercial aluminum alloys.

There are two principal kinds of wrought alloys: (1) heat-treatable alloys—those strengthened primarily by solution heat treatment or solution heat treatment and artificially aging (precipitation hardening) and (2) non-heat-treatable alloys—those which depend primarily on cold work for strengthening. Alloy designations are a continuation of the four-digit system noted for aluminum, followed with a letter to designate the temper or condition of the alloy: F (as-fabricated condition), O (annealed), H (strain-hardened), W (solution heat-treated and unstable; that is, the alloy is prone to natural aging in air at room temperature), and T (heat-treated to a stable condition). Numerals following T and H designations further distinguish between tempers or conditions. T3, for example, refers to alloys that have been solution heat-treated, cold-worked, and naturally aged to a substantially stable condition. T6 denotes alloys that have been solution heat-treated and artificially aged. H designations are followed by two or three digits. The first (1, 2, or 3) indicates a specific sequence of operations applied. The second (1 to 8) refers to the degree of strain hardening (the higher the number, the greater the amount of strain, or hardening, 8 corresponding to the amount induced by a cold reduction of about 75%). The third (1 to 9) further distinguishes between mill treatments.

The **aluminum alloy 2XXX series** is characterized by copper (2.3 to 6.3%) as the principal alloying element. Most of these alloys also contain lesser amounts of magnesium and manganese, and some may contain small amounts of other ingredients, such as iron, nickel, titanium, vanadium, zinc, and zirconium. 2XXX alloys are strengthened mainly by solution heat treatment, sometimes by solution heat treatment and artificial aging. Among the more common, especially for structural aircraft applications, are **aluminum alloys 2014, 2024,** and **2219,** which can be heat-treated to tensile yield strengths in the range of 40,000 to 60,000 lb/in^2 (276 to 414 MPa). A recent alloy for auto body panels is **aluminum alloy 2036,** which in the T4 temper provides a tensile yield strength of about 28,000 lb/in^2 (193 MPa). Aluminum alloy 2XXX series is not as corrosion resistant as other aluminum alloys and thus is often clad with a thin layer of essentially pure aluminum or a more corrosion-resistant aluminum alloy, especially for aircraft applications. **Aluminum alloy 2011,** which nominally contains 0.4% bismuth and 0.4 lead, is the most easily machinable wrought aluminum alloy and is widely used for screw-machine products. **Toolrite 2011,** of Alcoa, is cold-finished by proprietary processing to increase tool life in screw-machine operations.

Manganese (0.5 to 1.2%) is the distinguishing alloying element in the **aluminum alloy 3XXX series. Aluminum alloy 3003** also contains 0.12% copper. **Aluminum alloys 3004** and **3105** also contain 1 and 0.5% magnesium, respectively. Strengthened by strain hardening, 3XXX alloys provide maximum tensile yield strengths in the range of 27,000 to 36,000 lb/in^2 (186 to 248 MPa) and are used for chemical equipment, storage tanks, cooking utensils, furniture, builders' hardware, and residential siding.

The **aluminum alloy 4XXX series** is characterized by the addition of silicon: about 12%, for example, in **aluminum alloy 4032** and 5% in **aluminum alloy 4043.** Aluminum alloy 4032, which also contains 1% magnesium and almost as much copper and nickel, is heat-treatable, providing a tensile yield strength of about 46,000 lb/in^2 (315 MPa) in the T6 temper. This, combined with its high wear resistance and low thermal expansivity, has made it popular for forged engine pistons. **Deltalloy 4032,** of Alcoa, is cold-finished 4032 alloy screw-machine stock. Aluminum alloy 4043, which is alloyed only with silicon, is a strain-hardenable alloy used for welding rod and wire. Some aluminum-silicon alloys are also used for brazing, others for architectural applications. Their appeal for architectural use stems from the dark gray color they develop in anodizing.

The principal alloying element in the **aluminum alloy 5XXX series** is magnesium, which may range from about 1 to 5% and is often combined with lesser amounts of manganese and/or chromium. Like 3XXX alloys, the 5XXX are hardenable only by strain hardening, and all are available in a wide variety of H tempers. Tensile yield strengths range from less than 10,000 lb/in^2 (69 MPa) in the annealed condition to more than 40,000 lb/in^2 (276 MPa) in highly strained conditions. **Aluminum alloys 5083, 5154, 5454,** and **5456** are widely used for welded structures, pressure vessels, and storage tanks; others for more general applications, such as appliances, cooking utensils, builders' hardware, residential siding, auto panels and trim, cable sheathing, and hydraulic tubing.

The **aluminum alloy 6XXX series** is characterized by modest additions (0.4 to 1.4%) of silicon and magnesium and can be strengthened by heat treatment. Except for auto sheet **aluminum alloys 6009** and **6010,** which are typically supplied and used in the T4 temper, the alloys are strengthened by solution heat treatment and artificial aging. As a class, these alloys are intermediate in strength to aluminum alloy 2XXX and aluminum alloy 7XXX but provide good overall fabricability. The auto sheet alloys are relatively new. More traditional among the dozen or so alloys of this kind are **aluminum alloy 6061,** which is used for truck, marine, and railroad-car structures, pipelines, and furniture, and **aluminum alloy 6063** for furniture, railings, and architectural applications. Other applications include complex forging, high-strength conductors, and screw-machine products. **Aluminum alloy 6013-T8,** in cold-finished bar and rod from Alcoa, has typical tensile yield strengths of 58,000 to 62,000 lb/in^2 (400 to 427 MPa) and 11 to 12% elongation. Compared with 6061-T6, it is stronger, more fatigue-resistant, and more machinable, though comparable in weldability, corrosion resistance, and anodizing response.

Zinc is the major alloying element in the **aluminum alloy 7XXX series,** and it is usually combined with magnesium for strengthening by heat treatment. An exception is **aluminum alloy 7072,** which is alloyed only with 1% zinc, is hardenable by strain hardening, and is used for fin stock or as a clad for other aluminum alloys. The other alloys, such as **aluminum alloys 7005, 7049, 7050, 7072, 7075, 7175, 7178,** and **7475,** contain 4.5 to 7.6% zinc, 1.4 to 2.7% magnesium, and, in some cases, may also include copper, manganese, silicon, titanium, or zirconium. These heat-treatable alloys are the strongest of aluminum alloys, with tensile yield strengths of some exceeding 70,000 lb/in^2 (483 MPa) in the T6 temper. They are widely used for high-strength structures, primarily in aircraft. And **aluminum alloy X7093,** a PM alloy of Alcoa, provides greater resistance to stress corrosion than 7075 without reducing tensile yield and ultimate strengths or fracture toughness.

Aluminum-lithium alloys are a significant recent development in high-strength wrought aluminum alloys. Because of the very low density of lithium, every 1% of this alkali metal can provide a 3% reduction in density and a 10% increase in stiffness-to-density ratio relative to conventional 2XXX and 7XXX alloys. These alloys, such as **aluminum alloys 2090** and **2091** (there are several others, including proprietary ones), are comparable in strength to some of the strongest traditional alloys. Some are ingot-cast products, others PM wrought alloys. Alcoa's **aluminum alloy C155** is 7% more lightweight and has greater fracture toughness and resistance to fatigue-crack growth than 7075-T651 and 2024-T351. Reynolds Metals' **Weldalite 2195** is 5% more lightweight and 30% stronger than 2219 alloy and has good fracture toughness at cryogenic temperatures. It is a candidate to replace 2219 for the external fuel storage tanks of the Space Shuttle orbiter. **VacLite** aluminum-lithium alloys, of Comalco Aluminum Ltd. of Australia, are vacuum-refined alloys having low amounts of hydrogen and alkali-metal impurities. Their high purity is believed to account for improved toughness, greater resistance to stress corrosion, and superior weldability, or less tendency for weld cracking, than other aluminum-lithium alloys. One of the alloys, with 3.3% lithium, has a tensile yield strength of 73,000 lb/in^2 (504 MPa) and a tensile modulus of 12.1×10^6 lb/in^2 (83,400 MPa). Aluminum-lithium alloys were developed primarily for aircraft applications. They are used in the horizontal stabilizer of two Airbus models and in the cargo floor of the C-17 military transport.

Wrought superplastic alloys, such as Superform Metals' **Supral** alloys and fine-grain **aluminum alloy 7475** or modifications thereof, permit far greater tensile elongation than conventional alloys at low strain rates and temperatures in the 700 to 980°F (371 to 527°C) range, permitting extremely complex shapes to be formed at low gas pressure [100 to 300 lb/in^2 (0.69 to 2.1 MPa)].

For structural aircraft applications, high-strength alloys such as 2024 and 7075 are generally limited to temperatures below 400°F (204°C). Developmental **dispersion-strengthened aluminum alloys** may extend the range to 600°F (315°C). Key elements of these PM wrought systems include iron and cerium; iron and molybdenum; iron, vanadium, and silicon; or chromium, zirconium, and manganese. All provide tensile yield strengths greater than 30,000 lb/in^2 (270 MPa) at 600°F (315°C), the strongest being an aluminum-iron-vanadium-silicon alloy. At room temperature, this alloy has an ultimate tensile strength of 90,600 lb/in^2 (625 MPa), a yield strength of 82,400 lb/in^2 (568 MPa), and 6% elongation. At 600°F, these values are 45,000 lb/in^2 (310 MPa), 44,000 lb/in^2 (300 MPa), and 7%, respectively.

Aluminum PM parts are made mainly from **aluminum alloys 201 AB** and **601 AB.** Both are copper, silicon, and magnesium compositions that can be heat-treated to the T4 or T6 temper after sintering. 201 AB is the stronger, having 47,000 lb/in^2 (324 MPa) tensile yield strength for 95% dense material in the T6 temper compared to about 35,000 lb/in^2 (241 MPa) for 601 AB for these conditions.

A four-digit system, based on principal alloying elements, is also used to designate **aluminum casting alloys**; but the third and fourth digits are separated by a decimal point, and a capital letter before the first digit designates alloy modification. Essentially pure aluminum (99% minimum) is designated aluminum IXX.X. Aluminum alloys 2XX.X are aluminum-copper alloys; 3XX.X are aluminum-silicon alloys with copper and magnesium; 4XX.X are aluminum-silicon alloys; 5XX.X are aluminum-magnesium alloys; 6XX.X is an unused series; 7XX.X are aluminum-zinc alloys; and 8XX.X are aluminum-tin alloys. 2XX.X alloys may, however, contain other alloy elements, and whether an aluminum-copper-silicon alloy is a 2XX.X or 3XX.X alloy depends on which of the alloying elements is present in the greater quantity. In the IXX.X designations, the second and third digits indicate the degree of minimum aluminum content beyond 99.00% to the nearest 0.01%, and the fourth digit denotes product form (0 for castings or I for ingot). In the other designations, the second and third digits have no numerical significance and serve only to distinguish one alloy from another. The fourth digit again indicates product form: 0, castings; 1, standard ingot; 2, ingot having a composition within but narrower than that of the standard.

Of some 40 or so standard casting alloys, 10 are **aluminum die-casting alloys.** The others are **aluminum sand casting alloys** and/or **aluminum permanent-mold casting alloys.** Some of the latter also are used for plaster-mold casting, investment casting, and centrifugal casting. Although the die-casting alloys are not normally heat-treated, those for sand and/or permanent-mold casting often are, usually by solution heat treatment and artificial aging. **Aluminum alloy 380.0** and its modifications constitute the bulk of die-casting applications. Containing 8.5% silicon, 3.5 copper, and as much as 2 iron, it provides a tensile yield strength of about 24,000 lb/in^2 (165 MPa) and good corrosion resistance, is quite fluid and free from hot shortness, and thus is readily castable. Engine cylinder heads, type-writer frames, and various housings are among its applications. **Aluminum alloy 390.0,** a high-silicon (17%) copper-magnesium-zinc alloy, is the strongest of the die-casting alloys, providing a tensile yield strength of about 35,000 lb/in^2 (240 MPa) as cast and 38,000 lb/in^2 (260 MPa) in the T5 temper. Like high-silicon wrought alloys, it also features low thermal expansivity and excellent wear resistance. Typical applications include auto engine cylinder blocks, brake shoes, compressors, and pumps requiring abrasion resistance. **Aluminum alloys 383.0, 384.0, 413.0,** and **A413.0** provide the best die-filling capacity and have excellent resistance to hot cracking and die stick-ing. **Aluminum alloy 518.0** provides the best corrosion resistance, machinability, and polishability, but is more susceptible to hot crack-ing and die sticking than the other alloys.

Sand and/or permanent-mold casting alloys are encompassed in each of the **aluminum alloy 2XX.X to 8XX.X series.** As-cast, tensile yield strengths range from about 14,000 lb/in^2 (97 MPa) for 208.0 to 29,000 lb/in^2 (200 MPa) for **aluminum alloy A390.0.** In various solu-tion-treated and aged conditions, several alloys can provide tensile yield strengths exceeding 40,000 lb/in^2 (280 MPa). Strongest of the alloys—60,000 lb/in^2 (415 MPa) in the T7 temper and 63,000 lb/in^2 (435 MPa) in the T6—is **aluminum alloy 201.0,** which contains 0.7% silver and 0.25% titanium in addition to 4.6% copper and small amounts of magnesium and manganese. The T7 temper is suggested for applications requiring stress-corrosion resistance. Applications for 201.0 include aircraft and ordnance fittings and housings; engine cylinder heads; and pistons, pumps, and impellers. Other more com-monly used high-strength alloys are **aluminum alloys 354.0, 355.0,** and **356.0,** which are alloyed primarily with silicon, copper, magne-sium, manganese, iron, and zinc and which are used for auto and air-craft components. **Premium-quality aluminum-alloy castings** are those guaranteed to meet minimum tensile properties throughout the casting or in specifically designated areas.

Aluminum alloys also constitute a major class of **metal-matrix composites. Duralcan composites,** from Duralcan USA of Alcan Aluminum Ltd., comprise **alumina particulates** in a wrought matrix alloy and **silicon-carbide particulates** in a cast matrix alloy. The amount of particulate pertains to the volume percent. The composites are made by mixing the particulate in a molten alloy, which is then poured into foundry ingot, extrusion billet, rolling bloom, or rolling ingot. These forms are then processed conventionally by casting, extruding, rolling, or forging. Matrix alloys are similar in composition to standard wrought and cast alloys, and all the composites are heat-treatable. The composites, however, are stronger, more rigid, harder, and far more wear- and abrasion-resistant. On the other hand, their ductility and fracture toughness are less, and machining typically requires the use of polycrystalline diamond cutting tools for efficient production. In general, alumina decreases electrical and thermal conductivity and the coefficient of thermal expansion whereas silicon carbide increases thermal conductivity while decreasing electrical conductivity and thermal expansivity.

Here are some typical properties of extruded bar and rod in the T6 temper: With 15% Al_2O_3, **aluminum alloy composite 2014 (Duralcan W2A)** has a tensile yield strength of 73,000 lb/in^2 (500 MPa), 2% elongation, and a tensile modulus of 13.6 × 10^6 lb/in^2 (93,800 MPa). Although yield strength is only 6% greater than that of 2014-T6 at room temperature, it is 16 to 67% greater in the 200 to 700°F (93 to 371°C) range. **Aluminum alloy composite 2618 (Duralcan W2F)**, with 20% Al_2O_3, has a yield strength of 61,000 lb/in^2 (420 MPa), 2% elongation, and a modulus of 15.1 × 10^6 lb/in^2 (104,000 MPa). At 400°F (204°C), the yield strength is 47,000 lb/in^2 (320 MPa). With this alumina content, Duralcan W2F has a density of 0.109 lb/in^3 (3,017 kg/m^3).

Cast Duralcan alloys are based on matrix alloys 339 and 359 for gravity casting and 360 and 380 alloys for die casting. The 359 is for general room-temperature applications and the 339 is for elevated-temperature applications. Alloy 380 is also for general use, and 360 is for uses requiring additional corrosion resistance. In the T6 temper, **aluminum alloy composite 359 (Duralcan F3S)** with 20% SiC has a tensile yield strength of 49,000 lb/in^2 (340 MPa), 0.4% elongation, and a modulus of 14.3 × 10^6 lb/in^2 (98,600 MPa). Rockwell B hardness is 77, compressive yield strength is also 49,000 lb/in^2 (340 MPa), and the modulus in compression is 14.6 × 10^6 lb/in^2 (101,000 MPa). **Aluminum alloy composite 339 (Duralcan F3K)** with 10% SiC has a yield strength of 52,000 lb/in^2 (360 MPa) and a modulus of 12.7 × 10^6 lb/in^2 (87,500 MPa). In the T5 temper, Rockwell B hardness is 63, and the tensile yield strength is 36,000 lb/in^2 (248 MPa) at room temperature, 20,000 lb/in^2 (138 MPa) at 400°F (204°C), and 10,000 lb/in^2 (69 MPa) at 600°F (316°C). In the T5 temper, **aluminum alloy composite 380 (Duralcan F3D)** with 10% SiC has a Rockwell B hardness of 84, a tensile yield strength of 48,000 lb/in^2 (330 MPa),

0.7% elongation, and a tensile modulus of 13.6×10^6 lb/in^2 (93,800 MPa). It has a density of 0.100 lb/in^3 (2,768 kg/m^3).

Composite castings also are made by casting aluminum alloys about selectively placed **fiber preforms.** Honda has cast auto engine blocks with alumina and graphite preforms along the cylinder-bore surfaces, eliminating the need for cast-iron cylinder liners. **Silicon-carbide whiskers** and preforms have potential for squeeze-cast aluminum composites, especially for military vehicles. **Arall,** of Alcoa, is a laminate composite comprising thin layers of aramid-reinforced epoxy and wrought high-strength aluminum alloy. Besides being more lightweight than aluminum alloys, the laminate has high fatigue resistance.

ALUMINUM BRASS. A wrought copper (76 to 79%) and zinc alloy containing 1.8 to 2.5% aluminum and 0.02 to 0.06% arsenic. The aluminum addition markedly improves the alloy's resistance to impingement attack in high-velocity seawater and the arsenic inhibits dezincification. **Aluminum-brass C68700** is commonly used for condensers and heat exchangers in marine applications.

ALUMINUM BRONZE. A series of wrought and **cast copper-aluminum alloys,** most of which also contain several other elements, principally iron. Aluminum content ranges from 4 to 15%, depending on the alloy, and iron, if present, from less than 1 to 6%. Alloy designations for **wrought aluminum bronzes** are C60XXX, C61XXX, C62XXX, C63XXX, and C64XXX; cast alloys are C95XXX. The alloys are also referred to by their nominal aluminum content, such as 5% aluminum bronze, or by other distinguishing alloying ingredients, such as **nickel aluminum bronze.** Aluminum bronzes are noted for their moderate to high strength, and corrosion resistance in potable, brackish, and sea waters, industrial and marine atmospheres, and various chemicals. They are attacked, however, by oxidizing acids. Because of their pale gold look and amenability to lustrous finishing, some of the alloys are also used for their decorative appeal. Aluminum bronzes are available in a full range of wrought products, and the cast alloys are amenable to various casting methods. Typical applications include condenser and heat-exchanger tubing and tube sheets, fasteners, shafts, cams, gears, bearings, bushings, valves and valve seats, deep-drawn parts, wear plates, pickling equipment, impellers and agitators, nonsparking hardware, and decorative ware.

Single-phase (alpha) alloys containing only copper and aluminum can only be strengthened by strain hardening during cold working. The 4 to 7% **aluminum bronze C60600** alloy in the form of 0.5-in (13-mm) plate has a tensile yield strength of 17,000 lb/in^2 (115 MPa) annealed and 24,000 lb/in^2 (165 MPa) in the H04, or "hard," cold-worked temper. In these same conditions, 1-in (25-mm) diameter rod of **aluminum (6 to 8.5%) bronze C61000** has a tensile yield strength of 30,000 lb/in^2 (205 MPa) and 55,000 lb/in^2 (380 MPa), respectively. Two-phase aluminum bronzes, such as those alloyed with iron or iron and nickel, are typically strengthened by quenching and then are tempered. One such casting alloy, **aluminum bronze C95500** (nominally 81% copper, 11 aluminum, 4 iron, 4 nickel), has a tensile yield strength of 60,000 lb/in^2 (415 MPa) in the quenched and tempered (TQ50) condition, 50% greater than as cast.

ALUMINUM PALMITATE. One of the important **metallic soaps.** A yellow, massive salt, or a fine white powder of composition $Al(C_{16}H_{31}O_2)_3 \cdot H_2O$, made by heating a solution of aluminum hydroxide and palmitic acid. It is soluble in oils, alkalies, and benzol, but insoluble in water, and is used in waterproofing fabrics, paper, and leather and in paints as a drier. In finishing leather and paper it adds to the gloss. It is also used to increase viscosity in lubricating oils. Another material of the same class is **aluminum resinate,** $Al(C_{14}H_{63}O_5)_3$, a brown mass made by heating rosin and aluminum hydroxide. **Aluminum oleate,** $Al(C_{18}H_{33}O_2)_3$, is a white salt of oleic acid used as a drier. **Aluminum stearate,** $Al(C_{18}H_{35}O_2)_3$, is a salt of stearic acid. It is repellent to water and is valued for waterproofing fabrics and as a drier, and in waterproofing concrete and stucco. It is also used to give adherence to dyes and as a flux in soldering compounds. It is a white, fluffy powder of 200 mesh, soluble in oils and in turpentine. Grades high in free fatty acid, up to 22%, do not gel readily and are useful as a flatting agent and suspending medium in paints. Grades low in free fatty acid have a thickening effect on solvents. Those with 5 or 6% are used in lubricating grease, and those with about 8% are used in paints. **Aluminum monostearate** and **aluminum distearate** have similar uses.

ALUMINUM POWDER. Aluminum powder is produced by drawing a stream of molten metal through an atomizing nozzle and impinging that stream with compressed air or inert gas, solidifying and disintegrating the metal into small particles, which are then drawn into a collection system and screened, graded, and packaged. Particle sizes range from -325 mesh (fine) to $+200$ mesh (granules). In the process, the metal reacts with oxygen in the air and moisture, causing a thin film of aluminum oxide to form on the surface of the particles. Oxide content, which increases with decreasing particle size, ranges from 0.1 to 1.0% by weight. The reactivity and combustibility of aluminum powder require special handling precautions. Dust clouds, having a low ignition temperature—less than 1110°F (600°C)—and low explosive limits, need little oxygen (less than 3%) to be ignited by an electric spark or other ignition sources. Although coarse particles are difficult to ignite, particles in the -200 to $+325$ range and finer can be quite explosive, even in small concentrations. First used to produce **aluminum flake** by ball milling for paint pigments, aluminum powder now finds many other uses: ferrous and nonferrous metals production, powder metal (PM) parts, PM wrought aluminum alloy mill products, coatings for steel, asphalt roof products, spray coatings, and vacuum metallizing. Other applications include rocket fuels and explosives, incendiary bombs, pyrotechnics, signal flares, and heat and magnetic shields. It is also used in permanent magnets, high-temperature lubricants, industrial cements, chalking compounds, printing inks, and cosmetic and medical products. **Aluminum powder alloy A201AB** is used to produce cam bearing caps for dual-overhead-cam auto engines.

AMALGAM. A combination of a metal with mercury. The amalgams have the characteristic that when slightly heated, they are soft and easily workable, and they become very hard when set. They are used for filling where it is not possible to employ high temperatures. A native **silver amalgam** found in South America contains 26 to 95% silver. Native **gold amalgams** are found in California and Colombia, and they contain about 40% gold. Although native amalgams are chemical combinations of the metals, some of the artificial amalgams are alloys and others are compounds. **Dental amalgams** are prepared by mixing mercury with finely divided alloys composed of varying proportions of silver, tin, and copper. A **silver-tin alloy,** developed at the National Institute of Standards and Technology, is mercury-free and less brittle than standard amalgams.

   Cadmium amalgam was formerly employed for filling holes in metals and was called **Evans' metallic cement.** It is a silvery-white compound of composition Cd_5Hg_8, with about 74% mercury, the excess mercury separating out on standing. It softens at about 100°F (38°C) and can be kneaded like wax, remaining soft for a considerable time and then becoming hard and crystalline. Tin or bismuth may be added. Amalgams with an excess of cadmium are ductile and can be hammered into sheet. **Bismuth amalgams** are lustrous, very fluid combinations of mercury and bismuth, used for silvering mirrors. They are also added to white bearing metals to make them more plastic and to fusible alloys to lower the melting point. **Crilley metal** was a self-lubricating bearing alloy containing bismuth amalgam. The binary amalgams of mercury and bismuth are usually too fluid for ordinary use.

   The usual quaternary alloy has equal parts of bismuth, mercury, tin, and lead, with the proportion of mercury increased to give greater fluidity. A bismuth amalgam containing bismuth, lead, and mercury was used for lead pencils. A **thallium amalgam,** with 8.5% thallium, which freezes at $-76°F$ ($-60°C$), is used for thermometers for low readings. **Mackenzie's amalgam** is a two-part amalgam in which each part is a solid but becomes fluid when the parts are ground together in a mortar at ordinary temperatures. One part contains bismuth and mercury, and the other part contains lead and mercury. **Sodium amalgam** contains from 2 to 10% sodium. It is a silvery-white mass which decomposes water and can be used for producing hydrogen. **Potassium amalgam,** made by mixing sodium amalgam with potash, is a true chemical compound, used for amalgamating with other metals.

AMARANTH. One of the few broad-leafed vegetable plants that also produces edible grain. It was a mainstay in the diet of the Aztecs, Incas, and other Native Americans prior to the conquest of those empires by Spain in the 1500s. It is currently grown in Mexico, India, and China.

The brilliantly colored plant grows vigorously in a wide range of climates, resisting drought, heat, and pests. Its tiny seeds contain 16% protein compared to 12 to 14 for wheat and 7 to 10 for rice. The protein contains as much of the amino acid lysine as does milk protein, making it an excellent complement to traditional grains, which are low in this essential nutrient. The grains have a good taste. They can be popped like corn and eaten as a breakfast cereal, used as breading for meat and vegetables, or mixed with honey and compressed into blocks as a sweet snack. Other fast-growing species of amaranth produce spinachlike greens that are high in vitamin A.

AMBER. A fossil resin found buried in the countries along the Baltic Sea and in Malagasy. It is employed for making varnishes and lacquers and for ornaments. The original German name for the material was **Glassa,** and in early writings it is referred to by the Greek word **elektron** and the Persian name **karaba.** It was called **vernice** by the Italian painters who used it as a varnish resin. Amber came from a coniferous tree, *Pinus succinifera,* now extinct. It is hard, brittle, and tasteless but with an aromatic odor, and it dissolves in acids. It is sometimes transparent, but usually semitransparent or opaque with a glossy surface, yellow or orange. It takes a fine polish. When rubbed, it becomes electrically charged. In fact, the word **electricity** evolved from **elektron,** the Greek name for amber. Amber contains succinic acid in a complex form, and the finest specimens are known as **succinite,** although this is the name of an amber-colored garnet.

The amber of Malagasy was prized for necklaces and pipe mouth-pieces. It is semitransparent, wave-streaked, and honey-colored. Synthetic amber is plasticized phenol formaldehyde or other synthetic resin. **Amberoid** is reclaimed scrap amber pieces compressed into a solid, sometimes mixed with pieces of copal or other resin. It has the same uses as amber. **Amber oil,** distilled from scrap amber, is a mixture of terpenes and is used in varnish. **Succinic acid,** $(CH_2COOH)_2$, a solid melting at 361°F (183°C), is obtained by distilling scrap amber, but is now made by fermentation of tartaric acid, or synthetically from benzene and called **butane diacid.** It is used in foods as an acidifier and taste modifier, and **succinic aldehyde** is used for making plastics. The small, highly polar molecule gives crystalline resins of high strength. The dioctyl sodium sulfo salt of succinic acid is used in the foodstuffs industry to prevent agglomeration of colloids. **Complemix,** of American Cyanamid Co., is this material. **Maleic acid,** $(HCCO_2H)_2$, formed by heating succinic acid, is made commercially from benzene and is used widely for plastics production. It can be obtained by the dehydration of **malic acid,** which is a hydroxysuccinic acid. **Pomalius acid,** of Allied-Signal Inc., is a food grade of malic acid used to replace acetic acid as a more powerful acidulating agent, stabilizer, and flavor enhancer.

AMBERGRIS.   A solid, fatty, inflammable, grayish to black substance found in the intestines of some sperm whales or found floating in the ocean. It softens at 140°F (60°C) and is composed of 80 to 85% **ambrein (triterpenic tricyclic alcohol)** and 12 to 15% **ambergris oil,** an active ingredient. Ambergris is becoming scarce due to excessive whaling by fishing fleets. **White amber,** formed by aging in the ocean, is the finest quality. When dried, it pulverizes to a fine dust. Ambergris has a peculiar sweet fragrance and is highly valued as a fixative in perfumes. In Asia it is used as a spice and in Egypt for scenting cigarettes. Lumps weigh from a few ounces up to 600 lb (272.2 kg). Synthetic ambergris, of the Swiss company Firmenick & Co., is a gamma-dihydro ionone, $C_{13}H_{22}O$, which gives the odor, with also **ambreinolide,** a gamma lactone, $C_{17}H_{28}O$, which has fixative power. **Ambropur** is a German synthetic ambergris. When the terpene alcohol **manool,** of the wood of *Dacrydium* trees, is oxidized, it yields an acetal which also has the ambergris odor.

AMETHYST.   A violet or purple transparent quartz. The color is due to manganese and iron oxides, and it becomes yellow on heating. It has a specific gravity of 2.65 and Mohs hardness of 7. Amethyst is composed of alternate right- and left-hand crystals and breaks with a rippled fracture instead of the conchoidal fracture of ordinary quartz.

The crystals are doubly refractive. One of its uses is for making pivot bearings for instruments. It is the most esteemed of the quartzes for cutting into gem stones, but only deep and uniformly colored stones are used as gems and they are not common. Any large amethyst of deep and uniform color is likely to be synthetic. The chief production of natural amethyst is in Brazil and Uruguay.

AMINES.   A large group of chemicals derived from ammonia, NH_3, in which one or more of the hydrogen atoms have been replaced by an organic radical. A **primary amine,** such as **methylamine,** CH_3NH_2, has one ammonia hydrogen substituted. It is used in the tanning industry for unhairing skins and as a catalyst and solvent in the manufacture of synthetic resins. It is a gas and, like ammonia, is soluble in water and handled in water solution. It is flammable, a 40% solution having a flash point of 20°F (-7°C), and the vapors are explosive in air. A **secondary amine,** dimethylamine, $(CH_3)_2NH$, is more effective for unhairing. In water solution it forms a hydrate, $(CH_3)_2NH \cdot 7H_2O$, which has a low freezing point, 1.76°F (-16.8°C). The **tertiary amine,** trimethylamine, or **secaline,** $(CH_3)_3N$, is a gas liquefying at 37.2°F (2.87°C). The methylamines are used widely as a source of nitrogen. **Isopropylamine,** $(CH_3)_2CNNH_2$, is used as a replacement for ammonia in many chemical processes and as a solvent for oils, fats, and rubber. It is a clear liquid of specific gravity 0.686, boiling point 89°F (31.9°C), freezing point -150°F (-101°C), and flash point -15°F (-25°C). The **Sipenols,** of Alcolac Chemical Co., are a group of amines used as textile lubricants and softeners. They are clear to light yellow liquids and are dimethyl ethanol amines or dibutyl ethanol amines.

Changes in characteristics are obtainable in the amines by altering the position of the **amino group,** NH_2, in the carbon chain. **Hodag Amine 50,** of Hodag Chemical Corp., is amino ethyl ethanolamine, $HO \cdot CH_2 \cdot CH_2 \cdot NH \cdot CH_2 \cdot CH_2 \cdot NH_2$, a clear liquid with specific gravity of 1.027. It has both primary and secondary amino groups which give versatility for production of plastics, plasticizers, agricultural chemicals, and textile wetting agents.

Fatty acid amines are used as flotation agents, oil additives to prevent sludge, and rubber-mold release agents. **Alamine,** of General Mills Co., is **lauryl amine, palmityl amine,** or other amines of composition RNH_2, where R is the fatty acid radical. **Delamin,** of Hercules Inc., is a series of fatty acid amines produced from tall oil. The **Armeens,** of Akzo Chemicals Inc., are fatty acid amines with a wide range of uses.

Amino acids are organic compounds with both amino and carboxylic functional groups. They occur free and in combined states in nature. In the combined state, they serve as the monomers that form the carbon skeleton of protein polymers. Proteins are natural **polyamides** of various amino acids linked together by amide groups, also called **peptide linkages.** An **amide** is formed by replacing the hydroxyl group, OH, of an organic acid by an amino group to form $CONH_2$. The ease with which even complex radicals can be attached to introduce nitrogen into compounds makes the amines useful for the production of plastics, dyes, pharmaceuticals, explosives, detergents, and other chemicals.

Acetamide, also called **acetic acid amine** and **ethanamide,** is a grayish-white crystalline solid with a melting point of 171 to 178°F (77 to 81°C), specific gravity of 1.139, composition $CH_3 \cdot CO \cdot NH_2$, and slight mousy odor. It is soluble in water and in alcohol. It is used as a liquid flux for soldering on painted or oily surfaces, as an antacid in lacquers and explosives, as a softening agent in glues and leather coatings, and as a nonhazing plasticizer in cellulose nitrate and acetate films. Its ability to dissolve starch and dextrine makes it useful in adhesives for waxy papers. With added corrosion inhibitors, it is used as an antifreeze, a 50% solution in water having a freezing point of $-17.5°F$ ($-27.5°C$).

Aramids are synthetic fibers produced from long-chain polyamides (nylons) in which 85% of the amide linkages are attached directly to two aromatic rings. The fibers are exceptionally stable and have good strength, toughness, and stiffness, which are retained well above 300°F (150°C). Two aramids are **Nomex** and **Kevlar,** from Du Pont Co., and **Twaron,** from Akzo NV. They have high strength and intermediate stiffness and are suitable for cables, ropes, webbings, and tapes. Kevlar 49 fiber, with high strength and stiffness, is used as a reinforcement in polymer-matrix composites and personnel armor. Nomex, a **meta-aramid** best known for its excellent flame and abrasion resistance, is used for protective clothing, air filtration bags, and electrical insulation. **Korex,** a **para-aramid** of Du Pont, is a stronger, more rigid honeycomb-core material. **Aeroweb HMX** and **HMX-20,** of Ciba Composites, use Nomex in paper form.

AMMONIA. A gas of the formula NH_3, originally called **alkaline air** and **volatile alkali** and later in water solution called **spirits of hartshorn.** Ammonia is predominantly made by direct synthesis at high temperatures and pressures of hydrogen and nitrogen, derived by steam re-forming hydrocarbons. It is a major building block for making fertilizers, nitric acid, plastics, and explosives. Ammonia is readily absorbed by water, which at 60°F (16°C) takes up 683 times its own volume of the gas, forming the liquid commonly called ammonia, but which is **ammonium hydroxide,** a colorless, strongly alkaline, and pungent liquid of composition NH_4OH with a boiling point of 100°F (38°C). At 80°F (27°C) it contains 29.4% ammonia in stable solution. It is also known as **ammonium hydrate** and **aqua ammonia,** and it is used for the saponification of fats and oils, as a deodorant, for cleaning and bleaching, for etching aluminum, and in chemical processing.

Ammonia gas is used in refrigeration, in nitriding steels, and in the manufacture of chemicals. Chlorine unites with it to form **chloramines** which are used as solvents, chlorinating agents, and disinfectants. The gas does not burn in air, but a mixture of ammonia and oxygen explodes when ignited. **Anhydrous ammonia** is the purified gas liquefied under pressure, marketed in cylinders. At 68°F (20°C) the liquid has a vapor pressure of 122.1 lb/in² (0.84 MPa). The anhydrous ammonia used for controlled atmospheres for nitriding steel, bright annealing, and sintering metals contains 90% NH_3 and is oxygen-free. When dissociated by heat, each pound yields 45 ft³ (1.3 m³) of hydrogen and 11 ft³ (0.31 m³) of nitrogen.

Smelling salts, sometimes referred to as **ammonia,** and in solution as **aromatic spirits of ammonia,** is **ammonium carbonate,** $(NH_4)_2CO_3 \cdot H_2O$, forming in colorless or white crystals. It was also called **hartshorn salts. Ammonium bicarbonate,** NH_4HCO_3, or **acid ammonium carbonate,** is a water-soluble, white, crystalline powder used as a source of pure ammonia and carbon dioxide and to decrease density in organic materials by creating voids, such as for making foamed rubber and in the food-baking industry. It gasifies completely at 140°F (60°C). **Ammonium gluconate,** $NH_4C_6H_{11}O_7$, is a water-soluble, white, crystalline powder used as an emulsifier for cheese and mayonnaise and as a catalyst in textile printing.

The **ammonium radical,** $—NH_4$, has the chemical reaction of an alkali metal and forms many important chemicals. **Ammonium nitrate** is made by the action of nitric acid on ammonium hydroxide. It is a colorless to white, crystalline, hygroscopic powder of composition NH_4NO_3, specific gravity 1.725, melting point 338°F (170°C), decomposing at 410°F (210°C), and soluble in water, alcohol, and alkalies. It is also used in fertilizers, pyrotechnics, dental gas, insecticides, freezing mixtures, and explosives. For use as a slow-burning propellant for missiles, it is mixed with a burning-rate catalyst in a synthetic-rubber binder and pressed into blocks. **Riv,** a vapor-phase rust inhibitor, is ammonium nitrate. The explosive **amatol** is a mixture of ammonium nitrate and TNT, which explodes violently on detonation. The 50-50 mixture can be melted and poured, while the 80-20 mixture is like brown sugar and was used for filling large shells. **ANFO,** a mixture of ammonium nitrate and fuel oil formed into porous prills, is one of the largest-selling explosives. **Macite,** for tree trunk blasting, is ammonium nitrate coated with TNT, with a catalyst to make it more sensitive. **Akremite** is ammonium nitrate and carbon black, used as an explosive in strip mining. **Nitramex 2H,** of Du Pont Co., is **TNT-ferrosilicon-ammonium nitrate** in a metal can.

Ammonium perchlorate is another explosive made by the action of perchloric acid on ammonium hydroxide. It is a white crystalline substance of composition NH_4ClO_4 and specific gravity 1.95, is soluble in water, and decomposes on heating. **Nitrogen trichloride,** NCl_3, which forms in reactions of chlorine and ammonia when there is an excess of chlorine, is a highly explosive yellow oil of specific gravity 1.653. **Ammonium sulfate,** $(NH_4)_2SO_4$, is a gray, crystalline, water-soluble material used as a fertilizer and for fireproofing. It is made by reacting sulfuric acid with ammonia by-product from coal distillation, and, increasingly, from sulfuric acid and synthetic ammonia, from caprolactam production, or from gypsum.

Ammonium chloride, or **sal ammoniac,** NH_4Cl, is a white crystalline powder of specific gravity 1.52, used in electric batteries, in textile printing, as a soldering flux, and in making other compounds. Many salts and metallic soaps are also formed in the same manner as with the alkali metals. **Ammonium vanadate,** NH_4VO_3, is a white to yellow crystalline powder used as a paint drier, in inks, as a mordant for textiles, and in pottery mixes to produce luster. The specific gravity is 2.326, and it decomposes at 410°F (210°C). **Ammonium chromate,** $(NH_4)_2CrO_4$, is a bright-yellow, water-soluble, granular powder used as a textile mordant, in inks, and for the insolubilization of glues. On boiling, the water solution liberates ammonia. At 356°F (180°C) the powder decomposes to the dichromate. **Ammonium lactate,** $CH_3CHOHCOONH_4$, is a yellowish syrupy liquid with a slight odor of ammonia, used in leather finishing. **Ammonium stearate** is obtainable as a tan-colored waxlike solid, melting at 74°F (23°C). It can be dispersed in hot water, but above 190°F (88°C) it decomposes to ammonia and stearic acid.

AMORPHOUS METALS. Also known as **metallic glasses,** amorphous metals are produced by rapid quenching of molten metal-metalloid alloys, resulting in a noncrystalline grain-free structure. They are extremely strong and hard, yet reasonably ductile and quite corrosion-resistant. Perhaps of greatest interest is the magnetic performance of several complex iron-base compositions, called **Metglas,** developed by Allied Signal Corp. and produced in the form of ribbon or narrow strips. Because of their very low hysteresis and power losses, they can markedly reduce the size of transformer cores traditionally made of silicon steels. Several nickel-alloy compositions also are available for use as brazing foils. One such foil, **Metglas MBF60A,** contains 11% phosphorus and 0.1% maximum carbon and meets **BNi-6 brazing filler** specification of the American Welding Society. For aircraft turbines, it is used as a preform to braze **fiber metal** to 430 stainless steel for abradable seals.

Amorcor hardfacing alloys, of Amorphous Metals Technologies, are iron-chromium alloys of Vickers hardness 625 to 675 in their crystalline form as applied. After abrasion in service or by surface grinding, the two-phase crystalline structure transforms to an amorphous one of Vickers hardness 1,050 to 1,350. There are three basic alloys: alloy M for general corrosion resistance plus resistance to carburization, oxidation, and sulfidation at temperatures up to 1700°F (927°C); alloy C, for resistance to bleach, chlorides, and medium-strength sulfuric acid and to wear, with a maximum use temperature of 1600°F (871°C); and alloy T for especially severe wear applications. The alloys have been used on down-hole drilling pipe and drill bits, engine valve guides, and steel-making, coal-mining, agricultural, and earth-moving equipment.

AMYL ALCOHOL. A group of monohydroxy, or simple, alcohols, which are colorless liquids and have the general characteristic of five carbon atoms in the molecular chain. **Normal amyl alcohol,** $CH_3(CH_2)_4OH$, called also **fusel oil, grain oil, pentanol,** and **fermentation amyl alcohol,** has a specific gravity of 0.82 and boiling point of 279°F (137°C). It is only slightly soluble in water. It is used as a solvent for oils, resins, and varnishes; in the manufacture of amyl acetate; and in rubber vulcanization. **Secondary amyl alcohol** has a differently arranged molecule, $CH_3CHOH(C_3H_7)$. The specific gravity is 0.82 and flash point 80°F (27°C). It is used in the manufacture of secondary amyl acetate for lacquers and in chemical manufacture. **Tertiary amyl alcohol** has the formula $(CH_3)_2C(OH)C_2H_5$ and a camphor-like odor. The specific gravity is 0.81 and boiling point 216°F (102°C). It is highly soluble in water and soluble in alcohol and ether. It is used as a flavor and as a plasticizer in paints, varnishes, and cellulose plastics. **Isoamyl alcohol,** or **isobutyl carbinol,** $(CH_3)_2CHCH_2CH_2OH$, has a flash point above 80°F (27°C). It is used in pharmaceutical manufacture. **Amyl acetate,** $CH_3COOC_5H_{11}$, called **banana oil** because of its odor of bananas, is an ester made by the action of acetic acid on amyl alcohol. It is a colorless oily liquid of specific gravity 0.896 and boiling point 286°F (141°C). It is insoluble in water but soluble in alcohol. It is a good solvent and plasticizer for cellulose plastics and is used in cellulose lacquers and adhesives. It is also used in linoleum and oilcloth and as a banana flavor. **Amyl xanthate** is a common collector for sulfides in mineral flotation.

ANILINE. Also known as **aminobenzene, phenylamine, aminophen,** and **aniline oil** and, when first made, **krystallin** and **kyanol.** A yellowish, oily liquid of composition $C_6H_5 \cdot NH_2$, boiling at 364°F (184.4°C), freezing at 20.8°F (−6.2°C), and soluble in alcohol, benzene, and hydrochloric acid. The specific gravity is 1.022. It turns brown in air, finally oxidizing into a resin. The vapor is toxic, and it is poisonous when in contact with skin, requiring protective handling. Its largest uses are in the making of dyes and rubber chemicals, but it is also used for the production of plastics, drugs, explosives, perfumes, and flavors. With nitric acid as an oxidizer it has been used as a rocket fuel. **Aniline salt** is **aniline hydrochloride,** $C_6H_5NH_2HCl$, coming in white crystalline plates of specific gravity 1.2215, melting at 388°F (198°C), and soluble in alcohol.

ANNATTO. One of the chief **food colors.** It is a salmon-colored dye made from the pulp of the seeds of the tree *Bixa orellana* of the West Indies and tropical America and Africa. It contains **bixin,** $C_{25}H_{30}O_4$, a dark-red crystalline carotenoid carboxylic acid, and **bixol,** $C_{18}H_{30}O$, a dark-green oily alcohol. It is more stable than carotene and has more coloring power. Annatto is sometimes called **bixine,** and in West Africa it is called **rocou.** It is soluble in oils and in alcohol. **Annatto paste** is used as a food color especially for butter, cheese, and margarine, but has a tendency to give a slightly mustardy flavor unless purified. It is also used as a stain for wood and silk. Water-soluble colors are made by alkaline extraction, giving orange to red shades. For coloring margarine yellow, a blend of annatto and turmeric may be used. **Anattene** is a microcrystalline powder produced from annatto, giving a range of colors from light yellow to deep orange. It comes either oil-soluble or water-soluble.

A substitute for annatto for coloring butter and margarine, having the advantage that it is rich in vitamin A, is **carrot oil** obtained from the common carrot. The concentrated oil has a golden-yellow color and is odorless and tasteless. **Carex** is a name for carrot oil in cottonseed oil solution used for coloring foods. Many of the fat-soluble coloring matters found in plant and animal products are terpenes that derive their colors from conjugated double bonds in the molecule. The yellow **carotene** of carrots and the red **lycopene** of tomatoes both have the formula $C_{40}H_{56}$ and are tetra terpenes containing 8 isoprene units but with different molecular structures. **Beta carotene,** produced synthetically from acetone by Hoffmann-La Roche, is identical with the natural food color.

A beautiful water-soluble yellow dye used to color foods and medicines is **saffron,** extracted from the dried flowers and tips of the saffron crocus, *Crocus sativas,* of Europe, India, and China. It is expensive, as about 4,000 flowers are required to supply an ounce of the dye.

Saffron contains **crocin,** $C_{44}H_{70}O_{28}$, a bright-red powder soluble in alcohol. Both red and yellow colors are obtained from the orange thistlelike heads of the safflower, which are dried and pressed into cakes.

ANODE METALS. Metals used for the positive terminals in electroplating. They provide in whole or in part the source of the metal to be plated, and they are as pure as is commercially possible, are uniform in texture and composition, and have the skin removed by machining. They may be either cast or rolled, with their manufacture controlled to obtain a uniform grade and to exclude impurities, so that the anode will corrode uniformly in the plating bath and not polarize to form slimes or crusts. In some plating, such as for white bronze, the anode efficiency is much higher than the cathode efficiency, and a percentage of steel anodes is inserted to obtain a solution balance. In other cases, as in chromium plating, the metal is taken entirely from the solution, and insoluble anodes are employed. Chromium-plating anodes may be lead-antimony, with 6% antimony, or tin-lead, with 7% tin. In addition to pure single metals, various alloys are marketed in anode form. The usual brass is 80% copper and 20 zinc, but other compositions are used, some containing 1 to 2 tin. **Brass anodes** are called **platers' brass. Copper anodes** for metal plating are usually hot-rolled oval bars, 99.9% pure, while those for electrotype deposits may be hot-rolled plates, electrodeposited plates, or cast plates. **Copper ball anodes** are forged instead of cast to give a finer and more even grain. **Zinc anodes** are 99.99% pure. **Nickel anodes** are more than 99% pure, rolled or cast in iron molds, or 97% sand-cast. **Bright nickel** anodes may have 1% or more of cobalt. **Lead anodes** have low current-carrying capacity and may be made with a sawtooth or multiple-angled surface and ribs, to provide more area and give greater throwing power. Anodes of other metals are also made with sections gear-shaped, fluted, or barrel-shaped to give greater surface area and higher efficiency. **Rhodium anodes** are made in expanded-mesh form. **Platinum anodes,** also made in mesh form, have the platinum clad on tantalum wire. Special anode metals are marketed under trade names, usually accenting the color, hardness, and corrosion resistance of the deposited plate.

ANTHRACITE. Also called **hard coal.** A variety of mineral coal found in Wales, France, and Germany, but in greatest abundance in an area of about 500 mi² (1,295 km²) in northeastern Pennsylvania. It is distinguished by its semimetallic luster, high carbon content, and high specific gravity, which is about 1.70. The carbon content may be as high as 95%, but the usual fixed carbon content is from 78 to 84%. It should give 13,200 Btu/lb (30,700 kJ/kg). In theory, the best grades of anthracite should have 90% carbon, 3 to 4.5% hydrogen, 2 to 5.5% oxygen and nitrogen, and only 1.7% ash. Anthracite, when pure and dry, burns without smoke or smell and is thus preferred to bituminous coal for household furnaces. But coal will absorb a high proportion of water, and commercial coal may be wetted down to add to the weight, thus lessening its efficiency. Hard coal is graded as anthracite and **semianthracite,** depending upon the ratio of fixed carbon to volatile matter. When the ratio is 10:1, it is anthracite.

The commercial gradings of anthracite are chiefly by size, varying from three sizes of very fine grains called silt, rice, and buckwheat, to the large size of furnace, or lump, coal. Standard ASTM sizes for anthracite are as follows: broken, 4.375 to 3.25 in (11.1 to 8.3 cm); egg, 3.25 to 2.4375 in (8.3 to 6.2 cm); stove, 2.4375 to 1.625 in (6.2 to 4.1 cm); chestnut, 1.625 to 0.8125 in (4.1 to 2.1 cm); pea, 0.8125 to 0.5625 in (2.1 to 1.4 cm); No. 1 buckwheat, 0.5625 to 0.3125 in (1.4 to 0.8 cm); No. 2 buckwheat (rice), 0.3125 to 0.1875 in (0.8 to 0.5 cm); No. 3 buckwheat (barley), 0.1875 to 0.09375 in (0.5 to 0.2 cm). As the coal comes from the breaker, the proportions are about 8% silt, 9 rice, 15 buckwheat, 10 pea, 24 chestnut, 23 stove, and 8 egg.

ANTIFREEZE COMPOUNDS. Materials employed in the cooling systems and radiators of internal-combustion engines to ensure a liquid circulating medium at low temperatures to prevent damage from the formation of ice. The requirements are that the compound give a freezing point below that likely to be encountered without lowering the boiling point much below that of water, that it not corrode the metals or deteriorate rubber connections, that it be stable up to the boiling point, and that it be readily obtainable commercially. Calcium chloride was early used for automobile radiators but corroded the metals. It is still used in fire tanks, sodium chromate being added to retard corrosion. Oils were also used, but the high boiling points permitted overheating of the engine, and the oils softened the rubber. Denatured **ethyl alcohol** may be used, but **methanol** is less corrosive and less expensive. A 30% solution of ethyl alcohol in water has a freezing point of about 5°F (−15°C), and a 50% solution freezes at −24°F (−31°C). Alcohol, however, must be renewed frequently because of loss by evaporation.

Glycerol is also used as an antifreeze, a 40% solution in water lowering the freezing point to about 0°F (−18°C), and a 50% solution to −15°F (−25°C). It has the disadvantage of high viscosity, requiring forced circulation at low temperatures, but it does not evaporate easily. **Ethylene glycol** lowers the freezing point to a greater extent than alcohol and has a high boiling point so that it is not lost by evaporation, but it has a higher first cost and will soften ordinary natural rubber connections. **Acetamide** in water solution may also be used as an antifreeze. Antifreezes are sold under various trade names. **Zerone,** of Du Pont, has a methanol base, while **Zerex** has a base of ethylene glycol. **Prestone,** marketed by Union Carbide Corp., is ethylene glycol antifreeze. **Pyro** is an antifreeze of U.S. Industrial Chemicals, Inc., with a low freezing point. **Ramp** is ethylene glycol with anticorrosion and antifoam agents added. **Antifreeze PFA55MB,** of Phillips Petroleum Co., used in jet engine fuels, is ethylene glycol monoethyl ether with 10% glycerin. **Dowtherm 209,** of Dow Chemical Co., is an antifreeze material of inhibited methoxypropanol which boils off without forming gum. **Sierra,** a proplyene-glycol-based antifreeze of Safe Brands Corp., performs as well as ethylene glycol compounds but is less toxic. A 50-50 blend with water freezes at $-26°F$ ($-32°C$).

ANTIMONY. A bluish-white metal, symbol Sb, having a crystalline scalelike structure. It is brittle and easily reduced to powder. It is neither malleable nor ductile and is used only in alloys or in its chemical compounds. Like arsenic and bismuth, it is sometimes referred to as a **metalloid,** but in mineralogy it is called a **semimetal.** It does not have the free cloudlike electrons that occur in metal atoms, and thus it lacks plasticity and is a poor conductor of electricity.

The chief uses of antimony are in alloys, particularly for hardening lead-base alloys. The specific gravity of the metal is 6.62, melting point 824°F (440°C), and Brinell hardness 55. It burns with a bluish light when heated to redness in air. Antimony imparts hardness and a smooth surface to soft-metal alloys; and alloys containing antimony expand on cooling, thus reproducing the fine details of the mold. This property makes it valuable for type metals. When alloyed with lead, tin, and copper, it forms the **babbitt metals** used for machinery bearings. It is also much used in white alloys for pewter utensils. Its compounds are used widely for pigments. **Antimony red** is the common name of **antimony trisulfide,** Sb_2S_3, also known as **antimony sulfide** and **antimony sulfuret,** found in the mineral stibnite, but produced by precipitation from solutions of antimony salts. It comes in orange-red crystals and has a specific gravity of 4.56 and a melting point of 1015°F (545°C). It is used as a paint pigment, for coloring red rubber, and in safety matches. **Antimony pentasulfide,** Sb_2S_5, an orange-yellow powder, was once used for vulcanizing rubber, coloring the rubber red. It breaks down when heated, yielding sulfur and the red pigment antimoney trisulfide.

ANTIMONY ORES. The chief ore of the metal antimony is **stibnite,** an impure form of antimony trisulfide, Sb_2S_3, containing theoretically

71.4% antimony. The usual content of the ore is 45 to 60%, which is concentrated to an average of 92% for shipment as matte. Sometimes gold or silver is contained in the ore. Stibnite occurs in slender prismatic crystals of a metallic luster and lead-gray color with a hardness of 2 Mohs. The metal is obtained by melting the stibnite with iron, forming FeS and liberating the antimony, or by roasting the ore to produce the oxide, which is then reduced with carbon. For pyrotechnic uses, stibnite is liquated by melting the mineral and drawing off the metal, which on cooling and solidifying is ground. Stibnite comes from China, Mexico, Japan, West Germany, Bolivia, Alaska, and the western United States.

Senarmontite, found in Mexico, Nevada, and Montana, is **antimony oxide,** Sb_2O_3, occurring in cubic crystals with a yellow color. The specific gravity is 5.2, Mohs hardness 2.5, and theoretical metal content 83.3%. **Valentinite,** also found in the same localities, has the same theoretical formula and antimony content as senarmontite, but has a rhombic crystal structure, a hardness of 3, and a specific gravity of 5.5. These oxides are used as opacifiers in ceramic enamels. **Cervantite,** found in Mexico, Nevada, and Montana, is **antimony tetraoxide,** Sb_2O_4. It is grayish yellow, has a specific gravity of 5 and hardness of 4.5, and contains theoretically 79.2% antimony. **Stibiconite,** from the same area, is a massive pale-yellow mineral, $Sb_2O_4 \cdot H_2O$, with specific gravity 5.1, hardness 4.5, and antimony content 71.8%.

Kermesite, known as **red antimony** or **antimony blend,** found in Mexico and Italy, is a mineral resulting from the partial oxidation of stibnite. The composition is Sb_2S_2O, and when pure, it contains 75% antimony and 20 sulfur. It occurs in hairlike tufts, or radiating fibers of a dark-red color and metallic luster, with hardness 1.5 and specific gravity 4.5. Another sulfide ore of antimony is **jamesonite,** $Pb_2Sb_2S_5$, found in Mexico and the western United States. It has a dark-gray color, specific gravity 5.5, and hardness 2.5 and contains 20% antimony. When the ore is silver-bearing, it can be worked profitably for antimony. **Stephanite** is classified as an ore of silver, but yields antimony. It is a **silver sulfantimonite,** Ag_5SbS_4, containing 68.5% silver and 15.2 antimony. It occurs massive or in grains of an iron-black color with a Mohs hardness of 2 to 2.5 and specific gravity of 6.2 to 6.3. It is found in Nevada, Mexico, Peru, Chile, and central Europe. Much antimony is in lead ores and is left in the lead as hard lead. Antimony is marketed in flat cakes or in broken lumps. The highest grade of pure refined antimony is known as **star antimony** because of the glittering, spangled appearance on the surface, but starring can be done with lower grades of antimony by special cooling of the ingots. **Crude antimony** is not antimony metal, but is beneficiated ore, or ore matte, containing 90% or more of metal. High-grade antimony is +99.8% pure, Standard grade is 99 to 99.8% pure, and Chinese is 99% pure.

ANTIOXIDANT. A material used to retard oxidation and deterioration of vegetable and animal fats and oils, rubber, or other organic products. Antioxidants embrace a wide variety of materials, but in general for antioxidant activity the hydroxy groups must be substituted directly in an aromatic nucleus. In the phenol group of antioxidants, the hydrogen atoms must be free. In the naphthol group, the alpha compound is a powerful antioxidant. Usually, only minute quantities of antioxidants are used to obtain the effect. **Ionol,** an antioxidant, or **oxidation inhibitor,** of Shell Chemical Co., is a complex butyl methyl phenol used in gasoline, oils, soaps, rubber, and plastics. It is an odorless, tasteless, nonstaining granular powder, insoluble in water, melting at 158°F (70°C). In gasoline the purpose of an antioxidant is to stabilize the diolefins that form gums. **Norconidendrin,** an antioxidant for fats and oils, is produced from the high-phenol **confidendrin,** obtained from hemlock pulp liquor. The **Tenox antioxidants,** of Eastman Chemical Co., used for meats and poultry, are mixtures of butyl hydroxyanisole, butyl hydroxytoluene, propyl gallate, and citric acid, in solution in corn oil, glyceryl monooleate, or propylene glycol. **Tenox 2** contains 20% BHA, 6 propyl gallate, 4 citric acid, and 70 propylene glycol. **Tenox 4** contains 20% BHA, 20 BHT, and 60 corn oil. **Tenox HQ,** used to prevent rancidity in margarine, dried milk, and cooking fats, is a purified hydroquinone. **Tenamene,** of the same company, used in rubber, is a complex phenylenediamine. Most of the antioxidants for rubber and plastics are either phenols or aromatic amines. **Lead diamyldithiocarbamate (LDADC),** long used to prolong the life of hydrocarbon-based lubricants, can also be used to inhibit asphalt's cracking with age and exposure to varying climatic conditions. **Metilox,** a phenolic intermediate of Ciba Geigy, is used to produce antioxidants for plastics using a metal-hydroxide catalyst.

A **synergist** may be used with an antioxidant for regeneration by yielding hydrogen to the antioxidant. Synergists are acids such as citric or maleic, or they may be ferrocyanides. The presence of small quantities of metallic impurities in oils and fats may deactivate the antioxidants and nullify their effect. **Phytic acid** not only is an antioxidant for oils and foodstuffs, but also controls the metallic contaminations. It does not break down as citric acid does or impart a taste to edible oils as phosphoric acid does. It occurs in the bran of seeds as the salt **phytin,** $CaMg(C_2H_6P_2O_9)_2$, and is obtained commercially from corn steep liquor. Chemicals used to control metallic ions and stabilize the solutions are called **sequestering agents. Pasac** is such an agent. It is **potassium acid saccharate,** $KHC_6H_8O_8$, in the form of a water-soluble white powder. **Sequelene,** of A. E. Staley Mfg. Co., for treating hard and rusty waters, is a **sodium glucoheptonate.**

Since odor is a major component of flavor, and the development of unpleasant odors in edible fats arises from oxidation, the use of antioxidants is generally necessary, and in such use they are called **food stabilizers.** But degradation of some organic materials may not be a simple oxidation process. In polyvinyl chloride plastics, the initial stage of heat degradation is a dehydrochlorination with hydrogen chloride split out of the molecular chain to give a conjugated system subject to oxidation. Materials called **stabilizers** are thus used to prevent the initial release. Traces of iron and copper in vegetable oils promote rancidity, and citric acid is used as a stabilizer in food oils to suppress this action. **Densitol,** of Abbott Laboratories, for stabilizing citrus-fruit beverage syrups, is a brominated sesame oil. It also enhances the flavor, although it has no taste.

Light stabilizers may be merely materials such as carbon black to screen out the ultraviolet rays of light. Most commercial antioxidants for foodstuffs are mixtures, and all the mixtures are synergistic with the total antioxidant effect being greater than the sum of the components. **Sustane 3** is a mixture of butylated hydroxyanisole, propyl gallate, citric acid, and propylene glycol. **Inhibitors** for controlling color in the chemical processing of fats and oils are usually organic phosphates, such as the liquids **triisooctyl phosphate** and **chloroethyl phosphate.** They are mild reducing agents and acid acceptors, and they complex with the metal salts. **Ultraviolet absorbers,** to prevent yellowing and deterioration of plastics and other organic materials, are substituted hydroxybenzophenones. The photons of the invisible ultraviolet rays of sunlight have great energy and attack organic materials photochemically. Ultraviolet absorbers are stable in this light and absorb the invisible rays. **Sorbalite,** an ultraviolet-absorbing acrylic polymer of Monsanto Co., can be dispersed in water as a latex to form thin, clear ultraviolet-resistant coatings on polypropylene and polyester films. **Antirads** are antioxidants that increase the resistance of rubber or plastics to deterioration by gamma rays. Such rays may break the valence bonds and soften a rubber, or cross-link the chains and harden the rubber.

The term **corrosion inhibitors** usually refers to materials used to prevent or retard the oxidation of metals. They may be elements alloyed with the metal, such as columbium or titanium incorporated in stainless steels to stabilize the carbon and retard intergranular corrosion; or they may be materials applied to the metal to retard oxygen attack from the air or from moisture. Many paint undercoats, especially the phosphate and chromate coatings applied to steel, are corrosion inhibitors. They may contain a ferrocyanide synergist. **Propargyl alcohol,** C_2H_4CO, a liquid boiling at 239°F (115°C), is used in strong mineral acid pickling baths to prevent hydrogen embrittlement and corrosion of steel. **VPI 260,** of Shell Chemical Co., is **dicyclohexylamine nitrite,** a white crystalline powder which sublimes to form a shield on steel or aluminum to passivate the metal and make it resistant to moisture corrosion. VPI means **vapor-phase inhibitor. VPI paper** is wrapping paper impregnated with the nitrite, used for packaging steel articles.

Inhibitors are also added to process water to inhibit corrosion of containment equipment. Oxygen is a major contributor to boiler corrosion, and until it was marked as a suspected carcinogen, **hydrazine** was the principal **purging agent.** Besides scavenging oxygen, it passivates metal surfaces, further inhibiting corrosion. **Mekor,** or **methyl ethyl ketoxime** of Drew Industrial Div. of Ashland Chemical, behaves similarly and is not a suspected carcinogen. Other alternatives are Grace Dearborn's **diethyl hydroxyl amine** (DEHA) and Nalco Chemical's **Eliminox** and **Surgard.** Still others are **carbohydrazide, erythorbate, hydroquinone,** and **sodium sulfate.** However, although they will protect the boiler, they are not volatile as hydroxine is and thus will not protect the entire system. Boilers and process equipment also can be protected with FMC Corp.'s **hydroxy phosphine carboxylic acid** (HPCA), an organic metal-free additive, and small doses of **sodium silicate.**

ANTISLIP METALS. Metals with abrasive grains cast or rolled into them, used for floor plates, stair treads, and car steps. They may be of any metal, but are usually iron, steel, bronze, or aluminum. The abrasive may be sand, but it is more usually a hard and high-melting-point material such as aluminum oxide. In standard cast forms, antislip metals are marketed under trade names. **Alumalun** is the name of an aluminum alloy cast with abrasive grains. **Bronzalum** is a similar product made of bronze. **Algrip steel** is steel plate 0.125 to 0.375 in (0.32 to 0.95 cm) thick, with abrasive grains rolled into one face. It is used for loading platforms and ramps.

ANTLER. The bony, deciduous horns of animals of the deer family, used for making handles for knives and other articles but now replaced commercially by plastic moldings. Antlers are true outgrowths of bone and are not simply hardenings of tissue, as are the horns of other animals. Unlike horn, antlers are solid and have curiously marked surfaces. They come in various shapes and sizes and are usually found on the male during the mating season, although both sexes of reindeer and American caribou possess them. They grow in 3 to 4 months and are shed annually.

ARGENTITE. An important ore of silver, also called **silver glance.** It has composition Ag_2S, containing theoretically 87.1% silver. It usually occurs massive, streaked black and lead gray, with a metallic luster and a Mohs hardness of 2 to 2.5. It is found in Nevada, Arizona, Mexico, South America, and Europe. **Argyrodite** is another silver ore found in Bolivia, and it is a source of the rare metal germanium. When pure, it has composition $4Ag_2S \cdot GeS_2$ and contains 5 to 7% germanium. A similar mineral, **canfieldite,** found in Bolivia, has 1.82% germanium and some tin.

ARGOLS. Also called **wine lees.** A reddish crust or sediment deposited from wine, employed for the production of tartaric acid, cream of tartar, and rochelle salts. It is crude **potassium acid tartrate,** or **cream of tartar,** $KH(C_4H_4O_6)$. When grape fermentation is complete, the wine is drawn off and placed in storage tanks where the **lees** settle out. The amount of tartrate varies in different types of wine, from 0.1 to 1.0 lb/gal (0.01 to 0.12 g/cm^3) of cream of tartar. From wines clarified by refrigeration, as much as 1 to 3 lb/gal (0.12 to 0.36 g/cm^3) of tartrate crystallizes out. Cream of tartar is also obtained from **grape pomace,** which is the residue skins, seeds, and pulp, containing 1 to 5% tartrate. **Wine stone** is cream of tartar, 70 to 90% pure, which crystallizes on the walls of wine storage tanks. Purified cream of tartar is a colorless to white crystalline powder of specific gravity 1.956, soluble in water, and used in baking powders.

    Tartaric acid is a colorless crystalline product of composition $HOOC(CHOH)_2COOH$, which has a melting point of 338°F (170°C) and is soluble in water and in alcohol. It has a wide variety of uses in pharmaceuticals, in effervescent beverages, and as a mordant in dyeing. The pods of the tamarind tree, *Tamarindus indica,* of India, contain 12% tartaric acid and 30 sugars. They are used in medicine and for beverages under the name of **tamarind. Rochelle salts** is **potassium sodium tartrate,** $KNa(C_4H_4O_6) \cdot 4H_2O$, a colorless to bluish-white crystalline solid of specific gravity 1.79 and melting point 167°F (75°C), which is soluble in water and in alcohol. It is used in medicines and in silvering mirrors. Like quartz, it is doubly refractive and is used in piezoelectric devices where water solubility is not a disadvantage.

ARSENIC. A soft, brittle, poisonous element of steel-gray color and metallic luster, symbol As. The melting point is 1562°F (850°C), and specific gravity is 4.8. In atomic structure it is a semimetal, lacking plasticity, and is used only in alloys and in compounds. When heated in air, it burns to **arsenious anhydride** with white odorous fumes. The bulk of the arsenic used is employed in insecticides, rat poisons, and weed killers, but it has many industrial uses, especially in pigments. It is also used in poison gases for chemical warfare. The white, poisonous powder commonly called *arsenic* is **arsenic trioxide,** or **arsenious oxide,** As_2O_3, also known as **white arsenic.** When marketed commercially, it is colored pink to designate it as a poison. White arsenic is marketed as Refined, +99% pure; High-grade, 95 to 99%; and Low-grade, −95%. Refined arsenic trioxide is used as a decolorizer and fining agent in the production of glass, and for the production of arsenic compounds. **Monosodium methylarsonate, disodium methylarsonate,** and **methane arsenic acid,** also called **cacodylic acid,** are used for weed control. Arsenic is added to antimonial lead alloys and white bearing metals for hardening and to increase fluidity, and to copper to increase the annealing temperature for such uses as radiators. It is also used in lead shot to diminish cohesion, and small amounts are used as negative electron carriers in rectifier crystals.

Arsenic acid is a white crystalline solid of composition $(H_3AsO_4)_2 \cdot H_2O$, produced by the oxidation of white arsenic with nitric and hydrochloric acids. It is soluble in water and in alcohol, has a specific gravity of 2 to 2.5, and a melting point of 95.9°F (35.5°C). Arsenic acid is sold in various grades, usually 75% pure, and is used in glass manufacture, printing textiles, and insecticides. The **arsines** comprise a large group of alkyl compounds of arsenic. They are **arsenic hydrides,** AsH_3, a colorless gas. The primary, secondary, and tertiary arsines are not basic, but the hydroxides are strongly basic. The arsines are easily oxidized to **arsonic acid,** $RAsO_3H_2$, and related acids. **Arsenic disulfide,** also known as **ruby arsenic, red arsenic glass,** and **red orpiment,** is an orange-red, poisonous powder with specific gravity 3.5 and melting point 585°F (307°C), obtained by roasting arsenopyrite and iron pyrites. The composition is As_2S_2. It is employed in fireworks, as a paint pigment, and in the leather and textile industries. Another arsenic sulfur compound used as a pigment is **orpiment,** found as a natural mineral in Utah, Peru, and central Europe. It is an **arsenic trisulfide,** As_2S_3, containing 39% sulfur and 61 arsenic. The mineral has a foliated structure, a lemon-yellow color, and a resinous luster. The specific gravity is 3.4, Mohs hardness 1.5 to 2, and melting point 572°F (300°C). Artificial arsenic sulfide is now largely substituted for orpiment and is referred to as **king's yellow.**

ARSENIC ORES. Arsenopyrite, also called **mispickel,** is the most common ore of arsenic. It is used also as a source of white arsenic, and directly in pigments and as a hide preservative. The composition is FeAsS. It occurs in crystals or massive forms of a silvery-white to gray-black color and a metallic luster. The specific gravity is 6.2, and Mohs hardness 5.5 to 6. Arsenic is usually not a primary product from ores, but is obtained as a by-product in the smelting of copper, lead, and gold ores. A source of white arsenic is the copper ore **enargite,** $Cu_2S \cdot 4CuS \cdot As_2S_3$, theoretically containing 48.3% copper and 19.1 arsenic. It occurs in massive form with a hardness of 3 and specific gravity of 4.45 and is gray, with a pinkish variety known as **luzonite.** The mineral is commonly intertwined with **tennantite,** $5Cu_2S \cdot 2(CuFe)S \cdot 2As_2S_3$, a gray to greenish mineral. **Realgar,** known also as **ruby sulfur,** is a red or orange arsenic disulfide, As_2S_2, occurring with ores of lead and silver in monoclinic crystals. The hardness is 1.5, and specific gravity is 3.55. It is used as a pigment. Another ore is **smaltite,** or **cobalt pyrites,** $CoAs_2$, occurring in gray masses of specific gravity 6.5 and Mohs hardness 5.5. It occurs with ores of nickel and copper. It may have nickel and iron replacing part of the cobalt, and it is a source of cobalt, containing theoretically 28.1% cobalt.

ASBESTOS. A general name for several varieties of fibrous minerals, the fibers of which are valued for their heat-resistant and chemical-resistant properties, and which have been made into fabrics, paper, insulating boards, insulating cements, fireproof garments, curtains, shields, brake linings, shingles, pipe coverings, and molded products. During the past 20 years or so, great concern has developed over the effects of asbestos, especially dust, on human health and strict regulations regarding its uses have been imposed in many countries, markedly reducing consumption. For example, U.S. consumption declined from 881,058 tons (800,962 metric tons) in 1973 to 56,650 tons (51,500 metric tons) in 1989. The original source of asbestos was the mineral actinolite, but the variety of **serpentine** known as **chrysotile** later furnished most of the commercial asbestos. **Actinolite** and **tremolite,** which furnished some of the asbestos, belong to a great group of widely distributed minerals known as **amphiboles,** which are chiefly metasilicates of calcium and magnesium, with iron sometimes replacing part of the magnesium. They occur as granules, in crystals, compact such as **nephrite,** which is the jade of the Orient, or in silky fibers such as in the iron amphibole asbestos. This latter type is more resistant to heat than chrysotile. Its color varies from white to green and black.

    Jade occurs as a solid rock and is highly valued for making ornamental objects. Jade quarries have been worked in Khotan and Upper Burma for many centuries, and large pebbles are also obtained by divers in the Khotan River. The most highly prized in China was white speckled with red and green and veined with gold. The most valued of the **Burma jade** is a grass-green variety called **Ayah kyauk.** Most jade is emerald green, but some is white and others are yellow, vermilion, and deep blue. This form of the mineral is not fibrous.

Asbestos is a hydrated metal silicate with the metal and hydroxyl groups serving as lateral connectors of the molecular chain to form long crystals which are the fibers. The formula for **chrysotile** is given as $Mg_6Si_4O_{11}(OH)_6 \cdot H_2O$. Each silicon atom in the Si_4O_{11} chain is enclosed by a tetrahedron of four oxygen atoms so that two oxygen atoms are shared by adjacent tetrahedra to form an endless chain. When the crystal orientation is perfect, the fibers are long and silky and of uniform diameter with high strength. When the orientation is imperfect, the Si_4O_{11} chain is not parallel to the fiber axis and the fibers are uneven and harsh. In chrysotile the metal connector is magnesium with or without iron, but there are at least 30 other different types of asbestos.

Chrysotile fibers are long and silky, and the tensile strength is 80,000 to 200,000 lb/in² (552 to 1,379 MPa). The color is white, amber, gray, or greenish. The melting point is 2770°F (1521°C), and specific gravity is 2.4 to 2.6. Chrysotile has been mined chiefly in Vermont, California, Quebec, Arizona, Turkey, and Zimbabwe. Only about 8% of the total mined is long spinning fiber, the remainder being too short for fabrics or rope. The Turkish fiber is up to 0.75 in (1.9 cm) in length. Asbestos produced in Quebec is chrysotile occurring in serpentized rock in veins 0.25 to 0.50 in (0.64 to 1.27 cm) wide, though veins as wide as 5 in (12.7 cm) occur. The fibers run crosswise to the vein, and the width of vein determines the length of fiber. **Calidria asbestos** is short-fiber chrysotile from California and has about 14% water of crystallization. At temperatures near 1800°F (980°C), it loses its water, and the dehydration has a cooling effect.

Blue asbestos, from South Africa, is the mineral **crocidolite,** $NaFe(SiO_3)_2 \cdot FeSiO_2$. The fiber has high tensile strength, averaging 600,000 lb/in² (4,080 MPa), is heat resistant to 1200°F (650°C), and is resistant to most chemicals. The fibers are 0.125 to 3 in (0.32 to 7.6 cm) long with diameters from 0.06 to 0.1 in (0.15 to 0.25 cm). It is compatible with polyester, phenolic, and epoxy resins.

The classes of **cape asbestos** from South Africa are chrysotile, **amosite,** and **Transvaal blue.** Amosite has a coarse, long, resilient fiber, and it has been used chiefly in insulation, being difficult to spin. It comes in white and dark grades, and the fibers are graded also by length from 0.125 to 6 in (0.32 to 15.2 cm). It has a chemical resistance slightly less than that of crocidolite and a tensile strength of 200,000 lb/in^2 (1,379 MPa). The name *amosite* was originally a trade name for South African asbestos, but now refers to this type of mineral. Transvaal blue is a whitish, iron-rich, **anthophyllite,** (MgFe)SiO$_3$, noted for the length of its fiber. The best grades are about 1.5 in (3.8 cm) long. The fibers are resistant to heat and to acids, and the stronger fibers are used for making acid filter cloth and fireproof garments. This type of asbestos is also found in the Appalachian range from Vermont to Alabama. Canadian, Vermont, and Arizona asbestos is chrysotile; that from Georgia and the Carolinas is anthophyllite.

Canadian asbestos is graded as crude, mill fibers, and shorts. Crudes are spinning fibers 0.375 in (0.95 cm) or longer. Mill fibers are obtained by crushing and screening. Shorts are the lowest grades of mill fibers. **Rhodesian asbestos** comes in five grades. Kenya asbestos is anthophyllite, and that from Tanzania is largely amphilbole. **Nonspinning asbestos** is graded as shingle stock, 0.25 to 0.375 in (0.38 to 0.95 cm); paper stock, 0.125 to 0.250 in (0.32 to 0.38 cm); and shorts, 0.0625 to 0.125 in (0.16 to 0.32 cm). In England this material is known as **micro asbestos.**

Caposite is rope 0.5 to 2 in (1.3 to 5.1 cm) in diameter made of twisted rovings of long-staple asbestos covered with a braided jacket of asbestos yarn. Uses have included pipe, valve, joint insulation and furnace door packing. **Asbestos felt,** also for insulation, can be made by saturating felted asbestos with asphalt, although synthetic rubber or other binder may be used.

Asbestos shingles and boards have been made of asbestos fibers and portland cement formed under hydraulic pressure. Another type of asbestos for some insulation is **paligorskite,** known as **mountain leather,** found in Alaska. It is a complex mineral which may be an alteration product of several asbestos minerals. It absorbs moisture and is thus not suited to the ordinary uses of asbestos, but it can be reduced to a smooth pulp and molded with a resistant plastic binder into a lightweight insulating board.

ASH. The wood of a variety of species of ash trees valued for uses where strength, hardness, stiffness, and shock resistance are important. Most of the species give dense, elastic woods that polish well, but they do not withstand exposure well. The color is yellowish, which turns brown on exposure. The woods from the different species vary in their qualities and are likely to be mixed in commercial shipments, but the general quality is high. Ash is used for quality cooperage such as tubs, flooring, veneer, vehicle parts, tool handles, bearings, and trim lumber. **American ash** and **Canadian ash,** also called **cane**

ash, white ash, and **Biltmore ash,** come chiefly from the tree *Fraxinus americana* which grows over a wide area east of the Mississippi River. **Arkansas ash** is from *F. platycarpa;* **Japanese ash,** also called **tamo,** is from *F. mandschurica;* and European ash is from *F. excelsior.* **European ash** is heavier than American ash and is tough and elastic. It is valued for hockey sticks, tennis rackets, and tool handles. Japanese ash is a close-grained wood, but browner.

White ash has a density of 41 lb/ft^3 (657 kg/m^3) dry; **red ash,** *F. pennsylvanica,* 39 lb/ft^3 (625 kg/m^3); and **green ash,** *F. pennsylvanica lanceolata,* also called **water ash** and **swamp ash,** 44 lb/ft^3 (704 kg/m^3). This latter tree grows over the widest area throughout the states east of the Rockies, and it is commercially abundant in the southeast and Gulf states. It is a hardy tree, and it has been used for farm windbreaks in the Great Plains area. All these woods vary in tensile strength from 11,000 to 17,000 lb/in^2 (76 to 117 MPa). White ash has a compressive strength perpendicular to the grain of 2,250 lb/in^2 (15.5 MPa). **Mountain ash** and **black ash,** *F. nigra,* are also species of American ash. The latter, also called **brown ash** and **hoop ash,** is a northern tree and was formerly used in aircraft construction. It has a specific gravity of 0.53 when oven-dried, a compressive strength perpendicular to the grain of 1,260 lb/in^2 (8.7 MPa), and a shearing strength parallel to the grain of 1,050 lb/in^2 (7.2 MPa). **Oregon ash,** *F. oregona,* is somewhat lighter and not as strong as white ash. It grows along the west coast of Canada. **Blue ash,** *F. quadrangular,* grows in the central states. **Pumpkin ash,** *F. profunda,* grows in the lower Mississippi Valley and in Florida. A wood that has similar uses to ash, for handles, levers, and machine parts, but is harder than ash, is **hornbeam.** It is from the tree *Ostrya virginiana* of the eastern United States. The wood is very hard, tough, and strong, but is available only in limited quantities.

ASPEN. The wood of the aspen tree, *Populus tremula,* used chiefly for match stems and for making excelsior, but also for some inside construction work. The color is yellowish, and it is tough and close-grained. The tree is native to Europe. The American aspen is from the tree *P. tremuloides,* called also **American poplar,** and from the **large-tooth aspen,** *P. grandidentata.* Both species are also called *poplar,* and the lumber may be mixed with poplar and cottonwood. The trees grow in the lake and northeastern states and in the west. The heartwood is grayish white to light brown with a lighter-colored sapwood. It is straight-grained with a fine and uniform texture, but is soft and weak. It has a disagreeable odor when moist. The wood is used for excelsior, matches, boxes, and paper pulp. The pulp is easily bleached. **Salicin** is extracted from the bark.

ASPHALT.   A bituminous, brownish to jet-black substance, solid or semisolid, found in various parts of the world. It consists of a mixture of hydrocarbons and is fusible and largely soluble in carbon disulfide. It is also soluble in petroleum solvents and in turpentine. The melting points range from 90 to 100°F (32 to 38°C). Large deposits occur in Trinidad and Venezuela. Asphalt is of animal origin, as distinct from coals of vegetable origin. Native asphalt usually contains much mineral matter; and crude **Trinidad asphalt** has a composition of about 47% bitumen, 28 clay, and 25 water. **Artificial asphalt** is a term applied to the bituminous residue from coal distillation mechanically mixed with sand or limestone. Asphalt is used for roofings, road surfacing, insulating varnishes, acid-resistant paints, and cold-molded products.

   Bitumen refers to asphalt clean of earthy matter. It is obtained at Athabasca, Canada, in tar sands which are strip-mined. In general, bitumens have the characteristics that they are fusible and are totally soluble in carbon disulfide, as distinct from the **pyrobitumens,** albertite, elatarite, and coals, which are infusible and relatively insoluble in carbon disulfide. **Pyrogenous asphalts** are residues from the distillation of petroleum or from the treatment of wurtzilite. **Asphaltite** is a general name for the bituminous asphaltic materials which are fusible with difficulty, such as gilsonite and grahamite. It is thought that **benzopyrene,** a constituent of coal-tar pitch and asphalt, will produce cancer in living tissues. This material also occurs in shale oil, soot, and tobacco smoke.

   Rock asphalt, or **bituminous rock,** is a sandstone or limestone naturally impregnated with asphalt. The asphalt can be extracted from it, or it may be used directly for paving and flooring. **Kyrock** is a rock asphalt from Kentucky consisting of silica sand of sharp grains bound together with a bituminous content of about 7%. The crushed rock is used as a paving material. **Albertite** is a type of asphalt found originally in Albert County, New Brunswick, and first named **Albert coal.** It belongs to the group of asphalts only partly soluble in carbon disulfide, infusible, and designated as **carboids,** although they are true asphalts and not of vegetable origin. The commercial albertite is a type called **stellarite** from Nova Scotia. It is jet black, brittle, contains 22 to 25% fixed carbon, and yields oil and coke when distilled. It is easily lighted with a match and burns with a bright, smoky flame, throwing off sparks. The albertite found in Utah is called **nigrite** and contains up to 40% fixed carbon. A species found in Angola is called **libollite.** These materials are weathered asphalts. **Ipsonite** is a final stage of weathered asphalt. It is black, infusible, and only slightly soluble in carbon disulfide; contains 50 to 80% fixed carbon; and is very low in oxygen. It is found in Oklahoma, Arkansas, Nevada, and various places in South America. The **rafaelite** found in large beds on the eastern slopes of the Andes Mountains in Argentina is a form of ipsonite.

Cutback asphalt is asphalt liquefied with petroleum distillates, used for cementing down floor coverings and for waterproofing walls. Protective coatings based on asphalt cutback form economical paints for protection against salts, alkalies, and nonoxidizing acids at temperatures up to 110°F (43°C). They are black but may be pigmented with aluminum flake. They are often marketed under trade names such as **Atlastic** and **Protek-Coat.** Many corrosion-resistant coatings for chemical tanks and steel structures are asphalt solutions compounded with resins and fillers. **Perfecote,** for steel and concrete, contains an epoxy resin. The color is black, but it will accept a cover coat of colored plastic paint. High-temperature [500°F (260°C)] **asphaltic membranes** are applied to carbon-steel ducts and bypass stacks of incinerators for corrosion protection from acid condensate.

Modified asphalt, for laminating paper and for impregnating flooring felts, is asphalt combined with a rosin ester to increase the penetration, tack, and adhesion; but asphalt for paints and coatings may also be modified with synthetic resins. **Emulsified asphalt** is an asphalt emulsion in water solution, used for floor surfacing, painting pipes, and waterproofing concrete walls. Emulsified asphalts may be marketed under trade names such as **Elastex** and **Ebontex.** **Thermotex** is an emulsified asphalt mixed with asbestos fibers, used for painting steam pipes. **Brunswick black** is a mixture of asphaltite with fatty acid pitch in a volatile solvent, used for painting roofs. **Amiesite** is asphalt mixed with rubber latex or is a premixed asphalt with an aggregate employed for road filling. Rubbers are sometimes incorporated into paving asphalts to give resilience. The natural or synthetic rubber is mixed into the asphalt either in the form of powder or as a prepared additive. **Catalyzed asphalt** is asphalt treated with phosphoric anhydride, P_2O_5, used for road construction to resist deterioration of the pavement from weathering. An asphalt mix developed by Shell Chemical Co. for aircraft runways to resist the action of jet fuels is petroleum asphalt with an epoxy resin and a plasticizer. **Flooring blocks** and **asphalt tiles** are made in standard shapes and sizes from mixtures of asphalt with fillers and pigments. They are sold under many trade names, such as **Elastite** and **Accotile.**

Oil asphalt, petroleum asphalt, petroleum pitch, or **asphalt oil** is the heavy black residue left after removal of the tar tailings in the distillation of petroleum. It contains 99% bitumen, is not soluble in water, and is durable. As it adheres well to metals, wood, or paper and forms a glossy surface, it is used in roofings or is mixed with natural asphalt for paints and coatings. It is also used for roads. **Vanadiset** is a series of resin fractions of petroleum asphalt with small amounts of vanadium pentoxide, varying from semisolids to a brittle solid. They are used as softeners for rubber and in bitumen paints.

AVOCADO OIL.   An oil obtained from the ripe, green, pear-shaped fruit of the avocado, *Persea americana,* a small tree of which more than 500 varieties grow profusely in tropical America. The oil is also called **alligator pear oil.** In California, where the fruit is grown for market, it is also known as **Calavo.** The fruits weigh up to 3 lb (1.4 kg), and the seeds are 8 to 26% of the fruit. The fresh pulp contains 71% water, 20 oil, and 2.37 proteins. The seeds contain about 2% of an oil, but the avocado oil is extracted from the fruit pulp, the dehydrated pulp yielding 70% oil. In Central America the oil is extracted by pressing in bags, and the oil has been used by the Mayans since ancient times for treating burns and as a pomade. It contains 77% oleic acid, 10.8 linoleic, 6.9 palmitic, and 0.7 stearic, with a small amount of myristic and a trace of arachidic acid. It is also rich in lecithin, contains **phytostearin,** and is valued for cosmetics because it is penetrating, as lanolin is. It also contains **mannoketoheptose,** a highly nonfermentable sugar. The oil has good keeping qualities and is easily emulsified. The oil-soluble vitamins are absorbed through the skin, and the oil for cosmetics is not wintered in order to retain the sterols. The specific gravity is 0.9132. Another oil used in cosmetics and for lubricating fine mechanisms is **ben oil,** a colorless to yellow oil obtained from the seeds of trees of the genus *Moringa,* notably *M. aptera, M. oleifera,* and *M. pterygosperma,* of Arabia, Egypt, India, and the Sudan. The latter species is also grown in Jamaica. The seeds contain 25 to 34% oil varying from a liquid to a solid, with specific gravity of 0.898 to 0.902 and saponification value of 179 to 187.

BABASSU OIL.   An oil similar to coconut oil obtained from the kernels of the nut of the palm tree *Attalea orbignya* which grows in vast quantities in northeastern Brazil. There are two to five long kernels in each nut, the kernel being only 9% of the heavy-shelled nut, and these kernels contain 65% oil. A bunch of the fruits contains 200 to 600 nuts. The oil contains as much as 45% lauric acid and is a direct substitute for coconut oil for soaps, as an edible oil, and as a source of lauric, capric, and myristic acids. The melting point of the oil is 72 to 79°F (22 to 26°C), specific gravity 0.868, iodine value 15, and saponification value 246 to 250. **Tucum oil,** usually classified with babassu but valued more in the bakery industry because of its higher melting point, is from the kernels of the nut of the palm *Astrocaryum tucuma* of northeastern Brazil. The oil is similar but heavier with melting point up to 95°F (35°C), and it consists of 49% lauric acid. In Colombia it is called **guere palm.**

Another similar oil is **murumuru oil,** from the kernels of the nut of the palm *A. murumuru,* of Brazil. The name is a corruption of the two Carib words *marú* and *morú,* meaning bread to eat. The oil contains as much as 40% lauric acid, with 35% myristic acid, and some palmitic, stearic, linoleic, and oleic acids. It is usually marketed as babassu oil. The **awarra palm,** *A. janari,* of the Guianas, yields nuts with a similar oil. **Cohune oil** is a white fat from the kernels of the nut of the palm *Attalea cohune* of Mexico and Central America. It is a small tree yielding as many as 2,000 nuts per year. The oil has the appearance and odor of coconut oil, and it contains 46% lauric acid, 15 myristic, 10 oleic, with stearic, capric, and linoleic acids. All these oils yield a high proportion of glycerin. Cohune oil has a melting point of 64 to 68°F (18 to 20°C), saponification value 252 to 256, iodine value 10 to 14, and specific gravity 0.868 to 0.971. The cohune nut is much smaller than the babassu but is plentiful and easier to crack. **Curua oil** is from the nut of the palm *A. spectabilis* of Brazil. It is similar to cohune oil and is used for the same purposes in soaps and foods. **Mamarron oil** is a cream-colored fat with the odor and characteristics of coconut oil, obtained from another species of *Attalea* palm of Colombia. Another oil high in lauric acid, and similar to babassu oil, is **corozo oil,** obtained from the kernels of the nuts of the palm *Corozo oleifera* of Venezuela and Central America. **Macanilla oil** is a similar oil from the kernels of the nuts of the palm *Guilielma garipaes* of the same region. **Buri oil** is from the nuts of the palm *Diplothemium candescens* of Brazil.

BABBITT METAL. The original name for tin-antimony-copper white alloys used for machinery bearings, but the term now applies to almost any white bearing alloy with either tin or lead base. The original babbitt, named after the inventor, was made by melting together 4 parts by weight of copper, 12 tin, and 8 antimony, and then adding 12 parts of tin after fusion. It consisted, therefore, of 88.9% tin, 7.4 antimony, and 3.7 copper. This alloy melts at 462°F (239°C). It has a Brinell hardness of 35 at 70°F (21°C) and 15 at 212°F (100°C). As a general-utility bearing metal, the original alloy has never been improved greatly, and makers frequently designate the tin-base alloys close to this composition as **genuine babbitt.**

Commercial **white bearing metals** now known as *babbitt* are of three general classes: tin-base, with more than 50% tin hardened with antimony and copper, and used for heavy-duty service; intermediate, with 20 to 50% tin, having lower compressive strength and more sluggish as a bearing; and lead-base, made usually with antimonial lead with smaller amounts of tin together with other elements to hold the lead in solution. These **lead-base babbitts** are cheaper and serve to conserve tin in times of scarcity of that metal, but they are suitable only for light service, although many ingenious combinations of supplementary alloying elements have sometimes been used to give hard, strong bearings with little tin. The high-grade babbitts, however, are usually close to the original babbitt in composition. **SAE Babbitt 11,** for connecting-rod bearings, has 86% tin, 5 to 6.5% copper, 6 to 7.5% antimony, and not over 0.50% lead. A babbitt of this kind will have a compressive strength up to 20,000 lb/in^2 (138 MPa) compared with only 15,000 lb/in^2 (103 MPa) for high-lead alloys.

Copper hardens and toughens the alloy and raises the melting point. Lead increases fluidity and raises antifriction qualities, but softens the alloy and decreases its compressive strength. Antimony hardens the metal and forms hard crystals in the soft matrix, which improve the alloy as a bearing metal. Only 3.5% of antimony is normally dissolved in tin. In the low-antimony alloys, copper-tin crystals form the hard constituent; and in the high-antimony alloys, antimony-tin cubes are also present. Alloys containing up to 1% arsenic are harder at high temperatures and are fine-grained, but arsenic is used chiefly for holding lead in suspension. Zinc increases hardness but decreases frictional qualities, and with much zinc the bearings are inclined to stick. Even minute quantities of iron harden the alloys, and iron is not used except when zinc is present. Bismuth reduces shrinkage and refines the grain, but lowers the melting point and lowers the strength at elevated temperatures. Cadmium increases the strength and fatigue resistance, but any considerable amount lowers the frictional qualities, lowers the strength at higher temperatures, and causes corrosion. Nickel is used to increase strength but raises the melting point. The normal amount of copper in babbitts is 3 or 4%, at which point the maximum fatigue-resisting properties are obtained with about 7% antimony. More than 4% copper tends to weaken the alloy and raises the melting point. When the copper is very high, tin-copper crystals are formed and the alloy is more a bronze than a babbitt. All the SAE babbitts contain some arsenic, ranging from 0.10% in the high-tin **SAE Babbitt 10** to about 1% in the high-lead **SAE Babbitt 15.** The first of these contains 90% tin, 4.5 antimony, 4.5 copper, and 0.35 lead, while babbitt 15 has 82% lead, 15 antimony, 1 tin, and 0.60 copper.

Because of increased speeds and pressures in bearings and the trend to lighter weights, heavy cast babbitt bearings are now little used despite their low cost and ease of casting the alloys. The alloys are used mostly as **antifriction metals** in thin facings on steel backings, the facing being usually less than 0.010 in (0.03 cm) thick, in order to increase their ability to sustain higher loads and dissipate heat.

Babbitts are marketed under many trade names, the compositions generally following the SAE alloy standards but varying in auxiliary constituents, the possibilities for altering the physical qualities by composition rearrangement being infinite. Some of the trade names that have been used for babbitt-type alloys marketed in ingots are **Leantin** and **Cosmos metal** for high-lead alloys, **stannum metal** for high-tin alloys, and **Lubeco metal** and **Lotus metal** for medium-composition alloys. **Hoo Hoo metal** and **nickel babbitt** are high-tin alloys containing nickel, while **Silver babbitt** has no tin but contains a small amount of silver to aid retention of the lead and to give hardness at elevated temperatures. **Glyco** is the name of a group of lead-base alloys of Joseph T. Ryerson & Son, Inc. **Satco,** of NL Industries, Inc., is a high-melting-point alloy for heavy service. It melts at 788°F (420°C). **Tinite** is a tin-base metal hardened with copper. **Ajax bull** contains 76% lead, 7 tin, and 17 antimony, modified with other elements.

BAGASSE. The residue left after grinding sugarcane and extracting the juice, employed in making paper and fiber building boards. In England it is called **megass.** The fiber contains 45% cellulose, 32 pentosan, and 18 lignin. It is marketed as dry- and wet-separated, and as dry fiber. The dry-separated fibers bulk 4.5 lb/ft^3 (72 kg/m^3), with 62 to 80% passing a 100-mesh screen. The dry fiber bulks 6 to 8 lb/ft^3 (96 to 128 kg/m^3) and is about 14 mesh. The fibers mat together to form a strong, tough, light, absorptive board. The finer fibers in Cuba and Jamaica are soaked in molasses and used as a cattle feed under the name of **molascuit.** **Celotex** is the trade name of the Celotex Corp. for wallboard, paneling, and acoustic tile made from bagasse fibers. **Ferox-Celotex** is the material treated with chemicals to make it resistant to fungi and termites. **Celo-Rock** is the trade name for Celotex-gypsum building boards. **Acousti-Celotex** is Celotex perforated to increase its sound-absorbing efficiency. In India, the Philippines, and some other countries where sugarcane is plentiful, paper is made from the bagasse. Newsprint is made from a mixture of mechanical and chemical bagasse pulp, and writing papers may be made by delignifying the bagasse and digesting with soda. **Aconitic acid,** $HOOCCH:C(COOH)CH_2COOH$, occurs in bagasse and is extracted from Louisiana cane. The acid is esterified for use as a plasticizer for vinyl resins, or sulfonated for use as a wetting agent. This acid is also produced as a white powder of melting point 383°F (195°C) by the dehydration of citric acid.

BALATA. A nonelastic rubber obtained chiefly from the tree *Manilkara bidentata* of Venezuela, Brazil, and the Guianas. It is similar to gutta percha and is used as a substitute. The material contains a high percentage of gums and is more tacky than rubber, but it can be vulcanized. It differs from rubber in being a transisomer of isoprene with a different polymerization. Balata has been used principally for transmission and conveyor belts and for golf ball covers. For conveyer belts, heavy duck is impregnated with balata solution and vulcanized. The belts have high tensile strength, good flexibility, and wear resistance. The wood of the balata tree is used for cabinetwork and for rollers and bearings. It is called **bulletwood** in the Guianas, but this name is also applied to the wood of the gutta-percha trees of Asia. The wood is extremely hard and durable and has a density of 66 lb/ft^3 (1,057 kg/m^3). It has a deep-red color and a fine, open grain.

BALSA WOOD.   The wood of large and fast-growing trees of the genus *Ochroma* growing from southern Mexico to Ecuador and northern Brazil. It is the lightest of the commercial woods and combines also the qualities of strength, stiffness, and workability. It is about one-fourth the weight of spruce, with a structural strength half that of spruce. The crushing strength is 2,150 lb/in^2 (14 MPa). The wood is white to light yellow or brownish and has a density of about 8 lb/ft^3 (128 kg/m^3) from a 4-year-old tree. Wood from a 6-year-old tree has a density of 10 to 12 lb/ft^3 (160 to 192 kg/m^3). Its peculiar cellular structure makes it valuable as an insulating material for refrigeration. It is also used for life preservers, buoys, floats, paneling, vibration isolators, insulating partitions, and inside trim of aircraft. The small pieces are used for model airplanes. **Balsa sawdust** may be used as a lightweight filler for plastics.

Much of the commercial wood is from the tree *O. grandiflora* of Ecuador. **Barrios balsa,** *O. concolor,* grows from southern Mexico through Guatemala and Honduras. **Limos balsa** is from the tree *O. limonensis* of Costa Rica and Panama, and **Santa Marta balsa** is *O. obtusa* of Colombia. **Red balsa** is from *O. velutina* of the Pacific Coast of Central America. The balsa known in Brazil as **Sumaúma** is from a kapok tree *Ceiba pentandra.* It is used for life preservers and rafts and is quite similar to balsa. A Japanese lightweight wood used for floats, instruments, and where lightness is required is **Kiri,** from the tree *Paulownia tomentosa.* It has a density of 14 to 19 lb/ft^3 (224 to 304 kg/m^3), has a coarse grain, but is strong and resists warping. Grown as a shade tree since 1834 under the names **paulownia** and **empress tree,** it is now common in the United States, and the wood is used as a lightweight crating lumber.

BALSAM FIR.   The wood of the coniferous tree *Abies balsamea* of the northeastern United States and Canada. It is brownish white and soft and has a fine, even grain. It is not strong and not very durable, and it is used chiefly for pulpwood and for packing boxes and light construction. The density is 26 lb/ft^3 (417 kg/m^3). Liquid pitch comes from blisters on the outer bark. It was formerly used as a transparent adhesive. **Canada balsam,** or **Canada turpentine,** is a yellowish, viscous oleoresin liquid of pleasant odor and bitter taste, obtained from the buds of the tree. The specific gravity is 0.983 to 0.997. It is a class of turpentine and is used as a solvent in paints and polishes, in leather dressings, adhesives, and perfumes. It is also referred to as **balm of Gilead** for medicinal and perfumery use, but the original balm of Gilead, marketed as buds, was from the small evergreen tree *Balsamodendron gileadense* of the Near East. *Southern balsam fir* is **Frazer fir,** from the tree *A. fraseri* of the Appalachian Mountains. The wood is similar to balsam fir.

BAMBOO. A genus of gigantic treelike grasses, of the order *Graminaceae,* of which the *Bambusa arundinacea* is the most common species. It grows most commonly in Indonesia, the Philippines, and southern Asia, but many species have been brought to Latin America and to the southern United States. The stems of bamboo are hollow and jointed and have an extremely hard exterior surface. They sometimes reach more than 1 ft (0.3 m) in diameter and are often 50 ft (15 m) high, growing in dense masses. Nearly 1,000 species are known. The *B. spinosa* of the Philippines grows as much as 10 ft (3 m) in one week. Bamboo is a material which has had innumerable uses. The stalks are used for making pipes, buckets, baskets, walking sticks, fishing poles, rug-winding poles, lance shafts, window blinds, mats, arrows, and for building houses and making furniture. The density is about 22 lb/ft^3 (352 kg/m^3). **Tonkin bamboo** is strong and flexible and is used for making fishing poles. **Tali bamboo** of Java, *Gigantochloa apus,* is used for construction. **Betong bamboo,** *G. asper,* is one of the largest species. **Giant bamboo,** *Dendrocalamus gigantea,* of Sri Lanka, grows to a height of 100 ft (30 m). The fast-growing **eeta bamboo** is used in India as a source of cellulose for rayon manufacture.

BARITE. Sometimes spelled **baryte,** and also called **heavy spar,** and in some localities known as **tiff.** A natural **barium sulfate** mineral of the theoretical composition of $BaSO_4$, used chiefly for the production of lithopone, in chemical manufacture, and in oil-drilling muds. Mixed with synthetic rubber, it is used as a seal coat for roads. For chemicals it is specified 90 to 95% pure $BaSO_4$, with not more than 1% ferric oxide. Prime white and floated grades are used for coating paper. **Baroid,** of NL Industries, Inc., used in oil wells, is barite ore crushed, dried, and finely ground. **Artificial barite, permanent white,** and

blanc fixe are names for white, fine-grained precipitated paint grades. **Micronized barite,** for rubber filler, is a fine white powder of 400 to 1,000 mesh. Barite is widely distributed and especially associated with ores of various metals or with limestones. It occurs in crystals or massive form. It may be colorless, white, or light shades of blue, red, and yellow, and transparent to opaque. Its hardness is Mohs 3 to 3.5, and its specific gravity is 4.4 to 4.8. It is insoluble in water. The mineral is produced in the western United States and from Virginia to Georgia. The barite of Cartersville, Georgia, contains 96% $BaSO_4$, 0.6 iron, with silica, alumina, and traces of calcium, strontium, and magnesium. Large deposits of high-grade barite occur in Nova Scotia. In the west, much ground crude barite is used as a drilling mud in oil wells. The white pigment marketed by American Zinc Sales Co. under the name of **Azolite** is 71% barium sulfate and 29% zinc sulfide in 325-mesh powder. **Sunolith,** of Wishnick-Tumpeer, Inc., is a similar product. A substitute for barite for some filler uses is **witherite,** an alteration mineral of composition $BaCO_3$, which is **barium carbonate,** found associated with barite. Precipitated barium carbonate is a white, tasteless, but poisonous powder used in rat poisons, optical glass, ceramics, and pyrotechnics; as a flatting agent in paints; and as a filler for paper. With ferric oxide it is used for making ceramic magnets. **Barium oxide,** BaO, of 99.99% purity, is made by the reduction of barite. It is used as an additive in lubricating oils.

BARIUM. A metallic element of the alkaline earth group, symbol Ba. It occurs in combination in the minerals witherite and barite, which are widely distributed. The metal is silvery white and can be obtained by electrolysis from the chloride, but it oxidizes so easily that it is difficult to obtain in the metallic state. Powdered or granular barium is explosive when in contact with carbon tetrachloride, fluorochloromethanes, and other halogenated hydrocarbons. Its melting point is 1562°F (850°C) and its specific gravity 3.78. The most extensive use of barium is in the form of its compounds. The salts which are soluble, such as sulfide and chloride, are toxic. An insoluble, nontoxic barium sulfate salt is used in radiography. Barium compounds are used as pigments, in chemical manufacturing, and in deoxidizing alloys of tin, copper, lead, and zinc. Barium is introduced into lead-bearing metals by electrolysis to harden the lead. When barium is heated to about 392°F (200°C) in hydrogen gas, it forms **barium hydride,** BaH_2, a gray powder which decomposes on contact with water and can be used as a source of nascent hydrogen for life rafts. Barium is also a key ingredient in ceramic superconductors.

BARIUM CHLORIDE. A colorless crystalline material of composition $BaCl_2 \cdot 2H_2O$, or in anhydrous form without the water of crystallization. The specific gravity is 3.856, and the melting point 1760°F (960°C). It is soluble in water to the extent of 25% at 68°F (20°C) and 37% at 212°F (100°C). In the mechanical industries it is used for heat-treating baths for steel, either alone or mixed with potassium chloride. The molten material is free from fuming and can be held at practically any temperature within the range needed for tempering steels. It is also used for making boiler compounds, for softening water, as a mordant in dyeing and printing inks, in tanning leather, in photographic chemicals, and in insecticides. Two of the most widely used red colorants, **Lithol Red** and **Red Lake,** are made by treating a hot, aqueous solution of barium chloride with the appropriate diazo dye. **Barium chlorate,** $Ba(ClO_3)_2 \cdot H_2O$, is a colorless crystalline powder, soluble in water. The melting point of the anhydrous material is 777°F (414°C). It is used in explosives as an oxygen carrier and in pyrotechnics for green-colored light. **Barium fluoride,** BaF_2, is used in crystal form for lasers. When "doped" with uranium, it has an output wavelength of 8,530 nft (2,600 nm). Doping with other elements gives diffused wavelengths for different communication beams. **Barium cyanide,** $Ba(CN)_2$, is a poisonous, colorless, crystalline material melting at 1112°F (600°C). It is marketed by Koppers Co. as a 30% water solution for adding to cyanide plating baths, in which it removes carbonates and increases the current efficiency.

BARIUM NITRATE. Also called **nitrobarite.** A white crystalline powder of composition $Ba(NO_3)_2$, with specific gravity of 3.24, melting at 1098°F (592°C), and decomposing at higher temperatures. It is a barium salt of nitric acid obtained by roasting barite with coke, leaching out the precipitated barium sulfide, precipitating as a carbonate by the addition of soda ash, and then dissolving in dilute nitric acid. It has a bitter metallic taste and is poisonous. Barium nitrate is used in ceramic glazes, but its chief use is in **pyrotechnics.** It gives a pale-green flame in burning and is used for green signals and flares, and for white flares in which the delicate green is blended with the light of other extremely luminous materials. It is also used as an oxygen carrier in **flare powders** and to control the time of burning of the aluminum or magnesium. **Sparklers** are composed of aluminum powder and steel filings with barium nitrate as the oxygen carrier. The steel filings produce the starlike sparks. **Barium nitrite,** $Ba(NO_2)_2$, decomposes with explosive force when heated. **Barium oxalate,** BaC_2O_4, is used in pyrotechnics as a combustion retarder.

BARLEY. The seed grains of the annual plant *Hordeum vulgare* of which there are many varieties. It is one of the most ancient of the cereal grains. The plant is hardy, with a short growing season, and can be cultivated in cold latitudes and at high altitudes, giving high yields per acre. The grains grow in a dense head with three spikelets, and the six-row variety has a high protein content, but has low gluten, thus making a poor breadstuff. **Pearl barley** is the husked and polished grain. When used for cattle feed, barley produces lean meats. The chief industrial use is for making **malt,** for which the two-rowed varieties with low protein and thin husk are used. Malt is barley that has been germinated by moisture and then dried. Malting develops the diastase enzyme, which converts the insoluble starch to soluble starch and then to sugars. It is used for brewing beer and for malt extracts. **Caramel malt** is browned with high-temperature drying and is used for the dark-colored **bock beer. Barley straw** is employed in Europe and Asia for making braided plaits for hats. In the United States it is used for packing material, especially for glassware.

BASALT. A dense, hard, dark-brown to black igneous rock, consisting of feldspar and augite and often containing crystals of green olivine. It occurs as trap or as volcanic rock. The specific gravity is 2.87 to 3, and it is extremely hard. Masses of basalt are frequently found in columns or prisms, as in the celebrated basalt cliffs of northern Ireland. It differs from granite in being a fine-grained extrusive rock and in having a high content of iron and magnesium. Basalt is used in the form of crushed stone for paving, as a building stone, and for making rock wool. A Russian cast basalt used for electrical insulators is called **angarite.** In Germany cast basalt has been used as a building stone, for linings, and for industrial floors. It is made by melting the crushed and graded basalt and then tempering by slow cooling. The structure of the cast material is dense with needlelike crystals, and it has a Mohs hardness of 8 to 9. **Basalt glass** is not basalt, but pumice.

BASSWOOD. The wood of several species of lime trees, *Tilia americana, T. heterophylla, T. glabra,* and *T. pubescens,* all native to the United States and Canada. The European **limewood,** from the tree *T. cordata,* is not called basswood. The wood of *T. glabra,* called in the eastern states the **lime tree** and the **linden,** and also **white basswood,** *T. heterophylla,* is used for containers, furniture, and such millwood as blinds. It is soft and lightweight and has a fine, even grain, but is not very strong or durable on exposure. The white sapwood merges gradually with the yellow-brown heartwood. The specific gravity is 0.40 when oven-dried, and the compressive strength perpendicular to the grain is 620 lb/in^2 (4.3 MPa).

BATE. Materials used in the leather industry to remove lime from skins and to make them soft and flaccid before tanning by bringing the collagen into a flaccid or unswollen condition. Since ancient times, dung has been used for this purpose, and until recent years the U.S. tanning industry imported dog dung from Asia Minor for bating leather. **Artificial bates** are now used because of their greater uniformity and cleanliness. **Boric acid** is sometimes used for deliming, and it gives a silky feel to the leather, but most bates have both a deliming and an enzyme action. **Trypsin** is a group of enzymes from the pancreatic glands of animals, and its action on skins is to dissolve protein. They are generally used with ammonium chloride or other salt. **Oropon,** of Rohm & Haas Co., is this material carried in wood flour and mixed with a deliming salt. **Sulfamic acids** are also used as bates. The lime compounds used for dehairing are called **depilating agents.**

BAUXITE. A noncrystalline, earthy-white to reddish mineral, massive or in grains, having composition $Al_2O_3 \cdot 2H_2O$, theoretically containing 74% alumina. It is the most important ore of aluminum, but is also used for making aluminum oxide abrasives, for refractories, white cement, and decolorizing and filtering.

Bauxite is graded on the Al_2O_3 content. High-grade bauxite, Grade A, contains a minimum of 55% alumina and a maximum of 8% silica. Grade B contains a minimum of 50% alumina with a silica content from 8 to 16%. Chemical grades should have less than 2.5% Fe_2O_3. Grades appearing in price quotations with up to 84% alumina content are calcined and are based on the dehydrated alumina content of the ore.

Bauxite has a high melting point, 3308°F (1820°C), and can be used directly as a refractory. Cement-making **white bauxite** from Greece ranks very high in alumina content. Brazilian, Arkansas, and Indian ores also contain some titanium oxide, and the Surinam ore has as high as 3% TiO_2. Two kinds of **red bauxite** are found in Italy, a dark variety containing 54 to 58% Al_2O_3, and only 2 to 4 SiO_2, but having 22 to 26% Fe_2O_3, and 2 to 3 TiO_2, and a light variety containing 60 to 66% Al_2O_3, 5 to 9 silica, 10 to 16 iron oxide, and 3 to 5 titanium oxide. The best French white bauxite contains 66 to 74% alumina, 6 to 10 silica, 2 to 4 iron oxide, and 3 to 4 titanium oxide. It is preferred for ceramic and chemical purposes, while the best grade of the red variety is used for producing aluminum, and the inferior grade for refractories and for cement manufacture. Malayan and Indonesian bauxite averages 57 to 60% Al_2O_3, 6.7 Fe_2O_3, 3 to 5 SiO_2, and 0.9 to 1 TiO_2. The large deposits on Ponape and other Pacific islands average 50 to 52% alumina, 3 to 6 silica, and 10 to 20 Fe_2O_3, but the bauxite of

Hawaii contains only 35% alumina with up to 15% silica. **Phosphatic bauxite,** from the island of Trauhira off the coast of Brazil, is a cream-colored porous rock containing 31.5% alumina, 25.2 P_2O_5, 7.3 iron oxide, 6.8 silica, and 1.3 titania. **Diaspore,** $Al_2O_3 \cdot H_2O$, mined in Missouri, and **gibbsite,** $Al_2O_3 \cdot 3H_2O$, from the Guianas, are bauxites also used for refractories. Gibbsite is also called **wavellite. Filter bauxite,** or **activated bauxite,** is bauxite that has been crushed, screened, and calcined, and it is usually in 20- to 60- and 30- to 60-mesh grades. It may be sold under trade names such as **Porocel** and **Floride.** It is preferred to fuller's earth for oil-refinery filtering because it can be revivified indefinitely by calcining. **Calcined bauxite** for the abrasive industry is burned bauxite and contains 78 to 84% alumina. **Laterite,** or **ferroginous bauxite,** has been used in Europe to produce alumina and iron. The laterite of Oregon contains 35% alumina, about 35 iron oxide, and about 7 silica. Low-alumina, high-silica bauxites can be lime-sintered to release the sodium aluminate which goes back into the process while the silicate goes out with the calcium, thus giving high alumina recovery with low soda loss. **Anorthosite,** an abundant **aluminum silicate** mineral containing up to 50% silica, is also used to produce aluminum. The anorthosite of Wyoming is sintered with limestone and soda ash and calcined to yield alumina and a by-product portland cement base of dicalcium silicate.

BEARING MATERIALS. A large variety of metals and nonmetallic materials in monolithic and composite (laminate) form are used for bearings. **Monolithic ferrous bearings** are made of gray cast iron, pressed and sintered iron and steel powder, and many wrought steels, including low- and high-carbon plain-carbon steels, low-alloy steels, alloy steels, stainless steels, and tool steels. Most **cast-iron bearings** are made of gray iron because it combines strength with the lubricity of graphitic carbon. **Pressed and sintered bearings** can be made to controlled porosity and impregnated with oil for lubricity. Because of its wide use in ball and roller bearings, one of the best-known bearing steels is **AISI 52100 steel,** a through-hardening 1% carbon and 1.3 to 1.6% chromium alloy steel. Many steels, however, are simply surface-hardened for bearing applications. In recent years, the performance of **bearing steels** has been markedly improved by special melting practices that reduce the presence of nonmetallic inclusions.

   Monolithic nonferrous bearings include **copper-zinc bronze, leaded bronzes, unleaded bronzes,** and an **aluminum-tin alloy,** containing about 6% tin as the principal alloying element. The bronze and aluminum alloy provide similar load-bearing capacity and fatigue resistance, but the bronze is somewhat better in resistance to corrosion by fatty acids that can form with petroleum-based oils. It is also less prone to seizure and abrasion from mating shafts; more able to embed foreign matter and thus prevent shaft wear; and more tolerant of shaft misalignment. The load-bearing capacity of tin bronzes depends on the lead content. Low-lead and lead-free tin bronzes have the highest load capacity, about 5,000 lb/in^2 (34 MPa), and fatigue strength. Applications include auto engine starter-motor bearings, or bushings, for the copper-zinc bronze; auto engine connecting-rod bearings for the aluminum alloy; and various bearings in motors, machine tools, and earthmoving equipment for the tin bronzes.

Monolithic bearings are also made of cemented tungsten and chromium carbides, plastics, carbon-graphite, wood, and rubber. Plastics provide good combinations of inherent lubricity, corrosion resistance, and adequate strength at room to moderately elevated temperatures. Thermal conductivity and other performance features that may be required can be provided by metal and other fillers. **Plastic bearings** can be made of acetal, nylon, polyester, ultrahigh-molecular-weight polyethylene, polytetrafluoroethylene, polysulfone, polyphenylene sulfide, polyimide, polybenzimidazole, and polyamide-imide. **Carbon-graphite bearings** are more heat-resistant but rather brittle, thus limited to nonimpact applications. **Wood bearings** are made of maple and the hard lignum vitae. **Rubber bearings,** usually steel-backed, are used for applications requiring resilience.

Nonferrous metals are widely used in dual- or trimetal systems. **Dual-metal bearings** comprise a soft, thin, inner liner metallurgically bonded to stronger backing metal. Steel lined with bronze containing 4 to 10% lead provides the highest load-bearing capacity—8,000 lb/in^2 (55 MPa), or about twice that of the bronze alone—and fatigue strength. However, the aluminum alloy with a steel backing provides the best corrosion resistance and only moderately less load-bearing capacity. Tin and lead babbitt linings excel in surface qualities conducive to free-sliding conditions and are used with steel, bronze, or aluminum-alloy backings; load-bearing capacities range from 1,500 to 7,000 lb/in^2 (10 to 48 MPa). Dual-metal systems cover a gamut of bearings for motors, pumps, piston pins, camshafts, and connecting rods.

Trimetal bearings, all with steel backings, have an inner liner of tin or lead babbitt and an intermediate layer of a more fatigue-resistant metal, such as leaded bronze, copper-lead, aluminum-tin, tin-free aluminum alloys, silver, or silver-lead. Load-bearing capacity ranges from 1,500 to 12,000 lb/in^2 (10 to 83 MPa). The silver bearing systems provide the best combination of load-bearing capacity, fatigue and corrosion resistance, and compatibility to mating materials; but a lead babbitt, medium-lead bronze and steel system is a close second, sacrificing only a moderate reduction in corrosion resistance but at a reduction in cost. Applications include connecting-rod, camshaft, and main bearings in auto engines and reciprocating aircraft engines.

Hybrid bearings comprise silicon-nitride balls, a fiber-reinforced polyimide separator, and a steel race. The balls are much more lightweight than steel balls, markedly reducing centrifugal force, and ride on the low-friction separator, increasing wear life and running speed. They also have high fatigue resistance, increasing service life. These bearings are used in medical instruments and machine-tool spindles.

BEECH. The wood of several species of beech trees, *Fagus atropunicea, F. ferruginea,* and *F. grandifolia,* common to the eastern parts of the United States and Canada. The wood is strong, compact, fine-grained, durable, and light in color, similar in appearance to maple. The density is 47 lb/ft^3 (753 kg/m^3). It is employed for tool handles, shoe lasts, gunpowder charcoal, veneer, cooperage, pulpwood, and small wooden articles such as clothespins. The beech formerly used for aircraft, *F. grandifolia,* has a specific gravity, oven-dried, of 0.66, a compressive strength perpendicular to the grain of 1,670 lb/in^2 (12 MPa), and a shearing strength parallel to the grain of 1,300 lb/in^2 (9 MPa). The wood may be obtained in large pieces, as the tree grows to a height of 100 ft (30.5 m) and a diameter of 4 ft (1.2 m). It grows from the Gulf of Mexico northward into eastern Canada. **White beech** refers to the light-colored heartwood. **Red beech** is from trees with dark-colored heartwood. The sapwood of beech is white tinged with red and is almost indistinguishable from the heartwood. The wood is noted for its uniform texture and its shock resistance.

Antarctic beech, *F. antarctica,* known locally as **rauli,** grows extensively in southern Chile. It is commonly called by the Spanish word **roble,** or oak, in South America, and is used for cooperage to replace oak. It has a coarser grain than American beech. **European beech,** *F. sylvatica,* is reddish; has a close, even texture; is not as heavy as American beech; but is used for tools, furniture, and small articles. **New Zealand beech,** known as *red beech* and **tawhai,** is from the very large tree *Nothofagus solandri.* The wood has a density of 44 lb/ft^3 (705 kg/m^3), is brown, and has high strength and durability. **Silver beech,** of New Zealand, is *N. menziesii.* The trees grow to a height of 80 ft (24.3 m) and a diameter of 2 ft (0.61 m). The wood is light-brown, straight-grained, and strong and has a density of 34 lb/ft^3 (545 kg/m^3). It is used for furniture, implements, and cooperage.

BEEF. The edible meat from full-grown beef cattle, *Bos taurus.* The meat from the younger animals that have not eaten much grass is called **veal** and is lighter in color and softer. The production of beef and beef products is one of the great industries of the world. In the industrial countries, much of the beef is prepared in organized packing plants, but also the production from city slaughterhouses is important. After slaughter and preparation of the animal, the beef is marketed in animal quarters either chilled or frozen. Fresh-killed beef from local slaughterhouses is also chilled to remove animal heat before marketing. The amount of marketable beef averages 55 to 61% of the live weight of the animal. The hide is from 5 to 7%, the edible and inedible fat and tallow are 3.5 to 7.5%, and the bones, gelatin, and glue material are 2.8 to 4.9%. From 10 to 17% of the live weight may be shrinkage and valueless materials, although the **tankage,** which includes entrails and scraps, is sold as fertilizer. **Offal** includes tongues, hearts, brains, tripe (stomach lining), livers, tails, and heads, and may be from 3 to 5.5% of the live animal. The **glands** are used for the production of insulin. **Lipid** is the name for a yellow waxy solid melting at 212°F (100°C), extracted from beef spinal cord after removal of cholesterol. It contains phosphatides and complex acids and is used in medicine as an emulsifier and anticoagulant. **Cortisone,** used in medicine, is a steroid produced from ox bile, but now it is made synthetically.

Canned beef, which includes **corned beef, canned hash** (beef mixed with potatoes), and various **potted meats,** is not ordinarily made from the beef of animals suitable for sale as chilled or frozen beef, but is from tough or otherwise undesirable meat animals, or from animals rejected by government inspectors as not suitable for fresh beef. In the latter case, the beef canned is held at high temperature for a sufficient time to destroy any bacteria likely to be in the fresh meat. Federal specifications for canned corned beef require freedom from skin, tendons, and excessive fat, and a maximum content of not more than 3.25% salt and 0.2 saltpeter. Government inspection of beef for health standards is rigid, but the federal grading of beef is little more than a rough price evaluation.

Beef extract was first made by Prof. Justus von Liebig in 1840 as a heavy concentrated paste that could be kept indefinitely. It is now made on a large scale in both paste and cubes, and it is used for soups and hot beverages, but much of the extract marketed in bouillon cubes is highly diluted with vegetable protein. The so-called *nonmeat* beef extract is made with corn and wheat hydrolysates and yeast. Pure **nonfat beef extract** is used in the food processing industry for soups, gravies, and prepared dishes. The extract of International Packers, Ltd., is a paste of 17% moisture content. It contains thiamine, niacin, riboflavin, pyrodoxine, pantothenic acid, vitamins B_1 and B_{12}, purine, creatine, and the nutrient proteins found only in meat. **Dehydrated**

beef is lean beef dried by mechanical means into flake or powder form. It is semicooked, and when it is wet with water, it resumes its original consistency but has a somewhat cooked taste. Its advantage is the great saving in shipping space. Beef is also marketed in the form of **dried beef,** usually sliced and salted. **Jerked beef,** or **tasajo,** is beef that has been cut into strips and dried in the sun. It is used in some Latin American countries, but has a strong taste.

BEESWAX.   The wax formed and deposited by the honey bee, *Apis mellifera.* The bees build combs for the reception of the honey, consisting of two sheets of horizontal, six-angled prismatic cells formed of wax. Between 1.5 and 3 lb (0.56 to 1.1 kg) of wax can be obtained from 10 combs when they are scraped. After the extraction of the honey, the wax is melted and molded into cakes. New wax is light yellow, but turns brown with age. It may be bleached with sunlight or with acids. It is composed largely of a complex long-chain ester, **myricil palmitate,** $C_{15}H_{31}COOC_{30}H_{61}$, and **cerotic acid,** $C_{25}H_{51}COOH$. The specific gravity is 0.965 to 0.969 and the melting point 145°F (63°C). It is easily colored with dyes, and the Germans marketed powdered beeswax in various colors for compounding purposes. Beeswax is used for polishes, candles, leather dressings, adhesives, cosmetics, molded articles, as a protective coating for etching, and as a filler in thin metal tubes for bending. It is frequently adulterated with paraffin, stearin, or vegetable waxes, and the commercial article may be below 50% pure. Standards for the Cosmetics, Toiletry, and Fragrance Assoc. require that it contain no carnauba wax, stearic acid, paraffin, or ceresin and show no more than 0.01% ash content. Beeswax is produced in many parts of the world as a by-product of honey production from both wild and domesticated bees, the honey being used as a sweetening agent or for the making of alcoholic beverages. **Honey** varies greatly in flavor owing to the different flowers upon which the bees feed, but the chemical properties of both the honey and the wax vary little. Honey is composed largely of fructose. In the food industry, small proportions are added to the sugar to enhance the flavor of cookies and bakery products. Honey, normally 82% solids, is also dehydrated to a free-flowing **honey powder** used in confectionery. Sugar may be added to raise the softening temperature and make the powder more resistant to caking. West Africa produces much wax from wild bees. Abyssinia is a large producer of beeswax, where the honey is used for making **tej,** an alcoholic drink. The ancient drink known as **mead** was a fermented honey solution. **Scale wax** is produced by removing the combs from the hives, thus forcing production of wax which is dropped in scales or particles by the bees and prevented from being picked up by a screen.

BELL METAL. A bronze used chiefly for casting large bells. The composition is varied to give varying tones, but the physical requirements are that the castings be uniform, compact, and fine-grained. The standard is 78% copper and 22 tin. The alloy has a density of 0.312 lb/in^3 (8,636 kg/m^3), is yellowish red, has a fine grain, is easily fusible, and gives a clear tone. Increasing the copper slightly increases the sonorous tone. Large bells of deeper tone are made of 75% copper and 25 tin. Big Ben, at Westminster Abbey, cast in 1856, contains 22 parts copper and 7 tin. Another bell metal, containing 77% copper, 21 tin, and 2 antimony, is harder, giving a sharper tone. An alloy for fire-engine bells contains 20% tin, 2 nickel, 0.1 silicon for deoxidation, and the balance copper. The nickel reduces the tendency to embrittlement from pounding. One bell metal contains 80% copper and 20 tin, deoxidized with phosphorus. **Silver bell metal,** for bells of silvery tone, is a white alloy containing 40% copper and 60 tin. This type of alloy, with tin contents up to as high as 60%, is also used for valves and valve seats in food machinery.

BENTONITE. A **colloidal clay** which has the property of being hydrophilic, or water-swelling, with some clay absorbing as much as 5 times its own weight in water. It is used in emulsions, adhesives, for oil-well drilling, to increase plasticity of ceramic clays, and as a bonding clay in foundry molding sands. In combination with alum and lime, it is used in purifying water as it captures the fine particles of silt. Because of its combined abrasive and colloidal properties, it is much used in soaps and washing compounds. It is also used as an absorbent in refining oils, as a suspending agent in emulsions, and in lubricants.

Bentonite occurs in sediment deposits from a few inches to 10 ft (3 m) thick. It is stated to have been formed through the devitrification and chemical alteration of glassy igneous materials such as volcanic ash, and it is a secondary mineral composed of deposits from the mineral **leverrierite,** $2Al_2O_3 \cdot 5SiO_2 \cdot 5H_2O$, crystallizing in the orthorhombic system, though some of the bentonite marketed may be montmorillonite. The finely powdered bentonite from Wyoming was originally called **wilkinite.** Wyoming bentonite is characterized by a very sticky nature and soapy feel when wet, and it is highly absorbent. Bentonites are usually light in color, from cream to olive green. Some have little swelling property, and others are gritty. The material from Otay, California, has been called **otaylite.** It is brownish and not as highly colloidal as **Wyoming bentonite.** Analyses of bentonites from various areas vary from 54 to 69% silica, 13 to 18 alumina, 2 to 4 ferric oxide, 0.12 to 3.5 ferrous oxide, 1 to 2.2 lime, 1.8 to 3.6 magnesia, 0.1 to 0.6 titania, 0.5 to 2 soda, and 0.14 to 0.46 potash.

The material known as **hectorite** from California is lower in silica and alumina and higher in magnesia and lime. In general, the highly colloidal bentonites contain the highest percentages of soda which have been adsorbed by the clay particles. Most crude bentonites contain impurities, but are purified by washing and treating.

Bentonites are marketed under various trade names such as **Volclay** of American Colloid Co., **Refinite,** and **Eyrite,** of Baroid Division, NL Industries, Inc. **Bentone,** produced in various grades by the latter is purified montmorillonite. It is a fine white powder of 200 mesh and is used as a gelling agent for emulsion paints, adhesives, and coatings. **Bentone 18-C** is an organic compound of the material used for gelling polar organic materials such as cellulose lacquers and vinyl solutions.

BENZENE. Also called **benzol,** although the term is now reserved for the less pure grades. A colorless, highly inflammable liquid of composition C_6H_6. It is an aromatic hydrocarbon obtained as a by-product of coke ovens or in the manufacture of gas, and also made synthetically from petroleum. Its molecular structure is the closed **benzene ring** with six CH groups in the linkage, which forms a convenient basic chemical for the manufacture of styrene and other chemicals. It is also an excellent solvent for waxes, resins, rubber, and other organic materials. It is employed as a fuel or for blending with gasoline or other fuels. Industrially pure benzene has a distillation range of 172.6 to 179.8°F (78.1 to 82.1°C), a specific gravity of 0.875 to 0.886, and a flash point below 60°F (15.5°C). The pure nitration grade, used for nitrating and for making organic chemicals, has a 1.8°F (1°C) boiling range starting not below 174.6°F (79.2°C) and a specific gravity of 0.882 to 0.886. Benzene has a characteristic odor, is soluble in alcohol but insoluble in water, and all its combinations are toxic. The terms **aromatic chemicals** and **aromatics** refer to all the chemicals made from the benzene ring.

Nitrobenzene, $C_6H_5NO_2$, is a highly poisonous and inflammable liquid made by the action of nitric and sulfuric acids on benzene, used in soaps and cosmetics. It is called **myrbane oil** as a perfuming agent. The nitrated derivative called **benzedrine,** or **amphetamine,** originally used by wartime pilots to combat fatigue, is **phenylaminobenzine,** $C_6H_5 \cdot CH_2 \cdot CH(NH)_2 \cdot CH_3$. It is used in medicine to control obesity, but it is a stimulant to the central nervous system and is habituating. The isomer dextroamphetamine is d-phenylaminopropane sulfate, commonly called **Dexedrine.** It causes a rise in blood pressure and stimulates cerebral activity which lasts several hours, but it has a depressant effect on the intestinal muscles, causing loss of appetite and delayed activity of the stomach with other side effects.

Diphenyl carbonate, $(C_6H_5)_2CO_3$, is much used for the manufacture of chemicals where two benzene rings are desired. It is a white crystalline water-insoluble solid melting at 172°F (78°C). **Benzyl alcohol,** $C_6H_5CH_2OH$, is a colorless liquid soluble in water, having a boiling point of 401.4°F (205.2°C) and a freezing point of 4.5°F (−15.3°C). It is also called **phenylcarbinol** and is used as a solvent for resins, lacquers, and paints. **Benzyl chloride,** $C_6H_5CH_2Cl$, is a colorless liquid of specific gravity 1.103 and boiling point 354°F (179°C), which was used as a lachrymatory gas and is employed in the production of plastics. **Benzyl cellulose** is a thermoplastic of ICI Americas Inc., produced by the action of benzyl chloride and caustic soda on cellulose. The plastic is nonflammable and resistant to acids, can be molded easily, and is produced in various grades by different degrees of benzylation. **Benzyl dichloride,** $C_6H_5CH \cdot Cl_2$, is a liquid heavier than benzyl chloride and has a higher boiling point, 414°F (212°C), but was also used as a war gas. It is also called **benzylidene chloride** and is used for producing dyestuffs.

BERGAMOT OIL. An essential oil contained in the fresh peel of the fruit of *Citrus aurantium Linn.,* subspecies *Bergamia,* a small, spiny tree cultivated principally in southern Calabria in Italy. In Turkey, the sour, pear-shaped fruit is called **beg-armudi,** or **Bey's pear.** The yellowish-green oil is expressed by pressing the fruit between two corrugated disks. The oil, filtered to remove any rind, has an aromatic, agreeable odor, but a bitter and unpleasant taste. It is used principally in perfumery to make eau de colognes and to mask unpleasant odors. About 36 to 45% of the oil is *l*-**linalyl acetate,** about 6% *l*-**linalool,** *d*-**limonene,** and **dipentene. Lavender oil** also has significant quantities of *l*-linalyl acetate. The white, hard crystals that separate out from an alcohol solution of bergamot oil are known as **bergaptene.**

BERYLLIUM. This steel-gray lightweight metal, symbol Be, was formerly known as **glucinum** and is used mainly for its excellent physical properties rather than its mechanical properties. Except for magnesium, it is the most lightweight of common metals, having a density of 0.067 lb/in³ (1,855 kg/m³). It also has the highest specific heat [0.45 Btu/(lb · °F), 1,833 J/kg · K)] and a melting point of 2354°F (1290°C). It is nonmagnetic, has about 40% the electrical conductivity of copper, a thermal conductivity of 110 Btu · ft/(h · ft² · °F) [190 W/(m · K)], high permeability to X-rays, and the lowest neutron cross section of any metal having a melting point above 932°F (500°C). Also, its tensile modulus [42 × 10⁶ lb/in² (28.9 × 10⁴ MPa)] is far greater than that of almost all metals. Mill forms, such as block or

billet, are made from powder by hot pressing and can be thermomechanically processed to extruded billet and cross-rolled sheet.

Ultimate tensile strength ranges from 33,000 to 100,000 lb/in² (228
to 690 MPa) and tensile elongation from 1 to 40%, depending on the
mill form. Thus, because of its low density, beryllium excels in specific
strength, especially in specific stiffness. However, tensile properties,
especially elongation, are extremely dependent on grain size and orientation and are highly anisotropic, so that results based on uniaxial
tensile tests have little significance in terms of useful ductility in fabrication or fracture toughness in structural applications. From these
standpoints, the metal is considered to be quite brittle. Ductility, as
measured by elongation in tensile tests, increases with increasing
temperature to about 750°F (400°C), then decreases above about
930°F (500°C). Although resistant to atmospheric corrosion under
normal conditions, beryllium is attacked by oxygen and nitrogen at
elevated temperatures and certain acids, depending on concentration,
at room temperature.

The metal is also very expensive. It is toxic if inhaled or ingested,
necessitating special precautions in handling. Most beryllium parts
are made by machining from block, and machining leaves a damaged
surface layer that must be removed by etching for stressed applications. Most applications are quite specialized and stem largely from
the metal's good thermal and electrical properties. Uses include precision mirrors and instruments, radiation detectors, X-ray windows,
neutron sources, nuclear reactor reflectors, aircraft brakes, and rocket nozzles. It is also used as an alloying element to produce beryllium-
aluminum, beryllium-copper, and beryllium-nickel alloys.

BERYLLIUM-ALUMINUM. A 62% beryllium 38% aluminum alloy
developed by Lockheed Aircraft in the 1960s and called **Lockalloy.**
Noted primarily for its lightweight and high tensile modulus, thus
high specific rigidity, it was used in the form of extrusions for missile
skin stiffeners and in rolled sheet on the YF-12 reconnaissance aircraft. The alloy has since become available as a powder-metallurgy
product from Brush Wellman Inc. as **AlbeMet AM162** rolled sheet
and extruded bar and **AlbeMet AM162H** hot isostatically pressed
billet, which contain by weight 60 to 64% beryllium, 36 to 40 aluminum, and as much as 1 oxygen, 0.1 carbon, and 0.2 each of other
metallics. The density is 0.0748 to 0.0767 lb/in³ (2,070 to 2,123
kg/m³). Typical minimum tensile properties of rolled sheet are
55,000 lb/in² (379 MPa), 40,000 lb/in² (279 MPa) yield strength, and
5% elongation. For extruded bar, the minimum properties are 52,000
to 55,000 lb/in² (358 to 379 MPa), 40,000 lb/in² (279 MPa), and 6 to
7% elongation, respectively, depending on the extrusion ratio. For

the hot isostatically pressed billet, typical minimums are 38,000 lb/in² (262 MPa), 28,000 lb/in² (193 MPa), and 2%, respectively. These properties pertain to material heat-treated for 24 h at 1100°F (593°C).

BERYLLIUM-COPPER. Beryllium-coppers, as these copper-base alloys are commonly called, are among the hardest, strongest, and most wear-resistant of copper alloys. They also feature good electrical and thermal conductivity and corrosion resistance in various environments, are nonmagnetic and nonsparking, and can be readily fabricated. Wrought **beryllium-coppers C17000** to **C17300** contain 1.6 to 2.0% beryllium with much smaller amounts of iron, nickel, cobalt, silicon, and aluminum; **beryllium-coppers C17400** to **C17700** contain less beryllium, 0.7% maximum, and, depending on the alloy, greater amounts of nickel or cobalt, sometimes with other ingredients, such as magnesium and zirconium (C17520), silver (C17600), or tellurium (C17700). Casting alloys, **beryllium-coppers C81300** and **C81400** contain only 0.02 to 0.10% beryllium, with 0.6 to 1.0 cobalt in the former and this range of chromium in the latter. **Beryllium-coppers C81700** to **C82200** contain 0.3 to 0.8% beryllium, with generally larger amounts of cobalt, nickel, and, in some cases (C81700 and C81800), silver. **Beryllium-coppers C82400** to **C82800** contain 1.65 to 2.75% beryllium with smaller amounts of other alloying elements, such as cobalt, nickel, silicon, and iron. All the alloys, wrought or cast, are 94% or more copper. The general corrosion resistance of the alloys is similar to that of deoxidized copper and other high-copper alloys. Some of the alloys have excellent resistance to stress corrosion in many environments. At elevated temperatures, however, the alloys can form intergranular oxidation, causing surface deterioration and abrasion of fabrication tools. The beryllium oxide film can be removed, however, mechanically or by pickling. Because beryllium is toxic, precautions are required in many fabricating operations.

Mechanical properties vary widely depending on solution treatment or annealing, cold working and precipitation, or age hardening. For example, the tensile yield strength and elongation of **beryllium-copper C17000** strip ranges from 25,000 to 35,000 lb/in² (170 to 240 MPa) and 35 to 60%, respectively, in the annealed condition to 140,000 to 195,000 lb/in² (965 to 1,345 MPa) and 2 to 5% in hard tempers. **Beryllium-copper C17200** strip is somewhat stronger and generally less ductile in all tempers. The hardness of both alloys ranges from roughly Rockwell B 45 to 78 and Rockwell C 39 to 45, respectively. Electrical conductivity increases with increasing strength and hardness, from about 18 to 25% that of copper, respectively. **Beryllium-copper C17500** strip, though not as strong or hard, is more conductive—25 to 30% (annealed) to as much as 52% (hard)—relative to copper.

Of the casting alloys, **high-beryllium C82400** to **C82800** provide the greatest strength and hardness. As sand-cast, yield strengths range from 37,000 lb/in^2 (255 MPa) for **beryllium-copper C82400** to 55,000 lb/in^2 (379 MPa) for **beryllium-copper C82800** and hardness from Rockwell B 78 to 85, respectively. Following solution heat treatment and precipitation hardening, yield strength and hardness increase to 135,000 lb/in^2 (931 MPa) and Rockwell C 38 for C82400, and to 155,000 lb/in^2 (1,069 MPa) and Rockwell C 43 for **beryllium-copper C82600.** In the fully heat-treated condition, electrical conductivity is on the order of 18 to 25% that of copper, **beryllium-copper C82400** being the most conductive. Alloys containing less beryllium are not nearly as strong. Yield strengths after full heat treatment range from 36,000 lb/in^2 (248 MPa) for **beryllium-copper C81300** to 75,000 lb/in^2 (517 MPa) for **beryllium-copper C81800, C82000,** and **C82200.**

Typical applications for wrought alloys include diaphragms, bellows, fasteners, bushings, washers, springs, electrical and instrument parts, valves, pump parts, tools and dies, connectors, and welding equipment. Casting alloys are used for resistance-welding tips and equipment, soldering irons, casting and molding dies and equipment, electrical and thermal conductors, valves, gears, cams, bearings, and pump parts.

BERYLLIUM-NICKEL. Wrought beryllium-nickel contains about 2% beryllium, 0.5 titanium, and the balance nickel. Casting alloys contain a bit more beryllium (2 to 3%) and, in one alloy, 0.4 carbon. As in the case of beryllium-copper alloys, mechanical properties vary widely, depending on temper condition—from 45,000 to 230,000 lb/in^2 (310 to 1,586 MPa) in tensile yield strength and Rockwell B 70 to Rockwell C 55 in hardness at room temperature. The alloys retain considerable yield strength at high temperature: 130,000 to 170,000 lb/in^2 (896 to 1,172 MPa) at 1000°F (538°C). They also have good corrosion resistance in general atmospheres and reducing media. Because beryllium is toxic, special precautions are required in many fabricating operations. The wrought alloy is used for springs, bellows, electrical contacts, and feather valves; and the casting alloys for molding plastics and glass, pump parts, seal plates, and metal-forming tools.

BERYLLIUM ORES. Beryllium is widely distributed in possible recoverable quantities in more than 30 minerals, but the chief ore is **beryl,** $3BeO \cdot Al_2O_3 \cdot 6SiO_2 \cdot H_2O$. This mineral is usually in pale-yellow rhombic crystals in pegmatic dikes. The crystals are 0.25 to 0.5 in

(0.64 to 1.27 cm) in diameter, with a specific gravity of 2.63 to 2.90, and a Mohs hardness of 7.5 to 8. The ore is resistant to acid attack and requires calcining to make it reactive, although the ore of Utah, called **vitroite,** is of simpler composition and can be acid-leached. Beryl ore may contain up to 15% beryllium oxide, but most ore averages below 4%. The Indian ore contains a minimum of 12% BeO, and the beryl of Ontario has 14% BeO, or 5% metallic beryllium. The ore of Nevada contains only 1% BeO, but can be concentrated to 20%.

The secondary ores of beryllium—**bertrandite, herderite,** and **beryllonite**—usually have only small quantities of BeO disseminated in the mineral. But Utah clay from Topaz Mountain, Utah, in which the bertrandite is associated with pyrolusite, fluorspar, opal, and mixed with montmorillonite and other clays, is concentrated by flotation, acid-leached, and chemically processed to 97% BeO. Other ores of beryllium are **chrysoberyl,** $BeO \cdot Al_2O_3$, and **phenacite,** which is a **beryllium silicate,** Be_2SiO_4. **Helvite,** $(MnFe)_2(Mn_2S)Be_2(SiO_2)_3$, is in cubic crystals of various colors from yellow through green to dark brown, associated with garnet and having the appearance of garnet. The specific gravity is 3.3, and the Mohs hardness is 6.5.

Choice crystals of beryl, colored with metallic oxides, are cut as gemstones. **Alexandrite,** a gem variety of chrysoberyl, is emerald green in natural light but red in transmitted or artificial light. Allied Corp. produces a synthetic alexandrite for use as a tunable solid-state laser. The **emerald** is a flawless beryl-colored green with chromium. High-grade natural emeralds are found in Colombia, but occur in the United States only in North Carolina. The rose-pink, rose-red, and green beryl crystals of Malagasy, called **morganite,** are cut as gemstones, and the dark-blue stone is made by heating the green crystals. The pale blue-green crystals are **aquamarines,** and the **heliodor** is **golden beryl** from southwest Africa. But the yellowish-green gemstone of Brazil, called **brazilianite,** is not beryl, but is a hydrous sodium-aluminum phosphate and is softer. Lemon-yellow crystals of chrysoberyl found in Brazil are valued as gemstones. Synthetic emerald of composition $3BeO \cdot Al_2O_3 \cdot 6SiO_2$ was first made in Germany by heat and pressure under the name of **Igmerald.** Synthetic emeralds are now grown from high-purity alumina, beryllia, and silica, with traces of Cr_2O_3 and Fe_2O_3 to give the green color. Synthetic beryl is used for bearings in watches and instruments.

BERYLLIUM OXIDE. A colorless to white crystalline powder of composition BeO, also called **beryllia,** and known in mineralogy as **bromellite.** It has a specific gravity of 3.025, a high melting point, about 4680°F (2585°C), and a Knoop hardness of 2,000. It is used for polishing hard metals and for making hot-pressed ceramic parts. Its high heat resistance and thermal conductivity make it useful for crucibles, and its high dielectric strength makes it suitable for high-frequency insulators. Single-crystal **beryllia fibers,** or whiskers, developed by National Beryllia Corp., have a tensile strength above 1×10^6 lb/in^2 (6,895 MPa).

Ceramic parts with beryllia as the major constituent are noted for their high thermal conductivity, which is about 3 times that of steel, and second only to that of the high-conductivity metals (silver, gold, and copper). They also have high strength and good dielectric properties. Properties of typical grades of **beryllia ceramics** are: tensile strength, 14,000 lb/in^2 (96 MPa); compressive strength, 300,000 lb/in^2 (2,068 MPa); hardness (micro), 1,300 Knoop; maximum service temperature, 4350°F (2400°C); dielectric strength, 5.8 V/mil (0.23 $\times$ 10^6 V/m). Beryllia ceramics are costly and difficult to work with. Above 3000°F (1650°C) they react with water to form a volatile hydroxide. Also, because beryllia dust and particles are toxic, special handling precautions are required. Beryllia parts are used in electronic, aircraft, and missile equipment. A more recent application has been beryllia's use as thermocouple insulators in vacuum furnace equipment operating below 3000°F (1650°C).

Beryllia is used in ceramics to produce gastight glazes. Thin films of the oxide are used on silver and other metals to protect the metal from discoloration. Very thin films are invisible, but heavier films give a faint iridescence. Two other beryllium compounds used especially in chemical manufacturing are **beryllium chloride,** $BeCl_2$, a water-soluble white powder melting at 824°F (440°C), and **beryllium fluoride,** BeF_2, melting at 1472°F (800°C). Another beryllium compound, useful for high-temperature, wear-resistant ceramics, is **beryllium carbide,** Be_2C. The crystals have a Mohs hardness of 9, and the compressed and sintered powder has a compressive strength above 100,000 lb/in^2 (690 MPa). **Berlox,** of National Beryllia Corp., is a beryllium oxide powder in particle sizes from 80 to 325 mesh for flame-sprayed, heat- and wear-resistant coatings.

BESSEMER STEEL.   Steel made by blowing air through molten iron. The original pneumatic process, which for the first time made possible the production of steel on a large scale, involved blowing air through molten pig iron held in a bottom-blown vessel lined with acid (siliceous) refractories, and thus is commonly referred to as the **acid Bessemer process.** It was developed independently by Henry Bessemer of England, whose U.S. patent was issued in 1856, and by William Kelly, Eddyville, Kentucky, who didn't apply for a patent until 1857 but proved that he had been working on the process as early as 1847. Bessemer built a steel works in Sheffield, England, and began operating in 1860. In the United States, where both men shared rights to the process, Kelly Pneumatic Process Co. was formed in 1863. By 1871, some 55% of total U.S. steel production was made by this process, and it remained the significant steelmaking process for many years until it was eventually replaced worldwide by the **open-hearth process,** which, in turn, was replaced by the **basic-oxygen process.**

The acid Bessemer process was the major steelmaking process until 1908. Among the product forms made were free-machining bars, flat-rolled stock, seamless and welded tubing, wire, and castings. Fully killed (deoxidized) acid Bessemer steel was first used by U.S. Steel for making steel pipe, and the dephosphorized steel was used extensively in the production of welded pipe and galvanized sheet. The **basic Bessemer process,** or **Thomas process,** patented in 1879 by Sidney G. Thomas in England, involved use of a basic lining and flux in the converter, making it possible to use the pneumatic process to refine pig iron smelted from high-phosphorus ores common in Europe. The process was never used in the United States, and production developed more widely in other European countries than in England. Steel produced by the process was called **Thomas steel.** In the acid Bessemer process, ferromanganese and sometimes steel scrap were added to the steel when pouring into the ladle in order to regulate the content. In the blowing process, the chemical action between the oxygen of the air and the molten mass increases the temperature, and air then forms the chief fuel as carbon is oxidized and driven off. The blowing required only a few minutes, and carbon was reduced to 0.04% or less. The carbon desired in the steel was then regulated by the addition of carbon to the melt. The two processes, acid and basic, differ in the type of refractories employed for lining the converters, and there is a difference in the resulting steel since the acid process does not remove as much sulfur and phosphorus.

BIOMATERIALS. Materials used to repair, restore, or replace damaged or diseased tissue, or those that comprise parts of artificial organs, artificial tissues, or prostheses. The use of biomaterials dates back to antiquity. Hair, cotton, animal sinew, tree bark, and leather have been used as natural suture materials for almost 4,000 years. Synthetic biomaterials are composed of metals, ceramics, polymers, and their composites, and they are often called **biomedical materials** to differentiate them from natural ones. Gold plates for skull repair were in use in 1000 B.C., and gold-wire sutures as early as 1550. Besides the mechanical properties of strength, elasticity, and durability, biomaterials need to be nontoxic, sterilizable, and biocompatible. Biocompatibility implies that the material will be inert when in contact with the body and not activate the immune system or lead to blood coagulation. Biomaterials are also categorized as bioabsorbable or nonbioabsorbable. Among the former are the peptides (collagen, fibrin, albumin, and gelatin), hemiacetals (starch, hyaluronic acid, chitin), esters (poly-ß-hydroxybutyrate and polymalic acid), and phosphates (DNA and RNA). Synthetic polymers that are bioabsorbable are polylactides, polylactones, polycarbonates, poly-α-cyanoacrylates, polyphosphazenes, and polyanhydrides.

No natural suture material has been so prevalent as **catgut,** derived from the small intestines of animals, usually the outer serosal layer in cattle or the submucosal layer in sheep. For **surgical catgut,** or **gut,** the intestinal tracts of animals are slit lengthwise; the resulting ribbons are twisted into bonded strands and then sterilized by electron-beam irradiation. Catgut becomes too stiff to handle when dry, so it is packed in aqueous alcohol. When the material is treated with a chromic salt solution to prolong in vivo strength, it is called **chromic catgut.** For a more uniform surface, chromic catgut is treated with glycerine to form **glycerine catgut.** Catgut varies widely in quality and elicits relatively severe tissue reactions. It is absorbed by the body in 90 days.

A more uniform suture material has been prepared by wet-spinning dispersions of **purified collagen** into strands. Three kinds of synthetic, absorbable polyester sutures are now marketed. The first is a **glycolide homopolymer,** also known as **poly(glycolic acid)** or **PGA,** obtained by ring-opening polymerization of cyclic diester monomers at high vacuum using tin catalysts. The suture is formed by melt extrusion spinning, followed by hot drawing to give high orientation and crystallinity, and annealed to improve dimensional stability. It is braided into a multifilament and can be dyed or coated. It is absorbed by the body in 90 days. **Dexon** is such a material from Davis and Geck Div. of American Cyanamid Co., and **Medifit** is a competitive product from Japan Medical Supply Co. **Vicryl** is a copolymer of 90 to 92% glycolide and 8 to 10% lactide, with properties similar to PGA. It is made by the Ethicon division of Johnson & Johnson. Also from the same company is **PDS,** an absorbable suture made from **polydioxanone.** It is a smooth monofilament that can reduce tissue trauma. PDS takes twice as long as Vicryl to be absorbed by the body. Among nonabsorbable sutures, the most common are silk, linen, cotton, polyester, polyamide (nylon 6 or nylon 6,6), polypropylene, and steel. **Mersilene** is a braided polyester from Ethicon, which also produces **Prolene,** a monofilament polypropylene, and **Ethilon,** a monofilament nylon. **Surgilon** and **Ticron** are braided nylon and polyester products, respectively, that have also been silicon-treated. They are from Davis and Geck. Thermoplastic polyurethane products of Polymedica Industries include wound dressings and **Spyroflex** films that can be stretched over surgical incisions to close them, replacing sutures or staples.

Polymer matrixes are receiving attention for cell transplant devices to regenerate human tissue in wound healing and reconstruction surgery. For skin tissue, Type I **collagen,** a protein that strengthens tissue and is available from animals, has been combined with **chondroitin sulfate,** a carbohydrate polymer, to form largely porous insoluble matrixes with controlled rate of degradation. Concern over adverse biological reactions of the collagen, however, has prompted development of matrixes based on synthetic polymers, such as biodegradable polyester, which is widely used for absorbable sutures. Also being studied are hybrids of a degradable polymer such as **hyaluronic acid,** an anionic polysaccharide of the skin, covalently modified by a small **peptide ligand** to control biological response. For cartilage tissue, polyesters in the family of **polylactides, polyglycolides,** and their copolymers are of interest. They degrade mainly by hydrolysis to yield natural metabolic intermediates, most of which then convert to carbon dioxide and water. These polymers are sufficiently strong to make plates and screws for setting bone fractures.

Biofix screws, from Bioscience of Finland, are made of either poly (L-lactide) (PLLA) semicrystalline polyesters or PGA thermoplastic resins and are used to reassemble broken bones. The screws, which are strengthened with sutures of the same materials, hold bone fragments in place during healing, then break down and are absorbed by the body.

Joint replacement is one of the largest uses of biomaterials. The first implants, artificial hips, were made from 316L low-carbon stainless steel, followed by **chromium-cobalt,** and in the 1970s by titanium. Initially, metal femurs with ball-shaped ends were inserted into the acetabulum, the hip socket, as a direct replacement for the natural femur. Now, the acetabulum itself may be made of metal cups coated with a self-lubricating plastic with good wear properties, such as high-molecular-weight (HMW) polyethylene. Europeans commonly use alumina for femoral balls and acetabular cups, and in the United States, Richards Medical is marketing an alumina-capped femur. Astro Met is developing a **partially stabilized zirconia** for the same application. Osteonics, of Stryker Corp., uses wrought **cobalt alloy CMM** for the femoral cap of hip joints. The cap fits over the top of a hip stem made of either the cast or wrought alloy or a **titanium alloy.** The cap is within an acetabular cup made of **ultrahigh-molecular-weight polyethylene** that is mounted in a hemispherical titanium shell. The combination serves as the articulating couple of the hip joint. To reduce friction and wear of the cup, nitrogen ions are implanted in the cap surface. LifeCore Biomedical is testing graphite-polysulfone composites, and Zimmer is sheathing titanium in polysulfone or polyetheretherketone (PEEK). **Hyaluronic acid** is a viscoelastic liquid used to lubricate the traumatized joints in arthritic patients.

Thermoplastic aromatic polyetherurethanes are used for implants because of their resistance to flexural fatigue, self-bonding characteristics, and good tensile strength. Some grades, however, are degraded by enzymes. **Corethane** thermoplastic elastomers have similar properties and are believed to resist enzymes. They are being studied by Corvita Corp., a small-device manufacturer. The company is also developing a copolymer of polyurethane and polycarbonate that is free of polyether segments responsible for degradation in the body. Potential uses include insulators for pacemaker leads, long-term in-dwelling catheters, and pump diaphragms for artificial hearts. A composite of the copolymer and a silane is seen as a potential replacement for silicone in breast implants. Being stronger, the composite is expected to have less tendency to rupture. Also, unlike silicone, it would permit X-ray examination. **Chronoflex,** from Polymedica Industries, is another urethane believed to resist enzyme degradation. It is being developed for artificial hearts, vascular parts, and grafts.

A **zirconium-palladium-ruthenium alloy** developed at Paffenbarger Research Center at the National Institute of Standards and Technology, to extend the life of medical and dental implants, holds promise for hip and knee implants. Extremely wear-resistant, it may reduce the amount of wear debris generated by other implant materials, which, even if inert, can degrade surrounding tissue. A dentin-regenerative protect based on **OP-1,** a protein that appears to stimulate uncommitted cells to lay down dentin, may preclude root-canal operations. Developed by Creative Biomolecules, Inc., it may also be useful in bone reconstruction and in treating osteoporosis. **Bioglass,** of US Biomaterials, consists of silicon, sodium, and natural bone ingredients—calcium and phosphorus. It can be used to replace middle-ear bone and tooth roots. **Perioglas,** a granulated form, is used to fill gum pockets formed by periodontal disease.

To anchor artificial bones in place, they are layered with a porous coating that encourages the growth of natural bone tissue and blood vessels. Because these materials interact with the body, they are called **bioactive materials.** A porous coating of cobalt-chromium or titanium beads on metal femurs is bioactive. **Calcium phosphate ceramics** and coatings are osteophilic (bone-loving); their porosity provides the templates on which new bone growth can form for natural healing. **HTR** polymer from HTR Sciences, a division of U.S. Surgical, is a porous plastic bead made from polyhydroxyethylmethacrylate and polymethylmethacrylate. It has a negative charge that attracts cell formation in bones and is hydrophilic, so it stays where surgeons place it. It is derived from a bacterium by LifeCore Biomedical. **Hydroxyapatite,** a ceramic similar to bone and which bonds to bone naturally, is used to coat bioactive implants. It is applied by plasma spraying, although high-velocity oxy-fuel deposition also may be suitable and advantageous due to its lower processing temperature and ease of control. In coating titanium, a common implant material, good results were achieved with HVOF at the Interdisciplinary Research Center in Biomaterials at Queen Mary College, London. Artificial blood is bioactive. One of the earliest and best known is **Fluosol, a perfluorocarbon blood** substitute made of **perfluorodecalin** and **perfluorotripropylamine.** Under testing are other perfluorocarbons with larger oxygen-carrying capacity than Fluosol, such as **perfluorobutyltetrahydrofurane.** Under testing are other perfluorocarbons with larger oxygen-carrying capacity than Fluosol, such as **perfluorobutyltetrahydrofurane.** A biodegradable skin substitute, **Stage 1 skin,** that stimulates new skin growth is being developed. It is a mixture of bovine collagen fibers and **chondroitan-6-sulfate,** a polysaccharide from shark cartilage, that lowers the rate at which collagen degrades in the body.

Pyrolytic carbon, a pure, glassy carbon developed originally to encapsulate fuel for nuclear reactors, is used for making heart valves because it is inert and extremely wear-resistant. It is also compatible with blood. The valves in the **Jarvik-7** artificial heart are made from this material. Medtronics produces highly polished titanium for heart valves and for insulin delivery pumps.

PMMA lenses are implants made of polymethylmethacrylate and are commonly used for patients suffering from severe cataracts. Recent improvements include lenses made of **silicone rubber** or of **hydrogel,** a viscous, jellylike polymer. These require a much smaller incision in the eye during insertion, because they are "foldable." Silicone has been widely used for breast implants, a controversial use that has led to lawsuits by women, alleging that leaks from the devices have injured their health.

Elastomeric polypeptides that will do work in response to changes in the body's chemical potential have been developed at the School of Medicine, University of Alabama. **Polypentapeptide** is the parent of this class of biomaterials. They are prepared by solution- and solid-phase peptide synthesis. The elastomeric polypeptides can match the compliance of natural biological tissues and can be modified to obtain desirable tissue reactions to the cellular and enzymatic levels. They can be made to contract and relax in response to changes in chemical potential analogous to the production of motion in living organisms. Some of their possible biomedical applications are synthetic arteries and ligaments, burn cover materials, and targeted drug delivery systems. In many other product areas, the materials can function as sensors, based on their thermomechanical and chemomechanical transduction properties. **OncoScint,** of Cytogen Corp., is a genetically formulated, injectable, diagnostic imaging agent that tests for colorectal and ovarian cancer.

BIRCH. The wood of birch trees, of which more than 15 varieties grow in the northeastern and Lake states of the United States and in Canada, and other varieties in Europe and north Asia. The birch of north Europe is called **Russian maple.** The wood of the American birches has a yellow color; is tough, strong, hard, and close-textured; and polishes well. It has a fine wavy grain, sometimes beautifully figured, and can be stained to imitate cherry and mahogany. Birch is used in construction work for trim and paneling, for furniture, and for turned articles such as handles, shoe pegs, clothespins, toys, and woodenware. The lumber usually includes the wood of several species. It has a specific gravity, oven-dried, of 0.68, a compressive strength perpendicular to the grain of 1,590 lb/in^2 (10 MPa), and a shearing strength parallel to the grain of 1,300 lb/in^2 (9 MPa). **Yellow birch,** *Betula lutea,* highly prized for furniture, is now getting scarce. It is also called **silver birch** and **swamp birch.** The commercial wood includes that from the **gray birch,** *B. populifolia.* **Sweet birch,** *B. lenta,* ranks next in importance. It is called **black birch, cherry birch,** and **mahogany birch,** and it may be marketed together with yellow birch. Sweet birch may also include **river birch,** *B. nigra,* but sweet birch is a heavier and stronger wood. **Paper birch,** *B. papyrifera,* is the variety known as **canoe birch** because the silvery-white flexible bark was used by the Indians in making canoes. It is also referred to as *silver birch* and is much used for pulpwood. It is similar to and mixed with **white birch,** *B. alba,* which is strong, elastic, and uniform and is much used in Vermont and New Hampshire for making spools, bobbins, handles, and toys. Yellow birch of the Canadian border reaches a height of 60 to 80 ft (18.3 to 24.4 m) and a diameter up to 2 ft (0.61 m). A 50-year-old tree has a diameter of about 15 in (0.38 m) and a height of 40 ft (12 m).

Birch oil is a viscous, yellowish, poisonous oil of specific gravity 0.956, with a characteristic birch odor, obtained by distilling **birch tar,** a product of the dry distillation of the wood of the white birch. It contains phenols, cresol, and xylenol and is used in disinfectants and in pharmaceuticals. It is also called **birch tar oil,** and in pharmacy it is known as **oil of white birch. Sweet birch oil,** also called **betula**

oil, is a lighter volatile oil distilled from the steeped bark of *B. lenta,* or sweet birch. It contains methyl salicylate and is used as a flavoring agent, in perfumes, in dressing fancy leathers, in cleaning solutions and soaps, and as a disinfectant to neutralize odors of organic compounds.

BISMUTH. An elemental metal, symbol Bi, sometimes occurring native in small quantities. American bismuth is obtained chiefly as a by-product in the refining of lead and copper. Foreign bismuth comes largely from the mineral bismuthinite. The metal is grayish white with a reddish tinge, is very brittle, and powders easily. It is highly crystalline in rhombohedral crystals. It has few uses in its pure state. Specific gravity is 9.75, melting point 520°F (271°C), and hardness Brinell 73. Thermal conductivity is less than that of any other metal except mercury, and it is the most diamagnetic of all the metals. It is one of the few metals that increases in volume upon solidification. It expands 3.32% when changing from the liquid to the solid state, which makes it valuable in type-metal alloys and in making small castings where sharp impressions of the mold are needed. The metal imparts to lead and tin alloys hardness, sonorousness, and luster and lowers the melting point. By regulating the amount of bismuth, it is possible to cast the alloys to fill the mold without expansion or contraction on cooling. It is used in white alloys for molds for casting plastics, and because it lowers the melting point, it is valued in fusible alloys and soft solders. Very fine **bismuth wire** used for thermocouples is drawn in glass tubes, and extruded bismuth wire in diameters of 0.003 to 0.039 in (0.0062 to 0.0991 cm) is ductile enough to be wound very tightly.

Small amounts of bismuth are used to enhance the machinability of metals, especially steel, a use fostered by concern about the toxicity of lead, a traditional free-machining additive for this purpose. Bismuth has been used both as an additive to resulfurized, rephosphorized, and leaded **AISI 1214 steel** and as a substitute for lead in **AISI 1214** and **1215 steels.** As a substitute, it is believed to improve machinability in a manner similar to lead. As an additive, it alloys with lead, providing more effective internal lubrication than lead alone, further improving machinability. Its advantages over lead also have been attributed to its lower density, which permits more uniform dispersion in the steel matrix. **Aluminum alloy 6262,** which was developed specifically for improved machinability, contains about equal amounts (0.6%) of bismuth and lead. Bismuth is also used in amalgams and is employed in the form of its salts in pigments, in pharmaceuticals as an antacid, and in many chemicals. For medicinal purposes bismuth must be completely free of traces of arsenic. The paint pigment known as **pearl white** is **bismuth oxychloride,** BiOCl, a white crystalline powder of specific gravity 7.717, insoluble in water. Another bismuth pigment is **bismuth chromate,** $Bi_2O_3 \cdot 2CrO_3$, a water-insoluble orange-red powder. The material known in medicine as bismuth is **bismuth phosphate,** $BiPO_4$, a white powder insoluble in water. **Bismuth telluride,** Bi_2Te_3, is a semiconductor, used mainly as a thermoelectric material.

BISMUTHINITE.   An ore of the metal bismuth, found in Bolivia, Peru, central Europe, Australia, and the western United States. It is **bismuth trisulfide,** Bi_2S_3, theoretically containing 81.3% bismuth. The richest Bolivian ores contain more than 25% bismuth, and concentrates from northwest Argentina contain 40 to 48% bismuth. The mineral has a massive foliated structure with a metallic luster, a lead-gray streaked color, and a Mohs hardness of 2. The concentrated ore is roasted and smelted with carbon, and the resulting impure bismuth is refined by an oxidizing fusion. Other bismuth ores are **bismite,** or **bismuth ocher,** $Bi_2O_3 \cdot 3H_2O$, containing theoretically 80.6% bismuth, and **bismutite,** $Bi_2O_3CO_3 \cdot H_2O$, containing theoretically 78.3% bismuth, both of which are widely distributed minerals.

BITUMINOUS COAL.   Also called **soft coal.** A variety of coal with a low percentage of carbon, and easily distinguished from anthracite by the property of losing moisture and breaking up into small pieces. Because of its cheapness, it is the coal used most exclusively for industrial fuel, but is not preferred for household use because of its smoke and odor. However, considerable fine or powdered coal called **slack coal** is used in making fuel briquettes. Bituminous coal is widely distributed in many countries and is found in 28 states of the United States. Much bituminous coal is used for the production of coke, coal tar, liquid fuels, and chemicals. The bituminous coals vary in quality from near lignite to the hard grades near anthracite, called **semibituminous coal,** depending upon their geologic age. They are not true bitumens. The specific gravity of clean bituminous coal is 1.75 to 1.80. The best steam coals are the semibituminous grades from West Virginia, Virginia, Pennsylvania, and some parts of the midwest. The latter is very compact and is extracted in large blocks, called **block coal.** Good coals for industrial use should give 13,500 to 14,500 Btu/lb (30,400 to 33,700 kJ/kg) and should have from 55 to 60% fixed carbon and 30 to 37% volatile matter. The best grades are in lumps. **Coking coals** are the higher-carbon grades low in sulfur, and with capacity to leave the residue coke in large firm lumps. **Sea coal** is a name for finely ground bituminous coal used in sand mixtures for molds for cast iron to prevent fusing of the sand to the castings.

BLACKFISH OIL. A pale-yellow oil extracted from the pilot whale, porpoise, or blackfish, *Globicephala melas,* found off the North Atlantic Coast as far south as New Jersey, and the *G. ventricosa* of other seas. The blackfish averages 15 to 18 ft (4.6 to 5.5 m) in length, with a weight of about 1,000 lb (454 kg). The oil has a saponification value of 290, iodine value 27, and specific gravity 0.929. **Dolphin oil,** of the common dolphin, *Delphinus delphis,* of all seas, is also classified as blackfish oil, as is the oil of the killer whale, *Grampus orca,* of all seas. The oil is used as a lubricant for fine mechanisms, in cutting oils, and for treating leather. The product from the head and jaw is of the best quality and is known as **jaw oil,** although the best grade of sperm oil is also called jaw oil. The jaw oil from blackfish does not oxidize easily and is free-flowing at low temperatures, having a pour point of $-20°F$ $(-29°C)$. It consists of 71% mixed acids, of which 86% is **valeric acid,** $C_2H_5(CH_3)CHCOOH$, and 13% oleic and palmitic acids. Normal valeric acid is methyl ethyl acetic acid, and it can be made by the oxidation of amyl alcohol. It is produced in the human system by the action of enzymes on amino acids. **Arginine,** of General Mills, Inc., used for treatment of shock and for reducing the toxic effects of ammonia in the blood, is a **guanidine valeric acid.** A variant of valeric acid is **levulinic acid,** a liquid boiling at $473°F$ $(245°C)$, used for making synthetic resins. It is produced as a by-product in the production of furfural from corncobs, and it has composition $CH_3COCH_2CH_2COOH$ with reactive methylene groups that undergo condensation readily. It can also be produced from starches and sugars.

BOILERPLATE. Originally, a high-grade plain iron or steel plate 0.1406 in (0.36 cm) thick or thicker, used for making steam boilers; however, the term came to mean plate of this kind for any purpose, and plates 0.25 in (0.64 cm) thick or thinner are referred to as *sheet.* Actually, boilerplates for boilers, tanks, and chemical equipment, and **flange plates** for dished ends, are now made in various alloy steels to incorporate high strength, corrosion resistance, and creep resistance, and also in clad steels. Ordinary boilerplate is divided into firebox, flange, and extra-soft. A plain steel **firebox plate** contains not more than 0.30% carbon, 0.30 to 0.50 manganese, and not more than 0.40 each of sulfur and phosphorus. Flange plates contain less carbon. The tensile strength is about 60,000 lb/in^2 (414 MPa). **Boiler-tube steel** may be carbon steel or alloy steels and may be hot-rolled or cold-drawn. But the tubes are given an internal hydrostatic test and marked with the test pressure used.

BONES. The dried bones from cattle and from Asiatic buffalo form an important item in international commerce. In general, organized packing plants do not ship much bone, but utilize it for the production of glue, gelatin, bone meal, and fertilizer. The bones shipped from packing plants are called **packer bones,** but the source of much commercial bone is slaughterhouses, local retail meat shops, and the bones from farms and fields known as **prairie bones** and **camp bones,** the latter name being from the Argentine word *campo,* meaning field. U.S. imported bone comes from Argentina, Canada, Uruguay, Brazil, South Africa, and India. **Raw bone meal** was the term for the scrap pieces and sawdust from the manufacture of these articles.

Case-hardening bone, for carbonizing steel, and **bone meal,** for animal feed, are processed to remove fats and prevent rancidity and are steam-sterilized. **Fertilizer bone** is bone meal that is of too fine a mesh for carbonizing use or for making bone black, but it is not cooked or processed and has an analysis of about 45% ammonia and 50 phosphate of line. Knucklebones, and the hard shinbones, are preferred for gelatin manufacture. **Dissolved bone,** used for fertilizer, is ground bone treated with sulfuric acid, or the residue bone after dissolving out the gelatin with acid. The **steamed bone meal** produced as a residue by-product of the glue factories and soap factories is suitable only for fertilizer, but it contains only 1% ammonia. The steamed bone meal for animal feed is merely steam-sterilized raw bone.

Bone black, also called **animal charcoal** and **bone char,** is charred bone ground to a fine silky powder for use as a pigment, or as a decolorizing agent for sugar and oils. It has a deep, dense, bluish-black color valued for engraving inks of depth and tone, and to give a dull velvety-black finish to coated paper. Its covering power, however, is inferior to that of carbon black, as it has only about 10% carbon, is largely calcium phosphate, and has a very high ash content. The best blacks may be treated with acid to remove the lime salts. Federal specifications for bone black for pigment require that 97.5% pass through a 325-mesh screen. For filtering, the material used is from 4 to 16 mesh. The specific gravity is 2.6 to 2.8. Bone black is made by calcining ground, fat-free, dried bones in airtight retorts.

Drop black is the spent bone black from the decolorizing of sugar, which has been washed and reground for pigment use. **Ivory black,** having the same uses as bone black, is made by heating the refuse of ivory, working in a closed retort, and then grinding to a fine powder. **Aquablak No. 1,** of Binney & Smith Co., is a water dispersion of bone black used to give a velvety-black color to inks, water paints, and leather finishes.

Bone oil, used in sheep dips and disinfectants, and in insecticides and fungicides, is a chemically complex oil with a pungent, disagreeable odor derived as a by-product in the destructive distillation of bones. It contains nitrides, pyrroles, pyridine, and aniline.

BORAX. A white or colorless crystalline mineral used in glass and ceramic enamel mixes, as a scouring and cleansing agent, as a flux in melting metals and in soldering, as a corrosion inhibitor in antifreeze liquids, as a constituent in fertilizers, in the production of many chemicals and pharmaceuticals, and as a source of boron. Borax is a hydrous **sodium borate,** or tetraborate, first obtained from Tibet under the Persian name **borak,** meaning white. **Tincal** is the natural borax, $Na_2O \cdot 2B_2O_3 \cdot 10H_2O$, obtained originally through Iran from Tibet, and later found in quantity in the western United States. The specific gravity is 1.75, Mohs hardness 2 to 2.5, and melting point 1125°F (608°C). It contains 47.2% water and is readily soluble in water. The borax from California and Nevada known as **colmanite** is **calcium borate,** $2CaO \cdot 3B_2O_3 \cdot 5H_2O$. That known as **kernite,** or **rasorite,** is another variety of sodium borate of composition $Na_2O \cdot 2B_2O_3 \cdot 4H_2O$. **Ulexite,** or **boronatrocalcite,** is found in the western United States, Chile, Bolivia, and Peru, and it has composition $Na_2O \cdot 2CaO \cdot 5B_2O_3 \cdot 16H_2O$; but the ulexite of Chile and Bolivia is also mixed with sodium and calcium sulfates and sodium and magnesium chlorides. The dried mineral contains 45 to 52% boric acid. **Priceite,** or **pandermite,** chiefly from Asia Minor, is $5CaO \cdot 6B_2O_3 \cdot 9H_2O$. All these minerals are **boron ores** suitable for producing boron.

Borax is found in great quantities in the desert regions of the western United States, and in the Andean deserts of South America, where the borate deposits are in land-locked basins at altitudes above 11,000 ft (3,353 m). The colmanite is usually associated with ulexite, shale, and clay and is concentrated to 40% B_2O_3. Kernite ore from Kern County, California, contains about 29% B_2O_3 and is concentrated to 45%. Federal specifications for borax call for not less than 99.5% hydrous sodium borate in grades from large crystals to fine white powder. **Pyrobar** is a crystalline anhydrous borax of American Potash & Chemical Co., produced from the brine of Searles Lake. **Puffed borax,** of this company, is an expanded borax powder with bulking density as low as 2 lb/ft³ (32 kg/m³) which is very soluble and is used for detergents.

Boric acid, also called **boracic acid** and **orthoboric acid,** is a white, crystalline powder of composition $B_2O_3 \cdot 3H_2O$, derived by adding hydrochloric or sulfuric acid to a solution of borax and crystallizing. It is also obtained from the boron ores of California, Chile, Bolivia, and Peru as one of a number of chemicals. It occurs naturally in volcanic fissures in Italy as the mineral **sassolite,** $B_2O_3 \cdot 3H_2O$. The specific gravity is 1.435 and melting point 1050°F (562°C). It is soluble in water and in alcohol. Boric acid is used as a preservative and weak antiseptic in glass and pottery mixes, in the tanning industry for deliming skins by forming calcium borates soluble in water, and as a flux in soldering and brazing. **Anhydrous boric acid** is nearly pure **boric oxide.** It is used in glass and ceramic enamels.

BORNEO TALLOW. A hard, brittle, yellowish-green solid fat obtained from the seed nuts of trees of the family *Dipterocarpaceae* of Borneo, Java, Sumatra, and Malaya. They include the *Shorea aptera, S. robusta, S. stenoptera, Hopea aspera,* and *Pentacme siamensis,* which trees also produce copals and valuable woods. The kernels of the nuts of *S. stenoptera* contain 45 to 60% fat. It contains 39% stearic acid, 38 oleic, 21.5 palmitic, and 1.5 myristic acid. The specific gravity is 0.852 to 0.860, saponification value 185 to 200, iodine value 29 to 38, and melting point 93 to 102°F (34 to 39°C). The seeds are exported as **Sarawak illipé** nuts or **Pontianak illipé,** the "black" grades being of higher quality and richer in fat than the "brown" ones. They are not the same as the **illipé nuts** and **illipé tallow** from the species of *Bassia* of India. **Illipé butter,** also called **mowrah butter** and **bassia butter,** is from the seed nuts of the trees *B. longifolia* and *B. latifolia,* of India. The oil content of the seeds is 50 to 60%. The crude fat is yellow to green, but the refined product is colorless with a pleasant taste. It is used as a food and in soaps and candles. **Phulwa butter,** or **Indian butter,** is from the seed nuts of *B. butyracea* of India. It is white, stiffer than lard, and is used as a food. **Siak tallow** is from the *Palanquium oleosum* of Malaya and various species of *Dipterocarpaceae* of Indonesia. The nuts are smaller than the illipé and are known as **calam seeds** in Sumatra. The fat is yellow to greenish, and was used as a substitute for cocoa butter. **Jaboty fat** is a similar heavy oil from the kernels of the fruit of the tree *Erisma calcaratum* of Brazil. It contains 44% palmitic acid, 28 myristic acid, and 4 stearic acid, and it has a melting point of 104 to 115°F (40 to 46°C). The tropical forests of Brazil, Africa, and southern Asia abound in trees that produce a great variety of nuts from which useful fats and oils may be obtained.

BORON. A metallic element, symbol B, closely resembling silicon. When pure, it is in the form of red crystals or a brownish amorphous powder. Crystalline boron in lumps of 99.5% purity is extremely hard, has a metallic sheen, is not hygroscopic and resists oxidation, and is a semiconductor. It has a specific gravity of 2.31, a melting point of about 4172°F (2200°C), and a Knoop hardness of 2,700 to 3,200, equal to a Mohs hardness of about 9.3. At 112°F (600°C) boron ignites and burns with a brilliant green flame. Obtained by the electrolysis of fused boric oxide from the mineral borax, it is used mostly in its compounds, especially borax and boric acid. Minute quantities of boron are used in steels for case hardening by the nitriding process to form a boron nitride, and in other steels to increase hardenability, or depth of hardness. In these **boron steels** as little as 0.003% is beneficial, forming an iron boride, but with larger amounts the steel becomes brittle and susceptible to hot-short unless it contains titanium or some other element to stabilize the carbon. In cast iron, boron inhibits graphitization and also serves as a deoxidizer. It is added to iron and steel in the form of **ferroboron.**

Boron 10, an isotope that occurs naturally in boron in the proportion of 10 to 12%, is 10 times more effective than lead in stopping neutrons and is used for neutron shielding, but it does not stop gamma rays. **Boron 11,** which comprises 88 to 90% of natural boron, is transparent to neutrons. **Boron steel foil** and sheet for neutron shielding, called **Bo-Stan,** has the composition of a stainless steel with 2% boron evenly distributed. It is made from metal powders, compacted under pressure, sintered below the melting point to bond the metals, and hot-rolled. A similar foil called **Binal** is aluminum with 2% boron.

Boron fibers for reinforced structural composites are continuous fine filaments which are, themselves, composites. They are produced by vapor deposition of boron on a tungsten substrate. Their specific gravity is about 2.6, and they range from 4 to 6 mils (0.10 to 0.15 mm) in diameter. They have tensile strengths around 500,000 lb/in² (3,448 MPa), and a modulus of elasticity of nearly 60×10^6 lb/in² (0.41×10^6) MPa). The fibers are used chiefly in aluminum or epoxy matrixes. Unidirectional **boron-aluminum composites** have tensile strengths ranging from 110,000 to over 200,000 lb/in² (758 to over 1,379 MPa). Their strength-to-weight ratio is about 3 times greater than that of high-strength aluminum alloys.

Boron compounds are employed for fluxes and deoxidizing agents in melting metals and for making special glasses. Boron, like silicon and carbon, has an immense capacity for forming compounds, although it has a different valence. The boron atom appears to have a lenticular shape, and two boron atoms can make a strong electromagnetic bond, the B_2 acting as carbon does but with a double ring.

Typical of a complex boron compound is the natural mineral **danburite** which occurs in orthorhombic crystals that will fuse to a colorless glass. The composition of the mineral is given as $CaB_2(SiO_4)_2$, but each silicon atom is at the center of a group of four oxygen atoms, two such groups having one oxygen atom in common. The eighth oxygen atom forms another tetrahedral group with one oxygen atom from each of the three Si_2O_7 groups, and in the center of such groups are the boron atoms. Thus, even tiny amounts of boron can affect alloys greatly. Boron is used in specialty refractories, electronics, borane fuels, and high-temperature carborane plastics.

Boron trichloride, BCl_3, is used as a catalyst. Above 55°F (12.5°C) it is a gas, but it is also used as the dihydrate, a fuming liquid. **Boron trifluoride,** BF_3, is a gas used for polymerizing epoxy resins, usually in the solid form of **boron ethyl amine,** $BF_3 \cdot C_2H_5NH_2$, which releases the BF_3 at elevated temperatures. **Boron tribromide,** BBr_3, is also a highly reactive compound, used to produce boron hydrides as sources of hydrogen. **Hydrazine diboride,** $BH_3 \cdot NH_2 \cdot NH_2 \cdot BH_3$, used as a source of hydrogen in rocket fuels, is a white, crystalline, free-flowing powder. **Trimethyl boroxine,** $(CH_3O)_3B \cdot B_2O_3$, is a liquid used for extinguishing metal fires. Under heat it breaks down, and the molten coating of boric acid smothers the fire.

Carborane plastics are produced from the boron molecule, a boron hydride of composition $B_{10}H_{14}$, by replacing the four terminal hydrogens with two carbon atoms. The monomer, $C(HB_2H)_5C$, polymerizes at high heat and pressure to form rubbery solids which will withstand temperatures above 600°F (315°C); but the commercial plastics are usually copolymers with vinyls, silicones, or other plastics.

BORON CARBIDE. A black crystalline powder of high hardness used as an abrasive, or pressed into wear-resistant products such as drawing dies and gages, or into heat-resistant parts such as nozzles. The composition is either B_6C or B_4C, the former being harder but usually containing an excess of graphite difficult to separate in the powder. It can be used thus as a deoxidizing agent for casting copper and for lapping, since the graphite acts as a lubricant. **Boroflux,** of General Electric Co., is boron carbide with flake graphite, used as a casting flux.

The boron carbide marketed by Norton Co. under the name of **Norbide** is B_4C, over 99% pure. It is used as a hard abrasive and for molding. Parts molded without a binder have a compressive strength of 400,000 lb/in² (2,758 MPa), a tensile strength of 22,500 lb/in² (155 MPa), and a Knoop hardness of 2,800. The coefficient of expansion is only one-third that of steel, and the specific gravity is 2.51, or less than that of aluminum. The melting point is 4450°F (2454°C), but above 1800°F (982°C) it reacts with oxygen. The material is not resistant to fused alkalies. **Boron carbide powder** for grinding and lapping comes in standard mesh sizes to 240, and in special finer sizes to 800. The usual grinding powders are 220 and 240 mesh, and the finishing powder is 320 mesh. Boron carbide powders are also added to molten aluminum and then rolled into sheets for use in shielding against neutrons. **Boral,** of Aluminum Co. of America, is this sheet containing 35% by weight of carbide and clad on both sides with pure aluminum. A 0.25-in (0.64-cm) sheet is equal to 25 in (0.64 m) of concrete for neutron shielding.

BORON NITRIDE. A light, fluffy, white powder produced by the reaction of boric oxide with ammonia results in the composition BN, used as a lubricant for high-pressure bearings and for compacting into mechanical and electrical parts. Its X-ray pattern and platy crystal structure are almost identical with those of graphite, and it is called **white graphite.** Boron nitride has a very low coefficient of friction; but unlike carbon, it is a nonconductor of electricity, and it is attacked by nitric acid. It sublimes at 5432°F (3000°C). It reacts with carbon at about 3632°F (2000°C) to form boron carbide. It is used for heat-resistant parts by molding and pressing the powder without a binder to a specific gravity of 2.1 to 2.25.

Sintered parts have an ivorylike appearance, a tensile strength of 3,500 lb/in² (24 MPa), compressive strength of 45,000 lb/in² (310 MPa), and dielectric strength of 1,000 V/mil (39 × 10⁶ V/m). They are soft, with a hardness of Mohs 2, and can be machined easily.

When compressed at very high pressure and heat, the hexagonal crystal structure changes to a cubic structure of greater strength and hardness. **Cubic boron nitride** (CBN), such as General Electric's **Borazon,** has a Knoop hardness of 4,700, second only to diamond, a compressive strength of 700,000 lb/in² (4,830 MPa), twice the thermal conductivity of copper, and is thermally stable to about 3500°F (1925°C). It is also chemically stable with iron and oxidation-resistant in air to about 3600°F (1982°C), a temperature far exceeding that at which diamond oxidizes. Thus CBN is valued as a bonded abrasive powder for cutting tools used to machine materials that cannot be readily cut with more common materials.

Boron nitride fibers are produced in diameters as small as 197 to 276 μin (5 to 7 μm) and in lengths to 15 in (0.38 m). The fibers have a tensile strength of 200,000 lb/in² (1,378 MPa). They are used for filters for hot chemicals and as reinforcement to plastic lamination. **Boron nitride HCJ,** of Union Carbide, is a fine powder, 99% pure, used as a filler in encapsulating and potting compounds to add thermal and electric conductivity. Boron nitride (BN) is basically the same as BC and can only be densified by hot pressing, Hot-pressed hexagonal BN is easy to machine and has a low density but must be kept free of boron oxide or the boron oxide must be stabilized with CaO-containing additives so that BN is not destroyed by hydration during heating.

Another boron compound used for the production of high-temperature ceramic parts by pressing and sintering is **boron silicide,** B_4Si. It is a black, free-flowing, crystalline powder. The powder is microcrystalline, with particles about 2,950 μin (75 μm) in diameter, and the free silicon is less than 0.15%. This compound normally reacts at 2190°F (1200°C) to form B_6Si and silicon; but when compacted and sintered, the ceramic forms a boron silicate oxygen protective coating, and the parts have a serviceable life in air at temperatures to 2550°F (1400°C). Molded parts have high thermal shock resistance and can be water-quenched from 2000°F (1093°C) without shattering. **Boralloy** pyrolytic boron nitride, a dielectric, and **pyrolytic graphite** are combined in **Boralectric PBN/PG** resistance-heating elements of Union Carbide Coating Services Corp. The elements, which operate at temperatures up to 3632°F (2000°C), are electrically and dimensionally stable and resistant to thermal shock and many chemicals.

BOTANICALS. Known also as **plant extracts** and **crude drugs,** they are preparations, usually in concentrated form, obtained by chemically treating plant tissue from roots, leaves, stalks, flowers, or bark to remove odiferous, flavorful, or nutritive substances of the tissue for use in medicines, cosmetics, insecticides, and foodstuffs. Their use goes back to ancient times when they were used for herb healing and aromatic bath preparations. Over thousands of years, plant extracts have been the main ingredient in many folk remedy drugs.

Today botanicals find use in sophisticated drugs and in an immense variety of cosmetics. It is estimated that there are approximately 750,000 different species of higher plants on earth. Of these, only about 1% have been studied for possible use as botanicals.

Among the plant extracts used in cosmetics are those from cornflowers, marigolds, wild pansies, hawthorn blossoms, sage, horse chestnut, ivy, juniper, hops, mint, and chamomile. Chamomile is a popular botanical in cosmetics, such as shampoo and skin care products.

Many of the most valuable **drug plants** contain **alkaloids** which have complex arrangements of carbon, hydrogen, oxygen, and nitrogen. Most of the alkaloids are violent poisons even in small quantities, and with few exceptions, such as quinine, they are not employed as medicines except under the direction of trained physicians or pharmacists. The actual amount to produce sufficient effect without injury must be understood prior to use of a drug. **Pain-killing drugs** for example, do not kill pain but diminish the conductivity of the nerve fibers and thus prevent the brain from recording the pain. The alkaloid **hyoscine, or scopolamine,** a heavy liquid of composition $C_{17}H_{21}NO_4$, causes loss of part of the normal inhibition control, and it is known as **truth serum.** But what is expressed as truth by a mentally disturbed person may be only the product of imagination, and the drug should be given only by trained operators.

Drug plants are almost unlimited in number. **Belladonna,** used to relieve pain, to check perspiration, and as a dilatant, consists of the leaves, roots, flowers, and small stems of the perennial herb *Atropa belladonna,* cultivated chiefly in Yugoslavia and Italy, but also grown in the United States. The plant is called **banewort** and **deadly nightshade.** The leaves are dried in the shade to retain the green color. They contain two alkaloids, **atropine** and **hyoscyamine.** Atropine is used as an antidote for military nerve gas. **Henbane** is the dried leaves and flowering tops of the ill-smelling herb *Hyoscyamus niger,* a plant of the nightshade family containing several alkaloids. It is used as a sedative. The plant is grown in southern Europe and Egypt and to some extent in the United States. It is harvested in full bloom and dried in the shade.

Stramonium is the dried leaves, flowers, small stems, and seeds of the **thorn apple** or **jimson weed,** *Datura stramonium,* an annual of the nightshade family which grows as a common weed in the United States. It is shipped from Ecuador under the name **chamico.** It produces the alkaloids atropine and hyoscyamine and is used as a nerve sedative, hypnotic, and antispasmodic. Under cultivation, the yield is 1,000 to 1,500 lb (453 to 680 kg) of dry leaf or 500 to 2,000 lb (227 to 906 kg) of seed per acre. In general, the yield of drug plants in cultivation is high, so that only small acreages are needed. **Aconite** is the dried root of **monkshood,** a perennial plant, *Aconitum nepullus.* It is grown in Europe and the United States. The root contains the colorless, crystalline, extremely poisonous alkaloid **aconitine,** $C_{34}H_{47}O_{11}N$. It is used as a cardiac sedative and diaphoretic and to relieve pain and fever. **Matrine,** obtained from the plant *Sophora angustifolia,* is a lupine alkaloid having two piperidine rings. Combined with phenol sulfonate salts under the name of **dysentol,** it is used for amoebic dysentery.

Digitalis, an important heart stimulant, consists of the leaves of the perennial **foxglove plant,** *Digitalis purpurae,* native to Europe but also grown in New England and the Pacific northwest. The leaves contain the bitter glucoside **digitoxin,** $C_{34}H_{54}O_{11}$. **Peyote** is the root of a **cactus,** *Lophophora williamsii,* growing wild in the desert region of Mexico. The buttonlike tops of the root, called **mescal buttons,** contain several alkaloids, and were used by the Indians to produce a sense of well-being with visions. **Peyotina hydrochloride,** made from the buttons, is used in medicine as a heart stimulant similar to digitalis and as an anesthetic to the nervous system. The **glucosides,** unlike the alkaloids, contain no nitrogen. They are ethers of single sugars. The glucoside **rutin,** used for treatment of hemorrhage and for high blood pressure, is a yellow nontoxic powder extracted from flue-cured tobacco and from **Tartary buckwheat** or **rye buck-**

wheat. The active ingredient, **quertin,** is extracted and used to prevent hemorrhage in hypertensive persons. **Ipecac** is the rhizome of
the shrub *Cephalis ipecacuamha,* of tropical South America, chiefly
Brazil. It contains **emetine,** $C_{30}H_{40}O_5N_2$, and other alkaloids and is
used as an emetic tonic and expectorant. **Gentian,** also known as **bitter root,** is the dried root of the perennial herb *Gentiana lutea* of central Europe. It contains glucosides and is used in tonics and in cattle
feeds. **Coca** is a South American shrub from which the drug and narcotic **cocaine** is produced. The shrub, indigenous to Peru and Bolivia,
is produced by cultivation in Java. The dried leaves, when chewed,
have a mild narcotic effect.

   Ginseng is the dried root of the plant *Panex ginseng* of China and
P. quinquefolium of North America. It is used as a stimulant and
stomachic. **Dog-grass root,** known in medicine as **triticum,** is the
powdered dried rhizome of **Scotch grass,** *Agropyron repens,* which
abounds in the meadows of the northern United States. It has a slight
aromatic odor and sweet taste and is used as a diuretic. **Lobelia** is
the dried leaves and tops of the small annual plant *Lobelia inflata,*
known in the northeastern United States as **Indian tobacco.** It contains the alkaloid **lobinine** and is used as an emetic and antispasmodic. **Erigeron** is the herb and seeds of the **fleabane herb,**
Erigeron canadensis, which grows wild in the north central and western states. The herb yields 0.35 to 0.65% **erigeron oil,** used as an
astringent and tonic. **Hamamelis** is the dried leaves of the witch
hazel shrub, *Hamamelis virginiana,* of the eastern United States. It
is used as a tonic and sedative, and the water extract of the leaves
and twigs with 14% alcohol and 1% active ingredient is known as
witch hazel. It is used as an external astringent. It is fragrant and
contains also tannic acid from the bark of the twigs. **Ephedrine,** or
ephedra, is an alkaloid extracted from the dried twigs of the
mahuang, *Ephedra sinica,* and other species of mountain shrubs of
China and India. It is also grown in South Dakota. It is used as an
adrenaline substitute to raise blood pressure and in throat medicines
and nasal sprays. **Hoarhound** is an extract from the leaves and flowering tops of the small perennial herb *Marrubium vulgare* of North
America, Europe, and Asia. It is used in preparations for colds and as
a flavor in confections.

   Cascara is the bark of the small **buckthorn** tree, *Rhamnus pur-
shiana,* of the northwestern United States and Canada. It was also
called **chittem bark.** It was used by the Indians and given the name
cascara sagrada by the Spaniards. The word *cascara* simply means
bark; the word *sagrada* is not translated as *sacred* but is a term
applied to medicinal botanicals. Cascara is used as a laxative and
tonic. The **European buckthorn,** *R. frangula,* also yields cascara.

Aloe, used in purgative medicines, is the dried resinous juice from the leaves of the bush *Aloe vulgaris* of the West Indies and *A. perryi* of western Africa. The cut leaves are placed in troughs where the juice exudes. It is then evaporated to a viscous black mass which hardens. Both aloe and cascara contain the glucoside **emodin,** occurring also in senna and in **rhubarb.** The latter, the stalks of which are much used as a food, is known in medicines as **rheum** and is employed as a laxative and stomach tonic. **Aloin,** used in skin creams for radiation and sunburns, is an extract from *A. vera,* a plant of the lily family growing in Florida. **Aletris root** is a botanical drug from **stargrass,** *Aletris farinosa,* growing in the eastern United States. It is also known as **colic root** and is used as a tonic and uterine stimulant. **Senega,** used as an emetic and stimulant, is the dried root of the **snakeroot** or **milkwort,** *Polygala senega,* a small perennial herb grown in the eastern United States. **Calumba root** is the yellow root of the woody climbing plant *Coscinium fonestratum* of Sri Lanka. It is used as a cure for tetanus. The wood, which is bright yellow, is used locally as a dye. **Tanacetum** is the dried leaves and tops of the **silverweed,** *Tanacetum vulgaris,* of Michigan and Indiana, used as a vermifuge. The green herb yields about 0.2% **tansy oil,** or **tanacetum oil,** which contains a terpene—**tanacetene**—and borneol. It is used as an anthelmintic.

Ergot, used to stop hemorrhage, is the dried sclerotium of the fungus *Claviceps purpurea,* which develops on rye and some grasses. The purple structure that replaces the diseased rye grain contains the alkaloid **clavine,** $C_{11}H_{22}O_4N_2$. Ergot is also used to produce the alkaloids **ergotamine** and **ergonovine,** used for treating high blood pressure and migraine headache and used for mental-disease research. It requires 1,000 lb (453 kg) of ergot to yield 1.5 oz (43 g) of ergonovine, but it is produced synthetically by Eli Lilly & Co. from **lysergic acid,** which is synthesized from **indole propionic acid,** $C_8H_6N(CH_2)_2COOH$. **Ergot oil,** used in medicinal soaps, is obtained by extraction from dry ergot. It was the original source of **ergosterol,** one of the most important **sterol alcohols** from plants, known as **phytosterols,** which, when irradiated, yield vitamin D. Ergot is produced chiefly in Spain, and its commercial production has not been encouraged in the United States because it is an undesirable disease on grain. Ergot also contains the amino acid **thiozine** found in blood.

Araroba, or **Goa powder,** is a brownish, bitter, water-soluble powder scraped out of the split logs of the tree *Andira araroba* of Brazil. The Carib Indians used the powder for skin diseases, and it is now employed for eczema and skin infections. It contains **chrysarobin,** a complex mixture of reduction products of a complex acid contained in the wood. **Labdanum** is an oleoresin obtained in dark-brown or greenish lumps from the branches of the **rock rose,** *Cistus ladani-ferus,* and other species of the Mediterranean countries. It yields a volatile oil with a powerful sweet characteristic odor, and is used as a stimulant and expectorant and as a basis for lavender and violet perfumes. It was originally obtained in Greece by combing from the fleece of sheep that browsed against the bushes. **Buchu** is the dried leaves of the South African herb *Barosma betulina,* used as an anticatarrhal. **Serpasil,** of Ciba Pharmaceutical Products, Inc., is an extract from the root of the plants *Rauwolfia serpentina* of India and *R. heterophylla* of Central America. It is used as a sedative in hypertension cases and belongs to the class of **tranquilizing agents. Rauwolfia extract** contains the two alkaloids **reserpine** and **rescinnamine,** both having hypotensive activity. They are produced by Riker Laboratories as **Serpiloid** and **Rauwiloid,** and the product of Squibb is called **Raudixin.** A synthetic material, **meprobamate** is a complex **propanediol dicarbamate,** and **Singoserp,** of Ciba, is a synthetic reserpine.

BOXWOOD. The wood of the **Turkish boxwood** tree *Buxus semper-virens,* native to Europe and Asia but also grown in America. It is used for rulers, instruments, engraving blocks, and inlay work. The wood is light yellow and hard and has a fine grain and a dense structure that does not warp easily. The density is about 65 lb/ft^3 (1,041 kg/m^3). **African boxwood,** or **cape boxwood,** comes from the tree *B. macowani,* of South Africa, and is similar to boxwood but softer. **Kamassi wood** is a hard, fine-grained wood from the tree *Gonioma kamassi,* of South Africa, sometimes substituted for boxwood, but it does not have the straight grain of boxwood. It is valued for loom shuttles. **Coast gray boxwood,** or the **Gippsland boxwood** of New South Wales, is from the tree *Eucalyptus bosistoana.* It is a durable wood of uniform texture, but it has an interlocking grain. **Maracaibo boxwood,** or **zapatero,** which comes chiefly from Venezuela, is from the tree *Casearia praecox.* It comes in straight knotless logs 8 to 10 ft (2.4 to 3 m) long and 6 to 10 in (15.2 to 25.4 cm) in diameter. The light-yellow wood has a fine uniform texture and straight grain. It replaces Turkish boxwood for all purposes except wood engravings. The **ginkgo wood** used in China for making chess pieces and chessboards is from the large tree *Ginkgo biloba* which appears to be the sole survivor of a family of trees with fernlike leaves once very abundant. The wood is white to yellowish, lightweight, fine-textured, and easy to work. Ginkgo extracts obtained from the tree's leaves are used widely in medicines in Asia and much of Europe. The synthesis recently in the United States of the compound **ginkgolide B** may lead to the development of ginkgo-based drugs in this country.

BRASS. **Copper-zinc** alloys whose zinc content ranges up to 40%. If the copper crystal structure is face-centered cubic, there will be up to 36% of zinc present. This solid solution, known as the *alpha phase,* or **alpha brass,** has good mechanical properties, combining strength with ductility. Corrosion resistance is very good, but electric conductivity is considerably lower than that for copper. When 30 to 36% of the alloy is zinc, a body-centered-cubic crystal structure is formed, known as the *beta phase,* or **beta brass.** This phase is relatively brittle and high in hardness compared to the alpha phase. However, ductility increases at elevated temperatures, thus providing good hot-working properties. **Gamma brass,** with zinc above 45%, is not easily worked, either hot or cold.

The mechanical properties of brasses vary widely. Strength and hardness depend on alloying and/or cold work. Tensile strengths of annealed grades are as low as 30,000 lb/in² (207 MPa), although some hard tempers approach 90,000 lb/in² (621 MPa). Although brasses are generally high in corrosion resistance, two special problems must be noted. With alloys containing a high percentage of zinc, dezincification can occur. The corrosion product is porous and weak. To prevent dezincification, special inhibitors—antimony, phosphorus, or arsenic—in amounts of 0.02 to 0.05% can be added to the alloy. The other problem is stress corrosion, or seasonal cracking, which occurs when moisture condenses on the metal and accelerates corrosion.

Brass is annealed for drawing and bending by quenching in water from a temperature of about 1000°F (538°C). Simple copper-zinc brasses are made in standard degrees of temper, or hardness. This hardness is obtained by cold-rolling after the first anneal, and the degree of hardness depends on the percentage of cold reduction. When the thickness is reduced about 10.9%, the resulting sheet is ¼-hard. The other grades are ½-hard, hard, extra hard, spring, and finally extra spring, which corresponds to a reduction of about 68.7% without intermediate annealing. Degrees of softness in annealed brass are measured by the grain size, and annealed brass is furnished in grain sizes from 0.00039 to 0.0059 in (0.010 to 0.150 mm). The ASTM standard grain sizes are 0.00059 to 0.00098 in (0.015 to 0.025 mm) for light anneal, 0.0014 in (0.035 mm) for drawing or rod anneal, 0.002 in (0.050 mm) for intermediate anneal, 0.0028 in (0.07 mm) for soft anneal, and 0.0047 in (0.120 mm) for deadsoft anneal. Brasses with smaller grain sizes are not as ductile as those with larger grain sizes, but they have smoother surfaces and require less polishing.

Even slight additions of other elements to brass alter the characteristics drastically. Slight additions of tin change the structure, increasing the hardness but reducing the ductility. Iron hardens the alloy and reduces the grain size, making it more suitable for forging but difficult to machine. Manganese increases strength, increases the solubility of iron in the alloy, and promotes stabilization of aluminum, but makes the brass extremely hard. Slight additions of silicon increase strength, but large amounts promote brittleness, loss of strength, and danger of oxide inclusion. Nickel increases strength and toughness, but when any silicon is present, the brass becomes extremely hard and more a bronze than a brass.

There are hundreds of brasses with a bewildering array of names—some misleading, such as *commercial bronze, jewelry bronze,* and *manganese bronze,* which are brasses. But most brasses can be grouped into a few major classes. The **straight brasses** constitute by far the largest and most widely used group. They are binary copper-zinc alloys with zinc content ranging from 5 to about 40%. Some of the common names of these alloys are **gilding metal** (5% zinc), **commercial bronze** (10 zinc), **jewelry bronze** (12.5 zinc), **red brass** (15 zinc), **yellow brass** (35 zinc), and **muntz metal** (40 zinc). As the zinc content increases in these alloys, the melting point, density, electric and thermal conductivity, and modulus of elasticity decrease while the coefficient of expansion, strength, and hardness increase. Work hardening also increases with zinc content. These brasses have a pleasing color, ranging from the red of copper for the low-zinc alloys through bronze and gold colors to the yellow of high-zinc brasses. The color of jewelry bronze closely matches that of 14-karat gold, and this alloy and other low brasses are used in inexpensive jewelry.

The low-zinc brasses have good corrosion resistance along with moderate strength and good forming properties. Red brass, with its exceptionally high corrosion resistance, is widely used for condenser tubing. The high brasses (**cartridge brass** and **yellow brass**) have excellent ductility and high strength and are widely used for engineering and decorative parts fabricated by drawing, stamping, cold heading, spinning, and etching. Muntz metal, primarily a hot-working alloy, is used where cold working is not required.

Another group is the **leaded brasses.** These alloys have essentially the same range of zinc content as the straight brasses. Lead is present, ranging from less than 1 to 3.25%, to improve machinability and related operations. Lead also improves antifriction and bearing properties. Common leaded brasses include **leaded commercial bronze** (0.5% lead), **medium-leaded brass** (1 lead), **high-leaded brass** (2 lead), **free-cutting brass** (3.25 lead), and **hardware bronze** (1.75 lead). Free-cutting brass provides optimum machinability and is ideally suited for screw-machine parts.

Another group, the **tin brasses,** is copper-zinc alloys with small amounts of tin. The tin improves corrosion resistance. Pleasing colors are also obtained when tin is added to the low brasses. Tin brasses in sheet and strip form, with 80% or more copper, are used widely as low-cost spring materials. **Admiralty brass** is a standard alloy for heat-exchanger and condenser tubing. **Naval brass** and **manganese bronze** are widely used for products requiring good corrosion resistance and high strength, particularly in marine equipment.

Casting brasses are usually made from brass ingot metal and are seldom plain copper-zinc alloys. In melting brass for casting, any overheating causes loss of zinc by vaporization, thus lowering the zinc content. Small amounts of antimony or some arsenic are used to overcome this dezincification. The casting brasses are roughly divided into two classes as **red casting brass** and **yellow casting brass,** which are various compositions of copper, tin, zinc, and lead to obtain the required balance of color, ease of casting, hardness, and machining qualities.

Many brasses are now designated by alloy number and grouped into several families of standard alloys. There are three families of wrought brass: **copper-zinc brasses** (C20500 to C28580); **copper-zinc-lead brasses,** or **leaded brasses** (C31200 to C38590); and **copper-zinc-tin brasses,** or **tin brasses** (C40400 to C49080). Casting alloys comprise four families; **copper-tin-zinc red brasses** and **copper-tin-zinc-lead brasses,** or **leaded red brasses** (C83300 to C83810); **copper-zinc-tin-lead brasses, semired brasses,** and **leaded semired brasses** (C84200 to C84800); **copper-zinc-lead-tin yellow brasses, leaded yellow brasses,** and **manganese-bearing high-strength yellow brasses** (C85200 to C85800); and **copper-zinc-silicon brasses,** or **silicon brasses** (C87300 to C87900). Wrought alloys are available in most mill forms, and casting alloys are amenable to most casting processes, including sand, permanent-mold, investment, plaster, and centrifugal. A few alloys can be die-cast.

Mechanical properties vary depending on the alloy, grain size, and temper condition. Thin, annealed flat products of **red brass C23000,** for example, have tensile yield strengths of 10,000 to 18,000 lb/in^2 (69 to 124 MPa), with strength increasing and ductility decreasing (48 to 42% elongation) with decreasing grain size. Cold work increases yield strength to 39,000 lb/in^2 (269 MPa) in the ¼-hard temper and to 63,000 lb/in^2 (434 MPa) in the spring temper. Sand-cast red brass (C83300) has a minimum tensile yield strength of 60,000 lb/in^2 (414 MPa) and 12% elongation as cast.

BRASS INGOT METAL. Commercial ingots made in standard composition grades and employed for casting various articles designated as brass and bronze. They are seldom true brasses, but are composition metals intermediate between the brasses and the bronzes, and their selection for any given purpose is based on a balance of the requirements in color, strength, hardness, ease of casting, and machinability. Brass ingot metal is usually made from secondary metals, but, in general, the grading is now so good that they will produce high-grade uniform castings. In producing the ingot metal there is careful sorting of the scrap metals, and the impurities are removed by remeltings. An advantage of ingot metal over virgin metals is the ease of controlling mixtures in the foundry. The ATSM designates eight grades for brass ingot metal. Number 1 grade, the highest in copper, contains 88% copper, 6.5 zinc, 1.5 lead, and 4 tin, with only slight percentages of impurities. The No. 8 grade contains 63.5% copper, 34 zinc, 2.5 lead, and no tin. **Yellow ingot,** for plumbing fixtures, contains 65% copper, 1 tin, 2 lead, and the balance zinc. The most widely used ingot metal is **ASTM alloy No. 2,** which is the 85:5:5:5 alloy known as **composition metal.** Yellow brass, or **yellow casting brass,** is frequently cast from **ASTM alloy No 6,** which contains 72% copper, 22 zinc, 4 lead, and 2 tin. It has a tensile strength of 20,000 to 25,000 lb/in^2 (138 to 172 MPa), elongation 15 to 20%, and Brinell hardness 40 to 50. It is yellow and makes clean, dense castings suitable for various machine parts except bearings.

BRAZIL NUTS. Also called **Pará chestnut.** The nuts of the large tree *Bertholletia excelsa* growing wild in the Amazon Valley. The trees and nuts are called **tacarí** in Brazil and **toura** in French Guiana. The tree begins to bear in 8 years and yields up to 1,000 lb (453 kg) of large, round fruit pods containing 18 to 24 hard-shelled kernels which are the commercial nuts. The shelled kernels are several times the size of the peanut, and they have a pleasant nutty flavor. The kernels contain 67% of a pale yellow oil of specific gravity 0.917, saponification value 192 to 200, and iodine value 98 to 106. The oil contains 51% oleic acid, 19 linoleic, 2 myristic, 3 stearic, and 12 palmitic. It is a valuable food oil and is also a good soap oil, but is normally too expensive for this purpose, as the nuts are more valued for eating.

BRAZILWOOD. The wood of the trees *Caesalpinia brasiliensis, C. crista,* and *C. echinata,* of tropical America, and *C. sappan,* of Sri Lanka, India, and Malaya. The latter species is called **sappanwood.** Brazilwood formerly constituted one of the most valuable exports from Brazil to Europe as a dyewood. It produces purple shades with a chrome mordant and crimson with alum. **Brazilwood extract** is still valued for silk dyeing, wood staining, and inks. The wood is prized for such articles as violins and fine furniture. It has a rich bright-red color and takes a fine, lustrous polish.

BRAZING METAL. A common name for high-copper brass used for the casting of such articles as flanges that are to be brazed on copper pipe. Federal specifications for brazing metal call for 84 to 86% copper and the balance zinc. **Brazing brass,** of American Brass Co., has 75% copper and 25 zinc. Some alloys also contain up to 3% lead for ease of machining. It also makes the metal easier to cast.

The term *brazing metal* is also applied to **brazing rods** of brass or bronze, used for joining metals. A common brazing rod is the 50–50 brass alloy with a melting point of 1616°F (880°C). The SAE designates this alloy as **spelter solder.** The joints made with it are inclined to be brittle. **Brazing wire,** of Chase Brass & Copper Co., contains 59% copper and 41 zinc, while the **brazing solder,** for brazing high-zinc brasses, has 51% copper and 49 zinc. **Phos-copper** is a phosphor copper which gives joints 98% the electric conductivity of copper. It flows at 1382°F (750°C). **Tri-metal** is a brazing metal consisting of sheet brass with a layer of silver rolled on each side. It is used for brazing carbide tips to steel tools. The silver ensures a tightly brazed joint, while the brass center acts as a shock absorber for the cutting tool.

Brazing rods for brazing brasses and bronzes are usually of a composition similar to that of the base metal. For brazing cast iron and steel, various bronzes, naval brass, manganese bronze, or silicon bronze may be used. Brass rods may contain some silicon. Small amounts of silver added to the high-copper brazing metals give greater fluidity and better penetration into small openings.

BRICK. The most ancient of all artificial building materials, consisting of clay molded to standard shape, usually rectangular, and burned to a hard structure. In some areas bricks, known as **adobe bricks,** are still made by baking in the sun, and a modern adaptation of sun-baked brick, called **bitudobe,** has a binder of emulsified asphalt. Commercial bricks in the United States and in Europe are all hard-burned clay. They are used for buildings, walls, and paving and are classified apart from the bricks of fireclays used for refractories. **Brick clays** are of two general classes. The first consists of noncalcareous clays or shales composed of true clay with sand, feldspar grains, and iron compounds, which when fired, become buff or salmon in color. The second class comprises calcareous clays containing up to 40% calcium carbonate, called **marls.** When fired, they are yellowish. Brick clays of the first type are widely distributed. Iron oxide in them varies from 2 to 10%, and the red color of common brick depends largely on this content. In practice, the composition of bricks varies widely, but much of the difference is in the burning as well as in the method of pressing. **Pressed brick** is a stiff mud brick made under high pressure. It is homogeneous and has increased density and strength. **Building brick,** made by machine of ground and tempered clay, has great uniformity of strength and color. Such brick is made by pressing soft, stiff, or dry. The burning is done in kilns at temperatures of 1652 to 2282°F (900 to 1250°C). The calcareous clays require a temperature of 2192°F (1200°C) to bring about chemical combination. The bricks are sorted according to hardness and color, both largely resulting from their position in the kiln. **Paving brick** is usually a hard-burned common brick. **Floor brick** is highly vitrified brick. The common hard brick for building has a crushing strength of 5,000 to 8,000 lb/in^2 (34 to 55 MPa) and a density of 125 lb/ft^3 (2,025 kg/m^3). The common standard for building-brick size is 8.25 by 4 by 2.5 in (21 by 10.2 by 6.4 cm); other sizes are also used, especially 8 by 3.875 by 2.25 in (20.3 by 9.8 by 5.7 cm). Specially sized paving bricks are 8.5 by 4 by 3 or 8.5 by 4 by 3.5 in (21.6 by 10.2 by 7.6 or 21.6 by 10.2 by 8.9 cm). **Sand-lime bricks** for fancy walls are of sand and lime pressed in an atmosphere of steam. They are not to be confused with the sand-lime bricks used for firebrick, which are of refractory silica sand with a lime bond. Ceramic glazes and semiglazes are used on some building bricks, especially on the yellow.

BRISTLES. The stiff hairs from the back of the hog, used chiefly in making brushes. The very short bristles, rejects, and scrap pieces are used for filler in plastics. The best brush bristles do not include hair from the sides of the animals or the product from the fat-meat animals killed in slaughterhouses. They come mostly from types of semiwild swine grown in cool climates, notably northern China and Russia. Bristles are in form similar to a tiny tube outwardly covered with microscopic scales and filled with a fatty substance. The so-called *flag,* or split end, gives the valuable paint-carrying characteristic for brushes. The taper of the bristle gives the brush stiffness at the base and resiliency toward the end. Quality varies according to the type of animal, climate, and feeding. The colors are white, yellow, gray, and black. They are graded by locality, color, and length; and in normal times the name of the place at which they are graded, such as Tsingtao, Hankow, and Chungking, is an indication of the grade. The best fibers are more than 3 in (7.6 cm) in length. The Chinese natural black bristles are sometimes sold at a premium. The U.S. bristles from Chester hogs are light in color and of high quality. Bristles from the Duroc hog are bronze in color, stiff, and superior to most Chinese grades. Those from the Poland China hogs are black and stiff, but they have a crooked flag and are of poor quality. **Artificial bristles** are made from various plastics, the **nylon bristles** being of high quality and much used. **Exton,** of Du Pont, was one of the original nylon monofilament nontapered bristles. **Tynex,** of this company, now comes both tapered and level. They are more durable than natural bristles. **Casein bristles** are made by extruding an acid solution of casein, stretching the fiber, and insolubilizing with formaldehyde or other chemicals. They have good paint-carrying capacity and good wear resistance, but are dissolved by some paint solvents. **Keron bristle,** of Rubberset Co., is produced from the protein extracted from chicken feathers. It is nearly identical in composition to natural bristle.

BROMINE. An elementary material, symbol Br. It is a reddish-brown liquid having a boiling point of 138°F (59°C). It gives off very irritating fumes and is highly corrosive. It is one of four elements called **halogens,** a name derived from Greek words meaning *salt producer.* They are fluorine, chlorine, iodine, and bromine. They are all chemically active, combining with hydrogen and most metals to form **halides.** Bromine is less active than chlorine but more so than iodine. It is moderately soluble in water. It never occurs free in nature, and it is obtained from natural bromide brines by oxidation and steaming, or by electrolysis. It occurs in seawater to the extent of 65 to 70 parts per million and is extracted. It is marketed 99.7% minimum purity with specific gravity not less than 3.1, but dry elemental bromine, Br_2, is marketed 99.8% pure for use as a brominating and oxidizing agent. For these uses, also, bromine is available as a crystalline powder as **dibromodimethyl hydantoin,** containing 55% bromine. **Brom 55** is this material. Bromine is also used as a flame retardant in plastics, although its use as such has been questioned due to environmental concerns.

A pound of bromine is obtained from 2,000 gal (7,570 L) of seawater. It is also produced as a by-product from the brine wells of Michigan and from the production of chemicals at Searles Lake, California, where the bromine concentration is 12 times that of seawater. It is used in the manufacture of agricultural chemicals, dyes, photographic chemicals, poison gases for chemical warfare, pharmaceuticals, disinfectants, and many chemicals. It is also employed in the extraction of gold. Bromine's major end use, as ethylene dibromide for scavenging lead antiknock compounds in gasoline, is decreasing as these environmentally hazardous additives are phased out.

BRONZE. The term *bronze* is generally applied to any copper alloy that has as the principal alloying element a metal other than zinc or nickel. Originally the term was used to identify copper-tin alloys that had tin as the only, or principal, alloying element. Some brasses are called bronzes because of their color, or because they contain some tin. Most commercial copper-tin bronzes are now modified with zinc, lead, or other elements.

The **copper-tin bronzes** are a rather complicated alloy system. The alloys with up to about 10% tin have a single-phase structure. Above this percentage, a second phase, which is extremely brittle, can occur, making plastic deformation impossible. Thus high-tin bronzes are used only in cast form. Tin oxide also forms in the grain boundaries, causing decreased ductility, hot workability, and castability. Additions of small amounts of phosphorus, in the production of phosphor bronzes, eliminate the oxide and add strength. Because tin additions increase strength to a greater extent than zinc, the bronzes as a group have higher strength than brasses—from around 60,000 to 105,000 lb/in^2 (414 to 724 MPa) in the cold-worked high-tin alloys. In addition, fatigue strength is high.

Bronzes containing more than 90% copper are reddish; below 90% the color changes to orange-yellow, which is the typical bronze color. Ductility rapidly decreases with increasing tin content. Above 20% tin the alloy rapidly becomes white and loses the characteristics of bronze. A 90% copper and 10% tin bronze has a density of 0.317 lb/in^3 (8,775 kg/m^3); an 80–20 bronze has a density of 0.315 lb/in^3 (8,719 kg/m^3). The 80–20 bronze melts at 1868°F (1020°C), and a 95–5 bronze melts at 2480°F (1360°C).

The family of **aluminum bronzes** is made up of alpha-aluminum bronzes (less than about 8% aluminum) and alpha-beta bronzes (8 to 12% aluminum) plus other elements such as iron, silicon, nickel, and manganese. Because of the considerable strengthening effect of aluminum, in the hard condition these bronzes are among the highest-strength copper alloys. Tensile strength approaches 100,000 lb/in^2 (690 MPa). Such strengths plus outstanding corrosion resistance make them excellent structural materials. They are also used in wear-resistance applications and for nonsparking tools. **Phosphor bronzes** have a tin content of 1.25 to 10%. They have excellent mechanical and cold-working properties and a low coefficient of friction, making them suitable for springs, diaphragms, bearing plates, and fasteners. Their corrosion resistance is also excellent. In some environments, such as salt water, they are superior to copper. **Leaded phosphor bronzes** provide improved machinability. **Silicon bronzes** are similar to aluminum bronzes. Silicon content is usually between 1 and 4%. In some, zinc or manganese is also present. Besides raising strength, the presence of silicon sharply increases electrical resistivity. Aluminum-silicon bronze has exceptional strength and corrosion resistance and is particularly suited to hot working.

Gear bronze may be any bronze used for casting gears and worm wheels, but usually means a tin bronze of good strength deoxidized with phosphorus and containing some lead, to make it easy to machine and to lower the coefficient of friction. A typical gear bronze contains 88.5% copper, 11 tin, 0.25 lead, and 0.25 phosphorus. It has a tensile strength of up to 40,000 lb/in^2 (276 MPa), elongation 10%, and Brinell hardness of 70 to 80, or up to 90 when chill-cast. The density is 0.306 lb/in^3 (8,470 kg/m^3). This is **SAE bronze No. 65.** A **hard gear bronze,** or **hard bearing bronze,** contains 84 to 86% copper, 13 to 15 tin, up to 1.5 zinc, up to 0.75 nickel, and up to 0.5 phosphorus. Hard and strong bronzes for gears are often silicon bronze or manganese bronze.

In a modified 90–10 type of bronze, the zinc is usually from 2 to 4%, and the lead up to 1%. A cast bronze of this type will have a tensile strength of about 40,000 lb/in^2 (276 MPa), an elongation of 15 to 25%, and a Brinell hardness of 60 to 80, those high in zinc being the stronger and more ductile, those high in lead being the weaker. Bronzes of this type are much used for general castings and are classified as **composition metal** in the United States. In England they are called **engineer's bronze.**

Architectural bronze, or **art bronze,** is formulated for color and is very high in copper. One foundry formula for art bronze of a dull-red color calls for 97% copper, 2 tin, and 1 zinc. For ease of casting, however, they are more likely to contain lead, and a **gold bronze** for architectural castings contains 89.5% copper, 2 tin, 5.5 zinc, and 3 lead. In leaded bronze, the hard copper-tin crystals aid in holding the lead in solution. These bronzes are resistant to acids and are grouped as **valve bronze,** or as **bearing bronze** because of the hard crystals in a soft matrix. Federal specifications for bronze give 10 grades in wide variations of tin, zinc, and lead. The ASTM designates five grades of **bronze casting alloys. Alloy No. 1** contains 85% copper, 10 tin, and 5 lead; **Alloy No. 5** contains 70% copper, 5 tin, and 25 lead. The British **coinage copper** is a bronze containing 95.5% copper, 3 tin, and 1.5 zinc.

Many of these bronzes are designated by alloy number and grouped into several families of standard alloys. There are four principal families of wrought bronze: **copper-tin-phosphorus alloys,** or **phosphor bronzes** (C50100 to C52400); **copper-tin-phosphorus-lead alloys,** or **leaded phosphor bronzes** (C53200 to C54800); **copper-aluminum alloys,** or **aluminum bronzes** (C60600 to C64400); and **copper-silicon alloys,** or **silicon bronzes** (C64700 to C66100). A few, such as two **manganese bronzes,** are included in the copper-zinc family of copper alloys. The aluminum bronze family is the largest, containing nearly three dozen standard compositions.

Casting alloys comprise five main families: **copper-zinc-man-**

ganese alloys, or **manganese bronzes** (C86100 to C86800); **copper-tin alloys,** or **tin bronzes** (C90200 to C91700); **copper-tin-lead alloys,** or **leaded and highly leaded tin bronzes** (C92200 to C94500); **copper-tin-nickel alloys,** or **nickel-tin bronzes** (C94700 to C94900); and **copper-aluminum alloys,** or **aluminum bronzes** (C95200 to C95900). **Copper-silicon bronzes** are included in the C87300 to C87900 family of copper-silicon alloys.

Flat products 0.12 in (3 mm) thick of **aluminum bronze C61400** have tensile yield strengths ranging from 45,000 lb/in^2 (310 MPa) in the annealed (O60) temper to 60,000 lb/in^2 (414 MPa) after cold working to the H04, or hard, temper, with ductility decreasing from 40 to 32% elongation, respectively. All casting alloys can be sand-cast, many can be centrifugally cast, and some can be permanent-mold, plaster-, and investment-cast but not die-cast. Among the strongest is **manganese bronze C86300,** which, as cast in sand molds, provides a minimum tensile yield strength of 60,000 lb/in^2 (414 MPa) and at least 12% elongation. Some bronzes, such as **nickel-bearing silicon bronzes** and aluminum bronzes, especially those containing more than 9% aluminum, can be strengthened by age-hardening.

BRONZE POWDER. Pulverized or powdered bronze made in flake form by stamping from sheet metal. It is used chiefly as a paint pigment and as a dusting powder for printing. In making the powder, the sheets are worked into a thin foil which becomes harder under the working and breaks into small flakes. Lubricant keeps the flakes from sticking to one another. Usually stearic acid is used, but in the dusting powder, hot water or nonsticky lacquers are used. The powder is graded in standard screens and is then polished in revolving drums with a lubricant. This gives it the property of leafing, or forming a metallic film in the paint vehicle. The leaf is also called **composition leaf,** or **Dutch metal leaf,** when used as a substitute for gold leaf. **Flitters** are made by reducing thin sheets to flakes, and they are not as fine as bronze powder. **Alpha Bronze,** a prealloyed 91% copper, 10% tin powder from Makin Metal Powders of England for powder-metal parts, provides greater green strength and less wear and noise in bearing applications than a premixed 90–10 composition.

The compositions of bronze powder vary, and seven alloys form the chief commercial color grades from the reddest, called *pale gold,* which has 95% copper and 5 zinc, to the rich gold which has 70% copper and 30 zinc. Colors are also produced by heating to give oxides of deep red, crimson, or green-blue. The powder may also be dyed in colors, using tannic acid as a mordant, or treated with acetic acid or copper acetate to produce an antique finish. The color or tone of bronze powders may also be adjusted in paints by adding a proportion of mica powder. A **white bronze powder** is made from aluminum bronze, and the silvery colors are obtained with aluminum powder. The bronze powder of 400 mesh used for inks is designated as extra fine. The fine grade, for stencil work, is 325 mesh. Medium fine, for coated paper, has 85% of the particles passing through a 325-mesh screen and 15% retained on the screen. Near mesh, for paint pigment, has 30% passing through a 325-mesh screen. A 400-mesh powder has 500 million particles per gram. The old name for bronze powder is **gilding powder.** It is also called *gold powder* when used in cheap gold-colored paints, but bronze powders cannot replace gold for use in atmospheres containing sulfur, or for printing on leather where tannic acid would corrode the metal. **Gold pigments** used in plastics are bronze powders with oxygen stabilizers.

BROOMCORN. A plant of the sorghum family, *Holcus sorghum,* grown in the southwest, in Illinois and Kansas, and in Argentina and Hungary. It is used for making brushes and brooms and for the stems of artificial flowers. The jointed stems of the dwarf variety grown in semiarid regions are 12 to 24 in (0.30 to 0.61 m) long, but the standard brush corn is up to 30 in (0.76 m) long. The fibers are yellow and, when dry, are coarse and hard. They are easily cleaned and readily dyed. As a brush material, they have the disadvantage of breaking easily and are therefore unsuited for mechanical brushes for hard service. **Broom root,** or **rice root,** is similar to broomcorn and is suitable for mixing with it or for coarse brushes. It is from a type of grass, *Epicampes macroura,* of Mexico and Guatemala. The fiber is from the tough, crinkly, yellowish roots. After removal of the outer bark, the dry root is treated with the fumes of burning sulfur to improve the color. The fibers are 8 to 18 in (0.20 to 0.46 m) long. In Mexico it is called *raiz de Zacaton,* or **Zacaton root,** and its American name, *rice,* is a corruption of the Spanish word for root.

BRUSH FIBERS. Industrial brushes are made from a wide variety of fibers, varying from the fine and soft camel's hair to the hard, coarse, and brittle broomcorn. Bristles are the most commonly used, but tampico and piassava fibers are important for polishing brushes. The vegetable fibers used for brushes are tough and stiff compared to the finer, flexible and cohesive fibers used for twine and for fabrics. They may, however, come from the same plant, or even from the same leaf, as the textile fibers, but be graded out for stiffness. **Palmetto fiber** is from the **cabbage palm** tree, *Sabal palmetto,* of Florida. Whiskbrooms and brushes are made from the young leafstalks and stiff floor sweeps from the leaves.

A fiber finer than palmetto is obtained from the twisted roots of the

scrub palmetto, *S. megacarpa.* **Arenga fiber** is a stiff, strong fiber from the stems of the **aren palm** tree, *Arenga saccharifera,* of Indonesia. The finest grades resemble horsehair. **Kittool** is a similar strong, elastic fiber from the large leaves of the palm tree *Caryota urens,* of India and Sri Lanka. It is very resistant and is valued for machine brushes. **Gomuti fiber** and **Chinese coir** are fibers from other species of this palm. **Bass,** or **raphia,** is a coarse fiber used for hard brushes and brooms. The heavier piassava fibers are also known as bass, but bass is from the leaves of the palm tree *Raphia vinifera,* of West Africa. **Darwin fiber,** used for brooms and scrubbing brushes in Australia, is from the *Gahnia trifida.* **Crin** is from the leaves of the palm tree of Algeria, although the word *crin* originally referred to horsehair. **Crin vegetal,** or **vegetable crin,** is fiber from the leaf of the **yatay palm,** *Diplothemium littorale,* of Corrientes Province, Argentina. **Horsehair,** from the manes and tails of horses, is used for some paintbrushes.

    Red sable hair is used for fine-pointed and knife-edged brushes for show-card and watercolor use. It is from the tail of the **kolinsky,** *Mustela siberica,* of Siberia, and the pale red hair has strength and resiliency and very fine points. **Russian sable hair,** used for artists' brushes, is stronger than red sable hair, but is less pointed and not as elastic for water painting. It is from the tail of the **fitch,** *Putorius putorius,* of central Asia, but the so-called **fitch hair** used for ordinary flowing brushes is usually **skunk tail hair.** It is stiffer and coarser than fitch hair. **Badger hair,** also used for flowing brushes, is a resilient hair with fine points and is from the back of the badger of Turkey and southern Russia. **Black sable hair,** used for signwriter brushes, is not from a sable, but is the trade name for mixtures of marten hair, bear hair, and some other Siberian hairs.

    Vegetable and animal fibers are not resilient to alkalies or acids and cannot be wetted with them. The artificial fibers of plastics such as nylon are resistant to many chemicals. For hard-service mechanical brushes, and for resistance to strong chemicals, brush fibers are of steel, brass, or aluminum wire. **Brush wire** for rotary-power brushes for metal brushing is soft- to hard-drawn steel wire usually 0.005 in (0.013 cm) in diameter. Finer wire for soft rotary brushes is a soft-drawn steel wire 0.0025 in (0.006 cm) in diameter.

BUFFING COMPOSITIONS.   Materials used for buffing or polishing metals, originally consisting of dolomitic lime with 18 to 25% saponifiable grease as a bond. The lime acts as the abrasive, and in some compositions is partly replaced by other abrasives such as emery flour, tripoli, pumice, silica, or rouge. Harsher abrasives are used in the compositions employed for the cutting-down or buffing operations.

Abrasive grains are selected for combinations of hardness, toughness, and sharpness, from the soft iron oxide to the hard and sharp aluminum oxide. Buffing compositions are usually sold under trade names for definite uses rather than by composition. **Metal polishes** for hand use are now usually liquids. The pastes, formerly known as **Putz cream** and **brash polish,** contained tripoli or pumice with oxalic acid and paraffin. The liquid polishes now generally contain finer abrasives such as pumicite or diatomite, in a detergent, together with a solvent, and sometimes pine oil or an alkali.

BUILDING SAND. Selected sand used for concrete, for mortar for laying bricks, and for plastering. Early specifications called for sand grains to be sharp, but rounded grains are now preferred because there are fewer voids in the mixture. Building sand is normally taken from deposits within a reasonable haul of the site of building, and is not usually specified by analysis, but should be a hard silica sand that will not dissolve. Pure white sand for finish plaster is made by grinding limestone. Building sand is required to be clean, with not more than 3% clay, loam, or organic matter. ASTM requirements are that all grains pass through a 0.375-in (0.95-cm) sieve, 85% through a No. 4 sieve, and not more than 30% through a No. 50 sieve. For brick mortar, all the sand should pass through a 0.25-in (0.64-cm) sieve. For plaster, not more than 6% should pass through a No. 8 sieve. **Flooring sand** for mastic flooring is a clean sand passing through a No. 3 sieve, with 7% passing through a No. 100 sieve. **Roofing sand** is a fine, white silica sand. **Paving sand** is divided into three general classes: for concrete pavements, for asphaltic pavements, and for grouting.

The U.S. Bureau of Public Roads requires that all sand for concrete pavements pass through a 0.25-in (0.64-cm) sieve, 5 to 25% should be retained on a No. 10 sieve, from 50 to 90% on a No. 50 sieve, and not more than 10% should pass through a No. 100 sieve. Not more than 3% of the weight should be matter removable by elutriation. For asphaltic pavements small amounts of organic matter are not objectionable in the sand. All should pass through a 0.25-in (0.64-cm) sieve, 95 to 100% through a No. 10 sieve, and not more than 5% through a No. 200 sieve. **Grouting sand** should all pass through a No. 20 sieve, and not more than 5% through a No. 200 sieve. **Chat sand,** used for concrete pavements, is a by-product of zinc and lead mines. It is screened through a 0.375-in (0.95-cm) sieve.

BUILDING STONE. Any stone used for building construction may be classified as building stone. **Granite** and **limestone** are among the most ancient of building materials and are extremely durable. Two million limestone and granite blocks, totaling nearly 8 million long tons (8,128 million kg), were used in the pyramid of Giza, built about 2980 B.C., the granite being used for casing. Availability, or a near supply, may determine the stone used in ordinary building, but for public buildings stone is transported long distances. Some sandstones, such as the **red sandstone** of the Connecticut Valley, weather badly and are likely to scale off with penetration of moisture and frost. Granite will take heavy pressures and is used for foundation tiers and columns. Limestones and well-cemented sandstones are employed extensively above the foundations. Nearly half of all the limestone used in the United States in block form is **Indiana limestone.** Marble has a low crushing strength and is usually an architectural or facing stone.

Crushed stone is used for making concrete, for railway ballast, and for road making. The commercial stone is quarried, crushed, and graded. Much of the crushed stone used is granite, limestone, and **trap rock.** The last is a term used to designate basalt, gabbro, diorite, and other dark-colored, fine-grained igneous rocks. Graded crushed stone usually consists of only one kind of rock and is broken with sharp edges. The sizes are from 0.25 to 2.5 in (0.64 to 6.35 cm), although larger sizes may be used for massive concrete aggregate. Screenings below 0.25 in (0.64 cm) are employed largely for paving. **Granite granules** for making hard terrazzo floors are marketed in several sizes, and in pink, green, and other selected colors. **Roofing granules** are graded particles of crushed rock, slate, slag, porcelain, or tile, used as surfacing on asphalt roofing and shingles. Granules have practically superseded gravel for this purpose. Black **amphibole ryolite** may be used, or gray basalt may be colored artificially for granule use. The **suzorite rock** of Quebec contains feldspar, pyroxenite, apatite, and mica, and it is treated to remove the mica. **Ceramic granules** are produced from clay or shale, fired and glazed with metallic salts. They are preferred because the color is uniform.

BULK MOLDING COMPOUND. BMC is a puttylike mixture of thermosetting polyester resin, additives, and fillers, often extruded into easily handled shapes for compression, transfer, or injection molding. Glass-fiber-reinforced grades provide tensile strengths of 3,000 to 10,000 lb/in^2 (21 to 69 MPa), flexural strengths of 11,000 to 24,000 lb/in^2 (76 to 165 MPa), and notched-Izod impact strengths of 2 to 13 ft · lb/in (107 to 694 J/m). Washtubs, trays, and equipment housings are common applications. **Corelyn,** of Bulk Molding Compounds, Inc., has a density of 0.067 lb/in^3 (1,855 kg/m^3), a notched-Izod impact strength of 6 to 8 ft · lb/in (320 to 427 J/m), and resists heat up to 400°F (204°C). **Nu-Stone,** of Industrial Dielectrics Inc., is a BMC that looks like granite.

BURLAP. A coarse, heavy cloth made of plain-woven jute, or jutelike fibers, and used for wrapping and bagging bulky articles, for upholstery linings, and as a backing fabric for linoleum. Finer grades are used for wall coverings. The standard burlap from India is largely from jute fibers, but some hibiscus fibers are used. For bags and wrappings, the weave is coarse and irregular, and the color is the natural tan. The coarse grades such as those used for wrapping cotton bales are sometimes called *gunny* in the United States, but **gunny** is a general name for all burlap in Great Britain. Dundee, Scotland, is the important center of burlap manufacture outside of India, but considerable quantities are made from native fibers in Brazil and other countries. Burlap is woven in widths up to 144 in (3.6 m), but 36, 40, and 50 in (0.91, 1.02, and 1.27 m) are the usual widths. **Hessian** is the name of a 9.5-oz (269-g), plain-woven finer burlap made to replace an older fabric of the same name woven from coarse and heavy flax fibers. When dyed in colors, it is used for linings, wall coverings, and upholstery. **Bithess** was a name for Hessian fabric coated with bitumen, used in India to spread over soft-earth areas as a seal for a top coating to form airplane runways. **Brattice cloth** is a very coarse, heavy, and tightly woven jute cloth, usually 20 oz (567 g) used for gas breaks in coal mines; but a heavy cotton duck substituted for the same purpose is called by the same name. Most burlap for commercial bags is 8, 9, 10, and 12 oz (226, 255, 283, and 340 g), feed bags being 8 oz (226 g) and grain bags 10 oz (283 g).

BUTADIENE. Also called **divinyl, vinyl ethylene, erythrene,** and **pyrrolylene.** A colorless gas of composition $CH_2{:}CH \cdot CH{:}CH_2$ used in the production of neoprene, nylon, latex paints, and resins. Butadiene has a boiling point of 26.6°F (−3°C) and a specific gravity of 0.6272. Commercial butadiene is at least 98% pure.

Butadiene is primarily obtained as an ethylene coproduct during the steam cracking of naphtha or gas oil. It is also made by oxidation dehydrogenation of *n*-butenes, the dehydrogenation of butanes, and conversion of ethyl alcohol. The largest use for butadiene is the production of elastomers, such as **polybutadiene, styrene-butadiene, polychloroprene,** and **acrylonitrile-butadiene,** or **nitrile rubbers.**

Three types of polybutadiene are available: high-*cis* (97%), medium-*cis* (92%), and low-*cis* (40%). The high-*cis* rubber is made by polymerization with a cobalt or nickel catalyst to keep the detrimental vinyl content below 1%. The medium-*cis,* the most popular grade, employs a titanium catalyst, and the low-*cis* product uses an alkyl-lithium initiator. The rubbers have less resilience and a higher heat buildup than natural rubber, but they also give much greater wear life, low-temperature flexibility, and increased groove-cracking resistance in automotive tire treads and sidewalls, and bias truck-tire body plies. Polybutadiene is almost always used in blends with other rubbers. In tire treads, the concentration is 25 to 35% by weight; the other components include corubbers, **carbon black, extending oils,** and **zinc stearate** activator. Polybutadiene is also used as a raw material for making **hexamethylenediamine,** the precursor for **nylon 6,6** and **acrylonitrile-butadiene-styrene plastics.**

BUTTER. An edible fat made from cow's milk by curdling with bacterial cultures and churning. The production of butter is one of the large industries of the Western nations, with an annual production exceeding 10 billion lb (4.5 billion kg), 30% of which is made in the United States. Other important producers are Germany, Holland, the Scandinavian countries, Australia, New Zealand, Canada, Ireland, and Argentina. Butter is an important raw material in the bakery and confectionery industries. Federal regulations require that creamery butter be made exclusively from milk or cream, with or without salt and coloring matter, and contain not less than 80% by weight of milk fat, not over 15% moisture, and not over 2.5% salt. Butter varies greatly in color and flavor according to the feed of the animal, the processing, and the storage. The natural color is whitish in winter and yellow in summer, when the animal feeds on green pasturage. Commercial butter is usually brought to a uniform yellow by coloring with annatto. Musty, garlicky, and fishy flavors may be caused by noxious weeds eaten by the animal; cheesy or yeasty flavors may be from stale cream; metallic, greasy, scorched, or alkaline flavors may be from improper processing. **Whipped butter** has 50% greater volume in the same weight and has greater plasticity for spreading.

United States grades for creamery butter range from 93 score for the best butter of fine flavor and body down to 85 score for the lowest grade having pronounced obnoxious weed flavor and defects in body, color, or salt. The grading, or scoring, of butter is done by experts. The flavor is determined by the senses of taste and smell. The flavor, body, color, and salt are rated independently, and points, or scores, are subtracted for defects. Body and texture of the butter are determined by the character of the granules and their closeness. The most common body defects are gumminess, sponginess, crumbliness, and stickiness. The most common defect in color is lack of uniformity, with waves or mottles. Defects in salting are excessive salt and undissolved salt grains. Butter held in storage at improper temperatures is likely to develop rancid or unpleasant flavors and acidity due to chemical changes, or it may absorb flavors from surrounding products. High-grade butter can be held in well-regulated cold storage for long periods without appreciable deterioration.

An important substitute for butter is **margarine. Oleomargarine** is a term still retained in old food laws, but the product is no longer manufactured. It was a compound of mutton fat with vegetable tallows and fats, invented by the French chemist Mege-Mouries. Margarine is made from a mixture of about 80% vegetable oils and 20 milk in the same manner as butter. It has a slightly lower melting point than butter, 72 to 81°F (22 to 27°C), but the melting point and a desired degree of saturation of the fatty acids can be regulated by hydrogenation of the oils. Margarine of lower melting points is used in the bakery industry, and grades with higher melting points are for table use. From 2.5 to 4% salt is used, together with vitamins A and D, lecithin, annatto coloring, and sometimes phosphatides to prevent spattering when used for frying. **Biacetyl,** $C_4H_6O_2$, a colorless, pungent, sweet liquid which gives the characteristic flavor to butter, is also added. The food value is, in general, higher than that of butter; but because of the competition with butter, various federal and state regulations restrict its use. **Soya butter** is made from emulsified soybean; and when fortified with butyric acid, the characteristic acid of butter, it is practically indistinguishable from butter. It is, however, subject to restrictive regulations. **Butter flavors** are used in confectionery and bakery products. **Butter-Aid** is made by extracting and concentrating the esters of natural butter. It is used as a high-strength flavor in foodstuffs in the form of powder or liquid emulsion. **Butta-Van** is a butter flavor with vanilla. It contains butyric acid, ethyl butyrate, coumarin, vanillin, and glycerin in water solution. **Ghee butter,** used in India, is made from buffalo milk, sometimes mixed with cow's milk. It is clarified and the moisture removed by boiling and slow cooling and separating off the opaque white portion. It is light in color and granular.

Cheese is an important solid food product made from whole or skim milk. It contains all the food value of milk, including the proteins of the casein. The biotics used in the manufacture produce *n*-butyric acid, also with caproic, caprylic, and capric acids in varying amounts which produce the flavor of the various types of cheese. In the same manner, **lipase enzymes** from the glands of calves and lambs are used for enhancing the flavor of food products containing milk or butterfat. The enzymes hydrolyze the butyric or other short-chained fatty acids into the glycerides. **Lipolyzed butter,** of Marschall Dairy Laboratories, Inc., is made by treating natural butterfat with enzymes. It gives intensity and uniformity of flavor to margarine and bakery products.

CADMIUM. A silvery-white crystalline metal, symbol Cd. It has a specific gravity of 8.6, is very ductile, and can be rolled or beaten into thin sheets. It resembles tin and gives the same characteristic cry when bent, but is harder than tin. A small addition of zinc makes it very brittle. It melts at 608°F (320°C) and boils at 1409°F (765°C). Cadmium is employed as an alloying element in soft solders and in fusible alloys, for hardening copper, as a white corrosion-resistant plating metal, and in its compounds for pigments and chemicals. It is also used for nickel-cadmium batteries and to shield against neutrons in atomic equipment; but gamma rays are emitted when the neutrons are absorbed, and these rays require an additional shielding of lead. The metal is marketed in small, round sticks 12 in (0.31 m) long, in variously shaped anodes for electroplating, and as foil. **Cadmium foil** is 99.95% pure cadmium and is as thin as 0.0005 in (0.013 mm). It is used for neutron shielding and for electronic applications requiring high corrosion resistance. Electrolytic cadmium is 99.95% pure. It is obtained chiefly as a by-product of the zinc industry by treating the flue dust and fumes from the roasting of the ores. Flue dust imported from Mexico averages 0.66 ton (600 kg) of cadmium per ton (metric ton) of dust. About half the world production is in the United States. Other important producers are West Germany, Belgium, Canada, and Poland. The only commercial ore of the metal is **greenockite,** CdS, which contains theoretically 77.7% cadmium. This mineral occurs in yellow powdery form in the zinc ores of Missouri. Cadmium occurs in sphalerite to the extent of 0.1 to 1%.

Most of the consumption of cadmium is for electroplating. For a corrosion-resistant coating for iron or steel, a cadmium plate of 0.0003 in (0.008 mm) is equal in effect to a zinc coat of 0.001 in (0.025 mm). The plated metal has a silvery-white color with a bluish tinge, is denser than zinc and harder than tin, but electroplated coatings are subject to hydrogen embrittlement, and aircraft parts are usually coated by the vacuum process. **Cadmium plating** is not normally used on copper or brass since copper is electronegative to it; but when these metals are employed next to cadmium-plated steel, a plate of cadmium may be used on the copper to lessen deterioration.

Small amounts of cadmium added to copper give higher strength, hardness, and wear resistance, but decrease the electrical conductivity. Copper containing 0.5 to 1.2% cadmium is called **cadmium copper** or **cadmium bronze. Hitenso** is a cadmium bronze of American Brass Co. It has 35% greater strength than hard-drawn copper and 85% the conductivity of copper. The cadmium bronze known in England as **conductivity bronze,** used for electric wires, contains 0.8% cadmium and 0.6 tin. Tensile strength, hard-drawn, is 85,000 lb/in^2 (586 MPa), and conductivity is 50% that of copper. **Cadmium nitrate,** $Cd(NO_3)_2$, is a white powder used for making cadmium yellow and fluorescent pigments, and as a catalyst. **Cadmium sulfide** CdS, is used as a yellow pigment, and when mixed with **cadmium**

selenide, CdSe, a red powder, it gives a bright-orange pigment. The sulfide is used for growing **cadmium sulfide crystals** in plates and rods for semiconductor uses. Crystals grown at 1922°F (1050°C) are nearly transparent, but those grown at higher temperatures are dark amber. Cadmium, a carcinogen, can be extremely toxic, and caution is required not to create dust or fumes.

CAFFEINE. An alkaloid which is a white powder when it has the composition $C_8H_{10}N_4O_2$ and occurs in crystalline flakes when it has one molecule of water of crystallization. The melting point is 459°F (237°C). It is soluble in chloroform and slightly soluble in water and alcohol. It is the most widely used of the purine compounds, which are found in plants. Caffeine stimulates physically to lessen fatigue, but in large amounts is highly toxic. Its prime use is in medicine, but most of the production is used in soft drinks. Caffeine does not normally break down in the human body, but passes off in the urine, and the effect is not cumulative; but sarcosine, which occurs in muscles, is a decomposition product of caffeine, though it normally comes from nitrogen metabolism. Caffeine is obtained from coffee, tea waste, kola nuts, or guarana by solvent extraction, or as a by-product in the manufacture of noncaffeine coffees, or in the processing of coffee for the production of oil and cellulose. It is made synthetically from **dimethyl sulfate,** a volatile toxic liquid of composition $H(CH_2)O(SO_2)O(CH_2)H$, also used for making codeine and other drugs. Synthetic caffeine is made from urea and sodium cyanoacetate and is equal chemically to natural caffeine.

Less than 1% caffeine is obtained from coffee, about 2 from **tea waste,** and 1.5 from kola nuts. In tea it is sometimes called **theine.** Cocoa waste contains **theobromine,** from which caffeine may be produced by adding one more methyl group to the molecular ring. The name is a deception, as there is no bromine in the molecule. Theobromine is a more powerful stimulant than caffeine. It is a bitter white crystalline powder of composition $C_7H_8N_4O_2$, also called **dimethyl xanthine** and used in medicine. **Guarana** contains the highest percentage of caffeine of all the beverage plants, about 3%. It comes from the seeds of the woody climbing plant *Paullinia cupana,* of the Amazon Valley. The Indians grind the seeds with water and mandioca flour and dry the molded paste with smoke. For use it is grated into hot water. **Kola nuts** are the seeds of the fruit of the large spreading tree *Kola acuminata,* native to West Africa and cultivated also in tropical America, and *K. nitida* of West Africa. The nuts of the latter tree contain the higher percentages of theobromine and caffeine. The white nuts are preferred to the pink or red varieties. A similar product consists of caffeine and **sodium benzoate.** Both formulations are far more soluble in water than caffeine. **Citrated caffeine,** used in pharmaceuticals, is a white powder produced by the action of citric acid on caffeine, and it contains about equal quantities by weight of anhydrous caffeine and citric acid.

CAJEPUT OIL. A greenish essential oil distilled from the leaves of the tree *Melaleuca leucadendron,* growing chiefly in Indonesia. It contains the cineole of eucalyptus oil and the **terpinol** which is characteristic of the lilac. It has a camphorlike odor. It is used in medicine as an antiseptic and counterirritant, and in perfumes. **Naouli oil** is a similar oil from the leaves of the tree *M. viridi* of New Caledonia. **Cajeput bark,** from the same tree, is used as an insulating material in place of cork. The bark, up to 2 in (5.08 cm) thick, is soft, light, resistant, and a good insulator.

CALCITE. One of the most common and widely diffused materials, occurring in the form of limestones, marbles, chalks, calcareous marls, and calcareous sandstones. It is a **calcium carbonate,** $CaCO_3$, and the natural color is white or colorless, but it may be tinted to almost any shade with impurities. The specific gravity is about 2.72 and Mohs hardness 3. Calcite is usually in compact masses, but **argonite,** formed by water deposition, develops in radiating flowerlike growths often twisted erratically. **Iceland spar,** or **calc spar,** is the name for the perfectly crystallized, water-clear, flawless calcite crystals of optical grade used for the manufacture of **Nicol prisms** for polarizing microscopes, photometers, calorimeters, and polariscopes. It comes from Iceland, Spain, South Africa, and New Mexico, and some crystals have been found as large as 17 lb (7.7 kg). The common **black calcite,** containing manganese oxide, often also contains silver in proportions high enough to warrant chemical extraction of the metal.

CALCIUM. A metallic element, symbol Ca, belonging to the group of alkaline earths. It is one of the most abundant materials, occurring in combination in limestones and calcareous clays. The metal is obtained 98.6% pure by electrolysis of the fused anhydrous chloride. By further subliming, it is obtained 99.5% pure. **Calcium metal** is yellowish white. It oxidizes easily, and when heated in air, burns with a brilliant white light. It has a density of 0.056 lb/in^3 (1,550 kg/m^3), a melting point of 1540°F (838°C), and a boiling point of 2625°F (1440°C). Its strong affinity for oxygen and sulfur is utilized as a cleanser for nonferrous alloys. As a deoxidizer and desulfurizer, it is employed in the form of lumps or sticks of calcium metal or in ferroalloys and calcium-copper. For the reduction of light-metal ores, it is used in the form of the hydride. **Crystalline calcium** is also used in the form of a very reactive free-flowing powder of 94 to 97% purity and containing 2.5% of calcium oxide with small amounts of magnesium and other impurities. The specific gravity of the powder is 1.54, and the melting point is 1562°F (851°C). Natural calcium compounds, such as **dolomite,** are used directly as a flux in melting iron. Calcium is also used to harden lead, and **calcium silicide** is used in making some special steels to inhibit carbide formation.

Many compounds of calcium are employed industrially, in fertilizers, foodstuffs, and medicine. It is an essential element in the formation of bones, teeth, shells, and plants. **Oyster shells** form an important commercial source of calcium for animal feeds. They are crushed, and the fine flour is marketed for stock feeds and the coarse for poultry feeds. The shell is calcium carbonate. **Edible calcium,** for adding calcium to food products, is **calcium lactate,** a white powder of composition $Ca(C_3H_5O_3)_2 \cdot 5H_2O$, derived from milk. **Calcium lactobionate** is a white powder that readily forms chlorides and other double salts and is used as a suspending agent in pharmaceuticals. It contains 4.94% available calcium. **Calcium phosphate,** used in the foodstuffs industry and in medicine, is marketed in several forms. **Calcium diphosphate,** known as **phosphate of lime,** is $CaHPO_4 \cdot 2H_2O$, or in anhydrous form. It is soluble in dilute citric acid solutions and is used to add calcium and phosphorus to foods, and as a polishing agent in toothpastes. **Calcium monophosphate** is a stable, white, water-soluble powder, $CaH_4P_2O_8 \cdot H_2O$, used in baking as a leavening agent. The anhydrous **monocalcium phosphate,** $CaH_4(PO_4)_2$, for use in prepared flour mixes, is a white powder with each particle having a coating of a phosphate that is soluble only with difficulty, to delay solution when liquids are added. **Calcium triphosphate,** $Ca_3(PO_4)_2$, is a white, water-insoluble powder used to supply calcium and phosphorus to foods, as a polishing agent in dentifrices, and as an antacid. **Calcium sulfite,** $CaSO_3 \cdot 2H_2O$, is a white powder used in bleaching paper pulp and textiles, and as a disinfectant. It is only slightly soluble in water, but it loses its water of crystallization and melts at 212°F (100°C). **Calcium silicate,** $CaO \cdot SiO_2$, is a white powder used as a reinforcing agent in rubber, as an absorbent, to control the viscosity of liquids, and as a filler in paints and coatings. It reduces the sheen in coatings. **Silene EF** is a precipitated calcium silicate for rubber. **Micro-Cal,** of Manville Corp., is a synthetic calcium silicate with particle size as small as 0.79 μin (0.02 μm). It will absorb up to 6 times its weight of water, and 3 lb (1.36 kg) will absorb 1 gal (0.0038 m^3) of liquid and remain a free-flowing powder.

Calcium metasilicate, $CaO \cdot SiO_3$, is found in great quantities as the mineral **wollastonite** near Willsboro, New York, mixed with about 15% andradite. The thin, needlelike crystals are easy to crush and grind, and the impurities are separated out. The ground material is a brilliant white powder in short fibers, 99.5% passing a 325-mesh screen. It is used in flat paints, for paper coatings, as a filler in plastics, for welding-rod coatings, and for electrical insulators, tile, and other ceramics. **Calcium acetate,** $Ca(C_2H_3O_2)_2 \cdot H_2O$, is a white powder used in liming rosin and for making metallic soaps and synthetic resins. It is also called **lime acetate, acetate of lime,** and **vinegar salts.**

CALCIUM CARBIDE.   A hard, grayish-black, crystalline substance used chiefly for the production of acetylene gas for welding and cutting torches and for lighting. It was discovered in 1892 and was widely employed for theater stage lighting and for early automobile headlights. It is made by reducing lime with coke in the electric furnace, at 3632 to 3992°F (2000 to 2200°C). It can also be made by heating crushed limestone to a temperature of about 1832°F (1000°C), flowing a high-methane natural gas through it, and then heating to 3092°F (1700°C). The composition is CaC_2, and the specific gravity is 2.26. It contains theoretically 37.5% carbon. When water is added to calcium carbide, **acetylene** gas is formed, leaving a residue of slaked lime. Pure carbide will yield 5.83 ft³ (0.16 m³) of acetylene per 1 lb (0.45 kg) of carbide, but the commercial product is usually only 85% pure. Federal specifications require not less than 4.5 ft³ (0.13 m³) of gas per l lb (0.45 kg). Although calcium carbide is principally used for making acetylene, this market is shrinking as acetylene is recovered increasingly as a by-product in petrochemical plants. A growing application for calcium carbide is desulfurization and deoxidation of iron and steel. It is also a raw material for production of **calcium cyanamide.**

CALCIUM CHLORIDE.   A white, crystalline, lumpy or flaky material of composition $CaCl_2$. The specific gravity is 2.15, the melting point is 1422°F (772°C), and it is highly hygroscopic and deliquescent with rapid solubility in water. The commercial product contains 75 to 80% $CaCl_2$, with the balance chiefly water of crystallization. Some is marketed in anhydrous form for dehydrating gases. It is also sold in water solution containing 40% calcium chloride. Calcium chloride has been used on roads to aid in surfacing, absorb dust, and prevent cracking from freezing. It is used for accelerating the setting of mortars, but more than 4% in concrete decreases the strength of the concrete. It is also employed as an antifreeze in fire tanks, for brine refrigeration, for storing solar energy, as an anti-ice agent on street pavements, as a food preservative, and in textile and paper sizes as a gelling agent. In petroleum production, it is used in drilling muds, cementing operations, and workover or completion fluids. Calcium chloride is obtained from natural brines and dry lake beds, after sodium chloride, bromide, and other products are extracted. The **magnesium-calcium brine** remaining is marketed for dust control or purified into calcium chloride. It is a by-product of sodium bicarbonate production via the Solvay process and is made in small quantities by neutralizing waste hydrochloric acid with lime or limestone.

CALCIUM-SILICON. An alloy of calcium and silicon used as a deoxidizing agent for the elimination of sulfur in the production of steels and cast irons. Steels deoxidized or treated with calcium or calcium and silicon can have better machinability than those deoxidized with aluminum and silicon. It is marketed as low-iron, containing 22 to 28% calcium, 65 to 70 silicon, and 5 maximum iron, and as high-iron, containing 18 to 22% calcium, 58 to 60 silicon, and 15 to 20 iron. It comes in crushed form and is added to the molten steel. At the temperature of molten steel, all the calcium passes off and leaves no residue in the steel. **Calcium-manganese-silicon** is another master alloy containing 17 to 19% calcium, 8 to 10 manganese, 55 to 60 silicon, and 10 iron.

CAMEL'S HAIR. The fine, tough, soft hair from the mane and back of the camel, *Camelus bactrianus,* used for artists' brushes and industrial stripping brushes. Most of the hair is produced in central Asia and Iran, and the grades preferred for brushes are from the crossbred Boghdi camel. The hair from the dromedary, also called *djemel,* or camel, is of poor quality. Much of the camel hair is not cut, but is molted in large patches and is picked up along the camel routes. The plucked beard hair and the coarse outerguard hair obtained in combing are the brush fibers. They are tough, silky, and resilient. The length is 5 to 8 in (12.7 to 20.3 cm). The fine body hair, or **camel wool,** which constitutes about 90% of the total fiber, is 1.5 to 2 in (3.8 to 5.1 cm) long, has a fine radiance, a pale tan color, and a downy feel. It is the textile fiber. The beard hair from the Cashmere goat is very similar to camel hair and is used for brushes. Various other hairs are used for making camel's-hair brushes, including ox-ear hair, badger hair, and sable hair.

CAMPHOR. The white resin of *Cinnamomum camphora,* an evergreen tree with laurellike leaves, reaching a height of 100 ft (30 m). The tree occurs naturally in China and southern Japan, and is also grown in Florida. Taiwan is the center of the industry. Camphor, $C_{10}H_{16}O$, has a specific gravity of 0.986 to 0.996 and melts at 356°F (180°C). It is insoluble in water, but soluble in alcohol or ether. Camphor is used for hardening nitrocellulose plastics, but it is also used in pharmaceuticals, disinfectants, and explosives and chemicals. It is obtained from the trunks, roots, and large branches by steam distillation. From 20 to 40 lb (9.1 to 18.1 kg) of chips produces 1 lb (0.5 kg) of camphor. Crude camphor is pressed to obtain the **flowers of camphor** and **camphor oil.** The crude red camphor oil is fractionated into white and brown oils; the white oil is used in soaps, polishes, varnishes, cleaners, and pharmaceuticals; and the brown oil is used in perfumery. **White camphor oil** is a colorless liquid with a camphor odor and a specific gravity of 0.870 to 1.040, and it is soluble in ether or chloroform. Camphor oil may also be distilled from the twigs. **Camphor sassafrassy oil** is a camphor-oil fraction having a specific gravity of 0.97. It is a sassafras tone and is used for scenting soaps and sprays.

Borneo camphor, or **borneol,** is a white, crystalline solid obtained from the tree *Dryobalanops camphora* of Borneo and Sumatra. It is used as a substitute for camphor in cellulose plastics. It has composition $C_{10}H_{17}OH$ and a specific gravity of 1.01, is soluble in alcohol, and sublimes at 414°F (212°C). The wood of this tree, known as **Borneo camphorwood,** or **kapur,** is used for cabinetwork. It has a density of 50 lb/ft^3 (801 kg/m^3), an interlocking grain, and a scent of camphor. It is also known as **camphorwood.**

Artificial camphor is **bornyl chloride,** $C_{10}H_{17}Cl$, a derivative of the pinene of turpentine. It has a camphor odor and the same industrial uses as camphor, but is optically inactive and is not used in pharmaceuticals. A compound derived from natural camphor, **10-camphor-sulfonic acid,** is used extensively in the optical resolution of amines. **Synthetic camphor,** made from turpentine, in refined form is equal to the natural product for medicinal use, and the technical grade is used in plastics. The camphor substitute **Lindol,** of Hoechst Celanese Corp., is **tricresyl phosphate,** or **tolyl phosphate,** $(CH_3C_6H_4)_3PO_4$, a colorless, odorless viscous liquid which solidifies at −4°F (−20°C). Like camphor, it hardens cellulose nitrate and makes it nonflammable. Tricresyl phosphate is also used as an additive to gasoline to prevent buildup of carbon deposits on the spark plugs and in the engine, thus increasing power by preventing predetonation. Other uses are as a plasticizer for synthetic resins, as a hydraulic fluid, and as an additive in lubricants. It is made from petroleum and from the cresylic acid from coal. **Triphenyl phosphate,** $(C_6H_5)_3PO_4$, is also used as a substitute for camphor in cellulose nitrate and for making coating compounds nonflammable. It is a colorless solid, melting at 120°F (49°C). **Dehydranone,** of Union Carbide Corp., Chemicals Div., is **dehydracetic acid,** $C_8H_8O_4$, a white, odorless solid with some of the properties of camphor, used in nitrocellulose and vinyl resins. **Cyclohexyl levulinate,**

$CH_3CO(CH_2)_2COOC_6H_{11}$, is used as a substitute for camphor in nitro-cellulose and in vinyl resins and chlorinated rubber. It is a liquid of specific gravity 1.025, boiling point 509°F (265°C), and freezing point -94°F (-70°C). **Adamantane** has the odor of camphor and turpentine. It is obtained from the crude petroleum of Moravia as a stable, crystalline solid, melting at 514°F (268°C). It has the empirical formula $C_{10}H_{16}$, and the molecule has four transcyclohexane rings. **Camphorene,** $C_{20}H_{32}$, is made from turpentine by polymerizing two myrcene molecules. It is a raw material for producing geraniol and linalol.

CAMWOOD. The wood of the tree *Baphia nitida,* native to West Africa, used for tool handles and for machine bearings. It will withstand heavy bearing pressures. The wood is exceedingly hard, has a coarse, dense grain, and has a density of 65 lb/ft³ (1,041 kg/m³). It contains a red coloring matter known as **santalin** and was once valued as a dyewood for textiles. **Barwood,** from the tree *Pterocarpus santalinus,* of West Africa, is a similar reddish hardwood containing the same dye and used for the same purposes.

CANAIGRE. A tanning material extracted from the roots of the low-growing plant *Rumex hymenosepalus* of northern Mexico and the arid southwest of the United States. The plant is known locally as **sour dock,** and the roots contain up to 40% tannin. The cultivated plant yields as much as 20 tons/acre (4.8 kg/m²) of root. **Canaigre extract** contains 30% tannin. It produces a firm, orange-colored leather. Canaigre was the tanning agent of the Aztec Indians, and is still extensively cultivated.

CANARY SEED. The seeds of the **canary grass,** *Phalaris canariensis,* native to the Canary Islands, but now grown on a large scale in Argentina for export and in Turkey and Morocco for human food and for export. In international trade it is known by the Spanish name **alpiste.** It is valued as a bird food because it contains phosphates, iron, and other minerals and is rich in carbohydrates. It is, however, low in proteins and fats and is usually employed in mixtures. **Birdseed** is an extensive item of commerce, but the birdseed that reaches the market in the United States is usually a blend of canary seed and millet, with other seeds to give a balanced food. Canary seed is small, pale yellow, and convex on both sides. The term **Spanish canary seed** is applied to the choice seed regardless of origin. **Niger seed,** also valued as a birdseed, is from the plant *Guizotia abyssinica,* of the thistle, or *Compositae* family, grown in India, Africa, Argentina, and Europe. It is also known as **inga seed, rantil, kala til seed,** and

black sesame. It is called **gingelli** in India, although this name and **til** are more properly applied to sesame. The seed is high in proteins and fats.

CANDELILLA WAX. A yellowish amorphous wax obtained by hot water or solvent extraction from the stems of the shrubs *Pedilanthus pavonis* and *Euphorbia antisyphilitica,* growing in the semiarid regions of Texas and Mexico. The plants grow to a height of 3 to 5 ft (0.9 to 1.5 m) and consist of a bundle of stalks without leaves. The stems yield 3.5 to 5% wax that consists of, unusually for a vegetable wax, about 55% hydrocarbons, principally **hentriacontane,** and less than 30% esters. The wax has a specific gravity of 0.983, melting point of 153 to 158°F (67 to 70°C), iodine value of 37, and saponification value of 45 to 65. The refined grade is purified by remelting and contains not more than about 1% water. It is soluble in turpentine and is used for varnishes, polishes, and leather finishes; as a substitute for carnauba wax; or to blend with carnauba or beeswax. About half the production goes into furniture and show polishes, but it does not have the self-polishing characteristics of carnauba wax. It is also used in electrical insulators, candles, and sound records.

CANNEL COAL. A variety of coal having some of the characteristics of petroleum, valued chiefly for its quick-firing qualities. It consists of coallike matter intimately mixed with clay and shale, often containing fossil fishes, and probably derived from vegetable matter in lakes. It is compact in texture, dull black, and breaks along joints, often having an appearance similar to black shale. It burns with a long, luminous, smoky flame, from which it derives its old English name, meaning candle. On distillation, cannel coal yields a high proportion of illuminating gas, up to 16,000 ft^3/ton (450 m^3/ton), leaving a residue consisting mostly of ash. At low temperatures it yields a high percentage of tar oils. The proportion of volatile matter may be as high as 70%. It is found in Great Britain and in Kentucky, Ohio, and Indiana. Cannel coal from Scotland was originally called **parrot coal,** and **boghead coal** was a streaky variety.

CARBOHYDRATES. The most abundant class of organic compounds, constituting about three-fourths of the dry weight of the plant world. They are distinguished by the fact that they contain the elements carbon, hydrogen, and oxygen, and no others. Many chemical compounds, such as alcohols and aldehydes, also have these elements only, but the term *carbohydrate* refers only to the starches, sugars, and cellulose, which are more properly called **saccharides.** Their properties vary enormously. Sugars are soluble, crystalline, and sweet; starches form pastes and are colloidal; celluloses are insoluble. They are best known for their use as foodstuffs, as carbohydrates compose more than 50% of all U.S. food, but they are also used in many industrial processes. The digestible carbohydrates are the sugars and the starches. The indigestible carbohydrates are cellulose and hemicellulose, which form the chief constituents of woods, stalks, and leaves of plants, the outer covering of seeds, and the walls of plant cells enclosing the water, starches, and other substances of the plants. Much cellulose is eaten as food, especially in the leaves of vegetables and in bran; but it serves as bulk rather than as food and is beneficial, if not consumed in quantity. The digestible carbohydrates are classified as **single sugars, double sugars,** and **complex sugars,** chemically known as **monosaccharides, disaccharides,** and **polysaccharides.** The single sugars—glucose, fructose, and galactose—require no digestion and are readily absorbed into the bloodstream. The double sugars—sucrose, maltose, and lactose—must be broken down by enzymes in the human system. **Lactose,** produced from milk solids, is a nonhygroscopic powder. It is only 16% as sweet as sugar and not as soluble, but it enhances flavor. It digests slowly. It is used in infant foods, dairy drinks, and ice cream to improve low-fat richness, in bakery products to decrease sogginess and improve browning, and as a dispersing agent for high-fat powders. **Galactose** is derived from lactose by hydrolysis. **Multisugars** are mixed sugars with the different sugars interlocked in the crystals. They dissolve rapidly to form clear solutions.

The complex sugars are the starches, dextrins, and glycogen. These require digestion to the single stage before they can be absorbed in the system. The common starches are in corn, wheat, potatoes, rice, tapioca, and sago. **Animal starch** is the reverse food of animals stored in the liver and muscles. It is **glycogen,** a sweet derivative of glycolic acid. It is not separated out commercially because it is hygroscopic and quickly hydrolyzed. **Dextran,** related to glycogen, is a **polyglucose** made up of many molecules of glucose in a long chain. It is used as an extender of blood plasma. It can be stored indefinitely and, unlike plasma, can be sterilized by heat. It is produced commercially by biotic fermentation of common sucrose sugar.

The **hemicelluloses** are agar-agar, algin, and pectin. They differ chemically from cellulose and expand greatly on absorbing water. The hemicelluloses of wood, called **hexosan,** consist of the **wood sugars,** or **hexose,** with six carbon atoms, $(C_6H_{10}O_5)_n$. They are used to make many chemicals. The water-soluble hemicellulose of Masonite Corp., known as **Masonex** in water solution and **Masonoid** as a powder, is a by-product of the steam-exploded wood process. It is used to replace starch as a binder for foundry cores and for briquetting coal, and for emulsions. It contains 70% wood sugars, 20 resins, and 10 lignin. **Lichenin,** or **moss starch,** is a hemicellulose from moss and some seeds.

The **pentosans** are gums or resins occurring in nutshells, straw, and the cell membranes of plants. They may be classified as hemicellulose and on hydrolysis yield **pentose,** or **pentaglucose,** a sugar containing five carbon atoms. **Pectin** is a yellowish, odorless powder soluble in water and decomposed by alkalies. It is produced by acid extraction from the inner part of the rind of citrus fruits and from apple pomace. In east Africa it is obtained from sisal waste. **Flake pectin** is more soluble and has a longer shelf life than the powdered form. It is produced from a solution of apple pomace containing 5% pectin by drying on steam-heated drums, and the thin film obtained is flaked to 40 mesh. Another source is **sugar-beet pulp,** which contains 20 lb (9.7 kg) of pectin per ton (0.91 metric ton).

Pectin has a complex structure, having a lacturonic acid with methanol in a glucoside chain combination. It is used for gelling fruit preserves, and the gelling strength depends on the size of the molecule, the molecular weight varying from 150,000 to 300,000. It is also used as a blood coagulant in treating hemorrhage, and for prolonging the effect of some drugs by retarding their escape through the body. **Sodium pectate** is used for creaming rubber latex, and in cosmetics and printing inks. Hemicellulose and pectin are valuable in the human system because of their ability to absorb and carry away irritants, but they are not foods in the normal sense of the term. **Oragen** is a pectin-cellulose complex derived from orange pulp. It is used in weight-reduction diets, increasing bulk and retaining moisture, thus suppressing the desire for excess food. Each of the saccharides has distinctive characteristics of value in the system, but each also in excess causes detrimental conditions.

CARBON. A nonmetallic element, symbol C, existing naturally in several allotropic forms and in combination as one of the most widely distributed of all the elements. It is quadrivalent and has the property of forming chain and ring compounds, and there are more varied and useful compounds of carbon than of all other elements. Carbon enters into all organic matter of vegetable and animal life, and the great branch of **organic chemistry** is the chemistry of carbon compounds. The black amorphous carbon has a specific gravity of 1.88; the black crystalline carbon known as *graphite* has a specific gravity of 2.25; the transparent crystalline carbon, as in the diamond, has a specific gravity of 3.51. **Amorphous carbon** is not soluble in any known solvent. It is infusible, but sublimes at 6332°F (3500°C), and is stable and chemically inactive at ordinary temperatures. At high temperatures it burns and absorbs oxygen, forming the simple oxides CO and CO_2, the latter being the stable oxide present in the atmosphere and a natural plant food. Carbon dissolves easily in some molten metals, notably iron, exerting great influence on them. Steel, with small amounts of chemically combined carbon, and cast iron, with both combined carbon and graphitic carbon, are examples of this. **Volatile organic compounds** (VOCs) are carbon compounds, readily passed off by evaporation, that react to form ground-level ozone, a primary component of smog. They pertain to many solvents, degreasers, paints, and chemicals, and great efforts have been made in recent years to reduce their emission.

Carbon occurs as hydrocarbons in petroleum, and as carbohydrates in coal and plant life, and from these natural basic groupings an infinite number of carbon compounds can be made synthetically. Carbon, for chemical, metallurgical, or industrial use, is marketed in the form of compounds in a large number of different grades, sizes, and shapes; or in master alloys containing high percentages of carbon; or as activated carbons, charcoal, graphite, carbon black, coal-tar carbon, petroleum coke; or as pressed and molded bricks or formed parts with or without binders or metallic inclusions. Natural deposits of graphite, coal tar, and petroleum coke are important sources of elemental carbon. Charcoal and activated carbons are obtained by carbonizing vegetable or animal matter.

Carbon 13 is one of the isotopes of carbon, used as a tracer in biologic research where its heavy weight makes it easily distinguished from other carbon. **Carbon 14, or radioactive carbon,** has a longer life. It exists in air, formed by the bombardment of nitrogen by cosmic rays at high altitudes, and enters into the growth of plants. The half-life is about 6,000 years. It is made from nitrogen in a cyclotron. **Carbon fullerenes,** such as C_{60}, are a new form of carbon, discovered in the mid-1980s, with considerable potential in diverse applications.

Carbon fibers are made by pyrolysis of organic precursor fibers in an inert atmosphere. Pyrolysis temperatures can range from 2012 to 5432°F (1000 to 3000°C); higher process temperatures generally lead to higher-modulus fibers. Only three precursor materials—**rayon, polyacrylonitrile** (PAN), and pitch—have achieved significance in commercial production of carbon fibers. The first high-strength and high-modulus carbon fibers were based on a rayon precursor. These fibers were obtained by being stretched to several times their original length at temperatures above 5072°F (2800°C). The second generation of carbon fibers is based on a PAN precursor and has achieved market dominance. In their most common form, these carbon fibers have a tensile strength ranging from 350,000 to 450,000 lb/in^2 (2,413 to 3,102 MPa), a modulus of 28×10^6 to 75×10^6 lb/in^2 (193,000 to 517,000 MPa), and a shear strength of 13,000 to 17,000 lb/in^2 (90 to 117 MPa).

This last property controls the traverse strength of composite materials. The high-modulus fibers are highly graphitic in crystalline structure after being processed from PAN at temperatures in excess of 3600°F (1982°C). Higher-strength fibers obtained at lower temperatures from rayon feature a higher carbon crystalline content. There are also carbon and graphite fibers of intermediate strength and modulus. The third generation of carbon fibers is based on **pitch** as a precursor. Ordinary pitch is an isotropic mixture of largely aromatic compounds. Fibers spun from this pitch have little or no preferred orientation and hence low strength and modulus. Pitch is a very inexpensive precursor compared with rayon and PAN. High-strength and high-modulus carbon fibers are obtained from a pitch that has first been converted to a mesophase (**liquid crystal**). These fibers have a tensile strength of more than 300,000 lb/in^2 (2,069 MPa) and a Young's modulus ranging from 55×10^6 to 75×10^6 lb/in^2 (379,000 to 517,000 MPa). The average filament diameter of continuous yarn is 0.0003 in (0.008 mm). Pitch-based carbon and graphite fibers are expected to see essentially the same applications as the more costly PAN- and rayon-derived fibers, e.g., ablative, insulation, and friction materials and in metals and resin matrixes. **Thornel,** of Union Carbide Corp., is a yarn made from these filaments for high-temperature fabrics. It retains its strength to temperatures above 2800°F (1538°C). **Carbon yarn** is 99.5% pure carbon. It comes in plies of 2 to 30, with each ply composed of 720 continuous filaments of 0.0003-in (0.008-mm) diameter. Each ply has a breaking load of 2 lb (0.91 kg). The fiber has the flexibility of wool and maintains dimensional stability to 5700°F (3150°C). **KIIOOX** fiber, from Amoco Performance Products, Inc., is a pitch-carbon fiber for **prepreg** used to produce composites for thermal management systems in space satellites. **Ucar,** of Union Carbide Corp., is a conductive **carbon fabric** made from carbon yarns woven with insulating glass yarns with resistivities from 0.2 to 30 Ω for operating temperatures to 550°F (288°C). **Carbon wool,** for filtering and insulation, is composed of pure-carbon fibers made by carbonizing rayon. The fibers, 197 to 1,970 μin (5 to 50 μm) in diameter, are hard and strong and can be made into rope and yarn, or the mat can be activated for filter use. **Avceram RS,** of FMC Corp., is a composite rayon-silica fiber made with 40% dissolved sodium silicate. A highly heat-resistant fiber, **Avceram CS** is woven into fabric and then pyrolyzed to give a porous interlocked mesh of **carbon silica fiber,** with a tensile strength of 165,000 lb/in^2 (1,138 MPa). **Dexsan,** of C. H. Dexter & Sons Co., for filtering hot gases and liquids, is a **carbon filter paper** made from carbon fibers pressed into a paperlike mat, 0.007 to 0.050 in (0.18 to 0.127 mm) thick, and impregnated with activated carbon.

Carbon brushes for electric motors and generators and **carbon**

electrodes are made of carbon in the form of graphite, petroleum coke, lampblack, or other nearly pure carbon, sometimes mixed with copper powder to increase the electrical conductivity, and then pressed into blocks or shapes and sintered. Carbon-graphite brushes contain no metals but are made from **carbon-graphite powder** and, after pressing, are subjected to a temperature of 5000°F (2760°C), which produces a harder and denser structure, permitting current densities up to 125 A/in^2 (1,538 A/m^2). **Carbon brick,** used as a lining in the chemical processing industries, is carbon compressed with a bituminous binder and then carbonized by sintering. If the binder is capable of being completely carbonized, the bricks are impervious and dense. **Graphite brick,** made in the same manner from graphite, is more resistant to oxidation than carbon bricks and has a higher thermal conductivity, but it is softer. The binder may also be a furfural resin polymerized in the pores. **Karbate No. 1** is a carbon-base brick, and **Karbate No. 2** is a graphite brick. Karbate has a crushing strength of 10,500 lb/in^2 (72 MPa) and a density of 110 to 120 lb/ft^3 (1,762 to 1,922 kg/m$_3$). **Impervious carbon** is used for lining pumps, for valves, and for acid-resistant parts. It is carbon- or graphite-impregnated with a chemically resistant resin and molded to any shape. It can be machined. **Karbate 21** is a phenolic-impregnated graphite, and **Karbate 22** is a modified phenolic-impregnated graphite. Molded impervious carbon has a specific gravity of 1.77, tensile strength of 1,800 lb/in^2 (12.4 MPa), and compressive strength of 10,000 lb/in^2 (69 MPa). **Impervious graphite** has a higher tensile strength, 2,500 lb/in^2 (17.2 MPa), but a lower compressive strength, 9,000 lb/in^2 (62 MPa). The thermal conductivity is 8 to 10 times that of stainless steel. **Graphitar,** of U.S. Graphite Co., is a strong, hard carbon molded from amorphous carbon mixed with other forms of carbon. It has high crushing strength and acid resistance and is used for sealing rings, chemical pump blades, and piston rings. **Porous carbon** is used for the filtration of corrosive liquids and gases. It consists of uniform particles of carbon pressed into plates, tubes, or disks without a binder, leaving interconnecting pores of about 0.001 to 0.0075 in (0.025 to 0.190 mm) in diameter. The porosity of the material is 48%, tensile strength 150 lb/in^2 (1 MPa), and compressive strength about 500 lb/in^2 (3.5 MPa). **Porous graphite** has graphitic instead of carbon particles, and is more resistant to oxidation but is lower in strength.

Carbon/carbon composites, which comprise carbon fibers in a carbon matrix, are noted for their heat resistance, high-temperature strength, high thermal conductivity, light weight, low thermal expansivity, and resistance to air/fuel mixtures. However, they are costly to produce. Also, they react with oxygen at temperatures above 800°F (427°C), necessitating oxygen-barrier coatings. **Silicon carbide,**

0.005 to 0.007 in (0.127 to 0.178 mm) thick, serves as such a coating for applications in the nose cone and wing leading edges of the Space Shuttle. Other uses include the brakes of large commercial aircraft, clutches and brakes of Formula 1 race cars, and rocket nozzles.

Carbon films, usually made by chemical vapor deposition (CVD) at 2012°F (1100°C), can strengthen and toughen **ceramic-matrix composites** but are not readily adaptable to coating fibers, platelets, or powder. The Japanese have developed what is said to be a more economical method using silicon carbide and other ceramics. Nanometer- to micrometer-thick films are formed on these forms, including silicon carbide single crystals, by treating them with water under pressure at 572 to 1472°F (300 to 800°C). This treatment transforms the surface layer to carbon.

The so-called carbons used for electric-light **arc electrodes** are pressed from coal-tar carbon, but are usually mixed with other elements to bring the balance of light rays within the visible spectrum. Solid carbons have limited current-carrying capacity, but when the carbon has a center of metal compounds such as the fluorides of the rare earths, its current capacity is greatly increased. It then forms a deep positive crater in front of which is a flame 5 times the brilliance of that with the low-current arc. The **sunshine carbon,** used in electric-light carbons to give approximately the same spectrum as sunlight, is molded coal-tar carbon with a core of cerium metals to introduce more blue into the light. **Arc carbons** are also made to give other types of light, and to produce special rays for medicinal and other purposes. **B carbon,** of National Carbon Co., Inc., contains iron in the core and gives a strong emission of rays from 9,055 to 12,598 nin (230 to 320 nm), which are the antirachitic radiations. The light seen by the eye is only one-fourth the total radiation since the strong rays are invisible. **C carbon** contains iron, nickel, and aluminum in the core and gives off powerful lower-zone ultraviolet rays. It is used in light therapy and for industrial applications. **E carbon,** to produce penetrating infrared radiation, contains strontium. **Electrode carbon,** used for arc furnaces, is molded in various shapes from carbon paste. When calcined from petroleum coke, the electrodes contain only 0.2% moisture, 0.25 volatile matter, and 0.3 ash and have a specific gravity of 2.05. The carbon is consumed in the production of light and of furnace heat. For example, from 1,100 to 1,320 lb (500 to 600 kg) of carbon is consumed in producing 1 ton (0.91 metric ton) of aluminum.

CARBON BLACK. An amorphous powdered carbon resulting from the incomplete combustion of a gas, usually deposited by contact of the flame on a metallic surface, but also made by the incomplete combustion of the gas in a chamber. The carbon black made by the first process is called **channel black,** taking the name from the channel iron used as the depositing surface. The modern method, called the *impingement process,* uses many small flames with the fineness of particle size controlled by flame size. The air-to-gas ratio is high, giving oxidized surfaces and acid properties. No water is used for cooling, keeping the ash content low. The supergrade of channel black has a particle size as low as 512 μin (13 μm) and a pH of 3 to 4.2. Carbon black made by other processes is called **soft black** and is weaker in color strength, not so useful as a pigment. **Furnace black** is made with a larger flame in a confined chamber with the particles settling out in cyclone chambers. The air-to-gas ratio is low, and water cooling raises the ash content. The particle surface is oily, and the pH is high.

Carbon black from clean artificial gas is a glossy product with an intense color, but all the commercial carbon black is from natural gas. To remove H_2S, the sour gas is purified and water-scrubbed before burning. **Thermotomic black,** a grade made by the thermal decomposition of the gas in the absence of oxygen, is preferred in rubber when high loadings are employed because it does not retard the vulcanization; but only a small part of the carbon black is made by this process. This thermal process black has large particle size, 5,906 μin (150 μm), and a pH of 8.5. It gives a coarse oily carbon.

The finer grades of channel black are mostly used for color pigment in paints, polishes, carbon paper, and printing and drawing inks. The larger use of carbon black is in automotive tires to increase the wear resistance of the rubber. The blacker blacks have a finer particle size than the grayer blacks, hence have more surface and absorptive capacity in compounding with rubber. Channel black is valued for rubber compounding because of its low acidity and low grit content. The high pH of furnace black may cause scorching unless offsetting chemicals are used, but some furnace blacks are made especially for tire compounding. In general, the furnace black with particle sizes from 1,100 to 3,350 μin (28 to 85 μm) and a pH from 8 to 10, and the channel blacks with particle size of about 1,140 μin (29 μm) and pH of 4.8, are used for rubber. **Micronex EPC,** an impingement channel black of Binney & Smith Co., has a particle diameter of 1,140 μin (29 μm) and a pH of 4.8, while **Thermax MT,** a thermal process black of Cancarb Ltd., has a particle size of 10,800 μin (274 μm) and a pH of 7.

In rubber compounding, the carbon black is evenly dispersed to become intimately attached to the rubber molecule. The fineness of the black determines the tensile strength of the rubber, the structure of the carbon particle determines the modulus, and the pH determines the cure behavior. Furnace blacks have a basic pH which activates the accelerator, and delaying-action chemicals are thus needed, but fine furnace blacks impart abrasion resistance to the rubber.

Furnace black made with a confined flame with limited air has a neutral surface and a low volatility. Fineness is varied by temperature, size of flame, and time. Carbonate salts raise the pH. Most of the channel black for rubber compounding is made into dustfree pellets less than 0.125 in (0.3 cm) in diameter with a density of 20 to 25 lb/ft^3 (320 to 400 kg/m^3). **Color-grade black** for inks and paints is produced by the channel process or the impingement process. In general, carbon black for reinforcement has small particle size, and the electrically conductive grades, **CF carbon black** and **CC carbon black,** conductive furnace and conductive channel, have large particle sizes.

Carbon black from natural gas is produced largely in Louisiana, Texas, and Oklahoma. About 35 lb (15.9 kg) of black is available per 1,000 ft^3 (28 m^3) of natural gas, but only 2.2 lb (1 kg) is recovered by the channel process and 10 lb (4.5 kg) by the furnace method. By using gas from which the natural gasoline has been stripped, and by controlled preheating and combustion, as much as 27 lb (12.2 kg) can be recovered. **Acetylene black** is a carbon black made by heat decomposition of acetylene. It is more graphitic than ordinary carbon black with colloidal particles linked together in an irregular lattice structure and has high electrical conductivity and high liquid-absorption capacity. Particle size is intermediate between that of channel black and furnace black, with low ash content, nonoiliness, and a pH of 6.5. It is valued for use in dry cells and lubricants. **Ucet,** of Union Carbide Corp., is in the form of agglomerates of irregular fine crystals. The greater surface area gives higher thermal and electrical conductivity and high liquid absorption.

For electrically **conductive rubber,** the mixing of the black with the rubber is regulated so that carbon chain connections are not broken. Such conductive rubber is used for tabletops, conveyor belts, and coated filter fabrics to prevent static buildup. Carbon blacks are also made from liquid hydrocarbons, and from anthracite coal by treatment of the coal to liberate hydrogen and carbon monoxide and then high-temperature treatment with chlorine to remove impurities. The black made from anthracite has an open-pore structure useful for holding gases and liquids.

Carbon-black grades are often designated by trade names for particular uses. **Kosmovar** is a black with a slight bluish top tone used as a pigment for lacquers. The specific gravity is 1.72, and mesh is 325. **Gastex** and **Pelletex** are carbon blacks used for rubber compounding. **Statex** is a colloidal furnace black for synthetic rubber compounding. **Kosmos 60** is a furnace black of high density and structure, while **Continex FF** is a finely divided furnace black. Both are used in rubber compounding, the first giving easier extrusion of the rubber and the second giving better abrasion resistance.

Aquablak H, of Binney & Smith Co., is a colloidal water dispersion of channel black to give a jet-black color. **Aquablak M** is a water dispersion of furnace black to give a blue-gray tone. They are used as pigments in casein paint, inks, and leather finishes. **Liquimarl-Black** is a stable colloidal dispersion of pure food-grade carbon black for use in coloring confectionery and for modifying food colors in bakery products.

CARBON DIOXIDE.   Also called **carbonic anhydride,** and in its solid state, **dry ice.** A colorless, odorless gas of composition CO_2, which liquefies at $-85°F$ ($-65°C$) and solidifies at $-108.8°F$ ($-78.2°C$). Release of CO_2 into the atmosphere by the burning of fossil fuels is said to be causing global warming by the process known as the *greenhouse effect.* It is recovered primarily as a by-product of the steam reforming of natural gas to make hydrogen or synthesis gas in petroleum and fertilizer plants. Smaller quantities are obtained by purifying flue gases generated from burning hydrocarbons or lime, and from distilleries. Its biggest uses are captive, as a chemical raw material for making urea and in enhanced oil recovery operations in petroleum production. Merchant CO_2 is more than 99.5% pure, with less than 500 ppm (parts per million) of nonvolatile residues. In liquid form it is marketed in cylinders and is used in fire extinguishers, in spray painting, in refrigeration, for inert atmospheres, for the manufacture of **carbonated beverages,** and in many industrial processes. It is also marketed as dry ice, a white, snowlike solid used for refrigeration in transporting food products. **Cardox** is a trade name of Cardox Corp. for liquid carbon dioxide in storage units at 30 lb/in^2 (0.21 MPa) pressure for fire-fighting equipment. Other uses include hardening of foundry cores, neutralization of industrial wastes, and production of salicylic acid for aspirin. Carbon dioxide is a key lasing gas in **carbon dioxide lasers** and is also used as a shielding gas in welding and as a foaming agent in producing plastic foam products. It can behave as a supercritical fluid, in which state it can be used to foam plastics and extract hazardous substances in waste treatment processes and in soil remediation.

CARBON MONOXIDE.   CO is a product of incomplete combustion and is very reactive. It is one of the desirable products in synthesis gas for making chemicals, the synthesis gas made from coal containing at least 37% CO. It is also recovered from top-blown oxygen furnaces in steel mills. It reacts with hydrogen to form methanol, which is then catalyzed by zeolites into gasoline. Acetic acid is made by methanol carbonylation, and acrylic acid results from the reaction of CO, acetylene, and methanol. CO forms a host of neutral, anionic, and cationic

carbonyls, with such metals as iron, cobalt, nickel, molybdenum, chromium, rhodium, and ruthenium. Pressure Chemical Co. and Strem Chemicals Inc. make **molybdenum carbonyl, chromium carbonyl,** and other complexes for olefin carbonylation and isomerization, and carboxylation reactions. Carbon monoxide is an intense poison when inhaled and is extremely toxic even in the small amounts from the exhausts of internal-combustion engines.

CARBON STEEL.   The **wrought carbon steels** covered here are sometimes termed **plain carbon steels.** The old shop names of **machine steel** and **machinery steel** are still used to mean any easily worked low-carbon steel. By definition, plain carbon steels are those that contain up to about 1% carbon, not more than 1.65 manganese, 0.60 silicon, and 0.60 copper, and only residual amounts of other elements, such as sulfur (0.05% maximum) and phosphorus (0.04% maximum). They are identified by means of a four-digit numerical system established by the American Iron and Steel Institute (AISI). The first digit is the number 1 for all carbon steels. A 0 after the 1 indicates nonresulfurized grades, a 1 for the second digit indicates resulfurized grades, and 2 for the second digit indicates resulfurized and rephosphorized grades. The last two digits give the nominal (middle of the range) carbon content in hundredths of a percent. For example, for grade 1040, the 40 represents a carbon range of 0.37 to 0.44%. If no prefix letter is included in the designation, the steel was made by the basic open-hearth, basic oxygen, or electric furnace process. The prefix B stands for the acid Bessemer process, which is obsolete, and the prefix M designates merchant quality. The letter L between the second and third digits identifies leaded steels, and the suffix H indicates that the steel was produced to hardenability limits.

For all plain carbon steels, carbon is the principal determinant of many performance properties. Carbon has a strengthening and hardening effect. At the same time, it lowers ductility, as evidenced by a decrease in elongation and reduction of area. In addition, increasing carbon content decreases machinability and weldability, but improves wear resistance. The amount of carbon present also affects physical properties and corrosion resistance. With an increase in carbon content, thermal and electrical conductivity decline, magnetic permeability decreases drastically, and corrosion resistance is less.

Carbon steels are available in most wrought mill forms, including bar, sheet, plate, pipe, and tubing. Sheet is primarily a low-carbon-steel product, but virtually all grades are available in bar and plate. Plate, usually a low-carbon or medium-carbon product, is used mainly in the hot-finished condition, although it also can be supplied heat-treated. Bar products, such as rounds, squares, hexagonals, and flats (rectangular cross sections), are also mainly low-carbon and medium-carbon products and are supplied hot-rolled and cold-finished. Cold finishing may be by drawing (cold-drawn bars are the most widely used); turning (machining) and polishing; drawing, grinding, and polishing; or turning, grinding, and polishing. Bar products are also available in various quality designations, such as merchant quality (M), cold-forging quality, cold-heading quality, and several others. Sheet products have quality designations as noted in low-carbon steels, which follow. Plain carbon steels are commonly divided into three groups, according to carbon content: low carbon, up to 0.30%; medium carbon, 0.31 to 0.55; and high carbon, 0.56 to 1.

Low-carbon steels are the grades **AISI 1005** to **1030.** Sometimes referred to as **mild steels,** they are characterized by low strength and high ductility and are nonhardenable by heat treatment except by surface-hardening processes. Because of their good ductility, low-carbon steels are readily formed into intricate shapes. These steels are also readily welded without danger of hardening and embrittlement in the weld zone. Although low-carbon steels cannot be through-hardened, they are frequently surface-hardened by various methods (carburizing, carbonitriding, and cyaniding, for example) which diffuse carbon into the surface. Upon quenching, a hard, wear-resistant surface is obtained.

Low-carbon sheet and strip steels (1008 to 1012) are widely used in cars, trucks, appliances, and many other applications. Hot-rolled products are usually produced on continuous hot strip mills. Cold-rolled products are then made from the hot-rolled products, reducing thickness and enhancing surface quality. Unless the fully work-hardened product is desired, it is then annealed to improve formability and temper-rolled to further enhance surface quality. Hot-rolled sheet and strip and cold-rolled sheet are designated *commercial quality* (CQ), *drawing quality* (DQ), *drawing quality special killed* (DQSK), and *structural quality* (SQ). The first three designations refer, respectively, to steels of increasing formability and mechanical property uniformity. SQ, which refers to steels produced to specified ranges of mechanical properties and/or bendability values, do not pertain to cold-rolled strip, which is produced to several tempers related to hardness and bendability. Typically, the hardness of CQ hot-rolled sheet ranges from Rockwell B (RB) 40 to 75, and tensile properties range from ultimate strengths of 40,000 to 68,000 lb/in^2 (276 to 469 MPa), yield strengths of 28,000 to 48,000 lb/in^2 (193 to 331 MPa), and elongations of 14 to 43%. For DQ hot-rolled sheet: RB 40 to 72; 40,000 to 60,000 lb/in^2 (276 to 414 MPa); 27,000 to 45,000 lb/in^2 (186 to 310 MPa); and 28 to 48%, respectively. For CQ cold-rolled sheet: RB 35 to

60; 42,000 to 57,000 lb/in^2 (290 to 393 MPa); 23,000 to 38,000 lb/in^2 (159 to 262 MPa); and 30 to 45%. And for DQ cold-rolled sheet: RB 32 to 52; 38,000 to 50,000 lb/in^2 (262 to 345 MPa); 20,000 to 34,000 lb/in^2 (138 to 234 MPa); and 34 to 46%.

Special (modified) low-carbon sheet steels may contain small amounts of other alloying elements. Nitrogen in quantities of 0.010 to 0.018% or phosphorus (0.03 to 0.15) permits increasing strength without decreasing ductility as much as traditional amounts of carbon and manganese. Thus, their use has increased appreciably in recent years, especially in the auto industry. As supplied, these steels have tensile yield strengths of 35,000 to 50,000 lb/in^2 (241 to 345 MPa) and tensile elongations of 28 to 32%. **Nitrogenized steels** exhibit substantial strain aging—to 70,000 lb/in^2 (483 MPa) or greater—during cold forming. Although such strengthening may occur naturally, a brief low-temperature age [15 to 30 min at 350°F (177°C)], such as in auto paint-bake cycles, is sometimes recommended. The most formable, however, because of their metallurgical cleanliness, are the **interstitial-free steels,** typified by Armco's **I-F steel.** Produced by aluminum deoxidation and vacuum decarburization deoxidation, the carbon content is only 0.004 to 0.010% and nitrogen 0.004 or less. Columbium (0.08 to 0.12%) or columbium and vanadium serve as carbide and nitride formers. The drawability of the steel exceeds that of traditional DQSK grades, but its tensile yield strength is 2,000 to 8,000 lb/in^2 (14 to 55 MPa) less. The formability of low-carbon sheet steels also can be enhanced by **inclusion-shape control,** which was initially implemented for high-strength low-alloy steels. This involves small additions of zirconium, titanium, or rare-earth elements and special mill practices to alter the shape of nonmetallic inclusions from stringerlike to small, dispersed globules. The strongest of the sheet steels are Inland Steel's low- and medium-carbon **MartINsite** grades. Produced by rapid water quenching after cold rolling, they provide tensile yield strengths of 130,000 to 220,000 lb/in^2 (896 to 1,517 MPa) but little ductility, 4 to 2% elongation, respectively.

Low-carbon steels 1018 to **1025** in cold-drawn bar 0.625 to 0.875 in (16 to 22 mm) thick have minimum tensile properties of about 70,000 lb/in^2 (483 MPa) ultimate strength, 60,000 lb/in^2 (413 MPa) yield strength, and 18% elongation. Properties decrease somewhat with increasing section size to, say, 55,000 lb/in^2 (379 MPa), 45,000 lb/in^2 (310 MPa), and 15%, respectively, for 2- to 3-in (50- to 76-mm) cross sections.

Medium-carbon steels are the grades **AISI 1030** to **1055.** They usually are produced as killed, semikilled, or capped steels and are hardenable by heat treatment. However, hardenability is limited to thin sections or to the thin outer layer on thick parts. Medium-carbon steels in the quenched and tempered condition provide a good balance of strength and ductility. Strength can be further increased by cold work. The highest hardness practical for medium-carbon steels is about Brinell 550 (Rockwell C 55). Because of the good combination of properties, they are the most widely used steels for structural applications, where moderate mechanical properties are required. Quenched and tempered, their tensile strengths range from about 75,000 to over 150,000 lb/in^2 (517 to over 1,034 MPa).

Medium-carbon steel 1035 in cold-drawn bar 0.625 to 0.875 in (16 to 22 mm) thick has minimum tensile properties of about 85,000 lb/in^2 (586 MPa) ultimate strength, 75,000 lb/in^2 (517 MPa) yield strength, and 13% elongation. Strength increases and ductility decreases with increasing carbon content to, say, 100,000 lb/in^2 (689 MPa), 90,000 lb/in^2 (621 MPa), and 11%, respectively, for **medium-carbon steel 1050.** Properties decrease somewhat with increasing section size to, say, 70,000 lb/in^2 (483 MPa), 60,000 lb/in^2 (414 MPa), and 10%, respectively, for 1035 steel 2- to 3-in (50- to 76-mm) thick.

High-carbon steels are the grades **AISI 1060** to **1095.** They are, of course, hardenable with a maximum surface hardness of about Brinell 710 (Rockwell C 64) achieved in the 1095 grade. These steels are thus suitable for wear-resistant parts. So-called **spring steels** are high-carbon steels available in annealed and pretempered strip and wire. Besides their spring applications, these steels are used for such items as piano wire and saw blades. Quenched and tempered, high-carbon steels approach tensile strengths of 200,000 lb/in^2 (1,378 MPa).

Free-machining carbon steels are low- and medium-carbon grades with additions usually of sulfur (0.08 to 0.13%), sulfur-phosphorus combinations, and/or lead to improve machinability. They are **AISI 1108** to **1151** for sulfur grades, and **AISI 1211** to **1215** for phosphorus and sulfur grades. The latter may also contain bismuth and be lead-free. The presence of relatively large amounts of sulfur and phosphorus can reduce ductility, cold formability, forgeability, weldability, as well as toughness and fatigue strength. **Calcium deoxidized steels** (carbon and alloy) have good machinability and are used for carburized or through-hardened gears, worms, and pinions.

Low-temperature carbon steels have been developed chiefly for use in low-temperature equipment and especially for welded pressure vessels. They are low- to medium-carbon (0.20 to 0.30%), high-manganese (0.70 to 1.60%), silicon (0.15 to 0.60%) steels, which have a fine-grain structure with uniform carbide dispersion. They feature moderate strength with toughness down to $-50°$F ($-46°$C).

For grain refinement and to improve formability and weldability, carbon steels may contain 0.01 to 0.04% columbium. Called **columbi-**

um steels, they are used for shafts, forgings, gears, machine parts, and dies and gages. Up to 0.15% sulfur, or 0.045 phosphorus, makes them free-machining, but reduces strength.

Rail steel, for railway rails, is characterized by an increase of carbon with the weight of the rail. Railway engineering standards call for 0.50 to 0.63% carbon and 0.60 manganese in a 60-lb (27-kg) rail, and 0.69 to 0.82% carbon and 0.70 to 1.0 manganese in a 140-lb (64-kg) rail. Rail steels are produced under rigid control conditions from deoxidized steels with phosphorus kept below 0.04% and silicon 0.10 to 0.23%. Guaranteed minimum tensile strength of 80,000 lb/in^2 (551 MPa) is specified, but it is usually much higher.

Sometimes a machinery steel may be required with a small amount of alloying element to give a particular characteristic and still not be marketed as an alloy steel, although trade names are usually applied to such steels. **Superplastic steels,** developed at Stanford University, with 1.3 to 1.9% carbon, fall between high-carbon steels and cast irons. They have elongations approaching 500% at warm working temperatures of 1000 to 1200°F (538 to 650°C) and 4 to 15% elongation at room temperature. Tensile strengths range from 150,000 to over 200,000 lb/in^2 (1,034 to over 1,378 MPa). The extra-high ductility is a result of a fine, equiaxed grain structure obtained by special thermomechanical processing.

CARBON TETRACHLORIDE. A heavy, colorless liquid of composition CCl_4, also known as **tetrarchloromethane,** which is one of a group of chlorinated hydrocarbons. It is an important solvent for fats, asphalt, rubber, bitumens, and gums. It is more expensive than the aromatic solvents, but it is notable as a nonflammable solvent for many materials sold in solution and is widely used as a degreasing and cleaning agent in the dry-cleaning and textile industries. Since the fumes are highly toxic, it is no longer permitted in compounds for home use. It is used as a chemical in fire extinguishers such as **Pyrene**; but when it falls on hot metal, it forms the poisonous gas phosgene. It is also used as a disinfectant, and because of its high dielectric strength has been employed in transformers. It was first produced in 1839 and used in Germany as a grease remover under the name **Katharin.** Carbon tetrachloride is obtained by the chlorination of carbon bisulfide. The specific gravity is 1.595, boiling point 169°F (76°C), and the freezing point 73°F (23°C). **Chlorobromomethane,** $Br \cdot CH_2 \cdot Cl$, is also used in fire extinguishers, as it is less corrosive and more than twice as efficient as an extinguisher. It is a colorless, heavy liquid with a sweet odor, a specific gravity 1.925, boiling point 153°F (67°C), and a freezing point −85°F (−65°C). It is also used as a high-gravity flotation agent.

CARNAUBA WAX. A hard, high-melting lustrous wax from the fanlike leaves of the palm tree *Copernicia cerifera* of the arid region of northeastern Brazil, sometimes referred to as **Brazil wax,** or **ceara wax.** It is composed largely of **ceryl palmitate,** $C_{25}H_{51}COOC_{30}H_{61}$. The trees grow up to 60 ft (18 m) with leaves 3 ft (1 m) long. The wax comes in hard, vitreous, yellowish cakes or lumps that melt at about 185°F (85°C) and have a specific gravity of 0.995. It is soluble in alcohol and in alkalies. **Olho wax** is the wax from young yellow leaves and is whitish gray. **Palha wax,** from the older, green leaves, is a deeper grayish yellow. In melting, water is added to the palha to make the **chalky wax.** No. 3 chalky contains up to 10% water. Olho wax without water yields the prime yellow wax. **Flora wax** is the highest quality and is clear yellow. Fully 70% of the production of carnauba goes into the manufacture of floor waxes and carbon paper. It has the property of being self-polishing in liquid floor waxes. In carbon paper it is nongreasy and nonsmearing. Other uses are in shoe polishes, in leather finishes, in cosmetics, and for blending with other waxes in coating compounds. **Burnishing wax,** in the shoe industry, is carnauba wax blended with other waxes.

A wax quite similar to carnauba is **guaruma,** or **cauassu wax,** from the leaves of *Calathea lutea,* a small plant with large leaves like those of the banana, growing in the lower Amazon Valley. Its melting point is 176°F (80°C). Another similar wax is from the trunk of the **wax palm** *Ceroxylon andicola,* growing on the Andean slopes. A wax that is very similar to carnauba in properties and is more plentiful, but which contains the green leaf coloring difficult to bleach out, is **ouricury wax.** The name is also spelled *urucury* (*uru,* the Carib name for a shell; *o* means leaf). The wax is from the leaves of the palm tree *Syagrus coronata,* or *Cocos coronata,* of northeastern Brazil. Ouricury wax has a melting point of about 185°F (85°C), acid number 10.6, iodine value 16.9, and saponification value 78.8. It has the same uses as carnauba where color is not important, or it is used to blend with carnauba to increase the gloss. The nuts of the tree are called **licuri nuts,** and they are used to produce **licuri oil** employed in soaps. The name **licuri wax** is sometimes erroneously given to ouricury.

Cotton wax, which occurs in cotton fiber to the extent of about 0.6%, is very similar to carnauba wax. It is a combination of C_{28} to C_{32} primary alcohols with C_{24} to C_{32} fatty acids. It has not been produced commercially. **Sugarcane wax** is a hard wax similar to carnauba occurring on the outside of the sugarcane stalk. A ton of cane contains 2 to 3 lb (1 to 1.4 kg) of wax, which concentrates in the filter press cake after clarification of the cane juice. The filter cake contains as high as 21% wax, which is solvent-extracted, demineralized with hydrochloric acid, and distilled to remove the low-molecular-weight constituents. It is used in floor and furniture polishes. The wax has a tan color, a melting point at about 176°F (80°C), and acid number 23 to 28. **Duplicane wax,** of Warwick Wax Co., Inc., is a grade of sugarcane wax for carbon paper, and **Technicane wax** is a grade for polishes. Sugarcane wax is miscible with vegetable and petroleum waxes and has greater dispersing action than carnauba wax. **Henequen wax,** extracted from the waste pulp of the henequen plant, has a melting point of 185°F (85°C) and is similar to carnauba. **Moss wax,** used for polishes, is extracted from Spanish moss which contains up to 4% wax. **Spanish moss** is the fiber from the plant *Tillandsia usneoides,* which grows throughout tropical and subtropical America and along the southeastern coast of the United States, hanging from branches of trees. It is used for packing fragile articles and for mattresses.

CARNOTITE.   A mineral found in Utah and Colorado and employed as a source of uranium, radium, and vanadium. It is a vanadate of uranium and potassium, $V_2O_5 \cdot 2U_2O_3 \cdot K_2O \cdot 3H_2O$. It is found as a powder with other sands and gives them a pale-yellow color. The ore may contain 2 to 5% uranium oxide and up to 6 vanadium oxide, but it usually runs 2% V_2O_5. The vanadium is produced by roasting the ore, leaching, precipitating the oxide with acids, and sintering. The production of radium from the residue ore is a complex process, and 400 tons (362,800 kg) of ore produces only 0.0022 lb (1 g) of radium. **Patronite,** mined in Peru as a source of vanadium, is a greenish mineral, V_2S_9, mixed with pyrites and other materials. Carnotite ore may contain up to 2,500 parts per million of selenium and is a source of this metal.

CAROA.   Pronounced car-o-áh. The fiber from the leaves of the plant *Neoglaziovia variegata* of northeastern Brazil. It is more than twice as strong as jute and is lighter in color and in weight, but is too hard to be used alone for burlap. It is employed as a substitute for jute in burlap when mixed with softer fibers and also for rope, and in mixtures with cotton for heavy fabrics and suitings. Some suiting is made entirely of the finer caroa fibers. **Fibrasil** is a trade name in Brazil for fine, white caroa fibers used for tropical clothing.

CARTRIDGE BRASS.   Basically a 70% copper, 30% zinc wrought alloy, designated **brass alloy C26000,** which may also contain as much as 0.07% lead and 0.05 iron. Besides *cartridge brass,* a name resulting from its use in munitions, notably cartridge cases, it has been known as **brass alloy 70–30** brass, **spinning brass, spring brass,** and

extraquality brass. Physical properties include a density of 0.308 lb/in³ (8,525 kg/m³), a melting-temperature range of 1680 to 1750°F (915 to 954°C), a specific heat at 68°F (20°C) of 0.09 Btu/lb · °F 375 J/kg · K, a thermal conductivity at 68°F (20°C) of 70 Btu/ft · h · °F [120 W/(m · K)], and an electrical conductivity at 68°F (20°C) of 28% that of copper. Typical tensile properties of thin, annealed, flat products range from ultimate strengths of 44,000 to 53,000 lb/in² (300 to 365 MPa), yield strengths of 11,000 to 22,000 lb/in² (75 to 150 MPa), and elongations of 68 to 54%. In the ¼-hard to extrahard cold-worked temper conditions, the tensile properties of these products range from 54,000 to 86,000 lb/in² (370 to 595 MPa), 40,000 to 65,000 lb/in² (275 to 450 MPa), and 43 to 5%, respectively. Besides flat products, the alloy is available in bar, rod, wire, tubing, and, for cartridge cases, cups. It has excellent cold-forming characteristics and a machinability about 30% that of free-cutting brass. It is also readily brazed and soldered and can be welded by oxyfuel and resistance methods. Its weldability by gas-metal-arc methods, however, is limited, and other welding methods are not advisable. Although corrosion-resistant in various waters and chemical solutions, the alloy may be susceptible to dezincification in stagnant or slow-moving, brackish waters and salt or slightly acidic solutions. Also, it is prone to stress-corrosion cracking, particularly in ammonia environments. Besides munition applications, it is used for various stamped, spun, or drawn shapes, including lamp fixtures, shells and reflectors, auto radiator cores, locks, springs, fasteners, cylinder components, plumbing fixtures, and architectural grille work.

CASE-HARDENING MATERIALS. Materials for adding carbon and/or other elements to the surface of low-carbon or medium-carbon steels or to iron so that upon quenching a hardened case is obtained, with the center of the steel remaining soft and ductile. The material may be plain charcoal, raw bone, or mixtures marketed as **carburizing compounds.** A common mixture is about 60% charcoal and 40 barium carbonate. The latter decomposes, giving carbon dioxide, which is reduced to carbon monoxide in contact with the hot charcoal. If charcoal is used alone, action is slow and spotty. Coal or coke can be used, but action is slow, and the sulfur in these materials is detrimental. Salt is sometimes added to aid the carburizing action. By proper selection of the carburizing material, the carbon content may be varied in the steel from 0.80 to 1.20%. The carburizing temperature for carbon steels typically ranges from 1550 to 1750°F (850 to 950°C) but may be as low as 1450°F (790°C) or as high as 2000°F (1095°C). The articles to be carburized for case hardening are packed in metallic boxes for heating in a furnace, and the process is called **pack hard-**

ening, as distinct from the older method of burying the red-hot metal in charcoal.

Steels are also case-hardened by the diffusion of carbon and nitrogen, called **carbonitriding,** or nitrogen alone, called **nitriding.** Carbonitriding, also known as **dry cyaniding, gas cyaniding, liquid cyaniding, nicarbing,** and **nitrocarburizing,** involves the diffusion of carbon and nitrogen into the case. Nitriding also may be done by gas or liquid methods. In carbonitriding, the steel may be exposed to a carrier gas containing carbon and as much as 10% ammonia, the nitrogen source, or a molten cyanide salt, which provides both elements. Ammonia, from gaseous or liquid salts, is also the nitrogen source for nitriding. Although low- and medium-carbon steels are commonly used for carburizing and carbonitriding, nitriding is usually applied only to alloy steels containing nitride-forming elements, such as aluminum, chromium, molybdenum, and vanadium. In **ion nitriding,** or **glow-discharge nitriding,** electric current is used to ionize low-pressure nitrogen gas. The ions are accelerated to the workpiece by the electric potential, and the workpiece is heated by the impinging ions, obviating an additional heat source. All three principal case-hardening methods provide a hard, wear-resistant case. Carburizing, however, which gives the greater case depth, provides the best contact-load capacity. Nitriding provides the best dimensional control, and carbonitriding is intermediate in this respect.

The principal liquid-carburizing material is sodium cyanide, which is melted in a pot that the articles are dipped in, or the cyanide is rubbed on the hot steel. **Cyanide hardening** gives an extremely hard but superficial case. Nitrogen as well as carbon is added to the steel by this process. Gases rich in carbon, such as methane, may also be used for carburizing, by passing the gas through the box in the furnace. When ammonia gas is used to impart nitrogen to the steel, the process is not called carburizing but is referred to as **nitriding. Tufftriding,** of Kolene Corp., is a nitriding process using molten potassium cyanate with a small amount of sodium ferrocyanide in titanium-lined melting pots.

Case-hardening compounds are marketed under a wide variety of trade names. These may have a base of hardwood charcoal or of charred bone, with sodium carbonate, barium carbonate, or calcium carbonate. **Char** is a carburizing material in which the particles of coal-tar carbon are surrounded by an activator and covered with a carbon coating. **Accelerated Salt WS,** of Du Pont, for heat-treating baths, has a content of 66% sodium cyanide, with graphite to minimize fuming and radiation losses. For selective case hardening on steel parts, a stiff paste of carburizing material may be applied to the surfaces where a carbon impregnation is desired. **Carburit** is a carburizing paste of this kind. **Aerocarb** and **Aerocase,** of American Cyanamid Co., are mixtures of sodium and potassium nitrates and nitrides for use in carburizing baths at a temperature up to 1850°F (1010°C).

Chromized steel is steel surface-alloyed with chromium by diffusion from a chromium salt at high temperature. The reaction of the salt produces an alloyed surface containing about 40% chromium. **Plasmaplate** was a name given by the former Linde Div. of Union Carbide to protective coatings of tungsten or molybdenum, deposited by a plasma torch which gives a concentrated heat to 30,000°F (16,650°C); but the refractory metals can now be deposited at lower temperatures by decomposition of chemical compounds. **Molybdenum pentachloride,** $MoCl_5$, is a crystalline powder which deposits an adherent coating of molybdenum metal when heated to 1652°F (900°C).

Metalliding is a **diffusion coating** process involving an electrolytic technique similar to electroplating, but done at higher temperatures [1500 to 2000°F (816 to 1093°C)]. Developed by General Electric, the process uses a molten fluoride salt bath to diffuse metals and metalloids into the surface of other metals and alloys. As many as 25 different metals have been used as diffusing metals, and more than 40 as substrates. For example, **boride coatings** are applied to steels, nickel-base alloys, and refractory metals. **Beryllide coatings** can be applied to many different metals by this process. The coatings are pore-free and can be controlled to a tolerance of 0.001 in (0.025 mm).

CASEIN. A whitish to yellowish, granular or lumpy protein precipitated from skim milk by the action of a dilute acid, or coagulated by rennet, or precipitated with whey from a previous batch. The precipitated material is then filtered and dried. Cow's milk contains about 3% casein. It is insoluble in water and in alcohol, but soluble in alkalies. Although the casein is usually removed from commercial milk, it is a valuable food accessory because it contains **methionine,** a complex mercaptobutyric acid which counteracts the tendency toward calcium hardening of the arteries. This acid is also found in the ovalbumin of egg white. Methionine, $CH_3 \cdot S \cdot CH_2CH_2CHNH \cdot COOH$, is one of the most useful of the amino acids, and it is used in medicine to cure protein deficiency and in dermatology to cure acne and falling hair. It converts dietary protein to tissue, maintains nitrogen balance, and speeds wound healing. It is now made synthetically for use in poultry feeds. Some casein is produced as a by-product in the production of lactic acid from whole milk, the casein precipitating at a pH of 4.5. It is treated with sodium hydroxide to yield **sodium caseinate.**

Most of the production of casein is by acid precipitation, and this casein has a moisture content of not more than 10% with no more than 2.25% fat and not over 4 ash. The casein made with rennet has up to 7.5% ash content, less than 1 fat, and is less soluble in alkalies. It is the type used for making plastics. **Rennet** used for curdling cheese is an extract of an enzyme derived from the stomachs of calves and lambs and is closely related to pepsin. Rennet substitutes produced from pepsin and other vegetable sources are only partial replacements and often have undesirable off-flavors. But **Sure-Curd,** of Pfizer and Co., is derived from a strain of *Endothia parasitica* and is similar to true rennet in coagulating and proteolytic properties. **Whey** is the thin, sweet, watery part separated out when milk is coagulated with rennet. Whey solids are used in prepared meats and other foods to enhance flavor and in pastries to eliminate sogginess. **Tekniken** is a dry whey for use in margarine, chocolate, and cheese. **Orotic acid,** NH(CO · NH · CO · CH):C · COOH, produced synthetically, is identical with the biotic *Lactobacillus bulgaricus* of **yogurt,** the fermented milk whey used as food. It is a vitaminlike material.

Argentina and the United States are the most important producers of casein. France, Norway, and Holland are also large producers. Casein is employed for making plastics, adhesives, sizing for paper and textiles, washable interior paints, leather dressings, and as a diabetic food. **Casein glue** is a cold-work, water-resistant paste made from casein by dispersion with a mild base such as ammonia. With a lime base it is more resistant but has a tendency to stain. It is marketed wet or dry, the dry powder being simply mixed with water for application. It is used largely for low-cost plywoods and in water paints, but is not waterproof. Many gypsum wallboard cements are fortified with casein. Concentrated **milk protein,** available as **calcium caseinate** or **sodium caseinate,** is for adding proteins and for stabilizing prepared meats and bakery products. It contains eight amino acids and is high in lysine. **Sheftene** is this material.

CASEIN PLASTICS. A group of thermoplastic molding materials made usually by the action of formaldehyde on rennet casein. The process was invented in 1885, and the first commercial casein plastic was called **Galalith,** meaning **milkstone.** Casein plastics are easily molded, machine easily, are nonflammable, withstand temperatures up to 300°F (150°C), and are easily dyed to light shades. But they are soft, have high water absorption (7 to 14%), and soften when exposed to alkalies. They are thus not suitable for many mechanical or electrical parts. They are used for ornamental parts, buttons, and such articles as fountain-pen holders. The specific gravity of the material is 1.34, and the tensile strength is 8,000 lb/in² (55 MPa). **Casein fiber** is made by treating casein with chemicals to extract the albumen and salts, forcing it through spinnerets, and again treating it to make it soft and silklike. The fiber is superior to wool in silkiness and resistance to moth attack, but is inferior in general properties. It is blended with wool in fabrics and in hat felts.

CASHEW SHELL OIL. An amber-colored, poisonous, viscous oil obtained by extraction from the by-product shells of the cashew nut industry of India and Brazil. The cashew nut grows on the distal end of the fruit of the tree *Anacardium occidentale.* The thin-skinned, yellow, pear-shaped fruit may be eaten or used in preserves. It is also distilled into a spirit in Mozambique and India. The kernel of the seed nut, known as the **cashew nut,** is roasted and widely used as an edible nut or in confections. The kernel is crescent-shaped, and the nuts are graded by sizes from 200 per lb (0.45 kg) to 400 to 500 per lb (0.45 kg). On crushing, the nuts produce 45% of an edible oil, but the nuts are more valuable as a confection than for oil, and there is no commercial production of **cashew nut oil.** One pound of shells yields 0.335 lb (0.152 kg) of cashew nut shell oil, which contains 90% **anacardic acid,** a carboxypenta-dica-dienyl phenol, very blistering to the skin. It is used for the production of plastics, drying oils, and insulating compounds. The oil reacts with formaldehyde to give a drying oil. With furfural it produces a molding plastic. Reacted with other chemicals, it forms rubberlike masses used as rubber extenders and in electrical insulating compounds. The other 10% of cashew nut shell oil is **cardol,** a dihydroxypenta-dica-dienyl benzene. When decarboxylated, the anacardic acid yields **cardanol,** a light oil liquid of composition $C_6H_4 \cdot OH(CH_2)_6CH{:}CH(CH_2)_6CH_3$, with boiling point of 680°F (360°C) and freezing point of about -4°F (-20°C). Cardanol polymerizes with formaldehyde to form a heat-resistant, chemical-resistant, flexible resin of high dielectric strength valued for wire insulation. Small amounts of this resin also improve the chemical and electrical properties of the phenol resins. **Cardolite** is a high-molecular-weight, straight-chain bisphenol derived from cashew nut shell oil. It is used for making flexible epoxy resins, supplanting about half the normal amount of epichlorhydrin used in the resin.

CASHMERE. A fine, soft, silky fabric made from the underhair of the Cashmere goat raised on the slopes of the Himalayas in Asia. The hair is obtained by combing the animals, not by shearing, and only about 3 oz (0.09 kg) is obtained from a goat. The hair is straight and silky, but not lustrous, and is difficult to dye. The fabrics are noted for warmth, and the production now goes mostly into the making of shawls and fine ornamental garments. **Cotton cashmere** is a soft, loosely woven cotton fabric made to imitate cashmere, or it may be a cotton-and-wool mixture, but it lacks the fineness of true cashmere. **Cashmere hair,** used for fine paintbrushes, is from the beard of the Cashmere goat. It is similar to camel hair. **Qiviut,** the underwool of the musk ox of northern Canada, is a finer and longer fiber than cashmere, and about 6 lb (2.7 kg) may be obtained from each animal. It is shed in May or June. One pound (0.45 kg) of qiviut will make a 40-strand thread 26 mi (44 km) long. It dyes easily and does not shrink, even when boiled. It is used for fine gloves and sweaters.

CASSITERITE.   Also called **tin stone.** It is the only commercial **tin ore** and is a **tin dioxide,** SnO_2, containing theoretically 78.6% tin. It is a widely distributed mineral, but is found on a commercial scale in only a few localities, notably Malaya, East Indies, Bolivia, Cornwall (England), Nevada, Isle of Youth, and Australia. The mineral occurs granular massive with a specific gravity of 6.8 to 7.1, a Mohs hardness of 6 to 7, and a brown to black color. It is present in the ore usually in amounts of 1 to 5% and is found in veins, called **lode tin,** or in placer deposits. The concentrated ore averages 65 to 70% tin oxide. It is roasted to eliminate sulfur and arsenic and then smelted in reverberatory furnaces.

CAST IRON.   The generic name for a broad family of materials comprised basically of carbon, silicon, and iron, but which may also contain small or large amounts of alloying elements. The principal kinds are gray iron, ductile (or nodular) iron, malleable iron, white iron, and alloy irons. The borderline between steel and cast iron is 2% carbon, cast irons having more than this amount, and at least 1% silicon, usually 1 to 3. Carbon is present in two forms: graphite, often referred to as *free carbon,* and iron carbide (cementite).

Each of the five major types differs in the form in which carbon is present. High carbon content makes molten iron fluid, easing castability. Precipitation of graphite during solidification counteracts metal contraction as it cools, producing sound castings. Graphite also provides excellent machinability, damping qualities, and lubricity on wear surfaces. When most of the carbon is combined with iron in the form of carbides, as in white iron, it provides excellent wear resistance. Silicon serves to promote graphite formation and provide desired metallurgical structures.

The matrix structures of cast irons, where any graphite present is embedded, vary widely depending not only on casting practice and cooling rate but also on the shape and size of casting. Furthermore, it is possible to have more than one kind of matrix in the same casting. Also, the matrix structure can be controlled by heat treatment, but once graphite is formed, it is not changed by subsequent treatments. The matrix can be entirely ferritic. It differs from the ferrite found in wrought carbon steels because the relatively large amount of silicon produces a structure that makes the iron free-machining. Addition of alloys can produce an acicular (needlelike) matrix. Hardening treatments yield a martensitic matrix. Other possible matrix structures are pearlite and ledeburite. Because the same composition in a cast iron can produce several different types of structure, cast irons are seldom specified by composition. Within each major type, standard grades are classified by minimum tensile strength.

Cast iron is usually made by melting pig iron and scrap in a cupola in contact with the fuel, which is normally coke. Pouring temperature, which varies with the analysis, is important, especially to prevent cold shut, which is a discontinuity in the structure caused by two streams of metal meeting and failing to unite. With an electric furnace, scrap iron may be employed alone with carbon without pig iron, and the furnace may be operated continuously. The product is called **synthetic cast iron.**

Gray iron, which contains graphite in flake form and usually contains 2 to 4% carbon and 1 to 3 silicon, is noted primarily for its ability to dampen vibrations, withstand moderate thermal shock, and provide moderate strength: ultimate tensile strengths of 20,000 to 60,000 lb/in^2 (138 to 414 MPa). In general, the greater the strength, the lower the damping capacity and thermal-shock resistance, and the less amenability to be cast in thin sections. Machinability also decreases with increasing strength, although high-strength grades can be machined to finer finishes. Although the various grades are designated by tensile strength, compressive strength is often a major design selection factor. Compressive strengths corresponding to the above tensile strength range are about 80,000 to 185,000 lb/in^2 (552 to 1,276 MPa). Modulus of elasticity in tension also increases with increasing strength, ranging from about 9.6×10^6 to 23×10^6 lb/in^2 (66,000 to 159,000 MPa). Although gray iron can be strengthened and toughened by heat treatment, these requirements are usually met by adjusting composition. Quenching from elevated temperature is done more commonly to increase wear resistance by increasing hardness, with tempering used to enhance toughness. Gray cast iron is widely used in the auto, truck, and off-highway equipment industries for engine blocks, gearboxes, brake drums, camshafts, and many other components.

Ductile iron, also known as **nodular iron** or **spheroidal-graphite iron** because of the shape of the graphite particles, is noted primarily for its high strength and toughness. Though made from the same basic materials as gray iron, a small amount of magnesium, or magnesium and trace amounts of cerium, is inoculated during casting to control the shape and distribution of the graphite. Tensile properties range from 50,000 to 120,000 lb/in^2 (345 to 827 MPa) ultimate strength, 25,000 to 90,000 lb/in^2 (172 to 621 MPa) yield strength, and 20 to 2% elongation. Most ductile iron castings are used as cast, but subsequent heat treatment can be beneficial. Annealing, which provides a ferritic structure, maximizes toughness at the expense of strength. Normalizing, often followed by tempering, induces a pearlitic structure, providing intermediate strength and toughness. And a martensitic structure, induced by quenching, usually in oil, provides the highest strength and hardness, but the least toughness. The modulus of elasticity of ductile iron—22 $\times$ 10^6 to 25 $\times$ 10^6 lb/in^2 (152,000 to 172,000 MPa)—is typically greater than that of gray iron, as is its high-temperature oxidation resistance, but its machinability is about the same. Ductile-iron castings are widely used in the automotive industry for crankshafts, camshafts, steering knuckles, pinions, gears, and many other components. They are also used for a variety of machinery applications, marine components, and equipment used in the paper and glass industries.

Compacted graphite cast iron, also known as **CGI** and **vermicular iron,** is characterized by coarser, more rounded graphite than the flake graphite in gray iron. It is produced by adding a small but precise amount of magnesium, in a process similar to making ductile iron, and resulting mechanical properties are generally intermediate to those of gray and ductile irons. In some cases, however, properties may be superior to either of the two more common cast irons. CGI is about equal to gray iron in thermal conductivity and damping quality but can be twice as strong. It is similar to ductile iron in strength and rigidity. The Backerud process for casting CGI is patented by the Swiss firm SinterCast S.A. and named after its inventor, Lennart Backerud.

Malleable iron is white cast iron that is heat-treated to transform the carbon phase from iron carbide to a nodular form of graphite called *temper carbon.* The resulting structure can be ferrite with dispersed nodules (**ferritic malleable iron**); pearlitic, which also contains combined carbon; or **martensitic malleable iron,** which is produced by quenching and tempering **pearlitic malleable iron.** The nodules are more irregular than those of ductile iron, but otherwise the structure and mechanical properties are roughly comparable to standard nodular iron. Malleable iron has a slight advantage in modulus of elasticity—25 $\times$ 10^6 to 28 $\times$ 10^6 lb/in^2 (172,000 to 193,000 MPa)—and a definite advantage in amenability to casting thin-section components. Ductile iron shrinks less on solidifying and has the advantage in casting thick sections because, in making malleable iron castings, there is a limit to the section thickness that can be cast completely as white cast iron.

In **white cast iron,** the carbon is not transformed to graphite but remains combined with iron, usually in the form of large carbides. High hardness, thus high wear resistance, is its principal advantage. Unalloyed white iron contains a small amount of silicon and has a pearlitic structure. Alloy grades contain small amounts of carbide-stabilizing elements, such as chromium, molybdenum, and vanadium, and have a bainitic or martensitic structure and can provide a hardness of Brinell 700. **Chilled iron** combines white iron and gray iron. Iron or graphite chills are used in select areas of the mold to increase the solidification rate and form white iron while the rest of the casting solidifies at a slower rate and forms gray iron.

Although some of the four major classes of cast irons—gray, ductile, malleable, and white—may contain small amounts of alloying elements, **alloy cast irons** may contain appreciable amounts. Their purpose is to increase strength, hardness, hardenability, abrasion resistance, heat resistance, corrosion resistance, or combinations of these properties. Among alloy cast irons are **abrasion-resistant white irons** that may contain 1 to 5% nickel, 1 to 28 chromium, 0.5 to 3.5 molybdenum, and, sometimes, 1.2 to 2.5 copper, which may supplant or be an addition to nickel. **Corrosion-resistant cast irons** include **high-silicon** (14 to 17%) **irons,** such as **Duriron, Durichlor 51,** and **Superchlor,** which also contain 5% chromium, 1 molybdenum, and 0.5 copper; **nickel-chromium gray irons,** such as **Ni-Resist austenitic iron,** having 13.5 to 36% nickel, 1.5 to 6 chromium, 7 copper, and 1 molybdenum; and **nickel-chromium ductile iron,** such as **Ni-Resist austenitic ductile iron,** having 18 to 36% nickel, 1 to 5.5 chromium, and 1 molybdenum. **Heat-resistant gray irons** include 4 to 7% silicon grades, such as **Silal; high-chromium iron** (15 to 35% chromium, 5 nickel); **nickel-chromium irons,** such as **Ni-Resist austenitic iron; nickel-chromium-silicon irons,** such as **Nicrosilal,** having 13 to 43% nickel, 10 copper, 5 to 6 silicon, 1.8 to 5.5 chromium, and 1 molybdenum; and **high-aluminum** (20 to 25%) **iron,** which also contains 1.3 to 6% silicon. **Heat-resistant ductile irons** include **medium-silicon ductile iron** (2.5 to 6% silicon, 1.5 nickel) and **nickel-chromium ductile iron** (18 to 36% nickel, 1.75 to 3.5 chromium, 1.75 to 5.5 silicon, and 1 molybdenum).

Austempered ductile iron, alloyed ductile iron having a structure of ferrite and carbon-rich austenite, has been known for many years but seldomly used because of the finesse required to induce this structure by heat treatment. Because of the exceptional strength and toughness possible with careful control of heat treatment, however, it has recently emerged as a promising material, especially for auto and truck applications. The alloying elements are nickel, copper, or molybdenum, or combinations of these, and their purpose is to increase hardenability. These elements delay pearlite formation, permitting the casting to be cooled from austenitizing temperatures to the austempering transformation range without forming pearlite or other high-temperature transformation products during quenching.

Heat treatment involves (1) heating to austenitizing temperature and holding at this temperature until the structure has transformed to face-centered-cubic austenite and this austenite is saturated with carbon; (2) quenching to a temperature above the martensite start temperature [450 to 750°F (232 to 399°C)] usually in molten salt or a medium capable of providing a quenching rate sufficient to avoid pearlite formation, and holding at this temperature for sufficient time to transform the austenite to a structure of acicular ferrite and carbon-rich austenite; and (3) cooling to room temperature. No subsequent tempering is necessary. The bainitic reaction temperature, commonly called the *austempering temperature,* determines mechanical properties. High austempering temperatures promote ductility, fatigue strength, and impact strength, but reduce hardness. Low austempering temperatures increase strength and hardness. Tensile yield strength can range from 80,000 to 180,000 lb/in^2 (552 to 1,240 MPa), with corresponding ultimate strengths of 125,000 to 230,000 lb/in^2 (860 to 1,585 MPa), elongations of up to 10%, and hardness from the range of Brinell 269 to 321 to Brinell 444 to 555. Impact strength is about 75 ft · lb (102 J) for 80,000 lb/in^2 (552 MPa) yield-strength material, and 30 ft · lb (41 J) for the 140,000 lb/in^2 (965 MPa) material.

Gun iron, formerly used for casting cannons, was a fine-grained iron of uniform texture, low in sulfur and in total carbon, made with charcoal in an air furnace.

Graphite is a weakening element in cast iron, and the high-graphite irons are desired only because of their ease of casting and machining. The lower the carbon, the stronger the cast iron. To obtain this result, steel scrap is used in the mix. Low-carbon steel of known chemical content, such as plate and rod ends and rail croppings, is used. The amount of steel varies from 15 to 60%, and the product resulting from the larger additions is called **semisteel.** Tensile strengths as high as 40,000 lb/in^2 (276 MPa) can be obtained without great reduction in the casting and machining qualities of the cast iron. Semisteel castings can be softened and made more ductile by annealing at a temperature of about 800°F (427°C), but they then lose 25 to 35% tensile strength.

Many trade names have been used to designate cast irons. **Pomoloy** is an unalloyed cast iron with a tensile strength of 40,000 lb/in^2 (276 MPa) and hardness Brinell 215. **DeLavaud metals** are made by a centrifugal process in rotating steel molds. After annealing, the pipe has an outer layer of malleable iron, a center layer resembling steel, and an inner surface of gray iron. **Hi-Tem iron** is a corrosion-resistant cast iron used for processing vessels. **Hi-Tem S** is a high-manganese iron used for retorts.

High-test cast iron was originally cast iron that was superheated in the melting for pouring, poured in chilling molds, and then heat-treated, the only change in composition being to keep the silicon and manganese high. The term now means **high-strength irons** that are processed to give a careful balance of ferrite, pearlite, cementite, and carbon by the treatment, by additions of steel scrap, and by additions of nickel, chromium, and other elements that give strength to the metal by balancing the structure, but are not in sufficient quantities to classify the iron as an alloy cast iron. Tensile strengths above 50,000 lb/in^2 (345 MPa) are obtained, and all the high-test irons are fine-grained, not spongy like gray iron. Steel scrap gives a stronger and finer structure; nickel aids in the chilling and eases machining; chromium gives hardness and resistance to growth; molybdenum raises the combined carbon and adds strength and hardness. **Oxygenized iron** is high-test cast iron made by blowing air through a part of the metal and then returning the blown metal to the cupola. There is no sharp dividing line between some of these processed irons and steel, and when the combined carbon is high and the graphitic carbon is well distributed in even flakes, the metal is called **graphitic steel.**

High-test cast irons are used for brake drums, cams, rolls, and high-strength parts. In many cases they are substitutes for malleable iron. They are marketed under many trade names. **Ermal** is a **pearlitic cast iron** with a tensile strength up to 70,000 lb/in^2 (483 MPa). **Perlit** is another pearlitic cast iron. **Armite** is a synthetic cast iron, and **Jewell alloy** is the name of a group of high-strength and heat-resistant irons. **Ermalite** and **Wearloy** are high-strength, wear-resistant cast irons. **Gunite** is a graphitic steel which, when quenched to a hardness of Brinell 477, has a compressive strength of 200,000 lb/in^2 (1,379 MPa). **Arma steel** is a graphitic steel, or arrested malleabilized iron, of high strength and shock resistance, used for connecting rods, gears, and camshafts where both high strength and bearing properties are required. **Meehanite metal** is made in a wide range of high-strength, wear-resisting, corrosion-resisting, and heat-resisting castings for dies, hydraulic cylinders, brake drums, pump parts, and gears. Tensile strengths range from 35,000 to 55,000 lb/in^2 (241 to 379 MPa), compressive strengths from 135,000 to 175,000 lb/in^2 (931 to 1,207 MPa), and hardness from Brinell 193 to 223. **Cylinder iron** is a general term for cast iron for engine and compressor cylinders, but is also used for a variety of mechanical parts. The iron must be easily cast into a dense structure without hard spots or blowholes. Combined carbon must be sufficient to give wear resistance without brittleness, and the content of free graphite must be high enough to give a low coefficient of friction without great loss of strength.

CASTOR OIL.   A light-yellow to brownish viscous oil obtained from the seed beans of the castor plant, *Ricinus communis.* In the tropics the plant grows to the proportions of a sturdy tree, but in temperate climates it is small with a poor yield. Besides its original use as a purgative in medicine, castor oil is one of the most widely used industrial vegetable oils. When pure and fresh, the oil is nearly colorless and transparent. The hot-pressed oil is brownish. It has a characteristic acrid, unpleasant taste. The specific gravity is 0.960 to 0.970, iodine value 82 to 90, saponification value 180, and solidifying point 14°F (-10°C). The oil is chiefly composed of the glyceride or **ricinoleic acid,** which has a complex double-bonded molecular structure that can be polymerized easily. It is used for making alkyd resins for surface coatings and in plasticizers, perfumes, and detergents. **Castor seeds** have the appearance of mottled colored beans and are enclosed in hard husks which are removed before crushing. The chief commercial production has been in Brazil, where two types are grown. The large Zanzibar type has seeds 0.63 in (16 mm) long containing 30 to 35% oil, and the sanguineous type has seeds 0.39 in (10 mm) long containing up to 60% oil. They are usually mixed in shipments, and the average yield is calculated as 0.45 lb (0.20 kg) of oil from 1 lb (0.45 kg) of beans. In the southwestern United States, dwarf disease-resistant hybrid varieties are grown that give high oil yields. Cold-pressed oil is used in medicine and lubricants, but the industrial oil is usually hot-pressed. Castor oil is used in paints, as a hydraulic oil, for treating leather and textiles, in soaps, and for making urethane resins. It increases the lathering power of soaps and their solubility in cold water. In lubricating oils and in cutting oils, it has excellent keeping qualities and does not gum on exposure.

When castor oil is chemically dehydrated by removing the hydroxyl groups in the form of water by means of a catalyst, a double bond is formed, giving an oil of heavy viscosity, light color, and with iodine value 116, acid value 3.5, and saponification value 191. **Dehydrated castor oil** gives a better gloss in varnishes than tung oil with a softer and less brittle film, but it has less alkali resistance than tung oil, unless it is mixed with synthetic resin. **Sulfonated castor oil,** known as **Turkey red oil** in the textile industry, is made by treating crude hot-pressed castor oil with sulfuric acid and neutralizing with sodium sulfate. It is miscible with water and lathers as a solution of soap. It is used for the preparation of cotton fibers to be dyed, and it gives clearer and brighter colors. It is also employed in soaps and cutting compounds. Sulfonated dehydrated castor oil is used in nonalkaline water-washable skin ointments. It has a softening point of 86°F (30°C) and an SO_2 content of 10%. **Synthenol,** of Spencer Kellogg, is a dehydrated castor oil for paints and varnishes. **Castung** and **Isoline** are dehydrated castor oils. **Copolymer 186** is a polymerized dehydrated castor oil which adds flexibility and improved general qualities to paints and outside enamels. **Mannitan drying oil** is an ester of dehydrated castor oil that dries faster than linseed oil and has better resistance in paints.

Hydrogenated castor oil is a hard, nongreasy, white solid melting at 180°F (82°C), used as an extender for waxes in coating compositions and as a hard grease for making resistant lithium-type lubricating greases. Hydrogenated castor oil is odorless and tasteless and is valued for coatings. **Castorwax, Emery S-751-R,** and **Cenwax G** are hydrogenated castor oil. In general, these materials are white, nongreasy, waxlike solids melting at about 185°F (85°C). **Primawax** is a flaked form of hydrogenated castor oil used as a plasticizer in vinyl and cellulose plastics. The destructive distillation of castor oil yields **cognac oil,** a mixture of **undecylenic acid** and **heptaldehyde,** also known as **oenanthaldehyde.** All are important intermediates in perfumes; heptaldehyde is the basis of synthetic jasmine perfumes.

The hydrogenated ricinoleic acid, known as **hydroxystearic acid,** may also be separated out and used for making waxy esters for pharmaceutical ointments, or for reacting with amines to make white, waxy solids useful as water repellents. By reacting castor oil with sodium hydroxide under heat and pressure, **sebacic acid,** $HO_2C(CH_2)_8CO_2H$, is produced. It is a powder melting at 264°F (129°C) and is a versatile raw material for alkyd resins, fibers, and heat-resistant plasticizers. It is also used for making nylon polymers and for **sebacate esters** for cold-weather lubricants, although the lower-cost azelaic and adipic acids may be substituted. Both sebacic acid and isosebacic acid are now produced synthetically from butadiene. **Isosebacic acid** is a mixture of sebacic acid with the isomers of this acid, **diethyl adipic acid** and **ethyl suberic acid.** It can replace sebacic acid for resin manufacture. Also similar in chemical properties to the ricinoleic acid of castor oil is dimorphecolic acid, obtained naturally from **daisy oil** from the seeds of the Cape marigold, of the genus *Dimorphotheca,* grown in California.

A substitute for castor oil in medicine is **croton oil,** a yellow-brown oil obtained from the dried ripe seeds of the small tree *Croton tiglium* of India and Sri Lanka. It has a burning taste and unpleasant odor and is a more violent purgative than castor oil. The leaves and flowers of the tree are used like derris to kill fish. **Curcas oil** is a yellowish oil from the kernels of the seeds of *Jatropha curcas* which grows in Central America. The kernels yield 50% oil with a specific gravity 0.920, iodine value 98 to 104, and saponification value 192. It is also a good soap oil but has an unpleasant odor. The ethyl and methyl esters of **crotonic acid** are used as monomers for flexible plastics for coatings. The acid with composition $CH_3CH:CHCOOH$ is now made synthetically from acetylene and aldol.

CAST STEEL. Low-carbon (less than 0.20%), medium-carbon (0.20 to 0.50), high-carbon (more than 0.5), and low-alloy (less than 8 total alloy content) steels that have been cast in sand, graphite, metal, ceramic, or other molds to produce finished or semifinished products. **Steel castings** having greater alloy content are commonly identified by other terms, such as **heat-resistant castings** or **corrosion-resistant castings.** Cast and wrought steels of equivalent composition respond similarly to heat treatment and have fairly similar properties. A major difference, however, is that cast steels are more isotropic in mechanical properties because, for wrought steels, these properties generally vary with respect to grain direction, that is, the direction of hot or cold working. For example, the impact strength of wrought steels is typically greater than that of cast steels, but the values reported for the wrought steels usually pertain only to the longitudinal grain direction. Values transverse to grain are lower. Impact strength of cast steels is generally intermediate to that of wrought steels in the longitudinal and transverse directions.

   Low-carbon cast steels and **medium-carbon cast steels** generally contain 0.5 to 1.20% manganese, as much as 0.8 silicon, and small amounts of phosphorus and sulfur. Low-carbon grades used for electrical equipment are restricted to 0.20% manganese to enhance magnetic properties. As-cast, tensile properties of a 0.19% carbon, 0.74% manganese grade are about 64,000 lb/in^2 (441 MPa) ultimate strength, 35,000 lb/in^2 (241 MPa) yield strength, and 33% elongation. Annealing markedly improves impact strength without appreciably affecting tensile properties. Surface-hardening methods, such as carburizing, are often used to increase wear resistance. Besides electrical equipment, low-carbon cast steels are used for railroad components, auto and truck parts, and heat-treating equipment. Medium-carbon grades, the most widely used, are almost always heat-treated by annealing, normalizing, normalizing and tempering, or quenching and tempering after casting. Depending on the grade, tensile properties range from 65,000 to 175,000 lb/in^2 (448 to 1,207 MPa) ultimate strength, 35,000 to 145,000 lb/in^2 (241 to 1,000 MPa) yield strength, and 24 to 6% elongation. **High-carbon cast steels** are less frequently used, and tensile properties are markedly influenced by carbon content. The ultimate tensile strength of one annealed grade, for example, ranges from about 94,000 lb/in^2 (648 MPa) to 126,000 lb/in^2 (869 MPa) as carbon content increases from 0.50 to 1.00%. The steels also may be normalized and tempered or quenched and tempered.

Low-alloy cast steels are generally medium-carbon grades containing chromium, nickel, molybdenum, and vanadium. Compared with the plain-carbon cast steels, they provide better hardenability, toughness (at greater strength levels), wear resistance, and/or heat and corrosion resistance. For example, the cast grades similar to wrought grades 41XX, 43XX, and 86XX can provide 50% greater tensile yield strength and equivalent impact strength to plain-carbon cast steels. Although they can provide ultimate tensile strengths exceeding 200,000 lb/in^2 (1,379 MPa), specified strength levels are generally less. **ASTM A487 cast steels,** for example, are normally specified for ultimate strengths of 70,000 to 145,000 lb/in^2 (483 to 1,000 MPa) and yield strengths of 30,000 to 100,000 lb/in^2 (207 to 690 MPa) in the normalized and tempered or quenched and tempered conditions. Corresponding elongations range from 24 to 14%. Applications include auto, truck, steam-turbine, and earthmoving equipment parts, machine tools, valves, marine hardware, and processing equipment of many kinds.

CATALYST. A material used to cause or accelerate chemical action without itself entering into the chemical combination. Catalysts are chosen for selectivity as well as activity, mechanical strength, and life. They should give a high yield of product per unit and be capable of regeneration whenever possible for economy. Small amounts of cocatalysts or promoters increase activity measurably. In the cracking of petroleum, activated carbon breaks the complex hydrocarbons into the entire range of fragments; activated alumina is more selective, producing a large yield of C_3 and C_4; and silica-alumina-zirconia is intermediate. **Contact catalysts** are the ones chiefly used in the chemical industry, and they may be in various forms. For bed reactors the materials are pelleted. **Powdered catalysts** are used for liquid reactions such as the hydrogenation of oils. **Chemical catalysts** are usually liquid compounds, especially such acids as sulfuric and hydrofluoric.

Various metals, especially platinum and nickel, are used to catalyze or promote chemical action in the manufacture of synthetics. Nitrogen in the presence of oxygen can be "fixed" or combined in chemicals at ordinary temperatures by the use of ruthenium as a catalyst. Acids may be used to aid in the polymerization of synthetic resins. Mineral soaps are used to speed up the oxidation of vegetable oils. Cobalt oxide is used for the oxidation of ammonia. Cobalt and thorium are used for synthesizing gasoline from coal. All these are classified as **inorganic catalysts.** Sometimes more complex chemicals are employed, silicate of soda being used as a catalyst for high-octane gasoline. In the use of **potassium persulfate,** $K_2S_2O_8$, as a catalyst in the manufacture of some synthetic rubbers, the material releases 5.8% active oxygen, and it is the nascent oxygen that is the catalyst. **Sodium methylate,** also called **sodium methoxide,** $CH_3 \cdot O \cdot Na$, used as a catalyst for ester-exchange reactions in the rearrangement of edible oils, is a white powder soluble in fats but violently decomposed in water. Transition-metal complexes, dispersed uniformly in solution, are called **homogeneous catalysts.** The most common ones are organometallic complexes, such as the carbonyls. They are more resistant to poisoning than solid **heterogeneous catalysts,** and they are highly active, specific, and selective.

Magnetite, a magnetic iron ore, is used as a catalyst in the synthesis of ammonia. In a system from M. W. Kellog Co., ruthenium, supported on a proprietary graphite structure, is more active, increasing ammonia production by 12 to 16% over magnetite. **Metallocenes,** organometallic coordination compounds obtained as cyclopentadienyl derivatives of a transition metal or metal hylide, are recent catalysts in the production of various plastics. Also referred to as **single-site catalysts,** they allow closer control of molecular weight and comonomer distribution, permitting monomers and comonomers previously considered incompatible to be combined. They also allow production of plastics in isotactic and syndiotactic forms and have been applied to polyethylene, polypropylene, and polystyrene.

Aluminum chloride, $AlCl_3$, in gray granular crystals which sublime at 1742°F (950°C), is used as a catalyst for high-octane gasoline and synthetic rubber and in the synthesis of dyes and pharmaceuticals. **Antimony trichloride,** $SbCl_3$, is a yellowish solid, melting at 164°F (73.4°C), used as a catalyst in petroleum processing to convert normal butane to isobutane. This chemical is also used for antimony plating and as a cotton mordant. **Bead catalysts** of activated alumina have the alumina contained in 0.1-in (3-mm) beads of silica gel. **Catasil** is alumina adsorbed on silica gel, used for polymerization reactions.

Vocat 350, of Salem Engelhard, can be used to reduce chlorinated hydrocarbon emissions in industrial processes, soil remediation, and groundwater cleanup. The catalyst operates between 437 and 886°F (225 and 475°C) and achieves up to 99% oxidation of chlorocarbons in the feed stream. It has greater activity treating aliphatic compounds than aromatics, forming carbon dioxide and hydrogen chloride gas when 1.5% or more water is present, simplifying treatment relative to the use of precious-metal catalysts.

Molecular sieve zeolites are crystalline **aluminosilicates** of alkali and alkali-earth metals. The aluminum and silicon atoms form regular tetrahedral structures that have large voids interconnected by open three-dimensional channels. The micropores may amount to 50% of the volume, resulting in crystals with some of the highest internal surface areas. The alkaline cations are mobile and may be ion-exchanged with metals with catalytic properties. Only reactants of the right molecular size may enter the channels and be catalyzed by the metal cations in the voids. Zeolites occur naturally in volcanic or basaltic rocks, the most important industrially being **faujasite, erionite, clinoptilolite,** and **mordenite.** Synthetic **zeolite X** and **zeolite Y,** with structures similar to faujasite, are made by Union Carbide Corp. The firm's **zeolite A** has no natural analog. In the production of gasoline, a petroleum-cracking catalyst consists of a crystalline aluminosilicate zeolite for breaking long-chain molecules, **kaolin** for strength and density, and a binder or gel to hold the two together. **Reduxion,** of Englehard Corp., is a line of fluid catalysts for precracking longer hydrocarbon molecules in petroleum refining before they are released for cracking in zeolites. Mobil Corp. markets **zeolites ZSM-5** and **ZSM-11,** which have been used for reacting methanol into gasoline. W. R. Grace & Co.'s Davison Chemicals Division **XP series,** Engelhard Corp.'s **Precision** line, **Katalistiks,** International's **LZ-210,** and Akzo Chemicals Inc.'s **Vision** are all targeted for cracking oil into high-octane gasoline, an application where they have largely replaced alumina.

Catalyst carriers are porous inert materials used to support the catalyst, usually in a bed through which the liquid or gas may flow. Materials used are generally alumina, silicon carbide, or mullite, and they are usually in the form of graded porous granules or irregular polysurface pellets. High surface area, low bulk density, and good adherence of the catalyst are important qualities. Pellets are bonded with a ceramic that fuses around the granules with minute necks that hold the mass together as complex silicates and aluminates with no trace elements exposed to the action of the catalyst or chemicals. Catalyst carriers are usually bonded to make them about 40% porous. The pellets may be 50 mesh or finer, or they may be in sizes as large as 1 in (2.5 cm). Platinum, palladium, and rhodium supported on activated alumina carriers are used in the catalytic converters of automobiles to clean up exhaust gases. A catalyst of precious metals supported on zeolite removes hydrocarbons, carbon monoxide, and nitrogen oxides from auto exhaust gases even in the presence of excess oxygen, as is the case for lean-burn engines. Developed by Mazda Motor of Japan, it could improve fuel efficiency of such engines by 5 to 8%. **NC-300** catalyst, of Norton Chemical Process Co., has a homogeneous zeolite composition and is used to reduce nitrogen oxide emissions from power-generating equipment. Reliable at temperatures exceeding 1004°F (540°C), it could be used in coal-fired boilers and gas turbines. Refractory filters known as **porous media,** used for filtering chemicals and gases at high temperatures, are essentially the same materials as catalyst carriers with ceramic bonds fired at about 2282°F (1250°C); but they are usually in the form of plates or tubes, and the porosity is usually about 35%. They may be used directly as filters, or as underdrain plates for filter powders.

A catalyst of palladium, cobalt, molybdenum, potassium, and a bromide compound, developed by Sumiken Chemical of Japan, is used in the production of 2,6-naphthalene dicarboxylic acid by air oxidation of 2,6-diisopropyl naphthene. A metal oxide catalyst of molybdenum, nickel, cobalt, and aluminum is effective for off-site activation of hydrotreatment catalysts for hydrogenation, denitrogenation, and desulfurization in a process developed by Leuna AG of Germany and Exxon Chemical of Brussels. Normally such catalysts are activated by sulfiding them in situ, necessitating reactor downtime and considerable emission of hydrogen sulfide. **Chiral** catalysts, of Regis Technologies Inc., are made from binuclear rhodium compounds with bridging ligands. They are applied in carbenoid reactions for the production of cyclopropanes, lactones, and lactems. Substrates for these reactions are diazoacetates, which can be prepared from various alcohols. The catalysts promote loss of nitrogen by the substrate to form an intermediate metal carbenoid. Potential applications include production of optically pure pharmaceuticals and agricultural chemicals.

Sunlight or ultraviolet rays are also used as catalysts in some reactions. For example, chlorine and hydrogen combine very slowly in the dark, but combine with great violence when a ray of sunlight is turned on. **Biologic catalysts** are the **enzymes,** which are **organic catalysts** that are a form of life. They are sensitive to heat and light and are destroyed at 212°F (100°C). Enzymes are soluble in water, glycerin, or dilute saline solutions, and water must always be present for enzyme action. Their action may be simulated or checked by other substances. When dehydrated vegetables lose their flavor by destruction of the enzymes, the flavor may be restored by adding small percentages of enzymes from the same or similar vegetables.

Enzymes have various actions. **Diastase,** found in the seeds of barley and other grains, converts starch to maltose and dextrin. **Diastase 73,** of Rohm & Haas Co., is an enzyme chemical for converting gelatinized starches to dextrose. It is **amyloglucosidase** modified to remove the bitter taste. One pound (0.45 kg) will convert 100 lb (45 kg) of starch. **Cytase,** found in seeds and fruits, decomposes cellulose to galactose and mannose. **Zymose,** found in yeast, hydrolyzes glucose to alcohol. **Thiaminase,** an enzyme which occurs in small amounts in salmon, cod, rockfish, and some other fish, destroys the vitamin thiamine; and if taken in high concentration in the human diet, it causes ventritional polyneuritis. **Rhozyme LA,** of

Rohm & Haas Co., is a diastatic enzyme concentrate in liquid form for desizing textiles. **Bromelin,** an enzyme used in breweries, is produced from pineapples by alcohol precipitation from the juice. **Fermcozyme** is a liquid glucose-oxidase-catalase used in carbonated beverages to remove dissolved oxygen which would combine with glucose to form gluconic acid, resulting in loss of color and flavor. It is also used in egg powders to remove undesirable glucose.

Fermenting agents comprise a wide range of yeasts, bacteria, and enzymes which break down molecules to form other products. **Yeasts** are important in foodstuffs manufacture. A yeast is a fungus, and the life organisms produce carbon dioxide gas to raise doughs. These are called **leavening yeasts. Fermenting yeasts** produce alcohols by action on sugars. Many of the yeasts are high in proteins, vitamins, and minerals, and as dry, inactive powders they are used to raise the nutritional values of foodstuffs. **Torula yeast,** *Torulopsis utilis,* used as an additive in processed foods, is a by-product of the sulfite paper mills, growing on the 5- and 6-carbon wood sugars. It contains more than 50% proteins and has 10 different vitamins and 15 minerals. The dry powder is inactive and does not cause rising in baked foods. **Prostay,** of St. Regis Paper Co., is this material.

CATECHU.   An extract obtained from the heartwood and from the seed pods of the tree *Acacia catechu* of southern Asia. It is used in tanning leather and as a dyestuff, giving brown, drab, and khaki colors. It is used in medicine as an astringent for diarrhea and hemorrhage. The name is sometimes applied to gambier, which also contains catechu tannin, $C_{15}H_9(OH)_5$. Catechu, or **cutch,** comes either as a liquid which is a water solution or as brownish, brittle, glossy cakes. The liquid contains 25% tannin; and the solid, 50%. A ton (0.91 metric ton) of heartwood yields, by hot-water extraction, 250 to 300 lb (113 to 136 kg) of solid cutch extract. It is a powerful astringent. When used alone as a tanning agent, the leather is not high-quality, being of a dark color, spongy, and water-absorbent. It is normally employed in mixtures. **Burma cutch** is from *A. catechuoides.* **Indian cutch** is from *A. sundra.* The latter is frequently adulterated with starch, sand, and other materials. **Wattle** is an extract from Australian and east African acacia, *A. dealbata,* and other species. The wattle tree is called **mimosa** in Kenya. **Wattle bark** contains 40 to 50% tannin. It gives a firm, pinkish leather and is employed for sole leathers. The solid extract contains 65% tannin. **Golden wattle,** used for tanning in New Zealand, is the tree *A. pycantha.* Much wattle extract is produced in Brazil from the **black wattle. Turwar bark,** or **avarem,** used in India for tanning cattle hides, is from the tree *Cassia auricula* and is similar to wattle.

CATGUT. String made from the intestines of sheep, used for violin strings and for tough, durable cords for rackets and other articles. After cleansing and soaking in an alkali solution, the intestines are split, drawn through holes in a plate, cured in sulfur or other material, and graded according to size. Sheep intestines are also used for making surgical sutures, but for this purpose they are not called catgut, but simply **gut.** The sutures are encased in tubes and bombarded by electron-beam radiation for sterilization. In the meat-packing industry, the intestines of sheep and goats are referred to as **casings** and are employed as the covering of sausage and other meat products. They are graded by diameter, freedom from holes, strength, color, and odor. Intestines of hogs and beef cattle are also used as casings, but they are not as edible as those from sheep.

CEDAR. A general name that includes a great variety of woods. The true cedars comprise trees of the natural order of *Coniferae,* genus *Cedrus,* of which there are three species: **Lebanon cedar,** *Cedrus libani*; **Himalayan cedar,** *C. deodora*; and **Atlas cedar,** *C. atlantica.* The differences are slight, and all the species are sometimes classed as *C. libani.* The Himalayan cedar is also known as **deodar.** All are mountain trees and are native to southern Europe, Asia, and northern Africa. The true cedar is yellow and fragrant, takes a beautiful polish, and is very durable. It is used in construction work, and timbers in temples in India more than 400 years old are still perfectly preserved. The wood has a density of about 36 lb/ft^3 (576 kg/m^3). Numerous species of *Cedrela* occur in tropical America, Asia, and Africa, and they are also called cedar, but the wood has greater resemblance to mahogany. In the United States and Canada, the name *cedar* is applied to woods of species of *Thuya, Juniperus,* and *Cupressus,* more properly classified as thuya, juniper, and cypress.

   Spanish cedar, or **Central American cedar,** used in the United States as a substitute for mahogany in patternmaking, and for cigar boxes, furniture, carving, cabinetwork, and interior trim, is a softwood from numerous species of *Cedrela,* called in Spanish America by the name **Cedro.** It has a light-red color sometimes beautifully figured with wavy grain, has an agreeable odor, is easily worked, seasons well, and takes a fine polish. The density is 28 to 33 lb/ft^3 (449 to 529 kg/m^3). The trees grow to a large size, logs being available 40 in (1.02 m) square. The imports come chiefly from Central America and the West Indies, but the trees grow as far south as northern Argentina. Paraguayan cedar is the wood of the tree *C. braziliensis,* of Paraguay, Brazil, and northern Argentina, employed locally for cabinetwork, car building, and interior building work. It is similar in appearance to Spanish cedar but is denser, harder, and redder. The wood known as

southern white cedar, and called *juniper* in the Carolinas, is from the tree *Chamaecyparis thyoides,* growing in the coastal belt from Maine to Florida. The heartwood is light brown tinged with pink, and the thin sapwood is lighter in color. The wood is lightweight, straight-grained, durable, and fragrant. The more plentiful **white cedar** of the west coast, known also as **Port Orford cedar, Oregon cedar, ginger pine,** and in England as **Lawson cypress,** is from the tree *C. lawsoniana* of California and Oregon, mostly from a narrow coastal strip in Oregon to an altitude of about 5,000 ft (1,524 m). Mature trees reach a height of 160 ft (49 m) and a diameter of 6 ft (1.8 m). The wood is white with a yellow tinge and a trace of red. It is rather hard and tough, with a fine, straight grain, and is very durable. It has an agreeable aromatic odor and is free from pitch. The wood is used for doors, sashes, boats, matches, patterns, and where a light, strong straight-grained wood is required. **Toon,** the wood of the tree *Cedrela toona* of India, southeast Asia, and Australia, resembles Spanish cedar but is somewhat harder, denser [35 lb/ft^3 (560 kg/m^3)], and deeper red. It has a beautiful grain and takes a high polish. It is durable, does not warp, and is used for furniture and cabinetwork.

CELLULOSE.   The main constituent of the structure of plants, which, when extracted, is employed for making paper, plastics, and many combinations. Cellulose is made up of long-chain molecules in which the complex unit $C_6H_{10}O_5$ is repeated as many as 2,000 times. It consists of glucose molecules with three **hydroxyl groups** for each glucose unit. These OH groups are very reactive, and an almost infinite variety of compounds may be made by grafting on other groups, either repetitively or intermittently, such as reaction with acetic or nitric acids to form acetates or nitrates, reaction with ethylene oxide to form hydroxyethel cellulose, reaction with acrylonitrile to form cyanoethylated cellulose, or reaction with vinyls. Cellulose is the most abundant of the nonprotein natural organic products. It is highly resistant to attack by the common microorganisms, but the enzyme **cellulase** digests it easily, and this organism is used for making paper pulp, for clarifying beer and citrus juices, and for the production of citric acid and other chemicals from cellulose. Cellulose is a white powder insoluble in water, sodium hydroxide, or alcohol, but it is dissolved by sulfuric acid. The highly refined insoluble cellulose with all the sugars, pectin, and other soluble matter removed is called **alpha cellulose,** or **chemical cellulose,** used for the production of chemicals. It was formerly made only from cotton linters, but is now largely made from wood pulp. It is a white crystalline powder for use in foodstuffs to give body and gel stability to such products as peanut butter, cheese spreads, and prepared puddings. It forms a firm gel in water and absorbs oils easily. It is odorless and tasteless and has no calorie content.

One of the simplest forms of cellulose used industrially is **regenerated cellulose,** in which the chemical composition of the finished product is similar to that of the original cellulose. It is made from wood or cotton pulp digested in a caustic solution. The viscous liquid is forced through a slit into an acid bath to form a thin sheet, which is then hardened and bleached. **Cellophane,** of Du Pont, is a regenerated cellulose in thin sheets for wrapping. It is transparent, dyed in colors, or embossed. It is up to 0.0016 in (0.041 mm) thick with tensile strengths from 8,000 to 19,000 lb/in^2 (55 to 130 MPa). It chars at about 375°F (190°C). The thinnest sheets, 0.0009 in (0.023 mm) in thickness, have 21,500 in^2/lb (30.8 m^2/kg). The three-digit gage system used for cellophane indicates the total film yield. Thus, 180 gage has a film yield of 18,000 in^2/lb (25.8 m^2/kg). The waterproofed material is coated with a thin film of cellulose lacquer, or the cellophane may be laminated with a film of a synthetic resin. Cellophane has greater transparency than polyethylene, but is not as strong or as chemically resistant. For food packaging, the printing is done on the reverse side of the cellophane before laminating.

A highly purified and bleached cellulose produced from wood pulp is used for making high-grade writing papers. Nearly pure cellulose is used in plastics or for carbonizing. It is a buff-colored, odorless powder or granular material with residual ash content of 1.6%. Some cellulose is obtained from potatoes as a by-product in the production of starch. It is pure white and is used in plastics. **Solka-Floc,** of Gretco, Inc., is 99.5% pure wood cellulose in the form of tough, white fibers 39 to 79 μin (1 to 2 μm) in diameter and 1,378 to 6,496 μin (35 to 165 μm) long, bulking 9 to 34 lb/ft^3 (144 to 544 kg/m^3). It is used as a filler for plastics requiring a fine surface finish and dimensional stability, such as buttons, knobs, trays, and vinyl floor tile. It is also used in welding rod coatings, in adhesives, and for cellulose chemicals. Water-soluble cellulose, or **cellulose gum,** used as a substitute for gum arabic and carob-bean flour as a stabilizer, thickener, or emulsifier, is **sodium cellulose glycollate,** or **sodium carboxymethyl cellulose,** in powder form. It is also used to increase the effectiveness of detergents. Water-soluble film is also made from this material. **Carbose,** of BASF Corp., and **Cellocel S,** of Dow Chemical Co., are sodium carboxymethyl cellulose. **Carboxymethyl cellulose** is used as a temporary binder for ceramic glazes. It burns out in the firing. A purified grade of this gum is used as a stabilizer in pharmaceuticals and low-acid foodstuffs. **Cellocel A** is **aluminum cellulose glycollate,** a water-soluble, brownish powder used for waterproofing paper. **Natrosol,** of Hercules, Inc., is **hydroxyethyl cellulose,** a white powder used for textile finishes and as a thickener for water-base paints. Hydroxyethyl cellulose, with a low degree of substitution of ethylene oxide in the molecular chain, is insoluble in water, but is alkali-soluble. It is used in paper coating to add gloss and water resistance. **Cellosize QP-4400,** of Union Carbide Corp., is a hydroxyethyl cellulose powder easily soluble in water but nongelling. It is used as a thickener in latex paints, inks, cosmetics, and pharmaceuticals. Alkali-soluble **cellulose ether** is marketed as a white fibrous powder. When dissolved in a water solution of caustic soda, it forms a viscous liquid used for sizing textiles. **Sodium cellulose sulfate** is a water-soluble granular powder used as a thickener in emulsion paints, foods, and cosmetics and for sizing paper and textiles. It produces a clear, tough, greaseproof coating. It is the sodium salt of cellulose acid sulfate produced by sulfuric acid treatment of wood pulp, with the sulfate groups in ester-type linkages on the cellulose chain.

Ethyl cellulose is a colorless, odorless ester of cellulose resulting from the reaction of ethyl chloride and cellulose. The specific gravity is 1.07 to 1.18. It is nonflammable, very flexible, stable to light, and forms durable alkali-resistant coatings. It is used as a thin wrapping material, for protective coatings, as a hardening agent in resins and waxes, and for molding plastics. **Ethyl cellulose plastics** are thermoplastic and are noted for their ease of molding, light weight, and good dielectric strength, 400 to 520 V/mil (15 to 20.5×10^6 V/m), and retention of flexibility over a wide range of temperature from -70 to $150°F$ (-57 to $66°C$), the softening point. They are the toughest and the lightest and have the lowest water absorption of the cellulosic plastics. But they are softer and lower in strength than cellulose-acetate plastics.

Ethocel is ethyl cellulose of Dow Chemical Co. Solutions of ethyl cellulose are used for dipping automotive and aircraft replacement parts or other metal products to form a thin, waterproof protective coating to prevent corrosion. The coating strips off easily when the part is to be used. The same material is marketed by a number of other companies for the same purpose under a variety of trade names. **Methyl cellulose** is a white, granular, flaky material, which is a strong emulsifying agent and is used in soaps, floor waxes, shoe cleaners; in emulsions of starches, glues, waxes, and fats; and as a substitute for gum arabic. It gives colorless, odorless solutions resistant to fermentation. It dissolves in cold water, but is stable to alkalies and dilute acids. In soaps it lowers the surface tension of the water and aids lathering. It is also used for tree-wound dressings and as a moisture-conserving soil conditioner. **Methocel HB,** of Dow Chemical Co., is a hydroxybutyl methyl cellulose for use in paint removers. **Cyanoethylated cellulose** is a white, fibrous solid used to produce thin transparent sheets for insulating capacitors and as carriers for luminescent phosphors. It has a high dielectric constant and low dissipation factor. A 0.002-in (0.051-mm) film has a tensile strength of 5,300 lb/in^2 (37 MPa) and is flexible.

CELLULOSE ACETATE. An amber-colored, transparent material made by the reaction of cellulose with acetic acid or acetic anhydride in the presence of sulfuric acid. In Germany it was made by treating beechwood pulp with acetic acid in the presence of an excess of zinc chloride. It is employed for lacquers and coatings, molding plastics, rayon, and photographic film. Cellulose acetate may be the triacetate $C_6H_7O_2(OOCCH_3)_3$, but may be the tetracetate or the pentacetate, or mixtures. It is made in different degrees of acetylation with varying properties. Unlike nitrocellulose, it is not flammable, and it has better light and heat stability. It has a refractive index of 1.47 to 1.50, and a sheet 0.125 in (0.32 cm) thick will transmit 90% of the light. The specific gravity is 1.27 to 1.37, Brinell hardness 8 to 15, tensile strength 3,500 to 8,000 lb/in^2 (24 to 55 MPa), compressive strength up to 20,000 lb/in^2 (138 MPa), elongation 15 to 80%, dielectric strength 300 to 600 V/mil (12×10^6 to 24×10^6 V/m), and softening point 122 to 205°F (50 to 96°C). It is thermoplastic and is easily molded. The molded parts or sheets are tough, easily machined, and resistant to oils and many chemicals. In coatings and lacquers, the material is adhesive, tough, and resilient, and it does not discolor easily. **Cellulose acetate fiber** for rayons can be made in fine filaments that are strong and flexible, nonflammable, mildewproof, and easily dyed. Standard cellulose acetate for molding is marketed in flake form. **Cellulose triacetate,** with 60 to 61.5% combined acetic acid, is more insoluble, has higher dielectric strength, and is more resistant to heat and light than other types. It is cast into sheets and is also used for resistant coatings and textile fibers. **Cellulose acetate film,** used for wrapping, is somewhat more lightweight than regenerated cellulose, giving 14,500 in^2/lb (20 m^2/kg) for the 0.0015-in (0.0381-mm) film. **Tenite** is a cellulose-acetate molding material of Eastman Chemical Products, Inc. **Estron** is a name adopted by this company to designate cellulose ester yarns and staple fiber. **Protectoid** is Lumarith in the form of nonflammable motion-picture film.

   Cellulose acetate lacquers are acetate in solvents with plasticizers and pigments. Cellulose propionate plastic for injection molding has high impact resistance, and requires less plasticizer than cellulose acetate. Cellulose acetate molding powder produces moldings with tensile strengths from 4,000 to 7,000 lb/in^2 (28 to 48 MPa) and elongations from 14 to 22%. **Avcocel,** used as a filler in plastics to increase the impact strength, is a by-product of cellulose acetate production. It contains 50% cellulose acetate and 50% white cotton.

Cellulose acetate butyrate is made by the esterification of cellulose with acetic acid and butyric acid in the presence of a catalyst. It is particularly valued for coatings, insulating types, varnishes, and lacquers. Commonly called **butyrate** or **CAB,** it is somewhat tougher and has lower moisture absorption and a higher softening point than acetate. Special formulations with good weathering characteristics plus transparency are used for outdoor applications such as signs, light globes, and lawn sprinklers. Clear sheets of butyrate are available for vacuum-forming applications. Other typical uses include transparent dial covers, television screen shields, tool handles, and typewriter keys. Extruded pipe is used for electric conduits, pneumatic tubing, and low-pressure waste lines.

Cellulose acetate propionate is similar to butyrate in both cost and properties. Some grades have slightly higher strength and modulus of elasticity. Propionate has better molding characteristics, but lower weatherability than butyrate. Molded parts include steering wheels, fuel filter bowls, and appliance housings. Transparent sheeting is used for blister packaging and food containers. Cellulose acetate butyrate also is used for cable coverings and coatings. It is more soluble than cellulose acetate and more miscible with gums. It forms durable and flexible films. A liquid cellulose acetate butyrate is used for glossy lacquers, chemical-resistant fabric coatings, and wire-screen windows. It contains 17% butyl with one hydroxyl group per four anhydroglucose units. It transmits ultraviolet light without yellowing or hazing and is weather-resistant.

CELLULOSE NITRATE. Materials made by treating cellulose with a mixture of nitric and sulfuric acids, washing free of acid, bleaching, stabilizing, and dehydrating. For sheets, rods, and tubes it is mixed with plasticizers and pigments and rolled or drawn to the shape desired. The cellulose molecule will unite with from one to six molecules of nitric acid. The lower nitrates are very inflammable, but they do not explode as the high nitrates do; and they are the ones used for plastics, rayons, and lacquers, although their use for clothing fabrics is restricted by law. The names **cellulose nitrate** and **pyroxylin** are used for the compounds of lower nitration, and the term **nitrocellulose** is used for the explosives. **Collodion** is a name given to the original solution of cellulose nitrate in a mixture of 60% ether and 40 alcohol for making fibers and film, and the name is still retained in pharmacy. The name **soluble cotton** is used to designate batches of cellulose nitrate wet with alcohol for storing for the production of lacquers, but the soluble cotton gauze, used for surgical dressings, is cotton oxidized with nitrogen dioxide.

Cellulose nitrate was first used as a plastic in England in 1855 under the name **Parkesine.** It consisted of nitrocellulose mixed with camphor and castor oil for hardening and making it nonexplosive. Later, in 1868, an improved cellulose nitrate and camphor plastic was called **Celluloid,** now the trade name of Hoechst Celanese for cellulose nitrate plastics. **Xylonite** was the name used in England for the nitrocellulose hardened with camphor made by Daniel Spill in 1868. Cellulose nitrate is the toughest of the thermoplastics. It has a specific gravity of 1.35 to 1.45, tensile strength of 6,000 to 7,500 lb/in^2 (41 to 52 MPa), elongation of 30 to 50%, compressive strength of 20,000 to 30,000 lb/in^2 + (138 to 207 MPa), Brinell hardness 8 to 11, and dielectric strength of 250 to 550 V/mil (9.9×10^6 to 21.7×10^6 V/m). The softening point is 160°F (71°C), and it is easy to mold and easy to machine. It also is readily dyed to any color. It is not light-stable and is therefore no longer used for laminated glass. It is resistant to many chemicals, but has the disadvantage that it is inflammable. The molding is limited to pressing from flat shapes. It burns with a smoky flame, and the fumes are poisonous. Methyl or amyl alcohols are the usual solvents for the material, and various plasticizers are used, some of which aid in reducing the flammability. Camphor is the usual hardener and plasticizer, from 24 to 30% being the usual amount.

CEMENT. A material, generally in powder form, that can be made into a paste usually by the addition of water and, when molded or poured, will set into a solid mass. Numerous organic compounds used for adhering, or fastening materials, are called cements, but these are classified as adhesives, and the term *cement* alone means a construction material. The most widely used of the construction cements is **portland cement.** It is a bluish-gray powder obtained by finely grinding the clinker made by strongly heating an intimate mixture of calcareous and argillaceous minerals. The chief raw material is a mixture of high-calcium limestone, known as **cement rock,** and clay or shale. Blast-furnace slag may also be used in some cements. United States specifications call for five types of portland cement. Type I, for general concrete construction, has a typical analysis of 63.2% CaO, 21.3 SiO_2, 6 Al_2O_3, 2.7 Fe_2O_3, 2.9 MgO, and 1.8 SO_3. Type III, for use where high early strength is required, has 64.3% CaO, 20.4 SiO_2, 5.9 Al_2O_3, 3.1 Fe_2O_3, 2 MgO, and 2.3 SO_3. The color of the cement is due chiefly to iron oxide. In the absence of impurities, the color would be white, but neither the color nor the specific gravity is a test of quality. The specific gravity is at least 3.10. Good cement is always ground fine, with 98.5% passing a 200-mesh screen.

    White cement is from pure calcite limestone, such as that found in eastern Pennsylvania. It is ground finer and used for a better class of work, but the physical properties are similar to those of ordinary cement. A typical analysis of white cement is 65% CaO, 25.5 SiO_2, 5.9 Al_2O_3, 0.6 Fe_2O_3, 1.1 MgO, and 0.1 SO_3. The white cements of France and England are made from the chalky limestones and have superior working qualities, as they are usually ground finer. White cement is also made from inferior iron-bearing limestone by treatment with fluorspar.

Aluminous cement, or **aluminate cement,** sometimes referred to as **high-speed cement,** will set to high strength in 24 h and is thus valued for laying roads or bank walls. It is made with bauxite and contains a high percentage of alumina. A typical analysis is 39.8% Al_2O_3, 33.5 CaO, 14.6 Fe_2O_3, 5.3 SiO_2, 1.3 MgO, and 0.1 SO_3. **Lumnite cement** is a cement of this type. **Accelerated cements** are intermediate cements that will set hard in about 3 days. The raw mixture for making portland cement is controlled to give exact proportions in the final product, and some quartz or iron ore may be added to balance the mix. The temperature of the rotary kiln is raised gradually to about 2650°F (1454°C). The burned clinker is then ground with a small amount of gypsum, which controls the set.

There are a number of other construction cements not classified as portland cement. **Natural cement** is made by heating to complete decarbonation, but not fusion, a highly argillaceous soft limestone. This is the most ancient of the manufactured cements, and it is still called **Roman cement.** It is low-cost and will set more quickly than portland cement, but is softer and weaker. It is sometimes called **hydraulic lime.** When used for laying brick and stone, it is called **masonry cement**; and ordinary **mortar** for laying brick is not this product, but is slaked lime and sand. **Cement mortar** is made with portland cement, sand, water, and sometimes lime to aid spreading.

Oxychloride cement, or **Sorel cement,** is composed of **magnesium chloride,** $MgCl_2$, and calcined magnesia. It is strong and hard and, with various fillers, is used for floors and stucco. **Magnesia cement** is magnesium oxide, prepared by heating the chloride or carbonate to redness. When mixed with water, it sets to a friable mass but of sufficient strength for covering steam pipes or furnaces. It is usually mixed with asbestos fibers to give strength and added heat resistance. The term *85% magnesia* means 85% magnesia cement and 15% asbestos fibers. The cement will withstand temperatures up to 600°F (316°C).

Keene's cement, also known as **flooring cement** and **tiling plaster,** is a **gypsum cement.** It is made by burning gypsum at about 1100°F (593°C), to drive off the chemically combined water, grinding to a fine powder, and adding alum to accelerate the set. It will keep better than ordinary gypsum cement, has high strength, is white, and takes a good polish. **Parian cement** is similar, except that borax is used instead of alum. **Martin's cement** is made with potassium carbonate instead of alum. These cements are also called **hard-finish plaster,** and they will set very hard and white. They are used for flooring and to imitate tiling. An ancient natural cement is **pozzuolana cement.** It is a volcanic material found near Pozzuoli, Italy, and in several other places in Europe. It is a volcanic lava modified by steam or gases so that it is powdery and has acquired hydraulic properties. The chief components are silica and alumina, and the color varies greatly, being white, yellow, brown, or black. It has been employed as a construction cement since ancient times. **Trass** is a similar material found in the Rhine district of Germany. **Santorin** is a light-gray volcanic ash with somewhat similar characteristics from the Greek island of Santorin. Artificial pozzuolana cements and trass cements are made in the United States by intergrinding pumicite, tufa, or shale with portland cement. **Slag cement** is made by grinding blast-furnace slag with portland cement. **Pozzolans** are siliceous materials which will combine with lime in the presence of water to form compounds having cementing properties. **Fly ash** is an artificial pozzolan composed principally of amorphous silica with varying amounts of the oxides of aluminum and iron and traces of other oxides. It is a fine, dark powder of spheroid particles produced as the by-product of combustion of pulverized coal, and collected at the base of the stack. As an admix, it improves the workability of concrete, and in large amounts its pozzolanic action adds to the compressive strength. A **fire-resistant cement,** developed by Arthur D. Little, Inc., is made of magnesium oxychlorides and magnesium oxysulfates. This inorganic resin foam cement contains 40 to 50% bond water that is released when the material is exposed to high temperatures and absorbs heat. It is said not to burn, smoke, or produce poisonous fumes when subjected to a direct flame.

CERAMICS. Ceramics, one of several major materials families, are crystalline compounds of metallic and nonmetallic elements. The ceramic family is large and varied, including such materials as refractories, glass, brick, cement and plaster, abrasives, sanitaryware, dinnerware, artware, porcelain enamel, ferroelectrics, ferrites, and dielectric insulators. There are other materials which, strictly speaking, are not ceramics, but which nevertheless are often included in this family. These are carbon and graphite, mica, and asbestos. Also intermetallic compounds, such as aluminides and beryllides, which are classified as metals, and cermets, which are mixtures of metals and ceramics, are usually thought of as ceramic materials because of similar physical characteristics to certain ceramics.

A broad range of metallic and nonmetallic elements are the primary ingredients in ceramic materials. Some of the common metals are aluminum, silicon, magnesium, beryllium, titanium, and boron. Nonmetallic elements with which they are commonly combined are oxygen, carbon, or nitrogen. Ceramics can be either simple, one-phase materials, composed of one compound, or multiphase, consisting of a combination of two or more compounds. Two of the most common are **single oxide ceramics,** such as alumina (Al_2O_3) and magnesia (MgO), and **mixed oxide ceramics,** such as cordierite (magnesia alumina silica) and forsterite (magnesia silica). Other newer ceramic compounds include borides, nitrides, carbides, and silicides. Macrostructurally there are essentially three types of ceramics: crystalline bodies with a glassy matrix; crystalline bodies, sometimes referred to as holocrystalline; and glasses.

The specific gravity of ceramics ranges roughly from 2 to 3. As a class, ceramics are low-tensile-strength, relatively brittle materials. A few have strengths above 25,000 lb/in^2 (172 MPa), but most have less than that. Ceramics are notable for the wide difference between their tensile and compressive strengths. They are normally much stronger under compressive loading than in tension. It is not unusual for compressive strength to be 5 to 10 times the tensile strength. Tensile strength varies considerably depending on composition and porosity.

One of the major distinguishing characteristics of ceramics, as compared to metals, is their almost total absence of ductility. They fail in a brittle fashion. Lack of ductility is also reflected in low impact strength, although impact strength depends to a large extent on the shape of the parts. Parts with thin or sharp edges or curves and with notches have considerably lower impact resistance than those with thick edges and gentler, curving contours.

Ceramics are the most rigid of all materials. A majority are stiffer than most metals, and the modulus of elasticity in tension of a number of types runs as high as 50×10^6 to 65×10^6 lb/in^2 (0.3×10^6 to 0.4×10^6 MPa) compared with 29×10^6 lb/in^2 (0.2×10^6 MPa) for steel. In general, they are considerably harder than most other materials, making them especially useful as wear-resistant parts and for abrasives and cutting tools.

Ceramics have the highest known melting points of materials. Hafnium and tantalum carbide, for example, have melting points slightly above 7000°F (3870°C), compared to 6200°F (3424°C) for tungsten. The more conventional ceramic types, such as alumina, melt at temperatures above 3500°F (1927°C), which is still considerably higher than the melting point of all commonly used metals. Thermal conductivities of ceramic materials fall between those of metals and polymers. However, thermal conductivity varies widely among ceramics. A 2-order-of-magnitude variation is possible between different types, or even between different grades of the same ceramic. Compared to metals and plastics, the thermal expansion of ceramics is relatively low, although as with thermal conductivity, it varies widely between different types and grades. Because the compressive strengths of ceramic materials are 5 to 10 times greater than the tensile strength, and because of relatively low heat conductivity, ceramics have fairly low thermal-shock resistance. However, in a number of ceramics, the low thermal expansion coefficient succeeds in counteracting to a considerable degree the effects of thermal conductivity and differences between tensile and compressive strengths.

Practically all ceramic materials have excellent chemical resistance, being relatively inert to all chemicals except hydrofluoric acid and, to some extent, hot caustic solutions. Organic solvents do not affect them. Their high surface hardness tends to prevent breakdown by abrasion, thereby retarding chemical attack. All technical ceramics will withstand prolonged heating at a minimum of 1830°F (999°C). Therefore atmospheres, gases, and chemicals cannot penetrate the material surface and produce internal reactions which normally are accelerated by heat. Paintable ceramic coatings, a specialty of Zyp Coatings, Inc., combine corrosive resistance with heat resistance to 2000°F (1093°C).

Unlike metals, ceramics have relatively few free electrons and therefore are essentially nonconductive and considered to be dielectric. In general, dielectrical strengths, which range between 200 and 350 V/mil (7.8 × 10^6 and 13.8 × 10^6 V/m), are lower than those of plastics. Electrical resistivity of many ceramics decreases rather than increases with an increase in impurities, and is markedly affected by temperature.

Larsenite, of Blasch Precision Ceramics, Inc., is a ceramic composite of alumina and silicon carbide. It is more resistant to thermal shock than alumina and resists oxidation at higher temperatures [over 3000°F (1649°C)] than the carbide. It is made by firing alumina and a particular grain size of silicon carbide, which then forms a lattice and improves the thermal shock resistance of the alumina. The composite has been used instead of fused silica for nozzles used in atomizing metals into powder. **Sulfide ceramics,** developed at Argonne National Laboratory, hold promise for effective bonding of difficult-to-join materials, such as ceramics to metals. Because they form at lower temperatures than traditional welds, joints are stronger and less brittle. Materials having coefficients of thermal expansion differing by as much as 200% have been joined. The ceramics are candidates for use in lithium-iron sulfide batteries being developed for battery-powered cars.

CERMETS. A composite material made up of ceramic particles (or grains) dispersed in a metal matrix. Particle size is greater than 39 μin (1 μm), and the volume fraction is over 25% and can go as high as 90%. Bonding between the constituents results from a small amount of mutual or partial solubility. Some systems, however, such as the metal oxides, exhibit poor bonding between phases and require additions to serve as bonding agents. Cermet parts are produced by powder-metallurgy (PM) techniques. They have a wide range of properties, depending on the composition and relative volumes of the metal and ceramic constituents. Some cermets are also produced by impregnating a porous ceramic structure with a metallic matrix binder. Cermets can be used in powder form as coatings. The powder mixture is sprayed through an acetylene flame and is fused to the base material.

Although a great variety of cermets have been produced on a small scale, only a few types have significant commercial use. These fall into two main groups: oxide-based and carbide-based cermets. The most common type of **oxide-based cermets** contains aluminum-oxide ceramic particles (ranging from 30 to 70% volume fraction) and a chromium or chromium-alloy matrix. In general, oxide-based cermets have a specific gravity of 4.5 to 9.0 and a tensile strength of 21,000 to 39,000 lb/in^2 (145 to 269 MPa). Modulus of elasticity ranges from 37 $\times$ 10^6 to 50 $\times$ 10^6 lb/in^2 (255,000 to 345,000 MPa) and the hardness is Rockwell A 70 to 90. The outstanding characteristic of oxide-based cermets is that the metal or ceramic can be either the particle or the matrix constituent. The 6 MgO–94 Cr cermets reverse the roles of the oxide and chromium; that is, MgO is added to improve the fabrication and performance of the chromium. Chromium is not ductile at room temperature. Adding MgO not only permits press forging at room temperature but also increases oxidation resistance to 5 times that of pure chromium. Of the cermets, the oxide-based alloys are probably the simplest to fabricate. Normal PM or ceramic techniques can be used to form shapes, but these materials can also be machined or forged. The oxide-based cermets are used for high-speed cutting tools for difficult-to-machine materials. Other uses include thermocouple-protection tubes, molten-metal-processing equipment parts, mechanical seals, gas-turbine flameholders (resistance to flame erosion), and flow control pins (because of chromium-alumina's resistance to wetting and erosion by many molten metals and to thermal shock).

There are three major groups of **carbide-based cermets:** tungsten, chromium, and titanium. Each of these groups is made up of a variety of compositional types or grades. **Tungsten-carbide cermets** contain up to about 30% cobalt as the matrix binder. They are the heaviest type of cermet (specific gravity is 11 to 15). Their outstanding properties include high rigidity, compressive strength, hardness, and abrasion resistance. Modulus of elasticity ranges between 65×10^6 to 95×10^6 lb/in^2 (448,000 to 655,000 MPa), and hardness is about Rockwell A 90. Structural uses of tungsten carbide–cobalt cermets include wire-drawing dies, precision rolls, gages, and valve parts. Higher-impact grades can be applied where die steels were formerly needed to withstand impact loading. Combined with superior abrasion resistance, the higher impact strength results in die-life improvements as high, in some cases, as 5,000 to 7,000%. Most **titanium-carbide cermets** have nickel or nickel alloys as the metallic matrix, which results in high-temperature resistance. They have relatively low density combined with high stiffness and strength at temperatures above 2200°F (1204°C). Typical properties are specific gravity, 5.5 to 7.3; tensile strength, 75,000 to 155,000 lb/in^2 (517 to 1,069 MPa); modulus of elasticity, 36×10^6 to 55×10^6 lb/in^2 (248,000 to 379,000 MPa); and Rockwell hardness A 70 to A 90. Typical uses are integral turbine wheels, hot-upsetting anvils, hot-spinning tools, thermocouple protection tubes, gas-turbine nozzle vanes and buckets, torch tips, hot-mill-roll guides, valves, and valve seats. **Chromium-carbide cermets** contain from 80 to 90% chromium carbide, with the balance being either nickel or nickel alloys. Tensile strength is about 35,000 lb/in^2 (241 MPa), the tensile modulus about 50×10^6 to 56×10^6 lb/in^2 (345,000 to 386,000 MPa), and hardness about Rockwell A 88. They have superior resistance to oxidation, excellent corrosion resistance, and relatively low density (specific gravity is 7.0). Their high rigidity and abrasion resistance make them suitable for gages, oil-well check valves, valve liners, spray nozzles, bearing seal rings, bearings, and pump rotors.

Other cermets are **barium-carbonate-nickel** and **tungsten-thoria,** which are used in higher-power pulse magnetrons. Some proprietary compositions are used as friction materials. In brake applications, they combine the thermal conductivity and toughness of metals with the hardness and refractory properties of ceramics. **Uranium-dioxide cermets** have been developed for use in nuclear reactors. Cermets play an important role in sandwich-plate fuel elements, and the finished element is a **siliconized silicon carbide** with a core containing uranium oxide. Control rods have been fabricated from boron carbide–stainless steel and rare-earth oxides–stainless steel. Other cermets developed for use in nuclear equipment include **chromium-alumina cermets, nickel-magnesia cermets,** and **iron-zirconium-carbide cermets.** Nonmagnetic compositions can be formulated for use where magnetic materials cannot be tolerated.

CESIUM.   Also spelled **caesium.** A rare metal, symbol Cs, obtained from the mineral pollucite, $2Cs_2O \cdot 2Al_2O_3 \cdot 9SiO_2 \cdot H_2O$, of southwest

Africa and Canada. The metal resembles rubidium and potassium, is silvery white and very soft. It oxidizes easily in the air, ignites at ordinary temperatures, and decomposes water with explosive violence. It can be contained in vacuum, inert gas, or anhydrous liquid hydrocarbons protected from oxygen and air. The specific gravity is 1.903, melting point 83.3°F (28.5°C), and boiling point 1238°F (670°C). It is used in low-voltage tubes to scavenge the last traces of air. It is usually marketed in the form of its compounds such as **cesium nitrate,** $CsNO_3$, **cesium fluoride,** CsF, or **cesium carbonate,** Cs_2CO_3. In the form of **cesium chloride,** CsCl, it is used on the filaments of radio tubes to increase sensitivity. It interacts with the thorium of the filament to produce positive ions. In photoelectric cells, cesium chloride is used for a photosensitive deposit on the cathode, since cesium releases its outer electron under the action of ordinary light, and its color sensitivity is higher than that of other alkali metals. The high-voltage rectifying tube for changing alternating current to direct current has cesium metal coated on the nickel cathode and has cesium vapor for current carrying. The cesium metal gives off a copious flow of electrons and is continuously renewed from the vapor. Cesium vapor is also used in the infrared signaling lamp, or **photophone,** as it gives infrared waves without visible light. **Cesium 137,** recovered from the waste of atomic plants, is a gamma-ray emitter with a half-life of 33 years. It is used in teletherapy, but the rays are not as penetrating as cobalt 60, and twice as much is required to produce equal effect.

CHALK. A fine-grained limestone, or a soft, earthy form of **calcium carbonate,** $CaCO_3$, composed of finely pulverized marine shells. The natural chalk comes largely from the southern coast of England and the north of France, but high-calcium marbles and limestones are the sources of most U.S. chalk and precipitated calcium carbonate. Chalk is employed in putty, crayons, paints, rubber goods, linoleum, calcimine, and as a mild abrasive in polishes. **Whiting** and **Paris white** are names given to grades of chalk that have been ground and washed for use in paints, inks, and putty. **French chalk** is a high grade of massive talc cut to shape and used for marking. Chalk should be white, but it may be colored gray or yellowish by impurities. The commercial grades depend on the purity, color, and fineness of the grains. The specific gravity may be as low as 1.8.

Precipitated calcium carbonate is the whitest of the pigment extenders. **Kalite,** of Diamond Alkali Co., is a precipitated calcium carbonate of 39-μin (1-μm) particle size, and **Suspenso, Surfex,** and **Nonferal** are grades with particle sizes from 197 to 394 μin (5 to 10 μm). **Whitecarb RC,** of Witco Corp., for rubber compounding, is a fine-grained grade, 2.56 μin (0.065 μm), coated to prevent dusting and for easy dispersion in the rubber. **Purecal SC** is a similar material. **Limeolith, Calcene,** of PPG Industries, and **Kalvan,** of R. T. Vanderbilt Co., Inc., are precipitated calcium carbonates. A highly purified calcium carbonate for use in medicine as an antacid is **Amitone.**

CHAMOIS. A soft, pliable leather originally made from the skins of the chamois, *Antilopa rupicapra,* a small deer inhabiting the mountains of Europe but now nearly extinct. The leather was a light-tan color, with a soft nap. All commercial chamois is now made from the skins of lamb, sheep, and goat or from the thin portion of split hides. The Federal Trade Commission limited the use of the term *chamois* to oil-dressed sheepskins mechanically sueded, but there are no technical precedents for such limitation. The original **artificial chamois** was made by tanning sheepskins with formaldehyde or alum, impregnating with oils, and subjecting to mechanical sueding; but chamois is also made by various special tannages with or without sueding. Those treated with fish oils have a distinctive feel. Chamois leather will withstand soaking in hot water and will not harden on drying. It is used for polishing glass and plated metals. **Buckskin,** a similar pliable leather, but heavier and harder, was originally soft-tanned, oil-treated deerskin, but is now made from goatskins.

CHARCOAL. An amorphous form of carbon, made by enclosing billets in a retort and exposing them to a red heat for 4 or 5 h. It is also made by covering large heaps of wood with earth and permitting them to burn slowly for about a month. Much charcoal is now produced as a by-product in the distillation of wood, a retort charge of 10 cords of wood yielding an average of 2,650 gal (10,030 L) of pyroligneous liquor, 11,000 lb (4,950 kg) of gas, and 6 tons (5.4 metric tons) of charcoal. **Wood charcoal** is used as a fuel, for making black gunpowder, for carbonizing steel, and for making activated charcoal for filtering and absorbent purposes. **Gunpowder charcoal** is made from alder, willow, or hazelwood. Commercial wood charcoal is usually about 25% of the original weight of the wood and is not pure carbon. The average composition is 95% carbon and 3 ash. It is an excellent fuel, burning with a glow at low temperatures and with a pale-blue flame at high temperatures. Until about 1850, it was used in blast furnaces for melting iron, and it produces a superior iron with less sulfur and phosphorus than when coke is used. **Red charcoal** is an impure charcoal made at a low temperature that retains much oxygen and hydrogen.

CHAULMOOGRA OIL. A brownish, semisolid oil from the seeds of the fruit of the tree *Taraktogenos kurzii* and other species of Thailand,

Assam, and Indonesia. It is used chiefly for skin diseases and for lep-
rosy. A similar oil is also obtained from other genera of bushes and
trees of the family *Flacourtiaceae*; and that obtained from some
species of *Hydnocarpus,* called **lukrabo oil** or **krabao oil,** is superior
to the true chaulmoogra oil. The tree *H. anthelminthica,* native to
Thailand, is cultivated in Hawaii. This oil consists mainly of chaulmoogric and hydnocarpic acids, which are notable for their optical
activity. **Sapucainha oil,** from the seeds of the tree *Carpotroche
brasiliensis,* of the Amazon Valley, contains chaulmoogric, hydnocarpic, and **gorlic acids** and is a superior oil. **Gorliseed oil,** from
the seeds of the tree *Onchoba echinata* of tropical Africa, and cultivated in Costa Rica and Puerto Rico, contains about 80% chaulmoogric
acid and 10 gorlic acid. **Dilo oil** is from the kernels of the nuts of the
tree *Calophyllum inophyllum* of the South Sea Islands. In Tahiti it is
called **tamanu.** The **chaulmoogric acids** are cyclopentenyl compounds, $(CH)_2(CH)_2CH(CH_2)_x COOH$, made easily from cyclopentyl
alcohol.

CHEESECLOTH.   A thin, coarse-woven cotton fabric of plain weave, 40
to 32 count, and of coarse yarns. It was originally used for wrapping
cheese, but is now employed for wrapping, lining, interlining, filtering, as a polishing cloth, and as a backing for lining and wrapping
papers. The cloth is not sized and may be either bleached or
unbleached. It comes usually 36 in (0.91 m) wide. The grade known as
beef cloth, originally used for wrapping meats, is also the preferred
grade for polishing enameled parts. It is made of No. 22 yarn or finer.
For covering meats the packing plants now use a heavily napped
knitted fabric known as **stockinett.** It is made either as a flat fabric
or in seamless tube form, and it is also used for covering inking and
oiling rolls in machinery. Lighter grades of cheesecloth, with very
open weave, known as **gauze,** are used for surgical dressings and for
backings for paper and maps. **Baling paper** is made by coating
cheesecloth with asphalt and pasting to one side of heavy kraft or
Manila paper. **Cable paper,** for wrapping cables, is sometimes made
in the same way but with insulating varnish instead of asphalt.
Buckram is a coarse, plain-woven open fabric similar to cheesecloth
but heavier and highly sized with water-resistant resins. It is usually
made of cotton, but may be of linen, and is white or in plain colors. It
is used as a stiffening material, for bookbindings, inner soles, and
interlinings. **Cotton bunting** is a thin, soft, flimsy fabric of finer
yarn and tighter weave than cheesecloth, used for flags, industrial
linings, and decorations. It is dyed in solid colors or printed. But usually the word **bunting** alone refers to a more durable, nonfading,
lightweight, worsted fabric in plain weave.

CHELATING AGENTS.   Also called **chelants** and used to capture undesirable metal ions in water solutions, increase color intensity in organic dyes, treat waters and organic acids, and preserve quality of food products and pharmaceuticals. In 1993, **ethylene diaminetetraacetic acid** (EDTA) was the principal industrial chelant in the United States, followed by **citric acid, phosphates, phosphonates, and gluconates.** Phosphates, the leading chelant in the late 1980s, have been severely restricted for environmental reasons, especially in household detergents. EDTA has been implicated for raising metal concentrations in rivers, by remobilizing metals in sludge, so that its position as the leading chelant may be threatened. **Citrates,** which are biodegradable, are being used increasingly as substitutes for phosphates in liquid laundry detergents, increasing the use of citric acid and its sodium and potassium salts. **Nitrilotrioacetic acid,** listed as a suspected carcinogen in the United States, is a biodegradable member of EDTA and has largely replaced phosphates in detergents in Canada. Phosphonates, used in industrial cleaners and in treating water, are comparable to EDTA in stability and capture a greater variety of metals than the calcium and magnesium for which phosphates are best suited. **Zeolites,** though not chelants, serve as phosphate substitutes in detergents but are not as effective in removing magnesium. **Polyelectrolytes,** lightweight polymers of **acrylic acid** and **maleic anhydride,** reduce scale formation by dispersing calcium as fine particles.

CHEMICAL INDICATORS.   Dyestuffs that have one color in acid solutions and a different color in basic or alkaline solutions. They are used to indicate the relative acidity of chemical solutions, as the different materials have different ranges of action on the acidity scale. The materials are mostly weak acids, but some are weak bases. The best known is **litmus,** which is red below a pH of 4.5 and blue above a pH of 8.3 and is used to test strong acids or alkalies. It is a natural dye prepared from several varieties of lichen, *Variolaria,* chiefly *Rocella tinctoria,* by alllowing them to ferment in the presence of ammonia and potassium carbonate. When fermented, the mass has a blue color and is mixed with chalk and made into tablets of papers. It is used also as a textile dye, wood stain, and food colorant. **Azolitmin,** $C_7H_7O_4N$, is the coloring matter of litmus and is a reddish-brown powder. **Orchil,** or **cudbear,** is a red dye from another species. **Alkanet,** also called **orcanette, anchusa,** or **alkanna,** is made from the root of the plant *Alkanna tinctoria* growing in the Mediterranean countries, Hungary, and western Asia. The coloring ingredient, **alkannin,** is soluble in alcohol, benzene, ether, and oils, and is produced in dry extract as a dark red, amorphous, slightly acid powder. It is also used for coloring fats and oils in pharmaceuticals and in cosmetics, for giving an even red color to wines, and for coloring wax.

Some coal-tar indicators are **malachite green,** which is yellow below a pH of 0.5 and green above 1.5; **phenolphthalein,** which is colorless below 8.3 and magenta above 10.0; and **methyl red,** which is red below 4.4 and yellow above 6.0. A **universal indicator** is a mixture of a number of indicators that gives the whole range of color changes, thereby indicating the entire pH range. But such indicators must be compared with a standard to determine the pH value.

The change in color is caused by a slight rearrangement of the atoms of the molecule. Some of the indicators, such as **thymol blue,** exhibit two color changes at different acidity ranges because of the presence of more than one chromophore arrangement of atoms. These can thus be used to indicate two separate ranges on the pH scale. **Curcumin,** a crystalline powder obtained by percolating hot acetone through turmeric, changes from yellow to red over the pH range of 7.5 to 8.5, and from red to orange over the range of 10.2 to 11.8. **Test papers** are strips of absorbent paper that have been saturated with an indicator and dried. They are used for testing for acidic or basic solutions, and not for accurate determination of acidity range or hydrogen-ion concentration, such as is possible with direct use of the indicators. **Alkannin paper,** also called **Boettger's paper,** is a white paper impregnated with an alcohol solution of alkanet. The paper is red, but it is turned to shades from green to blue by alkalies. **Litmus paper** is used for acidity testing. **Starch-iodide paper** is paper dipped in starch paste containing potassium iodide. It is used to test for halogens and oxidizing agents such as hydrogen peroxide.

CHERRY. The wood of several species of cherry trees native to Europe and the United States. It is brownish to light red, darkening on exposure, and has a close, even grain. The density is about 40 lb/ft^3 (641 kg/m^3). It retains its shape well and takes a fine polish. The annual cut of commercial cherry wood is small, but it is valued for instrument cases, patterns, paneling, and cabinetwork. **American cherry** is mostly from the tree *Prunus serotina,* known as the **black cherry,** although some is from the tree *P. emarginata.* The black cherry wood formerly used for airplane propellers has a specific gravity of 0.53 when oven-dried, compressive strength perpendicular to the grain of 1,170 lb/in^2 (8.1 MPa), and shear strength parallel to the grain of 1,180 lb/in^2 (8.1 MPa). This tree is thinly scattered throughout the eastern part of the United States. The wood is light to dark reddish with a beautiful luster and silky sheen, but has less figure than mahogany. **English cherry** is from the trees *P. cerasus* and *P. avium.*

CHESTNUT.   The wood of the tree *Castanea dentata,* which once grew plentifully along the Appalachian range from New Hampshire to Georgia, but is now very scarce. The trees grow to a large size, but the wood is inferior to oak in strength, though similar in appearance. It is more brittle than oak; has a coarse, open grain often of spiral growth; and splits easily in nailing. The color is light brown or yellowish. It was used for posts, crossties, veneers, and some mill products. The wood contains from 6 to 20% tannin, which is obtained by soaking the chipped wood in water and evaporating. **Chestnut extract** was valued for tanning leather, giving a light-colored strong leather. The seed nuts of all varieties of chestnut are used for food and are eaten fresh, boiled, or roasted. The **European chestnut,** *C. sativa* and *C. vesca,* also called the **Spanish chestnut** and the **Italian chestnut,** has large nuts of inferior flavor. The wood is also inferior. The **horse chestnut** is a smaller tree, *Aesculus hippocastanum,* grown as a shade tree in Europe and the United States. The nut is round and larger than the chestnut. It is bitter but is rich in fats and starch, and when the saponin is removed, it produces an edible meal with an almondlike flavor used in confections in Europe. The nuts of the American horse chestnut, **buckeye,** or **Ohio buckeye,** *A. glabra,* and the **yellow buckeye,** *A. octandra,* are poisonous. The trees grow in the central states, and the dense, white wood is used for furniture and artificial limbs.

CHICLE.   The coagulated latex obtained from incisions in the trunk of the evergreen tree *Achras zapota* and some other species of southern Mexico, Guatemala, and Honduras. The crude chicle is in reddish-brown pieces and may have up to 40% impurities. The purified and neutralized gum is an amorphous white to pinkish powder insoluble in water, which forms a sticky mass when heated. The commercial purified gum is molded into blocks of 22 to 26 lb (10 to 12 kg) for shipment. It contains about 40% resin, 17 rubber, and about 17 sugars and starches. Under the name of **txixtle,** the coagulated latex was mixed with asphalt and used as **chewing gum** by the Aztec Indians, and this custom of chewing gum has been widely adopted in the United States. Chicle is used chiefly as a base for chewing gum, sometimes diluted with gutta gums. For chewing it is compounded with polyvinyl acetate, microcrystalline wax, and flavors.

CHITIN.   A celluloselike polysaccharide, it holds together the shells of such crustaceans as shrimp, crab, and lobster; and it is also found in insects, mollusks, and even some mushrooms. It ranks after cellulose as nature's most abundant polymer. Deacylation of chitin, a poly-*N*-acetyl glucose amine, yields **chitosan,** a cationic electrolyte that finds occasional use as a replacement for some cellulosic materials. Chitosan may serve as a flocculant in wastewater treatment, thickener or extender in foods, coagulant for healing wounds in medicine, and coating for moistureproof films. Chitin is insoluble in most solvents, whereas chitosan, although insoluble in water, organic solvents, and solutions above pH 6.5, is soluble in most organic acids and dilute mineral acids. Since only 17 to 25% of the live weight of crustaceans is edible, the remaining shell consisting of calcium carbonate (40 to 55%), protein (25 to 40%), and chitin (5 to 35%) poses a disposal burden for seafood processors. Dried waste shells are ground and treated with a dilute alkaline solution to dissolve protein; the residue is reacted with hydrochloric acid to convert the calcium carbonate to calcium chloride brine and carbon dioxide. The remaining material, chitin, can be treated in a 40 to 50% caustic solution to remove actyl groups, to form chitosan. Yield is about 75%. Norway's Protan A/S is one of the principal manufacturers of chitosan. Canada's Nova Chem Ltd. produces a water-soluble form, **N,O-carboxymethyl chitosan** (NOCC), by reacting chitosan with monochloroacetic acid under alkaline conditions. Aqueous solutions of NOCC are used for coating fruits and vegetables, the coating acting as a barrier to limit the passage of oxygen into the product. For removing heavy metals from wastewater, Manville Corp. immobilizes bacteria on diatomaceous earth and then coats the complex with chitosan; the bacteria degrade organic material, and the chitosan absorbs heavy metals, such as nickel, zinc, chromium, and arsenic.

CHLORIDE OF LIME.   A white powder, a **calcium chloride hypochlorite,** of composition $CaCl(OCl)$, having a strong chlorite order. It decomposes easily in water and is used as a source of chlorine for cleaning and bleaching. It is produced by passing chlorine gas through slaked lime. Chloride of lime, or **chlorinated lime,** is also known as **bleaching powder,** although commercial bleaching powder may also be a mixture of calcium chloride and calcium hypochlorite, and the term **bleach** is used for many chlorinated compounds. The dry bleaches of the FMC Corp. are chlorinated **isocyanuric acids,** the CDB-85 being a fine white powder of composition $CINCO_3$, containing 88.5% available chlorine. **Perchloron,** of Pennsylvania Salt Mfg. Co., is **calcium hypochlorite,** $Ca(OCl)_2$, containing 70% available chlorine.

CHLORINATED HYDROCARBONS.   A large group of materials that have been used as solvents for oils and fats, for metal degreasing, dry cleaning of textiles, as refrigerants, in insecticides and fire extinguishers, and as foam-blowing agents. They are hydrocarbons in which hydrogen atoms were replaced by chlorine atoms. They range from the gaseous methyl chloride to the solid **hexachloroethane,** CCl_3CCl_3, with most of them liquid. The increase in the number of chlorine atoms increases the specific gravity, boiling point, and some other properties. They may be divided into four groups: the methane group, including methyl chloride, chloroform, and carbon tetrachloride; the ethylene group, including dichlorethylene; the ethane group, including ethyl chloride and dichlorethane; and the propane group. All these are toxic, and the fumes are injurious when breathed or absorbed through the skin. Some decompose in light and heat to form more toxic compounds. Some are very inflammable, while others do not support combustion. In general, they are corrosive to metals. Some have been implicated in the depletion of ozone in the stratosphere. For example, on a scale of 1.0 (high ozone depletion potential) to 0 (no such potential), **chlorofluorocarbon** CCl_3 **(CFC-11)** is rated 1.0, **hydrochlorofluorocarbon** $CHClF_2$ **(HCFC-22)** is rated 0.055, **hydrochlorofluorocarbon** $CHCl_2CF_3$ **(HCFC-123)** is rated 0.02, and **hydrochlorofluorocarbon** CCl_2FCH_3 **(HCFC-1416)** is 0.11.

Chloroform, or **trichloromethane** or **methenyl trichloride,** is a liquid of composition $CHCl_3$, boiling point 142.2°F (61.2°C), and specific gravity 1.489, used industrially as a solvent for greases and resins and in medicine as an anesthetic. It decomposes easily in the presence of light to form phosgene, and a small amount of ethyl alcohol is added to prevent decomposition. **Ethyl chloride,** also known as **monochlorethane, kelane,** and **chelene,** is a gas of composition CH_3CH_2Cl, used in making ethyl fuel for gasoline, as a local anesthetic in dentistry, as a catalyst in rubber and plastics processing, and as a refrigerator in household refrigerators. It is marketed compressed into cylinders as a colorless liquid. The specific gravity is 0.897, freezing point −221.4°F (−140°C), and boiling point 54.5°F (12.5°C). The condensing pressure in refrigerators is 12.4 lb (5.6 kg) at 6°F (−14°C), and the pressure of vaporization is 10.1 lb (4.6 kg) at 5°F (−15°C). Its disadvantage as a refrigerant is that it is highly inflammable, and there is no simple test for leaks. **Methyl chloride** is a gas of composition CH_3Cl, which is compressed into cylinders as a colorless liquid of boiling point −10.65°F (−23°C) and freezing point −144°F (−98°C). Methyl chloride is one of the simplest and cheapest chemicals for methylation. In water solution it is a good solvent. It is also used as a catalyst in rubber processing, as a restraining gas in high-heat thermometers, and as a refrigerant. **Monochlorobenzene,** C_6H_5Cl, is a colorless liquid boiling at 269.6°F (132°C), not soluble in water. It is used as a solvent for lacquers and resins, as a heat-transfer medium, and for making other chemicals. **Trichlor cumene,** or **isopropyl trichlorobenzene,** is valued as a hydraulic fluid and dielectric fluid because of its high dielectric strength, low solubility in water, and resistance to oxidation. It is a colorless liquid, $(CH_3)_2CHC_6H_2Cl_3$, boiling at 500°F (260°C) and freezing at -40°F (-40°C). **Halane,** used in processing textiles and paper, is dichlorodimethyl hydantoin, a white powder containing 66% available chlorine.

CHLORINATED POLYETHER. A high-priced, high-molecular-weight thermoplastic used chiefly in the manufacture of process equipment. Crystalline in structure, it is extremely resistant to thermal degradation at molding and extrusion temperatures. The plastic has resistance to more than 300 chemicals at temperatures up to 250°F (120°C) and higher, depending on environmental conditions.

Along with the mechanical capabilities and chemical resistance, chlorinated polyether has good dielectric properties. Loss factors are somewhat higher than those of polystyrenes, fluorocarbons, and polyethylenes, but are lower than many other thermoplastics. Dielectric strength is high, and electrical values show a high degree of consistency over a range of frequencies and temperatures.

The material is available as a molding powder for injection-molding and extrusion applications. It can also be obtained in stock shapes such as sheet, rods, tubes, or pipe, and blocks for use in lining tanks and other equipment, and for machining gears, plugs, etc. Rods, sheet, tubes, pipe, blocks, and wire coatings can be extruded on conventional equipment and by normal production techniques. Parts can be machined from blocks, rods, and tubes on conventional metalworking equipment.

Sheet can be used to convert carbon steel tanks to vessels capable of handling highly corrosive liquids at elevated temperatures. Using a conventional adhesive system and hot gas welding, sheet can be adhered to sandblasted metal surfaces.

Coatings of chlorinated polyether powder can be applied by several coating processes. Using the fluidized-bed process, pretreated, preheated metal parts are dipped in an air-suspended bed of finely divided powder to produce coatings, which after baking are tough, pinhole-free, and highly resistant to abrasion and chemical attack. Parts clad by this process are protected against corrosion both internally and externally.

CHLORINATED RUBBER. An ivory-colored or white powder produced by the reaction of chlorine and rubber. It contains about 67% by weight of rubber and is represented by the empirical formula $(C_{10}H_{13}Cl_7)_x$, although it is a mixture of two products, one having a CH_2 linkage instead of a CHCl. Chlorinated rubber is used in acid-resistant and corrosion-resistant paints, in adhesives, and in plastics.

The uncompounded film is brittle, and for paints chlorinated rubber is plasticized to produce a hard, tough, adhesive coating, resistant to oils, acids, and alkalies. The specific gravity of chlorinated rubber is 1.64 and bulking value 0.0735 gal/lb (0.045 L/kg). The tensile strength of the film is 4,500 lb/in² (31 MPa). It is soluble in hydrocarbons, carbon tetrachloride, and esters, but insoluble in water. The unplasticized material has a high dielectric strength, up to 2,300 V/mil (90.6 × 10⁶ V/m). **Pliofilm,** of Goodyear Tire & Rubber Co., is a rubber hydrochloride made by saturating the rubber molecule with hydrochloric acid. It is made into transparent sheet wrapping material which heat-seals at 221 to 266°F (105 to 130°C), or is used as a coating material for fabrics and paper. It gives a tough, flexible, water-resistant film. **Pliolite,** of this company, is a cyclized rubber made by highly chlorinating the rubber. It is used in insulating compounds, adhesives, and protective paints. It is soluble in hydrocarbons, but is resistant to acids and alkalies. **Pliowax** is this material compounded with paraffin or ceresin wax. **Pliolite S-1** is this material made from synthetic rubber. Resistant fibers have also been made from chlorinated rubbers. **Betacote 95** is a maintenance paint for chemical processing plants which is based on chlorinated rubber. It adheres to metals, cements, and wood and is rapid-drying; the coating is resistant to acids, alkalies, and solvents.

Cyclized rubber can be made by heating rubber with sulfonyl chloride or with **chlorostannic acid,** $H_2SnCl_6 \cdot 6H_2O$. It contains about 92% rubber hydrocarbons and has the long, straight chains of natural rubber joined with a larger, ring-shaped structure. The molecule is less saturated than ordinary natural rubber, and the material is tougher. It is thermoplastic, somewhat similar to gutta percha or balata, and makes a good adhesive. The specific gravity is 1.06, softening point 176 to 212°F (80 to 100°C), and tensile strength up to 4,500 lb/in² (31 MPa). It has been used in adhesives for bonding rubbers to metals and for waterproofing paper.

CHLORINE. An elementary material, symbol Cl, which at ordinary temperatures is a gas. It occurs in nature in great abundance in combinations, in such compounds as common salt. A yellowish-green gas, it has a powerful suffocating odor and is strongly corrosive to organic tissues and to metals. During World War I, it was used as a poison gas under the name **Bertholite.** An important use for liquid chlorine is for bleaching textiles and paper pulp, but it is also used for the manufacture of many chemicals. It is a primary raw material for chlorinated hydrocarbons and for such inorganic chemicals as titanium tetrachloride. Chlorine is used extensively for treating potable, process, and wastewaters. Its use as a biocide has declined due to toxicological and safety issues. A key issue is the chlorinated organics, such as **trichalomethanes** (THMs), that form when chlorine reacts with organics in water. One alternative to chlorine biocides for process waters is FMC Corp.'s **tetra alkyl phosphonium chloride,** a strong biocide containing a surface-active agent that cleans surfaces fouled by biofilm. Another is Dow Chemical's 2,2-dibromo-3-nitrilopropionamide (DBNPA). This nonoxidizing biocide remains active only for a few hours, quickly destroying unwanted constituents, then breaks down into naturally occurring products believed to be harmless. It is available in slow-release tablets in water-soluble bags for periodic addition to water. Use of chlorine in fluorocarbons also has decreased as chlorofluorocarbons have been replaced with non-ozonedepleting compounds. Its use in **chlorofluorocarbons,** such as CFC-11 and CFC-12, is decreasing, as these are replaced with more environmentally acceptable refrigerants. Chlorine's use in bleach also has declined. For bleaching, it has been widely employed in the form of compounds easily broken up. The other two oxides of chlorine are also unstable. **Chlorine monoxide,** or **hypochlorous anhydride,** Cl_2O, is a highly explosive gas. **Chlorine heptoxide,** or **perchloric anhydride,** Cl_2O_7, is an explosive liquid. The **chlorinating agents,** therefore, are largely limited to the more stable compounds. Dry chlorines are used in cleansing powders and for detinning steel, where the byproduct is tin tetrachloride.

Chlorine may be made by the electrolysis of common salt. The specific gravity of the gas is 3.214, or 2.486 times heavier than air. The boiling point is 28.5°F (-33.6°C), and the gas becomes liquid at atmospheric pressure at a temperature of -24.48°F (-31°C). The vapor pressure ranges from 39.4 lb (17.9 kg) at 32°F (0°C) to 602.4 lb (273.2 kg) at 212°F (100°C). The gas is an irritant and not a cumulative poison, but breathing large amounts destroys the tissues. Commercial chlorine is produced in making caustic soda, by treatment of salt with nitric acid, and as a by-product in the production of magnesium metal from seawater or brines. The chlorine yield is from 1.8 to 2.7 times the weight of the magnesium produced.

CHLOROPHYLL. A complex chemical which constitutes the green coloring matter of plants and the chief agent of their growth. It is obtained from the leaves and other parts of plants by solvent extraction and is used as food color and as a purifying agent. When extracted from alfalfa by hexane and acetone, 50 tons (45.4 metric ton) of alfalfa yields 400 lb (181 kg) of chlorophyll. A higher yield is obtained in California from the cull leaves of lettuce. It is one of the most interesting of chemicals and is a sunlight-capturing, food-making agent in plants. It has the empirical formula $C_{55}H_{72}O_5N_4Mg$, having a complex ring structure with **pyrrole,** $(CH:CH)_2NH$, as its chief building block and a single magnesium atom in the center. It is designated as a **magnesium-porphyrin** complex. The **iron-porphyrin** complex **hematin,** of blood, is the same structure with iron replacing magnesium. The **vanadium-porphyrin** complex of fishes and cold-blooded animals, found also in petroleum, is the same thing with vanadium replacing magnesium. Under the influence of sunlight and the pyrrole complex, carbon dioxide unites with water to produce formaldehyde and oxygen and enables plant and animal bodies to produce carbohydrates and proteins. Failure of the pyrrole ring to link up with NH, connecting with sulfur instead, completely suspends the functioning of the blood.

Chlorophyll is obtained as a crystalline powder soluble in alcohol and melting at 361°F (183°C). It combines with carbon dioxide of air to form formaldehyde which is active for either oxidation or reduction of impurities existent in the air, changing such gases to methanol, formic, acid, or carbonic acid. It is thus used in household air-purifying agents. In plants, some of the formaldehyde is given off to purify the air, but most of it is condensed in the plant to form **glycolic aldehyde,** $HOCH_2CHO$, the simplest carbohydrate, and **glyceric aldehyde,** another simple carbohydrate. Although chlorophyll is used as an odor-destroying agent in cosmetics and foods, its action when taken into the human body in quantity in its nascent state is not fully understood, and the magnesium in the complex is capable of replacing the iron in the blood complex.

The **porphyrins,** each having a nucleus of four pyrrole rings and a distinctive metal such as the magnesium of the chlorophyll of plants, are termed *pigments* in medicine, and the disease of unbalance of porphyrin in human blood is called *porphyria*. In addition to photosynthesis, they have catalytic and chelating actions and may be considered as the chief growth agents in plant and animal life. For example, the **zinc porphyrin** of the eye is formed in the liver, and the lack of supply to the fluid of the eye may cause loss of vision.

Pyrrole can be obtained from coal tar and from bone oil, or it can be made synthetically, and is used in the production of fine chemicals. **Pyrrolidine,** used as a stabilizer of acid materials and as a catalyst, is a water-soluble liquid, $(CH_2CH_2)_2NH$, made by the hydrogenation of pyrrole, or by treating tetrahydrofuran with ammonia. **Polyvinyl pyrrolidine,** $H_2C \cdot H_2C \cdot NH \cdot CH_2T \cdot CH_2$, is a cyclic secondary amine made from formaldehyde and acetylene. It is used as a supplementary blood plasma and for making fine chemicals. Small amounts are added to fruit beverages such as prune juice, as a color stabilizer. It combines with the phenols which cause the oxidation, and the combination can be filtered off.

CHROMIC ACID. A name given to the red, crystalline, strongly acid material of composition CrO_3 known also as **chromium trioxide** or as **chromic anhydride.** It is, in reality, not the acid until dissolved in water, forming a true chromic acid of composition H_2CrO_4. It is marketed in the form of porous lumps. The specific gravity is 2.70, melting point 385°F (196°C). It is produced by treating sodium or potassium dichromate with sulfuric acid. The dust is irritating and the fumes of the solutions are injurious to the nose and throat because the acid is a powerful oxidizing agent. Chromic acid is used in chromium-plating baths, for etching copper, in electric batteries, and in tanning leather. **Chromous chloride,** $CrCl_2$, is used as an oxygen absorbent and for chromizing steel. **Chromic chloride,** $CrCl_3$, is a volatile white powder used for tanning and as a mordant, for flame metallizing, and in alloying steel powders.

Chrome oxide green is a **chromic oxide** in the form of dry powder or ground in oil, used in paints and lacquers and for coloring rubber. It is a bright-green crystalline powder of composition Cr_2O_3, with specific gravity 5.20 and melting point 3614°F (1990°C), insoluble in water. The dry powder has a Cr_2O_3 content of 97% minimum and is 325 mesh. The paste contains 85% pigment and 15 linseed oil. Chrome oxide green is not as bright in color as chrome green but is more permanent.

CHROMITE. An ore of the metal chromium, called **chrome ore** when used as a refractory. It is found in the United States, chiefly in California and Oregon, but most of the commercial production is in South Africa, Zimbabwe, Cuba, Turkey, the Philippines, Greece, and New Caledonia. The theoretical composition is FeO[mdt]Cr_2O_3, with 68% chromic oxide, but pure **iron chromate** is rare. Part of the iron may be replaced by magnesium, and part of the chromium by aluminum. The silica present in the ore, however, is not a part of the molecule. Chromite is commonly massive granular, and the commercial ores contain only 35 to 60% chromic oxide. The hardness is Mohs 5.5, and the specific gravity 4.6. The color is iron black to brownish black, with a metallic luster. The melting point is about 3900°F (2149°C), but when it is mixed with binders as a refractory, the fusion point is lowered. New Caledonia ore has 50% chromic oxide, Turkish ore averages 48 to 53%, Brazilion ore runs 46 to 48%, and Cuban ore averages only 35%. The high-grade Guleman ore of Turkey contains 52% Cr_2O_3, 14 Al_2O_3, 10.4 FeO, 4.4 Fe_2O_3, 16 magnesia, and 2.5 silica. Most of the domestic ore in the United States is low-grade.

Cuban ore is rich in spinel and deficient in magnetite, and this type is adapted for refractory use even when the chromic oxide is low. Ore from Baluchistan is also valued for refractory use, as are other hard lumpy ores high in Al_2O_3 and low in iron. For chemical use the ores should have more than 45% chromic oxide and not more than 8 silica, and should be low in sulfur. Metallurgical ore should have not less than 49% chromic oxide, and the ratio of chromium to iron should not be less than 3:1. Chromite is used for the production of chromium and ferrochromium, in making **chromite bricks** and refractory linings for furnaces, and for the production of chromium salts and chemicals. For bricks the ground chromite is mixed with lime and clay and burned. Chromite refractories are neutral and are resistant to slag attack. A chrome-ore high-temperature cement marketed by General Refractories Co. under the name **Grefco** has a fusion point of 3400°F (1871°C).

CHROMIUM. An elementary metal, symbol Cr, used in stainless steels, heat-resistant alloys, high-strength alloy steels, electrical-resistance alloys, wear-resistant and decorative electroplating, and, in its compounds, for pigments, chemicals, and refractories. The specific gravity is 6.92, melting point 2750°F (1510°C), and boiling point 3992°F (2200°C). The color is silvery white with a bluish tinge. It is an extremely hard metal, the electrodeposited plates having a hardness of Mohs 9. It is resistant to oxidation, is inert to nitric acid, but dissolves in hydrochloric acid and slowly in sulfuric acid. At temperatures above 1500°F (816°C) it is subject to intergranular corrosion.

Chromium occurs in nature only in combination. Its chief ore is chromite, from which it is obtained by reduction and electrolysis. It is marketed for use principally in the form of master alloys with iron or copper. The term **chromium metal** usually indicates a pure grade of chromium containing 99% or more of chromium. A grade marketed by Sheldalloy Corp. has 99.25% minimum chromium, with 0.40 maximum iron and 0.15 maximum silicon. **High-carbon chromium** has 86% minimum chromium and 8 to 11% carbon with no more than 0.5% each of iron and silicon. **Isochrome** is a name given by Battelle Memorial Institute to chromium metal, 99.99% pure, made by the reduction of chromium iodide. Chromium metal lacks ductility and is susceptible to nitrogen embrittlement, and it is not used as a structural metal. **Chromium plating** is widely used where extreme hardness or resistance to corrosion is required. When plated on a highly polished metal, it gives a smooth surface that has no capillary attraction to water or oil, and chromium-plated bearing surfaces can be run without oil. For decorative purposes, chromium plates as thin as 0.0002 in (0.0006 cm) may be used; for wear resistance, plates up to 0.050 in (0.127 cm) are used. Increased hardness and wear resistance in the plate are obtained by alloying 1% molybdenum with the chromium. Ultrathin and dense electroplated chromium coatings, developed by the U.S. Air Force, improve the corrosion resistance and wear resistance of aircraft turbine bearings. **Alphatized steel,** also known as **chromized steel,** is steel coated with chromium by a diffusion process. The deposited chromium combines with the iron of the steel and forms an adherent alloy rather than a plate. Less penetration is obtained on high-carbon steels, but the coating is harder. **Securacoat GPX 9160,** of Securamax International of Canada, is a plasma-sprayed chromium oxide coating with high resistance to oxidation, corrosion, and wear. It is applied to stainless steel and titanium ball valves used in the separation of gold from sulfide ore slurries by autoclave processing in the mining industry.

CHROMIUM COPPER. A name applied to master alloys of copper with chromium used in the foundry for introducing chromium into nonferrous alloys or to **copper-chromium alloys,** or **chromium coppers,** which are high-copper alloys. A chromium-copper master alloy, **Electromet chromium copper,** contains 8 to 11% chromium, 88 to 90 copper, and a maximum of 1 iron and 0.50 silicon.

    Wrought chromium coppers are designated C18200, C18400, and C18500 and contain 0.4 to 1.0% chromium. C18200 also contains as much as 0.10% iron, 0.10 silicon, and 0.05 lead. C18400 contains as much as 0.15% iron and 0.10 silicon, and several other elements in small quantities. C18500 is iron-free and contains as much as 0.015% lead and several other elements in small quantities. Soft, thus ductile, in the solution-treated condition, these alloys are readily cold-worked and can be subsequently precipitation-hardened. Depending on such treatments, tensile properties range from 35,000 to 70,000 lb/in^2 (241 to 483 MPa) ultimate strength, 15,000 to 62,000 lb/in^2 (103 to 427 MPa) yield strength, and 15 to 42% elongation. Electrical conductivity ranges from 40 to 85% that of copper. Chromium coppers are used for resistance-welding electrodes, cable connectors, and electrical parts.

CHROMIUM-MOLYBDENUM STEEL. Any alloy steel containing chromium and molybdenum as key alloying elements. However, the term usually refers specifically to steels in the **AISI 41XX series,** which contain only 0.030 to 1.20% chromium and 0.08 to 0.35 molybdenum. Chromium imparts oxidation and corrosion resistance, hardenability, and high-temperature strength. Molybdenum also increases strength, controls hardenability, and reduces the tendency to temper embrittlement. **AISI 4130 steel,** which contains 0.30% carbon, and 4140 (0.40) are probably the most common and can provide tensile strengths well above 200,000 lb/in^2 (1,379 MPa). Many other steels have greater chromium and/or molybdenum content, including high-pressure boiler steels, most tool steels, and stainless steels. **Croloy 2,** of Babcock & Wilcox Co., used for boiler tubes for high-pressure superheated steam, contains 2% chromium and 0.50 maximum molybdenum and is for temperatures to 1150°F (621°C). **Croloy 5** has 5% chromium and 0.50 maximum molybdenum, for temperatures to 1200°F (649°C) and higher pressures. **Croloy 7** has 7% chromium and 0.50 molybdenum.

ASTM A 387 steels, used as plate for pressure vessels, include 10 grades based on chromium content. Five often used grades are Grade 5 (1% chromium), Grade 11 (1.25), Grade 12 (2.25), Grade 22 (5), and Grade 91 (9). Of these, Grades 11 and 12 are the most widely used. Grade 11 also contains 0.05 to 0.17% carbon, 0.40 to 0.65 manganese, 0.50 to 0.80 silicon, and 0.45 to 0.65 molybdenum. Grade 12 contains slightly less carbon, manganese, and silicon but 0.90 to 1.1% molybdenum. In recent years, toughness has been improved by changes in steelmaking practice to yield finer-grain steels with sulfur contents of less than 0.005%, and by calcium treatments for inclusion-shape control. Preheat and postheat treatments are required to preclude or minimize hydrogen embrittlement during welding. The steels are typically used at temperatures of 600 to 1100°F (315 to 595°C). Grade 91, which is used mainly for vessel components, contains 0.18 to 0.25% vanadium, 0.06 to 0.10 columbium, and 0.03 to 0.07 nitrogen. These additional elements enhance mechanical properties, including notch toughness. This steel is resistant to hydrogen embrittlement in welding and less susceptible than Grade 22 to stress-relief cracking.

CHROMIUM STEEL. Any steel containing chromium as the predominating alloying element may be termed chromium steel, but the name usually refers to the hard, wear-resisting steels that derive the property chiefly from the chromium content. **Straight chromium steels** are low-alloy steels in the **AISI 50XX, 51XX,** and **61XX** series. Chromium combines with the carbon of steel to form a hard chromium carbide, and it restricts graphitization. When other carbide-forming elements are present, double or complex carbides are formed. Chromium refines the structure, gives deep hardening, increases the elastic limit, and gives a slight red-hardness so that the steels retain their hardness at more elevated temperatures. Chromium steels have great resistance to wear. They also withstand quenching in oil or water without much deformation. Up to about 2% chromium may be included in tool steels to add hardness, wear resistance, and nondeforming qualities. When the chromium is high, the carbon may be much higher than in ordinary steels without making the steel brittle. Steels with 12 to 17% chromium and about 2.5 carbon have remarkable wear-resisting qualities and are used for cold-forming dies for hard metals, for broaches, and for rolls. However, chromium narrows the hardening range of steels unless it is balanced with nickel. Such steels also work-harden rapidly unless modified with other elements. The high-chromium steels are corrosion-resistant and heat-resistant but are not to be confused with the high-chromium stainless steels which are low in carbon, although the nonnickel 4XX stainless steels are very definitely chromium steels. Thus, the term is indefinite but may be restricted to the high-chromium steels used for dies, and to those with lower chromium used for wear-resistant parts such as ball bearings.

Chromium steels are not especially corrosion-resistant unless the chromium content is at least 4%. Plain chromium steels with more than 10% chromium are corrosion-resistant even at elevated temperatures and are in the class of stainless steels, but are difficult to weld because of the formation of hard, brittle martensite along the weld.

Chromium steels with about 1% chromium are used for gears, shafts, and bearings. One of the most widely used bearing steels is **AISI 52100,** which contains 1.3 to 1.6% chromium. Many other chromium steels have greater chromium content and, often, appreciable amounts of other alloying elements. They are used mainly for applications requiring corrosion, heat, and/or wear resistance.

CHROMIUM-VANADIUM STEEL. Alloy steel containing a small amount of chromium and vanadium, the latter having the effect of intensifying the action of the chromium and the manganese in the steel and controlling grain growth. It also aids in formation of carbides, hardening the alloy, and in increasing ductility by the deoxidizing effect. The amount of vanadium is usually 0.15 to 0.25%. These steels are valued where a combination of strength and ductility is desired. They resemble those with chromium alone, with the advantage of the homogenizing influence of the vanadium. A chromium-vanadium steel having 0.92% chromium, 0.20 vanadium, and 0.25 carbon has a tensile strength of 100,000 lb/in^2 (690 MPa), and when heat-treated, has a strength up to 150,000 lb/in^2 (1,034 MPa) and elongation 16%. Chromium-vanadium steels are used for such parts as crankshafts, propeller shafts, and locomotive frames. High-carbon chromium-vanadium steels are the mild-alloy tool steels of high strength, toughness, and fatigue resistance. The chromium content is usually about 0.80%, with 0.20 vanadium, and with carbon up to 1%.

Many high-alloy steels also contain some vanadium, but where the vanadium is used as a cleansing and toughening element and not to give the chief characteristics to the steel, these steels are not classified as chromium-vanadium steel.

A high-strength steel, developed by Sumitomo Metal Industries of Japan for boiler and heat-exchanger tubes, contains 2.25% chromium, 1.5 tungsten, 0.25 vanadium, 0.06 carbon, and 0.05 columbium. It is weldable without preheat or postheat and provides a stress rupture strength of 15,370 lb/in^2 (106 MPa) at 1067°F (575°C).

CINCHONA. The hard, thick, grayish bark of a number of species of evergreen trees of genus *Cinchona,* native to the Andes from Mexico to Peru but now grown in many tropical countries chiefly as a source of **quinine.** The small tree *Remijia pendunculata* also contains 3% quinine in the bark, and quinine occurs in small quantities in other plants and fruits, notably the grapefruit. **Cinchona bark** was originally used by the Quechua Indians of Peru in powdered form and was called **loxa bark.** It derives its present name from the fact that in 1630 the Countess of Cinchon was cured of the fever by its use. In Europe, it became known as **Peruvian bark** and **Jesuits' bark.** Quinine is one of the most important drugs as a specific for malaria and as a tonic. It is also used as a denaturant for alcohol, as it has an extremely bitter taste. Metallic salts of quinine are used in plastics to give fluorescence and glow under ultraviolet light. Quinine is a colorless crystalline alkaloid of composition $C_{20}H_{24}O_2N_2 \cdot 3H_2O$. It is soluble with difficulty in water and is marketed in the form of the more soluble **quinine sulfate,** a white powder of composition $(C_{20}H_{24}O_2N_2)H_2SO_4 \cdot 2H_2O$. **Quinine bisulfate** has the same composition but with seven molecules of water. During the Second World War **quinine hydrochloride** was preferred by the Navy. It contains 81.7% quinine compared with 74% in the sulfate and is more soluble in water but has a more bitter taste. **Synthetic quinine** can be made, but is more expensive. **Atabrine,** of I. G. Farbenindustrie, is **quinacrine hydrochloride.** It is not a complete substitute, is toxic, and is a dye that colors the skin when taken internally. **Primaquine,** of Winthrop-Stearns, Inc., is an 8-aminoquinoline, and as an antimalarial it is less toxic than other synthetics. In Germany, copper arsenite has been used as an effective substitute for quinine. The **maringin** of grapefruit is similar to quinine, and in tropical areas where grapefruit is consumed regularly, the incidence of malaria is rare.

The bark of the tree *C. ledgeriana* yields above 7% quinine, but it is not a robust tree and in cultivation is grafted on the tree *C. succirubra* which is hardy but yields only 2 to 3% quinine. Ledgeriana trees on plantations in Mindanao and in Peru yield as high as 13% total alkaloids from the bark. Most of the world production is from *C. officinalis* and *C. calisaya,* which are variations of *C. ledgeriana,* or **yellow bark.** The **red bark,** *C. succirubra,* is grown in India. The peak gathering of bark is 10 years after planting of the 2-year seedlings, and the trees are uprooted to obtain bark from both trunk and root. An 8-year-old tree yields 8.8 lb (4 kg) of bark, and a 25-year-old tree yields 44 lb (20 kg) but of inferior quality. The bark is dried and ground to powder for the solvent extraction of the alkaloids. Besides quinine the bark contains about 30 other alkaloids, chief of which are **cinchonidine, quinidine,** and **cinchonine. Totaquina** is the drug containing all the alkaloids. It is cheaper than extracted quinine, is effective against malaria, and is a better tonic. Quinidine has the same formula as quinine but is of right polarization instead of left. It is used for heart ailments. The **gluconate** and **polygalacturonate** are available for oral use. Cinchonine, $C_{19}H_{22}ON_2$, has right polarization and is 13 times more soluble in water than quinine sulfate. Cinchonidine has the same formula, but has left polarization. **Australian quinine,** or **alstonia,** is not true quinine. It is from **dita bark,** the bark of the tree *Alstonia scbolaris* of Australia, and is used as a febrifuge. It contains the water-soluble alkaloid **ditaine,** $C_{22}H_{28}O_4N_2$, and the water-insoluble alkaloid **ditamine,** $C_{16}H_{19}O_2N$. **Fagarine,** used as a substitute for quinidine for heart flutter, is extracted from the leaves of the tree *Fagara coco* of northern Argentina. **Chang shan,** used as an antimalarial in China, is the root of the plant *Dichroa febrifuga*. It contains the alkaloid **febrifugine.**

CINNABAR. The chief ore of the metal mercury. As a pigment it was originally called **minium,** a name now applied to red lead. It is a **mercuric sulfide,** HgS, which when pure, contains 86.2% mercury. The ores are usually poor, the best ones containing only about 7% mercury, and the average Italian ore having only 1.1% Hg and American ore yielding only 0.5%.

The chief production is in Italy, Spain, Mexico, and the United States. Cinnabar has a massive granular structure with a Mohs hardness of 2 to 2.5, a specific gravity of about 8, and usually a dull, earthy luster. It is brownish red, from which it derives the name **liver ore. Chinese cinnabar** is ground as a fine scarlet pigment for inks. Cinnabar is not smelted, the extraction process being one of distillation, made possible by the low boiling point of the metal. Another ore of mercury found in Mexico is **livingstonite,** $2Sb_2S_3 \cdot HgS$. It is a massive, red-streaked mineral of specific gravity 4.81 and hardness 2. **Calomel,** a minor ore in Spain, is a white crystalline mineral of composition Hg_2Cl_2 with a specific gravity of 6.5. It is also called **horn mercury.** It is used in medicine as a purgative, but is poisonous if retained in the system. The ore found in Colorado and known as **coloradoite** is a **mercuric telluride,** HgTe. It has an iron-black color and a specific gravity of 8. **Tiemanite,** found in California and Utah, is a **mercuric selenide,** HgSc, having a lead-gray color and a specific gravity of 8.2. There are more than 20 minerals classified as **mercury ores.**

CINNAMON. The thin, yellowish-brown, highly aromatic bark of the tropical evergreen laurel tree *Cinnamomum zeylanicum,* of Sri Lanka and southeast Asia. It is used as a spice and as a flavor in confectionery, perfumery, and medicine. The bark is marketed in rolls or sticks packed in bales of 112 lb (51 kg). **Cassia** is the bark from the *C. cassia* of South China and is less expensive than cinnamon. **Saigon cinnamon,** *C. loureirii,* is cinnamon, but is not as thin or as smooth a bark, and it does not have as fine an aroma and flavor. **Cassia buds** are small, dried flowers of the *C. cassia,* used ground as a spice or for the production of oil. They resemble cloves in appearance and have an agreeable, spicy odor and sweet, warm taste. **Cinnamon oil, cinnamon leaf oil,** and **cassia oil** are essential oils distilled, respectively, from the bark, leaf, and bud. They are used in flavoring, medicine, and perfumery. The bark contains between 0.5 and 3% cinnamon oil, which consists of about 70% **cinnamic aldehyde,** 8 to 10% cugenol, and also pinene and linalol. The specific gravity is 1.03, and the refractive index 1.565 to 1.582. The pale-yellow color darkens with age. Cinnamic aldehyde is also made synthetically. **Flasolee,** of J. Hilary Herchelroth Co., is **amyl cinnamic aldehyde,** redistilled to remove the unpleasant odor of heptyl aldehyde, for use in perfumes. The leaf oil is used as a substitute for clove oil. About 1.9% oil is obtained from cassia buds, but it lacks the delicate fragrance of cinnamon oil. **Nikkel oil,** a bright-yellow liquid with an odor of lemon and cinnamon, is distilled from the leaves and twigs of the tree *C. laureirii* of Japan. It contains citral and cincol and is used in perfumery. Some of the cinnamon marketed in the United States is **Padang cassia,** from the tree *C. burmannii* of Indonesia. It does not have the delicate aroma of true cinnamon.

CITRIC ACID. $C_6H_8O_7$, produced from lemons, limes, and pineapples, is a colorless, odorless, crystalline powder of specific gravity 1.66 and melting point 307°F (153°C). It is also produced by the fermentation of blackstrap molasses. It is used as an acidulent in effervescent salts in medicine, and in jams, jellies, and carbonated beverages in the food industry. **Acetyl tributyl citrate** is a vinyl resin plasticizer. It is also used in inks, etching, and as a resist in textile dyeing and printing. It is a good antioxidant and stabilizer for tallow and other fats and greases, but is poorly soluble in fats. **Tenox R,** of Eastman Chemical Products, Inc., a soluble antioxidant, consists of 20% citric acid, 60 propylene glycol, and 20 butylated hydroxyanisol. Citric acid is also used as a preservative in frozen fruits to prevent discoloration in storage. Its salt, **sodium citrate,** is a water-soluble crystalline powder used in soft drinks to give a nippy saline taste, and it is also used in plating baths. Citric acid is a strong chelant and finds use in regenerating ion-exchange resins, recovering metals in spent baths, decontaminating radioactive materials, and controlling metal-ion catalysis.

CLAD METALS. Two or more metals or alloys metallurgically bonded together to combine the characteristic properties of each in composite form. **Copper-clad steel,** for example, is used to combine the electrical and thermal characteristics of copper with the strength of steel. A great variety of metals and alloys can be combined in two or more layers, and they are available in many forms, including sheet, strip, plate, tubing, wire, and rivets for application in electrical and electronic products, chemical processing equipment, and decorative trim, including auto trim. **Clad strip** is probably the most common form, and it is available from most clad-metal producers: American Clad Metals, Art Wire/Doduco, Cook-Horton, Cyclops, Engelhard, GTE Metal Laminates, Handy & Harmon, Heraeus Volkert, Hood & Co., Olin, Pfizer, Polymetallurgical, Revere Copper & Brass, Technical Materials, Texas Instruments, and Scientific Metals. Revere is also a major sheet supplier; Du Pont and Lukens Steel are major **clad-plate** suppliers. **Clad wire** is produced by Cook-Horton, Copperweld Bimetallics, Engelhard, GTE, Handy & Harmon, and Texas Instruments. **Clad rivets** are made by Art Wire/Doduco and Gibson Electric, and **clad welding tapes** by American Clad Metals, Art Wire/Doduco, and Heraeus Volkert. **Clad-plate transition joints** are made by Du Pont.

Laminated metals were used very early in the jewelry and silverware industries, and silver-clad iron was made by the Gauls by brazing together sheets of silver and iron for lower-cost products as substitutes for the Roman heavy silver tableware. An early French **duplex metal** called **doublé,** for costume jewelry, had a thin facing of a noble metal on a brass or copper base, and **Efkabimetal** was a German name for this material. **Gold shell,** used for costume jewelry, is a duplex metal with gold rolled on a rich low brass. **Abyssinian gold, talmi gold,** and other names were used for these duplex metals in traders' jewelry. **Inter-Weld metal,** of American Silver Co., has a base metal of brass to which is silver-soldered a sheet of nickel over which is welded the gold sheet. When rolled, the gold is extremely thin, but the nickel prevents the color of the base metal from bleeding through.

Composite tool steel, used for shear blades and die parts, is not a laminated metal. The term refers to bar steel machined along the entire length and having an insert of tool steel welded to the backing of mild steel. **Clad steels** are available regularly in large sheets and plates. They are clad with nickel, stainless steel, Monel metal, aluminum, or special alloys, on one or both sides of the sheet. Where heat and pressure are used in the processing, there is chemical bonding between the metals. For some uses the cladding metal on one side will be 10 to 20% of the weight of the sheet. A composite plate having an 18–8 stainless-steel cladding to a thickness of 20% on one side saves 144 lb (65 kg) of chromium and 64 lb (29 kg) of nickel per 1,000 lb (454 kg) of total plate. The clads may also be extremely thin.

Many trade names have evolved over the years. **Pluramelt,** of Allegheny Ludlum Steel Corp., is composite steel with various types of stainless steels integrally bonded to a depth of 20% by a process of intermelting. **Ingaclad** consists of stainless steel bonded to carbon-steel plate. **Silver-Ply** is a **stainless-steel-clad steel** with the stainless 10 or 20% of the thickness of the plate, combined with the mild-steel backing by hot rolling. **Silver-clad sheet,** with silver rolled onto a cheaper nonferrous metal, is used for food processing equipment. It resists organic acids but not products containing sulfur. **Silver-clad steel,** used for bearings, shims, and reflectors, is rolled with pure silver bonded to a steel billet. The silver-clad stainless steel of American Cladmetals has the silver rolled onto one side for electrical conductivity. **Permaclad** has stainless steel bonded to one side of carbon steel. **SuVeneer steel** has a veneer of stainless steel bonded to spring steel. **Bronze-clad steel** is sheet steel with high-tensile-strength and corrosion-resistant bronze rolled on one or both sides. The cladding is from 0.031 in (0.079 cm) up to 40% of the thickness of the sheet. It is used for tanks and chemical equipment. **Hortonclad** has the stainless steel or other cladding joined to the steel baseplate by a process of heating the assembly of base metal, cladding metal, and brazing material together under vacuum. Since there is no rolling, the clad thickness is uniform, and there is no migration of carbon from the steel plate to the surface of the cladding.

Titanium-clad steel, of Lukens Steel Co., is produced by sandwich-pack rolling without the use of any interlayer foil between the plate and the cladding. An atmosphere of argon gas is used during the heating, and there are no impurities that normally make the titanium brittle. **Nickel-clad flange steel** is also produced by this company. **Niclad** has the nickel deposited on the steel by a continuous welding process. The duplex metal called **Bronze-on-steel** that is used for bearings is made by sintering a homogeneous alloy powder of 80% copper, 10 tin, and 10 lead, to strip steel in a hydrogen atmosphere, and then rolling the strip and forming it into bearings and bushings. **Nifer,** of Texas Instruments, is **nickel-clad steel** with the nickel bonded to both sides of a carbon steel, while **Alnifer** has nickel on one side and aluminum on the other. It comes in thin gages, up to 0.010 in (0.0254 cm) for electronic uses. The company's aluminum (5052 alloy)-clad steel (1008) can serve as an insert material to prevent galvanic corrosion, as when placed between aluminum and steel parts to be resistance-welded or otherwise joined.

Stainless-clad copper is copper sheet with stainless steel on both sides, used for making cooking utensils and food processing equipment. With stainless steel alone, heat remains localized and causes sticking and burning of foodstuffs. Copper has high heat conductivity, is corroded by some foods, and has an injurious catalytic action on milk products. Thus, the stainless-clad copper gives the conductivity of copper with the protection of stainless steel. The internal layer of copper also makes the metal easier to draw and form. **Rosslyn metal,** of American Clad Metals Co., is this material. **Ferrolum** is sheet steel clad with lead to give protection against sulfuric acid in tanks and chemical equipment. **Copper-clad steel** usually has a cladding of copper equal to 10% of the total thickness of the sheet on each side of a soft steel. But **Conflex,** of Texas Instruments, has the copper laminated to a hardenable carbon steel so that spring characteristics can be given by heat treatment of the finished parts. The electrical conductivity is 30% that of solid copper.

Brass-clad steel, used for making bullet jackets and shell cases, consists of 90–10 brass on one side of a low-carbon-steel sheet, with the brass equal to 20% of the weight of the sheet. **Bronco metal,** of Metals & Controls, is copper-strip-coated on both sides with 25% by weight of phosphor bronze. The bronze gives good resiliency for springs, and the material has an electrical conductivity 55% that of solid copper.

Coppered steel wire is produced by wet-drawing steel wire which has been immersed in a copper sulfate or copper-tin sulfate solution. The tin gives a brass finish or a white finish, depending on the proportion of tin. **Fernicklon,** of Kenmore Metals Corp., is nickel-coated wire for instrument use, made by nickel-plating steel or copper rod and then drawing into wire. **Copper-clad steel wire,** marketed by Copperweld Steel Co., for line wires, screens, and staples, has an electrical conductivity 40% that of an equal section of pure copper and a tensile strength 250% higher than that of copper. **Copperply wire,** of National-Standard Co., has either 5 or 10% by weight of copper electroplated on hard-drawn or annealed steel wire in 5 to 36 B&S gage. The conductivity of the 10% coated wire is 20% that of copper wire, or 23% when low-carbon soft wire is used. It is employed for electrical installations where high strength is needed. **Nickel-clad copper wire** is used where an electrical conductor is required to resist oxidation at high temperatures. It is made by inserting a copper rod into a nickel tube and drawing. **Kulgrid,** of GTE Corp., is a nickel-clad copper wire for lead-in wires. The cladding is 28% of the total weight, and the electrical conductivity is 70% that of solid copper. The tensile strength of the hard-drawn wire is 85,000 lb/in^2 (586 MPa), and it resists oxidation at high temperatures.

Feran, a German duplex metal, was made by passing strips of aluminum and iron through rolls at a temperature of 662°F (350°C) and then cold-rolling to sheet. **Alclad** is an aluminum-clad aluminum alloy, with the exposed pure aluminum giving added corrosion resistance and the aluminum-alloy base metal giving strength. The German **Lautal** with pure aluminum rolled on is called **Allautal.** **Zinnal** is a German aluminum sheet with tin cladding on both sides, while **Cupal** is a copper-clad aluminum sheet. **Copper-clad aluminum** is regularly available in sheet, strip, and tubing. The copper is rolled on to 5% of the total thickness on each side, or 10% on one side, with a minimum thickness of copper of 0.001 in (0.003 cm). It gives a metal with good working characteristics and high electrical and heat conductivity. **Alcuplate,** of Texas Instruments is aluminum with copper bonded to both sides, used for stamped and formed parts where good electrical conductivity and easy soldering in combination with light weight are desired. **Alsiplate,** of this company, has silver bonded to both sides of aluminum sheet. **Alfer,** of Texas Instruments is **aluminum-clad steel.** The aluminum cladding is 10% of the total thickness on each side. It comes in strips of thin gages. **Aliron,** of this company, is a five-ply metal in very thin gages for radio-tube anode plates. It has a core of copper amounting to 40% of the thickness, with a layer of iron and a layer of aluminum on both sides. The copper gives good heat dissipation, and the iron-aluminum compound formed when the metal is heated makes it highly emissive. **Aluminum-clad wire** for electric coils is copper wire coated with aluminum to prevent deterioration of the enamel insulation caused by copper oxide.

CLAY.   Naturally occurring sediment produced by chemical actions resulting during the weathering of rocks. Often clay is the general term used to identify all earths that form a paste with water and harden when heated. The **primary clays** are those located in their place of formation. **Secondary clays** are those that have been moved after formation to other locations by natural forces, such as water, wind, and ice. The U.S. Department of Agriculture distinguishes clay as having small grains, less than 0.00008 in (0.002 mm) in diameter, as distinct from **silt** with grains from 0.00008 to 0.002 in (0.002 to 0.05 mm). Most clays are composed chiefly of silica and alumina. Clays are used for making pottery, tiles, brick, and pipes, but more particularly the better grades of clays are used for pottery and molded articles not including the fireclays and fine porcelain clays. **Kaolins** are the purest forms of clay. The clayey mineral in all clays is kaolinite, or minerals closely allied, such as **anauxite,** $Al_2O_3 \cdot 3SiO_2 \cdot 2H_2O$, and **montmorillonite,** $Al_2O_3 \cdot 4SiO_2 \cdot 2H_2O$, the latter having an expanding lattice molecular structure which increases the bond strength of ceramic clays. When the aluminum silicates are in colloidal form, the material is theoretically true clay, or **clayite.** Some clays, however, derive much of their plasticity from colloids of organic material, and since all clays are of secondary origin from the weathering or decomposition of rocks, they may vary greatly in composition. Hardness of the clay depends on the texture as well as on the cohesion of the particles. Plasticity involves the ability of the clay to be molded when wet, to retain its shape when dry, and to have the strength to withstand handling in the green or unfired condition. The degrees of plasticity are called *fat, rich, rubbery,* and *waxy;* or the clays may be termed *very plastic,* which is waxy; *sticky plastic, medium plastic,* and *lean,* which is nonplastic. Clays that require a large amount of water for plasticity tend to warp when dried. Those that are not easily worked may be made plastic by **ceramic binders** such as alkaline starch solutions, ammonium alginate, or lignin. For making pressed or cast whiteware, methyl cellulose is used as binder for the clay. It gives good binding strength, and it fires out of the ceramic with an ash residue of only 0.5%.

Clays with as much as 1% iron burn red, and titanium increases this color. Yellow ochers contain iron as a free hydrate. Most clays contain quartz sand and sometimes powdered mica. **Calcareous clays** are known as **marls.** Pyrites burn to holes in the brick bordered by a ring of magnetic iron oxide, and a clay should be free of this mineral. Limestone grains in the clay burn to free lime which later slakes and splits the ceramic. Most of the common brick clays are complex mixed earths likely to have much undesirable matter that makes them unsuitable for good tile, pipe, or pottery. **Kingsley clay** of Georgia, used for artware, wall tile, dishes, and refractories, has only 0.4% iron oxide, 0.15 Na_2O, 0.1 K_2O, and 0.05 CaO. It contains about 45% silica, 40 alumina, and 1.15 titanium oxide. The **seito ware** of Japan is made with the **Gaerome clay** found near Nagoya. It is a granite with quartz particles, and when used with a high percentage of zirconium oxide, it produces ceramics of close density and brilliant whiteness. Alumina clay of western Idaho contains on a dry basis 28.7% alumina, 5.6 iron oxide, and a high percentage of titania.

CLOVES. The dried flower buds of the evergreen tree *Caryophillus aromaticus,* grown chiefly in Zanzibar, but also in Malagasy, East Africa, and Indonesia. The buds yield 15 to 19% of a pungent, yellowish essential oil, **clove oil,** also called **caryophil oil** and **amboyna.** It contains 85% **eugenol** and the terpene **clovene,** $C_{15}H_{24}$. Clove oil is used in medicine as an antiseptic, in toothpastes, in flavoring, and for the production of artificial vanilla. Eugenol is a viscous, phenol-type liquid. It is also the basis for carnation-type perfumes. **Clove buds** are chiefly valued as a highly aromatic spice. Lower-grade **Zanzibar cloves** containing only about 5% oil are used in the strootjes cigarettes of Indonesia, in a mixture of 75% tobacco and 25 cloves. Ground clove is also an efficient antioxidant and is sometimes used in lard and pork products. The clove tree attains a height up to 40 ft (12.2 m), bearing in 7 or 8 years, and continuing to bear for a century, yielding 8 to 10 lb (3.6 to 4.5 kg) of dried cloves annually. Clove stems are also aromatic, but contain only 5 to 6% oil of interior value. Clove was one of the most valued spices of medieval times. It grew originally only on five small islands, the Moluccas, in a volcanic-ash soil, and was carried by Chinese junks and Malayan outriggers to India from whence the Arabs controlled the trade, bringing the tree also to Zanzibar. The *Victoria* of Magellan's fleet returned to San Lucar with 26 tons (24 metric tons) of cloves, enough to pay for the loss of the other four ships and the expenses of the voyage around the world.

COAL. A general name for a black mineral formed of ancient vegetable matter, and employed as a fuel and for destructive distillation to obtain gas, coke, oils, and coal-tar chemicals. Coal is composed largely of carbon with smaller amounts of hydrogen, nitrogen, oxygen, and sulfur. It was formed in various geological ages and under varying conditions, and it occurs in several distinct forms. **Peat** is the first stage, followed by lignite, bituminous coal, and anthracite, with various intermediate grades. The mineral is widely distributed in many parts of the world. The value of coal for combustion purposes is judged by its fixed carbon content, volatile matter, and lack of ash. It is also graded by the size and percentage of lumps. The percentage of volatile matter declines from peat to anthracite, and the fixed carbon increases. A good grade of coal for industrial powerplant use should contain 55 to 60% fixed carbon and not exceed 8% ash. The heating value should be 13,500 to 14,000 Btu/lb (31,400 to 33,700 kJ/kg). Finely ground coal, or **powdered coal,** is used for burning in an air blast like oil, or it may be mixed with oil. Coal in its natural state absorbs large amounts of water and also, because of impurities and irregular sizes, is not so efficient a fuel as the **reconstructed coal** made by crushing and briquetting lignite or coal and waterproofing with a coating of pitch. **Anthracite powder** is used as a filler in plastics. **Carb-O-Fil,** of Shamokin Filler Co., is powdered anthracite in a range of particle sizes used as a carbonaceous filler. It has a plasticizing effect. It can also be used to replace carbon black in phenolic resins.

Low-sulfur coal burns cleaner than regular coal, but its heating value is much less so that it is uneconomical as a fuel. A conversion process developed by SGI International Inc., however, can raise the heating value of a 8,300 Btu/lb (19,000 kJ/kg) low-sulfur coal to about 12,000 Btu/lb (28,000 kJ/kg). The process involves crushing the coal, removing its moisture, drying, and pressurizing at 1000°F (538°C). Pressurizing at this temperature releases volatile gaseous material, which can be condensed to coal liquids and sold as industrial fuel.

Increasing amounts of coal are being used for the production of gas and chemicals. By the hydrogenation of coal much greater quantities of phenols, cresols, aniline, and nitrogen-bearing amines can be obtained than by means of by-product coking, and low grades of coal can be used. The finely crushed coal is slurred to a paste with oil, mixed with a catalyst, and reacted at high temperature and pressure. **Synthesis gas,** used for producing gasoline and chemicals, is essentially a mixture of carbon monoxide and hydrogen. It is made from low-grade coals. The pulverized coal is fed into a high-temperature reactor with steam and a deficiency of oxygen, and the gas produced contains 40% hydrogen, 40 carbon monoxide, 15 carbon dioxide, 1 methane, and 4 inert materials. It is made by passing steam through a bed of incandescent coke to form a **water gas** of about equal proportions of carbon monoxide and hydrogen. It is made from natural gas.

COATED FABRICS. The first coated fabric was a rubberized fabric produced in Scotland by Charles Macintosh in 1823 and known as **Mackintosh cloth** for rainwear use. The cloth was made by coating two layers of fabric with rubber dissolved in naphtha and pressing them together, making a double fabric impervious to water. **Rubberized fabrics** are made by coating fabrics, usually cotton, with compounded rubber and passing between rollers under pressure. The vulcanized coating may be no more than 0.003 in (0.008 cm) thick, and the resultant fabric is flexible and waterproof. But most coated fabrics are now made with synthetic rubbers or plastics, and the base fabric may be of synthetic fibers, or a thin plastic film may be laminated to the fabric.

Coated fabrics now have many uses in industrial applications, and the number of variations with different resins and backing materials is infinite. They are usually sold under trade names and are used for upholstery, linings, rainwear, bag covers, book covers, tarpaulins, outerwear, wall coverings, window shades, gaskets, and diaphragms. Vinyl-type resins are most commonly used, but for special purposes other resins are selected to give resistance to wear, oils, or chemicals.

The coated fabric of Reeves Bros., Inc., called **Reevecote,** for gaskets and diaphragms, is a Dacron fabric coated with Kel-F fluorocarbon resin. An industrial sheeting of Auburn Mfg. Co. is a cotton fabric coated with urethane rubber. It is tough, flexible, and fatigue-resistant, and it gives 10 times better wear resistance than natural rubber.

Vinyl-coated fabrics are usually tough and elastic and are low-cost, but unless specially compounded are not durable. Many plastics in the form of latex or emulsion are marketed especially for coating textiles. **Rhoplex WN-75** and **WN-80,** of Rohm & Haas Co., are water dispersions of acrylic resins for this purpose. Coatings cure at room temperature, have high heat and light stability, give softness and flexibility to the fabric, and withstand repeated dry cleaning. A water emulsion of a copolymer of vinyl pyrrolidone with ethyl acrylate forms an adherent, tough, and chemical-resistant coating. **Geon latex,** of Geon Co., is a water dispersion of polyvinyl chloride resin. Polyvinyl chloride of high molecular weight is resistant to staining, abrasion, and tearing and is used for upholstery fabrics. The base cloth may be of various weights from light sheetings to heavy ducks. They may be embossed with designs to imitate leather. The **Boltaflex cape vinyl,** of DiversiTech General, is a rayon fabric coated with a vinyl resin embossed with a leatherlike grain. It has the appearance, feel, and thickness of a split leather and, when desired, is impregnated with a leather odor.

One of the first upholstery fabrics to replace leather was **Fabrikoid,** of Du Pont. It was coated with a cellulose plastic and came in various weights, colors, and designs, especially for automobile seating and book covers. **Armalon** is twill or sateen fabric coated with ethylene plastic for upholstery. For some uses, such as for draperies or industrial fabrics, the fabric is not actually coated, but is impregnated, either in the fiber or in the finished cloth, to make it water-repellent, immune to insect attack, and easily cleaned. **Tontine,** of Stauffer Chemical Co., is a plastic-impregnated fabric for window shades. The **Fairprene fabrics,** also of Du Pont, are cotton fabrics coated with chloroprene rubber or other plastics. **Corfam,** of the same company, used as a leather substitute, is a nonwoven sheet of urethane fibers reinforced with polyester fibers, with a porous texture. The fabric can be impregnated or coated.

Terson voile, of Athol Mfg. Co., for umbrellas, rainwear, and industrial linings, is a sheer-weight rayon coated with a vinyl resin. It weighs 2 oz/yd^2 (0.07 kg/m^2). Coated fabrics may also be napped on the back, or coated on the back with a flock, to give a more resilient backing for upholstery.

Impregnated fabrics may have only a thin, almost undetectable surface coating on the fibers to make them water-repellent and immune to bacterial attack, or they may be treated with fungicides or with flame-resistant chemicals or waterproofing resins. **Stabilized fabrics,** however, are not waterproofed or coated, but are fabrics of cotton, linen, or wool that have been treated with a water solution of a urea formaldehyde or other thermosetting resin to give them greater resiliency with resistance to creasing and resistance to shrinking in washing. **Shrinkproof fabrics** are likewise not coated fabrics, but have a light impregnation of resin that usually remains only in the core of the fibers. The fabric retains its softness, texture, and appearance, but the fibers have increased stability. Various resin materials are marketed under trade names for creaseproofing and shrinkproofing fabrics, such as **Lanaset,** a methylomelamine resin of American Cyanamid Co., and **Synthrez,** a methylourea resin of Synthron, Inc.

Under the general name of **protective fabrics,** coated fabrics are now marketed by use characteristics rather than by coating designation since resin formulations vary greatly in quality. For example, the low-cost grades of vinyl resins may be hard and brittle at low temperatures and soft and rubbery in hot weather, and thus unsuitable for all-weather tarpaulins. Special weaves of fabric are used to give high tear strength with light weight, and the plastic may be impregnated, coated on one side or both, bonded with an adhesive or electronically bonded, or some combination of all these. Flame resistance and static-free qualities may also be needed. Many companies have complete lines to meet definite needs. The **Coverlight** fabrics of Reeves Bros., Inc., which come in many thicknesses and colors, are made with coatings of neoprene, Hypalon, or vinyl chloride resin, with weights from 6 to 22 oz/yd^2 (0.18 to 0.67 kg/m^2) and widths up to 72 in (1.8 m). The **H.T.V. Coverlight** is a high-tear-resistant nylon fabric with specially formulated vinyl coating. The 22-oz (0.62-kg) grade for such heavy-duty, all-weather uses as truck-trailer covers and concrete-curing covers remains flexible at temperatures down to $-50°F$ ($-46°C$).

COBALT AND COBALT ALLOYS. A white metal, Co, resembling nickel but with a bluish tinge instead of the yellow of nickel. It is rarer and costlier than nickel, and its price has varied widely in recent years. Although allied to nickel, it has distinctive differences. It is more active chemically than nickel. It is dissolved by dilute sulfuric, nitric, or hydrochloric acid and is attacked slowly by alkalies. The oxidation rate of pure cobalt is 25 times that of nickel. Its power of whitening copper alloys is inferior to that of nickel, but small amounts in nickel-copper alloys will neutralize the yellowish tinge of the nickel and make them whiter. The metal is diamagnetic like nickel, but has nearly 3 times the maximum permeability. Like tungsten, it imparts red-hardness to tool steels. It also hardens alloys to a greater extent than nickel, especially in the presence of carbon, and can form more chemical compounds in alloys than nickel.

Cobalt has a specific gravity of 8.756, a melting point of 2723°F (1495°C), Brinell hardness 85, and an electrical conductivity about 16% that of copper. The ultimate tensile strength of pure cast cobalt is 34,000 lb/in^2 (234 MPa), but with 0.25% carbon it is increased to 62,000 lb/in^2 (427 MPa). Strength can be increased slightly by annealing and appreciably by swaging or zone refining. The metal is used in tool-steel cutters, in magnet alloys, in high-permeability alloys, and as a catalyst; and its compounds are used as pigments and for producing many chemicals. The metal has two forms: a close-packed hexagonal crystal form, which is stable below 782°F (417°C), and a cubic form stable at higher temperatures to the melting point. Cobalt has valences of 2 and 3, while nickel has only a valence of 2.

The natural cobalt is **cobalt 59,** which is stable and nonradioactive, but the other isotopes from 54 to 64 are all radioactive, emitting beta and gamma rays. Most have very short life, except **cobalt 57** which has a half-life of 270 days, **cobalt 56** with a half-life of 80 days, and **cobalt 58** with a half-life of 72 days. **Cobalt 60,** with a half-life of 5.3 years, is used for radiographic inspection. It is also used for irradiating plastics and as a catalyst for the sulfonation of paraffin oils, since gamma rays cause the reaction of sulfur dioxide and liquid paraffin. Cobalt 60 emits gamma rays of 1.1- to 1.3-MeV energy, which gives high penetration for irradiation. The decay loss in a year is about 12%, the cobalt changing to nickel.

Cobalt metal is marketed in rondels, or small cast slugs, in shot and anodes, and as a powder. Powders with low nickel content for making cobalt salts and catalysts are in particle sizes down to 39 μin (1 μm). About one-quarter of the supply of cobalt is used in the form of oxides and salts for driers, ceramic frits, and pigments. **Cobalt carbonyls** are used for producing **cobalt powder** for use in powder metallurgy, as catalysts, and for producing cobalt chemicals. **Dicobalt octacarbonyl,** $Co_2(CO)_8$, or **cobalt tetracarbonyl,** is a brownish powder melting at 123°F (51°C) and decomposing at 140°F (60°C) to **tetracobalt dodecacarbonyl,** $(CoCO_3)_4$, a black powder which oxidizes in the air.

The best-known **cobalt alloys** are the **cobalt-base superalloys** used for aircraft-turbine parts. The desirable high-temperature properties of low creep, high stress-rupture strength, and high thermal-shock resistance are attributed to cobalt's allotropic change to a face-centered cubic structure at high temperatures. Besides containing 36 to 65% cobalt, usually more than 50%, most of these alloys also contain about 20 chromium for oxidation resistance and substantial amounts of nickel, tungsten, tantalum, molybdenum, iron and/or aluminum, and small amounts of still other ingredients. Carbon content is in the 0.05 to 1% range. These alloys include **L-605; S-816; V-36; WI-52; X-40; J-1650; Haynes 21** and **151; AiResist 13, 213,** and **215;** and **MAR-M 302, 322,** and **918.** Their 1,000-h stress-rupture strengths range from about 40,000 lb/in^2 (276 MPa) to 70,000 lb/in^2 (483 MPa) at 1200°F (649°C) and from about 4,000 lb/in^2 (28 MPa) to 15,000 lb/in^2 (103 MPa) at 1800°F (982°C). Cobalt is also an important alloying element in some nickel-base superalloys, other high-temperature alloys, and alloy steels. Besides tool steels, the **maraging steels** are a good example. Although cobalt-free grades have been developed, due to the scarcity of this metal at times, most maraging steels contain cobalt, as much as 12%. Cobalt is also a key element in **magnet steels,** increasing residual magnetism and coercive force, and in nonferrous-base magnetic alloys.

An important group of cobalt alloys is the **Stellites.** These alloys include the relatively low-carbon **Stellite 21** with 28% chromium, 5.5 molybdenum, 2.5 nickel, 2 iron, 2 silicon, 1 manganese, and 0.25 carbon; and **Stellite 306** with 25 chromium, 6 columbium, 5 nickel, 2 tungsten, and 0.4 carbon. There are also high-carbon (1 to 3.3) alloys **Stellite 1, 3, 6, 12, 190,** and **F,** which contain 25 to 31% chromium, 4 to 14.5 tungsten, 3 iron, 2.5 to 3 nickel (22 in Stellite F), 1 to 1.5 molybdenum, 1 to 1.4 manganese, and 0.7 to 2 silicon. Stellite 3 also has 0.1% boron. These alloys excel in resistance to abrasion, corrosion, and heat and are used for weld overlays, or hardfacings, and cast parts in the power-generating, steel-producing, chemical processing, and petroleum industries.

The interesting properties of cobalt-containing permanent, soft, and constant-permeability magnets are a result of the electronic configuration of cobalt and its high curie temperature. In addition, cobalt in well-known Alnico magnet alloys decreases grain size and increases coercive force and residual magnetism.

Cobalt is a significant element in many **glass-to-metal sealing alloys** and **low-expansion alloys.** One iron-base alloy containing 31% nickel and 5 cobalt provides a lower coefficient of thermal expansion than the iron–36% nickel alloy called **Invar** and is less sensitive to variations in heat treatment. **Cobalt-chromium alloys** are used in dental and surgical applications because they are not attacked by body fluids. Alloys named **Vitallium** are used as bone replacements and are ductile enough to permit anchoring of dentures on neighboring teeth. They contain about 65% cobalt. **CMM alloy,** of Carpenter Technology, is a wrought version of the cast **ASTM F75 cobalt alloy** and is used for surgical implants. It is a vacuum-melted and electroslag-remelted product containing 26% chromium, 6 molybdenum, 1 iron, 1 manganese, 1 nickel, 1 silicon, 0.5 copper, 0.5 tungsten, 0.18 nitrogen, 0.05 carbon, 0.015 phosphorus, and 0.015 sulfur.

Cobalt is a necessary material in human and animal metabolism, and it is used in fertilizers in the form of **cobaltous carbonate,** $CoCO_3$, in which form it is easily assimilated. This form occurs in nature in the mineral **cobalt spar** and is mixed with magnesium and iron carbonates. **Cobaltous citrate,** $Co(C_6H_5O_7) \cdot 2H_2O$, is a rose-red powder soluble in water, used in making pharmaceuticals. **Cobaltous fluorosilicate,** $CoSiF_6 \cdot H_2O$, is an orange-red, water-soluble powder used in toothpastes. It furnishes fluorine and silica as well as cobalt. **Cobaltous hydroxide,** $Co(OH)_2$, has a high cobalt content, 61.25%, is stable in storage, and is used for paint and ink driers and for making many other compounds. **Cobaltous chloride,** $CoCl_2$, a black powder, is an important cobalt chemical. It is used as a **humidity indicator** for silica gel and other desiccants. As the **desiccant** becomes spent, the blue of the cobaltous chloride changes to the pink color of the hexahydrate; but when the material is regenerated by heating to drive off the moisture, the blue reappears.

Cobalt metal may be obtained from the sulfur and arsenic ores by melting and then precipitating the cobaltous hydroxide powder which is high in cobalt, has high stability in storage, and is readily converted to the metal or the oxide or used directly for driers and other applications. The chief **cobalt ores** are cobalite and smaltite. **Cobalite,** or **cobalt glance,** from Ontario and Idaho, is a sulfarsenide, CoAsS, and occurs with **gersdorffite,** NiAsS. Another sulfide is **linnaeite,** Co_3S_4, containing theoretically 58% cobalt, but usually containing also nickel and iron. Cobalt is also found with pyrites as the mineral **bieberite,** which is **cobaltous sulfate,** $CoSO_4 \cdot 7H_2O$, but combined with iron sulfate. Some cobalt is extracted from the iron pyrites of Pennsylvania, the concentrated pyrite containing 1.41% cobalt, 42 iron, and 0.28 copper. **Erythrite** is a hydrous cobalt arsenate occurring in the smaltite deposits of Morocco. **Skutterudite** also occurs in Morocco. It is a silvery-gray, brittle mineral of composition $(CoNiFe)AS_3$, with a specific gravity of 6.5 and Mohs hardness of 6.

Asbolite, an important ore in Shaba and New Caledonia, is a soft mineral, hardness Mohs 2, consisting of varying mixtures of cobaltiferous manganese and iron oxides. A number of minerals classified as **heterogenite,** black and containing only cobalt and copper, occur in copper deposits, especially in Shaba. Among these are **mindigite,** $2Co_2O_3 \cdot CuO \cdot 3H_2O$, and **trieuite,** $2Co_2O \cdot CuO \cdot 6H_2O$. **Carrollite,** $CuS \cdot Co_2S_3$, a steel-gray mineral with a specific gravity of 4.85 and hardness of 5.5, is an important ore in Zimbabwe. The copper ores of Zaire and Zimbabwe form one of the chief sources of commercial cobalt. Some of the metal is exported as **white alloy,** containing 40% cobalt, 9 copper, and the balance iron. Cobalt occurs naturally in many minerals, and the metal may be considered as a by-product of other mining. Small quantities are produced regularly as a by-product of zinc production in Australia, although the cobalt content of the concentrate is only 0.015%. Some cobalt is obtained from the lead and zinc ores of Missouri. Its relative scarcity is a matter of cost of extraction.

High-purity cobalt can be produced from lower-grade cobalt, such as that containing copper, iron, and zinc impurities, by an electrolytic process developed by the U.S. Bureau of Mines. The lower-grade cobalt is dissolved at the anode, generating a cobalt-chloride anolyte, while the high-purity metal plates out at the cathode. An ionic double membrane in the cell allows only chloride ions to migrate to the cathode. The anolyte is continuously removed, impurities are separated by cementation and solvent extraction, and the purified solution flows to the cathode side of the cell. The process is aimed at upgrading lower-grade material in the U.S. stockpile to **Grade A cobalt,** which is at least 99.85% pure.

COBALT OXIDE. A steel-gray to blue-black powder employed as a base pigment for ceramic glazes on metal, as a colorant for glass, and as a chemical catalyst. It gives excellent adhesion to metals and is valued as an undercoat for vitreous enamels. It is the most stable blue, as it is not changed by ordinary oxidizing or reducing conditions. It is also one of the most powerful colorants for glass, 1 part in 20,000 parts of a batch giving a distinct blue color. Cobalt oxide is produced from the cobalt-nickel and pyrite ores, and the commercial oxide may be a mixture of the three oxides. **Cobaltous oxide,** CoO, is called **gray cobalt oxide** but varies from greenish to reddish. It is the easiest to reduce to the metal, and it reacts easily with silica and alumina in ceramics. **Cobaltic oxide,** Co_2O_3, occurs in the mixture only as the unstable hydrate, and it changes to the stable **black cobalt oxide,** or **cobalto-cobaltic oxide,** Co_3O_4 on heating. Above about 1652°F (900°C) this oxide loses oxygen to form cobaltous oxide.

Cobalt dioxide, CoO_2, does not occur alone, but the dioxide is stable in combination with other metals. The blue-black powder called **lithium cobaltite,** $LiCoO_2$, is used in ceramic frits to conserve cobalt, since the lithium adds fluxing and adherent properties. The pigment known as **smalt,** and as **royal blue** and **Saxon blue,** is a deep-blue powder made by fusing cobalt oxide with silica and potassium carbonate. It contains 65 to 71% silica, 16 to 21 potash, 6 to 7 cobalt oxide, and a little alumina. It is used for coloring glass and for vitreous enameled signs, but does not give good covering power as a paint pigment. **Thenaud's blue** is made by heating together cobalt oxide and aluminum oxide. **Rinmann's green** is made by heating together cobalt oxide and zinc oxide.

COCAINE.   An alkaloid derived from the leaves of the **coca** shrub. It is used as a local anesthetic and as a narcotic. It is habit-forming. In small and moderate doses it is stimulating and increases physical energy. Depression usually follows. Continued heavy use of cocaine has debilitating effects on the nervous system and can lead to insanity. Cocaine crystallizes from alcohol and is readily soluble in ordinary solvents except water. In the manufacture of cocaine, the alkaloids of coca leaves are hydrolyzed to **ecgonine.**

COCHINEAL.   A dyestuff of animal origin, which before the advent of coal-tar dyes was one of the most important coloring materials. Cochineal is the female of the *Coccus cacti,* an insect that feeds on various species of cactus, *Nopalea coccinellifera,* of Mexico. The insects have no wings, and at the egg-laying season they are brushed off the plants, killed by boiling, and dried; or they are bagged in linen and dried in an oven, preserving a peculiar white down covering the insect. They are dark reddish brown. Cochineal contains 10 to 20% pure coloring matter, carminic acid, mostly in the eggs, from which the **carmine red,** $C_{11}H_{12}O_7$, is obtained by boiling with mineral acid. Carmine red produces brilliant lake colors of various hues with different metals. Commercial cochineal may be adulterated with starch, kaolin, red lead, or chrome lead. The brilliant red pigment known as **carmine lake** is made by precipitating a mixture of cochineal and alum, and a fiery scarlet is obtained by treating with stannous and stannic chlorides. Salmonella-free cochineal in water solution is now used in foods to give a reddish-purple color. A species of cochineal insect that feeds on the leaves of the **tamarisk** tree, *Tamarix manifera,* produces **manna,** a viscous, white, sweet substance composed mostly of sugars. It forms in small balls and falls usually in May to July. When dry, it is hard and stable and is a good food. It is native to the Near East.

COCOA BEANS.   The seed beans from the large fruit pods of the **cacao tree,** *Theobroma cacao,* native to Mexico, and *T. leiocarpum,* native to Brazil. The tree was cultivated in Mexico from ancient times, and the beans were used by the Aztecs to produce a beverage called **choclatl** which contained the whole substance of the fermented and roasted bean flavored with vanilla. Cocoa beans are now produced in many countries, and the United States imports them from about 40 countries. Ghana, Nigeria, and Brazil are noted producers.

The flavor and aroma vary with soil and climate, and differences in curing methods also produce differences in the beans, so that types and grades are best known by the shipping ports and districts in which they grow. **Mico coca** is wild cocoa of Central America. The beans are smaller and are noted for fine flavor. Cocoa beans are shipped dried but not roasted. They are roasted just before use to develop the flavor, to increase the fat content, and to decrease the tannin content. The hard shells are removed, and the roasted seeds are ground and pressed to produce **bitter chocolate,** generally known as **chocolate liquor. Sweet chocolate** is made by adding sugar and flavoring, usually vanilla. **Cocoa,** for beverage purposes, is made by removing about 60% of the fatty oil from chocolate by hydraulic pressing and powdering the residue, to which is usually added ground cocoa shells. The removed fatty oil is **cocoa butter,** used for bakery products, cosmetics, and pharmaceuticals. A hundred pounds of cocoa beans yields 48 lb (21.8 kg) of **chocolate powder,** 32 lb (14.5 kg) of cocoa butter, and 20 lb (9.1 kg) of waste. Also an artificial cocoa butter is made by fractionating palm kernel oil. **Pakena,** a substitute cocoa butter, contains 53% lauric acid, 21.5 myristic, 12 palmitic, 8 oleic, 3.5 stearic, and 2 capric acids. Besides fat, chocolate contains much starch and protein and has high food value, but is not as stimulating as the cocoa since the alkaloid is largely contained in the waste and shells. These contain 1 to 1.5% theobromine and are used for the synthetic production of caffeine. The chocolate is used in the manufacture of confectionery, chocolate bars, bakery products, and flavoring syrups. **Microfine cocoa,** used for bakery products, is ground to 325 mesh and contains from 9 to 16% cocoa butter. **Postonal** is a German substitute for cocoa butter for pharmaceuticals. It is a polymerized ethylene oxide containing chemically combined castor oil.

Cocoa powder, used in the United States for beverages and for adding chocolate flavor to foodstuffs, as distinct from the sweet chocolate used in Latin countries for beverages, was originally made from the shells, but is now made from the residue cake after extraction of the chocolate liquor and the pressing out of the cocoa butter. It is widely used as a flavor for cakes and confectioneries. Sugar makes the powder easily soluble in water; **instant cocoa** is cocoa powder processed with about 70% sugar and sometimes with nonfat milk powder. The fat content of commercial cocoa powders ranges from 6 to 22% with a color range from light brown to reddish black. **Breakfast cocoa** is the high-fat grade. Cocoa powder is usually acidic with the pH as low as 3.3, but **Dutch cocoa,** for nonacid foods, is **stabilized cocoa** with the pH raised to as high as 9.0 by treatment with solutions of sodium or potassium carbonate.

COCOBOLA. The wood of the hardwood tree *Dalbergia retusa,* of Central America, also known as **Honduras rosewood.** It is a beautiful wood, extremely hard, and very heavy with a density of 75 to 85 lb/ft^3 (1,202 to 1,362 kg/m^3). It has orange and red bands with dark streaks and takes a fine polish. The thick sapwood is hewn off before shipment, and the heartwood logs are usually not more than 18 in (45.7 cm) in diameter. The wood is used for canes, turnery, inlaying, scientific-instrument cases, and knife handles. **Cocos wood,** also called **cocoawood** and **West Indian ebony,** used chiefly for inlaying, is from the tree *Brya ebenus* of tropical America. The sapwood is light yellow, and the heartwood is brown, streaked with yellow. The grain is dense and even, and the wood is hard and tough.

COCONUT OIL. The oil obtained from the thick kernel or meat adhering to the inside of the shell of the large nuts of the palm tree *Cocos nucifera,* growing along the coasts of tropical countries. The tree requires salt air, and inland trees do not bear fruit unless supplied with salt. The name *coco* is the Carib word for palm. **Copra** is the dried meat of the coconut from which the oil is pressed and alkali refined and bleached. Dried copra contains 60 to 65% oil. It is an excellent food oil and is valued as a shortening for crackers, but its use for margarine has declined. It is also valued for soaps because of its high lathering qualities due to the large percentage of lauric and myristic acids, although these acids are irritating to some skins. It is also employed as a source of lauric acid, but lauryl alcohol is now made synthetically. Coconut oil was once the chief illuminating oil in India, and the oil for burning was exported under the name **Cochin oil.** This oil was cold-pressed and filtered and was water-clear. Coconut oil has a melting point of 81 to 90°F (27 to 32°C), specific gravity 0.926, saponification value 251 to 263, and iodine value 8 to 9.6. It contains 45 to 48% lauric acid, 17 to 20 myristic, 10 capric, 5 to 7 palmitic, up to 5 stearic, and some oleic, caprylic, and caproic acids.

In sun-drying coconut meat to make copra, there is a loss of some of the sugars and other carbohydrates, and some proteins. The oil from copra contains more free fatty acid than that from fresh-dried coconut and is rancid, requiring neutralization, decolorization, and deodorization. The meal and cake are also dirty and rancid but are useful for animal feed or fertilizer. **Dehydrated coconut** meat gives a better yield of oil and is not rancid. The **copra cake** of India is called **poonac.** The chief production of copra and coconut oil is in southern Asia, Indonesia, the Philippines, and in the South Sea Islands. About 5,000 coconuts are required to produce 1 metric ton of copra, and the average yield of crude oil is 63%. The stearine separated from crude coconut oil by the process of wintering, to remove the more-liquid glycerides, is known as **coconut butter** and is used in confectionery. It has a melting point of 81 to 90°F (27 to 32°C) and saponification value of 250 to 260. Hydrogenated coconut oil is a soft solid with a melting point of 113°F (45°C). **Desiccated coconut,** produced by oven-drying or dehydration of the fresh coconut meat, is used shredded as a food and also powdered in many bakery products as a food and stabilizer. It has high food value, containing not less than 60% oil, 15 carbohydrates, 14 cellulose, 6 to 7 protein, various mineral salts, and considerable vitamin B. It is easily digested and has antitubercular value, but its characteristic coconut flavor is not universally liked and its use is largely confined to confections.

COFFEE. The seed berries, or beans, of the **Arabian coffee** tree, *Coffea arabica,* the **Liberian coffee,** *C. liberica,* and the **Congo coffee,** *C. robusta,* of which the first species furnishes most of the commercial product. The coffee bean contains the alkaloid caffeine used in medicine as a stimulant and in soft drinks, but most of the commercial coffee beans are used for the preparation of the beverage coffee, with small quantities for flavoring. The alkaloid is stimulating and is harmless in small amounts as it does not break down in the system and is easily soluble in water and thus carried off rapidly; but in large quantities at one time it is highly toxic. Coffee contains niacin, and rubidium and other metallic salts useful in small quantities in the human system.

The Arabian coffee plant is a small evergreen tree first introduced to Europe through Arabia. The first plants were brought to America in 1723, and the trees are now grown in most tropical countries. It requires a hot, moist climate, but develops best at higher altitudes. There are numerous varieties, and the coffee beans also vary in aroma and taste with differences in climate and cultivation. The Liberian and Congo species, grown on the west coast of Africa, are hardier plants, but the coffee is different in aroma and is used only for blending. **Mocha coffee** and **Java coffee** are fragrant varieties of Arabian coffee. The fruits are small fleshy berries containing two greenish seeds. They are dried in the sun, or are pulped by machine and cleaned in fermenting baths and dried in ovens or in the sun. After removal of the skin from the dried beans, they are graded and shipped as green beans. The general grades are by shipping ports or regions with numbered grades or qualities. Coffee is always roasted for use. This consists in a dry distillation with the formation of new compounds which produce the flavor and aroma. The **caffeic acid** in coffee is a complex form of cinnamic acid which changes readily to a complex coumarin. **Coffee-Captan,** of Cargille Scientific, Inc., is alpha furfuryl mercaptan, one of the essential constituents in the aroma of freshly roasted coffee. It is a water-white liquid used in masking agents and is a vulcanizer for rubber. **Coffee flavor,** made synthetically for adding to coffee blends, is furfural mercaptan. The **mercaptans** are thioalcohols, or sulfur alcohols, which have compositions resembling those of the alcohols but react differently to give **mercaptals** with aldehydes and **mercaptols** with ketones and produce various flavors from offensive to pleasant.

   Brazilian coffee is the base for many blends, though the average quality is not high. In blends, **Medellin coffee** from Colombia is used for rich flavor, Mexican **Coatepec** for winey flavor, El Salvadoran for full body, Costa Rican for fragrance, and Arabian mocha for distinctive flavor. Some coffees, such as Guatemalan, which have a full body and rich flavor are used without blending, though trade-named coffees are usually blends because of the lack of quantity of superior types. **Powdered coffees,** commonly known as **instant coffee,** are produced by evaporating coffee brew. To drink, it is only necessary to add hot water. **Chicory,** which is used extensively in Europe for blending with coffee, is the dried, roasted, and ground root of the perennial plant *Cichorium intybus,* native to Europe. From 5 to 40% chicory may be used in some blends of coffee. It gives a taste preferred by some. **Caffeine-free coffee** brands have the alkaloid removed by solvent extraction and the tannic acid neutralized to improve digestibility. **Postum,** a noncaffeine coffee substitute now of Kraft Foods, was introduced in 1895 by Charles W. Post.

COIR.   A fiber by-product of the coconut industry. The **fiber** is retted from the outer husks, hammered with wooden mallets, and then combed and bleached. The coarse and long fibers are used for brushmaking; the finer and curly fibers are spun into coir yarn used for mats, cordage, and coarse cloths. In the West Indies it is mixed with sisal and jute to make coffee-bag cloth. In the Philippines it has been used with cement to make a hard-setting, lightweight board for siding. In India **coir fiberboard** is made by bonding with shellac, pressing, and baking. The boards are hard and have a good finish. Coir is easily dyed. The Sri Lankan **coir yarn** is sold in two quality grades, **Kogalla** and **Colombo,** with subdivisions according to the thickness and texture. The yarn is properly called *coir,* and the harsh brush fiber is best known as **coconut fiber.** Coir yarn averages 491 ft/lb (330 m/kg). The Indian yarn is in 450-yd (411-m) lengths tied into bundles. A hundred nuts yield 17 or 18 lb (7.7 or 8.2 kg) of fiber. **Coconut shell,** a by-product of the copra industry, is used for making activated charcoal and for **coconut shell flour** used as a filler in molded plastics. It has a composition similar to walnut shell, being chiefly cellulose with about 30% lignin, 17 pentosan, and 5 methoxyl.

COKE.   The porous, gray, infusible residue left after the volatile matter is driven out of bituminous coal. The coal is heated to a temperature of 2192 to 2552°F (1200 to 1400°C), without allowing air to burn it, and the volatile matter expelled. The residue, which is mainly fixed carbon and ash, is a cellular mass of greater strength than the original coal. Its nature and structure make it a valuable fuel for blast furnaces, burning rapidly and supporting a heavy charge of metal without packing. Soft, or bituminous, coals are designated as coking or noncoking, according to their capacity for being converted to coke. Coal low in carbon and high in ash will produce a coke that is friable and not strong enough for furnace use, or the ash may have low-melting-point constituents that leave glassy slag in the coke. Coke is produced in the beehive and by-product ovens, or is a by-product of gas plants. One ton (907 kg) of coal will yield an average of 0.7 ton (635 kg) of coke, 11,500 ft^3 (325 m^3) gas, 12 gal (45 L) tar, 27 lb (12 kg) ammonium sulfate, 50 gal (189 L) benzol, 0.9 gal (3.4 L) toluol and naphtha, and 0.5 lb (0.2 kg) naphthalene, but the product yield varies with the temperature. When steel production is low and coking ovens are run at lower temperature with a longer cycle, the yield of naphthalene is low.

The fixed carbon of good coke should be at least 86%, and sulfur not more than 1%. The porosity may vary from 40 to 60%, and the apparent specific gravity should not be less than 0.8. **Foundry coke** should have an ignition point of about 1000°F (538°C), with sulfur below 0.7%, and the pieces should be strong enough to carry the burden of ore and limestone. Coke suitable for foundry use is also made from low-grade coals by reducing them to a semicoke, or char, and briquetting, but **semicoke** and **smokeless fuel** are generally coals carbonized at low temperatures and briquetted for household use. These fuels are sold under trade names such as **Coalite** and **Carbolux,** and they are really by-products of the chemical industry since much greater quantities of liquids and more lighter fractions in the tar are obtained in the process.

Pitch coke, made by distilling coal tar, has a high carbon content, above 99%, with low sulfur and ash, and is used for making carbon electrodes. **Petroleum coke** is the final residue in the distillation of petroleum and forms about 5% of the weight of the crude oil. With the sand and impurities removed, it is about 99% pure carbon and is used for molded carbon products. **Calcined coke** is petroleum coke that has been calcined at 2400°F (1316°C) to remove volatile matter. It is used for electrodes. **Carbonite** is a natural coke found in England

and in Virginia. It is a cokelike mineral formed by the baking action of igneous rocks on seams of bituminous coal.

COLD-MOLDED PLASTICS. This is the oldest group of plastic materials, and they were introduced into the United States in 1908. The materials fall into two general categories: inorganic or refractory materials, and organic or nonrefractory materials.

Inorganic cold-molded plastics consist of asbestos fiber filler and either a silica-lime cement or portland cement binder. Clay is sometimes added to improve plasticity. The silica-lime materials are easier to mold although they are lower in strength than the portland cement types.

In general, advantages of these materials include high arc resistance, heat resistance, good dielectric properties, comparatively low cost, rapid molding cycles, high production with single-cavity molds (thus low tool cost), and no need for heating of mold. On the other hand, they are relatively heavy, cannot be produced to highly accurate dimensions, are limited in color, and can be produced only with a relatively dull finish. They have been used generally for arc chutes, arc barriers, supports for heating coils, underground fuse shells, and similar applications.

Organic cold-molded plastics consist of asbestos fiber filler materials bound with bituminous (asphalt, pitches, and oils), phenolic, or melamine binders. The binder materials are mixed with solvents to obtain proper viscosities and then thoroughly mixed with the asbestos, ground, and screened to form molding compounds. The bituminous-bound compounds are lowest in cost and can be molded more rapidly than the inorganic compounds; the phenolic and melamine-bound compounds have better mechanical and electrical properties than the bituminous compounds and have better surfaces as well as being lighter in color. Like the inorganic compounds, organic compounds are cold-molded, followed by oven curing.

Compounds with melamine binders are similar to the phenolics, except that melamines have greater arc resistance and lower water absorption, are nontracking, and have higher dielectric strength.

Major disadvantages of these materials, again, are relatively high specific gravity, limited colors, and inability to be molded to accurate dimensions. Also they can be produced only with a relatively dull finish.

Compounds with bituminous binders are used for switch bases, wiring devices, connector plugs, handles, knobs, and fuse cores. Phenolic and melamine compounds are used for similar applications where better strength and electrical properties are required.

An important benefit of cold-molded plastics is the relatively low tooling cost usually involved for short-run production. Most molding is done in single-cavity molds, in conventional compression-molding presses equipped for manual, semiautomatic, or fully automatic operation.

The **water-fillable plastics** used to replace wood or plaster of paris for ornamental articles, such as plaques, statuary, and lamp stands, and for model making are thermoplastic resins that cure to closed-cell lattices that entrap water. The resin powders are mixed with water and a catalyst and poured into a mold without pressure. They give finer detail than plasters, do not crack or chip, and are lightweight, and the cured material can be nailed and finished like wood. Water content can be varied from 50 to 80%.

COLD-ROLLED STEEL. Flat steel products produced by cold-rolling hot-rolled products. The hot-rolled product is cleaned of oxide scale by pickling and passed through a cold-reduction mill to reduce and more uniformly control thickness and enhance surface finish. Cold rolling also increases hardness, reducing ductility. Although the steel is sometimes used as rolled, it is often subsequently annealed to improve formability and then temper-rolled or roller-leveled for flatness. Cold-rolled steels are available in carbon and alloy grades as well as high-alloy grades, such as stainless steels. For plain carbon steels, carbon content is usually 0.25% maximum, often less. Quality designations include **commercial-quality (CQ) steel,** which is produced from rimmed, capped, or semikilled steel; **drawing-quality (DQ),** which is made from specially processed steel and is more ductile and uniform in forming characteristics; and **drawing-quality special-killed (DQSK) steel,** which is still more ductile and more uniform in forming characteristics. **Cold-rolled structural-quality (SQ) steel** refers to cold-rolled steel produced to specific mechanical properties. Bar and rod products are often cold-drawn through dies and called **cold-drawn bar steel,** or cold-finished in other ways and called **cold-finished bar steel.**

COLUMBITE. An ore of the metal columbium. Its composition varies and may be $FeO \cdot Cb_2O_5$ or $(FeMn)Cb_2O_6$, or it may also contain tungsten and other metals. It is produced chiefly in Nigeria and marketed on the basis of its Cb_2O_5 content. But columbium occurs more usually in combination with tantalum. Concentrates generally average 44 to 70% Cb_2O_5 and 0.4 to 7% Ta_2O_5. The combined mineral known as **columbotantalite,** mined in South Dakota, Idaho, and the Congo, is marketed on the basis of the total $Ta_2O_5 \cdot Cb_2O_5$ content, and as the tantalum increases and the specific gravity increases, the mineral is called *tantalite*. The black mineral is associated with pegmatite, and some crystals are up to a ton in weight. Columbite concentrates contain about 60% **columbium pentoxide,** Cb_2O_5.

COLUMBIUM AND COLUMBIUM ALLOYS. One of the basic elements, columbium (Cb) is also known as **niobium (Nb)** and occurs in the minerals columbite and tantalite. A refractory metal, it closely resembles tantalum, is yellowish-white, has a specific gravity of 8.57, a melting point of 4474°F (2468°C), and an electrical conductivity of 13.2% relative to copper. Columbium has a body-centered-cubic crystal structure, a coefficient of thermal expansion at room temperature of 3.9×10^{-6}/°F (7.1×10^{-6}/°C), a ductile-to-brittle transition temperature of −255°F (−160°C), and a superconducting transition temperature of −433°F (−264°C). It is quite ductile when pure or essentially free of interstitials and impurities, notably nitrogen, oxygen, and hydrogen, which are limited to very small amounts. Tensile properties depend largely on purity, and columbium, having a total interstitial content of 100 to 200 ppm (parts per million), provides about 40,000 lb/in^2 (276 MPa) ultimate strength, 30,000 lb/in^2 (207 MPa) yield strength, 30% elongation, and 15.2×10^6 lb/in^2 (105,000 MPa) elastic modulus. Drawn wire having an ultimate tensile strength of 130,000 lb/in^2 (896 MPa) has been produced. The metal is corrosion-resistant to many aqueous media, including dilute mineral and organic acids, and to some liquid metals, notably lithium, sodium, and sodium potassium. It is strongly attacked, however, by strong dilute alkalies, hot concentrated mineral acids, and hydrofluoric acid. At elevated temperatures, gaseous atmospheres attack the metal primarily by oxidation even if the oxygen content is low, attack being especially severe at 750°F (399°C) and higher temperatures, necessitating the use of protective coatings. Columbium tends to gall and seize easily in fabrication. Sulfonated tallow and various waxes are the preferred lubricants in forming, and carbon tetrachloride in machining. **Ferrocolumbium** is used to add the metal to steel. Columbium is also an important alloying element in nonferrous alloys.

Columbium alloys are noted mainly for their heat resistance at temperatures far greater than those that can be sustained by most metals, but protective coatings are required for oxidation resistance. Thus, they find use for aircraft-turbine components and in rocket engines, aerospace reentry vehicles, and thermal and radiation shields. **Columbium-tin** and **columbium-titanium alloys** have found use as superconductors, and **Cb-1Zr,** a **columbium–1% zirconium alloy,** has been used for high-temperature components, liquid-metal containers, sodium or magnesium vapor-lamp parts, and nuclear applications. It has a tensile yield strength of about 37,000 lb/in^2 (255 MPa) at 70°F (21°C) and 24,000 lb/in^2 (165 MPa) at 2000°F

(1093°C). Thin cold-rolled sheet of **columbium alloy C-103,** which contains 10% hafnium and 1 titanium, has a tensile yield strength of 94,000 lb/in^2 (648 MPa) at 70°F and 25,000 lb/in^2 (172 MPa) at 2000°F (1093°C). After recrystallization at 2400°F (1315°C), however, yield strength drops to 50,000 lb/in^2 (345 MPa) at 70°F and 18,000 lb/in^2 (124 MPa) at 2000°F (1093°C). The room-temperature tensile properties of the 10% tungsten, 10 hafenium, 0.1 yttrium columbium alloy, known as **columbium alloy C-129,** are 90,000 lb/in^2 (620 MPa) ultimate strength, 75,000 lb/in^2 (517 MPa) yield strength, 25% elongation, and 16×10^6 lb/in^2 (110,000 MPa) elastic modulus. Its strength falls rapidly with increasing temperatures, tensile yield strength declining to about 34,000 lb/in^2 (234 MPa) at 1832°F (1000°C). Other columbium alloys and their principal alloying elements are **Cb-752** (10% tungsten, 2.5 zirconium), B-66 (5 molybdenum, 5 vanadium, 1 zirconium), **Cb-132M** (20 tantalum, 15 tungsten, 5 molybdenum, 1.5 zirconium, 0. 12 carbon), **FS-85** (28 tantalum, 10 tungsten, 1 zirconium), and **SCb-291** (10 tantalum, 10 tungsten). Typical tensile properties of **columbium alloy B-66** at room temperature and 2000°F (1093°C), respectively, are 128,000 lb/in^2 (882 MPa) and 65,000 lb/in^2 (448 MPa) ultimate strength, 108,000 lb/in^2 (745 MPa) and 58,000 lb/in^2 (400 MPa) yield strength, 12 and 28% elongation, and 15.3×10^6 lb/in^2 (105,500 MPa) and 12×10^6 lb/in^2 (82,700 MPa) elastic modulus. B-66 contains 5% molybdenum, 5 vanadium, and 1 zirconium.

Columbium alloys can be categorized in terms of strength and ductility. Cb-1Zr and C-103 are low-strength, high-ductility alloys. Other such alloys and their ingredients are **columbium alloys B-3 and D-14,** each with 5% zirconium, and **columbium alloy D-36,** (10 titanium and 5 zirconium). B-66, FS-85, C-129, Cb-752, and SCb-291 are moderate in strength and ductility. Others in this group are **columbium alloy AS-55** (10% tungsten, 1 zirconium, and 0.06 yttrium), **columbium alloy D-43** (10 tungsten, 1 zirconium, and 0.1 carbon), **columbium alloy PWC-11** (1 zirconium and 0.1 carbon), and **columbium alloy SU-16** (10 tungsten, 3 molybdenum, and 2 hafnium). Cb-132M is noted for its high strength. Others in this group are **columbium alloy B-88** (28% tungsten, 2 hafnium, and 0.07 carbon), **columbium alloy Cb-1** (30 tungsten, 1 zirconium, and 0.05 carbon), **columbium alloy F-48** (15 tungsten, 5 molybdenum, 1 zirconium, and 0.05 carbon), **columbium alloy F-50** (15 tungsten, 5 molybdenum, 5 titanium, 1 zirconium, and 0.05 carbon), and **columbium alloy SU-31,** (17 tungsten, 3.5 hafnium, 0.12 carbon, and 0.05 silicon).

Columbium selenide, CbSe$_2$, is more electrically conductive than graphite and forms an adhesive lubricating film. It is used in powder form with silver, copper, or other metal powders for self-lubricating bearings and gears. Columbium also comes in the form of **columbi-**

um oxide, Cb_2O_5, a white powder melting at 2768°F (1520°C), and as **potassium columbate,** $4K_2O \cdot 3Cb_2O_5 \cdot 16H_2O$. **Columbium ethylate,** $Cb(OC_2H_5)_5$, has a melting point of 43°F (6°C). It is used for producing thin dielectric films and for impregnating paper for dielectric use. Other such **metal alcoholates** are **columbium methylate,** $Cb(OCH_3)_5$, with a melting point of 127°F (53°C, and the **tantalum alcoholates** of the same formula. **Columbium carbide,** CbC, is an extremely hard crystalline powder, which can be molded with a metal binder and sintered for use in cutting tools. The melting point is about 6872°F (3800°C). It is made by sintering columbium powder and carbon in a hydrogen furnace.

COMPOSITES. In the broadest sense, materials comprising at least two distinct intended materials, providing superior performance or lower cost than that of the constituent materials alone. Many materials more commonly designated by other terms are indeed composites, including clad, coated, and plated metals and filled or reinforced plastics. The term was established in the aerospace industry and caught on elsewhere, perhaps because it became sort of a buzzword symbolic of "high performance." In the auto industry and others, it is now often used to refer to reinforced plastics, which have been used for many years and referred to as such or, simply, as plastics. To distinguish such routinely used materials from the aerospace kind, the term **advanced composites** also has been used to designate the latter.

In the aerospace industry, composites have come to be categorized by the matrix material, which contains the reinforcing elements. Thus there are **polymer-matrix composites,** or **PMCs,** the most mature and widely used; and the emerging **metal-matrix composites,** or **MMCs; ceramic-matrix composites,** or **CMCs;** and **intermetallic-matrix composites,** or **IMCs.** There are also **carbon/carbon composites,** or **CCCs,** containing the same basic material for both reinforcement and matrix. These are sometimes referred to as **graphite/graphite composites.**

The matrix material generally governs the service temperature. **Epoxy,** the most widely used for PMCs, allows service temperatures up to about 300°F (149°C). **Bismaleimide,** which has replaced epoxy to some extent in military aircraft applications, permits use to about 350°F (177°C). **Polyimide,** with a maximum service temperature of at least 500°F (260°C), is used to a much more limited extent. The principal load-bearing elements, however, are the fibers, typically continuous, contained by the matrix. These include **aramid, Kevlar** mainly, **boron, glass,** and **graphite.** PMCs are lightweight, strong, and rigid, thus providing high strength-to-weight ratios (specific strength) and high rigidity-to-weight ratios (specific stiffness).

Another advantage is that fiber direction can be oriented to suit applied load direction. Such composites are made by manual or automatic layup of thin [0.010-in (0.254-mm)] **prepreg** plies or by filament winding, followed by curing in autoclaves or presses. Prepreg is a partially cured and somewhat tacky fiber-reinforced resin, which must be kept in refrigerated storage to keep from spoiling. Filament winding involves winding a tow of fibers or a series of tows (band) around a mandrel of the shape of the part to be produced. In "dry winding," tows of pregreg are used. In "wet winding," the tows or bands are first drawn through a resin bath.

MMCs, like PMCs, were in use long before this term was coined. Examples include **cermets,** or ceramic-reinforced metals, such as **tungsten-carbide** particles in a cobalt matrix for cutting tools and **titanium-carbide** particles in steel for heat- and wear-resistant parts. MMCs may contain continuous or discontinuous fibers, particulates, or whiskers as the reinforcing constituent. As a class, they are far more heat-resistant than PMCs. Among the MMCs that have been made are aluminum, copper, cobalt, lead, and magnesium reinforced with graphite. Boron has served as a reinforcement for aluminum, magnesium, and titanium; silicon carbide for aluminum, titanium, and tungsten; and alumina for aluminum. Compared with PMCs, applications so far have been limited, and these are largely limited to aluminum. Aluminum reinforced with continuous boron fibers is used for struts in the Space Shuttle, and aluminum reinforced with continuous graphite fibers is used for the Hubble telescope masts. **Fiber preforms** have been used to selectively reinforce cast aluminum products. And **aluminum alloys** reinforced with **alumina** or **silicon-carbide** particulates are commercially available as wrought and foundry products.

CMCs and IMCs are largely developmental. Both are promising for still greater heat resistance, although the inherent brittleness of the CMCs may limit their use in structural applications. IMCs are seen as potential candidates for aircraft, aircraft-engine, and spacecraft components exposed to temperatures above 2000°F (1093°C). Promising matrix materials include **molybdenum disilicide** ($MoSi_2$), **nickel aluminides,** and **titanium aluminides.** Reinforcements include particles, whiskers, and continuous or discontinuous fibers of alumina or silicon carbide. $MoSi_2$, which excels in corrosion and oxidation resistance, has a brittle-to-ductile transition temperature of about 1832°F (1000°C), but alloying with tungsten disilicide (WSi_2) improves toughness at lower temperatures. Reinforced with 20% by volume silicon-carbide particles, $MoSi_2/WSi_2$ has a tensile yield strength of about 65,000 lb/in^2 (450 MPa) at 2192°F (1200°C). With silicon-carbide whiskers of this amount, the yield strength at this temperature is about 84,000 lb/in^2 (579 MPa). The nickel aluminide, Ni$_3$Al, with 0.5% boron and reinforced with alumina fibers, has a potential service temperature of 1500°F (816°C) or greater. For titanium aluminide, TiAl, reinforced with alumina, this temperature may approach 1900°F (1038°C), and for Ti$_3$Al with columbium, reinforced with silicon-carbide fibers, it is within the range of 1472 to 1562°F (800 to 850°C).

CCCs are noted for their good strength and low thermal expansion at temperatures to greater than 3600°F (2000°C). They are also tough and resistant to abrasion and corrosion, have high thermal and electrical conductivity at high temperatures, and have excellent resistance to thermal shock. However, they will react with oxygen at temperatures above 800°F (427°C), necessitating an **oxygen-barrier coating.** One method of manufacture is chemical vapor deposition, in which a mass of premolded carbon fibers is furnace-heated to high temperature while a hydrocarbon gas is fed into the furnace. The gas is thermally cracked to form carbon, which desposits on the mass. In another method, yarns or woven or nonwoven fabrics of carbon fiber with a **phenolic** or **epoxy** binder are shaped, then heated in inert atmosphere to carbonize the resin. With **silicon carbide** as the oxygen-barrier coating, CCCs serve as thermal-protection systems in the nosecone and wing leading edges of the Space Shuttle. They are also used in the brakes of large airliners, in the clutches and brakes of racecars, and for rocket nozzles.

COMPOSITION METAL. Also called **composition brass,** although it does not have the characteristics of a true brass. A general name for casting alloys, such as **copper alloy C83600,** that are in a midposition between the brasses and the bronzes. The most widely used standard composition metal is **ounce metal,** containing 85% copper, 5 zinc, 5 tin, and 5 lead, which derived its name from the fact that originally 1 oz (0.03 kg) each of the white metals was added to 1 lb (0.45 kg) of copper. It makes a good average bearing metal, and because it gives a dense casting that will withstand liquid pressures, it is also used for valves, pumps, and carburetor parts. It casts well, machines easily, and takes a good polish, so that it is widely employed for mechanical castings. It has about the same coefficient of expansion as copper and can thus be used for pipe fitting. **ASTM alloy No. 2** is this metal, and it may also contain up to 1% nickel and small amounts of iron, either as intentional additions to increase strength or as impurities. As-cast, tensile properties are 37,000 lb/in^2 (255 MPa) ultimate strength, 17,000 lb/in^2 (117 MPa) yield strength, 30% elongation, and 12×10^6 lb/in^2 (82,700 MPa) elastic modulus. Hardness is typically Brinell 60. This alloy also has been called **red**

casting brass, hydraulic bronze, and **steam brass,** and it has also been used for forgings, producing parts with a tensile strength of 33,000 lb/in² (227 MPa) and 25% elongation.

In the high-copper red casting-brass series, for any given content of copper and zinc, the higher the ratio of tin to lead, the stronger but less ductile the alloy. The higher the content of zinc, the more ductile the alloy. For cast pipe fittings, the alloy may have 80 to 86% copper, 4 to 15 zinc, 2 to 6 lead, and 3 to 6 tin. This type of alloy is called **valve bronze,** and when the copper content is higher, it is called **valve copper.** The **M bronze** of the U.S. Navy, for valves, contains 86 to 91% copper, 6.25 to 7.25 tin, 1.5 to 5 zinc, 1 to 2 lead, and not over 0.25 iron. It has a tensile strength of 34,000 lb/in² (234 MPa) and elongation of 17%. It withstands continuous temperatures up to 500°F (260°C), while the 85:5:5:5 bronze can be used for temperatures only to 400°F (204°C). **ASTM alloy No. 1,** designated as high-grade red casting brass for general castings, contains 85% copper, 6.5 tin, 4 zinc, and 1.5 lead. It has a tensile strength of 36,000 lb/in² (248 MPa), elongation 25%, and Brinell hardness 50 to 60.

Nickel is added to composition metals for hydraulic and steam castings to densify the alloy and make the lead more soluble in the copper. One company uses an alloy containing 84.5% copper, 7 zinc, 5 lead, 2.5 tin, and 1 nickel for casting injectors and lubricator parts. The nickel is added to the melt in the form of nickel shot which contains 5 to 7% silicon to deoxidize the metal and increase the hardness. For heavy high-pressure hydraulic castings, as much as 5% silicon may be added to alloys containing nickel, giving strengths above 40,000 lb/in² (275 MPa). The alloys for machinery bearings usually contain higher proportions of tin or lead, or both, and are classified as high-lead bronze, but **Johnson bronze No. 44,** for bearings, contains 88% copper, 4 tin, 4 lead, and 4 zinc. The **hardware bronze** used for casting hardware and automobile fittings to be highly polished and plated is likely to be a true copper-zinc brass or a leaded brass with only a small amount of lead. **Oreide bronze,** a term still used in the hardware industry, was the metal employed for carriage and harness hardware. It contains 87% copper and 13 zinc and polishes to a golden color. The hardware bronze of Chase Brass & Copper Co. contains 86% copper, 12.25 zinc, and 1.75 lead. Aluminum, even in small amounts, is not considered a desirable element in the red casting brasses as it decreases the ductility and requires more care in casting.

CONCRETE. A construction material composed of portland cement and water combined with sand, gravel, crushed stone, or other inert material such as expanded slag or vermiculite. The cement and water form a paste which hardens by chemical reaction into a strong, stone-like mass. The inert materials are called **aggregates,** and for economy no more cement paste is used than is necessary to coat all the aggregate surfaces and fill all the voids. The concrete paste is plastic and easily molded into any form or troweled to produce a smooth surface. Hardening begins immediately, but precautions are taken, usually by covering, to avoid rapid loss of moisture since the presence of water is necessary to continue the chemical reaction and increase the strength. Too much water, however, produces a concrete that is more porous and weaker. The quality of the paste formed by the cement and water largely determines the character of the concrete.

Proportioning of the ingredients of concrete is referred to as *designing* the mixture, and for most structural work the concrete is designed to give compressive strengths of 2,500 to 5,000 lb/in^2 (16 to 34 MPa). A rich mixture for columns may be in the proportion of 1 volume of cement to 1 of sand and 3 of stone, while a lean mixture for foundations may be in the proportion of 1:3:6. Concrete may be produced as a dense mass which is practically artificial rock, and chemicals may be added to make it waterproof, or it can be made porous and highly permeable for such use as filter beds. An air-entraining chemical may be added to produce minute bubbles for porosity or light weight. Normally, the full hardening period of concrete is at least 7 days. The gradual increase in strength is due to the hydration of the tricalcium aluminates and silicates. Sand used in concrete was originally specified as roughly angular, but rounded grains are now preferred. The stone is usually sharply broken. The weight of concrete varies with the type and amount of rock and sand. A concrete with traprock may have a density of 155 lb/ft^3 (2,483 kg/m^3). Concrete is stronger in compression than in tension, and steel bars or mesh is embedded in structural members to increase the tensile and flexural strengths. In addition to the structural uses, concrete is widely used in precast units such as block, tile, sewer and water pipe, and ornamental products.

Concrete blocks may be made from cement, sand, and gravel, or from cement and sand alone. For insulating purposes they may be made with cement and asbestos fibers. **Reinforced concrete** is a combination of concrete with a steel internal structure generally composed of rods or metal mesh. The strength of the concrete is thus greatly increased, and it is used for buildings, bridges, telegraph poles, roads, and fences. **Nonslip concrete,** for steps, is made by applying aluminum oxide grains, sizes 3 to 60 mesh, to the concrete before it hardens.

Insulating concrete and lightweight concretes are made by special methods or by the addition of spongy aggregates. Slag may be used for this purpose. **Aerocrete,** is a porous, lightweight concrete produced by adding aluminum powder to the cement. The reaction between the aluminum flakes and the lime in the cement forms hydrogen bubbles. **Durox,** produced as lightweight blocks, panels, and wall units, is a **foamed concrete** made from a mixture of sand, lime, cement, and gypsum, with aluminum powder which reacts to produce $3CaO \cdot Al_2O_3$ and free hydrogen, which generates tiny bubbles. The set material contains about 80% cells and has only about one-third the weight of ordinary concrete with a compressive strength of 1,000 lb/in² (6.9 MPa). **Acid-resistant concrete,** developed by the Dutch firm of Ocrietfabrick and called **Ocrete,** is made by passing the well-dried concrete products through a treatment tunnel containing **silicon tetrafluoride** gas, SiF_4, which converts the free lime to calcium fluoride. In the center of the concrete parts where moisture still remains, silicic acid is formed and fills the pores. The parts have increased density and are more wear-resistant than the original concrete.

Many prepared aggregates are used for special-purpose concretes. **Haydite** is a lightweight aggregate made by kiln-burning shale to produce a material of expanded cellular structure. Haydite concrete has a density of less than 100 lb/ft³ (1,602 kg/m³), but is not as strong as gravel concrete. **Superock** and **Waylite** are trade names for expanded aggregate made by treating molten slag with water or steam. **Microporite** is a German aggregate made by steam-treating ground silica and lime. **Calicel** is a lightweight spongy aggregate made by fusing silicates of lime and alumina and cooling to produce a stone of cellular structure. **Fluftrok** is a lightweight aggregate made by heating obsidian in a kiln. The rock expands to 16 times its original volume, forming a porous material. Mixed with about 10% portland cement, it is made into building blocks that are light and strong. A **conductive concrete,** known as **Marconite,** produced by Marconi Communication Systems, England, can be used for radio-frequency grounding of TV, radio, and computer systems. The special aggregate can be added to the concrete mix to provide predetermined resistivity values.

A new fast-drying and hard concrete mix, **Pyrament,** is now available. It is an alkali-activated alumina silicate hydrate, which, due to alumina, requires less water in the mix. Therefore, it dries in only 4 h versus a week for regular cement. An application found to date is for runway repairs (pour cement in California when the jet leaves New York, and the pavement is ready for landing).

CONDUCTIVE (ELECTRICAL) POLYMERS AND ELASTOMERS. Polymers made electrically conductive by the addition of carbon-black nickel, silver, or other metals. Volume resistivities of plastics and rubbers, which normally are in excess of 2.5×10^8 Ω/in (10^8 Ω/cm) can be lowered to between 0.25 Ω/in (10^{-1} Ω/cm) and 2.5×10^6 Ω/in (10^6 Ω/cm) by addition of conductive materials. **Carbon black** is the most widely used filler. The relationship of carbon black loading and volume resistivity is not proportional. With up to a 25% loading, conductivity significantly increases, but it falls off sharply thereafter. Generally, the addition of carbon black lowers the polymer's mechanical properties. However, the use of carbon fibers to enhance conductivity improves mechanical properties.

Polyethylene and **polyvinyl chloride** resins loaded with carbon black are perhaps the most widely used conductive plastics. Plastics often made conductive by adding up to 30% carbon fiber are **polysulfone, polyester, polyphenylene sulfide, nylon 6/6, ethylene tetrafluoroethylene,** and **vinylidene fluoride-polytetrafluoroethylene.**

While **silicone** is the most widely used base polymer for **conductive rubber,** other rubbers frequently used in compounding conductive elastomers include SBR, EPDM, TPR, and neoprene.

Another type of **electrically conductive polymers** is materials that are doped with either electron acceptors, such as alkali metal ions and iodine, or electron donors, such as arsenic trifluoride. Also referred to as **organic conductors,** their conductivity is about one-hundredth that of copper. The most widely used are **polyacetylene, polyparaphenylene,** and **polyparaphenylene sulfide.** Polyacetylene, used in the form of foil for battery electrodes, has an energy storage density comparable to that of a lead-acid automobile battery, but can deliver 20 to 25 times the current. By stretching the foil, the fibers of which the foil is composed conduct electricity preferentially in one direction. Environmental stability, especially water sensitivity, is a problem with these materials. It can be improved by encasing them in other plastics. Another problem is that these polymers are difficult to form. Polyacetylene is insoluble and infusible, polyparaphenylene can be formed only by sintering, while polyparaphenylene can be melt-processed. **Phthalocyanines** can also be made electrically conductive by doping them with an electron acceptor, such as iodine, bromine, and charge-transfer salts.

Electrorheological (ER) fluids are suspensions of fine particles, usually polymers, in nonconducting oils or other liquids. When an electric current is passed through them, they turn from liquid to gel-like solids or vice versa in 0.001 to 0.0001 s. With the amount of applied voltage governing the degree of solidity, the fluid itself can perform various control functions, such as damping shock and vibration. Particles include **alumino-silicate zeolites** and **polyacene quinones.** The most common fluids are **silicone oils, gasoline,** and **kerosene.** Advanced Fluid Systems, Ltd., of England, supplies a 20 to 40% solution of **lithium-polymethacrylate** particles in chlorinated paraffins, silicone, or mineral oil depending on property requirements. ER Fluid Development Ltd., of England, has a series of ERFs that use crosslinked lithium polymethacrylate particles in low-molecular-weight **fluorosilicone** fluids. Asahi Chemical Industries, of Japan, has developed specially plated and coated **nylon spheres** in **dimethyl silicone.**

CONDUCTORS. A term usually applying to materials, generally metals, used to conduct electric current, though heat conductors and sound conductors have important uses. Good conductors of electricity tend to be good conductors of heat, too. Silver is the best conductor of electricity, but copper is the most commonly used. The conductivity of pure copper is 97.6% that of silver. The electrical conductivity of metals is often expressed as a percentage of the electrical conductivity of copper, which is arbitrarily set at 100%. Tough-pitch copper is the standard conductivity metal, and it is designated as the International Annealed Copper Standard (IACS).

Because of the low conductivity of zinc, the brasses have low current-carrying capacity, but are widely used for electrical connections and parts because of their workability and strength. The electrical conductivity of aluminum is only 63% that of copper, but it is higher than that of most brasses. **Copper wire** for electrical conductors in high-temperature environments has a plating of heat-resistant metal. **Aluminum wire,** usually with a steel core, is used for power transmission because of the long spans possible. Steel has a conductivity only about 12% that of copper, but the current in a wire tends to travel near the surface, and the small steel core does not reduce greatly the current-carrying capacity. Aluminum is now much used to replace brass in switches and other parts. Aluminum wire for electrical equipment is usually commercially pure aluminum with small amounts of alloying elements such as magnesium which give strength without appreciably reducing the conductivity. Plastics, glass, and other nonconductors are given conductive capacity with coatings of transparent lacquer containing metal powder, but **conductive glass** usually is made by spraying on at high temperature an extremely thin, invisible coating of tin oxide. Coated glass panels are available with various degrees of resistivity.

CONVERSION COATINGS. Surface transformations formed naturally or by chemical or electrochemical methods on ferrous and nonferrous metals and alloys. **Natural coversion coatings,** usually oxides, include **rust,** the ferric oxide and hydroxide that forms on iron and plain carbon steels in air or moisture, and the adherent and protective oxides that form on aluminum, copper, and other metals. **Chemical conversion coatings** are mainly phosphates, chromates, and oxides induced by the reaction of specific chemicals with metal surfaces. **Electrochemical conversion coatings** are formed by anodic oxidation, or anodizing, in an electrolytic cell in which the metal being treated is the anode.

Phosphate conversion coatings are formed on iron, steel, galvanized steel, and aluminum by chemicals containing phosphoric acid and its salts. **Zinc phosphates,** one of three major types, are applied by immersion or spray in a wide variety of coating weights and crystal sizes. The microcrystalline type enhance paint adhesion, minimize paint consumption, and improve bonding to plastics and rubber. The heavy kind are quite absorbent, thus capable of retaining forming lubricants and rust-preventive oils. They also reduce friction and enhance wear resistance and corrosion resistance. **Iron phosphates,** similarly applied, are produced from alkali-metal phosphate solutions. When amorphous or of fine crystal size, they are used mainly to improve paint adhesion and to increase resistance to paint flaking from impact or flexing. **Manganese phosphates,** applied only by immersion, are used to retain oil so as to facilitate part break-in and prevent galling of mating surfaces. Phosphate coatings have specific colors, depending on the chemicals used and the metal to which they are applied. Special colors can be developed by pretreatments and post-treatments.

Chromate conversion coatings are formed on aluminum, cadmium, copper, magnesium, silver, zinc, and their alloys by immersion or spray using aqueous solutions of chromic acid, chromium salts such as sodium or potassium chromate or dichromate, and hydrofluoric, phosphoric, or other mineral acids. Generally, the basic ingredients are **hexavalent chromium** and sufficient acid for the desired pH. Solutions of **trivalent chromium** are used for clear coatings on electroplated cadmium and zinc. Chromate-phosphate mixtures are used to produce combination coatings on aluminum. Chromates are typically amorphous, porefree, and gellike initially but, on drying, harden and become hydrophobic, less soluble, and more abrasion-resistant. They are used primarily to increase corrosion resistance, especially in marine, humid, and tropical atmospheres; but they also serve as a good base for paint. Various colors can be provided, depending on the particular solution and posttreatments.

Oxide chemical conversion coatings are the bluish, black, or brown oxides formed on iron or steel with hot caustic or alkaline solutions, and the black or brown oxides formed on cadmium, copper, iron, steel, or zinc alloys with acidic solutions. Although they are used mainly for abrasion resistance and aesthetics, some add a modest degree of corrosion protection. Alkaline chromate solutions and fused-salt solutions, for example, impart corrosion resistance and abrasion resistance to iron and steel.

Electrochemical conversion coatings, or **anodic coatings,** pertain mainly to aluminum alloys and magnesium alloys, although several other metals are also anodized. **Anodized aluminum** is produced primarily in aqueous solutions of sulfuric, chromic, or oxalic acid for a variety of reasons: to improve corrosion resistance, abrasion resistance, paint adhesion, adhesive bonding, or electroplating; to provide decorative finishes, including color; and to impart an electrically insulative surface or a base for photographic or lithographic emulsions. **Hard anodic coatings** on aluminum alloys, for abrasion resistance, are typically thicker than those for corrosion protection. **Anodized magnesium** is produced in aqueous solutions of ammonium bifluoride, sodium dichromate, and phosphoric acid or in aqueous solutions of potassium and aluminum hydroxide, trisodium phosphate, potassium fluoride, and potassium manganate or permanganate. Thin coatings serve mainly as a base for paint, thick ones for corrosion resistance and abrasion resistance. Other nonferrous metals are anodized primarily to increase corrosion resistance. **Anodized zinc** is produced by immersion in an alkali-carbonate solution and then in a silicate solution, or in a single alkaline solution of sodium silicate, borate, and tungstate. **Anodized beryllium** is made in an aqueous solution of nitric and chromic acids, and **anodized titanium** and anodized thorium are made in mixtures of glacial acetic and nitric acids. **Anodized zirconium** also has been made in such mixtures, although, for nuclear applications, it and **anodized hafnium** are made in aqueous solutions of ethanol, glycerine, and lactic, phosphoric, and citric acids, followed by autoclaving. **Amodized columbium** and **anodized tantalum** are produced in solutions of ammonium citrate or borate, with ammonium hydroxide for basicity.

Conversion coatings are also known by various terms and trade names. Among the latter by Allied-Kelite, of Witco Corp., are **Keycote phosphate coatings, Iridite chromate coatings, Iridize zinc anodic coatings,** and **Irilac clear coatings.** The clear coatings can be applied to the anodic for additional corrosion protection. For anodizing aluminum, Alumilite, Oxydal, Anodal, and Anoxal refer to sulfuric acid baths; Bengough-Stuart is the original chromic acid system; Oxal and Eloxal GX are oxalic acid systems; and Ematal is the titanium-salt triacid system. Alumilite also pertains to a sulfuric and oxylic acid system for hard anodizing. Other terms and key ingredients applicable to hard-anodizing aluminum are Martin Hard Coat (sulfuric acid); Alcanodoz, Hardas, and Lasser (oxalic acid); Sanford (sulfuric acid with organic additives); and Kalcolor (sulfosalicylic and sulfuric acids). **Magnaplate HCR,** of General Magnaplate, is a surface treatment to improve the hardness, corrosion resistance, and abrasion resistance of aluminum and aluminum alloys. **Nituff,** of Nimet Industries Inc., is Teflon-impregnated hard-anodized aluminum having a Rockwell C hardness of 60 to 70. Combining such hardness and lubricity markedly increases release properties and wear resistance. One application is chemical polishing tubs used to make sapphire optics.

COOLANTS. Liquids used to quench metals in heat treating, used to cool and lubricate cutting tools and workpieces in machining, or applied to forming tools and workpieces to assist in forming operations. In the case of machining, they are also called **cutting fluids,** and in the case of forming, **forming lubricants.** When water is used for the normal water-hardening steels, it may be modified with soda or other material to give a less drastic and more uniform cooling. A water bath containing 5% sodium hydroxide gives uniform, rapid cooling. Oils are used in cooling or quenching baths to provide a more moderate cooling effect. **Quenching oils** are usually compounded, although fish oils alone are sometimes employed. Fish oils, however, have offensive odors when heated. Vegetable oils alone are likely to oxidize and become gummy. Animal oils become rancid. Lard and palm oils give low cooling rates, while cottonseed, neatsfoot, and fish oils give more rapid cooling. Mineral oils compounded with fish, vegetable, or animal oils are sold under trade names and vary considerably in their content. **Oil-quenching baths** are usually kept at a temperature of not over 150°F (66°C) by providing cooling pipes. **Tempering oils** differ from quenching oils only in that they are compounded to withstand temperatures up to about 525°F (274°C).

Coolants for machining are classified into five groups: straight oils, soluble oils, chemical coolants, synthetics, and semisynthetics. **Straight oils,** which contain no water, are petroleum or mineral oils with or without additional compounding. Without further compounding, they are suitable for light- to moderate-duty cutting on readily machined metals. For more severe machining, they are typically compounded with up to 20% fatty oils, sulfur, chlorine, phosphorus, or combinations thereof. Sulfur, chlorine, and phosphorus are commonly called **extreme-pressure additives** (EP additives). For the most severe applications, compounding, mainly with chlorine and sulfurized fatty oils, may exceed 20%. **Soluble oils,** such as emulsified sulfonated mineral oils, are also suitable for light- and moderate-duty applications. Although they do not match the straight oils in lubricity, they, like water-dilutable fluids in general, are better heat dissipators. Because of their water content, they are usually formulated with additives to prevent corrosion of the workpiece and to resist microbial degradation and souring, necessitating maintenance in service to retain these characteristics. Heavy-duty soluble oils are suitable for most applications for which straight oils are used. **Chemical coolants** were originally **amine nitrites,** but **amine borates** are commonly used now because nitrites in contact with amine form **nitrosamine,** a suspected carcinogen. They are noted for excellent cooling capacity, inhibiting corrosion, and resisting microbial degradation and souring. They have limited lubricity, however, and are confined to light-duty operations, mainly light grinding. **Synthetic coolants,** which have been likened to soluble oils without oil, are water-dilutable systems designed for high-cooling capacity, lubricity, corrosion prevention, and easy maintenance. These synthesized materials are chemically similar to mineral-oil derivatives, but can be dispersed in water and are suitable for more severe operations than chemical coolants. They tend to defat human skin, however, causing dermatitis, thus necessitating that workers adhere to prescribed methods of personal hygiene. **Semisynthetic coolants** contain small dispersions of oil in an otherwise water-dilutable system, are almost transparent, are more broadly applicable than soluble oils, and are easier to maintain.

Straight oils, soluble oils, and synthetics are also used as forming lubricants. They also may contain a **wetting agent,** or **polarity agent,** such as animal fats, fatty acids, long-chain polymers, emulsifiers, and EP additives. The straight oils are the most varied in formulation and the most widely used. The soluble oils, however, can match their performance in many applications and, because of their superior cooling effect, sometimes provide better performance. The synthetics, which have been improved in recent years, also feature excellent cooling capacity as well as cleanliness and lubricity, and they have replaced both the straight and soluble oils in many applications. Because of their cleanliness, they are especially useful in forming precoated metals.

COPAL. A general name for fossil and other hard resins found in nearly all tropical countries and used in making varnishes and lacquers, adhesives, and coatings, though now largely replaced by synthetic resins. Copals are distinguished by their solubility in chloral hydrate. All the copals are also soluble in alcohol, linseed oil, and turpentine. The hardest varieties come from Africa. **Zanzibar copal,** from the tree *Trachylobium verrucosum,* or from species no longer existent, is one of the hardest of the varnish resins, with a melting point of 464 to 680°F (240 to 360°C), compared with 356 to 392°F (180 to 200°C) for **Congo copal** from Guinea. **Madagascar copal** is from the tree *Hymenaca verrucosa* and is darker than Zanzibar. **Gum benguela** is a semifossil resin from the tree *Gulbourtia copaifera* of West Africa. The melting point is 338°F (170°C). Many species of trees of the genus *Hymenaca* of tropical America furnish copals. **Animi gum,** or **gum Zanzibar,** is from the stem of the plant *H. coubarii* of Zanzibar and East Africa. It belongs to the group called **East African copals,** but is distinguished from other copals by its solubility in alcohol. The specific gravity is about 1.065, and melting point 473°F (245°C). The **Brazilizan copal** known as **jutahycica resin** is from the jatahy tree which is plentiful in the Amazon Valley. **Jatabó** and **trapucá resins** are fossil copals from species of *Hymenaca* of the state of Bahia, Brazil. **Congo gum,** chiefly from the tree *Copaifera demensi,* is the most insoluble of the natural resins, but after thermal processing, it is soluble in a wide range of solvents. The specific gravity of copals is from 1.04 to 1.13. The colors vary from white through yellow, red, brown, to brownish black; generally speaking, the harder the copal, the greater the value.

The commercial copals are classified in five groups: East African, West African, Manila, East Indian, and South American. The name *copal* is applied in Indonesia to the resin of the tree *Agathis alba,* closely related to the kauri pine. The types include **Manila copal, Loba,** and **Boea.** In Malaya the tree has been classified as *Dammara orientalis,* and the copal is known as **white dammar.** In the Philippines the tree is called **almacido,** and the gum, Manila copal. There are seven grades of Manila copal, from No. 1 pale, scraped chunks, to the No. 7 dust. Hard copal is harder than dammar, and has a higher melting point, but the hardness of the resin depends greatly upon the seasoning time in the ground. The semihard and soft copals are produced directly from the trees by tapping. The melting point of copal from *A. alba,* collected 1 day after tapping, averages 185°F (85°C), compared with 221°F (105°C) when collected 3 months after tapping. **Fossil copal,** or **copalite,** or **copaline,** of high quality, is obtained by separation from the low-grade coals of Utah, which contain about 5%. The copal has an amberlike appearance of light yellow to red color, with a specific gravity of 1.02 to 1.06, melting point of 329°F (165°C), and hardness about the same as that of Congo gum. In London, England, where copalite occurs as irregular fragments in blue clay, it is called **Highgate resin.**

COPPER. One of the most useful of the metals, and probably the one first used by humans. It is found native and in a large number of ores. Its apparent plentifulness is only because it is easy to separate from its ores and is often a by-product of silver and other mining. Copper has a face-centered-cubic crystal structure. It is yellowish red, tough, ductile, and malleable; gives a brilliant luster when polished; and has a disagreeable taste and a peculiar odor. It is the best conductor of electricity next to silver, having a conductivity 97% that of silver. Copper, Cu, refers to the metal at least 99.3% pure. Standard wrought grades number more than 50, many of which are more than 99.7% pure. They are represented by the C10XXX to C15XXX series of copper and copper alloy numbers of the Copper Development Association. These include **oxygen-free coppers, oxygen-free-with-silver coppers,** and **oxygen-bearing coppers** (C10100 to C10940); **electrolytic-tough-pitch coppers** and **tough-pitch-with-silver coppers** (C11100 to C11907); **phosphorus-deoxidized coppers, fire-refined tough-pitch coppers,** and **fire-refined tough-pitch-with-silver coppers** (C12000 to C13000); and certain coppers distinguished by very small amounts of specific ingredients such as **cadmium copper** (not to be confused with the high-copper alloys having a greater cadmium content), **tellurium-bearing copper, sulfur-bearing copper, zirconium copper,** and **aluminum-oxide-bearing coppers** (Cl4XXX to Cl5XXX). The highest-purity grade, **oxygen-free-electronic copper,** is at least 99.99% pure. There are seven standard **cast coppers** (C80XXX to C81100), and their minimum purity percentage ranges from 99.95 (C80100) to 99.70 (C81100).

Oxygen-free coppers C10100 and C10200 have a melting point of 1980°F (1082°C), a density of 0.323 lb/in^3 (8,941 kg/m^3), an electrical conductivity of 101%—or slightly greater than the 100% for electrolytic-tough-pitch copper (C11100) used as the International Annealed Copper Standard (IACS) for electrical conductivity—a thermal conductivity of 226 Btu/(ft · h · °F) [391 W/(m · K)], and a specific heat of 0.092 Btu/(lb · °F) [385 J/(kg · K)]. Typical tensile properties of thin, flat products and small-diameter rod and wire having an average grain size of 0.002 in (0.050 mm) are 32,000 lb/in^2 (220 MPa) ultimate strength, 10,000 lb/in^2 (69 MPa) yield strength, 45 to 50% elongation, and 17 × 10^6 lb/in^2 (117,000 MPa) elastic modulus. Hardness is about Rockwell F. 40. These properties are fairly typical of other wrought coppers as well. Strength increases appreciably with cold work, yield strengths reaching 50,000 lb/in^2 (345 MPa) in the spring and hard-drawn conditions. Zirconium copper, which may be heat-treated after cold working, can provide yield strengths of 50,000 to 70,000 lb/in^2 (345 to 483 MPa) in rod and wire forms and retains considerable strength at temperatures to 800°F (426°C). The aluminum-oxide-bearing coppers are high-strength **dispersion-strengthened coppers.** They are designated C15710 to C15760, and their oxide content ranges from a nominal 0.2% to 1.1. They are used mainly for the tips of spot-welding electrodes. Cast coppers are suitable for sand, plaster, permanent-mold, investment, and centrifugal castings as well as for continuous casting. Regardless of grade, typical tensile properties are 25,000 lb/in² (172 MPa) ultimate strength, 9,000 lb/in² (62 MPa) yield strength, and 40% elongation. Hardness is typically Brinell 44.

Lake copper, from the Lake Superior region, is a silver-bearing copper having varying amounts of silver up to about 30 oz (0.9 kg) per ton (907 kg). Coppers are generally corrosion-resistant to rural, marine, and industrial atmospheres. **Copper C11000,** for example, corrodes at rates ranging from 0.005 mil/yr (0.13 μm/yr) in dry, rural regions to 0.055 mil/yr (1.40 μm/yr) in industrial regions. They also resist corrosion by various waters, saline solutions, soils, nonoxidizing mineral and organic acids, and caustic solutions, but are attacked by oxidizing acids, such as nitric, moist ammonia, and halogens, sulfides, and solutions containing ammonium ions. Wrought coppers are among the most formable of metals. Forgeability is about 65% that of C37700 forging brass, but machinability is only about 20% that of C36000 free-cutting brass. And they are readily joined by welding, brazing, and soldering. Coppers are used for a great variety of applications: bus bars, commutators, terminals, waveguides, electric wire, power transmission lines, motor windings, printed circuits, springs, water pipe and tubing, heat exchangers, building products, gaskets, and fasteners of many kinds. **Roofing copper** is soft, hot-rolled copper sheet. **Cornice copper** is cold-rolled to a hard temper. **Braziers' copper** refers to heavy sheet, 1.5 to 6 lb/ft² (7.3 to 29 kg/m²), used for coppersmiths' work. **Coppersmiths' copper** is hot-rolled, soft-temper, heavy sheet up to 0.5 in (13 mm) thick. **Copper foil** is less than 0.005 in (1.3 mm) thick. **Free-cutting copper** is deoxidized copper containing up to 0.7% tellurium in rod form for making screw-machine parts. **Pyralux AP,** of DuPont, is a copper-polyimide-copper laminate for flexible printed-circuit boards. **Narloy Z** is a copper and silver alloy composite developed in the program for the U.S. national aerospace airplane and intended for engine inlets and wing leading edges.

COPPER ACETATE. Also known as **crystals of Venus.** A dark-green, crystalline, poisonous powder of composition $Cu(CH_3COO)_2 \cdot H_2O$, of specific gravity 1.882 and melting point 239°F (115°C). It is soluble in water and in alcohol. It is used as a pigment in paints, lacquers, linoleum, and inks and for making artificial verdigris or patina on copper articles. It is used as a catalyst in making phthalic anhydride plastics. When used for mildew-proofing cotton cloth, the copper precipitates out to form the **waxate,** or copper soap coating. **Verdigris** is an old name for basic copper acetate as a blue-green pigment, but the name is now usually applied to the bluish-green corrosion crust on copper. The greenish-brown crust known as **patina,** formed on bronze, is esteemed as a characteristic of antiquity. It is a basic sulfate of copper, usually with oxides of tin, copper, and lead. Another green copper paint pigment is **copper carbonate,** also called **artificial malachite.** It is a poisonous powder of composition $CuCO_3 \cdot Cu(OH)_2$, made by adding sodium carbonate to a solution of copper sulfate. The specific gravity is 3.7. It is insoluble in water. As a pigment it is also named **mineral green, Bremen green,** and **mountain green.**

COPPER ALLOYS. Copper serves as the base metal for a great variety of wrought and cast alloys, details of which are included in other sections under their common names, such as brass, bronze, and beryllium copper. The major wrought alloys and their designations, or alloy numbers, are **high-copper alloys** (C16200 to C19750), which include **cadmium copper, beryllium copper,** and **chromium copper**; **copper-zinc brasses** (C20500 to C28580); **copper-zinc-lead leaded brasses** (C31200 to C38590); **copper-zinc-tin alloy** or **tin brasses** (C40400 to C49080); **copper-tin phosphor bronzes** (C50100 to C52400); **copper-tin-lead bronzes** or **leaded phosphor bronzes** (C53200 to C54800); **copper-phosphorus alloys** and **copper-silver-phosphorus alloys** (C55180 to C55284); **copper-aluminum alloys** or **aluminum bronzes** (C60600 to C64400); **copper-silicon alloys** or **silicon bronzes** (C64700 to C66100); miscellaneous **copper-zinc alloys** (C66400 to C69950); **copper-nickel alloys** (C70100 to C72950); and **copper-nickel-zinc alloys** or **nickel silvers** (C73150 to C79900). All told, there are about 300 standard wrought alloys, and many have cast counterparts (C81300 to C99750). There are about 140 standard cast alloys. Many copper alloys are also available in powder form for powder-metal parts. There are also **memory alloys,** or **shape-memory alloys,** which can be deformed and then revert to their original shape when heated to their transformation temperature. Reusable locknuts, from Memry Corp., are one application. With copper-aluminum-zinc or copper-aluminum-nickel alloy used for an insert, the insert can be deformed to lock the nut in place. When the nut is removed and heated to the alloy's transformation temperature, the insert returns to its original shape so that the nut can be reused.

COPPER-NICKEL ALLOYS. A series of wrought and cast copper alloys containing nickel as the main alloying element. **Copper-nickel wrought alloys** are designated C70100 to C72950; cast alloys, C96200 to C96800. The alloys also have been referred to as **cupronickels,** copper-nickel 20% (or whatever the percentage of nickel), and 80–20 (or whatever the percentage of copper and nickel). Nickel content may be as low as 2 to 3% (C70200) or as high as 43 to 46

(C72150), but intermediate amounts, nominally 10 (C70600 and C96200), 20 (C71000 and C96300), and 30 (C71500 and C96400), are the most common. Most of the alloys also contain small to moderate amounts of iron, zinc, manganese, and other alloying ingredients, and some contain substantial amounts of tin (1.8 to 2.8% in C72500; 7.5 to 8.5 in C72800).

The 10, 20, and 30% nickel wrought alloys (C70600, C71000, and C71500, respectively) are available as flat products, rod, bar, forgings, pipe, and tubing; and their cast counterparts are amenable to sand and centrifugal casting, and some to investment and continuous casting as well. These alloys are noted for their outstanding resistance to aqueous corrosion, the 30% nickel alloy being the best of all major copper alloys in this respect, although the 10% nickel alloy is more widely used because of its lower cost. All three alloys, however, are widely used for condenser and heat-exchanger tubing in recirculating-steam systems. They are also superior to coppers and many other copper alloys in their resistance to acid solutions. And they are highly resistant to stress-corrosion cracking. Other applications include condenser plates, tube sheets, distiller tubes, salt-water piping systems, marine components, water boxes, and springs.

Each of the three alloys has a density of 0.323 lb/in^3 (8,941 kg/m^3) and a specific heat of 0.09 Btu/(lb · °F) [380 J/(kg · K)]. Electrical conductivity decreases with increasing nickel content: 9, 6.5, and 4.6%, respectively, relative to copper. Thermal conductivity ranges from 23 Btu/(ft · h · °F) [40 W/(m · K)] for C70600 to 17 Btu/(ft · h · °F) [29 W/(m · K)] for C71500. In the annealed condition, tensile properties of C70600 and C71000 thin, flat products having 0.002-in (0.050-mm) average grain size are 51,000 lb/in^2 (352 MPa) ultimate strength, 13,000 lb/in^2 (90 MPa) yield strength, and 35% elongation. Elastic modulus in tension is 20×10^6 lb/in^2 (138,000 MPa), and hardness is Rockwell B 25. Cold working increases strength and decreases ductility appreciably to, say, 85,000 lb/in^2 (586 MPa) ultimate strength, 79,000 lb/in^2 (545 MPa) yield strength, and 3% elongation in the extra spring temper. C71500 rod of 1-in (0.98-mm) diameter provides 55,000 lb/in^2 (379 MPa), 20,000 lb/in^2 (138 MPa), and 45%, respectively, in the annealed condition, and 75,000 lb/in^2 (517 MPa), 70,000 lb/in^2 (483 MPa), and 15% in the half-hard temper. As cast, the 10, 20, and 30% nickel cast alloys (C96200, C96300, and C96400, respectively) have minimum tensile properties of 45,000 to 75,000 lb/in^2 (310 to 517 MPa) ultimate strength, 25,000 to 55,000 lb/in^2 (172 to 379 MPa) yield strength, and 10 to 20% elongation. C96600, which contains 0.5% beryllium and responds to solution heat treatment and precipitation hardening, has typical tensile properties of 110,000 lb/in^2 (758 MPa), 70,000 lb/in^2 (483 MPa), and 7%, respectively.

C71900, which contains 28 to 33% nickel and 2.2 to 3.0 chromium, hardens by spinodal decomposition. Spinodal structures form in alloys having a miscibility gap and in which atoms of the component metals are sufficiently mobile at heat-treating temperatures. After such **spinodal alloys** are heated to a temperature above this gap, they are rapidly cooled to a temperature within the gap and are held there until spinodal decomposition, at a rate governed by the diffusion rate of the component metals, has been completed. In the case of C71900, the spinodal structure is induced by heating the alloy to 1650 to 1850°F (900 to 1000°C), rapid cooling to 1400°F (760°C), and then slow cooling through the 1400 to 850°F (760 to 425°C) range. Resulting tensile properties are 78,000 lb/in^2 (538 MPa) ultimate strength, 47,000 lb/in^2 (324 MPa) yield strength, 25% elongation, and 22×10^6 lb/in^2 (152,000 MPa) elastic modulus.

Several copper-nickel alloys are also part of the family of **electrical-resistance alloys.** These include **radio alloys,** which contain 78 to 98% copper and 2 to 22 nickel; the **manganins,** 83 to 85 copper, 10 to 13 manganese, and 4 nickel; and the **constantans,** 55 to 57 copper and 43 to 45 nickel.

COPPER ORES. There are about 15 copper ores of commercial importance, and these are widely distributed in almost all parts of the world. More than 40 countries produce copper on a commercial scale. The average copper content of ores, however, is usually low, and copper would be an expensive metal if it were not for the valuable byproducts: silver, gold, nickel, and other metals. About 80% of the ores in the United States contain only 1.17 to 1.57% copper and are concentrated before smelting. The direct smelting ores average from 4.3 to 6.2% copper. The most important ore of copper is **chalcopyrite,** also known as **copper pyrite** and **yellow copper ore.** It occurs widely distributed, associated with other minerals, and may carry gold and silver. It is the chief copper ore in many parts of the United States, Canada, Chile, Africa, England, and Spain. Chalcopyrite is a sulfide of copper and iron, $CuFeS_2$, containing theoretically 34.5% copper. It usually occurs massive, with Mohs hardness 3.5 and a specific gravity of 4.2. The color is brass yellow, with greenish-black streaks. To obtain the copper, first the ore is smelted with enough sulfur to combine with all the copper, producing a matte which is a mixture of CuS_2 and FeS together with impurities. Then air is blown through the molten matte in a reverberatory furnace, converting the iron sulfide to oxide and the sulfur to sulfur dioxide. The remaining copper is cast into pigs, which are called **blister copper,** owing to its blistered appearance. Blister copper contains 96 to 99% copper, with various metals and arsenic and sulfur. It is not used commercially, but is refined in furnaces or electrolytically. The **cement copper** shipped from Cyprus contains about 51% copper.

Chalcocite is another important ore found in Montana, Arizona, Alaska, Peru, Mexico, and Bolivia. It is a **cuprous sulfide,** Cu_2S, containing theoretically 79.8% copper. It usually occurs massive, but crystals are also found. The hardness is 2.5 to 3, and the specific gravity 5.5. It has a shining lead-gray color. But the emerald-green platy mineral **chalcolite** is a **copper-uranium mica,** $CuO \cdot 2UO_3 \cdot P_2O_5 \cdot 8H_2O$, with a high percentage of uranium oxide, U_3O_8. **Tennantite,** or **gray copper ore,** found in Colorado, Wyoming, and Montana, has composition $3Cu_2S \cdot As_2S_3$, with iron and antimony. When much of the arsenic is replaced by antimony, it is called **tetrahedrite. Azurite,** also called **blue copper carbonate** and **chessylite,** is found with other copper ores. It is a basic carbonate of copper, $Cu(OH)_2 \cdot 2CuCO_3$, occurring in azure-blue crystals. **Malachite,** or **green copper ore,** is an important carbonate ore, $Cu(OH)_2 \cdot CuCO_3$, containing theoretically 57.4% copper. It has a bright-green color, specific gravity 3 to 4, and Mohs hardness 3.5 to 4. **Cuprite,** or **red copper ore,** is a **cuprous oxide,** Cu_2O, containing theoretically 88.8% copper. It occurs usually massive, but sometimes in crystals. The specific gravity is 6, and the hardness 3.5 to 4. The color may be various shades of red, with an adamantine luster in the clear crystalline form or a dull, earthy luster in the massive varieties. Cuprite is found in the copper deposits in Arizona and is one of the ores in Chile, Peru, and Bolivia.

Bornite, also known as **horseflesh ore, peacock ore,** and **variegated ore,** is an important ore of copper widely distributed and mined in Chile, Peru, Canada, and the United States. It occurs in massive form, having a bronze color that turns purple on exposure. The composition is Cu_5FeS_4, having theoretically 63.3% copper. It has a metallic luster and a hardness of 3. **Chrysocolla** is a highly refractory ore of copper occurring in the oxidized parts of copper veins of Arizona and New Mexico. It is a hydrous **copper silicate** of composition $CuSiO_3 \cdot 2H_2O$. It occurs in compact masses with a specific gravity of 2 to 2.4 and a hardness of 2 to 4. The color is green to bluish. It was used as a green pigment by the ancient Greeks. Large reserves of this ore occur in Gambia and other copper regions of Africa, and it is treated by high-temperature methods to obtain the copper. **Atacamite** is an ore found in Bolivia, Arizona, and Australia. It is a copper chloride with copper hydroxide, $CuCl_2 \cdot 3Cu(OH)_2$, generally found in confused crystalline aggregates, fibrous or granular. The hardness is 3 to 3.5, specific gravity 3.75, and the color may be various shades of green. The unique copper ores of Japan, called **kuromono,** are complex sulfide-sulfate replacement minerals.

Much **native copper** metal occurs in the Lake Superior region, particularly in Michigan, but it occurs irregularly and not in continuous veins. The Ontonagon boulder of native copper in the National Museum, weighing 3 tons (2.7 metric tons), came from Michigan. A mass of native copper found in 1847 was 10 ft (3 m) long and weighed 6 tons (5.4 metric tons). The largest ever found weighed 18 tons (16.3 metric tons).

COPPER OXIDE. There are several oxides of copper, but usually the term refers to **red copper oxide,** or **cuprous oxide,** Cu_2O, a reddish crystalline powder formed by the oxidation of copper at high temperatures. It also occurs naturally in cuprite ore. The specific gravity is 6.0, and the melting point 2255°F (1235°C). It is insoluble in water but soluble in acids and alkalies. It is used in coloring glass and ceramics red, in electroplating, and in alternating-current rectifiers. **Rextox,** of Westinghouse Electric Corp., is copper upon which a layer of copper oxide has been formed. Electric current will flow easily from the oxide to the copper, but only with difficulty from the copper to the oxide. It may be used for transforming alternating current to pulsating direct current. **Black copper oxide,** or **cupric oxide,** CuO, is a brownish-black amorphous powder of specific gravity 6.4 and melting point 1949°F (1065°C). It is used for coloring ceramics green or blue. In its natural ore form, it is called **tenorite.** Together with the red oxide, it is used as a copper paint for ships' bottoms. **Copper hydroxide,** formed by the action of an alkali on the oxides, is a poisonous blue powder of composition $Cu(OH)_2$ and specific gravity 3.37. It is used as a pigment.

COPPER STEEL. Steel containing up to 0.25% copper and very low in carbon, employed for construction work where mild resistance to corrosion is needed and where the cost of the more resistant chromium steels is not warranted. It is employed in sheet form for culverts, ducts, pipes, and such manufacturing purposes as washing-machine boilers. The copper-bearing grade specified for culverts by the ASTM contains not less than 0.20% copper and not more than 0.10 carbon, manganese, phosphorus, sulfur, and silicon as impurities. The alloy steels containing considerable copper for special purposes are not classified as copper steels. The copper neutralizes the corroding influence of the sulfur in the steel and aids in the formation of a fine-grained oxide that retards further corrosion. Copper is not added to unalloyed high-carbon steels because it causes brittleness and hot-shortness. Since the carbon content of copper steel is usually very low, the material is more a **copper iron.** Unless balancing elements, especially nickel, are present, more than 0.2% copper in steel may cause rolling defects. Molybdenum in small quantities may also be added to give additional corrosion resistance, and the percentage of carbon may be raised to 0.40% when about 0.05% molybdenum is added. **Toncan iron** has this composition and has a tensile strength of 40,000 to 48,000 lb/in² (276 to 331 MPa), elongation of 32 to 40%, and density of 0.283 lb/in³ (7,833 kg/m³).

COPPER SULFATE.   Also called **bluestone, blue copperas,** and **blue vitriol.** An azure-blue, crystalline, lumpy material of composition $CuSO_4 \cdot 5H_2O$ and specific gravity 2.286. It is soluble in water and insoluble in alcohol. When heated, it loses its water of crystallization and melts at 302°F (150°C). In its natural form, called **chalcanthite,** it is a rare mineral found in arid regions and deposited from the water in copper mines. It is produced as a by-product in copper refineries, or by the action of sulfuric acid on copper or copper oxide. A major market for copper sulfate is agriculture, where it is used in fungicides, micronutrients for fertilizers and animal feeds, and seed treatment. In chemical processes, it is used as an algicide in water treatment, for separating sulfide ores, in electroplating, in froth flotation, in leather tanning and hide preservation, and as a raw material for other salts and dyes. It is a component of **chromated copper arsenate,** a mixture of **potassium dichromate,** copper sulfate, and **arsenic pentoxide,** a major wood preservative that is being phased out of commercial use due to its carcinogenic properties.

CORAL.   A shiny, hard, calcareous material valued for jewelry, buckles, beads, and novelties. It is a growth composed of the skeletons of *Corallium nobile* and other species of aquatic protozoa. The structures are built up by these creatures into forms like leafless trees or shrubs, fans, mushrooms, or cups. **White coral** is common and is not used commercially. The most valuable is **red coral,** a twiglike species that grows about 12 in (30 cm) high with thin stems. **Pink coral** and **black coral** are also valued. Red and pink corals come from the Indian Ocean and off the coast of northeastern Africa. Black coral is from southeastern Asia. The red and black varieties are very hard and take a beautiful polish. The pink is softer, with a more delicate appearance, and is used for beads. The rate of growth of coral is very slow. The gleaming white sand of tropical beaches called **coral sand** is usually not coral, but consists of the disintegrated limy skeletons of the seaweed *Halimeda opuntia.*

CORDAGE.   A general term for the flexible string or line of twisted fibers used for wrapping, baling, power transmission, and hauling. **Cordage fibers** are any materials used for making ropes, cables, twine, and cord. In general, cordage fibers are hard compared with those used for weaving into fabrics, but cotton and some other soft fibers are used for cord. **Twine** is cordage less than 0.1875 in (0.48 cm) in diameter and is composed of two or more rovings twisted together. **Rope** is cordage made by twisting several yarns into strands and then twisting the strands into a line. A **cable** is a strong rope, usually referring to the large sizes of special construction. **Cord** is an indefinite term for twine but is, more specifically, the soft cotton twines used for wrapping. The term **string** is applied to the weak cotton cords used for wrapping light packages. **Seaming twines** are made of flax fibers. **Seine twine** is a three-strand cotton twine with 2 to 56 plies per strand. Most of the **binder twine** is made from sisal, but **Indian twine** is made from jute. Ramie fiber is used for marine twines. **Binder twine** has 15 turns per foot (49 turns per meter) and 500 ft/lb (336 m/kg). **Baler twine,** for heavier work, has 12 turns per foot (39 turns per meter) and 125 ft/lb (84 m/kg). Before the advent of synthetics, about half of American strong cordage was from Manila hemp and about 30% from sisal. Manila hemp is very resistant to seawater. Sisal is used for the cheaper grades of rope, but it absorbs water easily. True hemp is considered a superior fiber for strong ropes. Untarred **hemp rope** is used for elevator cables, and tarred hemp is employed for ship cables. **Marine rope,** used by the Navy, was formerly true hemp, then Manila hemp, and is now often synthetic fiber. Most industrial rope has at least three strands, each strand having at least two yarns, and may be hard lay, medium lay, or soft lay. Twisting may be S twist or Z twist, conforming approximately to the shape of these letters. Cable twist has the twists alternating in each successive operation. Hawser twist, to give greater strength and resilience, has the plies twisted SSZ.

Cordage fibers are also obtained from a wide variety of plants. Generally, after the fibers are retted, the softer and finer fibers are separated for use in weaving into fabrics and the harder and coarser fibers are marketed as cordage fibers. **New Zealand hemp,** or **New Zealand flax,** is a strong cordage fiber obtained from the leaves of the swamp lily *Phormium tenax,* grown in New Zealand and Argentina. The fibers are white, soft, and lustrous. One variety of the plant reaches a height of 16 ft (4.9 m) and the other variety 6 ft (1.8 m). **Olona fiber,** grown in Hawaii and used locally for fishnets, is from the nettle plant *Touchardia latifolia.* The bast fibers of the bark of the slender branches are soft and flexible, are very water-resistant, and have a tensile strength 3 times that of Manila hemp. **Gravatá** is a Brazilian name for the very long and resistant fibers from the leaves of the pineapple plant *Ananas sagenaria.* The leaves of this species are up to 7 ft (2.1 m) in length. The fiber known as **widuri** of Indonesia is bast fiber from the tree *Calotropis gigantea* which yields the madar kapok. It has great strength and is resistant to seawater. It is used for ropes and fishnets. **Agel fiber** is from the stems and leaves of the **gebang palm** of the Celebes where the various grades are used for sailcloth, rope, and fishnets; the coarser fibers are woven into Bangkok hats. The fibers from the leafstalks are fine and white. **Caraguatá** is a strong, highly resistant fiber from the plant *Bromelia balansea* of Paraguay. It is employed by the Indians for making hammocks, and is now used for cordage and burlap fabrics.

Synthetic fibers are also used for cordage. **Nylon rope** is about twice as strong as **Manila rope,** is lighter, and because of its property of stretching rapidly but recovering slowly, it makes a desirable rope for lifting and towing, giving a smooth, shock-absorbing pull. Nylon ropes are used for pulling airplane gliders and for tugboat lines. **Mylar rope,** is made by slitting Mylar film and stretching and spinning the strands. A three-strand rope of 1-in (2.5-cm) diameter has a breaking strength of 18,000 lb/in^2 (124 MPa), compared to 9,000 lb/in^2 (62 MPa) for Manila rope of the same size. Moisture absorption is less than 0.3%. Elongation at 50% of breaking strength is about 4.75%. **Saran rope,** for chemical-plant use, is formed of three strands of vinylidene chloride monofilament. The breaking strength is 70% that of Manila rope, and it is flexible and chemical-resistant, but it is not recommended for temperatures above 170°F (77°C). **M-cord** is a strong wrapping twine made with a core of Manila fiber wrapped with a tough, smooth paper. Nylon and some other plastics have a tendency to fray in cordage and may be coated with polyvinyl butyral to give abrasion resistance. **Chemclad** is **rayon cordage** coated with polyvinyl chloride. Nylon rope is steel-wire rope with an extruded coating of nylon in various colors, used for automotive brake cables, aircraft control cables, and luggage handles. **Glass rope,** woven from continuous filaments of glass fiber, is used for chemical and electrical applications where resistance to chemicals or electrical insulation is needed. It is strong, but is expensive and has low flexing strength. It comes in diameters from 0.25 to 0.75 in (0.64 to 1.90 cm). **Fiberglas cordage,** of Owens-Corning Fiberglas Corp., is marketed in diameters from 0.0156 to 0.125 in (0.04 to 0.32 cm) and made of continuous filament or staple glass fibers. The 0.125-in (0.32-cm) untreated continuous-filament cord has a breaking strength of 258 lb (116.5 kg). **Newbroc** is chemical-resistant and heat-resistant thread and cord made with continuous-filament glass fiber impregnated with Teflon plastic, in diameters from 0.0046 to 0.076 in (0.12 to 0.19 cm). It remains flexible at subzero temperatures and is used for lacings and for sewing canvas. The 0.020-in (0.05-cm) fiber has a tensile strength of 70 lb (31.6 kg). Cordage made with high-modulus polyethylene fiber has high tensile strength and elasticity and is used for tugboat hawsers.

CORE OILS.   Liquid binders used for sand cores in foundry work. The binder should add strength to the core, should bake to a dry bond, should not produce much gas, and should burn out after the metal is poured, so that the sand core will collapse. Linseed oil is considered one of the best binders, but it is usually expensive and may be mixed with cheaper vegetable oils or mineral oil. In some cases fish oil or rosin is also used. Molasses, dextrin, or sulfite liquor may be included in prepared core oils. The specifications of the American Foundrymen's Society call for 50% raw linseed oil, 25 H grade rosin, and 25 water-white kerosene, with no fish oil. A good core oil should have a specific gravity of 0.9368 maximum, flash point 165 to 200°F (73 to 93°C), Saybolt viscosity 155 minimum, and iodine number 154, and should be of light color. However, any drying oil or semidrying oil can be used to replace all or part of the linseed oil. Perilla and corn oils are used, and core oils of linseed and soybean oil mixtures have good strength. The liquor from sulfite pulp mills contains lignin and is used as a core binder. **Glutrin** is a core oil with sulfite liquor. **Truline** is a resinous binder in a powder form marketed by Hercules Inc. **Uformite 580,** of Rohm & Haas Co., is a **core binder** especially for aluminum sand cores. It is a modified urea-formaldehyde resin which bakes in the core at 325 to 375°F (162 to 190°C), and it will break down in the core at temperatures above 450°F (232°C). **Cycor 191,** of American Cyanamid Co., is a urea-formaldehyde resin in water solution for sand cores for short-cycle baking in an electronic oven. **Dexocor** and **Kordex** are dextrin binders.

CORK.   The thick, spongy bark of a species of oak tree, *Quercus suber,* grown in Spain, Portugal, Italy, Algeria, Morocco, Tunisia, and to a limited extent the United States. It is used for bottle stoppers, insulation, vibration pads, and floats for rafts and nets. The scrap cuttings are used for packing for the transportation of fruits and the manufacture of linoleum and pressed products. When marketed as **granulated cork,** this material usually comes in sizes of 0.5 in (1.27 cm) and No. 8 mesh. Cork is also used natural or in the form of pressed composition for gaskets, oil retainers, roll coverings, polishing wheels, and many other articles. The material has a cellular structure with more than 50% of the volume in air cells. The cell structure is peculiar, and each cell is in contact with 14 neighboring cells, and because of lack of capillarity it does not absorb moisture. When dried, cork is light, porous, easily compressed, and very elastic. It is one of the lightest of solid substances, the specific gravity being 0.15 to 0.20. It also has low thermal conductivity. Charring begins at 250°F (121°C), but it ignites only with difficulty in contact with flame. The cork tree grows to a height of about 30 ft (9 m). After it has attained the age of about 25 years, it can be barked in the summer, and this barking is repeated every 8 or 10 years. The quality of the bark improves with the age of the tree, and with proper barking, a tree will live for 150 years or more. The thickness of the bark varies from 0.5 to 2 in (1.27 to 5.08 cm). **Cork bark** is shipped in bales of 170 lb (77.1 kg), and **cork wastes** in bales of 148 lb (67.1 kg).

Brazilian cork is the bark of the tree *Angico rayado,* called **pao santo bark,** and also the trees *Piptadenia incuriale* and *P. colomurina.* The bark has a cellular structure and, when ground, has the appearance of a low grade of true cork, but is softer. It is suitable for insulation. A substitute for cork for insulation packings and acoustical panels is **Palmetex.** It is the compressed pith from the internal fibers of the sawtooth palm *Cerano repens,* of the eastern Gulf states. It has lower conductivity than cork, but without a binder it is more friable. **Corkboard** is construction board made by compressing granulated cork and subjecting it to heat so that the particles cement themselves together. It is employed for insulating walls and ceilings against heat and cold and as a sound insulator. **Cork tile** is corkboard in smaller, regularly shaped blocks for the same purposes. The natural gum in the cork is sufficient to bind the particles, but other binders may be used.

CORN. One of the most important food grains of the world for both human and animal consumption, but also used industrially for the production of starch, glucose, alcohols, alcoholic beverages, and corn oil. Corn was unknown to Europe before the discovery of America, where it was one of the chief foods of the Indians from Canada to Patagonia. In Europe and in foreign trade, it is known by its original name **maize,** and the Incan name **choclo** still persists in South America for the grains on the cob. In Great Britain, corn means all hard grains including wheat, and the U.S. term *corn* is an abbreviation of the name **Indian corn.** In South Africa, it is called **mealies.** Corn is the seed grain of the tall leafy plant *Zea mays,* of which there are innumerable varieties of subspecies. It grows in temperate climates and in the high elevations of the tropics where there is a warm growing season without cold nights; but high commercial yields are limited to areas where there is a combination of well-drained friable soil, plenty of moisture, few cloudy days, and no night temperatures below 66°F (19°C) during the growing season of 4 months. Corn is an unnatural plant, with seeds not adapted for natural dispersal; it does not revert to a wild species. It is a product of long selection. No wild plants have ever been found, but it is believed to have been a cultivated selection from the grass **teosinte** of Mexico. About half the world production of corn is in the United States and Argentina, but large amounts are also grown in southern Europe and northern India.

Confectionery flakes, used as an additive and conditioner in candy, cookies, and pastries, is a bland, yellowish, flaky powder made from degerminated yellow corn. It contains 8% protein and is pregelatinized to require no cooking. The **pregelatinized corn flour** of General Foods Corp., used to improve texture, binding qualities, and flavor of bakery products, is a cream-colored powder which hydrates in cold water and needs no cooking. It contains 82% starch, 9 protein, 1 corn oil, and 8 moisture, and it is a food ingredient rather than an additive, although it may replace 10% of the wheat flour. In the corn belt of the United States, 40% of the corn grown is used for hog feed, while in the dairy belt the hogs are fed on skim milk, buttermilk, and whey, and most of the corn is fed to poultry or shipped commercially.

Corn grains grow in rows on a cob enclosed by leafy bracts. They are high in starch and other food elements, and they form a valuable stock feed especially for hogs and poultry. Nearly 90% of the commercial corn in the United States is for animal feed. But corn is one of the cheapest and easiest sources of starch, and much of the Argentine corn is used for starch and glucose.

Sweet corn is a type of **soft corn,** *Z. saccharata,* cultivated for direct eating and for canning. There are about 70 varieties grown widely on farms, but not cultivated for industrial applications. **Popcorn,** *Z. everta,* has very hard, small, elongated oval grains which, when heated, explode into a white, fluffy, edible mass without further cooking. It was used by the Indians as a food for journeys and is now grown for food and confections. The corns cultivated for stock feeding and for starch and glucose are varieties of **flint corn,** *Z. indurata,* and **dent corn,** *Z. indentata.* Flint corn has long, cylindrical ears with hard, smooth grains of various colors. Dent corn has larger and longer ears which are tapering, with white or yellow grains. About 300 varieties of dent corn are grown in the corn belt of the United States, while the Argentine corn is largely flint varieties which yield high starch. Much of the corn grown in the United States is **hybrid corn.** This is not a species, but consists of special seed stocks produced by crossing inbred strains. It is resistant to disease and gives high yields. The **waxy corn** grown in Iowa produces a starch comparable with the root starches. In the wet milling of corn for the production of cornstarch, the germ portion of the grain is separated as a by-product and used for the extraction of **corn oil,** or **maize oil.** The germ contains 50% oil which is a bright-yellow liquid of specific gravity 0.920 to 0.925, iodine value 123. It contains 56% linoleic, 7 palmitic, 3 stearic, and the balance mainly oleic acid. About

1.75 lb (0.80 kg) of oil per bushel of corn is obtained by crushing the germ, and another 1.4% is obtained by solvent extraction. About 1% of oil remains in the **corn oil meal** marketed as feed. Corn oil is used as an edible oil as a substitute for olive oil and in margarine, and also in soaps, belt dressings, core oils, and for vulcanizing into factice. Corn syrups and glucose are produced directly from the starchy corns. **Zein** is a protein extracted from corn. It is dissolved in alcohol to form a lacquerlike solution which will dry to a hard, tough film. It is used as a substitute for shellac and is more water-resistant than shellac. **Zein G210** is a water solution of prolamine protein extracted from corn gluten, used to produce hard, tough, grease-resistant coatings and for formulating polishes and inks. **Corn tassels** are used for livestock and poultry feed. They are a rich source of vitamins. About 270 lb (122 kg) of dry tassels is produced per acre. **Cornstalks** contain up to 11% sugars, usually about 8% sucrose, and 2 other sugars, but little sugar is produced commercially from this source, the stalks being used as cattle feed. **Corncobs** are used to produce **cob meal** for feeds and are processed to produce lignin, xylose, furfural, and dextrose. **Korn-Kob** is granular corn cob used as an abrasive material for finishing metal parts in tumbling barrels. It is tougher than maple and will not absorb water as wood granules do.

Kafir corn is a variety of sorghum grass not related to true corn. The plant is a tall annual with a stalk similar to corn but with smaller leaves and long, cylindrical, beardless heads containing small, round seed grains. It is widely grown in tropical Africa, and a number of subvarieties are grown on a limited scale in Kansas, Texas, and Oklahoma. The grain is similar in composition to corn, but has a peculiar characteristic flavor. It is used as flour in bread mixtures and in biscuit and waffle flour.

CORROSION-RESISTANT CAST ALLOYS.   In general, these are the cast counterparts to 3XX and 4XX wrought stainless steels and, thus, are also referred to as **cast stainless steels.** Designations of the Alloy Casting Institute of the Steel Founders Society of America and the wrought designations to which they roughly correspond (compositions are not identical) include CA-15 (410), CA-40 (420), CB-30 (431), CC-50 (446), CE-30 (312), CF-3 (304L), CF-3M (316L), CF-8 (304), CF-8C (347), CF-8M (316), CF-12M (316), CF-16F (303), CF-20 (302), CG-8M (317), CH-20 (309), and CK-20 (310). There are also other alloys that do not correspond to wrought grades. The cast alloys corresponding to 3XX wrought grades have chromium contents in the range of 17 to 30% and nickel contents in the range of 8 to 22%. Silicon content is usually 2.00% maximum (1.50 for CE-8M), manganese 1.50 maximum, and carbon 0.08 to 0.30 maximum, depending on the alloy.

Other common alloying elements include copper and molybdenum. Those corresponding to 4XX grades may contain as much chromium but much less nickel: 1 to 5.5%, depending on alloy. Manganese and silicon contents are also generally less, and carbon may be 0.15 to 0.50%, depending on the alloy. All the alloys are **iron-chromium-nickel alloys,** and the most widely used are CF-8 and CF-8M, which limit carbon content to 0.08%. CN-7M and CN-7MS contain more nickel than chromium and, thus, are referred to as **iron-nickel-chromium alloys.**

The alloys are noted primarily for their outstanding corrosion resistance in aqueous solutions and hot, gaseous, and oxidizing environments. Oxidation resistance stems largely from the chromium. Nickel improves toughness and corrosion resistance in neutral chloride solutions and weak oxidizing acids. Molybdenum enhances resistance to pitting in chloride solutions. Copper increases strength and permits precipitation hardening to still greater strength. After a 900°F (482°C) age, for example, the room-temperature tensile properties of CB-7Cu are 187,000 lb/in^2 (1,290 MPa) ultimate strength, 160,000 lb/in^2 (1,100 MPa) yield strength, 10% elongation, and 28.5×10^6 lb/in^2 (196,500 MPa) elastic modulus. Hardness is Brinell 412 and impact strength (Charpy V-notch) 7 ft · lb (9.5 J). At 800°F (426°C), yield strength approaches 120,000 lb/in^2 (827 MPa). Higher aging temperatures, to 1150°F (621°C), decrease strength somewhat but markedly increase impact strength. The alloys are widely used for pumps, impellers, housings, and valve bodies in the power-transmission, marine, and petroleum industries; and for chemical, food, pulp and paper, beverage, brewing, and mining equipment.

CORUNDUM. A very hard crystalline mineral used chiefly as an abrasive, especially for grinding and polishing optical glass. It is **aluminum oxide,** Al_2O_3, in the alpha, or hexagonal, crystal form, usually containing some lime and other impurities. It is found in India, Burma, Brazil, and in states of Georgia and the Carolinas, but most of the commercial production is in South Africa. The physical properties are theoretically the same as for synthetic alpha alumina, but they are not uniform. The melting point and hardness are generally lower because of impurities, and the crystal structure also varies. The hexagonal crystals are usually tapered or barrel-shaped, but may be flat with rhombohedral faces.

The Hindu word *corundum* was originally applied to gemstones. The ruby and the sapphire are corundum crystals colored with oxides. **Oriental topaz** is yellow corundum containing ferric oxide. **Oriental emerald** is a rare green corundum, but it does not have the composition of the emerald, and the use of the name is discouraged in the jewelry industry. The clear-colored crystals are sorted out as gemstones, and the premium ore is the large-crystal material left after sorting. Some material is shipped in grain. The crude ore is washed, crushed, and graded. There are four grades of abrasive corundum shipped from South Africa: Grade A is over 92% Al_2O_3, Grade B is 90 to 92%, Grade C is 85 to 90%, and Grade D is under 82%. In the United States most of the natural corundum used for optical-glass grinding is in sizes from 60 to 275 mesh, while the grain sizes for coarse grinding and snagging wheels are 8 to 36 mesh. Corundum is now largely replaced by the more uniform, manufactured aluminum oxide, and even the name **synthetic corundum,** or the German name **Sintercorund,** is no longer used.

COSMETICS. Substances applied to the outer surface of the body for enhancing appearance and/or for improving the condition of the skin. Most cosmetics also contain odorants and perfume oil. **Face powders** are composed of white pigments having high covering power, such as titanium oxide and zinc oxide; pigments, such as iron oxide and talc (hydrated magnesium silicate), to import slip; and adhesion-promoting ingredients, such as zinc or magnesium stearate. **Rouges** for the face, which contain many of the ingredients present in face powders, are produced in pressed powder or paste form. The coloring agents are usually water-insoluble, bright red lakes, and the binder is an oil, lanolin, or gum tragocanth. The ingredients of lipstick are principally a vehicle of castor oil and a mixture of waxes, such as beeswax, carnauba wax, candililla wax, lanolin, butyl stearate, and spermaceti. A great variety of other substances are used for special effects. The color ingredients are usually lakes.

Mascaras, used on eyelashes, are made of an oil-soluble soap base, such as triethanolamine stearate; waxes; and color pigments, such as carbon blacks, iron oxide, and ultramarine blue.

Nail polishes, or **nail lacquers,** are made of a nitrocellulose, gum resins, and plasticizers dissolved in a mixture of solvents. For color and opacity, lakes and a substance like titanium oxide are also present.

Although produced in great variety, most **skin creams,** or **cold creams,** are emulsions composed of oils, water, beeswax, and borax. A typical cold cream contains spermaceti, beeswax, oil of lemon or mineral oil, borax, and rose water. **Handcreams** and **hand lotions** for protection against chapping are emulsions formed from a soap, an oil, and glycerine. Other ingredients that can be present include water.

The active ingredients in **astringents,** sold by the name of **skin bracers** or **aftershave lotions,** are witch hazel or alcohol. Often they contain 50% water by volume. **Refiners** are astringents containing aluminum salts that when applied to the skin cause slight swelling, which in turn causes the pores to look smaller for a brief period of time. **Clarifiers** are liquids containing such chemicals as bromelin, resorcinol, or a salicylate, which remove the skin's top layer of dead cells and give the skin a fresher appearance. **Facial masks,** consisting of various "clay" minerals, such as bentonite and kaolin, produce a tight film over the skin upon drying, causing the skin pores to become smaller. Paint-on–peel-off masks use polyvinyl alcohol or vinyl pyrrolidone to form the dry film.

Suntan lotions are formulated to protect the skin against damage from excessive exposure to sunlight. They generally are composed of ingredients similar to those in other skin creams. In addition, however, substances that screen out ultraviolet radiation are present.

Deodorants are of two different types. **Antiperspirants** use zinc and/or aluminum salts that have an astringent action to block the pores through which perspiration is secreted. Other deodorants prevent the bacterial decomposition of the perspiration that produces unwanted odors. These **antibacterial deodorants** contain germicides, such as hexachlorophene.

Bath salts are generally composed of sodium sesquicarbonate or sodium phosphates dissolved in alcohol along with some color and perfume oil. **Bubble bath preparations** contain foaming agents such as sulfated alcohols or sulfated glyceryl monolaurate. In one type of **bath oil** perfume oils are mixed with an agent such as polyoxyethylene sorbitan monolaurate, which disperses the oil in the water. In another type of bath oil, the perfume is dissolved in a low-viscosity oil.

Shampoos for washing hair are composed of one or more detergent materials. **Soaps** derived from coconut oil are the most widely used because they are high in detergency, are excellent foaming agents, and are resistant to precipitation by hard water. In recent years increasing use has been made of **synthetic detergents,** such as sulfated castor oil, sulfated lauryl alcohol, and sulfated glyceryl monolaurate.

Hair rinses and **hair conditioners** are intended to restore the hair to its natural condition after shampooing or the use of various treatments. The **acid rinses** remove scum left by the shampoo and restore the hair's acid pH to its previous level. The **conditioning rinses,** which restore the hair's natural oily coating, contain stearalkonium chloride. Also included may be such ingredients as an alkali, an emollient of oil or fatty substance, thickeners, humectants, and fragrances.

Hair sprays coat the hair with a film that makes the hair strands stick together. Available as lotions, gels, and sprays, they contain a synthetic resin such as vinyl pyrrolidone dissolved in alcohol and water.

COTTON. The white to yellowish fiber of the calyx, or blossom, of several species of plants of the genus *Gossypium* of the mallow family. It is a tropical plant, and the finest and longest fibers are produced in hot climates, but the plant grows well in a belt across southeastern United States and as far north as Virginia. It requires a growing season of about 200 days with an average summer temperature of about 75°F (24°C) and a dry season during the time of ripening and picking. Cotton was used in India and China in most ancient times, was described in Greece as a vegetable wool of India, but was not used in Europe until the early Middle Ages. All the Asiatic species are short-staple, and the long-staple cottons are from species cultivated by the American Indians. Cotton has a wide variety of uses for making fabrics, cordage, and padding, and for producing cellulose for plastics, rayon, and explosives.

There are many species and varieties of the plant, yielding fibers of varying lengths, coarseness, whiteness, and silkiness. Cotton fiber contains 88 to 96% cellulose (dry weight), together with protein, pectin, sugars, and 0.4 to 0.8% wax. Ordinary treatment does not remove the wax. When the wax is removed by ether extraction, the fiber is stronger but is harsh and difficult to spin. The most noted classes are Sea Island, Egyptian, American upland, Brazilian, Arabian, and Nanking. **Sea Island cotton,** *G. barbadense,* was native to the West Indies, and named when brought to the islands off the American coast. It is grown best in hot, moist climates, and it is the longest, finest, and silkiest of the fibers. Its length varies from 1.25 to 2.5 in (3.18 to 6.35 cm), but it is cream-colored. **Egyptian cotton,** grown in Egypt and the Sudan, came originally from Peruvian seed. **Peruvian cotton,** *G. acuminatum,* is long-staple, silky, has strength and firmness, but is brownish. The **tanguis cotton** from Peru is valued for fine English fabrics. Egyptian cotton, or **maco cotton,** is next in quality to Sea Island. The long staple is from 1.125 to 1.375 in (2.86 to 3.49 cm), and the extra-long staple is over 1.375 in. It has a fine luster and great strength. It also has a remarkable twist, which makes a strong, fine yarn. It is used chiefly in yarns for the production of fine fabrics, thread, and automobile-tire fabrics. **American-Egyptian cotton** is grown in Arizona. The fiber has an average length of 1.625 in (4.13 cm), and it has the same uses as the Egyptian. **Upland cotton,** *G. hirsutum,* is the species originally grown by the Aztecs of Mexico. It is whiter than Egyptian or Sea Island cotton and is the easiest and cheapest to grow. There are 1,200 named varieties of this plant. The short-staple upland has a fiber under 1.125 in in length, and it can be spun only into coarse and medium yarns, but it is the most widely grown of cottons in the United States. Long-staple upland is from 1.125 to 1.375 in in length. The common grades of cotton fiber in the United States vary in diameter from 0.0006 to 0.0009 in (0.0152 to 0.0229 mm). Sea Island cotton fiber is as fine as 0.0002 in (0.005 mm), compared with 0.001 in (0.025 mm) for the coarse Indian cotton. The cotton of India, China, and the Near East is from *G. herbaceum,* and the fiber is short, 0.375 to 0.75 in (0.95 to 1.91 cm), but strong.

Cotton linters removed from the cottonseed after ginning are from 0.04 to 0.6 in (0.10 to 1.5 cm) long. The first cuts, or longer fibers, are used for upholstery and for mattresses, and amount to 20 to 75 lb (9 to 34 kg) per ton (907 kg) of seed. The second-cut short fibers vary from 125 to 180 lb (57 to 82 kg) per ton (907 kg) of seed, and are called **hull fiber.** The No. 1 grade of long linters is spinnable and can be used for mixing with cotton for yarns. This grade is also used for making absorbent cotton. The short hull fiber is cleaned and processed to produce **chemical cotton,** which is a pure grade of alpha cellulose used for making rayon, nitrocellulose, and plastics. Chemical cotton is marketed as loose pulp in bales and as sheet pulp with the sheet stacked in bales of 200 or 400 lb (91 or 181 kg), or with the continuous sheet in rolls. Formerly, cotton linters were considered the only source of pure cellulose for making nitrocellulose explosives, but pure alpha cellulose from wood is now used for this purpose.

Chaco cotton, grown in Argentina, is from Louisiana seed, and probably 70% of total world cotton is now grown from U.S. upland seed although it varies in characteristics because of differences in climate and soil. Cotton is shipped in bales of 478 lb (216 kg) each. **Cotton yarn** is put up in 840-yd (768-m) hanks, and the number, or count, of cotton yarn indicates the number of hanks to the pound. Number 10 cotton yarn, for example, has 10 hanks, or 8,400 yd/lb (16,933 kg/m).

Mercerized cotton, developed in 1851 by John Mercer, is prepared by immersing the yarn in a stretched condition in a solution of sodium hydroxide, washing, and neutralizing with dilute sulfuric acid. Mercerized yarns have a silky luster resembling silk, are stronger, have less shrinkage, and have greater affinity for dyes. The fabrics are used as a lower-cost substitute for silk, or the yarns are mixed with silk.

Absorbent cotton is cotton fiber that has been thoroughly cleaned and has had its natural wax removed with a solvent such as ether. It is very absorbent and will hold water. It is marketed in sterilized packages for medical use. **Cotton batting** is raw cotton carded into matted sheets and usually put up in rolls to be used for padding purposes. **Cotton waste,** used in machine shops for wiping under the general name of waste, is usually in mixed colors, but the best grades are generally all white, of clean soft yarns and threads without sizing. It is very oil-absorbent. **Comber waste** consists of the lengths of fiber up to 1 in (2.5 cm) and is not sold with the waste from yarns, but is sent to mills that produce cheap fabrics. **Cotton fillers,** used as reinforcing materials in molding plastics to replace wood flour or other fibers, are made by cutting cotton waste or fabric pieces into short lengths. **Filfloc** is cotton flock for this purpose; **Fabrifil** is cotton fabric cut into small pieces; and **Cordfil** is cotton cord cut into very short pieces. These fillers give greater strength to the molded product than wood flour. **Acetylated cotton** is a mildewproof cotton made by converting part of the fiber to cellulose acetate by chemical treatment of the raw fiber. **Aminized cotton** is produced by reacting the raw cotton with aminoethyl sulfuric acid in an alkaline solution. Amino groups are chemically combined with the cellulose of the fiber, which gives ion-exchange properties and good affinity for acid wool dyes, and absorption of metallic waterproofing agents. **Cyanoethylated cotton** is produced by treating the fibers with acrylonitrile, and caustic and acetic acid. The acrylonitrile reacts with the hydrogen of the hydroxyl groups, forming cyanoethyl ether groups in the fiber. The fibers retain the original feel and appearance, but have increased heat strength, better receptiveness to dyes, and strong resistance to mildew and bacterial attack. Another method of adding strength, chemical resistance, and dyeing capacity to cotton fibers is by treating them with anhydrous monoethylamine. It forms an amine-cellulose complex instead of the hydrogen bond. Since cotton is nearly pure cellulose, many chemical variations can be made, and even some dyes may alter the fiber.

COTTON FABRICS.   Cotton cloth is made in many types of weave and many weights, from the light, semitransparent **voile,** made of two-ply, hard-twisted yarn, and **batiste,** a fine, plain-woven fabric, to the coarse and heavy canvas and duck. They may have printed designs, as in **calico,** which is highly sized; or yarn-dyed plain stripes, plaids, or checks, as in **gingham**; or woven figures, as in **madras. Muslin,** a plain white fabric widely used for garments, filtering, linings, and polishing cloths, has a downy nap on the surface. The full-bleached cloth is usually of finer yarns than the unbleached. Cheaper grades are usually heavily sized, and the sizing is removed in washing. **Crinoline** is an open-weave fabric of coarse cotton yarn and is heavily sized to give stiffness. It was originally made as a dress fabric of horsehair and linen. It is now used for interlinings and as a supporting medium where a stiff, coarse fabric is needed. **Wigan** is similar to crinoline, but is more closely woven. **Percale** is a softer fabric similar to calico but with a higher yarn count. **Swiss** is a plain-woven, fine, thin muslin, stiff and crisp. **Dotted Swiss** is a very thin, transparent, plain-woven cotton with colored swivel or lappet woven dots. It is sized stiff and crisp. **Dimity** is a plain-woven, sheer fabric with ribs in the form of corded checks or stripes. It comes in white or colors. **Organdy** is a plain-woven, thin, transparent, crisp fabric stiffened with shellac or gum, usually in delicate color shades. All of these are plain-woven. **Poplin** is a lateral-ribbed fabric, often mercerized. It is heavier than broadcloth. **Rep** has a rib produced by heavy warp yarns. **Crash** is a rough-texture fabric with effects produced by novelty yarns. **Charmeuse** in the cotton industry designates a soft, fine, satin-weave fabric of Egyptian cotton used industrially as a lining material. **Chambray** is a plain-woven, lightweight cotton similar to gingham but with no pattern and a dyed warp and white filling. It is used for linings, shirtings, and dresses. **Cotton damask** is a type of jacquard-figured fabric having warp sateen figures in a filling sateen ground, or vice versa. The surface threads of the figures lie at right angles to those in the ground so that the light is diffusely reflected, causing them to stand out in bold relief. The fabric is usually of coarse or medium yarns, 15s to 30s, bleached and finished to imitate linen. **Cotton crepe** is a cotton fabric having a pebbled surface. The pebble is produced with sulfonated oil, lauric acid ester oil, or other soluble oil which is washed off after the treatment. When the word *crepe* is used alone, it usually signifies silk crepe. **Domet** is a warp-stripe cotton fabric similar to flannel, used for apparel linings. **Venetian** is a highly mercerized, stout, closely woven fabric with the yarn in reverse twist. It is used as a lining for hats, pocketbooks, and luggage. **Cottonade** is a coarse, heavy cotton fabric made to look like woolens and worsteds in weave and finish, and it is used for men's suit linings. **Eiderdown** is a cotton fabric of knitted soft-spun yarns, heavily napped on one or both sides. It is used for shoe and glove linings. **Tarlatan** is a thin cotton fabric with a net weave, heavily sized, used for linings. **Cambric** was originally a fine, thin, hard-woven linen but is now a strong cotton fabric of fine weave and hard-twist yarn. It was used as **varnished cambric** and **varnished cloth** with a coating of insulating varnish or synthetic resin. The strength exceeded that of the older **varnished silk** but was less than that of **varnished rayon.** A 0.003- to 0.008-in (0.076- to 0.203-mm) thick fabric made from high-tenacity rayon has a dielectric strength of 1,000 V/mil (39.4×10^6 V/m).

Strex, developed by Uniroyal, Inc., is an elastic, full-cotton fabric that has 100% elongation without the use of rubber. It is made from yarn that has a twisting like a coiled spring. The fabric is used for surgical bandages, gloves, and wearing apparel. **Glass cloth** is a name given to cotton fabric made of smooth, hard-twisted yarns which do not lint. It is used for wiping glass, but is now largely replaced by silicone-treated soft papers. It may be of the type known as **sponge cloth,** which is a twill fabric of **nub yarn** or honeycomb effect, or it may be of **terry cloth,** which has a heavy loop pile on one or both sides. Another **wiping cloth** for glass and instruments where a lint-free characteristic is important is made with a cotton warp and a high-tenacity rayon filling. It is strong, soft, and absorbent. For polishing glass and fine instruments, a nonwoven fabric is made by binding the cotton fibers with a plastic.

Twill is a fabric in which the threads form diagonal lines. **Tackle twill,** used for football uniforms, is also used in olive-drab color for army parachute troop uniforms. It is a strong, snag-resistant fabric having a right-hand twill with a rayon warp and combed cotton filling. It is 8.5 oz/yd² (0.29 kg/m²), 180-lb (82-kg) warp, and 80-lb (36-kg) filling. **Cavalry twill** is not a cotton cloth, but is of worsted or rayon twill woven with a diagonal raised cord. It is similar to **gabardine** except that gabardine has a single cord and cavalry twill has a double cord. **Bedford cord** has the cord running lengthwise, and the cord is more pronounced than in cavalry twill. These three are usually woolen fabrics, but **parade twill** is a mercerized cotton fabric of combed two-ply yarns, with the fabric vat-dyed in tan. It is employed for work clothing. **Byrd cloth** is a wind-resistant fabric made originally for Antarctic use. It has a close-twill weave with about 300 threads per inch. It is soft and strong and comes in light and medium weights. **Sateen** is fabric made with a close-twill weave of mercerized cotton in imitation of satin. The wind-resistant sateen used for military garments is a 9 oz/yd² (0.30 kg/m²) cotton fabric in satin weave with two-ply yarn in warp and filling. The thread count is 112 ends per inch, 68 picks per inch. The fabric is singed, mercerized, and given a water-repellent finish. **Foulard** is a highly mercerized twill-woven cotton with a silky feel. It is plain or printed and is used for dresses or sportswear. **Cotton duvetyn** is a twill-woven, mercerized cotton fabric with a fine nap that gives it a soft, velvety feel. It is much used for apparel linings and pocket linings. **Brilliantine** is a lightweight fabric with a cotton warp and a twilled worsted filling, yarn-dyed. It is used for apparel linings.

Balloon cloth is a plain-woven cotton fabric used originally as a base material in making coated fabrics for the construction of balloons, but now used in many industries under the same name. The various grades differ in weight, thread count, and strength. Grade HH, having 120 threads per inch in each direction, is most widely used. A Navy fabric has a weight of 2.05 oz/yd² (0.07 kg/m²) and a tensile strength of 38 lb/in^2 (0.26 MPa) in each direction. When several layers are built up and rubberized or plastic-coated, they may be on the bias, and the outside layer is coated with aluminum paint to reduce the heat absorption. **Gas cell fabric** is a single-ply, coated balloon cloth. **Airplane cloth,** formerly used for fabric-covered training planes, is a plain-woven cotton fabric of two-ply combed yarns mercerized in the yarn. It is usually 4 oz/yd^2 (0.14 kg/m^2), but wide fabrics may be 4.5 oz/yd^2 (0.15 kg/m^2). The cotton is 1.5 in minimum staple, and the threads per inch are 80 to 84.

COTTONSEED OIL. One of the most common vegetable oils, used primarily as a food oil in salad oils, margarine, cooking fats, and for sardine packing. It also has a wide industrial use in lubricants, cutting oils, soaps, quenching oils, and paint oils, although soybean oil is used as a more abundant substitute. The hydrogenated oil is widely used as a cooking grease. Its food value is lower than that of lard, but it is often preferred because it is odorless and does not scorch. A new market is in the formulation of pesticides. Here it is preferred over petroleum and mineral oils as a carrier for pesticides, because it is natural, safer for plants, and easily available. Cottonseed oil is expressed from the seed of the cotton plant, *Gossypium,* and is entirely a by-product of the cotton industry, its production depending upon the cotton crops. The yield of seed is 890 lb (403 kg) per 478-lb (217-kg) bale of cotton, and 100 lb (45 kg) of seed yields 15.5 lb (7 kg) of oil. When the seeds are crushed whole, the oil is dark in color and requires careful refining. U.S. practice is to hull the seeds before crushing. The oil is colorless and nearly odorless and has a specific gravity of 0.915 to 0.921. Upland cottonseed contains about 25% oil, which has 40% linoleic, 30 oleic, and 20 palmitic acids. The residue is caked and sold as **cottonseed meal** for cattle feed and fertilizer. About 900 lb (408 kg) of meal and from 450 to 620 lb (204 to 281 kg) of hulls are obtained per short ton (0.9 metric ton) of seed, the yield of hulls varying inversely with the yield of linters. The U.S. oil has an iodine value up to 110 and a saponification value of 192 to 200. Egyptian and Indian oils are inferior in color, and the Indian oil has a fishy odor and a fluorescence. **Cotton seed stearin** is the solid product obtained by chilling the oil and filtering out the solid portion. It has an iodine value between 85 and 100 and consists largely of palmitin. It is used for margarine, soap, and as a textile size. Winter-yellow cottonseed oil is the expressed oil after the stearin has been removed.

COTTONWOOD. The wood of the large trees *Populus monilifera, P. deltoides,* and other species of the United States and Canada. It is a soft wood of a yellowish-white color and a fine, open grain. It is sometimes called **poplar,** or **Carolina poplar,** and **whitewood.** The density is about 30 lb/ft^3 (480 kg/m^3). The wood is easy to work, but is not strong and warps easily. It is used for packing boxes, paneling, and general carpentry. The *P. deltoides,* or **eastern cottonwood,** used in paneling, has a specific gravity when kiln-dried of 0.43, a compressive strength perpendicular to the grain of 650 lb/in^2 (4.5 MPa), and a shearing strength parallel to the grain of 660 lb/in^2 (4.6 MPa). This wood comes from the lower Mississippi Valley. **Black cottonwood** is from the large tree *P. trichocarpa,* of the Pacific coast. The wood is used for boxes, excelsior, and pulpwood. It has a light color, uniform texture, and fairly straight grain. **Swamp cottonwood,** *P. heterophylla,* also called **river cottonwood,** grows in the Mississippi and Ohio river valleys. **Balsam poplar** is from the tree *P. balsamifera,* of the northeastern states. It is a soft, weak wood used chiefly for containers and for making excelsior. The tree also goes under the Algonquin name of **tacamahac.** The wood may be marketed as cottonwood even when mixed with aspen. It is an excellent paper-pulp material. The name cottonwood is also applied to the wood of the tree *Bombax malabaricum,* native to India, which produces kapok. The wood is white and soft and has a density of about 28 lb/ft^3 (448 kg/m^3). It is much softer than cottonwood.

COUMARONE. A colorless, oily liquid of composition C_8H_6O, used chiefly in making synthetic resins. It occurs in the fractions of naphtha between 329 and 347°F (165 and 175°C). It has a specific gravity of 1.096, is insoluble in water, and is easily oxidized. Another similar product is **indene,** C_9H_{10}, a colorless liquid of specific gravity 0.993, boiling at about 360°F (182°C), obtained from coal tar. When oxidized, it forms phthalic acid, and with sulfuric acid it polymerizes readily. It is a bicyclic ring compound with an active double bond and methylene group in the five-membered ring fused to the benzene nucleus. It can be reacted with butadiene to form an **indene-butadiene rubber** of superior properties. All the **cumenes** are variants of benzene.

The **indene resins** are classified with the coumarone resins, but they are lighter in color and are used in varnishes. The simple polymer, or di-indene resin, is a crystalline solid melting at about 136°F (58°C). The polyindene resins are made by polymerizing indene with ultraviolet light and oxygen. The courmarone **resins,** which are polymers of $C_6H_4 \cdot O \cdot CH{:}CH$, made by the action of sulfuric or phosphoric acid on coumarone, are very soluble in organic solvents and are used in lacquers, waterproofing compounds, molding, and adhesives. The specific gravity of the molded resins is 1.05 to 1.15. They have high dielectric strength. **Paracoumarone,** also called **paraindene** and **cumar gum,** is a synthetic resin which is a copolymer of coumarone and indene. The grades vary from a soft gum to a hard, brown solid, with melting points 41 to 284°F (5 to 140°C). Varnishes made with it are resistant to alkalies. **Nevindene,** of the Neville Co., is a coumarone-indene resin of specific gravity 1.08 and melting point 50 to 320°F (10 to 160°C), used for compounding with rubber and synthetics. **Nevilloid C-55** is a **coumarone-indene resin** in water emulsion for coatings. It forms cohesive translucent films of slightly tacky nature. Blended with melamine resin, it forms a clear and hard film. **Cumar** is the name of a coumarone-indene resin of Barrett Co., but the name *cumar* has been applied to a range of pale-yellow to reddish-brown coal-tar resins which are polymers of indene, coumarone, and other compounds, with melting points of 113 to 320°F (45 to 160°C). They are used in rubber compounding to increase tensile strength and tear resistance. **Piccoumarone resins** of Pennsylvania Industrial are paracoumarone-indene thermoplastic resins produced by the polymerization of unsaturates in coal-tar oils. They vary from light liquids to tacky solids with melting points of 50 to 248°F (10 to 120°C). The colors vary from pale yellow to reddish brown. They are resistant to alkalies and are used in paints and waterproofing for concrete and in adhesives for floor tile.

CREOSOTE. Also called **dead oil** and **pitch oil.** A yellowish, poisonous oily liquid obtained from the distillation of coal tar. It has the odor of carbolic acid, a specific gravity of 1.03 to 1.08, and a boiling point of 392 to 572°F (200 to 300°C). The crude **creosote oil** is used as a wood preservative and as a harsh disinfectant, but its use in these applications is expected to decrease because it has been recently classified as a possible carcinogen. Other applications include use as a fluxing oil for coal-tar pitch and bitumen, production of carbon black, and use in sprays for dormant fruit plants. Creosote is also obtained in the distillation of pinewood tar and is then a yellowish liquid with a smoky odor, a mixture of phenols and derivatives. Creosote oil contains **acridine,** a dibasic pyridine, used as an insecticide, and is also the source of other complex heterocyclic ring compounds. The distillation of wood also produces charcoal, gas, and **methyl acetate,** a sweet-smelling liquid of composition $CH_3COO \cdot CH_3$, and boiling point 129°F (54°C), used as a solvent.

 Cresol, also known as **cresylic acid** and as **methyl phenol,** obtained in the distillation of coal tar, is a mixture of three isomers of cresol, $CH_3 \cdot C_6H_4 \cdot OH$, and **xylenol,** $(CH_3)_2 \cdot C_6H_3 \cdot OH$. The crude material is a brownish-yellow liquid solidifying at 52°F (11°C). It is used for making plastics, in ore flotation, in refining petroleum, in soap-emulsion cutting oils as a disinfectant, and in medicine as a strong antiseptic such as **Lysol,** which is a 50% solution of cresols in liquid soap. It is also used in the production of other chemicals. Commercial cresols are mixtures of **orthocresol, metacresol,** and **paracresol,** or just of the latter two, and are defined as phenolic mixtures in which 50% of the material boils below 399°F (204°C). Cresol and xylenol mixtures in which 50% of the mixture boils above this temperature are called **cresylic acid,** while refined cresylic acid contains higher amounts of **xylenol,** including some higher-boiling-point phenolic tar acids. Sherwin-Williams Co. produces high-purity p-cresol by toluene sulfonation. A 60% m-cresol–40% p-cresol is made from **cymene,** obtained by alkylating toluene by propylene, by Mitsui Petrochemical Industries and Sumitomo Chemical Co. (both of Japan). **Orthocresol** is a colorless solid with a melting point of 86°F (30°C) and a boiling point of 376.7°F (191.5°C). It is soluble in alcohol, but only slightly soluble in water. It is used in the manufacture of cumerones, disinfectants, and fumigants, and as a plasticizer. It is a component of specialty phenolic resins and is employed as an intermediate in the manufacture of the herbicides **MCPA, MCPB, MCPP,** and **DNOC. Metacresol** is a yellow liquid freezing at 54°F (12°C) and boiling at 397°F (202.8°C). It is used in the manufacture of photographic developers, nitrocresols, disinfectant soaps, printing inks, paint, and varnish removers; as a preservative in leathers, glues, and pastes; in the reclaiming of rubber; and in making synthetic resins, perfumes, and pharmaceuticals. Metacresol is used for making **Thymol,** an ingredient in cold and cough syrups. A growing application is synthetic pyrethroid insecticides, for which high-purity metacresols are required. **Paracresol** is a colorless solid melting at 97°F (36°C) and boiling at 397°F (202.5°C). It is the least soluble of the cresols. It is used in the manufacture of cresotinic acid dyes, disinfectants, and pharmaceuticals. A major application of paracresol is for **butylated hydroxytoluene,** or **BHT,** which is used primarily as an antioxidant in rubber and plastics and, to a lesser extent, in food. Non-BHT antioxidants are also produced via paracresol.

   Tricresyl phosphate (TCP) and **cresyl diphenyl phosphate (CDP)** are major cresol-derived phosphate esters, but are being replaced by **isopropyl** and **butylated phenolic phosphates** in plasticizer uses. Production of **cresyl diphenyl phosphate,** also used as a plasticizer, has decreased substantially.

CRYOLITE.   A mineral of composition Na_3AlF_6, found in commercial quantities in Greenland and used as a flux in the electrical production of aluminum, in the making of special glasses and porcelain, as a binder for abrasive wheels, and in insecticides. One ton (907 kg) of cryolite is used for flux for 40 tons (36,280 kg) of aluminum. For glass batches 30 lb (14 kg) of cryolite is equivalent to 22.7 lb (10 kg) soda ash, 16.3 lb (7 kg) fluorine, and 11 lb (5 kg) aluminum hydrate. It acts as a powerful flux because of its solvent power on silicon, aluminum, and calcium oxides. In opal and **milky glasses,** it forms a complex AlF_6 anion, retaining the alumina and preventing loss of the fluorine. Cryolite occurs in masses of a vitreous luster, colorless to white, with a Mohs hardness of 2.5. It fuses easily. **Kryolith** is cryolite of 98 to 99% purity, and **Kryocide** is a grade of 90% purity. The latter is the dust from the natural ore and is used as an insecticide. Synthetic cryolite is made by reacting fluorspar with boric acid to form fluoroboric acid, and then reacting with hydrated alumina and sodium carbonate to form cryolite and regenerate boric acid.

CRYPTOSTEGIA RUBBER. Rubber obtained from the leaves of two species of perennial vines native to Malagasy, *Cryptostegia grandiflora* and *C. madagascariensis.* The former was grown in India, and the rubber was known as **palay rubber.** It was brought to Mexico and Florida as an ornamental plant and now grows extensively in Mexico and the West Indies. The maximum rubber content is found in the leaves 3.5 months old, at which time it is 2 to 3% of the dry weight of the leaf. There is also about 8% resin in the leaf, which must be separated from the rubber because it makes the rubber soft and tacky. The *C. madagascariensis* contains less rubber, but the leaves of hybrid plants grown from both species give increased yields of rubber. The hybrid does not come true to type from seed, and it is propagated from cuttings. When extracted and separated from the resin, cryptostegia has the same uses as ordinary hevea rubber.

Another plant that yields rubber from the leaves is the desert **milkweed,** *Asclepias erosa, A. subulata,* and other species growing in the dry regions of southwest United States. The short and slender leaves are produced only on the young stems, and the gathering season is short. The dry leaves are ground, and the rubber is obtained by solvent extraction. The average rubber content is about 2%, but as much as 12% has been obtained from some species of wild plants. As with guayule and cryptostegia, a considerable amount of resin is extracted with the rubber. **Goldenrod rubber** is extracted similarly from the leaves of the goldenrod, the dry leaves containing as much as 7% rubber mixed with resin. The species which contains the most rubber is *Solidago leavenworthii.* It does not occur in the plant as a latex, but is in isolated globules in the cells, mostly in the leaf. The **milk bush,** *Euphorbia tirucalli,* of Cuba and Jamaica, also produces rubber of good elasticity, but the crude latex from the bush causes skin blisters, and the extraction requires special treatment.

Dandelion rubber is the gum latex extracted from the roots of the Russian dandelion, which, when separated from the contained resin, has practically the same characteristics as the rubber from the hevea tree. Dandelion rubber, from various species of the genus *Taraxacum,* chiefly the plants known as **kok sagyz, tau sagyz,** and **crim sagyz,** native to Turkmen, is produced in Russia. The plant is grown only on a small scale in the United States and Canada. The roots, which extend 15 to 20 in (38 to 51 cm) into the ground, contain up to 10% rubber after the plant has passed the first-year flowering period. The normal yield is about 6% rubber with considerable resin. The dry roots also contain a high percentage of inulin.

CURUPAY.   The wood of the tree *Piptadenia cebil,* native to Argentina, Paraguay, and Brazil. In northern Argentina and Paraguay, it is also known under the Guarani name **cevil.** The wood is very hard and heavy, having a density of 74 lb/ft^3 (1,185 kg/m^3), and it has a reddish color and a handsome, wavy grain. It is used as an ornamental hardwood and is much employed locally for construction. Another wood of the same order is angico, from the *Angico rigida* of Brazil, also known as **queenwood**; the lighter-colored wood is called **angico vermelho,** or **yellow angico.** It is very hard, with a dense close grain, a reddish-brown color, and density of 70 lb/ft^3 (1,121 kg/m^3). It is employed where a heavy hardwood is required, and in cabinetmaking.

CUTTING ALLOYS.   Usually of complex Co-Cr-W-Fe-Si-C composition, used for lathe and planer tools for cutting hard metals. They form a class distinct from the cemented carbides, which are not true alloys; from the refractory hard metals, which are chemical compounds; and from the cobalt high-speed steels, which are high in iron and usually have less carbon. The hardness is inherent in the alloy and is not obtained by heat treatment, as with the tool steels. Cutting alloys are cast to shape and are usually marketed in the form of tool bits and shear blades. Complex alloys, however, may have heat-transition points at which the metal complexes change structure, limiting the range of use.

Since the development of balanced high-speed steels and cermet-type cutting tools, these alloys with a high proportion of the scarcer cobalt have lost their importance as cutting alloys and, because of their high corrosion, heat, and wear resistance, are used chiefly for weld-facing rods and heat-corrosion applications. One of the earliest of the alloys, called **Cooperite,** was based on nickel. The first of the commercial cobalt cutting alloys was **Stellite,** of Haynes Stellite Co., in various composition grades and with trade names, such as **J-metal** and **Star J-metal.** The hardest alloy, with a Rockwell C hardness to 68, contained about 45% cobalt, 32 chromium, 17 tungsten, 1.5 iron, 1.5 silicon, and up to 2.7 carbon. The tensile strength is above

100,000 lb/in^2 (689 MPa), and compressive strength is about 325,000 lb/in^2 (2,240 MPa). It is silvery white. **Delloy** is of somewhat similar composition. Other similar alloys were **Speedaloy, Rexalloy, Crobalt,** and **Borcoloy,** the last two containing also boron for added wear resistance. This type of alloy is now also used in **surgical alloys** for surgical tools and dental plates since they are not attacked by body acids and set up no electromotive currents. To make them more workable for this purpose, they usually contain a higher content of cobalt, 60% or more, with a smaller amount of molybdenum instead of tungsten, and with less carbon and silicon.

CYPRESS. A number of different woods are called cypress, but when the name is used alone, it is likely to refer to the wood of the **Italian cypress,** *Cupressus sempervirens,* native to the Mediterranean countries but now grown in the Gulf states and in California. The wood is lightweight, soft, and light brown and has a pleasant aromatic odor. It is very durable and is used for furniture, chests, doors, and general construction. **Citrus wood,** or **citron board,** is the wood from which the massive dining tables of ancient Rome were made. Heavy plates of the wood of this tree were cut across the trunk near the roots to show a variegated grain. The wood was cut in Mauritania. **Arizona cypress,** *C. arizonica,* is a smaller tree, and the wood is used chiefly for fence posts. The wood, usually referred to in the eastern United States as *cypress,* and also as **marsh cypress, red cypress, bald cypress, yellow cypress, gulf cypress,** and **southern cypress,** is from the coniferous tree *Taxodium distichum*; the **pond cypress** is from *T. ascendens,* of the southeastern states. Southern cypress grows along the coast from Delaware to Mexico, especially in Florida and the lower Mississippi Valley. The red cypress is along the coast, and the yellow is inland, the coastal types being darker in color. The trees are sometimes very old, reaching a height of 120 ft (37 m) in 800 years. The wood is yellowish red or pink and is moderately hard with an open grain. The density is about 32 lb/ft^3 (513 kg/m^3). It is very durable and is valued for shingles, tanks, boatbuilding, or construction where resistance to weather exposure is needed. The wood called *yellow cypress* on the west coast, also known as **Sitka cypress, Alaska cedar,** and **yellow cedar,** is from the tree *Chamaecyparis nootkaensis,* or *Cupressus sitkaensis,* growing on the Pacific coast from Alaska to Oregon. The trees reach 6 ft (2 m) in diameter and 120 ft (37 m) in height in 500 years. The heartwood is bright yellow, and the sapwood slightly lighter. The wood has a fine, uniform, straight grain and is lightweight, moderately hard, easily worked and polished, shock-resistant, and durable. It is used for furniture, boatbuilding, and interior finish. **Monterey cypress,** *C. macrocarpa* of

California, is one of the chief trees planted on reforestation projects in New Zealand.

DAMMAR. Also written **damar.** The resin from various species of trees of genera *Shorea, Balanocarpus,* and *Hopea,* but the name is also applied to the resins of other trees, especially from the *Agathis alba,* the source of Manila copal. There is no dividing line between the dammars and the copals, and dammar may be considered as a recent or nonfossil **copal,** the Malay word *damar* meaning simply a gum. The best and hardest dammars are from deposits at the bases of the trees, which are then the seasoned or fossil resins like the copals. Dammar is obtained by tapping the trees and collecting the solidified gum after several months. It is used in varnishes, lacquers, adhesives, and coatings. The usual specific gravity is 1.04 to 1.12, and the melting point is up to 248°F (120°C). The average grade of dammar does not have a melting point much higher than 212°F (100°C). Dammar is a spirit varnish resin, gives a flexible film, but is softer and less durable than the copals. It is noted for its complete solubility in turpentine. It is also soluble in alcohol, and the Batavia and Singapore dammars are soluble in chlorinated compounds and in hydrocarbons. Dammar is classified according to color and size, the best grades being colorless and in large lumps. The high-grade pale-colored dammars from Batavia and Sumatra, including the **cat's-eye dammar,** are from species of *Hopea.* Most of the **white dammar** equivalent to Manila copal comes from Malagasy. It is semihard to hard and is used in paints where resistance to wear is required, as in road-marking paints, but is not as hard as Congo copal. In general, the true dammars are from the *Shorea* and *Balanocarpus,* and they are inferior in hardness to the fossilized resins approaching the copals. The *Shorea* resins are usually dark in color. The **Malayan black dammar, dammar hitam,** is from a species of *Balanocarpus.* The plentiful **dammar penak** is from the Malayan tree *B. heimii,* which also yields the important wood known as **chengal** used for furniture and boatbuilding. **Black dammar** is from the tree *Canarium strictum,* of India, and comes in black, brittle lumps, easily ground to powder. The reddish **dammar sengai** is also from a species of *Canarium.* These are types of elemi. **Dewaxed dammar,** for making colorless, glossy lacquers, is highly purified dammar in xylol solution.

DEGRADABLE PLASTICS. Plastics that are decomposed by any of three mechanisms—biodegradation, solubility, and photodegradation. **Biodegradable plastics** are those that are susceptible to being assimilated by microorganisms, such as fungi and bacteria, through enzyme action. The assimilating action requires heat, oxygen, and moisture. For all practical purposes, almost all synthetic polymers are immune to enzyme attack. Only **aliphatic polyesters** and **urethanes** derived from **aliphatic ester diols** and low-molecular-weight (under 500) unbranched **polyethylene** derivatives can be assimilated. Certain mutant soil microorganisms, when inoculated into resistant types of polymers in waste disposal areas, have increased the degradability of the polymers. Union Carbide Corp. has formulated **polycaprolactone resins** which are biodegradable in contact with a nutrient soil environment. They are not attacked by airborne spores. Cargill makes the **Ecopla** line of polylactic-acid degradable bioplastics. A biodegradable plastic developed by Takassago International Corp. of Japan and marketed by Zeneca of England is a copolyester of poly-3-hydroxybutyrate and poly-3-hydroxyvalerate, synthesized by bacteria. **Biopol,** a bacteria-synthesized polyester introduced by ICI, is also marketed by Zeneca.

The solubility of **water-soluble plastics** varies with formulations, molecular weight, and temperature. **Hydroxypropyl cellulose** is insoluble in water above 115°F (46°C). Below this temperature, when immersed in water, it quickly forms a slippery gel on the outer surface. The gel layer must dissolve and wash away before further dissolving takes place. **Polyethylene oxides** are soluble in water above 150°F (66°C). They are nontoxic, eatable but nonnutritive, and nonchloric, and they wash through plumbing without damage or clogging. They are resistant to grease, oil, and petroleum hydrocarbons. Water-soluble and/or compostable **EnviroPlastic resins,** of Planet Polymer Technologies, include polyester-, polyethylene-, and cellulose-based resins. The solubility of **polyvinyl chloride** depends on the degree of alcoholization. Thus, completely alcoholized grades are hot-water-soluble and cold-water-soluble. Partially alcoholized types (about 87%) are soluble in both hot and cold water.

Photodegradable plastics are sensitive to ultraviolet light. Energy in the form of photons breaks down the bonds between the carbon and hydrogen atoms, and oxygen-reactive free radicals are formed. The free radicals react with oxygen in the environment to produce peroxide and hydroperoxides that decompose further to produce carbonyl groups, hydroxyl groups, water, and carbon dioxide. The best photodegradable materials are the linear, nonaromatic, molecular structured plastics. Unvulcanized **syndiotactic polybutadiene** is typical. It is degradable under direct sunlight in periods ranging from one week to more than one year. Additives such as pigments, ultraviolet accelerators, and promoters, and ultraviolet absorbers and antioxidants promote ultraviolet degradation in **polyethylenes, polystyrenes, polypropylenes, polybutadienes, polybutylenes, ABS,** and **polyvinyl chloride.**

DENATURANTS. Materials used chiefly for mixing with ethyl alcohol to be employed for industrial purposes to prevent the use of the alcohol as a beverage and to make it tax-free under the Tax Free Industrial Alcohol Act. The qualities desired in a denaturant are that its boiling point be so close to that of the alcohol that it is difficult to remove by ordinary distillation, and that it be bad tasting. Some of the denaturants are poisonous and cause death if the alcohol is taken internally. The usual denaturants are **methyl alcohol, pyridine, benzene, kerosene,** and **pine oil.** One or several of these may be employed, but denaturants must be approved by the Bureau of Internal Revenue. Completely **denatured alcohol** is a term used to designate alcohol containing poisonous denaturants, and these are employed only for antifreeze, fuels, and lacquers, but not in contact with the human body. Special denatured alcohol is alcohol containing denaturants authorized for special uses, such as pine oil for hair tonics. Many approved denaturants are marketed under trade names. **Denol** is the name of a mixture of primary and secondary aliphatic higher alcohols. **Agadite** is a compounded petroleum product. **Hydronol** is a hydrogenated organic product. Denaturants are also used in imported oils that are permitted entry at lowered tax rates for industrial use so that they cannot be diverted for edible use. Rapeseed oil, for example, is denatured with brucine.

DENDRIMERS. **Dendritic polymers,** or dendrimers, consist of highly branched globular molecules grown from a core molecule and formed in stages, allowing the molecules to be built with specific diameters, weights, and surface characteristics for improved processibility. Dendritech Inc.'s **polyamidoramine dendrimers,** called **Paman,** begin with an ammonia molecule, which is reacted with methylacrylate and ethylenediamine. This results in a molecule with three branches, each ending in an amino group. As the process repeats, the dendrimer grows in layers, with each amino group reacting with two ethylenediamine molecules so that the new molecule has six branches ending in an amino group. Each successive reaction doubles the number of branches. The molecular weight of Paman varies only by as little as 0.005% in contrast with as much as 5% for straight-chain polymers. DSM's (of the Netherlands) **polypropyleneamine dendrimer** begins with diaminobutane, which has four amino branches. After acrylonitrile is added to the amino groups, the molecule is hydrogenated, making eight branches. Then the process repeats. These dendrimers have higher glass transition temperatures and lower viscosity than analogous straight-chain polymers.

DERRIS. The root of various species of vines of the bean family, *Derris uliginosa, D. elliptica,* and *D. trifoliata,* growing in Indonesia.

It is imported as crude root and marketed as a fine powder of 200 mesh for use as an insecticide diluted with dusting clay to a rotenone content of 1%, or as a spray in kerosene or other liquid. The root contains **rotenone,** a colorless, odorless, crystalline solid poison of complex composition, $C_{22}H_{22}O_6$, and melting point 325°F (163°C). The value of rotenone as an insecticide is that it is highly toxic to cold-blooded animals, including insects and worms, and nonpoisonous to warm-blooded animals. It is widely used as an agricultural insecticide as it is harmless to birds. It is about 30 times more toxic to cutting worms than lead arsenate and is more potent than pyrethrum. Besides rotenone, other insecticidal constituents of derris root are **deguelin, tephrosin,** and **toxicarol.**

Rotenone is also found in many other plants, and when separated has the same toxic power. **Cubé** is the root of the vine *Lonchocarpus utilis,* of Peru, containing rotenone and used for the same purposes as derris. **Timbó,** also known as **urucú** and as **tingi** and **conambi,** is the root of the vine *L. urucu,* of Brazil, also containing rotenone and used in the same manner. **Barbasco** is a name applied to timbó and all other fish-killing plants of the Orinoco Valley. The Caribs used the root either in shredded or in extract form for catching and killing fish. A cubic foot of root will poison an acre of water without harming the fish as food. The tubers of the wild yam called barbasco yield **diosgenin,** a steroidal used in the synthesis of steroids, which are oxidized to produce cortisone. Other plants of the same family are **nicou, nekoe,** and **haiari** of the Guianas, and rotenone sometimes goes under the name of **nicouline.** The high yield of rotenone from Indonesian derris, up to 12%, is due to careful selection and propagation in cultivation, the semiwild roots of South America sometimes containing only about 2%. The Brazilian government standard for timbó is 4% rotenone content. From 1 to 4% rotenone is also obtained from the long, leathery shoots of the perennial weed *Tephrosia virginiana,* known as **devil's shoestring,** growing in Texas. Piperonyl butoxide is sometimes mixed with rotenone to give greater insect-killing power.

DETERGENTS. Materials which have a cleansing action like soap. Although soap itself is a detergent, as are the sodium silicates and the phosphates, the term usually applies to the synthetic chemicals, often referred to as **detergent soaps** or **soapless soaps,** which give this action. The detergents may be the simple sulfonated fatty acids such as turkey-red oil; the **monopole soaps,** or highly sulfonated fatty acids of general formula $(SO_2OH)_x R \cdot COONa$; or the **gardinols,** which are sulfonated fatty alcohols.

All the synthetic detergents are **surface-active agents,** or **surfactants,** with unsymmetric molecules which concentrate and orient at the interface of the solution to lower interfacial tension. They may be **anion-active agents,** with a positive-active ion; **cation-active agents,** with a negative-active ion; or **nonionic agents.** The anions and cations are sometimes called **gengenions.** Most of the household detergents are anion-active and are powders. Most of the nonionics are liquids and are useful in textile processing since they minimize the difference in dye affinity of various fibers. The cationics have lower detergency power and are usually skin irritants, but they have disinfectant properties and are used in washing machines and dairy cleansers. They are called **invert soaps** by the Germans. The synthetic detergents do not break down in the presence of acids or alkalies, and they do not form sludge and scum, or precipitate salts in hard waters as soap does. They do not form quantities of suds as some soaps do, but suds contribute little to cleansing and are not desirable in automatic washing equipment. **Textile softeners** are different from surface-active agents. They are chemicals that attach themselves molecularly to the fibers, the polar, or charged end, of the cation orienting toward the fiber, with the fatty tails exposed to give the softness to the fabric. **Arquad 2HT** is a distearyldimethyl ammonium chloride for this purpose. A special-purpose surfactant used for dispersing oil slicks on the sea is **Dispersol,** of ICI Americas, Inc. It is a polyethanoxy dissolved in isopropyl alcohol. It is soluble in oil but not in water. It agglomerates the oil into small blobs that are scattered by the winds and eventually destroyed by marine organisms.

Synthetic detergents have now largely replaced soaps for industrial uses. They are employed in textile washing, metal degreasing, paper-pulp processing, and industrial cleansing. They are also used in household cleansers, soapless shampoos, and toothpastes. **Biodegradable detergents** are those which can be chemically disintegrated by bacteria so that the discharged wastes do not contaminate the groundwaters. **Millox** is a group of biodegradable detergents made by the reaction of sucrose and fatty acids with a linking of ethylene oxide. This type of detergent is more powerful than petroleum-based detergents. **Millox 120** is made from the fatty acids of coconut oil, and **Millox 180** is from tallow. The detergents produced from straight-chain paraffinic hydrocarbons derived from petroleum cracking are **alkyl aryl sulfonates,** $R \cdot Ar \cdot SO_3Na$, **alkylbenzene sulfonates,** or **dodecylbenzene sulfonates.** These detergents do not break down in wastes and therefore do tend to contaminate the groundwaters. The detergent characteristics vary with the number of carbon atoms in the alkyl chain and the arrangement of atoms in the chain. Detergency increases to a maximum at 12 to 15 atoms and then decreases. These detergents are 10 times as bulky as soda ash, but can be mixed with alkaline or phosphate cleaners.

The detergents are more efficient than toilet soaps, but tend to leave the skin with an alkaline hardness. Lecithin may be used in detergent bars to reduce tackiness, and starch may be used for hardening. **Nytron** is a **sodium sulfonate** derived from petroleum hydrocarbons. It is a buff-colored powder. **Surfax 1288,** of E. F. Houghton & Co., is an **aryl sulfopropionate** with only slight detergent power, used in textile processing for rewetting and as a leveling agent for dye baths. **Clavenol,** of Dexter Chemical Corp., is a **polyethylene glycol** condensate of the nonionic class.

Ultrawets, Kamenol D, Oronite, Kreelon, Parnol, Wicamet, and Monsanto's **Santomerse** are alkyl aryl sulfonates. This type of chemical is available in powder, bead, and paste forms, and one molecule in 40,000 molecules of water gives good detergency. It is effective in hard water or in acid and alkaline solutions. **Sulframin E** is this material in liquid form.

Superonyx is a modified **sodium alkyd sulfate** and is a neutral detergent and dye assistant for processing textiles. **Maprosyl 30** is called a **modified soap.** It has the detergent and emollient properties of soap but does not form scum as soap does, and does not cause skin irritation as many detergents do. Unlike soap, it is soluble in highly alkaline solutions, and unlike most detergents, it has high foaming qualities. It is a **sodium lauroyl sarcosinate** produced from fatty acids, and it may also be in the form of stearoyl, linoleyl, or derivatives of other fatty acids. The **sarcosine** is **methyl glycine,** $CH_3NHCH_2CO_2H$, an amino acid occurring in small amounts in animal muscle, but now made synthetically. It is a decomposition product of caffeine. **Lauryl pyridium chloride** is also a soaplike detergent. It is a tan-colored semisolid with a soapy feel and with germicidal properties. It is used for textile washing.

The **Pluronics** are nonionic detergents produced from polyoxypropylene glycol, ethylene oxide, and ethylene glycol. When the ethylene oxide content is 70%, the detergent is a solid which can be flaked. It is formulated with alkyl sulfonate and sodium carboxymethyl cellulose for laundry work. Somewhat similar chemicals to the detergents are used as **dispersing agents** for latex, paper coatings, dyestuffs, and agricultural sprays. **Daxad 11** is a polymerized salt of alkyl naphthalene sulfonic acid. Its action is to impart an electric charge to each particle, giving a repelling action to space the particles and to prevent agglomeration or settling. It increases fluidity and permits a higher solids content in dispersions without increasing the viscosity.

To reduce package size, laundry detergents have become increasingly concentrated. **Phosphates,** once widely used, have lost favor because the water discharges can be environmentally damaging.

Aluminum silicate zeolites are preferred but require additives for alkalinity, water softening, and equivalent cleaning. Containing water, however, they limit detergent concentration, and although they are effective in removing calcium, they only remove some of the magnesium, another hard-water constituent. **SKS-6,** a layered, crystalline **sodium silicate** of Hoechst Celanese, is water-free, thus easier to concentrate. Also, it removes both calcium and magnesium and provides sufficient water alkalinity. **Varisoft 475,** a **water softener** of Witco Chemicals, can also be used as a base for concentrates. It suits cooler-water washes, an energy conservation trend. **Chlorine bleach,** a whitening product, is another water pollutant. **Sodium perborate** is less environmentally offensive but, due to the presence of high-temperature activators, works best at 140 to 160°F (60 to 71°C). Although such wash temperatures are common outside the United States, 110°F (43°C) is typical in the United States. **Sodium nonanoyloxybenzene sulfonate,** of Proctor & Gamble, is a low-temperature activator used in the company's Tide With Bleach detergent.

DEXTRIN.   Also called **amylin.** A group of compounds with the same empirical formula as starch $(C_6H_{10}O_5)_x$, but with a smaller value of x. The compounds have strong adhesive properties and are used as pastes, particularly for envelopes, gummed paper, and postage stamps; for blending with gum arabic; in pyrotechnic compositions; and in textile finishing. Dextrin is a white, amorphous, odorless powder with a sweetish taste. It dissolves in water to form a syrupy liquid and is distinguished from starch by giving violet and red colors with iodine. Dextrin is made by moistening starch with a mixture of dilute nitric and hydrochloric acids and then exposing to a temperature of 212 to 257°F (100 to 125°C). Dextrin varies in grade chiefly owing to differences in the type of starch from which it is made. **British gum** is a name given to dextrins that give high tack for paste use and are products containing partially converted starch. **Feculose** is obtained by treating starch with acetic acid; it forms clear, flexible films and can be used as a textile finish. A wood adhesive, commonly called **vegetable glue,** is prepared by heating starch with caustic. **Cartonite** is a liquid solution of a converted dextrin used as an adhesive in box-sealing machines. It is also marketed as a brown water-soluble powder. **Koldrex,** of A. E. Staley Mfg. Co., is a formulated dextrin which dissolves easily in cold water to produce stable liquid adhesives of uniform viscosity. It is produced by combining dextrin with borax, preservatives, and defoamers and then spray-drying the mixture into powder. The **borated dextrin** of National Starch and Chemical Corp., for automatic packaging machines, has high initial tack and good adhesion. It gives 400 sealings per minute.

DIAMOND. A highly transparent and exceedingly hard crystalline stone of almost pure carbon, 99% of the isotope carbon 12. When pure, it is colorless, but it often shows tints of white, gray, blue, yellow, or green. It is the hardest known substance and is 10 on the Mohs hardness scale. But the **Mohs scale** is only an approximation, and the hardness of the diamond ranges from Knoop 5,500 to 7,000 compared with 2,670 to 2,940 for boron carbide, which has a Mohs hardness of 9.

The diamond always occurs in crystals in the cubic system and has a specific gravity of 3.521 and a refractive index of 2.417. Carbon is normally quadrivalent in flat planes, but in the diamond the carbon atoms are arranged in face-centered lattices forming interlocking tetrahedrons and also hexagonal rings in each cleavage plane.

The diamond has been valued since ancient times as a gemstone, but it is used extensively as an abrasive, for cutting tools, and for dies for drawing wire. These **industrial diamonds** are diamonds that are too hard or too radial-grained for good jewel cutting. **Jewel diamonds** have the formation in regular layers, while industrial diamonds are grown in all directions. Technically these are called **feinig** and **naetig. Ballas diamonds,** valued for industrial drilling, are formed with the crystallization starting from one central point. The stones thus formed do not crack in the tool as easily as those with layer formation. Stones for diamond dies are examined in polarized light to determine the presence of internal stresses. They are then drilled normal to the rhombic dodecahedron plane with cleavage planes parallel to the die-hole axis to obtain the greatest die-service life. The stones for industrial purposes are also the fragments and the so-called **bort** which consists of the cull stones from the gem industry including stones of radiating crystallization that will not polish well. Bort also includes a cryptocrystalline variety of diamond in brown, gray, or black, known as **black diamonds, carbonados,** or **carbons,** found in Brazil in association with gem diamonds. The carbons have no cleavage planes, are compact, and thus offer greater resistance to breaking forces. The carbons vary greatly in quality and hardness. Some rare natural diamonds of South America contain small amounts of aluminum and other elements which give stability to the crystal above the normal disintegrating temperature. These diamonds are not suitable as gemstones but are efficient semiconductors.

The value of diamonds is based on the gem value and is determined by color, purity, size, and freedom from flaws. The weight is measured in carats. Diamond splinters as small as $\frac{1}{500}$ carat may be cut and faceted. Small diamonds are sieved into straight sizes, and the tinted stones are separated. Then each stone is examined for cut, brilliance, and degree of perfection, and diamond merchants who sell by grade are meticulously careful of their reputation for uniform judgment.

The most valued gems are blue-white. A faint straw color detracts from the value, but deep shades of yellow, red, green, or blue are prized. The largest diamond found in Brazil, the **Vargas diamond,** was a flawless stone weighing 726.6 carats. It was cut into 23 stones. The famous **Kohinoor diamond** weighed originally 793 carats, and the **Jonkers diamond** from South Africa was a blue-white stone weighing 726 carats. The **Cullinan diamond,** or Star of Africa, measured 4 by 2.5 by 2 in (10 by 6 by 5 cm) and weighed 3,106 carats. The annual world production of natural diamonds reaches as high as 28 million carats, or about 6 tons (5.4 metric tons), of which 5 tons (4.5 metric tons) is industrial diamonds. An average of 250 tons (228 metric tons) of ore is processed to obtain 1 carat. In Angola the average find is 0.004 carat/ft^3 (0.14 carat/m^3) of ore.

Most of the diamonds come from South Africa, Brazil, India, and Zaire. The average diamond content of the Bushimaie deposits of Zaire is 16.4 or 19.7 carat/ft (5 or 6 carat/m). The diamonds are associated with pebbles of flint, jasper, agate, and chalcedony, but diamonds usually occur in **kimberlite,** an intrusive rock with the appearance of granite but with a composition similar to basalt plus much olivine. It occurs in South Africa, North Carolina, and Arkansas. Diamonds are formed at very high pressures and heat, and since at ordinary pressure the diamond disintegrates into graphite at 1600°F (871°C), the natural diamonds could not have been released until the temperature of the rock was below that point. The stones found in the beach sands of southwest Africa and in sandstone in Brazil are not native to the sand, but were washed into it after scattering from the exploded rock. Diamonds have been found irregularly in Arkansas since their discovery in 1906. The average weight of the Arkansas diamonds is less than 1 carat, with the largest 40.22 carats. Some diamonds are found in the Appalachian region, the largest from West Virginia, weighing 34.46 carats. Few of the U.S. diamonds are of gem quality, but they are of full hardness.

Synthetic diamond was first produced from graphite at pressures from 800,000 to 1.8×10^6 lb/in^2 (5,512 to 12,402 MPa) and temperatures from 2200 to 4400°F (1204 to 2427°C) by General Electric in the early 1950s. A molten metal catalyst of chromium, cobalt, nickel, or other metal is used, which forms a thin film between the graphite and the growing diamond crystal. Without the catalyst, much higher pressures and temperatures are needed. The shape of the crystal is controllable by the temperature. At the lower temperatures cubical shapes predominate, and at the upper limits octahedra predominate; at the lower temperatures the diamonds tend to be black, while at higher temperatures they are yellow to white. The diamonds ranged up to 0.01 carat in size, with quality comparable with natural diamond powders. The powder has been used on saws and grinding and polishing wheels to cut and finish hard materials. By 1990, GE had combined chemical vapor deposition (CVD) with the high-pressure, high-temperature technology to create diamond of 99.9% carbon 12 or 99% carbon 13. The carbon 13 is produced using methane enriched with carbon 12 to vacuum-deposit a polycrystalline sheet of the material. The sheet is then crushed into powder, which serves as the carbon source for the high-pressure, high-temperature process. The resulting crystals are said to be the best thermal conductors: 50% better than natural diamond and 850% better than copper. GE also has synthesized gem-quality **jadeite,** a rare gem that has been used in jewelry and sculptures for 3,000 years. In this case, the source materials for the high-pressure, high-temperature process are the oxides of sodium, aluminum, and silicon. Either white jadeite or, using additives, color jadeites can be made.

DuPont synthesizes diamond for abrasive powder polishing applications by underground explosive shocking of graphite at pressures of 2 $\times$ 10^6 to 7 $\times$ 10^6 lb/in^2 (13,800 to 48,000 MPa). Then a series of mechanical and chemical operations extract the diamond as a fine powder which is cleaned, shaped, and graded to particle sizes of about 3.9 to 2,362 μin (0.1 to 60 μm). Polycrystalline and gray to black because of trace elements, the powder resembles the natural carbonado diamond. First used for polishing synthetic sapphire for watch stones, it is also used for precision finishing other semiprecious gemstones, alumina and ferrite electronic ceramics, alumina and cemented-carbide wear parts, and composite coatings for special uses. Single-crystal and polycrystalline synthetic and natural diamond powders of Warren Diamond Powder Co. are used as abrasives for grinding, precision machining, honing, lapping, and polishing. Free-flowing in the dry state, the powders mix well with resin or liquid carriers and can be formulated with special water- or oil-soluble bases. **Polycrystalline diamond (PCD)** has a Vickers hardness of about 8,000 and is valued as a cutting-tool material for machining materials that quickly wear out more common cutting-tool materials, such as tungsten carbide (WC), which has a Vickers hardness of about 1,800. Also, although PCD and WC particles are combined in the substrate with a cobalt alloy, WC is cemented by the alloy whereas the PCD is fused with the aid of a cobalt-rich catalyst for greater strength. Most cutting-tool manufacturers buy the PCD from General Electric Superabrasives, DeBeers of South Africa, or Sumitomo of Japan, which supply various grades, such as fine, medium, and coarse. **Compax PCD** and **Stratapax PCD** are GE trade names for the material.

CVD diamond coatings date back to the late 1940s when a Union

Carbide researcher used a hydrocarbon gas in a low-pressure reaction chamber and energized the gas with a plasma or microwave heat source. Too much graphite remained mixed with the diamond, necessitating lengthy and costly removal procedures. Russian researchers added hydrogen to the gas to remove the graphite, but the results were not taken seriously until confirmed by the Japanese in the 1980s. Current CVD techniques usually include directing a high-energy beam at the substrate to accelerate surface reactions that produce the free carbon necessary for true diamond coatings. Coatings containing graphite or other impurities are called **diamondlike carbon coatings.** These coatings, or films, can be beneficial because of their extreme hardness, low coefficient of friction, outstanding thermal conductivity, excellent optical transmissivity, and high electrical resistivity. The first U.S. company to offer a commercial product—a diamond-coated X-ray window—was Crystallume in 1989. One commercial use by the Japanese—Sony Corp.—is for loudspeakers. The diamond film imparts harness and stiffness to the substrate, improving reproduction of high-frequency sound. The Japanese have also introduced diamond-coated carbide and silicon-nitride cutting-tool inserts. Horton Diamond Film has found two applications. Because of their hardness and low friction and wear resistance, the coatings are used on the measuring faces of a line of micrometers from L. S. Starret Co. High thermal conductivity led to the use of CVD diamond substrates, replacing aluminum oxide, for microwave frequency divider circuits in fighter-aircraft test equipment. Because the diamond has 50 times the thermal conductivity of the oxide, its use simplified package design and improved performance.

At Lockheed Missile & Space Co., diamond films are made by burning a mixture of acetylene gas and oxygen at temperatures of 4941 to 6741°F (2727 to 3727°C) and energy levels of 1,290 to 1,935 W/in^2 (200 to 300 W/cm^2). This method is said to be far more productive than CVD. Films as thick as 0.02 in (500 μm) have been deposited. One application is infrared-sensor windows on missiles, which stem from the film's hardness, thermal conductivity, and ability to transmit infrared and visible light.

DIATOMACEOUS EARTH. A class of compact, granular, or amorphous minerals composed of hydrated or **opaline silica,** used as an abrasive, for filtering, in metal polishes and soaps, as a filler in paints and molding plastics, for compacting into insulating blocks and boards, and in portland cement for fine detail work and for waterproofing. It is formed of fossil diatoms in great beds and is not earthy. In mineralogy it is called **diatomite,** and an old name for the ground powder is **fossil flour.** Tripoli and kieselguhr are varieties of crystalline diatomite.

The U.S. production of the mineral is mainly in Oregon, California, Washington, Idaho, and Nevada. After mining, the material is crushed and calcined. When pure, it is white; with impurities it may be gray, brown, or greenish. The powder is marketed by fineness and chemical purity. The density is usually 12 to 17 lb/ft^3 (192 to 272 kg/m^3). Its high resistance to heat, chemical inertness, and dielectric strength, and the good surface finish it imparts make it a desirable filler for plastics. For insulating purposes, bricks or blocks may be sawed from the solid or molded from the crushed materials, or it may be used in powdered form. **Diatomite block** has a porosity of 90% of its volume and makes an excellent filter. **Celite,** of Celite Corp., is a 325-mesh, uncalcined, amorphous diatomaceous earth for portland-cement mixtures, paper finishes, and use as a flatting agent in paints. **Sil-O-Cel** is diatomaceous earth in powder or in insulating block to withstand temperatures to 1600°F (871°C). **Superex** is calcined diatomite powder bonded with asbestos fibers to resist temperatures up to 1900°F (1038°C). **Dicalite,** of Philip Carey Co., is a fine diatomite powder having a density of 8 to 8.5 lb/ft^3 (128 to 136 kg/m^3) loose, and 15 to 17 lb/ft^3 (240 to 272 kg/m^3) tamped, used for heat-insulating cement or as insulation for walls. **Compressible insulation,** to absorb the expansion stresses known as drag stress, between the firebrick and the steel shell of metallurgical furnaces, may be of diatomaceous silica. **Superex SG,** for blast furnaces, is a composite block with Superex on the hot side and a blanket of fine spun-glass fibers on the cold side. It recovers to 97% of its original thickness in cooling from a temperature of 1900°F (1038°C) and a compression of 10%.

DIE-CASTING METAL. Any alloy employed for making parts by casting in metal molds, or dies, in pressure casting machines as distinct from other permanent-mold casting methods where little or no pressure is used. The pressures may be as high as 25,000 lb/in^2 (172 MPa) to give a uniformly dense structure and smooth finish to castings of intricate design and varying section thickness. The cost of equipment, including heat-resistant dies, limits the process economically to high production quantities of nonferrous metals only. A characteristic of the castings, also, is that they must have a draft of at least 2° on all sides to give rapid ejection from the die.

Zinc alloys and aluminum alloys are the most common die-casting metals. However, **magnesium alloys,** a limited number of **copper alloys,** as well as **lead alloys** and **low-melting alloys** of lead, zinc, tin, and bismuth, are also die-cast. **Slush castings,** in which excess metal is poured out after a skin of metal on contact with the die has set, leaving a hollow casting, are cast without pressure and are classified as permanent-mold castings rather than die castings, although the composition of the alloys may be essentially the same.

DIE STEELS.   Any of the various types of **tool steels** used for cold- and hot-forming dies, including forging, casting, and extrusion dies; stamping and trim dies; piercing tools and punches; molds; and mandrels. In general, all the major families of tool steels except the high-speed types are used for dies, including the hot-work, cold-work, shock-resisting, mold, special-purpose, and water-hardening types. The high-speed types, however, which are typically used for cutters, are also used for punches.

DISINFECTANTS.   Materials used for killing germs, bacteria, or spore, and thus eliminating causes of disease or bad odors in factories, warehouses, or in oils and compounds. The term **antiseptic** is employed in a similar sense in medicine, and the term **germicide** is often used for industrial disinfectants. Some disinfectants are also used as preservatives for leather and other materials, especially chlorine and chlorine compounds. **Phenol** is one of the best-known disinfectants, and the germ-killing power of other chemicals is usually based on a comparison with it. Practically all bacteria are killed in a few minutes by a 3% solution of phenol in water, but phenol has the disadvantage of being irritating to skin. Industrial disinfectants are usually sold as concentrates to be diluted to the equivalent of a 3 to 5% solution of phenol.

Too large a proportion of disinfectants in oils, solutions, or the air may be injurious to workers, so the advice of health officials is ordinarily obtained prior to general use. Creosote oil and cresylic acid are employed in emulsions in disinfecting sprays and dips, but continuous contact with creosote may be injurious. **Formaldehyde** has high germicidal power and is used for hides and leather, and some air sprays may contain chemicals such as chlorophyll which unite with moisture in the air to produce formaldehyde. But formaldehyde is not generally recommended for odor control, as it is an **anesthetizer.** It desensitizes the olfactory receptors so that the individual is no longer able to detect the odor. **Masking agents,** which introduce a stronger, more pleasant odor, are likewise not a recommended method of disinfecting. They do not destroy the undesirable odor and may permit raising the total odor level to unhealthy proportions. Elimination of odors requires chemicals that neutralize or destroy the cause of the odors without causing undesirable effects.

The **silver** ion is an effective cleanser of water that contains bacteria which produce sulfur-bearing enzymes, and silver sterilization is done with silver oxides on activated carbon, or with organic silver compounds. The safe limit of silver in water for human consumption is specified by the U.S. Health Service as 10 parts per billion, and as with many other disinfectants, the use requires competent supervision. Among other metals, mercury is effective as an antibacterial agent in the form of **mercuric chloride.** Several organic mercuricals are used as antiseptics, such as **mercurochrome, Metaphen,** of Abbott Laboratories, and **Merthiolate** of Eli Lilly & Co. Antiseptic atmospheres may be produced by spraying chloramine T, iodine, or argyrol. **Chloramine T** is a white crystalline powder of composition $CH_3C_6H_4SO_2NClNa \cdot 3H_2O$, soluble in water and in organic solvents. Besides its use as an antiseptic and germicide, it is employed as an oxidizing and chlorinating agent. **Dichloramine T, Halazone, chlorinated cyanuric acid** derivatives, and **trichloroisocyanuric acids** are other halogenated compounds. Inorganic chlorinated compounds include **Pittside,** of Columbia Chemical, used as an industrial germicide. It is a stabilized calcium hypochlorite, available as water-soluble granulates. **Iodine** is a strong bactericidal antiseptic and is commonly used as a 2% tincture. Mixtures of iodine and nonionic solubilizing surfactants are called **iodophors.** Purdue Frederick Co. markets an iodine and **polyvinylpyrrolidinone** combination under the name **Betadine.** Disinfectants sold under trade names are usually complex chemicals and may be chlorinated or fluorinated phenyl compounds not harmful to skin.

A 3 to 5% solution of **hydrogen peroxide** is used as an antiseptic, although the chemical can cause corrosive burns at concentrations exceeding 25%. Compounds of various metals, such as iron, manganese, and cobalt, enhance peroxide's bactericidal action. Hydrogen peroxide is often trapped in **urea,** forming a solid containing up to 35% peroxide. **Ozone** generators are becoming popular for disinfecting swimming pools. **Zinc peroxide** and **benzoyl peroxide** are especially effective as antiseptic dressings.

Hexachlorophene is an antibacterial additive that does not lose its activity in soaps, as do phenolics. **G-11** is from the Swiss firm Givaudan Corp., and **pHisoHex** is produced by Winthrop Laboratories. Hexachlorophene exhibits synergistic antibacterial action with **trichlorocarban,** an antigermicidal and deodorizing compound from Monsanto Chemical Co.

Thymol is used as a disinfectant in ointments, mouthwashes, soaps, and solutions. Condensation products of thymol with other materials are also used. **Thymoform,** $C_{21}H_{28}O_2$, made by condensing thymol with formaldehyde, is a yellow powder used as an antiseptic dusting powder. **Thymidol** is an antiseptic made by condensing thymol with menthol. **Dihydroxyacetic acid,** $CH(OH)_2COOH$, and its sodium salt, both white powders, are used in cosmetics, pharmaceuticals, and coatings for food-wrapping papers, as they are nontoxic and do not irritate skin. **Hexylresorcinol,** $CH_3(CH_2)_5C_6H_3(OH)_2$, is a more powerful antiseptic than phenol and is not injurious to skin or tissues. **Caprokol,** of Merck, is hexylresorcinol. The antiseptic throat lozenge of this company, known as **Sucret,** has a base of sugar and glucose, with hexylresoricinol and a flavor. **Pinosylvine** is a natural antiseptic extracted from the heartwood of the pine tree, where it protects the tree against decay and insects. It is related chemically to resorcinol, and its germ-killing power is 30 times that of phenol. **Ceresan M,** of Du Pont, is a powder designated as **ethyl mercury toluene sulfonanilide,** used for disinfecting seeds to protect against soil-borne plant diseases.

DISPERSION-STRENGTHENED METALS. Particular composites in which a stable material, usually an oxide, is dispersed throughout a metal matrix. The particles are less than 39 μin (1 μm) in size, and the particle volume fraction ranges from only 2 to 15%. The matrix is the primary load bearer while the particles serve to block dislocation movement and cracking in the matrix. Therefore, for a given matrix material, the principal factors that affect mechanical properties are the particle size, the interparticle spacing, and the volume fraction of the particle phase. In general, strength, especially at high temperatures, improves as interparticle spacing decreases. Depending on the materials involved, dispersion-hardened alloys are produced by powder-metallurgy, liquid-metal, or colloidal techniques. They differ from **precipitation-hardened alloys** in that the particle is usually added to the matrix by nonchemical means. Precipitation-hardened alloys derive their properties from compounds that are precipitated from the matrix through heat treatment.

There are a rather wide range of dispersion-hardened-alloy systems. Those of aluminum, nickel, and tungsten, in particular, are commercially significant. **Tungsten thoria,** a lamp-filament material, has been in use for more than 30 years. **Dispersion-hardened aluminum** alloys, known as **SAP alloys,** are composed of aluminum and aluminum oxide and have good oxidation and corrosion resistance plus high-temperature stability and strength considerably greater than that of conventional high-strength aluminum alloys. Another dispersion-hardened metal, **TD nickel,** has dispersion of thoria in a nickel matrix. It is 3 to 4 times stronger than pure nickel at 1600 to 2400°F (871 to 1316°C). **TD-nickel-chromium** also has been produced for increased resistance to high-temperature oxidation. Other metals that have been dispersion-strengthened include copper, lead, zinc, titanium, iron, and tungsten alloys. The copper is used for resistance-welding (spot-welding) tips.

DIVI-DIVI. The dried seed pods of the tree *Caesalpinia coriaria,* native to tropical America, employed in tanning leather. Most of the divi-divi is produced in Colombia, the Dominican Republic, and Venezuela. It is used chiefly in blends with other tannins to increase acidity, to give a light color to the leather, and to plump and soften the leather. The pods are about 3 in (7.6 cm) long and contain up to 45% pyrogallol tannin, which consists of **ellagitannin** and **ellagic acid.** They must be kept from fermentation, which develops a red coloring matter. The best pods are the thickest and lightest in color, and they are used to replace gambier, valonia, and myrobalans. The commercial extract contains 25% tannin. **Algarobilla,** from the pods of the *C. brevifolia,* of Chile, is a similar tanning agent. **Cascalote** is from the pods of the tree *C. cacolaco* of Mexico and is the standard tanning material of Mexico. It is also used to replace quebracho for oil-well-drilling mud. **White tan,** or **tari,** is from the pods of the *C. digyna* of the Far East. **Tara,** or **Bogotá divi-divi,** also called **cevalina,** is from the pods of the tree *C. tinctoria* of Colombia and Peru. The pods contain 32% tannin, and 1,000 lb (454 kg) of tara pods produces 500 lb (227 kg) of **tara powder.** The material makes a soft leather and is used to replace sumac.

DOGWOOD. A heavy hardwood noted for its ability to stay smooth under long-continued rubbing. Its outstanding use is for shuttles for weaving. The texture is fine and uniform. Other uses of the wood are for small pulleys, golf-club heads, mallet heads, jewelers' blocks, skate rollers, and bobbins. There are 17 known varieties of the plant in the United States, only four of which grow to tree size. The **white dogwood** is *Cornus florida*; the **Pacific dogwood** is *C. nuttalli*; **rough-leaf dogwood** is *C. asperfolia*; and **blue dogwood** is *C. alternifolia.* Dogwood grows widely throughout the eastern states. **Turkish dogwood** was formerly imported for shuttles, as was also the **Chinese dogwood,** or **kousa,** *C. kousa.*

DOLOMITE. A type of limestone employed in making cement and lime, as a flux in melting iron, as a lining for basic steel furnaces, for the production of magnesium metal, for filtering, and as a construction stone. It is a carbonate of calcium and magnesium of composition $CaCO_3 \cdot MgCO_3$, differentiated from limestone by having a minimum of 45% $MgCO_3$. It occurs widely distributed in coarse, granular masses or in fine-grained compact form known as **pearl spar.** The specific gravity is 2.8 to 2.9 and Mohs hardness 3.5 to 4. It is naturally white, but may be colored by impurities to cream, gray, pink, green, or black. For furnace linings it is calcined, but for fluxing it is simply crushed. The raw dolomite, marketed by Basic Refractories, Inc., for open-hearth steel making, is washed crushed stone in 0.625-in (1.6-cm) size. When calcinated at a temperature of about 3100°F (1704°C), dolomite breaks down to MgO and CaO, and it is limited to about 3000°F (1649°C) as a refractory. **Calcined dolomite** used in Germany as a water-filter material under the name of **magno masse** is in grain sizes 0.02 to 0.2 in (0.5 to 5.0 mm). Dolomite for the production of magnesia, some of which is cut as building marble, contains 10 to 20% magnesia, 27 to 33 lime, 1 to 12 alumina, 40 to 46 carbonic acid, 1 to 5 silica, and 0 to 3 iron oxide. The dolomite found in huge deposits in Oklahoma contains 30.7% CaO, 21.3 MgO, and only very small amounts of silica, alumina, and iron oxide. For the production of magnesium metal, calcined dolomite and ferrosilicon are brought to a high temperature in a vacuum, and the magnesium is driven off as a vapor. In the ceramic industry, dolomite is sometimes called **bitter spar** and **rhombic spar.**

DOUGLAS FIR. The wood of the tree *Pseudotsuga taxifolia,* of the northwestern United States and British Columbia. It is sometimes called **Oregon pine, Douglas pine, Douglas spruce, red fir, fir, yellow fir,** and **Puget Sound pine.** The wood of young trees with wide growth rings is reddish brown and is the type called *red fir,* though the true red fir is from the large tree *Abies magnifica* of California and Oregon, the lumber of which is called **golden fir,** and the wood of which is used also for paper pulp. The wood of older trees of slower growth with narrow rings is usually yellowish brown and is called *yellow fir.* Both woods may come from the same tree. The narrow-ringed wood is stronger and heavier. Douglas fir averages below longleaf pine in weight, strength, and toughness, but above loblolly pine in strength and toughness, though below it in weight. The grain is even and close, with resinous pores less pronounced than in pitch pine. It is a softwood and is fairly durable. The density is 34 lb/ft³ (545 kg/m³). The compressive strength perpendicular to the grain is 1,300 lb/in² (9 MPa); the shearing strength parallel to the grain is 810 lb/in² (5.5 MPa).

Douglas fir is used for general construction and millwork, plywood, boxes, flooring, and where large timbers are required. It is also used for pulping and yields kraft paper of high folding endurance but low bursting strength. The fibers are large. The trees grow to great heights, the average being 80 to 100 ft (24 to 30 m). The stand is estimated at more than 450 billion bd ft (1 billion m³), or about one-fourth of all timber in the United States. **Douglas fir bark** contains from 7.6 to 18.3% of a catechol tannin, the bark of young trees yielding the higher percentages. It is suitable for tanning heavy leathers and yields a pliable, light-colored leather. **Silvacon 383,** of

Weyerhaeuser Co., is Douglas fir bark in flaky, corklike granules used in flooring and acoustical tile. **Silvacon 490** is the bark as a reddish powder used in dusting powders and paints. **Silvacon 508** is hard, spindle-shaped small fibers from the tissue of the bark, used as a filler for plastics and in asphalt and fibrous paints. **Douglas-fir bark wax** is a hard, glossy wax extracted from the bark of the Douglas fir and is a partial replacement for carnauba wax. A ton of bark yields 150 lb (68 kg) of wax by solvent extraction with 150 lb (68 kg) of tannin and 10 lb (4.5 kg) of quercetin as by-products.

DRIERS. Materials used for increasing the rapidity of the drying of paints and varnishes. The chief function of driers is to absorb oxygen from the air and transfer it to the oil, thus accelerating its drying to a flexible film. They are in reality catalyzers. Excessive use of driers will destroy the toughness of the film and cause the paint to crack. Solutions of driers are called *liquid driers*; it is in this form that **paint driers** are most used. Certain oils, such as tung oil, have inherent drying properties and are classified as drying oils but not as driers. Driers may be oxides of metals, but the most common driers are metallic salts of organic acids. **Manganese acetate,** $(CH_3COO)_2Mn \cdot 4H_2O$, is a common paint drier. It is a pinkish, crystalline powder soluble in water and in alcohol and is used in strengths of 6, 9, or 12% metal. **Sugar of lead,** used as a drier, is **lead acetate,** $Pb(CH_3COO)_2 \cdot 3H_2O$, a white, cyrstalline powder with a faint acetic acid odor, also used as a mordant in textile printing. It is known as **plumbous acetate** and **Goulard's powder. Lead oleate,** $Pb(C_{18}H_{33}O_2)_2$, is a drier made by the action of a lead salt on oleic acid. It is used for thickening lubricants. **Lead linoleate,** $Pb(C_{18}H_{31}O_2)_2$, is a drier made by adding litharge to linseed oil and heating. Lead and manganese compounds together act more effectively as driers than either alone. **Lead resinate** adds toughness of film as well as drying power. Because of antilead laws, this metal is being replaced by zinc, cobalt, calcium, and zirconium compounds. **Zinar** is a zinc resinate with 5.6% zinc content. **Cobalt octoate,** which has about 12% cobalt in combination with hexoic acid, is used as a drier. **Cobalt driers** are twice as rapid in drying power as manganese driers, but too rapid drying often makes a wrinkled film which is desirable for some finishes but not for others.

Naphthenate driers are metallic salts made with naphthenic acids instead of fatty-oil acids. They are usually more soluble in paint solvents, and since the naphthenic acids can be separated into a wide range of molecular weights by distillation, a wider variety of characteristics can be obtained. **Sodium naphthenate,** with 8.6% metal content, and **potassium naphthenate,** with 13.1%, are powders that are good bodying agents and emulsifiers as well as driers. **Tin naphthenate,** with 20% tin, may be added to lubricating oils as an antioxidant. **Mercuric naphthenate,** with 29% mercury, retards the growth of bacteria and mold when added to finishes. **Barium naphthenate,** with 22.6% barium, has binding and hardening properties and is used in adhesives and in linoleum. **Uversols** are naphthenic acid salts of aluminum, calcium, cobalt, lead, manganese, or zinc, in liquid form for use as paint driers, wetting agents, and catalysts. **Octoic driers,** of Witco Corp., are metallic salts made with ethylhexoic acid, and the metal content is lower than that of driers made with naphthenic acids. They are light in color, have no odor, and have high solubility. The **Octasols** are ethylhexoic acid metal salts. **Drying agents** for resin coatings and inks may act by oxidation or other chemical reaction. **Sulfur dichloride,** S_2Cl_2, speeds the drying action of coatings and inks formulated with alkyd, urea, or melamine resins, and such inks dry almost instantly.

DRILL ROD. Tool-steel round rod made to a close degree of accuracy, generally not over or under 0.0005 in (0.0127 mm) the diameter size, and usually polished. It is employed for making drills, taps, reamers, punches, or for dowel pins, shafts, and rollers. Some mills also furnish square rods to the same accuracy under the name of drill rod. Common drill rod is of high-carbon steel hardened by quenching in water or in oil. The usual commercial sizes are from 1.5 in (3.8 cm) in diameter down to No. 80, which is 0.0135 in (0.343 cm) in diameter. The usual lengths are 1 to 3 ft (0.3 to 0.9 m). The sizes are by the standard of drill gages, with about 200 different diameters. The carbon content is usually from 0.90 to 1.05%, with 0.25 to 0.50 manganese, 0.10 to 0.50 silicon, and a maximum of 0.04 phosphorus or sulfur. It also comes in high carbon with from 1.50 to 1.65% carbon and 0.15 to 0.35 manganese. Drill rod can be obtained regularly in high-speed steels and in special alloy steels for dowel pins. **Needle wire** is round tool-steel wire used for making needles, awls, and latch pins. It comes in coils, in diameters varying by gage sizes from 0.010 to 0.105 in (0.025 to 0.267 cm). **Needle tubing** for surgical instruments and radon implanters is stainless-steel tubing 0.014 to 0.203 in (0.036 to 0.516 cm) in diameter in 6-ft (1.8-m) lengths. **Hypodermic tubing** is hard-drawn stainless-steel tubing 0.008 to 0.120 in (0.020 to 0.304 cm) in outside diameter, with wall thicknesses from 0.004 to 0.012 in (0.010 to 0.304 cm), in 2-ft (0.6-m) lengths, with a fine finish. **Capillary tubing** is also stainless steel, but comes in lengths to 200 ft (61 m), with outside diameters from 0.060 to 0.125 in (0.152 to 0.318 cm). The inside bore can be had in various diameters from 0.006 to 0.025 in (0.015 to 0.064 cm) for the 0.060-in (0.152-cm) tubing and from 0.010 to 0.024 in (0.025 to 0.061 cm) for the 0.125-in (0.318-cm) tubing. **Stud steel** is an English name for round bar steel made to close limits and hardened and descaled, used for heavy pins and studs. **Pin bar** is small-diameter rod of case-hardened steel used for dowel pins. **Drill steel,** for mine and quarry drills, comes in standard rounds, octagons, squares, and cruciform bars, solid or hollow, usually in carbon steel.

DRYING OILS. Vegetable oils which are easily oxidized by exposure to air and thus suitable for producing a film in paints and varnishes, known as **paint oils.** The use of drying oils as the sole or main binder in alkyd coatings is steadily decreasing with the advent of water-based latex paints. Currently, it is limited to solvent-thinned exterior house paints and some metal paints. The oils are also used in oleoresinous varnishes and in the manufacture of synthetic resins for coating binders, epoxy ester resins, and oil-modified urethane resins. Small amounts are used in printing inks, linoleum, putty and caulking compounds, core oils, and hardboard. The best drying oils are those which contain the higher proportions of unsaturated acids, in which oxidation causes polymerization of the molecules. The drying of an oleoresinous varnish takes place in two stages. First, the reducer or solvent evaporates, leaving a continuous film composed of gums and drying oil. The drying oil is then oxidized by exposure, leaving a tough, hard skin. This oxidation is hastened by driers, but the drying oil itself is responsible for the film. The drying power of oils is measured by their **iodine value,** as their power of absorbing oxygen from the air is directly proportional to their power of absorbing iodine. Drying oils have typical iodine values about 140, semidry oils above 120, and nondrying oils are below 120. Linseed oil is the most common of the drying oils, though tung oil and oiticica oil are faster in drying action. Linseed oil alone will take about 7 days to dry, but can be quickened to a few hours by the addition of driers. Linseed oil and other oils may be altered chemically to increase the drying power.

 Conjugated oils are oils that have been altered catalytically by nickel, platinum, palladium, or carbon to give conjugated double bonds in place of isolated double bonds in the molecules of the fatty acids. **Conjulinol** is a drying oil of this class made from linseed oil. The iodine value is 180, and the drying time is greatly reduced. Normally, soybean oil is not classed as a drying oil although it may be blended with drying oils for paint use. But by chemical alteration and, lately, by mixing with synthetic resins, it can be given good drying power. **Conjusoy** is a drying oil made by conjugation of soybean oil. The iodine value is 128, and the drying time is about half that of boiled linseed oil.

Castor oil, which has poor drying properties, is dehydrated to form a good drying oil. Other methods are used to alter oils to increase the drying power, notably polymerization of the linoleic and some other acids in the oils; or oils may be fractionated and reblended to increase the percentage of acids that produce drying qualities. The **Admerols,** of the Archer-Daniels-Midland Co., comprise a series of drying oils made by treating linseed or soybean oil with butadiene, styrene, or pentaerythritol. **Kel-X-L oil,** of Spencer-Kellogg, is a modified linseed oil with an iodine value up to 170, used as a substitute for tung oil in quick-drying varnishes. **Kellin,** of the same company, is a quick-drying blended oil with a linseed-oil base, while **Kellsoy** is a similar oil with a soybean-oil base. **Cykelin,** of the same company, is a quick-drying oil made by treating linseed oil with **cyclopentadiene,** $(CH:CH)_2 \cdot CH_2$, a low-boiling liquid obtained from coal tar or from cracking petroleum. **Cykelsoy** is another drying oil made by treating soybean oil with cyclopentadiene. **Dorscolene** is a drying oil made from fractionated and blended fish oils. The German substitute drying oil known as **Resinol** was a liquid obtained by the distillation of the heavy fractions of the benzolated oils derived from scrubbing coke-oven gas. **Resigum** is the final residue in the distillation of tar-oil benzol which has been washed with sulfuric acid, caustic soda, and water. It contains a maximum naphthalene content of 5%. It is miscible with resins or copals, and with vegetable oils, and makes a good paint without other drying oils. Synthetic drying oil is **glycerol allyl ether** derived from propylene gas obtained in cracking petroleum. **C oil** is a heavy, sticky liquid with a butadiene base. In paints it gives high adhesion to metals and masonry and produces a smooth, hard, glossy coating with good chemical resistance.

Although the great volume of drying oils is produced from linseed, soybean, tung, oiticica, castor, and fish oils, many other oils have drying properties and are used in varying quantities. **N'gart oil** is from the seed nuts of a climbing plant of Africa and is equal in drying power to linseed oil. **Lallemantia oil,** obtained from the seeds of *Lallemantia iberica,* of southeastern Europe and Asia, resembles linseed oil in physical properties. **Isano oil,** obtained from the kernel of the nut of *Ongokea klaineana* of tropical Africa, is a pale-yellow viscous oil that has little drying power, but when heat-treated sets up an exothermic action to produce a varnish oil. **Anda-assu oil,** also used in Brazil for paints, is from the seeds of the plant *Joannesia princeps.* The seeds yield 22% of a clear yellow oil with an iodine value of 142 which is bodied by heating. **Manketti oil** is a varnish oil with about two-thirds the drying power of linseed oil. It is a light-yellow viscous oil from the seed nuts of the tree *Ricinodendron rautanenii,* of southwest Africa.

Chia-seed oil is a clear amber-colored oil extracted from the seeds of the plant *Salvia hispanica* of Mexico. It has a higher drying value than linseed oil. The seeds yield about 30% oil, which contains 39% linolenic acid, 45 linoleic, 5 palmitic, 2.7 stearic, with some arachidic, oleic, and myristic acids. The specific gravity is 0.936, iodine value 192, and acid value 1.4. The seeds scatter easily from the pods and are difficult to collect.

DUCK. A strong, heavy cotton fabric employed for sails, awnings, tents, heavy bags, shoe uppers, machine coverings, and where a heavy and durable fabric is needed. It is woven plain, but with two threads together in the warp. It is made in various weights and is designated by the weight in ounces per running yard 22 in (0.6 m) wide. It is marketed unbleached, bleached, or dyed in colors, and there are about 30 specific types with name designations usually for particular uses such as **sailcloth.** When woven with a colored stripe, it is called **awning duck. Russian duck** is a fine variety of **linen duck.** Large quantities of cotton duck are used for making laminated plastics and for plastic-coated fabrics, and it is then simply designated by the weight. **Belt duck,** for impregnated conveyor and transmission belts, is made in loosely woven, soft ducks and in hard-woven, fine-yarn hard fabric. The weights run from 28 to 36 oz (0.80 to 1.02 kg). **Conveyor belting** for foodstuff plants is usually of plastic fabric for cleanliness. **Transilon,** of Extremultus, Inc., is a belting of good strength and flexibility to operate over small-diameter rollers. It is made of nylon fabric faced on both sides with polyvinyl chloride sheet. It may have a variety of surface finishes such as tetrafluoroethylene.

Hose duck, for rubber hose, is a soft-woven fabric of plied yarns not finer than No. 8, made in weights from 10 to 24 oz (0.28 to 0.68 kg). The grade of duck known as **elevator duck** for conveyor belts is a hard-woven 36-oz (1.02-kg) fabric. **Plied-yarn duck** is used for army tents instead of flat duck as it does not tear easily and does not require sizing before weaving. **Canvas** is duck of more open weave. The term is used loosely in the United States to designate heavy duck used for tarpaulins, bags, sails, and tents. But more properly it is a heavy duck of square mesh weave more permeable than ordinary duck, such as the canvas used for paintings and for embroidery work. The word *duck* is from the Flemish *doeck,* meaning cloth, originally a heavy linen fabric. The word *canvas* is from the Latin *cannabis,* originally a coarse, heavy hempen cloth for tents. **Osnaburg cloth** is a heavy, coarse, plain-woven fabric used for wrapping and bailing and for inside sacks for burlap flour bags. It is made from lower grades of short-staple cotton and from waste. In colored checks and stripes it is used for awnings.

Drill is a stout, twilled cotton fabric used for linings and where a strong fabric lighter than duck is required. It differs from duck also in that it has a warp-flush weave that brings more warp than filling to the face of the cloth. It comes unbleached, bleached, or piece-dyed, or it may be yarn-dyed. It is made in various weights and is designated in ounces per yard, the same as duck. Tan-colored drill is called **khaki. Denim** is a heavy, twill-woven, warp-flush fabric usually lighter in weight than drill. The warp is yarn-dyed. The filling is made with one black and one white yarn. It is much used for workers' clothing, and the light weights for sportswear are called **jean.** Denim is also used industrially where a tough fabric is needed. **Art denim,** in plain colors or woven with small figures, is used for upholstery.

DYESTUFFS.   Materials, also called **colorants,** used to color textiles, paper, leather, wood, or other products. They may be either natural or artificial. Many chemicals will stain and color other materials, but a product is not considered a dye unless it will impart a distinct color of some permanence to textiles. The natural dyestuffs may be mineral, animal, or vegetable, but the artificial dyes are derived mainly from coal-tar bases. Almost all naturally extracted dyes have been replaced by synthetic counterparts for commercial use; an exception is **logwood,** a Central American tree extract, known as **natural black 1, CI 75290. Tyrian purple,** from various Mediterranean snails, was in ancient times the most noted of the animal dyestuffs. **Cochineal** and **kermes** are other animal dyes. One of the earliest metallic or mineral dyestuffs was called **iron buff.** It was made by allowing pieces of iron to stand in a solution of vinegar to corrode. Fabrics that had been dipped in this solution were rinsed in a solution of wood ashes. **Mineral dyes** now include ocher, chrome yellow, and Prussian blue. **Vegetable dyes** may be water solutions of woods, barks, leaves, fruits, or flowers. The buff and brown textile colors of early New England were made by boiling fresh green butternuts in water, while a dark-red dye was made by boiling the common red beet in water. The yellow to red colors known by the Algonquin name of **puccoon** were from the orange-red juice of the root of the **bloodroot,** a perennial of the poppy family. Vegetable dyes now include brazilwood, barwood, sappanwood, fustic, logwood, madder, henna, saffron, annatto, indigo, and alkanet. The **camphire** of the ancients mentioned in the Bible and Koran was a reddish-orange dyestuff made by grinding to a paste the red, sweet-scented spikes of the small cypress tree *Lawsonia inermis,* of Egypt and the Near East. It was used by Eastern and Roman women to stain fingernails, and is now used under the name **henna** for dyeing leather and hair. It gives various shades from yellowish to red or brown. **Argol,** a brilliant red used extensively until replaced by synthetic dyes, is from the **orchilla,** a lichen found in the Canaries and Near East. It was used to produce the brilliant colors of the medieval **Florentine cloth. Chinese green, buckthorn bark,** or **lokao** is the powdered bark of the buckthorns, *Rhamorus globosa* and *R. utilis,* of China and Russia. It is used in dyeing silk and cotton. **Weld,** from the plant *Reseda luteola* of Europe, produces a very bright-yellow color with an alum mordant. With indigo it produces green. **Woad** is the dried fermented leaves of the plant *Isatis tinctoria* of Europe. It gives a blue color, but is now little cultivated. Ecolor dyes, of Allegro Natural Dyes, are derived from plants, such as the *Maclura pomifera* (Bodark tree) and the insect, cochineal. No heavy metals are required as mordants to pretreat fibers to accept the dyes, and no toxics or solvents, other than water, are used in dyeing. More than 100 colors exist for cotton fibers.

Synthetic dyes are mostly coal-tar or aniline colors. They are more intense, brighter, faster, and generally cheaper than natural dyestuffs. The dyes are complex chemicals, but they usually contain characteristic groups of atoms so that the color or change in color can be predicted. The Colour Index (CI) from the U.K.'s Society of Dyers and Colourists is a new system for classifying dyes. It assigns a number defining the chemical class while a generic name identifies the application; in industry, the trade name is also appended. **Benzene-azo-*m*-phenylenediamine hydrochloride** is sold as **Chrysoidine Y** and classified as **Azo** or **Basic Dye, CI 11270, Basic Orange 2.** The Colour Index recognizes 26 types of dyes by chemical classification. The **azo dyes,** with an $\cdot$N:N$\cdot$ linkage, constitute about half of the production and are produced by diazotization of primary arylamines, followed by reaction with aromatic amines, phenols, and enolizable ketones. **Azobenzene,** $C_6H_5 \cdot N{:}N \cdot H_5C_6$, is made from nitrobenzene and is in crystalline red plates melting at 154°F (68°C). This may be converted to **hydrazobenzene,** $C_6H_5NHNHC_6H_5$, a solid of camphorlike odor melting at 268°F (131°C). **Substituted azo dyes** constitute a class containing OH and NH groups, made by coupling amines or phenols with the salts. The azo dyes are in general poisonous, but are sometimes used in restricted quantities to color foodstuffs. Some are poisonous in contact with skin, such as **xylyazonaphthol** and the sodium salt of **sulfophenylazo,** designated by the Food and Drug Administration as **red No. 32** and **orange No. 1,** and proscribed for use in coloring lipstick and oranges.

Three other important classes are **anthraquinone dyes, indigoid dyes,** and **thioindigoid dyes,** the latter being **sulfur dyes.** The sulfur dyes may be made by treating the organic compounds with sodium sulfide. They are fast to washing and to light, but the range of color is limited, and their use is generally limited to fibers where a strongly alkaline bath is tolerable.

Some of the synthetic dyes will color animal fibers well and not vegetable fibers, or vice versa, while some will color all fibers. As a result, it is possible to divide the bulk of the dyes into six classes, including **azoic dyes.** The **direct dyes** can be dyed directly, while others require a mordant. Some are permanent, or fast, while others are water-soluble and will fade when the fabric is washed, or some may not be light-fast and will fade when exposed to light. Direct dyes usually have a weak OH bond between the nitrogen in the dye and the fiber, usually cotton. In **reactive dyes,** the dye reacts with the fiber to produce both an OH and an oxygen linkage, the chlorine combining with the hydroxyl to form a strong ether linkage. Such dyes are fast and very brilliant and are used for cotton, rayon, and nylon. **Acid dyes** contain a carboxylic or sulfonic acid group and operate best in an acid bath. They are used for drying protein fibers, such as wool, silk, and nylon, and sometimes for leather and paper. They are usually azo, **triaryl methane,** or anthroquinone complexes. **Basic dyes** are commonly amino and substituted amino compounds, such as triaryl methane or **xanthenes.** They are used for dyeing cotton with a mordant. **Vat dyes** are insoluble and are applied in the soluble colorless form and then reduced or oxidized to color. They usually have an anthraquinone or **indanthrene** structure and are solubilized by the reducing agent, a hydroxyl group, OH, diffusing into the fiber where it is fixed. The best known example is **indigo,** the dye synonymous with the color of bluejeans, which has become one of the most important colorants because of the popularity of **denim** garments. Originally derived from plants, synthetic indigo dye now dominates the market. Synthesis, first commercialized by BASF of Germany more than 100 years ago, uses aniline formaldehyde and sodium cyanide as the raw materials. The process generates toxic wastewater. A biosynthesis process, of Genecor International, Inc., however, is said to be environmentally benign. Pioneered by Amgen, Inc., and the University of Texas, it uses no petrochemical feedstocks, and biomass is the only waste product. Genecor produces recombinant *Escherichia coli* bacteria, which contain an enzyme capable of converting **indole** to indigo. Indole is a naturally occurring by-product of an enzyme within *E. coli.*

Bifunctional reactive dyes, such as the **Sumifix Supra** line of Sumitomo, of Japan, combine several reactive groups in a single molecule. Each group compensates for the other during changing process conditions, improving color fixation and reproducibility. The **Levafix** line, of Miles Inc., combines vinyl sulfone, monofluorotriazine, and fluorochloropyrimidine to increase color fixation rates to 75 to 90% and higher in contrast with the 50 to 70% rates typical of conventional reactive dyes. **Cibacron C** bifunctional reactives, of Ciba Geigy, are said to provide color fixation rates of 95%. Increasing the fixation rate reduces the amount of dye discharged in textile wastewater. To reduce sulfur effluents in nonreactive sulfur dyes, **Sandozol RDT,** of Sandoz Chemical Corp., contains nonsulfide-reducing agents, which cut sulfide consumption in half.

Color carriers, used to aid adherence of dyes to synthetic fibers, are usually chemicals that act as swelling agents to open the fiber structure, such as phenylphenol, benzoic acid, or dichlorobenzoic acid. **Ketosol 75,** of Union Carbide, is 75% methylphenyl carbinol and 25 acetophenone. Monochlorobenzene, used as a color carrier for Dacron fiber, acts to promote a concentrated layer of dye solution around the fiber. **Ring-dyed fiber** is a synthetic fiber not receptive to dyes that has been passed through a bath of a receptive plastic before dyeing. The dye then adheres to the coated surface and encases the fiber.

EBONY. A hard, black wood valued for parts subject to great wear, and for ornamental inlaying. It is the wood of various species of trees of the ebony family, *Ebenaceae,* although the name is also applied to some woods of the genus *Dalbergia,* family *Leguminosae.* **Black ebony,** from the tree *Diospyros dendo* of West Africa, and ebony, from the tree *D. melanoxylos,* of India, are the true ebonies. Black ebony has a black heartwood with brownish-white sapwood. It is next to lignum vitae in hardness and has a fine, open grain and a density of 78 lb/ft^3 (1,250 kg/m^3). It is used for inlaying, piano keys, and turnery. The ebony of India is also extremely hard, with a fine, even grain. The heartwood is black with brownish streaks. **Marblewood,** or **Andaman marblewood,** is an ebony from the tree *D. kurzii* of India and the Andaman Islands. The wood is black with yellowish stripes. It has a close, firm texture, is hard, takes a fine polish, and has a density of 65 lb/ft^3 (1,041 kg/m^3). **Marble ebony** is another species from Malagasy. The ebony from Japan, called **kaki,** is from the tree *D. kaki.* It has a black color streaked with gray, yellow, and brown. The grain is close and even, and the wood is very hard, but the density is less than that of African ebony. Ebony wood is shipped in short billets and is graded according to the color and the source, as Niger, Macassar, or Cameroon. **Green ebony** is a name sometimes given to the cocoswood of the West Indies. **Artificial ebony,** formerly composed of asphaltic compounds, is now usually molded plastics. **Partridgewood,** a heavy blackish wood used for fine inlay work, is **acapau,** from the large tree *Voucapapoua americana* of the Amazon Valley. It is valued in Brazil for furniture because of its resistance to insect attack.

ELASTOMERS. **Synthetic rubbers,** often referred to as **rubbers,** are hydrocarbon polymeric materials similar in structure to plastic resins. The difference between plastics and elastomers is largely one of definition based on the property of extensibility, or stretching. The American Society for Testing and Materials defines an *elastomer* as "a polymeric material which at room temperature can be stretched to at least twice its original length and upon immediate release of the stress will return quickly to approximately its original length." Some grades of plastics approach this rubberlike state, for example, certain of the polyethylenes. Also, a number of plastics have elastomer grades, such as the olefins, styrenes, fluoroplastics, and silicones. As indicated above, the major distinguishing characteristic of elastomers is their great extensibility and high-energy storing capacity. Unlike many metals, for example, which cannot be strained more than a fraction of 1% without exceeding their elastic limit, elastomers have usable elongations up to several hundred percent. Also, because of their capacity for storing energy, even after they are strained several hundred percent, virtually complete recovery is achieved once the stress is removed.

Up until World War II, almost all rubber was natural. During the war, synthetic rubbers began to replace the scarce **natural rubber,** and since that time, production of synthetics has increased until now their use far surpasses that of natural rubber. There are thousands of different elastomer compounds. Not only are there many different classes of elastomers, but also individual types can be modified with a variety of additives, fillers, and reinforcements. In addition, curing temperatures, pressures, and processing methods can be varied to produce elastomers tailored to the needs of specific applications.

In the raw-material or crude stage, elastomers are thermoplastic. Thus **crude rubber** has little resiliency and practically no strength. By a vulcanization process in which sulfur and/or other additives are added to the heated crude rubber, the polymers are cross-linked by means of covalent bonds to one another, producing a thermosetlike material. The amount of cross-linking which occurs between the sulfur (or other additive) and the carbon atoms determines many of the elastomer's properties. As cross-linking increases, resistance to slippage of the polymers over one another increases, resiliency and extensibility decrease, and the elastomer approaches the nature of a thermosetting plastic. For example, hard rubbers, which have the highest cross-linking of the elastomers, in many respects are similar to phenolics. In the unstretched state, elastomers are essentially amorphous because the polymers are randomly entangled and there is no special preferred geometric pattern present. However, when stretched, the polymer chains tend to straighten and become aligned, thus increasing in crystallinity. This tendency to crystallize when stretched is related to an elastomer's strength. Thus, as crystallinity increases, strength also tends to increase.

There are roughly 20 major classes of elastomers; we cannot do much more here than identify them and highlight the major characteristics of each group. Two basic specifications provide a standard nomenclature and classification system for these classes. The ASTM standard D1418 categorizes elastomers into compositional classes. A joint ASTM-SAE specification (ASTM D2000/SAE J200) provides a classification system based on material properties. The first letter indicates specific resistance to heat aging, and the second letter denotes resistance to swelling in oil.

Styrene-butadiene elastomers, sometimes also called **Buna S, SBR,** and **GR-S,** are copolymers of butadiene and styrene. They are similar in many ways to the natural rubbers, and were the first widely used synthetics. They top all elastomers in volume of use, chiefly because of their low cost and use in auto tires. A wide range of property grades are produced by varying the relative amounts of styrene and butadiene. For example, styrene content varies from as low as 9% in low-temperature resistant rubbers to 44% in rubbers with excellent flow characteristics. Those grades with styrene content above 50% are by definition considered plastics. Carbon black is sometimes added also as it substantially improves processing and abrasion resistance. SBR elastomers are similar in properties to natural rubber. They are non-oil-resistant and are generally poor in chemical resistance. Although they have excellent impact and abrasion resistance, they are somewhat below natural rubber in tensile strength, resilience, hysteresis, and some other mechanical properties. The largest single use is in tires. Other applications are similar to those of natural rubber. **Styrene-butadiene latex,** typically a 70% SBR emulsion in water, is used mainly for coatings and adhesives. **Carbon tetrachloride,** CCl_4, long used as a weight modifier in some of the latex, is being phased out for environmental reasons. Japan Synthetic Rubber, that country's largest producer of latex, has developed a nonhalogen hydrocarbon-based substitute for all latex grades formerly using CCl_4. New solution polymers of Goodyear Chemical Co., based on **styrene-isoprene-butadiene rubber,** improve wet traction of tires over standard emulsion SBR. This is also true of **Li polymers,** so called because of the lithium-based catalyst used during polymerization in hexane or other organic solvent.

Neoprene, also known as **chloroprene,** was developed in the 1930s, and it has the distinction of being the first commercial synthetic rubber. It is chemically and structurally similar to natural rubber, and its mechanical properties are also similar. Its resistance to oils, chemicals, sunlight, weathering, aging, and ozone is outstanding. Also, it retains its properties at temperatures up to 250°F (121°C), and it is one of the few elastomers that does not support combustion, although it is consumed by fire. In addition, it has excellent resistance to permeability by gases, having about one-fourth to one-tenth the permeability of natural rubber, depending on the gas. Although it is slightly inferior to natural rubber in most mechanical properties, neoprene has superior resistance to compression set, particularly at elevated temperatures. It can be used for low-voltage insulation, but is relatively low in dielectric strength. Typical products made of chloroprene elastomers are heavy-duty conveyor belts, V belts, hose covers, footwear, brake diaphragms, motor mounts, rolls, and gaskets. **Butyl rubbers,** also referred to as **isobutylene-isoprene elastomers,** are copolymers of isobutylene and about 1 to 3% isoprene. They are similar in many ways to natural rubber and are one of the lowest-priced synthetics. They have excellent resistance to abrasion, tearing, and flexing. They are noted for low gas and air permeability (about 10 times better than natural rubber), and for this reason they make a good material for tire inner tubes, hose, tubing, and diaphragms. Although butyls are non-oil-resistant, they have excellent resistance to sunlight and weathering and generally have good chemical resistance. They also have good low-temperature flexibility and heat resistance up to around 300°F (149°C); however, they are not flame-resistant. They generally have lower mechanical properties, such as tensile strength, resilience, abrasion resistance, and compression set, than the other elastomers. Because of their excellent dielectric strength, they are widely used for cable insulation, encapsulating compounds, and a variety of electrical applications. Other typical uses include weather stripping, coated fabrics, curtain wall gaskets, high-pressure steam hoses, machinery mounts, and seals for food jars and medicine bottles.

 Isoprene is **synthetic natural rubber.** It is processed as natural rubber, and its properties are quite similar, although isoprene has somewhat higher extensibility. Like natural rubber, its notable characteristics are very low hysteresis, low heat buildup, and high tear resistance. It also has excellent flow characteristics and is easily injection-molded. Its uses complement those of natural rubber. And its good electrical properties plus low moisture absorption make it suitable for electrical insulation. **Polyacrylate elastomers** are based on polymers of butyl or ethyl acrylate. They are low-volume-use, specialty elastomers, chiefly used in parts involving oils (especially sulfur-bearing) at elevated temperatures up to 300°F (149°C) and even as high as 400°F (204°C). A major use is for automobile transmission seals. Other oil-resistant uses are gaskets and O rings. Mechanical properties such as tensile strength and resilience are low. And, except for recent new formulations, they lose much of their flexibility below −10°F (−23°C). The new grades extend low-temperature service to $-40°F$ $(-40°C)$. Polyacrylates have only fair dielectric strength, which improves, however, at elevated temperatures.

Nitrile elastomers, or **NBR rubbers,** known originally as **Buna N,** are copolymers of acrylonitrile and butadiene. They are principally known for their outstanding resistance to oil and fuels at both normal and elevated temperatures. Their properties can be altered by varying the ratio of the two monomers. In general, as the acrylonitrile content increases, oil resistance, tensile strength, and processability improve while resilience, compression set, low-temperature flexibility, and hysteresis characteristics deteriorate. Most commercial grades range from 20 to 50% acrylonitrile. Those at the high end of the range are used where maximum resistance to fuels and oils is required, such as in oil-well parts and fuel hose. Low-acrylonitrile grades are used where good flexibility at low temperatures is of primary importance. Medium-range types, which are the most widely used, find applications between these extremes. Typical products are flexible couplings, printing blankets, rubber rollers, and washing-machine parts. Nitriles as a group are low in most mechanical properties. Because they do not crystallize appreciably when stretched, their tensile strength is low, and resilience is roughly one-third to one-half that of natural rubber. Depending on acrylonitrile content, low-temperature brittleness occurs at from -15 to $-75°F$ $(-26$ to $-60°C)$. Their electrical insulation quality varies from fair to poor.

Hydrogenated nitrile rubber, of Bayer AG of Germany, has good heat stability, abrasion resistance, and dynamic-load capacity. It is used for synchronous belts in auto applications. **Zeptol elastomers,** of Zeon Chemicals, Inc., are hydrogenated **nitrile-butadiene rubbers** for service temperatures of 30 to $302°F$ $(-1$ to $150°C)$. They have good tensile strength and resistance to lubricants. **Phosphonitrile elastomers** have high elasticity and high-temperature resistance. They are derived from **chlorophosphonitrile,** or **phosphonitrilic chloride,** $P_3N_3Cl_3$, which has a hexagonal ring of alternating atoms of phosphorus and nitrogen with chlorine atoms attached. (In **phosphonitrile plastics,** the chlorine atoms are replaced by other groups.)

Polybutadiene elastomers are notable for their low-temperature performance. With the exception of silicone, they have the lowest brittle or glass transition temperature, $-100°F$ $(-73°C)$, of all the elastomers. They are also one of the most resilient, and have excellent abrasion resistance. However, resistance to chemicals, sunlight, weathering, and permeability by gases is poor. Some uses are shoe heels, soles, gaskets, and belting. They are also often used in blends with other rubbers to provide improvements in resilience, abrasion resistance, and low-temperature flexibility.

Polysulfide elastomer is rated highest in resistance to oil and gasoline. It also has excellent solvent resistance, extremely low gas permeability, and good aging characteristics. Thus, it is used for such products as oil and gasoline hoses, gaskets, washers, and diaphragms. Its major use is for equipment and parts in the coating production and application field. It is also widely applied in liquid form in sealants for the aircraft and marine industries. Its mechanical properties, including strength, compression set, and resilience, are poor. Although it is poor in flame resistance, it can be used in temperatures up to 250°F (121°C). **Ethylene-propylene elastomers,** or **EPR rubber,** are available as copolymers and terpolymers. They offer good resilience, flexing characteristics, compression-set resistance, and hysteresis resistance, along with excellent resistance to weathering, oxidation, and sunlight. Although they are fair to poor in oil resistance, their resistance to chemicals is good. Their maximum continuous service temperature is around 350°F (177°C). Typical applications are electrical insulation, footwear, auto hose, and belts.

Urethane elastomers are copolymers of diisocyanate with polyester or polyether. Both are produced in solid gum form and viscous liquid. With tensile strengths above 5,000 lb/in^2 (34 MPa) and some grades approaching 7,000 lb/in^2 (48 MPa), urethanes are the strongest available elastomers. They are also the hardest, and have extremely good abrasion resistance. Other notable properties are low compression set, and good aging characteristics and oil and fuel resistance. The maximum temperature for continuous use is under 200°F (93°C), and their brittle point ranges from −60 to −90°F (−51 to −68°C). Their largest field of application is for parts requiring high wear resistance and/or strength. Typical products are forklift truck wheels, airplane tail wheels, shoe heels, bumpers on earth-moving machinery, typewriter damping pads, and a variety of seals. For seal applications, maximum recommended deflection limits are 50% for Shore A hardness 40 to 60, 30% for 70, 20% for 80, 10% for 95, and 5% for Shore D 70. **Isoloss** is a series of specially formulated urethanes, of E-A-R Specialty Composites Div. of Cabot Safety Corp., for damping noise, vibration, and shock. **Chlorosulfonated polyethylene elastomer,** commonly known as **Hypalon,** contains about one-third chlorine and 1 to 2% sulfur. It can be used by itself or blended with other elastomers. Hypalon is noted for its excellent resistance to oxidation, sunlight, weathering, ozone, and many chemicals. Some grades are satisfactory for continuous service at temperatures up to 350°F (177°C). It has moderate oil resistance. It also has unlimited colorability. Its mechanical properties are good but not outstanding, although abrasion resistance is excellent. Hypalon is frequently used in blends to improve oxidation and ozone resistance. Typical uses are tank linings, high-temperature conveyor belts, shoe soles and heels, seals, gaskets, and spark plug boots. **Epichlorohydrin elastomers** are noted for their good resistance to oils and excellent resistance to ozone, weathering, and intermediate heat. The homopolymer has extremely low permeability to gases. The copolymer has excellent resilience at low temperatures. Both have low heat buildup, making them attractive for parts subjected to repeated shocks and vibrations.

Fluorocarbon elastomers, like their plastic counterparts, excel in resistance to oxidation, chemicals, oils, solvents, and heat. They are also quite costly. Many have continuous-use service temperatures as high as 400°F (204°C), some can withstand higher temperatures, and they will not support combustion. Most are brittle, however, at -10°F (-23°C), and their mechanical properties are only moderate. **Viton,** of Du Pont, is a vinylidene fluoride-hexafluoropropylene-tetraflouroethylene elastomer. 3M's **Fluorel** elastomers, with 65 to 71% fluorine, are either dipolymers of VF_2 and HFP or terpolymers of VF_2, HFP, and TFE. Depending on the grade, their density is 0.065 to 0.069 lb/in^3 (1799 to 1910 kg/m^3), tensile strength 1,460 to 2,560 lb/in^2 (10 to 18 MPa), elongation 180 to 330%, Shore A hardness 70 to 84, compression set for 70 h at 392°F (200°C) 9 to 45%, and the continuous-use temperature 0 to 392°F (-17.8 to 200°C). The properties of **Fluorel II,** a VF_2-TFE-propylene elastomer, are within these ranges except for density [0.058 lb/in^3 (1605 kg/m^3)]. **Aflas,** a 3M dipolymer of TFE and propylene with 57% fluorine, has a density of 0.056 lb/in^3 (1550 kg/m^3) and a service temperature range of 35 to 392°F (2 to 200°C). Depending on the grade, tensile strength is 1,690 to 2,440 lb/in^2 (11.7 to 17 MPa), elongation 220 to 325%, Shore A hardness 72 to 73, and compression set for 70 h at 392°F (200°C) of 42 to 50%. Except for its higher low-temperature service temperature and the greater acid resistance of peroxide-cured Fluorel, the overall environmental resistance of Alfas is superior to that of Fluorel and Fluorel II. A **phosphonitrilic fluoroelastomer,** developed by Firestone Tire and Rubber Co., retains flexibility at temperatures as low as -70°F (-57°C), sustains temperatures as high as 350°F (177°C), and is especially resistant to oils and solvents over this temperature range. **Kalrez perfluoroelastomer,** of Du Pont, has the highest continuous-use service temperature of any fluorocarbon elastomer: 550°F (288°C). This most costly of elastomers can withstand short-term temperatures to 650°F (343°C) and is also resistant to a variety of solvents, bases, and fuels.

Chlorinated polyethylene elastomers are produced by substitution of chlorine for hydrogen on a high-density polyethylene chain, resulting in a fully saturated structure with no double or triple bonds. The elastomer requires the catalytic reaction of a peroxide for curing.

Thus, most molded parts are black. Five grades of **CPE polymers** are produced, differing principally in chlorine content. The higher-chlorine-content grades have best oil and fuel resistance, tear resistance, gas impermeability, and hardness. Those with lower chlorine content have lower viscosities, better low-temperature properties, and improved resistance to heat and compression set. **Silicone elastomers** are polymers composed basically of silicone and oxygen atoms. There are four major elastomer composition groups. In terms of application, silicone elastomers can be divided roughly into the following types: general-purpose, low-temperature, high-temperature, low-compression-set, high-tensile–high-tear, fluid-resistant, and room-temperature vulcanizing. All silicone elastomers are high-performance, high-price materials. The general-purpose grades, however, are competitive with some of the other specialty rubbers and are less costly than the fluorocarbon elastomers. Silicon elastomers are the most stable group of all the elastomers. They are outstanding in resistance to high and low temperatures, oils, and chemicals. High-temperature grades have maximum continuous service temperatures up to 600°F (316°C); low-temperature grades have glass transition temperatures of $-180°F$ ($-118°C$). Electrical properties, which are comparable to the best of the other elastomers, are maintained over a temperature range from $-100°F$ ($-73°C$) to over 500°F (260°C). However, most grades have relatively poor mechanical properties. Tensile strength runs only around 1,200 lb/in^2 (8 MPa). However, grades have been developed with much improved strength, tear resistance, and compression set. **Fluorosilicone elastomers** have been developed which combine the outstanding characteristics of the fluorocarbons and silicones. However, they are expensive and require special precautions during processing. A unique characteristic of one of these elastomers is its relatively uniform modulus of elasticity over a wide temperature range and under a variety of conditions. Silicone elastomers are used extensively in products and components where high performance is required. Typical uses are seals, gaskets, O rings, insulation for wire and cable, and encapsulation of electronic components.

ELECTRICAL-CONTACT MATERIALS. These are materials used to make and break electrical contact, thus make-and-break electric circuits, or to provide sliding or constant contact. Both require high electrical conductivity to ease current flow, high thermal conductivity to dissipate heat, high melting point or range to inhibit arc erosion and prevent sticking, corrosion and oxidation resistance to prevent formation of films that impede current flow, high hardness for wear resistance, and amenability to welding, brazing, or other means of joining. The sliding-contact types also require low friction, and a lubricant is always required between the sliding materials to prevent seizing and galling.

The materials used range from pure metals and alloys to composites, including those made by powder-metallurgy methods. **Copper** is widely used but requires protection from oxidation and corrosion, such as by immersion in oil, coating, or vacuum sealing. **Copper-tungsten alloys** or mixtures of **copper-graphite** increase resistance to arcing and sticking, and some copper alloys provide greater hardness, thus greater wear resistance, and better spring characteristics. **Silver** is more oxidation-resistant in air and, pure or alloyed, is the most widely used metal for make-and-break contacts for application at currents to 600 A. **Silver-copper alloys** provide greater hardness but less conductivity and oxidation resistance; **silver-cadmium alloys** increase resistance to arc erosion and sticking; and **silver-platinum alloys, silver-palladium alloys,** and **silver-gold alloys** increase hardness, wear resistance, and oxidation resistance. All alloying elements, however, decrease conductivity. **Gold** has outstanding oxidation and sulfidation resistance but, being soft and prone to wear and arc erosion, is limited to low-current (0.5-A maximum) applications. To enhance these properties, **gold alloys,** such as **gold-silver, gold-copper, gold-silver-platinum, gold-silver-nickel,** and **gold-copper-platinum-silver,** are more commonly used. **Platinum** and **palladium** are also used for contacts but, again, in alloy form more than as pure metals. Among the most common ones are **platinum-iridium, platinum-ruthenium, platinum-palladium-ruthenium, palladium-ruthenium, palladium-copper,** and **palladium-silver.** A **palladium-silver-platinum-gold alloy** for brushes and sliding contacts is noted for its exceptional modulus of elasticity and high proportional limit. **Aluminum, tungsten,** and **molybdenum** are also used for electrical contacts but mainly in composite form. Aluminum used for contacts provides an electrical conductivity of about 60% that of copper, but is prone to oxidation and thus clad or plated with silver, tin, or copper. The refractory metals, though providing excellent resistance to wear and arc erosion, are poor conductors and oxidize readily.

The principal metals made in composite form by PM methods are the **refractory metals,** including those in carbide form, and copper-base and silver-base metals. The refractory metals, notably tungsten and molybdenum or their carbides, usually serve as a base for infiltrating with copper or silver, thus combining electrical conductivity and resistance to wear and arc erosion in a single material. Many such composites are common, including **tungsten-copper, tungsten-silver, tungsten carbide-silver, tungsten carbide-copper, tungsten-graphite-silver,** and **molybdenum-silver.** The amount of conductive metal may exceed or be less than that of the refractory metal or refractory-metal carbide. A common silver composite is **silver-cadmium oxide,** which, for a given amount of silver, provides greater conductivity than a silver-cadmium alloy as well as greater hardness and resistance to sticking. Others include **silver-graphite, silver-nickel,** and **silver-iron.** The silver-graphite composites are used mainly for sliding or brush contacts.

ELECTRICAL INSULATORS.   Any materials that retard the flow of electricity and are used to prevent the passage or escape of electric current from conductors. No materials are absolute nonconductors; those rating lowest on the scale of conductivity are therefore the best insulators. An important requirement of a good insulator is that it not absorb moisture which would lower its resistivity. Glass and porcelain are the most common line insulators because of low cost. Pure silica glass has an average dielectric strength of 500 V/mil (20×10^6 V/m), and glass-bonded mica about 450 V/mil (17.7×10^6 V/m), while ordinary porcelain may be as low as 200 V/mil (8×10^6 V/m), and steatite about 240 V/mil (9.4×10^6 V/m). Slate, steatite, and stone slabs are still used for panelboards, but now a great variety of insulating boards are made by compressing glass fibers, quartz, or minerals with binders, or standard laminated plastics of good dielectric strength may be used. **Vulcoid,** of Budd Co., is typical. For slots and separators, natural mica is still valued because of its heat resistance, but because of the irregular quality of natural mica and the difficulty of handling the small pieces, it has been largely replaced by synthetic mica paper, polyester sheet, or impregnated papers or fabrics. The impregnated fish paper called **Armite** comes in thicknesses down to 0.004 in (0.010 cm) and has a dielectric strength of 500 V/mil (20×10^6 V/m).

Synthetic rubbers and plastics have now replaced natural rubber for wire insulation, but some aluminum conductors are insulated only with an anodized coating of aluminum oxide. Wires to be coated with an organic insulator may first be treated with hydrogen fluoride, giving a coating of copper fluoride on copper wire and aluminum fluoride on aluminum wire. The thin film of fluoride has high dielectric strength and heat resistance. The AIEE classification for wire insulation is by heat resistance, from **Class O insulation,** for temperatures to 195°F (90°C), to **Class C insulation,** for temperatures above 355°F (179°C).

Insulating oils are mineral oils of high dielectric strength and high flash point employed in circuit breakers, switches, transformers, and other electrical apparatus. An oil with a flash point of 285°F (140°C) and fire point of 310°F (154°C) is considered safe. A clean, well-refined oil will have high dielectric strength, but the presence of as low as 0.01% water will reduce the dielectric strength drastically. The insulating oils, therefore, cannot be stored for long periods because of the danger of absorbing moisture. Impurities such as acids or alkalies also detract from the strength of the oil. Since insulating oils are used for cooling as well as insulating, the viscosity should be low enough for free circulation, and they should not gum. **Askarel** is an ASTM designation for **insulating fluids** which give out only nonflammable gases if decomposed by an electric arc. They are usually chlorinated aromatic hydrocarbons such as trichlorobenzene (TCB), but fluorinated hydrocarbons are also used. They have high dielectric strength, and a dielectric constant below 2. Askarel also refers to dielectric fluids containing **polychlorinated biphenyls** (**PCBs**), which had been widely used in transformers. These fluids, which may contain as much as 50% PCBs, have been replaced because of environmental concerns regarding PCBs. **Insulating gases** are used to replace air in closed areas to insulate high-voltage equipment. Sulfur hexafluoride for this purpose has a dielectric strength 2.35 times that of air. The insulating oil, fluids, and gases are generally referred to as **dielectrics,** although this term embraces any insulator.

    Insulation porcelain, or **electrical porcelain,** is not usually an ordinary porcelain except for common line insulators. For such uses as spark plugs they may be molded silica, and for electronic insulation they may be molded steatite or specially compounded ceramics, more properly called **ceramic insulators.** Insulation porcelains compounded with varying percentages of zirconia and beryllia have a crystalline structure and good dielectric and mechanical strengths at temperatures as high as 2000°F (1093°C). These porcelains may have some magnesia, but are free of silica. However, **zircon porcelain** is made from zirconium silicate, and the molded and fired ceramic is equal to high-grade steatite for high-frequency insulation. **Vitrolain,** of Star Porcelain Co., is an electrical porcelain of high strength and density with porosity of only 0.25%. **Thyrite,** of General Electric Co., is a porcelain that possesses the property of being an insulator at low potentials and a conductor at high potentials. It is used for lightning arresters. The German **Hartporzellan,** or **hard porcelains,** are specially compounded ceramics having good resistance to thermal shock. The material called **Nolex** by the Naval Surface Weapons Center (NSWC) is made by hot-molding finely powdered synthetic fluorine mica. The molded parts are practically pure mica. They can be machined, have high dimensional precision because they need no further heat treatment, and have high dielectric strength. Beryllia is a valued insulator for encapsulation coatings on heat-generating electronic devices as it has both high electrical resistivity and high heat conductivity.

Most **ceramics** are electrical insulators and are used widely for insulation of electric power lines. The applications range from structural power insulators to electronic packaging and substrates. Of particular interest is the use of ceramics as substrates. The ceramics serve as the structural and insulating base on which electronic components are deposited or attached. The requirement for good surface finish has led to the development of fine-grained alumina material that can be prepared with a very good finish. The requirement for high thermal conductivity in some applications has led to the use of **beryllia,** high-purity **alumina,** and more recently, **aluminum nitride** as substrate materials. Other materials are **silicon carbide** doped with beryllia to give electrical insulation and **glass-ceramics** which can easily be produced in the complex shapes often needed. They can also be produced with a tailored thermal expansion coefficient **multilayer ceramic** (MLC) substrate for high-speed computer processing modules.

ELECTRICAL-RESISTANCE METALS AND ALLOYS. This major family of metals, including alloys as well as pure metals, includes **resistance alloys** used in controls and instruments to measure or regulate electrical performance, **heating alloys** used to generate heat, and **thermostat metals** used to convert heat to mechanical energy. There are seven types of electrical-resistance alloys: (1) **radio alloys,** which contain 78 to 98% copper with the balance nickel; (2) **manganins,** 87% copper and 13 manganese or 83 to 85 copper, 10 to 13 manganese, and 4 nickel; (3) **constantans,** 55 to 57% copper and 43 to 45 nickel; (4) **nickel-chromium-aluminum alloys,** 72 to 75% nickel, 20 chromium, 3 aluminum, and either 5 manganese or 2 copper, iron, or manganese; (5) **iron-chromium-aluminum alloys,** 73 to 81% iron, 15 to 22 chromium, and 4 to 5.5 aluminum; (6) various other alloys, mostly **nickel-base alloys,** some of which contain substantial amounts of chromium or iron, chromium and iron, chromium and silicon, and, in some cases, manganese and aluminum; and (7) pure metals, notably **aluminum, copper, iron, nickel, precious metals, and refractory metals.**

Key characteristics of resistance alloys are uniform resistivity, stable resistance, reproducible temperature coefficients of resistance, and low thermoelectric potential compared to copper. Less critical, but also important, are the coefficient of thermal expansion; strength and ductility; corrosion resistance; and joinability to dissimilar metals by welding, brazing, or soldering. Heating alloys require high heat resistance, including resistance to oxidation and creep in particular environments, such as furnaces, in which they are widely used; high electrical resistivity; and reproducible temperature coefficients of resistance. Also desirable are high emissivity, low coefficients of thermal expansion, and low modulus to minimize thermal fatigue, strength, and resistance to thermal shock, and ductility for fabricability. Thermostat metals, two or more bonded materials of which one may be nonmetallic, are chosen based on different electrical resistivities and thermal expansivities so that applied heat can be converted to mechanical energy.

The electrical resistivity in ohm · circular mil/ft ($n\Omega \cdot m$) for the alloys encompassed by the seven groups range from 9.6 (16) for silver to 872 (1,450) for an alloy made up of 72.5% iron, 22 chromium, and 5.5 aluminum. Iron has the highest resistivity, 583 (970), of the pure metals, followed by tantalum, 81 (135); platinum, 64 (106); nickel, 48 (80); and tungsten, 33 (55). The radio alloys are in the 30 to 180 (50 to 300) range, the manganins 228 to 289 (380 to 480), the constantans 295 to 300 (490 to 500), and most of the nickel-chromium-aluminum, iron-chromium-aluminum, and various other alloys are in the 610 to 872 (1,015 to 1,450) range. Temperature coefficients of resistance in parts per million per degree Fahrenheit (Celsius) range from ±10 to ±15 at 59 to 113°F (15 to 45°C) for the manganins to +6,000 at 68 to 95°F (20 to 35°C) for pure nickel. Thermoelectric potentials versus copper in the μV/°F range from −43 to 77 to 221°F (25 to 105°C) for the 57 copper–43 nickel constantan to +12.2 at 32 to 167°F (0 to 75°C) for pure iron. The refractory metals have the lowest coefficients of thermal expansion, aluminum the highest, and most of the alloys are intermediate. Tensile strength and ductility also range widely depending on the alloy or metal. Maximum operating temperatures in air for the commonly used resistance-heating alloys range from 1700°F (927°C) for 43.5 Fe–35 Ni–20 Cr–1.5 Si alloy to 2505°F (1374°C) for 72.5 Fe–22 Cr–5.5 Al. For **platinum,** this temperature is 2750°F (1510°C). The refractory metals are suitable for still higher temperatures in vacuum and, in the case of **molybdenum** and **tungsten,** in select environments.

Electrical-resistance alloys are mainly wire products, and the alloys have been known by a multitude of trade names. The standard alloy for electrical-resistance wire for heaters and electrical appliances is **nickel-chromium,** but **nickel-manganese** and other alloys are used. For consumer products made in large quantities, cost and the relative scarcity of the alloying elements are important considerations. For high-temperature furnaces, tungsten, molybdenum, and alloys of the more expensive high-melting metals are employed. The much-used alloy with 80% nickel and 20 chromium resists scaling and oxidation to 2150°F (1177°C), but it is subject to an intergranular corrosion, known as *green rot,* which may occur in chromium above 1500°F (816°C) unless modified with other elements. The 80–20 alloy has a resistivity of $354 \times 10^{-8} \; \Omega \cdot$ ft ($108 \times 10^{-8} \; \Omega \cdot$ m). The tensile strength of the annealed wire is 100,000 lb/in^2 (689 MPa), with elongation of 35%, and the hardness is Rockwell B 80. The specific gravity is 8.412. In many appliances, high elongation is undesirable because it causes the wire to sag.

In times of nickel stringency, or for cost reduction, various **nickel-chromium-iron** alloys are used. An alloy of 60% nickel, 16 chromium, and 24 iron has a resistivity of 675 Ω with oxidation resistance to 1950°F (1066°C). **Tophet C** is this alloy. The alloy with 30% nickel, 20 chromium, and 50 iron is resistant to 1560°F (849°C). The resistivity of the low-nickel, chromium-iron alloys is high, and the heat resistance is ample for some types of appliances, but the strength is lower, with a tendency for the hot wire to sag. **Cromel AA,** of Hoskins Mfg. Corp., is an 80–20 nickel-chromium alloy for continuous service to 2150°F (1127°C). It is modified with small amounts of cobalt, manganese, columbium, silicon, and iron, the columbium stabilizing the chromium to prevent green rot. It is also resistant to carbon pickup which tends to make the chromium-iron alloys brittle. The resistance of the wire is $381 \times 10^{-8} \; \Omega \cdot$ ft ($116 \times 10^{-8} \; \Omega \cdot$ m).

The **chromium-aluminum-iron** alloys have high resistivity and high oxidation resistance, but have a tendency to become brittle. **Hoskins alloy 870** contains 22.5% chromium, 5.5 aluminum, 0.5 silicon, 0.10 carbon, and the balance iron. The resistivity is 466×10^{-8} $\Omega \cdot$ ft ($142 \times 10^{-8} \; \Omega \cdot$ m). It is used as wire or ribbon in furnaces to 2350°F (1288°C). The **Kanthal alloys** marketed as wire and ribbon by Kanthal Corp. have 20 to 25% chromium with some cobalt and aluminum, and the balance iron. **Kanthal A,** with 5% aluminum, will withstand temperatures to 2370°F (1299°C), has a resistivity of $456 \times 10^{-8} \; \Omega \cdot$ ft ($139 \times 10^{-8} \; \Omega \cdot$ m), and is resistant to sulfuric acid. The tensile strength is 118,000 lb/in^2 (813 MPa) with elongation of 12 to 16%. **Kanthal A-1,** for furnaces, has a resistance of 872 Ω and an operating temperature to 2505°F (1373°C). The **Nikrothal alloys** of this company are nickel-base modifications of Kanthal. They have higher tensile strength, up to 200,000 lb/in^2 (1,378 MPa), to permit rapid winding of tape without breakage. **Nikrothal 6** has 60% nickel, 15 chromium, and 25 iron. **Heating tape,** of Rogers Corp., designed for heating rocket batteries, is also used for panel heating and is operable at continuous temperatures up to 250°F (121°C). It has three flat wires of copper-nickel alloy encapsulated in Mylar tape 0.008 in (0.020 cm) thick and 0.375 in (0.953 cm) wide in lengths to 250 ft (76 m). The rating is 2 W/ft (6.6 W/m), and the dielectric strength is 2,400 V.

A series of alloys of Westinghouse Electric Corp., called **Hirox alloys,** contain 6 to 10% aluminum, 3 to 9 chromium, up to 4 manganese, with the balance iron except for small additions of boron and zirconium to reduce the size of the aluminum-iron grains and refine the structure. The alloy with 9% aluminum and 9 chromium has a resistivity of 850 Ω and a tensile strength of 118,000 lb/in^2 (813 MPa). At 1300°F (704°C) the tensile strength is 15,000 lb/in^2 (103 MPa) with elongation of 94%. Wire will give continuous service in air at 2350°F (1288°C) without failure.

Resistance alloys are generally specified for specific uses rather than by composition. Controlled resistivity over a temperature range instead of high heat resistance is desired for instrument use, while a definite coefficient of expansion is required for spark-plug wire and other uses where the wire is embedded. In some cases, good heat resistance with selected low resistivity is desired. **Oxalloy 28,** of GTE Corp., is copper wire clad with 28% by weight of chromium-iron alloy. It withstands continuous service at 1300°F (704°C). The resistivity at 1100°F (593°C) is 28×10^{-8} $\Omega \cdot$ ft (8.6×10^{-8} $\Omega \cdot$ m). **Neyoro G,** of J. M. Ney Co., used for fine resistance wire in potentiometers and electronic applications where high cost is not a factor, has a high gold content with platinum, silver, and copper. The drawn wire has a tensile strength of 185,000 lb/in^2 (1,275 MPa) and high corrosion resistance.

Copper-manganese alloys have high resistivity, an alloy with 96 to 98% manganese having a resistance of more than 16,400 $\mu\Omega$/in^3 (1,000 $\mu\Omega$/cm^3). But when the manganese content is high, they are brittle and difficult to make into wire. Addition of nickel makes them ductile, but lowers resistivity. A typical alloy contains 35% manganese, 35 nickel, and 30 copper. A resistance alloy developed by the National Bureau of Standards, called **Therlo,** contains 85% copper, 9.5 manganese, and 5.5 aluminum. **Fecraloy** has 15% chromium, 5 aluminum, and the balance iron. It is for temperatures to 1400°F (760°C). **Sparkaloy** is a **spark-plug wire** and is a manganese-nickel alloy. The spark-plug wire of Hoskins Mfg. Co., called **Hoskins alloy 667,** contains 4% manganese, 1 silicon, and the balance nickel. The resistance is 82×10^{-8} $\Omega \cdot$ ft (25×10^{-8} $\Omega \cdot$ m), specific gravity 8.4, and coefficient of expansion 15.1×10^{-6}/°F (27×10^{-6}/K). **Manganin** contains 80 to 85% copper, 2 to 5 nickel, and 12 to 15 manganese. It has a tensile strength of 70,000 lb/in^2 (482 MPa) and a resistance of 157×10^{-8} $\Omega \cdot$ ft (48×10^{-8} $\Omega \cdot$ m). It is used for coils and shunt wires in electrical instruments and in sheet form for instrument springs. **Tophet A** is a standard 80–20 nickel-chromium alloy. The tensile strength is 120,000 lb/in^2 (827 MPa) and resistance 354×10^{-8} $\Omega \cdot$ ft (108×10^{-8} $\Omega \cdot$ m). Electrical-resistance alloys, developed by Inco

Alloys International, contain 60 to 80% nickel plus chromium and iron. They are used for heater elements, resistors, rheostats, resistance thermometers, and in potentiometers.

   Calorite, of General Electric Co., has 65% nickel, 8 manganese, 12 chromium, and 15 iron. **Excello metal** contains 85% nickel, 14 chromium, and 0.5 each manganese and iron. It is used in electric heaters for temperatures up to 2000°F (1093°C). **Alumel,** of Hoskins Mfg. Co., intended for temperatures up to 2282°F (1250°C), has 94% nickel, 2.5 manganese, 0.5 iron, and small amounts of other elements. **Calido,** of Driver-Harris Co., contains 59% nickel, 16 chromium, and 25 iron. **Nichrome V,** of the same company, is the 80–20 alloy. **Nichrome S** contains 25% nickel, 17 chromium, and 2.5 silicon. It is marketed in sheet form for temperatures up to 1800°F (982°C). **Comet metal,** of the same company, used for rheostats, contains 30% nickel, 5 chromium, and the balance iron. It has high strength, up to 160,000 lb/in² (1,103 MPa), and a resistivity of 570 Ω. The Driver-Harris resistance alloy known as **Karma** contains 20% chromium, 3 iron, 3 aluminum, 0.30 silicon, 0.15 manganese, 0.06 carbon, and the balance nickel. Its melting point is 2552°F (1400°C), its resistivity 800 Ω, and the annealed wire has a tensile strength of 130,000 lb/in² (896 MPa) with elongation 25%. **Hytemco,** of the same company, is an iron-nickel alloy used for low-temperature wire. The resistance is 6.6×10^{-8} $\Omega \cdot$ ft ($2.0 \times 10^{-8}\Omega \cdot$ m). **Magno** is a 95% nickel, 5 manganese alloy of the same company; **Climax metal** has 74% iron, 25 nickel, and 1 manganese.

ELEMI.   A soft, sticky, opaque resin with a pleasant odor, obtained from the **pili tree,** *Canarium luzonicum,* of the Philippines and employed for giving body and elasticity to lacquers and in lithographic inks. In medicine it is used in ointments. It contains **dipentene,** $C_{10}H_{18}$, which is called **limonene** from its lemonlike odor, and is known as **cajeputene** when obtained from cajeput. **Limonene dioxide,** or **dipentene dioxide,** a colorless liquid of composition $C_{10}H_{12}O_2$, is a valuable synthetic chemical for making epoxy resins and for cross-linking acrylic and other resins. Elemi also contains a related terpinene oil, **phellandrene.** Substitute elemi resins are obtained from various trees of the family *Burseraceae* of tropical Africa and America. The pili trees are hacked or stripped, and the resin collects on the bark, a tree yielding about 5 lb (2.3 kg) per year. **West Indian elemi** is from the tree *Dacryodes hexandra* of the West Indies. **Nauli gum** is elemi from the tree *C. commune* of the Solomons. **Elemi oil,** obtained by distilling elemi, is a colorless liquid of specific gravity 0.87 to 0.91, used in perfumes and in medicines. It has an aniselike odor.

ELKSKIN. The commercial name for soft, pliable, and durable leather made from the bundled rawhides known as kips, or from overgrown calf by a special tanning process and impregnation with oils. It is used chiefly for children's shoe uppers and for pocketbooks. A heavier elkskin, or **elk leather,** for sport shoes and boots, is made from cowhides by the same treatment. Elkskin, like chamois, dries out to its original softness after wetting. **Smoked elk** is elk leather dyed cream-colored to imitate the original leather of elks, which was smoked over a wood fire.

ELM. The wood of several species of the elm tree, of the eastern United States and Canada and northern Europe. The wood of the **American elm,** or **white elm,** *Ulmus americana,* has a fine grain, has a density of about 40 lb/ft^3 (641 kg/m^3), is hard and tough, and is whitish brown. It is the best known commercially of the six species grown in the United States. The American elm is not a forest tree, but is grown as a shade and ornamental tree. It does not grow in the mountains. The trees sometimes reach a diameter of 6 ft (1.8 m) and a height of 100 ft (30 m). The tough, durable wood is valued for ax handles and for parts requiring a combination of strength, bending qualities, and ability to withstand rough use. The wood of this tree, and of the rock elm, was formerly used for superstructures of naval ships because it did not sharp-splinter like oak. It was also the favorite wood for hubs and spokes of heavy wagon wheels. **Rock elm,** or **hickory elm,** *U. thomasii,* is also native to the United States and Canada. It has a very fine, close grain and is slightly heavier. It is sometimes called **cork elm,** although this name applies to the **wahoo,** or **winged elm,** *U. alata,* of the southeastern states, because of the corky appearance of the twigs. The winged elm is grown as a shade tree, but the wood was valued for vehicle parts. **English elm,** *U. procera,* has a straight trunk and rounded crown more like the oaks. **Chinese elm,** *U. parvifolia,* has small leaves and is very resistant to disease. **Slippery elm** is a smaller forest tree, *U. fulva,* of the northeastern United States. Considerable lumber came from this tree under the name **red elm.** The inner bark of the tree is mucilaginous with a sweet taste and characteristic odor. It was used by the Indians as a chewing gum and as a poultice for skin infections. The dried and powdered bark is now used in medicine for skin infections and for the throat. It contains **ulmic acid,** or **geic acid,** $C_{20}H_{14}O_6$.

EMERY. A fine-grained, impure variety of the mineral corundum, with the fine crystals of aluminum oxide embedded in a matrix of iron oxide. It usually contains only 55 to 75% Al_2O_3. The specific gravity is 3.7 to 4.3 and Mohs hardness about 8. It occurs as a dark-brown, granular massive mineral. It is used as an abrasive either ground into powder or in blocks and wheels. In the natural block material, the grains are irregular, giving a varying grinding performance. The grains are graded in sizes from 220 mesh, the finest, to 20 mesh, the coarsest. **Emery paper** and cloth are usually graded from 24 to 120 mesh, and the grains are glued to one side of 9- by 11-in (23- by 28-cm) sheets. **Flour of emery** is the finest powder, usually dust from the crushing. **Emery cake** as made for buffing and polishing is not likely to be made of emery but a graded combination of aluminum oxide and iron oxide, with a higher percentage of the hard aluminum oxide for buffing, and higher iron oxide for polishing. It is furnished in various grades of fineness, with grains of 120 to 200 mesh, or flour sizes, F, FF, and FFF. Emery takes its name from Cape Emery, on the Island of Naxos.

EMULSIFYING AGENTS.   Materials used to aid in the mixing of liquids that are not soluble in one another, or to stabilize the suspension of nonliquid materials in a liquid in which the nonliquid is not soluble. The suspension of droplets of one liquid in another liquid in which the first liquid is not soluble is called an **emulsion.** The emulsion of oil and water, used in machine shops as a cutting lubricant and work coolant, may be made with soap as the emulsifying agent. The emulsifying agent protects droplets of the dispersed medium from uniting and thus separating out. The oil itself may be treated so that it is self-emulsifying. Sulfonated oils contain strong negatively charged ester sulfate groups in the molecule and do not react and conglomerate with the molecules of a weakly charged liquid. They will thus form emulsions with water without any other agent.

In emulsions of a powder in a liquid, an emulsifying agent called a **protective colloid** may be used. This is usually a material of high molecular weight such as gelatin, and such materials form a protective film around each particle of the contained powder. A photographic emulsion is a suspension of finely divided **silver halide** grains in gelatin. The gelatin serves as a binder, protective colloid, and 3 sensitizer for the silver halide. The emulsion consists of 35 to 40% silver halide, 60 to 65 gelatin, with small amounts of stabilizers, antifoggers, and hardeners. **Saponin** and **starches** are commonly used thus as **suspending agents.** For the suspension of drug materials in pharmaceutical mixtures, **gum arabic** or **tragacanth** may be used. Starches, egg albumin, and proteins are common emulsifying agents for food preparation. **Alginates** are among the best suspending agents for a wide range of emulsions because of the numerous repelling charges in the high molecular weight and the irregular configuration of the chain, but when added to protein-containing liquids such as many foodstuffs, the similar conditions of the algin and the protein molecules cause a neutralization reaction and a precipitation of the agglomerated particles. Suspending agents generally increase the viscosity of the liquid, and with high concentrations of some gums or resins, the water molecules may be completely encased in the resin lattice as a semisolid or water-filled plastic.

Sucrose esters, used as emulsifiers for foods, cosmetics, and drugs, are made from sugar and palmitic, lauric, or other fatty acids. The monoesters are soluble in water and in alcohol, and the diesters are oil-soluble. **Sorbester,** of Howards of Ilford, Ltd., for emulsifying fatty foodstuffs, is a diester of sucrose. **Sucrodet D-600,** of Millmaster Chemical Co. is a white, tasteless, and odorless powder made from sugar and palmitic acid. **Myrj 45,** of Atlas Chemical, is polyoxyethylene stearate. **Propylene laurate** is a light, high-boiling-point liquid that is self-emulsifying in water and is employed in foodstuffs and pharmaceuticals to stabilize the mixtures. The sodium salt of **ursolic acid** is a strong emulsifying agent for oil-in-water mixtures. The acid is a complex triterpene obtained from the skins of the cranberry. An ionic emulsifier has an organic lipophilic group (L) and a hydrophilic group (H), the balance between the two characterizing such classes of emulsifiers and surfactants. An anionic emulsifier is **triethanol stearate,** and a nonionic one is **P.E.G. 300 distearate**; both emulsify oil in water. Cold-cream emulsions are water in oil; the emulsifier is **sodium cerotate** produced by reacting borax with **cerotic acid.** The latter occurs freely in beeswax.

Some solid materials may be suspended indefinitely in liquids if ground to such a fineness that the electronegative mutually repelling force, or **zeta potential,** of the particles is greater than the force of gravity. Silica, for example, has only a feeble electronegativity, and if the particles are below about 39 μin (1 μm), they will give a permanent suspension in water. These finely ground solids are used as **thickening agents** for paints and coatings. Bentonite is thus used in adhesives and paints and for imparting nonnewtonian rheologies to drilling muds. Thickening agents may add other properties such as better adhesion or strengthening the film. Some long-chain chemicals used as emulsifying agents in cutting oils also give antirust properties. **Thickening agent ASE-95,** of Rohm & Haas Co., has a powerful thickening action on water-base emulsions which can be halted at any desired viscosity by neutralizing the acidity. It is an acrylic copolymer of 20% solids containing an organic acid with a pH of 3. When added to the emulsion to be thickened, the solids dissolve in minute particles, and the process is stopped at the desired viscosity by adding an alkali.

ENAMEL. A coating which upon hardening has an enameled or glossy face. **Pottery enamels, ceramic enamels,** or **ceramic coatings,**

and **vitreous enamels** are composed chiefly of quartz, feldspar, clay, soda, and borax, with saltpeter or borax as fluxes. The quartz supplies the silica, and such enamels are fusible glasses. In acid-resisting enamels, alkali earths may be used instead of borates. To make enamels opaque, opacifiers are used. They may be tin oxide for white enamel, cobalt oxide for blue, or platinum oxide for gray. Enamel-making materials are prepared in the form of a powder which is called **frit.** The frit-making temperature is about 2400°F (1316°C), but the enamel application temperatures are from 1400 to 1600°F (760 to 871°C). Each succeeding coat has a lower melting point than the one before it, so as not to destroy the preceding coat. It must also have about the same coefficient of expansion as the metal to prevent cracking. Enamels for aluminum usually have a high proportion of lead oxide to lower the melting point, and enamels for magnesium may be based on lithium oxide. Some enamels for low-melting-point metals have the ceramic frit bonded to the metal with monoaluminum phosphate at temperatures as low as 400°F (204°C).

The mineral oxide coatings fused to metals are often called **porcelain enamels,** but they are not porcelain, and the term **vitreous enamel** is preferred in the industry, although ceramic-lined tanks and pipe are very often referred to as **glass-lined steel.** The composition varies greatly, one company having more than 3,000 formulas. Vitreous enameled metals are used for cooking utensils, signs, chemical tanks and piping, clock and instrument dials, and siding and roofing. Ground coats are usually no more than 0.004 in (0.010 cm) thick, and cover coats may be 0.003 to 0.008 in (0.006 to 0.020 cm) thick. The hardness ranges from Knoop 150 to 500. Thick coatings on thin metals are fragile, but thin coatings on heavy metals are flexible enough to be bent. Standard porcelain-type enamel has a smooth, glossy surface with a light reflectance of at least 65% in the white color, but pebbly surfaces that break up the reflected image may be used for architectural applications.

High-temperature coatings may contain a very high percentage of zirconium and will withstand temperatures to 1650°F (899°C). **Refractory enamels,** for coating superalloys to protect against the corrosion of hot gases to 2500°F (1371°C), may be made with standard ceramic frits to which is added boron nitride with a lithium chromate or fluoride flux. Blue undercoats containing cobalt are generally used to obtain high adhesion on iron and steel, but some of the **enameling steels** do not require an undercoat, especially when a specially compounded frit or special flux is used. When **sodium aluminum silicate,** $Na_2O \cdot Al_2O_3 \cdot 6SiO \cdot xH_2O$, is used instead of borax, a white finish is produced without a ground coat. **Mirac** is a white enamel which gives good adhesion directly to steel. Enamels containing titanium oxide will adhere well to steels alloyed with a small amount of titanium. **Ti-Namel,** of Inland Steel Co., is an enameling steel containing titanium.

Many trade names are applied to vitreous enamels and to enameled metals. **Vitric steel** is an enameled corrugated sheet steel for construction. **Majolica** is an old name for marblelike enamels made by mixing enamels of different colors, but **mottled graywear** is made with cobalt oxide on steel that has a controlled misting on the surface. **Cloisonné enamel** is an ancient decorative enamel produced by soldering thin strips of gold on the base metal to form cells into which the colored enamel is pressed and fused into place. It requires costly hand methods and is now imitated in synthetic plastics under names such as **Enameloid.**

The word *enamel* in the paint industry refers to glossy **varnishes** with pigments or to paints of oxide or sulfate pigments mixed with varnish to give a glossy face. They vary widely in composition, in color and appearance, and in properties. As a class, enamels are hard and tough and offer good mar and abrasion resistance. They can be formulated to resist attack by the most commonly encountered chemical agents and corrosive atmospheres. Because of their wide range of useful properties, enamels are one of the most widely used organic finishes in industry and are especially used as household appliance finishes. **Japan** is a name applied to black baking enamels. Japan consists of a pigment, a gum, a drying oil, and a reducer, the same as any oil enamel. It is always baked, which drives off the solvent and fuses the gum into a uniform vitreous layer. Japans have now been replaced by synthetic baking finishes. The modified phenolmelamine and alkyd-melamine synthetic resins produce tough and resistant enamel coatings. Quick-drying enamels are the cellulose lacquers with pigments. **Fibrous enamel,** used for painting roofs, is an asphalt solution in which asbestos fibers have been incorporated. When of heavy consistency and used for caulking metal roofs, it is called **roof putty.**

EPOXY RESINS. A class of synthetic resins characterized by having in the molecule a highly reactive **oxirane** ring of triangular configuration consisting of an oxygen atom bonded to two adjoining and bonded carbon atoms. They are usually made by the reaction of epichlorohydrin with phenol compounds, but epoxidation is also done by the oxidation of a carbon-to-carbon double bond with an organic peracid such as peracetic acid. **Epichlorohydrin** is produced from allyl chloride and is a colorless liquid with a chlorine atom and an epoxide ring. The **dipoxy resins** made by the oxidation of olefins with peracetic acid have higher heat resistance than those made with bisphenol. Epoxidation is not limited to the making of plastic resins, and **epoxi-**

dized oils, usually epoxidized with peracetic acid, are used as paint oils and as plasticizers for vinyl resins.

Epoxy resins are generally more costly than many other thermosetting resins, but, because of their combinations of high mechanical and electrical properties, they are important, especially for such uses as adhesives, chemically resistant coatings, and encapsulation of electronic units. The resins are thermosetting and inert. For encapsulation, they cast easily with little shrinkage. They have very high adhesion to metals and nonmetals, heat resistance from 350 to 500°F (177 to 260°C), dielectric strength to 550 V/mil (22 V/m), and hardness to Rockwell M 110. The tensile strength may be up to 12,000 lb/in^2 (83 MPa), with elongation to 2 to 5%, but some resilient encapsulating resins are made with elongation to 150% with lower tensile strengths. The resins have high resistance to common solvents, oils, and chemicals.

An unlimited variety of epoxy resins are possible by varying the basic reactions with different chemicals or different catalysts, or both, by combination with other resins, or by cross-linking with organic acids, amines, and other agents. To reduce cost when used as laminating adhesives, they may be blended with furfural resins, giving adhesives of high strength and high chemical resistance. Blends with polyamides have high dielectric strength, mold well, and are used for encapsulating electrical components. By using a polyamide curing agent an epoxy can be made water-emulsifiable for use in water-based paints. An epoxy resin with 19% bromine in the molecule is flame-resistant. Another grade, with 49% bromine, is a semisolid, used for heat-resistant adhesives and coatings. **Epoxidized polyolefins** have five or more reactive epoxy groups along each molecule of the chain instead of the usual two terminal epoxy groups on each molecule. With dibasic acids or anhydrides they form strong, hard resins of high heat resistance; or resins of lower viscosity are made for laminating and casting. Epoxy resins made by the reaction of epichlorohydrin with a phenol-formaldehyde resin with an anhydride catalyst have heat distortion points of 570°F (300°C). As an adhesive for laminates, they give very high strength at elevated temperatures. Epoxies can be copolymerized with other resins. **Epoxy-acrylate resin,** used for glass-fiber laminates, combines the resistance and adhesiveness of the epoxy with the fast cure and strength of the acrylate. Epoxy resins can be made with cyclopentyl oxide terminal groups instead of diglycidyl ether. The yield strength at 392°F (200°C) is 18,200 lb/in^2 (123 MPa), and they have a heat deflection temperature of 434°F (223°C). Epoxy resins can be produced by a reaction of **hydantoin** with epichlorohydrin. Hydantoin is a nitrogen-containing heterocyclic compound. They have high mechanical properties, good dielectrical characteristics, and ultraviolet light resistance. They retain light transmission properties after thermal aging of several thousand hours at 302°F (150°C).

Epoxy has been the major matrix material of **polymer-matrix composites** for aircraft applications for many years. This is attributable to ease of processing (low-pressure, moderate-temperature autoclave or press curing), good mechanical properties, and low cost. The principal reinforcements are fibers of aramid (Kevlar), boron, glass, and graphite. In such applications, the composites are used for service temperatures up to about 300°F (149°C). In recent years, however, some aircraft manufacturers have replaced epoxy with **bismallimides,** which process much as epoxy does and can be used at service temperatures up to about 350°F (177°C). **Prepreg 977,** of ICI Fiberite, is an epoxy toughened with a proprietary thermoplastic elastomer so as not to sacrifice compression strength while increasing toughness. Unlike some elastomer-modified grades, the elastomer is an integral part of the resin so that it unites in the epoxy backbone on curing. Compression-after-impact strength and wet-service temperatures range from 47,000 lb/in^2 (324 MPa) and 180°F (82°C) for 977-1 to 30,000 lb/in^2 (207 MPa) and 250°F (121°C) for 977-3. Shell Chemical's **Epon HPT 1077** epoxy is an amine-based compound, which combines low viscosity with good mechanical properties and chemical and heat resistance. At 77°F (25°C), viscosity is about 3,500 cP, one-fourth that of **Epon 828.** It also has a glass transition temperature of 500°F (260°C), which is high for a low-viscosity resin. **PR-500,** of 3M, is a one-part compound that can resist temperatures up to 350°F (177°C). Reinforced with 50% glass fiber, it is used for resin-transfer-molded vent louvers of auxiliary power units on large commercial aircraft. **FR-4** is a halogenated epoxy compound widely used for printed-circuit boards.

Novoloids are fibers containing at least 85%, by weight, cross-linked **novalac epoxies. Kynol** is a novoloid noted for its exceptionally high temperature resistance. At 1920°F (1049°C) the fiber is virtually unaffected. The fiber also has high dielectric strength and excellent resistance to all organic solvents and nonoxidizing acids. Shell Chemical's **Epon HPT 1050** epoxy is a novalac compound in semisolid neat resin form or as a 75% by weight solution in acetone. **Epon 861** epoxy is a bisphenol F low-viscosity compound for resin-transfer molding or use as an adhesive. **Eposert,** of Ciba Geigy, is a line of epoxy syntactic inserts for reinforcing **honeycomb. SynSpand,** of Dexter Aerospace, is a line of epoxy-based, expandable syntactic films.

A family of one-component epoxy resins, named **Arnox,** was developed by General Electric Co. Suitable for compression, transfer, injection molding, filament winding, and pultrusion, they cure rapidly at temperatures of 250 to 350°F (121 to 177°C). The compression and transfer-molding grade is a black, mineral-filled compound. The injection-molding grade is a pelletized glass-fiber-reinforced compound with a shelf life of 9 to 12 months below 80°F (27°C).

ESSENTIAL OILS. Aromatic oils found in uncombined form in various parts of plants and employed for flavors, perfumes, disinfectants, medicines, and stabilizers; for masking undesirable odors; and as raw materials for making other products. They are usually the esters upon which the odiferous properties of the plants depend, and they are called essential oils because of their ease of solubility in alcohol to form essences. They are also called **volatile oils,** although this term is sometimes also applied to the light and volatile distillates from petroleum. The essential oils are of four general classes: the **pinenes** or **terpenes** of coniferous plants, containing carbon and hydrogen of the empirical formula $C_{10}H_{16}$, such as **oil of turpentine**; **oxygenated oils** containing carbon, hydrogen, and oxygen, such as **oil of cassia**; **nitrogenated oils** containing carbon, hydrogen, oxygen, and nitrogen, such as **oil of bitter almonds**; **sulfurated oils** containing carbon, hydrogen, and sulfur, such as **oil of mustard.**

Although fixed vegetable oils are obtained by expression, the essential oils are obtained by distilling the buds, flowers, leaves, twigs, or other parts of the plant. **Rose oil** is found only in the flowers. **Orange oil** and **lemon oil** are from the flowers and the fruits, but are of different compositions. **Sweet birch oil** and **cinnamon oil** are from the bark. **Valerian** and **calamus** are only in the roots, while **sandalwood oil** and **cedar oil** are only in the wood. Sometimes the essential oil is not in the plant, but is developed when the plant is macerated with water. The **alpha pinene** extracted from turpentine is used for paints and varnishes because it has a high evaporation rate. It is a water-white liquid of pleasant odor boiling at 325°F (163°C). It is also used in the synthesis of camphor. **Pinic acid** is a complex carboxycyclobutane acetic acid produced from alpha pinene. Its esters are used for synthetic lubricants. **Balsams** are solid or semisolid resinous oils and are mixtures of resins with cinnamic or benzoic acid, or both, with sometimes another volatile oil. They are obtained from a variety of trees and are used in antiseptics, perfumes, flavors, and medicine.

Some of the essential oils contain alkaloids which have a physiological effect. **Wormwood oil,** distilled from the dried leaf tops of the perennial herb *Artemisia absinthium,* native to southern Europe but also grown in the United States, is used in medicine for fevers, and for flavoring the liqueur absinthe. The drug **santonin,** used for worm treatment for animals, is an alkaloid extracted from the unopened flower heads of the **Levant wormseed,** *A. cina,* of the Near East, but **wormseed oil,** or **Baltimore oil,** used for the same purpose, is an essential oil containing the alkaloid ascoridole. It is distilled from the seeds and leaf stems of the annual plant *Chenopodium anthelminticum,* grown in Maryland.

ESTERS. Combinations of alcohols with organic acids, which form several important groups of commercial materials. The esters occur naturally in vegetable and animal oils and fats as combinations of acids with the alcohol glycerin. The natural fats are usually mixtures of esters of many acids, coconut oil having no less than 14 acids. Stearic, oleic, palmitic, and linoleic acid esters are the common bases for most vegetable and animal fats, and the esters of the other acids such as linolenic, capric, and arachidic give the peculiar characteristics of the particular fat, although the physical characteristics and melting points may be governed by the basic esters. Esters occur also in waxes, the vegetable waxes being usually found on the outside of leaves and fruits to protect them from loss of water. The waxes differ from the fats in that they are combinations of monacids with monohydric, or simple, alcohols, rather than with glycerin. They are harder than fats and have higher melting points. Esters of still lower molecular weights are also widely distributed in the essential oils of plants where they give the characteristic odors and tastes. All the esters have the characteristic formula ArCOOR or RCOOR, where R represents an **alkyl group,** and Ar an **aryl group,** that is, where R is a univalent straight-chain hydrocarbon having the formula C_nH_{2n+1} and Ar is a univalent benzene ring C_6H_5. In the esters of low molecular weight which make the odors and flavors, the combination of different alcohols with the same acid yields oils of different flavor. Thus the ester methyl acetate, CH_3COOCH_3, is **peppermint oil**; amyl acetate, $CH_3COOC_5H_{11}$, is **banana oil**; and isoamyl acetate, $CH_3COO(CH_2)_3(CH_3)_2$, is **pear oil.** Esters are used as solvents, flavors, perfumes, waxes, oils, fats, fatty acids, pharmaceuticals, and in the manufacture of soaps and many chemicals. Ester liquid lubricants have good heat and oxidation resistance at high temperatures and good fluidity at low temperatures. They are widely used in jet aircraft.

The natural esters are recovered by pressing or extraction, and steam distillation. Synthetic esters are prepared by reacting an alcohol with an organic acid in the presence of a catalyst, such as sulfuric acid or *para*-**toluenesulfonic acid.** The product is purified with an azeotrope, such as benzene or toluene. A range of **cellulose acetate esters** are made by esterification of cellulose with acetic anhydride. **Cellulose nitrate ester** is obtained by reacting cellulose with nitric acid, **cellulose sulfate** from **chlorosulfonic acid** in **pyridine** solvent, and **cellulose phosphate** from phosphoric acid in molten urea. **Alkoxysilanes** are silicon esters in which the silicon is connected to an organic group by oxygen. **Tetraethoxysilane,** a low-molecular-weight compound, is reactive and is used in binders, resins, and glasses and as a cross-linking agent. **Tetrabutoxysilane** is more stable and is used in lubricants and heat-transfer fluids.

Ester alcohols are intermediates that require less acid for esterification. **Texanol,** of Eastman Chemical Co., has both a hydroxy group and an ester linkage with the empirical formula $C_{12}H_{24}O_3$. It produces a wide range of chemicals and compounds with low, $-71°F$ $(-57°C)$, pour point.

ETCHING MATERIALS.   Chemicals, usually acids, employed for cutting into, or etching, the surface of metals, glass, or other material. In the metal industries they are called **etchants.** The usual method of etching is to coat the surface with a wax, asphalt, or other substance not acted upon by the acid; cut the design through with a sharp instrument; and then allow the acid to corrode or dissolve the exposed parts. For etching steel, a 25% solution of sulfuric acid in water or a ferric chloride solution may be used. For etching stainless steels, a solution of ferric chloride and hydrochloric acid in water is used. For high-speed steels, brass, or nickel, a mixture of nitric and hydrochloric acids in water solution is used, or nickel may be etched with a 45% solution of sulfuric acid. Copper may be etched with a solution of chromic acid. Brass and nickel may be etched with an acid solution of ferric chloride and potassium chlorate. For red brasses, deep etching is done with concentrated nitric acid mixed with 10% hydrochloric acid, the latter being added to keep the tin oxide in solution and thus retain a surface exposed to the action of the acid. For etching aluminum a 9% solution of copper chloride in 1% acetic acid, or a 20% solution of ferric chloride may be used, followed by a wash with strong nitric acid. Sodium hydroxide, ammonium hydroxide, or any alkaline solutions are also used for etching aluminum. Zinc is preferably etched with weak nitric acid, but requires a frequent renewal of the acid. Strong acid is not used because of the heat generated, which destroys the wax coating. A 5% solution of nitric acid will remove 0.002 in (0.005 cm) of zinc per minute, compared with the removal of over 0.005 in (0.013 cm) per minute in most metal-etching processes. Glass is etched with hydrofluoric acid or with **white acid.** White acid is a mixture of hydrofluoric acid and ammonium bifluoride, a white crystalline material of composition $(NH_4)FHF$. **Sodium chlorate** may be used as the electrolyte in producing **chemical finishes.** The process in which the metal is removed chemically to give the desired finish as a substitute for mechanical machining is called **chemical machining.**

To trace the electrical circuit pathways on silicon chips and printed-circuit boards, liquid etchants containing acids are used. **Buffered hydrofluoric acid** is a selective etchant for silicon dioxide in the presence of silicon. **Ammonium fluoride** is a common buffer, and its concentrations in the mixture range from 20% to more than 90. Formulations containing combinations of nitric, acetic, phosphoric, and sulfuric acids are called **mixed-acid etchants. Ammonium chloride, ammonium persulfate,** and **cupric chloride** are used for etching copper printed-circuit boards. **Ceric ammonium nitrate** is suited for etching silicon wafers. **Dry etching,** carried out in the gas phase, employs **silicon tetrafluoride** and **carbon tetrafluoride.**

ETHER. The common name for **ethyl ether,** or **diethyl ether,** a highly volatile, colorless liquid of composition $(C_2H_5)_2O$ made from ethyl alcohol. It is used as a solvent for fats, greases, resins, and nitrocellulose, and in medicine as an anesthetic. The specific gravity is 0.720, boiling point 93.6°F (34.2°C), and freezing point -177°F (-116°C). Its vapor is heavier than air and is explosive. Actually, ether is a more general term, and an ether is an **alkyl oxide** with two alkyl groups joined to an oxygen atom. The ethyl ether would thus be expressed as $C_2H_5 \cdot O \cdot C_2H_5$, and there are many ethers. **Butyl ether,** $(C_4H_9)_2O$, has a much higher boiling point, 284°F (140°C); is more stable; and is used as a solvent for gums and resins. **Isopropyl ether,** $(CH_3)_2CHOCH(CH_3)_2$, is a by-product in the manufacture of isopropyl alcohol from propylene. It has a higher boiling point than ethyl ether, 156°F (69°C); lower solubility in water; and is often preferred as an extractive solvent. **Methyl ether,** or **dimethyl ether,** also known as **wood ether,** is a colorless gas of composition $(CH_3)_2O$, with a pleasant aromatic odor. The boiling point is -10.3°F (-23.5°C). The specific gravity is 1.562 or, as a liquid compressed in cylinders, 0.724. It is used for fuel, as a welding gas, as a refrigerant, and for vapor-pressure thermometers. **Hexyl ether,** $C_6H_{13}OC_6H_{13}$, has a high boiling point, 439°F (226°C); very low water solubility; and a specific gravity of 0.7942. It is stable and not volatile, with a flash point of 170°F (77°C). It is used in foam breakers and in chemical manufacture where anhydrous properties are desired. A low-boiling-point chemical used as an extractive solvent and for plastics because of its stability in alkalies and its high water solubility is **methylal,** $CH_3OCH_2OCH_3$. It is a water-white liquid boiling at 108°F (42.3°C). Ether reacts slowly with the oxygen of air to form highly explosive and poisonous compounds, so that long-stored ether is dangerous for use as an anesthetic.

ETHYL ALCOHOL.   Also called **methyl carbinol,** and **ethanol** when
made synthetically. It is the common beverage alcohol, which when
denatured for nonbeverage purposes is called **industrial alcohol.**
About 90% of the ethyl alcohol used in the United States is denatured. Ethyl alcohol is a colorless liquid with a pleasant odor but
burning taste. The composition is CH_3CH_2OH, specific gravity 0.79,
boiling point 173.3°F (78.5°C), and freezing point −179°F
(−117.3°C). It mixes with water in all proportions and takes up moisture from the air. It burns with a bluish flame and high temperature,
yielding carbonic acid and water. The ignition temperature is 965°F
(518°C). It is one of the best solvents and dissolves many organic
materials such as gums, resins, and essential oils, making solutions
called **essences.**

Alcohol is sold by the proof gallon, with 100 proof containing 50%
alcohol by volume and having a specific gravity of 0.7939. The term
alcohol, alone, refers to 188 to 192 proof. High-purity alcohol, **grain
alcohol,** and pure ethyl alcohol are terms for 190 proof. **Absolute
alcohol,** or **anhydrous alcohol,** is 200 proof, free of water.
Methylated spirits is a term first used in England to designate the
excise-free mixture of 90% ethyl alcohol and 10 wood alcohol for
industrial use. Denatured ethyl alcohol, made unsuitable for beverage
purposes, may be marketed under trade names such as **Synasol** of
Union Carbide. **Solox** consists of 100 parts 190-proof alcohol, 5 ethyl
acetate, and 1 gasoline, used for lacquers, fuel, and as a solvent.
Neosol, of Shell Chemical Corp., is 190-proof ethyl alcohol denatured
with four parts of a mixture of tertiary butyl alcohol, methyl isobutyl
ketone, and gasoline.

Ethyl alcohol is used as a solvent in varnishes, explosives, extracts,
perfumes, and pharmaceuticals; as a fuel; as a preserving agent; as
an antifreeze; and for making other chemicals. Up to 15% of alcohol
can be used in gasoline motor fuels, called generically by the name
gasohol, without change in carburetion. **M85,** sold in the western
United States, is methanol with 15% alcohol. Brazil produces large
quantities of **Proalcohol,** which contains 22% anhydrous ethanol.
The German motor fuel **Monopolin** was a mixture of absolute alcohol
and benzene. Ethyl alcohol is classified as a poison when pure, but is
employed as a beverage in many forms. In small quantities it is an
exhilarant and narcotic. In all countries large amounts of **beverage
alcohol** are made from starches, grains, and fruits, retaining the
original flavor of the raw material and marketed directly as wines,
whiskies, and brandies. But **synthetic wines** are made by fermenting sugar and adding vegetable extracts to supply flavor and bouquet.
No methyl alcohol or fuel oil is produced in the process. Alcohol is produced easily by the fermentation of sugars, molasses, grains, and

starch. It is also made cheaply by directly or indirectly hydrating ethylene produced by the cracking of petroleum hydrocarbons. In Europe it is also made from the waste liquor of pulp mills by fermentation of wood sugar. **Sulfite pulp liquor** contains 1.8% fermentable hexose sugar. It is also made directly from wood waste by fermenting the wood sugar molasses. Ethanol is concentrated and purified by extractive distillation using an azeotrope, such as benzene.

A substitute for ethyl alcohol for solvent purposes and as a rubbing alcohol is **isopropyl alcohol,** or **isopropanol,** a colorless liquid of composition $(CH_3)_2CHOH$, boiling point 180°F (82°C), and produced by the hydration of propylene from cracked gases. It is also used as a stabilizer in soluble oils and in drying baths for electroplating. **Petrohol** is isopropyl alcohol. **Trichloroethanol,** $CCl_3 \cdot CH_2OH$, is a viscous liquid with an ether odor, boiling at 302°F (150°C) and freezing at 55°F (13°C), slightly soluble in water, used for making plasticizers and other chemicals. The spent grain from alcohol distilleries, called **stillage,** is dried and marketed as livestock feed and is a better feed than the original grain because of the high concentration of proteins and vitamins, with the starch removed. The **leaf alcohol** which occurs in fruits and many plants is a hexene alcohol. It is made synthetically for blending in synthetic flavors and for restoring full flavor and fragrance to fruit extracts.

ETHYL SILICATE. A colorless liquid of composition $(CH_2H_5)_4SiO_4$, used as a source of colloidal silica in heat-resistant and acid-resistant coatings and for moldings. The specific gravity is 0.920 to 0.950. It is a **silicic acid ester,** with a normal content of 25% available silica, although **tetraethyl orthosilicate** has 27.9% available silica, and **ethyl silicate 40** of Union Carbide has 40% silica. The latter is a brown liquid. Water hydrolyzes ethyl silicate to alcohol and **silicic acid,** H_4SiO_2, which dehydrates to an adhesive amorphous silica. For moldings, the ester is mixed with silica powder, and for such products as bearings, wood flour may be incorporated to absorb and retain the lubricating oil. Ethyl silicate solutions are employed for the surface hardening of sand molds and graphite molds for special casting. Silicic acid ester paints are used to harden and preserve stone, cement, or plaster, and for coating insulating brick. They are resistant to heat and to chemical fumes. **Kieselsol,** a German material for clarifying wine and fruit juices by precipitation of the albumin, is a 15% water solution of silicic acid.

ETHYLENE. Also called **ethene.** A colorless, inflammable gas, $CH_2:CH_2$, produced in the cracking of petroleum. Ethylene liquefies at −154.8°F (−68.2°C). It was first produced in Holland by dehydrating

ethyl alcohol with sulfuric acid, and is now made from cracking petroleum or by breaking down alcohol by catalytic action. It was originally
employed for enriching illuminating gas to give it a more luminous
flame, and it was called **olefiant gas** because it formed an oil, ethylene dichloride, called **Dutch liquid,** when treated with chlorine.
Ethylene is the largest-volume organic chemical produced today, and
it is the basic building block of the petrochemical industry.
Polymerization of ethylene is its largest use. When ethylene is reacted in the presence of transition-metal catalysts, such as **molybdenum oxide** or **chromium oxide,** at high pressures, it forms **low-
density polyethylene,** or **LDPE**; at lower pressures, **high-density
polyethylene,** or **HDPE,** is produced. Recently, low pressures have
been employed for making a new variant, **linear low-density polyethylene** (**LLDPE**). Ethylene is now used to produce ethyl alcohol,
acrylic acid, and styrene, and it is the basis for many types of reactive
chemicals. **Ketene,** for example, used as a reactant in connecting
polymers to improve physical properties of the plastics, has the basic
formula $H_2C:C:O$, which is ethylene modified by substituting oxygen
for two of the hydrogens. **Butyl ethyl ketene** of Eastman Chemical
Products, Inc., for modifying compounds with active double bonds or
active hydrogens, is a yellow liquid of specific gravity 0.826, having
composition $(C_4H_9)(C_2H_5):C:C:O$. **Calorene** is ethylene in pressure
cylinders for flame cutting. When burned with oxygen, it gives a
flame lower in temperature than acetylene, and it is more stable in
storage. For making resins and waxes, and for solvent use, it may be
employed in the form of **ethylene diamine,** $NH_2CH_2CH_2NH_2$, a colorless liquid of specific gravity 0.968, boiling at about 248°F (120°C).
Ethylene imene, C_3H_7N, is a very reactive chemical useful for making a wide range of products. It is a water-white liquid of specific
gravity 0.79, boiling at 151°F (66°C), soluble in water and in common
solvents. The **imene ring** in the molecule has two carbon atoms and
a nitrogen atom forming a triangle. The ring is stable with basic
chemicals, but is strongly reactive to acid compounds, opening at the
carbon-nitrogen bond to receive hydrogen. By acid catalyzation and
control with alkaline solutions to avoid violent simultaneous opening
of the two carbon bonds, the material can be polymerized or made to
receive other chemical groups.

   Trichlorethylene is a heavy colorless liquid of pleasant odor of
composition $CHCl:CCl_2$, also known as **westrosol.** Its boiling point is
189°F (87°C) and its specific gravity 1.471. It is insoluble in water
and is unattacked by dilute acids and alkalies. It is not flammable
and is less toxic than tetrachlorethane. Trichlorethylene is a powerful
solvent for fats, waxes, resins, rubber, and other organic substances
and is employed for the extraction of oils and fats, for cleaning fab-

rics, and for degreasing metals preparatory to plating. The freezing point is $-126°F$ ($-88°C$), and it is also used as a refrigerant. It is used in soaps employed in the textile industry for degreasing. **Tri-Clene** is a trade name of Du Pont for trichlorethylene, marketed for dry cleaning. **Triad** and **Perm-A-Clor** are trichlorethylene stabilized with a basic organic stabilizer that prevents breakdown of the solvent in degreasing metals.

Ethylene resins are a class of synthetic resins which range from greaselike liquids in the low molecular weights, to waxlike materials at molecular weights from about 4,000 to 10,000, to tough white solids at molecular weights above about 12,000, which are thermoplastic resins melting at 210 to 235°F (99 to 112°C). In the ethylene molecule, the two carbon atoms, each of which has two attached hydrogen atoms, are linked together with a straddle bond of the number 1 and 2 electrons of the carbons, which normally form the hexagonal carbon ring. This type of **double bond** is not double in a mechanical sense and is termed a **reactive bond,** that is, a bond that can be broken readily to receive other attachments.

Polyox resins, of Union Carbide, are white granular powders of water-soluble **ethylene oxide plastics** with a wide range of molecular weights for films, fibers, and molded articles. **Polyox film** has a tensile strength of 1,800 to 2,400 lb/in^2 (12.4 to 16.5 MPa), with elongation from 100 to 2,000%, and heat seals at temperatures from 170 to 265°F (77 to 129°C). It is used for packaging soaps, detergents, and chemicals to be added in measured amounts without removing the package. The plastic has high adhesive strength and is also used for adhesives where water solubility is wanted.

ETHYLENE GLYCOL. Also known as **glycol** and **ethylene alcohol.** A colorless syrupy liquid, CH_2OHCH_2OH, with a sweetish taste, very soluble in water. It has a low freezing point, $-13°F$ ($-25°C$), and is much used as an antifreeze in automobiles. A 25% solution has a freezing point of $-5°F$ ($-20.5°C$), without appreciably lowering the boiling point of the water. It has the advantage over alcohol that it does not boil away easily, and it permits the operation of engines at much higher temperatures than with water, giving greater fuel efficiency. **Prestone** from Union Carbide Corp. (Danbury, Conn.) is primarily ethylene glycol. It is also used for the manufacture of acrylonitrile fibers and as a solvent for nitrocellulose. It is highly toxic in contact with skin.

Diethylene glycol, $C_4H_{10}O_3$, is a water-white liquid boiling at 471°F (244°C), used as an antifreeze, as a solvent, and for softening cotton and wool fibers in the textile industry. A 50% solution of diethylene glycol freezes at $-18°F$ ($-28°C$). **Cellosolve,**

$C_2H_5OCH_2CH_2OH$, of Union Carbide, is the monoethyl ether of ethylene glycol. It is a colorless liquid boiling at 275°F (135°C) and is a powerful solvent used in varnish removers, cleaning solutions, and as a solvent for paints, varnishes, plastics, and dyes. **Carbitol,** of the same company, is an ether of diethylene glycol of composition $CH_3CH_2OCH_2CH_2OCH_2CH_2OH$, used as a solvent for oils, dyes, resins, and gums. The boiling point is 396°F (202°C) and freezing point −103°F (−75°C). **Propylene glycol,** or **propanediol,** $CH_3 \cdot CHOH \cdot CH_2OH$, is a colorless and odorless liquid boiling at 370°F (188°C), used in cosmetics and perfumes; in flavoring extracts as a humectant, wetting agent, and color solvent; and in baked foods to maintain freshness. It is also used in **Sierra antifreeze** of Safe Brands Corp. **Methyl carbitol,** with one less CH_2 group, is also a high-boiling-point solvent for gums and resins, and **carbitol acetate** is used as a high-boiling-point solvent for cellulose acetate. **Glycol diformate,** $HCOOCH_2CH_2OOCH$, used as a solvent for cellulose acetate and nitrocellulose, is a colorless liquid soluble in water, alcohol, and ether. It hydrolyzes slowly, liberating formic acid.

EUCALYPTUS. A tree genus of several hundred species native to Australia, but now grown in many parts of the world. It is known as **gumwood** in the southern United States. The **blue gum,** which attains a height of 300 ft (91 m), is grown on the west coast of the United States. The wood has a pale straw color and is hard and tough. It has a twisted grain and shrinks and warps easily, but is very durable. The density is about 50 lb/ft³ (801 kg/m³), greater than that of the southern gum. **Salmon gum,** from *E. salmonophloria,* has a salmon-red color, is dense and hard, and has a fine, open grain. It is superior and has a great variety of uses. The density is about 60 lb/ft³ (961 kg/m³). **Red gum,** from *E. calophylla,* has a yellowish-red color; is strong and tough; and has a density of about 45 lb/ft³ (721 kg/m³). The grain is fine but has gum veins intersecting. Other species of gumwood are marketed under the names **York gum, blackbutt, tuart,** and **Australian red mahogany.**

Three Australian timbers—**jarrah, karri,** and **ironbark**—are members of the *Eucalyptus* genus. Jarrah resembles karri so closely that it is difficult to distinguish one from the other. Both are dark red and similar in weight and appearance. Ironbark is heavier than either of these and is more gray. Also, it is nearly always severely surface-checked, a characteristic which does not detract significantly from its strength. It is very strong, having a modulus of rupture in bending of 27,100 lb/in² (187 MPa), whereas jarrah and karri run about 16,000 and 19,000 lb/in² (110 and 131 MPa), respectively. Ironbark and jarrah are rated "very durable" by the British Forest

Products Research Laboratory. Karri is rated only "moderately durable."

The **wandoo tree,** *E. redunca,* of western Australia, the wood of which is known as **redunca wood,** has a high percentage of pyrogallol tannin. The solid extract is called **myrtan,** and it produces a solid, firm sole leather lighter in color than that of chestnut. The wood of *E. saligna,* of South Africa, is hard and has a fine, even, interlocking grain which makes it strong in all directions. It is used in the United States for small turned articles, saw handles, and paintbrush handles. It has a reddish tinge. **Blackbutt** is from the trees *E. pilularis, E. patens,* and some other species native to Australia, but now grown in other countries. It is used as a substitute for oak, but tends to warp and crack.

Eucalyptus oil, obtained from the dried leaves of *E. globulus,* is used in pharmaceuticals for nose and throat treatment. It is the source of **cineole,** also called **eucalyptole.** From 3 to 4% oil is obtained from the leaves. It is a pungent, yellowish oil. This type of eucalyptus oil contains **phellandrine,** used in Australia as an antiknock agent in gasoline. **Eucalyptus dives oil,** from the leaves of the Australian tree *E. dives,* contains 92 to 94% **piperitone** and is used in the manufacture of menthol. The yield is about 50% **levomenthol** with a melting point of 91 to 95°F (33 to 35°C). It lacks the odor of USP methol of which only 15% can be produced from this oil. Much eucalyptus oil is produced in Chile. More than 300 species of eucalyptus trees are known, and each produces a different type of oil.

EXPANDED METAL. Sheet metal that has been slit and expanded to form a mesh, which is used for reinforced-concrete work or plaster wall construction, and for making grills, vents, and such articles as trays, where stiffness is needed with light weight. The expanded metal has greater rigidity than the original metal sheet and permits a welding of the concrete or plaster through the holes. It is made either with a plain diamond-shape mesh or with rectangular meshes. One type is made by slitting the sheet and stretching the slits into the diamond shape. The other variety is made by pushing out and expanding the metal in the meshes so that the flat surface of the cut strand is nearly at right angles to the surface of the sheet. Expanded metal is made from low-carbon steel, iron, or special metals, in sheets from 8 to 12 ft (2.4 to 3.7 m) in length and 3 to 6 ft (0.9 to 1.8 m) in width, in several thicknesses. It is also marketed as **metal lath,** usually 96 in (2.4 m) long and 14 to 18 in (0.4 to 0.5 m) wide. Expanded metal of U.S. Gypsum Co. is made of stainless steel and aluminum alloys in various thicknesses with openings from 0.5 to 1.5 in (1.27 to 3.81 cm). **Rigidized metal,** or **textured metal,** is thin sheet that is not perforated, but has the designs rolled into the sheet so that the rigidity of the sheet is increased 2 to 4 times. Thus, extremely thin sheets of stainless steel can be used for novelties, small mechanical products, and paneling. Rigidized steel, of Rigidized Metals Corp. and previously known as **Rigid-Tex steel,** is made in many ornamental designs and also comes in vitreous enameled sheets for paneling. **Crimp metal,** of American Nickeloid Co., has various embossed designs in either raised or depressed ridges rolled into the polished side of the metal. **Perforated metals** are sheet metals with the perforations actually blanked out of the metal. They are marketed in sheets of carbon steel, stainless steel, or Monel metal, with a great variety of standard designs. Those with round, square, diamond, and rectangular designs are used for screens and for construction. **Agaloy** is perforated metal made into tube form.

EXPANSIVE METAL. An alloy which expands on cooling from the liquid state. The expansive property of certain metals is an important characteristic in the production of accurate castings having full details of the mold such as type castings. The alloys are also used for proof-casting of forging dies, for sealing joints, for making duplicates of master patterns, for holding die parts and punches in place, and for filling defects in metal parts or castings. Antimony and bismuth are the metals most used to give expansion to the alloys. **Lewis metal,** one of the original expansive alloys, had one part of tin and one of bismuth, and melted at 280°F (138°C). **Matrix alloy** and **Cerromatrix,** of Cerro Metal Products, contain 48% bismuth, 28.5 lead, 14.5 tin, and 9 antimony. The melting point is 248°F (120°C), tensile strength 13,000 lb/in^2 (90 MPa), and Brinell hardness 19. **Cerrobase,** of this company, is another alloy balanced to give the exact impression of the mold without shrinkage or expansion in cooling. It is harder than lead and melts at 255°F (123°C).

EXPLOSIVE. A material which, upon application of a blow or by rise in temperature, is converted in a small space of time to other compounds more stable and occupying much more space. Commercial explosives are solids or liquids that can be instantaneously converted by friction, heat, shock, or spark to a large volume of gas, thereby developing a sudden rise in pressure which is utilized for blasting or propelling purposes. **Gunpowder** is the oldest form of commercial or military explosive, but this has been replaced for military purposes by more powerfully acting chemicals. **Smokeless powder** was a term used to designate nitrocellulose powders as distinguished from the smoky black gunpowder. **Blasting powders** are required to be relatively slow-acting to have a heaving or rending effect. **Military**

explosives used as propellants must not give instantaneous detonation, which would burst the gun, but are arranged to burn slowly at first and not reach a maximum explosion until the projectile reaches the muzzle. This characteristic is also required in explosives used for the explosive forming of hard metals. The more rapid-acting **high explosives** are generally used for bombs, torpedoes, boosters, and detonators. The **detonators** are extremely sensitive explosives, such as the fulminates, set off by a slight blow but too sensitive to be used in quantity as a charge. The **booster explosives** are extremely rapid but not as sensitive as the detonators. They are exploded by the detonators and in turn set off the main charge of explosive. Some explosives such as **nitroglycerin** can be exploded by themselves, while others require oxygen carriers or carbon carriers mixed with them. In combination with **nitrocellulose,** it is the principal component of powders and solid rocket propellants. Together with **nitroglycol,** it is the major constituent of gelatinous industrial explosives. Other requirements of explosives are that they not react with the metal container, be stable at ordinary temperatures, and not decompose easily in storage or on exposure to air.

Shaped charges of high explosive give a penetrating effect, known as the *Monroe effect,* used in armor-piercing charges. A solid mass of explosive spends itself as a flat blast; but with a conical hole in the charge, and having the open end facing the target, a terrific piercing effect is generated by the converging detonation waves coming from the sides of the cone. This effect drives a jet of hot gases through the steel armor. **Permissible explosives** are explosives that have been passed by the U.S. Bureau of Mines as safe for blasting in gaseous or dusty mines. Most of the permissible ones are of ammonium nitrate or gelatin base. **Wet-hole explosives,** for oil-well and mining operations, may be ammonium nitrate in plastic containers, or various combinations in containers. **Lox,** used in mines and quarries, is an explosive consisting of a paper cartridge filled with carbon black or wood pulp soaked in liquid air. It cannot be tamped, as it is very sensitive. It is fired by electric detonators. **Cardox,** an explosive used in coal mining, consists of liquid carbon dioxide in a steel cylinder with aluminum powder. The powder is fired by an electric spark, heating and gasifying the carbon dioxide. **Picric acid,** or **trinitrophenol,** $C_6H_2(OH)(NO_2)_3$, a lemon-yellow crystalline solid melting at 248°F (120°C), is a powerful explosive used in shells, and because of its persistent color also used as a dyestuff. It is called **melanite** by the French, **lyddite** by the English, and **schimose** by the Japanese. It is made by treating phenol with sulfuric and nitric acids, or it can be produced by treating acaroid resin with nitric acid. It reacts with metals to form dangerous explosive salts, so that the shells must be lacquered. **Cressylite,** used for shells, is a mixture of picric acid and trinitrocresol. It has a lower melting point.

Explosive D, or **dunnite,** made by the neutralization of picric acid with ammonium carbonate, is **ammonium picrate,** $C_6H_2(NO_2)_3ONH_4$. It forms orange-red needles that explode when heated to 572°F (300°C), but is not highly sensitive to friction. It is used as a bursting charge in armor-piercing shells. **Trinitrotoluene,** or **trinitrotoluol,** $C_6H_2(CH_3)(NO_2)_3$, also commonly known as **TNT** and also called **trotyl** and **tolite,** is the principal constituent of many explosives. It resembles brown sugar in appearance, it melts at 176°F (80°C), and the fumes are poisonous even when absorbed through skin. Its detonation velocity is 23,000 ft/s (7,010 m/s). It is thus not as powerful as picric acid, but it is stable, not hygroscopic, and does not form unstable compounds with metals. It is safe in handling because it does not detonate easily, but is exploded readily with mercury fulminate and is used for shrapnel, hand grenades, mines, and depth bombs. TNT is made by the nitration of toluol with nitric and sulfuric acids. The intermediate product, **dinitrotoluol,** is employed with hexanitrodiphenylamine for torpedoes. **Hexanitrodiphenylamine,** $(NO_2)_3C_6H_2 \cdot NH \cdot C_6H_2(NO_2)_3$, is a powder that explodes with great violence. It is highly poisonous and causes painful blisters and inflammation. The commercial explosive **sodatol** is made by mixing TNT with nitrate of soda.

Trinitroaniline, $(NO_2)_3C_6H_2NH_2$, commonly known as **TNA,** is derived from aniline by nitration and is one of the strongest of the high explosives. It is a yellowish-green crystalline powder melting at 419°F (215°C). It stains skin yellow but is not poisonous. It is more sensitive to shock than TNT and is more costly. **Trinitroanisol,** used in Japanese Baka planes, has composition $C_6H_2OCH_3(NO_2)_3$. It is about equal to TNT in power and has the advantage that it does not attack metals.

Tetryl, or **pyronite,** $(NO_2)_3C_6H_2N(NO_2)CH_3$, is a nitro derivative of benzene. It is a yellow crystalline powder that melts at 266°F (130°C) and explodes when heated to 367°F (186°C). It is more sensitive to shock than TNA and has a higher rate of detonation than TNT. It is too sensitive to be used as a shell filler, and it is employed as a booster and in commercial explosives to replace mercury fulminate for detonators. The high explosive **RDX** is **cyclotrimethylene trinitroamine,** or **Cyclonite,** and has a detonation velocity of 27,500 ft/s (8,382 m/s). It is used in bombs, torpedoes, mines, and rockets, but is very sensitive to shock and is mixed with waxes or plasticizers to reduce sensitivity. **HMX, cyclotetramethylene tetranitramine,** is superior to Cyclonite in high-power applications. It is also known as **Octogen,** or **homocyclonit. Octol** is a 75:25 mixture of Octogen and

TNT. **PETN** is **pentaerythritol tetranitrate,** with a detonation velocity of 26,500 ft/s (8,077 m/s). **Pentolite** is a 50–50 mixture of TNT and PETN with less sensitivity and a detonation velocity of 25,000 ft/s (7,620 m/s). It is used as a booster. When aluminum powder is added to high explosives, the brisance, or blast effect, is increased. A powerful explosive used during the Second World War contained 40% RDX, 40 TNT, and 20 aluminum powder. Various combinations of high explosives are now used in thin sheet form for **explosive welding** of laminated metals.

FABRICS. **Woven fabrics** and **knit fabrics** are composed of webs of fiber yarns. The yarns may be of either filament (continuous) or staple (short) fibers. In knit fabrics, the yarns are fastened to each other by interlocking loops to form the web. In woven fabrics, the yarns are interlaced at right angles to each other to produce the web. The lengthwise yarns are called the *warp,* and the crosswise ones are the *filling* (or *woof*) yarns.

The many variations of woven fabrics can be grouped into four basic weaves. In the **plain weave fabric,** each filling yarn alternates up and under successive warp yarns. With a plain weave, the most yarn interlacings per square inch can be obtained for maximum density, "cover," and impermeability. The tightness or openness of the weave, of course, can be varied to any desired degree. In **twill weave fabrics,** a sharp diagonal line is produced by the warp yarn crossing over two or more filling yarns. **Satin weave fabrics** are characterized by regularly spaced interlacings at wide intervals. This weave produces a porous fabric with a smooth surface. Satins woven of cotton are called **sateen.** In the **leno weave fabrics,** the warp yarns are twisted and the filling yarns are threaded through the twist openings. This weave is used for meshed fabrics and nets.

Because the variety of woven fabrics is endless, we can only briefly outline here the way woven textiles are characterized or specified. Generally, specifications include the type of weave; the thread count, in both warp and fillings; whether the yarn is filament or staple; the crimp, in percent; the twist per inch; and the yarn numbers for warp and fill. Over the years a rather unsystematic fabric designation system has evolved. For example, some fabrics, such as twills and sateens, are designated by width in inches, number of linear yards per pound, and number of warp and filling threads per inch. Other fabrics are identified by width, ounces per linear yard, and warp and filling count.

Exxaire, microporous polyolefin films of Exxon Chemicals, are used for surgical gowns. Highly breathable to air and water vapor but resistant to microorganisms, water, alcohol, and blood, they are used in composites of polyester and nonwoven polyolefins, competing with **Gore-Tex** and disposable fabrics, such as **Sontara.** Potential applications include wound dressings, diapers, and feminine hygiene products. The films, 0.001 to 0.0025 in (0.025 to 0.064 mm) thick, can be made of high- and low-density polyethylene, linear low-density polethylene, polypropylene, and polyolefin copolymers. Elf Atochem's **Pebax** line of amide-based thermoplastic elastomers is used for catheters in angiographic procedures and as transdermal drug-delivery patches. The elastomers, which also can be made breathable, have potential in the form of fabric composites as surgical drapes and wound dressings. Extruded as strong, flexible film as thin as 0.0005 in (0.013 mm), they also can be made hydrophilic.

While the largest single use of woven fabrics is, of course, for wearing apparel, they are used in many other areas: in mechanical applications such as machine and conveyor belting, for filtration, for packaging, and as reinforcement for plastics and rubber.

FAT LIQUORS.   Oil emulsions used in tanneries for treating tanned leather to lubricate the fibers, increase the flexibility, and improve the finish. Dyeing and fat liquoring are conducted in the same drum after tanned stock is aged, neutralized, and retanned to impart special properties. There are two general types of fat-liquor emulsions: acid and alkaline. The acid group includes sulfonated oils and some soluble-oil combinations. Alkaline types are emulsions of oils with soaps or alkalies. Leather may be treated first with an alkaline liquor and then with an acid, or borax or soda ash may be added to sulfonated oils to produce alkaline liquors. For suede and white leathers, egg-yolk emulsions may be used. The oils employed in emulsions may be sperm, cod, or castor oil, and those that are neutral have a neatsfoot-oil base. The soaps are usually special for the tannery trade. Prepared fat liquors are marketed under trade names. **Tanners' greases,** used for sponging or milling onto the leather, are also trade name mixtures of waxes, sulfonated oils, and soaps.

FATS.   Natural combinations of glycerin with fatty acids, so-called **triglycerides,** some fats having as many as 10 or more different fatty acids in the combination. At ambient temperature fats are solids; if liquid, they are normally called **fat oils.** Fats are also known as **lipids. Waxes** differ slightly in composition from fats and are mixed esters of polyhydric alcohols, other than glycerin, and fatty acids. **Animal fats** are **butter, lard,** and edible and inedible **tallows.** Fats contain less than 5% of phospholipids, pigments, vitamins, antioxidants, and sterols. They are derived from animal or vegetable sources, the latter source being chiefly the seeds or nuts of plants.

Fats in a pure state would be odorless, tasteless, and colorless, but the natural fats always contain other substances that give characteristic odors and tastes. Fats are used directly in foods and in the making of various foodstuffs. They are used in making soaps, candles, and lubricants, and in the compounding of resins and coatings. They are also distilled or chemically split to obtain the fatty acids. Crude fats are refined to remove nonglyceride impurities, including free fatty acids, phosphatides, and proteinaceous and mucilaginous matter, by treatment with strong caustic soda. The fatty acids are converted to oil-insoluble soaps known as **foots** or **soapstock.** Treatment with sulfuric acid produces **acidulated soapstock,** used for making soap and as an animal feed. Adding **citric acid** to refined fats prevents rancidity and flavor reversion. **Phosphoric acid** has a similar use in refined oils.

Fats are most important for food, containing more than twice the fuel value of other foods. They are also important carriers of glycerin necessary to the human system. Metabolism, or absorption of fats into the system, is not a simple process and is varied with the presence of other food materials. The fats with melting points above 45°C are not readily absorbed into the system. The heavy fats are called *tallow.* Lack of certain fats, or fatty acids, causes skin diseases, scaly skin, and other conditions. Some fatty acids are poisonous alone, but in the glyceride form in the fats they may not be poisonous but beneficial. Fats can be made synthetically from petroleum or coal. **Edible fats** were first made synthetically by the Germans in wartime by the hydrogenation of brown coal and lignite and then esterifying the C_9 to C_{16} fractions of the acids. But the world resources of natural fats are potentially unlimited, especially from tropical nuts, forming a cheap source of fatty acids in readily available form.

Margarine, shortening, confectionery fat, and other edible fats are made by hydrogenating a variety of semisolid or liquid fats. The hardening process converts unsaturated fatty glycerides to more saturated forms. About one cubic meter of hydrogen is needed per metric ton of oil to reduce the oil's iodine number one unit. The catalyst is nickel. **Accolade, Chiffon, Flair, Glen Eden, Golden Mist, Hollandale,** and **Log Cabin** are some trade names for margarines produced by Anderson Clayton Foods. **Crisco Oil,** a shortening from Procter & Gamble, is a partially hydrogenated, winterized, and deodorized soybean oil that has similar liquid properties as the unhydrogenated raw oil. Lard and **lard oil,** produced by rendering fat, are randomized by catalytic molecular rearrangement to produce shortenings.

Engineered, imitation, or **artificial fats** represent a new market for products that have low-calorie saturated fat, or cholesterol contents. They are predominantly mixtures of such hydrogenated vegetable oils as soybean, corn, peanut, palm, and cottonseed. **Simplesse** from The Nutrasweet Co. is a fat substitute consisting of the proteins from milk, whey-protein concentrate, egg whites, soy proteins, or a combination of these. It is made by blending these proteins under precisely controlled high shear at 175°F (79°C) for 20 s, the conditions normally needed for pasteurization. The product consists of 3.9- to 118-μin (0.1- to 3-μm) beads, has a calorie count of 50% fat, and can be used in ice creams and mayonnaise.

FATTY ACIDS. A series of **organic acids** deriving the name from the fact that the higher members of the series, the most common ones, occur naturally in animal fats, but fatty acids are readily synthesized, and the possible variety is almost infinite. All these acids contain the **carboxyl group** · COOH. The acids are used for making soaps, candles, and coating compounds; as plasticizers; and for the production of plastics and many chemicals. The hydrogen atom of the group can be replaced by metals or alkyl radicals with the formation of salts or esters, and other derivatives such as the halides, anhydrides, peroxides, and amides can also be made. The **neoacids,** in general, have the formula R(COOH) in which R is the substituted methyl or other groups. Some of the fatty acids can be polymerized to form plastics. Various derivatives of the acids are used as flavors, perfumes, driers, pharmaceuticals, and antiseptics. Certain fatty acids, such as oleic and stearic, are common to most fats and oils regardless of their source, while others, such as arachidic and erucic, are characteristic only of specific fats and oils.

    Saturated acids are acids that contain all the hydrogen with which they can combine, and they have the type formula $C_nH_{2n+1}COOH$. They have high melting points. **Unsaturated acids,** such as oleic, linoleic, and linolenic, are liquid at room temperature and are less stable than saturated acids. **Fatty acid glycerides** in the form of animal and vegetable fats form an essential group of human foods. Fats of the highly unsaturated acids are necessary in the metabolism of the human body, the glycerides of the saturated acids such as palmitic being insufficient alone for food.

    Polyunsaturated acids of the linoleic type with more than one double bond lower blood cholesterol, but saturated acids with no double bonds do not. **Arachidonic acid** with four double bonds lowers blood cholesterol greatly. It is manufactured in the body from linoleic acid if vitamin B_6 is present. **Linoleic acid,** $C_{18}H_{32}O_2$, the characteristic unsaturated food acid, has two double bonds. **Linolenic acid,** $C_{18}H_{30}O_2$, found in linseed oil, has three double bonds. A mixture of linoleic and **oleic acids** called **tall oil fatty acid,** or **TOFA,** is obtained by distilling crude **tall oil,** which contains, in addition,

rosin acids. TOFA is used for making soaps, detergents, ore-flotation chemicals, protective chemicals, and agrichemicals.

The names of the fatty acids often suggest their natural sources, though commercially they may be derived from other sources or made synthetically. **Butyric acid,** $CH_3CH_2CH_2 \cdot COOH$, is the characteristic acid of butter. Also called **butanoic acid** and **ethylacetic acid,** it is made synthetically as a colorless liquid with a strong odor and completely soluble in water. With alcohols it forms butyrates of pleasant fruity odors used as flavors. The cellulose esters of butyric acid are used in lacquers and have good water resistance and easy solubility in hydrocarbons. The acid is also used as a starting point for fluoro rubbers. **Goat fat** contains 6 to 10 carbon acids.

Some acids, such as linoleic, are found in greater amount in cold-climate products, while some others are found in greatest abundance in hot-climate products. **Lauric acid,** or **dodecanoic acid,** $CH_3(CH_2)_{10}COOH$, occurs in high percentage in the oil of the coconut and other kernels of tropical palm nuts. It is a saturated acid much lower in carbon and hydrogen than linoleic acid, and it is a semisolid melting at 111°F (44°C). It is one of the chief constituents of coconut oil that gives sudsing properties to soaps. It is also used for making detergents and plasticizers and as a modifier for waxes in coatings and polishes. **Neo-Fat 12,** of Armour & Co., is 95% pure lauric acid. The ester of lauric acid is used for treating cotton fabrics to give a pebbly surface. **Lauralene** is a lauric acid with an acid value of 324 and saponification number of 366. **Methyl laurate** is often preferred to lauric acid for all the uses. It is a stable, noncorrosive, water-white liquid. **Methyl esters** of other acids are similarly used. **Methyl stearate** is an economical compounding agent for rubbers, waxes, and textile coatings. **Myristic acid,** $CH_3(CH_2)_{12}COOH$, is a hard crystalline solid melting at 136°F (58°C), obtained from coconut oil. It is soluble in alcohol and is compatible with waxes and oils. It is used in cosmetics and will produce high-lathering soaps that are not irritating to skin, as are the coconut-oil soaps. **Neo-Fat 14,** of Armour Industrial Chemical Co., is myristic acid 94% pure.

Caprylic acid, $CH_3(CH_2)_6COOH$, obtained from coconut oil, has a melting point of 52°F (11°C), acid number of 382, and iodine value of 1. It is used in cosmetics, as a fungicide, and in the manufacture of pharmaceuticals. **Capric acid,** or **decanoic acid,** $CH_3(CH_2)_8COOH$, obtained from coconut oil, is a bad-smelling white crystalline solid melting at 89°F (31.5°C) with an acid value of 321. It is used for making esters for perfumes and flavors. **Neo-Fat 10** is capric acid 92% pure, containing 5% lauric acid and 3 caprylic acid. **Aliphat 2** and **Aliphat 3,** of General Mills Co., are caprylic acid and capric acid, respectively. **Caproic acid,** or **hexanoic acid,** $CH_3(CH_2)_4COOH$, occurs in coconut and palm kernel oils, but is produced synthetically on a large scale for the manufacture of hexylresorcinol, hexylphenols, flavors, and high-boiling-point plasticizers. It is a liquid boiling at 397°F (203°C) and has a goatlike odor from which it derives its name. **Oenanthic acid,** or **heptoic acid,** is a homolog of caproic acid with one more carbon atom. When polymerized with lactam, it gives a nylon stronger and more flexible than ordinary nylon 6. **AB fatty acid,** used for soaps, is composed of the acids from coconut oil distilled to remove most of the low fractions to improve color and odor. It contains 60% lauric acid, 18 myristic, 7 palmitic, 7 oleic, 3 linoleic, 3 capric, and 1 each of stearic and caprylic. Some fatty acids that occur only occasionally in small amounts in vegetable oils are made synthetically. **Undecylenic acid,** $CH_2:CH(CH_2)_8COOH$, is a highly reactive acid of this kind used for making synthetic resins, fungicides, and perfumes. The **Duomeens,** of Armour Industrial Chemical Co., are alkyl trimethylenediamines derived from fatty acids and are used as pigment dispersants, metal-working lubricants, and flotation agents. They have the general formula $RNHCH_2CH_2NHH$, where R is the alkyl group from the fatty acid. **Duomeen C** is from coconut oil, **Duomeen S** is from soybean oil, and **Duomeen O** is from oleic acid. The **lactams** and **lactones,** used in making plastics, form a wide range of amino-fatty acid ring compounds. They are produced from fatty acids.

FEATHERS.   The light, fluffy outgrowth or plumage of birds. The industrially important feathers are those from the duck, goose, chicken, and ostrich. Radiantly colored feathers from many other types of birds are used for ornamental and artistic purposes. An important featherwork art exists in Mexico as a development of the Aztec featherwork. **Down** is the soft feathers of young birds or the soft undergrowth of adult birds, used as a stuffing material. **Eiderdown,** from the eider duck, is highly valued as an insulation in sleeping bags. In Iceland the female duck plucks the down from her breast to line the nest, and this down is gathered commercially after the birds are hatched.

The midrib and quill of **chicken feathers** are made into protein plastic, and the fluffy barbs are used as stuffing, but many of the feathers are processed directly into protein. The inedible protein is used for making brush bristles and insulating fiber, or is split into edible proteins for poultry feed. **Ostrich feathers,** from the domesticated ostriches of Argentina, South Africa, and Australia, are used for ornamental purposes, hats, and dusting brushes. The ostrich has 24 feathers on each wing, some as long as 25 in (0.6 m), and the grade depends upon the color and the length. Male ostrich feathers are black. The female feathers are a soft gray, with white feathers in the wings and tail. The life of the ostrich is 50 to 75 years, and the feathers begin to be clipped at the age of 10 months. **Ostrich eggs,** which weigh 4 lb (1.8 kg) and are laid every other day, are a valuable food by-product.

FELDSPAR. A general name for a group of abundant minerals used for vitreous enamels, pottery, tile, and glass in fertilizers; in fluxes; for roofing granules; and as an abrasive in soaps and cleaning compounds. Ground feldspar is also used for extinguishing magnesium fires, as it melts and gives a smothering action. There are many varieties of feldspar, but those of greatest commercial importance are the **potash feldspars, orthoclase** or **microline,** $K_2O:Al_2O_3 \cdot 6SiO_2$, the **soda feldspar, albite,** $Na_2O:Al_2O_3 \cdot 6SiO_2$, and the **calcium feldspar, anorthite,** $CaO:Al_2O_3 \cdot 2SiO_2$. Orthoclase and microline have the same composition but different crystal structures. Anorthite crystals occur in many igneous rocks and are white, gray, or reddish. **Aplite,** used as a flux for ceramics, has more silica and less alumina. **Japanese aplite** has 77.6% silica, 12.8 alumina, 3.7 K_2O, and 3.9 Na_2O, with small amounts of calcia, magnesia, and iron oxide. Orthoclase is called **suntone. Adularia** is a pure form of orthoclase with only a little sodium. Pieces with an opalescent sheen are called **moonstone** and are used as gemstones. This stone is white with a bluish adularescence caused by the action of light on the laminations. Mohs hardness is 6 to 6.5, but the cleavage in two directions makes it fragile. The blue opalescent moonstone of New Mexico is **sanidine,** a quartz mineral. **Amazon stone,** or **amazonite,** is a beautiful green microline found in Italy, Malagasy, and Colorado and used as a gemstone. The Amazon stone of Virginia has bluish-green and white streaks, and was formerly shipped to Germany for cutting into ornamental objects. The colors of feldspar are from mineral oxides and impurities and are white, gray, yellow, pink, brown, and green. Albite is generally white; while microline is more often green.

All the chemical components of feldspar are glassmaking materials. In making glass about 150 lb (68 kg) is used to each 1,000 lb (454 kg) of sand. But the mineral in its natural occurrence varies widely in composition even in the same mine, and thus it must be controlled chemically to obtain uniform results in glass and ceramic enamels. It occurs in pegmatite dikes associated with quartz, mica, tourmaline, garnet, and spodumene. The mineral is ground to a uniform size, from 80 to 140 mesh, and shipped in bags. Crude unground feldspar is also marketed in bulk. The melting point varies from 2165 to 2714°F (1185 to 1490°C), but the preferred range is 2282 to 2462°F (1250 to 1350°C). Mohs hardness is 6 to 6.5, and the index of refraction is 1.518 to 1.588, the lowest being orthoclase and the highest anorthite. The specific gravity is 2.44 to 2.62 for orthoclase and microline and 2.6 to 2.8 for anorthite. Tennessee and North Carolina feldspar has about 70% SiO_2 and 17 Al_2O_3, with 9 to 11 K_2O, and 2 to 3 Na_2O. New England feldspar is lower in silica and higher in potash. **Potash spar** from New York and New Jersey has about 12% K_2O and is suited for glass and pottery. **Soda spar,** with about 7% Na_2O, is preferred for ceramic enamels. **Cornwall stone,** from England, is a kaolinized feldspar with about 2% CaO. A similar stone from North Carolina is called **Carolina stone.** Aplite is a ceramic fluxing stone found in Virginia and used chiefly to supplement feldspar to provide more alkalies. It is a white massive material of feldspars and other minerals, containing 60% silica, 24 alumina, 6 calcia, 6 sodium oxide, and 3 potassium oxide. Another feldspar material is **alaskite,** a feldspar and quartz mixture from North Carolina. It is classified as a pegmatitic granite. Ground feldspar for enamels is sometimes called **glass spar. Dental spar** is specially selected potash feldspar used in making artificial teeth.

FELT. A fabric of wool, fur, hair, or synthetic fibers made by matting the fibers together under pressure when thoroughly soaked or steam-heated. The matting may also be accomplished by blowing the wet fibers under a powerful air blast and then pressing. The animal fibers mat together, owing to minute scales on their surface. Cotton and other vegetable fibers do not have the property of felting, but a percentage of vegetable or synthetic fibers may be incorporated to vary the characteristics of the felt. So great is the felting property of wool that only 20% is needed in mixtures.

Wool felt can be composed of 100% virgin wool or a combination of synthetic fibers and reused wool. The top grade, which has a density of 0.0094 lb/in³ (260 kg/m³), is used where high strength, purity, and fineness are needed. It is made of the best grades of wool, which are usually white. Wool felts are produced in sheet and roll form. Standard widths are the minimum widths of trimmed felt or the width between pinholes of untrimmed felt. All widths of 60 to 72 in (152 and 183 cm) refer to felt made in roll form. All sizes of 36 by 36 in (91 by 91 cm) refer to felt made in sheet form and have a tolerance of 1.5 in (3.8 cm). Special sizes are available.

Synthetic fiber felts are composed of such fibers as polyester, nylon, Teflon, polypropylene, and acrylic. They are made by a needle loom process that simulates the natural entanglement of wool fiber.

New and reworked wool and noils are mixed with cotton, rayon waste, ramie, jute, casein fiber, and other fibers. Cotton decreases the density and prevents voids in the felt. **Kapok** gives lower thermal and sound conductance, and **insulating felt** may contain a high percentage of kapok. Felt is made of staple fibers of about 1.5 in (3.8 cm) in length, and noils of 0.75 to 1 in (1.9 to 2.5 cm). Longer fibers tend to mat. Shorter fibers lack depth of penetration to give necessary strength. Since most of the wool used is secondary or waste, all grades are employed, from the fines to the coarse carpet wools. Grading is by characteristic symbols. Thus, felt No. 26R1 is a felt with a specific gravity of 0.26 in roll form of first quality. But, although true felt is based on wool, most of the roll and sheet felt is now produced from synthetic fibers mechanically or chemically bonded, and they have the chemical resistance and physical properties of the particular synthetic fiber. **Needled felt** is a fabric made of natural or synthetic fibers physically interlocked by the action of a needle loom. It may also be treated chemically, or by heat or moisture, for special effects.

Felt is one of the earliest manufactured materials. It is now used for insulation, sound and vibration absorption, padding and lining in instrument cases, hats, roofing, and where a soft resilient fabric is needed. Although the best hat felts are made with nutria or beaver fur, vast quantities of rabbit furs or mixed furs and wool are used. **Hair felt** is made of cattle hair and is used for insulating cold-water pipes and refrigerating equipment and for cushioning and padding. The **Ozite felt** of American Felt Co. is an all-hair felt. Felt comes in thicknesses from 0.25 to 2 in (0.6 to 5.1 cm), the 0.25-in (0.6-cm) weighing 4 oz/ft^2 (18 g/cm^2). The **K felt** of this company is made to Army-Navy specifications and weighs 3.24 lb/(yd^2 · in) [0.69 kg/(m^2 · cm)] of thickness. It has a tensile strength of 12 lb/in^2 (0.08 MPa) and compressive strength of 3 lb/in^2 (0.02 MPa) at 50% deflection. It is for sound and thermal insulation. **Filtering felts,** for filtering gases and liquids, are usually made from various synthetic fibers to meet specific chemical-resistance requirements. **Teflon felt,** of American Felt Co., for filtering hot, strong acids and alkalies, is made from fluorocarbon fibers. Because of the high chemical and physical properties of this fiber, it is called **dragon fur** in the felting industry. It also has a low friction coefficient and is repellent to sticky materials, giving high filtering efficiency and easy cleaning. The **Scottfelt,** of Scott Paper Co., is a **foam filter** made from urethane foam compressed under heat and pressure to 0.067 of its original thickness. The tensile strength is increased from 35 to 270 lb/in^2 (0.2 to 1.9 MPa), and the porosity may be graduated up to 1,340/linear in (52,800/linear m).

Cattle- and goat-hair felts are also used for glass polishing. **Baize** is an old name for a thin, woolen felt used for desktops and tabletops, box linings, and bases of instruments. Its name is derived from the fact that it was originally bay, or brown, in color, but the industrial braize is now usually green. The name is now used to designate a plain-woven, loose, cotton or woolen fabric with a short, close nap, in plain colors for the same purposes. **Feltex** is an asphalt-saturated felt for roofing, and **Mica-kote** is a heavy felt coated with asphalt and finished with mica flakes, used for roofing. **Unisorb,** of Felters Co., is a heavy felt in blocks and sheets for isolation pads under machinery to absorb vibration. **Slaters' felt** is a tarred sheathing felt used in building construction, usually in 25- and 30-lb (11.3- and 13.6-kg) rolls. **Slatekote** is a heavy felt saturated with asphalt and coated with colored crushed slate, used for roofing. The term **roofing felt** is also applied to the thick asphalt-impregnated papers used for that purpose, and **papermakers' felt** is the woven wool or part-wool belting used in papermaking.

FERRIC OXIDE.   The red **iron oxide,** Fe_2O_3, also called **gamma ferric oxide,** found in abundance as the ore **hematite,** or made by calcining the sulfate. It has a dark-red color and comes in powder or lumps. The specific gravity is 5.20 and melting point about 2822°F (1550°C). It is used as a paint pigment under such names as **Indian red, Persian red,** and **Persian Gulf oxide.** In cosmetics and in polishing compounds it is called **rouge.** The Persian red oxide from the Island of Hormuz contains from 60 to 90% Fe_2O_3 and is marketed on a 75% basis. **Brown iron oxide** is made from ferrous sulfate and sodium carbonate and is not a pure oxide, though its chemical formula is given as Fe_2O_3. It is also called **iron subcarbonate** and is used in making green glass, paints, and rubber.

The names **metallic red** and **metallic brown** are applied to pigments from Pennsylvania ores containing a high percentage of **red iron oxide. Venetian red** is a name for red iron oxide pigments mixed with various fillers, most commonly an equal proportion of the pigment extender calcium sulfate. Commercially it is made by heating ferrous sulfate with quicklime in a furnace. Venetian red is a permanent and inert pigment that is generally used on wood. It cannot be used on many metals, including iron, because the calcium sulfate can cause corrosion. The **Tuscan red** pigments are red iron oxide blended with up to 75% of lakes, but may also be barium sulfate with lakes. **Ferric oxide pigments** make low-priced paints and are much used as base coats for structural steel work. The natural oxides come chiefly from Alabama, Tennessee, Pennsylvania, Iran, and Spain.

The **Mapico colors** of Binney & Smith Co. are **iron oxide pigments** refined under controlled conditions to give uniformity free of other mineral impurities. **Mapico red** and **Mapico crimson** contain 98% Fe_2O_3, the balance being almost entirely material lost on ignition or water-soluble impurity. The red oxide has a spheroidal particle shape, while the crimson has an acicular, or needle-shaped, particle. **Mapico lemon yellow** contains 87% Fe_2O_3, with 11.85% ignition loss. The particles are acicular and are only half the size of the crimson particles, being only 3.9 to 31.5 μin (0.1 to 0.8 μm). **Mapico brown** contains 93.1% Fe_2O_3 and 5 FeO. Its particles are cubic and of sizes from 7.9 to 15.7 μin (0.2 to 0.4 μm). **Mapico black** contains 76.3% Fe_2O_3 and 22.5 FeO, with a cubic particle shape. The **Auric brown** of Du Pont, used for giving light-fast shades to paper, is a hydrated ferric oxide ground to an extremely fine particle size.

Yellow iron oxide, known also as **ferrite yellow** and **Mars yellow,** used as a paint pigment, is $Fe_2O_3 \cdot 3H_2O$ plus from 2 to 12% calcium sulfate. It is made by precipitating ferrous hydroxide from iron sulfate and lime and then oxidizing to the yellow oxide. **Black ferric oxide, ferroferric oxide,** or **magnetic iron oxide** is a reddish-black amorphous powder, $FeO \cdot Fe_2O_3 \cdot H_2O$. It is used as a paint pigment, for polishing compounds, and for decarbonizing steel. The finely ground material used as a pigment is called **magnetic black,** and when used for polishing, it is called **black rouge. Hammer scale** is the iron oxide Fe_3O_4, formed in the hot rolling or forging of steel, and is used for decarbonizing steel by packing the steel articles in the scale and raising to a high temperature. It is very hard, Mohs 5.5 to 6.5, and is used as an abrasive.

Acicular gamma iron oxide is representative of a typical magnetic particle used in magnetic storage media. It is made from **goethite,** which is used in the paint industry as **yellow ocher.** When goethite is reduced in hydrogen, it forms **magnetite,** a strongly ferrimagnetic particle. It can be used as is in magnetic recording, but is usually reconverted to the acicular form by gradual oxidation in air to **maghemite,** because it is possible to orientate all the particles in a longitudinal direction by an external magnetic field. Recent trends are to dope the iron oxide with cobalt to get higher magnetic coercivity. **Acicular chromium oxide** from Du Pont Co. contains no iron. Hardened polyurethane, epoxy, and phenoxy resins are used to bind the iron oxide particles for high-grade tapes, such as for video and computer disks. For audio tapes, vinyl-based copolymers or tripolymers are employed. Hercules Inc. makes an iron oxide called **NP,** for nonpolar, that consists of smooth and regularly shaped particles. These remain suspended in the binder longer and therefore are evenly dispersed when the binder cures.

FERROCHROMIUM. A high-chromium iron master alloy used for adding chromium to irons and steel. It is also called **ferrochrome.** It is made from chromite ore by smelting with lime, silica, or fluorspar in an electric furnace. **High-carbon ferrochrome,** of Union Carbide,

contains 66 to 70% chromium in grades of 4.5, 5, 6, and 7% carbon. It
is used for making tool steels, ball-bearing steels, and other alloy
steels. It melts at about 2280°F (1250°C). It is marketed as crushed
alloy in sizes up to 2 in (5 cm) and as lump alloy in lumps up to about
75 lb (34 kg). **Low-carbon ferrochrome** of this company contains 67
to 72% chromium, in grades of 0.06, 0.10, 0.15, 0.20, 0.50, 1, and 2%
carbon. It is used for making stainless steels and acid-resistant steels.
Simplex ferrochrome, of Union Carbide, contains as little as 0.01%
carbon. It comes in pellet form to dissolve easily in the steel and is
used for making low-carbon stainless steels. Low-carbon ferrochrome
is also preferred for alloy steel mixtures where much scrap is used
because it keeps down the carbon and inhibits the formation of hard
chromium carbides. The various grades of ferrochromium are also
marketed as **high-nitrogen ferrochrome,** with about 0.75% nitrogen for use in making high-chromium cast steels which would normally have a coarse crystalline structure. The nitrogen refines the
grain and increases the strength. **Foundry-grade ferrochrome,** for
making cast irons, contains 62 to 66% chromium and 5 carbon.
Another has about 40% chromium, 18 silicon, and 9 manganese. It is
used for ladle additions to cast iron to give uniform structure and
increase the strength and hardness. Addition of 1% of the alloy to a
cast iron of 3.40% total carbon, with resultant balance of 1.30% silicon, 0.60 manganese, and 0.35 chromium, gives a dense iron of good
hardness.

FERROMANGANESE. A master alloy of manganese and iron used for
deoxidizing steels and for adding manganese to iron and steel alloys
and bronzes. Manganese is the common deoxidizer and cleanser of
steel, forming oxides and sulfides that are carried off in the slag.
Ferromanganese is made from the ores in either the blast furnace or
the electric furnace. Standard ferromanganese has 78 to 80% manganese. British ferromanganese contains about 7% carbon, but the
content in the U.S. alloy is usually 5 to 6.5%. **Low-carbon ferromanganese** is also marketed containing 0.10 to 1% carbon. Low-
phosphorus ferromanganese contains less than 0.10% phosphorus.
The alloys are marketed in lumps to be added to the furnace.
Spiegeleisen is a form of low-manganese ferromanganese with from
15 to 30% manganese and from 4.5 to 5.5 carbon. The German name,
meaning *mirror iron,* is derived from the fact that the crystals of the
fractured face shine as mirrors do. Spiegeleisen has the advantage
that it can be made from low-grade manganese ores, but the quantity
needed to obtain the required proportion of manganese in the steel is
so great that it must be premelted before it is added to the steel. It
was used for making irons and steels by the Bessemer process. Grade

A spiegeleisen has 19 to 21% manganese and 1 silicon; Grade B has 26 to 28% manganese and 1 silicon. The melting point is from 1950 to 2265°F (1066 to 1240°C).

FERROPHOSPHORUS. An iron containing a high percentage of phosphorus, used for adding phosphorus to steels. Small amounts of phosphorus are used in open-hearth steels to make them free-cutting, and phosphorus is also employed in tinplate steels to prevent the sheets from sticking together in annealing. Ferrophosphorus is made by melting phosphate rock together with the ore in making the pig iron. The phosphorus content is about 18% and is chemically combined with the iron. Another grade, made in the electric furnace and containing 23 to 25% phosphorus, is used for adding phosphorus to bronzes. A master alloy for adding selenium to steels, especially stainless steels, to give free-machining qualities, is **ferroselenium.** A typical ferroselenium contains about 52% selenium and 0.90 carbon.

FERROSILICON. A high-silicon master alloy used for making silicon steels, and for adding silicon to transformer irons and steels. It is made in the electric furnace by fusing quartz or silica with iron turnings and carbon. It is marketed in various grades with from 15 to 90% silicon. The silicon forms a chemical combination with the iron, but the alloys having more than about 30% silicon are fragile and unstable. The silicon also causes the carbon to be excluded in graphite flakes. The alloys of high silicon content are called **silicon metal.** One producer markets two grades, 15 and 45% silicon, while another has 15, 50, 75, 85, and 90% grades. Grades with silicon from 80 to 95% are marketed for use where small ladle additions are made for producing high-silicon steels, and for producing hydrogen by reaction with caustic soda. The alloy is marketed in lumps or crushed form. Silicon is often added to steels in combination alloys with deoxidizers or other alloying elements. **Ferrosilicon aluminum,** containing about 45% silicon and 12 to 15 aluminum, is a more effective deoxidizer for steel than aluminum alone. It is also used for adding silicon to aluminum casting alloys. **Silvaz** is a ferrosilicon aluminum containing also vanadium and zirconium. The alloy serves as a deoxidizer, fluxes the slag inclusions, and also controls the grain size of the steel. **Simanal** is a **deoxidizing alloy** containing 20% each of silicon, aluminum, and manganese. **Alsifier** contains 40% silicon, 20 aluminum, and 40 iron. The aluminum and silicon are in the form of an aluminum silicate which forms a slag that is eliminated during the teeming of the steel. **Alsimin** is a Swiss ferrosilicon aluminum with 50% aluminum. **Silicon aluminum** is a master alloy for adding silicon to aluminum alloys, and it does not contain iron. A 50–50 silicon aluminum has a melting point of 1920°F (1049°C), but is soluble in aluminum at 1275°F (690°C). It comes in pyramid waffle form for breaking into small lumps.

FERROTITANIUM. A master alloy of titanium with iron used as a purifying agent for irons and steel owing to the great affinity of titanium for oxygen and nitrogen at temperatures above 1472°F (800°C). The value of the alloy is as a cleanser, and little or no titanium remains in the steel unless the percentage is gaged to leave a residue. The **ferrocarbon titanium** is made from ilmenite in the electric furnace, and the carbon-free alloy is made by reduction of the ore with aluminum. Ferrotitanium comes in lumps, crushed, or screened. **High-carbon ferrotitanium** has 17% titanium and 7 carbon. It is used for ladle additions for cleansing steel. **Low-carbon ferrotitanium** has 20 to 25% titanium, 0.10 carbon, 4 silicon, and 3.5 aluminum. It is used as a deoxidizer and as a carbide stabilizer in high-chromium steels. **Graphidox** has 10% titanium, 50 silicon, and 6 calcium. It improves the fluidity of steel, increases machinability, and adds a small amount of titanium to increase the yield strength. The **Grainal alloys,** for controlling alloy steels, have various compositions. Grade No. 6 has 20% titanium, 13 vanadium, 12 aluminum, and 0.20 boron. **Tam alloy No. 78** contains 15 to 18% titanium, 7 to 8 carbon, with low silicon and aluminum. It is used in cast iron and steels. **Tam alloy No. 35** has 18 to 21% titanium and only 3.5 to 4.5 carbon. Its melting point is 2750°F (1510°C). Ferrotitaniums with 18 to 22% titanium are used for making fine-grained forging steels. **Carbotam** contains 16 to 17% titanium, 2.5 to 3 silicon, 6.5 to 7.5 carbon, 1.5 to 2 boron, and less than 1 calcium. It is used for cast steels to contain boron for high hardness. **Manganese titanium** is used as a deoxidizer for high-grade steels and for nonferrous alloys. A common grade contains 38% manganese, 29 titanium, 8 aluminum, 3 silicon, 22 iron, and no carbon. **Nickel titanium** is used for hard nonferrous alloys. The low-iron grade contains 15% titanium, 5 aluminum, 4 silicon, 1 iron, and 75 nickel. **Thermocol** is a **ferrocolumbium** for adding columbium to steel. It contains 53% columbium and 0.15 maximum carbon. It has an exothermic reaction which prevents chilling of the molten metal.

FERROUS SULFATE. Also called **iron sulfate** and **green vitriol.** It is a green crystalline material of composition $FeSO_4 \cdot 7H_2O$. It occurs naturally as the mineral **melanterite** and is a by-product of the galvanizing and tinning industries. The specific gravity is 1.898, the melting point is 147°F (64°C), and it is soluble in water. On exposure to the air it becomes yellowish because of the formation of basic iron sulfate, and on heating to 284°F (140°C) it becomes a white powder,

$FeSO_4 \cdot H_2O$, which also occurs as the mineral **szomolnokite.** Ferrous sulfate, under the name **copperas,** is an important salt in the ink industry to give color permanence to the inks. It is also employed in water purification, as a disinfectant, in polishing rouge, as a mordant in dyeing wool, and in the production of pigments. **Ferric sulfate** is a grayish amorphous powder of composition $Fe_2(SO_4)_3 \cdot 9H_2O$, or $Fe_2(SO_4)_3$. The specific gravity of the hydrous form is 2.1 and of the anhydrous 3.097. It is very soluble in water and is used as a pigment, as a mordant in dyeing, for etching aluminum and steel, and as a disinfectant. **Ferrisul,** of Monsanto Co., is anhydrous ferric sulfate used for speeding the action of metal pickling baths and for descaling boilers. In etching steel, the action of anhydrous ferric sulfate is 30 times more rapid than that of sulfuric acid.

FERTILIZERS. Materials added to the soil to supply plant food either directly or by chemical reaction with the soil. The preparation of fertilizers is now one of the major industries, and commercial fertilizers include nitrates, phosphates, potash salts, calcium salts, and mixtures. They may also include the materials which regulate the acidity of the soil for better plant production, such as lime, and the materials which act as **soil conditioners,** i.e., **synthetic mulches,** such as methyl cellulose or polymeric plasticlike organic chemicals. **Plant regulators** are fertilizers containing selected metals or minerals for specific plant foods, and they are applied either in the soil or to the plant.

Chemicals used as fertilizers must not be of such a nature as to kill earthworms. It is stated that at least 50,000 earthworms per acre are needed for invigorating and loosening the soil. Also, millions of bacteria are in every pound of good soil, and millions of ants, bugs, and invertebrates in every acre perform a tremendous pattern of interdependent chemical conversion. Thus, fertilizers should not contain drastic chemicals that make the soil sterile.

Much barnyard manure is employed as fertilizer, but does not enter the commercial mixed fertilizers except the dried and ground sheep and cow manures. Much local fertilization is also done by the planting and plowing under of legumes that bring nitrogen from the air and also serve as soil conditioners. Conditioning of the soil, for the retention of moisture and to prevent hard-caking so that plants may take deep root and have the needed elements readily available, is a necessary part of fertilization. Decayed vegetable matter, or peat moss, may thus be added to the soil as **humus.** These materials also often add plant foods to the soil. **Fersolin** is such a material produced by heating sawdust with a catalyst below the charring temperature to convert the cellulose to lignin and humus. It is usually mixed with fertilizers to give greater plant yields. **Merloam,** a soil conditioner of Monsanto Co., is a vinyl acetate–maleic acid compound.

Chilean nitrate, also known as **Chile saltpeter, phosphate rock,** and **potash,** is the chief natural mineral used as fertilizers. Nitrogen is needed in most soils, and phosphorus is a necessary ingredient in soils. Large quantities of **muriate of potash** are used in fertilizers to supply K_2O, while vast quantities of hydrated lime are employed to supply MgO and to reduce the acidity of some soils. Potassium, calcium, and sodium are also supplied in combination forms especially with ammonia to yield nitrogen. Ammonium sulfate yields both nitrogen and sulfur. **Ground gypsum** is a source of sulfur trioxide for cotton, tobacco, grapes, and some other crops. It also helps to liberate soluble potash and stimulates growth of nitrogen-fixing bacteria in the soil. **Calcium cyanamid** is employed as a fertilizer to yield nitrogen and calcium. **Guanylurea sulfate** is a 37% nitrogen product made by direct acidification of calcium cyanamid by Japan's Nitto Chemical Industry.

The composition of commercial fertilizers is expressed as grade or units: A 10-10-10 fertilizer contains 10% by weight of N, P_2O_5, and K_2O; in metric units, a unit is 10 kg, and a 10-10-10 fertilizer has 100 kg each of the three components. Nitrogen is the fertilizer nutrient used most abundantly, with almost all nitrogenous fertilizer being made from ammonia. Of nitrogenous fertilizers, urea is most widely used. **Crude urea** is now also used as a fertilizer and has the nitrogen in the same form as in the natural guanos and manures. **Ureaform,** developed by the U.S. Department of Agriculture, is a hygroscopic powder made by reacting urea with a small amount of formaldehyde and crushing. It may also be mixed with ammonium nitrate. These products are called **controlled-release fertilizers** because the solubility of urea-aldehyde condensation products is low, enabling them to release their nutrients over long periods. **Nitroform** is a urea-formaldehyde product (N-P-K composition of 38-0-0) made by Nor-Am Chemical Co. It is available in two forms: **Blue Chip,** a granular material, and **Powder Blue,** a sprayable powder. O. M. Scott & Sons makes a comparable product, **Hi-Tech** (38-0-0). The firm also markets a line of granulated mixed fertilizers containing **methylene ureas,** including **Burpee, Scotts, ProTurf,** and **ProGrow.** A methylene diurea product, **Nitro-26 CRN** (26-0-0), is available as a solution from C. P. Chemical Co. **Fluf** (18-0-0) is microcrystalline dispersion of soluble and insoluble urea reaction products produced by W. A. Cleary Chemical Corp. A Fluf product (16-2-4) is also available.

Isobutylidene diurea (IBDU) is a condensation product of urea and isobutyraldehyde and is made in Germany by Farbwerk Hoechst

Aktiengesellschaft and in Japan by Mitsubishi Chemical Co. Fertilizer-grade IBDU contains 31% nitrogen. Urea reacts with acetaldehyde under the influence of an acid catalyst to produce **crotonylidene diurea.** It is marketed as **CDU** by Chisso-Asahi Fertilizer Co. in Japan and as **Crotodur** by BASF in Germany. **FAN** is a urea-acetaldehyde solution (20-0-0) from W. A. Cleary Chemical Corp. that has a storage life of up to 3 years, compared with 3 months of urea-formaldehyde solutions. **Melamine, or triaminortriazine,** is a slow-release product made from urea by Melamine Chemicals, Inc. The **Super 60** variety has a 60-0-0 composition. **Oxamide,** $(CONH_2)_2$, made by Ube Industries in Japan contains 32% nitrogen. **Fertilizer pellets** that resist the bleaching action of rains and release nitrogen slowly are made in granules from ammonium sulfate with a binder of asphalt and wax. **UAN fertilizers** are aqueous solutions of urea, ammonia, and ammonium nitrate.

Superphosphate, or **phosphate fertilizer,** is made by treating the phosphate rock with sulfuric or nitric acid, reacting with ammonia to neutralize the acid and add nitrogen, and then adding potash salts. The final ground product contains 12% each of nitrogen, phosphoric acid, and potash. Or it may be produced by digesting the rock with ammonium sulfate, yielding ammonium phosphate and gypsum. The ammonium phosphate is then treated with sulfuric acid to yield 70% phosphoric acid and ammonium sulfate. The German fertilizer **Nitrophoska** is a nitrate-phosphate-potash made by treating phosphate rock with nitric acid, neutralizing with ammonia, and then granulating with potassium salts. The **calcium nitrate tetrahydrate** which is precipitated off is also used as fertilizer. The calcium nitrate used in Europe is produced by treating phosphate rock with nitric acid. It is highly alkaline and efficient in the release of nitrogen, but is very hygroscopic and sets up in lumps. Fish meal, castor pomace, cottonseed meal, soybean meal, copra cake, and other residues from oil pressing are used as commercial fertilizers. Tankage from the meat-packing plants is also an important fertilizer material. **Whale guano,** from South Georgia and Newfoundland, was a mixture of whale-meat meal and bone meal. Ground bonemeal is used in fertilizers to give phosphorus, calcium, and other mineral salts to the soil. Some plants require boron, and borax is applied as a fertilizer to some soils. Many vegetable products obtain their coloring and some characteristic properties from small quantities of copper, manganese, rubidium, iodine, and other elements that do not occur in all soils. Boron is necessary for sunflower growth, iron is necessary for pineapples, molybdenum is needed for cauliflower, and cobalt oxide is necessary in the soil to prevent salt sickness in cattle. Lack of manganese in the soil also causes yellow spot on leaves of tomatoes and citrus fruits. For use as a fertilizer, **manganese sulfate,** $MnSO_4$, comes as a water-soluble powder of porous spherical particles. Most plants require minute quantities of zinc to promote formation of **auxin,** a complex butyl-cyclopentene ring compound needed for growth of plants. However, plants require balanced feeding, and indiscriminate use of fertilizers is often injurious. Too much manganese in the soil, for example, may cause necrosis, or inner bark rot, on apple trees, or excess of some common fertilizers may cause abnormal growth of stalk and leaves in plants.

FIBERBOARD. Heavy sheet material of fibers matted and pressed or rolled to form a strong board, used for making containers and partitions and for construction purposes. Almost any organic fiber may be used, with or without a binder. The softboards are made by felting wood pulp, wood chips, or bagasse, usually without a binder. **Masonite** is produced from by-product wood chips reduced to the cellulose fibers by high steam pressure. The long fibers and the lignin adhesive of the wood are retained, and no chemicals are used in pressing the pulp into boards. **Masonite quarter board,** for paneling, is made in boards 0.25 in (0.64 cm) thick. **Presdwood** is a grainless grade made by compressing under hydraulic pressure and is dense and strong.

These types belong to the class known as **hardboard,** in the processing of which the carbohydrates and soluble constituents of the original wood are dissolved out and the relative proportion of lignin is increased, resulting in a grainless, hard, stiff, and water-resistant board free from shrinkage. The specific gravity of most hardboards is greater than 1.0, and the modulus of rupture is from 5,000 to 15,000 lb/in^2 (34 to 103 MPa). The lignin acts as a binder for the fibers, but some hardboards are made harder and more resistant by adding a percentage of an insoluble resin. The usual density is 50 to 65 lb/ft^3 (801 to 1,041 kg/m^3), but with added resin binder the weight may be as much as 70 lb/ft^3 (1,121 kg/m^3). **Densified hardboard,** made with high pressure, is 85 lb/ft^3 (1,362 kg/m^3) or greater. Hardboards have uniform strength in all directions and have smooth surfaces. Tensile strengths are up to 7,700 lb/in^2 (53 MPa), and compressive strength is 26,000 lb/in^2 (179 MPa).

Irradiated wood is natural wood impregnated with resins of low molecular weight and irradiated with gamma rays from cobalt 60 which cross-links the resin molecules and binds them to the fibers of the wood. The resins add strength and hardness without changing the grain structure and color of the wood. Maple, impregnated with 0.5% by weight of methyl methacrylate and irradiated, is 3 times as hard as the natural wood but can be worked with ordinary tools.

Particleboards, made with wood particles, have lower density, about 40 lb/ft^3 (641 kg/m^3), and have greater flexibility but lower strength than hardboard. The process is not limited to the making of boards. **Wood particles** are also used for low-cost molded parts, with up to 90% wood particles and the balance urea, phenolic, or melamine resin. Birch or maple particles are preferred. These **Granuplast moldings** are made with low heat and pressure to densities of 45 to 85 lb/ft^3 (721 to 1,362 kg/m^3).

Wood molding powder is made by the same method of treating wood fibers with steam pressure and hydrolyzing the hemicellulose, leaving the lignin free as a binder. Hardboard is used for countertops, flooring, furniture, and jigs and templates. **Forall,** of Forest Fiber Products Co., is a light-colored hardboard in thicknesses from 0.375 to 0.75 in (1 to 1.9 cm), made by compressing Douglas fir free of bark. It is grain-free and will not split or splinter. **Presdply,** of Masonite Corp., has surfaces of grainless hard Presdwood and a core of soft plywood that will hold screws.

Hardwood is a hardboard made from hardwood waste compressed into sheets under heat and hydraulic pressure. The surface is hard with a high polish. A hardboard, developed by the Scottish Cooperative Wholesale Society and called **heatherwood,** is made by pulping heather and pressing into boards with a synthetic resin binder. **Heather,** or **heath,** is a small flowering shrub, *Ericaceae tetralix* and *E. cinerea,* which grows profusely in Great Britain. **Wonderwood,** of Wonderwood Corp., is a development of the **Novopan** made in Switzerland. It is made by pulping chipped waste wood and compressing with a resin binder. **Tensilite 300** is made of pulp combined with nitrile rubber and a phenolic resin and pressed into sheets. The specific gravity is 1.35, dielectric strength 600 V/mil (23.6 × 10^6 V/m), and compressive strength 32,000 lb/in^2 (220 MPa). It is suitable for mechanical and electrical applications as well as for paneling. **Forest hardboard** of Forest Fiber Products Co. is made of chipped wood that is pulped and mixed with synthetic resin and wax and then is hydraulically pressed. The board has a smooth glossy face and is suitable for making furniture and toys and for paneling. **Prespine** is a paneling board of lower hardness and density made by mixing 5 to 15% phenolic resin with sawdust and wood chips and pressing at only 200 lb/in^2 (1.4 MPa). These hardboards are used for many construction parts, but the lighter and less dense fiberboards are preferred for insulation and some construction uses. **Granite board,** of National Starch and Chemical Corp., is a strong, nonsplintering **building board** made from fine particles of eastern white pine molded under pressure with a resin binder to a density about equal to that of natural wood. It has an acoustical value higher than that of natural wood. **Kimflex board,** of Kimberly-Clark Corp., is a lightweight, pliable fiberboard used for shoe counters. It is made from balsawood pulp, using rubber latex as a binder. Wood fiberboard of high strength and high dielectric strength is used for electrical panels. **Temlok** is a fiberboard made from pinewood fibers impregnated with resin and compressed into building boards and tiles. **Temwood** is a lightweight board of wood fibers hydraulically pressed into grainless boards in hard and semihard grades. **Temboard** is a decorative wood fiberboard used for interior paneling. **Veneer fiberboard** is made by cutting veneer waste into fibers of 0.010- to 0.015-in (0.025- to 0.038-cm) thickness with strand lengths from 1 to 8 in (2.54 to 20.32 cm). The flat side of the strand is edge-grained, and when felted, the broad surface lies parallel to the faces of the board. From 10 to 20% phenolic resin is used as a fiber binder. **Thermax** is an insulating board made of shredded wood fibers with a fire-resistant cement.

FIBER-REINFORCED PLASTICS. A broad group of composite materials composed of fibers embedded in a plastic resin matrix. The short designation for these materials is **FRP.** In general, they have relatively high strength-to-weight ratios and excellent corrosion resistance compared to metals. They can be formed economically into virtually any shape and size. In size, FRP products range from tiny electronic components to large boat hulls. In between these extremes, there are a wide variety of FRP gears, bearings, bushings, housings, and parts used in all product industries.

Glass is by far the most used fiber in FRPs. **Glass-fiber-reinforced plastic** is often referred to as **GFRP** or **GRP. Asbestos fiber** has some use, but is largely limited in applications where maximum thermal insulation or fire resistance is required. Other fibrous materials used as reinforcements are paper, sisal, cotton, nylon, and the aramid Kevlar. For high-performance parts and components, more costly fibers, such as boron, carbon, and graphite, can be specified.

Although a number of different plastic resins are used as the matrix for reinforced plastics, thermosetting **polyester resins** are the most common. The combination of polyester and glass provides a good balance of mechanical properties as well as corrosion resistance, low cost, and good dimensional stability. In addition, curing can be done at room temperature without pressure, thus making for low processing equipment costs. For high-volume production, special **sheet-molding compounds** are available in continuous-sheet form. Resin mixtures of thermoplastics with polyesters have been developed to produce high-quality surfaces in the finished molding.

Prepregs are partially cured thermoset-resin reinforced fabrics in roll or sheet. The prepregs (short for *preimpregnated*) can be laid or wrapped in place and then fully cured by heat.

Other glass-reinforced thermosets include **phenolics** and **epoxies. GR phenolics** are noted for their low cost and good overall performance in low-strength applications. Because of their good electrical resistivity and low water absorption, they are widely used for electrical housings, circuit boards, and gears. Since epoxies are more expensive than polyesters and phenolics, **GR epoxies** are limited to high-performance parts where their excellent strength, thermal stability, chemical resistance, and dielectric strength are required.

Initially, GRP materials were largely limited to thermosetting plastics. Today, however, more than 1,000 different types and grades of **reinforced thermoplastics** or **GR+P** are commercially available. Leaders in volume use are **nylon** and the **styrenes.** Unlike thermosetting resins, GR+P parts can be made in standard injection-molding machines. The resin can be supplied as pellets containing chopped glass fibers. As a general rule, a GR+P with chopped fibers at least doubles the plastic's tensile strength and stiffness. Glass-reinforced thermoplastics are also produced as sheet materials for forming on metal-stamping equipment and compression-molding machines.

Hy-bor, of Textron Specialty Materials, is a **prepreg** of boron and graphite fibers in an epoxy or polyimide matrix. With 70 to 80% by volume fibers in epoxy, it provides twice the flexural strength and rigidity of graphite-epoxy prepreg; 40% greater compression, flexural, and tensile strengths than boron-epoxy prepreg; and 30% greater rigidity than boron-epoxy prepreg. **PIX** is a carbon-fiber thermoplastic polyimide prepreg of Mitsui Toatsu of Japan. **Towflex,** of Custom Composite Materials Inc., is a prepreg of fiber bundles, or tows.

FIBERS. By definition (ASTM), a **fiber** has a length at least 100 times its diameter or width, and its length must be at least 0.2 in (0.5 cm). Length also determines whether a fiber is classified as staple or filament. **Filaments** are long and/or continuous fibers. **Staple fibers** are relatively short and, in practical applications, range from under 1 to 6 in (2.5 to 15.2 cm) long (except for rope, where the fibers can run to several feet). Of the natural fibers, only silk exists in filament form, while synthetics are produced as both staple and filaments.

The internal, microscopic structure of fibers is basically no different from that of other polymeric materials. Each fiber is composed of an aggregate of thousands of polymer molecules. However, in contrast to bulk plastic forms, the polymers in fibers are generally longer and aligned linearly, more or less parallel to the fiber axis. Thus fibers are generally more crystalline than bulk forms.

Also in contrast to bulk forms, fibers are used not alone, but either in assemblies or aggregates such as yarn or textiles or as a constituent with other materials, such as in composites. Also, compared with other materials, the properties and behavior of both fibers and textile forms are more critically dependent on their geometry. Hence fibers are sometimes characterized as tiny microscopic beams, and as such, their structural properties are dependent on such factors as cross-sectional area and shape, and length. The cross-sectional shape and diameter of fibers vary widely. Glass, nylon, Dynel, and Dacron, for example, are essentially circular. Some other synthetics are oval, while others are irregular and serrated round. Cotton fibers are round tubes, and silk is triangular.

Fiber diameters range from about 394 to 1,575 μin (10 to 40 μm) in diameter. Because of the irregular cross section of many fibers, it is common practice to specify diameter or cross-sectional area in terms of *fineness,* which is defined as a weight-to-length or linear density relationship. One exception is wool, which is graded in micrometers. The common measure of linear density is the *denier,* which is the weight in grams of a 29,530-ft (9,000-m) length of fiber. Another measure is the *tex,* which is defined in grams per kilometer. A millitex is in grams per 1,000 km.

Of course, the linear density, or denier, is also directly related to fiber density. This is expressed as the denier/density value, commonly referred to as *denier per unit density,* which represents the equivalent denier for a fiber with the same cross-sectional area and a density of 1.

The cross-sectional diameter or area generally has a major influence on fiber and textile properties. It affects, for example, yarn packing, weave tightness, fabric stiffness, fabric thickness and weight, and cost relationships. Similarly, the cross-sectional shape affects yarn packing, stiffness, and twisting characteristics. It also affects the surface area, which in turn determines the fiber contact area, air permeability, and other properties.

Plastic optical fibers, with polymethylmethacrylate as the core and, generally, a fluoropolymer coating, are used in illuminating signs and displays. Metal fibers of stainless steel, titanium alloy, and iron-chromium-aluminum-yttrium are made into **Feltmetal,** or **Fibermetal,** sheet by Technetics Corp. The fibers, 0.0004 to 0.010 in (0.01 to 0.25 mm) thick, are bonded by sintering at all contact points.

FILTER FABRICS.   Any fabric used for filtering liquids, gases, or vapors, but, because of the heat and chemical resistance usually required, generally synthetic or metal fibers. Weave is an important consideration. Plain weave permits maximum interlacings, and a tight weave gives high impermeability to particles. Twill weave has lower interlacings in sharp diagonal lines and gives a more selective porosity for some materials. Satin weave has fewer interlacings, is spaced widely but regularly, and is used for dust collection and gaseous filtration.

Fibers are chosen for their particular chemical resistance, heat resistance, and strength. Dacron has good acid resistance except for concentrated sulfuric or nitric acid. It can be used to 325°F (162°C). High-density polyethylene has good strength and abrasion resistance, and its smooth surface minimizes clogging of the filter, but it has an operating temperature only to 230°F (110°C). Polypropylene can be used to 275°F (134°C). Nylon gives high strength and abrasion resistance. It has high solvent resistance, but low acid resistance. Its operating limit is about 250°F (121°C). Teflon is exceptionally resistant to a wide variety of chemicals. It can be operated above 400°F (204°C), and its waxy, nonsticking surface prevents clogging and makes it easy to clean, but the fiber is available only in single-filament form.

FILTER SAND. A natural sand employed for filtration, especially of water. Much of the specially prepared filter sand comes from New Jersey, Illinois, and Minnesota and is from ocean beaches, lake deposits, and sandbanks. The specifications for filter sand require that it be of fairly uniform size, free from clay and organic matter, and chemically pure, containing not more than 2% combined carbonates. The most common grain sizes are 0.014 to 0.026 in (0.35 to 0.65 mm). Very fine sand clogs the filter. **Greensand,** produced from extensive beds in New Jersey, is used as a water softener. It is a type of marl classed as **zeolite** and consists largely of **glauconite,** which is a greenish granular mineral containing up to 25% iron, with a large percentage of silica and some potash and alumina. Synthetic zeolite is a **sodium alumina silicate,** $Na_2O \cdot Al_2O_3 \cdot 6SiO_2 \cdot xH_2O$, made by reacting caustic soda with bauxite to form sodium aluminate and then reacting with sodium silicate. In addition to filtering, the greensand softener extracts the calcium and magnesium from the water. It is regenerated for further use by passing common salt brine through it. **Zeolex** of J. M. Huber Corp. is **sodium silicoaluminate** in extremely fine powder form. Up to 2% is added to such products as dried egg-yolk powder to prevent caking and to keep dry ingredients in automatic food processing equipment free-flowing.

Molecular sieves are synthetic crystalline zeolites whose molecules are arranged in a crystal lattice so that there are a large number of small cavities interconnected by smaller pores of uniform size, the network of cavities and pores being up to 50% of the volume of the crystal. The sieves consist of three-dimensional frameworks of SiO_4 and AlO_4 tetrahedra. Electrovalance of each tetrahedron is balanced by the inclusion in the crystal of a metal cation of Na, Ca, or Mg. Firing in a kiln drives out the water, and by exchanging the sodium ion for a smaller or larger ion, the pore openings can be varied from 7.8 to 47 nin (0.2 to 1.2 nm). For gasoline upgrading, 15.7-nin (0.4-nm) openings are used, while 39-nin (1.0-nm) openings serve for removing oil vapor or hydrogen sulfide from gas. The mean path required for oxygen and nitrogen molecules is about 3.9 μin (0.1 μm). **Filter plates,** for filtering acids and oils, are porous fused alumina with pore diameters 0.004 to 0.012 in (0.09 to 0.30 mm). **Zeolon,** of Norton Co., is a zeolite with a crystal structure known as **nordenite.** The pore diameter is about 39 nin (1.0 nm). **Zeolite 4A,** of Linde, for chemical separations, will pass molecules no larger than 15.7 nin (0.4 nm).

FIRECLAY.   Clays that will withstand high temperatures without melting or cracking have been used for lining furnaces, flues, and for making firebricks and lining tiles. Common fireclays are usually **silicate of alumina.** Theoretically these clays contain 45.87% alumina and 54.13 silica, but in general they contain considerable iron oxide, lime, and other impurities. Most U.S. clays are from New Jersey, Kentucky, Pennsylvania, Ohio, and Missouri. They are largely $Al_2O_3 \cdot SiO_2$, with CaO, Fe_2O_3, and TiO_2. Those low in iron oxide, lime, and magnesia, and alkalies are chosen. The clays are grouped as low-duty, intermediate-duty, high-duty, and super-duty. The low-duty has low alumina and silica with high impurities and is limited to a temperature of 1600°F (871°C). Standard types are good for temperatures of 2400 to 2700°F (1316 to 1482°C), and the super-duty to temperatures of 2700 to 3000°F (1482 to 1649°C). Kiln-burned clay should have a balanced proportion of coarse, intermediate, and fine grain sizes. Clays with an excess of silica are also used. The German **Klingenberg clay** used for crucibles has about 60% silica. The term **refractory clay** embraces nearly all clays having a melting point above 320°F (160°C). But the clays alone are likely to shrink and crack, and they may be mixed with other clays, sand, or graphite. **Firebrick** is made in various shapes and sizes and is usually white or buff color. Common firebrick from natural clays will melt from 2800 to 3100°F (1538 to 1704°C).

   Insulating firebrick is made with fireclay and a combustible material, such as sawdust, which burns out to leave a porous structure. The weight is 1.25 to 4 lb (0.57 to 1.81 kg) per brick compared with 8 lb (3.6 kg) for regular firebrick. Firebrick containing more than 47.5% alumina is not classified as fireclay brick but as **alumina brick. SL firebrick** of Manville Corp. is a **kaolin firebrick** with 62% silica, and it can be used up to 2300°F (1260°C). Spalling is a common failure of fireclay brick, but it can be reduced under long soaking at 1650°F (899°C) or alternating periods of heating and cooling at higher temperatures. **Alamo brick** and **Varnon brick,** of Harbison-Walker Refractories Co., are high-duty firebrick. **Kaosil firebrick,** of the same company, is designated as a **semisilica firebrick.** It is made from low-alkali siliceous kaolin of Pennsylvania, rotary-fired at high temperature. The nominal composition is 75.6% silica, 21.8 alumina, 1.7 titania, 0.5 iron oxide, 0.27 magnesia, 0.15 lime, and 0.10 alkalies. The brick can be used in a soaking heat of 2700°F (1482°C), is resistant to spalling and to fluxing by alkali slags, and has high load-carrying ability. **Korundal,** of the same company, is a corundum-mullite brick for temperatures to 3425°F (1880°C). It contains 91% corundum alumina, 8 silica, and less than 1 iron oxide, lime, magnesia, and alkalies. It melts at 3668°F (2020°C), converting all the mullite to corundum, but slow cooling returns the brick to the original mixture.

Some other materials used in making firebrick are chromite, bauxite, diatomaceous earth, and magnesite, or the artificial materials silicon carbide and aluminum oxide, but brick made of these is designated by the name of the material or by trade names. **Chromite brick** will withstand temperatures up to 3700°F (2038°C), and **magnesia brick** up to 3900°F (2149°C), while silicon-carbide brick without a clay binder will withstand heats to 4000°F (2204°C). **Firecrete,** of Manville Corp., is a lightweight refractory consisting of calcined high-alumina clay used for furnace doors and floors. It will withstand temperatures of 2400°F (1316°C) in continuous operation. **Insuline,** of Quigley Co., is a calcined fireclay in small cellular particles. In insulating brick it is called **Insulbrix,** and as a lightweight concrete it is known as **Insulcrete. Allmul firebrick,** of Babcock & Wilcox Co., for glass furnaces, is mullite with no free silica.

FIRE EXTINGUISHERS. Materials used for extinguishing fires, usually referring to chemicals in special containers rather than the materials, like water, used in quantity for cooling and soaking the fuel with a noncombustible liquid. There are three general types of fire extinguishers: those for smothering, such as carbon dioxide; those for insulating the fuel from the oxygen supply, such as licorice and protein foams, which also include mineral powders that melt and insulate metallic fires; and chemicals which react with the combustion products to terminate the chain reaction of combustion, such as **bromotrifluoromethane,** $CBrF_3$, a nontoxic colorless gas liquefied in cylinders. **Freon FE 1301,** of Du Pont, is this chemical, while **Freon 13B1** of the same company is **monobromotrifluoromethane** gas pressurized with nitrogen. The relative effectiveness of extinguishers varies with the type of fuel in the fire, but on average, with bromotri-fluoromethane taken as 100%, dibromodifluoromethane would be about 67, the dry chemical sodium hydrogen carbonate 66, carbon tetrachloride 34, and carbon dioxide about 33. Others in the family include the **Halons,** halogenated hydrocarbons. **Halon 1211 is bromochlorodifluoromethane, Halon 1301 is bromotrifluoromethane, Halon CTFE is polychlorotrifluoroethylene,** and **Halon TFE is polytetrafluoroethylene.** The brominated compounds destroy stratospheric ozone even more drastically than do the **chlorofluorocarbons,** but since smaller quantities are released, their use is not being curtailed as severely as the CFCs.

Protein foams are made by hydrolyzing fish meal, feather meal, and horn-and-hoof meal. A **fluoroprotein foam** includes a fluorocarbon surfactant. For confined spaces, high-expansion foams, or **Hi-Ex foams,** are made by bubbling air through a porous material soaked in surfactant. Dry chemicals used for extinguishing fires include **sodium bicarbonate, potassium bicarbonate, monoammonium phosphate,** and **potassium chloride.** The potassium products impart a purple color to the flames and so are known as **Purple K. Sodium chloride** mixed with **vinylidene dichloride copolymer** is called **Met L-X** and is used to fight metal fires. **G-1** is a **graphite-** based product used for the same purpose. **Fire-Trol,** of Arizona Agrochemical Co., is ammonium sulfate and **attapulgic clay** for fighting forest fires. It is sprayed in a water solution, and the slurry mixture coats the trees to stop fire spread.

Flame-retardant and **smoke-suppression chemicals** are used in plastics, wood paper, and textiles. **Alumina trihydrate** is the dominant flame retardant in plastics, followed by **halogenated phosphates.** The situation is reversed for wood and paper, where phosphorus is used extensively as a coating. **Cyclic phosphonate esters** are the basis for **Antiblaze 19,** a flame retardant for polyesters developed by Mobil Corp. For cotton blends, **Caliban P-44,** an **antimony oxide**—based organic oxide, is available from White Chemical Co. **Zinc borate, molybdenum, nitrogen,** and **magnesium carbonates** and **hydrates** find use in niche markets. **Saytex** is a series of flame retardants from Ethyl Chemicals. **Saytex VBR** is **vinyl bromide** mixed in monomers and modacrylic fibers. **Saytex 102** is **decabromodiphenyl oxide** used in such plastics as high-impact polystyrene and thermoplastic polyesters. **Reoflam PB-460,** of FMC Corp., is a bromine and phosphate flame retardant for thermoplastics and blends. Processible at low temperatures, it facilitates production of thin-wall, highly filled moldings. In recent years, halogen-type flame retardants for plastics have become an environmental concern.

FISHERY PRODUCTS. Fisheries constitute one of the largest industries of the world. In addition to its use as food, fish is important as the source of fatty oils, animal feeds, fertilizer, vitamin products, fish flour, protein powders, pearl essence, and skins. More than half of all species of vertebrates (animals with backbones) are fish, and more than 40,000 kinds of fish have been classified, varying from the small goby, weighing less than 0.01 oz (0.28 kg), to the whale shark, sometimes weighing more than 20 tons (18 metric tons). Extreme shapes vary from the snakelike eel to the sea horse, but the **marine herring,** *Clupea herengus,* is designated by the U.S. Fish and Wildlife Service as the typical **fish** because of its abundance; its lack of extremes in form, size, and structure; and its water efficiency due to its streamlined shape and fin arrangement.

Ocean perch, *Sebastes marinas,* is the chief commercial fish caught in the New England area, and **menhaden,** *Brevoortia tyrannus,* constitutes the chief catch of the Atlantic coasts. The red **snapper,** *Lutianus aya,* and the red **grouper,** *Epinephelus morio,* constitute the major part of the catch in the Gulf of Mexico. **Smelt,** *Oamerus mordax,* planted in the Great Lakes in 1912, constituted 97% of the commercial catch of fish in Lake Erie before the recent pollution of the lake.

One-third of all fish and frozen packaged fish in the United States is ocean perch, or **rosefish,** compared with 24% haddock. **Rockfish,** very low in oil and sodium and high in protein, is important in the frozen-fish industry of the west coast, reaching a tonnage about half that of the total salmon. With these processed fish, only about 25% of the total weight of the fish is packaged, the remainder being used for oils, feeds, and fertilizer. The variety of **tuna fish** known as **skipjack,** *Katsuwonus pelamis,* is the most important commercial fish of the Pacific. It is migratory, ranging from California to Japan and the Philippines. Known as **aku** in Hawaii, it abounds in that area, but is also caught in warm areas of the Atlantic. The **albacore tuna** likewise ranges over the Pacific and Indian oceans and into the Atlantic.

The usual types of fish processed for marketing as **smoked fish** are mackerel, mullet, sturgeon, catfish, and flounder. The **sturgeon** of the Caspian Sea attains a length of 30 ft (9.1 m) and a weight of 4,000 lb (1,814 kg). It is valued for meat, liver vitamins, isinglass, oils, skins, and **caviar,** the latter being the **roe,** or eggs, also obtained from shad and some other large fish. In the processing of frozen fish, an average of only 33% of the whole fish is shipped as edible packaged fish, but the amount of fillets taken from cleaned fish may be as high as 55% for pollock and 70% for large, 8-lb (3.61-kg) steelhead trout. **Industrial fish,** for the production of oil, animal feeds, and fertilizer, consists usually of mixed small fish, but in some areas there is no sorting and fish inedible to humans may be included.

The **fish scales** which are removed from processed edible fish are high in edible proteins and are used in animal feeds. Scales from the pollock contain 70% protein. **Fish solubles** usually consist of a 50% solids concentration of the residue liquor, known as **stickwater,** from processing plants and canneries. In addition to protein, it is rich in vitamins B, G, and B_{12}. It is mixed with alfalfa leaf meal for animal feed. **Fishskins** have a close texture and are impervious, but sharkskin is the only fishskin of commercial importance.

Fish meal is produced from whole fish or from the residue of processed food fish. Whole fish is ground and cooked below 212°F (100°C) to avoid loss of protein, and the oil solvent is extracted. About 2% oil is retained in the meal, but for the manufacture of fish flour, this residue oil is removed by alcohol extraction. Prior to the extraction of the oil, which may be up to 16%, fish meal contains up to about 23% protein and up to about 30% minerals, including calcium, phosphorus, iron, and copper. The proteins have all the essential amino acids to supplement cereal foods for poultry raised by commercial methods where the birds lack access to normal feeding.

Fish flour is a low-cost source of protein and is used for enriching flours, baby foods, sauces, and prepared soups, and for adding proteins to breads, cakes, pastries, and other bakery products. It is prepared from fish meal by refining and deodorizing. It is an additive rather than a flour; it does not thicken soups, and in bakery products it does not have the elastic and extensible properties inherent in cereal flours. In the food industries it is called **animal protein concentrate,** and the high-protein grade contains 95% animal protein. It is also high in calcium and phosphorous, and it contains thiamin, niacin, and riboflavin. An edible **fish-protein concentrate** may now be made directly from finely ground industrial fish. Oils, fats, fatty acids, and lipid-containing materials are extracted by a solvent such as isopropyl alcohol. About 20% by weight of raw fish is recovered as a dry, odorless, tasteless powder which can be mixed with a variety of foods to upgrade the human diet. Fish flour has negligible carbohydrate content and no more than 0.4% fat, but **Viking egg white,** an odorless gray powder made in Germany from whitefish, is a soluble albumin used as a substitute for egg white in bakery goods.

Fish oils are obtained by boiling the fish and skimming off the oil, or by solvent extraction from the fish meal. The crude oil has a brownish color and an offensive odor, but it is usually decolorized and deodorized. Oil content of fish varies from 0.5 to about 16%, depending on the type of fish, the season, and the area. Fish in cold waters tend to have more oil than those in warm waters. There is only a small difference in the composition of oils from different species. They usually contain 20 to 30% of saturated acids and 70 to 80% unsaturated acids. The average specific gravity is about 0.930. Much of the commercial oil is from the cod, herring, menhaden, sardine, and salmon. **Japan fish oil** consists of a mixture of sardine and herring oils. Fish oil is of the nondrying class and is used for lubricants, leather dressings, soaps, and heat-treating oils, but is also used for blown oils or for fractionating for use in paints and in plastics.

FLAX. A fiber obtained from the flax, or linseed, plant, *Linum usitatissimum,* used for making the fabrics known as linens and for thread, twine, and cordage. It is valued because of its strength and durability. It is finer than cotton and very soft, and the fibers are usually about 20 in (50.8 cm) long. Flax consists of the **bast fibers,** or those in the layer underneath the outer bark, which are of fine texture. The plants are pulled up by the roots, retted, or partly decayed, scraped, and the fibers combed out and bleached in the sun. For the best European flax, the preparation is entirely by hand. The important centers of flax preparation are in Russia, central Europe, Italy, Ireland, France, and Egypt. Some flax is also grown in the United States. The plants that are grown for the oil seed yield a poor fiber and are not employed to produce flax.

FLINT. An opaque variety of **chalcedony** or nearly pure amorphous **quartz** which shows no visible structure. It is deposited from colloidal solution and is an intimate mixture of quartz and opal. It contains 96 to 99% silica and may be colored to dull colors by impurities. Thin plates are translucent. When heated, it becomes white. Flint is finely crystalline. It breaks or chips with a convex, undulating surface. The hardness is Mohs 7, and the specific gravity is 2.6. It was the prehistoric utility material for tools, and was later used with steel to give sparks on percussion. **Gun flints** are still made from a type of flint mined at Brandon, England, for special uses. **Lydian stone,** or **touchstone,** was a cherry flint used for testing gold. Flint is now chiefly used as an abrasive and in pottery and glass manufacture. **Flint paper** for abrasive use contains crushed flint in grades from 20 to 240 mesh, usually coated on one side of 70- or 80-lb (32- or 36-kg) paper. Flint is also used in the form of grinding pebbles. **Potters' flint,** used for mixing in ceramics to reduce the firing and drying shrinkage and to prevent deformation, is ground flint of about 140 mesh made from white **French pebbles. Bitstone** is a name used in the ceramic industry for calcined flint chips ground to the size of wheat, employed for sprinkling on the bottom of the saggers so that the ware will not stick in firing. **Hornstone** is a flint with chalcedony inclusions. It splinters rather than chips and is not used for abrasives.

FLUORINE.   An elementary material, symbol F, which at ordinary temperatures is an irritating pale-yellow gas, F_2. Fluorine gas is obtained by the reduction and electrolysis of fluorspar and cryolite. It has a specific gravity of 1.69 and a boiling point of $-305°F$ ($-187°C$), and it solidifies at $-369°F$ ($-223°C$). It is used in the manufacture of fluorine compounds. It combines violently with water to form hydrofluoric acid, and it also reacts strongly with silicon and most metals. **Liquid fluorine,** at temperatures below $-367°F$ ($-221°C$), is used as an oxidizer for liquid rocket fuels. In combustion, a pound of fluorine produces a pound of hydrogen fluoride which is highly corrosive. Fluorine is one of the most useful of the halogens.

The gas **sulfur hexafluoride,** SF_6, resembles nitrogen in its inactivity. It is odorless, colorless, nonflammable, nontoxic, and 5 times as heavy as air. It is used as a refrigerant, as a dielectric medium in high-voltage equipment, as an insecticide propellant, and as a gaseous diluent. It remains stable to 1472°F (800°C). **Aluminum fluoride,** AlF_3, is a white crystalline solid used in ceramic glazes and for fluxing nonferrous metals. Other metallic fluorides are marketed for special purposes. **Silver difluoride,** AgF_2, is a blackish powder used as a fluorinating agent; it contains about 26% fluorine. **Chlorofluorine gas** is a violent fluorinating agent and is used for the fluorination of some metals that are otherwise difficult to separate, such as uranium.

Chlorine- and bromine-free **perfluorocarbon fluids,** of 3M, are intended for spotfree drying of metal parts, replacing ozone-depleting chlorofluorocarbons. At a boiling temperature of 121°F (50°C), the fluids remove deionized water applied after an aqueous or semiaqueous wash and remove contaminants trapped in the water residue. They also can be used as heat-transfer media and for solubilizing perfluoropolyether and chlorotrifluoroethylene oils, lubricants, and greases. However, because they absorb infrared energy and thus may contribute to global warming, the company recommends that the fluids be used in enclosed systems. The fluids are derived from common organic compounds by replacing all carbon-bound hydrogen atoms with fluorine atoms so that the fluids are fully fluorinated.

FLUOROCARBONS.   Compounds of carbon in which fluorine instead of hydrogen is attached to the carbon atoms. They range from gases to solids. When not less than two fluorine atoms are attached to a carbon atom, they are very firmly held, and the resulting compounds are stable and resistant to heat and chemicals. Fluorocarbons may be made part hydrocarbon and part fluorocarbon, or they may contain chlorine. The fluorocarbons used as plastic resins may contain as much as 65% fluorine and also chlorine, but are very stable. **Liquid fluorocarbons** are used as heat-transfer agents, hydraulic fluids, and fire extinguishers. Benzene-base fluorocarbons are used for solvents, dielectric fluids, and lubricants and for making dyes, germicides, and drugs. Synthetic lubricants of the fluorine type consist of solid particles of a fluorine polymer in a high-molecular-weight fluorocarbon liquid. Chlorine reacts with fluorocarbons to form **chlorofluorocarbons,** commonly referred to as **CFCs.** CFC 11 is used as a foam-blowing agent, and CFC 12 is employed as a refrigerant. CFC 113 is a degreasant in semiconductor manufacturing. Because they are strong depletants of stratospheric ozone, the use of CFCs as aerosol propellants has been banned in the United States since 1978, and is being phased out in Europe. Alternatives to CFCs are being sought for other applications by partially substituting the chlorine with other elements. FC 134a is a fluorocarbon from Du Pont Co. CFC 22, which has 95% less ozone-depleting capacity than CFC 12, is a potential candidate from Allied-Signal. Pennwalt Corp. offers a CFC blend called **Isotron 142b/22** that can replace CFC 11 and CFC 12 in specific applications. The firm is also testing CFC 141b for use in foam insulation.

FLUOROPLASTICS. Also termed **fluoropolymers, fluorocarbon resins,** and **fluorine plastics.** A group of high-performance, high-price plastics. They are composed basically of linear polymers in which some of or all the hydrogen atoms are replaced with fluorine, and they are characterized by relatively high crystallinity and molecular weight. All fluoroplastics are natural white and have a waxy feel. They range from semirigid to flexible. As a class, they rank among the best of the plastics in chemical resistance and elevated-temperature performance. Their maximum service temperature ranges up to about 500°F (260°C). They also have excellent frictional properties and cannot be wet by many liquids. Their dielectric strength is high and is relatively insensitive to temperature and power frequency. Mechanical properties, including tensile creep and fatigue strength, are only fair, although impact strength is relatively high.

There are three major classes of fluoroplastics. In order of decreasing fluorine replacement of hydrogen, they are **fluorocarbons, chlorotrifluoroethylene,** and **fluorohydrocarbons.** There are two fluorocarbon types: **tetrafluoroethylene (PTFE** or **TFE)** and **fluorinated ethylene propylene (FEP).** PTFE is the most widely used fluoroplastic. It has the highest useful service temperature, 500°F (260°C), and chemical resistance. FEP's chief advantage is its low-melt viscosity, which permits it to be conventionally molded. **Zymaxx** composite, of Du Pont, is a carbon-fiber-reinforced PTFE produced in stock shapes and custom-machined parts for applications in the chemical, glass, and petroleum processing industries. The material combines resistance to chemicals, compressive creep, and wear with toughness and dimensional stability.

Teflon, of Du Pont, is a tetrafluoroethylene of specific gravity up to 2.3. It was discovered by Du Pont scientist Dr. Roy J. Plunkett in 1938 and has since become a household word, primarily because of its "nonstick" surface. The tensile strength is up to 3,500 lb/in^2 (23.5 MPa), elongation 250 to 350%, dielectric strength 1,000 V/mil (39.4 × 10^6 V/m), and melting point 594°F (312°C). It is water-resistant and highly chemical-resistant. **Teflon S** is a liquid resin of 22% solids, sprayed by conventional methods and curable at low temperatures. It gives a hard, abrasion-resistant coating for such uses as conveyors and chutes. Its temperature service range is up to 400°F (204°C). **Frelon** is a Teflon compound used to line aluminum bearings. **Teflon fiber** is the plastic in extruded monofilament, down to 0.01 in (0.03 cm) in diameter, oriented to give high strength. It is used for heat- and chemical-resistant filters. **Teflon tubing** is also made in fine sizes down to 0.10 in (0.25 cm) in diameter with wall thickness of 0.01 in (0.03 cm). **Teflon 41-X** is a collodial water dispersion of negatively charged particles of Teflon, used for coating metal parts by electrode-position. **Teflon FEP** is **fluorinated ethylenepropylene** in thin film, down to 0.0005 in (0.001 cm) thick, for capacitors and coil insulation. The 0.001-in (0.003-cm) film has a dielectric strength of 3,200 V/mil (126 × 10^6 V/m), tensile strength of 3,000 lb/in^2 (20 MPa), and elongation of 250%.

Chlorotrifluoroethylene (**CTFE** or **CFE**) is stronger and stiffer than the fluorocarbons and has better creep resistance. Like FEP and unlike PTFE, it can be molded by conventional methods.

The fluorohydrocarbons are of two kinds: **polyvinylidene fluoride** (**PVF$_2$**, or **PVDF**) and **polyvinyl fluoride** (**PVF**). While similar to the other fluoroplastics, they have somewhat lower heat resistance and considerably higher tensile and compressive strength. Like PTFE, PVDF is used for coatings and linings of chemical processing equipment. The fully fluorinated PTFE is more chemically inert and thermally stable than the partially fluorinated PVDF. However, PVDF is easier to process and has better mechanical properties. PVDF of Solvay Polymers has a tensile strength of 7,250 lb/in^2 (50 MPa) and 12% elongation. The flexural strength is 13,650 lb/in^2 (90 MPa), and flexural modulus is 478,500 lb/in^2 (3,296 MPa). The melting point is about 352°F (178°C), and the continuous-use temperature is 302°F (150°C). **Kynar** is Pennwelt's PVF$_2$ film, and **Tedlar** is Du

Pont's PVF film. **Ultra-Cap,** of Geon Co., is an alloy based on PVF_2 that can be coextruded into thin capstocks.

Except for PTFE, the fluoroplastics can be formed by molding, extruding, and other conventional methods. However, processing must be carefully controlled. Because PTFE cannot exist in a true molten state, it cannot be conventionally molded. The common method of fabrication is by compacting the resin in powder form and then sintering. **Parofluor,** a fluorinated polymer of Parker Hannifin Corp., combines the chemical resistance of PTFE with the flexibility of fluorocarbon elastomers and is used for O rings and gaskets.

The following three fluoroplastics are melt-processible: **Perfluoroalkoxy** (**PFA**) can be injection-molded, extruded, and rotationally molded. Compared to FEP, PFA has slightly greater mechanical properties at temperatures over 300°F (150°C) and can be used up to 500°F (260°C). PFA is sold under the trade names **Teflon** (PFA) and **Neoflon.**

Ethylene-chlorotrifluoroethylene (**ECTFE**) copolymer resins also are melt-processible with a melting point of 464°F (240°C). Their mechanical properties—strength, wear resistance, and creep resistance in particular—are much greater than those of PTFE, FEP, and PFA, but their upper temperature limit is about 330°F (165°C). ECTFE also has excellent property retention at cryogenic temperatures. ECTFE is sold under the trade name **Halar.**

Ethylene-tetrafluoroethylene (**ETFE**) copolymer resin is another melt-processible fluoroplastic with a melting point of 518°F (270°C). It is an impact-resistant, tough material that can be used at temperatures ranging from cryogenic up to about 355°F (179°C). ETFE is sold under the trade names **Tefzel, Halon ET,** and **Neoflon.**

Fluorocarbon powder of 39-μin (1-μm) particle size is used as a dry lubricant or for incorporation into rubbers, plastics, and lubricating greases. Tetrafluoroethylene powder of fine particle size, 984 to 1,181 μin (25 to 30 μm), is used to mold parts that have tensile strengths to 6,500 lb/in^2 (44.5 MPa).

Fluorothene plastic has the formula $(CF \cdot CFCl)_n$, differing from Teflon in having one chlorine atom on every unit of the polymer chain, replacing the fourth fluorine atom. It is transparent, and molded parts have a specific gravity of 2.1, a tensile strength of 9,400 lb/in^2 (65 MPa), and high dielectric strength; and it will withstand temperatures to 300°F (149°C). **KEL-F,** of 3M, is **chlorotrifluoroethylene** used for moldings, gaskets, seals, liners, diaphragms, and coatings. The molded parts have high chemical resistance. The compressive strength is 30,000 lb/in^2 (210 MPa), but it can be heat-treated to increase the compressive strength to 80,000 lb/in^2 (560 MPa). The tensile strength of the molded material is 5,000 lb/in^2 (34

MPa), but oriented fibers have tensile strength to 50,000 lb/in^2 (344 MPa). **Fluorocarbon rubber** produced by this company for tubing, gaskets, tank linings, paints, and protective clothing has a tensile strength of 3,000 lb/in^2 (21 MPa), elongation of 600%, heat resistance to 400°F (204°C), and high resistance to oils and chemicals. It is a saturated fluorocarbon polymer containing 50% fluorine. **Aclar,** of Allied-Signal, Inc., is chlorotrifluorethylene transparent packaging film, which is exceptionally resistant to oils and chemicals, has a moisture-barrier efficiency 400 times that of polyethylene film, has good strength to 390°F (199°C), and retains its flexibility to −300°F (−184°C). It is also used for wire covering.

FLUORSPAR.   Also called **fluorite.** A crystalline or massive granular mineral of composition CaF_2, used as a flux in the making of steel, for making hydrofluoric acid, in opalescent glass, in ceramic enamels, for making artificial cryolite, as a binder for vitreous abrasive wheels, and in the production of white cement. It is a better flux for steel than limestone, making a fluid slag and freeing the iron of sulfur and phosphorous. About 5.54 lb (2.5 kg) of fluorspar is used per ton (0.91 metric ton) of basic open-hearth steel.

Fluorspar is mined in Illinois, Kentucky, Nevada, and New Mexico. United States ore usually runs 35 to 75% CaF_2, but high-grade ore from Spain and Italy contains up to 98%. The specific gravity is 3.18, Mohs hardness 4, and the colors light green, yellow, rose, or brown. When ground, the color is white. The melting point is 1650°F (899°C). The usual grades for fluxing are smaller than 0.5 in (1.27 cm) and contain 85% minimum CaF_2, with 5 maximum SiO_2. High-grade fluorspar for ceramic frit has 95 to 98% CaF_2, 3 maximum SiO_2, and 0.12 maximum Fe_2O_3 and is known as No. 1 ground. **Acid spar** is a grade used in making hydrofluoric acid. It contains over 98% CaF_2 and 1 maximum SiO_2 and is produced by flotation. It is also used for making refrigerants, plastics, and chemicals and for aluminum reduction. **Optical fluorspar** is the highest grade but is not common. **Fluoride crystals** for optical lenses are grown artificially from acid-grade fluorspar. Pure **calcium fluoride,** Ca_2F_6, is a colorless crystalline powder used for etching glass, in enamels, and for reducing friction in machine bearings. It is also used for ceramic parts resistant to hydrofluoric acid and most other acids. **Calcium fluorite** has silicon in the molecule, $CaSiF_6 \cdot 2H_2O$, and is a crystalline powder used for enamels. The clear rhombic fluoride crystals used for transforming electric energy to light are **lead fluoride,** PbF_2.

FLUX.   A substance added to a refractory material to aid in its fusion. A secondary action of a flux, which may also be a primary reason for its use, is as a reducing agent to deoxidize or decompose impurities and remove them as slags or gases. In soldering, a flux may serve to remove oxides from the surface to be soldered. Materials such as charcoal or impure boron carbide used to cover baths of molten metals may also be considered as fluxes. **Fluxing stone** is a common term for the limestone or dolomite used in the melting of iron. About 900 lb (408 kg) of limestone is employed for every long ton (0.91 metric ton) of pig iron produced in the blast furnace. If iron ore were reduced without a basic flux, the silica and alumina would unite with the iron oxides to form double silicates of iron and alumina, and there would be a heavy loss of iron. With the addition of limestone, the silica and alumina, having strong affinity for the lime and magnesia, form compounds that contain very little iron. These compounds form a liquid slag which floats on the surface of the molten iron and can be removed readily. The flux also removes sulfur and phosphorus from the iron. Some iron ores contain sufficient lime carbonate to be almost self-fluxing. Lime is more effective as a flux than limestone, but is more expensive. The action of the blast furnace is first to convert the limestone to lime. Upon being heated to 1525°F (829°C), limestone breaks down to lime, which then begins fusion with silica to form the slag at about 2600°F (1427°C). Limestone for use as flux must be fairly pure, or additional undesirable compounds will be formed. For brass, bronze, or soft white metals, resins may be used, and the covering flux may be charcoal, salt, or borax. Cryolite is a flux for aluminum and for glass.

Fluxing alloys for brasses and bronzes are phosphor tin, phosphor copper, or silicon copper. They deoxidize the metals at the same time that alloying elements are added. For tinning steel, palm oil is used as a flux. For ordinary soldering, zinc chloride is a common flux. Tallow, rosin, or olive oil may also be used for soldering. Acetamide is used for soldering painted metals. For silver braze filler metals, borax is a common flux. For soldering stainless steel, borax is mixed with boric acid, or pastes are made with zinc chloride and borax. Borax may also be used as a welding flux. **White flux** is a mixture of sodium nitrate and nitrite and is a strong oxidizer used for welding. Other fluxes used in brazing contain potassium chloride, lithium fluoride, boric acid, borates, and fluoroborates.

Welding fluxes for high-temperature welding are usually coated on the rod and contain a deoxidizer and a slag former. **Lithium fluoride,** LiF, is a powerful flux with the fluxing action of both lithium and fluorine, and it gives a low-melting-point liquid slag. Deoxidizers may be ferromanganese or siliconmanganese. **Slag formers** are titanium dioxide, magnesium carbonate, feldspar, asbestos, or silica. Soluble silicate is a binder, while cellulose may be used for shielding the arc.

FOAM MATERIALS. Materials with a spongelike, cellular structure. They include the well-known sponge rubber, plastic foams, glass foams, refractory foams, and a few metal foams. Ordinary chemically blown **sponge rubber,** is made up of interconnecting cells in a labyrinthlike formation. When made by beating latex, it may show spherical cells with the porous walls perforated by the evaporation of moisture. It is also called **foam rubber.** Special processes are used to produce celltight and gastight **cellular rubber** which is nonabsorbent. **Unicel ND,** of Du Pont, used as a blowing agent for sponge rubber, is dinitropentamethyl tetramine. It is mixed into the rubber, and in the presence of the rubber acids it is decomposed, liberating gas during the vulcanization to form small cells. The cellular rubber of Uniroyal, Inc., produced in sheets of density 3.5 to 12 lb/ft^3 (56 to 192 kg/m^3) for refrigeration insulation, is made with a chemical that releases nitrogen gas to produce innumerable microscopic cells during the molding. **Rubatex** of Rubatex Corp. is this type of cellular rubber. It comes in sheets of any thickness for gaskets, seals, weather stripping, vibration insulation, and refrigerator insulation.

Nitrogen-filled rubber is used in the form of insulation board. The 5.5 lb/ft^3 (88 kg/m^3) board has a crushing strength of 33 lb/in^2 (0.23 MPa). Under the name of **Royal insulation board** it is marketed in thicknesses from 0.75 to 1.5 in (1.9 to 3.8 cm), but most of the so-called sponge rubbers are not made of natural rubber but are produced from synthetic rubbers or plastics and may be called by a type classification, such as **urethane foam,** or marketed under trade names. Some of these foams may be made by special processes, and some of the materials are marketed in liquid form for use as foamed-in-place insulation. The urethane foams expanded with fluorocarbon gas are 95% gas in closed microscopic pores. They are rigid, have low weight, and are used in sheets with metal, paper, felt, or other facings for wallboard and roofing, and in thermal insulation. The contained gas has very low thermal conductivity. **Phenolic foam** is made by incorporating sodium bicarbonate and an acid catalyst into liquid phenol resin. The reaction liberates carbon dioxide gas, expanding the plastic.

Cellular cellulose acetate is expanded with air-filled cells to densities from 4 to 9 lb/ft^3 (64 to 144 kg/m^3) for use as insulation and as a buoyancy materials for floats. It is tough and resilient. **Strux,** of Strux Corp., is **cellulose acetate foam,** made by extruding the plastic mixed with barium sulfate in an alcohol-acetone solvent. When the pressure is removed, it expands into a light, cellular structure. **Foamex** is a foam rubber made from synthetic latex in several density grades. It is stronger than that made from natural rubber and is flexible at very low temperatures. **Polystyrene foam** is widely used for packaging and for building insulation. It is available as prefoamed board or sheet or as beads that expand when heated. Densities range from 1 to almost 5 lb/ft^3 (16 to 80 kg/m^3).

Styrofoam, of Dow Chemical USA, is polystyrene expanded into a multicellular mass 42 times the original size. It has only one-sixth the weight of cork, but will withstand hot water or temperatures above 170°F (77°C), as it is thermoplastic. It is used for cold-storage insulation and is resistant to mold. **Pyrofoam,** composed of expanded obsidian, a dark, hard, glassy volcanic rock, is slightly heavier than styrofoam and is extremely fire-resistant. It has a closed-cell structure and a low and stable rate of thermal expansion, and it can be formed into heat-resistant panels. **Polyethylene foam** and **polypropylene foam** also are available. Compared to expanded polystyrene, expanded polyolefins have greater toughness and can be molded more easily. Current uses include automotive bumper cores, sunvisor cores, and electronic packaging. Although many plastic foams require additives for flame retardancy, **polyimide** foams are inherently flame-retardant. Imi-Tech's **Solimide** has a broad temperature range, retaining flexibility at −300 to 500°F (−184 to 260°C).

Ensolite, of Uniroyal, and **Vinylaire** are **foamed vinyl plastisols. Expanded polyvinyl chloride** molded to densities from 2 to 12 lb/ft^3 (32 to 192 kg/m^3) is used for floats, buoys, and insulation. Vinyl plastisol is sprayed on and cured by heat into a foamed texture with an increase in volume of 400%. It is white and has an insulating K value of 0.36, or about the same as felt. It is used for gaskets, seals, and refrigerator doors. **Disperseplast 1150,** of Byk-Chemie USA, is a **dispersing agent** for blending with fillers and pigments used in polyvinyl chloride plastisols. The compound is a polar acidic ester of a long-chain alcohol, resists yellowing, and is especially effective for dispersing azodicarbonamide, a blowing agent for PVC foam. **Polyester foam** is odorless, flame-resistant, and resistant to oils and solvents. It has only half the weight of foamed rubber with greater strength and high resistance to oxidation. It is used for upholstery and insulation.

Vinyl foams are widely used. They are made from various types of vinyl resins with the general physical properties of the resin used. Open-cell vinyl foam contains interconnecting voids and is very flexible. It is made by mechanical foaming by absorption under pressure of an inert gas in a vinyl plastisol. It is used for furniture and transport seating. Closed-cell vinyl foam contains separate, discrete voids. It is made by chemical foaming, using chemical blowing agents. It is used for impermeable insulation and marine floats.

Silicone foam, of Dow Corning Corp., used for insulation, is silicone rubber foamed into a uniform unicellular structure of 8 to 24 lb/ft^3 (128 to 384 kg/m^3) density. It will withstand temperatures above 600°F (316°C). For structural sheets the rubber is foamed between two sheets of silicone glass laminate.

Epoxy foam comes as a powder consisting of an epoxy resin mixed with diaminodiphenyl sulfone. When the powder is placed in a mold or in a cavity and heat is applied, it foams to fill the space and cures to a rigid foam. The foam has a density of 16 lb/ft^3 (256 kg/m^3), tensile strength of 360 lb/in^2 (2.5 MPa), and compressive strength of 710 lb/in^2 (4.9 MPa); and it will withstand temperatures to 500°F (260°C). Pour-in-place urethane foams expand with a fluorocarbon to a density of 2 lb/ft^3 (32 kg/m^3). The insulation K factor is 0.13 Btu (137 J), and they retain their properties at subzero temperatures when used for freezer insulation. Polyether-based urethanes expand to a density of 2 lb/ft^3 (32 kg/m^3) into stable, rigid foam of good strength for refrigerator insulation. **Polymethylene polyphenyl isocyanate foam** is infusible and withstands temperatures to 900°F (482°C) before beginning to carbonize. It is used in aircraft and missile sandwich structures. Urethane-type foams with a high chlorine content are fire-resistant. Their thermal conductivity K factor is 0.10, and they are used for refrigerator and building insulation.

As for **foaming agents,** also called **blowing agents,** use of chlorofluorocarbons is being phased out due to their potential for depleting the stratosphere of protective ozone. Substitutes include less-ozone-depleting hydrochloroflurocarbons (HCFCs), non-ozone-depleting hydrofluorocarbons, pentane, cyclopentane, carbon dioxide, water, and other agents. Du Pont's **Formacel Z-2** is HCFC-152. Methylene chloride, once widely used as an auxiliary blowing agent for flexible polyurethane foam, has been banned or restricted as a suspected carcinogen. Acetone, another such agent, is more efficient, but there are concerns regarding air pollution and fire hazard. Despite the need for explosionproof equipment, pentanes have become the choice in Europe for rigid polyurethane foam in refrigerators and freezers. In the United States, however, pentanes do not meet regulations regarding emission of volatile organic compounds. Water is the blowing agent for **Waterlily** polyurethane foam, a slabstock product of TCT Polyurethanes in Belgium for seating applications. With 10% melamine-type flame retardant, it meets stringent ignition tests. This foam is based on diphenylmethane diisocyanate, and its molecular weight is about 10 times that of low-density toluene-diisocyanate-based foam. Also, it takes but 1 to 2 h to cure versus 1 to 2 days for the TDI foam, no measurable isocyanate vapors are emitted, and no ozone-depleting agents are used. Dow Plastics uses 0.5 to 6% carbon dioxide to extrude polystyrene foam sheet; no other additives are required. Axiomatics Corp uses supercritical carbon dioxide to foam plastics. In this state, the carbon dioxide permits rapid dissolution of the polymer in the melt phase, yielding foams of high cell density.

Several plastics are available as expandable **foam beads:** polystyrene, polypropylene, and blends of polyphenylene oxide/polystyrene and polyphenylene ether/polystyrene (PPE/PS). **Gecet,** of General Electric Plastics, is a line of PPE/PS with heat-distortion temperatures of 220 to 290°F (104 to 143°C). Foron dkk-Scharfenstein of Germany uses polystyrene pellets foamed with steam as an alternative to polyurethane foamed with CFCs for the wall insulation of small refrigerators. **Polycor 267,** a foaming agent of Environmental Products available in pellets and intended for polypheylene oxide foam, combines the good surface properties of an endothermic chemical agent with the high-pressure and rapid degassing characteristics of exothermic agents. **Excerol LBA-39,** a liquid endothermic/exothermic blend of Hanley Div. of Boehringer Ingleheim, is for olefinics and styrenics. Other LBA grades are for acrylonitrile butadiene styrene, polycarbonate, and other plastics.

Many thermoplastics can be used to make **structural foams:** thick [0.25-in (6.4-mm) or thicker] moldings having thin, rigid, solid skins, or facings, encasing a rigid cellular core. They are noted for high rigidity-to-weight ratios and better thermal and sound insulation than solid moldings. Applicable plastics include acrylonitrile butadiene styrene, modified phenylene oxide, nylon, polycarbonate, polyetherimide, high-density polyethylene, polyester, polypropylene, polystyrene, and polyurethane. Mechanical properties depend mainly on the base resin, density, and amount of reinforcement. Foamed to a 10% by weight reduction relative to solid moldings and 0.25 in (6.4 mm) thick, General Electric Plastics' Lexan polycarbonate can provide flexural moduli of 280,000 to 1×10^6 lb/in^2 (1,900 to 6,900 MPa), tensile yield strengths of 7,000 to 10,500 lb/in^2 (48 to 72 MPa), and unnotched Izod impact strengths of 8 to 50 ft · lb/in (427 to 2,670 J/m).

Glass foam is used as thermal insulation for buildings, industrial equipment, and piping. **Ceramic foams** of alumina, silica, and mullite are used principally for high-temperature insulation. **Aluminum foam** is a **metal foam** that has found appreciable industrial use as a core material in sandwich composites. **Foamed zinc,** of Foamalum Corp., is a lightweight structural metal with equal strength in all directions, made by foaming with an inert gas into a closed cell structure. It is used particularly for shock and vibration insulation. Foaming agents for metals are essentially the same as those used for plastics. They are chemical additives that release a gas to expand the material by forming closed bubbles. Or they may be used to cause froth as in detergents or fire-fighting foams.

FOIL. Very thin sheet metal used chiefly for wrapping, laminating, packaging, insulation, and electrical applications. **Tinfoil** is higher in cost than some other foils, but is valued for wrapping food products because it is not poisonous; it has now been replaced by other foils such as aluminum. Ordinary tinfoil is made in thicknesses from 0.00024 in (0.006 mm) to 0.00787 in (0.200 mm), the former having 16,037 in^2/lb (22.7 m^2/kg) and the latter 432 in^2/lb (0.612 m^2/kg). Tinfoil for radio condensers has 14,500 in^2/lb (20.6 m^2/kg). An English modified tinfoil, which has greater strength and is nontoxic in contact with foods, contains 8.5% zinc, 0.15 nickel, and the balance tin. It can be rolled to thinner sheets.

Lead foil, used for wrapping tobacco and other nonedible products, is rolled to the same thickness as tinfoil, but because of its higher specific gravity gives less coverage. The thinnest has 10,358 in^2/lb (14.7 m^2/kg), and the thickest 279 in^2/lb (0.4 m^2/kg). Lead foil has a dull luster, but it may be modified with some tin and other elements to give it a brighter color. **Stainless steel foil** is produced in thicknesses from 0.002 to 0.015 in (0.005 to 0.038 cm) for laminating and for pressure-sensing bellows and diaphragms. Type 302, for facing, comes in a thickness of 0.003 in (0.008 cm) highly polished, in rolls 24 in (61 cm) wide. **Steel foil** is carbon steel coated with tin in sheets as thin as 0.001 in (0.003 cm) with widths to 40 in (102 cm). It is easily formed and soldered and is used for strong packaging and laminating.

Aluminum foil has high luster, but is not as silvery as tinfoil. The thin foil usually has a bright side and a matte side because two sheets are rolled at one time. Aluminum foil also comes with a satin finish, or in colors or embossed designs. It is made regularly in 34 thicknesses, from 0.0002 in (0.0006 mm), having 43,300 in^2/lb (61 m^2/kg), to 0.0078 in (0.200 mm), having 1,169 in^2/lb (1.7 m^2/kg), but can be made as thin as 60,000 in^2/lb (85 m^2/kg). The most used thickness is 0.00035 in (0.0089 mm), with 29,300 in^2/lb (42 m^2/kg). For electrical use the foil is 99.999% pure aluminum, but foil for rigid containers is usually **aluminum alloy 3003,** with 1 to 1.5 manganese, and most other foil is of **aluminum alloy 1145,** with 99.45% aluminum. The tear resistance of thin aluminum foil is low, and it is often laminated with paper for food packaging. **Trifoil** is aluminum foil coated on one side with Teflon and on the other wide with an adhesive. It is used as coatings for tables and conveyors or liners in chemical and food plants. Since polished aluminum reflects 96% of radiant heat waves, this foil is applied to building boards or used in crumpled form in walls for insulation. **Alfol** is a name applied to crumpled aluminum foil for this purpose by the British National Physical Laboratory. Aluminum foil cut in tiny strips has been used for scattering from aircraft to confuse radar detection. A bundle of 6,000 such strips weights only 6 oz (0.2 kg) and scatters widely.

Aluminum yarn, for weaving ribbons, draperies, and dress goods, is made from aluminum foil, 0.001 to 0.003 in (0.003 to 0.008 cm) thick, by gang-slitting to widths from 0.0125 to 0.125 in (0.0317 to 0.3175 cm) and winding the thread on spools. **Gold foil** is called **gold leaf** and is not normally classified as foil. It is used for architectural coverings and for hot-embossed printing on leather. It is made by hammering in books and can be made as thin as 0.0000033 in (0.0000083 cm), 0.0022 lb (1 g) of gold covering 5.184 in^2 (3.4 m^2). Usually, gold leaf contains 2% silver and copper for hardening.

Metal film, or **metal foil,** for overlays for plastics and for special surfacing on metals or composites, comes in thicknesses from flexible foils as thin as 0.002 in (0.005 cm) to more rigid sheets for blanking and forming casings for intricately shaped parts. **Hot stamping foils** are decorative foils on a disposable carrier film applied to parts by means of a heated die. The carrier is usually polyester or cellophane. There are many types of these foils, including **metallic pigment foils, printed foils,** and **vacuum metallized foils. Composite metal films** come in almost any metal or alloy, such as film of tungsten carbide in a matrix of nickel alloy for wear-resistant overlays. They are made by rolling and sintering 325-mesh powder. The foils may be cemented to the substrate with an epoxy resin, but for operating temperatures above 400°F (204°C), brazing or welding is needed. **Metal overlays** give a smoother and more uniform surface than is usually obtained by flame-spray or chemical deposition.

FOLIC ACID. **Sodium folate** ($C_{19}H_{18}N_7NaO_6$), a yellow to yellow-orange liquid used in medicine for folic acid deficiency. It is also known as **folic acid sodium salt, sodium pteroylglutamate, folacin** and, in its biologically active form, **folate.** Discovered in 1941 in green, leafy vegetables, its name derives from the Latin word for leaf, *folium.* It is a B vitamin known to prevent spina bifida and anencephaly, devastating birth defects, and may be effective in preventing cardiovascular diseases and common cancers. Foods rich in the nutrient include chicken liver and various grains, cereals, beans, vegetables, nuts, and fruit juices.

FOODSTUFFS. A great group of materials employed for human consumption, while those employed for feeding animals are called **feeds.** Foodstuffs are derived mainly from vegetable and animal life, but some, like common salt, are produced from mineral sources, and others may be entirely synthesized. Foodstuffs are intended primarily for the maintenance and growth of the body, and technically they could include drugs which are taken primarily for their physiological effects, and cosmetics that feed the skin and hair. Tobacco, used also for its physiological effect, is classified with foodstuffs in government statistics.

Proteins, carbohydrates, and fats are called the *essential foodstuffs,* and about 1.5 lb (0.7 kg) total of these is considered as needed daily for the human system. But these alone are inadequate for metabolism. Almost every element is required directly, or indirectly as catalysts, in the building up and maintenance of the innumerable highly complex, chemical compounds in the human body, and the form in which they are taken into the body is of great importance. Iodine, for example, is an essential element in the thyroid gland, with minute amounts also required in every cell of the body; but if it is taken directly into the system in even small amounts, it is an intense poison.

Some complex chemical compounds, required for proper health, cannot be synthesized in the human body and must be taken in through the eating of foodstuffs containing them. First-class proteins are essential for full health, but they are not synthesized in the human body and are not available in vegetable products, so they must be obtained from the eating of meats from animals that synthesize them. About one-third of the protein required daily should come from animal sources. Fish and shellfish can supply the essential amino acids, but abnormally large quantities would be necessary.

Concentrated foods are now used in vast quantities for military supplies and for prepared food mixes. They may be spray-dried powders, comminuted dehydrated vegetables or fruits, or freeze-dried cooked foods. For example, dehydrated onions can be handled in automatic metering equipment for adding to soup mixes or meat dishes, and 1 lb (0.45 kg) of the dried onions equals 8 lb (3.6 kg) of fresh onions. For mixes, cooking times are balanced by partial-pressure cooking or puffing. Celery is cross-cut in fine flakes and dehydrated. Rehydration gives 25-to-1 volume. For use uncooked, it requires only immersion in cold water. **Concentrated wine,** used in flavoring foods, is made by evaporating sherry or other wines. One ounce (0.03 kg) is equal to 8 oz (0.23 kg) of wine, but it contains no alcohol. **Banana crystals,** for flavoring bakery products and milk-based beverages, are a light-tan crystalline powder made by vacuum dehydration of bananas, the fruit of the tall, treelike herbaceous plant *Musa paradisiaca,* native to southern Asia but now grown in all tropical countries. There are more than 300 varieties of the plant. The powder contains 50% banana solids and 50 corn-syrup solids, while the natural banana contains 25 solids and 75 water. The powder crystals are easily soluble in water, reconstituting to 80% banana solids and 20 corn-syrup solids, giving a true banana flavor and the natural food value. For food manufacturers bananas are also marketed in canned form, either sliced or mashed.

For efficient high-production processing of bakery, confectionery, and other food products, most types of fruits and flavors are now available in the form of dry powders, nuggets, pastes, solutions, or cooked or semicooked sections. Advantages are economy in shopping and storing, ease of handling in automatic equipment, and uniformity of quality. For example, **fruit nuggets** consisting of foamed particles are easily metered in automatic machines. Nuggets of berries contain no seeds. For dry-mix packaging the flavors may be in capsule.

Caloric value is only an imperfect measurement of food value. The trace percentages of mineral compounds and such chemical compounds as vitamins are of vital importance to proper metabolism and health. **Therapeutic foods** are mixtures, usually in powder form to be mixed in water, to give a balance of proteins, fats, and carbohydrates for food satisfaction without overeating. They may be designed for reducing excessive weight, for adding weight, or as a soft diet for invalids or babies, and they are usually intended to contain sufficient caloric value for a rounded diet. But it is extremely difficult to obtain all the essential ingredients in a synthetic food. Iron is a common essential, readily absorbed from many natural foods, but as an additive for deficient systems it must be in chemical compounds that yield ferrous iron to the system. Trace quantities of zinc in enzymatic proteins are necessary to catalyze the reaction of carbonic anhydrase to sustain life, and they are also needed in the liver and in the eyes. Other metals, such as cobalt and rubidium, are also necessary, and the highly complex processes are as yet only imperfectly understood.

Frozen foods are frozen by convection or conduction. Convection, used for irregularly shaped products, such as whole poultry, and loosely packaged products, involves blowing cold air over the packaged products as they move on a conveyor. The conduction method, developed by Charles Birdseye, is used for foods packaged in straight-side boxes. The boxes are slightly compressed by hollow metal plates through which a refrigerant runs.

Disease-resistant or herbicide-tolerant strains of food crops can be made by inserting genetic material into **plant cells.** Methods include microinjection with micromanipulators; ballistically firing **gold particles** coated with the genetic material; and brief agitation by a mixer of a test tube containing some 1,000 plant cells, the genetic material, and about 10,000 **silicon carbide whiskers.** In the latter method, developed by Zeneca Plc's Seed Research Center and the Seed Research Station in the United Kingdom, collisions between the cells and whiskers create pores through which the genetic material diffuses into the cells. **Disease-resistant corn** has been produced this way.

FORMALDEHYDE. Also called **methylene oxide.** A colorless, poisonous gas of composition HCHO, boiling at $-5.8°F$ ($-21°C$). It is very soluble in water and is marketed as a 40% solution by volume, 37 by weight, under the name of **formalin.** The commercial formalin is a clear, colorless liquid with a specific gravity of 1.075 to 1.081. When shipped by tank cars, it contains 11 to 12% methanol, or 6 to 7 when shipped in drums, as a stabilizer to prevent precipitation of polymerized formaldehyde. The material is obtained by oxidation from methyl alcohol. It is used in making plastics, as a reducing agent, as a disinfectant, and in the production of other chemicals, such as ethylene glycol, hexamethylenetetramine, pentaerythritol, and butadiene. **Trioxane** is polymerized formaldehyde, or a ring compound of anhydrous formaldehyde, $(HCHO)_3$. It is marketed as colorless crystals of a pleasant ether-alcohol odor, with a specific gravity of 1.17, melting point of $144°F$ ($62°C$), and boiling point of $239°F$ ($115°C$). It is used as a source of dry formaldehyde gas, as a tanning agent, and as a solvent. It ignites at $235°F$ ($113°C$), burns with a hot, odorless flame, and is used in tablet form to replace solidified alcohol for heating.

Glyoxal, of Union Carbide, is **dialdehyde,** or **ethanedial,** CHO · CHO, marketed as a water solution as a substitute for formaldehyde for resin manufacture and as a hardening and preserving agent, and for treating rayon fabrics. It is faster-acting and has less volatility and odor than formaldehyde. It is a light-yellow liquid boiling at $122°F$ ($50°C$). **Paraformaldehyde,** $(CH_2O)_3$, also called **paraform,** is a white amorphous powder used instead of formaldehyde where a water solution is not desirable. It is used as a catalyst and hardener for the resorcinol and some other synthetic resins, and as an antiseptic. **Formamide,** $H(CO)NH_2$, is a clear, viscous, water-soluble liquid with a faint odor of ammonia, boiling at $410°F$ ($210°C$). It is used as a solvent for metal chlorides and inorganic salts, and for lignin, glucose, or cellulose, and as a softener for glues. **Dimethylformamide,** $HCON(CH_3)_2$, is a colorless, polar liquid used as a solvent for spinning acrylic fibers, polyurethane and polyamide coatings, and dyes. **Formol** is a trade name for a solution of formaldehyde in methanol and water, used as an antiseptic. **Formcel,** of Hoechst Celanese Corp., is a solution of formaldehyde in either methanol or butanol, to replace formalin where a waterfree solution is desired. **Glutar aldehyde,** $O:HC(CH_2)_3 · CH:O$, made from acrolein, has a reaction similar to formaldehyde, and is used as a cross-linking agent in plastics and for insolubilizing starches, casein, and gelatin.

Hexamine is a white, crystalline powder used chiefly for the manufacture of synthetic resins in place of formalin and its sodium hydroxide catalyst. It is formed by the action of formaldehyde and ammonia. It is hexamethylene tetramine, $(CH_2)_6N_4$, melting at $536°F$

(280°C), and is very stable when dry. It is readily soluble in water and in alcohol. It is also known as **formin, ammonio formaldehyde, urotropin, crystogen, aminoform,** and **cystamine.** In pharmacy it is called **methenamine.** In the presence of an acid, it yields formaldehyde and is used in medicine as an internal antiseptic. It is also used as an accelerator for rubber.

FORMIC ACID. Also called **methanoic acid** and **hydrogen carboxylic acid.** Formic acid is the simplest of the organic acids, with composition HCOOH, and was originally distilled from red ants, receiving its name from the Latin name for ants. It is made synthetically or is obtained from the black liquor of sulfite paper mills where it occurs as a sodium salt. It is a pungent, colorless liquid of specific gravity 1.22, boiling point 214°F (101°C), and freezing point 47°F (8.4°C), soluble in water and in alcohols. It is an easily oxidized reducing agent, and small amounts will blister skin and give the stinging sensation of ant or bee bites or nettle stings, all of which are caused by formic acid. The acid has greater reducing action than acetic acid and is thus used in textile furnishing, especially in chrome dyeing of wool. It is also used in leather processing as a dye-bath exhausting agent. Other uses are as a food preservative, in electroplating, as a germicide, as a fermentation assistant in brewing, and as a coagulant for rubber. **Methyl formate,** $HCOOCH_3$, is a white, volatile liquid boiling at 89.2°F (31.8°C), with a pleasant ester odor, soluble in water. It hydrolyzes to form methanol and formic acid and is used as an intermediate, but it is also a good solvent for cellulose esters and for acrylic resins. **Protan,** of Hercules Inc., is a **sodium formate.**

FUEL BRIQUETTES. Also termed **coal briquettes.** Various-shaped briquettes made by compressing powdered coal, usually with an asphalt or starch binder, but sometimes as **smokeless fuel** without a binder. They are sometimes also made waterproof by coating with pitch or coal tar. They have the great advantage over raw coal that they do not take up large amounts of water, as coal does, and thus have uniformity of firing. Fuel briquettes are made from anthracite screenings usually mixed with bituminous screenings, as the bituminous coals require no binders. The usual forms of the briquettes are pillow-shaped, cubic, cylindrical, ovoid, and rectangular, and the usual size is not over 5 oz (0.14 kg). The term **packaged fuel** is used for cube-shaped briquettes wrapped in paper packages, used for hand firing in domestic furnaces. The **coal briquettes** of Blaw-Knox Co. are made with anthracite dust and a small amount of Pocahontas-type bituminous coal with an asphalt binder. The cakes are 3 by 3 by

3 in (7.6 by 7.6 by 7.6 cm) wrapped in kraft paper. **Coal logs** are briquetted, smokeless fuel which consists of high-temperature coke
made by carbonizing Utah low-grade coal, yielding 20 to 40 gal (76 to
151 L) of tar as a by-product per ton (0.91 metric ton) of coal.
Charcoal briquettes for home fuels are charcoal powders pressed
with a starch binder. **Fuelettes,** solid fuel of BioFuels Corp. made
from the low-grade nonrecyclable portion of mixed-paper waste, are
intended for cofiring with coal in industrial boilers. Heating value
ranges from 7,100 to 8,000 Btu/lb (16,500 to 18,600 kJ/kg), and they
are best suited for solid-fuel and fluidized-bed boilers. Because they
are sulfur-free, their use reduces emissions of sulfur dioxide.

FUEL OIL. Distillates of petroleum or shale oil used in diesel engines
and in oil-burning furnaces. True fuel oils are the heavier hydrocarbons in kerosene, but the light or distillate oils are used largely for
heating and the heavy or residual oils for industrial fuels. In some
cases only the light oils, naphtha and gasoline, are distilled from
petroleum, and the residue is used for fuel oil, but this is wasteful of
the lighter oils. Commercial grades of **furnace oil** for household use
and **diesel oil** for trucks may be low grades of kerosene. Fuel oil of
low specific gravity requires preheating to obtain complete atomization. At 10°Bé the minimum temperature to atomize fuel oil is 300°F
(149°C), but at 40°Bé a temperature of only 40°F (4°C) is required.
Fuel oils used in oil burners are 28 to 32°Bé and have a heating value
of 142,000 Btu/gal (537,170 Btu/L), completely atomizing at 90°F
(32°C). **Gas oil,** which receives its name from its use to enrich fuel
gas and increase the luminance of the flame, is also used as a fuel oil
in engines. It is the fraction of petroleum distilling off after kerosene,
or above about 572°F (300°C). It is brownish and has a specific gravity of about 0.850. Ignition temperatures of crude oils vary from 715 to
800°F (379 to 427°C) in air at atmospheric pressure. **Bunker C oil,**
for diesel engines, is a viscous black oil of specific gravity 1.052 to
0.9659 with a flash point above 15°F (−9°C). The National Bureau of
Standards lists six grades of fuel oils with flash points from 100 to
200°F (38 to 93°C). For general comparison of oil and coal, with comparable fire and boiler equipment, 4.5 bbl of Bunker C oil equals 1.1
ton (1 metric ton) of low-ash, highly volatile West Virginia coal.

FUELS. The term normally covers a wide range, since innumerable
organic materials can be used as fuel. Coal, oil, or natural gas or
products derived from them are the basic industrial fuels, but other
materials are basic in special situations, such as sawdust in lumbering areas and bagasse in sugarcane areas. Fuel conversion factors for
industrial fuels are based on the relation to the fuel value of 1.1 ton (1

metric ton) of coal having a caloric value of 12,000 Btu/lb (27,900 kJ/kg). Thus 1.1 ton (1 metric ton) of lignite equals 0.3 ton (0.3 metric ton) of coal; 1.1 ton (1 metric ton) of crude oil equals 1.4 ton (1.3 metric ton) of coal; 1,100 ft^3 (31 m^3) of natural gas equals 1.4 ton (1.3 metric ton) of coal; and 1,000 kWh of hydroelectricity equals 0.4 ton (0.4 metric ton) of coal.

But modern technical reference to fuels generally applies to **high-energy fuels** for jet engines, rockets, and special-use propellants; and the comparisons of these fuels are in terms of **specific impulse,** which is the thrust in pounds per pound of propellant per second. The molecular weight of the products produced by the reaction of a fuel must be extremely low to give high specific impulse, that is, above 400. Hydrogen gives a high specific-impulse rating, but it has very low density in the liquid state and other unfavorable properties, so that it is usually employed in compounds. The initial impulse of a rocket is proportional to the square root of the combustion temperature of the fuel. **Hydrogen fuels** reacted with pure oxygen produce temperatures above 5000°F (2760°C), and some fuels may react as high as 9000°F (4982°C). Aluminum powder or lithium added to hydrogen increases the efficiency. **Boron fuels** in general release 50% more thermal energy than petroleum hydrocarbons. The first Saturn space rocket had kerosene–liquid oxygen in the first stage and liquid hydrogen–oxygen in the following three stages. **Solid rocket fuels,** designed for easier handling, have a binder of polyurethane or other plastic. **Fuel oxidizers,** for supplying oxygen for combustion, may be ammonium, lithium, or potassium perchlorates. In solid fuels, oxidizers make up as much as 80% of the total.

A **monopropellant,** high-energy fuel is a chemical compound which, when ignited under pressure, undergoes an exothermic reaction to yield high-temperature gases. Examples are nitromethane, methyl acetylene, ethylene oxide, and hydrogen peroxide. Gasoline oxidized by hydrogen peroxide gives a specific impulse of 248, while pentaboranes under pressure and oxidized with hydrogen peroxide give a specific impulse of 363. **ASTM fuel A,** for jet engines, is isooctane, and **ASTM fuel B** is isooctane and toluene. **Turbine jet fuel,** or **JP fuel,** has been the naphtha-based **JP-4** for military aircraft and the kerosene-based **JP-8** for commercial aircraft, with JP-4 having a lower flash point, or ignition temperature. However, military aircraft are switching to JP-8 in the interest of greater safety on impact. Besides its higher flash point, flame spread of a pool of JP-8 is much slower. The **naphthalenes,** such as **decahydronaphthalene,** have high thermal stability, and they have a high density which gives high thermal energy per unit volume. **Naphthene,** C_nH_{2n}, is a general term for nonaromatic cyclic hydrocarbons in petroleum.

Hydrazine with liquid oxygen has a specific impulse of 282, and with liquid fluorine as an additive the specific impulse is 316. The rocket fuel **hydyne** is a 60–40 mixture of unsymmetric dimethylhydrazine and diethylene triamine. Other liquid fuels may be this dimethylhydrazine with nitric acid, or with nitrogen tetroxide, or with liquid oxygen. **Alkyl boranes** are used in rockets, but they leave a deposit and they exhaust boric acid in a dense cloud. **Diborane,** B_2H_6, is the simplest compound of boron and hydrogen. It is a gas which burns with high flame speed and high heat. It is used to produce the **boron hydrides** such as **decarborane,** $B_{10}H_{14}$, or $HH(H \cdot BB \cdot H)_5HH$, employed for boron fuels and for making boron plastics.

Sodium boron hydride, a white, crystalline solid of composition $NaBH_4$, made by reacting sodium hydride with methyl borate, is also used to produce the boranes for fuels. **Triethyl borane,** $(C_2H_5)B$, used for jet fuels and as a flame-speed accelerator, is a colorless liquid. It is spontaneously flammable, its vapors igniting with oxygen. Any element or chemical which causes spontaneous ignition of a rocket fuel is called a **hypergolic material.**

Chemical radicals are potential high-energy fuels, as the recombining of them produces high specific impulses. But chemical radicals normally exit only momentarily and are thus not stable materials and, in general, are not commercial fuels. **Ion propellants** operate on the principle that like charges repel each other, and the fuel is an ion-plasma jet actually formed outside the engine. The original fuel is a metal such as cesium from which electrons can be stripped by passing the vapor through a hot screen, leaving positive cesium ions, which are formed into a beam and exhausted from the jet thrust to be electronically neutralized in the ionized plasma.

FULLERENES. Introduced by researchers at Rice University in the mid-1980s as large carbon molecules having 60 or more carbon atoms arranged in cagelike pseudospheres similar in shape to Buckminster Fuller's geodesic dome and thus also called **bucky balls. Carbon 60,** or C_{60}, molecules are a new form of carbon with potential for use in steel, catalysts, lubricants, superconductors, diamond synthesis, monocrystalline film, and as building blocks for high-strength polymers.

Closely related to carbon fullerenes are **carbon nanotabules,** synthesized by graphite vaporization at NEC Corp. of Japan. They consist of concentric graphitic carbon capped with fullerenelike hemispheres that curve by incorporating five-member rings. The tips of the tabules can be opened by an oxidizing agent, creating nanoscale test tubes useful in studying catalysis and conducting experiments.

At AT&T Bell Laboratories, researchers adding alkali metals to **carbon fullerenes** created $(NH_3)_4Na_2CsC_{60}$, a compound which converts from insulator to superconductor at $-405°F$ ($-243°C$). **Noncarbon fullerenes,** created at Iowa State University (Ames), comprise a nearly round outer cage of 70 or 74 indium atoms and successively smaller sodium and indium cages, with a single nickel, palladium or platinum atom in the center. By computer simulation, researchers at Japan's National Institute of Materials and Chemical Research and Nissan Motor Co. have determined the feasibility of creating fullerenelike nitrogen with 60 or 70 nitrogen atoms. **Nitrogen 60,** or N_{60}, would resemble the C_{60} structure but not be stable. It could serve as a source for nitrogen gas and could be useful for rocket fuels and explosives.

Rhondite and **DiaSteel,** commercial products of MicroMet Technology, are described as **metallofullerite composites** of iron encapsulated in a fullerene structure in the form of helix-wound, cablelike crystals. Rhondite 3591 contains 1.12% carbon, 0.89 silicon, 0.15 chromium, 0.12 manganese, and the balance iron. It has a modulus of elasticity of about $33×10^6$ lb/in^2 (228,000 MPa) and, as cast, a hardness of Rockwell C 27 and a compressive strength of 109,600 lb/in^2 (756 MPa). Hardening increases hardness to Rockwell C 62 and compressive strength to 338,900 lb/in^2 (2,337 MPa). The composite features exceptional wear resistance and exhibits self-healing, which improves with wear. DiaSteel is made from Rhondite by a process that converts a high percentage of the fullerenes to diamond. The diamond-embedded steellike structure has a hardness of Rockwell C 65 and is devoid of internal stresses. Further processing can remove the iron, leaving diamond crystals nanometers to a few micrometers in size, which can be used as abrasives or for electronic applications.

FULLER'S EARTH. A soft, opaque clay with a greasy feel used as a filtering medium in clarifying and bleaching fats, greases, and mineral and vegetable oils. It absorbs the basic colors in the organic compounds. It is also used as a pigment extender and a substitute for talcum powder. It was formerly much used in the textile industry as a fuller for woolen fabrics, cleansing them by absorbing oil and grease. It is a hydrated compound of silica and alumina. It may contain 75% silica, 10 to 19 alumina, 1 to 4 lime, 2 to 4 magnesia, and sometimes ferric oxide. The usual color is greenish white to greenish brown. The rose-colored fuller's earth from Florida is a variation of **montmorillonite,** $(MgCa)O \cdot Al_2O_3 \cdot 4SiO_2$. Florida's fuller's earth marketed by the Floridian Co. under the name of **Floridin** is a grayish-white material graded by sizes from B, which is 16 to 30 mesh, to XXX, which will pass 90% through a 200-mesh screen. **Florex** is this material processed by extrusion to increase the absorption capacity. A typical analysis gives 58.1% silica, 15.43 alumina, 4.95 iron oxide, 2.44 magnesia, with small amounts of CaO, N_2O, and K_2O. The specific gravity is 2.3. **Florigel** is the hydrated material which forms viscous suspensions in water, and it is used to replace bentonite as a filler for soaps, and for clarifying liquids. **Diluex** is finely powdered Florida fuller's earth used as a diluent or a carrier for insecticides. **Activated clay,** or **bleaching clay,** for bleaching oils, may be acid-treated fuller's earth, or it may be bauxite or kaolin. **Florite,** of Floridian Co., is activated bauxite. It is a red, granular, porous material of 20 to 60 mesh.

FULMINATES. Explosives used in percussion caps and detonators because of their sensitivity. They may be called **cap powder** in cartridge caps and detonators when used for detonating or exploding artillery shells. **Mercury fulminate,** $Hg(CNO)_2$, a gray or brown, sandy powder, is the basis for many detonating compositions. It is made by the action of nitric acid on mercury and alcohol and is 10 times more sensitive than picric acid. It may be mixed with potassium chlorite and antimony sulfide for percussion caps. The fulminates may be neutralized with a sodium thiosulfate. The **azides** are a group of explosives containing no oxygen. They are compounds of hydrogen or a metal and a monovalent N_3 radical. **Hydrogen azide,** HN_3, or **axoic acid,** and its sodium salt are soluble in water. **Lead azide,** $Pb(N_3)_2$, is used as a substitute for fulminate detonators. It is much more sensitive than mercury fulminate and in large crystals is subject to spontaneous explosion, but it is precipitated as a 93% pure product to suppress crystal formation and to form a free-flowing powder less sensitive to handling. Lead azide detonators for use in coal mining have **copper detonators**; all other blastings employ **aluminum caps. Lead bromate,** $Pb(BO_3)_2 \cdot H_2O$, is in colorless crystals which will detonate if mixed with lead acetate. For primer caps, a substitute for mercury fulminate is a mixture of lead styphnate, lead triagoacetate, lead nitrate, and lead sulfocyanate. **Lead styphnate,** used as a detonator, is made by the sulfonation and nitration of resorcinol to form **styphnic acid,** or **trinitroresorcinol,** $(NO_2)_3C_6H(OH)_2$. This powder is treated with magnesia and with a lead nitrate solution to form the lead styphnate powder. **Azoimide,** or **iminazoic acid,** $HN \cdot N_2$, is an extremely explosive colorless gas liquefying at 99°F (37°C) which can be used in the form of its salts. **Silver azoimide,** $AgN \cdot N_2$, is highly explosive, and **barium azoimide,** BaN_6, explodes with a green flash. **Cyanuric triazide** is a powerful explosive made by reacting cyanuric chloride with sodium azide. **Lead nitrate,** a white, crystalline, water-soluble powder of

composition $Pb(NO_3)_2$, is used in match heads and explosive compositions. It is also employed as a mordant in dyeing and printing, and in
paints. **Nitrostarch,** a primary high explosive, is used as a detonator
and in caps. **Mannitol hexanitrate** is a colorless crystal that is used
in blasting caps as a replacement for nitrostarch.

FURFURAL. Also known as **furfuraldehyde, furol,** and **pyromucic
aldehyde.** A yellowish liquid with an aromatic odor, having composition $C_4H_3O \cdot CHO$, specific gravity 1.161, boiling point 323°F
(161.7°C), and flash point 132°F (56°C). It is soluble in water and in
alcohol but not in petroleum hydrocarbons. On exposure it darkens
and gradually decomposes. Furfural occurs in different forms in various plant life and is obtained from complex carbohydrates known as
pentosans, which occur in such agricultural wastes as cornstalks,
corncobs, straw, oat husks, peanut shells, bagasse, and rice. Furfural
is used for making synthetic plastics, as a plasticizer in other synthetic resins, as a preservative, in weed killers, and as a selective solvent
especially for removing aromatic and sulfur compounds from lubricating oils. It is also used for the making of butadiene, adiponitrile, and
other chemicals.

Various derivatives of furfural are also used, and these, known col-
lectively as **furans,** are now made synthetically from formaldehyde
and acetylene which react to form butyl nedole. This is hydrogenated
to butanediol, then dehydrated to tetrahydrofuran. Furan, or **tetrol,**
C_4H_4O, used for plastics manufacture, is a colorless liquid boiling at
90°F (32°C). **Methyl furan,** or **sylvan,** $C_4H_3O \cdot CH_3$, is a colorless
liquid boiling at 144°F (62°C). **Tetrahydrofuran,** $(CH_2)_4O$, is a
water-white liquid boiling at 151°F (66°C), having strong solvent powers on resins. It reacts with carbon monoxide to form adipic acid and
is also an intermediate for the production of other chemicals. Quaker
Oats converts tetrahydrofuran to **polytetramethylene glycol,**
which is used for producing **Spandex** fibers, polyurethane elastomers, and other polymers. **Polytetramethylene ether glycol,** Du
Pont's Terathane, is a precursor of the company's **Lycra** Spandex
fiber and **Hytrel** copolyether-ester elastomers. **Furfuryl alcohol,**
$C_4H_3O \cdot CH_2OH$, a yellow liquid with a brinelike odor, boiling at
349°F (176°C) with a flash point at 167°F (75°C), is used as a solvent
for nitrocelulose and for dyes, and for producing synthetic resins. It is
made by hydrogenation of furfural. **Furfuryl alcohol resins,** made
by reacting with an acid catalyst, are liquid materials that are low-
cost and highly chemical-resistant. They are much used for protective
coatings, tank linings, and chemical-resistant cements. They are dark
in color. **Alkor cement** is a furfuryl alcohol solution which produces
coatings resistant to chemicals and to temperatures to 380°F (193°C).

Furacin is a nitro-furfural semicarbazone, a yellow crystalline powder made by nitrating furfural and reacting it with hydrazine hydrate. It is used as a bacterial treatment for wounds and burns. **Furoic acid,** or **pyromucic acid,** $C_4H_3O \cdot COOH$, is a colorless crystalline powder soluble in water. It is **furan carboxylic acid** used for making pharmaceuticals, flavors, and resins.

Tetrahydrofurfuryl alcohol, $(CH_2)_3OCH \cdot CH_2OH$, is the usual starting point for making furfuryl esters, ethers, and straight-chain compounds, and it is also a high-boiling solvent for gums, resins, and dyes. It is a liquid of specific gravity 1.064 boiling at 352°F (177.5°C), and is soluble in water. **Furfural acetone,** $C_4H_3O \cdot CH:CHCOCH_3$, is a reddish-brown liquid boiling at 444°F (229°C). Furfural, when treated with aniline at 302°F (150°C), forms an insoluble black **furfural-aniline resin** used in resistant protective coatings and enamels. **Furfural-acetone resin,** or **furfuracetone,** is a transparent elastic resin made by the reaction of furfural and acetone in the presence of an alkali. Furfural also polymerizes with phenol to form **furfuralphenol resins** that are self-curing. They have high heat, chemical, and electrical resistance and excellent adhesion to metals and other materials, making them adaptable for chemical and electrical coatings. The resins have high gloss, but a very dark color. The **Tygon resins** of U.S. Stoneware Co. are furfural resins used for brush application as protective coatings for such purposes as chemical tank linings. They cure by self-polymerization, will withstand temperatures to 350°F (177°C), and are resistant to acids, alkalies, alcohols, and hydrocarbons. **Furafil,** of Quaker Oats Co., is a by-product material containing modified cellulose, lignin, and resins, used as an extender for phenolic plywood glues, as an additive for phenolic molding resins, and as a binder for foundry sand molds. Under the name of **Fur-Ag,** it is used as a conditioner and anticaking agent in fertilizer mixtures. The material is a dark-brown, absorbent powder.

Furane plastics have high adhesion and chemical resistance, but they do not have high dielectric strength, and are black or dark in color. They are used for pipe, fittings, and chemical equipment parts and for adhesives and coatings. **Eonite** is produced from **Durez 16470,** a furfural alcohol resin of Hooker Chemical. The pipe will resist hot acids and alkalies to 300°F (149°C), is strong, and does not sag in long lengths. **Furfural-ketone resin** is used to blend with epoxy laminating resins to reduce cost and improve the properties.

FUSIBLE ALLOYS. Alloys having melting points below the boiling point of water 212°F (100°C). They are used as binding plugs in automatic sprinkler systems, for low-temperature boiler plugs, for soldering pewter and other soft metals, for tube bending, and for casting patterns and many ornamental articles and toys. They are also used for holding optical lenses and other parts for grinding and polishing. They consist generally of mixtures of lead, tin, cadmium, and bismuth. The general rule is that an alloy of two metals has a melting point lower than that of either metal alone. By adding still other low-fusing metals to the alloy, a metal can be obtained with almost any desired low melting point. The original **Newton's alloy** contains 50% bismuth, 31.25 lead, and 18.75 tin. **Newton's metal,** used as a solder for pewter, contains 50% bismuth, 25 cadmium, and 25 tin. It melts at 203°F (95°C) and will dissolve in boiling water. **Lipowitz alloy,** another early metal, contains 3 parts cadmium, 4 tin, 15 bismuth, and 8 lead. It melts at 158°F (70°C), is very ductile, and takes a fine polish. It was employed for casting fine ornaments, but now has many industrial uses. A small amount of indium increases the brilliance and lowers the melting point 34°F (19°C) for each 1% of indium up to a maximum of 18. **Wood's alloy,** or **Wood's fusible metal,** was patented in 1860 and was the first metal used for automatic sprinkler plugs. It contains 7 to 8 parts bismuth, 4 lead, 2 tin, and 1 to 2 cadmium. It melts at 160°F (71°C), and this point was adopted as the operating temperature of sprinkler plugs in the United States; in England it is 155°F (68°C). The alloy designated as **Wood's metal** by Cerro Metal Products Co. contains 50% bismuth, 25 lead, 12.5 tin, and 12.5 cadmium. It melts at 158°F (70°C). An early alloy for tube bending contains 50% bismuth, 16.7 lead, 13.3 tin, and 10 cadmium. It melts at 158°F (70°C) and can be easily removed from the tube after bending by dipping in boiling water or by applying steam.

    Cerrobend, or **Bendalloy,** of Cerro Metal Products Co., is a fusible alloy for tube bending which melts at 160°F (71°C). **Cerrocast** is a **bismuth-tin alloy** with pouring range of 280 to 338°F (138 to 170°C) and shrinkage of only 0.0001 in/in (0.0025 cm/cm), used for making pattern molds. **Cerro-safe,** or **Safalloy,** is a fusible metal used for toy-casting sets because the molten metal will not burn wood or cause fires. Alloys with very low melting points are sometimes used for this reason for pattern and toy casting. A fusible alloy with a melting point at 140°F (60°C) contains 26.5% lead, 13.5 tin, 50 bismuth, and 10 cadmium. These alloys expand on cooling and make accurate impressions of the molds. **Boiler-plug alloys** have been made under a wide variety of trade names with melting points usually ranging from 212 to 342°F (100 to 172°C). **D'Arcet's alloy,** melting at 200°F (93°C), contains 50% bismuth, 25 tin, and 25 lead. **Lichtenberg's alloy,** melting at 198°F (92°C), contains 50% bismuth, 30 lead, and 20 tin. **Guthrie's alloy** has 47.4% bismuth, 19.4 lead, 20 tin, and 13.2 cadmium. **Rose's alloy** contains 35% lead, 35 bismuth, and 30 tin. **Homberg's alloy,** melting at 251°F (121°C), contains 3 parts lead, 3 tin, and 3 bismuth. **Malotte's metal,** melting at 203°F (95°C), has 46% bismuth, 20 lead, and 34 tin. The variation of these different alloys is largely due to the relative cost of the different alloying metals at various times. Fusible metals have also been used in strip form to test the temperature of steels for heat-treating. The **Temperite alloys** are for this purpose with melting points between 300 and 625°F (149 and 329°C) in steps of 25°F (13.9°C). The **Tempil pellets** of Tempil Big Three Industries, Inc., are alloy pellets made with melting points in steps of 12.5, 50, and 100°F (6.9, 27.8, and 55.6°C) for measuring temperatures of 113 to 2500°F (45 to 1371°C). The **Semalloy metals** of Semi-Alloys Inc. cover a wide range of fusible alloys with various melting points. **Semalloy 1010** with a melting point of 117°F (47°C) can be used where the melting point must be below that of thermoplastics. It contains 45% bismuth, 23 lead, 19 indium, 8 tin, and 5 cadmium. **Semalloy 1280,** for uses where the desired melting is near the boiling point of water, melts at 204°F (96°C). It contains 52% bismuth, 32 lead, and 16 tin.

FUSTIC.   Known also as **Cuba wood.** The wood of the tree *Chlorophora tinctoria* of tropical America, used for cabinetmaking and as a dyewood. It is yellow and very hard and has a fine, open grain. The density is about 41 lb/ft^3 (657 kg/m^3). The liquid extract of the wood produces the yellow dyestuff **morin,** $C_{15}H_{11}O_7$, and the red dye **morindone,** $C_{15}H_{11}O_5$. **Fustic extracts** are mordant dyes and give colors from yellow to olive with various mordants. Morin is used also as an indicator to detect aluminum, with which it develops a green fluorescence. **Young fustic,** or **Hungarian yellowwood,** is a yellow dyewood from *Rhus cotinus.* **Osage orange,** called **bois d'arc,** is the bright orangewood of the bush *Maclura pomifera* growing in the swamplands of Texas and Oklahoma. It has a high tannin content and is used in the textile and leather industries for orange-yellow and gold colors and to blend with greens. As a tanning agent, it may be blended with quebracho and chestnut extracts.

GALLIUM.   An elementary metal, symbol Ga, originally called **austrium.** It is silvery white, resembling mercury in appearance but having chemical properties more nearly like those of aluminum. It melts at 85.6°F (30°C) and boils at 4359°F (2403°C), and this wide liquid range makes it useful for high-temperature thermometers. Like bismuth, the metal expands on freezing, the expansion amounting to about 3.8%. Pure gallium is resistant to mineral acids and dissolves with difficulty in caustic alkali. It forms many salts at different valences. The weight is only about half that of mercury, having a specific gravity of 5.9. Commercial gallium has a purity of 99.9%. In the molten state it attacks other metals, and small amounts have been used in tin-lead solders to aid wetting and decrease oxidation, but it is expensive for this purpose. **Gallium-tin alloy** has been used when a low-melting-point metal was needed. It is also used as an electron carrier in silicon semiconductors, and crystals of **gallium arsenide,** GaAs, are used as semiconductors. Gallium arsenide can be used in rectifiers to operate to 600°F (316°C). The material has high electron mobility. This material in single-crystal bars is produced by Monsanto for lasers and modulators. GaAs is also used for lenses in CO_2 laser systems. Made under certain conditions, the material exhibits superconductivity at $-440°F$ ($-262°C$). **Gallium selenide,** GaSe, **gallium triiodide,** GaI_3, and other compounds are also used in electronic applications.

Gallium exists in nature in about the same amount as lead, but it is widely dissipated and not found concentrated in any ore. It is found in small amounts associated with zinc ores and is recovered from smelter flue dust. In Germany it is produced as a by-product of copper smelting. It is also a minor constituent in the mineral sphalerite to the extent of 0.01 to 0.1%, and it occurs in almost all aluminum ores in the ratio of 0.11 to 0.22 lb (50 to 100 g) of gallium per ton of aluminum. In the United States it is a by-product of aluminum production. About 1 oz (0.03 kg) of gallium is obtained commercially per ton (0.91 metric ton) of bauxite.

GALLS. Tanning materials obtained from the **nutgalls,** or **gall nuts,** from the oaks of Europe and the Near East and from the sumac of China and Japan. Nutgalls are plant excrescences caused by the punctures of insects. They contain 50 to 70% tannins and are the richest in tannin of all the leather-tanning materials. The tannin is also valued for ink making and in medicine for treating burns. **Green galls, Turkey galls,** or **Aleppo galls** are obtained from the twigs of the **Aleppo oak,** *Quercus infectoria,* a shrub of the Near East. Those of blue color are the best quality, with green second, and white of inferior grade. **Chinese galls,** from the species of *Rhus,* are in the form of irregular, roundish nuts which enclose the insect. They show no vegetable structure but have a dense resinous fracture and are very high in tannin.

The product known as **gall** in the pharmaceutical industry is an entirely unrelated material. It is **beef gall,** or **ox bile,** a bitter fluid from the livers of cattle. It is used for steroid production and in the textile industry for fixing dyes and in soaps for washing dyed fabrics. **Steroids,** or **hormones,** made from ox bile, have a great number of possible combinations that have an influence on the behavior of the human cellular system. They are based on a four-ring, 17-carbon cyclopentamorphen anthrene nucleus, and arranging the side group in different ways gives compounds with distinct physiological properties. **Cortisone,** made by moving the oxygen atom of the steroid nucleus from the 12th to the 11th position, is one of the many steroids. Steroids are not synthesized from the more plentiful cholesterol, from the **stigmastrol** of vegetable oils, or from the **sapogenins** of plants. **Corticosteroids,** inhaled from bronchodilators, are antiinflammatory medications for treating asthma.

GALVANIZED STEEL AND IRON.   Galvanizing is the process of coating irons and steels with zinc for corrosion protection. The zinc may be applied by immersing the substrate in a bath of the molten metal (**hot-dip galvanizing**), by electroplating the metal on the substrate (**electrogalvanizing**), or by spraying atomized particles of the metal onto otherwise finished parts. The zinc protects the substrate in two ways: (1) as a barrier to atmospheric attack and (2) galvanically; that is, if the coating is broken, exposing the substrate, the coating will corrode sacrificially, or in preference to the substrate. From the standpoint of barrier protection alone, a coating weight of 0.82 lb/ft^2 (4 kg/m^2) on sheet steel will provide a service life of about 30 years in rural atmosphere and about 5 years in severe industrial atmosphere.

Both hot-dip galvanizing and electrogalvanizing are continuous processes applied in the production of galvanized steel, and the coating may be applied on one or both sides of the steel. In the case of **hot-dipped galvanized steel,** the zinc at the steel face alloys with about 25% iron from the steel. Iron alloying decreases progressively to a region that is 100% zinc. The amount of coating is commonly indicated by designations G235 to G01 for regular zinc coatings and A60 to A01 for zinc-iron coatings. The numbers in the designations denote the minimum total coating weight on both sides of the sheet as determined by dissolving the zinc from samples of the sheet, usually at three locations (triple-spot test), and weighing it. For example, G210 refers to a total coating weight of 2.10 oz/ft^2 (640 g/m^2), that is, 1.05 oz/ft^2 (320 g/m^2) per side or, in thickness, 0.0017 in (0.04 mm) per side. The lower the number in these designations, the lesser the amount of zinc. **Electrogalvanized steel** typically has a more homogeneous but thinner coating of pure zinc and is somewhat more formable than the hot-dipped variety.

A spangled surface has long been characteristic of traditional hot-dipped galvanized steel. Although that effect can be minimized, concern by automakers that the spangles might show through on painted external body panels gave rise to the development of **Zincrometal** in the early 1970s and increased use of electrogalvanized steel. Zincrometal, developed by Diamond Shamrock and now a product of

Metal Coatings International, is one-side-coated sheet steel produced on coil-coating lines. The coating consists of a chromate base coat and a special zinc-rich top coat. Total coating thickness is about 0.0005 in (0.00002 mm). However, it serves only as a barrier coating and does not provide sacrificial, or galvanic, protection. Through the 1970s, it was produced under license by virtually all major sheet-steel producers serving the auto industry, but its use has declined since the late 1970s as steel producers developed hot-dipped galvanized steels having more uniform surface appearance.

The American Iron and Steel Institute lists eight types of galvanized steel for auto applications in addition to Zincrometal. Five are of the hot-dipped variety: (1) regular and minimum spangle; (2) fully alloyed zinc-iron coated; (3) differentially zinc-coated (both sides coated but one side having a substantially lower weight, or thickness, of zinc than the other); (4) differentially zinc-iron-coated (same as differentially zinc-coated except that the side with the thin coating is heat-treated or wiped to produce a fully alloyed zinc-iron coating); and (5) one-side-coated (one side is zinc-free). The three types of electrogalvanized sheet steels are (1) electrolytic flash-coated [0.10 to 0.20 oz/ft^2 (30 to 60 g/m^2) on both sides for minimal corrosion protection]; (2) electroplated zinc-coated [coated on one or both sides, the latter with equal or differential coating weights, with as much as 0.65 oz/ft^2(200 g/m^2) total coating], and (3) electroplated iron-zinc-coated (coated on one or both sides, the latter with equal or differential coating weights, by simultaneous electroplating of zinc and iron to form an alloy coating).

Among the recently developed hot-dipped galvanized steels having improved surface quality is National Steel's **Unikote,** a one-side-coated steel made by first differentially coating cold-rolled steel on a hot-dip galvanizing line and then electrolytically treating the steel to remove the minimal coating on one side while simultaneously depositing an equal amount on the thicker-coated side. Another "one-side" is Armco's **Zincgrip O.S.,** which is coated by a meniscus roll precisely positioned just above the molten bath in a chamber. The chamber, sealed in the bath, contains high-purity nitrogen. When the steel approaches the bath surface, a meniscus forms across its width, supported only by surface tension. There is no tendency for the zinc to move onto the backside unless the steel is immersed substantially below the bath surface. The nitrogen, from a finishing jet within the chamber, impinges on the steel while the zinc is still molten, regulating coating thickness and uniformity. Because of the protective atmosphere, the steel is virtually free of surface defects. **Galva-One,** a one-side electrogalvanized steel developed by U.S. Steel, has a zinc coating of 0.32 oz/ft^2 (97 g/m^2) or 0.00054 in (0.014 mm).

Bethlehem Steel's **Automotive Jetcoat** is an iron-zinc **galvan-**

nealed sheet steel coated with less than normal amounts of zinc. It is said to enhance paint appearance and resistance welding but at a sacrifice in corrosion protection. It consists of a light iron-zinc alloy coating and a thin overlay of zinc on both sides. Inland Steel's **Paint-Tite B,** which has a 0.32 oz/ft^2 (97 g/m^2) coating on one side and a 0.00015-in (0.0038-mm) zinc-iron-alloy coating on the other, provides good corrosion resistance and paint appearance. Armco's latest two-side galvanized steel, **Zincgrip Ultrasmooth,** like Zincgrip O.S., is coated in a nitrogen atmosphere for improved surface quality. The steel is then sprayed with a steam mist containing ammonium phosphate to produce a spangle free finish and is temper-rolled to minimize strain effects and further condition the surface. **Galfan,** another two-side-coated hot-dipped galvanized steel, was developed by International Lead Zinc Research Organization and is produced under license by many companies. The coating on Galfan, which is 95% zinc and 5 aluminum and mischmetal, is said to provide superior corrosion resistance compared to conventional galvanized steels in rural, marine, and severe-marine atmospheres.

GAMBIER.   Also spelled **gambir.** A tanning and dyeing material extracted from the leaves and twigs of the shrubs *Uncaria gambier, U. dacyoneuro,* and other species of India, Malaysia, and the East Indies. It is similar to catechu and is also called **white cutch, yellow cutch, cube cutch,** and **tara japonica.** The cubes of extract are brittle and have a dull-gray color and a bitter astringent taste. The material contains **catechin,** $C_{15}H_{14}O_6$, a yellow astringent dye also found in catechu and kino resin. Catechin is soluble in hot water and in alkaline solutions and is neutral with no acid properties. Gambier also contains **catechutannic acid,** a reddish tannin. The liquid extract contains 25% tannin, and the cube gambier has 30 to 40%. Gambier is used in tanning leather and in dyeing to give yellow to brown colors. It produces the **cutch brown** on cotton fabrics. It has excellent fastness as a dye and is used in shading logwood and fustic, or as a mordant for fixing basic dyes. In tanning it give a soft, porous leather, has less astringency than other tannins, and is employed for tanning coat leathers and for blending with other tannins. Gambier is also used in boiler compounds and in pharmaceuticals. **Plantation gambier,** or **Singapore cube,** is refined clear gambier in small, square cubes, while ordinary gambier may have dark patches or a black color with a fetid odor. **Gambier bulat** is round gambier, and **gambier papu** is in long, black sticks, both used locally for chewing.

GARNET.   A general name for a group of minerals varying in color, hardness, toughness, and method of fracture, used for coating abrasive paper and cloth, for bearing pivots in watches, for electronics, and the finer specimens for gemstones. Garnets are trisilicates of alumina, magnesia, calcia, ferrous oxide, manganese oxide, or chromic oxide. The general formula is $3R''O \cdot R_2'''O_3 \cdot 3SiO_2$, in which R'' is Ca, Mg, Fe, or Mn, and R''' is Al, Cr, or Fe. There are thus 12 basic types of garnets, but sodium and titanium may also occur in the crystals, replacing part of the silicon, and there are also color varieties. Asterism in the garnet, because of the isometric crystal, appears as 4-, 8-, or 12-ray stars instead of the 6 rays of the ruby.

Hardness of garnet varies from the Mohs 6 of grossularite to 7.5 of almandite and rhodolite, a hard garnet having a Knoop hardness of 1,350. The specific gravity is 3.4 to 4.3, with a melting point about 2372°F (1300°C). The color is most often red, but it may be brown, yellow, green, or black. **Spanish garnet** is pale pink. High-iron garnets have the lowest melting points and fuse to a dark glass, while the high-chromium garnets have high melting points. Garnets occur in a wide variety of rocks in many parts of the world. The immense alluvial schist deposits of the Emerald Creek area of Idaho contain about 10% garnet. **Almandite,** $3FeO \cdot Al_2O_3 \cdot 3SiO_2$, forms crystals of a fine, deep-red color with a hardness of about 7.5, which fuses to a glassy mass at 2399°F (1315°C). It is the most common garnet and is produced chiefly in New York State. It is crushed and graded for coating abrasive paper and cloth. The choice crystals are used as gemstones. **Cape ruby,** from South Africa, is a red almandite. **Pyrope** has composition $3MgO \cdot Al_2O_3 \cdot SiO_2$. It is deep red to nearly black and has hardness of 6.5 to 7.5. The nearly transparent or translucent crystals are selected for gemstones. **Rhodolite,** a pale-rose to purple variety of garnet, is a mixture of pyrope and almandite. These two garnets are found in the eastern states. The garnet of North Carolina is a by-product of kyanite mining, the ore containing 10% garnet, 15 kyanite, and 30 mica.

Andradite has composition $3CaO \cdot Fe_2O_3 \cdot 3SiO_2$; and the colors are yellow, green, and brown to black, with a hardness of 6.5. The orange-yellow variety is called **hessenite**; the yellow-green is **topazolite**; the green is **demantoid**; and the black variety is called **melanite. Uralian emeralds** are choice green crystals of demantoid from the Ural Mountains. **Uvarovite** has composition $3CaO \cdot CR_2O_3 \cdot 3SiO_2$ and is emerald green. **Grossularite,** $3CO \cdot Al_2O_3 \cdot 3SiO_2$, or $Ca_3Al_2(SiO_4)_3$, may be white, yellow, or pale green. **Succinite** is an amber variety, **romanzovite** is brown, and **wiluite** is pale green. **Spessartite,** $3MnO \cdot Al_2O_3 \cdot 3SiO_2$, is brown to red, the yellowish-brown variety being **rothoffite.** The sodium garnet is **lagoriolite,** and one variety of calcium spessartite is called **essonite.**

The best **abrasive garnets** come from almandite, although andradite and rhodolite are also used. Garnet is crushed, ground, and separated and graded in settling tanks and sieves. Hornblende is a common impurity and is difficult to separate, but good-quality abrasive garnet should be free of this softer mineral. **Garnet-coated paper** and cloth are preferred to quartz for the woodworking industries, because garnet is harder and gives sharper cutting edges; but aluminum oxide is often substituted for garnet. The less expensive quartz is sometimes colored to imitate garnet. The grades of garnet grains used on **garnet paper** and cloth range from No. 5, the coarsest, which is about 15 mesh, to 7/0, the finest, which is about 220 mesh. The paper used as a backing is a kraft of 50- to 70-lb (23- to 32-kg) weight, or a manila stock. The usual size is 9 by 11 in (23 by 28 cm). The cloth is usually in two weights, the lightweight being used as a flexible rubbing-down material. Some garnet is made into wheels by the silicate process, but vitrified wheels are not made because of the low melting point of garnet.

Synthetic garnets for electronic application are usually **rare-earth garnets** with the general structure of grossularite, but with a rare-earth metal substituted for the calcium, and iron substituted for the aluminum and the silicon. **Yttrium garnet** is thus $Y_3Fe_2(FeO_4)_3$. **Yttrium-aluminum garnets** of 0.118-in (3-mm) diameter are used for lasers. **Gadolinium garnet** is for microwave use. **Gadolinium gallium garnet** made from gadolinium oxide and gallium oxide is used for computer bubble memories.

GASKET MATERIALS.   Any sheet material used for sealing joints between parts to prevent leakage, but gaskets may also be in the form of cordage or molded shapes. A usual requirement is that the material not deteriorate by the action of water, oils, or chemicals. Metals for gaskets include copper, brass, bronze, iron, steel, nickel, and various alloys. Nonmetallics include cork, paper, natural and synthetic rubbers or elastomers, various polymers, glass and aramid fibers, asbestos, carbon fibers, and graphite. Sometimes metals and nonmetallics are combined. Combination gaskets include corrugated-metal, jacketed, and spiral-wound types. Soft material is bonded to the face of the corrugated-metal type or is used as a filler in the spiral-wound. In the metal-jacketed kind, a soft, compressible filler is enclosed in a metal jacket. The spiral-wound type with flexible graphite probably sees the widest use. Upon bolting, the conformable filler flows into flange-surface irregularities; the metal spiral provides strength, resiliency, and blowout resistance. Gasket materials are usually marketed under trade names. **Felseal** consists of sheets of paper or fiber, 0.010 to 0.125 in (0.025 to 0.318 cm) thick, coated with Thiokol to withstand oils and gasoline. **Corbestos,** which resists high heat and pressure, consists of sheet metal coated with graphited asbestos, the sheet metal being punched with small tongues to hold the asbestos. **Chrome lock** is felt impregnated with zinc chromate to prevent corrosion and electrolysis between dissimilar metal surfaces.

Foamed synthetic rubbers in sheet form, and plastic impregnants, are widely used for gaskets. Some of the speciality plastics, selected for heat resistance or chemical resistance, are used alone or with fillers, or as binders for fibrous materials. **Haveg 16075,** of Haveg Industries, Inc., a gasketing sheet to withstand hot oils and superoctane gasolines, is based on **Viton,** of Du Pont, a copolymer of vinylidene fluoride and hexafluoropropylene. It contains about 65% fluorine, has a tensile strength of 2,000 lb/in^2 (14 MPa) with elongation of 400%, and withstands operating temperatures to 400°F (204°C) with intermittent temperatures to 600°F (316°C). Another synthetic of this company, used for gaskets of high chemical resistance at temperatures to 350°F (177°C), is **Hypalon CSM-60,** a chlorosulfonated polyethylene, made by reacting polyethylene with sulfur and chlorine. It has a tensile strength of 2,250 lb/in^2 (15.5 MPa), elongation of 30%, and Durometer hardness of 60. The square rope-type packing called **Graphlon C** has a core of braided asbestos fibers impregnated with Teflon, over which is a braided jacket of Teflon fiber, with an outer jacket of graphite fabric. It has a low coefficient of friction, is self-lubricating, has high chemical resistance, and withstands service temperatures to 675°F (357°C).

Since the decline in the use of asbestos as a gasket material, several high-temperature alternatives have been developed. Aramid fibers, such as Akzo Fibers' **Twaron** and Du Pont's **Kevlar,** contained within a nitrile-butadiene binder, provide gaskets that will withstand temperatures up to 450°F (232°C). These products are also available coated with tetrafluoroethylene for use as braided packings. **Carbon-fiber gaskets** with the same binder will resist temperatures to 800 to 900°F (427 to 482°C). For temperatures to 1100 to 1200°F (593 to 649°C), gaskets of graphite flakes compressed without a binder are used. Japan Gore-Tex Inc. offers a laminate of expanded-graphite sheets and tetrafluorethylene that can contain steam, hot oils, acids, alkalies, and solvents at temperatures of −400 to 572°F (−240 to 300°C). **Grafoil** gaskets, of Industrial Gasket & Shim Co., are made of **flexible graphite.** Intended for high-temperature uses, they also resist fire, acids, alkalies, salt solutions, halogens, and various organics. The **G-9900** gaskets of Garlock Mechanical Packing Div. rely on graphite fibers for heat resistance to 1004°F (540°C). They also resist saturated steam, hot oils, gasoline, aliphatic gases, and hydrocarbons. Palmetto Packings offers **Flexi-Braid 5000** gaskets, based on graphite ribbon braided yarn. They are used for packings in pumps and valves.

GASOLINE. Known in England as **petrol.** A colorless liquid hydrocarbon obtained in the fractional distillation of petroleum. It is used chiefly as a motor fuel, but also as a solvent. Ordinary gasoline consists of the hydrocarbons between C_6H_{14} and $C_{10}H_{22}$, which distill off between 156 and 345°F (69 and 174°C), usually having the light limit at **heptane,** C_7H_{16}, or **octane,** C_8H_{18}. The octane number is the standard of measure of detonation in the engine. **Motor fuel,** or the general name *gasoline,* before the wide use of high-octane gasolines obtained by catalytic cracking, meant any hydrocarbon mixture that could be used as a fuel in an internal combustion engine by spark ignition without being sucked in as a liquid and without being so volatile as to cause imperfect combustion and carbon deposition. These included mixtures of gasoline with alcohol or benzol.

Gasanol, used in the Philippines, contained only 20% gasoline, with 5 kerosene and 75 ethyl alcohol, while the German **Dynakol** contained 70% gasoline with alcohol and benzol. The **Dynax motor fuel** contained a methanol-benzol blend. In Brazil, alcohol produced from excess sugar is mixed with gasolines, and at times all commercial automotive fuels there contain alcohol. But gasolines containing as much as 15% alcohol require special carburetion. In the United States, alcohol was not normally used in gasoline fuels until the oil shortages of the 1970s, after which its use increased. **Gasohol** is the trade name of the University of Nebraska Agriculture Products Industrial Utilization Committee for a blend of 90% unleaded gasoline and 10 (by volume) anhydrous ethanol. The **ethanol** is derived mostly from corn. Ethanol from corn is also used by Aquinas Technology to make windshield washer fluid for cars. Most the world's ethanol, however, comes from fermenting sugarcane. **Lignocellulosic** biomass, from agricultural, forestry, and municipal wastes, is a major potential source. So are woody and grassy crops, **sweet sorghum,** and **black locust,** or *Robina pseudoacacia,* which grows rapidly.

The common commercial gasolines in the United States had an upper limit at 437°F (225°C) and an average specific gravity of 0.75, with the **aromatic-free gasolines** having a specific gravity of 0.718. **Aviation gasoline** formerly had a boiling range below 302°F (150°C), but aviation gasolines are now the high-octane cracked and treated gasolines. Straight-run distillation yields gasoline octane numbers from about 40 to 60, and the yield is 20 to 30% of the petroleum. In the heat-cracking process, the heavy hydrocarbons of petroleum are fragmented and converted to lighter hydrocarbons of gasoline range. The octane number of the gasoline then ranges from 55 up to 70, and the yield is higher. With catalytic cracking the octane number can be brought up to 100 or higher, and the yield proportionately increased.

Cracked gasolines may be rich in the olefins, C_nH_{2n}, or ethylene series, which have antiknock properties but which polymerize and form resins with high heat. They are stabilized with antioxidants. Filtering straight-run gasoline with bauxite removes sulfur impurities that cause knocking and raises the octane number. The octane number and the antiknock qualities are improved by slight additions of tetraethyl lead, but high-quality gasolines that do not need lead additions are produced by catalytic cracking. Small amounts of tricresyl phosphate may also be added to motor gasoline to give more uniform combustion and to eliminate knock. All lead compounds are toxic, and the exhaust fumes from **leaded gasoline** may remain in the atmosphere for long periods, causing injury to animals and vegetation. The federal government's Clean Air Act, passed in 1970, requires the use of **unleaded gasoline** in all motor vehicles built in 1975 or later. To attain the minimum of 91 octane in unleaded gas, as required by the federal government and automobile manufacturers, a manganese compound, called **MMT,** and the high-octane boosters, reformates, and alkylates replace the formerly used **tetraethyl lead.**

Alkylated gasoline is made by adding **neohexane,** which is produced by combining isobutane and ethylene, or by adding other alkylates. **Alkylates** are produced by reacting butylenes plus propylenes and amylenes with isobutane and an acid catalyst. They are clean-burning fuels with high octane rating. As much as 35% may be added to a premium grade of gasoline to improve the sensitivity and raise the aviation performance number to above 115.

Auto emission standards imposed by 1990 amendments to the Clean Air Act gave rise to **oxygenated gasoline,** also called **reformulated gasoline,** containing by weight 2% oxygen and no more than 1 benzene and 25 aromatics. The leading **oxygenate** is **methyl *tert*-butyl ether** (**MTBE**), which has an octane rating of 110. MTBE's ethanol-derived counterpart, **ethyl *tert*-butyl ether** (**ETBE**), has an octane rating of 105 and certain performance advantages, including a higher blending octane and lower vapor pressure. Other oxygenates include methanol-derived ***tert*-amyl methyl ether** (**TAME**) with an octane rating of 112 and the ethanol-derived ***tert*-amyl ethyl ether** (**TAEE**) with 100, both also having lower vapor pressure than MTBE. Blending oxygenates into gasoline is now required during winter months in many states with high ambient levels of carbon monoxide emissions. Using gasoline with 2.7% oxygen by weight reduces these emissions by 15 to 20%. The 2.7% is typically met by blending 7.7% ethanol or 15 MTBE by volume. California, however, limits oxygen to 2% so as not to increase emissions of nitrogen oxides. Arco Product's **EC-1** gasoline, introduced in 1989, is a reformulated gasoline for older vehicles operating on leaded gasoline.

It then removed the lead, reduced aromatics from over 30% to 20%, added 5.5% MTBE and more alkylate, and reduced olefin, sulfur, and benzene contents. To meet still stricter emission regulations, it has since developed a 15% MTBE grade.

Synthetic gasoline was first produced in Germany by the Bergius process of hydrogenation of powdered lignite at high temperatures and pressures to produce gasoline, an intermediate oil, and a heavy oil. But gasolines are also produced from the high-volatile bituminous coal. The low-grade fuel bituminous coals contain a high proportion of **anthraxylon,** or **vitrain,** which is that part of bituminous coal consisting of undisintegrated parts of trees and plants. This structure of coal has less carbon and can be liquefied more easily to produce gasoline and oil. A low-grade bituminous, such as Vigo No. 4 vein, with 65% anthraxylon, gives a liquefaction to gasoline and oils of more than 96%, while a Pennsylvania medium-volatile fuel coal gives only 79% liquefaction and requires a greater consumption of hydrogen. **Natural gasoline** consists of the liquid hydrocarbons from C_5 upward, extracted from natural gas and separated from propane and other higher fractions. It has high vapor pressure and is used for blending with motor fuels. However, under controlled production, gasolines from the various sources are the same.

Gasoline gel, used in incendiary bombs, is gasoline made into a thick gel with a chemical thickener. It adheres to the surface where it strikes and will produce a temperature of 3000°F (1649°C) for 10 min. **Kerosene gel,** similarly made, is used to dissolve tight formations and increase the flow in oil wells. The gel known as **napalm** is sodium palmitate, but may also be an aluminum soap made with oleic, naphtenic, and coconut fatty acid mixture. The same principle is used for making quick-firing gels for commercial power boilers, but diesel oil is used instead of gasoline. The motor fuel known as **triptane** is **trimethyl butane.** In automotive engines it is free from knock so that higher compressions may be employed. In blends with 100-octane gasoline, it increases the power output about 25% in aviation engines.

GELATIN. A colorless to yellowish, water-soluble, tasteless colloidal hemicellulose obtained from bones or skins and used as a dispersing agent, sizing medium, coating for photographic films, and stabilizer for foodstuffs and pharmaceutical preparations. It is also flavored for use as a food jelly, and it is a high-protein, low-calorie foodstuff. While albumin has a weak, continuous molecular structure that is cross-linked and rigidized by heating, gelatin has an ionic or hydrogen bonding in which the molecules are brought together into large aggregates, and it sets to a firmer solid. Gelatin differs from glue only in the purity. **Photographic gelatin** is made from skins. **Vegetable gelatin** is not true gelatin, but is algin from seaweed.

Collagen is the gelatin-bearing protein in bone and skins. The bone is dissolved in hydrochloric acid to separate out the calcium phosphate and is washed to remove the acid. The organic residue is called **osseine** and is the product used to produce gelatin and glue. About 25% of the weight of the bone is osseine, and the gelatin yield is about 65% of the osseine. One short ton (0.9 metric ton) of green bones, after being degreased and dried, yields about 300 lb (136 kg) of gelatin. When skins are used, they are steeped in a weak acid solution to swell the tissues so that the collagen may be washed out. The gelatin is extracted with hot water, filtered, evaporated, dried, and ground or flaked.

GERMANIUM. A rare elemental metal, symbol Ge. It is a grayish-white crystalline metal of great hardness: Mohs 6.25. Its specific gravity is 5.35 and melting point 1720°F (937°C). It is resistant to acids and alkalies. The metal is trivalent and will form chain compounds like carbon and silicon. It gives greater hardness and strength to aluminum and magnesium alloys, and as little as 0.35% in tin will double the hardness. It is not used commonly in alloys, however, because of its rarity and great cost. It is used chiefly in the form of its salts to increase the refraction of glass, and as metal in rectifiers and transistors. A **gold-germanium alloy,** with about 12% germanium, has a melting point of 680°F (359°C) and has been used for soldering jewelry.

Germanium is obtained as a by-product from flue dust of the zinc industry, or it can be obtained by reduction of its oxide from the ores, and it is marketed in small irregular lumps. Metal of 99.9+ purity for electronic use is made by passing an ingot slowly through an induction heater so that the more soluble impurities pass along through the molten zones and are cut off at the end of the ingot. **Germanium crystals** are grown in rods up to 1.375 in (3.49 cm) in diameter for use in making transistor wafers. High-purity crystals are used for both P and N semiconductors. They are easier to purify and have a lower melting point than other semiconductors, specifically silicon.

The chief ore of the metal is **germanite,** which is a copper ore found in southwest Africa. Germanite contains no less than 20 elements. Together with about 45% copper and 30 sulfur, it contain 6 to 9% germanium and 1 gallium, with various amounts of iron, zinc, lead, arsenic, silica, titanium, tungsten, molybdenum, nickel, cobalt, manganese, cadmium, and carbon. **Renierite,** found in the Congo, is a **germanium sulfide** containing up to 7.8% germanium in the ore. The rare lead-silver ore, **ultrabasite,** found in central Europe, also contains germanium. Many other metal ores, such as lepidolite, sphalerite, and spodumene, contain small amounts of germanium, so that it has an indirect use, especially in ceramics. As much as 1.6% **germanium oxide,** GeO, occurs in some English coals.

GILSONITE.   A natural asphalt used for roofing, paving, floor tiles, storage-battery cases; in coatings; and for adding to heavy fuel oils. It is also referred to as **Utah coal resin.** It is a lustrous, black, almost odorless, brittle solid, having a specific gravity of 1.10. Gilsonite is one of the purest asphalts and has high molecular weight. It is soluble in alcohol, turpentine, and mineral spirits. The mineral was named for Samuel Gilson, who discovered it around 1875. The very pure gilsonite called **uintahite** takes its name from the Indian word *uintah,* meaning mountain. The mineral in Utah occurs in vertical veins up to 20 ft (6.1 m) wide and 1,400 ft (427 m) deep, sometimes 8 mi (13.5 km) long. The selects come from the center of the veins, are very pure, have high solubility in naphtha, and have fusing points from 270 to 310°F (132 to 154°C). But most commercial gilsonite is now melted and regraded. **Elaterite** and **wurtzilite** are similar asphalts found in Utah, used chiefly for acid-resisting paints. **Grahamite,** or **glance pitch,** is another pure asphalt found in large deposits in Oklahoma, and in Mexico, Trinidad, and Argentina. It is used in insulation and molding materials and in paints. **Gilsonite dust,** used for foundry cores, is a by-product of Utah mining. **Manjak** is a variety of grahamite from Cuba, Barbados, and Trinidad, used for insulation and varnishes. It is the blackest of the asphalts, has a higher melting point than gilsonite, but is usually not as pure. **Millimar,** of R. T. Vanderbilt Co., Inc., is processed gilsonite in 20-mesh powder used for rubber compounding. **Gilsulate,** of American Gilsonite Co., is an insulation grade of gilsonite. **Millex,** of Cary Chemicals, Inc., is a dark-brown gilsonite melting at 250°F (121°C), used for blending in synthetic rubbers and vinyl resins to improve processing. **Gilsonite coke** is a high-grade, low-sulfur coke produced by American Gilsonite Co. from Utah gilsonite, 630 tons (571 metric tons) of gilsonite yielding 250 tons (227 metric tons) of coke, 1,300 bbl of gasoline, and 330 bbl of fuel oil. **Insulmastic** is a solution of gilsonite with mica flakes and asbestos fibers, used for protective coatings.

GINGER.   The most important spice obtained from roots, and one of the first Oriental spices known in Europe. It is the prepared root of the perennial herb *Zingiber officinale,* grown in India, China, Indonesia, and Jamaica. The roots are pale yellow and contain starch, an essential oil, and a pungent oleoresin, **gingerin. Ginger oil** is light yellow, is not pungent to taste, and occurs in yields of 1 to 3%. The first crop is the best; the product from the regrowth is called **rhatoon ginger** and is inferior. Ginger is employed as a spice and condiment, in flavoring beverages and confections, and in medicine as a digestive stimulant and carminative. It has a cooling effect on the body. **Preserved ginger** is made by peeling off the thick scaly skin of the boiled roots and boiling in a sugar solution. **Dried ginger** is prepared by drying the peeled roots in the sun. When the roots have been boiled in lime water before peeling, **black ginger** is produced. **White ginger** is made by bleaching the roots.

Other plants of the ginger family are also used as spices and flavors. **Angelica** is from the perennial herb *Angelica archangelica,* of Syria and Europe. All parts of the plant are aromatic. **Angelica oil,** distilled from the fruit, is used in perfumes, flavors for vermouth and bitters, and in medicine. **Candied angelica** consists of the stems steeped in syrup. It is bright green, has an aromatic taste, and is used for decorating confections. **Galangal** is from the roots of the perennial herb *Languas officinarum,* of China. It has an aromatic odor and pungent taste similar to a mixture of ginger and pepper. It is used chiefly as a flavor, but also in medicine. **Turmeric,** highly popular in India and Malaya as an ingredient of curries, is from the rhizomes of the perennial plant *Curcuma longa,* of southeast Asia and Indonesia. When used as a dye, it is called **India saffron.** The roots are cleaned and dried in the sun. It is very aromatic, with a pungent bitter taste. **Curry powders** are not turmeric alone, but are mixtures of turmeric with pepper, cumin, fenugreek, and other spices. Turmeric is used to flavor and color foodstuffs, and as a dyeing agent for textiles and leather and in wood stains. The natural dye is reddish brown and gives a yellowish color to textiles and foods. It is also used as a chemical indicator, changing color with acidity or alkalinity. **Zedoary** consists of the dried roots of the perennial plant *C. zedoaria,* grown in India. **Zedoary oil** is a viscid liquid of reddish color with an odor resembling ginger and camphor. It is used in flavoring, medicine, and perfumery.

GLASS. An amorphous solid made by fusing silica (silicon dioxide) with a basic oxide. Its characteristic properties are its transparency, its hardness and rigidity at ordinary temperatures, its capacity for plastic working at elevated temperatures, and its resistance to weathering and to most chemicals except hydrofluoric acid. The major steps in producing glass products are melting and refining, forming and shaping, heat-treating, and finishing. The mixed batch of raw materials, along with broken or reclaimed glass, called **cullet,** is fed into one end of a continuous-type furnace where it melts and remains molten at around 2730°F (1499°C). Molten glass is drawn continuously from the furnace and runs in troughs to the working area, where it is drawn off for fabrication at a temperature of about 1830°F (999°C). When small amounts are involved, glass is melted in pots.

There are a number of general families of glasses, some of which have many hundreds of variations in composition. It is estimated that there are over 50,000 glass formulas. The **soda-lime glasses** are the oldest, lowest in cost, easiest to work, and most widely used. They account for 90% of the glass used in the United States. They are composed of silica, sodium oxide (soda), and calcium oxide (lime). These glasses have only fair to moderate corrosion resistance and are useful at temperatures up to about 860°F (460°C) annealed and 480°F (249°C) in the tempered condition. Thermal expansion is high, and thermal shock resistance is low compared with other glasses. These are the glass of ordinary windows, bottles, and tumblers.

Bottle glass is a simple soda-lime glass, the greenish color being due to iron impurities. The lime acts as a flux, and the calcium silicate produced gives the glass a chemical stability which quartz and fused silica do not have. The brilliance and sparkle in glass for bottles for food and drug containers are obtained by adding a small amount of barium. Small amounts of cerium will absorb ultraviolet light without distorting the passage of visible light. Cerium also acts as a **glass decolorizer** by changing the crystal structure of the iron impurities, energizing it from the normal divalent which is bluish to the trivalent which is colorless. **Crown glass** for windows is a hard, white soda-lime glass high in silica, a typical composition being 72% SiO_2, 13 CaO, and 15 Na_2O. It derives its name from the circular crowning method of making the sheets, but it is highly transparent and will take a brilliant polish. **Hard glass,** or **Bohemian glass,** for brilliant glassware, is a potash-lime glass with a high silica content, the potash glasses in general not being as hard as the soda glasses and having lower melting points. The artistry is largely responsible for the quality of Bohemian glass.

Lead glasses or **lead-alkali glasses** are produced with lead contents ranging from low to high. They are relatively inexpensive and are noted for high electrical resistivity and a high refractory index. Corrosion resistance varies with lead content, but they are all poor in acid resistance compared with other glass. Thermal properties also vary with lead content. The coefficient of expansion, for example, increases with lead content. High-lead grades are the heaviest of the commercial glasses. As a group, lead glasses are the lowest in rigidity. They are used in many optical components, for neon-sign tubing, and for electric lightbulb stems.

Borosilicate glasses, which contain boron oxide, are the most versatile of the glasses. They are noted for their excellent chemical durability, for resistance to heat and thermal shock, and for low coefficients of thermal expansion. There are six basic kinds. The low-expansion type is best known as **Pyrex** or **Kimax.** The low-electrical-loss types have a dielectric loss factor second only to fused silica and some grades of 96% silica glass. Optical grades, which are referred to as *crowns,* are characterized by high light transmission and good corrosion resistance. Ultraviolet-transmitting and laboratory apparatus grades are the two other borosilicate glasses.

Because of this wide range of types and compositions, borosilicate glasses are used in such products as sights and gages, piping, seals to low-expansion metals, telescope mirrors, electronic tubes, laboratory glassware, ovenware, and pump impellers.

Aluminosilicate glasses are roughly 3 times more costly than the borosilicate types, but are useful at higher temperatures and have greater thermal shock resistance. Maximum service temperature in the annealed condition is about 1200°F (649°C). Corrosion resistance to weathering, water, and chemicals is excellent, although acid resistance is only fair compared with other glasses. Compared with 96% silica glasses, which they resemble in some respects, they are more easily worked and are lower in cost. They are used for high-performance power tubes, traveling-wave tubes, high-temperature thermometers, combustion tubes, and stovetop cookware.

Vicor glass is a silica glass made from a soft alkaline glass by leaching in hot acid to remove the alkalies and then heating to 2000°F (1093°C) to close the pores and shrink the glass. The glass will withstand continuous temperatures to 1600°F (871°C) without losing its strength or clarity. It will withstand temperatures to 1800°F (982°C) but becomes cloudy and opaque. The glass has high thermal shock resistance. **Filter glass** is flat, porous glass sheets or disks to replace high-alloy metals for filtering chemicals. Filter disks come in five porosities, from A, with pore diameters of 5,710 to 6,890 μin (145 to 175 μm), to E, with pore diameters of 157 to 315 μin (4 to 8 μm).

Phosphate glass, developed by American Optical Co., will resist the action of hydrofluoric acid and fluorine chemicals. It contains no silica, but is composed of P_2O_5 with some alumina and magnesia. It is transparent and can be worked like ordinary glass, but it is not resistant to water. **Fluorex glass** is a phosphate glass containing 75% P_2O_5 and less than 0.5 silica. It is decomposed by alkalies.

Sodium aluminosilicate glasses that are chemically strengthened are used in premium applications, such as aircraft windshields. Molten salt baths are used in the strengthening process. **Corning 0315 glass,** a chemically strengthened glass, has the highest modulus of rupture of any commercially available glass.

Industrial glass is a general name usually meaning any glass molded into shapes for product parts. The lime glasses are the most frequently used because of low cost, ease of molding, and adaptability to fired colors. For such uses as light lenses and condenser cases, the borosilicate heat-resistant glasses may be used. **Glass flake** is produced by cooling rapidly very thin sheets of glass which shatter into fine flakes 0.0003 in (0.0008 cm) thick. It is used for reinforcement in plastics to permit higher filler loadings than with glass fiber. It also gives higher strength and rigidity, and dielectric strengths as high as 3,000 V/mil (118×10^6 V/m) in plastics with 65% glass flake filler. The flake is used as a substitute for mica flake in paints and plastics comes in particles 39 to 156 μin (1 to 4μm) thick and 0.03125 in (0.079 cm) in diameter. In paints it produces a tilelike appearance. **Carboglas 1600** is a coating material of a chemically resistant polyester solution containing glass flake. When sprayed, a 0.030-in (0.076-cm) coating will contain 120 layers of the thin glass flake. **Filmglas** is glass flake in tiny platelets. The 10- to 18-mesh flake is 78 μin (2 μm) thick. **Glass powder,** used as a filler in plastics and coatings, is made by grinding broken scrap glass.

Cer-Vit, of Owens-Illinois Inc., consists of tiny ceramic crystals in a vitreous matrix and is used for molded electronic parts. The thermal expansion can be varied from zero to small positive or negative values. **Glass spheres** and **ceramic spheres** are thin-walled, hollow balls of glass or oxide ceramics used as fillers and strengthening agents in plastics and lightweight composites.

Fused silica glass is 100% silicon dioxide. If it occurs naturally, the glass is known as **fused quartz** or **quartz glass.** There are many types and grades of both glasses, depending on the impurities present and the manufacturing method. Because of its high purity level, fused silica is one of the most transparent glasses. It is also the most heat-resistant of all glasses; it can be used at temperatures up to 1650°F (899 °C) in continuous service and to 2300°F (1260°C) for short-term exposure. In addition, it has outstanding resistance to thermal shock, maximum transmittance to ultraviolet light, and excellent resistance to chemicals. Unlike most glasses, its modulus of elasticity increases with temperature. However, because fused silica is high in cost and difficult to shape, its use is restricted to such specialty applications as laboratory optical systems and instruments and crucibles for crystal growing. Because of a unique ability to transmit ultrasonic elastic waves with little distortion or absorption, fused silica is also used in relay lines in radar installations.

Flint glass for windows is a highly transparent, soda-lime quartz glass. **Lustraglas** is a highly transparent, flat-drawn quartz glass. **Thermopane** is a heat- and sound-insulating window glass made with two panes of glass separated by a metal bonded to the edges of the glass, leaving an insulating layer of dehydrated air between the two glass sheets. The highly refractive flint glass used in the **rhinestones** for cheap jewelry and for the **paste diamonds,** which derive their name from their softness compared with the diamond, contains lead. It has an index of refraction of 1.67, but lacks the double refraction and regular molecular arrangement of true gem crystals. Lead is also used in the **crystal glass** for cut glassware, and the brilliance from the lead is higher in the potash glasses than the soda. **English crystal,** which is a potash glass, contains as high as 33% lead oxide. It has a high clarity and brilliancy, but is soft. Leaded glasses are heavy, a crystal glass having a specific gravity of 6.33 compared with 2.25 for a borate glass.

Ninety-six percent silica glasses are similar to fused silica. Although less expensive than fused silica, they are still more costly than other glasses. Compared to fused silica, they are easier to fabricate, and they have a slightly higher coefficient of expansion, about 30% lower thermal stress resistance, and a lower softening point. They can be used continuously up to 1470°F (799°C). Uses include chemical glassware and windows and heat shields for space vehicles.

The boric oxide glasses, or **borax glasses,** are transparent to ultraviolet rays. The so-called **invisible glass** is a borax glass surface treated with a thin film of sodium fluoride. It transmits 99.6% of all visible light rays, thus casting back only slight reflection and giving the impression of invisibility. Ordinary soda and potash glasses will not transmit ultraviolet light. Glass containing 2 to 4% ceric oxide absorbs ultraviolet rays and is also used for X-ray shields. Glass capable of absorbing high-energy X-rays or gamma rays may contain tungsten phosphate, while the glass used to absorb slow neutrons in atomic energy work contains cadmium borosilicate with fluorides. The shields for rocket capsule radio antennas are made of 96% silica glass. It is transparent to radio signals and will withstand temperatures above 900°F (482°C). **Fluorescent glass** for mercury-vapor discharge tubes contains ceric oxide. **Kromex glass,** used for gasoline-dispensing pumps, is a glass made to stop the passage of ultraviolet rays.

Optical glass is a highly refined glass; it is usually a flint glass of special composition, or made from rock crystal, used for lenses and prisms. It is cast, rolled, or pressed. In addition to the regular glass-making elements, silica and soda, optical glass contains barium, boron, and lead. The high-refracting glasses contain abundant silica or boron oxide. A requirement of optical glass is transparency and freedom from color. Traces of iron make the glass greenish, while manganese causes a purple tinge. First-quality optical glass should contain a minimum of 99.8% SiO_2. Borax is used in purifying and in increasing the strength and brilliance of the glass. Besides the control of chemical composition, careful melting and cooling are necessary to obtain fine transparency, followed by intense polishing. The pouring temperature is about 2192°F (1220°C). The best optical glass has a transparency of 99%, compared with 85 to 90% for ordinary window glass. A **borate glass** with lanthanum and tantalum oxides but no silica is used for airplane camera lenses and for eyepieces for wide-angle field glasses. It has a high refractive index and a low dispersion. The **telescope mirror glass** of Owens-Illinois, called **Cer-Vit optical glass,** has near-zero expansion and no deflection of light rays. **Beryllium fluoride glass** of American Optical Co. is made by substituting beryllium fluoride for silicon dioxide. It has a low refractive index with low color dispersion, and light travels faster through it than through ordinary optical glass. It has the disadvantage that it is hygroscopic. Wound fibers of quartz optical glass are used in fiber-optic cables to transmit the light beam of neodymium-doped, yttrium-aluminum-garnet (YAG) lasers in manufacturing operations. Optical-glass lenses are used for optics in YAG laser systems.

Plate glass is any glass that has been cast or rolled into a sheet and then ground and polished. But the good grades of plate glass are, next to optical glass, the most carefully prepared and the most perfect of all the commercial glasses. It generally contains slightly less calcium oxide and slightly more sodium oxide than window glass, and small additions of agents to give special properties may be added, such as agents to absorb ultraviolet or infrared rays, but inclusions that are considered as impurities are kept to a minimum. The largest use of plate glass is for storefronts and office partitions. Plate glass is now made on a large scale on continuous machines by pouring on a casting table at a temperature of about 1832°F (1000°C), smoothing with a roller, annealing, and then setting rigidly on a grinding table and grinding to a polished surface. Normally, the breaking stress of a glass with a ground surface is much less than that of blown or pressed glass, but highly polished plate glass with the surface flaws removed may have double the breaking strength of average pressed glass. The chief advantage of plate glass, however, is that the true parallelism of the ground surfaces eliminates distortion of objects seen through the glass. **Herculite** is a glass of PPG Industries that withstands temperatures up to 650°F (343°C) without cracking. **Carrara structural glass** of this company is made like plate glass and ground to true plane surfaces. It is made in many colors in thicknesses from 0.34375 to 1.5 in (0.86 to 3.81 cm) or laminated to give various color effects. It is used for storefronts, countertops, tiling, and paneling. It does not craze, check, or stain like tile. **Spandrelite,** of this company, is an ornamental structural plate glass made by fusing ceramic color to a plate glass. It is used for cladding buildings.

Conductive glass, employed for windshields to prevent icing and for uses where the conductive coating dissipates static charges, is plate glass with a thin coating of stannic oxide produced by spraying glass, at 900 to 1300°F (482 to 704°C), with a solution of stannic chloride. Coating thicknesses are 164 to 1,804 nft (50 to 550 nm) and will carry current densities of 6 W/in^2 (9,300 W/m^2) indefinitely. The coatings are hard and resistant to solvents. The light transmission is 70 to 88% that of the original glass, and the index of refraction is 2.0, compared with 1.53 for glass. **Electrapane** is a conductive glass, as is **Nesa,** of PPG Industries.

Transparent mirrors are made by coating plate glass on one side with a thin film of chromium. The glass is a reflecting mirror when the light behind the glass is less than in front, and is transparent when the light intensity is higher behind the glass. **Photosensitive glass** is made by mixing submicroscopic metallic particles in the glass. When ultraviolet light is passed through the negative on the glass, it precipitates these particles out of solution; and since the shadowed areas of the negative permit deeper penetration into the glass than the highlight areas, the picture is in three dimensions and in color. The photograph is developed by heating the glass to 1000°F (538°C). **Photochromic glass** becomes dark under ultraviolet light and regains full transparency when the rays are removed. In sunlight, rearrangement of the oxides forms dark spots that impede the light rays.

Metal salts are used in glass for coloring as well as controlling the characteristics. Manganese oxide is added to most glass to neutralize iron oxide, but an excess colors the glass violet to black. **Jena blue glass** gets its color and fluorescence from a mixture of cobalt oxide and ceric oxide. **Ruby glass,** of a rich red color, is produced with selenium and cadmium sulfide, or with copper oxide. It is also produced with **purple of cassius,** or **gold-tin purple,** a brown powder which is a mixture of the yellow **gold chloride,** $AuCl_3$, and the dark-brown tin oxide. **Copper-ruby glass** has a greenish tinge and is suitable for automobile taillights, but signal lens glass is made with selenium, cadmium sulfide, and arsenious oxide, which give a distinct ruby color with heat treatment. **Amber glass** is made with controlled mixtures of sulfur and iron oxide that give tints varying from pale yellow to ruby amber. Amber glass is much used for medicine bottles to prevent entry of harmful light rays. The **actinic glass** used for skylights and factory windows has a yellow tint that softens and diffuses the light and impedes passage of heat rays. **Neophane glass** is glass containing neodymium oxide which reduces glare and is used in yellow sunglasses, or small amounts may be used in windshield glass. **Opticolor glass,** containing neodymium oxide for television tube filters, transmits 90% blue, green, and red light rays and only 10% of the less desirable yellow rays. The pictures are bright with improved color contrasts. **Opalescent glass,** or **opal glass,** used for light shades, tabletops, and cosmetic jars, has structures that cause light falling on them to be scattered, and thus they are white or translucent. They owe their properties to tiny inclusions with different indices of refraction, such as fluorides, sulfides, or oxides of metals. **Alabaster glass** has inclusions of larger dimensions in lower numbers per unit than opal glass, and it shows no colors, whereas opalescent glass appears white by reflected light but shows color images through thin sections. Opal glass may contain lipidolite, a mineral which contains various metals.

Monax glass is a white diffusing glass for lamp shades and architectural glass. **Glass blocks** for translucent units in factory walls are made from types of opalescent glass. **Insulex,** of Owens-Illinois, is a translucent glass brick of this kind. **Cellulated glass,** or **foamed glass,** is **expanded glass** in the form of blocks and sheets used for thermal insulation of walls and roofs. It is made by heating pulverized glass with a gas-forming chemical at the flow temperature of the glass. The glass expands and forms hollow cells which may comprise up to 90% of the total volume. The density is usually about 10 lb/ft^3 (160 kg/m^3). It has a compressive strength of 100 lb/in^2 (0.69 MPa) and a flexural strength of 75 lb/in^2 (0.5 MPa) and will retain rigidity to 800°F (427°C). **Low-melting-point glass,** for encapsulating electronic components, is made by adding selenium, thallium, arsenic, or sulfur to give various melting points from 260 to 660°F (127 to 349°C). These glasses can be vaporized and condensed as thin films. Some are insulators, others are semiconductors, and all are chemical-resistant.

Polarized glass, for polarizing lenses, is made by American Optical Co. by adding minute crystals of tourmaline or peridot to the molten glass and stretching the glass while still plastic to bring the axes of the crystals into parallel alignment. **Florentine glass** does not refer to a mixture but to an ornamental glass made by casting on an embossed bed, or by rolling with a roll upon which the designs are cut.

Porous glass is made of silica sand, boric acid, oxides of alkali metals (sodium, potassium, etc.), and a small amount of zirconia. This glass has 500 times better resistance to alkali solution than has been achieved to date. The glass is made by first melting, heat treatment at 1202 to 1472°F (650 to 800°C), then separating into two phases, one composed mainly of silica and the other containing boric acid and alkali metal oxides; this is known as *phase splitting*. The boric acid–alkali oxides phase elutes in heat treatment, producing small holes that give porous glass its name. However, porous glass of this type easily dissolves in water or alkali and, lacking durability, has little practical application. A technique has been devised of adding a large quantity of alkaline earth metal (calcium) oxide, i.e., quicklime, to the raw materials along with zirconia. This enables most of the zirconia to move into the silica phase at the time of phase splitting, thus becoming the silica-zirconia phase and producing porous glass of high durability. The size of pores can be controlled within a range of 3.28 nft (nm) to 79 μin (2 μm) by changing the conditions of heat treatment. The glass is heat-resistant, transmits gas, and permits adhesion of many substances to its surface. It is, therefore, used at high temperatures as a gas-separating membrane and as a carrier of various substances.

An **oxycarbide glass** has been developed in which substituting carbon for oxygen or even nitrogen can create a whole new category of **high-strength glasses.** In a 1.0% C glass system of Mg-Si-Al-O, Vickers hardness was increased significantly as well as the glass transition temperature. The oxycarbide glass was prepared by firing SiO_2, Al_2O_3, MgO, and SiC in a molybdenum crucible at 3272°F (1800°C) for 2 h. Oxycarbide glass systems based on Si-metal-O-C and Al-metal-O-C are likely to exist, and could potentially be used to produce refractory glasses. Controlled recrystallization of oxycarbides could lead to stable glass-ceramic matrices for ceramic-reinforced composites.

Retroreflective glass beads, with trihedral cube corners, are the optical elements used to reflect light rays along paths parallel to those of the incident rays in 3M's **Diamond Grade sheeting.** The sheeting, used for more-visible traffic signs, truck markings, and pavement marking, has a brightness of 800 to 1000 candela/ft candle/ft^2 (candela/lux/m^2), or cpl, compared with 300 cpl for conventional reflective materials. The beads, with sheeting color incorporated, are formed on the rear surface of dimensionally stable **plastic film,** covered with a transparent protective film, and sealed with a white film in an interlocking pattern to the rear of standard film. The sealing film is backed with a pressure-sensitive adhesive and protected by a flexible product liner. This arrangement results in encapsulated areas where only air contacts the beads so that light will be totally reflected internally by the bead facets, obviating reflective coatings.

Reinforcing glass and glass ceramics with continuous **silicon carbide fibers** or **carbon fibers** markedly increases strength, ductility, and toughness. In a system used by Schott Glaswerk of Germany, the fibers are coated with a slurry of fine **glass powder,** which becomes the matrix, and an organometallic **sol-gel,** which tailors the composition, controls fiber-to-matrix bonding, and transforms to an inorganic material when the glass is dried and tempered. To produce glass ceramics, nucleating agents are added. The coated fibers are then wound in wide bands to form tape, which can be cut and stacked to make plate shapes or wound to form annular shapes. The mass is then pressed at 725 lb/in² (5 MPa) and 1832°F (1000°C).

GLASS CERAMICS. A family of fine-grained crystalline materials made by a process of controlled crystallization from special glass compositions containing nucleating agents. They are sometimes referred to as **nucleated glass, devitrified ceramic,** or **vitro ceramics.** Since they are mixed oxides, different degrees of crystallinity can be produced by varying the composition and heat treatment. Some of the types produced are cellular foams, coatings, adhesives, and photosensitive compositions.

Glass ceramics are nonporous and generally either opaque white or transparent. Although not ductile, they have much greater impact strength than commercial glasses and ceramics. However, softening temperatures are lower than those for ceramics, and they are generally not useful above 2000°F (1093°C). Thermal expansion varies from negative to positive values depending on the composition. Excellent thermal shock resistance and good dimensional stability can be obtained if desired. These characteristics are used to advantage in "heatproof" skillets and rangetops. Like chemical glasses, these materials have excellent corrosion and oxidation resistance. They are electrical insulators and are suitable for high-temperature, high-frequency applications in the electronics field. They are also used for dental restorations.

Pyroceram, of Corning Inc., is a hard, strong, opaque-white nucleated glass with a flexural strength to above 30,000 lb/in² (206 MPa), a density of 2.4 to 2.62, a softening point at 2460°F (1349°C), and high thermal shock resistance. It is used for molded mechanical and electrical parts, heat exchanger tubes, and coatings. **Macor,** by the same company, is a machinable glass ceramic. Axles for mechanisms that provide power in pacemakers have been made of Macor due to chemical inertness and light weight. It is also used in welding fixtures and welding nozzles. The company's **MemCor,** which is also machinable, serves as a substrate for a magnetic memory-storage coating on small hard-disk drives for computers. Compared with aluminum for such applications, it is stronger, more rigid, and more shock-resistant and thus can be used in thinner sections without nickel coating. **Pyroceram balls** are used to replace steel balls in bearings and valves. They have the hardness of hardened tool steel with only one-third the weight, are corrosion-resistant, have a low coefficient of expansion, and withstand temperatures to 2200°F (1204°C). **Nucerite,** for lining tanks, pipes, and valves, is nucleated glass. It has about 4 times the abrasion resistance of a hard glass, withstands sudden temperature differences of 1200°F (649°C), and has high impact resistance. It also has high heat-transfer efficiency.

Polychromic glass is a glass ceramic that reproduces colors much like photographic film in a thin layer or in thick sections of glass up to 1 in (2.5 cm) thick. Before processing, the glass is colorless and transparent. The colored images are produced when activated by ultraviolet light and heat.

GLASS FIBER. Fine, flexible fibers made from glass are used for heat and sound insulation, fireproof textiles, acid-resistant fabrics, retainer mats for storage batteries, panelboard, filters, and electrical insulating tape, cloth, and rope. Molten glass strings out easily into threadlike strands, and this **spun glass** was early used for ornamental purposes; but the first long fibers of fairly uniform diameter were made in England by spinning ordinary molten glass on revolving drums. The original fiber, about 0.001 in (0.003 cm) in diameter, was called **glass silk** and **glass wool,** and the loose blankets for insulating purposes were called **navy wool.** The term *navy wool* is still used for the insulating blankets faced on both sides with flameproofed fabric, employed for duct and pipe insulation and for soundproofing.

Glass fibers are now made by letting the molten glass drop through tiny orifices and blowing with air or steam to attenuate the fibers. The usual composition is that of soda-lime glass, but it may be varied for different purposes. The glasses low in alkali have high electrical resistance, while those of higher alkali are more acid-resistant. They have very high tensile strengths, up to about 400,000 lb/in^2 (2,757 MPa).

The standard glass fiber used in glass-reinforced plastics is a borosilicate type known as **E-glass.** The fibers spun as single glass filaments, with diameters ranging from 0.0002 to 0.001 in (0.0005 to 0.003 cm), are collected into strands that are manufactured into many forms of reinforcement. E-glass fibers have a tensile strength of 500,000 lb/in^2 (3,445 MPa). Another type, **S-glass,** is higher in strength, about one-third stronger than E-glass, but because of higher cost, 18 times more costly per pound than E-glass; its use is limited to advanced, higher-performance products. **Hollex glass fibers** of Owens-Corning Fiberglas Corp. are **hollow fibers** of S-2 glass. They have a dielectric constant of 3.6, versus 3.4 for quartz fibers, which they can replace at lower cost for reinforcing electrically transparent plastics used for radomes and other components that protect aircraft radar and communications equipment.

Staple glass fiber is usually from 0.00028 to 0.00038 in (0.0007 to 0.0009 cm) in diameter, is very flexible and silky, and can be spun and woven on regular textile machines. **Glass fiber yarns** are marketed in various sizes and twists, in continuous or staple fiber, and with glass compositions varied to suit chemical or electrical requirements. **Vitron yarn** is a plied, low-twist yarn for braided insulation for wire. **Glass sewing thread** has a high twist. The minimum breaking strength of the 0.014-in (0.036-cm) thread is 12 lb (5 kg).

Halide glass fibers are composed of compounds containing fluorine and various metals such as barium, zirconium, thorium, and lanthanum. They appear to be promising for fiber-optic communication systems. Their light-transmitting capability is many times better than that of the best silica glasses now being used.

Glass fiber cloth and **glass fiber tape** are made in satin, broken twill, and plain weaves, the satin-wear cloth being 0.007 in (0.02 cm) thick and weighing 7 oz/yd^2 (0.24 kg/m^2). **Lagging cloth** for high-temperature pipe insulation is 20 oz/yd^2 (0.68 kg/m^2). **Glass insulating sheet,** for electrical insulation, is a tightly woven fabric impregnated with insulating varnish, usually in thicknesses from 0.005 to 0.012 in (0.013 to 0.030 cm). **Glass fabric** for varnishing is 0.001 in (0.003 cm) thick and weighs 0.81 oz/yd^2 (0.03 kg/m^2).

Glass cloth of plain weave of either continuous fiber or staple fiber is much used for laminated plastics. The usual thicknesses are from 0.002 to 0.023 in (0.005 to 0.058 cm) in weights from 1.43 to 14.7 oz/yd^2 (0.05 to 0.50 kg/m^2). Cloth woven of monofilament fiber in loose rovings to give better penetration of the impregnating resin is also used. **Glass mat,** composed of fine fibers felted or intertwined in random orientation, is used to make sheets and boards by impregnation and pressure. **Fluffed glass fibers** are tough, twisted glass fibers. For filters and insulation the felt withstands temperatures to 1000°F (538°C). **Chopped glass,** consisting of glass fiber cut to very short lengths, is used as a filler for molded plastics. Translucent corrugated building sheet is usually made of glass fiber mat with a resin binder. All these products, including chopped fiber, mat, and fabric preimpregnated with resin, and the finished sheet and board, are sold under a wide variety of trade names. Glass fiber bonded with a thermosetting resin can be made for pipe and other insulation coverings. **Glass fiber block** is also available to withstand temperatures to 600°F (316°C). **Glass filter cloth** is made in twill and satin weaves in various thicknesses and porosities for chemical filtering. **Glass belting,** for conveyor belts that handle hot and corrosive materials, is made with various resin impregnations. Many synthetic resins do not adhere well to glass, and the fiber is sized with vinyl chlorosilane or other chemical.

Four major principles should be recognized in using glass fibers as composite reinforcement. Mechanical properties depend on the combined effect of the amount of glass-fiber reinforcement used and its arrangement in the finished composite. The strength of the finished object is directly related to the amount of glass in it. Generally speaking, strength increases directly in relation to the amount of glass. A part containing 80% by weight glass and 20 resin is almost 4 times stronger than a part containing the opposite amounts of these two materials. Chemical, electrical, and thermal performance is influenced by the resin system used as the matrix. Materials selection, design, and production requirements determine the proper fabrication process to be used. Finally, the cost-performance value achieved in the finished composite depends upon good design and judicious selection of raw materials and processes.

GLASS SAND. Sand employed in glassmaking. Glass sands are all screened, and usually washed, to remove fine grains and organic matter. The grain standards of the American Ceramic Society specify that all should pass through a No. 29 screen, between 40 and 60% should remain on a No. 40 screen, between 30 and 40% should remain on a No. 60 screen, between 10 and 20% on a No. 80 screen, and not more than 5% should pass through a No. 100 screen. Sand for first-quality optical glass should contain 99.8% SiO_2 and maximums of 0.1 Al_2O_3 and 0.02 Fe_2O_3. Third-quality flint glass may contain only 95% SiO_2 and as high as 4 Al_2O_3. Only in the eighth- and ninth-quality amber glasses is the content of Fe_2O_3 permitted to reach 1%. **Potters' sand** is usually a good grade of glass sand of uniform grain employed for packing to keep the wares apart.

GLUCOSE. A syrupy liquid of composition $CH_2OH(CHOH)_4CHO$, which is a monosaccharide, or simple sugar, occurring naturally in fruits and in animal blood, or made by the hydrolysis of starch. It is also produced as a dry, white solid by evaporation of the syrup. Glucose is made readily from cornstarch by heating the starch with dilute hydrochloric acid, which is essentially the same process as occurs in the human body. Commercial glucose is made from cornstarch, potato starch, and other starches, but in Japan it is also produced from wood. Glucose is only 70% as sweet as cane sugar and has a slightly different flavor. It is used in confectionery and other foodstuffs for blending with cane sugar and syrups to prevent crystallization on cooling and because it is usually cheaper than sugar. It is used in tobacco and inks to prevent drying and in tanning as a reducing agent. The name *glucose* is usually avoided by the manufacturers of edible products, because of prejudices against its substitution for sugar; but in reality it is a simple form of sugar easily digested. It is used in medicine as a blood nutrient and to strengthen heart action, and it may be harmful only in great excess. When free from starch, it is called **dextrose.** It is also marketed as corn syrup, but corn syrup is not usually pure glucose, as it contains some dextrine and maltose. The **maltose, or malt sugar,** in the combination has the empirical formula $C_{12}H_{22}O_{11}$. When hydrolyzed in digestion, it breaks down easily to glucose. It is produced from starch by enzyme action. When purified, it is transparent and free of malt flavor. It is not as sweet as the sucrose of cane sugar, but is used in confectionery and as an extender of cane sugar. **Dry corn syrup** is in colorless, glasslike flakes. It is made by instantaneous drying and quick cooling of the syrup. **Sweetose,** of A. E. Staley Mfg. Co., is a crystal-clear enzyme-converted corn syrup used in confectionery to enhance flavor and increase brightness.

Glucose derived from grapes is called **grape sugar.** The glucose in fruits is called **fruit sugar, levulose,** or **fructose.** This is dextroglucose, and when separated out, it is in colorless needles which melt at 219°F (104°C). It is used for intravenous feeding and is absorbed faster than glucose. It is also used in low-calorie foods, and in honey to prevent crystallization. It is normally expensive, but is made synthetically. It can be made from corn and is superior to corn syrup as a sweetening agent. **Isomerose** is a corn syrup that is treated with an enzyme to convert part of the dextrose to fructose. It contains 14% fructose and has twice the sweetness of sucrose. Generically known as **high-fructose corn syrup,** or **HFCS,** the product for the most part has replaced sugar in soft drinks. **Maple syrup** is prepared by concentrating sap from the maple tree. **Molasses** is a by-product of cane sugar manufacture. **Corn sugar** is also a solid white powder, being glucose with one molecule of water crystallization. When the refined liquor is cooled, the corn sugar crystallizes in a mother liquor known as **hydrol, or corn-sugar molasses.** This molasses is screened, washed off, and marketed for livestock feed. The crystalline monohydrate sugar is known as **cerelose. Methyl glucoside,** made from corn glucose, is a white, crystalline powder melting at 327°F (164°C). It has composition $C_7H_{13}O_6$, with four esterifiable hydroxyls, and is used in making tall oil esters and alkyd resins. **Ethyl glucoside,** $C_8H_{16}O_6$, is marketed as a colorless syrup in water solution with 80% solids and a specific gravity of 1.272. It is noncrystallizing and is used as a humectant and plasticizer in adhesives and sizes.

GLUE. A cementing material usually made from impure gelatin from the clippings of animal hides, sinews, horn and hoof pith; from the skins and heads of fish; or from bones. The term **animal glue** is limited to **hide glue, extracted bone glue,** and **green bone glue.** Fish glue is not usually classified with animal glue, nor is casein glue. The vegetable glues are also misnamed, being classified with the mucilages. Synthetic resin glues are more properly classified with adhesive cements. Animal glues are **hot-work glues** which are applied hot and bind on cooling. Good grades of glue are semitransparent, free from spots and cloudiness, and not brittle at ordinary temperatures. **Bone glue** is usually light amber; the strong hide and sinew glues are light brown. The stiffening quality of glue depends upon the evaporation of water, and it will not bind in cold weather. Glues made from blood, known as **albumin glues,** and from casein are used for some plywood. However, they do not have the strength of the best grades of animal glue and are not resistant to mold or fungi. **Marine glue** is a glue insoluble in water, made from solutions of rubber or resins, or both. The strong and water-resistant plywoods are now made with synthetic resin adhesives.

Animal glue has been made since ancient times, and is now employed for cementing wood, paper, and paperboard. It will not withstand dampness, but white lead or other material may sometimes be added to make it partly waterproof. Casein glues and other protein glues are more water-resistant. **Soybean glue** is made from soybean cake and is used for plywood. It is marketed dry. It has greater adhesive power than other vegetable glues, or pastes, and is more water-resistant than other vegetable pastes. Hide glue is used in the manufacture of furniture, abrasive papers and cloth, gummed paper and tape, matches, and print rollers. The bone glues are used either alone or blended in the manufacture of cartons and paper boxes. Green bond glue is used chiefly for gummed paper and tape for cartons. In making bone glue the bones are crushed, the grease is extracted by solvents, and the mineral salts are removed by dilute hydrochloric acid. The bones are then cooked to extract the glue. Glues are graded according to the quality of the raw material, method of extraction, and blend.

There are 16 grades of hide glue and 15 grades of bone glue. Those with high viscosity are usually the best. Most glue is sold in ground form, but also as flake or pearl. Glues for such uses as holding abrasive grains to paper must have flexibility as well as strength, obtained by adding glycerin. The **animal protein colloid** of Swift & Co. is a highly purified bone glue especially adapted for use as an emulsifier, and for sizing, water paints, stiffening, and adhesives. Hoof and horn pith glue is the same as bone glue and is inferior to hide glue. **Fish glue** is made from the jelly separated from fish oil or from solutions of the skins. The best fish glue is made from Russian isinglass. Fish glues do not form gelatin well and are usually made into **liquid glues** for photographic mounting, gummed paper, household use, and use in paints and sizes. Liquid glues are also made by treating other glues with a weak acid. Pungent odors indicate defective glue. Glues made from decomposed materials are weak. Preservatives such as sulfur dioxide or chlorinated phenol may be used. The melting point is usually about 140°F (60°C).

GLYCERIN. A colorless, syrupy liquid with a sweet, burning taste, soluble in water and in ethyl alcohol. It is the simplest **trihydroxy alcohol,** with composition $C_3H_5(OH)_3$. It has a specific gravity of 1.26, a boiling point of 554°F (290°C) and a freezing point of 68°F (20°C). It is also called **glycerol,** and was used as a lotion under the name of **sweet oil** for more than a century after its discovery in 1783.

Glycerin occurs as **glycerides,** or combinations of glycerin with fatty acids, in vegetables and animal oils and fats, and is a by-product in the manufacture of soaps and in the fractionation of fats, and is also made synthetically from propylene. Coconut oil yields about 14% glycerin, 11 palm oil, 10 tallow, 10 soybean oil, and 9 fish oils. It does not evaporate easily and has a strong affinity for water, and it is used as a moistening agent in products that must be kept from drying, such as tobacco, cosmetics, foodstuffs, and inks. As it is nontoxic, it is used as a solvent in pharmaceuticals, as an antiseptic in surgical dressings, as an emollient in throat medicines, and in cosmetics. Since a different type of group can replace any one of or all three **hydroxyl groups,** (OH), a large number of derivatives can be formed, and it is thus a valuable intermediate chemical, especially in the making of plastics. Commercially, the most important are the **alkyd resins.** It is also used as a plasticizer in resins, and to control flexibility in adhesives and coatings. An important use is in nitroglycerin and dynamite. In water solutions the freezing point is lowered, reaching −60°F (−51°C), the lowest point, at 37% of water, and it is thus valuable as an antifreeze. **Glycerogen,** a German substitute made by the hydrogenation of wood hexose, is not pure glycerin, but also contains glycols and other hexyl alcohols.

GOLD. A soft, yellow metal, known since ancient times as a precious metal on which all material trade values are based. It is so chemically inactive that it is found mostly in the native state. It is found widely distributed in all parts of the world. It is used chiefly for coinage, ornaments, jewelry, and gilding. Gold is extracted by crushing the ores and catching the metal with quicksilver. About 25% of the gold produced in the United States is placer gold, and about 5% is a by-product of the copper industry. The average gold recovered from ore in the western United States is 0.2 oz/ton (0.006 kg/metric ton) of ore;

that in Alaska is 0.044 oz/ton (0.001 kg/metric ton); and the low-grade carbonaceous ore of Nevada contains about 0.3 oz/ton (0.009 kg/metric ton). **Native gold** is usually alloyed with silver, placer gold being the purest. The natural alloy of gold and silver was known as **electrum,** and under the Egyptian name of **Asem** it was thought to be an elementary metal until produced as an alloy by the Romans.

Among most civilized nations gold has always been the standard upon which trade values were set, even when gold itself was not used. For more than 25 centuries, until the extensive use of precious metals for industrial purposes, gold retained a 15.5-to-1 or 16-to-1 value with silver, the only other metal meeting the tests for a **coinage metal.** These tests are: that it have an intrinsic value to the people as a whole, as for ornamentation; that it be readily workable but highly corrosion-resistant and permanent; and that while reasonably scarce, it be available in all parts of the world so that it cannot be monopolized by any nation or group of nations. Thus, the value of gold (and silver) is regulated by law in all countries, and only gold and silver pieces are true coinage, those of other metals being merely **tokens** and paper money being merely promissory notes.

Gold is the most malleable of metals, and it can be beaten into extremely thin sheets. A gram (0.0022 lb) of gold can be worked into leaf covering 6 ft^2 (0.6 m^2), and only 0.0000033 in (0.0000084 cm) thick, or into a wire 1.5 mi (2.5 km) in length. **Cast gold** has a tensile strength of 20,000 lb/in^2 (138 MPa). The specific gravity is 19.32, and the melting point is 1948°F (1063°C). It is not attacked by nitric, hydrochloric, or sulfuric acid, but is dissolved by aqua regia, or by a solution of azoimide, and is attacked by sodium and potassium cyanide plus oxygen. The metal does not corrode in air, only a transparent oxide film forming on the surface. Its reflectivity of ultraviolet and visual light rays is low, but it has high reflectivity of infrared and red rays and is thus valued for plating some types of reflectors. For radiation-control coatings for spacecraft, **gold flake** is used. Gold of 14-karat purity is used in the form of tiny, laminar platelets with overlap to form a film. Gold-plated grid wires in electronic tubes give high conductivity and suppress secondary emission. **Gold-gallium** and **gold-antimony** alloys of J. M. Ney Co., for electronic uses, come in wire as fine as 0.005 in (0.013 cm) in diameter and in sheet as thin as 0.001 in (0.003 cm). The maximum content of antimony in workable gold alloys is 0.7%. A gold-gallium alloy with 2.5% gallium has a resistivity of 90 Ω/cir mil ft (15 $\times$ 10^{-8} $\Omega \cdot$ m), a tensile strength of 55,000 lb/in^2 (379 MPa), and 22% elongation. **Gold powder** and **gold sheet,** for soldering semiconductors, are 99.999% pure. The gold wets silicon easily at a temperature of 700°F (371°C). Chemically reduced powder is amorphous and comes in particle sizes of 39 to 315 μin (1 to 8 μm). **Atomized gold powder** has spherical free-flowing particles from 30 to 400 mesh. The "gold" powder used in some paints and plastics is bronze powder. The gold coating on spacecraft for radiation control must be hard for resistance to meteorite particles. The metallizing powder contains up to 20% platinum or palladium with a glass frit.

Because of its softness, gold is almost always alloyed with other metals, usually copper, silver, or nickel, and graded on the basis of degrees of fineness in 1,000 parts, or on the basis of karat gold value, pure gold being 24 karats. **Green gold,** used in making jewelry, is an alloy of gold, silver, and copper, graded from 14 to 18 karats. The 18-karat green gold contains 18 parts gold and 6 silver, with no copper. The 15-karat grade contains 15 parts gold, 8 silver, and 1 copper. The 14-karat grade contains 14 parts gold, 8.25 silver, and 1.75 copper. **Coinage gold** in the United States is 90% gold and 10 copper. In England it is 91.66% gold and 8.33 copper, and this alloy is called **standard gold.** In Australia 8.33% silver is used instead of the copper, and the **gold-silver alloy** is called **Australian gold. Dental gold** is a term for a wide range of wrought and cast alloys with usually from 65 to 90% gold, 5 to 12 silver, and frequently platinum and sometimes palladium. A very small amount of iridium may also be used for hardening. Colors vary from white to yellow, and such alloys are also used for jewelry and acid-resistant plates. Gold is easily electroplated on other metals from cyanide solutions in controlled thicknesses from 0.000005 to 0.005 in (0.000013 to 0.013 cm). **Sodium gold cyanide,** used for gold-plating solutions, is a water-soluble, yellow powder of composition $NaAu(CN)_2$, containing 46% gold. **Gold plate** is thus much used for jewelry, ornaments, and for chemical resistance, but the gold is so soft that thin plates wear off easily. But small amounts of other metals harden the gold and give a color range from red to pink gold to lemon yellow. For plating electrical contacts, alloys with 1 to 6% nickel are used. The hardness of gold doubles with each 2% increase in nickel content, but the electrical resistance increases. Hard gold plate is a **gold-indium plate.** The process of Indium Corp. of America is to plate the gold and indium successively and then alloy by diffusion at 330°F (166°C). Precipitation hardening takes place, giving a hard, wear-resistant coating.

South Africa is the most important producer of gold. Other important producers are Canada, the United States, and Australia, but gold is also produced in 90 other countries. The ore of gold known as **calaverite,** found in Colorado, California, and west Australia, is a **gold telluride,** $AuTe_2$, occurring in monoclinic crystals, while the variety known as **krennerite** is Au_8Te_{16} and occurs in orthorhombic crystals. They are found with the **gold-silver telluride** called **syl-**

vanite, (AuAg)Te$_2$, a silvery-white granular material of specific gravity 9.35 and Mohs hardness 2.5, which is easily fusible.

GRANITE. A coarse-grained, igneous rock having an even texture and consisting largely of quartz and feldspar with often small amounts of mica and other minerals. There are many varieties. Granite is very hard and compact, and it takes a fine polish, showing the beauty of the crystals. It is the most important building stone and is also used as an ornamental stone. An important use is for large rolls in pulp and paper mills. **Granite surface plates,** for machine-shop layout work, are made in sizes up to 30 by 72 in (76 by 183 cm) and 10 in (25 cm) thick, ground and highly polished to close accuracy. It is extremely durable, and since it does not absorb moisture, as limestone and sandstone do, it does not weather or crack as these stones do. The colors are usually reddish, greenish, or gray. **Rainbow granite** may have a black or dark-green background with pink, yellowish, and reddish mottling; or it may have a pink or lavender background with dark mottling. The density is 170 lb/in^3 (2,723 kg/m^3), the specific gravity 2.72, and the crushing strength 23,000 to 32,000 lb/in^2 (158 to 220 MPa). The most notable granite quarries are in northern New England. **Mount Airy granite** from North Carolina is light gray and is a **biotite** containing feldspar, quartz, and mica. It is somewhat lighter in weight and of lower crushing strength than **Maine granite.** The hard composite igneous rock **diabase,** called **black granite,** is used for making precision parallels for machine-shop work. **Unakite** is a granite of Virginia and North Carolina with a mosaic of red, green, and other colors. It is used as an ornamental building stone. **Balfour pink granite** of North Carolina contains 72% silica, 14.1 alumina, 2 soda, 6 potash, 1.2 iron oxide, 0.12 titanium oxide, 0.20 manganese oxide, and 0.36 lime. The granite known as **pegmatite,** of which there are vast quantities, contains beryllium in the form of beryl as a minor constituent.

GRAPHITE. A form of carbon. Also called **plumbago.** It was formerly known as black lead, and when first used for pencils, it was called **Flanders' stone.** A natural variety of elemental carbon having a grayish-black color and a metallic tinge.

Natural graphite comes chiefly from Mexico, India, Sri Lanka, and Malagasy. In the United States it is found in Alabama. It occurs in veins in rocks and always contains some impurities. The high-grade lump graphite of Sri Lanka, often preferred for crucibles, is found in closely foliated masses in underground veins in gneiss interbedded with limestone. It occurs in two forms: foliated and amorphous. **Foliated graphite** is used principally for crucibles and lubricants, and amorphous for lead pencils, foundry facings, electric brush carbons, molded parts, and paint pigments. It is infusible, subliming at 6700°F (3704°C), but oxidizing above 1112°F (600°C). It is a good conductor of heat and electricity, is resistant to acids and alkalies, and is readily molded. **Molded graphite** is usually made by mixing calcined petroleum coke with a binder of coal-tar pitch, pressing, baking in an inert atmosphere, and then heating to above 3500°F (1927°C) to promote crystal growth. Molded parts increase in strength with increasing temperature up to about 4500°F (2482°C). The specific gravity of graphite is 2 to 2.5; Mohs hardness is 1 to 2, sometimes less than 1; and it has a decidedly greasy feel. It is a good lubricant, especially when mixed with grease; but the natural graphite contains silica and other abrasive materials so that artificial graphite is preferred for lubricants.

Graphite is marketed in grades by purity and fineness. Number 1 flake should contain at least 90% graphitic carbon. Mexican **amorphous graphite** contains 80%. The best Malagasy graphite contains 95% carbon, while the powders may be as low as 75%. **Crystalline graphite** and **flake graphite** are synonymous terms for material of high graphite content, as distinguished from amorphous. Some natural graphite, useful for paints, contains as little as 35% graphite carbon; but high-grade graphite, suitable for crucibles and nuclear reactors, is made from low-grade ores containing 20% carbon by flotation, purification at high heat, and pressing into blocks. Ultrapure graphite, for nuclear reactors, is graphitized at temperatures to 5400°F (2982°C) to free it of silicon, calcium, aluminum, and manganese, and it is treated with a Freon gas to eliminate boron and vanadium. In general, **artificial graphite** made at high temperatures in the electric furnace is now preferred for most uses because of its purity.

Recrystallized graphite is produced by a proprietary hot-working process which yields recrystallized or "densified" graphite of specific gravity 1.85 to 2.15, compared with 1.4 to 1.7 for conventional graphites. The material's major attributes are a high degree of quality reproducibility, improved resistance to creep, a grain orientation that can be controlled from highly anisotropic to relatively isotropic, lower permeability than usual, absence of structural macroflaw, and ability to take a fine surface finish.

Graphite fibers are produced from organic fibers. One line of development used **rayon** as the precursor, and the other used **polyacrylonitrile (PAN)**. Although the detailed processing conditions for converting cellulose or PAN to carbon and graphite fibers differ in detail, they both consisted fundamentally of a sequence of thermal treatments to convert the precursor to carbon by breaking the organic compound to leave a "carbon polymer." The fibrous carbon formed by the controlled pyrolysis of organic precursor fiber was viscous rayon or acrylonitrile. **Carbon fibers** produced by the rayon-precursor method have fine-grained, relatively disordered microstructure, which remains even after treatment at temperatures up to 5432°F (3000°C). Graphite crystallites with a long-range three-dimensional order do not develop. In both the rayon and PAN processes, a high degree of preferred crystal orientation was responsible for the high elastic modulus and tensile strength.

Although the names *carbon* and *graphite* are used interchangeably when related to fibers, there is a difference. Typically, **PAN-based carbon fibers** are 93 to 95% carbon by elemental analysis, whereas graphite fibers are usually 99+%. The basic difference is the temperature at which the fibers are made or heat-treated. PAN-based carbon is produced at about 2400°F (1316°C), while higher-modulus-of-elasticity graphite fibers are graphitized at 3450 to 5450°F (1899 to 3010°C). This also applies to carbon and graphite cloths. Unfortunately, with only rare exceptions, none of the carbon fibers are ever converted to classic graphite regardless of the heat treatment.

When used in composites, fibers are generally made into yarn containing some 10,000 fibers. Depending on the precursor fiber, tensile strength ranges from 200,000 to nearly 500,000 lb/in^2 (1,379 to 3,448 MPa) and modulus of elasticity from 28×10^6 to 75×10^6 lb/in^2 (193,060 to 517,125 MPa).

Graphite-reinforced graphite composites, also called **graphite-graphite composites** and **carbon-carbon composites,** can be used at temperatures to about 6332°F (3500°C). Most comprise 99.5 to 99.9% carbon, and there are no reinforcement-to-matrix compatibility problems because both the reinforcing elements, such as fibers, and the matrix are graphite. The matrix is produced by pyrolytic decomposition of polymeric systems in which the reinforcing elements are originally embedded. Many of the matrix starting materials are proprietary, but include phenolic, furfuryl ester, and epoxy resins. Even those produced of low-modulus reinforcements are up to 20 times stronger than unreinforced carbon and graphite, and their high purity provides good chemical and corrosion resistance. Typical properties include a tensile strength of 8,200 lb/in^2 (57 MPa), a flexural strength of 11,000 lb/in^2 (76 MPa), a compressive strength of 40,000 lb/in^2 (276 MPa), and a modulus of 2.5×10^6 lb/in^2 (172,375 MPa). And having a density of 0.034 to 0.05 lb/in^3 (941 to 1,384 kg/m^3), they provide high strength-to-weight and rigidity-to-weight ratios. They also have good thermal conductivity and erosion resistance to molten metals and perform well in reducing environments, but require protective coating in air or oxidizing atmospheres under certain conditions. For example, an oxygen-barrier coating is necessary for exposure in air at temperatures above 800°F (427°C). These composites are used mainly, however, for their exceptional thermal stability at very high temperatures such as in select regions of aerospace vehicles.

Graphite-reinforced plastics also provide high strength and rigidity and are used in a greater variety of applications. Graphite-reinforced epoxy constitutes the most widely used polymer-matrix composite for structural applications on aircraft. In recent years, bis-maleimide has replaced epoxy to some extent in the interest of somewhat greater heat resistance. Many other plastics, including thermoplastics, are also reinforced with graphite. Sports equipment, such as golf club shafts and tennis rackets, is a common application.

Extremely fine particles of pure artificial graphite, or **colloidal graphite,** will remain in suspension indefinitely, and are marketed in distilled water, oils, or solvents under trade names. **Mexacote,** of U.S. Graphite Co., is colloidal graphite powder to be mixed with water for spraying on sand molds. When a solution of colloidal graphite in alcohol is sprayed on machine bearings, the alcohol evaporates leaving a thin coating of graphite as a lubricant. **Prodag,** of Acheson Colloids Co., is a solution in water for foundry facings. **Dag dispersion 154,** of the same company, is colloidal graphite in ethyl silicate used to produce black coatings on glass. **Dag 440** is **graphite powder** in silicone resin for use as a resistance coating for continuous temperatures up to 500°F (260°C). A 0.002-in (0.005-cm) coating has a volume resistivity of 100 Ω/in^2 (0.065 Ω/m^2). **Grafita** and **Grafene** are grease and oil solutions of colloidal graphite for producing lubricating films. **Acheson 1127,** of Acheson Colloids, is a **release agent** for spraying on molds in aluminum die-casting machines. It is a water solution of graphite powder. It prevents adherence of the metal to the die and gives better flow of metal because of reduced friction.

For making lead pencils, amorphous graphite is mixed with clay and fired, the amount of clay determining the hardness of the pencil. **Flexicolor pencils,** of Koh-I-Noor, Inc., have a plastic binder to give pliable strength to the pencil. **Polo Graphite** is a graphite powder with a maximum particle size of 0.001 in (0.003 cm), used for molded parts. The parts have a smoother surface and a tensile strength to 6,000 lb/in^2 (41 MPa). **Graphite carbon raiser** is a term given to graphite powder added to molten steel to raise the carbon content. **Kish,** a waste product generated in steelmaking, is a fine powder of graphite and iron traditionally disposed of as landfill. In a process developed by the U.S. Bureau of Mines and commercialized by Inland Steel, synthetic flake graphite of 95% purity can be recovered by separating the two materials by wet centrifuging and leaching off the oxide with hydrochloric acid. Molded **graphite brushes** for motors and generators may have metal powders mixed with the graphite to regulate the conductivity or may be molded from carbon-graphite powder that is specially heat-treated.

Graphited metals, used for bearings and bushings, are made by molding metal powders with graphite and sintering, and they may contain up to about 45% graphite evenly dispersed in the metal matrix to act as a lubricant. Or powdered oxides of the metals may be used, and these oxides are reduced in the sintering to give greater porosity for oil retention. **Genelite,** of General Electric Co., is a porous **graphitized bronze** made with oxides of copper, tin, and lead with graphite powder. It will absorb about 20% by volume of oil. It has a compressive strength of 50,000 lb/in^2 (345 MPa), and a tensile strength of 8,000 lb/in^2 (55 MPa). **Durex bronze,** of General Motors Corp., made by reducing the oxides with graphite under pressure, will take up 29% by volume of oil. Graphited metals, with the matrix of bronze or of babbitt, are marketed in the form of rods and bushings under a variety of trade names, such as **Gramix, Graphex** of Wakefield Corp., and **Ledaloyl. Iron-bonded graphite,** developed by Ford Motor Co. for oilless bearings, is made by powder metallurgy with a content of 40 to 90% graphite and the balance iron powder and a calcium-silicon powder. The calcium-silicon overcomes the normal low compatibility of the iron and graphite, resulting in a strong, nonbrittle material.

Supergraphite, used for rocket casings and other heat-resistant parts, is recrystallized molded graphite. It withstands temperatures to 5500°F (3038°C). **Pyrolytic graphite,** developed by General Electric Co., is an oriented graphite. It has high density, with a specific gravity of 2.22; has exceptionally high heat conductivity along the surface, making it very flame-resistant; is impermeable to gases; and withstands temperatures to 6700°F (3704°C). It is made by deposition of carbon from a stream of methane on heated graphite, and the growing crystals form with thin planes parallel to the existing surface. The structure consists of close-packed columns of graphite crystals joined to each other by strong bonds along the flat planes, but with weak bonds between layers. This weak and strong electron bonding is characteristic of a semimetal, providing a laminal structure. The material conducts heat and electricity many times faster along the surface than through the material. The flexural strength is 25,000 lb/in^2 (172 MPa) compared to less than 8,000 lb/in^2 (55 MPa) for the best conventional graphite. At 5000°F (2760°C) the tensile strength is 40,000 lb/in^2 (276 MPa). Sheets as thin as 0.001 in (0.003 cm) are impervious to liquids or gases. It is used for nozzle inserts

and reentry parts for spacecraft. **Boron Pyralloy,** for atomic shielding, is this material with the addition of boron.

GRAVEL. A natural material composed of small, usually smooth,
rounded stones or **pebbles.** It is distinguished from sand by the size
of the grain, which is usually above 0.25 in (0.64 cm); but gravel may
contain large stones up to 3 in (7.62 cm) in diameter and some sand.
It will also contain pieces of shale, sandstone, and other rock materials. Gravel is used in making concrete for construction and as a loose
paving material. Commercial gravel is washed to remove the clay and
organic material and is screened. **Pea gravel** is screened gravel
between 0.25 and 0.50 in (0.64 and 1.27 cm) in diameter. It is used for
surfacing with asphalt or for roofing. Gravel is sold by the cubic yard
(cubic meter) or by the ton (metric ton) and is shipped by weight.
Bank-run gravel, with both large and small material, has a density
of about 3,000 lb/yd^3 (1,779 kg/m^3).

GREENHEART. The wood of the tree *Octotea rodioei* of Guyana, especially valued for shipbuilding, dock timbers, planking, and lock gates
because of its resistance to fungi and termites. It also goes under the
name of **Demerara greenheart** and under its native name of
bibiru. The tree grows to a height of 120 ft (37 m), with a diameter of
2 to 3 ft (0.6 to 0.9 m), clear of branches for 50 to 70 ft (15 to 21 m).
The wood is very strong and hard with good wearing qualities. The
average density is 62 lb/ft^3 (993 kg/m^3) with 12% moisture. The specific gravity, oven-dried, is 0.80. The heartwood is light olive to nearly
black, and the sapwood pale yellow to greenish.

   Surinam greenheart is the wood **bethabara,** from the tree
Tecoma leucoxylon of Surinam and French Guiana. It is distinguished
from greenheart by yellow deposits in the pores and is not as resistant as greenheart. **Manbarklak,** the wood of the tree *Eschweilera
corrugata,* of Surinam, is reddish brown, has a density of 76 lb/ft^3
(1,218 kg/m^3), is equal to greenheart for marine construction, but is
scarcer. **Angelique,** the wood of the tree *Dicorynia paraensis* of the
Guianas and lower Amazon, is very resistant to fungi and insect
attack and is used as a substitute for greenheart in marine construction. It is hard but not as heavy as greenheart, having a density of 50
lb/ft^3 (801 kg/m^3). It has an olive-brown color with reddish patches.
African greenheart is **dahoma,** the wood of the large tree
Piptadena africana of the west coast of Africa. It is yellowish brown,
also has a density of 50 lb/ft^3, but is not as resistant to marine borers
as greenheart. The wood of a species of tonka bean tree of Panama,
known as **almendro,** *Coumarouna panamensis,* is very resistant to
marine borers and is used as a substitute for greenheart. It is yellow-

ish brown and very heavy. The almondlike seeds of the tree are used as food.

GRINDING PEBBLES. Hard and tough, rounded small stones, usually of flint, employed in cylindrical mills for grinding ores, minerals, and cement. Pebbles from Greenland, marketed usually through Denmark and known as **Danish pebbles,** are very hard and tough. Newfoundland also supplies these pebbles. Quantities of **flint pebbles** also come from Denmark for use in tube mills. They are smooth, round pebbles formed by the washing of the sea on the chalk cliffs, and they come from the islands off the Danish coast. Danish pebbles are graded in seven sizes, according to French standards, from No. 0 which is 1.5 to 1.75 in (3 to 4 cm) to size E which is 4 to 5 in (10 to 13 cm). Small pebbles, 0.5 in (1.27 cm) in diameter, are used for polishing iron castings by tumbling. The U.S. grinding pebbles are chiefly from Minnesota, Ohio, Nevada, and the beaches of California. **Quartzite pebbles** are produced in Nova Scotia and Saskatchewan. **Quartz pebbles** from Alabama are 99% silica and low in iron, but they do not wear as well as true flint pebbles. Granite, rhyolite, and andesite pebbles are also used for grinding. The **granite pebbles** produced in North Carolina from feldspar or other mineral processing are broken into cubes and wet-milled to remove the edges and corners. They are graded in sizes from 1.5 in (3.8 to 12.7 cm). Because of greater uniformity of size and hardness, manufactured abrasive materials are now generally preferred to natural pebbles.

The **tumbling abrasives,** for use in tumbling barrels, come in aluminum oxide or silicon carbide preformed balls, cubes, triangles, or cylinders of various sizes to conform to the parts being tumble-polished. Typical are the **aluminum oxide balls,** 0.75 to 2 in (1.9 to 5.1 cm) in diameter, pressed to a uniform density and marketed under the name **Starrlum** by American Refractories & Crucible Corp. **Porcelain grinding balls** are made of high-grade resistant porcelain and are marketed in stock sizes for grinding and polishing. They have the advantage over flint pebbles of greater uniformity. The material is dense and is 80% heavier and 50% harder than flint.

GRINDSTONES. **Sandstones** employed for grinding purposes. Grindstones are generally used for the sharpening of edged tools, and do not compete with the hard emery, aluminum oxide, and silicon carbide abrasive wheels which are run at high speeds for rapid cutting. Grindstones are quarried from the sandstone deposits and made into wheels usually ranging from 1 to about 6 ft (0.3 to about 1.8 m) in diameter and up to 16 in (41 cm) in thickness. They are always operated at low speeds because of their inability to withstand high centrifugal stresses. The grades vary from coarse to fine. Good grind-stones have sharp grains, without an excess of cementing material that will cause the stone to glaze in grinding. The texture must also be uniform so that the wheel will wear evenly. The hard silica grains are naturally cemented together by limonite, clay, calcite, quartz, or mixtures. Too much clay causes crumbling, while too much calcite results in disintegration in the atmosphere. An excess of silica results in a stone that is too hard.

GUAIACUM OIL. An essential oil distilled from the wood of the guaya-can tree of Paraguay, used in medicine, soaps, and perfumes. It is light gray and is solid at temperatures below 45°C. The odor is that of a combination of tea leaves and roses. It is also called **guaiacwood oil.** The wood yields 5 to 6% of the oil. **Guaiac gum,** also called **gua-iacum,** is gum resin of the true lignum vitae trees for use in varnish-es, as a chemical indicator, and to prevent rancidity and loss of flavor in preserved and dehydrated foods. It is an effective antioxidant, although its chief use is in medicine as a stimulant and laxative. The resin comes in greenish-brown tears, and a good-quality product is 90% soluble in alcohol. **Azulene** is a blue dye extracted from gua-iacum oil, from eucalyptus oil, and from some balsams. The azulenes derive names from the source, as **guaiazulene,** but they all have the empirical formula $C_{15}H_{18}$, the molecule having two rings and five dou-ble bonds. The synthetic material is known as **vetivazulene.** It comes in cobalt-blue crystals melting at 194°F (90°C).

GUANIDINE NITRATE. A white granular powder of composition $(H_2N \cdot CNH \cdot NH_2) \cdot HNO_3$, produced from cyanamide and ammonium nitrate and employed in making flashless propellants and pharmaceu-ticals, for coating blueprint paper to speed up development time, and in photographic fixing baths. It melts at 403 to 414°F (206 to 212°C) and is strongly acidic. **Guanidine hydrochloride** is an intermediate in drug production, such as for **sulfadiazine** and **sulfamerazine,** both antibacterials. **Aminoguanidines** are dye intermediates. **Dodeylguanidine** salts are sold under the trade name **Cyprex,** as fungicides. **Guanidine carbonate** is a white nonhygroscopic granu-lar powder of composition $(H_2N \cdot CNH \cdot NH_2)_2 \cdot H_2CO_3$. It decomposes at 338°F (198°C). It is soluble in water and has an alkaline reaction. Guanidine carbonate is used in the manufacture of pharmaceuticals, as an emulsifier in soaps, and in photographic developing solutions. Other derivatives of guanidine, such as **diphenyl guanidine,** are used as plasticizers and accelerators for rubber.

   Guanidine, also called **carbamidine,** is a white crystalline hygro-scopic powder of composition $HN{=}C(NH_2)_2$, produced from ammonium thiocyanate, chloropicrin, or cyanogen chloride for use in plastics, pharmaceuticals, and other chemicals. It can also be made by reacting dicyandiamide with ammonium nitrate in the presence of ammonia. It is a strong base and forms salts with acids even as weak as carbonic. It is classified generally with the single-carbon group of chemicals based on **cyanogen,** $N{\equiv}CC{\equiv}N$, which itself is a poisonous, colorless gas. The chemistry of this vast interrelated group is highly complex, including urea and the amino acids, and the **purines,** which are the basic cyclic organic compounds of which **uric acid** is the best known. The purines also include caffeine and **xanthine,** found in plants, and **guanine,** found in guano and in the fish scales used for imitation pearl. **Kelzan,** of Kelco Co., used as a thickening and stabilizing agent, is synthetic **Xantham gum** made by fermentation of glucose. It is nontoxic and soluble in water. Xantham gum is a polysaccharide of **xanthic acid** occurring in many plants.

    Guanidine oxalate, guanidine polyacrylate, and **guanidine salts** of fatty acids can be used to produce high-purity ceramics better able to withstand temperatures over 2190°F (1199°C) and corrosion than those made traditionally using alkali metal hydroxides, such as sodium hydroxide and potassium hydroxide. Their superior performance stems from the absence of sodium or potassium residues.

GUAR GUM.   Used as a replacement for carob bean gum and gum arabic in foods and pharmaceuticals, it is a water-soluble, odorless, and tasteless white powder obtained from the endosperm of the seed of the guar plant, *Cyamopsis tetragonoloba,* grown in Pakistan as cattle feed and cultivated in Texas and Arizona. The gum is a polysaccharide with a straight grain of mannose united having one galactose group on every other mannose unit. There is 3:1 ratio of mannose to galactose compared with a 4:1 ratio in **locust bean gum.** It also contains about 6% protein, about the same as in locust bean gum. Guar gum and its derivatives are used to increase petroleum production by cracking open hydrocarbon-bearing zones; as a thickener in slurry explosives and water gels; as a viscosifier in ice cream; and as a wet-end additive in papermaking. The carboxymethyl and hydroalkyl ethers of guar are dye-solution thickeners in textile printing applications. **Guar flour** is a nearly white guar gum. Its high swelling properties in cold water make it suitable as a disintegrating agent in medical tablets. It has more than 6 times the thickening power of starch and is used for upgrading starches. **Jaguar gum** is guar gum. **Jaguar 315** and **Guartec,** of General Mills, Inc., are derivatives of guar gum that gel in either acid or alkaline solutions and form stiff gels with as much as 99% water content. They are polymers of mannose and galactose.

GUAVA. The fruit of the tree *Psidium guajava* of the myrtle family to which the clove, allspice, and eucalyptus belong. Because of its content of acids, sugars, pectin, and vitamins, it is valued in the food industries for blending in jellies, preserves, and beverages. The pulp is also used in ice-cream manufacture. Vast quantities of guava are used for hard jellies in Latin America. The fruit has more than 10 times the vitamin C content of the orange and retains it better. It also contains vitamins A and B_1 and 11.6% carbohydrates. The tree is native to America from Mexico and Peru and is also grown extensively in Brazil and the West Indies. It was one of the original Aztecan and Incan fruits, and is still known under the Carib name of **guayaba** in Brazil and Argentina. There are about 150 varieties, and the large, seedy fruit has a fragrant aroma and distinctive sweet flavor.

GUAYULE. A perennial plant, *Parthenium argentatum,* of the *Compositae* family, grown in the semiarid regions of northern Mexico and southern California as a source of rubber. The plants are hardy, woody shrubs that mature into the highest rubber content in 7 years. They contain in the dry state up to 22% guayule rubber, in all parts except the leaves. The plant is uprooted in 3 to 5 years and is crushed and pulverized in mills, and the rubber extracted by flotation. The guayule rubber contains from 20 to 25% resin so that it is suitable only for blending or for cements, unless deresinated. Natural crude guayule is softer than hevea rubber, owing to the content of natural resins which act as plasticizers. In the low-sulfur compounds, it remains permanently tacky and is thus valued for use as a coating adhesive for the permanently tacky binding tape known as Scotch tape. When deresinated by extraction of the resin with acetone or other solvent, the rubber is suitable for all the uses of hevea rubber. The by-product **guayule resin** is used in plastics. From 400 to 1,000 lb (181 to 454 kg) of rubber is produced per acre (4,047 m²) under cultivation.

GUM. The wood of the tree *Liquidambar styraciflua,* of the United States and Mexico. It is called **red gum** and **sweet gum.** In England it is known as **California red gum,** although the gumwood of California is from a eucalyptus tree. In Europe, also, the term **satin walnut** is used for the heartwood and **hazel pine** for the sapwood. Gum has a reddish-brown color; is soft with a fine, close grain; and has a density of about about 40 lb/ft³ (641 kg/m³). It is used for furniture, veneer, inside trim, cooperage, and the making of pulp for book paper. The timber is cut mostly in the southern states, especially in Louisiana, Mississippi, and Arkansas. The trees reach a height of 80 to 100 ft (24 to 30 m) and average 1.5 to 3 ft (0.5 to 0.9 m) in diameter with a straight, clear trunk. Red gum is from the heartwood of mature trees and is reddish brown. **Sap gum** comes from the outer portion of logs or from young trees and is white tinged with pink. Nearly 25% of all the hardwood used in the United States is red gum. It has an interlocking grain which gives a fine appearance in veneers, but has a tendency to warp.

Gum is graded according to standards of the National Hardwood Lumber Association from firsts through selects to No. 3B common. Local names for red gum are **southern gum, sycamore gum, bilsted,** and **starleafed gum. Cotton gum,** or **tupelo,** of Louisiana, is from the tree *Nyssa aquatica.* It is also known as **water tupelo, tupelo gum, swamp gum, sour gum. Black gum,** or **black tupelo,** also of the southern states, is *N. sylvatica.* **Swamp tupelo,** also called **water gum** and **swamp black gum,** is from the tree *N. biflora.* **Ogeche tupelo** is from the tree *N. ogecha* and is not common. It is called **gopher plum, wild limetree,** and **sour tupelo.** Black tupelo grows from New Hampshire to central Texas, but water tupelo is found chiefly along the coasts and river valleys of the south. The shipments of woods are usually mixed. Tupelo woods from the various species of *Nyssa* are fine-textured but with large pores. The heartwood is brownish gray, and the sapwood is grayish white. They are tough and difficult to split, having an interlocking grain, and find wide use for such articles as mallets, toilet seats, furniture, and bottle cases.

GUM ARABIC. Also called **acacia gum.** The gum exudation of the small tree *Acacia arabica* and various other species of acacia trees of Africa. **Kordofan gum,** or **hashab gum,** is a variety from the Red Sea area and forms the chief export of the Sudan. It is obtained by tapping the wild tree *A. verek* and is of high quality. **Senaar gum** is gum arabic exported from Arabian ports on the Red Sea. **Gum senegal,** from *A. senegal,* comes from the dry regions of northwest Africa. **Gum talha, talco gum,** or **talh gum** is a brittle and low grade of gum arabic from the north African acacia *A. stenocarpa.* It is unique in that it forms solutions of greater than 50% concentration. Gum arabic is used for adhesives, for thickening inks, in textile coatings, and in drug and cosmetic emulsions. As a binder in pharmaceutical tablets, the powder acts as a **disintegrating agent** to make the tablets easily soluble in water. In confectionery glazes it prevents crystallization of the sugar. It is a foam stabilizer in beer and a fixative in spray-dried flavors. IT is used as a glaze in confectionery because it prevents sugar crystallization.

To obtain the gum, the trees are wounded and the sap is allowed to run out, forming yellowish, transparent lumps. It is also marketed as a white powder of 120 mesh, soluble in water, but insoluble in alcohol.

Gum arabic is a mixture of calcium, magnesium, and potassium salts of **arabic acid,** in a complex of the saccharides **arabinose, galactose, rhamnose** or **mannomethylose,** and the open-chain **glucuronic acid.** It has a molecular weight of 240,000 and an acid reaction. For drug uses gums are selected, blended, and ground to a powder of uniform characteristics. **Arabasan** is a spray-dried, blended gum of this kind. It is a white powder, colorless in solution. **Cloud gum** is a combination of gum arabic and vegetable fat, spray-dried to a uniform product. It is used in imitation fruit drinks. **Larch gum,** or **galactan gum,** leached from chips of the western larch, is a copolymer of arabinose and galactose and is very similar to gum arabic. **Stractan** is this material in white powder form. It is used as a stabilizer and binder in coatings, adhesives, pharmaceuticals, and low-calorie foodstuffs. **Tamarind seed gum** is from the pod beans of the tamarind tree, *Tamarindus indica,* of India. It is a white to tan, water-soluble powder used as a low-cost alternate for gum arabic. It is a polysaccharide but differs from arabic chemically by containing 12% tartaric acid and 30% sugars. It is used also in medicine and for beverages under the name *tamarind.*

Various synthetic water-soluble gums and emulsifiers are now used as replacements for gum arabic in drugs, cosmetics, adhesives, and foodstuffs. **Bemul** is a glyceryl monostearate, soluble also in alcohol. The water-soluble **Kelzan gum** of Kelco Co. is made from glucose. The wood of the gum arabic tree is the **satinwood** of the Near East, valued since ancient times for its great durability. It is lightweight, hard, and close-grained and has an orange-brown color that darkens with age. But the satinwood of Brazil, known also as **setim,** is a yellowwood from the large tree *Aspidosperma eburneum,* used for inlays.

GUNMETAL. The name for a **casting bronze,** C90500, containing 88% copper, 10 tin, and 2 zinc. It was originally used for small cannons, but is now used where the golden color and strong, crystalline structure are desired. It casts and machines well and is suitable for making steam and hydraulic castings, valves, and gears. It has a tensile strength of 32,000 to 45,000 lb/in^2 (221 to 310 MPa), with elongation 15 to 30%. The specific gravity is 8.7, density 0.314 lb/in^3 (8,719 kg/m^3), and Brinell hardness 65 to 74. This alloy is similar to **G bronze** (C90300), which contains 88% copper, 8 tin, and 4 zinc. In England it is called **admiralty gunmetal** and is specified as **BES No. 383** for sand castings. **Gunmetal ingot** may have the zinc replaced by 2% lead. Such an alloy is easier to machine but has less strength. **Modified gunmetal** contains lead in addition to the zinc. It is used for gears and bearings. A typical modified gunmetal contains 86% copper, 9.5 tin, 2.5 lead, and 2 zinc. It has a tensile strength up to 40,000 lb/in^2 (276 MPa), elongation 15 to 25%, Brinell hardness 63 to 72, and density 0.31 lb/in^3 (8,580 kg/m^3).

GUNPOWDER. Also known as **black powder.** An explosive extensively used for blasting purposes and for fireworks. It was introduced into Europe prior to 1250 and was the only propellant used in guns until 1870. It is now superseded for military uses by high explosives. Black powder deteriorates easily in air from the absorption of moisture. It is a mechanical mixture of potassium nitrate, charcoal, and sulfur, in the usual proportions of 75%, 15, and 10. More saltpeter increases the rate of burning; additional charcoal decreases the rate. A slow-burning powder for fireworks rockets may have only 54% saltpeter and 32 charcoal. Commercial black powder comes in grains of graded sizes and is glazed with graphite. The grain sizes are known as **pebble powder,** large-grain, fine-grain, **sporting powder, mining powder, Spanish spherical powder,** and **cocoa powder.** The potential energy of gunpowder is estimated at 500 ft · ton/lb (305,000 kg · m/kg), but the actual gun efficiency is less than 10% of this. A temperature of about 3712°F (2100°C) is produced by the explosive. Gunpowder is the slowest-acting of all the explosives, and it has a heaving, not a shattering, effect. Hence, it is effective for blasting and breaking up stone. **Blasting powder** is divided by Du Pont into two grades: A and B. The **A powder** contains saltpeter; the **B powder** contains nitrate of soda. The saltpeter concentration varies from 64 to 74% in commercial formulations. The other ingredients are the usual sulfur and charcoal. The B powder is not so strong or water-resistant as A powder, but is cheaper and is extensively used. **Pellet powder** is blasting powder made up in cylindrical cartridges for easier use in mining. **White gunpowder** is a powder in which the saltpeter is replaced by potassium chlorate. It is very sensitive and explodes with violence. It is used only for percussion caps and fireworks. **Lesmok powder,** used in 22-caliber cartridges, is composed of 15% black powder and 85 nitrocellulose.

GURJUN BALSAM. Also known as **wood oil** and sometimes called **East Indian copaiba.** An oleoresin obtained from various species of the *Dipterocarpus* tree, about 50 varieties of which grow in India, Burma, Sri Lanka, and the Malay Peninsula. It is a clear liquid with a greenish fluorescence. The specific gravity is 0.955 to 0.965. It is soluble in benzene. Gurjun balsam is used in lacquers and varnishes that are capable of resisting elevated temperatures. The Burmese trees form two groups, yielding products known as kanyin and in. **Kanyin oils** are brown while the **in oils** are whitish and heavier. Gurjun balsam may consist of either or both of these products.

Commercial **gurjun oil** is obtained by steam distillation of the balsam and has a specific gravity of 0.900 to 0.930. It is soluble in alcohol. **Copaiba balsam** is a resin obtained from the **copaifera tree,** a species of *Dipterocarpus,* of South America. **Maracaibo copaiba** and **Paracopaiba** are the principal varieties. They are dark yellow or brown and are soluble in alcohol. The resin is used as a plasticizer, in varnishes, tracing paper, and pharmaceuticals. The specific gravity is 0.940 to 0.990.

GUTTA PERCHA.   A gum obtained by boiling the sap of species of trees of the order *Sapotaceae,* chiefly *Palaquium gutta* and *P. oblongifolia,* native to Borneo, New Guinea, and Malaya. It is grayish white, very pliable, but not elastic as rubber is. It is harder and a better insulator than rubber. Gutta percha, like rubber, will vulcanize with sulfur and form a hard material. It is used for mixing with rubber, but its chief use was in the covering of electric cables. It has a greater pliability than rubber for the given hardness required in cable insulation. This property with its greater resistance to water makes it valuable for golf balls and dental fillings. It molds easily at 180°F (82°C). It is also employed like balata for impregnating driving belts, for washers and valve seats, and in adhesives. Gutta percha is often imported as mixtures with inferior guttas from other trees. **Gutta soh,** for example, is a mixture from Singapore often colored red artificially. **Gutta siak** is a low-grade gutta from Sumatra. It has a reddish color and is lightly elastic. **Gutta sundik** is from the tree *Payena leerii* of Malaya and Indonesia. It is white and is mixed with gutta percha. Another gutta used to adulterate gutta percha is **gutta hangkang,** from the *Palaquium leiocarpum* of Borneo. It is slightly reddish. **Gutta jangkar** is a low-grade red gutta from Sarawak. **Gutta susu** of Indonesia is a gray-colored material, slightly elastic. With increasing use of synthetic resin insulating materials, gutta percha has become of only secondary importance.

GYPSUM.   A widely distributed mineral which is a hydrated **calcium sulfate,** $CaSO_4 \cdot 2H_2O$, used for making building plaster, wallboard, tiles, as an absorbent for chemicals, as a paint pigment and extender, and for coating papers. Natural gypsum of California, containing 15 to 20% sulfur, is used for producing ammonium sulfate for fertilizer. Gypsum is also used to make sulfuric acid by heating to 2000°F (1093°C) in an air-limited furnace. The resultant calcium sulfide is reacted to yield lime and sulfuric acid. Raw gypsum is also used to mix with portland cement to retard the set. Compact massive types of the mineral are used as building stones. The color is naturally white, but it may be colored by impurities to gray, brown, or red. The specific gravity is 2.28 to 2.33, and the Mohs hardness 1.5 to 2. It dehydrates when heated to about 374°F (190°C), forming the hemihydrate $2CaSO_4 \cdot H_2O$, which is the basis of most gypsum plasters. It is called **calcined gypsum,** or when used for making ornaments or casts, **plaster of paris.** When mixed with water, it again forms the hydrated sulfate that will solidify and set firmly owing to interlocking crystallization. Theoretically, 18% of water is needed for mixing, but actually more is necessary. Insufficient water causes cracking. Water solutions of synthetic resins are mixed with gypsum for casting strong, waterproof articles. **Palestic** is gypsum mixed with a urea-formaldehyde resin and a catalyst. It expands slightly on hardening and thus gives a good impression of the mold. The tensile strength of the molded material is 1,100 lb/in^2 (7.6 MPa), and the compressive strength is 12,000 lb/in^2 (83 MPa).

Calcium sulfate without any water of crystallization is used for paper filler under the name of **pearl filler,** but is not as white as the hydrated calcium sulfate called **crown filler.** The paint pigment known as **satin spar** is a fibrous, silky variety of gypsum and is distinct from the pigment called **satin white,** made by precipitation of aluminum sulfate with lime. **Terra alba** is an old name for ground gypsum as a paint filler. The anhydrous calcium sulfate in powder and granular forms will absorb 12 to 14% of its weight of water, and is used as a drying agent for gases and chemicals. It can then be regenerated for reuse by heating. **Drierite,** of W. A. Hammond Drierite Co., is this material. The mineral mined at Billingham, England, as **anhydrite** is **anhydrous calcium sulfate.** It is used for producing sulfur, sulfur dioxide, and ammonium sulfate. **Hydrocal** of U.S. Gypsum Co., used as a filler for paints and plastics, is $CaSO_4 \cdot 0.5H_2O$. Much calcined gypsum, or plaster of paris, is used as **gypsum plaster** for wall finish. For such use it may be mixed in lime water or glue water, and with sand. Because of its solubility in water it cannot be used for outside work. **Neat plaster,** for walls, is the plaster without sand. When the term **plaster** alone is used, it generally refers to gypsum plasters, but **calcium plaster** made from spent fuller's earth has greater workability and better water resistance than gypsum plaster.

Plasterboard, or **gypsum wallboard,** consists of sheets or slabs of gypsum mixed with up to 15% fibers, employed as a fire-resistant material for walls, ceilings, or partitions, but most of the **wallboard** used in dwelling houses is gypsum board faced on both sides and edged with paper. It usually comes 0.5 in (1.27 cm) thick, 4 by 8 ft (1.2 by 2.4 m), weighing 2 lb/ft^2 (10 kg/m^2). **Macoustic,** of National Gypsum Co., is a lightweight gypsum acoustical plaster. **Grain board** is a fireproof gypsum board with an imitation wood-grain surface used for walls. The hard-finish plasters for **flooring plaster** may contain alum or other materials. **Scott's cement** is a plaster made by grinding lime with calcined gypsum. It sets rapidly. **Mack's cement** is a hard cement made of dehydrated gypsum to which is added a small percentage of calcined sodium sulfate and potassium sulfate. It sets quickly and has good adhesion. It is used for walls, and for floors when mixed with sand.

Patterns, models, and molds of plaster of paris have their strength raised and are made water-resistant by impregnating the dried and hardened plaster with a synthetic resin, particularly the furane resins which cure at low temperatures without pressure. The ordinary casting plaster of 100 parts solids and 60 parts water, called **Hydrocal,** is slightly acid and must be treated with an acid-catalyzing resin, while the low-expansion plaster of 100 parts solids and 45 water, called **hydrostone,** is alkaline and treated with alkaline-catalyzed resin. Impregnation of plaster-of-paris castings with resin raises the compressive strength from 2,000 up to 9,000 lb/in^2 (13 to 62 MPa). Special liquid resins are marketed for this purpose. **Plaspreg** is a furane solution to which a catalyst is added before use.

A crystalline variety of gypsum known as **selenite** occurs in transparent crystals and usually splits in thin laminations. A fine-grained, marblelike variety called **alabaster** is employed in ornamental building work and for lamps, vases, and novelties. Much alabaster is produced in Colorado. **Travertine,** which resembles alabaster but is grained like wood, is a water-deposited calcium carbonate. The **Italian travertine** is notable as a decorative building stone, but is also found in Georgia, Montana, and California. It is an ancient building material, and the great Colosseum at Rome was built of this stone.

Gypsum produced at flue gas desulfurization plants in Germany is converted to **foam** useful in sound-insulating building materials. The gypsum is mixed with water and an isocyanate prepolymer, which react and release carbon dioxide, producing a polyurea network entrapping the gypsum. The release gas causes the material to expand 10-fold.

HAFNIUM.   An elementary metal, symbol Hf. It occurs in nature in about the same amount as copper, but is sparsely disseminated and is costly to extract. All zirconium minerals contain some hafnium, but the two metals are so similar chemically that separation is difficult. All zirconium chemicals and alloys may contain some hafnium, and hafnium metal usually contains about 2% zirconium. The melting point, 4032°F (2222°C), is higher than that of zirconium, and heat-resistant parts for special purposes have been made by compacting hafnium powder to a density of 98%. The metal has a close-packed hexagonal structure. The electric conductivity is about 6% that of copper. It has excellent resistance to a wide range of corrosive environments. Because it has a high thermal-neutron-capture cross section and excellent strength up to 1000°F (538°C), hafnium is useful in unalloyed form in nuclear reactors.

Hafnium oxide, or **hafnia,** HfO_2, is a better refractory ceramic than zirconia, but is costly. The inversion of the crystal from the monoclinic to the tetragonal occurs at 3100°F (1704°C) with an expansion of only 3.4%, compared with 2000°F (1093°C) and an expansion of 7.5% for zirconia. **Hafnium carbide,** HfC, produced by reacting hafnium oxide and carbon at high temperature, is obtained as a loosely coherent mass of blue-black crystals. The crystals have a hardness of Vickers 2,910 and a melting point of 7520°F (4160°C). It is thus one of the most refractory materials known. Heat-resistant ceramics are made from **hafnium titanate** by pressing and sintering the powder. The material has the general composition $_x(TiO_2) \cdot n(HfO_2)$, with varying values of x and n. Parts made with 18% titania and 82 hafnia have a density of 0.26 lb/in^3 (7,197 kg/m^3), a melting point at about 4000°F (2204°C), a low coefficient of thermal expansion, good shock resistance, and a rupture strength above 10,000 lb/in^2 (69 MPa) at 2000°F (1093°C).

HAIR. The fibrous covering of skins of various animals, used for making coarse fabrics and for stuffing purposes. It is distinguished from wool in having no epidermal scales. It cannot be spun readily, although certain hairs, such as camel hair, are noted for great softness and can be made into fine fabrics. **Horsehair** is from the manes and tails and is used as a brush fiber and for making haircloth. It is largely imported from China and Argentina, cleaned and sorted. The imported hair from live animals is more resilient than domestic hair from dead animals. **Brushhair** is usually cut to 3 to 5.5 in (7.6 to 14 cm) long, but tail hair for making **curled hair** for weaving comes in lengths up to 30 in (76 cm). **Cattle hair** is taken from slaughtered animals in packing plants. The body hair is used as a binder in plaster and cements, for hair felt, and for stuffing. The tail hair is used for upholstery, filter cloth, and stuffing. The **ear hair** is used for brushes. It has a strong body and fine, tapered point suitable for poster brushes. In the brush industry it is known as **ox hair. Camox** is a mixture of squirrel hair and ox hair to combine the fineness of squirrel hair with the springiness of ox hair for one-stroke brushes and flowing brushes.

Artificial horsehair, or **monofil,** was a single-filament cellulose acetate fiber, used for braids, laces, hairnets, rugs, and pile fabrics.

Crinex is a German artificial horsehair made of cuproammonium filament and used as a brush fiber. **Haircloth** is a stiff, wiry fabric with a cotton or linen warp and a filling of horsehair. It is elastic and firm and is used as a stiffening and interlining material. The colors are black, gray, and white. The fabric is difficult to weave and disintegrates easily, as the hairs cannot be made into a single strand and must be woven separately. **Press cloth,** used for filtering oils, was made from human hair, which has high tensile strength, resiliency, and resistance to heat. The hair came from China, but filter fabrics are now made from synthetic fibers. **Rabbit hair** from Europe and Australia is used for making felt hats and is referred to as **rabbit fur,** although it does not felt as wool does. The white rabbit hair known as **Angora wool** is from the Angora rabbit of France and Belgium, called **Belgian hare.** The hairs are clipped or plucked four times a year when they are up to 3 in (8 cm) long. They are soft and lustrous, dye easily to delicate shades, and are used for soft wearing apparel. Because of its fluffiness and hairy characteristics, the wool is difficult to spin and is usually employed in mixtures.

HARD-SURFACED POLYMERS. Polymers, developed at the U.S. Department of Energy's Oak Ridge National Laboratories, which combine the flexibility and corrosion resistance of plastics with the strength and durability of metals. They are made by high-energy ion-beam irradiation, which displaces some atoms while ionizing others, forming new bonds and creating a highly cross-linked microstructure of much greater hardness, wear resistance, and abrasion resistance. Chemical resistance and oxidation resistance are also improved.

HEAT INSULATORS. Materials having high resistance to heat rays, or low heat conductivity, used as protective insulation against either hot or cold influences. The materials are also called **thermal insulators.** Insulators for extremely high external temperatures, as on aerospace vehicles, are of ablative materials. Efficiency of heat insulators is measured relatively in Btu/(h · ft · °F) [W/(m · K)], known as the K factor. The thermal conductivity of air and gases is low, and the efficiency of some insulators, especially fibrous ones, is partly due to the airspaces. On the other hand, the thermal conductivity of a porous insulator may be increased if water is absorbed into the spaces.

A wide variety of materials are used as thermal insulators in the form of powder or granules for loose fill, blanket batts of fibrous materials for wall insulation, and in sheets or blocks. Although metals are generally high heat conductors, the polished white metals may reflect as much as 95% of the heat waves and make good **reflective insulators.** But for this purpose the bright surface must be exposed to airspace. Aluminum has a high K factor, up to 130, but crumpled aluminum foil is an efficient thermal insulator as a fill in walls. Wool and hair, either loose or as felt, with a K factor of 0.021 are among the best of the insulators, but organic materials are usable only for low temperatures, and they are now largely replaced by mineral wool or ceramic fibers. Mineral wool has a low K factor, 0.0225. **Tipersul,** of Du Pont, is a **potassium titanate fiber** used loose or in batts, blocks, or sheets. Its melting point is 2500°F (1371°C), and it withstands continuous temperatures to 2200°F (1204°C). Another ceramic fiber, called **Fibrox,** for the same purpose, is a **silicon oxycarbide,** SiCO, in light, fluffy fibers.

Magnesia or asbestos, or combinations of the two, is much used for insulation of hot pipelines, while organic fibrous materials are used for cold lines. **High-heat insulators,** for furnaces and boilers, are usually made of refractory ceramics such as chromite. For intermediate temperatures, expanded glass, such as **Foamglass** of Pittsburgh-Corning Corp., may be used. Foamed glass blocks withstand heats to 1000°F (538°C), and the blocks have a crushing strength of 150 lb/in^2 (0.7 MPa). Some rigid materials of good structural strength serve as structural parts as well as insulators. **Roofinsul,** used for roof decks, is a lightweight board compressed from wood fibers. **Ludlite board,** of Allegheny Ludlum Steel Div. of Allegheny Ludlum Corp., for paneling, is thin stainless steel backed with a magnesia-asbestos composition. Insulators in sheets, shapes, and other forms are sold under a great variety of trade names. **Dry-Zero,** for refrigeration insulation, consists of kapok batts enclosed in fiberboard. **Aura,** of Owens-Corning, consists of glass-fiber mats sealed in steel panels and is intended for refrigerator walls, ovens, and water heaters. The French cold-storage insulation known as **Isotela** consists of pads of matted coir. **Balsam wool** is wood fibers chemically treated to prevent moisture absorption.

HEAT-RESISTANT CAST ALLOYS. Cast alloys suitable for use at service temperatures to at least 1000°F (538°C) and, for some alloys, to 2000°F (1093°C). Included are the alloys designated as the H-series by the Alloy Casting Institute of the American Steel Founders Society as well as **nickel-base alloys** and **cobalt-base alloys.** Most of the nickel-base and cobalt-base alloys are also known as **superalloys** because of their exceptional high-temperature stress-rupture strength and creep resistance as well as corrosion and oxidation resistance.

The **H-Series cast alloys** include iron-chromium, iron-chromium-nickel, and iron-nickel-chromium alloys also containing 0.20 to 0.75% carbon, 1 to 2.5 silicon, and 0.35 to 2 manganese. A letter (A to X) following the H is used to more closely distinguish alloy compositions.

The **iron-chromium cast alloys** (HA, HC, and HD) contain as much as 30% chromium and as much as 7% nickel. The **iron-chromium-nickel cast alloys** (HE, HF, HH, HI, HK, and HL) contain as much as 32% chromium and 22 nickel. And the **iron-nickel-chromium cast alloys** (HN, HP, HP-50WZ, HT, HU, HW, and HX) contain as much as 68% nickel (HX) and 32 chromium (HN), so that some of these alloys are actually nickel base instead of iron base.

All are noted primarily for their oxidation resistance and ability to withstand moderate to severe temperature changes. Most are heat-treatable by aging, room-temperature tensile properties in the aged condition ranging from 73,000 lb/in^2 (503 MPa) to 115,000 lb/in^2 (793 MPa) in terms of ultimate strength, 43,000 lb/in^2 (297 MPa) to 80,000 lb/in^2 (552 MPa) in yield strength, and 4 to 25% in elongation. Hardness of the aged alloys ranges from Brinell 185 to 270. Applications include heat-treating fixtures, furnace parts, oil refinery and chemical processing equipment, gas-turbine components, and equipment used in manufacturing steel, glass, and rubber.

Both the nickel-base and cobalt-base alloys are probably best known for their use in aircraft turbine engines for disks, blades, vanes, and other components. The nickel alloys contain 50 to 75% nickel and usually 10 to 20 chromium; substantial amounts of cobalt, molybdenum, aluminum, and titanium; and small amounts of zirconium, boron, and, in some cases, hafnium. Carbon content ranges from less than 0.1 to 0.20%. Because of their complex compositions, they are best known by trade names, such as **B-1900**; **Hastelloy X**; **IN-100, -738X, -792**; **Rene 77, 80, 100**; **Inconel 713C, 713LC, 718, X-750**; **MAR-M 200, 246, 247**; **Udimet 500, 700, 710**; and **Waspaloy.** The high-temperature strength of most of these alloys is attributed to the presence of refractory metals, which provide solid-solution strengthening; the presence of grain-boundary-strengthening elements, such as carbon, boron, hafnium, and zirconium; and, because of the presence of aluminum and titanium, strengthening by precipitation of an $Ni_3(Al,Ti)$ compound known as *gamma prime* during age hardening. Many of these alloys provide 1,000-h stress-rupture strengths in the range of 100,000 lb/in^2 (690 MPa) to 110,000 lb/in^2 (759 MPa) at 1200°F (649°C), and 8,000 lb/in^2 (55 MPa) to 18,000 lb/in^2 (124 MPa) at 1800°F (982°C).

The cobalt alloys contain 36 to 65% cobalt, usually more than 50, and usually about 20 chromium; substantial amounts of nickel, tungsten, tantalum, molybdenum, iron, and/or aluminum; and small amounts of still other ingredients. Carbon content is 0.05 to 1%. Though not generally as strong as the nickel alloys, some may provide better corrosion and oxidation resistance at high temperatures. These alloys include **L-605**; **S-816**; **V-36**; **WI-52**; **X-40**; **J-1650**;

Haynes 21, 151; AiResist 13, 213, 215; and **MAR-M 302, 322, 918.** Their 1000-h stress-rupture strengths range from about 40,000 lb/in^2 (276 MPa) to 70,000 lb/in^2 (483 MPa) at 1200°F (649°C) and from about 4,000 lb/in^2 (28 MPa) to 15,000 lb/in^2 (103 MPa) at 1800°F (982°C).

HEAT-TRANSFER AGENTS. Liquids or gases used as intermediate agents for the transport of heat or cold between the heat source and the process, or for dissipating heat by radiation. Water, steam, and air are the most common heat-transfer agents, but the term is usually applied only to special materials. Air can be used over the entire range of industrially important temperatures, but it is a poor heat-transfer medium. Water can be used only between its freezing and boiling points, unless high pressures are employed to keep the water liquid. A liquid agent should have a wide liquid range, be noncorrosive and nontoxic, and have low vapor pressure to minimize operational loss.

Gallium, with a freezing point at 85.6°F (29°C) and boiling point of 3600°F (1982°C), offers an exceptionally wide liquid range; but it is too costly for ordinary use, and the liquid metal also attacks other metals. Mercury is used for heat transfer, but is costly and toxic, and at 1200°F (649°C) it exerts a vapor pressure of 500 lb/in^2 (3.4 MPa). Among commercial products, heat-transfer fluids span the range in chemical structure from **alkylated benzenes, alkylated biphenyls, alkylated naphthalenes, unhydrogenated polyphenyls, benzylated aromatics, diphenyl-diphenyl oxide eutectics, polyalkylene glycols, dicarboxylic acid esters, polymethyl siloxanes, mineral oils,** and inorganic nitrate salts. Union Carbide's **Ucon HTF 500** is based on polyalkylene glycol; it has a flash point of 540°F (282°C) and a fire point of 600°F (316°C). Dow Chemical's **Dowtherm A,** a mixture of biphenyl and diphenyl oxide, is stable up to 750°F (399°C), but it freezes at 53.6°F (12°C). **Dowtherm Q,** from the same firm, is another aromatic, mostly diphenylethane. **Syltherm XLT,** a polydimethylsiloxane from Dow Corning Corp., is nontoxic, odorless, and nonreactive and has an operating range from −100°F (−73°C) to 500°F (260°C). **Syltherm 800,** also a silicone fluid, can operate at −40 to 750°F (−40 to 399°C). A low-temperature fluid that works between −100°F (−73°C) and 600°F (316°C) is **MCS-2313** from Monsanto Co., but the fluid is used almost exclusively in the −20 to 20°F (−29 to 7°C) range. **Therminol 59,** also from Monsanto, is an alkylated benzene that is 5 times more stable than mineral oil at 600°F (376°C). **Therminol 66** operates to 650°F (343°C). **Marlotherm fluids,** of Huls America, Inc., are based on mixtures of isomeric dibenzyltoluenes and benzyltoluenes. **Syntrel 350** is an alkylated aromatic, a **phenyl tetrahydronaphthyl ethane,** from Exxon Chemical, and it can be used from 0 to 660°F (12 to 349°C). **Anisol** is a methyl phenyl ether of composition $C_6H_5OCH_3$. It freezes at $-35°F$ ($-37°C$) and boils at 309°F (154°C), has low vapor pressure, and is used for heat transfer although its chief use is as a solvent for plastics and for recrystallization processes. **Aroclor 1248,** of Monsanto, can be used for temperatures up to 550°F (288°C) and, like anisol, is easily pumped at room temperature. It is a chlorinated biphenyl and is noncombustible.

Brine solutions of sodium or calcium chlorides are used for heat transfer for temperatures down to $-6°F$ ($-21°C$), but are corrosive to metals. Molten sodium and potassium salts are used for temperatures from 1112 to 2552°F (600 to 1400°C), but are corrosive to metals. The **sodium-potassium salt,** NaK, called **Nack** is also highly corrosive. The salt known as **Hitec,** which is a 50–50 mixture of sodium nitrite and potassium nitrate, melts at 282°F (139°C), remains liquid at high temperatures, and has no appreciable vapor pressure at 1200°F (649°C). **Tetraryl silicate** remains liquid between -40 and 700°F (-40 and 371°C), but is costly for most uses.

HEAVY ALLOY. A name applied to **tungsten-nickel alloy** produced by pressing and sintering the metallic powders. It is used for screens for X-ray tubes and radioactivity units, for contact surfaces for circuit breakers, and for balances for high-speed machinery. The original composition was 90% tungsten and 10 nickel, but a proportion of copper is used to lower the sintering temperature and give better binding as the copper wets the tungsten. Too large a proportion of copper makes the product porous. In general, the alloys weigh nearly 50% more than lead, permitting space saving in counterweights and balances, and they are more efficient as gamma-ray absorbers than lead. They are highly heat-resistant, retain a tensile strength of about 20,000 lb/in² (138 MPa) at 2000°F (1093°C), have an electric conductivity about 15% that of copper, and can be machined and brazed with silver solder.

An alloy of 90% tungsten, 7.5 nickel, and 2.5 copper has a tensile strength of 135,000 lb/in² (930 MPa), compressive strength 400,000 lb/in² (2,758 MPa), elongation 15%, Rockwell C hardness 30, and density 0.61 lb/in³ (16,885 kg/m³). **Kenertium,** of Kennemetal, Inc., has this composition. **Hevimet,** of General Electric Co., and **Mallory 1000** are similar metals. **Mallory 3000** is in the form of rolled sheet for radiation shielding. The tensile strength of the sheet is 195,000 lb/in² (1,345 MPa), with a Rockwell A hardness of 63. **Fansteel 77 metal,** of Fansteel, Inc., contains 89% tungsten, 7 nickel, and 4 copper. Specific gravity is 16.7, tensile strength 85,000 lb/in² (586 MPa), elongation 17%, and Brinell hardness to 280. The coefficient of expansion is low, $3.6 \times 10^{-6}/°F$ ($6.5 \times 10^{-6}/K$). **Heavy metal powder,** of Astro Alloys Corp., for making parts by powder metallurgy, is prealloyed with the tungsten in a matrix of copper-nickel to prevent settling out of the heavy tungsten.

HELIUM. A colorless, odorless, elementary gas, He, with a specific gravity of 0.1368, liquefying at $-452°F$ ($-268.9°C$), freezing at $-458°F$ ($-272.2°C$). It has a valence of zero and forms no electron-bonded compounds. It has the highest ionization potential of any element. The lifting power of helium is only 92% that of hydrogen, but it is preferred for ballons because it is inert and nonflammable, and is used in weather balloons. It is also used instead of air to inflate large tires for aircraft to save weight. Because of its low density, also, it is used for diluting oxygen in the treatment of respiratory diseases. Its heat conductivity is about 6 times that of air, and it is used as a shielding gas in welding, and in vacuum tubes and electric lamps. Because of its inertness helium can be used to hold free chemical radicals which, when released, give high energy and thrust for missile propulsion. When an electric current is passed through helium, it gives a pinkish-violet light and is thus used in advertising signs. Helium can be obtained from atmospheric nitrogen, but comes chiefly from natural gas, the gas of Texas yielding 0.94%, with some gases yielding as much as 2%. It also occurs in the mineral **cleveite.** Helium is transported as a liquid in trucks and tube trailers. Helium marketed by Matheson Gas Products Co. for use in semiconductor production and where noncontaminating atmospheres are needed is 99.9999% pure. In carbon dioxide lasers, helium is used to cool the laser cavity. It is also used, singly or in mixtures with argon or argon and carbon dioxide, as a **shielding gas** in welding.

HEMLOCK. The wood of the coniferous tree *Tsuga mertensiana,* of the northeastern United States. This species is also called **mountain hemlock** and is now scarce. **Eastern hemlock,** *T. canadensis,* was formerly a tree common from eastern Canada to northern Alabama. In the southern area it is called **spruce pine,** and in the northern area **hemlock spruce.** The wood is coarse with an uneven texture, splintering easily. The trees are up to 80 ft (24 m) in height and up to 3 ft (0.9 m) in diameter. It is used for paper pulp, boxes and crates, and inferior lumber. **Western hemlock,** *T. heterophylla,* is a wood produced in abundance from Alaska to northern California. It is known also as **West Coast hemlock, hemlock spruce, Prince Albert fir, gray fir, Alaskan pine, and western hemlock fir.** Trees 100 years old are about 20 in (0.5 m) in diameter and 140 ft (43 m) high. The wood is light in color, with a pinkish tinge, lightweight, moderately soft, and straight-grained. It is nonresinous and free from resin ducts, but black knots are frequent. The select grades of the lumber are free from knots and suitable for natural and paint finishes. The wood is used for general construction, boxes, woodenware, and pulpwood. The lumber often comes mixed with Douglas fir. It is easy to work but does not plane smooth as pine does. It has frequent dark streaks from heart rot, common in old trees. **Hemlock-bark extract** is obtained from the bark of the eastern hemlock, and it is an important tanning material. **Western hemlock bark** is not in general use for tanning, but the bark contains 22% tannin. The extract is used with resorcinol-formaldehyde or other resins as **cold-setting adhesives** for plywood. They are strong and water-resistant. **Adhesive HT-120** is hemlock-bark extract modified with a phenol resin.

HEMP. A fiber from the stalk of the plant *Cannabis sativa,* valued chiefly for cordage, sacking, packings, and as a fiber for plastic filler. In normal times it is grown principally in southern Russia, central Europe, and Mediterranean countries, and Asia, but during the Second World War it was extensively cultivated in the United States. The fiber, which is obtained by retting, is longer than that of the flax plant, up to 75 in (2 m), but is coarser and not suitable for fine fabrics, although the finest and whitest fibers are sorted out in Europe and used in linen fabrics. It is also more difficult to separate the fiber and to bleach. It is stronger, more glossy, and more durable than cotton, and it has been used for toweling and coarse fabrics to replace the heavy linen fabrics. It is high in alpha cellulose, containing abut 78%. **Hemp rope** was once the chief marine cordage, but it has been replaced largely by rope of abaca which is lighter and more water-resistant. Hemp contains a toxic alkaloid, and in India the stalks are chewed for the narcotic effect. The drug, known in medicine as **cannabis,** is called **marijuana** when smoked in cigarettes. Cannabis is an exhilarator and painkiller and is used in medicine as a depressive antidote, but in excess the drug causes hallucinations. The plant's resin, which in a fully ripe cultivated plant covers the flowers and top leaves, contains the active ingredients. The least potent grade is **bhang,** derived from the tops of uncultivated plants. **Ganja** is the product of select, cultivated plants. The most potent and highest-grade version of the drug, called **charas,** is made from the resin alone and is the only grade which may properly be referred to as **hashish.** **Synthetic cannabis,** or **synhexyl,** is a pyrahexyl more powerful in action than the natural material. **Hempseed oil,** used in paints and varnishes, is made by pressing the seed of the hemp plant. It has a specific gravity of 0.926 and an iodine value of 148. **Oakum,** used for seam caulking, is made from old hemp ropes pulled into loose fiber and treated with tar, usually blended with some new tow. Some grades may have sunn or jute fibers. It comes in balls or in rope form. **Marine oakum** is made entirely from new tow fibers.

HERRING OIL. A fish oil obtained by extraction from several species of fish of the herring family, *Clupeidae.* The **sardine** is the smaller fish of this family. The **Norwegian herring,** *Clupea harengus,* or **sea herring,** is the sardine of Maine, eastern Canada, and the North Sea. The herring is an abundant fish, but it is objectionable as a food because of the quantity of sharp bones. In the very small sardines the bones are soft and edible when cooked. In Norway the oil is produced by boiling the whole fish, pressing, and separating the oil from the water centrifugally. A process used in the United States is to grind the whole fish into liquid form, remove the oil, and condense the remaining solution until it is 50% solids, which is marketed as homogenized, condensed fish for use as poultry feed.

In California and western Canada, the sardine is a much larger fish, the **pilchard,** *Sardinia coerulea,* usually about 8 in (20 cm) long. The pilchard, or **California sardine,** once constituted about 25% of the entire fish catch of the United States by weight, but since 1948 the number of sardines in California waters has decreased greatly. The oil yield is about 30 gal/ton (125 L/metric ton) of fish, but much of the sardine oil is a by-product of the canning industry. The oil content of herring is 10 to 15% of the total weight of the fish, being low in the 1-year-old fish and reaching a peak in the third year. The fish builds up its oil in the summer. In winter the herring tends to stay close to the bottom or at great depths and uses up much of its oil. Commercially, the yield of herring oil is from 3 to 5 gal (1.4 to 19 L) per 250-lb (113-kg) bbl of raw fish. Much of the fish oil of South Africa is from the pilchard, *C. sagax.* In France, Spain, and Portugal the **European pilchard,** *C. pilchardus,* and *C. sardinus* are used. The oil from the latter has a high iodine value. In Norway, the **sprat,** *C. sprattus,* is also used. The **Japanese herring** is *C. pallasi.* Herring oil, or **sardine oil,** is employed as a quenching oil in heat-treating, either alone or mixed with other oils; in soaps, printing inks, and lubricants; and for finishing leather. It is also fractionated to use in blends for paint oils. Herring oil contains 25% **clupanodonic acid,** $C_{21}H_{35}COOH$, 20 **arachidonic acid,** $C_{19}H_{31}COOH$, 18 **palmitoleic acid,** $C_{15}H_{29}COOH$, 13 linoleic, 9 oleic, 8 palmitic, and 7 myristic. The specific gravity is 0.920 to 0.933, iodine value 123 to 142, and saponification value 179 to 194. It can be made clear and odorless by hydrogenation. Sardine oil is richer in hydrocarbons than most marine oils; it also contains some stearic acid, higher percentages of palmitic and linoleic acids, and less of the other acids. **Pilchard oil** is quite similar, but has less oleic acid. Both oils contain about 15% **tetracosapolyenoic acid,** a 24-carbon acid, also occurring in herring oil.

HICKORY.   The wood of the **shagbark hickory** tree, *Carya ovata,* and other species of the walnut order. It is prized as a wood for ax, pick, and other tool handles and for other items where resiliency and shock absorption are necessary. The color of the thick sapwood is white, and the heartwood is reddish brown. It has a fine, even, straight grain and is tough and elastic, having 30% greater strength than white oak and twice the shock resistance, although it is not as durable. The density is 45 to 52 lb/ft^3 (721 to 833 kg/m^3). The chief producing states are Arkansas, Louisiana, Mississippi, Tennessee, and Kentucky, but the trees grow from New Hampshire to Texas. A mature shagbark tree 200 to 300 years old averages 100 ft (30 m) high and over 2 ft (0.6 m) in diameter. For handle manufacture, the white wood and the red wood are considered equal in physical properties, and both possess the smooth feel required for handles. The average specific gravity when kiln-dried is 0.79, compressive strength perpendicular to the grain 3,100 lb/in^2 (21 MPa), and shearing strength parallel to the grain 1,440 lb/in^2 (10 MPa). There are more than 30 species of hickory, including the pecan trees. Besides *C. ovata* three other species are important for the commercial wood: the **shellbark hickory,** *C. laciniosa,* also called **kingnut**; the **pignut hickory,** *C. glabra,* also called **black hickory** and **bitternut**; and the **mockernut hickory,** *C. alba,* also called **ballnut, hognut,** and **white hickory.** The kernels of the nuts of all species are edible, although some are bitter and astringent. The pecan hickories include the **pecan,** *C. illinoensis,* **water hickory,** *C. aquatica,* and **nutmeg hickory,** *C. myristicaeformis.* The pecan trees are cultivated widely in the southern states for the nuts. **Pecan nuts** are widely used in confectionery and bakery products, but they become off-taste rapidly unless sprayed with an antioxidant.

HOLLY.   The wood of the tree *Ilex aquifolium* and several other species of *Ilex,* or holly tree, native to Europe and the tree *I. opaca,* of southeastern United States. It is valued as a wood for inlaying because of its white color; its fine, close grain; and its ease of staining to imitate ebony. It is hard, and the density is 47 lb/ft^3 (753 kg/m^3). It is also used for scientific and musical instruments, model boats, and sporting goods.

HORN.   The excrescent growth, or horns, from the heads of certain animals, notably beef cattle. Horn is used for making handles and various articles. The quality depends largely upon the size and age of the animal from which it comes, the No. 1 grade being the large steer horns and the No. 2 those below 40 lb (18 kg) per 100. Horns occur on the head in pairs and are hollow, growing on a core of pithy bone. The horns are split by saws, soaked to make them flexible, and then flattened under pressure. **Horn meal,** made from bone refuse, is sold largely as fertilizer. **Horn pith,** extracted by boiling the horns, is used for glue and for gelatin.

HYDRAZINE. NH:NH is a colorless liquid boiling at 236°F (113.5°C) and freezing at 36°F (2°C). It is used as a propellant for rockets, yielding exhaust products of high temperature and low molecular weight. With a nickel catalyst it decomposes to nitrogen and hydrogen. It is a strong reducing agent and has been used in soldering fluxes, for corrosion control in boilers, in metal plating, in noble-metal catalysts, and in organic syntheses. However, since it has been declared a suspected carcinogen, its use has declined. It is a starting material for antioxidants and herbicides. Reacted with citric acid, it produces the antituberculosis drug **cotinazin,** which is isonicotinic acid hydrazine. It is also used as a blowing agent for foamed rubber and for the production of plastics. For industrial applications it may be used in the form of **dihydrazine sulfate,** $(N_2H_4)_2 \cdot H_2SO_4$, a white, crystalline, water-soluble flake decomposing at 356°F (180°C) and containing 37.5% available hydrazine. **Hydrazine hydrate,** $N_2H_4 \cdot H_2O$, is a colorless, water-miscible liquid boiling at 248°F (120°C) and freezing at −61°F (−51.7°C). A **monomethyl hydrazine** is also available from Olin Chemicals. Hydrazine is made by reacting chlorine and caustic soda and treating with ammonia.

HYDROCHLORIC ACID. Also called **muriatic acid** and originally called **spirits of salt.** An inorganic acid used for pickling and cleaning metal parts; producing of glues and gelatin from bones; manufacturing chlorine, chlorine dioxide, pharmaceuticals, dyes, and pyrotechnics; recovering zinc from galvanized iron scrap; making high-fructose corn syrup; tanning, etching, and reclaiming rubber; and treating oils and fats. It is a water solution of **hydrogen chloride,** HCl, and is a colorless or yellowish fuming liquid, with pungent, poisonous fumes. The specific gravity of the gas is 1.269, the solidifying point −170°F (−112°C), and boiling point −117°F (−83°C). It is made by the action of sulfuric acid on sodium chloride, or common salt. The commercial acid is usually 20°Bé, equaling 31.45% HCl gas, and has a specific gravity of 1.16. Other grades are 18 and 22°Bé. **Fuming hydrochloric acid** has a specific gravity of 1.194 and contains about 37% hydrogen chloride gas. Reagent-grade hydrochloric acid is usually of this high strength and is clear and colorless, unlike the impure fuming variety. Hydrochloric acid is shipped in glass carboys. Anhydrous hydrogen chloride gas is also marketed in steel cylinders under a pressure of 1,000 lb/in² (6.9 MPa) for use as a catalyst. The boiling point is 185°F (85°C). The acid known as **aqua regia,** used for dissolving or testing gold and platinum, is a mixture of 3 parts hydrochloric acid and 1 nitric acid. It is a yellowish liquid with suffocating fumes.

HYDROCYANIC ACID. Also called **prussic acid, formonitrile,** and **hydrogen cyanide.** A colorless, highly poisonous gas of composition HCN. The specific gravity is 0.697, the liquefying point is 79°F (26°C), and it is soluble in water and in alcohol. It is usually marketed in water solutions of 2 to 10%. It is used for the production of acrylonitrile and adiponitrile and for making sodium cyanide. It is also employed as a disinfectant and fumigant, as a military poison gas, and in mining and metallurgy in the cyanide process. It is so poisonous that death may result within a few seconds after it is taken into the body. It was used as a poison by the Egyptians and Romans, who obtained it by crushing and moistening peach kernels. It is produced synthetically from natural gas. The French war gas known as **vincennite** was hydrocyanic acid mixed with stannic chloride. **Manganite** was a mixture with arsenic trichloride. **HCN discoids,** of American Cyanamid Co., are cellulose disks impregnated with 98% hydrocyanic acid, used for fumigating closed warehouses.

HYDROFLUORIC ACID. A water solution of **hydrogen fluoride,** HF. It is a colorless, fuming liquid, highly corrosive and caustic. It dissolves most metals, except gold and platinum, and glass, stoneware, and organic material. The choking fumes are highly injurious. It is widely used in the chemical industry, for etching glass, and for cleaning metals. In cleaning iron castings it dissolves the sand from the castings. The specific gravity of the gas is 0.99, and the boiling point is 67°F (19.5°C). Hydrofluoric acid is made by treating calcium fluoride or fluorspar with sulfuric acid. It is marketed in solution strengths of 30, 52, 60, and 80%. The anhydrous material, HF, is used as an alkylation catalyst. **Hydrobromic acid,** HBr, is a strong acid which reacts with organic bases to form bromides that are generally more reactive than chlorides. The technical 48% grade has a specific gravity of 1.488.

HYDROGEN. A colorless, odorless, tasteless elementary gas. With an atomic weight of 1.008, it is the lightest known substance. The specific gravity is 0.0695, and its density ratio to air is 1:14.38. It is liquefied by cooling under pressure, and its boiling point at atmospheric pressure is $-423°F$ ($-252.7°C$). Its light weight makes it useful for filling balloons, but because of its flammable nature, it is normally used only for signal balloons, for which use the hydrogen is produced easily and quickly from hydrides. Hydrogen produces high heat and is used for welding and cutting torches. For this purpose it is used in atomic form rather than the usual H_2 molecular form. Its high thrust value makes it an important rocket fuel. It is also used for the hydrogenation of oil and coal, for the production of ammonia and many other chemicals, and for **water gas,** a fuel mixture of hydrogen and carbon monoxide made by passing steam through hot coke.

Hydrogen is so easily obtained in quantity, by the dissociation of water and as a by-product in the production of alkalies by the electrolysis of brine solutions, that it appears as a superabundant material; but its occurrence in nature is much less than that of many of the other elements. It occurs in the atmosphere to the extent of only about 0.01%, and in the earth's crust to the extent of about 0.2%, or about half that of the metal titanium. However, it constitutes about one-ninth of all water, from which it is easily obtained by high heat or by electrolysis.

Hydrogen has three isotopes. **Hydrogen 2,** called **deuterium,** occurs naturally in ordinary **hydrogen 1** to the extent of one part in about 5,000. Deuterium has one proton and one neutron in the nucleus, with one orbital electron. A gamma ray will split off the neutron, leaving the single electron revolving about a single proton. The physicist's name for hydrogen 1 is **protium.** Deuterium is also called **double-weight hydrogen. Deuterium oxide** is known as **heavy water.** The formula is H_2O, but with the double-weight hydrogen the molecular weight is 20 instead of 18 for ordinary water.

Heavy water is used for shielding in atomic reactors, as it is more effective than graphite in slowing down fast neutrons. It is also made with oxygen 17 and oxygen 18. Chemicals for special purposes are also made with hydrogen 2. The **deuterated benzene** of Ciba, Ltd., is made with deuterium, and the formula is expressed as C_6D_6. **Hydrogen 3** is **triple-weight hydrogen** and is called **tritium.** It has two neutrons and one proton in the nucleus and is radioactive. It is a by-product of nuclear fission reactors, and most commercial production is from this source. It is a beta emitter with little harmful secondary ray emission, which makes it useful in self-luminous phosphors. It is a solid at very low temperatures. **Liquid hydrogen** for rocket fuel is made from ordinary hydrogen. It is required to be within 0.00001% of absolute purity. This material has a boiling point of $-423°F$ ($-253°C$), and the weight is 0.6 lb/gal (0.07 kg/L). With a chamber pressure of 300 lb/in^2 (2 MPa), the specific impulse is 375. **Hydrides** are metals that contain hydrogen in a reduced state and as a solid solution in their lattice. **Titanium hydride** and **zirconium hydride** have catalytic activity. **Lanthanum** and **cerium** react with hydrogen at room temperature, forming hydrides, and are used for storing hydrogen. **Sodium borohydride** is used commercially as a reducing agent, for removing trace impurities from organic chemicals, in the synthesis of pharmaceuticals, for wood-pulp bleaching, for brightening clay, and for recovering trace metals in effluents. **Silicon hydride,** also known as **silane,** is a gas used for manufacturing ultrapure silicon for fabrication into semiconductors. The **Hydripills** of Metal Hydrides, Inc., used for producing small quantities of hydrogen, are tablets of a mixture of sodium borohydride and **cobalt chloride,** $CoCl_2$. When water is added, the chloride reacts to produce hydrogen from both the borohydride and the water. In Japan, Chuo Denki Kogyo Co. is building hydride-based heat pumps for water-cooling and hot water systems. **Gelled hydrogen** for rocket fuel is liquid hydrogen thickened with silica powder.

HYDROGEN PEROXIDE.   A liquid that readily yields oxygen for bleaching and oxidizing purposes. The C.P. grade of hydrogen peroxide is a colorless liquid with 90% H_2O_2 and 10 water. The specific gravity is 1.39. It contains 42% active oxygen by weight, and 1 volume yields 410 volumes of oxygen gas. Grades for oxidation and bleaching contain 27.5 and 35% H_2O_2. It is also used as an oxidizer for liquid fuels. A variety of chemicals are used for providing oxygen for chemical reactions. These are known as **oxidizers** or **oxidants,** and they may be peroxides or superoxides which are compounds with the oxygen atoms singly linked. They break down into pure oxygen and a more stable reduced oxide. Sodium peroxide is used in submarines to absorb carbon dioxide and water vapor and to give off oxygen to restore the air. To provide oxygen in rockets and missiles, **lithium nitrate,** $LiNO_3$, with 70% available oxygen, and **lithium perchlorate,** $LiClO_4 \cdot 3H_2O$, with 60% available oxygen, are used. Another rocket fuel oxidizer which is liquid under moderate pressure and is easily stored is **perchloryl fluoride,** ClO_3F, normally boiling at $-52°F$ ($-47°C$).

Albone, of Du Pont, is hydrogen peroxide, and **Solozone** is sodium peroxide with 20% available oxygen. **Tysul** and **Perone** are other Du Pont peroxides. **Peroxygen** is a peroxide line from Interox America. **Ingolin** was a German name for hydrogen peroxide used in rockets. **Liquid air** was used in the first V-2 rockets, with alcohol, potassium permanganate, and hydrogen peroxide. Liquid air is used in the chemical industry and for cold-treating. It is atmospheric air liquefied under pressure, and it contains more than 20% free oxygen. The boiling point is $-310°F$ ($-190°C$), and 1 ft^3 (0.028 m^3) makes 792 ft^3 (22 m^3) of free air.

Tetrabutyl hydroperoxide, an organic peroxide, is a powerful oxidizing agent used as an accelerator in curing rubbers; as a drying agent in oils, paints, and varnishes; and as a combustion aid for diesel fuel oils. The commercial 60% solution in water has a boiling point of 180°F (82°C) and specific gravity of 0.859. **Urea peroxide,** $(CO \cdot NH_2)_2O_2$, is a white crystalline material with 16% by weight of active oxygen, used in bleaching, polymerization, and oxidation processes. **Magnesium peroxide,** MgO_2, **calcium peroxide,** CaO_2, and **zinc peroxide,** ZnO_2, are stable white powders insoluble in water, containing, respectively, 14.2, 13.6, and 7.4% active oxygen. They are used where oxidation is required to be at high temperatures. **Uniperox,** of Union Oil Co. of California, is a peroxide of composition $C_7H_{13}OOH$, made from petroleum fractions. At low temperatures it is stable, but at 230°F (110°C) the decomposition is exothermic and rapid. It is used as a diesel fuel additive to raise the cetane number and as a polymerization catalyst for synthetic resins.

Oxidizers and **reducers,** or **reductants,** are used in solutions for water disinfection, bleaching, cyanide destruction, chromium reduction, and metal etching. Common oxidizers include **chlorine, bromine, ozone, sodium hypochlorite,** and **hydrogen peroxide.** Well-known reductants include **sodium bisulfate,** or **sodium metabisulfate,** and **sulfur dioxide.** The oxidation reduction potential, or capacity, of several oxidants to oxidize or reduce another material is 1.07 V for bromine, 1.36 for chlorine, 1.57 for **chlorine dioxide,** 1.68 for **potassium permanganate,** 1.78 for hydrogen peroxide, 2.07 for ozone, and 3.05 for **fluorine.** Thus, ozone is about twice as effective as bromine.

HYDROGENATED OILS. Vegetable or fish oils that have been hardened or solidified by the action of hydrogen in the presence of a catalyst. Partial hydrogenation also clarifies and makes odorless some oils. The solidifying process is carried on to any desired extent, and these oils have a variety of uses. For mechanical uses they are employed in cutting oils, and in place of palm oil in tinplate manufacture. By hydrogenation the fatty acids, such as oleic acid, are converted to stearic acid. Peanut oil, coconut oil, and cottonseed oil can thus be made to have the appearance, taste, and odor of lard, or they can be made like tallow. **Lard compound,** prior to the passage of the Food and Drug Act of 1906, was cottonseed oil mixed with oleostearin from beef tallow. It was later sold under trade names, but has now been replaced by hydrogenated oils under the general name **shortenings** and under trade names such as **Crisco.** Hydrogenated oils have lower iodine values and higher melting points than the original oils.

IMPREGNATED WOOD. Also called **compressed wood** or **densified wood.** Many types are forms of laminated wood. **Compreg,** developed by U.S. Forest Products Laboratory, consists of many layers of 0.0625-in (0.16-cm), rotary cut, yellow-birch plies bonded with about 30% resin under a pressure of 600 to 1,500 lb/in² (4 to 10 MPa). The specific gravity is 1.22 to 1.37, and tensile strength 43,000 to 54,000 lb/in² (296 to 372 MPa), depending upon the resin and the molding pressure. **Impreg,** developed by this laboratory for use in making patterns and models, is produced from 0.0625-in (0.16-cm) laminations of mahogany impregnated with a low-molecular-weight phenolic resin and bonded under pressure into a uniform solid of good dimensional stability. **Flaypreg,** another member of this group of wood products, is hard, strong, dense, and low-cost. It is made from wood flakes, usually fir or spruce, impregnated with resin, machine-felted, and pressed. The specific gravity is 1.39, and the water absorption is only 0.44% compared with 1.46 for Compreg. It is used for making gears, cams, patterns, and tabletops. **Delwood** is molded of wood chips, chopped glass fiber, and a binder of polyester resin. It has the strength of hard maple and takes nails and screws better than wood. It is used for shoe lasts, picture frames, and furniture. **Pregwood,** of Formica Corp., is a wood laminate impregnated with a phenolic resin and cured into a hard sheet. But **Impreg weldwood** has the wood plies impregnated only to a short depth before compressing, so that it remains a true plywood. It has higher strength and is more resistant than an ordinary resin-bonded plywood. **Sprucolite,** for bearings, rolls, gears, and pulleys, is cross-laminated like plywood with thin sheets of western spruce, but is **plywood block** impregnated with resin and subjected to high pressure to make it dense and hard. Its weight is about 35% that of cast iron. A similar English laminated wood, called **Hydulignum,** consists of thin birch veneers impregnated with vinyl formal resin and compressed into a dense board with a specific gravity of 1.31 and a tensile strength of 45,000 lb/in² (310 MPa). Still another type of building board, **Dylite,** of Koppers Co., has a core of polystyrene plastic jacketed on both sides with plywood or gypsum.

The material developed by U.S. Forest Products Laboratory under the name **Staypak** is made by compressing veneered softwood containing no resin except that used to bond the veneers. It has a smooth, satiny finish, a specific gravity of 1.3 to 1.4, and about double the tensile and flexural strengths of birch. The color is darker than the original wood because of the flow of lignin. **Hiden,** of Parkwood Corp., is a synthetic hardwood of about the same density and hardness as lignum vitae. It is made of birch veneers impregnated with phenolic resin and compressed to 30% of the original thickness and cured. The boards are used for tabletops, cutlery handles, sheet-metal forming dies, and textile picker sticks.

Wood impregnated with **polyethylene glycol** is known as **Peg.** This treatment is used for walnut gunstocks for high-quality rifles and for tabletops. This impregnant can be used to reduce checking of green wood during drying. Wood can also be vacuum-impregnated with certain liquid vinyl monomers and then treated by radiation or catalyst heat systems, which transform the vinyl to a plastic. **Methyl methacrylate,** or **acrylic,** is a common resin used to produce this type of product, known as **wood-plastic combinations,** or **WPCs.** A principal commercial use of this modified wood is as parquet flooring and for sporting goods such as archery bows. It is produced in squares about 5.5 in (14 cm) on a side from strips about 0.875 in (2.2 cm) wide and 0.3125 in (0.8 cm) thick. It has a specific gravity of 1.0. WPC material resists indentation from rolling, concentrated, and impact loads better than white oak. This is largely due to improved hardness, which is increased 40%. Abrasion resistance is no better than that of white oak.

INDIGO. Once the most important of all vegetable dyestuffs and valued for the beauty and permanence of its color. It is widely used to color **denim** for clothing. Commercial blue indigo is obtained from the plants *Indigofera tinctoria* and several other species, of India and Java, and the plant *Isatis tinctoria,* of Europe, by steeping the freshly cut plants in water, and after decomposition of the glucoside **indican,** $C_{14}H_{17}O_6N$, the liquid is run into beating vats where the indigo separates out in flakes which are pressed into cakes. About 4 oz (0.1 kg) of indigo is produced from 100 lb (45 kg) of plants. **Indigo red,** or **indirubin,** $C_{16}H_{10}N_2O_2$, is a crimson dyestuff obtained in the proportion of 1 to 5% in the manufacture of indigo by extraction in organic solvents. **Indigo brown** is an impurity that occurs during the manufacture of indigo, but has little influence in the dyeing process. Other constituents are **indigo yellow** and **indigo gluten. Indigo white** is obtained by reducing agents and an alkali. Another product obtained in the manufacture of synthetic indigo is **indole,** a white, crystalline solid with a melting point of 126°F (52°C). In concentrations it has a powerful, disagreeable odor, but in extreme dilution has a pleasant floral odor and is used in many perfumes. It occurs naturally in oils of jasmine, neroli, orange blossom, and others, and it is made synthetically as **benzopyrrole. Skatole** is made by adding a methyl group to the No. 3 position of the indole ring. It is a solid melting at 203°F (95°C) and is found as a decay product of albumin in animal excrement. It has an overpowering fecal odor, and the synthetic material is used as a fixative in fine perfumery. **Oxindole,** or **hydroxyindole,** is a lactam of aminophenyl acetic acid, easily made synthetically, and is the basis for the production of a wide variety of chemicals.

INDIUM. A silvery-white metal with a bluish hue, whiter than tin. It has a specific gravity of 7.31, tensile strength of 15,000 lb/in^2 (103 MPa), and elongation of 22%. It is very ductile and does not work-harden, as its recrystallization point is below normal room temperature, and it softens during rolling. The metal is not easily oxidized, but above its melting point, 315°F (157°C), it oxidizes and burns with a violet flame.

Indium was first found in zinc blende, but is now obtained as a by-product from a variety of ores. Because of its bright color, light reflectance, and corrosion resistance, it is valued as a plating metal, especially for reflectors. It is softer than lead, but a hard surface is obtained by heating the plated part to diffuse the indium into the base metal. It has high adhesion to other metals. When added to chromium plating baths, it reduces the brittleness of the chromium.

In spite of its softness, small amounts of indium will harden copper, tin, or lead alloys and increase the strength. About 1% in lead will double the hardness of the lead. In solders and fusible alloys, it improves wetting and lowers the melting point. In lead-base alloys, a small amount of indium helps to retain oil film and increases the resistance to corrosion from the oil acids. Small amounts may be used in gold and silver dental alloys to increase the hardness, strength, and smoothness. Small amounts are also used in silver-lead and silver-copper aircraft engine bearing alloys. **Lead-indium alloys** are highly resistant to corrosion and are used for chemical-processing equipment parts. **Gold-indium alloys** have high fluidity, a smooth, lustrous color, and good bonding strength. An alloy of 77.5% gold and 22.5 indium, with a working temperature of about 932°F (500°C), is used for brazing metal objects with glass inserts. **Silver-indium alloys** have high hardness and a fine silvery color. A silver-indium alloy of Westinghouse Electric Corp., used for nuclear control rods, contains 80% silver, 15 indium, and 5 cadmium. The melting point is 1375°F (746°C), tensile strength 42,000 lb/in^2 (290 MPa), and elongation 67%, and it retains a strength of 17,600 lb/in^2 (121 MPa) at 600°F (316°C). It is stable to irradiation and is corrosion-resistant in high-pressure water up to 680°F (360°C). The thermal expansion is about 6 times that of steel.

Indium sulfate, used for plating, has three forms. The normal sulfate is $In_2(SO_4)_3 \cdot 9H_2O$; the acid salt is $In_2(SO_4)_3 \cdot H_2SO_4 \cdot 7H_2O$; and the basic salt is $In_2O(SO_4)_2 \cdot 6H_2O$. **Indium oxide** is an amorphous yellow powder of composition In_2O_3 and specific gravity 7.179, used to give a beautiful yellow color to glass. The color may be varied from light canary to dark tangerine-orange. **Indium monoxide,** InO, is black and is not stable.

A mirrorlike **indium coating,** deposited by vacuum metallization on primed soft-polymer substrates in the form of discontinuous microscopic islands, permits repeated flexing without cracking. Developed by Davidson Instrument Panels Textron, it is used to metallize thermoplastic polyurethane elastomers for auto radiator grilles and is adaptable to other polymers and applications.

INDUSTRIAL JEWELS. Hard stones, usually **ruby** and **sapphire,** used for bearings and impulse pins in instruments and for recording needles. **Ring jewels** are divided into large and small. The large rings are about 0.050 in (0.127 cm) in diameter and 0.012 in (0.030 cm) thick with holes above 0.006 in (0.015 cm) in diameter. Ring jewels are used as pivot bearings in instruments, timepieces, and dial indicators. From 2 to 14 are used in a watch. **Vee jewels** are used in compasses and electrical instruments. **Cup jewels** are used for electric meters and compasses. **End stones** are flat, undrilled stones with polished faces to serve as end bearings. **Pallet stones** are rectangular impulse stones for watch escapements. **Jewel pins** are cylindrical impulse stones for timepiece escapements. In making **bearing jewels,** the synthetic sapphire boules are split in half, secured to wooden blocks, and then sawed to square blanks. These are then rounded on centerless grinding machines and flat-ground to thickness by means of copper wheels with diamond powder. **Quartz bearings** are made from fused quartz rods. A notch is ground in the end of the rod, then polished and cut off, repeating the process for each bearing. Quartz has a Mohs hardness of only 7 while the synthetic ruby and sapphire have a Mohs hardness of 8.8, but quartz has the advantage of low thermal expansion.

The making of industrial jewels was formerly a relatively small, specialized industry, and a national stockpile of cut jewels was maintained for wartime emergencies. But the process of slicing and shaping hard crystals for semiconductors and other electronic uses is essentially the same, and stones of any required composition and cut to any desired shape are now regularly manufactured.

INGOT IRON. Nearly chemically pure **iron** made by the basic open-hearth process and highly refined, remaining in the furnace 1 to 4 h longer than the ordinary time and maintained at a temperature of 2900 to 3100°F (1593 to 1704°C). In England, it is referred to as **mild steel,** but in the United States the line between iron and steel is placed arbitrarily at about 0.15% carbon content. Ingot iron has as low as 0.02% carbon. It is obtainable regularly in grades 99.8 to

99.9% pure iron. Ingot iron is cast into ingots and then rolled into plates or shapes and bars. It is used for construction work where a ductile, rust-resistant metal is required, especially for tanks, boilers, enameled ware, and galvanized culvert sheets. The tensile strength, hot-rolled, is 48,000 lb/in^2 (331 MPa), elongation 30%, and Brinell hardness 82 to 100. Dead soft, the tensile strength is 38,500 lb/in^2 (265 MPa), elongation 45%, and Brinell hardness 67. **Armco ingot iron,** of Armco Steel Corp., is 99.94% pure, with 0.013% carbon and 0.017 manganese. It is used as a rust-resistant construction material, for electromagnetic cores, and as a raw material in making special steels. The specific gravity is 7.858 and melting point 2768°F (1530°C). **Enamelite** is a sheet iron especially suited for vitreous enameling. Ingot iron may also be obtained in grades containing 0.25 to 0.30% copper, which increases the corrosion resistance. Iron of very low carbon content may also be used for molds and dies which are to be hobbed. The iron is quite plastic under the hob and is then hardened by carburizing. **Plastiron** is such an iron.

INK.   Colored liquid or paste for writing, drawing, marking, and printing. Black writing inks usually contain gallotannate of iron which is obtained by adding an infusion of nutgalls to a solution of ferrous sulfate. Good **writing ink** is a clear, filterable solution, not a suspension. It must flow easily from the pen without clogging; give a smooth, varnishlike coating; and adhere to inner fibers of the paper without penetrating through the paper. It must have an intense color that does not bleach out. Ink is essentially a pigment in a liquefying and adhesive medium, but the **iron-gallotannate** writing inks develop their full color by chemical action and become insoluble in the outer fibers of the paper. For the proper development of the black color in gallotannate inks, a high percentage of iron is needed, and this requires a liberal use of acid, which will tend to injure the paper. It is thus usually the practice to reduce the amount of iron and bring up the color with dyes or pigments. Gums or adhesive materials may also be added.

    Carbon inks are composed of lampblack or carbon black in solutions of gums or glutenous materials. **India ink** is a heavy-bodied drawing ink. The original India ink, or **Chinese ink,** was made with a jet-black carbon pigment produced by burning tung oil with insufficient air. The pigment was imported into Europe in compressed sticks known as **indicum.** India ink was originally only black ink, but the name is now used also for colored, heavy drawing inks made with various mineral pigments. **Marking inks** are usually solutions of dyes that are fast to laundering, but they may also be made with silver salts which develop full color and stability by the application of heat.

Fountain pen ink is not a special-composition ink, but is a writing ink free of sediment and tendency to gum. It usually contains tannic, gallic, and hydrochloric acids with a pH above 2 to avoid corrosion. **Permanent inks** contain dissolved iron, not over 1%, to avoid sludge. **Ballpoint ink** is usually a paste and is a true solution with 40 to 50% dye concentration. It must be stable to air, noncorrosive, and a good lubricant. An **encaustic ink** is a special writing ink that will penetrate the fibers of the paper and set chemically to make erasure difficult, but an **indelible ink** for textiles is a marking ink. **Invisible-writing inks,** or **sympathetic inks,** are inks that remain invisible until the writing is brought out by the application of heat or with another chemical which develops the color. They are made with sal ammoniac or salts of metals. **Magnetic ink** for use on bank checks to permit mechanical processing contains 50 to 70% of a magnetic powder with smaller than 197-μin (5-μm) particle size. The powder may be hydrogen-reduced iron, carbonyl iron, or electrolytic iron.

Printing inks are in general made with carbon black, lampblack, or other pigment suspended in an oil vehicle, with a resin, solvent, adhesive, and drier. But there are innumerable modifications of printing inks to meet different conditions of printing and varieties of surfaces. The oil or chemical vehicles are innumerable, and the pigments, resin strengtheners and gloss formers, adhesives, tackifiers, and driers vary greatly to suit the nature and surface coating of the base material. Oils may dry by oxidation, polymerization, absorption, or solvent evaporation; and resins may be used to add gloss, strength, hardness, and color fastness, or to increase the speed of drying. It is estimated that there are about 8,000 variables in an ink, and thus printing ink is a prescription product for any given job. They are not normally purchased on composition specification, but on ability to meet the requirements of the printing. **Aremco-Mark 530,** of Aremco Products, Inc., is a black **ceramic ink** that withstands temperatures to 2000°F (1093°C). These inks are ceramic solutions that must be heat-cured. **Azo, triphenylmethane, vat, anthraquinone,** and **phthalocyanine** dyes are common dye families used in ink manufacture. **Flexographic ink** and **rotogravure ink** may be made with cellulose-acetate-propionate ester resin which is soluble in alcohol and in other resins. When used with urea, it cross-links to form a permanent thermoset film.

INSECTICIDES. Chemicals, either natural or synthetic, used to kill or control insects, particularly agricultural pests. They are also referred to as **pesticides.** Of about 800,000 known species of insects, one-half feed directly on plants, retarding growth of the plant and causing low yields and inferior crops. The production of insecticides is now one of the important branches of the chemical industry, and increasing quantities are used; but the specification and use of insecticides require much skill because of the cumulative effect on the earth and animal or plant life. Indiscriminate use may destroy honeybees and other useful insects, produce sterility of soils by killing worms and anaerobic life, and poison the waters of lakes and streams. DDT, for example, is highly valuable for the control of malaria and other insect-borne diseases, but its uncontrolled use as an insecticide has been disastrous to wildlife.

Insecticides are generally classified as **stomach poisons** and **contact poisons.** Stomach poisons include **calcium arsenate,** a white powder of composition $Ca_3(AsO_4)_2$, which constitutes about one-half of all insecticides used, and also paris green, lead arsenate, and white arsenic. **Cryolite, or sodium fluoroaluminate,** and **sodium fluoride** are used occasionally. An **antimetabolite** is not a direct poison, but acts on the insect to stop the desire for food so that the insect dies from starvation. **Dimethyltriazinoacetamilide,** used against corn-ear worms, is such a chemical. Contact poisons include **rotenone dust, sulfur dust,** and **nicotine sulfate** solution. **Pyrethroids** are favored for consumer uses because they are fast-acting.

A **larvicide** is a chemical, such as chloropicrin, used to destroy fungi and nematodes in soils, and insect eggs and organisms in warehouses. Chemicals used against fungi and bacteria are called **fungicides** and **bactericides,** and those used to control plant diseases caused by viruses are called **viricides.** None of these are properly classified as insecticides, but are often used with them. **Herbicides** are used to kill weeds, usually by overstimulating cell growth. **Glysophate** is used as an aerial spray to destroy **coca** leaf, the raw material for producing **cocaine. Amionotriazole,** used as a weed killer, may cause cancer in animal life. Most of the pest control chemicals are cumulative toxic poisons. **Benzene hexachloride** destroys bone marrow, and all the chlorinated hydrocarbons affect the liver. **Deodorants** may have an insect-kill action, but are usually chemicals such as chlorophyll which combine with impurities in the air to eliminate unpleasant odors.

Fungi are being pursued as natural pesticides against grasshoppers and locust. Two strains that proved effective in field trials are **Metarhizium flavoviride** and **Beauveria bassiana. Resveratrol** is a natural pesticide produced by grapes to fight fungi. It has been found to reduce fat and cholesterol in the blood of animals, portending medicinal value. **Azadirachtin,** a chemical from the neem tree of India, kills insects in their larva stage. After the larva sheds one coat, the chemical prevents it from growing another. Azadirachtin products are marketed by AgriDyne Technologies and W. R. Grace & Co.

Bacteria, if cultivatable, can serve as pesticides and for environmental purposes. Cattle feed laced with bacteria from the stomachs of sheep and goats seems to detoxify tansy ragwort, a weed that kills cattle. Bacteria from the forestomach of whales may prove useful in cleaning up oil spills. **White-rot fungi,** such as **Phanerochaete chrysosporium,** which grow by degrading wood, are being used to clean up soils contaminated with dioxins and polychlorinated biphenyls. Unlike bacteria, fungi break down contaminants extracellularly, using excreted enzymes, and can attack more toxic compounds.

Insecticides may be solids or liquids, and the solids may be applied as a fine powder, usually in dilution in a dusting powder, or the powder may be suspended in a liquid carrier. Usually, the proportion of poison mixed with a mineral powder is no more than 5%. The mineral carrier, or **dusting powder,** for this purpose should be gritless and inert to the insecticide. Ordinary **dusting clay** is a light, fluffy, air-floated kaolin, or it may be finely ground, soft limestone. **Sevin,** a **carbamate** insecticide, also known as **carbaryl,** is carried on synthetic clays or talc. **Affirm,** a macrocyclic lactone isolated from the soil organism *Streptomyces avermitilis,* is carried on pregelled, defatted corn grit. It is available from Merck Sharp & Dohme Research Laboratories. **Sodium fluorosilicate,** Na_2SiF_6, comes as a fine white powder for this purpose. Calcium sulfate is also used as a carrier but has itself a poisoning effect. One of the oldest of solid insecticides still used, either dry or in liquid sprays, is **Bordeaux mixture,** made by reacting copper sulfate with lime, giving a product with an excess of hydrated lime. Liquid carriers for insecticides may be kerosene or other petroleum hydrocarbons, or they may be liquid chemicals that have toxic properties; but they must be chosen to avoid deleterious effects, such as the yellowing of papers and organic materials in warehouses or archives, or the injuring of plants from active chlorine in some chemicals.

Some materials, such as citronella oil, used as mosquito repellents in households, have little or no value as insecticides for the eradication of mosquitoes in important applications such as at military sites or mining and lumbering camps. The **aerosol bomb** employed during the Second World War contained 3% DDT, 2 to 20 pyrethrum concentrate, 5 cyclohexanone, 5 mineral oil, and the balance a carrier gas. **Dimethyl phthalate,** a liquid of composition $C_6H_4(CO_2CH_3)_2$, is a mosquito repellent having an effect lasting 1.5 h in the open air. **Thiourea** is used to kill mosquito larvae in water and is harmless to fish. **Permethrin** repels and kills mosquitoes on contact. Developed by U.S. Army scientists, it is a modification of a pesticide used to keep moths away from carpets and woolens. Sprayed on clothing, the treatment lasts through at least 25 launderings.

The insecticide called **DDT** is dichlorodiphenyltrichloroethane, $C_6H_3Cl_2(C_6H_4 \cdot CH_2CCl_3)$, used effectively during the Second World War against flies, mosquitoes, body lice, and agricultural pests. It has no noxious odor, but it is cumulative and in concentration is toxic to humans and other warm-blooded animals. Oil paint containing 0.5 to 5% DDT kills flies on walls painted with it. Because it is highly resistant to degradation, DDT is now limited to essential uses. **Aldrin** and **heptachlor** are widely used as soil insecticides. Aldrin is a chlorinated cyclodiene sold under the trade names **Aldrex, Bangald, Drinox, Octalene,** and **Seedrin,** among others. **Heptachlor** is from the same chemical family as aldrin. Velsicol Chemical Corp.'s product is **Velsicol 104**; others are **Heptalube, Heptagran,** and **Gold Crest H-60. Chlordane** is a liquid of composition $C_{10}H_6Cl_8$. It is a powerful insecticide.

Sabadilla is used in cotton sprays. It is also known as **cevadilla,** or **Indian barley,** and consists of the dried, ripe, poisonous seeds of the plant *Veratrum sabadilla* of the lily family growing in Central America, of which there are about 20 known species in Central and South America. The seeds contain **veratric acid,** from which is derived **veratraldehyde,** or **vetraldehyde,** a crystalline solid of composition $(CH_3O)_2C_6H_3 \cdot CHO$, which gives the heliotrope flavor to the vanilla of Samoa and to some synthetic vanilla from coniferin. The alkaloid poison is extracted with a hydrocarbon solvent, and when the extract is used as an insecticide in combination with a synthetic, it gives greatly increased toxicity. The powdered seeds are also used as an agricultural insecticide dust which has greater staying power than pyrethrum. The cresols in various forms are also used as insecticides. **Dinitroorthocresol,** a yellow, crystalline material melting at 182°F (83.5°C), is used in fruit-tree sprays. Sodium antimony lactophenate, known as **salp,** is an effective insecticide against chewing insects. Some insecticides are sprayed on the ground or on the foliage to be absorbed into the plant, poisoning the insect that feeds on the plant.

A **microencapsulated insecticide,** consisting of pyrethroids inside a plyurea shell, is produced by 3M. The pyrethroids become active when a biological synergist permeates the shell wall. Unprotected, pyrethroids normally decompose in several minutes to several hours.

Sodium fluoride, or **fluorol,** NaF, is a water-soluble, white powder used as a wood preservative as well as an insecticide and vermin poison, although this material is better known as an industrial chemical. Single crystals of it are used for windows for ultraviolet and infrared equipment, as it transmits these rays. When wood is treated with an alkaline water solution of acrylonitrile ethylates, the cellulose fibers are cyanoethylated and the wood becomes resistant to the attack of enzymes and fungi. Wood treated with pyradine and acetic anhydride is given dimensional stability as well as resistance to insect and fungi attack.

A **fumigant** is a liquid, powder, or gas which kills insects, worms, or burrowing animals by toxic fumes. For general use a fumigant should not be injurious to grains or stored foodstuffs. **Repellents** are fumigants used for driving out, rather than killing, insects. However, some repellents contain naphthalene, rotenone, or other materials having toxic properties, and these are insecticides rather than fumigants. **Methyl bromide,** or **bromomethane,** CH_3Br, a gas with a liquefying point at 40°F (4.6°C), is an effective fumigant not injurious to grains. Methyl bromide is also used for fumigating clothing warehouses and does not shrink or wrinkle woolen fabrics. **Dihydroacetic acid** is used on dried fruits in storage to prevent decomposition. It acts as a fungicide.

IODINE. A purplish-black, crystalline, poisonous elementary solid, chemical symbol I, best known for its use as a strong antiseptic in medicine, but also used in many chemical compounds and war gases. In tablet form it is used for sterilizing drinking water, and it has less odor and taste than chlorine for this purpose. It is also used in cattle feeds. Although poisonous in quantity, iodine is essential to proper cell growth in the human body and is found in every cell in a normal body, with larger concentration in the thyroid gland.

The Chilean production of iodine is a by-product of the nitrate industry. In Scotland, Norway, and Japan it is produced by burning seaweed and treating the ashes. A ton of seaweed produces about a pound of iodine. It is also produced from salt brines and from seawater, and in California from the wastewaters of oil wells, the brine containing 65 parts of iodine per million. The lump iodine from this source is 99.9% pure. As much as 1,000 tons (907 metric tons) of iodine is present in 1 mi^3 (42×10^8 m^3) of seawater. The specific gravity of iodine is 4.98, melting point 238°F (114.2°C), and boiling point 363°F (184°C). It is insoluble in water, but is soluble in alcohol, ether, and alkaline solutions. **Tincture of iodine,** a 7% alcohol solution of iodine in a 5% solution of potassium iodide, is used in medicine as a caustic antiseptic. As an antiseptic, iodine has the disadvantage that it burns and stains the skin. **Vodine** is a 2% oil-and-water emulsion of iodine containing also lecithin. It does not burn, and the faint stain washes off easily. An **iodophor** is a chemical containing iodine which is released on contact with organic material. **I-O-Dynamic** is a detergent containing iodine. **PVP iodine** of GAF Corp. is iodine combined with polyvinyl pyrrolidine to give a product that retains the germ-killing properties of iodine without the toxic and burning effects. **Wescodyne** is another nonburning and nonstaining iodine. **Clearodine** is a water-soluble iodine in powder form for disinfectant purposes. In water solution it releases a colorless **hypoiodous acid,** IHO. It has a higher bacteriological effect than ordinary iodine and does not stain or irritate. **Iodine cyanide,** ICN, an extremely poisonous, colorless, crystalline material soluble in water, is used as a preservative for furs and museum specimens. **Iodic acid,** HIO_3, and a stable **iodine pentoxide,** I_2O_5, are also marketed. The iodine atom is very regular with a valence of 7, but having three spheron pairs in opposite polarity which can be broken to give valences of 1, 3, 5, and 7. A wide range of compounds are made for electronic and chemical uses. Iodine is also a chemical reagent, being used for reducing vanadium pentoxide and zirconium oxide into high-purity metals.

IRIDIUM. A grayish-white metal of extreme hardness, symbol Ir. It is insoluble in all acids and in aqua regia. The melting point is 4436°F (2447°C), and the specific gravity is 22.50. The annealed metal has a Brinell hardness of 172. Iridium is found in the nickel-copper ores of Canada, pyroxinite deposits of South Africa, and platinum ores of Russia and Alaska. It occurs naturally with the metal osmium as an alloy, known as **osmiridium,** 30 to 60% osmium, used chiefly for making fountain pen points and instrument pivots. Iridium is employed as a hardener for platinum, the jewelry alloys usually containing 10%. With 35% iridium the tensile strength of platinum is increased to 140,000 lb/in² (965 MPa). **Iridium-rhodium alloys** are used for high-temperature thermocouples. Electrodeposition is difficult, but coatings have been obtained with a fused sodium-cyanide electrolyte or with an iridium salt and organic compound in solution and, after application, volatilizing the vehicle and reducing the compound to the metal. Iridium-clad molybdenum and tungsten are produced by swaging an iridium tube onto the base metal and drawing at 1112°F (600°C). Iridium plating is used on molybdenum to protect against oxidation at very high temperatures. Above 1112°F (600°C) iridium tarnishes, and above 1832°F (1000°C) it forms a volatile oxide. It gives a bright ductile plate with a Vickers hardness of 170. It is also used as a catalyst. It is resistant to most molten metals except copper, aluminum, zinc, and magnesium. **Iridium wire** is used in spark plugs as it resists attack of lead aviation fuels. **Iridium-tungsten alloys** are used for springs operating at temperatures to 1472°F (800°C). Iridium intermetallic compounds such as Cb_3Ir_2, Ti_3Ir, and $ZrIr_2$ are superconductors. Iridium is multivalent with most of its compounds formed in its trivalent state, as $IrCl_3$. **Iridium-columbium** intermetallic compound has been identified as a potential material for high-temperature aircraft-turbine parts because of its high melting temperature and room-temperature ductility. It has a specific gravity of 15.2 and melts at 3452°F (1900°C).

IRON. One of the most common commercial metals. It has been in use since the most remote times, but it does not occur native except in the form of meteorites, and early tools of Egypt were apparently made from nickel irons from this source. The common iron ores are magnetic pyrites, magnetite, hematite, and carbonates of iron. To obtain the iron, the ores are fused to drive off the oxygen, sulfur, and impurities. The melting is done in a blast furnace directly in contact with the fuel and with limestone as a flux. The latter combines with the quartz and clay, forming a slag which is readily removed. The resulting product is crude pig iron which requires subsequent remelting and refining to obtain commercially pure iron. A short ton (0.9 metric ton) of ore, with about 1,000 lb (454 kg) of coke and 600 lb (272 kg) of limestone, produces an average of 1,120 lb (508 kg) of pig iron. Sintered iron and steel are also produced without blast-furnace reduction by compressing purified iron oxide in rollers, heating to 2200°F (1204°C), and hot-strip rolling. The final cold-rolled product is similar to conventional iron and steel.

Originally, all iron was made with charcoal, but because of the relative scarcity of wood and the greater expense, charcoal is now seldom used in the blast furnace. **Charcoal iron** has less sulfur and phosphorus than iron made with coke, and cast iron made from it has a dense structure and a tendency to chill. **Elverite** is a charcoal type of cast iron which gives a hard chill with a soft, gray-iron core. **Charcoal pig iron** was formerly imported from Sweden and Norway and was used for such purposes as car wheels, magnet cores, and making high-grade steels for boiler tubes. **Stora** was a name for Swedish charcoal iron used for making malleable iron.

Iron is a grayish metal, which until recently was never used pure. It melts at 2778°F (1525°C) and boils at 4442°F (2450°C). Even very small additions of carbon reduce the melting point. It has a specific gravity of 7.85. All commercial irons except ingot iron and electrolytic iron contain perceptible quantities of carbon, which affect its properties. Iron containing more than 0.15% chemically combined carbon is termed *steel.* When the carbon is increased to above about 0.40%, the metal will harden when cooled suddenly from a red heat. Iron, when pure, is very ductile, but a small amount of sulfur, as little as 0.03%, will make it **hot-short,** or brittle at red heat. As little as 0.25% of phosphorus will make iron **cold-short,** or brittle when cold. Iron forms carbonates, chlorides, oxides, sulfides, and other compounds. It oxidizes easily and is also attacked by many acids.

Because pure iron is allotropic, it can exist as a solid in two different crystal forms. From subzero temperatures up to 1670°F (910°C), it has a body-centered cubic structure and is identified as alpha (α) iron. Between 1670 and 2552°F (910 and 1400°C), the crystal structure is face-centered cubic. This form is known as gamma (γ) iron. At 2552°F (1400°C) and up to its melting point of 2804°F (1540°C), the structure again becomes body-centered cubic. This last form, called delta (δ) iron, has no practical use. The transformation from one allotropic form to another is reversible. Thus, when iron is heated to above 1670°F (910°C), the alpha body-centered cubic crystal changes into face-centered cubic crystals of gamma iron. When cooled below this temperature, the metal again reverts to a body-centered cubic structure. These allotropic phase changes inherent in iron make possible the wide variety of properties obtainable in ferrous alloys by various heat-treating processes.

Electrolytic iron is a chemically pure iron produced by the deposition of iron in a manner similar to electroplating. Bars of cast iron are used as anodes and dissolved in an electrolyte of ferrous chloride. The current precipitates almost pure iron on the cathodes which are hollow steel cylinders. The deposited iron tube is removed by hydraulic pressure or by splitting and then is annealed and rolled into plates. The iron is 99.9% pure and is used for magnetic cores and where ductility and purity are needed. Highly refined, nearly pure iron, designated **Puron,** is used for spectroscopic and magnetic standards. It contains 99.95% iron, with only 0.005 carbon, 0.003 sulfur, and less than 0.001 phosphorus. By high-temperature hydrogen annealing, the carbon can be reduced to 0.001%, bringing the purity to 99.99%. **Iron whiskers** are single-crystal, pure iron fibers, 0.00004 in (0.0001 cm) in diameter, for electronic uses. The tensile strength is as high as 500,000 lb/in^2 (3,448 MPa).

Iron powder, as originally produced in Sweden, is made by reducing iron ore by the action of carbon monoxide at a temperature below the melting point of the iron and below the reduction point of the other metallic oxides in the ore. In the United States it is made by the reduction of iron oxide mill scale, by electrolysis of steel borings and turnings in an electrolyte of ferric chloride, or by atomization. Iron powders are widely used for pressed and sintered structural parts, commonly referred to as **powder-metal (PM) parts.** Pure iron powders are seldom used alone for such parts. Small additions of carbon in the form of graphite and/or copper are used to improve performance properties.

Iron and carbon powders containing up to 1% graphite are commonly combined singly or with alloying elements to produce steel PM parts by pressing and sintering. As with all PM parts, strength and other mechanical properties increase with increasing density. As the density approaches the theoretical density, these properties approach those of wrought steel. As with wrought steels, strength and hardness can be significantly increased with subsequent heat treatment. Copper, a common alloying element, is used to increase strength, hardness, and corrosion resistance. It is also often used as an infiltrant to fill porous PM parts and increase density. Sometimes, however, porosity is advantageous, such as for filters or for filling with oil for surface lubrication. **PM forgings,** with density often approaching the theoretical, are made by hot-forming specially formed, medium-density **PM preforms.** Phosphorus copper in a carbon-steel matrix is used for the PM main bearing caps of some auto engines. Iron–2% copper, used for PM connecting rods in auto engines, provides the following tensile properties: 120,000 lb/in^2 (827 MPa) ultimate strength, 79,000 lb/in^2 (545 MPa) yield strength, 18% elongation, and 29×10^6 lb/in^2 (20×10^5 MPa) modulus. Auto applications, largely steel, account for about 65% of PM parts production in North America.

Soft **electrolytic iron powder** for making PM parts has 99% minimum iron, 0.04 maximum carbon, and 0.1 maximum of other impurities. **Magnetic iron powder,** used for electrical cores of high permeability, is made by reducing iron oxide with hydrogen. For this purpose it must be free of carbon and sulfur. A 390- to 1,575-μin (10- to 40-μm) powder with an apparent density of 0.099 lb/in^3 (2,740 kg/m^3) will compress to a density of 0.247 lb/in^3 (6,837 kg/m^3) at 150,000 lb/in^2 (1,034 MPa). **Carbonyl iron powder,** used for magnetic cores for high-frequency equipment and for medical application of iron, is metallic iron of extreme purity and in microscopic spherical particles. It is made by the reaction of carbon monoxide on iron ore to give liquid **iron carbonyl,** $Fe(CO)_5$, which is vaporized and deposited on the surface as a powder. **Micro-powder iron,** of International Specialty Products, is made by chemical decomposition of **iron pentacarbonyl.** This fine and uniform, high-purity, spherical powder comes in grades for producing PM parts by traditional methods and in grades for producing such parts by injection molding. The carbonyl iron powder of BASF Corp. for these methods is made by thermal decomposition of iron pentacarbonyl.

HVA iron powder, produced electrolytically in Germany for making strong, high-density parts by powder metallurgy, contains a maximum of 0.02% silicon, 0.05 carbon, 0.06 manganese, 0.01 phosphorus, and 0.01 sulfur. The standard powder has up to 35% of the grains under 0.0024 in (0.06 mm), up to 30% from 0.0039 to 0.0024 in (0.10 to 0.06 mm), and a maximum of 5% over 0.0079 in (0.20 mm). The sponge grade, for making porous parts, has a greater percentage of smaller particles, with up to 50% under 0.0024 in (0.06 mm). Parts made with electrolytic iron powder to a density of 7.5 have a tensile strength of about 40,000 lb/in^2 (276 MPa) with elongation of 30%.

Reduced iron, used for special chemical purposes, is a fine, gray, amorphous powder made by reducing crushed iron ore by heating with hydrogen. **Nu-Iron** is iron powder made by reducing Fe_2O_3 with hydrogen at 1300°F (704°C) to FeO, and then at 1100°F (593°C) to iron to prevent sintering of the particles. **H-iron** is a **hydrogen-reduced iron.** Iron powder made by this process is usually pyrophoric and is made nonpyrophoric by heating in a nonoxidizing atmosphere. **Ferrocene,** used as a combustion-control additive in fuels, and in lubricants and plastics for heat stabilization and radiation resistance, is a **dicyclopentadienyl iron,** $(C_5H_5)_2Fe$. It is a double-ring compound with the iron atom between parallel planes and symmetrically bound to the five carbon atoms of each ring. It is an orange-yellow powder melting at 343°F (173°C), but resisting pyrolysis to 878°F (470°C). **Iron 55** is a very pure radioactive iron produced in minute quantities by cyclotronic bombardment of manganese and used in medicine for increasing the red blood pigment, **hemoglobin,** in the human system. **Busheled iron** is an inferior grade of iron or steel made by heating bundles of scrap iron or steel in a furnace and then forging and rolling into bars. It is not uniform in composition or in the welding, and it is used only in isolated instances or in wartime.

IRON CARBIDE.   A hard, brittle intermetallic compound of iron and carbon, having the approximate composition Fe_3C. It is also called **cementite,** which dominates **white cast iron,** a type of cast iron formed by rapid solidification of molten iron. Iron carbide has a density of 0.276 lb/in^3 (7,640 kg/m^3), which is slightly less than that of pure iron. It is extremely hard, thus exceptionally wear-resistant, retains high hardness up to red heat, has high compressive strength, and exhibits a white crystalline fracture. With suitable compositions it can be induced to form in select regions of otherwise softer irons by the use of chills in these regions of a casting mold to effect more rapid solidication.

By using a high-temperature fluid-bed reactor developed by Procedyne Corp., **iron ore** can be converted to iron carbide using natural gas. The process, licensed to Nucor Corp. by Carbide Holdings Inc., produces the carbide for use as a partial replacement for steel scrap in the production of **sheet steels.**

IRON ORES.   Iron-bearing minerals from which iron can be extracted on a commercial scale. The chief iron ores in order of importance are hematite, magnetite, limonite, and siderite. More than 90% of the iron ores mined in the United States are **red hematite,** Fe_2O_3, containing theoretically 70% iron, but usually not over 60%. The districts include the Lake Superior region and northern Alabama. It is also the ore from the Furness district in England and parts of Spain and Germany. The color is various shades of reddish brown, and the structure is usually earthy. The variety known as **kidney ore** is columnar with a fibrous appearance; **specular hematite** has a brilliant luster and foliated structure. The specific gravity is 4.8 to 5.3. **Brown hematites** contain from 35 to 55% iron. Ores containing more than 50% iron are considered high-grade. Lake Superior ores are now averaging only 52% iron, with 8 silica. This region also has large reserves of **taconite,** a ferroginous chert which is an alteration product from **greenalite,** or ferrous silicate. Taconite eventually leaches and becomes a hematite by the loss of silica, but the Mesaba taconite is a very hard, gray-green sedimentary chert in the form of a compact silica rock with 20 to 35% iron. It cannot be used directly in a blast furnace but is crushed to powder, concentrated to 65% iron, and pelletized for use. **Rubio iron ore** of Spain is classified both as limonite and as brown hematite. It is a **hydrated ferric oxide,** $Fe_2O_3 \cdot H_2O$, brown-streaked with a silica gangue. It contains 77.4% ferric oxide, 9.2 silica, 1.76 alumina, and 1.1 manganese oxide.

The **hematite ores** were preferred for the Bessemer process because of their freedom from phosphorous and sulfur. **Natural iron** is the percentage of iron in the ore before drying; **dry iron** is the percentage of iron in the ore after drying at 212°F (100°C). Low-grade hematite and limonite ores can be concentrated by passing the ground ore through a reducing gas at high temperature, which causes a part of the iron in each particle to become magnetic and thus capable of being separated. Low-grade ores of 30% iron are also ground to a fine powder and separated from the gangue by flotation, concentrated to 60% iron, and pelletized with a bentonite binder. **Self-fluxing iron ore** is concentrated iron ore combined with limestone and pelletized.

Magnetite, or **magnetic iron ore,** is found in northern New York, in New Jersey, and in Pennsylvania. It has composition $FeO \cdot Fe_2O_3$, containing theoretically 72.4% iron but usually only about 62. Magnetite may also contain some nickel or titanium. The specific gravity is 5.18, the melting point is 2804°F (1540°C), the color is iron black with a metallic luster, and the material is strongly magnetic. The natural magnet known as **lodestone** is magnetite.

Siderite and carbonate ores are used in Great Britain, Germany, and Russia. Much of these ores is not considered commercial in the United States, but the **dogger iron ore** of Germany contains as high as 35% iron. **Siderite** is the chief iron ore in Great Britain. It is found in Staffordshire, Yorkshire, and Wales, and in the United States in Pennsylvania and Ohio. It is an **iron carbonate,** $FeCO_3$, containing theoretically 48.2% iron. It usually occurs granular or compact and is earthy. Its specific gravity is 4.5 to 5, and the Mohs hardness is 3.5 to 4. The color is light to dark brown, with a vitreous luster. It often is impure, with a mixture of clay materials or forming stratified bodies with coal formations. It is also known as **carbonate ore, ironstone,** and **spathic iron ore.** Impure forms mixed with clay and sands are called **clay ironstones,** or **black band.** The ironstone and black-band ores are the important ores of England and Scotland, but there is only slight usage of carbonate ores in the United States. The **minette ore** of Luxembourg and Germany averages about 26% iron, 22 CaO, 11 silica, and 2 $CaOSiO_2$, with 0.54 phosphorus and 0.06 vanadium. The ore mined in Norway and Sweden is very pure, and it is the ore used for the **dannemora iron** made with charcoal as a fuel.

Limonite, also called **brown hematite, brown ore, bog iron ore,** and **shot ore** when in the form of loose, rounded particles, was the common iron ore of early New England. It is a mineral of secondary origin formed by the water solution of other iron minerals. Its composition is $2Fe_2O_33H_2O$, containing theoretically 59.8% iron, but usually 30 to 55%. It occurs earthy or in stalactitic forms of a dark-brown color. The specific gravity is 3.6 to 4. Limonite is found in Alabama, Tennessee, and Virginia and is also an English and German ore. **Goethite** is a minor ore of iron of composition $Fe_2O_3 \cdot H_2O$, found in the Lake Superior hematite deposits and in England. It is yellowish brown with a specific gravity of 4.3. It is also called **turgite.** The largest known deposits of high-grade iron ores are in Brazil and extend into Venezuela. The ore in southwestern Brazil is with deposits of manganese. The hematite of Minas Gerais contains 57 to 71% iron. Large deposits of high-grade iron ores are also found in the Labrador-Quebec regions of Canada, in India, and in the South Africa-Zimbabwe area. The nearest approach to a native iron is the iron-nickel mineral **awaruite,** $FeNi_2$, found in gravel in New Zealand and Alaska, and **josephinite,** $FeNi_3$, found in serpentine in Oregon.

IRON PYRITE. A mineral sometimes mined for the zinc, gold, or copper associated with it, but chiefly used for producing sulfur, sulfuric acid, and copperas. It is an **iron disulfide,** FeS_2, containing 53.4% sulfur. It often occurs in crystals, also massive or granular. It is brittle, with a Mohs hardness of 6 to 6.5 and specific gravity 4.95 to 5.1. The color is brass yellow, and it is called **fool's gold** because of the common error made in detection. In the Italian method of roasting pyrites, oxygen combines with sulfur to form sulfur dioxide, and with iron. In a yield of 100,000 tons (90,700 metric tons) of concentrated sulfuric acid there is obtained a by-product of 50,000 tons (45,350 metric tons) of 65% iron ore. **Pyrite** is found in rocks of all ages associated with different minerals. The pyrites mined in Missouri, known as **marcasite,** also used for gemstones, have the formula FeS, and the gem specimens have a yellow color with a greenish tinge. **Sulfur pyrites,** found in great quantity in Shasta County, California, are roasted to produce sulfuric acid, and the residue is used for making cement and ferric sulfate. The distillation of 370 tons (336 metric tons) of pyrite of Ontario yields 62 tons (56 metric tons) of sulfur and 252 tons (229 metric tons) of iron oxide sinter, the latter being used in steel mills. In Japan pyrite is chlorinated, producing sulfur, copper, gold, silver, and zinc, and the residue of sintered pellets of Fe_2O_3 is used in blast furnaces.

IRON SHOT. An abrasive material made by running molten iron into water. It is employed in tumbling barrels and in the cutting and grinding of stones. The round sizes between No. 30 and No. 20 are used for the shot peening of mechanical parts to increase fatigue resistance. **Peening shot** is marketed by SAE numbers from 6 to 157, the size numbers being the diameter in thousandths of an inch. Grit numbers in shot for grinding are from No. 10 to No. 200; the No. 10 is 0.0787 in (0.200 cm), and the No. 200 is 0.0029 in (0.007 cm) in diameter. **Steel grit** is made by forcing molten iron through a steam jet so that the metal forms into small and irregular pieces, and large globules are crushed into steel grit in sizes from No. 8 to No. 80. It is preferred to sand for sandblasting some materials. **Steelblast** is this material for tumbling and sandblasting.

Tru-Steel shot is steel shot which has been heat-treated to give toughness to prevent breakdown into fragments. **Cutwire shot** is made by cutting 1065 carbon steel wire into short pieces. **Pellets shot** is steel wire of Rockwell C hardness 50 cut to lengths equal to the diameter of the wire, ranging from 0.028 to 0.41 in (0.071 to 1.041 cm). **Kut-Steel** is a similar abrasive. **Permabrasive** is a treated **malleable iron shot** to give a soft, resilient body with a hard exterior. All these materials for metal cleaning are termed **blasting shot.**

IRONWOOD. The name for several varieties of wood, which may refer to any exceedingly hard wood that is used for making bearings, gears, tool handles, or parts of machinery. In the United States ironwood is most likely to refer to **hackia,** the wood of the hackia tree, *Ixora ferrea,* of the West Indies and of tropical South America, or it may refer to the wood of the quebracho tree. Hackia is brown, has a coarse, open grain, and is very hard and tough. The density is about 55 lb/ft^3 (881 kg/m^3). It is also used for furniture. The Burmese tree, *Mesua ferrea,* furnishes the wood **gangaw,** which is also known as ironwood. It is a tough, extremely hard wood of a rose-red color with a density of 70 lb/ft³ (1,121 kg/m³). Innumerable varieties of hardwoods occur in the tropical jungles, many unnamed or known only by native names. Many have rich colors, shading from yellow, orange, and red to purplish-black, and come in beautiful grains which can be varied by the angle of cutting. They are generally commercialized only on a small scale for use in furniture, cabinetwork, inlays, and panel facings.

Ekki, *Lophira alata procera,* sometimes called **African ironwood** or **azobe,** comes from West Africa. It is an excellent timber for piling, wharf and dock construction, bridge planking, ties, and all heavy timber structures. The wood is a chocolate-brown color, sometimes verging on dark red, and has a speckled surface caused by yellowish deposits in the pores. It is so heavy that the timbers will not float in water.

ISINGLASS. A gelatin made from the dried swimming bladders, or **fish sounds,** of sturgeon and other fishes. **Russian isinglass** from the sturgeon is the most valued grade and is one of the best of the water-soluble adhesives. It is used in glues and cements and in printing inks. It is also used for clarifying wines and other liquids. It is prepared by softening the bladder in water and cutting it into long strips. These dry to a dull-gray, horny, or stringy material. Isinglass is also known as **ichthyocolla. Ichthyol** is an entirely different material. It is a reddish-black, syrupy liquid with a peculiar odor and burning taste, distilled from an asphaltic shale found in the Austrian Tyrol. It is used as an antiseptic astringent. The name *isinglass* was also used for mica when employed as a transparent material for stove doors.

IVORY. The material that composes the tusks and teeth of the elephant. It is employed mostly for ornamental parts, art objects, and piano keys, although the latter are now usually a plastic which does not yellow. The color is the characteristic ivory white, which yellows with age. The specific gravity is 1.87. The best grades are from the heavy tusks weighing more than 55 lb (25 kg), sometimes 6 ft (1.8 m) long. The ivory from animals long dead is a gray color and inferior. The softer ivory from elephants living in the highlands is more valuable than the hard and more brittle ivory of the low marshes. The west coasts of Africa, India, and southern Asia are the chief sources of ivory. The tusks of the hippopotamus, walrus, and other animals, as well as the fossil mammoth of Siberia, also furnish ivory, although of inferior grades. **Odontolite** is ivory from fossil mammoths of Russia. Ivory can be sawed readily and is made into thin veneers for ornamental uses. It takes a fine polish. Artificial ivory is usually celluloid or synthetic resins.

Vegetable ivory, used for buttons and small articles, is from the **ivory nut,** called **tagua nut** in Columbia and Ecuador and **jarina** in Brazil. The ivory nut is the seed of the low-spreading palm tree *Phytelephas macrocarpa,* which grows in tropical America. The nuts are about 2 in (5 cm) in diameter, growing in clusters and encased in shells. They have a fine, white color and an even texture. They can be worked easily, and they harden on exposure to air. They take dyes readily and show fine, polished colors. A similar nut called the **dom nut** is used in Eritrea for making buttons and novelties.

JAPAN WAX. Sometimes called **sumac wax, China green tallow,** and **Japan tallow.** A vegetable fat used for extending beeswax and in candles and polishes. It occurs between the kernel and the outer skin of the berries of plants of the genus *Rhus,* which grow in Japan and California. The fat, which is misnamed *wax,* is extracted by steaming and pressing the berries and then refining. The crude wax is a coarse greenish solid, but the bleached wax is in cream-colored cakes which darken to yellow. Japan wax melts at 124°F (51°C). The specific gravity is about 0.975. It contains chiefly palmitic acid in the form of glyceryl palmitate with also stearic and oleic acids, and a characteristic acid, **japonic acid,** which is also found in catechu. The saponification value is 220, and the iodine value is 12. It is easily emulsified in dilute aqueous solutions of alkali carbonates, and so finds its chief use as an emulsifier in wax polishes and cosmetics. It is sometimes adulterated with common tallow. **Lac** is the gum exudation of the wood of the same sumac plants, notably *R. vernicifera,* of Japan and Korea. It is used as a drying oil in clear lacquers and baking enamels. Lac is distinct from shellac of the lac insect.

JEWELRY ALLOYS. An indefinite term which refers to the casting alloys used for novelties to be plated and to the copper and other nonferrous sheet and strip alloys for stamped and turned articles. The base-metal jewelry industry, as distinct from the precious-metal industry, produces costume jewelry, trinkets, souvenirs, premium and trade goods, clothing accessories known as notions, and low-cost religious goods such as medals and statuettes. The soft white alloys of this type are not now as important as they were before the advent of plastics, and articles now produced in metals are likely to be made from standard brasses, nickel brasses, and copper-nickel alloys, but laminated and composite sheet metals are much used to eliminate plating. When the base metals are clad or plated with gold, they are called **gold-filled metal** if the gold alloy is 10-karat or above and the amount used is at least 5% of the total weight. When the coating is less than 5%, it is called **rolled-gold plate.** Formerly, a great variety of trade names were used for jewelry alloys. **Argentine metal** was a silvery alloy for casting statuettes and small ornaments. It contained 85.5% tin and 14.5 antimony and produced silvery-white, hard, and clean-cut castings. An alloy known as **Alger metal** contained 90% tin and 10 antimony. A harder but more expensive white alloy known as **Warnes metal** contained 10 parts tin, 7 nickel, 7 bismuth, and 3 cobalt. **Fahlum metal,** used for stage jewelry, contained 40% tin and 60 lead. When faceted, it makes highly reflective brilliants. **Rosein** was a white metal for jewelry and ornamental articles containing 40% nickel, 30 aluminum, 20 tin, and 10 silver. Polished, it has a high white luster without plating. **Mock silver** was an aluminum alloy containing about 5% silver and 5 copper. **Argental** was a rich, low brass whitened and strengthened with about 5% cobalt. **Kuromi,** a Japanese white jewelry alloy, is copper whitened with tin and cobalt. **Argent français,** or **French silver,** is a copper-nickel alloy containing considerable silver. **Platinoid** was a nickel-silver-tungsten alloy with the small amount of tungsten added to the melt in the form of phosphor tungsten. **Proplatinum,** another substitute for platinum, was a nickel-silver-bismuth alloy. **Nuremberg gold,** with a nontarnishing golden color, contained 90% copper, 7.5 aluminum, and 2.5 gold.

JOJOBA WAX.   Also called **jojoba oil,** pronounced ho-ho-bah. A colorless, odorless liquid wax obtained by pressing or solvent extraction from the seed beans of the evergreen shrub *Simmondsia californica* growing in the semiarid region of the southwest United States and northern Mexico. Recently, commercial interest in jojoba has increased significantly, primarily because the seed consists of 50% liquid wax esters (97% pure); in contrast, other seed oils, such as corn, soy, sunflower, or olive, are rich in triglycerides. Resembling **sperm whale oil,** it remains liquid at room temperature and is used as a substitute in lubricants, leather dressing, and cosmetics. About 40,000 acres (162×10^6 m^2) is in cultivation in the southwestern United States, producing about 551 tons (500 metric tons) of oil. Processing is not unlike that for other oils, involving seed cleaning, drying, dehulling, pressing, and solvent extraction. Three grades are available: natural, refined, and modified. Single-pressed oils are premium natural grades and have less than 0.3% free fatty acids. Refined oil is treated by activated bleaching and is devoid of color, flavor, and odor, properties desirable in cosmetic formulations. The oil can be modified to become water-soluble through ethoxylation and propoxylation. Jojoba Growers and Processors, Inc.'s **Soluhoba E-80** and **Soluhoba E-120** have 80 and 120 mol of ethylene oxide per mole of jojoba oil. These derivatives are used as primary and secondary emulsifiers, wetting agents, and solubilizers for perfume oils. The melting point can be raised by isomerization with catalysts. Hydrogenation produces a solid, waxlike product similar to beeswax and candelilla wax. The fish lipids of **orange roughy, black oreo,** and **small-spined oreo** are comparable in composition to jojoba oil. **Jojoba alcohol,** extracted from the bean, is a complex alcohol used as an antioxidant to prevent deterioration of fats.

JUJUBE. The oval fruit of the small, spiny tree *Ziziphus jujuba* of the buckthorn family, growing in dry, alkaline soils. It is native to northern China, but is also grown in the Mediterranean countries and in the southwestern United States. It is also known as the **Chinese date,** or **tsao.** The fruit has high food value, being higher in sugars than the fig, 65 to 75%, and higher in protein than most fruits, 2.7 to 6%, but is very acidic, containing up to 2% acid. Some varieties develop butyric acid on ripening. Jujube is used for flavoring and in confections, preserves, and sweet pickles.

JUNIPER. The wood of the juniper tree *Juniperus virginiana* growing in the eastern United States from Maine to Florida. It is also called **red cedar, red juniper,** and **savin.** The heartwood has a bright to dull red color, and the thin sapwood is nearly white. The wood is lightweight, soft, weak, and brittle, but is durable. It is used for chests, cabinets, and closet lining because of its reputed value for repelling moths. It was formerly employed on a large scale for lead-pencil wood and was known as **pencil cedar**; but it is now scarce, and other woods are used for this purpose, notably **incense cedar,** *Libocedrus decurrens,* of California and Oregon, a close-grained brown wood with a spicy, resinous odor. **Rocky Mountain juniper** is from a medium-sized tree of the Rocky Mountain states, *J. scopulorum,* valued for fence posts and lumber. **African pencil cedar** is from the tree *J. procera,* of eastern Africa. It is harder and heavier and less fragrant than incense cedar. There are more than 40 species of juniper. The fruit or **juniper berries** of the common varieties are used in flavoring gin. **Cade oil** is a thick, brownish essential oil, of specific gravity 0.950 to 1.055, distilled from the wood of the **European juniper** and used in antiseptic soaps. It is also called **juniper tar oil.** The juniper oil known as **savin oil** is distilled from the leaves and tops of the **juniper bush,** *J. sabina,* of North America and Europe. It is used in medicine as a diuretic and vermifuge, and in perfumes. **Cedarwood oil,** used in perfumes and soaps, is distilled from the sawdust and waste of the eastern red cedar. The red heartwood contains 1 to 3% oil.

JUTE. A fiber employed for making burlap, sacks, cordage, ropes, and upholstery fabrics. It is obtained from several plants, of which *Corchorus capsularis* of India is the most widely cultivated, growing in a hot, steaming climate. This fiber in Brazil is called **juta indiana,** or **Indian jute.** Most of the commercial jute comes from Bengal. The plant grows in tall, slender stalks like hemp, and the fiber is obtained by retting and cleaning. The fiber is long, soft, and lustrous, but is not as strong as hemp. It also loses its strength when damp, but is widely used because of its low cost and because of the ease with which it can be spun. Its chemical composition is intermediate between those of hemp and kapok. It contains 60% alpha cellulose, 15 pentosan, 13 lignin, and 4.5 **uronic anhydride** which is also a constituent of kapok. The crude fiber may be as long as 14 ft (3.6 m), but the commercial fibers are from 4 to 8 ft (1.2 to 2.4 m). The butts, or short ends of the stalks, and the rough fibers are used for paper stock. **Jute paper,** used for cement bags, is a strong paper made of these fibers usually mixed with old rope and old burlap in the pulping. It is tan in color. **Jute fiber** is also used for machine packings. **Rel-Kol** is jute fiber treated with synthetic rubber and braided into square sections.

Brazilian jute is fiber from plants of the mallow family, *Hibiscus kitaibelifolius,* or *H. ferox,* cultivated in the state of Saõ Paulo, Brazil. It is also called **juta paulista.** The fiber is long, strong, resilient, and durable. It is used for making burlap for coffee bags. The fiber known as **pacopaco** is from *H. cannabinus,* of Bahia and Minas Gerais. It is used for cordage, burlap, and caulking. In Indonesia it is called **Java jute.** In India it is called **Bimlipatam jute** and **Deccan hemp,** and it is mixed with *Corchorus* fibers as jute and in burlap. As a substitute for hemp, it is known as **sunee,** or **brown Indian hemp.** The species known as **amaniurana** is from *H. furcellatus* of the Paraguay River Valley. The fibers are very soft and silky and are made into sacks and bags. The species *H. sabdariffa,* known as **vinagreira,** is used as a substitute for jute. In the East Indies it is called **roselle fiber** and has characteristics similar to India jute, but is lighter in color. In El Salvador it is called **kenaf fiber** and is used for coffee bags. **Brazilian hemp** is from *H. radiatus* and is stronger than true hemp. It is cultivated in Bahia. **Aramina fiber** is from the very long stalks of a plant *Urena lobata,* of the mallow family, of Brazil. In the north of Brazil it is known as **carrapicho.** The fiber is used for cordage, twine, and burlap fabrics. In Cuba this fiber is known as **white malva** and **bowstring hemp. Cuba jute** is from *Sida rhombifolia.* Both Cuba jute and aramina fiber belong to the mallow family, *Malvaceae,* and in Cuba and Venezuela are also known as **malva fiber.** The *U. lobata* is also grown in Zaire where it is known as **Congo jute.**

Pitafiber, used in Colombia and Central America for coffee bags, is from a plant of the pineapple genus, *Ananas magdalenae.* The fiber is a light cream color, lustrous, very long, and finer and more flexible than most hard fibers, so that it is useful for ropes, twines, and fabrics, although most of the fibers of this botanical class are brush fibers. It has excellent resistance to salt water. The word *pita* means yellow or reddish in the Indian language, and this name is also used for grades of yucca and other fibers. **Tucum fiber** is from the leaves of the oil palm *Astrocaryum tucuma* of Brazil. The fibers are long, flexible, water-resistant, and durable. They are used for ropes, hammocks, and marine cordage, but are not classified as burlap fiber. Another fiber of great length and noted for resistance to insect attack is **curana,** from the stalks of the plant *Bromelia sagenaria* of northeast Brazil. There are two chief grades: white and roxo or purple.

KAOLIN. Also called **China clay.** A pure form of hydrated **aluminum silicate clay.** There are three distinct minerals, **kaolinite, nicrite,** and **dickite,** all having similar composition. The formula for kaolin is usually given as $Al_2O_3 \cdot 2SiO_2 \cdot 2H_2O$, but is also expressed as $Al_2Si_2O_5(OH)_4$. It occurs in claylike masses of specific gravity 2.6 and a dull luster. Kaolin is used for making **porcelain** for **chinaware** and **chemical porcelain** for valves, tubes, and fittings; as a refractory for bricks and furnace linings; for electrical insulators; as a pigment and filler in paints; as a filler in plastics; and as an abrasive powder.

In firebricks kaolin resists spalling. Its melting point is 3200°F (1760°C), but this lowers with impurities. The color of all varieties is white, but inferior grades burn to a yellow or brownish color, and it should be free of iron. Porcelain made from kaolin is fired at about 2300°F (1260°C), but the upper service limit of the products is only about 500°F (260°C) since it has a low heat-transfer rate and low thermal shock resistance. Porcelain parts have a specific gravity of 2.4 to 2.9., a Mohs hardness of 7.5, and a compressive strength from 60,000 to 90,000 lb/in^2 (414 to 621 MPa).

When kaolin is employed as an inert colloidal pigment in paints, it is called **Chinese white.** The powder is hydrophobic and cannot be wet by water, but it has good compatibility in oils and many organic solvents. As an extender in plastics and rubbers, it reduces absorption of moisture and increases dielectric strength. Kaolin is a decomposition product of granite and feldspar, and its usual impurities are quartz, feldspar, and mica, which can be washed out. The **aluminum silicate RER-45,** of Georgia Kaolin Co., is purified kaolin ground to a fineness of 7.9- to 177-μin (0.2- to 4.5-μm) particle size. It is used in paints, coatings, and plastics. **Modified kaolin,** for rubber and plastics, has composition $(Al_2O_3)_3(SiO_2)_2$, and the particles are in thin, flat plates averaging 21.7 μin (0.55 μm).

The **Cornwall kaolin** of England and **Limoges kaolin** of France are the best known. English China clays contain little or no iron oxide, and the yellow clays contain only organic materials that can be bleached out. The best grade of English clay is used for coating and filling paper. **Cornish clay,** known as **China stone,** is used for the best grades of porcelain glazes. Cheaper grades of kaolin, called **mica clay,** are used for earthenware glazes and as an absorbent in oil purifying. The clay of Kentucky and Tennessee, known as **ball clay,** occurs in massive beds of great purity. It has high plasticity and good bonding strength and is light in color when fired. It is used for high-grade porcelain and for wall tiles. Impure varieties of kaolin, called **kaolonic earth,** are used for refractories. **Kaolin fiber** of extreme fineness, with average diameter of 118 μin (3 μm), is made from kaolin containing about 46% alumina, 51 silica, and 3 iron and titanium oxides. It withstands continuous temperatures to 2000°F (1093°C). **Kaowool,** of Babcock & Wilcox, is kaolin fiber in the form of insulating blankets of 0.25- to 3-in (0.64- to 7.62-cm) thickness. **Kaowool paper** is made from the fibers compressed to thicknesses up to 0.08 in (0.20 cm) with or without a binder. It withstands temperatures to 2000°F (1093°C) and is used for filters, separators, and gaskets. The fibers may also be compressed into **kaowool blocks** of 4-in (10-cm) thickness.

Halloysite has about the same composition as kaolinite but contains more alumina and water. It occurs with kaolinite. In association with alunite in Arkansas, it is called **newtonite.** Some varieties, such as **glossecollite,** are waxlike, and with an electron microscope the grains appear as tubular structures. A variety of halloysite is marketed in France as a fine, gray powder for use as a filler in rubber latex compounds. **Indianite,** or **allophane,** is an impure halloysite. It is a white, waxy clay found in Indiana and is used for pottery. The Indiana halloysite used for refractories is called **malinite. Bone clay** is a pure kaolin from feldspar and granite. It makes a strong porcelain. But **bone china** is a name given to high-grade English pottery made with China clay and 25% calcined bone. **Fresh clay** is formed from feldspar and quartz. It gives resilience to porcelain. Clays can be mixed to give desired characteristics.

Micronized clay is pure kaolin ground to a fineness of 400 to 800 mesh, used as a filter in rubber. **Dixie clay,** of R. T. Vanderbilt Co., is ground kaolin of 300 mesh, used as a stiffening or reinforcing agent in rubber and adhesives. **Osmose kaolin** is kaolin deposited by electroosmosis from an alkaline solution to eliminate iron and other impurities and to raise the alumina ratio. As a fine powder, it is used for making electrical insulators and synthetic mica, and for cosmetics. **Osmo** is a finely ground kaolin for cosmetics. **Alumina flake** is a white, flaky kaolinic clay from Missouri, ground for use in paints, adhesives, and rubber. The **tile kaolin** of Georgia Kaolin Co. is selectively mined and water-washed to obtain low iron and titania content. It has high plasticity and fires white with good translucency. The mean particle size is about 47 μin (1.2 μm) with 65% below 79 μin (2 μm). The pH is 4 to 6.5. **Quickcast kaolin** of the same company is a coarse powder of iron-free Georgia kaolin with 200-mesh particle size. It is used for mixing with other clays to increase the casting and drying times and to improve whiteness in tile and ceramic products.

KAPOK. A silky fiber obtained from the seed pods of the silk-cotton trees of genera *Ceiba, Bombax, Chorisia,* and *Ochroma,* now grown in most tropical countries. It is employed for insulation and fine padding work. It is extremely light and resilient. The chemical constituents are the same as those of jute, but the proportions are different. Kapok is low in alpha cellulose, 43%, and high in pentosans, 24, lignin, 15, and uronic anhydride, 6.6. Most of the commercial kapok normally comes from Java and is from *Ceiba pentandra.* The tree was brought originally from Brazil where it was known as **samaúma.** It grows to a height of 100 ft (30 m), with diameters to 10 ft (3 m), making the picking of the kapok pods difficult. The fibers are long, white, and silky, similar in appearance to cotton, but are too brittle for spinning. It is also known as **Java cotton** and **Illiani silk.** A silky cotton somewhat similar but inferior to kapok is **madar,** from the shrub *Calotropis gigantea* of India and the East Indies. Another species, **akund,** is from the *C. procera.* Both fibers are sometimes mixed with kapok, but are less resilient. They are known in the East Indies as **vegetable silk. Red silk cotton,** known as **simal,** is from the large tree *Bombax ceiba* or *B. malabaricum,* of India. The fiber is reddish. **Shilo fiber** is from the tree *B. ellipticum.* **White silk cotton** is from the tree *Cochlospermum gossypium* of India. It is quite similar to kapok. The kapok of Ecuador is from various species of the tree *Chorisia.* The fiber from the balsa trees, of the genus *Ochroma* of Central America, Colombia, and Ecuador, is dark in color. **Mexican kapok** is from the *Bombax palmeri* and from *Ceiba schottii* and *C. acuminata,* the fibers from the two latter being more buoyant than the Java fiber. The kapok from the lower Amazon region is from *C. samaúma,* and the **pochote fiber** of El Salvador is from *C. aesculifolia.* The **balsa fiber** of Central America is from the tree *O. velutina.* **Paina kapok** is from a ceiba tree of Brazil. **Kapok oil** is a semidrying oil obtained from the seeds of the kapok tree. It is used in margarine and for soaps.

A substitute for kapok for sound and heat insulation, and as stuffing for life jackets, cushions, furniture, and toys, is *typha,* or **cattail fiber.** It is the fluffy fiber from the cylindrical flowers of *Typha latifolia* which grows in swamps throughout the temperate climates of North America. As an insulating material, the fiber has about 90% the efficiency of wool. **Milkwood floss,** used also as a substitute for kapok, is the bluish-white, silky fiber from the seed pods of the common milkweed, *Asclepias syrica,* of the eastern United States. It is not grown commercially.

KARAKUL. The curly, lustrous fur pelts of newly born lambs of the karakul sheep originally of Afghanistan and Siberia, but now grown extensively in southwest Africa for the fur. The wool of the older sheep is clipped for carpet wool, but the fur pelts of the young animals are highly valued for coats and garment trim. The best qualities are those with small curl and medium curl to the fur. The fur skins are also graded by color, the gray and flora being most valued and the brown the least valued. **Persian lambskin** is a name for pelts of small, tight curl.

KAURI GUM. A fossil gum dug from the ground in New Zealand and New Caledonia, used in varnishes and enamels to increase the body, elasticity, and hardness. It is also used in adhesives, and the lower grades of chips are used in linoleum. It was first known as **New Zealand gum,** and first spelled *cowrie,* although **cowrie** is the name of a genus of mollusk shells found in the Indian Ocean and formerly used as money in China. Kauri is a product of kauri tree exudations buried for long periods, but it also comes from the conifer tree *Agathis australis.* There is little extraction of the gum from the present kauri forests, whose wood is employed for lumber, but some **bush gum** is obtained by collecting the deposits in the forks of branches. **Range gum** is found in clay deposits, and some is transparent. **Swamp gum** is brown and varies from hard to friable. The fossil gum has a specific gravity of 1.05, a melting point of 360 to 450°F (182 to 232°C), and is soluble in turpentine, benzol, and alcohol. The **kauri tree** grows to a height of 100 ft (30 m) and great diameters and yields a yellowish-brown, straight-grained wood free from knots and much prized as a useful softwood. The density is 36 lb/ft^3 (577 kg/m^3). Mottled and figured **kauri pine** is used as a cabinet wood.

KERMES. A brilliant red natural dyestuff similar in color to cochineal, having a beautiful tone and being very colorfast. It is one of the most ancient dyes, but is now largely replaced by synthetic dyestuffs. Kermes is an insect found on the **kermes oak** tree,

Quercus coccifera, of southern Europe and North Africa. The body of the animal is full of a red juice, and the coloring matter, **kermesic acid,** $C_{18}H_{12}O_9$, is separated out in brick-red crystals. The yield is 1.8 oz (50 g) from 110 lb (50 kg) of kermes. It has only about one-tenth the coloring power of cochineal. **Garouille** is the root bark of the kermes oak. It contains 18 to 32% tannin and is used for making sole leather. The color is darker than oak tannin.

KEROSENE. Originally the name of illuminating oil distilled from coal, and also called **coal oil.** It is now a light, oily liquid obtained in the fractional distillation of petroleum. It distills off after the gasoline and between the limits of 345 and 550°F (174 and 288°C). It is a hydrocarbon of composition $C_{10}H_{22}$ to $C_{16}H_{34}$, with a specific gravity between 0.747 and 0.775. Commercial kerosene may be as high a distillate as 617°F (325°C), with a corresponding higher specific gravity up to 0.850; but in states where it is distinguished from gasoline in the tax laws, it is more sharply defined. In Pennsylvania, kerosene is defined as having a flash point above 115°F (46°C), with not over 10% distillable at 347°F (175°C) and not over 45% at 392°F (200°C). Kerosene is employed for illuminating and heating purposes, as a fuel in internal combustion engines, and for turbine jet fuels. The heaviest distillate, known as **range oil,** is sufficiently volatile to burn freely in the wick of a heating range, but not so volatile as to be explosive. It is nearly free from odor and smoke. **Deodorized kerosene,** used in insect sprays, is a kerosene highly refined by treatment with activated earth or activated carbon. **Fialasol** is a **nitrated kerosene** used as a solvent for the scouring of wool. It is less flammable and has a slower rate of evaporation than kerosene, and it is odorless.

KHAYA. A class of woods from trees of the genus *Khaya,* growing chiefly in tropical West Africa and known commercially as **African mahogany.** The woods closely resemble mahogany, but they are more strongly figured than mahogany, are slightly lighter in weight and softer, and have greater shrinkage. The pores are larger, and the wood is coarser. The wood is used for furniture and store fixtures, musical instruments, and paneling. It is not as suitable for patterns as mahogany. African mahoganies are marketed under the names of the shipping ports, as the shipments from the various ports usually differ in proportion to the different species cut in the region. The chief wood of the genus, from which the native name *khaya* derives, is known commercially as **dry zone mahogany,** *K. senegalensis.* It is also known as **kail** and **oganwo.** It grows from Gambia to Angola on the west coast and eastward to Uganda. The heartwood is dark reddish brown, and the thick sapwood is grayish to pinkish red. The most favored commercial wood is that of the **red khaya,** or **red mahogany,** *K. ivorensis,* known locally as **dukuma** and **dubini.** This wood is highly figured, with interlocking grain, and when quartered shows a ribbon figure with alternate light and dark stripes. It comes chiefly from the Ivory Coast, Ghana, and Nigeria. **Sassandra mahogany** is chiefly this species. **Duala mahogany** is chiefly **white mahogany,** *K. anthoteca,* known also as **diala, krala,** and **mangona.** The wood is lighter in color but tinged with red. **Big-leaf mahogany,** *K. grandifolia,* has a reddish-brown color. Much African mahogany is cut into veneer, and the standard thickness for the face veneer is 0.036 in (0.09 cm).

Gaboon mahogany and **Port Lopez mahogany** are chiefly **okume wood,** from the tree *Aucoumea klaineana* of the Guinea coast. The tree belongs to the family *Meliaceae* to which khaya belongs, and the wood resembles African mahogany but is lighter in weight and softer. It is light pinkish brown. It is shipped chiefly to Europe where it is used for furniture, chests, boxes, and boats. **Cola mahogany,** from the Ivory Coast and Ghana, is **niangon,** *Tarrietia utilis.* The heartwood is light reddish brown, and the wood shows a herringbone figure on the quartered surface. It is heavier than khaya, and the pores are larger and more numerous. **Cherry mahogany,** or **makore,** is a plentiful wood on the Ivory Coast, Ghana, and Nigeria. It is from the tree *Minusops heckelii.* The wood is dark reddish brown without figure. It is heavier than khaya and is finer in texture.

KIESELGUHR. A variety of tripoli, or **infusorial earth,** obtained in Germany, and employed chiefly as an absorbing material. It is also used as an abrasive, as a heat insulator, for making imitation meerschaum, and as an absorbing material for nitroglycerin in making dynamite. Kieselguhr is very absorbent and will hold 75% of its own weight of sulfuric acid. It is insoluble in water. Its desirable characteristics as an insulator are closed cells and very high porosity, giving low density and low thermal conductivity. It is also used as a catalyst carrier in chemical processing. Kieselguhr from Oberhole, Germany, has 88% silica, 0.1 alumina, 8.4 water, and the remainder organic matter. **Randanite,** found at Clermont-Ferrand, France, is similar to kieselguhr but has a gray color. **Moler earth,** from the Jutland peninsula of Denmark, is similar to kieselguhr and is made into insulation bricks. **Nonpareil insulating brick** is made of pulverized kieselguhr mixed with ground cork, molded into brick form, and dried. The cork is burned out, leaving small air pockets to increase the insulating effect. The bricks withstand temperatures up to 1832°F (1000°C); the heat transmission is lower than that for natural kieselguhr.

KINO RESIN. Also known as **gum kino.** The red exudation of the tree *Pterocarpus marsupium* of India and Sri Lanka, and of *P. erinaceus,* of West Africa, formerly much used for colored varnishes and lacquers and used in throat medicines. **Bengal kino,** or **butea gum,** from the tree *Butea frondosa,* is now limited to medicinal use, as is the **Australian red kino** from species of eucalyptus. Kino belongs to the group of red resins known as **dragon's blood** when used in spirit varnishes for musical instruments and furniture, but it is now replaced by synthetic colors. The dragon's blood resins from the East Indies are from the ripe fruit of various species of the tree *Daemonorops.* The resin is separated out by boiling and is shipped in small, oval drops or long, cylindrical sticks. It is used in fine lacquers and varnishes. The dark-red dye known as **red sandalwood** is the boiled-down juice of another kino tree, *P. santelinus,* of India. Still another *Pterocarpus* tree of southern India produces the wood **padouk,** valued for furniture, cabinetwood, and veneer. The heartwood is red with black stripes. It is hard and takes a high polish.

KYANITE. A natural **aluminum silicate,** $Al_2O_3 \cdot SiO_2$, used as a refractory especially for linings of glass furnaces and furnaces for nonferrous metals. Molded or cast ceramic parts have a nearly zero thermal expansion up to 2300°F (1260°C). It is a common mineral but occurs disseminated with other minerals and is found in commercial quantities and grades in only a few places. Most of the world production has been in eastern India, but high-grade kyanite is now obtained from Kenya. It is also mined in North Carolina, Georgia, Virginia, and California. The related minerals sillimanite and dumortierite are mined in the western United States. The specific gravity is 3.56 to 3.67, and the Mohs hardness is 6 to 7. Kyanite of 97 to 98% purity is obtained by flotation, but gravity concentrates rarely exceed 90%. Low-grade kyanite ore from California, containing 35% kyanite and much quartz, is used for ceramics. Kyanite is usually marketed ground to finenesses from 35 to 325 mesh. Low-grade kyanite is used in glassmaking as a source of alumina to increase strength and chemical and heat resistance. Aluminum silicate materials are widely distributed in nature combined in complex forms, and the aluminum silicate extracted from them is called *synthetic.* **Kyanite powder,** produced from Florida beach sands, has round, single-crystal grains of kyanite and sillimanite in nearly equal proportions. The beach sands are screened and graded and marketed as granules of various sizes for compacting and sintering into refractory ceramic parts. The smooth, rounded grains give bonding properties superior to those of crushed kyanite. **Cerox ceramic,** of Babcock & Wilcox, is a synthetic aluminum silicate in molded shapes for electronic and furnace parts. It is marketed in grades containing 50 to 90% alumina. **Cerox 200,** with a melting point at 3290°F (1800°C), contains 64.3% alumina, 32.4 silica, 2.1 titania, and 1 iron oxide. It is a hard, white, nonporous ceramic which is spall- and wear-resistant.

LACQUER. Originally the name of an **Oriental finish** made with lac. The original **Chinese lacquer** was made with the juice of the **lac plant,** *Rhus vernicifera,* mixed with oils. The juice is milky, but it darkens with age, and the lacquer is a glossy black. Later, the name referred to transparent coatings made with shellac and to glossy, pigmented spirit varnishes. Still later, it referred to quick-drying finishes made with nitrocellulose or cellulose acetate. The word *lacquer* has now come to mean glossy, quick-drying finishes that are dried by evaporation of the solvents or thinners in which a resin (plastic) vehicle is dissolved.

The true lacquers were made with copals and other natural resins, a pigment, a softener or plasticizer, and one or more volatile solvents, with the time of drying controlled by the evaporation of the solvents. Too rapid drying may cause the cloudiness called *blushing,* because of the absorption of moisture from the cooling caused by the rapid evaporation. Various resins impart different characteristics. Dammar gives high gloss and hardness. Kauri gives hardness and wear resistance. The softeners are usually the amyl, ethyl, and butyl phthalates. The usual solvents are anhydrous alcohol, ethyl, acetate, benzol, and toluo. But modern lacquers may contain nitrocellulose or cellulose esters, or they may be made with acrylic, melamine, or other synthetic resins. **Shellac** is often referred to as a **spirit lacquer.**

For industrial work, lacquers are usually sprayed, and the feature is quick-drying. **Brushing lacquers** are also quick-drying, but they have a longer drying period to prevent streaks and lumps in the application. Lacquers are harder and tougher than enamels, but not as elastic, and they are more expensive. They are usually not as solvent-resistant or as weather-resistant, and are generally not suited for exterior work. A good lacquer requires no buffing and retains the original gloss well. The word *lacquer* is also used to describe a highly transparent varnish used to produce a thin protective film on polished or plated metals to preserve their luster. Lacquers are sold under a variety of trade names. Some early cellulose lacquers were marketed as **Duco, Agateen, Zapon, Brevolite,** and **Zeloctite,** but solutions of methyl acrylate or other crystal-clear resins may now be used. **Bronzing liquids** are the clear, lacquer-base media marketed ready for incorporation of bronze or aluminum powders. **Cable lacquers** are clear, black, or colored lacquers prepared from synthetic resins, and they are characterized by high dielectric strength, resistance to oils and heat, and ability to give tough, flexible coatings.

Chromate-protein films may be used instead of clear lacquers for the protection of metal parts. The parts are dipped in a solution of casein, albumin, or gelatin and, after drying, dipped in a weak chromic acid solution. The thin, yellowish film is hard and adherent and withstands temperatures to 300°F (149°C).

LAMPBLACK. A soot formed by the smudge process of burning oil, coal tar, resin, or other carbonaceous substances in an insufficient supply of air, the soot being allowed to settle on the walls or floors of the collecting chambers. Lampblack is practically pure carbon, but inferior grades may contain unburned oil. It is chemically the same as carbon black made from gas, but since it may contain as high as 2.5% oil, it is not generally used in rubber. The particle size is large, 2,559 to 3,937 μin (65 to 100 μm), and the pH is low, 3 to 3.5. However, some **amorphous carbon,** made from either tar oil or crude oil by spraying the oil and air into a closed retort at 3000°F (1649°C) to obtain partial combustion, is equal to carbon black for many uses. Lampblack is used in making paints, lead pencils, metal polishes, electric brush carbons, crayons, and carbon papers. Carbon brushes for use in electric machinery are made by mixing lampblack with pitch. Petroleum coke or graphite is added to impart special properties. For special arc lights, a mixture of lampblack, rare-earth oxides, fluorides, and coal as a binder is extruded. The combination of rare earths determines the type of radiation emitted. It is grayish black, flaky, and granular. The color is not as intensely black as carbon black. One pound (0.45 kg) occupies from 200 to 230 in^3 (3,278 to 3,770 cm^3). For use as a pigment for japan the powder should pass through a 325-mesh screen. **Lampblack oil** is a coal-tar product marketed for making lampblack.

LANCEWOOD. The wood of the tree *Guatteria virgata,* of tropical America. It is used as a substitute for boxwood, and for fine work where a uniform, tough, durable wood is needed. The wood is yellowish and has a fine, close, smooth grain. The density is 52 to 63 lb/ft^3 (833 to 1,009 kg/m^3). It is very hard and elastic. **Yaya** is a name given in the Honduras trade to lancewood. **Degami lancewood, or degami wood,** is a yellowish wood with a fine, dense grain, from *Calycophyllum candidissimum* of the West Indies. **Burma lancewood,** used in India for implements, is a strong, straight-grained, heavy wood from the large tree *Homalium tomentosum.* It has a light-brown color and a density of 60 lb/ft^3 (961 kg/m^3).

LANXIDES. A composite formed in the reaction between a molten metal and oxygen in air to some other vapor-phase oxidant. Normally, such a situation produces an unwelcome scum on the metal's surface. However, by controlling the molten metal's temperature and by adding traces of suitable dopant metals, a 1-in (2.5-cm) thick layer of a metal oxide composite can be grown on the liquid's surface. Composites up to 3.9 in (10 cm) thick and weighing up to 40 lb (18 kg) have been grown by this method, with no falling off in growth rate. Under the right conditions, a lanxide composite is considerably stronger than sintered alumina. This makes such a composite potentially useful for armor plating, rocket or jet engines, and other applications.

LARCH. The wood of the coniferous tree *Larix occidentalis,* of the northwestern United States and southwestern Canada. It is also called **western larch, western tamarack, mountain larch, Montana larch,** and **hackmatack.** The wood is heavier than most softwoods, having a specific gravity of 0.48 and a density of 30 lb/ft^3 (481 kg/m^3). It is fine-textured and straight-grained. In strength and hardness it ranks high among the softwoods, but it shrinks and swells more than most softwoods. The heartwood is reddish brown, and the narrow sapwood is yellowish. It finishes well, but splits easily. Butt logs contain galactan gum, which darkens the color. Larch is used for bridge timbers, flooring, paneling, and general construction. The trees reach a diameter of 5 ft (1.5 m) and a height of 200 ft (61 m). Shipments of western larch and Douglas fir mixed are known commercially as **larch fir.** Larch is also the name given to the **tamarack** tree of New England, *L. larincina.* **European larch,** *L. europea,* is an important wood in Russia and some other countries.

LARD. The soft, white fat from hogs. It is used chiefly as a shortening in bakery and food products and as a cooking grease. The inedible grades are used for the production of lard oil and soaps and for splitting into fatty acids and glycerin. The types of edible lard for use in the United States are defined in regulations of the Department of Agriculture. **Steam-rendered lard** is made by applying steam directly to the fats in a closed container. Open-kettle rendered lard is made by applying steam to the outside of the kettle. The **neutral lard** used for making margarine is produced by applying hot water in place of steam. **Rendered pork fat** is an edible material that does not meet the specifications for lard. Average production of edible lard is about 24 lb (11 kg) per animal.

Leaf lard is from around the kidneys and intestines and is the best edible grade. **White grease** is an inedible lard from the kidneys and the back. **Yellow grease** is the residue from the parts of the hog remaining after the parts yielding white grease are separated. USP grade of lard is called **adeps** in pharmacy. It is purified internal fat of the abdomen and is a soft, unctuous solid of bland taste. It is used in **benzoated lard** for ointments. **Modified lard,** of Armour Dairy & Food Oil Co., is a **plastic lard** which is plastic at lower temperatures and also retains its body at higher temperatures than ordinary hydrogenized shortenings do. It is made by treating with sodium methylate which re-forms the esters in random orientation to give the soft, plastic texture.

Lard oil is an oil expressed from lard by subjecting it to hydraulic pressure. Prime or first-grade lard oils are nearly colorless, or greenish, and have little odor. The commercial oils vary from the clear, sweet oil to the acidic and offensive-smelling brown oils. The oils contain oleic, stearic, and palmitic acids. They are used in cutting and in lubricating oils, sometimes in illuminating oils. They may be adulterated with cottonseed oil or blown oils. The flash point of pure lard oil is 480°F (249°C), saponification value 192, and specific gravity 0.915. **Mineral lard oil** is a mixture of refined mineral oil with lard oil, the fatty content being 25 to 30%. The flash point is about 300°F (149°C). **Petrofac** is a lard-oil substitute made entirely from petroleum. **Lardine** is an old name for blown cottonseed oil used in lubricants.

LATEX. The milklike juice of the rubber tree, now much used instead of the cured crude rubber for many rubber applications such as adhesives, rubber compounds, and rubber powder. The properties of latex vary with the type of tree, age of tree, method of tapping, and climate. Latex from young trees is less stable than that from old trees. Intensive tapping of the trees results in less rubber content, which may vary from 20 to 50%. For shipping, a preservative and anticoagulant are added to the latex, usually ammonia or sodium sulfate. Concentrated 60% latex is a stable liquid of creamlike consistency. **Latex foam** is a cellular sponge rubber made by whipping air into latex, pouring into molds, and vulcanizing. **Artificial latex** is a water dispersion of reclaimed rubber. Water dispersions of crude or reclaimed rubbers are produced by swelling and dissolving the rubber in an organic solvent, treating with an organic acid or with ammonia, and emulsifying. They resemble latex, but are softer and tackier and are used for adhesives. Latex foams are now usually made by incorporating a chemical which releases a gas to form the cells.

The term *latex* now also refers to water dispersions of synthetic rubbers and rubberlike plastics. **Neoprene latex** is a water dispersion of neoprene rubber, and it has dispersed particles smaller than those of natural latex, giving better penetration in coating paper and textiles. Butadiene-styrene latex with 68% solids is used for producing foamed rubber. **Latex water paints** are now usually made with synthetic rubber or plastic dispersions.

LAUAN. The wood of trees of several genera of the Philippines, Malaya, and Sarawak, known in the market as **Philippine mahogany.** The woods resemble mahogany in general appearance, weight, and strength, but the shrinkage and swelling with changes in moisture are greater than in the true mahoganies. The lauan woods are used for furniture and cabinet woods, paneling, and boatbuilding. The lauans belong chiefly to the genus *Shorea,* and the various species have local or common names. The so-called dark-red Philippine mahogany is **tangile,** S. *polysperma,* and **red lauan,** S. *negrosensis.* Tangile is also called **Bataan mahogany** and has the closest resemblance to true mahogany of all the species. The thick sapwood is light red, and the heartwood dark brownish red. It has greater tendency to warp than mahogany. Red lauan has larger pores, but is favored for boat construction because of the large sizes available. **Tiaong,** from S. *teysmanniana,* resembles tangile but is lighter and softer. **Almon,** from the tree S. *eximia,* is harder and stronger than red lauan or tangile, but is coarser in texture and less lustrous.

White lauan is from a different genus of the same family and is *Pentacme contorta.* It has about the same mechanical properties as tangile, but is gray with a pinkish tint. **Mindanao lauan,** P. *mindanensis,* is quite similar but is lighter and softer. **Mayapis,** from the tree *Shorea palosapis,* is coarser in texture than tangile and is subject to warping and checking, as is red lauan. It is intermediate in color, and the light-colored wood is marketed as white lauan, while the dark wood is sold as red lauan. **Yellow lauan** is from the trees **kalunti,** S. *kalunti,* **manggasinoro,** S. *philippensis,* and **malaanonang,** S. *polita.* Yellow lauan is yellowish and has lower strength and greater warpage than other lauans. **Bagtikan,** from the tree *Parashorea malaanonan,* is reddish gray, not lustrous, but is heavier and stronger than the other lauans. Sometimes mixed with Philippine mahogany is the wood known as **lumbayan,** from an entirely different family of trees. It is from the tree *Tarrietia javanica.* The thick sapwood is light gray and the heartwood reddish. The weight and strength are about equal to those of tangile, but the pores are larger. When marketed separately, it is a more valuable wood than the lauans for furniture manufacture. The reddish woods from Sarawak, Sumatra, and Malaya known as **meranti,** or **morenti,** are *Shorea* species of the lauan types. In the East Indies morenti is used for barrels, casks, and tanks for palm oil. Similar woods from North Borneo are called **seraya,** or known as **Borneo cedar** or **Borneo mahogany.** The *Shorea* trees yield Borneo tallow and dammar. **Merawan** is a wood from various species of trees of the genus *Hopea* of Malaya. It is valued for furniture and interior work. Much of the so-called mahogany normally shipped from the Philippines is **Apitong,** the wood of the tree *Dipterocarpus* *grandiflorus,* also grown in Borneo and Malaya. The wood has a density of 44 lb/ft^3 (705 kg/m^3).

LAVA. A name given to ceramic material used for molding gas-burner tips, electrical insulating parts, nozzles, and handles. It may be calcined talc, steatite, or other material. As produced by the American Lava Corp., it is molded from **magnesium oxide,** and it is hardened by heat treatment after shaping and cutting. It is baked at 2000°F (1093°C). The compressive strength is from 20,000 to 30,000 lb/in^2 (138 to 207 MPa). It resists moisture and has high dielectric strength. Rods as small as 0.020 in (0.05 cm) in diameter can be made. **Alsimag** is the trade name of lava produced by this company from ground talc and sodium silicate, but **Alsimag 602** is phosphate-bonded steatite talc, used for thyratron tubes and as a substitute for mica in receiving tube spacers. **Porcelava** is a similar ceramic for electric heating appliances. **Isolantite,** of Isolantite Manufacturing Co. of America, Inc., is a **steatite** ceramic molded with a binder and then vitrified. It can be machined or threaded before firing, and by allowing for the contraction, parts can be made with great accuracy. The specific gravity of the vitrified material is 2.5, Mohs hardness 8 to 9, crushing strength 80,000 to 120,000 lb/in^2 (552 to 827 MPa), and the dielectric strength of 0.125-in (0.32-cm) thickness is 40,500 V. **Lavalloy,** of Lava Crucible Refractories Co., is a high-strength ceramic made from a mixture of mullite and alumina. **Lavolain,** of Star Porcelain Co., is a strong, dense, heat-resistant ceramic for electrical insulation.

LEAD. A soft, heavy, bluish-gray metal, symbol Pb, obtained chiefly from the mineral galena. It surface-oxidizes easily, but is then very resistant to corrosion. It is soluble in nitric acid but not in sulfuric or hydrochloric, and it is one of the most stable of the metals. Its crystal structure is face-centered cubic. It is very malleable, but it becomes hard and brittle on repeated melting because of the formation of oxides. The specific gravity of the cast metal is 11.34, and that of the rolled is 11.37. The melting point is 621°F (327°C), and boiling point 3182°F (1750°C). The tensile strength is low, that of the rolled metal being about 3,600 lb/in^2 (25 MPa) with elongation of 52% at normal temperatures, but at low temperatures the strength is greatly increased. At −40°F (−40°C) it is about 13,000 lb/in^2 (90 MPa), with elongation of 30%. The coefficient of expansion is 0.0000183, and the thermal conductivity is 8.2% that of silver. The electric conductivity is only 7.8% that of copper. When used in storage batteries, the metal is largely returned as scrap after a period and is remelted and marketed as **secondary lead,** as is that from pipes and cable coverings. Lead is highly toxic and thus poses a health hazard. Inhalation of dust and fumes should be avoided, and it should not be used in contact with food or drink products.

Lead alloys easily with tin and zinc and forms many commercial alloys, including solders and bearing metals. Although the amount of lead going into chemicals other than white lead, red lead, and tetraethyllead is relatively small, there are about 200 lead chemicals that have a wide variety of industrial uses. Lead and its compounds are poisonous, and are not used in contact with skin or near foodstuffs. **Tetraethyllead,** also called **ethyl lead,** is a liquid of composition $Pb(C_2H_5)_4$, used in a blend with halogens as an additive in gasoline to increase octane number and resistance to knock. The compound is so volatile that it distributes widely in the atmosphere from exhaust fumes, and the lead is not recovered. It is highly toxic, and gasolines containing it are dyed for identification. Various other compounds, such as cyclopentadiene manganese carbonyl, can be used as antiknock agents. **Methyl lead,** $(CH_3)_4Pb$, a liquid boiling at 230°F (110°C), is more effective than tetraethyllead in some respects and is used in mixtures.

Commercial lead is sold in pigs weighing 100 lb (45 kg). Seven grades of pig lead are marketed. **Corroding lead,** 99.94% pure, is for making white lead. **Chemical lead,** 99.90% pure, with some silver and copper, is the **undesilverized lead** from Missouri ores. Soft, undesilverized lead is 99.93% pure and is used for storage batteries, sheet, and pipes and cable coverings. Chemical lead has a Rockwell B 80 hardness compared with 60 for the purer corroding lead. **Desilverized lead** is produced from the silver-lead ores of the Rocky Mountain states and is at least 99.85% pure. **Common lead A** is desilverized and is 99.85% pure, while **common lead B** is 99.73% pure. **Acid lead** is made by adding copper to fully refined lead, and **copper lead** is produced in the same way. These two grades are 99.90 and 99.85% pure, respectively. The percentages given for the seven grades of lead are ASTM minimums. **Work lead** is the pig lead from the blast furnaces before the silver is extracted. About 60% of the world production of lead is in the United States, Australia, Mexico, Germany, and Canada, though the ores are widely distributed and many other countries produce lead. **Arsenical lead** contains 0.15% arsenic, 0.10 tin, and 0.10 bismuth, and it is used for cable sheathing. There are several kinds of **calcium lead,** with calcium contents ranging from 0.06 to 0.09% and some also containing 0.3 to 1.3% tin. Calcium lead, containing 0.07% calcium and essentially tin-free, has an ultimate tensile strength of about 5,400 lb/in² (37 MPa) as-cast and air-cooled, and is used for cast grids in batteries. The **tin-bearing calcium leads** have tensile strengths of 6,000 to

10,000 lb/in^2 (41 to 69 MPa) and find similar applications. Some are for castings, others for sheet, strip, and wire products.

Lead wool is lead in a shredded form used for caulking. The strands are 0.005 to 0.015 in (0.013 to 0.038 cm) in diameter and come in ropes 0.625 to 0.75 in (1.59 to 1.90 cm) in diameter. **Blue lead** is a term meaning all lead products such as pipe and shot that have not been changed chemically in manufacture. Blue lead is also a name for a basic sulfate, a by-product of the smelting of lead ores obtained by collecting and filtering the smoke from the furnace. It is a mixture of the products of the partial combustion of the ore and coal. It consists of lead sulfate, lead sulfide, lead sulfite, lead oxide, zinc oxide, and carbon. It is used in base-coat paints for steel.

Sheet lead is produced by cold-rolling and is designated by weight. A sheet of lead 0.0156 in (0.38 cm) thick weighs approximately 1 lb/ft^2 (4.9 kg/m^2) and is called a 1-lb (0.45-kg) sheet. Thus 4-lb (1.8-kg) sheet lead is 0.0625 in (0.16 cm) thick, and 8-lb (3.6-kg) sheet is 0.125 in (0.32 cm) thick. As a sound barrier in building construction, a 0.008-in (0.02-cm) thickness of lead is equal to 0.75 in (1.9 cm) of fir plywood for sound absorption.

Lead has a high capacity for the capture of neutrons and gamma rays and is used for **radiation shielding** in the form of sheet lead or as metal powder in ceramic mortars and blocks, paints, and plastic composite structures. **DS Lead** is a **dispersion-strengthened lead** containing up to 1.5% lead monoxide evenly distributed throughout the structure. The oxide combines chemically with the lead, doubling the strength and stiffness of the metal, but increasing its brittleness. It is used for chemical piping and fittings. **Neoprene-lead fabric** is a neoprene fabric impregnated with lead powder. It has a radiation shielding capacity one-third that of solid lead sheet. It comes in thicknesses of 0.031 to 0.25 in (0.08 to 0.64 cm), and its flexibility makes it suitable for protective clothing and curtains. **Shielding cements** for X-ray and nuclear installation shielding are **metallic mortars** containing a high percentage of lead powder with ceramic oxides as binders and other elements for selective shielding. They are mixed with water to form plasters or for casting into sections and blocks. The formulation varies with the intended use for capture, attenuation, or dissipation of neutrons, gamma rays, and other radiation. **Shielding paints** are blended in the same manner. **Chemtree 82,** of Chemtree Corp., is a lead powder with an organic binder soluble in water for application as a paint or molded into brick. The cured material is 95% lead with linear attenuation about 60% that of lead. **Chemtree 1-6-18,** designed for neutron beam attenuation, has 78% lead and a high hydrogen content together with the neutron-capturing isotopes lithium 6 and boron 10.

Battery-plate lead for the grid plates of storage batteries contains 7 to 12% antimony, 0.25 tin, and small amounts of arsenic and copper. **Silvium alloy,** for positive-plate grids, contains silver. **Lead-coated copper,** used for roofing and acid-resistant tanks, is 16-oz (0.45-kg) copper sheet, coated on both sides with lead, made with either a rough or smooth finish. **Leadtex,** of Revere Copper & Brass, Inc., is a lead-coated sheet copper. **Roofloy,** of this company, is a strong, stiff, creep-resistant sheet lead used for flashings and gutters. It contains 99.7% lead, 0.20 tin, 0.015 calcium, and 0.012 magnesium. **Tea lead,** used for wrapping tea, is lead with 2% tin. **Tinsel lead** contains 1.5% antimony and up to 4 tin. **Shot lead** is lead-hardened with 2 to 6% antimony and a small amount of arsenic. **Frangible bullets,** which shatter on striking a target surface and are used for aerial gunner practice, are made of lead powder and synthetic resin powder pressed into shape.

Antimonial lead is an alloy containing up to 25% antimony with the balance lead, used for storage-battery plates, type metal, bullets, tank linings, pipes, cable coverings, bearing metals, roofing, collapsible tubes, toys, and small cast articles. The alloy is produced directly in the refining of some lead ores, but much is also made by adding antimony to soft lead. The alloy is also known as **hard lead,** and in England it is called **regulus metal.** Much of the lead used in the United States is in the form of antimonial lead. The antimony hardens the lead and increases the tensile strength. The usual alloy contains 4 to 8% antimony and has about twice the tensile strength of pure lead. Up to about 0.10% arsenic stabilizes and hardens the alloy, and from 0.25 to 1% tin may be added to improve the casting properties. Antimonial lead, of NL Industries, Inc., for chemical linings, contains 6 to 8% antimony, has a density of 0.398 lb/in^3 (11,016 kg/m^3), and melts at 475 to 555°F (245 to 290°C). An antimonial lead with 6% antimony has a tensile strength of 4,100 lb/in^2 (28 MPa) with elongation of 47%, as rolled. After heat treatment at 455°F (235°C) and aging one day, the tensile strength is 12,600 lb/in^2 (87 MPa) with elongation of 3%.

A hard lead with 10% antimony and 90 lead has a tensile strength of 8,800 lb/in^2 (61 MPa), elongation 17%, Brinell hardness 17, and melting point 486°F (252°C). **Cable lead,** or **sheathing lead,** used to cover telephone and power cables to protect against moisture and mechanical injury, contains about 1% antimony. The alloy for collapsible tubes usually contains 2% antimony. Antimonial lead may be used for machine bearings, but for this purpose it usually contains considerable tin and is classified with the babbitt metals. **Hoyt metal,** used for bearings, contains 6 to 10% antimony. Alloys containing from 70 to 90% lead, 5 to 20 antimony, and 2 to 20 tin have been used for railway car bearings under the name of **lining metal. Dandelion metal,** used for locomotive crosshead linings, had 72% lead, 18 antimony, and 10 tin. **Terne,** a lead alloy containing 8 to 12% tin, is used to plate steel for corrosion resistance. **Terneplate,** as the coated steel is called, is widely used for automobile gasoline tanks and also has been used for roofing on buildings.

LEAD ALKALI METALS. Lead hardened with small amounts of alkali metals used chiefly as bearing metals. Alloys of this type are also called **tempered lead** and **alkali lead.** The original alkali lead, known as **Bahnmetall,** was made under a German patent. It contains 0.73% calcium, 0.04 lithium, 0.55 sodium, and the balance lead, and it is made by electrolysis of the fused alkali salts, using a molten lead cathode. **Calcium-lead alloy** is made in the same manner and contains up to 1% calcium. The calcium forms a chemical compound, Pb_3Ca, with part of the lead, and the crystals of this compound are throughout the lead matrix. The Brinell hardness is 35, and the compressive strength 25,000 lb/in^2 (172 MPa), being thus superior in strength to the high-tin bearing metals. But the metal is difficult to melt without oxidation and is easily corroded. The melting point is 698°F (370°C), and it retains its bearing strength at more elevated temperatures than babbitts. **Calcium lead,** with about 0.04% calcium, is used as cable sheathing to replace antimonial lead, giving greater fatigue resistance. It is also used for grids in storage batteries and has a lower rate of self-discharge than antimonial lead. The small amount of lithium in Bahnmetall is intended to prevent the corrosion set up by the calcium. It also increases the compressive strength to about 30,000 lb/in^2 (207 MPa). **Mathesius metal,** another German alloy, contains calcium and strontium to form Pb_3Sr crystals. **Frary metal,** of NL Industries, Inc., is a lead-calcium-barium alloy containing about 1.5% of the alkali metals. These alloys give low friction loss at low temperatures, but they have higher coefficients of friction than the lead-tin alloys at temperatures above 149°F (65°C). A European lead-calcium-barium alloy known as **Ferry metal** contained 0.25% mercury to decrease the coefficient of friction. Another bearing alloy, **Noheet metal,** contained some antimony and tin in addition to the alkali metals. **B metal,** a German alloy, contains 0.73% calcium, 0.66 sodium, 0.05 lithium, and 0.03 potassium.

LEAD ORES. The chief ore of the metal lead is **galena,** a lead sulfide, PbS, containing theoretically 86.6% lead. The ore, however, contains many other minerals and usually carries only 4 to 11% lead. It is concentrated by gravity methods to contain 40 to 80% galena. Galena has a bright metallic luster, streaked gray color, a specific gravity of about

7.5, and Mohs hardness of 2.75. It frequently contains silver and sometimes cadmium, bismuth, and copper. The lead is obtained from the concentrated ore by roasting to remove the sulfur, and smelting. The **ingot lead** from the blast furnace contains silver, copper, zinc, and other impurities. It is refined and desilverized. Southern Missouri is the chief source of galena in the United States.

The abundant lead ores **cerussite** and **anglesite** are secondary minerals formed by the oxidation of galena. Cerussite is a lead carbonate, $PbCO_3$, found in crystals or in granular crystalline aggregate or massive. Its color is white to gray, transparent to opaque. The Mohs hardness is 3 to 3.5, and the specific gravity 6.55. Anglesite is usually found in the oxidized portions of lead veins associated with galena and other minerals. It is a common mineral occurring in many localities. Anglesite is a **lead sulfate** of composition $PbSO_4$, containing 68% lead. It occurs in crystals, massive or granular. Its hardness is 2.75 and specific gravity 6.12 to 6.39. The color may be white or pale shades of yellow or blue, or it may be colorless.

Basic lead sulfate, $PbSO_4 \cdot PbO$, called **lanarkite,** is in white crystals with a specific gravity of 6.92. It is formed by the action of heat and air on galena. It is used as a paint pigment combined with zinc oxide. **Alamosite** is **lead metasilicate,** $PbSiO_3$, and is in white crystals. It can be produced by fusing litharge and silica, and is used in glazing ceramics and in fireproofing textiles. **Barysilite** is **lead orthodisilicate,** $Pb_2Si_2O_7$, a white solid of specific gravity 6.6. Other rarer minerals are **clausthalite,** which is **lead selenide,** $PbSe$, occurring in lead-gray cubic crystals of specific gravity 8.1; **attoite,** which is **lead telluride,** $PbTe$, occurring in tin-white cubic crystals of specific gravity 8.16; and **lead tungstate,** $PbWO_4$, called **lead wolframite,** and occurring as **raspite** in monoclinic crystals, or as **stolzite** in colorless tetragonal crystals. Lead tungstate is manufactured as a yellow powder with a specific gravity of 8.235, and it is used as a paint pigment.

Vanadinite is a minor ore of lead and a source of vanadium. It is a mineral of secondary origin found in the upper oxidized parts of lead veins. It is found in Arizona, New Mexico, Mexico, and Spain. It has composition $9PbO \cdot 3V_2O_5 \cdot PbCl_2$, with phosphorus and arsenic sometimes replacing part of the vanadium. It occurs in reddish crystals and globular forms and as incrustations. The specific gravity is 7 and Mohs hardness 3.

LEAD PIGMENTS. Chemical compounds of lead used in paints to give color. They are to be distinguished from the lead compounds such as lead oleate, used as driers for paints. **Lead flake** is useful in exterior primers, where it exhibits excellent durability and rust inhibition.

But because of restrictions on the use of lead, it and many other lead compounds are being phased out as pigments. **White lead** is the common name for **basic lead carbonate,** the oldest and most important lead paint pigment, also used in putty and in ceramics. It is a white, amorphous, poisonous powder of composition $2PbCO_3 \cdot Pb(OH)_2$. It is insoluble in water and decomposes on heating. The specific gravity is 6.7. It is made from metallic lead and is marketed dry, or mixed with linseed oil and turpentine in paste form. **Lead carbonate,** $PbCO_3$, is used as a pigment in the same way as the basic compound, but it discolors more easily. Basic lead sulfate, called **sublimed white lead,** makes a fine white pigment. Commercial sublimed white lead contains 75% lead sulfate, 20 lead oxide, and 5 zinc oxide. Commercial white lead may be mixed with lithopone, magnesium oxide, antimony oxide, witherite, or other materials.

The basic **silicate white lead,** $3PbO \cdot 2SiO_2 \cdot H_2O$, is made by fusing silica sand with litharge and hydrating by ball-milling with water. It has corrosion-inhibiting properties and is used in metal-protective paints for underwater service. It is also used in ceramics and as a stabilizer in vinyl plastics. **Basic lead silicate** is made from silica and litharge with sulfuric acid as a catalyst. The material has a core of silica with a surface coating of basic lead silicate, giving a pigment of lower weight per unit of volume but retaining the activity of the silicate white lead. **Chrome yellow,** or **Leipzig yellow,** is **lead chromate** of composition $PbCrO_4$, and it comes in yellow, poisonous crystals. The specific gravity is 6.123. It is insoluble in water and decomposes at 1112°F (600°C). **Basic lead chromate,** $2PbO \cdot CrO_3$, is red and is used for anticorrosive base coats for steel. **Red lead** is a bright orange pigment used exclusively for corrosion inhibition; because of its poor opacity it is combined with iron oxide. **Basic lead silicochromate** is another rust inhibitor. **American vermilion,** also called **Chinese scarlet** and **chrome red,** is basic lead chromate made from white lead. **Orange mineral** is the red oxide, Pb_3O_4. **Cassel yellow,** or **Montpelier yellow,** is oxychloride of lead. Mixtures of white lead and heavy spar are known as **Venetian white. Dutch white** is composed of three parts of sulfate and one part of carbonate. **Lead thiosulfate,** PbS_2O_3, is a white insoluble powder used chiefly in matches. All the lead compounds are poisonous when absorbed through skin or taken internally. **Lead sulfide,** PbS, is used as a feeler in missiles as it is very sensitive to heat rays. **Molybdate oranges** are lead chromate, lead molybdate, and lead sulfate. Du Pont Co.'s **Kroler** is a heat-resistant variety suitable for employment in plastics.

Litharge is the yellow **lead monoxide,** PbO, also called **massicot.** It is a yellow powder used as a pigment and in the manufacture of glass, and for the fluxing of earthenware. An important use is as a filler in rubber. With glycerin, litharge is used as a plumber's cement. The specific gravity is 9.375. Litharge is produced by heating lead in a reverberatory furnace and then grinding the lumps. For storage-battery use the black oxide, or suboxide of lead, is now largely substituted. **Lead dioxide,** or **lead peroxide,** PbO_2, comes as a brown, crystalline powder and is used in making matches, dyes, and pyrotechnics, as a mordant, and as an oxidizing agent. The natural mineral is called **plattnerite.** Litharge, or **lead oxide,** converts to Pb_3O_4 at 800°F (427°C).

LEADED BRASS AND BRONZE.   Any of the copper-zinc, copper-zinc-tin, and copper-tin alloys containing generally 1 to 6% lead, sometimes more. The lead improves machinability, providing "free-machining" quality, but sacrifices ductility somewhat. Specific alloys are many, and they are available as wrought and cast alloys. Among the most common are the **leaded wrought brasses** (C31400 to C38600), **leaded cast brasses** (C83300 to C85800), **leaded phosphor bronzes** (C53200 to C54800), and **leaded tin bronzes** (C92200 to C92900).

LEADED STEEL.   A free-machining steel containing about 0.25% lead. Lead does not alloy with iron, but when a stream of finely divided lead is shot at the stream of molten steel to the ingot mold, the lead is distributed in the steel in tiny particles. The lead eases machining without imparting to the steel the unfavorable characteristics given by sulfur or phosphorus. There is little weakening of the physical properties of the steel. The lead forms a layer of liquid lubricant at the chip-tool interface, thus reducing the stress required to overcome friction. Also, by slightly embrittling the steel, lead reduces the deformation stress and serves to initiate microcracks to produce small chips. **Ledloy,** of Inland Steel Co., contains 0.15 to 0.30% lead in the regular SAE grades of steel. **La-Led steel** of LaSalle Steel Co. is a low-carbon leaded free-machining steel marketed in cold-finished rounds. **Super La-Led steel** contains 0.25% lead with up to 0.50 sulfur. **Rycut 40 steel,** of Joseph T. Ryerson & Son, Inc., is a chromium steel-molybdenum containing 0.15 to 0.35% lead. It can be machined 50% faster than unleaded steel of the same composition. The tensile strength is 129,000 to 200,000 lb/in^2 (889 to 1,379 MPa) with elongation of 12 to 21%, depending on the tempering. **Ledloy 170,** of the same company, is leaded, cold-drawn, seamless tubing. It is a 1015 analysis steel with 0.15 to 0.35% of lead added.

LEATHER.   The skin or hides of animals, cured by the chemical action of tannins. Leather is used for belting, gaskets, shoes, jackets, handbags, linings, coverings, and a variety of other products. The action of tannins precipitates the protein of the hide, changes its colloidal structure, and makes it more pliable and capable of resisting decay. The process of tanning hides consists essentially in soaking them in solutions of the tanning material after they have been unhaired in caustic lime. This soaking may be for weeks or months, after which the hides are washed, oiled, and rolled. Rapid tanning in strong solutions gives inferior leathers.

The quality of leather depends upon the type of animal, its physical condition, the care used in taking off the hide, the method of preserving the hide before tanning, and the care used in tanning. Leather is made from many kinds of skins, including sheep, goat, deer, alligator, seal, and shark, although the bulk of commercial leather is made from **cattle hides,** often referred to as **cowhide** or **steerhide.** Commercially, the term **hides** refers to the skins of full-grown beef cattle. Leathers from larger animals such as cowhide and steerhide are split into layers and divided into three grades. **Top-grain leather** is the top or outside layer of the hide and is the best quality. It has close fibers and is more flexible, durable, and attractive than the other top grades. Top grain is most often used natural and is seldom embossed. **First-split leather** comes from the layer next to the top layer and is second to top grain in quality. It is not as flexible or durable as top grain and is sometimes coated and embossed to resemble the better-grade leather. **Second-split leather,** the inner layer of the hide, is lowest in quality and generally considered waste.

Kip skins and **calfskins** are the names given to the hides of the younger animals. Calfskins, from animals that have not eaten grass, produce a softer leather than cattle hides. Hides of other animals besides the beef cattle, *Bos taurus,* are usually designated with the name of the animal, as in **horsehide.** The hides of smaller animals are designated as **skins,** either tanned or untanned, as in **pigskin.** Quantities of wild pigskins, known as **capivara skins,** are imported from Colombia.

Kid is leather made from the skins of young goats, but commercial **kidskin** leather is now from both young and old animals, and the term **goatskin** is not liked in the U.S. trade. It is thin and has a fine, close-grained texture, with tiny groups of pores, and the leather is soft and pliable but strong and nonscuffing. Kid is usually chrome-tanned and dyed to many colors. It has a natural lustrous surface which is heightened by glazing or polishing. The leather is sorted by grain or size of pores, weights, and sizes into 10 grades. India, Argentina, Brazil, South Africa, and northwest Africa are large producers of goatskins. About 70% of the goat and kid leather of the world is tanned in the United States, although this country produces less than 1% of the skins. Kidskin is used for shoes, gloves, pocket-books, jerkins, and pads and linings. The term **Vici kid,** used in the shoe industry, was the name originally given to chrome-tanned leather. The best kidskins come from arid regions, and these are used for the fine French kid leather. In the glove trade, the term **chevreau** is used to designate young goats that have never browsed, while **chevrette** refers to the small skins of the older kids that have eaten grass. **Capeskin** is goatskin from South Africa. **Glazed kid** is made by pressing a seasoning agent into the leather and then ironing the dry leather with a glass cylinder.

Buffalo hides are from the domesticated **water buffalo** of southern Asia and the Philippines. They are heavy, with a coarse grain, and are used to make heavy, rough leather for buffing wheels and heavy boots, and also for **rawhide** mallets and gears. Rawhide is a treated but untanned form of leather.

Belting leather is usually made from salted hides free from cuts and scratches, and is either oak- or chrome-tanned. It is then stuffed with oils or tallows. Belting leather has a density of about 0.035 lb/in³ (969 kg/m³) and has an ultimate strength of about 3,800 lb/in² (26 MPa). Belting is now sold by thickness instead of weight because the leather is easily weighted by the addition of heavy impregnations. But because of the use of direct drives and fabric belts impregnated with resins or rubbers, leather has lost its former importance for power transmission. **Lace leather,** for splicing machinery belting, was originally **porpoise leather** tanned with alum and impregnated with cod oil, making a tough, pliable, and strong leather, but it is now a semitanned, oil-treated cowhide or calfskin. Lacing for fastening sporting goods is now usually made of plastics, such as **Cotalace.** **Helvetia leather,** for light belts and lacings, was weak-tanned with gambier and oil-impregnated. **Belt dressings** are compounds of waxes and oils to maintain flexibility and rosin or chemicals to improve grip.

Leather was once used widely for packings and gaskets, but because it is dissolved by mineral oils and becomes brittle at high temperatures, it has now been largely replaced for such uses by compounded materials. **Fibrated leather,** for gaskets and packing, is shredded leather impregnated with a protein binder. It is resilient and resistant to oils and withstands temperatures to 300°F (149°C). It comes in sheets.

The **wax calf leather,** once popular for men's shoe uppers, was a leather heavily stuffed with grease and wax. It sheds water and is wear-resistant, but oily leathers are no longer desired in shoes. **Patent leather** is an old name for glossy, burnished black leather for shoes and handbags. It cracks easily, and plastic finishes are now usually employed. **Helios leather** has a base coat of linseed-oil gel on a chrome-tanned leather and a finish coat of urethane plastic, giving 20 times more flex life than patent leather. **Ostrich leather,** once valued for handbags, is marked with tiny rosettes where the quills are extracted. It is now imitated in embossed calfskin. **Kangaroo leather** is a strong, supple, and durable leather made from the skins of the Australian kangaroo, used chiefly for shoe uppers and gloves. The skins measure from 2 to 12 ft^2 (0.19 to 1.11 m^2) in area, and the small ones are known as **wallaby.** The fibers have an interwoven structure, and the leather does not scuff easily. It takes a brilliant polish. It is rated as the strongest shoe leather per unit of weight. **Cordovan** is a tough, smooth, close-grained leather made from the hind quarters of horsehides. It takes a beautiful polish and is used for boots and fancy articles.

Upholstery leather is very thin, finely finished leather once valued for upholstery, seats, and coverings for various articles, but now largely replaced by plastic-coated fabrics. It consists of split hides, tanned to a soft, even texture and usually dyed in colors. **Chrome-tanned leather** is softer and stronger than ordinary leather tanned with barks or quebracho. In splitting, the full hide thickness of about 0.25 in (0.64 cm) can be split into three or four thicknesses. After splitting, the leather is retanned and "nourished" with cod oil. Hand buffs are top grains with the top of the grain snuffed off. The second split of 0.047 in (0.11 cm) is called *deep buff* and has an artificial grain put on. The third split is called *No. 2 split.* What remains is called a *slab* and is unsuited for upholstery leather. Splitting with four cuts gives buffing, machine buff, No. 1 split, No. 2 split, and slab. Upholstery leather is finished by japanning, by coating with lacquer, by dyeing with aniline, or by combinations of the last with either of the first two methods. **Spanish leather,** used for upholstery, is made by tanning in strong quebracho liquor which draws the grain and gives a slightly wrinkled appearance.

Leather dust consists of the light, fluffy fibers blown from the buffing and sueding wheels in tanneries. It is used as a covering material for **artificial suede** and **sueded fabrics,** and as a filler in adhesives and caulking compounds. In Europe it is extensively collected and separated by grade and color, but the U.S. material generally lacks uniformity, as it is a mixture of many varieties of leather with different tannages and colors. **Leather flour** is ground and graded leather, uniform in color and in mesh, used as a filler and for sueded fabrics.

Pulped scrap leather is made into stiff board or flexible sheets and may be called **leather board, fiber leather, synthetic leather,** or **artificial leather.** Leather board and sheet are used for luggage, shoe heels, shoe linings, gaskets, and clutch linings. **Cotton-leather** is made from two to six plies of cotton fabric impregnated with a thermosetting resin, calendered, and surface-ground to resemble leather. It is flexible and resistant to oil and water, and it is used for shoe soles, buffing wheels, and machinery mountings. But for many of the former large uses of leather, including luggage and some types of shoes, leather has been replaced by impregnated laminated fabrics or fibers sheeted with a binder.

LECITHIN. A light-brown, soft, salvelike material of bland taste and odor obtained from soybeans by solvent and steam extraction. It is also obtained from cottonseed and occurs in small amounts in animal and plant cells. Lecithin is a **glyceride** of complicated chemical structure containing an amine, phosphoric acid, and choline. It bleaches to a faint gold color and is insoluble in water but soluble in oils and in alcohol. At about 150°F (66°C) it melts to an oil that will disperse readily in chocolate and other foods as an antioxidant to prevent spoilage, as a stabilizer and emulsifer, and as a dietetic. It is also employed in casein paints to regulate viscosity, in printing inks to impart flow, as a softener and wetting agent in some other products, and as a curing accelerator in synthetic rubbers. The **choline,** $CH_2OHCH_2N(CH_3)_3OH$, is used in medicine as a heart stimulant. Commercial grades are available as unbleached, single-bleached, and double-bleached lecithins in either fluid or plastic consistency. **Emultex R,** of A. E. Staley Mfg. Co., is a lecithin concentrate produced from soybeans. It is used as a stabilizer in water-based paints to prevent pigment migration and settling in storage. **Sta-Sol Plus,** of the same company, is a concentrated grade of soybean lecithin containing 68% active ingredients. It is a low-moisture emulsifier for coatings and paints.

LEMONGRASS OIL. A pale-reddish essential oil distilled from various species of *Cymbopogon* grass of the East Indies, Malaya, and tropical America. It contains about 75% **citral,** $C_9H_{15}CHO$, which is an aldehyde of several essential oils. The chief constituents are **citronellal,** $C_{10}H_{18}O$, and geraniol. It also contains some **ionone,** $C_{13}H_{20}O$, a terpene used as a base for violet perfumes. The oil has a powerful lemonlike odor. The plant is a tall, rapid-growing grass. Three tons (2.7 metric tons) of the East Indian grass yields 24 bottles of oil, each bottle being 25 oz (0.74 L) by volume. **Florida lemongrass** yields 6.3 lb (2.9 kg) of oil per ton of grass. The residue **lemongrass pulp** in Florida is dehydrated and mixed with cane molasses for cattle feed. The oil is soluble in 70% alcohol. It is used as a stabilizer and perfume in soaps and cosmetics. The two chief grades are **East Indian**

lemongrass, from *C. flexuosus,* and **West Indian lemongrass,** from *C. citratus.* The latter contains less citral and is less soluble, but partly owing to a difference in the method of distillation. **Citronella oil** is a yellow oil with a pleasant odor, distilled from the grass *C. nardus* of Sri Lanka, Java, and India. **Java citronella** is called **serah grass.** It contains high percentages of geraniol and citronellal. It is used as a stabilizer and perfume in soaps, as an insect repellent, and for the production of **synthetic menthol** and geraniol.

Java citronella oil is highest in citronellal, which yields about 20% levo menthol or about 12% USP menthol. Sri Lanka oil is highest in geraniol, being sold on the basis of 85%, and is used in soaps and insecticides. The rose perfume **rhodinal** is obtained from **Réunion geranium oil,** or **geranium Bourbon.** It is an isomer of **citronellol,** the alcohol of citronellal, and can be made synthetically by isomerizing citronellol with ultraviolet light. **Palmarosa oil,** called **Turkish geranium oil** because the original shipments were through Turkey, is from the **motia grass,** *C. martini,* of India and Java. It contains 75 to 90% geraniol. Another variety of *C. martini,* known as **sofia grass,** with *sofia* meaning mediocre, yields the product called **ginger grass oil,** inferior to palmarosa and containing less geraniol.

LEUCAENA.　A leafy, tropical evergreen found principally in Mexico. Its deep taproot enables it to live on moisture far below the surface. The plants also fix their own nitrogen from air in the soil. Because its foliage is rich in nitrogen, it can be used as fertilizer and as cattle feed. The trees yield high amounts of wood, and the stumps regrow rapidly.

LICORICE.　The sweet roots of a group of plants of the order Fabraeceae. The common licorice of the Near East and Spain is *Glycyrrhiza glabra,* and of Italy, *G. echinata.* Certain other species, all native to the Mediterranean countries, are also employed. The roots are crushed and boiled in water, and the juice is strained and evaporated. Finely ground root is added to aid in hardening, but starch or other adulterant may be added. In medicine it is known as **glycyrrhiza,** and it has an estrogenic, or female-hormone, action. The dark-colored extracted juice of the roots contains a glucoside, which will not ferment. **Glycyrrhizin** is an ammoniated form of the tri-terpenoid glycoside. It is a sweetening agent used in foodstuffs to reduce the amount of sugar and to intensify flavor. Licorice also contains a saponin, or froth-producing substance. In fire extinguishers it is sold under trade names. An insulating building board, made with a proportion of the residue fibers after extraction of the juice, is produced by National Gypsum Co. It is resistant to insect attack.

Licorice is employed in confectionery, medicines, and tobacco; as a flavoring; and in beverages as a frothing agent. But the saponins, in excess, are toxic.

LIGNIN. A colorless to brown crystalline product recovered from paper-pulp sulfite liquor, and used in furfural plastics, as an extender of phenol in phenolic plastics, as a corrosion inhibitor, in adhesives and coatings, as a natural binder for compressed-wood products, and for the production of synthetic vanilla when it contains coniferin. It is also used as a fertilizer, providing humus and organic material and some sulfur to the soil. For this use it may be mixed with phosphates. Lignin is coprecipitated with natural and synthetic rubbers to produce stronger and more lightweight products. When lignin is incorporated into nitrile rubber by coprecipitation, the rubber has higher tensile and tear strengths and greater elongation than with an equal loading of carbon black. **Indulin** is a pine wood lignin used as an extender for rubber latex to improve coating strength, and in phenolic resins to decrease cost. It is also used as a sequestering agent. About 100,000 tons (90,700 metric tons) of lignin occurs in the liquor from 1×10^6 tons (907,000 metric tons) of sulfate-process paper pulp.

The melting point of lignin is 482 to 527°F (250 to 275°C). It is a complex material, $C_{41}H_{32}O_6$, with four hydroxyl and four methoxyl groups, pine lignin containing 15% **methoxyl.** The contained **lignoceric acid** is a C_{24} fatty acid melting at 184°F (84.7°C). Lignin occurs in the waste liquor as a sodium compound and is precipitated by acidifying. For making laminates the lignin from soda-process aspen is preferred to spruce kraft lignin because of lower melting point and better flow. The lignin known as **tomlinite** is obtained from kraft pulp liquor by treating with CO_2, separating the black liquor, and acidifying with sulfuric acid to yield the water-insoluble brown powder. **Lignin binder,** for surfacing roads, is waste liquor from sulfite mills containing about 5% lignin. **Lignosol** is a concentrated lignin liquor or powder. **Spruce extract** is the liquor from sulfite spruce pulping used for mixing with quebracho for tanning leather, or in combination with syntans. The tanning agent in the liquor is **lignosulfonic acid. Alkali lignin** is produced by acidulation of the black liquor from the soda process. It is used as the organic expander in the negative plates of storage batteries, as an emulsifier for asphalt to reduce cold flow, and as a binder in paperboard products. It may also be hydrogenated to produce methane, methanol, and tar acids. **Dimethyl sulfide, dimethyl sulfoxide (DMSO),** and **methyl mercaptan** are important chemical derivatives of lignin. DMSO is an intermediate chemical, a solvent in spinning synthetic fibers, and a component of pharmaceuticals.

Lignosulfonates produced from the waste liquor of sulfate pulping of softwood are used to reduce the viscosity of slurries and oil-well muds; as foam stabilizers; as extenders in glues, cements, and synthetic resins; and as a partial replacement for quebracho in tanning leather. **Orzan A** is **ammonium lignosulfonate,** while **Orzan S** is **sodium lignosulfonate. Toranil** is **calcium lignosulfonate.** It comes as a tan powder or as a brown, viscous solution. These lignosulfonates have the wood sugars and the sulfur dioxide removed, and polymerize with heat.

LIGNITE. Also called **brown coal.** A variety of organic mineral of more recent age than coal, occurring in rocks of tertiary age, and intermediate in composition between wood and coal. It is widely distributed over Europe and found in many parts of the world. Freshly cut lignite often contains a large quantity of water, up to 40%, and is sometimes also high in ash. When dried, it breaks up into fine lumps and powder. Dry lignite contains 75% carbon, 10 to 30 oxygen, and 5 to 7 hydrogen. It kindles easily but burns with a low caloric power and a smoky flame. In retort gas production, lignite loses its gas in half the time required for gas removal from bituminous coal, with a temperature of 1270°F (688°C), compared with 1655°F (901°C) for bituminous. California and Arkansas lignites are processed to extract montan wax, **sap brown dye,** and **Van Dyke brown** pigment. The color of lignite varies from brown to black, and the lower grades of brown lignite show the woody structure. The **pitch coal** is brownish black, breaks with a pitchlike fracture, and shows no woody structure. Lignite is briquetted by crushing and pressing with a binder under heat. Belgian **lignite briquettes** have a binder of 8% asphalt, with 2 flour to assist binding. The reserves of lignite in the United States are estimated at 1000×10^9 tons (907×10^9 metric tons), and form a future fuel for use in powder form, or for distillation of oils and coal-tar products. Texas lignite is carbonized at 950°F (510°C) to produce a solid fuel known as **char.** From 7,000 tons (6,350 metric tons) of lignite the yield is 3,200 tons (2,902 metric tons) of char and 2,300 tons (2,091 metric tons) of by-product tar. **Lignite tar** is a mixture of hydrocarbons with nitrogen, oxygen, and sulfur complexes. The tar distillate contains 74% neutral oil, 23 tar acids, 3 tar bases, and 0.9 sulfur. The neutral oils are cracked for motor fuels. Other fractions are used for producing phenol, pyridine, and epoxy resins. **Jet** is a hard, compact, black lignite found in Colorado, Utah, and Yorkshire, England. The Utah jet occurs in the form of flattened-out trees, with the entire tree metamorphosed into jet. It was formerly much used for making buttons and ornamental articles, but is now largely replaced by plastics, and the less brittle black onyx is substituted for jewelry.

LIGNUM VITAE. The wood of the **guayacum** trees, *Guaiacum offinale* and *G. sanctum,* of tropical America, but the commercial shipments of lignum vitae are also likely to contain other species. The wood of the **guayacan** tree of Brazil and Paraguay is also called by this name. The best quality of the true lignin vitae comes in logs up to 18 ft (5.5 m) long and up to 3 ft (0.9 m) in diameter, and it originates mostly from Cuba, Haiti, Yucatan, Dominican Republic, and the west coast of Mexico and Central America, but species of the trees grow as far south as Paraguay. It is very hard, heavy, and tough. The color is brown to greenish black. The grain is very finely twisted, and the wood has a greasy feel, containing 3% of natural resin. The specific gravity is 1.14, density 80 lb/ft^3 (1,282 kg/m^3), and the crushing strength 10,500 lb/in^2 (72 MPa). The wood is used in places where extreme hardness is needed, such as for pulley blocks. It is also used for rollers, mallet heads, handles, novelties, bearings, and furniture. In machine bearings it withstands pressures up to 4,000 lb/in^2 (28 MPa), and is used for propeller-shaft bearings in steamships. **Palo santo,** *Bulnesia sarmenti,* known as guayacan, of Argentina and Paraguay, and also **vera,** *B. arborea,* of Colombia and Venezuela, are closely related to lignum vitae and were formerly classified as *Guaiacum.* The latter is called **Maracaibo lignum vitae,** and the former is known as **Paraguayan lignum vitae.** The trees are larger than true lignum vitae, but the wood is not as hard. **Philippine lignum vitae** is a hard, heavy wood with interwoven grain, from the tree *Xanthestemon verdugonianus* of the Philippines. **Yellow guayacan,** deriving its name from the yellow flowers, is from *Tecoma guayacan,* an important timber tree of Panama and Central America. The wood is very hard and strong, but neither this wood nor the Philippine wood has the self-lubricating bearing properties of the oily true lignum vitae. **Ibera vera,** of Argentina and Paraguay, is also called guayacan locally. It is from the tree *Caesalpinia melanocarpa,* and resembles lignum vitae in hardness, strength, and durability, but is lacking in the oily gum resin. It is a valuable ornamental construction wood.

LIME. A **calcium oxide,** CaO, chemically known as **calcia,** occurring abundantly in nature, chiefly in combination with carbon dioxide as calcium carbonate and in limestone, marble, chalk, coral, and shells. Lime is employed in mortars and cements, as a flux in iron melting, in many chemical processes, as an absorbent, in water treatment, in air pollution control, in glass manufacture, and for liming acid soils. It is obtained by heating limestone in a furnace or kiln to about 1000°F (538°C) to burn out the carbonic acid gas. The residue is called **quicklime** or **caustic lime.** Pure quicklime is white and amorphous or crystalline. The specific gravity is 3.2 and melting point 4660°F (2578°C).

Chemical lime, used in the chemical industries and for water softening, is a high-calcium lime with minimum impurities. A typical analysis is 97.9% CaO, 0.43 silica, 0.45 iron and aluminum oxides, and 0.52 magnesium oxide. The **Kemidol quicklime** marketed by USG Corp. for glass manufacture is a free-flowing, fine, powdered lime made from Ohio dolomite. When dolomite quicklime is hydrated, no more than 2 or 3% of the magnesium oxide is converted to the hydroxide, and the lime has a high neutralizing value, useful for neutralization of chemical solutions. **Oyster shells** have about the same calcium carbonate content as the best grade of high-calcium limestone, with low impurities, and they are used for making quicklime.

Commercial limes for building purposes contain about 94% calcium oxide, some calcium carbonate, and less than 0.50 magnesia. Water causes the lime to slake with much heat, leaving a white powder, $CaO \cdot H_2O$. High-calcium limes slake rapidly, expand greatly on slaking, and are the strongest. Limes with much magnesia slake slowly, but magnesia produces the slip that makes easier working. The **lean limes** contain considerable silica, alumina, and iron oxide and are slow-slaking and difficult to work. Lime is marketed in lumps or ground to 20 mesh, and as mill run. **Hydrated lime** is made by grinding quicklime, slaking the powder with water, and sifting to a fine powder. It is easier to handle and is a more reliable product than ordinary lime. High-grade hydrated lime will have a fineness so that 98% will pass through a 100-mesh sieve and will contain not over 2% magnesia. Some grades contain less than 0.50% magnesia, and 98% pass through a 200-mesh sieve. The pure hydrated lime for the chemical industry has 98.2% $Ca(OH)_2$, or 74.4 CaO, and is a fine air-float powder. This is **calcium hydroxide,** or **lime hydrate,** and it is used in paints, for dehairing hides, and in medicine. **Hydraulic lime** that will set underwater is a hydrated lime with more than 10% silica. The **Grappier cement** of France and Belgium is made from limestones that contain 20 to 22% silica and 2 alumina. **Lime mortar,** made of a mixture of hydrated lime, sand, and water, will have a compressive strength up to 400 lb/in² (3 MPa). **Soda lime,** used for freeing air of carbon monoxide, as a chemical purifying agent, and in gas masks, is a mixture of hydrated lime with sodium hydroxide.

In a system developed by Airpol Inc., lime particles and water are suspended in the exhaust gases of coal-fired power plants to gather sulfur, the source of acid rain, before the exhaust gases are released into the atmosphere.

LIME OIL. An essential oil expressed from the rind of the **lime,** the fruit of the small tree *Citrus aurantifolia,* native to India but now grown in many tropical and subtropical countries. The fruit is smaller than the lemon and is greenish yellow. Montserrat in the West Indies, Italy, and Florida are the chief commercial centers. **Lime juice,** from the pulp of the fruit, is used in flavoring and to give freshness to perfumes, in beverages, and was early used to prevent scurvy, but contains only about 25% as much vitamin C as orange juice. The seed of the lime yields up to 40% of a fatty oil, **lime seed oil,** used as an animal food supplement.

Of greater commercial importance is **lemon oil,** a bright-yellow oil of pleasant odor expressed from the rind of the lemon, the fruit of the tree *C. limonia,* grown commercially in the Mediterranean countries, California, Florida, and the West Indies. About 1,000 lemons are required to produce 1 lb (0.45 kg) of oil. The oil is high in citral with also limonene, terpineol, and citronellal. It is used in flavors, soaps, and perfumery. Synthetic lemon oil, used in confectionery and beverages, contains no terpenes. It has a much stronger flavor than the natural oil. **Lemon juice** is from the pulp. It contains 5% citric acid and is used in beverages and as a bleaching agent. **Lemon juice powder,** prepared by spray-drying from a solution that encases each particle, is a free-flowing, nonhygroscopic powder having long storage life. It is used for foodstuff flavoring.

LIMESTONE. A general name for a great variety of **calcite rocks.** Immense quantities of limestone are used as flux in the melting of iron, for the manufacture of lime, and as a building stone. In a broad sense, limestone includes dolomite, marble, chalk, or any mineral consisting largely of $CaCO_3$. When the proportion reaches 45% and the limestone is in the double carbonate, $CaCO_3 \cdot MgCO_3$, it is called **dolomite.** On calcination, limestone yields lime, CaO. **Portland stone,** of England, consists of fossils cemented together with lime, and the **coquina rock** of Florida is made up entirely of shells, mostly the tiny **coquina shell,** *Donax variabilis.*

The **Indiana limestone**, valued for building, is a noncrystalline, massive rock with aggregate filler, and matrix of 98% calcium carbonate in gray and buff colors. It is an **oolite** with small, round grains resembling fish roe. The density is 135 to 150 lb/ft^3 (2,163 to 2,403 kg/m^3), with compressive strength from 4,000 to 7,000 lb/in^2 (28 to 48 MPa). The mill blocks are from 8 to 12 ft (2.4 to 3.7 m) in length and are 3 ft 6 in (1 m) or 4 ft 4 in (1.3 m) square. **Bangor limestone,** from Alabama, is also an important binding stone. It is an oolite with 97% $CaCO_3$. **Dolomitic limestone** from Minnesota is also used for building. **Kasota stone** from Minnesota is a recrystallized variety resembling marble and coming in yellow and pink. It contains 49% $CaCO_3$ and 39 $MgCO_3$ with silica. Its crushing strength is more than double that of Indiana limestone. It is used largely for interior work.

Mankato stone contains slightly higher $CaCO_3$ with about 4.5% alumina in addition to silica. It has a density of 155 lb/ft^3 (2,483 kg/m^3) and a compressive strength of 13,500 lb/in^2 (93 MPa). Desirable buff and yellow coloring in limestones comes from small amounts of iron oxides, but iron sulfides produce weather staining and are not desired in building stone. **Sohnhofen stone, or lithographic stone,** produced at Sohnhofen, Germany, is a limestone with an exceptionally homogeneous fine grain, very low in impurities, and with a Mohs hardness of 3. It occurs in layers 2 to 6 in (5 to 15 cm) thick, with partings of gray clay. Limestone is widely distributed, and a very large proportion of all crushed and broken stone used for construction purposes in the United States is limestone. **Agricultural limestone,** known as **agstone,** used for liming soils, is stone of high calcium carbonate content ground to a fineness of 200 to 325 mesh. A grade for chemical use is specified with a minimum of 97.5% $CaCO_3$, and maximums of 0.30 silica, 0.20 alumina, and 0.07 iron.

LINEN. A general name for the yarns spun from the fiber of the variety of flax plant cultivated for its fiber, or for the cloth woven from the yarn. **Linen yarns** and fabrics have been made from the earliest times, and the ancient Egyptian linen fabrics were of exceeding fineness, containing 540 warp threads per inch, not equaled in Europe until the twentieth century. Ireland, Belgium, and France are the principal producers of linen. Linen yarns are used for the best grades of cordage, and linen fabrics are employed industrially wherever a fine, even, and strong cloth is required. **Linen fabrics** are sold under a wide variety of trade names. They are graded chiefly according to the fineness of the yarns and the class of weave and may contain some fine hemp fibers. **Lisle** was formerly a fine, hard linen thread, made at Lille, France, but is now a fine, smooth yarn made of long-staple cotton spun tightly in a moist condition. **Tow yarns** are the coarsest linen yarns, used for making **crash,** a coarse, plain-woven fabric used for towels or covers. **Chintz** is a plain-woven linen fabric in brightly covered designs. It is also made as **cotton chintz.** The name is from the Hindu word meaning color. **Damask** is a jacquard reversible woven linen fabric for table linen. It is now also made in cotton, silk, or rayon.

LINOLEUM. A general name for floor-covering material consisting of a mixture of ground cork or wood flour, rosin or other gum, blown linseed oil, pigments, with sometimes a filler such as lithopone, on a fabric backing of burlap or canvas, rolled under pressure to give a hard, glossy surface. It is pigmented or dyed in plain colors, or printed with designs. The cork is usually in about 50-mesh particles. Linoleum was patented in 1863 and displaced a more expensive floor covering known as **Kamptulicon,** which was made of rubber, cork, rosin, and oil. Because of its low cost, linoleum still sees wide use, but it is not durable and has a tendency to crack, and it has been largely replaced by sheeting compounded with synthetic resins marketed under trade names and by tile. **Battleship linoleum,** a very heavy grade in plain colors or mosaics, is described in federal specifications as made with oxidized linseed oil, fossil or other resins or oxidized rosin, and an oleoresinous binder, mixed with ground cork, wood flour, and pigments, processed on a burlap back. Because of its resiliency, battleship linoleum is still valued for high-grade flooring and is coated with an adherent, wear-resistant lacquer based on cellulose ester and alkyd resins.

LINSEED OIL.   This oil is the most common of the drying oils, and it is widely used for paints, varnishes, linoleum, printing inks, and soaps. It is obtained by pressure from the seeds of the flax plant, *Linum usitatissimum,* which is cultivated for oil purposes. It is sometimes referred to as **flaxseed,** though this name properly belongs only to the seed of the flax plant for producing flax fiber. The varieties producing linen fiber do not yield much seed. The seed contains up to 40% oil, and Argentine seed averages 38.5%, but the usual conversion factor is 34%. The linseed plant can stand high temperatures and drought, but is sensitive to cold, while the flax fiber plant needs a cool, humid climate. In normal times 80% of the world's linseed is grown in the Mesopotamia region of Argentina and Uruguay, and much of the remainder in northern India and southern Russia, while the flax plant is grown in northern Europe and the Baltic region. When prices are high, much linseed is grown in the United States and Canada. The commercial oil is hot-pressed and has a bitter taste, but a cold-pressed oil is used in Russia for food purposes. The residue, **linseed cake,** is used for cattle feed and fertilizer. The oil imported from Holland under the name of **Haarlem oil** is linseed oil mixed with sulfur and turpentine. It is used in the southern states as a diuretic medicine. Linseed oil contains about 48% linoleic and 34 linolenic acids. The specific gravity is 0.925 to 0.935. It is a yellowish, oily liquid with a peculiar odor and a bland taste. It is soluble in turpentine, ether, and benzene. It dries with a distinct gloss and makes a hard film.

The best **Baltic oil** is used as a standard in measuring the drying power of other oils. Genuine linseed oil has an iodine value of at least 170, and the best approaches 190. The linseed grown in cooler climates from the same type of plant generally yields oil of higher iodine value than that grown in warm climates. Baltic oil has an iodine value of 190 to 200, though this high value is from the type of plant as well as from the climate, the European plant being the flax plant yielding less seed. Oils from seeds grown in different areas vary widely in acid content. North Dakota oil contains 26 to 33% linolenic acid, while the **Punjab oil** of India has as high as 54% of this acid. The **puntis oil** used in Pakistan to extend linseed oil is a fish oil from the puntis, *Barbus stigma,* caught off the Indian coast.

For varnish use, linseed oil may be used as **boiled linseed oil,** prepared by heating to not over 600°F (316°C) in a closed container, or by heating with oxidizing driers such as the salts of lead, primarily litharge, or manganese. When prepared with driers, it is called **bung oil. Stand oil,** also known as **lithographic oil,** is linseed oil heated for several hours without blowing, at a temperature of 550 to 650°F (288 to 343°C). It has the consistency of honey and is used in oil enamel paints. Blown oils and boiled oils are not greasy like the original oil. **Linoxyn** is a trade name for blown linseed oil. The purity and adulteration of linseed oil for paint and varnish use are controlled by state laws. The law of the state of Ohio, which is typical, defines boiled linseed oil as prepared from pure, raw linseed oil heated to a temperature of 225°F (107°C) and incorporating no more than 4% by weight of drier, and with specific gravity at 60°F (16°C) of not less than 0.935 and not greater than 0.945. **Esskol** and **Solinox,** of Textron, Inc., are hydrogen-treated linseed oils used as substitutes for tung oil and castor oil. **Keltrol L,** of this company, is **styrenated linseed oil** made by reacting linseed oil with styrene. It is used for paints, in which it dries rapidly to a hard, tough, and alkali-resistant film. The heavy-bodied linseed oils are oxidized oils with specific gravities of 0.980 to 0.990 and iodine numbers from 210 to 230. **Linopol,** of Sherwin-Williams Co., is a polymerized linseed oil in water emulsion for use in latex paints.

LIQUID CRYSTALS. Nonisotropic materials—neither crystalline nor liquid—that are composed of long molecules parallel to each other in large clusters and that have properties between those of crystalline solids and liquids. They are also known as **liquid-crystal polymers (LCPs)**, can be processed by conventional thermoplastic-forming methods, and are suitable for intricate, thin-wall, precision parts and other applications. It is estimated that 1 in every 200 organic compounds has the capability of being produced in liquid-crystal form.

There are three principal types of liquid crystals, based on the arrangement of the molecules. In the smectic type, the molecules are parallel with their ends in line, forming layers that are usually curved or distorted, but are still capable of movement over one another. In the nematic type, the molecules are essentially parallel, but there is no regular alignment of their ends. The cholesteric type is formed by optically active compounds that have the capability for molecular organizations of the nematic type.

Liquid crystals have some of the properties of liquids, such as fluidity, and some of the properties of crystals, such as optical anisotropy. A major use of liquid crystals is for digital displays, which consist of two sheets of glass separated by a sealed-in, transparent, liquid-crystal material. The outer surface of the glass sheet is coated with a transparent conductive coating, with the viewing-side coating etched into character-forming segments. A voltage applied between the two glass sheets disrupts the orderly arrangement of the molecules, thus darkening the liquid to form visible characters. Self-frosting window glass comprises an LCP between clear, conductive-coated polyester films between glass panes. When low voltage is applied, the crystals align and the window is clear. When voltage is removed, the crystals unalign, frosting the window. Other typical applications of cholesteric liquid crystals are in skin thermography for tumor detection, in electronics for temperature mapping of circuits, and in nondestructive testing of laminates.

The nematic **Vectra LCP,** of Hoechst Celanese, features low-melting-point viscosity, permitting mold filling at injection pressures and cycle times about half those of semicrystalline resins. Depending on grade and reinforcement, tensile strength ranges from 10,000 to 35,000 lb/in^2 (69 to 241 MPa), flexural modulus from 1×10^6 to 4.8×10^6 lb/in^2 (68,950 to 330,960 MPa), notched Izod impact strength from 0.6 to 5.5 ft · lb/in (32 to 294 J/m), and heat deflection temperature at 264 lb/in^2 (1.8 MPa) from 410 to 655°F (210 to 346°C). Other features include low thermal expansion and moisture absorption, limited shrinkage in molding, inherent flame retardance, and resistance to various acids, chlorinated organic solvents, alcohols, and fuels. **Vectra LCP A130** is used for spacers in a color laser printer to maintain critical tolerances between the magnetic rolls and photoreceptive belt and between the rolls and trim bar in the developer housing.

LITHIUM AND LITHIUM ALLOYS.   This lightest of all metals, symbol Li, has a specific gravity of 0.534. It is found in more than 40 minerals, but is obtained chiefly from lepidolite, spodumene, and salt brines. The dried crude concentrate from the flotation cells contains about 20% Li$_2$O. It may be extracted by solar evaporation from the brines of underground lakes in Nevada and other locations. It occurs also in seawater in lesser concentrations. North Carolina has immense reserves of lithium ores.

Lithium melts at 356°F (186°C) and boils at 2444°F (1342°C). It is unstable chemically and burns in the air with a dazzling white flame when heated to just above its melting point. The metal is silvery white but tarnishes quickly in the air, and a **lithium nitride,** Li_3N, is formed. The metal is kept submerged in kerosene. Lithium resembles sodium, barium, and potassium, but has a wider reactive power than the other alkali metals. It combines easily with oxygen, nitrogen, and sulfur to form low-melting-point compounds which pass off as gases, and is thus useful as a deoxidizer and degasifier of metals. Lithium is also a key constituent in batteries for computers and camcorders and has potential for batteries being developed to power cars. In glass the small ionic radius of lithium permits a lithium ion coupled with an aluminum ion to displace two magnesium ions in the spinel structure. **Lithium cobaltite,** $LiCoO_2$, and **lithium zirconate,** Li_2ZrO_3, are also used in ceramics. **Lithium carbonate,** Li_2CO_3, is a powerful fluxing agent for ceramics and is used in low-melting-point ceramic enamels for coating aluminum. It is used in medicine to treat mental depression.

Lithium metal, 99.4% pure, is produced by the reduction of **lithium chloride,** LiCl. The salts of lithium burn with a crimson flame, and lithium chloride is used in pyrotechnics. It is also used for dehumidifying air for industrial drying and for air conditioning, as it absorbs water rapidly. It is employed in welding fluxes for aluminum and in storage batteries. The anode is lithium, the cathode is a **lithium-tellurium alloy,** and the electrolyte is a molten bath of lithium salts at 800°F (427°C). **Lithium ribbon,** of Foote Mineral Co., for high-energy battery use, is 99.96% pure metal in continuous-strip form, 0.02 in (0.05 cm) thick. It comes on spools packed dry under argon. An anhydrous form of **lithium hexafluoroarsenate** powder is used as the anode in dry batteries. **Aluminum-lithium alloys** are basically 2XXX and 7XXX aluminum alloys containing up to about 3% lithium. Because of lithium's extremely light weight, they provide higher stiffness-to-density and strength-to-density ratios than traditional structural aluminum alloys and thus have potential for aircraft applications.

Because of the low weight, lithium compounds give the highest content of hydrogen, oxygen, or chlorine. **Lithium perchloride,** $LiClO_4$, is a stable solid used as a source of oxygen in rockets and flares, with lithium chloride as a residue. One cubic foot (0.028 m^3) yields 91 lb (41 kg) of oxygen. On a volume basis, it has 29% more oxygen at normal temperature than liquid oxygen at its boiling point. **Lithium hydride,** LiH, a white or gray powder of specific gravity 0.82 and melting point 680°C, is used for the production of hydrogen for signal balloons and floats. A 1-lb (0.45-kg) can of hydride when immersed in water will liberate 45 ft^3 (1.3 m^3) of hydrogen gas. It is more stable to heat than sodium hydride, and it provides molecular hydrogen, not atomic hydrogen. **Lithium aluminum hydride,** or **lithium**

alanate, $LiAlH_4$, is used in the chemical industries for one-step reduction of esters without heat. Lithium metal is very sensitive to light and is also used in light-sensitive cells. **Lithium 7,** which comprises 94% of natural lithium, has low neutron absorption. The other isomer, **lithium 6,** has high neutron absorption and is used in nuclear reactors.

Lithium is soluble in most commercial metals only to a slight extent. It is a powerful deoxidizer and desulfurizer of steel, but no lithium is left in the **lithium-treated steel.** In stainless steels, it increases fluidity to produce dense castings. Cast iron treated with lithium has a fine-grain structure and increased density with high impact value. Not more than 0.01% remains in the casting when treated with lithium-copper. In magnesium alloys, the tensile strength is increased greatly by the addition of 0.05% lithium. The solid solubility of lithium in lead is not over 0.09%, but lithium refines the grain structure of the lead, increasing the strength, and it hardens the lead by the formation of a compound, Pb_3Li_2. **Lithium-treated lead** is called **alkali lead** and is used for machine bearings. Lithium up to 15% is added to magnesium to make alloys.

Lithium-copper master alloys consist of a group of foundry alloys containing usually 90, 95, or 98% copper with the balance lithium, used for deoxidizing and degasifying nonferrous alloys. Lithium combines easily in the molten bath with oxygen, hydrogen, nitrogen, sulfur, and the halides. The compounds formed are stable, of a nonmetallic nature, have low melting points, and volatilize easily so as to pass off as vapors at the pouring temperature of the metals. **Lithium copper** is a high-conductivity, high-density copper containing a minute quantity of residual lithium, 0.005 to 0.008%, made by treating copper with a 50–50 lithium-calcium master alloy. The conductivity of lithium copper is 101.5% IACS. The tensile strength is 31,500 to 36,500 lb/in^2 (217 to 252 MPa), with elongation 60 to 72%. The wrought metal is tougher than phosphorized copper, and it has exceptional deep-drawing properties. Lithium is an excellent desulfurizer for nickel alloys. From 1 to 7% calcium may also be included in lithium-copper master alloys. **Lithium-calcium alloys** usually contain 30 to 50% lithium, with the balance calcium. They are silvery white with a metallic luster and are hard and brittle. The melting range is 446 to 500°F (230 to 260°C). They must be kept in tight containers under kerosene. The alloys are used for treating steel, cast iron, or nickel where no residual copper is to be left. **Copper-manganese-lithium** contains 60 to 70% copper, 27 to 30 manganese, 0.5 to 5 lithium, and sometimes 5 to 7 calcium. **Copper-silicon-lithium** contains 80 to 84% copper, 10 to 11 silicon, 2.5 to 10 lithium, and sometimes 2.5 calcium.

LITHIUM ORES. One of the chief ores is **lepidolite** which also carries more rubidium than any other known mineral, containing from a trace to 3% **rubidium oxide,** Rb_2O. It may also have as much as 0.77% **cesium oxide.** It is a **lithia mica,** $LiF \cdot KF \cdot Al_2O_3 \cdot 3SiO_2$, occurring in small plates together with muscovite. It is the most widespread of the lithium minerals, being found in various parts of the United States, Canada, northern Zimbabwe, South Africa, India, China, Japan, Russia, and Germany. The Mohs hardness is 2.5 to 4 and specific gravity 2.8. It has a pearly luster and color of pink and lilac to grayish white. It is insoluble in acids. Lepidolite is employed as a source of lithium compounds and of the metals rubidium and cesium. It is also used in making opal and white glasses. **Glassmaker's lepidolite** contains 4% Li_2O; West African lepidolite usually contains 3.75%. **Amblygonite,** plentiful in Sweden, has the formula $Al_2O_3 \cdot 2LiF \cdot P_2O_5 \cdot Li_2O$ and contains 4.24 to 5.26% lithium. The amblygonite from South Dakota is sold on the basis of 8 to 9% Li_2O; and the ore from southern Zimbabwe contains 9% Li_2O, 48 P_2O_5, and 34 Al_2O_3. Other lithium ores are **lithiophilite,** $Li_2O \cdot 2MnO \cdot P_2O_5$, containing 4.6 to 5% lithium; **chryolithionite,** $3LiF \cdot 3NaF \cdot 2AlF_3$, with 5.35% lithium; **petalite,** $Li_2O \cdot Al_2O_3 \cdot 8SiO_2$, with 1.4 to 2.26% lithium; and **manandonite,** $7Al_2O_3 \cdot 2LiO:2B_2O_3 \cdot 6SiO_2$, with 2.13% lithium.

LITHOPONE. Also known under various trade names—**Ponolith, Sunolith, Beckton white, Zincolith, Sterling white,** and others. It is a white pigment consisting of barium sulfate and zinc sulfide and its use is dropping because of the superiority of titanium dioxide. A standard lithopone is 66% barium sulfate and 34 zinc sulfide. High-strength lithopones contain about 60% zinc sulfide, which is one of the whitest pigments. **Titanated lithopone** contains a percentage of titanium dioxide. **Tidolith** is a titanated lithopone of United Color & Pigment Co. having 85% lithopone and 15 titanium oxide. **Cadmolith** is the trade name of Glidden Co. for cadmium red and cadmium yellow lithopones used as pigments for plastics, as they are chemical-resistant and nonbleeding. Commercial lithopone is a fine, white powder used in the manufacture of paints, inks, oilcloth, linoleum, and rubber goods. For paints the powder should pass through a 325-mesh screen. The ground paste should contain 76 to 80% pigment and 20 to 24 linseed oil. As a paint pigment, lithopone has good hiding power and is lower in cost than other whites, but is not as durable for outside use as white lead or zinc white. It is one of the most used white pigments for interior work because it can be used in water-based formulations. **Albalith** is a 70–30 lithopone of New Jersey Zinc Co., also used in rubber goods, inks, and paper.

LOCUST. The wood of the locust tree, *Robinia pseudoacacia,* also known as **acacia, false acacia, black locust,** and **red locust.** The tree is native to North America, but is also grown in Europe. The wood is strong and durable, with a density of 43 to 52 lb/ft^3 (689 to 833 kg/m^3). Its hardness is about the same as that of ash, and the strength, flexibility, and shock resistance are greater than those of oak. The grain is coarse, but the surface is lustrous and satiny. Locust is used for furniture, wheel spokes, posts, cross ties, and in construction. **Honey locust** is a lighter and weaker wood from the tree *Gleditsia triacanothos.* The name *locust* is also applied to the wood of the tree *Hymensea courbaril* of tropical America. This wood has a brownish color, with an open grain, and takes a beautiful polish. The wood of the **Australian locust,** *Acacia melanoxylon,* known as **Australian blackwood** and **Tasmanian blackwood** and employed for cabinetwork, is reddish brown to black and beautifully grained. It is similar in durability and appearance to rosewood, but lighter in weight. Various species of true acacia trees furnish the tannins catechu and wattle. The **silver wattle** of New Zealand, *A. prominens,* used for ax handles and fruit cases, is hard and tough. The black locust has clusters of very fragrant, white flowers, and it is now widely grown as a shade tree and for shelterbelts in the eastern United States.

LOCUST BEAN GUM. Also called **locust bean flour,** and **carob flour.** A tasteless, odorless, white powder obtained by milling the bean kernels of the locust trees of tropical America, Africa, and the Mediterranean countries, notably *Ceratonia siliqua* of Cyprus, Syria, and Spain. The **carob bean,** or pod, contains 6 to 10 hard seeds, which are the **locust beans,** the bean in the pod averaging 9% of the weight. In Gambia the locust bean is called **netto.** When dissolved in water and boiled, it produces an adhesive, transparent jelly, which dries into a colorless, strong, elastic film. It contains galactose and mannose in a complex polymer and is a polysaccharide or complex sugar. It is used for coating textiles and as a thickener and binder in glues, pastes, and latex, in leather finishes, and in sizings for yarn and paper. It is used extensively together with guar gum as a wet-end additive during papermaking, to promote dry strength and improve sheet formation. The flour dissolves in cold water and swells in warm water. It is edible and is also used in jellies and bakery products. The dried pods are used in flavoring dog biscuit and tobacco. In the Near East they are eaten like candy and are also used as cattle and horse feed. The pods yield **tragasol** gum, which is used as a textile size and in leather tanning. The production of carob beans in Spain is large, and fully 90% of the crop is used locally for livestock feed.

The **algaroba tree** *Prosopis chilensis,* growing in semiarid regions of Mexico and Central and South America, called also **mesquite,** and in Hawaii called **keawe,** furnished an important stock feed from the pods and beans, which are similar to the locust. An acre of mesquite produces 4 times as much food for beef cattle as an acre of corn. The wood of the mesquite also contains up to 1% pyrogallol tannin, valuable for tanning leather.

LOGWOOD. An extract obtained from the wood of the tree *Haematoxylon campechianum,* of tropical America, used as a black dye or as a darkening agent in browns and grays. The wood yields 15% extract. The coloring matter, **hematine,** $C_{16}H_{12}O_6$, forms brownish-red crystals and is produced only in the aged wood or by oxidation of the white extract of fresh wood. **Logwood extract,** or hematine, is marketed in crystals, solid extract, or water extract.

LOOFA SPONGE. A yellowish, porous, skeletal, fibrous body obtained by retting the fruit or seed pods of the tropical edible cucumber, *Luffa cylindrica* and *L. acutangula,* obtained in India, Japan, and the Caribbean countries. It is also called **vegetable sponge,** although it is harder and coarser than sponge. The paper-thin skin is easily removed when the pod dries. The skinned sponge is washed to remove the slimy interior, and when redried, the seeds are shaken out. The seeds are used to produce a food oil similar to olive oil but colorless and tasteless. The sponges vary in size from 8 to 24 in (20 to 61 cm) long. The plants grow rapidly, giving four annual crops in Brazil, and a hill of 3 seeds can produce 30 sponges. Loofa sponges are used chiefly for filters in feed tanks of ships. A stiff, curly fiber used for making hats is obtained from the product by further retting.

LOW-ALLOY STEELS. Also known by other terms, including **alloy constructional steels,** they are generally limited to a maximum alloy content of 5%. One or more of the following elements may be present: manganese, nickel, chromium, molybdenum, vanadium, and silicon. Of these, nickel, chromium, and molybdenum are the most common. The steels are designated by a numerical code prefixed by AISI (American Iron and Steel Institute) or SAE (the Engineering Society for Advancing Mobility Land, Sea, Air, and Space). The last two digits show the nominal carbon content. The first two digits identify the major alloying element(s) or group. For example, 2317 is a nickel-alloy steel with a nominal carbon content of 0.17%.

Whereas surface hardness attainable by quenching is largely a function of carbon content, the depth of hardness depends in addition on alloy content. Therefore a principal feature of low-alloy steels is their enhanced hardenability compared to plain carbon steels. Like plain carbon steels, however, low-alloy steels' mechanical properties are closely related to carbon content. In heat-treated, low-alloy steels, the alloying elements contribute to the mechanical properties through a secondary hardening process that involves the formation of finely divided alloy carbides. Therefore, for a given carbon content, tensile strengths of low-alloy steels can often be double those of comparable plain carbon steels.

Low-alloy steels may be surface-hardening (carburizing) or through-hardening grades. The former are comparable in carbon content to low-carbon steels. Grades such as 4023, 4118, and 5015 are used for parts requiring better core properties than are obtainable with the surface-hardening grades of plain carbon steel. The higher-alloy grades, such as 3120, 4320, 4620, 5120, and 8620, are used for still better strength and core toughness.

Most through-hardening grades are medium in carbon content and are quenched and tempered to specific strength and hardness levels. These steels also can be produced to meet specific hardenability limits as determined by end quench tests. Identified as **H steels,** they afford steel producers more latitude in chemical composition limits. The **boron steels,** which contain very small amounts of boron, are also H steels. They are identified by the letter B after the first two digits.

A few low-alloy steels are available with high carbon content. These are mainly **spring-steel** grades 9260, 6150, 5160, 4160, and 8655 and **bearing steels** 52100 and 51100. The principal advantages of low-alloy spring steels are their high degree of hardenability and toughness. The bearing steels, because of their combination of high hardness, wear resistance, and strength, are used for a number of other parts, in addition to bearings.

High-strength low-alloy (HSLA) steels are low- to medium-carbon (0.10 to 0.30%) manganese (0.6 to 1.70) steels containing small amounts of alloying elements, such as aluminum, boron, chromium, columbium, copper, molybdenum, nickel, nitrogen, phosphorus, rare-earth metals, silicon, titanium, vanadium, and zirconium. Because of the small amount of some of these elements, these steels have been referred to as **microalloyed steels.** The chemical compositions and minimum mechanical properties of the steels are covered by about a dozen specifications of the American Society for Testing and Materials and the SAE's J410. The steels are commonly designated by minimum tensile yield strength, which ranges from about 35,000 lb/in^2 (241 MPa) to more than 80,000 lb/in^2 (552 MPa). They are available in most mill forms, hot-rolled sheet and plate probably being the most common, and they are typically used in the as-supplied condition. Thus, they provide high strength without heat treatment by users, and that is the principal reason for their use, which includes structural applications in cars and trucks, cargo vessels, railcars, and agricultural, earthmoving, and materials-handling equipment as well as office buildings and highway rails and bridges. **HSLA 100 steel,** with a tensile yield strength of 100,000 lb/in^2 (690 MPa), is used to a limited extent in naval vessels. **HY-100,** another steel of this strength, has been used in submarines.

HSLA steels are tougher than plain carbon steels, though not quite as formable, although sheet grades having yield strengths to 50,000 lb/in^2 (345 MPa) can be formed at room temperature to 1T (1 times thickness) to 2T bends, depending on thickness. The most formable are those produced with inclusion-shape control. That is, with the use of special alloying ingredients, such as rare-earth metals titanium or zirconium, and controlled-cooling practice, resulting inclusions are small, dispersed globules rather than stringerlike in shape. They are also readily welded by all common methods and can be brazed and soldered. Most of the steels are 2 to 8 times more resistant to atmospheric corrosion than plain carbon steels, and those commonly called **weathering steels** naturally acquire a deep purple-brown corrosion-inhibiting surface that precludes painting for corrosion protection. The color is considered attractive, especially in rural areas, and thus the steels have found considerable use for exposed building members and highway applications. **Dual-phase HSLA steel** has a deformable martensite phase in the ferrite matrix and exhibits a high rate of strain hardening during cold working. In the as-rolled condition in which it is supplied, it has a tensile yield strength of about 50,000 lb/in^2 (345 MPa) and thus the ductility (about 30% tensile elongation) and formability of conventional HSLA steels of this strength level. But strains of 2 to 3% during forming operations will increase yield strength in the strained regions to 80,000 lb/in^2 (552 MPa) or greater. Thus the steel provides the formability of medium-strength HSLA steels and the opportunity to achieve strength levels in selected regions equivalent to those of stronger, but less formable, as-supplied grades.

As contrasted to the HSLA steels, **quenched-and-tempered steels** are usually treated at the steel mill to develop optimum properties. Generally low in carbon, with an upper limit of 0.2%, they have minimum yield strengths from 80,000 to 125,000 lb/in^2 (552 to 862 MPa). Some two dozen types of proprietary steels of this type are produced. Many are available in three or four different strength or hardness levels. In addition, there are several special abrasion-resistant grades. Mechanical properties are significantly influenced by section size. Hardenability is chiefly controlled by the alloying elements. Roughly, an increase in alloy content counteracts the decline of strength and toughness as section size increases. Thus specifications for these steels take section size into account. In general, the higher-strength grades have endurance limits of about 60% of their tensile strength. Although their toughness is acceptable, they do not have the ductility of HSLA steels. Their atmospheric-corrosion resistance in general is comparable, and in some grades, it is better. Most quenched-and-tempered steels are readily welded by conventional methods.

Low-carbon ferritic steels, which were developed by Inco Alloys International, are low-alloy steels containing nickel, copper, and columbium. They are precipitation-hardened and have yield strengths from 70,000 to 100,000 lb/in² (483 to 690 MPa) in sections up to 0.75 in (1.9 cm). They possess excellent welding and cold-forming characteristics. A major use of these steels has been for vehicle frame members. Atmospheric corrosion resistance is roughly 3 or 4 times that of carbon steels.

LOW-EXPANSION ALLOYS. Alloys, mainly of iron and nickel, having low coefficients of thermal expansion, usually within a specific temperature range. Uses include precision-instrument parts requiring dimensional stability at various temperatures and glass-to-metal sealing applications, in which the thermal expansivity of the metal must closely match that of the glass. The best-known alloy is **Invar,** also known as **Nilvar,** an iron–36% nickel composition also containing (as impurities) minute amounts of carbon, manganese, and silicon. Developed in France, it has the lowest coefficient of thermal expansion of all metals in the −459 to 350°F (−273 to 177°C) range. In the annealed conditions, the alloy has a coefficient of thermal expansion ranging from about $0.8 \times 10^{-6}/°F$ ($1.44 \times 10^{-6}/K$) at 0 to 77°F (−17.8 to 25°C). At 300°F (149°C), the value is still only $1 \times 10^{-6}/°F$ ($1.8 \times 10^{-6}/K$). Expansivity is affected by heat treatment and cold work. Quenching from about 1526°F (830°C), for example, reduces the coefficient of thermal expansion below that of annealed material, as does cold forming. A combination of quenching and cold work can even result in zero or negative coefficients. Invar has a thermal conductivity of 6.4 Btu/(ft · h · °F) [11 W/(m · K)] from room temperature to 212°F (100°C) and is quite soft, having a Brinell hardness of about 160. Tensile properties are about 75,000 lb/in² (517 MPa) ultimate strength, 50,000 lb/in² (345 MPa) yield strength, and 35 to 40% elongation. The alloy is ferromagnetic at room temperature but becomes paramagnetic with increasing temperature. Because the alloy's thermal expansivity is rather constant within a specific temperature range, Invar is also known as a **controlled-expansion alloy.**

There are many other such alloys, each suited for specific coefficients of thermal expansion within certain temperature ranges. They include iron with 39% nickel, or **Fe–39 Ni, Fe–42 Ni (Dumet** and **Alloy 42), Fe–46 Ni, (Platinite), Fe–48.5 Ni, Fe–50.5 Ni, Fe–42 Ni–6 Cr, Fe–45 Ni–6 Cr, Fe–36 Ni–12 Cr (Elinvar), Fe–22 Ni–3 Cr,** and **Fe–42 Ni–5.5 Cr–2.5 Ti–0.40 Al (NiSpan C** and **Elinvar Extra).** Besides its low coefficient of thermal expansion, Elinvar is noted for its constant modulus of elasticity over a wide temperature range. Because of their titanium content, NiSpan C and Elinvar Extra can be solution heat-treated and aged to very high strength levels. Tensile yield strengths to 180,000 lb/in^2 (1,241 MPa), with a corresponding Rockwell C hardness of 42 have been reported. Some of these alloys are also **glass-to-metal sealing alloys.** Another is **Alloy 27** (Fe–28 Cr–0.6 Mn–0.5 maximum Ni–0.4 silicon–0.05 C), which grips the glass on cooling and is used in electronic and vacuum tubes as well as fluorescent and incandescent lamps. Annealed, it has a tensile yield strength of 55,000 lb/in^2 (379 MPa) and 25% elongation.

Cobalt in iron-nickel alloys increases the coefficient of thermal expansion at room temperature but enhances thermal stability over a wider temperature range. **Kovar** and **Fernico,** Fe–28 Ni–18 Co alloys, and **Fernichrome** (Fe–30 Ni–25 Co–8 Cr) are used for applications requiring vacuum sealing to glass. Kovar has a tensile yield strength of 59,500 lb/in^2 (410 MPa) and 25% elongation. A **Co 54–Fe 37–Cr 9 alloy** is noted for its near-zero and sometimes negative coefficient of thermal expansion in the 32 to 212°F (0 to 100°C) range. **Elgiloy** (40 Co–20 Cr–15.5 Ni–15.3 Fe–7 Mo–2 Mn–0.15 C–0.04 Be), originally a watch-spring alloy, has found many other spring applications, including seals, diaphragms, timers, instruments, surgical implants, and orthodontic fixtures. Produced in strip form by Elgiloy LP, it has an operating temperature range from the cryogenic to 900°F (480°C). Besides dimensional stability, the alloy is noted for its good fatigue strength, corrosion and heat resistance, and nonmagnetic characteristics. **Incoloy 903** (42 Fe–38 Ni–15 Co–3 Cb–1.4 Ti–0.7 Al), which is also heat-treatable, is noted for a near-constant coefficient of thermal expansion, about 4×10^{-6}/°F (7.2×10^{-6}/K) from 212 to 800°F (100 to 427°C) and a near-constant modulus of elasticity from −320 to 1200°F (−196 to 649°C). In the aged condition, the alloy has a yield strength of about 160,000 lb/in^2 (1,103 MPa).

Other low-expansion or controlled-expansion alloys that have been developed include **Nivar,** which contains 54% cobalt; the Swiss alloy **Nivarox** (Fe–37 Ni–8 Cr with small amounts of manganese, beryllium, silicon, and carbon) and **Contracid** (60 Ni–15 Cr–15 Fe–7 Mo–2 Mn and small amounts of beryllium and silicon); **Nicol** (40 Co–20 Cr–16 Fe–15 Ni–7 Mo–2 Mn and small amounts of beryllium and carbon); the French iron-nickel alloys **Dilvar** and **Adr**; **Super-Invar** from Japan, a 5% cobalt iron-nickel alloy; **Sylvania 4** (Fe–42 Ni–5.7 Cr with small amounts of manganese, silicon, carbon, and aluminum) and the similar **Sealmet HC-4**; **Niron 52** (52 Ni–48 Fe); **Rodar** (Fe–29 Ni–17 Co–0.3 Mn); and **Nicromet** (54 Fe–46 Ni).

LUBRICATING GREASE. Usually a compound of a mineral oil with a soap, employed for lubricating machinery where the speed is slow or where it would be difficult to retain a free-flowing oil. The soap is one that is made from animal or vegetable oils high in stearic, oleic, and palmitic acids. The lime soaps give water resistance, or a mineral soap may be added for this purpose. Aluminum stearate gives high film strength to the grease. All these greases are more properly designated as **mineral lubricating grease.** Originally, **grease** for lubricating purposes was hog fat or the inedible grades of lard, varying in color from white to brown. Some of these greases were stiffened with fillers of rosin, wax, or talc, which were not good lubricants. The stiffness of such a grease should be obtained with a mineral soap. ASTM specifications for heavy journal bearing grease require 45% soap content. About 2% calcium benzoate increases the melting point. Mineral lubricating grease may contain from 80 to 90% mineral oil and the remainder a lime soap. Federal specifications prescribe 85% mineral oil. Chemicals may be added to improve the physical properties of grease. **Oronite GA-10,** for example, is a sodium salt of terephthalic acid used as a gelling agent in high-temperature greases. It adds water resistance and stabilizes against emulsion. **Ortholeum 300,** marketed as a brown, flaky powder by Du Pont, is a mixture of complex amines, and small amounts added to a grease will give high heat stability. **Braycote 617,** of Bray Oil Co., is a synthetic grease for rockets subject to both heat and cold. It is a mixture of perfluorotrialkylamines gelled with tetrafluoroethylene. It comes as a translucent, lardlike semisolid with a boiling point at 446°F (230°C). The lubricating grease known as **trough grease,** used in food plants for greasing trays, tables, and conveyors, contains no mineral oil and is edible.

   Lime greases do not emulsify as readily as those made with a soda base and are thus more suitable for use where water may be present. **Hard grease** flows at a temperature of about 194°F (90°C); medium grease flows at 167 to 176°F (75 to 80°C). Paraffin wax, sometimes added, is an adulterant and not a lubricant. **Graphite grease** contains 2 to 10% amorphous graphite and is used for bearings, especially in damp places. Federal specifications call for 2 to 3% graphite. For large ball and roller bearings a low-lime grease is used, sometimes mixed with a small percentage of graphite. **Cylinder grease** is made of about 85% mineral oil or mineral grease and 15 tallow. Compounded greases are also marketed containing animal and vegetable oils, or are made with blown oils and compounded with mineral oils. The fatty acids in vegetable and animal oils, however, are likely to corrode metals. Tannin holds graphite in solution; in the gear grease sold under the name **Gredag** by Acheson Colloids Co., a graphite-tannin mixture is used. **Metaline** is a compound of powdered antifriction metal, oxide, and gums, which is packed in holes in the bearings to form self-lubricating bearings. **Lead-lube grease** has finely powdered lead metal suspended in the grease for heavy-duty lubrication.

Sett greases are mixtures of the calcium soaps of rosin acids with various grades of mineral oils. They are low-cost, semisolid greases used for lubricating heavy gears or for greasing skidways. Clay fillers may be added to improve the film strength, or copper or lead powders may be incorporated for heavy-load conditions. **Solidified oil** is also a name given to grease made from lubricating oil with a soda soap and tallow, used for heavy bearings. **Cup grease** is made with soda soaps and light lubricating oils. Greases made with potash and soda soaps tend to form soap fibers when water is present. A metallic soap that contains no fibers is called a **neat soap,** and it gives a smooth grease. Greases made with lithium stearate have good water and high-temperature resistance and have a buttery texture. **Alrania grease,** of Shell Oil Co., is a grease of this type. Fatty acids used for grease making may be hardened by hydrogenation to remove polyunsaturated acids. The greases have greater resistance to heat discoloration and do not gum or become rancid. **Lubrex 45** is a hardened fatty acid of this type. **Slushing oil,** for use in protecting machine parts from corrosion in shipping or storage, is usually a low-melting-point grease preferably compounded with a waxy fat such as lanolin. **Paralan** is such a slushing oil having a lanolin base.

Silicones and **fluorosilicones** are also used as greases. Although good at fast speeds, high temperatures, and in resisting washout by water, they are quite limited in load-carrying capacity. **BG 20,** of Dow Corning, uses polyol ester as the base fluid and can withstand loads up to about 585 lb (266 kg), compared with 270 lb (123 kg) for regular silicones, and temperatures of -50 to $360°F$ (-46 to $182°C$).

LUBRICATING OILS.   Oils used for lubricating the bearing parts of machinery. They are usually the heavy distillates following kerosene in the fractional distillation of petroleum, between 253 and $317°F$ (123 and $158°C$). They are separated into grades—light, medium, and heavy—depending upon the molecular weight. They are also classified as pale when yellow to reddish and dark when brownish black.

The flash point ranges from 300 to 600°F (149 to 316°C), and the specific gravity usually from 0.860 to 0.940. **Neutral oils** are light oils obtained by distillation without cracking, and they will not emulsify in contact with water, as do the paraffin oils. They are thus desirable for crankcase lubrication and in circulating systems. Lubricating oils may be bleached with acid, and they may be mixed with vegetable or animal oils. The ideal of lubrication is to obtain a full fluid film with little clearance between the moving surfaces so that the shaft rotates on a film of oil. Hydrodynamic lubrication with pressure gives this condition. Only a boundary lubrication with contact merely on the bottom is obtained when the clearance is too great, the oil viscosity too low, the load too heavy, or the speed is too slow, so that the film does not support the shaft.

Animal oils are greasier than mineral oils, but they are acidic. Vegetable oils are greasy and have more oiliness, but they oxidize easily and are also acidic. They are likely to gum in use unless an antioxidant is employed. Vegetable and animal oils add the property of adhesion to the lubricating oil, but in no case should any element be added to an oil that will cause emulsification. Federal specifications for marine engine oil call for 15 to 20% blown, refined rapeseed or peanut oil. This lubricating oil has a flash point of 350°F (177°C). **Steam cylinder oil** has 5 to 10% fatty acid vegetable oils, and the flash point is 450°F (232°C). **Absorbed oil** is a name of a combination oil of E. F. Houghton & Co. which acts as both film and lubricant. **Amlo** is a trade name of a mineral oil refined wax-free, used for low-temperature lubrication. The silicones are now often used to replace lubricating oils for very high and very low temperature conditions, but in general the lubricating value is not high.

Antioxidants used in oils to reduce oxidation and minimize sludging and acid formation are usually tin compounds such as tin dioxide, tin tetraphenyl, and tin ricinoleate. Tin dust alone also has an inhibitory action. Detergents are compounded in lubricating oils for internal combustion engines in order to prevent and break down carbon and sludge deposits. High percentages of animal or vegetable oils may be added to lubricating oils for use on textile machinery. They are called **stainless oils** for this purpose, since such oils wash out of the textile more easily than mineral oils do. They also give lower coefficients of friction. The high lubricating qualities of the vegetable oils without the disadvantage of gumming can be obtained with mineral oils by the addition of an **oiliness agent** such as **cetyl piperidine ricinoleate.** The **EP lubricants** (extreme pressure) for heavy-duty gear lubrication are made with a high-quality oil compounded with a lead-sulfonated soap.

For extreme high pressure and high temperatures where oils and greases oxidize, **molybdenum disulfide,** MoS_2, is used alone or mixed with oils or silicones. It is a fine, black powder, available in particle sizes as small as 30 μin (0.75 μm), which adheres strongly to metal surfaces, gives a low coefficient of friction, and permits operation up to 750°F (399°C), but it has an acid reaction and is corrosive to metals. Molybdenum disulfide resembles graphite but is twice as dense. The sulfur attaches itself with a weaker electron bond on one side than the other, forming laminal plates or scales in the molecular structure which tend to split off and give the sliding or lubrication action. Molybdenum disulfide may be used as a filler in nylon gears and bearings to reduce friction. It also increases the flexural strength of the plastic. **Molysulfide,** of Climax Molybdenum Co., is this material. **Tungsten disulfide** is also used as a lubricant in the same way as molybdenum disulfide. The electron bond of sulfur to tungsten is stronger than that to molybdenum, and it is thus more stable at high temperatures. The tungsten disulfide of GTE Corp. is a crystalline, gray-black powder with particle size from 39 to 79 μin (1 to 2 μm). **Liqui-Moly,** of Lockrey Co., and **Molykote** are molybdenum disulfide lubricants. **Dry-film lubricants** are usually graphite or molybdenum sulfide in a resin or volatile solution. They are sprayed on the bearing surface, and the evaporation of the solvent leaves an adherent thin film on the bearing. **Polytetrafluoroethylene,** which resists temperatures up to 500°F (260°C), is a dry-film lubricant. It is also used as an additive to lubricating oils. **Selenium disulfide,** SeS_2, will retain its lubricating qualities at temperatures to 2000°F (1093°C) and is useful for lubrication under vacuum because of its low emission of gas. Other materials used as **dry lubricants** are **tantalum disulfide,** TaS_2, **tantalum diselenide,** $TaSe_2$, **titanium ditelluride,** $TiTe_2$, and **zirconium diselenide,** $ZrSe_2$.

Hydraulic fluids for the operation of presses must lubricate as well as carry the pressure. They are mostly mineral oils, but chemicals are used where high temperatures are encountered, such as in die-casting machines. **Lindol HF-X,** of Hoechst Celanese Corp., is a flame-resistant hydraulic fluid with a tricresyl phosphate base. **Skydrol,** of Monsanto Chemical Co., for aircraft hydraulic systems, is an oily ester produced from petroleum gas. The ignition point is 1050°F (566°C), and it operates at temperatures as low as −40°F (−40°C). The **Fluorolube oils** of Hooker Chemical are polymers of trifluorovinyl chloride fractionated to provide grades from a colorless, low-viscosity oil to an opaque, heavy grease. They have high lubricating values, are resistant to acids and alkalies, and have an operating range from 572°F (300°C) down to very low subzero temperatures. **Hydraulic fluid QF-6-7009,** of Dow Chemical USA, for closed systems operating from −25 to 550°F (−32 to 288°C), is a diphenyl didodecyl silane. **Refrigeration oils,** for lubricating refrigerating machinery, are mineral oils refined to remove all moisture and wax. **Ansul oil,** of Ansul Chemical Co., is an oil of this class which remains stable at temperatures as low as $-70°F$ ($-57°C$). **Hydrolube HP-5046,** of Union Carbide, is a water-glycol hydraulic fluid that can be used at pressures up to 5,000 lb/in^2 (34 MPa). It has a pour point of $-81°F$ ($-63°C$) and a viscosity index of 170, and it can be used over a wide temperature range with minimal effect on the bulk fluid viscosity. Environmentally acceptable hydraulic fluids are being used increasingly in agricultural equipment and other environmentally sensitive applications. These fluids, such as Mobil Oil's **EAL 224H,** are highly biodegradable and relatively nontoxic to animal and fish life. Fluid life compares favorably with conventional hydraulic fluids, and they are compatible with seal and hose materials used with mineral oils.

The nature of the bearing metals often has an effect upon the action of the lubricating oil. In highly alloyed metals, some elements act as catalyzers to oxidize the oil, or the acids or moisture in the oils may act to break down the metal. In lead-bearing metals, free magnesium causes disintegration of the lead in contact with moisture. The alkali-lead metals also tend to dissolve in contact with animal or fish oils. Normally, however, none of the white bearing metals are attacked by the animal and vegetable oils used for lubrication unless there are perceptible amounts of a freely oxidizing element present. Graphite adds to the effectiveness of a lubricating oil and can be held in suspension with a tannin. **Graphite lubricants** are used where continuous lubrication is difficult, for running in, for springs, or for bearings where heavy films are desired. The **Dag Lubricants** and **Dag Dispersants** of Acheson Colloids Co. comprise a large group of lubricants, lubricant coatings, and mold partings consisting of graphite or molybdenum sulfide in oils, resins, or solvents, usually applied by spray. **Glydag** is a solution of 10% graphite in glycerin, **Castordag** is a graphite in castor oil, **Glydag B** is graphite in butylene glycol, and **Dag Dispersion 223** is molybdenum disulfide in an epoxy resin. **Neolube,** of Huron Industries, is graphite in alcohol. With these, the carrier liquid evaporates, leaving a film of graphite on the bearing. **Polyphenyl ether** lubricants are highly radiation-resistant. They lubricate after absorbing gamma-ray doses that solidify mineral oils. They are used as specialty lubricants under extreme high-temperature conditions.

There are five principal types of **synthetic lubricants: polyalphaolefins** (PAOs), **polyglycols, polyol esters, diesters,** and **phosphate esters,** the PAOs being the most widely used in the United States. Compared with mineral oils or petroleum-based lubricants, the synthetics can operate over a wider temperature range [−100 to 550°F (−73 to 288°C)], reduce friction better, and are more durable. They may also reduce the risk of fire in high-temperature applications. Phosphate esters containing 8 to 12% phosphorus are inherently fire-resistant. **Triaryl phosphate esters** have flash points from 460 to 485°F (238 to 251°C), compared with 300 to 350°F (149 to 177°C) for mineral oils. Their autoignition temperatures are between 950 and 1050°F (510 and 566°C), compared with 500 and 700°F (260 and 371°C) for the oils. However, the synthetics are far more costly. **Tricresyl phosphate ester,** once widely used, has lost appeal due to its neurotoxicity. New synthesis routes have been developed, however, to create other triaryl phosphate esters that are safer. Two such synthetics of Akzo Chemicals are **isopropyl phenyl phosphates** and *t*-**butyl phenyl phosphates.** *Durad-620B,* of FMC Corp., is a triaryl phosphate ester of superior hydrolytic and oxidative stability.

LUMINOUS PIGMENTS.   Pigments used in paints to make surfaces visible in the dark and in coatings for electronic purposes. They are used for signs, watch and instrument hands, airfield markings, and signals. They are of two general classes. The *permanent* ones are the **radioactive paints,** which give off light without activation, and the **phosphorescent paints,** or **fluorescent paints,** which require activation from an outside source of light. The radioactive paints contain a radioactive element that emits alpha and beta rays which strike the phosphors and produce visible light. Radium, sometimes used for paints for watch hands, gives a greenish-blue light, but it emits dangerous gamma rays. Also, the intense alpha rays of radium destroy the phosphors quickly, reducing the light. Strontium 90 gives a yellow-green light and has a long half-life of 25 years, but it emits both beta and gamma rays and is dangerous. **Tritium paints,** with a tritium isotope and a phosphor in the resin-solvent paint base, have a half-life of 12.5 years and require no shielding. The self-luminous phosphors for clock and instrument dials contain tritium, which gives off beta rays with only low secondary emission so that the glass or plastic covering is sufficient shielding. Other materials used are **krypton 85,** with a half-life of 10.27 years, **promethium 147,** with a half-life of 2.36 years, and **thallium 204,** with a half-life of 2.7 years.

Fluorescent paints depend upon the ability of the chemical to absorb energy from light and to emit it again in the form of photons of light. This variety usually has a base of calcium, strontium, or barium sulfide with traces of other metal salts to improve luminosity, and the vehicle contains a moistureproof gum or oil. Temporary luminous paints may be visible for long periods after the activating light is withdrawn. A paint activated by 5-min exposure to sunlight may absorb sufficient energy for 24 h of luminosity. **Luminous wall paints** used for operating rooms to eliminate shadows are made by mixing small amounts of zinc or cadmium sulfide into ordinary paints. After being activated with ultraviolet rays, they will give off light for 1.5 h.

Phosphorescent paints are lower in cost than radioactive paints and may be obtained in various colors. In general, the yellow and orange phosphorescent pigments are combinations of zinc and cadmium sulfides, the green is zinc sulfide, and the violet and blue pigments are combinations of calcium and strontium sulfides. They are marketed in powder form to be stirred into the paint or ink vehicle, since mixing by grinding lowers the phosphorescence. The natural minerals are not used, as the pigments must be of a high degree of purity, as little as a millionth part of iron, cobalt, or nickel killing the luminosity of zinc sulfide. These phosphorescent pigments are called **phosphors,** but technically they are incomplete phosphors; and copper, silver, or manganese is coprecipitated with the sulfide as an activator or to change the color of the emitted light. The metals that are used as activators are called **phosphorogens,** and their atoms diffuse into the lattice of the sulfide. For fluorescent screens the phosphors must have a rapid rate of extinguishment so that there will be no time lag in the appearance of the events. For television, electron microscope, and radar screens, the phosphors must cease to flow 0.02 s after withdrawal of excitation. They must also be of very minute particle size so as not to give a blurred image. For a white television screen, mixtures of blue zinc sulfide with silver and yellow zinc-beryllium silicates are used. For color television the screen is completely covered with a mixture of various colored phosphors, especially rare-earth metal combinations. For scintillation counters for gamma-ray-detection phosphors, the pulses should be of longer duration, and for this purpose crystals of cadmium or cadmium tungstate are used.

Fluorescent fabrics for signal flags and luminescent clothing are impregnated with fluorescent chemicals which can be activated by an ultraviolet light that is not seen with the eye. Some fluorescent paints contain a small amount of luminous pigment to increase the vividness of the color by absorbing the ultraviolet light and emitting it as visible color. **Fluorescein,** made from phthalic anhydride and resorcinol, has the property of fluorescence in a solvent. Since cellulose acetate will keep it in a permanently solvent state, acetate rayon is used as the carrier fabric. Signal panels are distinguishable from a plane at great heights even through a haze, and at night they give a brilliant glow when activated with ultraviolet rays. Fluorescent paints for signs may have a white undercoat to reflect the light passing through the semitransparent pigment. In passing through the color pigments the shorter violet and blue wavelengths are changed to orange, red, and yellow hues, and the reflected visible light is greater than the original light. **Uranine,** the sodium salt of fluorescein, is used by fliers to mark spots in the ocean. One pound (0.45 kg) of uranine will cover 1 acre (4,047 m^2) of water to a brilliant, yellowish green easily seen from the air. One part of uranine is detectable in 16×10^6 parts of water. **Luminous plastic** for aircraft markings is coated on the inside with radioactive material to give visibility in the dark.

The fluorescent pigments almost always consist of particles of a colorless resin containing a color-fluorescing dye. Two well-known dyes are **Potomac Yellow** and **Alberta Yellow** from Day-Glo Color Corp. BASF AG makes a series under the trade name **Rhodamine**; Sandoz Chemical markets **Xylene Red B**; and Bayer produces **Macrolex Fluorescent Yellow 10GN.** The pigment matrix is usually a mixed amino resin consisting of toluenesulfonamide, melamine, and formaldehyde resins. The **fluorescent plastic** of Rohm & Haas Co. is acrylic sheet containing a fluorescent dye. Lettering or designs cut from the sheet will glow brightly in the dark after exposure to light. It is used for direction signs and decorative panels. **Spot-Lite Glo** is a **phosphorescent frit** for incorporating into ceramics for luminous signs. It contains a zinc sulfide that is stable at high heats and has a long afterglow.

Whitening agents, optical whiteners, or **brightening agents,** used to increase the whiteness of paper and textiles, are fluorescent materials that convert some of the ultraviolet of sunlight to visible light. The materials are colorless, but the additional light supplied is blue, and it neutralizes yellow discolorations and enhances the whiteness. They were first developed in Germany and called **blankophors. Ultrasan,** the first of the German blankophors, was a 1,3,5-triazine derivative. The **M.D.A.C.,** of Carlisle Chemical Corp., is a methyl diethyl aminocoumarin of empirical formula $C_{14}H_{17}O_2N$. It comes in tan-colored granules melting at 158°F (70°C), soluble in water and in acid solutions. It gives a bright-blue fluorescence in daylight and adds whiteness to fabrics and makes colors more vivid. As little as 0.001% added to soaps, detergents, or starches is effective for wool and synthetic fibers, but it is not suitable for cotton. It is also used to overcome yellow casts in varnishes and plastics, and in oils and waxes. **Solium,** of Lever Bros., used in detergents, is a whitener of this type. The **DAS triazine,** triazinyl diaminostilbene disulfonic acid, used with naphthyl triazole, is effective for cotton and rayon.

Luminous materials also occur in nature as organic materials, with bioluminescence thought to be a form of chemiluminescence. Fireflies, bacteria, glow worms, and some luminous fish are capable of this feat. It occurs by the mixing of two substances present in the organism;

one is **luciferin,** which oxidizes the second, an enzyme known as **luciferase.** The reaction produces an excited form of luciferase, which emits light when it returns to its normal state.

LUTES. Adhesive substances, usually of earthy composition, deriving the name from the Latin *lutum,* meaning "mud." A clay cement was used by the Romans for cementing iron posts into stone. Although lutes often contain a high percentage of silica sand or clays, the active ingredient is usually sulfur. They may also contain other reactive ingredients such as lead monoxide or magnesium compounds. **Plumber's lutes** are used for pipe joints and seams and for coating pipes to withstand high temperatures. Plaster of paris mixed with a weak glue will withstand a dull-red heat. **Fat lute** is pipe clay mixed with linseed oil. **Spence's metal** is the name of an old lute for pipe jointing. It was made by introducing iron disulfide, zinc blende, and galena into melted sulfur. It melts at 320°F (160°C) and expands on cooling. It makes a good cast joint which is resistant to water, acids, and alkalies. It is not a metal, but is a mixture of sulfur with metallic oxides. **Sulfur cements,** or lutes, usually have fillers of silica or carbon to improve the strength. They are poured at about 235°F (113°C). They form a class of acidproof cements used for ceramic pipe connections. Modern lutes for very high heat resistance do not contain elemental sulfur. Industrial lutes are used for sealing in wires and connections in electrical apparatus, and are compounded to give good bonding to ceramics and metals. A lute cement for adhering knife blades to handles is composed of magnesium acid sulfates, calcined magnesia, with fine silica or powder. The term **sealant** generally refers to a wide range of mineral-filled plastics formulated with a high proportion of filler for application by troweling or air gun.

MADDER. Formerly the most important dyestuff with the exception of indigo. It is now largely replaced by the synthetic mauve dye alizarin. It was grown on a large scale in France and the Near East and was known by its Arabic name **alizari** and by the name **Turkey red.** Madder is the ground root of the plant *Rubia tinctorum,* which has been stored for a time to develop the coloring matter, the orange-red **alizarin,** $C_{14}H_{18}O_4$, which is a dihydroxyanthraquinone, a powder melting at 552°F (289°C). It occurs in a madder root as the glucoside, **ruberythric acid,** $C_{26}H_{28}O_{14}$, but is now made synthetically from anthracene. Its alkaline solution is used with mordants to give **madder lakes.** With aluminum and tin it gives **madder red,** with calcium it gives blue, and with iron it gives violet-black. **Purpurin,** $C_{14}H_{18}O_5$, is also obtained from madder, but is now made synthetically. Madder gives fast colors.

MAGNESIA. A fine, white powder of **magnesium oxide,** MgO, obtained by calcining magnesite or dolomite and refining chemically. It is used in pharmaceuticals, in cosmetics, in rubbers as a scorch-resistant filler, in soaps, and in ceramics. It requires 6.5 tons (5.9 metric tons) of dolomite to yield 1 ton (0.9 metric tons) of pure magnesia powder. Particle size of the powder is 19.7 μin (0.5 μm). For chemical uses it is 99.7% minimum purity with no more than 0.06% iron oxide and 0.08 calcium oxide, and the magnesia for electronic parts has a maximum of 0.03% iron oxide and 0.0025 boron. This powder is converted from magnesium hydroxide. **Maglite,** of Whittaker, Clark & Daniels, Inc., used for rubbers, is produced from seawater. **Magox magnesia,** of Basic Chemicals, is 98% pure MgO extracted from seawater. It comes in particle sizes to 325 mesh in high- and low-activity grades for rubber, textile, and chemical uses. A very pure magnesia is also produced by reducing magnesium nitrate.

Magnesia ceramic parts, such as crucibles and refractory parts, are generally made from magnesia that is usually electrically fused and crushed from the large cubic crystals. The crystals have ductility and can be bent. The particle size and shape are easily controlled in the crushing to fit the needs of the molded article. Pressed and sintered parts have a melting point of about 5070°F (2765°C) and can be employed to 4172°F (2300°C) in oxidizing atmospheres or to 3092°F (1700°C) in reducing atmospheres. The material is inert to molten steels and to basic slags. **Magnafrax 0340,** of Carborundum Co., is magnesia in the form of plates, tubes, bars, and disks. The material has a specific gravity of 3.3 and a thermal conductivity twice that of alumina. Its vitreous structure gives it about the same characteristics as a single crystal for electronic purposes. **Magnorite,** of Norton Co., is fused magnesia in granular crystals with a melting point of 5072°F (2800°C), used for making ceramic parts and for sheathing electric heating elements. **K-Grain magnesia,** of Kaiser Aluminum and Chemical Corp., is 98% magnesia, containing no more than 0.4 silica. The magnesia ceramic, of Corning Inc., is 99.8% pure. The cast, pressed, or extruded parts when high-fired have a fine-grained, dense structure with practically no shrinkage and a flexural strength of 15,500 lb/in^2 (107 MPa).

MAGNESITE. A white to bluish-gray mineral used in the manufacture of bricks for basic refractory furnace linings and as an ore of magnesium. The ground, burned magnesite is a light powder, shaped into bricks at high pressure and baked in kilns. Magnesite is a magnesium carbonate, $MgCo_3$, with some iron carbonate and ferric oxide. Magnesite releases carbon dioxide on heating and forms magnesia, MgO. When heated further, it forms a crystalline structure known as

periclase, which has a melting point of 5070°F (3076°C) and specific gravity of 3.58. The mineral periclase occurs in nature but is rare. A crystalline form is called **breunnerite.** The fused magnesia made in the arc furnace is actually synthetic periclase. The synthetic material is in transparent crystals up to 2 in (5 cm), which are crushed to powder for thermal insulation and for making refractory parts. Magnesite in compact, earthy form or granular masses has a vitreous luster, and the color may be white, gray, yellow, or brown. Mohs hardness is 3.5 to 4.5, and the specific gravity is about 3.1. The U.S. production of crude magnesite is in Nevada, Washington, and California.

The product known as **dead-burned magnesite** is in the form of dense particles used for refractories. It is produced by calcining magnesite at 2642 to 2732°F (1450 to 1500°C). **Caustic magnesite** is a product resulting from calcination at 1292 to 2192°F (700 to 1200°C), which leaves from 2 to 7% carbon dioxide in the material and gives sufficient cementing properties for use as a refractory cement. Beluchistan magnesite has 95 to 98% $MgCO_3$, with 0.5 to 1 iron oxide. Manchurian dead-burned magnesite has 90.9% magnesia with 4 silica, and some iron oxide and alumina.

Magnesite for use in producing magnesium metal should have at least 40% MgO, with not over 4.5 CaO and 2 FeO. **Brucite,** a natural hydrated magnesium oxide found in Ontario, contains a higher percentage of magnesia than ordinary magnesite and is used for furnace linings. Austrian magnesite has from 4 to 9% iron oxide, which gives it the property of fritting together more readily. Magnesite is a valued refractory material for crucibles, furnace brick and linings, and high-temperature electrical insulation because of its basic character, chemical resistance, high softening point, and high electrical resistance. Its chief disadvantage is its low resistance to heat shock. Magnesite brick and refractory products are marketed under a variety of trade names, such as **Ritex,** of General Refractories Co., and **Ramix.** It is also used as a covering for hot piping. The German artificial stone called **Kunststein** is magnesite.

MAGNESIUM. A silvery-white metal, symbol Mg, which is the lightest metal that is stable under ordinary conditions and produced in quantity. One of its chief uses is as an alloying element in aluminum, zinc, lead, and other nonferrous alloys. It is also used for cathodic protection of other metals from corrosion. It is the sixth most abundant element, and it was originally called **magnium** by Sir Humphry Davy. Specific gravity is 1.74, melting point 1202°F (650°C), boiling point about 2030°F (1110°C), and electrical conductivity about 40% that of copper. Ultimate tensile strengths are about 13,000 lb/in^2 (90 MPa) as cast, at least 23,000 lb/in^2 (159 MPa) for annealed sheet, and 26,000 lb/in^2 (179 MPa) for hard-rolled sheet, with corresponding elongations of about 4, 10, and 15%. The strength is somewhat higher in the forged metal. Magnesium has a close-packed hexagonal structure that makes it difficult to roll cold, and its narrow plastic range requires close control in forging. Repeated reheating causes grain growth. Sheet is usually formed at 300 to 400°F (150 to 200°C). It is the easiest of the metals to machine. Its heat conductivity is about half that of aluminum, and it has high damping capacity. Electrolytic magnesium is usually 99.8% pure, and the metal made by the ferrosilicon-hydrogen reduction process may be 99.95% pure.

Magnesium develops a corrosion-inhibiting film upon exposure to clean atmospheres and freshwater, but that film breaks down in the presence of chlorides, sulfates, and other media, necessitating corrosion protection in many applications. Many protective treatments have been developed for this purpose. It is also rapidly attacked by mineral acids, except chromic and hydrofluoric acids, but is resistant to dilute alkalies; aliphatic and aromatic hydrocarbons; certain alcohols; and dry bromine, chlorine, and fluorine gases. **Anodizing magnesium** improves its corrosion resistance.

Magnesium is valued chiefly for parts where light weight is needed. It is a major constituent in many aluminum alloys, and very light alloys have been made by alloying magnesium with lithium. **Photoengraving plates** made of commercially pure magnesium, or of slightly alloyed metal, are easier to etch than zinc, lighter in weight, and resistant to wear. It has also been used as a facing and shielding material in building construction; the light weight of magnesium gives high coverage, 1 lb (0.45 kg) of 0.005-in (0.013-cm) sheet covering 22.2 ft^2 (2 m^2).

The pure metal ignites easily, and even when it is alloyed with other metals, the fine chips must be guarded against fire. In alloying, it cannot be mixed directly into molten metals because of flashing, but is used in the form of master alloys. The metal is not very fluid just above its melting point, and casting is done at temperatures considerably above the melting point so that there is danger of burning and formation of oxides. A small amount of beryllium added to magnesium alloys reduces the tendency of the molten metal to oxidize and burn. The solubility of beryllium in magnesium is only about 0.05%. As little as 0.001% lithium also reduces fire risk in melting and working the metal. Molten magnesium decomposes water so that green-sand molds cannot be used, as explosive hydrogen gas is liberated. For the same reason, water sprays cannot be used to extinguish magnesium fires. The affinity of magnesium for oxygen, however, makes the metal a good deoxidizer in the casting of other metals.

Magnesium is produced commercially by the electrolysis of a fused chloride, or fluoride obtained either from brine or from a mineral ore, or it can be vaporized from some ores. Much of the magnesium produced in the United States is from brine wells of Michigan, whose brine contains 3% $MgCl_2$, and from seawater. From seawater the magnesium hydroxide is precipitated, filtered, and treated with hydrochloric acid to obtain a solution of magnesium chloride from which the metal is obtained by electrolysis. The magnesium ion content of seawater is 1,270 parts per million. One cubic mile (4.8 km^3) of seawater contains up to 12×10^6 lb (5.4×10^6 kg) of magnesium. In production, 1 lb (0.45 kg) of magnesium is obtained from 100 gal (379 L) of seawater. Magnesium is also obtained from dolomite by extracting the oxide by reacting the burned dolomite with crushed ferrosilicon in a sealed retort and filtering the vapor in a hydrogen atmosphere. The dolomite of Ohio averages 20% magnesium oxide, and the metal is obtained 99.98% pure in a solid, dense, crystalline mass which is then melted with a flux and poured into ingots. Magnesium metal and ferrosilicon are produced from the olivine of Washington state. In Russia, magnesium is produced from the mineral carnallite. In Germany it is produced from dolomite, carnallite, magnesite, and the end lyes of the potash industry. The metal can also be produced from serpentine, olivine, and other siliceous ores by heating the powdered ore in a vacuum retort and driving off the metal as a vapor which is condensed. Magnesium salt from the Dead Sea contains greater concentrations than found in normal seawater. Magnesium hydroxide made from the salt is used as a nonhalogen flame retardant, designated **FR-20,** for plastics.

Magnesium powder, for pyrotechnic and chemical uses, is made by reducing metallic magnesium into particles in the shape of curly shavings to give maximum surface per unit of weight. It is produced in four grades: cutting powder, standard powder, special specification, and fireworks powder. **Cutting powder** is finely cut shavings in a matted condition, made from magnesium of 99.8% purity. **Standard powder** is loose powder in fineness from 10 to 200 mesh. **Fireworks powder** is 100 mesh. The speed of ignition increases rapidly with decreasing particle size. A 200-mesh powder is used for **flashlight powder,** and a 30- to 80-mesh for more slowly burning flares. For flares, magnesium gives a brilliant light of high actinic value. **Incendiary powder,** for small-arms incendiary ammunition, is magnesium powder mixed with barium peroxide. **Ophorite** is an English name for magnesium powder and potassium perchlorate used as an igniter for incendiaries. The material known as **goop,** used in fire bombs, is a rubbery mixture containing magnesium powder coated with asphalt, gasoline, and chemicals.

A wide variety of **magnesium chemicals** are used for applications where the magnesium may be the desirable element or as the chemical carrier for another element. **Magnesium nitrate,** a colorless crystalline powder of composition $Mg(NO_3)_2 \cdot 6H_2O$, is made from magnesite and used in dry colors and pyrotechnics, and to produce magnesia. **Magnesium methoxide,** $Mg(OCH_3)_2$, is a white powder used for drying alcohol to produce absolute alcohol, and also for producing stable alcohol gels for use as solid alcohol fuels. The gels are made by adding water to an alcohol solution of the magnesium methoxide. **Magnesium fluoride,** MgF_2, or **sellaite,** is a pale-violet, crystalline powder that melts at 2545°F (1396°C). It has a very low refractive index and is used on lenses and instrument windows to eliminate reflection. To apply the coating, the fluoride is dissolved in **dimethyl formamide,** $(CH_3)_2NCHO$, and mixed with an essential oil to wet the glass surface. The applied coating is fired at 932°F (500°C) to leave a coating of pure magnesium fluoride about one-quarter light wavelength thick.

MAGNESIUM ALLOYS. Although magnesium alloys are moderate in strength and rigidity, they have high specific strength and rigidity because of their low density, which, being in the range of 0.064 to 0.066 lb/in^3 (1,772 to 1,827 kg/m^3), is the lowest of common metals. Modulus of elasticity in tension is typically 6.5×10^6 lb/in^2 (44,800 MPa), and ultimate tensile strengths range from 22,000 to 55,000 lb/in^2 (152 to 379 MPa), depending on the alloy and form. Applications are due mainly to the light weight, ease of casting, and superior machinability of the alloys and include auto parts, aerospace equipment, power tools, sporting goods, fixtures, and materials-handling equipment. Both wrought and cast alloys are available, the former in sheet, plate, rod, bar, extrusions, and forgings, and the latter for sand, permanent-mold, investment, and die castings. Alloys are designated by a series of letters and numbers followed by a temper designation. The first part of the alloy designation indicates by letters the two principal alloying elements (or one if the alloy contains only one alloying element): A for aluminum, E (rare-earth elements), H (thorium), K (zirconium), M (manganese), Q (silver), S (silicon), T (tin), and Z (zinc). The two (or one) numbers that follow indicate the amounts (in percent rounded off to whole numbers) of these elements, respectively. These numbers are followed by a letter to distinguish among alloys having the same amount of these alloying elements. The temper designations that follow are similar to those for aluminum alloys: F (as fabricated); O (annealed); H10 and H11 (slightly strain-hardened); H23, H24, and H26 (strain-hardened and partially annealed); T4 (solution heat-treated); T5 (artificially aged); T6 (solution heat-treated and artificially aged); and T8 (solution heat-treated, cold-worked, and artificially aged). Thus **AZ91C-T6** is the designation for an alloy containing 8.7% aluminum and 0.7 zinc as the major alloying elements. The letter C indicates that it is the third such alloy to be standardized, and, in this case, it is in the solution heat-treated and artificially aged temper. These designations, however, do not distinguish between wrought and cast alloys. And the **magnesium photoengraving alloy,** containing 3.3% aluminum and 0.7 zinc and made in special-quality sheet for photoengraving, goes simply by the designation **PE.**

Besides AZ91C, other **magnesium sheet and plate alloys** include AZ31B, HK31A, and HM21A, of which AZ31B is the strongest at room temperature and the most commonly used. In the H24 temper, it has an ultimate tensile strength of 42,000 lb/in^2 (290 MPa), a tensile yield strength of 32,000 lb/in^2 (221 MPa), 15% elongation, and can be used at service temperatures to about 200°F (93°C). The others, however, especially HM21A, are more heat-resistant and can sustain temperatures to about 600°F (315°C). For HM21A, the 100-h creep strength for 0.1% deformation is 12,500 lb/in^2 (86 MPa) at 400°F (214°C) and 7,500 lb/in^2 (52 MPa) at 600°F (316°C). None of the alloys are especially formable, minimum bend radii for AZ31B-O, the most formable, ranging from about 5 times thickness (5T) at room temperature to 2T at 500°F (260°C). Thus, heat is often required in forming operations, especially deep-drawing.

AZ31B is also widely used in the form of bar and other extruded shapes. The alloys for bar and extruded shapes are generally of two kinds: those alloyed principally with aluminum and zinc and those alloyed with zinc and a bit of zirconium. In the former, strength increases with increasing aluminum content, and in the latter with increasing zinc content. Some of each kind respond to artificial aging, providing in the T5 temper ultimate tensile strengths of 50,000 to 55,000 lb/in^2 (345 to 379 MPa). AZ31B and many of the alloys for bar and extrusions are also suitable for forging. The hot-working range may be as low as 450 to 700°F (232 to 371°C) or 560 to 1000°F (293 to 538°C).

There are several **magnesium die-casting alloys** but more than a dozen **magnesium sand-casting alloys** and **magnesium permanent-mold casting alloys.** One composition, AZ91, is available in four grades: AZ91A, B, C, and D. AZ91C, for sand- and permanent-mold castings, contains 8.7% aluminum, as opposed to 9 in the die-casting alloys (A and B). Each also contains 0.7% zinc and 0.13 manganese. As die-cast, AZ91A and AZ91B provide an ultimate tensile strength of 33,000 lb/in^2 (228 MPa), a tensile yield strength of 22,000 lb/in^2 (152 MPa), and 24% elongation. For AZ91C-T6, these values are

40,000 lb/in^2 (276 MPa), 21,000 lb/in^2 (145 MPa), and 6%, respectively. AZ91D is a high-purity version, containing extremely low residual contents of iron, copper, and nickel, which markedly improves corrosion resistance, precluding the need for protective treatments in certain applications, such as auto underbody parts. The benefits of high purity are apparently applicable to other alloys as well, such as AS41, a more heat-resistant die-casting alloy.

The sand- and permanent-mold casting alloys generally use either aluminum, zinc, and manganese or zinc, thorium, zirconium, and, in some cases, silver and rare-earth elements, in the way of alloying elements. Almost all these alloys respond to artificial aging or solution heat-treating and artificial aging. The strongest, ZE63A, in the T6 temper, and ZK61A, in the T5 or T6 temper, have an ultimate tensile strength of about 45,000 lb/in^2 (310 MPa), a tensile yield strength of about 28,000 lb/in^2 (193 MPa), and about 10% elongation. The alloys containing zirconium and rare-earth elements—ZE33A, ZE41A, and ZE63A—are more creep-resistant at higher temperatures than the aluminum-, zinc-, and manganese-bearing alloys, but are more difficult to cast. The **magnesium-silver-bearing alloys,** QE22A and QH21A, also are superior in elevated-temperature performance and have good castability and weldability, but are quite costly.

Like their base metal, magnesium alloys have outstanding machinability, providing faster cutting speeds and greater depths of cut at less power than all commonly machined metals. However, dust, chips, and turnings can pose a fire hazard, necessitating special precautions.

In general, media that are basic, neutral, or contain fluorine cause little or no corrosion, but those which are acidic attack magnesium. Thus, most acids, including fruit juices, attack the metal, although pure chromic, oleic, and dry stearic acids at room temperature are exceptions, as is 5 to 60% hydrofluoric acid. The metal also resists distilled water and acid-free rain but is attacked by boiling water, carbonated water, steam, and seawater. Pure compounds to which it is also generally resistant at room temperature include butyl, ethyl, isopropyl, and propyl alcohols; ethyl acetate, ethyl benzene, ethyl cellulose, ethyl chloride, and ethyl salicylate; methyl cellulose, methyl chloride, and methyl salicylate; and, in any concentration, sodium carbonate, sodium cyanide, sodium dichromate, sodium fluorine, sodium hydroxide, sodium phosphate (tribasic), and sodium silicate. Pure media at room temperature which attack the metal include methyl alcohol, ethyl bromide, most ammonium salts, milk, and nitrous gases. Pure chlorine and most chlorides in any concentration are corrosive, as are most heavy-metal salts, hydrogen peroxide, iodides, all nitrates, nitroglycerin, most sulfates, and tanning solutions.

Methylene chloride and vinylidene chloride, however, are not generally corrosive. Because magnesium is at the anodic end of the galvanic series, it will be corroded when coupled with many other metals, necessitating protective measures. Over the years, many protective coatings and treatments have been developed to improve the metal's suitability in many environments.

Although lithium is no longer used in magnesium alloys, a series of **magnesium-lithium alloys,** developed in the past for aerospace applications and ammunition containers, were noted for their extremely light weight and moderate ultimate tensile strengths— 14,500 to 36,000 lb/in^2 (100 to 248 MPa). One such alloy, LA141A, had a density of only 0.045 lb/in^3 (1245 kg/m^3), or three-quarters that of magnesium.

Magnesium-nickel is a master alloy of magnesium and nickel used for adding nickel to magnesium alloys and for deoxidizing nickel and nickel alloys. One such alloy contains about 50% of each metal, is silvery white in color, and is furnished in round bar form. **Magnesium-Monel** contains 50% magnesium and 50 Monel nickel-copper alloy. Alloys of magnesium with nickel, Monel, zinc, copper, or aluminum, used for deoxidizing nonferrous metals, are called **stabilizer alloys.**

MAGNESIUM CARBONATE.   A white, insoluble powder of composition $MgCO_3$, containing also water of crystallization. The specific gravity is 3.10. It is made by calcining dolomite with coke, slaking with water, saturating with carbonic acid gas, and crystallizing out the magnesium carbonate. It is employed as an insulating covering for steam pipes and furnaces, for making oxychloride cement, in boiler compounds, and as a filler for rubber and paper. **Montax,** of R. T. Vanderbilt Co., used as a filler, is a mixture of hydrated magnesium carbonate and silica powder. Magnesium carbonate is a good heat insulator because of the great number of microscopic dead-air cells in the material. The insulating material known as 85% magnesia has a density of 12 lb/ft^3 (192 kg/m^3). Hydrated magnesium carbonate is a fine, white powder called **magnesia alba levis,** slightly soluble in water, and used in medicine.

MAGNESIUM SULFATE.   A colorless to white, bitter-tasting material occurring in sparkling, needle-shaped crystals of composition $MgSO_4 \cdot 7H_2O$. The natural mineral is called **epsomite,** from Epsom Spa, Surrey, England. In medicine it is called **epsom salt.** It is used in leather tanning, as a mordant in dyeing and printing textiles, as a filler for cotton cloth, for sizing paper, in water-resistant and fireproof magnesia cements, and as a laxative. It is being sold in canisters as a heat-storage medium for solar energy by Pfizer Co. It can also be obtained in the anhydrous form, $MgSO_4$, as a white powder. The specific gravity of the hydrous material is 1.678 and of the anhydrous 2.65. It occurs naturally as deposits from spring waters and is also made by treating magnesite with sulfuric acid. In Germany it is produced from the mineral **kieserite,** $MgSO_4 \cdot H_2O$, which is abundant in the Strassfurt district, and is used as a source of sulfuric acid and magnesium. The magnesium sulfate from the waters at Seidlitz, Bohemia, was called **Seidlitz salt,** but **Seidlitz powder** is now a combination of Rochelle salt and sodium bicarbonate. **Synthetic kieserite** is made from the olivine of North Carolina. The mineral **langbeinite,** found in the potash deposits at Carlsbad, New Mexico, is a potassium-magnesium-sulfate containing 22% potassium oxide and 18 magnesia and is used in the production of potassium sulfate and magnesium metal. **Sulpomag** is langbeinite with the halite and clay washed out. When the magnesium of epsomite is replaced by zinc, the mineral is called **goslarite,** and when replaced by nickel, it is called **morenosite,** a green mineral occurring in nickel mines.

MAGNET WIRE. Insulated wire for the winding of electromagnets and for coils for transformers and other electrical applications. Since compactness is usually a prime consideration, **high-conductivity copper** is used in the wire, but where weight saving is important, aluminum wire may be used and has the advantage that an extremely thin, anodized coating of oxide serves as the insulation either alone or with a thin varnish coating. Square or rectangular wires may be used, but ordinary magnet wire is round copper wire covered with cotton and an enamel, in OD sizes from No. 40, AWG [0.0071 in (0.018 cm)], to No. 8 [0.1380 in (0.351 cm)]. **Vitrotex** is a magnet wire coated with a resin enamel and covered with alkali-free glass fiber. It withstands temperatures to 266°F (130°C), and the glass fibers dissipate heat rapidly. **Silotex,** of this company, has a silicon resin and glass-fiber insulation and withstands temperatures to 572°F (300°C). Various types of synthetic resins are used as insulation to give high dielectric strength and heat and abrasion resistance. Heat resistance is usually designated by the AIEE class standards. **Formvar magnet wire** has a coating of vinyl acetal resin with dielectric strength of 1,000 V/mil (39.4×10^6 V/m). The wire coating called **Alkanex** is modified glycerol terephthalate polyester resin for applications in which the operating temperatures are up to 311°F (155°C). **Bondar coating** is an epoxy-modified polyester amide for temperatures to 311°F (155°C). **Carthane 8063** is a liquid urethane resin for flexible and abrasion-resistant coating on magnet wire.

MAGNETIC MATERIALS. Metallic and ceramic materials that become magnetized when placed in a magnetic field. All magnetic materials can be classified into two broad groups—soft magnetic materials and hard magnetic materials. **Soft magnetic materials,** sometimes called **electromagnets,** do not retain their magnetism when removed from a magnetic field. **Hard magnetic materials,** sometimes referred to as **permanent magnets,** retain their magnetism when removed from a magnetic field. Cobalt is the major element used for obtaining magnetic properties in hard magnetic alloys.

Common soft magnetic materials are iron, iron-silicon alloys, and nickel-iron alloys. **Irons** are widely used for their magnetic properties because of their relatively low cost. Common **iron-silicon magnetic alloys** contain 1, 2, 4, and 5% silicon. There are about six types of nickel-irons, sometimes called **permeability alloys,** used in magnetic applications. For maximum magnetostriction the two preferred nickel contents are 42 and 79%. Additions of molybdenum give higher resistivities, and additions of copper result in higher initial permeability and resistivity.

Magnet steels, now largely obsolete, included plain high-carbon (0.65 or 1%) steels or high-carbon (0.7 to 1) compositions containing 3.5 chromium—**chromium magnet steels;** 0.5 chromium and 6 tungsten—**tungsten magnet steel;** or chromium, tungsten, and substantial cobalt (17 or 36)—**cobalt magnet steels.** They were largely replaced by ternary alloys of iron, cobalt, and molybdenum, or tungsten. **Comol** has 17% molybdenum, 12 cobalt, and 71 iron. **Indalloy** and **Remalloy** have similar compositions: about 20% molybdenum, 12 cobalt, and 68 iron. **Chromindur** has 28% chromium, 15 cobalt, and the remainder iron, with small amounts of other elements that give it improved strength and magnetic properties. In contrast to Indalloy and Remalloy, which must be processed at temperatures as high as 2280°F (1250°C), Chromindur can be cold-formed.

Some cobalt magnet steels contain 1.5 to 3% chromium, 3 to 5 tungsten, and 0.50 to 0.80 carbon, with high cobalt. **Alfer magnet alloys,** first developed in Japan to save cobalt, were **iron-aluminum alloys.** **MK alloy** had 25% nickel, 12 aluminum, and the balance iron, close to the formula Fe_2NiAl. It is age-hardening and has a coercive force of 520 Oe (41,340 A/m) and maximum energy product of 1.35×10^6 G · Oe (10,746 T · A/m). **Oerstit 400,** used by the Germans during the Second World War because it gave high coercive force in proportion to weight, contained 22% cobalt, 16 nickel, 8 aluminum, 4 copper, and the balance iron. **Cunife** is a nickel-cobalt-copper alloy that can be cast, rolled, and machined. It is not magnetically directional like the tungsten magnets and thus gives flexibility in design. The density is 0.300 lb/in³ (8,304 kg/m³), the electric conductivity is 7.1% that of copper, and it has good coercive force. **Cunife 1** contains 50% copper, 21 nickel, and 29 cobalt. **Cunife 2,** with 60% copper, 20 nickel, and 20 iron, is more malleable. This alloy, heat-treated at 1100°F (593°C), is used in wire form for permanent magnets for miniature apparatus. It has a coercive force of 500 Oe (39,750 A/m). **Hipernom,** of Westinghouse Electric Corp., is a high-permeability nickel-molybdenum magnet alloy containing 79% nickel, 4 molybdenum, and the balance iron. It has a Curie temperature of 860°F (460°C) and is used for relays, amplifiers, and transformers.

In the **Alnico alloys,** a precipitation hardening occurs with AlNi crystals dissolved in the metal and aligned in the direction of magnetization to give greater coercive force. This type of magnet is usually magnetized after setting in place. **Alnico 1** contains 21% nickel, 12 aluminum, 5 cobalt, 3 copper, and the balance iron. The alloy is cast to shape, is hard and brittle, and cannot be machined. The coercive force is 400 Oe (31,800 A/m). **Alnico 2,** a cast alloy with 19% nickel, 12.5 cobalt, 10 aluminum, 3 copper, and the balance iron, has a coercive force of 560 Oe (44,520 A/m). The cast alloys have higher magnetic properties, but the sintered alloys are fine-grained and stronger. **Alnico 4** contains 12% aluminum, 27 nickel, 5 cobalt, and the balance iron. It has a coercive force of 700 Oe (55,650 A/m), or 10 times that of a plain tungsten magnet steel. **Alnico 8,** of Crucible Steel Co., has 35% cobalt, 34 iron, 15 nickel, 7 aluminum, 5 titanium, and 4 copper. The coercive force is 1,450 Oe (115,275 A/m). It has a Rockwell C hardness of 59. The magnets are cast to shape and finished by grinding. **Hyflux Alnico 9,** of the same coercive force, has an energy product of 9.5×10^6 G · Oe (75,620 T · A/m). The magnets of this material, made by Indiana General Corp., are cylinders, rectangles, and prisms, usually magnetized and oriented in place. The **Alnicus magnets,** of U.S. Magnet & Alloy Corp., are Alnico-type alloys with the grain structure oriented by directional solidification in the casting which increases the maximum energy output. **Ticonal, Alcomax,** and **Hycomax** are Alnico-type magnet alloys produced in Europe. Various other alloys of high coercive force have been developed for special purposes. **Silmanal,** with 86.75% silver, 8.8 manganese, and 4.45 aluminum, has a coercive force of 6,300 Oe (500,850 A/m), but a low flux density. **Platinax,** with 76.7% platinum and 23.3 cobalt, has a coercive force of 2,700 Oe (214,860 A/m). **Bismanol,** developed by the Naval Ordnance Laboratory, is a **bismuth-manganese alloy** with 20.8% manganese. It has a coercive force of 3,600 Oe (286,560 A/m), but oxidizes easily. **Cobalt-platinum,** as an intermetallic rather than an alloy, has a coercive force above 4,300 Oe (341,850 A/m) and a residual induction of 6,450 G (0.645 T). It contains 76.8% by weight of platinum and is expensive, but is used for tiny magnets for electric wristwatches and instruments. **Placovar,** of Hamilton Watch Co., is a similar alloy that retains 90% of its magnetization flux up to 650°F (343°C). It is used for miniature relays and focusing magnets. **Ultra-mag** is a **platinum-cobalt** magnet material with a coercive force of 4,800 Oe (381,600 A/m). The Curie temperature is about 932°F (500°C), and it has only slight loss of magnetism at 662°F (350°C), whereas cobalt-chromium magnets lose their magnetism above 302°F (150°C). The material is easily machined. **Alloy 1751,** of Engelhard Corp., is a cobalt-platinum intermetallic with a coercive force of 4,300 Oe (341,850 A/m), or of 6,800 Oe (540,600 A/m) in single-crystal form. The metal is not brittle and can be worked easily. It is used for the motor and index magnets of electric watches.

Soft magnetic ceramics, also referred to as **ceramic magnets, ferromagnetic ceramics,** and **ferrites (soft),** were originally made of an iron oxide, Fe_2O_3, with one or more divalent oxides such as NiO, MgO, or ZnO. The mixture is calcined, ground to a fine powder, pressed to shape, and sintered. Ceramic and intermetal types of magnets have a square hysteresis loop and high resistance to demagnetization, and are valued for magnets for computing machines where a high remanence is desired. A ferrite with a square loop for switching in high-speed computers contains 40% Fe_2O_3, 40% MnO, and 20% CdO. Some intermetallic compounds, such as **zirconium-zinc,** $ZrZn_2$, which are not magnetic at ordinary temperatures become ferromagnetic with properties similar to ferrites at very low temperatures and are useful in computers in connection with subzero superconductors. Some compounds, however, are the reverse of this, being magnetic at ordinary temperatures and nonmagnetic below their transition temperature point. This transition temperature, or **Curie point,** can be arranged by the compounding to vary from subzero temperatures to above 212°F (100°C). **Chromium-manganese-antimonide,** $Cr_xMn_{2x}Sb$, is such a material. Chromium-manganese alone is ferromagnetic, but the antimonide has a transition point varying with the value of x.

Vectolite is a lightweight magnet made by molding and sintering ferric and ferrous oxides and cobalt oxide. The density is 0.114 lb/in^3 (3,156 kg/m^3). It has high coercive force and has such high electrical resistance that it may be considered a nonconductor. It is very brittle and is finished by grinding. **Magnadur** was made from barium carbonate and ferric oxide and has the formula $BaO(Fe_2O_3)_6$. **Indox** and **Ferroxdure** are similar. This type of magnet has a coercive force to 1,600 Oe (127,200 A/m), with initial force to 2,600 Oe (206,700 A/m), high electrical resistivity, high resistance to demagnetization, and light weight, with specific gravity from 4.5 to 4.9. **Ferrimag,** of Crucible Steel Co., and **Cromag** are ceramic magnets. Strontium carbonate is superior to barium carbonate for magnets but is more costly. **Lodex magnets,** of General Electric Co., are extremely fine particles of iron-cobalt in lead powder made into any desired shape by powder metallurgy.

Ceramic permanent magnets are compounds of iron oxide with oxides of other elements. The most used are **barium ferrite, oriented barium ferrite,** and **strontium ferrite. Yttrium-iron garnet (YIG)** and **yttrium-aluminum garnet (YAG)** are used for microwave applications.

Flexible magnets are made with magnetic powder bonded to tape or impregnated in plastic or rubber in sheets, strip, or forms. **Magnetic tape** for recorders may be made by coating a strong, durable plastic tape, such as a polyester, with a **magnetic ferrite powder.** For high-duty service, such as for spacecraft, the tape may be of stainless steel. For recording heads the ferrite crystals must be hard and wear-resistant. **Ferrocube** is manganese zinc. The tiny crystals are compacted with a ceramic bond for pole pieces for recorders. **Plastiform** is a barium ferrite bonded with rubber in sheets and strips. **Magnyl,** of Applied Magnetic Corp., is vinyl resin tape with the fine magnetic powder only on one side. It is used for door seals and display devices.

Rare-earth magnetic materials, used for permanent magnets in computers and signaling devices, have coercive forces up to 10 times those of ordinary magnets. They are of several types. **Rare-earth-cobalt magnets** are made by compacting and extruding the powders with a binder of plastic or soft metal into small precision shapes. They have high permanency. **Samarium-cobalt** and **cesium-cobalt magnets** are cast from vacuum melts and, as made by Bell Laboratories, are chemical compounds, $SmCo_5$ and $CeCo_5$. These magnets have intrinsic coercive forces up to 28,000 Oe (2.2×10^6 A/m). The **magnetooptic magnets** produced by IBM for memory systems in computers are made in thin wafers, often no more than a spot in size. These are ferromagnetic ceramics of **europium-chalcogenides.** Spot-size magnets of europium oxide only 157 μin (4 μm) in diameter perform reading and writing operations efficiently. Films of this ceramic less than a wavelength in thickness are used as memory storage media. **Neodymium-iron-boron magnets,** which are used in computer disk-drive systems, retain useful performance at temperatures up to 248°F (120°C). They also maintain their room-temperature energy at −452°F (−269°C). **Praseodymium-iron-boron magnets** can generate greater energy, appreciably greater at −452°F.

Magnetic fluids consist of solid magnetic particles in a carrier fluid. When a magnetic field is applied, the ultramicroscopic iron oxide particles become instantly oriented. When the field is removed, the particles demagnetize within microseconds. Typical carrier fluids are water, hydrocarbons, fluorocarbons, diesters, organometallics, and polyphenylene ethers. Magnetic fluids can be specially formulated for specific applications such as damping, sealing, and lubrication.

MAHOGANY.   A name applied to a variety of woods. All the true mahogany, however, comes from trees of the family Meliaceae, but of various genera and species. The tropical cedars, Spanish cedar and Paraguayan cedar, belong to this family. Mahogany, of the tree *Swietenia mahogani* and other species of *Swietenia,* is obtained from Mexico to as far south as northern Argentina. The Central American has the best reputation and is frequently referred to under the Spanish name **caoba.** The mahogany from Cuba and Santo Domingo has a close grain and beautiful color and is valued for furniture. The so-called **horseflesh maghogany** from Cuba is from the **sabicu tree. Baywood** is an English name originally applied to a superior, straight-grained mahogany from the shores of the Bay of Honduras. **Colombian mahogany** is the wood of the tree *Cariniana pyriformis* of northern South America. It resembles mahogany but is heavier and harder.

The wood of the mahogany tree is obtainable in large logs. It has a reddish color of various shades. The grain is often figured, and it has a high luster when polished. It seasons well, does not warp easily, and is prized for furniture and cabinetwork and for small patterns in foundry work. Density varies from 32 to 42 lb/ft^3 (513 to 673 kg/m^3), and the hardness and closeness also vary. The beautiful curled-grain woods are from selected forks of the trees. The mahogany formerly used for airplane propellers, and used also for small boats and boat trim, is either African *Khaya* or American *Swietenia,* with average specific gravity kiln-dried of 0.50. The compressive strength is up to 1,760 lb/in^2 (12 MPa) perpendicular to the grain, and the shearing strength 860 lb/in^2 (6 MPa) parallel to the grain.

Australian red mahogany is from the tree *Eucalyptus resinifera* of Australia. It is hard, durable, dark red, with a coarse, open grain. **Crabwood,** used as a substitute for mahogany, is the wood of the **carapa tree,** *Carapa guianensis,* of Brazil and the Guianas. It has a deep reddish-brown color with a coarse grain and a density of 40 lb/ft^3 (641 kg/m^3). This tree produces the seed nuts from which **carapa fat** is pressed and used for soap, candles, and as an edible fat. **Oleo vermelho,** from the tree *Myrospermum erythoxylum* of Brazil, is a fine-grained, reddish cabinet wood similar to mahogany. The specific gravity is 0.954. It has an agreeable odor. **Cameroon mahogany** is from the tree *Bassia toxisperma* of West Africa. The kernels of the nuts of the tree yield about 60% of a yellowish-white, semisolid oil known as

djave butter, or **adjab butter,** used in Europe for soapmaking. It is used locally for food by steaming off the traces of hydrocyanic acid.

MANGANESE. A metallic element, symbol Mn, found in the minerals manganite and pyrolusite, and with most iron ores and traces in most rocks. Manganese has a silvery-white color with purplish shades. **Distilled manganese,** with no iron, and with carbon and silicon not over 0.006% total, has a silvery-gray luster and is very resistant to corrosion. It is brittle but hard enough to scratch glass. The specific gravity is 7.42, melting point 2273°F (1245°C), and density 0.268 lb/in^3 (7,418 kg/m^3). It decomposes in water slowly. It is not used alone as a construction metal. The electrical resistivity is 100 times that of copper or 3 times that of 18–8 stainless steel. It also has a damping capacity 25 times that of steel and can be used to reduce the resonance of other metals.

Manganese is used in the steel industry as a deoxidizer and as a hardener, and nearly all steel now contains some manganese. For this purpose it is used largely in the form of ferromanganese. Manganese is also added to steel in considerable amounts for the production of wear-resistant alloy steels. **Manganese metal,** for adding manganese to nonferrous alloys, is marketed in crushed form containing 95 to 98% manganese, 2 to 3 maximum iron, 1 maximum silicon, and 0.25 maximum carbon, but for the controlled addition of manganese to nonferrous metals and to high-grade steels, high purity, 99.9% plus, electrolytic manganese metal is now used. Manganese metal has very high sound-absorbing properties, and **copper-manganese alloys** with high percentages of manganese are used as **sound-damping alloys** for thrust collars for jackhammers and other power tools.

Electrolytic manganese can be produced from low-grade ores by electrochemical methods and is 99.9% pure. The material produced from high-grade ores is designated **electromanganese.** It comes in chips about 0.0625 in (0.16 cm) thick in sizes larger than 1 in (2.54 cm) square. It is at least 99.97% pure, 150 to 325 mesh, and is employed for pyrotechnic and metallurgical uses. **Manganese tablets,** for use in steelmaking, are made by pressing electrolytically reduced powder in an inert atmosphere and then coating the tablets with ammonium chloride to prevent oxidation. **Manganese carbonyl,** used for vapor deposition of manganese coatings, is a yellow, crystalline solid of composition $Mn_2(CO)_{10}$, melting at 154°F (68°C) and soluble in common organic solvents.

MANGANESE ALLOY. A 72% manganese, 18 copper, 10 nickel alloy noted for its high coefficient of thermal expansion, electrical resistivity, strength, and vibration damping. It is used for rheostat resistors and electrically heated expansion elements.

MANGANESE-ALUMINUM. A **hardener alloy** employed for making additions of manganese to aluminum alloys. Manganese lowers the thermal conductivity of aluminum but increases its strength. Manganese up to 1.2% is used in aluminum alloys when strength and stiffness are required. One manganese-aluminum contains 25% manganese and 75 aluminum. **Manganese-boron** is used for deoxidizing and hardening bronzes. It contains 20 to 25% boron, with small amounts of iron, silicon, and aluminum. For deoxidizing and hardening brasses, nickel bronze, and copper-nickel alloys, **manganese copper,** or **copper manganese,** may be used. The alloys used contain 25 to 30% manganese and the balance copper. The best grades of manganese copper are made from metallic manganese and are free from iron. For nickel bronzes and nickel alloys, the manganese copper must be free of both iron and carbon, but grades containing up to 5% iron can be used for manganese bronze. Grades made from ferromanganese contain iron. Manganese copper is usually marketed in slabs with notched sections, or as shot. It has a lower melting point than metallic manganese and is thus more easily dissolved in the brass or bronze. The 30% alloy melts at about 1600°F (871°C).

MANGANESE BRONZE. Wrought and cast alloys of copper and zinc mainly, with lesser amounts of iron, aluminum, silicon, tin, and lead. The two standard alloys, C67000 (2.5 to 5.0% manganese) and C67500 (0.05 to 0.50 manganese), were formerly designated **manganese bronze B and A,** respectively. **Manganese bronze cast alloys** constitute the C86100 to C86800 Series, some of which also may contain as much as 5% manganese. Some manganese bronzes were formerly designated **high-strength yellow brasses** and **leaded high-strength yellow brasses.**

C67500, which contains 58.5% copper, 39 zinc, 1.4 iron, 1 tin, and 0.1 manganese, has an ultimate tensile strength of 65,000 lb/in^2 (448 MPa), a tensile yield strength of 30,000 lb/in^2 (207 MPa), and a tensile elongation of 33% in the annealed condition. In the half-hard temper, yield strength doubles and elongation reduces to 19%. Modulus of elasticity in tension is 15×10^6 lb/in^2 (103,000 MPa) and electrical conductivity is about 7% that of copper. The alloy is weldable, has good brazing and soldering characteristics, and has good resistance to corrosion in rural, industrial, and marine atmospheres. Available in rod and shapes, it is used in pumps, clutches, and valves.

Most of the cast alloys are castable by various methods, C86200 being the most versatile in this respect. As sand-cast, the alloys provide typical ultimate tensile strengths ranging from 65,000 to 115,000 lb/in² (448 to 793 MPa), and they are rather ductile, as indicated by tensile elongations of 15 to 30%. Specific gravity is in the range of 7.8 to 8.3, electrical conductivity is 7.5 to 22% that of copper, and modulus of elasticity in tension is 14×10^6 to 15×10^6 lb/in² (97,000 to 103,000 MPa). The alloys are not hardenable by heat treatment; weldability, including brazing and soldering, is generally poor or fair; and their machinability is 8 to 65% that of free-cutting brass, C86400 and C86700 being the best in this respect.

Turbadium bronze was an old name used by the British Admiralty for manganese bronze containing 50% copper, 44 zinc, 1 iron, 1.75 manganese, 2 nickel, and 0.5 tin, used for casting propellers and marine parts. The original **Turbiston's bronze** contained 55% copper, 41 zinc, 1 aluminum, 2 nickel, and 1 iron. **Manganese-tin alloy,** used in England as a substitute for nickel silver, is a white alloy containing 16% manganese, 8 tin, and the balance copper. The tensile strength is 57,000 lb/in² (392 MPa), with elongation of 48%, but when it is cold-worked, the strength is increased to 103,000 lb/in² (710 MPa) with elongation of 2%.

The alloys known as **manganese casting brass** are usually Muntz metal containing a small amount of manganese. The original **manganese brass,** patented in 1876 under the name of **Parsons' alloy,** contained 56% copper, 41.5 zinc, 1.2 iron, 0.7 tin, 0.1 manganese, and 0.46 aluminum. **Lumen manganese brass,** of Lumen Bearing Co., is a 60–40 brass with 3% of the copper replaced with 1% iron and some manganese, tin, and aluminum.

MANGANESE ORES. Manganese is a widely dispersed metal, occurring in many ores and in many parts of the world. The ores are used largely for producing ferromanganese, but some low-grade ores are reduced electrolytically to the metal, and the oxide ores are used directly in dry batteries, glassmaking, and the chemical industry. **Pyrolusite** is the most important manganese ore. It is a **manganese dioxide,** MnO_2, with a black color and a metallic luster. The specific gravity is 4.75 and Mohs hardness 2 to 2.5. It is mined in various parts of Europe, Australia, Brazil, Argentina, Ghana, Cuba, India, Canada, and the United States. It is valued for glass manufacture, and when used as a decolorizer for glass, pyrolusite has been called **glassmakers' soap.**

Some of the high-grade Montana and Ghana ores are used for batteries. **Battery-grade manganese** must be free of lead, copper, iron, and other impurities which are electronegative to zinc, and which would decrease the potential and the life of the dry cell. Battery manganese must also have the oxygen readily available, be poorly crystallized, and consist of the gamma oxide known as **cryptomelane,** or a black pseudoamorphous powder. Pyrolusite normally has an orthorhombic crystal structure, but also occurs pseudoamorphic, or as **psilomelane,** a colloidal form of the oxide, and it is this material which is separated as the battery grade. But the natural material is an alteration product in the ore veins and is irregular in quality. High-grade battery manganese of uniform purity is manufactured from low-grade ores by leaching the crushed ore with sulfuric acid and precipitating the heavy-metal sulfides with barium sulfide, aerating to oxidize the iron and sulfur, and electrolyzing the solution to obtain MnO_2 on the graphite anodes. For use as a dry-cell depolarizer, it is ground so that 65% passes a 100-mesh screen. Synthetic manganese dioxide made by electrolysis of the sulfate or by chemical reduction of the carbonate shows an irregularly shaped, amorphous structure under a microscope. It is more reactive and more uniform than the natural material, and gives a longer battery life with a smaller quantity. High-grade battery manganese is also made by reacting manganese sulfate with sodium chlorate at 200°F (93°C) in the presence of sulfuric acid, and the synthetic manganese oxides are now preferred for battery use.

The ore known as **bog manganese,** also called **wad,** is an impure mixture of MnO_2 and MnO, together with other oxides. It is a soft, friable mineral of black or brown color and is an impure psilomelane. The wad ore of Arkansas, used for making manganese sulfate, contains 15 to 50% manganese. **Manganblend,** or **alabandite,** is a natural **manganese sulfide,** MnS, and is an iron-black mineral with a specific gravity of 4 and a Mohs hardness of 3.5 to 4. This material in ground form is marketed by Foote Mineral Co. for making amber glass. It is stable and produces a clear amber color without muddiness.

Rhodochrosite, found in several parts of the United States and in central Europe, is a **manganese carbonate,** $MnCO_3$, with usually some iron replacing part of the manganese. It has a rose-red to dark-brown color, with a vitreous luster, specific gravity 3.5, and Mohs hardness 3.5 to 4.5. It usually has a massive cleavable structure. **Manganite,** found with other manganese minerals and with iron, is an iron-black mineral of composition $Mn_2O_3 \cdot H_2O$, containing theoretically 62.4% manganese. It is found in Germany, England, and the Lake Superior region of the United States. **Hausmannite,** found in Washington state, is another hydrated oxide, $Mn_3O_4 \cdot 2H_2O$, containing theoretically 62.26% manganese. It is used for coating welding rods. **Bementite,** from the same area, is a hydrated silicate, $8MnO \cdot 7SiO_2 \cdot 5H_2O$, containing 40.8% manganese. The high silica makes it difficult to use. **Rhodonite,** $MnO \cdot SiO_2$, found in Colorado, contains

35% manganese so tightly bound chemically that it is difficult to separate by ordinary methods. It is vaporized with a high-intensity arc, and the simple oxides, MnO_2 and SiO_2, then condense.

In general, an ore for ferromanganese should contain at least 35% manganese. Much of the U.S. ore contains only 5 to 10% manganese. Arkansas ores are low-grade, with as little as 18% manganese, and high-grade, with more than 70%. The low grades of Montana ore are concentrated by a nodulizing process up to 58% manganese. Manganese is also extracted from low-grade ores by a chemical process of leaching the pulverized ore with acid; treating with calcium chloride to remove calcium sulfate, and then with limestone; and filtering off the iron oxide. In Germany, low-grade ore is made into ferromanganese by first smelting to spiegeleisen and then treating part in an acid and part in a basic converter before mixing. Three grades of ore designated by manganese and iron contents are: High-grade, with 48% minimum manganese and 7 maximum iron; low-grade A, with 44% minimum manganese and 10 maximum iron; and low-grade B, with 40% minimum manganese and no maximum on iron. Chemical-grade manganese ore should have 80 to 90% MnO_2, equivalent to 51 minimum manganese, and not more than 2 iron. Indian ore is classed as first grade when it has 50% minimum manganese, second grade with 48 to 50, and third grade with 45 to 48. The battery-grade ore from Papua averages 86% manganese dioxide. The ore of Zaire averages 50% manganese and 4.5 minimum iron. By the Nossen process ores with as low as 11% Mn are converted to either metallurgical or battery-grade manganese dioxide. The process consists of leaching with nitric acid, evaporating the filtered manganese nitrate, and then decomposing in heated drums to form MnO_2, HNO, and some NO_2.

MANGANESE STEEL. All commercial steels contain some manganese which has been introduced in the process of deoxidizing and desulfurizing with ferromanganese, but the name was originally applied only to steels containing from 10 to 15% manganese. Steels with 1.0 to 1.5% manganese are known as **carbon-manganese steel, pearlitic manganese steel,** or **intermediate manganese steel. Medium manganese steels,** with manganese from 2 to 9%, are brittle and are not ordinarily used, but steels with 1 to 2% manganese and with or without small amounts of chromium and/or molybdenum are used for air-hardening and oil-hardening cold-work tool steels. The original **Hadfield manganese steel** made in 1883 contained 10 to 14.5% manganese and 1 carbon.

Manganese increases the hardness and tensile strength of steel. In the absence of carbon, manganese up to 1.5% has only slight influence on iron; as the carbon content increases, the effect intensifies. Air hardening becomes apparent in a 0.20% carbon steel with 1.5 manganese, and in a 0.35 carbon steel with 1.4 manganese. The manganese steels used for dipper teeth, tractor shoes, and wear-resistant castings contain 10 to 14% manganese, 1 to 1.4 carbon, and 0.30 to 1 silicon. The tensile strength is up to 125,000 lb/in^2 (862 MPa), elongation 45 to 55%, density 0.286 lb/in^3 (7,916 kg/m^3), and Brinell hardness, when heat-treated, of 185 to 200. Cold-working hardens this steel, and dipper teeth in service will work-harden to a Brinell hardness up to 550.

High-manganese steels are not commercially machinable with ordinary tools, but can be cut and drilled with tungsten carbide and high-speed steel tools. The austenitic steels, with about 12% manganese, are exceedingly abrasion-resistant and harden under the action of tools. They are nonmagnetic. The coefficient of expansion is about twice that of ordinary steel. Various trade names are used to designate the high-manganese steels. **Rol-man steel** contains 11 to 14% manganese and 1 to 1.4 carbon and has a tensile strength of 160,000 lb/in^2 (1,120 MPa) and elongation up to 50%. **Amsco steel** contains 12 to 13% manganese and 1.2 carbon. The tensile strength is 125,000 lb/in^2 (862 MPa), and it will work-harden to Brinell 500. **Tisco steel** has up to 15% manganese and is used for rails and crossovers where high resistance to abrasion is needed. **Timang** is a high-manganese steel made in the form of wire for rock screens. A German stainless type of steel, made without nickel, has 12.5% manganese and is called **Roneusil steel. High-manganese steels** are brittle when cast and must be heat-treated. For castings of thin sections or irregular shapes where the drastic water quenching might cause distortion, nickel up to 5% may be added. The **manganese-nickel steels** have approximately the same characteristics as the straight manganese steels. Nickel is also used in high-manganese steel wire, and the hard-drawn wire has strengths up to 300,000 lb/in^2 (2,069 MPa). **Manganal,** a hot-rolled plate steel of high strength and wear resistance, contains 11 to 13% manganese, 2.5 to 3.5 nickel, and 0.60 to 0.90 carbon. The tensile strength is 150,000 lb/in^2 (1,034 MPa). Pearlitic **nickel-manganese steel** contains only 1.25% manganese with 1.25 nickel. It has a high yield point and ductility.

A **manganese-aluminum steel,** developed by Ford Motor Co., has 30% manganese, 9 aluminum, 1 silicon, and 1 carbon. Its tensile strength is 120,000 lb/in^2 (827 MPa) with elongation of 18%, but it work-hardens rapidly; and when it is cold-rolled and heat-aged, the tensile strength is 300,000 lb/in^2 (2,069 MPa) with a yield strength of 290,000 lb/in^2 (2000 MPa). This alloy forms a special type of stainless steel, with high resistance to oxidation and sulfur gases to 1400°F (760°C).

Structural steels with 0.50% carbon and from 1 to 2 manganese have tensile strength above 90,000 lb/in² (621 MPa). **Martinel steel,** or **Martin elastic-limit steel,** was an early English steel of this type. **D-steel,** developed by the British Admiralty for warship construction, contains 1.1 to 1.4% manganese, 0.33 carbon, and 0.12 silicon. The tensile strength is 96,000 lb/in² (662 MPa) and elongation 17%. Penn Central rails have 1.30 to 1.60% manganese and 0.65 carbon. **Man-Ten steel** is a medium-carbon, medium-manganese structural steel, with corrosion resistance twice that of carbon steel. The tensile strength of the steel is 75,000 lb/in² (517 MPa), with elongation of 20%. Steels containing 1.30 to 1.90% manganese replace more expensive alloy steels for automotive parts. Most mills now list these steels as special alloy machinery steels; those containing about 0.10% sulfur are designated as **manganese screw stock.**

E.Z. Cut steel plate is a free-machining steel for molds, gears, and machine parts. It has 0.14 to 0.21% carbon, 1.15 to 1.4 manganese, and 0.17 to 0.23 sulfur. The tensile strength is 65,000 lb/in² (448 MPa) and elongation 30%; but when it is carbonized and water-quenched, the tensile strength is 100,000 lb/in² (690 MPa). **Max-El No. 4 steel** is a pearlitic manganese steel with a small amount of chromium and 0.75% carbon, used for spring collets and called **collet steel.** Slight amounts of chromium will increase the strength and hardness of the intermediate manganese steels. **Tank car steel** M-128 is a **manganese-vanadium steel** with 0.25% carbon, up to 1.5 manganese, and 0.02 or more vanadium. It has a minimum tensile strength of 81,000 lb/in² (558 MPa) with elongation of 18%. This type of steel with up to 1.75% manganese is used for forgings.

MANGROVE. An extract from the bark of the mangrove tree, *Rhizophora mangle,* of Venezuela and Colombia, the **red mangrove,** *R. racemosa,* of Nigeria, the **East African mangrove,** *R. mucronata,* and other species of Africa, the East Indies, southern Asia, and tropical America, used for tanning leather. In Java, it is called **baku bark.** The east African bark contains 22 to 38% tannin, and the Nigerian bark contains 15 to 29%, usually at the low level. The South American barks range in tanning content from 5 to 45%. In the **Brazilian mangrove,** the leaves contain 24%. The solid extract marketed in blocks contains 62 to 63% solids and 53 to 54 tannin. The liquid extract contains 25 to 35% tannin. Red mangrove contains a red coloring matter which is objectionable in tanning, but can be decolorized with albumin. **White mangrove** produces a pale, pinkish-brown leather, fairly soft and of firm texture. Mangrove from east Africa is called **mangrove cutch** and is sometimes erroneously referred to as *wattle.* The bark is sold in fibrous form and in pieces.

MANILA HEMP. A fiber obtained from the leaf stalks of the **abaca** plant, *Musa textilis,* a tree of the banana family growing in the Philippines, East Indies, and Central America. It is employed for rope and cordage and is the strongest of the vegetable fibers. The fibers are also very long, from 4 to 8 ft (1.2 to 2.4 m), and do not stiffen when wet. It is thus valued for marine cordage. The best grades are lightweight, soft and lustrous, and white. The finest fibers, called **lupis fibers,** are used in the East for weaving into cloth. The plant grows to a height of 20 to 30 ft (6.1 to 9.1 m), with huge leaves characteristic of the banana. Each successive layer of leaves toward the stalk yields fibers that are lighter in color, higher in strength, and of finer texture than those outside. There are 15 grades.

MAPLE. The wood of maple trees native to the United States and Canada which include 13 species in the United States. Of these the **sugar maple,** *Acer saccharum,* is the most plentiful and the most important. Other names for this tree and wood are **hard maple** and **rock maple.** The wood is tough and hard, close-grained, and does not splinter easily. The heartwood is light reddish brown, and the wide sapwood is white. The wood has an average specific gravity when kiln-dried of 0.67, compressive strength perpendicular to the grain of 2,170 lb/in^2 (15 MPa), and shearing strength parallel to the grain of 1,520 lb/in^2 (10 MPa). **Black maple,** *A. nigrum,* is similar to sugar maple and is marketed with it. The **broadleaved maple,** also known as **bigleaf maple** and **Oregon maple,** is *A. macrophyllum* and is the only species native to the western states. **Silver maple,** *A. saccharinum,* grows most extensively in the middle states. It is also called **soft maple, white maple, river maple,** and **swamp maple. Box elder,** *A. negundo,* grows over the northern states east of the Rocky Mountains. The **red maple** is *A. rubrum,* and the **vine maple** is *A. circinnatum.* The wood of the maples may be white or yellowish to brownish, and it is close-grained and hard. It often has a curly, twisted grain. The density is about 40 lb/ft^3 (641 kg/m^3). The wood of the soft maples is not as heavy or as strong as that of the sugar maple. Maple is used for furniture, cabinetwork, flooring, rollers, measuring rules, forming dies, shoe heels and lasts, and where a hard, fine-grained wood is needed. **Rose maple,** used in Australia for cabinetwork and paneling, is not a maple but is from the tree *Cryptocarya ethyroxylon.* The pinkish-brown wood has a wavy grain and density 45 lb/ft^3 (721 kg/m^3), is hard, and has a fragrant odor.

    Maple sugar, used in confectionery, and in sweetening agents as **maple syrup,** is the boiled-down sap of the sugar maple tree, harvested by tapping the tree in the early spring. The sugar contains the calcium salt of succinic acid. The ratio of sap to sugar is 40:1, and an average production is 2 lb (0.9 kg) per tree, or 20 qt (19 L) of sap from a 15-in (0.4-m) tree. Maple sugar is produced chiefly in Vermont, New Hampshire, New York, and Canada. **Maple flavor** is made artificially by the reaction of an alpha amino acid with a reducing saccharide.

MARAGING STEELS. A family of low-carbon high-alloy steels typically containing 12 to 18% nickel, 3 to 5 molybdenum, 0 to 12 cobalt, 0.2 to 1.6 titanium, and 0.1 to 0.3 aluminum (one cobalt-free grade also contains 5 chromium) and noted for their high strength and toughness, simple heat treatment, dimensional stability during heat treating, good machinability, and excellent weldability. The term *maraging* refers to the martensitic structure that forms during heat treatment, which is a precipitation-hardening, or aging, treatment usually at 900°F (482°C). The 18% nickel kinds, the most well known, are produced in four grades to provide tensile yield strengths of 200,000 lb/in^2 (1,379 MPa), 250,000 lb/in^2 (1,724 MPa), 300,000 lb/in^2 (2,069 MPa), or 350,000 lb/in^2 (2,413 MPa). Although the 18% nickel steels were originally developed for aerospace applications primarily, they also are now used for die-casting dies, cold-forming dies, and molds for forming plastics.

MARBLE. A compact, crystalline limestone used for ornamental building, for large slabs for electric-power panels, and for ornaments and statuary. In the broad sense, marble includes any limestone that can be polished, including **breccia, onyx,** and others. Pure limestone would naturally be white, but marble is usually streaked and variegated in many colors. **Carrara marble,** from Italy, is a famous white marble, being of delicate texture, very white, and hard. In the United States the marbles of Vermont are noted and occur in white, gray, light green, dark green, red, black, and mottled. A typical white **Vermont marble** slightly mottled with gray is 99% pure carbonates with only slight amounts of manganese and aluminum oxides and organic matter. But about 60% of American marble is quarried in Tennessee and Georgia. It is highly crystalline and is colored white, gray, bluish, or pink. The 56-ton (51-metric-ton) block in the tomb of the Unknown Soldier at Arlington, Virginia, is **yule marble** from Colorado. **Alabama marble** is a pure, low-porosity material of good statuary grade, mostly white. Much of the **Tennessee marble** is marked with stylolites of zigzag colored bands and is used for floor tile. The Victoria pink and Cumberland pink marbles of Tennessee have low porosity and high compressive strength, about 17,000 lb/in^2 (117 MPa). The marbles of southern Uruguay are famous for great variety of beautiful colors, and they occur in immense blocks.

Marble has a specific gravity of about 2.72 and a density of about 170 lb/ft^3 (2,723 kg/m^3) with compressive strengths from 8,000 to above 15,000 lb/in² (55 to 103 MPa). It ordinarily withstands heat up to 1200°F (649°C) without injury. **Translucent marble** is selected and processed marble, semitransparent to light. **Statuary marble** is always selected by experts who have had long experience in cutting. It must be of a single shade and free of hard or soft spots, iron inclusions, and other defects. **Marble chips** are irregular, small, graded pieces of marble marketed for making artificial stone. They are a byproduct of marble quarrying. **Marble flour,** or **marble dust,** is finely ground chips used as a filler or abrasive in hand soaps and for casting. **Marbelite** is an **artificial marble** used for casting statues and small ornamental articles. It is made by heating potassium alum in water, adding about 10% heavy spar, and then casting in rubber molds. Marble dust may be added. **Exsilite** is a synthetic onyx marble in slabs are large as 2 by 3 ft (0.6 to 0.9 m) and up to 3 in (7.6 cm) thick. It is made by fusion of pure silica with mineral colors incorporated at high temperatures.

MASTIC.   The gum exudation of the tree *Pistacia lentiscus,* called **Chios mastic,** and from *P. cabulica,* called **Bombay mastic,** both small evergreens native to the Mediterranean countries. It contains an ethereal oil (2%) that is mainly **pinene.** In ancient times the resin was highly valued for artists' paints and coating lacquers, adhesives, and for incense, dental cements, and as a chewing gum from which use it derives its name. Because of high cost, its use is now largely limited to fine art paints and lacquers and as an astringent in medicine.

Mastic is obtained by making an incision in the tree, a tree yielding 6 to 11 lb (2.7 to 5.0 kg) annually. There are two general grades, the purer resin adhering to the tree, and the resin collecting on the ground. It is easily soluble in turpentine but is more expensive than many other natural resins, and it is used for high-grade, quick-drying, and colorless varnish coatings on maps and watercolor paintings. The name *mastic* is also erroneously applied to asphalt when used in caulking or adhesive compounds.

MAURITIUS HEMP.   The fiber obtained from the fleshy leaves of the plant *Furcraea gigantea,* of Mauritius, Nigeria, and Ghana, used for rope and cordage. The product from west Africa is often erroneously termed *sisal.* The plant belongs to the lily family. Similar fibers are obtained from other species, notably *F. foetida,* of Brazil. The *F. gigantea* is also grown in Brazil under the name of **piteira.** Each plant yields about 40 leaves annually, from 10 to 12 ft (3.0 to 3.7 m) in length, each leaf giving 0.077 lb (35 g) of fiber. The plant *F. cabuya* produced the ancient cordage fiber of the Mayas. The term **cabuya,** which means cordage, is applied to the fibers of the several species growing through Central America, the West Indies, and northern South America to Ecuador. The fibers of *F. cabuya* of Costa Rica are up to 100 in (2.5 m) long. The leaves yield up to 3.5% of their green weight in dry fiber, which is coarser than henequen but is used for coffee-bag fabric. The fibers of the cabuya of Ecuador, *F. andina,* are not as long. They are used extensively for burlap for bagging. **Fique fiber,** of Colombia, used for rope and coffee bags, is from the leaves of *F. macrophylla.* The leaves are longer than those of henequen, and the fiber is finer and more lustrous.

MEERSCHAUM. A soft, white or gray, claylike mineral of composition $3SiO_2 \cdot 2MgO \cdot 2H_2O$, used for making pipes and cigar holders, but also employed for making various other articles, as it can be cut easily when wet and withstands heat. When fresh, the mineral absorbs grease and makes a lather; its German name means seafoam. It is used as a filler in soaps in Germany. The hardness is about Mohs 2 and the specific gravity 1.28. Most of the commercial meerschaum comes from Asia Minor; the mines at Eskisehir have been worked for 20 centuries. A little is produced in New Mexico and some in Spain. **Artificial meerschaum** is made from meerschaum shavings, kieselguhr, and from silicates of aluminum, calcium, and magnesium.

MELAMINE RESIN. A synthetic resin of the alkyd type made by reacting melamine with formaldehyde. The resin is thermosetting, colorless, odorless, and resistant to organic solvents. It is more resistant to alkalies and acids than to urea resins, has better heat and color stability, and is harder. The melamine resins have the general uses of molding plastics and also are valued for dishes for hot foods or acid juices, and they do not soften or warp when washed in hot water. **Melamine,** a trimer of cyanamide, has composition $(N:C \cdot NH_2)_3$.

It may be made by reacting urea with ammonia at elevated temperatures and pressure. It has a specific gravity of 1.56 and melting point of 669°F (354°C). Melamine alone imparts to other resins a high gloss and color retention. The melamine resins have good adhesiveness but are too hard for use alone in coatings and varnishes. They are combined with alcohol-modified urea-formaldehyde resins to give coating materials of good color, gloss, flexibility, and chemical resistance. Urea-modified melamine-formaldehyde resins are used for coatings and varnishes. Melamine-formaldehyde molding resin, with cellulose filler, has a tensile strength of 7,500 lb/in² (52 MPa) and dielectric strength 325 V/mil (12.8 × 10⁶ V/m). With a mineral filler, it has a dielectric strength of 400 V/mil (16 × 10⁶ V/m) and withstands temperatures to 300°F (149°C). Melamine-urea-formaldehyde resin with a lignin extender is used as an adhesive for water-resistant plywood. Phenol-modified melamine-formaldehyde resin solution is used for laminating fibrous materials. Highly translucent melamine-formaldehyde resin is used for molding high-gloss buttons. Methylol-melamine, made by alkylating a melamine-formaldehyde resin with methyl alcohol, is used for shrinkproofing woolen fabrics. **Resimene 812,** of Monsanto, is a colorless melamine-formaldehyde resin powder that can be dissolved in water or ethyl alcohol, for impregnating paper or fabrics or for laminating.

MENHADEN OIL. An oil obtained by steaming or boiling the fish *Brevoortia tyrannus,* caught along the Atlantic coast of the United States. It was first called **porgy oil,** the Maine name for the fish. Other names for the fish are **whitefish, fatback,** and **mossbunker.** The fish, when fully grown, are 12 to 15 in (30 to 38 cm) long, weighing about 1 lb (0.45 kg). They yield up to 15% oil, although fish from warm southern waters yield less oil. In May the fish migrate north to the New England coast, and they return south to below the Carolinas in November. An annual catch of 1.5 billion fish yields 10.2×10^6 gal (38.6×10^6 L) of oil and 103,000 tons (93,421 metric tons) of meal. Menhaden is not a desirable food fish because of its oily nature. The oil contains 27% oleic acid, 20 arachidonic, 16 clupanodonic, 17 palmitoleic, 7 myristic, and 1 stearic acid. It has an iodine number of 140 to 180 and a specific gravity of 0.927 to 0.933. The inedible oil is used for dressing leather, mixing in cutting oils, and making paint oils. It is also hydroxylated with acetic acid and used for making polyisocyanate and alkyd resins. Menhaden oil polymerizes easily, and the drying power is good, but it does not give an elastic film as do the vegetable oils. Its strong odor is due to the clupanodonic acid ester. The residue **fish meal** is sold for poultry feed and fertilizer. The meal is not as rich in vitamins A and D as that from some other fish, but as much as 15% can be used in poultry feed without producing a fishy taste in the eggs. Most edible oil is hydrogenated and blended in margarines and shortenings.

MERCURY. Also called **quicksilver.** A metallic element, symbol Hg. It is the only metal that is a liquid at room temperature. Mercury has a silvery-white color and a high luster. Its specific gravity is 13.596. The solidifying point is $-40°F$ ($-40°C$), and its boiling point at 1 atm is 674°F (357°C). It does not oxidize at ordinary temperatures, but when heated to near its boiling point, it absorbs oxygen and is converted to a red crystalline powder, **mercuric oxide,** HgO, used as a pigment in marine paints. Mercury is derived chiefly from the mineral cinnabar. Spain, Italy, Russia, Mexico, and the western United States are the chief producers. The metal is marketed in steel flasks holding 75 lb (34 kg). European flasks hold 76 lb (34.4 kg). It is used for separating gold and silver from their ores, for coating mirrors, as an expansive metal in thermometers, in mercury-vapor lamps, in tanning, in batteries, for the frozen-mercury molding process, in mercury-vapor motors, as a circulating medium in atomic reactors, in amalgams, and in its compounds for fungicides, pharmaceuticals, paint pigments, and explosives. The black **mercurous oxide,** Hg_2O, is used in skin ointments. **Mercuric chloride,** or **corrosive sublimate,** $HgCl_2$, is an extremely poisonous, white, crystalline powder soluble in water and in alcohol, used as a wood preservative, as an insecticide and rat poison, in tanning, as a mordant, and as a caustic antiseptic in medicine. **Vermilion red,** one of the oldest paint pigments, is red **mercury sulfide,** HgS, made directly by heating mercury and sulfur. It is a brilliant, water-insoluble, red powder of specific gravity 8.1. Because of its expense, it is often mixed with other red pigments. **Mercurochrome,** $C_{20}H_8O_6Na_2Br_2Hg$, is a green crystalline powder which gives a deep-red solution in water and is used as an antiseptic. Mercury forms a vast number of compounds, all of which are poisonous and some of which are explosive. **Mercury 203** is radioactive. A mercury distillation process, developed by Lumalampan AB of Switzerland and called the *MRT System,* recovers mercury from fluorescent lightbulbs, button batteries, amalgams, and electrical devices.

METALLIC MATERIALS. About three-quarters of the elements available can be classified as **metals,** and about half of these are of at least some industrial or commercial importance. Although the word *metal,* by strict definition, is limited to the pure metal elements, common usage gives it wider scope to include **metal alloys.** While pure metallic elements have a broad range of properties, they are quite limited in commercial use. Metal alloys, which are combinations of two or more elements, are far more versatile and for this reason are the form in which most metals are used by industry.

Metallic materials are crystalline solids. Individual crystals are composed of unit cells repeated in a regular pattern to form a three-dimensional crystal-lattice structure. A piece of metal is an aggregate of many thousands of interlocking crystals (grains) immersed in a cloud of negative-valence electrons detached from the crystals' atoms. These loose electrons serve to hold the crystal structures together because of their electrostatic attraction to the positively charged metal atoms (ions). The bonding forces, being large because of the close-packed nature of metallic crystal structures, account for the generally good mechanical properties of metals. Also, the electron cloud makes most metals good conductors of heat and electricity.

Metals are often identified by the method used to produce the forms in which they are used. When a metal has been formed or shaped in the solid, plastic state, it is referred to as a **wrought metal.** Metal shapes that have been produced by pouring liquid metal into a mold are referred to as **cast metals.**

There are two families of metallic materials—ferrous and nonferrous. The basic ingredient of all **ferrous metals** is the element iron. These metals range from cast irons and carbon steels, with over 90% iron, to specialty iron alloys, containing a variety of other elements that add up to nearly half the total composition.

Except for commercially pure iron, all ferrous materials, both irons and steels, are considered to be primarily **iron-carbon alloy** systems. Although the carbon content is small (less than 1% in steel and not more than 4 in cast irons) and often less than other alloying elements, it nevertheless is the predominant factor in the development and control of most mechanical properties.

By definition, metallic materials that do not have iron as their major ingredient are considered to be **nonferrous metals.** There are roughly a dozen nonferrous metals in relatively wide industrial use. At the top of the list is aluminum, which next to steel is the most widely used structural metal today. It and magnesium, titanium, and beryllium are often characterized as **light metals** because their density is considerably below that of steel.

Copper alloys are the second nonferrous material in terms of consumption. There are two major groups of copper alloys: **brass,** which is basically a binary-alloy system of copper and zinc, and **bronze,** which was originally a copper-tin alloy system. Today, the bronzes include other copper-alloy systems.

Zinc, tin, and lead, with melting points below 800°F (427°C), are often classified as **low-melting-point alloys.** Zinc, whose major structural use is in die castings, ranks third to aluminum and copper in total consumption. Lead and tin are rather limited to applications where their low melting points and other special properties are required. Other low-melting-point metals are bismuth, antimony, cadmium, and indium. A tin-bismuth alloy, which melts below 302°F (150°C), is used as the core in molding nylon air-intake manifolds for cars. After molding, the alloy is dissolved in a melt-out tank and pumped out for reuse.

Another broad group of nonferrous alloys is referred to as **refractory metals.** Such metals as tungsten, molybdenum, and chromium, with melting points above 3000°F (1649°C), are used in products that must resist unusually high temperatures. Although nickel and cobalt have melting points below 3000°F (1649°C), they serve as the base metal or as alloying elements of many heat-resistant alloys.

Finally, the **precious metals,** or **noble metals,** have the common characteristic of high cost. In addition, they generally have high corrosion resistance, many useful physical properties, and generally high density.

METALLIC SOAP. A term used to designate compounds of the fatty acids of vegetable and animal oils with metals other than sodium or potassium. They are not definite chemical compounds like the alkali soaps, but may contain complex mixtures of free fatty acid, combined fatty acid, and free metallic oxides or hydroxides. The name distinguishes the water-insoluble soaps from the soluble soaps made with potash or soda. Metallic soaps are made by heating a fatty acid in the presence of a metallic oxide or carbonate, and they are used in lacquers, leather and textiles, paints, inks, ceramics, and grease. They have the properties of being driers, thickening agents, and flattening agents. They are characterized by ability to gel in solvents and oils, and by their catalytic action in speeding the oxidation of vegetable oils.

When made with fatty acids having high iodine values, the metallic soaps are liquid, such as the oleates and linoleates, but the resinates and tungates are unstable powders. The stearates are fine, very stable powders. The fatty acid determines the physical properties, but the metal determines the chemical properties. Aluminum stearate is the most widely used metallic soap for colloid products. **Aluminum soaps** are used in polishing compounds, in printing inks and paints, for waterproofing textiles, and for thickening lubricating oils. The resinates, linoleates, and naphthanates are used as driers, the lead, cobalt, and manganese being the most common. **Calcium rosinate** in an insoluble soap of composition $Ca(C_{44}H_{62}O_4)_2$ produced by boiling rosin with lime water, and filtering. It is also called **calcium abiotate** and is a yellow powder of high molecular weight, 1349.06, used as a paint drier and for waterproofing. **Calcium linoleate** is a white, amorphous powder of composition $Ca(C_{18}H_{31}O_2)_2$ used in paints and in waterproofing. It is insoluble in water but soluble in alcohol. **Cobalt linoleate** is used as a drier in printing inks. **Calcium stearate,** $Ca(C_{18}H_{35}O_2)_2$, is a white, fluffy powder of 250 mesh. It is used as a flatting agent in paints, for waterproofing cements and stucco, as a lubricant for rubber and plastic molds, as a softening agent in lead pencils, and in drawing compounds for steelwire drawing. The calcium stearate of Mallinckrodt Chemical Works, used in food and pharmaceutical emulsions, is an air-float powder of 325 mesh.

Barium stearate, $Ba(C_{18}H_{35}O_2)_2$, is a waxy, nontacky, white powder with molecular weight 703 and melting point 284 to 302°F (140 to 150°C). It is used as a dry lubricant for molding plastics, greaseless bearings, wax compounding, and wire drawing. **Chemactant PFC-5,**

of Chemactant, Inc., is **barium lanolate,** a soft, waxy soap, 25% barium and 75 lanolin acids, used as an additive for paints and coatings to improve adhesion and pigment dispersion. **Strontium stearate,** $Sr(C_{18}H_{35}O_2)_3$, is a white powder with molecular weight 654 and melting point 266 to 284°F (130 to 140°C). It is partly soluble and gels in benzol, mineral spirits, and hydrocarbons. It is used in grease and wax compounding and in crimson flares and signals. **Chromium stearate** is a dark-green powder of composition $Cr(C_{18}H_{35}O_2)_3$ which melts at 203 to 212°F (95 to 100°C). It is used in ceramics and plastics, in plastic waxes and greases, and as a catalyst. **Manganese stearate** is a pink powder of composition $Mn(C_{18}H_{35}O_2)_2$, with a melting point 212 to 230°F (100 to 110°C). It is used in wax and grease compounding. **Cerium stearate,** $Ce(C_{18}H_{35}O_2)_2$, is a very inert, waxy, white powder melting at 212 to 230°F, used as a catalyst and in waterproofing compounds. **Nickel stearate** is a green powder, $Ni(C_{18}H_{35}O_2)_2$, with a melting point 302 to 320°F (150 to 160°C), soluble in aromatic hydrocarbons, and forming gels with petroleum oils. It is used in lubricants, waterproofing compounds, leather finishes, and as a flux in nickel welding.

Dibasic **lead stearate,** $2PbOPb(C_{18}H_{35}O_2)_2$, is a soft, white powder of specific gravity 2.02, insoluble in water or mineral spirits. It is used in greases, in cutting oils, and as a heat and light stabilizer in vinyl plastics. **Lithium stearate** is a white, odorless powder melting at 423°F (217°C). It is used in machinery greases and as an oil-soluble emulsifying and dispersing agent in cosmetics. The **lithium soap greases** are very adhesive to the bearings and are heat-resistant. The **Stanolith greases** of Standard Oil Co. of Indiana are lithium soap greases containing an oxidation inhibitor. **Uni-Temp grease,** of Texaco, is a lithium soap grease made with a synthetic hydrocarbon instead of an oil. It has uniform lubricating qualities between -148 and 572°F (-100 and 300°C).

Cobalt 2-ethylhexanoate, sometimes called **cobalt 2-ethylhexoate** or **cobalt octoate,** is mainly used as a primary or auxiliary paint drier. **Zirconium, calcium, lead, manganese,** and **zinc octoates** are also available commercially. **Aluminum octoate** is used to gel paints. **Barium** and **cadmium octoate** are components of complex heat stabilizers for polyvinyl chloride. The **stannous salts** and **dibutyltin salts** are catalysts in polyurethane foam production. **Lead naphthenate** is the predominant commercial compound from this family, being used as a paint drier. **Cobalt naphthenate** is a lubricating oil additive, a gelling agent, and a catalyst. For waterproofing, pigment suspending, and rubber and plastics compounding, **aluminum palmitate** and **zinc palmitate** formulations are used occasionally. Tallate salts are used almost exclusively as paint driers.

Lead, cobalt, and **manganese tallates** are the most common ones in use.

METHANE. Also known as **marsh gas,** in coal mines as **firedamp,** and chemically as **methyl hydride.** A colorless, odorless gas, CH_4, employed for carbonizing steel, in the manufacture of formaldehyde, and as a starting point for many chemical compounds. The molecule has no free electrons and is the only stable **carbon hydride,** though it reacts easily on the No. 1 and No. 2 electrons of the carbon to form the hexagonal molecule called the **benzene ring.** It may thus be considered the simplest of the vast group of **hydrocarbons** derived from petroleum, coal, and natural gas. Methane occurs naturally from the decomposition of plant and animal life and is one of the chief constituents of illuminating gas. It is made synthetically by the direct union of carbon or carbon monoxide with hydrogen. It is also produced by the action of water on **aluminum carbide,** a gray, massive substance of composition Al_4C_3. Methane has a specific gravity of 0.560, and since it is much lighter than air, it is easily diffused in it. In air the gas is highly explosive, although the gas alone is not explosive. It liquefies at $-258.6°F$ $(-161°C)$. **Pintsch gas,** once used widely for car lighting, contained up to 60% methane and was made by spraying petroleum oil into a hot retort. This type of gas, under the name of **oil gas,** or **carbohydrogen,** and containing as high as 85% hydrogen, gives a low-temperature flame used for flame-cutting torches. **Nitromethane** is nitrated methane. It is a liquid explosive more powerful than TNT. Methane is also the chief constituent of **natural gas** from oil fields. Natural gas contains usually at least 75% methane, although some Pennsylvania gas contains 98.8%, and some gas from Kentucky as low as 23%. A typical pipeline gas containing gas from several fields and freed of carbon dioxide, hydrogen sulfide, and water vapor has 78% methane, 13 ethane, 6 propane, 1.7 butane, 0.6 pentane, and some higher gases. Natural gas from the Po River in Italy has been used since 1935 to fuel Italian cars, buses, and trucks. Since 1990, to reduce air pollution, U.S. automakers have built cars, vans, and trucks fueled with natural gas for urban-area fleets. **Reformed gas,** used for copper refining, contains 86% methane and is free of H_2S and the higher homologs of CH_4. Natural gas has an energy value of 1,035 Btu/ft^3 (26,023 J/m^3), almost double that of manufactured gas from coal, but **synthetic gas,** or oil gas, made from crude oil, can be had with an energy value equivalent to that of natural gas. **Sour gas** is natural gas with more than 1 gr of H_2S per 100 ft^3 (2.8 m^3). This hydrogen sulfide is removed to eliminate the odor before it is piped. The propane, butane, and heavier hydrocarbons may also be removed for the production of chemicals. Much of the

American production of natural gas is in Texas and Louisiana, but there are large reserves in California and other areas and throughout the Canadian plains area. A new source of methane is a result of environmental cleanups. Methane is generated as a by-product during the biodegradation of polluted industrial waters in waste treatment plants. It is also obtained by tapping the gases produced as biomass matter degrades in municipal landfills.

MICA. Known originally as **Muscovy glass.** A group of **silicate** minerals with monoclinic crystals which break off easily into thin, tough scale, varying from colorless to black. **Muscovite** is the common variety of mica and is called **potash mica,** or **potash silicate,** $H_2KAl_3(SiO_4)_3$. It has superior dielectric properties and is valued for radio capacitors. **Phlogopite** is **magnesium mica,** $H_2KMgSi(SiO_4)_3$, and is distinguished from muscovite by its decomposition in sulfuric acid. It is also called **amber mica.** It is superior to muscovite in heat resistance, but is softer and has a brownish-yellow color. It comes from India, Canada, Malagasy, and Tanzania, while the chief producers of muscovite are India, Brazil, and Argentina. The peculiar crystal structure of phlogopite, and the almost infinite number of chemical combinations in which it can be produced, has made it an attractive mica for synthetic production which is generally replacing the natural product. The crystal structure is a repetition of four-layer units, and the layers are in conic thicknesses of indefinite extent. It consists of extremely thin sheets of strongly bonded silica tetrahedra with a weak ionic bond joining the sheets. **Dicor-MGC,** or **Dentsply,** is a two-phase micaceous glass used for dental restorations.

The **chromium mica** known as **fuchsite** comes only in small, emerald-green flakes. The rare greenish **vanadium mica** known as **roscoelite** is usually in fine scales, and where there are considerable amounts, as in the sandstones of Colorado, Utah, and Arizona, it is most valuable as a source of vanadium, as it contains 1.5 to 3.5% vanadium. **Margarite** is a yellow or purple **calcium mica,** $CaO(SiO_2)_2 \cdot (Al_2O_3) \cdot H_2O$, and is a transition product. The chief uses of mica are as an electrical insulator, a heat insulator, and a filler in plastics and insulating materials.

The value of sheet mica depends greatly upon the absence of staining, especially from iron inclusions which decrease the electrical insulating value. Most stains are black from the magnetite or other iron oxide. Reddish stains are usually red iron oxide. The brown micas containing much iron are valueless as electrical insulators. While selected high-grade mica is an excellent insulating material, natural mica is, in general, unsuited for economic high-production use because of the difficulty of handling the small irregular pieces and the great wastage of time and material in the selection of uncontaminated pieces. India has been the largest producer of the highest grades of phlogopite, and about 80% of this production comes from Bihar, but less than 1% of Bihar trimmed block mica is suitable for such uses as condenser film. Argentine mica runs 80% stained, 15 semiclear, and only 5 clear.

The specific gravity of mica is 2.7 to 3.1, and the Mohs hardness is 2 to 3. Dielectric strength is not an accurate measurement of the electrical quality of mica, and it is measured by the **Q value,** or reciprocal of the **power factor,** the power factor being the measure of the loss of electric energy in a capacitor in which mica is the dielectric. High-quality capacitor mica should have a minimum Q value of 2,500, or a power factor of 0.04% at a frequency of 1 mHz. **Ruby mica** is the finest grade of **Indian mica** for electric capacitor use. **Madras mica** is greenish and is not high-quality electric mica. Indian mica is graded as No. 5, first-quality and second-quality bookform; No. 6, loose with powder; extra loose, first-quality and second-quality; special loose, second-quality; and No. 6A loose, third-quality. **Argentine mica** has eight size grades from No. 6, in sheets of 1 to 3 in^2 (6.5 to 19.3 cm^2), to AA, or Special, in sheets of 48 in^2 (310 cm^2) or more.

Mica is marketed as cut or uncut block, sheet, splittings, and ground. **Block mica** is a deceptive term, since the pieces are not blocks. While Indian mica has been mined sometimes in blocks as large as 15 ft (4.6 m), and in sheets as large as 24 by 30 in (61 by 76 cm), the commercial block mica is usually no more than 0.030 in (0.076 cm) thick, and U.S. importers designate block mica as pieces not less than 0.007 in (0.018 cm) thick with a minimum usable area of 1 in^2 (6.5 cm^2). **Mica splittings** have a maximum thickness of 0.0012 in (0.003 cm) and a minimum usable area of 0.75 in (1.9 cm). About 85% of the mica imported into the United States is in thin splittings. **Film mica** is split from the best qualities of block mica to thicknesses from 0.0012 to 0.004 in (0.003 to 0.010 cm). Importers recognize 11 quality grades of block mica from the densely stained to the clear, all by visual inspection. Splittings, which are usually only 1 or 2 in (2.5 or 5.1 cm) in diameter, can be cemented together for use, but it is a costly operation. The original **micanite** made in India consisted of small pieces cemented together with shellac. **Built-up mica** is made by bonding the pieces with a synthetic resin and compressing at high temperature to give uniform thickness. Small pieces and the scrap from manufacture are made into mica powder for use as a filler in plastics and paints, or the powder is chemically cleaned and magnetically separated for making reconstituted sheet.

Because of the difficulty and cost of handling, natural mica sheet has now become almost obsolete for most uses. Sheets and strips made from mica flake can be handled in automatic machines, and are of a uniform quality not found in natural mica. **Reconstituted mica** without a binder is made by sheeting mica flakes by papermaking methods and submitting the sheet to high pressure and temperature to unite the flakes by recrystallization. Natural mica powder processed to give high purity may be used, but synthetic mica flake is preferred because of the ability to select a composition to suit the requirements for dielectric strength and heat resistance. In general, the absence of hydroxyl ions in synthetic mica gives higher heat resistance.

Mica sheet with a ceramic binder has been called **ceramoplastic insulation. Synthamica 202** is made with a heat-resistant grade of synthetic mica flake with a glass bond. It comes in continuous strips 3 in (7.6 cm) wide in thicknesses from 0.002 to 0.007 in (0.005 to 0.018 cm). The dielectric strength is 1,000 V/mil (39×10^6 V/m), operating temperature to 1800°F (982°C), and tensile strength 10,000 lb/in^2 (69 MPa). **Mykroy sheet** is glass-bonded mica in heavier sheets for panels and structural parts of electronic equipment. **Mica paper** and **mica mat** are usually made with an organic binder to give flexibility. One of the earliest mica papers, called **Watsonite,** was ground natural mica, dehydrated by heating, and sheeted with a resin binder. Many of the mica papers have superior dielectric strength, but the heat resistance is limited to that of the binder. **Crystal M,** of 3M, is a thin, flexible mica paper with a melting point of about 1900°F (1038°C). It is used for fire-resistant thermal insulation. Mica paper made with about 90% mica flake and 10 epoxy resin has a dielectric strength of 1,300 V/mil (51×10^6 V/m), power factor of 0.012, and flexural strength of 40,000 lb/in^2 (276 MPa). **Isomica** is mica paper with an epoxy binder. **Transformer-grade mica,** for Class H insulation, may be made with a silicone binder. **Flexi-Mica,** of Spruce Pine Mica Co., is mica bonded with a silicone resin. The 0.002-in (0.006-cm) sheet has a dielectric strength of 800 V/mil (31×10^6 V/m) and a tensile strength of 10,000 lb/in^2 (69 MPa). **Pyro-Putty,** of Aremco Products Inc., is stainless steel powder in an aqueous silicate ceramic binder. It has a thermal expansion coefficient similar to that of iron and steel, withstands continuous service temperatures up to 2000°F (1093°C), and is used to repair exhaust systems, other high-temperature parts, and castings.

The first **synthetic mica** as made by Siemens & Halske was produced by melting a mixture of 11.6% aluminum oxide, 32.6 magnesium oxide, 30.7 kieselguhr, and 25.1 K_2SiF_6. The synthetic mica developed by the Bureau of Mines is a **fluorine-phlogopite mica** produced by calcinating a mixture of quartz, bauxite, and magnesite to drive off the carbon dioxide, adding potassium fluorosilicate, and melting at 2552°F (1400°C). As the furnace cools, the mica crystals grow from a seed at the bottom of the crucible. The mica has composition $K_4Mg_{12}Al_3O_{40}F_8$ and is similar to a natural mica in which the hydroxyl radical has been replaced by fluorine. **Fluorine mica** has superior heat resistance and dielectric strength, but it is harder than and not as flexible as natural phlogopite mica. The number of combinations that can be produced in the fluorine type of phlogopite mica alone is very large. Potassium can be replaced by sodium, rubidium, calcium, cesium, strontium, barium, thallium, or lead. Magnesium can be replaced by iron, cobalt, nickel, manganese, titanium, copper, or zinc. Aluminum can be replaced by iron, manganese, vanadium, boron, or beryllium. **Boron phlogopite** as made synthetically is soft and flexible, with a melting point of 2102°F (1150°C). Almost any desired combination of characteristics of heat resistance, electrical resistivity, and chemical resistance can be obtained within wide ranges in synthetic micas by varying the composition. However, sheet mica made synthetically embraces the same costly procedures of splitting and handworking of the relatively small sheets as for natural mica, so that synthetic mica is normally made as flake for reconstituting into sheet and strip of uniform purity and thickness.

Plastics are often heavily filled with mica powder and marketed under trade names such as **Micabond** and **Lamicoid** in the form of sheets, tubes, and molded parts; but they are distinct from the **mica ceramics** molded with an inorganic binder and having generally higher physical properties. One of the first of these, called **Mycalex,** produced by Mycalex Corp. and General Electric Co., was composed of the ground mica and lead borate heated together to the softening point of the borate, 1247°F (675°C), and compressed while plastic. Part of the mica combines to form a lead borosilicate. Such a molding has good strength, water resistance, resistance to arcing, and a low coefficient of expansion. **Supramica 620BB** has a lead borate binder. The molded parts have a specific gravity of 3.8, a flexural strength of 12,000 lb/in² (83 MPa), Rockwell M hardness of 110, and operating temperatures to 1200°F (649°C). Mica ceramics are made in various grades. Molded parts made with **boron mica** have a power factor below 0.07, and those of **barium mica** have power factors as low as 0.03. The traverse strengths of the mica ceramics are as high as 10,000 lb/in² (69 MPa) at 752°F (400°C).

The first wartime German **mica substitute,** called **Glushartgewebe,** was made by impregnating a very fine, high-alkali glass-fiber fabric with an alcohol solution of osmose kaolin and a synthetic resin and compressing at high temperature. General Electric's **Terratex** was made from bentonite and asbestos fiber impregnated with ethyl silicate. Other substitute micas made from bentonite were **Alsifilm** and **Diaplex,** in thin, transparent hard sheets. **Tissuglas,** of

Pallflex Products Co., is made of extremely fine glass fibers matted on a papermaking machine into sheets 0.0006 to 0.012 in (0.0015 to 0.0304 cm) thick, in continuous rolls up to 38 in (97 m) wide. It withstands temperatures to 1200°F (649°C) and has higher dielectric strength than natural mica. **Fiberfilm,** of the same company, for such uses as capacitors and transformers, comes in thicknesses from 0.0008 to 0.0017 in (0.0020 to 0.0043 cm) and has a dielectric strength of 4,000 V/mil (158×10^6 V/m) and operating temperatures to 482°F (250°C). It is made with fine glass fiber bonded with tetrafluoroethylene resin. It is stiffer and stronger than ordinary plastic film. **Glass paper** developed by the Naval Research Laboratory is made from borosilicate glass flake bonded with alkyd, phenolic, or silicone resin. Where temperatures are not high, various plastic films are used for slot insulation. **Anilite** is a phenolic impregnate in sheets as thin as 0.004 in (0.010 cm). **Kynar,** of Atochem North America Inc., is vinylidene fluoride in thin sheet. The dielectric strength is 1,280 V/mil (50.4×10^6 V/m) and heat distortion temperature 340°F (171°C).

MICA POWDER. Mica in very small flakes used as a filler in plastics, in paints, in roofing and asphalt shingles, and for making glass-bonded mica. When produced by grinding the small scrap pieces from mica workings, it is known as **ground mica.** The ground mica from Canada is **phlogopite.** It comes in 20, 60, 120, and 150 mesh. Mica in paints helps to bond the film and prevent cracking, acting similarly to aluminum leaf. Mica powder for plastic filler and paints is produced directly from mica schists by froth flotation. The powder recovered averages more than 50% of sizes smaller than 200 mesh and 20% between 150 and 200 mesh. The recovered mica equals in every way the powder made by grinding mica scrap. **Sericite mica,** or **damourite,** from North Carolina kaolin workings, and in pockets in the fireclay deposits of Adelaide Co., Australia, is a type of potash mica related to muscovite but softer. It occurs as a finely divided powder with a talclike feel. Mixed with aluminum powder, it produces a finish paint superior to aluminum alone. It has some activating properties and is a substitute for zinc oxide in rubber. It is also used as a filler in plastics and replaces graphite as a foundry core and mold wash. The recovery from North Carolina clay is 10% of the gross weight of the clay. **Water-ground mica** is ground to a fineness of 90% through a 325-mesh screen. It is for paint and rubber use. The mica powder **Micalith G** is sericite mica from Pennsylvania washings with 0.5% graphite embedded in the mica crystals. The graphite improves the wetting and dispersion in paints. **Micronized mica** is a powder of a fineness of 400 to 1,000 mesh, used as a filler. **Mica flake,** used in the manufacture of shingles and roofing, is washed from pegmatite deposits, but the mica flake used for molding into mica ceramic electric insulators is ground phlogopite scrap, or from various compositions of synthetic mica.

MICROSPHERES. Spherical particles used in plastics and other materials as fillers and reinforcing agents. They are also used to encapsulate various materials. Those used as fillers or reinforcements are made of glass or ceramics or resins. There are two different kinds of **glass microspheres**—solid and hollow. Solid spheres, made of soda-lime glass, range in size from 157 to 197,000 μin (4 to 5,000 μm) in diameter and have a specific gravity of about 2.5. Hollow glass microspheres have densities ranging from 5 to 50 lb/ft^3 (80 to 801 kg/m^3) and diameters from 790 to 7,900 μin (20 to 200 μm). The strength of standard hollow glass spheres in terms of hydrostatic pressure required to reduce volume of the spheres by 10% is about 220 lb/in^2 (1.5 MPa). Stronger microspheres, of 3M, known as **B40BX spheres,** withstand a hydrostatic pressure of 2,200 lb/in^2 (15 MPa) for a 10% volume loss. The spheres in plastics improve tensile, flexural, and compressive strengths and lower elongation in water absorption. They also serve as thermal and sound insulators. Hollow, glass, **microballoon microspheres,** of Emerson & Cuming, can be as small as 200 μin (5 μm). They can be tempered to withstand 10,000 lb/in^2 (69 MPa), made almost alkali-free, and rendered heat-resistant up to 1650°F (899°C). **Plastic microspheres** are used mostly in the production of syntactic foams. **Polyvinylidene chloride microspheres** are excellent resin extenders and are used in sandwich construction of boat hulls. **Epoxy microspheres** are used as low-density bulk fillers for plastics and ceramics, and were developed for use in submerged deep-water floats. They can withstand hydrostatic pressures of 10,000 lb/in^2 (69 MPa). **Phenolic microspheres** filled with nitrogen are used for production of polyester foams and syntactic epoxy foams. **Polystyrene microspheres** are also used to produce syntactic foams.

In the pharmaceutical industry, microspheres are used to encapsulate active ingredients. Multilayered **liposome microspheres** are typically made of fatty acids or lipid materials. Liposome formation takes advantage of the amphipathic nature of fat molecules, which allows fats to form liposomes spontaneously in turbulent water. Water-soluble active materials are trapped in the fluid while fat-soluble materials are taken up in the liposome cell walls. For injectable drugs, nontoxic and biocompatible phospholipids, such as lecithin, cephalin, and phosphatidyl inositol, are used. For less demanding applications, liposomes can be made from fatty acids, alcohol derivatives, and amines. **Novasome microspheres,** of MicroVesicular Systems of IGI Inc., are used to encapsulate materials as diverse as

water-dispersible pigments and vaccine components. Using edible
waxy fats, M-Cap Inc. makes microspheres to encapsulate food ingredients, such as ascorbic and citric acids, flavors, salts, and nutrient
choline chloride. Monsanto offers a version of its herbicide **Alachlor**
encapsulated in polyurea shells. **Microsponges,** made from cross-
linked polymers by Advanced Polymer Systems, are used as time-
release delivery systems of process ingredients.

MILK. The secretion of the mammary glands of mammals.
Commercial milk in the United States is almost entirely from the cow.
Besides its use as a food for direct consumption and in bakery products, it is employed for making cheese, butter, casein, ice cream, lactic
acid, and lactose chemicals. The production of cow's milk in the
United States from registered dairy animals averages about 6,100 lb
(2,767 kg) per cow, but the output from cows is 8,000 lb (3,629 kg) per
year per animal. The composition of milk is 87.34% water, 3.75 butterfat, 3 caseinogen, 0.4 lactalbumin, and 0.75 mineral salts. It also
contains vitamins A, B, and C. The vitamin A content is low during
the winter months when the animals do not feed on green grasses.
The ultraviolet rays of sunlight may destroy the fat-digesting enzyme
and vitamins B and C in milk unless milk is protected in opaque bottles or in cartons.

Much of the U.S. production of milk is used in foodstuffs manufactured in the form of **dried milk,** or **milk powder,** which is made
largely from **skim milk,** that is, milk from which the fatty **cream**
has been separated. One gallon (3.8 L) of milk yields 1.2 lb (0.54 kg)
of dried milk. **Powdered milk** produced by spray drying in an
airstream has 90% of its particles larger than 2,953 μin (75 μm) and
will disperse more easily than a powder of fine particles. Milk powders processed above 180°F (82°C) may have a cooked taste, but powders with no off-taste are produced by evaporation at 95°F (35°C) and
a quick dry at 160°F (71°C). **Milk proteins** may be separated from
milk and used for enriching foodstuffs. **Crest 6S** is a milk protein
powder containing 60% soluble protein and 26 lactose, with calcium
and phosphorus. It also improves the whiteness and water-holding
ability of milk powders. **Condensed milk,** formerly used widely
where refrigeration was not available, is now used in confectionery to
replace cream. It is very sweet and contains 8.5% milk fats, 19.5 milk
protein solids, 42 sugar, and 30 water.

While milk has many useful applications in cookery and commercial baking, it is essentially a natural baby food to build up the original low-calcium soft bones and to provide calcium for bones and teeth.
Too much milk in the diet of adults may give an excess of calcium in
the arteries and also cause cirrhosis or fatty growth of the liver. It

may also cause an excess of lactic acid. **Savortex** milk powder, for use in sausage and other comminuted meat products, has sodium substituted for a part of the calcium by ion-exchange methods. It is less greasy and gives a smoother texture to meats.

Pasteurized milk is milk that has been heated to kill disease organisms. Sterilization at 285°F (140°C) for a few seconds leaves better flavor and color than heating at low temperature for long periods. **Homogenized milk** is milk that has been treated by sonic vibration to break up the fat globules and distribute them evenly in the liquid. In natural milk the fat and proteins are in colloidal solution. In most countries the handling of milk is regulated by laws, as it is easily contaminated.

Cow's milk for direct consumption and for industrial use is produced on a large scale in the United States, Canada, Europe, Argentina, New Zealand, and Australia. **Kumiss,** sweetened cow's milk fermented with yeast, is used as a nutrient. **Soy milk** is a water solution of soybean solubles combined with sugar, calcium phosphate, irradiated sesame oil, and vitamins. It has the approximate composition of cow's milk and is chemically almost indistinguishable from it. It is used in foodstuffs, but is not permitted to be marketed under the name of milk. **Sesa-Lac** is a **milk substitute** used by food processors in bakery products, creamed desserts, and beverages. It is made by crushing hulled sesame seed, homogenizing, and spray-drying into a cream-colored powder. It has good nutritional value and a pleasant, sweet flavor. Other **imitation milks,** known as **nondairy creamers,** are made from sodium caseinate, coconut oil, or other vegetable oils, and several other ingredients.

Lactic acid derives its name from the fact that it was originally obtained from milk. It has a wide industrial use and is now produced synthetically from sulfite pulp liquor, or by fermentation of carbohydrates. **Lactic acid** is a liquid of composition $CH_3CHOHCOOH$, but a carbon may attach to four different groups, and the acid can thus exist in four forms. The dextrorotary isomer which occurs in milk is optically active. This form, called **sarcolactic acid,** is also produced naturally in the human muscles and joints through exercise to give the normal feeling of tiredness and induce sleep, but large amounts of lactic acid taken into the body are injurious. The dextrorotary acid made by the fermentation of sucrose is used in beverages and foodstuffs to minimize undesirable odors, and in animal feeds to improve the efficiency of protein utilization. **Polylactic resins,** made by the heat reaction of lactic acid with castor oil or other fatty oils, are soft, elastic resins used to produce tough, water-resistant coatings. Many useful chemicals are made from lactic acid. **Lactonitrile,** $HO(CH_3)CHCN$, is a colorless liquid of specific gravity 0.9834, which has the reactions of an alcohol and those of a nitrile.

MILLET. The very small seeds of a number of grasses. It is one of the most ancient of food grains and is an important grain in Asia, being used as a food by one-third of the population of the world. Nearly 40×10^6 acres (162×10^9 m^2) is cultivated to this grain in India alone. The plant is drought-resistant, but will not withstand frost. The seeds are high in phosphate, protein, minerals, and oil. In the United States, millet seed is used as a birdseed, and the plant is grown for pasturage and silage. **German millet,** also known as **foxtail millet** and **Hungarian grass,** is from the grass *Setaria italica.* The seed contains phosphates, many minerals, and vitamins A and E. It is an important food in Europe and Asia, but in North America it is a forage crop. **White millet,** or **proso millet,** from the grass *Pancium miliaceum,* is one of the richest grains in food value, but is employed in the United States only as a birdseed. It is much used in Russia as a food. The millet from the plant *Sorghum vulgare* is a staple food in India. **Sanwa millet,** from *Echinochloa frumentacea,* is an important food in Japan. The plant will produce as many as eight forage crops per year.

MILLSTONE. Any stone employed for grinding paint, cement, grain, or minerals. Millstones are made from sandstone, basalt, granite, or quartz conglomerate. **Burrstone** is a millstone made from chalcedony silica of cellular texture, usually yellowish. The stone is used as a building stone. **Esopus stone** is a conglomerate of this type from Ulster County, New York. The most noted burrstones for grinding grains are the European stones cut in wedge-shaped blocks 6 in (15.2 cm) long, 3.5 to 4 in (8.9 to 10.2 cm) wide, and 2 to 2.5 in (5.1 to 6.4 cm) thick, bound together with iron hoops. The French stones are chalcedony quartz, creamy white, with a cellular structure, the cavities formerly occupied by fossil shells. German burrstones are basaltic lava. Millstones vary greatly in sharpness of grain and size of grain, and thus synthetic stones of even texture are often preferred. **Chaser stones** are very large stones run on edge in mills for grinding minerals.

Pulpstones are blocks of sandstone cut into wheels and used for grinding, chiefly for the grinding of wood pulp in paper manufacture. The U.S. pulpstones are produced in Ohio and West Virginia. The sandstones for pulpstones must be uniform in texture, have sharp grains, have medium hardness, and be composed of even quartz grains of which 85% will be retained on a 150-mesh screen, and 90% on a 200-mesh screen. The cementing material may be siliceous, calcareous, or argillaceous, but must be firm enough to hold the stone together when working under pressure, and soft enough to wear faster than the quartz grains and prevent glazing. The standard diameter of pulpstones is 54 in (137 cm) and width of face 27 in (69 cm). The stones are aged or seasoned from 1 to 2 years before use. Aging is quickened by heating the stones to about 180°F (82°C) in a closed room and cooling slowly. Large pulpstones for paper mills are now made of silicon carbide or aluminum oxide in fitted sections.

MINERAL WOOL. A fibrous material employed as a heat insulator in walls or as a sound insulator. It was first obtained as a natural product from volcanic craters in Hawaii and was known as **Pele's hair.** It is made by mixing stone with molten slag from blast furnaces and blowing steam through it. Slags from copper and lead furnaces are also used. **Slag wool** is made from slag without the rock. A lead slag containing 30 to 50% calcium and magnesium oxides makes a mineral wool that withstands temperatures to 1500°F (816°C) when made into blocks or boards. Mineral wool usually consists of a mass of fine, pliant, vitreous fibers, which are incombustible and nonconductors of heat. **Rock wool** is made by blowing molten rock in the same manner, and is more uniform than common mineral wool, with physical qualities depending, however, on the class of rock used. **Rock cork** is a name for a low-temperature insulating material made of rock wool molded in sheet form with a waterproof binder, used for walls in cold-storage rooms. **Mineral-wool board** is a moisture-resistant board in thicknesses of 1, 1.5, 2, 3, and 4 in (2.5, 3.8, 5.1, 7.6, and 10.2 cm) for cold-storage insulation. **Mono-Block** is rock wool made into standard blocks and slabs by a felting process. It is used for both cold and heat insulation to temperatures up to 1600°F (871°C). The density is 18 lb/ft^3 (288 kg/m^3).

Granulated mineral wool is the fiber milled into pellets of about 0.5-in (1.25-cm) diameter. The pellets can be poured into a space for insulation, giving a density of 4 to 6 lb/ft^3 (64 to 96 kg/m^3), or they may be pressed into insulating blocks. The block insulation of Harbison-Walker Refractories Co., for backing up firebrick, is mineral wool compressed with a mineral binder to a density of 21 lb/ft^3 (336 kg/m^3). The service temperature is 1900°F (1038°C). **Rockwool quilt** consists of felted fibers stitched between layers of kraft paper for wall insulation.

Selected minerals, to give various characteristics, may be used to make rock wools. **Wollastonite,** found in California and New York, is melted to produce a rock wool. It is also ground to a white acicular powder of 30 to 350 mesh for use as a filler in molded plastics. The grains are minute fibers. **Ceramic fiber** differs from mineral wool in that it is made of special composition. **Aluminum silicate fiber** is made by melting alumina and silica and passing the molten material through a stream of high-pressure air which produces a fluffy mass of extremely fine fibers up to 3 in (7.62 cm) long. The melting point of the fiber is about 3000°F (1649°C), but the maximum usable temperature is about 2300°F (1260°C), as the fibers devitrify at higher temperatures. The fibers are used for thin insulating paper, panelboard, filler for plastics, and chemical-resistant rope or are woven into fabrics. Aluminum silicate fabrics of 15 to 75 oz/yd (0.5 to 2.3 kg/m) are used for filters, gaskets, belting, insulation, and protective clothing. **Fiberfrax,** of Carborundum Co., is an aluminum silicate fiber.

MOHAIR.   The long, lustrous fleece of the **Angora goat,** *Capra angorensis,* important as a textile fiber because of its luster, length, strength, and spinning qualities. **Mohair fabrics** are used for upholstery material for hard service and are valued for summer wearing apparel, linings, and plushes. Army **necktie cloth,** notable for durability and not creasing, has a cotton warp and a mohair filling. Mohair commonly contains shorter fibers that are coarse and difficult to dye. These are known as **kemp** and sometimes comprise 18% of the fiber. They are removed by combing. Mohair has a natural curl but no crimp, and it does not felt as wool does. The **mohair fiber** used for fine paint brushes is usually from the kid and is white with a silklike luster and a slippery feel, and the scales are not sharply defined as in wool. The staple length is 5 to 8 in (12.7 to 20.3 cm), but the Turkish fiber is up to 10 in (25.4 cm) long. The U.S. production is mostly in Texas. The average weight of the clip is 5.3 lb (2.4 kg) per animal. South Africa produces much mohair.

MOLD STEEL.   A term that generally refers to the steels used for making molds for molding plastics. Mold steel should have a uniform texture that will machine readily with die-sinking tools. It must have no microscopic porosity and must be capable of polishing to a mirrorlike surface. When annealed, it should be soft enough to take the deep imprint of a hob, and when hardened, it must be able to withstand high pressure without sinking and have sufficient tensile strength to prevent breakage of thin mold sections. And it should be dimensionally stable during heat treatment and corrosion-resistant to the plastics being formed.

The principal mold steels are the **P-type mold steels** in the American Iron and Steel Institute family of tool steels, specifically P6, which contains 0.10% carbon, 3.50 nickel, and 1.50 chromium; P20 (0.35 carbon, 1.70 chromium, and 0.40 molybdenum); and P21 (0.20 carbon, 4.00 nickel, and 1.20 aluminum). P6 is a carburizing steel having moderate hardenability, P20 also has only moderate hardenability, but P21 is a deep-hardening steel. P20 and P21 are usually supplied hardened to Rockwell C 30 to 36, in which condition they can be machined to complex shapes. P21, a precipitation-hardening steel, can be supplied moderate hard, then machined and posthardened by a moderate-temperature aging.

MOLYBDENUM. A silver-white metal, symbol Mo, occurring chiefly in the mineral molybdenite but also obtained as a by-product from copper ores. About 90% of all molybdenum is produced in the United States, the remainder coming from Chile, Mexico, Peru, and Norway. The metal has a specific gravity of 10.2 and a melting point of 4730°F (2610°C). It is ductile, softer than tungsten, and readily worked or drawn into very fine wire. It cannot be hardened by heat treatment, only by working. The rolled metal has a tensile strength up to 260,000 lb/in^2 (1,792 MPa) with Brinell hardness 160 to 185. Above a temperature of 1400°F (760°C) the metal forms an oxide that evaporates as it is formed, but the resistance to corrosion is high, and molybdenum heating elements can be used to 3000°F (1649°C). The electrical conductivity is 34% that of copper. The thermal expansion is low, and the heat conductivity is twice that of iron. It is one of the few metals that has some resistance to hydrofluoric acid. While unalloyed molybdenum in the soft state has a tensile strength of 97,000 lb/in^2 (669 MPa) with elongation of 42%, small amounts of other elements will harden and strengthen it greatly. **Molybdenum-titanium alloy,** with only 0.5% titanium, has a tensile strength of 132,000 lb/in^2 (910 MPa) and at 1600°F (871°C) retains a strength of 88,000 lb/in^2 (607 MPa). This alloy and one modified with an addition of 0.8 zirconium are used for tubing which will retain its shape to near its melting point.

Molybdenum, 99.95% pure, is used for support members in radio and lightbulbs, heating elements for electric furnaces, arc-resistant electric contacts, and special high-temperature aerospace applications. Molybdenum is used as a flame-resistant and corrosion-resistant coating for other metals, and it may be arc-deposited for metallizing. But protective coatings of molybdenum can be deposited by vapor deposition of **molybdenum pentachloride** reduced with hydrogen chloride. The pentachloride melts at 381°F (194°C) and boils at 514°F (268°C), but a temperature of 1562°F (850°C) is used for deposition.

Molybdenum carbonyl, $MoCO_6$, may also be used for this purpose. Various molybdenum compounds are used for pigments, in catalysts, and in chemical manufacture. **Sodium molybdate** in the anhydrous form, Na_2MoO_4, is used as a dry powder in fertilizers, and the sulfides and selenides of molybdenum are used in lubricants.

Most of the molybdenum used is employed in alloys or in alloy steels. For the latter purpose it is used in the form of **ferromolybdenum,** which is made directly from the ore by reduction with carbon,

lime, and silicon, and adding iron. A standard grade of ferromolybdenum contains 50 to 60% molybdenum with up to 2.5 carbon, and it is
marketed on the basis of the contained molybdenum. **Calcium
molybdate,** $CaMoO_4$, is also used for adding molybdenum to steels.
It contains about 60.7% MoO_3, with the balance lime. The specific
gravity is 4.35. **Molyte,** of Molybdenum Corp., is calcium molybdate
with a flux. It is heavier and will sink in the molten steel. Also, briquettes of **molybdic oxide,** containing 70 to 75% molybdenum trioxide and 12 carbon, are used for alloying steel. **Molybdenum-chromi-
um,** produced by Shieldalloy Corp., for making nonferrous alloys,
contains 68 to 72% molybdenum, 28 to 32 chromium, with 0.50 maximum iron. **Climelt,** of Climax Molybdenum Co., is a **molybdenum-
tungsten alloy** for making high-temperature structural parts. It has
a melting point of 5100°F (2816°C). It contains 70% molybdenum and
30 tungsten. The tensile strength is 121,000 lb/in² (834 MPa) with
elongation of 26%, and it retains a strength of 65,000 lb/in² (448 MPa)
at 1800°F (982°C).

MOLYBDENUM ORES. Molybdenite is the chief ore of the metal
molybdenum. It is a **molybdenum disulfide,** MoS_2, containing 60%
molybdenum, occurring in granite, gneiss, and granular limestone.
Molybdenite resembles graphite in appearance, with a lead-gray
color, metallic luster, and greasy feel. The Mohs hardness is 1 and
specific gravity 4.75. It is infusible. The U.S. production of molybdenite is from Colorado, New Mexico, and Nevada, but about half of the
molybdenum is a by-product of copper mining. **Wulfenite,** another
important ore, is a **lead molybdate,** $PbMoO_4$, and occurs in lead
veins with other ores of lead. It is found in Utah, Nevada, Arizona,
and New Mexico. Wulfenite occurs in crystals and massive granular.
The specific gravity is 6.7 and Mohs hardness 3. Its color is yellow,
orange, gray, red, or white. **Molybdite,** another ore, is a hydrous ferric molybdate of composition $Fe_2O_3 \cdot 3MoO_3 \cdot 7H_2O$. It occurs either
crystalline massive or as an earthy powder. It is yellowish, with a specific gravity of 4.5 and a Mohs hardness of about 1.5. Molybdenum
ores are converted to ferromolybdenum or to calcium molybdate for
use in adding molybdenum to steel. **Molybdenum trioxide,** MoO_3,
the most important molybdenum compound for chemical manufacture, is made by heating molybdenite in air. It is a white crystalline
powder which melts at 1463°F (795°C). It is an electrical insulator,
but becomes a conductor when molten.

MOLYBDENUM STEEL. Next to carbon, molybdenum is the most effective hardening element for steel. It also has the property, like tungsten, of giving steel the quality of red-hardness, requiring a smaller

amount for the same effect. It is also used in hot-work steels, and to replace part of the tungsten in high-speed steels. It is added to heat-resistant irons and steels to make them resistant to deformation at high temperatures and to creep at moderate temperatures. It increases the corrosion resistance of stainless steels at high temperatures. Molybdenum in small amounts also increases the elastic limit of steel, reduces grain size, strengthens crystalline structure, and gives deep-hardening. It goes into solid solution, but when other elements are present, it may form carbides and harden the steel, giving greater wear resistance. It also widens the heat-treating range of tool steels. As it decreases the temper brittleness of aluminum steels, small amounts are added to nitriding steels.

Plain **carbon-molybdenum steels** are easier to machine than other steels of equal hardness. **Molybdenum structural steels** usually have from 0.20 to 0.75% molybdenum. **Steel 4130** and **steel 4140** contain about 1% chromium and 0.20 molybdenum and are high-strength forging steels for such uses as connecting rods. **Allenoy,** for hollow-head screws, is **steel 4150. Steels 4615 to 4650** have no chromium but contain about 1.75% nickel and 0.25 molybdenum. **Steel 4615** is used for molds and dies to be hobbed. It is easily worked. The alloying elements increase the rate of carbon pickup in carburizing, and the steel has a core Brinell hardness of 280 after hardening. The nickel-molybdenum **steel 4650** is used for form dies and, when hardened and drawn to a Brinell hardness of 435, has a tensile strength of 215,000 lb/in^2 (1,482 MPa). **Hyten M steel** and **Durodi steel** are high-strength nickel-chromium-molybdenum steels. Up to 3% molybdenum is used in stainless steels for cast parts for hot-oil and chemical equipment. **Lebanon 22-XM steel** has 19.5% chromium, 9 nickel, and 3 molybdenum. **Welmet** is a similar steel.

The old **Damascus steel** and **Toledo steel** were molybdenum steels, the molybdenum being in the original ore. **Damascene steel** is a name referring to the wavy marks made on blades by forging and was not necessarily a molybdenum steel. But the original **Wootz steel,** or **Indian steel,** of this type, contained small percentages of aluminum incorporated in some unknown manner. Wootz steel was made in the crucible, although the crucible method was not used in Europe until 1740.

MORDANT. A substance used in dyeing for fixing the color (the word means *bitter*). A mordant must have an affinity for the material being dyed and be chemically reactive; that is, the molecule must have free electrons that combine with the dyestuff. The vegetable fibers, such as cotton and linen, frequently require mordants. Basic aniline dyes require a mordant for cotton or rayon, and the water-soluble acid dyes need a mordant for vegetable fibers. The mordant may be applied first, usually in a hot solution, or simultaneously with the dye. **Mordant dyes** are dyes chemically able to accept and hold a metal atom, the metal being added in the mordant. The metal atom is held to the oxygen atoms and to one of the nitrogen atoms in each chromophoric group of the dye, forming highly insoluble materials known as **lakes.**

Besides fixing the color, mordants sometimes increase the brilliance of the dye, particularly when chelation occurs. Common mordants are **alum** and **sodium bichromate,** but salts of aluminum and other metals are used. **Sodium stannate,** used both as a mordant and for fireproofing textiles, is a gray-white, water-soluble powder of composition $Na_2SnO_3 \cdot 3H_2O$ made by treating tin oxide with caustic soda. **Chromium acetate,** used as a mordant for chrome colors and to produce khaki shades with iron solutions, is a grayish-green powder of composition $Cr(C_2H_3O_2)_2 \cdot H_2O$. It is soluble in water. In gilding, the term *mordant* is used to mean a viscous or sticky substance employed to make the gold leaf adhere, but such a material does not have the chemical action of a mordant.

MOTHER OF PEARL.   The hard, brilliant-colored internal layer of the **pearl oystershell** and of certain other marine shellfish. It is employed for knife handles, buttons, inlay work, and other articles. Large oysters of the Indian Ocean, especially off Sri Lanka and in the Persian Gulf, furnish much of the mother of pearl. Other producers are Australia and the lands bordering the Coral Sea. The large Hawaiian pearl oyster is *Pinctada galtsoff.* The iridescent appearance is due to the structure of the nacre coating, and the shells are also called **nacre.** Mother of pearl is brittle but can be worked with steel saws and drills using a weak acid lubricant. The shells are thick and heavy, and disks cut from them are split for buttons. A beautiful pink nacre occurs on the inner surface of the **conch shell,** *Strombus gigas,* a sea snail which grows in great abundance off the Caicos Island near Haiti. There are about 56 species, and they grow up to 12 in (30 cm) long and 8 in (20 cm) wide. The shells were formerly much used for making cameos. **Mussel shell,** from freshwater mussels of the Mississippi River, is also called **pearl shell,** but does not have the iridescence of mother of pearl. There is a large production in Iowa, used for buttons. The waste shell from the manufacture is crushed and marketed for poultry feed. **Pearl essence,** used for making **imitation pearls** and in plastics and lacquers, is a motley-silver compound extracted from the scales of fish, notably the sardine, herring, and alewife. Only a few types of fish have iridescent scales. The chief constituent is guanine, a chemical related to caffeine. One ton (0.907

metric ton) of scale is produced from 100 tons (90.7 metric tons) of Pacific herring, and this yields 1 lb (0.45 kg) of essence. **Nacromer,** of Mearl Corp., is pearl essence. It gives high luminosity and iridescence to lacquers. **Synthetic pearl essence** may be ground nacreous shells in a liquid vehicle, or it may be produced chemically. Compounds of basic lead carbonate and lead monohydrogen phosphate have multiple reflectivity and give an optical effect resembling that of mother of pearl. **H-scale,** of Hoechst Celanese, and **Ko-Pearl,** of Ultra Ray Pearl Essence Corp., are synthetic pearl essences.

MUCILAGE.   A sticky paste obtained from linseed and other seeds by precipitation from a hot infusion, and used as a light adhesive for paper and as a thickening agent. It contains arabinose, glucose, and galactose, and it is easily soluble in water. Mucilage as a general name also includes **water-soluble gums** from various parts of many plants and has the same uses. It is the stored reserve food of plants. There are two types: the cell-content mucilage, which acts as a disorganization product of some of the carbohydrates, and membrane mucilage, which acts as a thickening agent to the cell wall. Membrane mucilage occurs in the acacias, algae (seaweeds), linun (flax plants), ulmus (elms), and astragalus. When it is collected in the form of exudations from the trees, it is called a gum. **Cherry gum,** from *Prunus cerasus,* is this type of water-soluble gum, as is **medlar gum,** from a small tree *Mespilus germanica,* of the same family as the cherry, grown in Europe and the Near East.

MULLITE.   A mineral found originally in the Isle of Mull and employed as a refractory material for firebrick and furnace linings. The natural material occurs as fused argillaceous sediment inclusions in the mineral **buchite,** but it is rare and is produced artificially. It can be made by decomposing sillimanite or kyanite. **Artificial mullite,** or **synthetic mullite,** a ceramic material made by a prolonged fusing in the electric furnace of a mixture of silica sand or diasphoric clay and bauxite, has composition $3Al_2O_3 \cdot 2SiO_2$. It is valued as a refractory because it does not soften below its high melting point, 3290°F (1810°C), and its interlocked grain withstands continuous temperature changes without spalling. The bricks are resistant to flame and to molten ash, and they have a low, uniform coefficient of thermal expansion and a heat conductivity only slightly above that of fireclays. Normally, mullite has very fine crystals which change form and become enlarged after prolonged heating, making the product porous and permeable. For stable high-temperature refractories, the mullite is prefused to produce larger crystals. At very high temperatures mullite tends to decompose to form corundum and alkali-silicate minerals of lower heat resistance. Mullite is also made by burning silica clay and alumina at a very high temperature, and it is used for making spark plugs, chemical crucibles, and extruding dies. For spark-plug cores it is fired at a temperature of 1450°C and aged before use. The tensile strength is above 9,000 lb/in² (62 MPa), or double that of porcelain, and it has high dielectric strength. The Mohs hardness is 6 to 7.

Sillimanite, as well as **andalusite** found in California, has composition $Al_2O_3 \cdot SiO_2$, with the same melting point and specific gravity of 3.20. **Diaspore andalusite** from Mono County, California, is used for spark plugs. **Andalusite crystals** of fine brilliance and of brick-red to yellow-green colors, found in Brazil, are used as gemstones. Sillimanite is decomposed to mullite and silica when heated above 2822°F (1550°C). **Dumortierite,** produced in Nevada, has composition $8Al_2O_3 \cdot B_2O_3 \cdot 6SiO_2 \cdot H_2O$. Deep-blue crystals of this mineral from the Colorado River resemble lapis lazuli and are cut as gemstones. **Champion sillimanite,** an electric porcelain, is a mixture of andalusite and dumortierite, and **Shamra** is the trade name of a mullite refractory. **Mullfrax** is a mullite refractory of Carborundum Co. **Tervex,** of Du Pont, is **foamed mullite** with a uniformly latticed honeycomb structure for lightweight, heat-resistant structural parts.

MUNTZ METAL. A yellow brass containing 60% copper and 40 zinc, invented in 1832 by George F. Muntz. In England it is called **yellow metal.** It has also been known as **malleable brass.** Muntz metal is also modified with small amounts of iron and manganese. Iron above 0.35% forms a separate iron-rich constituent which is stable and gives hardness and high strength to the alloys. The addition of manganese helps to absorb the iron and hardens the alloy. These alloys were also called **high-strength brass** and were employed for such uses as hydraulic cylinders and marine forgings, but are now largely replaced by manganese bronze or aluminum bronze.

Current standard wrought alloys are C28000 (59 to 63% copper, as much as 0.30 lead, 0.07 iron, and the balance zinc); four leaded alloys—C36500, C36600, C36700, and C36800—each containing 58 to 61% copper, 0.25 to 0.70 lead, and as much as 0.15 iron and 0.25 tin, with the balance zinc except for 0.02 to 0.06 arsenic in C36600, 0.02 to 0.10 antimony in C36700, and 0.02 to 0.10 phosphorus in C36800. C37000 is a free-cutting alloy with a still greater lead content (0.8 to 0.15%). The alloys are available in most wrought forms, are hardenable only by cold-working, and are used for forged parts, heat-exchanger tubing and tube sheets, baffles, valve stems, and fasteners. The corrosion resistance of C28000 is said to be generally similar to that of copper and better than that of higher-copper alloys to sulfur-bearing compounds. Like the uninhibited leaded alloys (C36500), however, C28000 is susceptible to dezincification. Alloys inhibited with small amounts of arsenic, antimony, or phosphorus are resistant to this form of corrosion.

Density is 0.303 lb/in^3 (8,387 kg/m^3) for C28000 and only slightly greater [0.304 lb/in^3 (8,415 kg/m^3)] for the leaded alloys. Electrical conductivity is 27 to 28% that of copper and thermal conductivity is 69 to 71 Btu/(ft · h · °F) [120 to 123 W/(m · K)]. In the annealed condition, thin C28000 flat products have an ultimate tensile strength of 54,000 lb/in^2 (372 MPa), a tensile yield strength of 21,000 lb/in^2 (145 MPa), and a tensile elongation of 45%. In the half-hard temper, yield strength is 50,000 lb/in^2 (345 MPa) and elongation 10%. The alloy is not readily cold-worked, but is readily hot-worked, having 90% the forgeability of C37700 forging brass. Its machinability is about 40% that of C36000 free-machining brass. The leaded alloys also have good hot-working characteristics and superior machinability, to 70% of the free-machining brass in the case of C37000.

MUSIC WIRE. A high-grade, uniform steel originally intended for strings in musical instruments, but now employed for the manufacture of spiral springs. It is the highest grade of **spring wire** and is free from slag or dirt, and low in sulfur and phosphorus. For springs it is usually **steel 1085** reduced about 80% in 8 or 10 drawing passes, but piano wire may have as many as 40 draws from a No. 7 rod. The tensile strength of spring wire, when hard-drawn, is 225,000 to 400,000 lb/in^2 (1,551 to 2,758 MPa), but it should be formable enough to bend 180° flat without cracking, or wind into a close helix with inside diameter 1 to 1.5 times the diameter of the wire. The wire, which is 0.187 in (0.475 cm) in diameter, has an ultimate strength of 230,000 lb/in^2 (1,586 MPa); 0.015-in (0.038-cm) wire has a strength of 400,000 lb/in^2. The wire is usually marketed in gage sizes according to Washburn & Moen and the music-wire gages. Wire below 0.034 in (0.086 cm) in diameter (No. 15 gage) is furnished on reels. Larger sizes are in coils. Wire for springs is from No. 00, which is 0.0085 in (0.0216 cm) in diameter, to No. 36, which is 0.102 in (0.259 cm). The smallest size of music wire is 0.003 in (0.008 cm) in diameter. **Piano wire,** intended for piano strings but much used for springs, is a cold-drawn, high-quality wire formerly only drawn from Swedish billets or rods but now made from U.S. steels. Piano wire ranges in diameter from 0.03 to 0.065 in (0.076 to 0.165 cm). The tensile strength is from 350,000 to 400,000 lb/in^2 (2,413 to 2,758 MPa). **Supertensile steel wire,** for A strings of guitars, is a high-carbon steel made by long cold-working and heat treatment. It has a tensile strength of 460,000 lb/in^2 (3,172 MPa), permitting a vibrational load of 37.7 lb (17 kg) on a 0.010-in (0.025-cm) wire. **Tire cord wire** that is used for automobile tire plies and extra-thin conveyor belts has a tensile strength of 350,000 lb/in² (2,413 MPa). The 0.0059-in (0.0149-cm) wire is stranded into a 0.031-in (0.079-cm) yarn and coated with rubber to make the ply. **High tensile wire,** in diameters of 0.005 in (0.0125 cm) and finer, used as a reinforcement in high-strength plastics, contains 0.90% carbon and has a tensile strength of 575,000 lb/in² (3,965 MPa). **Stitching wire,** for wire-stitching machines, is not as hard-drawn and has a tensile strength of about 290,000 lb/in² (1,999 MPa).

Other spring steels and alloys include **austenitic stainless steels,** such as 302 and A286, an **iron-base superalloy,** containing about 25% nickel, 15 chromium, 2 titanium, 1.5 manganese, 1.25 molybdenum, 0.75 silicon, 0.3 vanadium, 0.08 maximum carbon, and not over 0.01 boron, and are used for springs in aircraft turbine engines. The tensile strength of the wire alloy, reduced 30%, is 198,000 lb/in² (1,365 MPa) at 1350°F (732°C). **Alloy 355,** for highly stressed springs in jet engines, has a nominal composition of 15.65% chromium, 4.38 nickel, 2.68 molybdenum, 1 manganese, 0.32 silicon, 0.12 copper, 0.124 nitrogen, and 0.142 carbon. The 0.004-in (0.010-cm) wire has a tensile strength of 500,000 lb/in² (3,450 MPa). The modulus of elasticity is 29.3 × 10⁶ lb/in² (202,000 MPa) at normal temperatures and 24.6 × 10⁶ lb/in² (170,000 MPa) at 800°F (427°C).

MUSTARD. A pungent, yellowish powder produced from the seeds of the **black mustard,** *Brassica nigra,* and the **white mustard,** *B. alba,* an annual plant of the turnip family grown in most temperate-climate countries. The ground seeds are treated with water, and the enzyme action yields a complex sulfur compound with a sharp taste and pungent action. It is used as a condiment, and in medicine as a counterirritant and emetic. It is one of the most ancient of the condiments and has a stimulating effect on the salivary glands. The product from the black mustard is more pungent than that from the white, and ground mustard may be a mixture of the two to give the desired pungency. **Mustard oil** is a volatile oil obtained from the seeds. It is a powerful vesicant and has a pungent aromatic odor. Also called **horseradish essence,** it is used in medicine and in flavors. Mustard seed which yields on expression the fixed **mustard-seed oil** used extensively in India as a food oil and in lubricants is *B. juncea.* It is closely related to rape oil, containing about 50% erucic acid. Another variety of *Brassica* of India yields **jamba oil,** which contains 46% erucic acid and is a substitute for rape oil for lubricants. Synthetic mustard oil is **allyl isothiocyanate.** It is a nasal irritant and lacrimator, and small quantities are added to solvent glues as a sniffing deterrent.

MYROBALAN EXTRACT. A liquid or solid extract made from the dried, unripe fruit of several species of *Terminalia,* especially *T. chebala* of India and China. *Phyllanthus emblica* of India also yields the fruit. The dried fruit, which resembles a plum, contains 30 to 40% tannin in the pulp. The fruit is graded and marketed as **myrobalans** chiefly on the basis of color, the lighter the color the higher the grade. The best grades of fruit are oval, pointed, and solid in structure. Inferior grades are round and spongy. The **bimlies** from Madras are rated best. Liquid myrobalan extract contains 25 to 30% tannin, and solid extract contains about 53%. It is used in tanning light leathers, and gives a quick tan. The natural acidity of myrobalan plumps the leather, but when used alone on heavy hides, it makes a porous leather. It is used with other tannins.

NANOPHASE MATERIALS. Emerging materials, often multilayered in composite form, produced in building blocks so small as to be measured in nanos, or 10^{-9} units, and having superior physical or mechanical properties compared with their conventional, or "bulk," counterparts. They are made by controlling the arrangement of matter on an atomic or molecular scale so as to create materials comprising particles or grains 39 to 3,900 nin (1 to 100 nm) wide or of layers or filaments of this thickness. Near-perfect materials, with critical flaw sizes less than the layer thickness, can be made, thereby creating materials approaching theoretical strengths. An early use, an extension of thin-film deposition, was in mirrors for soft X-rays. At Lawrence Livermore National Laboratory, sputtering techniques have produced some 75 natural elements in multilayers, or composites, including copper/nickel-copper (Monel), copper/304 stainless steel, copper/zirconium, 304 stainless steel/zirconium, and platinum/chromium. For the Cu/Zr Vickers hardness is 392,000 lb/in^2 (2,703 MPa) and tensile strength 152,000 lb/in^2 (1,048 MPa). For Pt/Cr hardness is 1.8 $\times$ 10^6 lb/in^2 (12,400 MPa) and tensile strength over 600,000 lb/in^2 (4,137 MPa).

As the grain size gets smaller, metals tend to get harder, stronger, and tougher. Also, traditionally brittle ceramics become ductile, even superplastic. Some nanophase materials exhibit unusual optical and magnetic behavior. One such material, being studied by Xerox for potential use in color copiers, is nanostructured iron oxide. Titanium dioxide powder, of Nanophase Technologies Corp., can be dispersed in sufficiently small size distributions to attenuate ultraviolet light without scattering visible wavelengths. The powder may be useful in sunblock lotions, fabrics, and optical instruments. The company also produces fine alumina, titania, and zirconia powder by oxidizing nanosize crystals rapidly made by condensing precursor vapors.

Nanodyne Inc. uses a solution-spray conversion process to make cobalt/tungsten carbide powder having a particle size of 787 to 1,181 nin (20 to 30 nm), which can be used for hard-facing by thermal spraying or for cutting tools. The process also can be used to produce **cermets,** such as chromium carbide/nickel and iron/titanium carbide as well as **oxide-dispersion-strengthened ceramics.** At BIRL, the industrial research laboratory of Northwestern University, magnetron vacuum reactive sputtering has led to nanophase **composite coatings** twice as hard as either constituent material. With thin alternating layers of columbium nitride/titanium nitride and vanadium nitride/titanium nitride, Vickers hardness values half that of diamond are achieved, portending potential applications for cutting tools and wear-resistant parts. Pratt & Whitney is exploring various metal/metal, metal/ceramic, and ceramic/ceramic coatings for greater heat, wear, or erosion resistance than conventional coatings can provide. In these combinations, ductile layers could mitigate cracking of hard and wear-resistant but brittle layers, or the various layers could be graded to better match the substrate's coefficient of thermal expansion.

NAPHTHA. A light, colorless to straw-colored liquid which distills off from petroleum between 158 to 194°F (70 and 90°C). The specific gravity is from 0.631 to 0.660, or slightly higher, ranging from C_6H_{14} to C_7H_{18}. The lightest of the distillates used as solvents for fats, rubber, and resins approaches petroleum ether; the heaviest distillates approach benzin and gasoline and are used for fuel. **Benzin** is a light distillate ranging from C_8H_{18} to C_9H_{20}, with specific gravity from 0.635 to 0.660, and boiling-point limits from 248 to 302°F (120 to 150°C). **Petroleum ether,** or **petroleum spirits,** ranges from C_5H_{12} to C_6H_{14} and is distilled off between 104 and 140°F (40 and 60°C). Benzine is used as a solvent, as a cleaner, and in lighters. **Varnoline** and **white spirit** are old names for petroleum ether used as a general-purpose solvent, especially for varnishes.

The name *naphtha* is also applied to heavier distillates from petroleum and to various grades of light oils obtained in the distillation of coal tar. They are widely used because of their low volatility and safety in handling. These include **solvent naphtha,** having a specific gravity of 0.862 to 0.872, with a boiling point below 320°F (160°C), and **heavy naphtha,** a dark liquid of specific gravity between 0.925 and 0.950 and boiling point between 320 and 428°F (160 and 220°C). **High-flash naphtha** for solvent purposes is a petroleum fraction with a specific gravity within the range of the gasolines and a boiling point from 302 to 392°F (150 to 200°C). **VM&P,** or varnish maker's and painter's naphtha, is a specialty product that is available from

Ashland Chemical Inc. Various trade names are given to the light petroleum distillates used as solvents and in paints and varnishes, such as **Naphtholite. Solvesso** is a hydrogenated distillate in various grades with specific gravities from 0.797 to 0.937. **Stoddard solvent** is a standardized fraction of petroleum, or naphtha, used in dry cleaning or as a solvent. It is water-white in color, has a flash point above 100°F (38°C), and consists of the distillation fraction not over 410°F (210°C) with 50% below 350°F (177°C). **HiSolve VM** is a petroleum naphtha with an initial distillation point at 212°F (100°C) and an end distillation at 275°F (135°C). It is used as a thinner in paints, as an ink solvent, and in paint removers. **HiFlash naphtha,** of the same company, is a coal-tar solvent for rubber, for dry cleaning, and for varnish thinning. The distillation range is 293 to 365°F (145 to 185°C), and the specific gravity is 0.860.

At high temperatures, and in the absence of air, naphtha is cracked to ethylene, the most widely used commodity organic. In this application, naphtha is replacing ethane and propane as the choice of feedstock.

NAPHTHALENE. Also called **tar camphor.** A white solid of composition $C_{10}H_8$, which is one of the heavy distillates from coal tar; it may also be obtained from petroleum. Crude naphthalene has a melting point of 158 to 173.3°F (70 to 78.5°C). The pure crystalline flakes melt at 174.9°F (79.4°C) and boil at 424°F (218°C), but vaporize slowly at room temperature. Naphthalene burns with a smoky flame, is soluble in benzene and in hot alcohol, but not in water. Refined naphthalene comes in balls, flakes, and pellets, largely for use as an insect repellent, but most of the material is sold in technical grades for use in making dyestuffs, synthetic resins, coatings, tanning agents, and Celluloid. The largest use is for production of phthalic anhydride. It is also the source of carbaryl, a **methyl carbamate** used extensively as a replacement for DDT. Union Carbide Corp.'s **Sevin** is such a product. **Naphthalene crystals** are very transparent to fluorescent radiation, and light produced in thick layers can escape and reach a photo surface. Low pulses can be obtained from the absorption of beta and gamma rays in naphthalene. It is thus used in photomultiplier tubes as a gamma-ray detector.

Halowax and **Halowax oil** of Koppers Co. consist of refined fractionated chlorinated naphthalene ranging from low-viscosity oils through waxlike solids to hard, resinlike solids. The melting range is 104 to 356°F (40 to 180°C). They are chemically stable, are resistant to acids and alkalies, have high dielectric strength, are nonflammable, and are used as solvents, for waterproofing and fireproofing fabrics, and for electric-cable coatings. **Monoamyl naphthalene,** used to give plasticity to synthetic resins at low temperatures, is an amber liquid, $C_{10}H_7C_5H_{11}$, of specific gravity 0.96 and boiling point 534°F (279°C). On hydrogenation, naphthalene yields the solvents Tetralin and Decalin. **Tetralin,** $C_{10}H_{12}$, of Du Pont, has a specific gravity of 0.975 and boiling point 403°F (206°C). It is a clear liquid that is a good solvent for fats, oils, and resins. It will burn with a bright flame and can also be used as a fuel. The flash point is 172°F (78°C). **Decalin,** $C_{10}H_{18}$, is a liquid with specific gravity 0.884 and boiling point 374°F (190°C).

NEATSFOOT OIL. A pale-yellow, inedible oil obtained by boiling the feet and shin bones of cattle in water, skimming the oil from the surface, and filtering. It was formerly highly valued for leather dressing and as a lubricating oil. For high-grade, cold-test lubricating oil for fine instruments, the stearin is pressed out. The oil contains 67% oleic acid, 17 palmitic, 9 palmitoleic, 3 stearic, 1 myristic, 1 myristoleic, and 2 arachidonic and clupanodonic. The specific gravity is 0.916, iodine value 70, and saponification value 197.

NICKEL. A silvery-white metal, symbol Ni, first isolated in 1751, but used in alloy form with copper since ancient times. Nickel has a specific gravity of 8.902, or a density of 0.322 lb/in^3 (8,913 kg/m^3), and is magnetic up to 680°F (360°C). Its electrical conductivity is 25% that of copper, and its thermal conductivity is 48 Btu/(ft · h · °F) [83 W/(m · K)]. The metal is highly resistant to atmospheric corrosion and resists most acids, although it is attacked by oxidizing acids, such as nitric. Sulfide ores and oxide ores are its principal sources, and its principal use is as an alloying element in stainless steels, alloy steels, and nonferrous metals. **Nickel-cadmium** and **nickel-metal hydride** are battery materials.

Nickel is also a common coating material, especially for other metals but also plastics. It can be applied electrolytically by electroplating or chemically by electroless or autocatalytic deposition. The principal ingredient for electroplating baths is **nickel sulfate, nickel chloride, nickel sulfamate,** or **nickel fluoborate.** The nickel sulfate bath also includes a lesser amount of nickel chloride, and all baths contain boric acid, antipitting agents, and other additives. Electroless-deposited nickel is actually nickel-phosphorus alloy with 1 to 13% phosphorus, nickel-boron alloy with 0.2 to 3 boron, nickel-cobalt (or cobalt-nickel) alloy with 20 to 80 cobalt, or nickel-thallium-boron alloy with 1 to 3.5 thallium and 4 to 5 boron. **Sodium hypophosphite,** a basic ingredient in electroless nickel plating, imparts corrosion resistance and wear resistance. Low-phosphorus formulations can provide a surface hardness of Rockwell C 58 to 60.

High-phosphorus formulations provide good resistance to gasoline-methanol fuel blends. **Niklad electroless nickel** systems, of Allied Kelite Div. of Witco Corp., are for plating onto ferrous and nonferrous metals, ceramics, and glass. There are also **composite electroless nickel coatings** of polytetrafluoroethane, alumina, silicon carbide, or diamond codeposited with nickel-phosphorus or nickel-cobalt alloys. Nickel and nickel-cobalt alloys are also used to produce electroformed parts. Nickel sulfamate and nickel fluoborate baths are used for electroforming.

There are several commercial **high-purity nickels,** containing at least 99% of the metal plus trace amounts of combined cobalt, and small amounts of other elements, such as carbon, copper, iron, manganese, silicon, and sulfur. They find use in electronic and aerospace applications, in chemical and food-processing equipment, and for anodes and cathodes, caustic evaporators, and heat shields. **Nickel 200,** also known as **commercially pure nickel,** may contain as much as 0.15% carbon, 0.25 copper, 0.40 iron, 0.35 manganese, 0.35 silicon, and 0.01 sulfur. It is especially resistant to caustics, high-temperature halogens and hydrogen halides, and salts, but not to oxidizing halides or ammonium hydroxide. Being susceptible to high-temperature embrittlement by carbon precipitation, however, it is restricted to a maximum service temperature of 600°F (316°C). **Nickel 201,** its low-carbon (0.02% maximum) counterpart, can be used at higher temperatures. **Nickel 270** is the highest-purity grade, being at least 99.97% nickel and containing no more than 0.02 carbon, 0.005 iron, and 0.001 each of other ingredients. Though similar to the other nickels in general corrosion resistance, it may be more prone to sulfur embrittlement under some conditions.

Annealed high-purity nickels have ultimate tensile strengths ranging from 50,000 to 85,000 lb/in^2 (345 to 586 MPa), tensile yield strengths of 15,000 to 50,000 lb/in^2 (103 to 345 MPa), and tensile elongations of 30 to 60%, depending on product form and section thickness. They can be hardened appreciably by cold-working. The tensile yield strength of Nickel 200 sheet in the hard temper, for example, is 3 to 4 times greater than that of annealed sheet. Annealed Nickel 200 to Nickel 201 retain essentially all of their room-temperature strength to temperatures as high as 600°F (316°C).

Dispersion-strengthened nickel, or **DS nickel,** contains about 2.2% thoria as the dispersion-strengthening phase and, thus, also has been called **TD nickel.** The thoria markedly increases high-temperature strength. Tensile strengths, about 70,000 lb/in^2 (483 MPa) ultimate and 48,000 lb/in^2 (331 MPa) yield at room temperature, are on the order of 17,000 lb/in^2 (117 MPa) at 2000°F (1093°C). The material also retains about 60% of its room-temperature tensile modulus, about 20×10^6 lb/in^2 (138×10^3 MPa), at 2000°F. DS nickel has poor oxidation resistance and, thus, requires coating for sustained use at high temperatures, such as for aircraft turbine components and furnace equipment. It has been alloyed with about 20% chromium to improve oxidation resistance, but still may require coating for prolonged use at high temperatures.

NICKEL ALLOYS. Any alloy containing nickel as the base metal, and there are several important families. Most of the alloys are wrought products, although many are also available for casting.

Permanickel 300 is a high-nickel alloy, containing at least 97% nickel and small amounts of titanium and magnesium and most of the other elements found in high-purity nickels. It is used for high-strength parts in electrical and electronics applications. It has a tensile modulus of 30×10^6 lb/in^2 (207,000 MPa) and, depending on form and condition, tensile yield strengths of 35,000 to 150,000 lb/in^2 (241 to 1,034 MPa). It is available in strip, rod, and wire and has an electrical resistivity of 6.2 $\mu\Omega \cdot$ in (0.157 $\mu\Omega \cdot$ m) and a thermal conductivity of 33 Btu/(ft $\cdot$ h $\cdot$ °F) [58 W/(m $\cdot$ K)].

Wrought **beryllium nickel** containing 97.5% nickel, 2 beryllium, and 0.5 titanium is used for springs, switches, bellows, diaphragms, and small valves. The age-hardenable alloy provides tensile yield strengths of 45,000 to 230,000 lb/in^2 (310 to 1,586 MPa), 130,000 lb/in^2 (896 MPa) at 1000°F (538°C), and has a good corrosion resistance in general atmospheres and to reducing media. Casting alloys, containing 2 to 3% beryllium and also age-hardenable, are about equally strong and are used for molds to form glasses and plastics and for metal-forming tools, aircraft fuel-pump impellers, and seal plates in aircraft engines. **Duranickel 301,** which is also age-hardenable and finds similar applications requiring high strength and corrosion resistance, contains 4 to 4.75% aluminum and 0.25 to 1.0 titanium. The wrought **nickel electrical-resistance alloys,** 80% Ni, 75 Ni, 70 Ni, and 60 Ni, are basically **nickel-chromium alloys** or **nickel-iron alloys.** The 70% nickel and 30 iron alloy, designated 70 Ni, has an electrical resistivity of 7.9 $\mu\Omega \cdot$ in (0.2 $\mu\Omega \cdot$ m) and is used in thermometer bulbs and resistance thermometers. The other alloys have resistivities in the 39 to 55 $\mu\Omega \cdot$ in (1 to 1.4 $\mu\Omega \cdot$ m) range and are used for resistors, rheostats, shunts, heating elements, and furnace and electronic equipment. The wrought **nickel permeability alloys,** which may have similar designations in terms of nickel content, include 50 to 80% nickel alloys containing iron as the predominant alloying element but also contain 4 to 5% molybdenum or 4 to 5 copper and 1.5 chromium. These alloys have resistivities of 18 to 26 $\mu\Omega \cdot$ in (0.45 to 0.65 $\mu\Omega \cdot$ m). The high-nickel (76 to 80%) alloys have an initial permeability in the range of 20,000 to 28,000, maximum permeabilities to as much as 500,000, saturation induction on the order of 7,000 to 7,500 G (700 to 750 mT), and coercive forces as low as 0.01 Oe (0.80 A/m). In the annealed condition, ultimate tensile strengths are 65,000 to 80,000 lb/in^2 (448 to 552 MPa), and tensile yield strengths are 18,000 to 25,000 lb/in^2 (124 to 172 MPa). Tensile elongation ranges from 25 to 65%. Applications include motor and transformer laminations and electrical and electronic devices. A **nickel-iron alloy** (50% nickel, 50 iron), noted for its low coefficient of thermal expansion [about $5 \times 10^{-6}/°F$ ($9 \times 10^{-6}/K$)], is used for glass-to-metal and ceramic-to-metal seals. **Nickel Alloy 42,** containing 42% nickel and the balance iron, is used for semiconductor lead frames.

Monel was originally a "natural" alloy produced from Canadian Bessemer mat by reduction of a nickel ore. For many years, Monels have been made by alloying nickel, copper, and other metals. **Mond metal,** containing 4% manganese, was an early Monel. Now there are four principal alloys: (1) **Monel 400**—63 to 70% nickel with as much as 2.5 iron, 2.0 manganese, 0.5 silicon, 0.30 carbon, and the balance copper; (2) **Monel R-405**—63% nickel (minimum), 28 to 34 copper, 0.025 to 0.060 sulfur, and as much as 2.5 iron, 2.0 manganese, 0.5 silicon, and 0.3 carbon; (3) **Monel K-500**—63% nickel (minimum), 27 to 33% copper, 2.30 to 3.15 aluminum, 0.35 to 0.85 titanium, and as much as 2.0 iron, 1.5 manganese, 0.5 silicon, 0.25 carbon, and 0.01 sulfur; and (4) **Monel 502**—63% nickel (minimum), 27 to 33 copper, 2.5 to 3.5 aluminum, and as much as 2.0 iron, 1.5 manganese, 0.5 silicon, 0.1 carbon, and 0.010 sulfur. For each alloy the nickel content may also contain some cobalt. Monel R-405 is a free-machining version of 400, and both are hardenable only by cold-working. Monel 502 is more machinable than K-500, and both are hardenable by heat treating.

Their crystal structure is face-centered cubic; 400 and R-405 have a density of 0.319 lb/in^3 (8,830 kg/m^3), and K-500 and 502 have a density of 0.305 lb/in^3 (8,442 kg/m^3). Depending on the alloy, electrical conductivity is in the range of 2.8 to 4.1% that of copper, and thermal conductivity at room temperature is 10 to 12.6 Btu/(ft $\cdot$ h $\cdot$ °F) [17.4 to 21.8 W/(m $\cdot$ K)]. In the annealed condition, 400 has a tensile yield strength of 25,000 to 55,000 lb/in^2 (172 to 379 MPa) and tensile elongations of 25 to 60%. Cold-worked sheet can provide yield strengths of about 100,000 lb/in^2 (690 MPa), but it is far less ductile. Still greater strengths are common in the spring temper. Monel K-500 and 502, which respond to solution heat treatment and age-hardening, can provide tensile yield strengths of 90,000 to 110,000 lb/in^2 (621 to 758 MPa) with elongations of 28 to 48%, and they maintain these strength levels to about 600°F (316°C).

There are also several casting alloys of roughly similar composition but greater silicon (1.25 to 4.5%), some of which are also hardenable by solution treating and aging. The age-hardenable alloys contain about 3.5% or more silicon. Increasing silicon content decreases weldability and toughness, but all the cast alloys remain tough to temperatures as low as $-320°F$ ($-195°C$). The cast Monels can provide tensile yield strengths ranging from 25,000 to 80,000 lb/in^2 (172 to 552 MPa), elongations of 10 to 25%, and Brinell hardness 125 to 300. The cast alloys are often used in part combinations with their wrought counterparts.

The Monels are used primarily for their excellent resistance in fresh, brackish, and seawaters; chlorinated solvents; numerous acids, including sulfuric and hydrofluoric solutions; and virtually all alkalies. Alloy 400 is said to be highly resistant to stress corrosion. Monel K-500 and 502, however, are more prone to stress corrosion in the age-hardened condition. Because of their corrosion resistance, the alloys are widely used for valves, pump parts, propellers, shafts, and fasteners in marine environments and for petroleum and chemical processing equipment.

Nickel-copper alloys are also used for making nonferrous alloys. A 50 nickel 50 copper composition has a melting point of 2330°F (1277°C) and dissolves readily. **Nickel-copper shot** of this composition is used for ladle additions to iron and steel. A 60% nickel, 33 copper, 3.5 manganese alloy containing as much as 3.5 iron is used for alloying high-strength bronzes.

Many **chemical-resistant nickel alloys,** both wrought and cast, are noted mainly for their resistance to chemicals that are aggressive toward other metals, and many of these also possess substantial strength at elevated temperature. Although trade names abound, the **Hastelloys** of Haynes International and the **Incoloys** and **Inconels** of Inco Alloys International Inc. are probably the most well known. Many of these alloys are resistant to normally aggressive acids, such as hydrofluoric and sulfuric, as well as to acetic and phosphoric acids, mixed acids, chlorides, solvents, and high-temperature oxidation and, thus, are widely used for chemical processing equipment. **Inconel 686,** for example, a wrought nickel, chromium, molybdenum, and tungsten alloy, features superior corrosion resistance to 80% sulfuric acid and 10 sulfuric, 5 hydrochloric acid mixtures at 176°F (80°C) and to boiling 2% hydrochloric acid. It also resists crevice corrosion in a mixture of 12% sulfuric acid, 1.3 hydrochloric acid, 1 ferric chloride, and 1 cupric chloride at 257 to 284°F (125 to 140°C).

Among the more common alloys and their principal ingredients are **Hastelloy B,** which contains 26 to 30% molybdenum, 4 to 6 iron, 2.5 cobalt, and 1 chromium; **Hastelloy B-2,** which contains the same amount of molybdenum and chromium but only 1 iron and 1 cobalt; **Hastelloy C,** having 15 to 18 molybdenum and about as much chromium plus 4 to 7 iron, 3 to 5 tungsten, and 2.5 cobalt; **Hastelloy C-4,** with molybdenum and chromium contents similar to those of Hastelloy C but no more than 3 iron and 2 cobalt; **Hastelloy D,** 8.5 to 10 silicon, 2 to 4 copper, and as much as 2 iron, 1.5 cobalt, and 1 chromium; **Hastelloy G,** 21 to 23 chromium, 18 to 21 iron, 5.5 to 7.5 molybdenum, 1.75 to 2.5 columbium plus tantalum, 1.5 to 2.5 copper, and as much as 2.5 cobalt; **Hastelloy S,** 14.5 to 17 chromium, 14 to 16.5 molybdenum, and as much as 3 iron and 2 cobalt; **Hastelloy C-276,** 15 to 17 molybdenum, 14.5 to 16.5 chromium, 4 to 7 iron, 3 to 4.5 tungsten, and as much as 2.5 cobalt. Hastelloy G provides greater high-temperature strength than Hastelloy C but is not as corrosion-resistant. Hastelloy C-276 is widely used in incinerator scrubbing systems used to dispose of chemical wastes, which, after combustion, form corrosive acidic wastestreams when absorbed in water. Other incinerator applications include mesh-type mist eliminators and draft-inducing fan wheels. **Hastelloy C-22** contains 20 to 22.5% chromium, 12.5 to 14.5 molybdenum, 2 to 6 iron, 2.5 to 3.5 tungsten, 2.5 maximum cobalt, and small amounts of other elements, including 0.15 maximum carbon. It is used for the quench body and variable venturi of incinerator systems. The alloy is brittle after welding, however, so welded components should be free from vibrations. **Incoloy 800,** though iron-base, is often grouped with these alloys. It contains 46% iron, 32.5 nickel, and 21 chromium. Nickel-base (42%) **Incoloy 825** contains 30% iron, 21.5 chromium, 3 molybdenum, and 2.3 copper.

Hastelloy X, which provides substantial strength and oxidation resistance at temperatures to about 2200°F (1204°C), contains 20 to 23% chromium, 17 to 20 iron, 8 to 10 molybdenum, 0.5 to 2.5 cobalt, and 0.2 to 1 tungsten. Solution-treated rapidly cooled sheet has room-temperature tensile properties of 114,000 lb/in^2 (786 MPa) ultimate strength, 52,000 lb/in^2 (359 MPa) yield strength, 43% elongation, and 28.5×10^6 lb/in^2 (197,000 MPa) modulus. At 1800°F (982°C), these properties are 22,500 lb/in^2 (155 MPa), 16,000 lb/in^2 (110 MPa), 45%, and 18.3×10^6 lb/in^2 (126,000 MPa), respectively. The alloy has a density of 0.297 lb/in^3 (8,221 kg/m^3), a coefficient of thermal expansion at 70 to 1600°F (21 to 871°C) of 9.1×10^{-6}/°F (16.3×10^{-6}/K), and a melting range of 2300 to 2470°F (1260 to 1354°C). The alloy is widely used for gas-turbine parts and other applications requiring heat and oxidation resistance. Though mainly a wrought alloy, it also can be investment cast.

There are a great variety of **high-temperature, high-strength nickel alloys,** called **superalloys** because of their outstanding strength, creep resistance, stress-rupture strength, and oxidation resistance at high temperatures. They are widely used for gas turbines, especially aircraft engines. Most of these alloys contain substantial chromium for oxidation resistance; refractory metals for solid-solution strengthening; small amounts of grain-boundary-strengthening elements, such as carbon, boron, hafnium, and/or zirconium; and aluminum and titanium for strengthening by precipitation of an Ni(Al,Ti) compound known as *gamma prime* during age-hardening. Among the well-known wrought alloys are **D-979; GMR-235-D; IN 102; Inconel 625, 700, 706, 718, 722, X750,** and **751; MAR-M 200** and **412; Rene 41, 95,** and **100; Udimet 500** and **700;** and **Waspaloy. Inconel 718SPF** is tailored for superplastic forming, as the letters in the alloy designation imply. Having an ultrafine grain size, ASTM 10 or less, it can be superplastically formed at a temperature of about 1740°F (950°C) at low strain rates. At these conditions, very little pressure, such as 300 lb/in^2 (2.1 MPa), is needed to achieve large deformation. Cast alloys include **B-1900; GMR-235-D; IN 100, 162, 738,** and **792; M252; MAR-M 200, 246,** and **421; Nicrotung; Rene 41, 77, 80,** and **100;** and **Udimet 500** and **700.** Some wrought alloys are also suitable for casting, primarily investment casting.

Controlled-expansion nickel superalloys have a nickel-iron-cobalt austenitic matrix optimized for minimal thermal expansion and are strengthened by gamma precipitation promoted by aluminum, columbium, and titanium additions. Specific alloys, of Inco Alloys International and Carpenter Technologies, respectively, include **Incoloy 903** and **Pyromet CTX-1, Incoloy 907** and **Pyromet CTX-3,** and **Incoloy 909** and **Pyromet CTX-909.** These alloys are used mainly in aircraft gas-turbine engines to maintain tight clearances between rotating and stationary components over a wide temperature range. Being chromium-free and, thus, lacking oxidation resistance in air, they must have coatings applied for service temperatures above 1000°F (538°C). Carpenter Technologies' **Thermo-Span alloy** contains 5.5% chromium, increasing oxidation resistance to 1300°F (704°C). The alloy matches the thermal expansivity of Pyromet CTX-909 at 200°F (95°C)—4.5 × 10^{-6}/°F (8.1 × 10^{-6}/K)—and at 400°F (205°C)—4.3 × 10^{-6}/°F (7.7 × 10^{-6}/K). At higher temperatures, however, its expansivity is greater, 18% greater at 1000°F.

In addition to the above families, there are **specialty nickel alloys** for glass sealing and other applications. **Paramagnetic alloys** called **Nitinol,** developed by the Naval Ordnance Laboratory, are intermetallic compounds of nickel and titanium rather than **nickel-titanium alloys.** The compound TiNi contains theoretically 54.5% nickel, but the alloys may contain Ti$_2$Ni and TiNi$_3$ with about 50 to 60% nickel. The TiNi and nickel-rich alloys are paramagnetic, with a permeability value of 1.002, compared with the unity value of a vacuum. A 54.5% nickel alloy has a tensile strength of 110,000 lb/in^2 (758 MPa) with elongation of about 15%, and Rockwell C hardness of 35. The alloys close to the TiNi composition are ductile and can be cold-rolled. The nickel-rich alloys are hot-rolled. They can be hardened by heat treatment to give Rockwell C hardnesses to 68 and tensile strengths to 140,000 lb/in^2 (965 MPa). This class of alloy can also be modified with small amounts of silicon or aluminum, forming complex intermetallic compounds that can be solution-treated.

The Nitinols, with nickel content ranging from 53 to 57%, are known as **memory alloys,** or **shape-memory alloys,** because of their ability to be deformed and then return to their original shape when heated to their transformation temperature. For example, a straight piece of Nitinol wire can be bent in multiple places and then straightened by simply applying heat to the bent regions. The alloys are ductile and have excellent fatigue resistance and damping capacity. Applications include fire-sprinkler actuators, tap water antiscalding devices, greenhouse window hinges, flow regulators, spacecraft solar-panel releases, various toys and novelties, and underwire brassieres (that return to shape at room temperature after warm machine washings).

Nickel alloy powders are used for flame-sprayed coatings for hard surfacing and corrosion resistance. **Metco 14E,** of Metco, Inc., is an alloy powder containing 14% chromium, 3.5 silicon, 2.75 boron, 4 iron, 0.60 carbon, with the balance nickel. The alloy is self-fluxing and gives an extremely hard coating. **Colmonoy 72,** of Wall Colmonoy Corp., is a similar alloy powder but with 13% tungsten. Coatings have a melting point of 1950°F (1066°C) and retain hardness and wear resistance at high temperatures. **Colmonoy 88,** with 17.3% tungsten, 15 chromium, and roughly similar iron, silicon, boron, and carbon contents, provides a Rockwell C hardness of 59 to 64 and somewhat greater abrasion resistance.

NICKEL BRONZE. A name given to bronzes containing nickel, which usually replaces part of the tin, producing a tough, fine-grained, and corrosion-resistant metal. A common nickel bronze containing 88% copper, 5 tin, 5 nickel, and 2 zinc has a tensile strength of 48,000 lb/in^2 (330 MPa), elongation 42%, and Brinell hardness 86 as cast. When it is heat-treated or age-hardened, the tensile strength is 87,000 lb/in^2 (599 MPa), elongation 10%, and Brinell hardness 196. Small amounts of lead take away the age-hardening quality of the alloy and lower the ductility. But small amounts of nickel added to bearing bronzes increase the resistance to compression and shock without impairing the plasticity. A bearing bronze of this nature contains 73 to 80% copper, 15 to 20 lead, 5 to 10 tin, and 1 nickel. In the **leaded nickel-copper,** which contains 1% nickel, 1 lead, 0.2 phosphorus, and the balance copper, a nickel phosphide is dispersed in the alloy by heat treatment, giving a machinability of 80% that of a free-cutting brass. The tensile strength is 85,000 lb/in^2 (586 MPa), elongation 5%, and electric conductivity 55% that of copper.

For decorative bronze parts, nickel is used to give a white color. In the hardware industry, the old name **Chinese bronze** was used for these white alloys. At least 10% nickel is needed to give a white color. This amount also gives corrosion resistance to the alloy. When more than 15% nickel is used, the bronzes are difficult to machine unless some lead is added. Hardware and plumbing fixtures of these alloys do not require plating.

NICKEL-CHROMIUM STEEL. Steel containing both nickel and chromium, usually in a ratio of 2 to 3 parts nickel to 1 chromium. The 2:1 ratio gives great toughness, and the nickel and chromium are intended to balance each other in physical effects. The steels are especially suited for large sections which require heat treatment because of their deep and uniform hardening. Hardness and toughness are the characteristic properties of these steels. Nickel-chromium steel containing 1 to 1.5% nickel, 0.45 to 0.75 chromium, and 0.38 to 0.80 manganese is used throughout the carbon ranges for case-hardened parts and for forgings where high tensile strength and great hardness are required. Low nickel-chromium steels, having more carbon, 0.60 to 0.80%, are used for drop-forging dies and other tools.

Nickel-chromium steels may have temper brittleness, or low impact resistance, when improperly cooled after heat treatment. A small amount of molybdenum is sometimes added to prevent this brittleness. A nickel-chromium **coin steel** used by the Italian government for coins, was a stainless-steel type cotaining 22% chromium, 12 nickel, and some molybdenum.

Low-carbon nickel-chromium steels are water-hardening, but those with appreciable amounts of alloying elements require oil quenching. Air-hardening steels contain up to 4.5% nickel and 1.6 chromium, but are brittle unless tempered in oil to strengths below 200,000 lb/in^2 (1,379 MPa). The alloy known as **Krupp analysis steel** contains 4% nickel and 1.5 chromium.

NICKEL-MOLYBDENUM STEEL. Alloy steels used mostly in compositions of 1.5% nickel and 0.15 to 0.25 molybdenum, with varying percentages of carbon up to 0.50%. These steels are characterized by uniform properties and are readily forged and heat-treated. Molybdenum toughens the steels, and in the case-hardening steels gives a tough core. Roller bearings are made of this class of steel. **Superalloy steel** is **3160 steel.** A 5%-nickel steel with 0.30% carbon and 0.60 molybdenum has a tensile strength of 175,000 to 230,000 lb/in^2 (1,207 to 1,586 MPa) with elongation 12 to 22%, depending on heat treatment. Molybdenum is more frequently added to the steels containing also chromium, the molybdenum giving air-hardening properties, reducing distortion, and making the steels more resistant to oxidation.

NICKEL ORES. Nickel occurs in minerals as sulfides, silicates, and arsenides, the most common being **pyrrhotite,** or **magnetic pyrites,** a sulfide of iron of formula $Fe_{1-x}S$, where x is between 0 and 0.2. When x is zero, the mineral is called **troilite.** Pyrrhotite has nickel associated with the iron sulfide. The ore of Copper Cliff, Ontario, is calcined to remove the sulfur, and the nickel is removed, leaving a fine magnetite which is pelletized and fired to give an iron concentrate of 68% iron. The chief sulfide ore deposit at Sudbury, Ontario, contains sulfides of iron, nickel, and copper, and small amounts of other elements; and some of the matte after removal of the iron and sulfur is used as Monel metal without separating the natural alloy. The extensive ore deposits at Lynn Lake, Manitoba, yield an ore averaging 1.74% nickel and 0.75 copper. The **garnierite,** or **noumeite,** of New Caledonia is a nickel silicate containing also iron and magnesium. It is amorphous and earthy, an apple-green color, with a specific gravity of 2.2 to 2.8, and Mohs hardness of 3 to 4. The ore contains about 5% nickel and is smelted with gypsum to a matte of sulfides of nickel and iron, the sulfur coming from the gypsum. This is then bessemerized, and the matte crushed, roasted to oxide, and reduced to nickel. The material exported from New Caledonia under the name of **fonte** is a directly smelted cast iron containing about 30% nickel.

A minor ore of nickel called **millerite,** occurring in Europe and in Wisconsin, is a **nickel sulfide,** NiS, containing theoretically 64.7% nickel. It is found usually in radiating groups of slender crystals with a specific gravity of 5.6, Mohs hardness 3.5, and of a pale-yellow color and metallic luster. **Nicolite,** NiAs, is a minor ore containing theoretically 43.9% nickel, usually with iron, cobalt, and sulfur. It is found in Canada, Germany, and Sweden. The mineral occurs massive, with a specific gravity of 7.5, Mohs hardness 5 to 5.5, and a pale-copper color. Nickel is also produced as a by-product from copper ores.

NICKEL SILVER. A name applied to an alloy of copper, nickel, and zinc, which is practically identical with alloys known in the silverware trade as **German silver. Packfong,** meaning **white copper,** is an old name for these alloys. The very early nickel silvers contained some silver and were used for silverware. **Wessell's silver** contained about 2%, and **Ruolz silver** about 20. **Baudoin alloy,** a French metal, contained 72% copper, 16 nickel, 1.8 cobalt, 2.5 silver, and the balance zinc, but the white jewelry alloys called **Paris metal** and **Lutecine** contained about 2% tin instead of silver. The English silver known as **Alpaca,** used as a base metal for silver-plated tableware, had about 65% copper, 20 zinc, 13 nickel, and 2 silver. Such an alloy takes a fine polish, has a silvery-white color, and is resistant to corrosion. **Lake copper,** sometimes classified as a nickel silver, is a silver-bearing copper with varying amounts of silver up to about 30 oz/ton (0.91 metric ton).

Nickel whitens brass and makes it harder and more resistant to corrosion, but the alloys are more difficult to cast because of shrinkage and absorption of gases. They are also subject to fire cracking and are more difficult to roll and draw than brass.

Some three dozen standard wrought alloys (C73150 to C79900) and four standard cast alloys (C97300 to C97800) are designated nickel silvers. Depending on the alloy, copper content of wrought alloys ranges from 48 to 80% and nickel content from about 7 to 25, with zinc the balance except for smaller quantities of other elements, mainly manganese, iron, and lead. The cast alloys range from about 55 to 65% copper, 12 to 25 nickel, 2.5 to 21 zinc, 2 to 10 lead, with lesser amounts of other elements.

The most common alloy, **nickel silver C75200,** nominally contains 65% copper and 18 nickel and, thus, is often referred to as **nickel silver 65–18.** The alloy's electrical conductivity is about 6% that of copper, and its thermal conductivity is 19 Btu/(ft · h · °F) [33 W/(m · K)]. Tensile properties for thin, flat products in the annealed condition are about 60,000 lb/in^2 (414 MPa) ultimate strength, 30,000 lb/in^2 (207 MPa) yield strength, and 30% elongation. Cold-working to the hard temper triples yield strength and markedly reduces ductility. Wire, which has similar tensile properties annealed, can be cold-worked to still greater tensile strength. Modulus of elasticity in tension is 18 × 10^6 lb/in^2 (124,000 MPa). All of the cast alloys are suitable for sand and investment casting and some also for centrifugal and permanent-mold casting. The strongest of these alloys, **nickel silver C97800,** has typical tensile properties of 55,000 lb/in^2 (379 MPa) ultimate strength, 30,000 lb/in^2 (207 MPa) yield strength, 15% elongation, and 19 × 10^6 lb/in^2 (131,000 MPa) modulus. Applications for wrought alloys include hollowware and tableware, watch and camera parts, hardware, dairy equipment, costume jewelry, nameplates, keys, fasteners, and springs. The cast alloys are used for fittings, valves, ornaments, pump parts, and marine equipment.

Over the years, nickel silvers have been known by a variety of names. **Benedict metal** originally had 12.5% nickel, with 2 parts copper to 1 zinc, but the alloy used for hardware and plumbing fixtures contains about 57% copper, 2 tin, 9 lead, 20 zinc, and 12 nickel. The cast metal has a strength of 35,000 lb/in^2 (241 MPa) with elongation of 15%. The white alloy known as **dairy bronze,** used for casting dairy equipment and soda-fountain parts, has 63% copper, 4 tin, 5 lead, 8 zinc, and 20 nickel. The higher-nickel alloys have a more permanent white finish for parts subject to corrosion. **Ambrac 854** is a wrought metal with 65% copper, 30 nickel, and 5 zinc. **Pope's Island white metal,** used for jewelry, has 67% copper, 19.75 nickel, and 13.25 zinc. **Victor metal,** an alloy of 50% copper, 35 zinc, and 15 nickel, is used for cast fittings. It is a white metal with a yellow shade. It casts easily and machines well.

For threaded parts and for casting metals, the nickel silvers usually contain some lead for easier machining. **White nickel brass,** for cast parts for trim, is a standard 18% nickel alloy with or without lead. **Silveroid,** an English alloy for this use, is a copper-nickel alloy without zinc. An English alloy for tableware, under the name of **Newloy,** contains 35% nickel, 64 copper, and 1 tin. The **stainless nickel** used for silverware by Viners, Ltd., has 30% nickel, 60 copper, and 10 zinc and is deoxidized with manganese copper, using borax as a top flux.

A number of other alloys of copper, nickel, and zinc are termed **nickel brass. Nickel-silicon brass** contains a very small percentage of silicon, usually about 0.60%, which forms a **nickel silicide,** Ni_2Si, increasing the strength and giving heat-treating properties. Rolled nickel-silicon brass, containing 30% zinc, 2.5 nickel, and 0.65 silicon, has a tensile strength of 114,000 lb/in^2 (786 MPa). **Imitation silver,** for hardware and fittings, is a nickel brass containing 57% copper, 25 zinc, 15 nickel, and 3 cobalt. The bluish color of the cobalt neutralizes the yellow cast of the nickel and produces a silver-white alloy. **Silvel** is a nickel brass containing 67.5% copper, 26 zinc, and 6.5 nickel, with sometimes a little cobalt. Nickel brass is an alloy used where white color and corrosion resistance are desired.

Seymourite, an alloy of 64% copper, 18 nickel, and 18 zinc produced by Seymour Mfg. Co., has a white color and corrosion resistance. **Nickeline,** used for hardware, is 58 to 60% copper, 16.5 nickel, 2 tin, and the remainder zinc. It has high strength, a white color, and casts well. **Nickelene** is an old name applied to nickel brass of various compositions, but an alloy patented in 1912 under this name had 55% copper, 12.5 nickel, 20.5 zinc, 10 lead, and 2 tin. Most of these alloys have good casting qualities, but do not machine easily unless containing some lead. Up to 2% lead does not affect the color or decrease strength greatly.

NICKEL STEEL. Steel containing nickel as the predominant alloying element. The first nickel steel produced in the United States was made in 1890 by adding 3% nickel in a Bessemer converter. The first nickel-steel armor plate, with 3.5% nickel, was known as **Harveyized steel.** Small amounts of nickel steel, however, had been used since ancient times, coming from **meteoric iron.** The nickel iron of meteorites, known in mineralogy as **taenite,** contains about 26% nickel.

Nickel added to carbon steel increases the strength, elastic limit, hardness, and toughness. It narrows the hardening range but lowers the critical range of steel, reducing danger of warpage and cracking, and balances the intensive deep-hardening effect of chromium. The nickel steels are also of finer structure than ordinary steels, and the nickel retards grain growth. When the percentage of nickel is high, the steel is very resistant to corrosion. At high nickel contents, the metals are referred to as **iron-nickel alloys** or **nickel-iron alloys.** The steel is nonmagnetic above 29% nickel, and the maximum permeability is at about 78% nickel. The lowest thermal expansion is at 36% nickel. The percentage of nickel in nickel steels usually varies from 1.5 to 5%, with up to 0.80 manganese. The bulk of nickel steels contain 2 and 3.5% nickel. They are used for armor plate, structural shapes, rails, heavy-duty machine parts, gears, automobile parts, and ordnance.

The standard ASTM **structural nickel steel** used for building construction contains 3.25% nickel, 0.45 carbon, and 0.70 manganese. This steel has tensile strength from 85,000 to 100,000 lb/in^2 (586 to 690 MPa) and a minimum elongation of 18%. An automobile steel contains 0.10 to 0.20% carbon, 3.25 to 3.75 nickel, 0.30 to 0.60 manganese, and 0.15 to 0.30 silicon. When heat-treated, it has a tensile strength up to 80,000 lb/in^2 (552 MPa) and an elongation 25 to 35%. Forgings for locomotive crankpins, containing 2.5% nickel, 0.27 carbon, and 0.88 manganese, have a tensile strength of 83,000 lb/in^2 (572 MPa), elongation 30%, and reduction of area 62%. A **nickel-vanadium steel,** used for high-strength cast parts, contains 1.5% nickel, 1 manganese, 0.28 carbon, and 0.10 vanadium. The tensile strength is 90,000 lb/in^2 (621 MPa) and elongation 25%. **Univan steel** for high-strength locomotive castings is a nickel-vanadium steel of this type. **Unionaloy steel** is an abrasion-resistant steel.

The federal specifications for 3.5% nickel carbon steel call for 3.25 to 3.75% nickel and 0.25 to 0.30 carbon. This steel has a tensile strength of 85,000 lb/in^2 (586 MPa) and elongation 18%. When oil-quenched, a hot-rolled, 3.5% nickel, medium-carbon steel, **Steel 2330,** develops a tensile strength up to 220,000 lb/in^2 (1,516 MPa) and Brinell hardness of 223 to 424, depending upon the drawing temperature. Standard 3.5 and 5% nickel steels are regular products of the steel mills, though they are often sold under trade names. Steels with more than 3.5% nickel are too expensive for ordinary structural use. Steels with more than 5% nickel are difficult to forge, but the very high-nickel steels are used when corrosion-resistant properties are required. **Nicloy,** used in fork tubing to resist the corrosive action of paper-mill liquors and oil-well brines, contains 9% nickel, 0.10 chromium, 0.05 molybdenum, 0.35 copper, 0.45 manganese, 0.20 silicon, and 0.09 maximum carbon. The heat-treated steel has a tensile strength of 110,000 lb/in^2 (758 MPa), with elongation 35%. The **cryogenic steels,** or **low-temperature steels,** for such uses as liquid-oxygen vessels, are usually high-nickel steels. **ASTM steel A-353,** for liquid-oxygen tanks at temperatures to $-320°F$ ($-196°C$), contains 9% nickel, 0.85 manganese, 0.25 silicon, and 0.13 carbon. It has a tensile strength of 95,000 lb/in^2 (655 MPa) with elongation of 20%. A **9% nickel steel,** for temperatures down to $-320°F$, contains 9% nickel, 0.80 manganese, 0.30 silicon, and not over 0.13 carbon. It has a minimum tensile strength of 90,000 lb/in^2 (621 MPa) and elongation of 22%.

NICKEL SULFATE. The most widely used salt for nickel-plating baths, and known in the plating industry as **single nickel salt.** It is easily produced by the reaction of sulfuric acid on nickel, and comes in pea-green, water-soluble crystalline pellets of composition $NiSO_4 \cdot 7H_2O$, of specific gravity 1.98, melting at about 212°F (100°C). **Double nickel salt** is **nickel ammonium sulfate,** $NiSO_4 \cdot (NH_4)_2 \cdot SO_4 \cdot 6H_2O$, used specifically for plating on zinc. To produce a harder and whiter finish in nickel plating, **cobaltous sulfamate,** a water-soluble powder of composition $Co(NH_2SO_3)_2 \cdot 3H_2O$, is used with nickel sulfate. **Nickel plate** has a normal Brinell hardness of 90 to 140, but by controlled processes file-hard plates can be obtained from sulfate baths. **Micrograin nickel,** with a grain diameter of 0.00002 in (0.00005 cm), is such a hard plate. In electroless plating, nickel sulfate, a reducing agent, a pH adjuster, and complexing and stabilizing agents are combined to deposit metallic nickel on an immersed object. General American Transportation Co. employs a hypophosphite reductant. The electroless nickel coating is comparable to electrolytic chrome.

Other nickel salts are also used for nickel plating. **Nickel chloride,** $NiCl_2 \cdot 6H_2O$, is a green crystalline salt which, when used with boric acid, gives a fine-grained, smooth, hard, strong plate. It requires less power, and the bath is easy to control. **Nickel carbonate,** $2NiCO_3 \cdot 3Ni(OH)_2 \cdot 4H_2O$, comes in green crystals not soluble in water, but soluble in acids and in solutions of ammonium salts.

Nickel carbonyl, $Ni(CO)_4$, used for nickel plating by gas decomposition, is a yellow volatile liquid. It is volatilized in a closed vessel with hydrogen as the carrier, and the nickel is deposited at about 350°F (177°C). It will adhere to glass and wood as well as to metals. The material is a strong reducing agent and is explosive when mixed with oxygen. **Nickel nitrate,** $(NiNO_3)_2{:}6H_2O$, used in electric batteries, comes in thin, flat flakes.

NITRIC ACID. Also called **aqua fortis** and **azotic acid.** A colorless to reddish, fuming liquid of composition HNO_3, having a wide variety of uses for pickling metals, in etching, and in the manufacture of nitrocellulose, plastics, dyestuffs, and explosives. It has a specific gravity of 1.502 (95% acid) and a boiling point of 187°F (86°C) and is soluble in water. Its fumes have a suffocating action, and it is highly corrosive and caustic. **Fuming nitric acid** is any water solution containing more than 86% acid and having a specific gravity above 1.480. Nitric acid is made by the action of sulfuric acid on sodium nitrate, or purified Chilean saltpeter, and condensation of the fumes. It is also made from ammonia by catalytic oxidation, or from the nitric oxide produced from air. The acid is sold in various grades depending on the amount of water. The strengths of the commercial grades are 38, 40, and 42°Bé, containing 67.2% acid. C.P., or reagent grade, is 43°Bé, with 70.3% acid, very low in iron, arsenic, or other impurities. It is usually shipped in glass carboys. **Anhydrous nitric acid** is a yellow fuming liquid containing the unstable anhydride **nitrogen pentoxide,** N_2O_5, It is violently reactive and is a powerful nitriding agent. The dark-red fuming liquid known as **nitrogen tetroxide,** N_2O_4, is really a concentrated water solution of nitric acid, as this oxide is an unstable polymer of NO_2. It is used as an oxidizer for rocket fuels, as it contains 70% oxygen. **Mixed acid,** or **nitrating acid,** is a mixture of nitric and sulfuric acids used chiefly in making nitrocellulose and nitrostarch. Standard mixed acid contains 36% nitric and 61 sulfuric acid, but other grades are also used.

NITRIDING STEELS. Low- and medium-carbon steels with combinations of chromium and aluminum or nickel, chromium, and aluminum.

Nitriding consists of exposing steel parts to gaseous ammonia at about 1000°F (538°C) to form metallic nitrides at the surface. The hardest coatings are obtained with aluminum-bearing steels. Nitriding of stainless steel is known as **Malcomizing.** After nitriding, these steels have extremely high surface hardnesses of about Rockwell N 92 to 95. The nitride layer also has considerable resistance to corrosion from alkalies, the atmosphere, crude oil, natural gas, combustion products, tap water, and still saltwater. Nitrided parts usually grow about 0.001 to 0.002 in (0.003 to 0.005 cm) during nitriding. The growth can be removed by grinding or lapping, which also removes the brittle surface layer. Most uses of nitrided steels are based on resistance to wear. The steels can be used at temperatures as high as 1000°F (538°C) for long periods without softening. The slick, hard, and tough nitrided surface also resists seizing, galling, and spalling. Typical applications are cylinder liners for aircraft engines, bushings, shafts, spindles and thread guides, cams, and rolls.

A composition range of **Nitralloy** steel is 0.20 to 0.45% carbon, 0.75 to 1.5 aluminum, 0.9 to 1.8 chromium, 0.4 to 0.70 manganese, 0.15 to 0.60 molybdenum, and 0.3 maximum silicon. Nitralloy is marketed by various steel companies. **Nitrard** is also the name of a nitriding steel. Nitralloy steel is used for tools, gages, gears, and shafts. Unlike the soft core of ordinary case-hardened steels, it will have a tough core with high hardness. **Nitralloy 135** contains 0.35% carbon, 0.55 manganese, 0.30 silicon, 1.20 copper, 1 aluminum, and 0.20 molybdenum, and has a tensile strength, hardened, of 138,000 lb/in^2 (952 MPa) with elongation of 20% and Brinell hardness of 280. **Nitralloy N** is similar but with about 3.5% nickel, higher chromium, and less carbon, providing a Brinell hardness of 415.

Carbonitrided steel is produced by exposing the steel at about 1500°F (816°C) in a carbon-nitrogen atmosphere and then quenching in oil. The depth of the case depends on the length of time of treatment. The surface is harder and more wear-resistant than carbon case-hardened steel.

NITROCELLULOSE. A compound made by treating cellulose with nitric acid, using sulfuric acid as a catalyst. Since cotton is almost pure cellulose, it was originally the raw material used, but alpha cellulose made from wood is now employed. The cellulose molecule will unite with from one to six molecules of nitric acid. **Trinitrocellulose,** $C_{12}H_{17}O_7(NO_3)_3$, contains 9.13% nitrogen and is the product used for plastics, lacquers, adhesives, and Celluloid. It is classified as cellulose nitrate. The higher nitrates, or **pyrocellulose,** are employed for making explosives. Dry nitrocellulose explodes with a detonation velocity of 4.5 miles/s (7.3 km/s), so it is always stored in a humid state. It was originally called **guncotton,** and the original U.S. government name for the explosive was **Indurite,** from the Indian Head Naval Powder Factory. It was called **cordite** in England. The nitrated cellulose is mixed with alcohol and ether, kneaded into a dough, and squeezed through orifices into long, multitubular strings which are cut into short, cylindrical grains. Solid grains become smaller as they burn, so that there would be high initial pressure and then a decreasing pressure of gases. When the multitubular grains burn, the surface becomes greater, and thus there is increasing pressure. **FNH powder,** or **flashless powder,** is nitrocellulose which is nonhygroscopic and which contains a partially inert coolant, such as potassium sulfate, to reduce the muzzle flash of the gun. **Ballistite** is a rapid-burning, double-base powder used in shotgun shells and as a propellant in rockets. It is composed of 60% nitrocellulose and 40 nitroglycerin, made into square flakes 0.005 in (0.013 cm) thick or extruded in cruciform blocks.

NITROGEN.   An element, symbol N, which at ordinary temperatures is an odorless and colorless gas. The atmosphere contains 78% nitrogen in the free state. It is nonpoisonous and does not support combustion. Nitrogen is often called an *inert* gas, and is used for some inert atmospheres for metal treating and in lightbulbs to prevent arcing, but it is not chemically inert. It is a necessary element in animal and plant life and is a constituent of many useful compounds. Lightning forms small amounts of nitric oxide from the air which is converted to nitric acid and nitrates, and bacteria continuously convert atmospheric nitrogen to nitrates. Nitrogen combines with many metals to form hard nitrides useful as wear-resistant metals. Small amounts of nitrogen in steels inhibit grain growth at high temperatures and increase the strength of some steels. It is also used to produce a hard surface on steels. Nitrogen has five isotopes, and **nitrogen 15** is produced in enrichments to 95% for use as a tracer.

Most of the industrial use of nitrogen is through the medium of nitric acid, obtained from natural nitrates or from the atmosphere. *Fixation of nitrogen* is a term applied to any process whereby nitrogen from the air is transferred into nitrogen compounds, or **fixed nitrogen,** such as nitric acid or ammonia. The first step is by passing air through an electric arc to produce **nitric oxide,** NO, a heavy, colorless gas, which oxidizes easily to form **nitrogen dioxide,** NO_2, a brown gas with a disagreeable odor. This oxide reacts with water to form nitric acid. Or, atmospheric nitrogen can be converted to the oxide by irradiation of the compressed heated air with uranium oxide. Vast quantities of nitrogen are reacted with hydrogen to make ammonia fertilizers. Nitrogen for these applications is obtained by liquefaction of air. A recent method is to separate air into its constituents by using polymeric membranes. Permea, Inc. separates air by using membranes, as do Generon Systems, and Air Products and Chemicals. In the Kryoclean process, nitrogen is used to remove volatile organic compounds (VOCs) from process emissions. The emissions are taken in a gaseous nitrogen stream to condensers where liquid nitrogen cools the stream to a temperature at which the VOCs condense. The liquefied VOCs are then recovered. Nitrogen is used to stimulate tertiary oil wells. Nitrogen gas is used in plasma-arc and laser cutting and as a shielding gas in welding. **Calcium cyanamide,** $CaCN_2$, made by reacting atmospheric nitrogen with calcium carbide, is used as a fertilizer and as a chemical raw material. The chemical radical **cyanamid,** or **hydrogen cyanamide,** $H_2N \cdot C \cdot N$, is marketed as a stable, colorless 50% aqueous concentrate. The nitrogen-containing gas **Drycolene,** of General Electric Co., used for furnace atmospheres for sintering metals, contains 78% N_2, 20 CO, and 2 H_2. It is produced by burning hydrocarbon gases and air, removing the moisture, and passing through incandescent charcoal to convert the CO_2 and residual moisture to CO and H_2. Nitrogen liquefies at about $-319°F$ ($-195°C$) and solidifies at about $-346°F$ ($-210°C$). Nitrogen gas occupies 696 times as much space as the **liquid nitrogen** used in surgery.

Cryogenic cooling with liquid nitrogen speeds extrusion and improves the quality of polyolefin pipe. Liquid-nitrogen–based atmospheres, such as blends of **nitrogen-hydrogen** and **nitrogen-methanol,** are used for brazing. **Purifire-BR** atmosphere systems, of Air Products, are low-cost alternatives for brazing carbon steel. They are used to produce gas atmospheres from on-site, noncryogenically generated nitrogen and natural gas, using a proprietary purification system. Brazed parts exhibit good braze flow, surface appearance, and joint strength. Nitrogen gas derived from the liquid gas eliminates sparks in soldering electronic components and acts as a safety curtain at the entrace and exit of hydrogen-atmosphere furnaces. Nitrogen gas is used as a blanket over volatile liquids in vapor-recovery systems to prevent emission of hazardous vapors in process vessels into the atmosphere during storage, handling, and processing. The gas reduces the oxygen content in the vapor space above the liquid, reducing fire and explosion hazards and preventing air, moisture, and other contaminants from entering. By maintaining a constant pressure in the vapor space, the vessels can breathe during pumping operations and during ambient temperature changes that cause the liquid to contract or expand.

Nitrogen oxide and nitrogen dioxide generated by the combustion of fossil fuels are air pollutants, contributing to the formation of ozone, or photochemical smog, and acid rain. Thus, regulations limiting their emission have been instituted. These nitrogen oxides, or **NO_x** compounds, can be reduced to nitrogen and water by selective catalytic reduction. This involves injecting ammonia into the flue gas of heaters, boilers, gas-turbine systems, and coal-fired steam plants, then passing the gas through a reactor housing the catalysts.

NITROGLYCERIN. A heavy, oily liquid known chemically as **glyceryl trinitrate** and having the empirical formula $C_3H_5(NO_3)_3$. It is made by the action of mixed acid (90% nitric and 25 to 30 oleum) on very pure glycerol in the presence of sulfuric acid. It is highly explosive, detonating upon concussion. Liquid nitroglycerin when exploded forms carbonic acid, CO_2, water vapor, nitrogen, and oxygen; 1 lb (0.45 kg) is converted into 156.7 ft^3 (4.4 m^3) of gas. The temperature of explosion is about 628°F (330°C). For use as a commercial explosive it is mixed with absorbents, usually kieselguhr or wood flour, under the name of **dynamite.** Cartridges of high density explode with greater shattering effect than those of low density. By varying the density and the mixture of the nitroglycerin with ammonium nitrate, which gives a heaving action, a great diversity in properties can be obtained. **Ethylene glycol dinitrate** (nitroglycol) and **diethylene glycol dinitrate** are also explosives. They are generally used to plasticize nitrocellulose.

Dynamites are rated on the percentage, by weight, of nitroglycerin that they contain. A 25% dynamite has 25% by weight of nitroglycerin and a rate of detonation of 11,800 ft/s (3,597 m/s). The regular grades contain from 25 to 60%. **Ditching dynamite** is the 50% grade. It has a rate of detonation of 17,400 ft/s (5,304 m/s), and will detonate sympathetically from charge to charge along a ditch line. **Extra dynamite** has half of the nitroglycerin replaced by ammonium nitrate. It is not so quick and shattering, and not as water-resistant, but is lower in cost. It is used for quarrying, stump and boulder blasting, and highway work. A 50% extra dynamite has a detonation rate of 10,800 ft/s (3,292 m/s). **Hercomite** and **Hercotol** are extra dynamites of Hercules, Inc., while **Durox** is an ammonium dynamite of Du Pont, and **Agritol,** a low-velocity dynamite also of Du Pont, is a low-density ammonium dynamite for stump blasting.

Gelatin dynamite is made by dissolving a special grade of nitrocotton in nitroglycerin. It has less fumes, it is more water-resistant, and its plasticity makes it more adaptable for loading solidly in holes for underground work. It is marketed as straight gelatin or as **ammonium gelatin,** called **gelatin extra.** The gelatin dynamites come in grades from 20 to 90%. All have a detonation rate of 8,500 ft/s (2,591 m/s), but modified high-pressure gelatin has rates to 19,700 ft/s (6,005 m/s). These, however, produce large amounts of fumes and are not for use in mines or confined spaces. **Blasting gelatin,** called **oil-well explosive,** is a 100% dense and waterproof gelatin with the appearance of crude rubber and having a detonation rate of 8,500 ft/s (2,591 m/s). **Gelamite** and **Hercogel** are gelatin **blasting dynamites** of Hercules, Inc., although **Bituminite,** of this company, is a slow permissible ammonium nitrate dynamite for coal mines.

Gelobel is a gelatin dynamite, and **Monobel** is an ammonium dyna-mite marketed by Du Pont for mine blasting. The **Gelodyn explosive** of Atlas Powder Co. is a combination of ammonium gelatin dynamite that is plastic, gives a shattering effect, and does not produce excessive fumes. It is used for construction blasting. **Amocol,** of this company, is a blasting explosive composed of grained ammonium nitrate mixed with ground coal. The double-base solid propellant for rockets, known as **ballistite,** is nitroglycerin-nitrocellulose. With potassium perchlorate as an oxidizer, it gives a specific impulse of 180 to 195. It leaves plumes of white smoke. Dynamite is also sometimes used for explosive metal forming, as it releases energy at a constant rate regardless of confinement, and produces pressures to 2×10^6 lb/in^2 (1,379 MPa). For bonding **metal laminates,** a thin sheet, or film, of the explosive is placed on top of the composite, and the progressive burning of the explosive across the film produces an explosive force downward and in vectors that produces a microscopic wave, or ripple, in the alloyed bond that strengthens the bond but is not visible on the laminated sheet.

NONMAGNETIC STEEL. Steel and iron alloys used where magnetic effects cannot be tolerated. **Manganese steel** containing 14% manganese is nonmagnetic and casts readily but is not machinable. **Nickel steels** and **iron-nickel alloys** containing high nickel are also non-magnetic. Many mills regularly produce nonmagnetic steels containing from 20 to 30% nickel. **Manganese-nickel steels** and **manganese-nickel-chromium steels** are nonmagnetic and may be formulated to combine desirable features of the nickel and manganese steels. One nonmagnetic steel with a composition of 10.5 to 12.5% manganese, 7 to 8 nickel, and 0.25 to 0.40 carbon has low magnetic permeability and low eddy-current loss, can be machined readily, and work-hardens only slightly. The tensile strength is 80,000 to 110,000 lb/in^2 (552 to 758 MPa), elongation 25 to 50%, and specific gravity 8.02. It is austenitic and cannot be hardened. The 18–8 austenitic chromium-nickel steels are also nonmagnetic. A nonmagnetic alloy used for watch gears and escapement wheels is not a steel but is a **copper-nickel-manganese** alloy containing 60% copper, 20 nickel, and 20 manganese. It is very hard, but can be machined with diamond tools.

NONSHATTERING GLASS. Also referred to as **shatterproof glass, laminated glass,** or **safety glass,** and when used in armored cars, it is known as **bulletproof glass.** A material composed of two sheets of plate glass with a sheet of transparent resinoid between, the whole molded together under heat and pressure. When subjected to a severe blow, it will crack without shattering. The first of these was a

German product marketed under the name of **Kinonglas,** which consisted of two clear glass plates with a cellulose nitrate sheet between, and it was first used for protective shields against chips from machines. Nonshattering glass is now largely used for automobile and car windows. The original cellulose nitrate interlining sheets had the disadvantage that they were not stable to light and became cloudy. Cellulose acetate was later substituted. It is opaque to actinic rays and prevents sunstroke but has the disadvantage of opening in cold weather, permitting moisture to enter between the layers. The acrylic resins are notable for their stability in this use; in some cases they are used alone without the plate glass, especially for aircraft windows. Polyvinyl acetal resins, as interlinings for safety glass, are weather-resistant and will not discolor. Polyvinyl butyral is much used as an interlayer, but in airplane glass at about 150°F (66°C) it tends to bubble and ripple. Silicone resins used for this purpose withstand heat to 350°F (177°C), and they are not brittle at subzero temperatures. **Silastic Type K,** of Dow Chemical Co., is such a silicone resin used as an interlayer. **Flexseal,** of PPG Industries, is a laminated plate glass with a vinyl resin interplate with an extension for sealing into the window frame. It withstands a pressure of 20 lb/in^2 (0.14 MPa), with a 0.125-in (0.32-cm) plastic interplate, and is used for aircraft windows. **Duplate** is the trade name of Duplate Canada Inc. for a nonshattering glass. Standard bulletproof glass is from 1.5 in (3.81 cm), 3 ply, to 6 in, 5 or more ply.

NONWOVEN FABRIC. In the most general sense, fibrous-sheet materials consisting of fibers mechanically bonded together by interlocking or entanglement, by fusion, or by an adhesive. They are characterized by the absence of any patterned interlooping or interlacing of the yarns. In the textile trade, the terms **nonwovens** and **bonded fabrics** are applied to fabrics composed of a fibrous web held together by a bonding agent, as distinguished from felts, in which the fibers are interlocked mechanically without the use of a bonding agent. There are three major kinds of nonwovens based on the method of manufacture. **Dry-laid nonwovens** are produced by textile machines. The web of fibers is formed by mechanical or air-laying techniques, and bonding is accomplished by fusion-bonding the fibers or by the use of adhesives or needle punching. Either natural or synthetic fibers, usually 1 to 3 in (2.5 to 7.6 cm) in length, are used. **Wet-laid nonwovens** are made on modified papermaking equipment. Either synthetic fibers or combinations of synthetic fibers and wood pulp can be used. The fibers are often much shorter than those used in dry-laid fabrics, ranging from 0.25 to 0.5 in (0.64 to 1.27 cm). Bonding is usually accomplished by a fibrous binder or an adhesive. Wet-laid nonwovens can also be produced as composites, for example, tissue-paper laminates bonded to a reinforcing substrate of scrim. **Spin-bonded nonwovens** are produced by allowing the filaments emerging from the fiber-producing extruder to form into a random web, which is then usually thermally bonded. These nonwovens are limited commercially to thermoplastic synthetics such as nylons, polyesters, and polyolefins. They have exceptional strength because the filaments are continuous and bonded to each other without an auxiliary bonding agent. Fibers in nonwovens can be arranged in a great variety of configurations that are basically variations of three patterns: parallel or unidirectional, crossed, and random. The parallel pattern provides maximum strength in the direction of fiber alignment, but relatively low strength in other directions. Cross-laid patterns (like wovens) have maximum strength in the directions of the fiber alignments and less strength in other directions. Random nonwovens have relatively uniform strength in all directions.

NUTMEG. The brown, round, wrinkled seed of the plumlike fruit of the evergreen tree *Myristica fragrans,* native to the Moluccas but now grown extensively also in Grenada. The bright-red aril covering of the seed is called **mace.** The trees average about 20 lb (9 kg) of kernels per year, but a large tree may bear as many as 10,000 nutmegs annually. The average yield in Grenada is taken as 1,500 lb (680 kg) of green nutmegs per acre (4,047 m^2) per year, giving 720 lb (327 kg) of dry sound nutmegs and 150 lb (68 kg) of mace per acre (4,047 m^2). The nutmeg tree grows best on tropical islands at a height of 500 to 1,500 ft (152 to 457 m) above sea level. It begins to bear at 6 years, and will bear for a century. The ripe fruit splits, and the seeds fall to the ground. Nutmeg is a delicately flavored spice for foodstuffs, but in large amounts is highly toxic. Mace has a finer but weaker flavor and is used as a savory, but **oleoresin mace** of Fritzsche Dodge & Olcott Inc., a dark-brown liquid produced from mace, gives a lasting spicy nutmeg flavor and is used as a substitute for nutmeg oil. **Nutmeg butter** is a solid yellow fat obtained from the rejected nutmegs of the spice trade. To obtain the fat, the kernels are roasted and ground before extraction. The nutmeg contains about 40% of the fat. It is used chiefly in ointments. **Nutmeg oil** is an essential oil extracted from nutmeg and used in medicine, flavoring tobacco, and dentifrices. It is also called **myristica oil** and is high in **myristicin,** a yellow poisonous oil of composition $C_3H_5 \cdot C_6H_2(O_2CH_2)OCH_3$. It is now synthesized from pine oil.

NUX VOMICA. The seeds of the ripe fruit of the deciduous tree *Strychnos nux vomica* of India, Ceylon, and Australia, used as the source of the alkaloids strychnine and brucine. The powdered seed may also be used. The fruits contain three to five hard, grayish seeds which yield 1 to 1.25% strychnine alkaloid and about the same amount of brucine. **Strychnine** is an odorless, crystalline, intensely bitter powder of composition $C_{21}H_{22}N_2O_2$ with a very complex multiring molecular structure. It is a spinal stimulant and in quantity is a violent convulsive poison. It is used in proprietary and prescription medicines of the tonic class, and in rat poisons. For medicinal use it is employed mostly in the form of *strychnine sulfate* which is easily soluble in water. **Brucine** is a bitter, crystalline alkaloid of composition $C_{23}H_{26}N_2O_4$ with similar characteristics but much less active. It is **dimethoxystrychnine.** It is also used as a denaturant for rapeseed oil and other industrial oils. The woody vine **woorali,** *S. toxifera,* of the Amazon and Orinoco valleys, from which the arrow poison **curare** was obtained, contains strychnine and **curine,** a benzyl isoquinoline alkaloid. Curare inactivates the motor nerves without affecting the sensory and central nervous system and is used in medicine as a local anesthetic. The synthetic **Mytolon** is used as a more potent and safer substitute. It is a complex diethylaminopropylaminobenzoquinone benzyl chloride in the form of red crystals.

NYLON. A group of **polyamide resins** which are long-chain polymeric **amides** in which the amide groups form an integral part of the main polymer chain, and which have the characteristic that when formed into a filament, the structural elements are oriented in the direction of the axis. Nylon was originally developed as a textile fiber, and high tensile strengths, above 50,000 lb/in² (345 MPa), are obtainable in the fibers and films. But this high strength is not obtained in the molded or extruded resins because of the lack of oriented stretching. When **nylon powder** that has been precipitated from solution is pressed and sintered, the parts have high crystallinity and very high compressive strength, but they are not as tough as molded nylon. Nylons are produced from the polymerization of a dibasic acid and a diamine. The most common one of the group is that obtained by the reaction of adipic acid with hexamethylenediamine.

All of the nylons are highly resistant to common solvents and to alkalies, but are attacked by strong mineral acids. Molded parts have light weight, with a specific gravity of about 1.14, good shock-absorbing ability, good abrasion resistance, very low coefficient of friction, and high melting point, up to about 482°F (250°C). A disadvantage is their high water absorption and the resulting dimensional changes in moldings in service. They are much used for such parts as gears, bearings, cams, and linkages. The electrical characteristics are about the same as those of the cellulosic plastics. As a wire insulation, nylon is valued for its toughness and solvent resistance. **Nylon fibers** are strong, tough, and elastic and have high gloss. The finer fibers are easily spun into yarns for weaving or knitting either alone or in blends with other fibers, and they can be crimped and heat-set. For making carpets, nylon staple fiber, lofted or wrinkled, is used to give the carpet a bulky texture resembling wool. **Tire cord,** made from Nylon 6 of high molecular weight, has the yarn drawn to 4 or 5 times its original length to orient the polymer and give one-half twist per inch. **Nylon film** is made in thicknesses down to 0.002 in (0.005 cm) for heat-sealed wrapping, especially for food products where tight, impermeable enclosures are needed. **Nylon sheet,** for gaskets and laminated facings, comes transparent or in colors in thicknesses from 0.005 to 0.060 in (0.013 to 0.152 cm). **Nylon monofilament** is used for brushes, surgical sutures, tennis strings, and fishing lines. Filament and fiber, when stretched, have a low specific gravity down to 1.068, and the tensile strength may be well above 50,000 lb/in^2 (345 MPa). Nylon fibers made by condensation with oxalic esters have high resistance to fatigue when wet.

Nylon 6 is made from **caprolactam,** which has the empirical formula $(CH_2)_5 NH \cdot C:O$, with a single six-carbon ring. Molded parts have a tensile strength of 11,700 lb/in^2 (79 MPa), elongation 70%, dielectric strength 440 V/mil (17.3×10^6 V/m), and melting point 420°F (216°C). **Nylon foam,** or **cellular nylon,** for lightweight buoys and flotation products, is made from Nylon 6. The foam is produced by Du Pont in slabs, rods, and sheets. Density ranges from 1 to 8 lb/ft^3 (16 to 128 kg/m^3). The low-density types are flexible, but the high-density material is rigid with a load-carrying capacity about the same as that of balsa wood. **Ultramid A3HG7,** a glass-fiber-reinforced Nylon 6/6 of BASF, and Du Pont's **Zytel 6/6** are used for auto engine air-intake manifolds for weight reduction over cast aluminum designs. **Nylatron GS-51,** of DSM Engineering Plastics, is a glass-reinforced and molybdenum-disulfide-filled Nylon 6/6 used for auto engine valve-lifter guides. **Lubriloys,** of LNP Engineering Plastics, are lubricated 6/6 blends. **Minlon 2C,** of Du Pont, is a glass- and mineral-reinforced 6/6. Nylon 6 and Nylon 6/6 are also used for a great variety of mechanical parts. **Nylon copolymers** of types 6 and 6/6 provide additional impact resistance, to temperatures as low as −40°F (−40°C), with good heat resistance. Nylon 6 or 6/6, in 420, 630, and 840 denier, is used for auto airbags. They are sometimes coated with neoprene for sealing and for protection from the heat of pyrotechnic inflators. **Nylon 6/10** is tough, relatively heat-resistant, and has a very low brittleness temperature. It absorbs about one-third as much moisture as type 6 and half as much as type 6/6. **Nylon 9** is made from soybean oil by reacting with ozone. It has better water resistance than other nylons and is used for coatings. **Nylon 11** is a polycondensation product of **aminoundecanoic acid** which is made by a complex process from the ricinoleic acid of castor oil. This type of nylon has superior dimensional stability and is valued for injection moldings. **Nylon 12** is a similar plastic. It is a lauro lactam synthesized from butadiene. It has a low water absorption and good strength and stability, and it is used for packaging film, coatings for metals, and moldings. **Nylon 4** is a polypyrrolidine used for textile fibers. The molecular chain has more amide groups than do the chains of other nylons, and its ability to absorb moisture is about the same as that of cotton. Fabrics made from it do not have the hot feel usual with other synthetic fibers, and they have better pressability and are free of static.

Grivory G21, of EMS-American Grilon, is an **amorphous polyamide** for extrusion into multilayer film, bottles, and tubes. It serves as a barrier to aroma, oxygen, and carbon dioxide. **Tepex,** of Du Pont, is a family of custom-made thermoplastic laminates, mostly nylon, combined with various fiber reinforcements.

OAK. The wood of a large variety of oak trees, all of the natural order Cupuliferae, genus *Quercus.* European oak, under various names, such as **Austrian oak** and **British oak,** is from two varieties of the tree *Q. robur.* The wood is light brown, with a coarse, open grain, firm texture, and density of about 45 lb/ft^3 (720 kg/m^3). **American red oak** is from the tree *Q. rubra* or *Q. falcata.* It is also called **black oak,** although black oak is from *Q. velutina,* and the **red oak** of the Lake states is *Q. borealis.* The heartwood is reddish brown, and the sapwood whitish. **Southern red oak** of the Gulf Coast, a valued wood for furniture and cabinetwork, is the **shumard oak,** *Q. shumardii,* also known as **Schneck oak** and **Texas oak. Nuttall oak,** *Q. nuttallii,* of the lower Mississippi Valley, is also called red oak. American **white oak** is from the tree *Q. alba* of the eastern states. The heartwood is brown, and the sapwood white. The grain of these species is coarse, but the texture is firm. **Post oak,** of the southern states, is *Q. stellata.* **Chestnut oak,** of the Appalachian range, is *Q. montana,* but this name is also applied to the **chinquapin oak,** *Q. muehlenbergii,* a large tree which grows profusely over a wide area of the eastern half of the United States, and was early valued for railroad ties and heavy construction timbers. **Overcup oak,** *Q. lyrata,* is an important tree from New Jersey to Texas. **Scarlet oak,** of Pennsylvania, is *Q. coccinea.* **Western white oak,** *Q. garryana,* has a more compact texture and straighter grain. **Spanish oak,** *Q. oblongifolia,* is native to California and New Mexico. The grain is finer and denser. American oaks are widely distributed in the United States and Canada. There are more than 400 varieties of oak on the North American continent. An enormous stand of oak in Costa Rica is made up of immense trees of **copey oak,** *Q. copeyensis,* the trees being up to 8 ft (2.4 m) in diameter with clean boles to 80 ft (24.4 m) to the first limb. The wood has a hardness between that of white and live oaks, and the bark has a high content of tannin.

Oak is used for flooring, furniture, cask staves, and where a hard, tough wood is needed. For cabinetwork the boards are variously sawed at angles and quarters to obtain grain effects known as **quartered oak. Fumed oak** is not a kind of oak, but a finish produced by the action of ammonia vapor. **Butt oak,** or **pollard oak,** also known as **burwood,** is the wood of the decapitated **European oak** trees, *Q. pedunculata* and *Q. sessiliflora,* of Great Britain. A pollard tree is one whose head has been cut for ornamental purposes. The growth in height is permanently arrested and innumerable branches shoot out from the trunk, which produce humps, or burrs, with the grain of the wood running in all directions. **Burr oak** is valued for ornamental work. Burr oak of the northern and central United States is not a pollard oak but is a name for the tree *Q. macrocarpa.* The commercial red and white oaks have an average specific gravity when kiln-dried of 0.69. The compressive strength perpendicular to the grain is 1,870 lb/in^2 (13 MPa) with shearing strength parallel to the grain of 1,300 lb/in^2 (9 MPa).

The woods often called oaks in the southern hemisphere are not true oaks. **Australian oaks** are from a variety of trees, and **Chilean oak** is from a species of beech. **Beef oak,** of Australia, is a hard, heavy, brownish wood from the tree *Grevillea striata.* It has an irregular grain. **She oak** is from the Australian tree *Casuarina stricta,* and **swamp oak** is from *C. suberosa.* These woods are lighter in weight than oak. **Silky oak,** used for cabinetwork, is a brownish wood that has a uniform texture and can be quartersawn to show attractive figuring. It is from the tree *Cardwellia sublimis* of Australia.

Oak extract, which is an important tanning material for the best grades of heavy leather, is chiefly from the bark of the **swamp chestnut oak,** *Q. prinus,* but also from the white oak and red oak. The **tanbark oak** of California is the tree *Lithocarpus densiflora.* The extract of the **scarlet oak,** *Q. coccinea,* is dark in color and is known as **quercitron extract.** The bark of the tanbark oak yields yields 10 to 14% tannin, but the extract contains 25 to 27% tannin. **Quercetin** is a complex phenyl benzyl pyrone derived from oak bark and from Douglas fir bark. It is an antioxidant and absorber of ultraviolet rays, and is used in rubber, plastics, and vegetable oils. It is also found in red grapes, red and yellow onions, broccoli, and yellow squash and is believed to be an anticarcinogen. **Valonia** consists of the acorn cups of the oak *Q. aegilops* of Asia Minor and the Balkans. **Smyrna valonia** contains 32 to 36% tannin which produces a light-colored, lightweight leather with a firm texture and bloom. When used alone, however, valonia makes a brittle leather and is thus always used in blends. Valonia is marketed as cups or as extract, the latter containing about 60% tannin.

OATS. An important grain which is the seed of the tall plant *Avena sativa.* The grain is surrounded by a hull and grows in many spikelets as a spreading or one-sided panicle inflorescence. It can be grown farther north than any other grain except rye, and on poor soils. Although it is one of the most nutritious of grains, most of the oats grown in the United States are used for animal feed. **Rolled oats** and **oatmeal** are used as cereal foods and for some bakery products, but the grain is not suitable for breadmaking. **Oat hulls** are used for the production of furfural and other chemicals. The largest production of oats is in the United States and Russia, but large quantities are produced in Canada, western Europe, and Argentina. It is the chief grain crop of Scotland. The yield per acre (4,047 m^2) in the United States is about 30 bu (1 m^3), but it is twice that figure in Great Britain. Oats are often called by the Spanish name **avena** in international trade. **Turkish oats,** cultivated in central Europe, are from the species *A. orientalis.* **Horse gram,** used as a substitute for oats in India, is from the plant *Dolichus bifloris.* The **gram,** from the *Cicer arientinum,* is an important food grain in India.

OCHRE. A compact form of earth used for paint pigments and as a filler for linoleum, also spelled **ocher.** It is an argillaceous and siliceous material, often containing compounds of barium or calcium, and owing the yellow, brown, or red colors to hydrated iron oxide. The tints depend chiefly upon the proportions of silica, white clay, and iron oxide. Ochres are very stable as pigments. They are prepared by careful selection, washing, and grinding in oil. They are inert and are not affected by light, air, or ordinary gases. They are rarely adulterated, because of their cheapness, but are sometimes mixed with other minerals to alter the colors. **Chinese yellow** and many other names are applied to the ochres. **Golden ochre** is ochre mixed with chrome yellow. **White ochre** is ordinary clay. A large part of the U.S. ochre is produced in Georgia. **Sienna** is a brownish-yellow ochre found in Italy and Cyprus. The material in its natural state is called **raw sienna. Burnt sienna** is the material calcined to a chestnut color. Indian red and Venetian red are hematite ochres.

   Vandyke brown is a deep-brown pigment made originally from lignitic ochre from Cassel, Germany. It was named after the Dutch painter Van Dyck, and is also called **Cassel brown, Cassel earth,** and **Rubens brown.** It contains up to 90% organic water, water and traces of iron oxides, and alumina. It is also obtained from low-grade coals of Oklahoma and California. Imitation Vandyke brown is made from a mixture of lampblack, yellow ochre, and iron oxide derived from copperas, ferrous sulfate. **Cologne earth** is a Vandyke brown made from U.S. clays which are mixtures of ochre, clay, and bituminous matter, roasted to make the color dark. **Yellow ochre** and **brown ochre** are limonite, but yellow iron oxide is made in Germany by the aeration of scrap iron in the presence of copperas. **Umber** is a brown siliceous earth colored naturally with iron oxides and manganese oxide. It comes chiefly from Italy and Cyprus. For use as a pigment it is washed with water and finely ground. It is inert and very stable. **Cyprus umber** is a rich, coffee-brown color and as a pigment has good covering qualities. It is a modified marl with impregnations of iron and manganese. **Burnt umber** is redder than umber and is made by calcining the raw umber. **Caledonian brown** and **Cappagh brown** are varieties of umber found in Great Britain.

OILCLOTH. A fabric of woven cotton, jute, or hemp, heavily coated with turpentine and resin compositions, usually ornamented with printed patterns, and varnished. It was employed chiefly as a floor covering, but a light, flexible variety having a foundation of muslin is used as a covering material. This class comes in plain colors or in printed designs. It was formerly the standard military material for coverings and ground protection, but has been replaced by synthetic fabrics. **Oilskin** is a cotton or linen fabric impregnated with linseed oil to make it waterproof. It was used for coverings for cargo and for waterproof coats, but has now been replaced by coated fabrics. **Oiled silk** is a thin silk fabric impregnated with blown linseed oil which is oxidized and polymerized by heat. It is waterproof, very pliable, and semitransparent. It was much used for linings, but has now been replaced by fabrics coated with synthetics.

OILS. A large group of fatty substances which are divided into three general classes: vegetable oils, animal oils, and mineral oils. The **vegetable oils** are either fixed or volatile oils. The fixed oils are present in the plant in combined form and are largely glycerides of stearic, oleic, palmitic, and other acids, and they vary in consistency from light fluidity to solid fats. They nearly all boil at 500 to 600°F (260 to 316°C), decomposing into other compounds. The volatile, or essential, oils are present in uncombined form and bear distillation without chemical change.

 Seed oils, or **oilseeds,** obtained from various plant seeds, are fatty acids of varying chain lengths containing hydroxy, keto, epoxy, and other functional groups. The oils are chemically very pure. Among important uses of these oils are for polymers, surface coatings, plasticizers, surfactants, and lubricants. The seeds of the **Chinese tallow tree** are coated with a semisolid fat. An oil similar to linseed oil is inside the kernel. The oil can be used as a substitute for cocoa butter and for fatty acids in cosmetics.

Fish oils are thick, with a strong odor. Vegetable and animal oils are obtained by pressing, extraction, or distillation. Oils that absorb oxygen easily and become thick are known as *drying oils* and are valued for varnishes, because on drying they form a hard, elastic, waterproof film. Unsaturation is proportional to the number of double bonds, and in food oils these govern the cholesterol depressant effect of the oil. Oils and fats are distinguished by consistency only, but waxes are not oils. **Mineral oils** are derived from petroleum or shale and are classified separately. The most prolific sources of vegetable oils are palm kernels and copra. About 2,500 lb^2 (1,134 kg) of palm oil is produced per acre (4,047 m^2) annually, and the yield of coconut oil per acre (4,047 m^2) from plantation plantings is 1,200 lb (544 kg). This compares with 350 lb (159 kg) of oil per acre (4,047 m^2) from peanuts and 200 lb (91 kg) per acre (4,047 m^2) from soybeans. Under comparable aggressive plantation work, from 10 to 20 times more palm and coconut oil can be produced per acre than peanut or soybean oil. Babassu oil is almost chemically identical with coconut oil, and vast quantities of babassu nuts grow wild in northeast Brazil.

Blown oils are fatty oils that have been oxidized by blowing air through them while hot, thereby thickening the oil. They are mixed with mineral oils to form special heavy lubricating oils, such as marine engine oil, or are employed in cutting oils. They are also used in paints and varnishes, as the drying power is increased by the oxidation. The flash point and the iodine value are both lowered by the blowing. The oils usually blown are rapeseed, cottonseed, linseed, fish, and whale oils. The **blown fish oils** of Archer-Daniels-Midland Co., used for paints, enamels, and printing inks, are preoxidized and destearinized, and have specific gravities from 0.980 to 1.025. **Crystol oils,** of this company, are kettle-boiled fish oils for paints.

OILSTONE. A fine-grained, slaty silica rock for sharpening edged tools. The bluish-white and opaque white oilstones of fine grain from Arkansas are called **novaculite,** and they received their name because they were originally used for razor sharpening. They are composed of 99.5% chalcedony silica and are very hard with a fine grain. Novaculite is a deposit from hot springs. It is fine-grained, and the ordinary grades are employed for the production of silica refractories. Arkansas oilstones are either hard or soft and have a waxy luster. They are shipped in large slabs or blocks, or in chips for tumbling barrel finishing. **Washita oilstone,** from Hot Springs, Arkansas, is a hard, compact, white stone of uniform texture. **Ouachita stones** come in larger and sounder pieces but are coarser than the Arkansas. **Water-of-Ayr stone,** also known as **Scotch hone,** is a fine sandstone used with water instead of with oil. **Artificial oilstones** are also produced of aluminum oxide. **India oilstone** was originally blocks of emery, but the name now may refer to aluminum oxide stones.

OITICICA OIL. A drying oil obtained from the kernels of the nuts of the tree *Licania rigida* of northeastern Brazil. The oil contains about 80% **licanic acid,** which, like the **eleostearic acid** of tung oil and isano oil, gives a greater drying power than is apparent from the iodine value. The specific gravity is 0.944 to 0.971, saponification value 187 to 193, and iodine number 142 to 155. The properties as a varnish oil are much like those of tung oil, both producing wrinkled films when applied pure and both lacking high gloss. **Cicoil** is a name for a treated oiticica oil with improved qualities. Treatment generally involves heating to above 437°F (225°C). Phenolic resins attain greater body with oiticica oil than with tung oil. The oiticica nuts are 1 to 2 in (2.5 to 5.1 cm) long with the kernel about 60% of the nut, yielding about 60% oil. The average yield per tree is 350 lb (159 kg) of nuts, but a full-grown tree may yield 10 times that amount. Another species of the tree, *L. crassifolia,* of Surinam, yields a similar oil. Mexican oiticica is from the nuts of another species and is called **cacahuanache oil.** The kernels yield 69% of light-colored heavy oil.

OLEFIN COPOLYMERS. A group of polyolefin plastics. They are thermoplastics and are also referred to as **polyolefin copolymers.** The principal olefin copolymers are the polyallomers, ionomers, and ethylene copolymers. The **polyallomers,** which are highly crystalline, can be formulated to provide high stiffness and medium impact strength, moderately high stiffness and high impact strength, or extrahigh impact strength. Polyallomers, with their unusually high resistance to flexing fatigue, have "hinge" properties better than those of polypropylenes. They have the characteristic milky color of polyolefins; they are softer than polypropylene but have greater abrasion resistance. Polyallomers are commonly injection-molded, extruded, and thermoformed, and they are used for such items as typewriter cases, snap clasps, threaded container closures, embossed luggage shells, and food containers.

Ionomers are nonrigid plastics characterized by low density, transparency, and toughness. Unlike polyethylenes, density and other properties are not crystalline-dependent. Their flexibility, resilience, and high molecular weight combine to provide high abrasion resistance. They have outstanding low-temperature flexural properties, but should not be used at temperatures above 160°F (71°C). Resistance to attack from organic solvents and stress-cracking chemicals is high. Ionomers have high melt strength for thermoforming and extrusion coating, and a broad temperature range for blow molding and interjection molding. Representative ionomer parts include injection-molded containers, housewares, tool handles, and closures; extruded film, sheet, electrical insulation, and tubing; and blow-molded containers and packaging.

Ethylene vinyl acetate, or **EVA, copolymers** approach elastomers in flexibility and softness, although they are processed as other thermoplastics are. Many of their properties are density-dependent, but in a different way from polyethylenes. Softening temperature and modulus of elasticity decrease as density increases, which is contrary to the behavior of polyethylenes. Likewise, the transparency of EVA increases with density to a maximum that is higher than that of polyethylenes, which become opaque when density increases above around 0.034 lb/in^3 (935 kg/m^3). Although EVA's electrical properties are not as good as those of low-density polyethylene, they are competitive with those of vinyl and elastomers normally used for electrical products. The major limitation of EVA plastics is their relatively low resistance to heat and solvents, the Vicat softening point being 147°F (64°C). EVA copolymers can be injection-, below-, compression-, transfer-, and rotationally molded; they can also be extruded. Molded parts include appliance bumpers and a variety of seals, gaskets, and bushings. Extruded tubing is used in beverage vending machines and for hoses for air-operated tools and paint spray equipment.

Ethylene ethyl acrylate, or **EEA, copolymer** is similar to EVA in its density-property relationships. It is also generally similar to EVA in high-temperature resistance, and like EVA, it is not resistant to aliphatic and aromatic hydrocarbons or chlorinated versions thereof. However, EEA is superior to EVA in environmental stress cracking and resistance to ultraviolet radiation. As with EVA, most of EEA's applications are related to its outstanding flexibility and toughness. Typical uses are household products such as trash cans, dishwasher trays, flexible hose and water pipe, and film packaging.

Ethylene n-butyl acrylate copolymers are closely related to EVA copolymers. The **EnBAs** are based on the comonomer n-butyl acrylate rather than vinyl acetate. Property advantages of the EnBAs include higher heat stability and greater low-temperature flexibility and oil resistance. Like EVAs, the EnBAs are used in hot melt adhesives and packaging film.

Two other ethylene copolymers are **ethylene hexene, or EH, copolymer** and **ethylene butene, or EB, copolymer.** Compared with the other two, these copolymers have greater high-temperature resistance, their useful service range being between 150 and 190°F (66 and 88°C). They are also stronger and stiffer, and therefore less flexible, than EVA and EEA. In general, EH and EB are more resistant to chemicals and solvents than the other two, but their resistance to environmental stress cracking is not as good.

Dow Plastics' **Affinity polyolefin plastomers** and **Engage polyolefin elastomers** are based on the company's "single-site" Insite metallocene catalysis, introduced in the early 1990s, which imparts very narrow molecular weight distribution and highly uniform comonomer distribution. Advantages claimed for the Affinity family of polymers include better low-temperature performance, thermal stability, and ease of processing than flexible polyvinyl chloride (f-PVC); better toughness, heat-sealing initiation, and taste and odor qualities than ionomers; better puncture, tear, and moisture resistance, clarity, and elasticity than EVA; and better clarity, puncture resistance, and elasticity than linear low-density polyethylene (LLDPE) and ultralow-density polyethylene, with a controllable range of lower and sharper melting points. They are also intended to compete with ethylene-propylene-diene-monomer-modified polypropylene and styrene block copolymers. Both polyolefins have a lower density than f-PVC— 0.031 to 0.032 lb/in^3 (858 to 886 kg/m^3) versus 0.044 to 0.047 lb/in^3 (1,218 to 1,301 kg/m^3)—and, thus, can provide about 40% more parts per given resin quantity. **Affinity SM 1250** is intended to replace PVC and EVA in appliances, toys, and siding. **Engage EG 8100** is a general-purpose grade, which with 24% octene comonomer, has a Mooney viscosity of 23, a melt index of 0.013 lb/h (0.006 kg/h), a tensile modulus of 350,000 lb/in^2 (2,413 MPa), an ultimate tensile strength of 1,500 lb/in^2 (10 MPa), 800% ultimate elongation, and a Shore A hardness of 75. Also based on metallocene catalysis—Exxpol technology—are the **Exact ethylene polymers** of Exxon Chemical for polyolefin modification. Among other possible applications, these are aimed at improving processability and performance of polyethylene by using standard LLDPE blown-film equipment. **Keldax,** of Du Pont, is a dense thermoplastic ethylene copolymer.

OLEIC ACID. Also called **red oil, elaine oil, octadecenoic acid,** and **rapic acid,** although the latter is a misnomer based on a former belief that it was the same as the crucic acid of rapeseed. It occurs in most natural fats and oils in the form of glyceride, and it is obtained in the process of saponification or by distillation. Much of this acid is obtained from lard and other animal fats, but **Emery 3758-R** is produced from soybean or other vegetable sources by hydrolysis of **glycerol trioleate.** It is an oily liquid with a specific gravity of 0.890, boiling at 547°F (286°C). Below about 57°F (14°C) it forms colorless needles. It is a complex acid of composition $CH_3(CH_2)_7CH{:}CH(CH_2)_7COOH$, and if heated to the boiling point of water, it reacts with oxygen to form a complex mixture of acids, including a small percentage of acetic and formic acids. When it is hydrogenated in food fats, it converts to stearic acid. When reacted with potassium hydroxide, it is converted to an acetate and a palmitate. It is also readily converted to pelargonic and other acids for making plastics. Oleic acid is a basic foodstuff in the form of the glyceride, and the acid has a wide use for making soaps, as a chemical raw material, and for finishing textiles. In soluble oils and cutting compounds it forms **sodium oleate,** $C_{17}H_{33}COONa$. The two commercial grades of oleic acid, yellow and red, are known as *distilled red oil* and *saponified.* They may be sold under trade names. **Alcholein 810** is a clear, distilled red oil used for textile treating. **Monoenoic acid** is a modified isomer of oleic acid which produces soaps that are nonirritating to skin. It is used in cosmetics. **Hydrofol C-18,** of Archer-Daniels-Midland Co., is this acid. **Aluminum oleate** is used for thickening lubricating oils.

OLIVE OIL. A pale-greenish, oily liquid extracted from the ripe fruit of the olive tree, *Olea europaea,* a small evergreen grown largely in the Mediterranean countries but also in California and Argentina. The fruits are eaten ripe (purple) and green. They are rich in oil, and vast quantities are crushed for oil. The oil contains 69 to 85% oleic acid, 7 to 14 palmitic acid, 4 to 12 linoleic acid, with some stearic, arachidic, and yristic acids. The specific gravity is 1.912, iodine value 85, and saponification value 190. The best grades of the oil result from the first cold-pressing step, and are used for food chiefly as a salad and cooking oil, and in canning sardines, but some are used in the manufacture of castile soaps. The industrial oil consists of the **olive oil foots** obtained in the third pressing or in the last extraction with carbon bisulfide, and is used for finishing textiles, degumming silk, and soaps. **Florence oil** is a grade of Italian olive oil. In Italy olive oil is also known as **Lucca oil. Synthetic olive oil,** or **olive-infused oil,** is used as a foodstuff. It is made from highly refined corn oil by infusing the corn oil with about 20% of a paste made of finely ground, partly dehydrated ripe olives ground with a small amount of corn oil. The olive-infused oil has the flavor of olive oil, and also contains carotene, or vitamin A, found in the olive pulp. Other fractionated oils reblended to give high oleic acid content are also used as substitutes for olive oil. **Olevene** is a sulfonated synthetic oil used instead of olive oil for treating textiles.

OLIVINE. A translucent mineral, usually occurring in granular form, employed as a refractory. The formula is usually given as $(Mg \cdot Fe)_2 \cdot SiO_4$, but it is a solid solution of **forsterite,** $2MgO \cdot SiO_2$, and **fayalite,** $2FeO \cdot SiO_2$. The fayalite lowers the refractory quality, but forsterite is not found alone. The mineral is also called **chrysolite,** and the choice green stones used as gems are called **peridot. Dunite** deposits in Washington and North Carolina carry up to 90% olivine which has only 5 to 15% fayalite. It is olive green in color, vitreous, with a Mohs hardness of 6.5 to 7 and a specific gravity of 3.3 to 3.5. As a refractory, it is neutral up to about 2912°F (1600°C) but may then react with silica. The fayalite fuses out at 2700°F (1482°C), making the material porous and subject to attack by iron oxide. Although the name *olivine* indicates a green color, not all is green. Dunite takes its name from Dun Mountain of New Zealand, *dun* being the Irish and Scottish word for reddish brown. The melting point of forsterite is 3470°F (1912°C). When it is used mixed with chrome ore, the low-fusing elements form a black glass which presents a nonporous face. Some refractory material marketed as forsterite may be olivine blended with magnesite, or may be serpentine treated with magnesite. **Forsterite firebrick** in the back walls of basic open-hearth steel furnaces gives longer life than silica brick but only two-thirds that of chrome-magnesite brick. Forsterite refractories are usually made from olivine rock to which MgO is added to adjust the composition to $2MgO \cdot SiO_2$. **Monticellite,** $CaMgSiO_4$, may also occur with forsterite. They are also made by synthetic mixtures of MgO and silica. The thermal expansion of olivine is lower than that of magnesite. **Olivine sand** is substituted for silica sand as a foundry sand where silica is expensive. There are large deposits of olivine in the Pacific northwest. When olivine is used as a foundry sand, it is noted that the heat-resisting qualities decrease with particle size. Olivine contains from 27 to 30% magnesium metal and is also used to produce magnesium by the electrolysis of the chloride. **Magnesium phosphate** fertilizer is made by fusing olivine with phosphate rock at 2912°F (1600°C), tapping off the iron, and spray-cooling and crushing the residue. It contains 20% citric-acid–soluble phosphate, 14 MgO, 29 CaO, and 23 SiO_2, and is useful for acid soils.

ONYX. A variety of chalcedony silica mineral differing from agate only in the straightness of the layers. The alternate bands of color are usually white and black, or white and red. Onyx is artificially colored in the same way as agate. It is used as an ornamental building stone, usually cut into slabs, and for decorative articles. **Onyx marble** is limestone with impurities arranged in banded layers. U.S. onyx comes largely from Arizona, California, and Montana. **Mexican onyx** is banded limestone obtained from stalactites in caves. These materials are cut into such articles as lamp stands. **Argentine onyx** is a dark-green or a green-yellow, translucent stone of great decorative beauty. In the United States it is called **Brazilian onyx** and is used for bookends, lamp bases, inkstands, and ornaments. **Opalized wood** is an onyxlike petrified wood from Idaho. It is cut into ornaments.

OPACIFIERS. Materials used in ceramic glazes and vitreous enamels primarily to make them nontransparent, but opacifiers may also enhance the luster, control the texture, promote craze resistance, or stabilize the color of the glaze. An opacifier must have fire resistance so as not to vitrify or decrease the luster. Tin oxide is a widely used white opacifier, and up to 3% also increases the fusibility of the glaze or enamel. Titanium oxide adds scratch hardness and high acid resistance to the enamel. It also increases the flow, making possible thinner coats which minimize chipping. Opacifiers may also serve as the pigment colors. Thus, cobalt oxide gives a blue color, and platinum oxide gives a gray. Lead chromate gives an attractive red color on glazes fired at 1652°F (900°C), but when fired at 1832°F (1000°C), the lead chromate decomposes and a green chromium oxide is formed. If the glaze is acid, the basic lead chromate is altered and the color tends toward green. **Lufax 77A,** of Rohm & Haas Co., is a crystalline zirconia which provides nuclei for the formation of zirconia crystals from the molten enamel, adding gloss and opacity and stabilizing the color on the blue side. Antimony oxide as an opacifier gives opaque white enamels of great brilliance but is expensive and poisonous. The **zirconium opacifiers** have a wide range of use from ordinary dishes to high-heat electrical porcelain and sanitary-ware enamels. The amount of zirconium oxide used is a minimum of 3%. The opacifiers may be in prepared form with lead oxide or other materials to give particular characteristics. **Opax** is a zirconium oxide with small percentages of silica, sodium oxide, and alumina. It is used for hard-glaze dinnerware and wall-tile glaze. **Zircopax** is zirconium silicate, $ZrSiO_4$, with 33.5% silica in the molecule. It gives color stability and craze resistance. **Superpax** is a finely milled zirconium silicate powder with an average particle size less than 197 μin (5 μm). In white ceramic glazes, very small amounts will give opacity. The **Ultrox opacifiers,** of M & T Chemicals, Inc., are refined zirconium silicates. **Ultrox 1000W,** for maximum opacity and whiteness, has 65% ZrO_2 and 35 SiO_2 with particle size of 20 μin (0.5

μm). Lead oxide is used to lower the melting point of a glaze. Matte effects are obtained by adding barium oxide, magnesia, or other materials to the opacifier.

OPEN-HEARTH STEEL.   Steel made by the process of melting pig iron and steel or iron scrap in a lined regenerative furnace, and boiling the mixture with the addition of pure lump iron ore until the carbon is reduced. The boiling is continued for 3 to 4 h. The process was developed in 1861 by Siemens in England. The furnaces contain regenerative chambers for the circulation and reversal of the gas and air. The fuels used are natural gas, fuel oil, coke-oven gas, or powdered coal. Both the acid- and the basic-lined open-hearth furnaces are used, but most steel made in the United States is basic open hearth. Ganister is used as a lining in the acid furnaces, and magnesite in the basic.

An advantage of the open-hearth furnace is the ability to handle raw materials that vary greatly and to employ scrap. Iron low in silicon requires less heating time. The duplex process consists in melting the steel in an acid Bessemer furnace until the silicon, manganese, and part of the carbon have been oxidized, and then transferring to a basic open-hearth furnace where the phosphorus and the remainder of the carbon are removed. Open-hearth steel is of uniform quality and is produced in practically all types.

OPIUM.   The dried fruit from the unripe capsules of the **poppy plant,** *Papaver somniferum,* cultivated extensively in China, India, and the Near East, but also growing wild in many countries. The opium poppy is an annual with white flowers. After the petals drop off, the capsules are cut and the juice exudes and hardens. The crude opium is a brownish mass. It contains about 20 alkaloids which are useful in medicine. Opium alone is a powerful narcotic, but the material is usually processed and the alkaloids are employed separately or in combinations for their particular effects.

Morphine, $C_{17}H_{19}NO_3 \cdot H_2O$, a white powder melting at 487°F (253°C), is the most important of the opium alkaloids. It is a powerful narcotic and painkiller. It has a complex five-ring molecular structure which can be synthesized from the three-ring **phenanthrene,** $C_{14}H_{10}$, an isomer of anthracene occurring in coal tar. **Codeine,** a white powder melting at 477°F (247°C), is a methyl ether of morphine and is a painkiller less powerful than morphine. It is much used in cough medicines. **Dionine** is **ethyl morphine** and is also an important drug. **Heroin** is **diacetyl morphine.** It is a powerful narcotic, but its use is prohibited in the United States. **Colchicine,** $C_{22}H_{25}NO_6$, is a complex three-ring alkaloid used as a gout remedy. Its action is to quicken the release of **heparin** from intestinal cells, which decomposes fat in the blood and prevents blood clotting. It is chemically similar to morphine, but has the acetyl amino group in a different position.

Laudanum is an alcohol solution of opium. **Amidone** is a German synthetic morphine. It is a diphenyl dimethylaminoheptanone, is stronger than morphine as a painkiller, and, like morphine, is an exhilarant and habit-forming. The English drug **Heptalgin** is a similar morphine substitute. **Poppy-seed oil** is a colorless to reddish-yellow liquid of specific gravity of about 0.925 and iodine number 157 used as a drying oil in artists' varnishes. The cold-pressed white oil is used locally as an edible oil. The very dark grades are used in soaps and in paints. The oil from the seed does not contain opium.

ORE. A metal-bearing mineral from which a metal or metallic compound can be extracted commercially. Earths and rocks containing metals that cannot be extracted at a profit are not rated as ores. Ores are named according to their leading useful metals. The ores may be oxides, sulfides, halides, or oxygen salts. A few metals also occur native in veins in the minerals. Ores are usually crushed and separated and concentrated from the **gangue** with which they are associated, and then shipped as **concentrates** based on a definite metal or metal oxide content. The metal content to make an ore commercial varies widely with the current price of the metal and with the content of other metals present in the ore. Normally, a sulfide copper ore should have 1.5% copper in the unconcentrated ore, but if gold or silver is present, an ore with much less copper is workable; or if the deposit can be handled by high-production methods, a mineral of very low metal content can be utilized as ore. Low-grade lead minerals can be worked if silver is recoverable, and low-grade manganese minerals become commercial when prices are high. Thus, the term *ore* is only relative, and under different economic conditions, minerals that are not considered ores in one country may be much used as ores in another.

OSMIUM. A platinum-group metal, symbol Os, noted for its high hardness, about Brinell 400. The heaviest known metal, it has a high specific gravity, 22.65, and a high melting point, 4890°F (2698°C). The boiling point is about 9900°F (5468°C). Osmium has a close-packed hexagonal crystal structure, and it forms solid-solution alloys with platinum, having more than double the hardening power of iridium in platinum. However, it is seldom used to replace iridium as a hardener except for fountain-pen tips where the alloy is called **osmiridium.** The name *osmium* comes from the Greek word meaning odor, and the tetroxide formed is highly poisonous. Osmium is not affected by the common acids and is not dissolved by aqua regia. It is practically unworkable, and its chief use is as a catalyst.

OXALIC ACID. Also known as **ethane diacid.** A strong organic acid of composition HO_2CCOOH, which crystallizes as the ortho acid $(HO)_3CC(OH)_3$. It reduces iron compounds and is thus used in writing inks, stain removers, and metal polishes. When it absorbs oxygen, it is converted to the volatile carbon dioxide and to water, and it is used as a bleaching agent, as a mordant in dyeing, and in detergents. Oxalic acid occurs naturally in some vegetables, notably **Swiss chard,** and is useful in carrying off excess calcium in the blood. The acid is produced by heating sodium formate and treating the resulting oxides with sulfuric acid, or it can be obtained by the action of nitric acid on sugar, or strong alkalies on sawdust. It comes in colorless crystals with a specific gravity of 1.653, containing about 71% of the anhydrous acid, melting at 215°F (101.5°C), and soluble in water and in alcohol. It is used in metal cleaning, dyeing, photography, and pulp bleaching. **Oxamide,** $(CONH_2)_2$, is a stable anhydrous derivative with a high melting point, 786°F (419°C). It is a white crystalline powder used in flameproofing and in wood treatment. **Potassium ferric oxalate,** $K_3Fe(C_2O_4)_3$, is stable in the dark, but is reduced by the action of light, and is used in photography.

OXYGEN. An abundant element, constituting about 89% of all water, 33% of the earth's crust, and 21% of the atmosphere. It combines readily with most of the other elements, forming their oxides. It is a colorless and odorless gas and can be produced easily by the electrolysis of water, which produces both oxygen and hydrogen, or by chilling air below −300°F (−184°C), which produces both oxygen and nitrogen. The specific gravity of oxygen is 1.1056. It liquefies at −171°F (−113°C) at 59 atm. **Liquid oxygen** is a pale-blue, transparent, mobile liquid. As gas, oxygen occupies 862 times as much space as the liquid. Oxygen is one of the most useful elements and is marketed in steel cylinders under pressure, although most of the industrial uses are in the form of its compounds. An important direct use is in welding and metal cutting, for which it should be at least 99.5% pure. Oxygen-enriched air is used in a number of oxidation and combustion processes in the steel, cement, glass, petrochemical, refining, and paper-and-pulp industries, and it has potential economic and environmental benefits in waste combustion. Oxygen enrichment improves overall combustion by raising oxygen partial pressure, thus increasing the combustion temperature and waste destruction.

Oxygen is the least refractive of all gases. It is the only gas capable of supporting respiration, but is harmful if inhaled pure for a long time. **Ozone** is an allotropic form of oxygen with three atoms of oxygen, O_3. It is formed in the air by lightning, or during the evaporation of water, particularly of spray in the sea. In minute quantities in the air it is an exhilarant, but pure ozone is an intense poison. It has a peculiar odor, which can be detected with 1 part in 20 million parts of air. Ozone is a powerful oxidizer, capable of breaking down most organic compounds and bleaching vegetable colors. Its use as an alternative to chlorine for treating water and wastewater has risen in recent years because it is safer and a stronger oxidizing agent. **Liquid ozone** explodes violently in contact with almost any organic substance. It is bright blue and is not attracted by a magnet, although liquid oxygen is attracted. Ozone absorbs ultraviolet rays, and a normal blanket in the upper ozonosphere at heights of 60,000 to 140,000 ft (18,288 to 42,672 m), with 1 part per 100,000 parts of air, shields the earth from excess shortwave radiation from the sun. The destruction of the stratospheric ozone layer is resulting in the cutting back in usage of chlorofluorocarbons and other inert chlorinated compounds. At ground levels, however, ozone concentrations are rising due to increased use of fossil fuels; such higher ozone levels are viewed as deleterious to health. As an oxidizer in the rubber industry, ozone is known as **activated oxygen.** It is used widely as a catalyst in chemical reactions. It is made commercially by bombardment of oxygen with high-speed electrons.

PAINT. A general name sometimes used broadly to refer to all types of **organic coatings.** However, by definition, paint refers to a solution of a pigment in water, oil, or organic solvent, used to cover wood or metal articles either for protection or for appearance. Solutions of gums or resins, known as *varnishes,* are not paints, although their application is usually termed *painting.* Enamels and lacquers, in the general sense, are under the classification of paints, but specifically the true paints do not contain gums or resins. **Stain** is a varnish containing enough pigment or dye to alter the appearance or tone of wood in imitation of another wood, or to equalize the color in wood. It is usually a dye rather than a paint.

　　Enamel paint is an intimate dispersion of pigments in either a varnish or a resin vehicle, or in a combination of both. Enamels may dry by oxidation at room temperature and/or by polymerization at room or elevated temperatures. They vary widely in composition, in color and appearance, and in properties. Although they generally give a high-gloss finish, some give a semigloss or eggshell finish, and still others give a flat finish. Enamels as a class are hard and tough and offer good mar and abrasion resistance. They can be formulated to resist attack by the most commonly encountered chemical agents and corrosive atmospheres, and they have good weathering characteristics.

　　Because of their wide range of useful properties, enamels are probably the most widely used organic coating in industry. One of their largest areas of use is as coatings for household appliances—washing machines, stoves, kitchen cabinets, and the like. A large proportion of refrigerators, for example, are finished with synthetic baked enamels. These appliance enamels are usually white, and therefore they must have a high degree of color and gloss retention when subjected to light and heat. Other products finished with enamels include automotive products; railway, office, sports, and industrial equipment; toys; and novelties.

House paint for outside work consists of high-grade pigment and linseed oil, with a small percentage of a thinner and drier. The volatile thinner in paints is for ease of application, the drying oil determines the character of the film, the drier is to speed the drying rate, and the pigment gives color and hiding power. Part of or all the oil may be replaced by a synthetic resin. Many of the newer house paints are water-base paints.

Paints are marketed in many grades, some containing pigments extended with silica, talc, barytes, gypsum, or other material; fish oil or inferior semidrying oils in place of linseed oil; and mineral oils in place of turpentine. Metal paints contain basic pigments such as red lead, ground in linseed oil, and should not contain sulfur compounds. Red lead is a rust inhibitor and is a good primer paint for iron and steel, though it is now largely replaced by chromate primers. White lead has a plasticizing effect which increases adhesion. It is stable and not subject to flaking. Between some pigments and the vehicle there is a reaction which results in progressive hardening of the film with consequent flaking or chalking, or there may be a development of water-soluble compounds. Linseed oil reacts with some basic pigments, giving chalking and flaking. Fading of a paint is usually from chalking. The composition of paints is based on relative volumes since the weights of pigments vary greatly, although the custom is to specify pounds of dry pigment per gallon of oil.

Bituminous paints are usually coal tar or asphalt in mineral spirits, used for the protection of piping and tanks and for waterproofing concrete. For line pipe, heavy pitch coatings are applied hot, but a bitumen primer is first applied cold. The **Bitumastic primer** of Koppers Co., Inc., for such purposes, is refined coal-tar pitch in a quick-drying solvent. The bituminous paints have poor solvent structure, but have high outdoor weathering resistance. **Battery paint** is usually asphalt or gilsonite in a petroleum solvent. It forms a heavy, acid-resistant, and water-resistant coating. Ordinary **aluminum paint** is made with aluminum flake in an oil varnish or in a synthetic lacquer. In lacquers the powder does not leaf, and the paint dries to a hard, metallic surface with a frosted effect. Aluminum paints will reflect 70% of light rays, and they are used for painting tanks, but where high resistance is needed, especially for industrial atmospheres, the paints have a synthetic resin base. For painting chimneys and ovens, aluminum paints consist of aluminum flake in a silicone resin, and they resist heats to 1000°F (538°C). When aluminum is used in asphalt paints for tanks and roofing, the aluminum pigment leaf comes to the surface to form a reflective shield. The so-called **heat-resistant paints** are usually aluminum pigment in a silicone resin. The heat resistance comes more from the reflective power of the aluminum than from the actual melting point of the resin. Sericite mica flake is sometimes mixed with aluminum flake to give a different color tone. The **Opal-Glo paint** of Sherwin-Williams Co. contains a small amount of opaque aluminum particles to give a three-dimensional opalescent glow without the metallic sheen of a flaked powder. Direct-to-metal **DTM Acrylic Gloss** of this company is acrylic-based and more flexible than alkyd paints. **Lumiclad paint,** for roofing, is aluminum flake and asbestos powder in an oil-resin vehicle. **Calibrite** is a borited aluminum powder which retains the silvery color of aluminum in the paint.

Lead powder may be incorporated in paints as a protection against gamma rays. **Leadoid paint** is an English paint of this kind. **Ceramic paints** are refractory oxides or carbides in a soluble silicate vehicle, but they are generally only temporary repair coatings. But the **Pyromark paint** of Tempil Big Three Industries, Inc., has the color pigment in a silicone vehicle which is converted by applied heat to an inorganic silica film which will withstand temperatures to 2500°F (1371°C). **Intumescent paints,** which bubble and swell to form an insulating barrier to protect the base material from fire damage, may contain borax or a percentage of an intumescent resin. **Masonry paints** may have a silicone resin base for water resistance, but they may also be made with synthetic rubbers and be designated for special purposes such as traffic paints, road-marking paints, and pool paints. Road-marking paints were formerly made with Manila copal, but they are now made with synthetics. **Imron,** of Du Pont, to withstand heavy traffic on industrial building floors, is based on a urethane resin. It dries quickly without a catalyst and is resistant to greases and cutting oils. **Pliolite AC,** of Goodyear Tire & Rubber Co., used in road-marking and pool paints, is a styrene-acrylate copolymer resin. It needs no catalyst for curing and has high adhesion to concrete. **Foliage paints** are made with a base of vinyl acrylic resin. They are used for coloring Zoyzia grass lawns in winter and for other turfs. **Vitalon,** of Mallinckrodt, Inc., is such a colorant.

Antifouling marine paints contain soaps of copper, arsenic, and mercury to inhibit action of marine organisms. A paint produced by International Paint Co. consists of a controlled-release biocide of organotin and copper in a copolymer base of methylmethacrylate and tributyltin, with oxylene as the solvent. Known as **SPC,** the paint film is made smoother by water friction, resulting in lower frictional resistance of a ship's hull as it passes through water.

Water-borne paints consist essentially of finely divided ingredients, including plastic resins, fillers, and pigments, suspended in water. An organic medium may also be involved. There are three types of water-borne coatings: **emulsion coatings** or **latexes, dispersion coatings,** and **water-soluble coatings.** Emulsions, or latexes, are aqueous dispersions of high-molecular-weight resins. Strictly speaking, latex coatings are dispersions of resins in water, whereas emulsion coatings are suspensions of an oil phase in water.

Emulsion and latex coatings are clear to milky in appearance and have low gloss, excellent resistance to weathering, and good impact resistance. Their chemical and stain resistance varies with composition. Dispersion coatings consist of ultrafine, insoluble resin particles present as a colloidal dispersion in an aqueous medium. They are clear or nearly clear. Their weathering properties, toughness, and gloss are roughly equal to those of conventional solvent paints.

Water-soluble types, which contain low-molecular-weight resins, are clear finishes, and they can be formulated to have high gloss, fair to good chemical and weathering resistance, and high toughness. Of the three types, they handle the flow most like conventional solvent coatings.

The simplest **water paints** consist of gypsum or whiting with some zinc oxide, with water as the vehicle and glue for adhesion. **Calcimine** is an old name for wall paint made with whiting and glue and some linseed oil and water colors. **Whitewash** may be merely quicklime and water, or may be slaked lime, salt, whiting, and glue. These materials are still used for interior painting of farm buildings where low cost is the prime factor. **Casein paints** consist of pigments and extenders in a casein solution. Interior paints and enamels are now mostly water paints with a vehicle of a latex water solution of a synthetic resin. The resin may be an acrylic emulsion, a styrene-acrylic, or water-dispersible polyesters and alkyds, or water-soluble epoxies. They can be applied to wet surfaces, and they cure rapidly to water- and chemical-resistant films. The acrylic resin emulsions are valued because of their ability to produce pastel shades and their good flow and leveling properties for one-coat application.

Paint removers, for removing old paint from surfaces before refinishing, are either strong chemical solvents or strong caustic solutions. In general, the more effective they are in removing the paint quickly, the more damaging they are likely to be to the wood or other organic material base. The hiding power of a paint is measured by the quantity which must be applied to a given area of a black and white background to obtain nearly uniform complete hiding. The hiding power is largely in the pigment, but when some fillers of practically no hiding power alone, such as silica, are ground to microfine particle size, they may increase the hiding power greatly. Paint making is a highly developed art, and the variables are so many and the possibilities for altering the characteristics by slight changes in the combinations are so great that the procurement specifications for paints are usually by usage requirements rather than by composition.

PALLADIUM. A rare metal, symbol Pd, found in the ores of platinum. It resembles platinum, but is slightly harder and lighter in weight and has a more beautiful, silvery luster. It is only half as plentiful but is less costly. The specific gravity is 12.10, and the melting point is 2826°F (1552°C). Annealed, the metal has a Brinell hardness of 40 and a tensile strength of 27,000 lb/in^2 (186 MPa). Drawing to 60% reduction increases Brinell hardness to 100 and tensile strength to 50,000 lb/in^2 (345 MPa). It is highly resistant to corrosion and to attack by acids, but, like gold, it is dissolved in aqua regia. It alloys readily with gold and is used in some white golds. It alloys in all proportions with platinum, and the alloys are harder than either constituent. About 1% by weight palladium increases the oxidation resistance of tungsten-chromium and tungsten-chromium-molybdenum alloys at 1830 to 2280°F (1000 to 1250°C).

Although palladium has low electrical conductivity, 16% that of copper, it is valued for its resistance to oxidation and corrosion. Palladium-rich alloys are widely used for low-voltage electrical contacts. **Palladium-silver alloys,** with 30 to 50% silver, for relay contacts, have 3 to 5% the conductivity of copper. A palladium-silver alloy with 25% silver is used as a catalyst in powder or wire-mesh form. A **palladium-copper alloy** for sliding contacts has 40% copper with a conductivity 5% that of copper. Many of the palladium salts, such as **sodium palladium chloride,** Na_2PdCl_4, are easily reduced to the metal by hydrogen or carbon monoxide and are used in coatings and electroplating. A **palladium-iridium alloy** with 20% iridium has a Brinell hardness of 140 and can be work-hardened to Brinell 260 and a tensile strength of 190,000 lb/in^2 (1,310 MPa). A **palladium-nickel alloy** with 20% nickel has a Brinell hardness of 200 and can be rolled to a Brinell hardness of 360 and a tensile strength of 170,000 lb/in^2 (1,172 MPa). **Palladium alloys** are also used for instrument parts and wires, dental plates, and fountain-pen nibs. Palladium is valued for electroplating as it has a fine white color which is resistant to tarnishing even in sulfur atmospheres. **Palladium leaf** is palladium beaten into extremely thin foil and used for ornamental work like gold leaf. Hydrogen forms solid solutions with palladium, forming **palladium sponge,** which has been used for gas lighters. **Palladium powder** is made by chemical reduction and has a purity of 99.9% with amorphous particles 12 to 138 μin (0.3 to 3.5 μm) in diameter. Atomized powder has spherical particles of 50 to 200 mesh and is free-flowing. The powders are used for coatings and parts for service temperatures to 2300°F (1260°C). **Palladium flake** has tiny laminar platelets of average diameter of 118 μin (3 μm) and thickness of 3.9 μin (0.1 μm). The particles form an overlapping film in coatings. For auto catalytic converters, a "palladium-only" catalyst developed by Allied Signal can reduce precious-metal cost by excluding platinum and rhodium.

PALM OIL. An oil obtained from the fleshy covering of the seed nuts of several species of palm trees, chiefly *Elaeis guineensis,* native to tropical Africa, but also grown in Central America. The tree attains a height of about 60 ft (18 m), and the nuts occur in large bunches similar to dates. The fruit is of an elongated ellipse shape, about 1.5 in (3.8 cm) long, enclosing a single kernel. The fleshy part carries about 65% oil, which is a semisolid fat. The iodine value is about 55, and the saponification value 205. West African palm oil has four grades: edible, with 11% maximum free fatty acid; soft, with 18% maximum; semihard, with 35% maximum; and hard, with more than 35%. The high-grade edible oil is from unfermented fruits. Fresh palm oil has an agreeable odor and a bright-orange color, but the oil often has a rancid stench and is of varying colors. The oil is used as a fluxing dip in the manufacture of tinplate; for soaps, candles, and margarine; and for the production of palmitic acid. About 10% by weight of the palm oil is recovered as by-product glycerin in making soaps or in producing the acid.

Palm oil contains 50 to 70% **palmitic acid,** $C_{15}H_{31}COOH$, which in the form of glyceride is an ingredient of many fats. When isolated, it is a white crystalline powder of specific gravity 0.866 and melting point 65°C, soluble in hot water. It is used in soaps, cosmetics, pharmaceuticals, and food emulsifiers and in making plastics. **Neo-Fat 16,** of Akzo Chemical Co., is 95% pure palmitic acid, with 4% stearic and 1 myristic acids. This is a powder with an acid value of 220 and a saponification value of 221. But **Greco 55L,** which is a white crystalline solid, for cosmetics and soaps, is 50% palmitic acid with the balance stearic acid.

The oil from the kernel of the palm nut, known as **palm kernel oil,** has different characteristics from palm oil. It contains about 50% lauric acid, 15 myristic acid, 16 oleic acid, and 7 palmitic acid, together with capric and caprylic acids found in coconut oil, while palm oil is very high in palmitic and oleic acids. The specific gravity is 0.873, iodine number 16 to 23, saponification value 244 to 255, and melting point 75 to 86°F (24 to 30°C). The American species of palm oil is from the dwarf tree *E. melanococca,* growing from Mexico to Paraguay and called **noli palm** in Colombia. The pulp of the nuts yields 30% of an oil similar to African palm oil. The tall **Paraguayan palm** *Acrocomia sclerocarpa* has the fruit also in bunches, and the pulp yields 60% of oil similar to palm oil.

PAPAIN.   The dried extract, or enzyme, obtained from the fruit and sap of the papaya tree, *Carica papaya,* of tropical America, east Africa, and Asia. It is marketed as a dry, friable powder and has a complex structure. It is a **proteolytic agent,** which splits proteins; it also contains a lipase which accelerates the hydrolysis of fatty acid glycerides; and it contains an antibacterial. The latex from the fruit is dried by low heat, since temperatures above 158°F (70°C) destroy the enzymes. Papain is used in beer and other beverages to remove protein haze, in medicine as a digestive aid and in combination with urea and chlorophyll to promote the healing of wounds, as a meat tenderizer, in degumming silk, and in treating textiles.

The **papaya** tree grows to a height of 25 ft (7.6 m) without branches and is crowned with large leaves. The melonlike fruit grows out from the trunk and has orange-colored flesh. It is eaten raw as a melon is, but as it spoils rapidly, it is not easily shipped. The papaya is called **pawpaw** in Florida. **Meat tenderizers** marketed by Papaya Industries, Inc., in powder and liquid forms, are papain with or without seasoning spices. They are applied to the meats before or during cooking. Papain is also injected into the beef animals 10 min before slaughtering. The enzyme spreads throughout the circulatory system of the animal, remaining in the meat and tenderizing it during the cooking cycle. It also inhibits discoloration of the meat in aging. **Pro-Ten** is a solution of papain for this purpose. **Augment,** of Calgon Corp., is a mixture of papain powder and sodium chloride for use in a dip solution for tenderizing meats.

PAPER.   The name given to cellulose made into paste form from plant sources and rolled into thin sheets, used as a material for writing, printing, and wrapping. It may be considered as a thin felting of fibers bonded by a water-soluble cellulose formed on the fiber surfaces, to which a coating material such as clay may be added with starch or other sizing material. Most papers are less than 0.006 in (0.152 cm) thick, but the dividing line between paper and **paperboard** is taken as 0.012 in (0.030 cm). Properties of paper are controlled by the following variables: (1) type and size of fiber, (2) pulp processing method, (3) web-forming operation, and (4) treatments applied after the paper has been produced.

The original **Egyptian paper,** known as **papyrus,** was made from the stems of the rush *Cyperus papyrus* growing along the Nile. It was made in sheets, sometimes as long as 130 ft (33 m). The Chinese process of papermaking from hemp and linen rags was brought to the Near East when the Arabs took Samarkand in A.D. 704. The papers used in medieval Europe were **charta damascena,** from the Arab factory at Damascus, and **charta bombycina,** from the factory at Bombyce near Antioch, both sold in reams (from the Arabic word *razmah*). **Greek parchment,** used in later medieval times, was made from cotton. **Aztec paper,** called **amatl,** was made from the inner bast fibers of species of wild fig trees, *Ficus.* The fibers were felted into sheets and beaten with a ribbed mallet. The thin white sheets used for writing were then polished with a curved stone celt which closed the pores and smoothed the sheets. Some very large sheets were made for folding into books. The fibers of the yellow fig tree, *F. petolaris,* were made into a yellow paper used for coloring for decorations.

There are many varieties and grades of paper, depending on the source of the cellulose and the method of manufacture. Wood is a lignified form of cellulose, and the wood is chipped and cooked with chemicals to dissolve out the lignin. The material so treated is known as **chemical wood pulp,** to distinguish it from **mechanical wood pulp** used for making wallboard and **newsprint paper,** the latter requiring some chemical pulp to give fiber and strength. There are four processes for producing chemical pulp: the sulfite, with calcium sulfite; the soda, with sodium carbonate; the sulfate, with sodium sulfate; and the magnesium bisulfite. Hardwoods are cooked in a soda-ash solution and sulfited. The bleaching of pulp is done with chlorine dioxide which oxidizes lignin to water-soluble colorless compounds without reducing the strength of the cellulose.

Cellulose fiber papers, made from wood pulp, constitute by far the largest number of papers produced. A great many of the engineering papers are produced from draft or sulfate pulps. The term **kraft paper** is used broadly today for all types of sulfate papers, although it is primarily descriptive of the basic grades of unbleached sulfate papers, where strength is the chief factor and cleanliness and color are secondary. Kraft paper can be altered by treatments to produce various grades of condenser, insulating, and sheathing papers.

Book paper is usually a mixture of sulfate and soda pulp, the latter process producing a bulky pulp. **Wrapping paper** is a strong, coarse paper made usually from mixed pulps. **Manila paper** is a strong wrapping paper originally made from Manila hemp, but the name is now applied to any strong chemical wood pulp or mixed paper of a slightly buff, or Manila, color. **Absorbent paper,** such as for **blotting paper** and **filter paper,** is made from spongy bulky fibers, such as poplar, or is loosely felted fiber. The **Kimtowels,** of Kimberly-Clark Corp., to replace cotton waste for machine cleaning, are made from bulky, specially treated pulp. The paper can be saturated with oil or solvent for cleaning purposes. **Netone filter paper** of National Filter Media Corp. is a 60-lb (27-kg) kraft paper impregnated with neoprene to give chemical resistance. This type of paper has 3 to 10 times the strength of cellulose papers, and it is also used for electrical insulation. But **Permalex paper,** for electrical insulation, is a kraft paper in which the cellulose fibers have been treated to replace hydroxyl groups with cyanoethyl groups. The paper has high tensile strength, is more heat-resistant, and has higher dielectric strength than ordinary paper. **X-Crepe paper,** of Cindus Corp., used as a substitute for burlap for bags, as a barrier paper, and as a reinforcement for laminated plastics, is a heavy, soft-texture kraft paper that is creped and cross-creped to produce a material that is stretchable in varying degrees from 15 to 60%. It has a bursting strength to 260 lb/in^2 (1.8 MPa). **Balancing paper,** used with the core material in structural plastic laminates to prevent warping, is heavy kraft paper impregnated with a phenol resin.

Cotton is nearly pure cellulose and makes an excellent paper material. Old cotton rags are thus scoured and used for papermaking. Linen rags are also used and produce a fine grade of **writing paper.** The best quality writing and **printing papers** have 50, 75, or 100% rag content. **Bond paper** is a hard-finished writing paper made from spruce, which has a long fiber. Highly rolled and coated printing papers are called **supercalendered papers.** They are used for printing fine-screen halftones. In England this paper is called **art paper.** **Offset paper** for offset printing is given the required porosity without affecting physical properties by coating with an alkali-swellable resin. Fine linen **ledger paper** is made with 100% white rags. Good-quality bond **typewriter paper** may have 80% white rags. These papers are sold by weight per ream, a ream usually consisting of 500 sheets of a specified size in inches. **Watermarked paper** can be made in various ways, but the simplest method is by printing the mark with a solution of castor oil in methyl alcohol. Papers are generally described in terms of basic weight, which is the weight in pounds per 3,000 ft^2 (279 m^2). Standard-weight papers are 90 to 105 lb (41 to 48 kg); lightweight papers are 60 to 65 lb (27 to 29 kg).

Carbon paper, used for duplicating typewriting, pencil, or pen writing, is made by coating the paper with a mixture of a pigment and a medium. The pigments include carbon black, Prussian blue, and organic red, or blue and green lakes. The medium is likely to be a blend of waxes and oils to give a composition of the desired consistency and melting point; but to make a good carbon paper that will not be gummy and will not smear, a proper proportion of a high-melting-point, nongreasy wax, such as carnauba, must be used. Papers of special texture, preferably rag papers, are employed. Smudgeproof carbon paper has a coating of plastic lacquer.

Drawing paper is a heavy paper, usually white or buff, employed for making drawings. For mechanical drawings the buff color is preferred as it is easier on the eyes and not so readily soiled. Drawing papers are smooth or rough, the smooth being hot-pressed. Good grades of drawing paper should permit considerable erasure without destroying the appearance. Buff detail paper for pencil use is made slightly rough or grained. High-grade paper for ink work is extra-hard-sized and coated. Drawing paper is marketed in rolls of widths from 30 to 72 in (76 to 183 cm) and in standard sheets varying from cap, 17 by 13 in (43 by 33 cm), to antiquarian, 52 by 31 in (132 by 79 cm). **Tracing cloth** is made from thin, fine cotton or linen fabric, of plain weave, heavily sized and glazed on one side. It is used for making tracings in ink and is quite transparent. It can also be obtained with the glaze on both sides. Tracing cloth is usually marketed in rolls of 24 yd (22 m). The sizing of ordinary tracing cloth is easily soluble in water, and will therefore not withstand wetting, but special grades are made with impervious resin coatings. Plastic-treated papers are now made that have high strength and better transparency than tracing cloth while retaining the drafting qualities of a fine paper. **Tracing paper** is usually a good grade of hard, transparent tissue paper in sheets and rolls, in white or buff.

Granite paper is made by the addition of colored fibers to the pulp or by adding several shades of dyed pulp to the regular stock. **Oatmeal paper,** used chiefly for wallpaper, has a flaky finish produced by washing a solution of wood flour over the sheet on the forming wire in the paper machine. The wood flour may be natural or dyed. **Cartridge paper** is 50- to 80-lb (23- to 36-kg) Manila paper, waxed on one side, originally used for muzzle-loading cartridges, but now employed where a stiff, waterproof material is needed. **Glassine** is a transparent thin paper used for envelope windows and for sanitary wrapping. It is made of sulfite pulp subjected to long-continued beating and supercalendered. **Glassoid** is a more highly finished transparent paper. **Onionskin paper** is a lightweight, highly finished, transparent writing paper made transparent by hydration of the pulp in the beaters. Transparent papers are now often made water-resistant and stronger by adding a synthetic resin to the pulp. **Albanene tracing paper,** of Keuffel & Esser Co., is a thin rag paper treated with a transparent synthetic resin. It takes ink well and erases easily. **Silicone tissue,** for wiping glass, is soft tissue paper treated with silicone resin. **Tissue paper** is a very thin, almost transparent paper. It may be loosely felted to give absorbent qualities, or it may have a hard, smooth surface for wrapping paper. **Detergent paper,** for washing windows, is a soft paper impregnated with a detergent.

Crepe paper has many consumer and industrial uses. Creping imparts stretch, strength, bulk, conformability, and texture similar to that found in fabrics. Creping consists of forming small pleats, or folds, with a blade. **Cross-creped paper** is made by creping in two opposite directions.

To make paper smooth-surfaced and resistant to the spreading of inks, adhesive sizing materials are used together with inert fillers such as China clay which give body, weight, opacity, and added strength to the paper. The usual coating adhesives are starches and proteins. The proteins, such as casein, are more uniform than starch, but give a more brittle film. Starch films are not water-resistant unless the treated starches are used. **Waterproof paper** was formerly paper treated with a copper-ammonium solution and hot-rolled, or was paper coated with rubber latex to which had been added a creaming agent such as a metallic soap; but various synthetic resins are now incorporated in the sizing or mixed in the pulp. High-styrene butadiene latex gives a flexible and glossy film for printing papers. Acrylic latex also gives a strong, glossy, and flexible coating. Polyvinyl acetate is also used for printing papers. **Scriptite 31,** of Monsanto Chemical Co., used to give a tough, water-impervious surface to offset papers, is a methylated methylol melamine resin which forms a molecular link with the protein of the coating. **Wet-strong paper** is usually specially processed paper in which the water resistance is due to the processing and interlocking of the fibers as well as to impregnation with a small amount of melamine, urea-formaldehyde, or other resin. It is used for maps, documents, and wrapping. The **Anti-adhesive paper** of Central Paper Co., for interleafing sticky materials and for box linings, is a kraft paper treated with a silicone resin. **Kastek paper** of Plastic Film Corp., for waterproof wrapping, is 30- to 100-lb (13.5- to 45-kg) paper with a very thin film of polyethylene or vinyl resin bonded to the surface.

Capacitor paper, used as a dielectric in capacitors, is made from Swedish spruce sulfate pulp, is highly purified, and is nearly transparent. It is extremely thin, 0.00015 to 0.004 in (0.00038 to 0.00101 cm), but is strong and tough. **Insulating paper,** commonly called **varnished paper,** is a standard material for insulation of electric equipment. It is usually bond or kraft paper coated on both sides with black or yellow insulating varnish. The thicknesses are 0.002 to 0.020 in (0.005 to 0.050 cm) with dielectric strengths of 500 to 2,000 V/mil (20 to 80 V/m). Special insulating varnishes of high dielectric strength are now marketed for this purpose. **Cyanoethylated paper,** used in condensers, is a thin paper treated with acrylonitrile which improves the electrical insulating properties. **Quinorgo,** of Johns-Manville, is a group of asbestos papers containing 80% chrysotile fibers bonded with organic resins. The resins are varied to give differing physical properties. The papers come in thicknesses from 0.003 to 0.015 in (0.007 to 0.038 cm) and have high dielectric strength. **Laminating paper,** for making laminated plastics, is a white or brown paper of uniform basis weight and uniform internal structure capable of having a controlled resin pickup. It usually comes in thicknesses from 0.004 to 0.020 in (0.010 to 0.050 cm). **Flameproof paper** is paper treated with ammonium sulfate and ammonium and sodium phosphates. Paper and nonwoven textile fabrics may be treated with a **fire-retardant agent. Metallized paper,** of Smith Paper, Inc., used for capacitors, is a lacquer-coated kraft paper with a thin layer of zinc deposited on one side. For packaging and as a barrier paper, **aluminum-coated paper** is produced by vacuum metallizing. It has the appearance of bright aluminum, but has the flexibility and physical properties of paper.

Building paper, used for sheathing houses, is a heavy kraft paper, plain or rosin-sized. Specially treated building papers are also marketed under trade names. **Barrier paper,** used for lining storage rooms, is kraft paper saturated with gilsonite, asphalt, and wax. It is odorless and black. **Copperskin** is an insulating construction material made by facing 1-oz (0.03-kg) electrosheet copper on one or two plies of heavy building paper impregnated with bitumen. **Fibreen** is a tough, strong, flexible, waterproof paper used for wrapping bundles of steel and other heavy products. It is made of two layers of kraft paper reinforced with two crossed layers of sisal fibers embedded in asphalt, and the whole combined under heat and pressure. **Brownskin,** a waterproof sheathing paper, is high-strength building paper impregnated with a bituminous compound and crimped to give it stretch and resiliency. **Burlap-lined paper,** for heavy wrapping, has 4- to 10-oz (0.11- to 0.28-kg) burlap laminated to heavy kraft paper with asphalt as the binder. It is waterproof. **Papier mâché** is comminuted paper made into a water paste with an adhesive binder and molded. It was formerly widely used for toys, dishes, and novelties, but dishes and novelties now made of paper stock are produced directly from the wood pulp and are more uniform and stronger. Kraft paper impregnated with phenolic resin laminated in blocks and heavy sheet is used for short-run tooling. **Laminated paperboard** is made by laminating together plies of paper about 0.0625 in (0.16 cm) thick. It is made in two general qualities, an interior and a weather-resistant quality. The main differences between these types are in the kind of bond used to laminate the layers together and in the amount of sizing used in the pulp stock from which the individual layers are made. Laminated paperboard is regularly manufactured in thicknesses of 0.1875, 0.25, and 0.375 in (0.48, 0.64, and 0.95 cm) for construction uses. For industrial uses, such as furniture and automotive liners, 0.125-in (0.32-cm) thickness is common.

There are three major types of **inorganic fiber papers: Asbestos papers,** the most widely used, are nonflammable, are resistant to elevated temperatures, and have good thermal insulating characteristics. They are available with or without binders and can be used for electrical insulation or for high-temperature reinforced plastics. **Fibrous glass papers** can be used to produce porous and nonhydrating papers. Such papers are used for filtration and thermal and electrical insulation and are available with or without binders. High-purity **silica glass papers** are also available for high-temperature applications. **Ceramic fiber (aluminum silicate) papers** provide good resistance to high temperatures, low thermal conductivity, and good dielectric properties, and they can be produced with good filtering characteristics.

Synthetic organic fiber papers consist of synthetic fibers, synthetic pulp, or plastic film. Plastics used include polyethylene, nylon, acrylic, and polypropylene. In general, synthetic papers have greater dimensional stability and tear resistance than conventional natural fiber papers.

PAPER PLANTS. Cellulose for papermaking is obtained from a wide variety of plant life, made directly into **paper pulp,** or obtained from old rags which were originally made from vegetable fibers. Animal fibers incorporated into some papers are fillers for special purposes, not papermaking materials. The papyrus of Egypt was made from a reed, but the **baobab** of India was from the bark of the tree *Adamsonia digitata.* **Rice paper** of China came from the *Tetrapanax papyriferum,* but the so-called rice paper used for cigarettes in the United States is made from flax fiber. **Cigarette paper** is also made from ramie and sunn hemp. The distinction between cigarette paper and the tissue paper used for wrapping is that it must be free of any substance that would impart a disagreeable flavor to the smoke; and it must be opaque and pure white, must burn at the same rate as tobacco, and must be tasteless.

Wood pulp is now the most important papermaking material. Spruce is the chief wood used for the sulfite process, but hemlock and balsam fir are also used. Aspen and other hardwoods are used in the soda process, and also southern pine. White fir is readily pulped by any process, but western red cedar is high in lignin content, about 30%, and reduced with difficulty by the sulfate process to a dark-colored pulp. It is pulped by the kraft process. Its fibers are fine and short, yielding a paper of high bursting strength. Normally, the **pulpwoods** of the west coast are western hemlock, white fir, and Sitka spruce, leaving the Douglas fir to the lumber mills. The same species of trees grow in Alaska and British Columbia as in Oregon and Washington, but Douglas fir decreases to the north and hemlock and spruce become more abundant but with smaller trees. A stand of spruce in Canada at the age of 80 years yields about 18 tons (16 metric tons) of pulp per acre (4,047 m²) while a stand of pine in the southern states at 24 years of age yields about the same amount. Western hemlock, balsam, and spruce are the chief pulpwoods of Canada. Pines are used extensively in the United States, especially for kraft paper, paperboard, and book paper. More than 50% of all pulpwood used in the United States is now from the southern states, and about 10% of this is salvage from lumber mills. But, in general, special methods are used for pulping pine since conventional sulfite liquor does not free the fibers as the phenolic compounds in the heartwood condense to form insoluble compounds.

Kraft paper is sulfate-pulped from a mixture of 50% western hemlock, 25 western red cedar, and 25 Douglas fir. The fir has a coarse fiber which gives high tear strength; cedar has a long, thin fiber which gives a smooth surface; and hemlock is abundant and used as a filler. Poplars are also used for pulpwood, and Scott Paper Co. uses fast-growing scrub alder. Newsprint made from hardwoods has a bursting strength 20% higher than that made from softwoods, and the brightness value is higher, but the pulping of hardwoods is usually a more involved chemical process.

In England fine printing papers are made by the soda process from **esparto** grass. It gives a soft, opaque, light paper, although the cellulose content is less than 50%. Esparto is the plant *Stipa tenacissima* of the dry regions of north Africa. In Tunisia it is called **alfa.** It grows to a height of about 3 ft (1 m), with cylindrical stem. The fine, light fibers, about 0.5 in (1.27 cm) long, are from the leaves. Some grades of cardboard and some newsprint are made from straw. **Deluwang paper** of the East Indies is made from the scraped and beaten bark of the **paper mulberry** tree, *Broussonetia papyrifera*. It is an ancient industry in Java, and the paper is used for lamp shades and fancy articles. Under the name of **tapa cloth,** the sheets were dyed and used as a muslinlike fabric by the Polynesians. The strips are welded together by overlapping and beating together the wet material.

Bagasse is of increasing importance as a papermaking material in the sugar-growing areas of the world.

PARAFFIN. A general name often applied to paraffin wax, but more correctly referring to a great group of hydrocarbons obtained from petroleum. Paraffin compounds begin with methane, CH_4, and are sometimes called the **methane group.** The compounds in the series have the general formula C_nH_{2n+2} and include the gases methane and ethane; the products naphtha, benzene, gasoline, lubricating oils, and jellies; and the common paraffin. The name *paraffin* indicates little affinity for reaction with other substances. In common practice, the name is limited to the waxes that follow petroleum jelly in the distillation of petroleum. These waxes melt at 104 to 140°F (40 to 60°C) and consist of the hydrocarbons between C_{22} and C_{27}; the refined waxes may range up to 194°F (90°C). They burn readily in air. Paraffin occurs to some extent in some plant products, but its only commercial source is from natural petroleum. **Vista C_{1416}** from Vista Chemical Co. is normal paraffin of high purity, low odor, and low viscosity. Uses include feedstocks for chlorinated paraffins, rolling oils, and as a clean fuel. **Chlorinated paraffin** is a pale to amber-colored, odorless, soft wax or viscous oil of specific gravity 0.900 to 1.50. It is flame-resistant and is used in treating paper and textiles. **Chlorowax,** of Diamond Shamrock Corp., is a chlorinated paraffin for adhesive, fire-resistant, and water-resistant compounds. It is a cream-colored powder containing 69 to 73% chlorine and is insoluble in water but soluble in organic solvents. A water-dispersible form of this wax is **Delvet 65,** a white, viscous liquid containing 65% solids. **Clorafin 42,** of Hercules, Inc., is an amber-colored, viscous, liquid chlorinated paraffin containing 42% chlorine. **Cereclor** is a similar product. **Clorafin 70** is a yellow solid containing 70% chlorine and softening at 194 to 212°F (90 to 100°C). The former is used as a plasticizer in resins and for coatings; the latter is for flameproofing and waterproofing textiles. **Kloro-Chek** from Ferro Corp. is employed in metalworking, as a flame retardant, and as a vinyl resin plasticizer. **Paraoil** is a liquid grade from Dover Chemical Corp.; **Chlorez 700-DD,** from the same firm, is a nonyellowing type, while **Chlorez 760** is a high-chlorine variety (74%).

   Paraffin oil is drip oil from the wax presses in the process of extracting paraffin wax from the wax-bearing distillate in the refining of petroleum. The oil is treated, redistilled, and separated into various grades of lubricating oils from light to heavy. They may be treated and bleached with sulfuric acid, and neutralized with alkali. When decolorized with acid and sold as filtered, they are brilliant liquids, but are not suitable in places where they may be in contact with water, since the sulfo compounds present cause emulsification. The specific gravities of paraffin oils are between 21 and 26°Bé. **Triton oil** is a 100% pure paraffin oil.

PARAFFIN WAX. The first distillate taken from petroleum after the cracking process is referred to as *wax-bearing,* and is put through a filter press and separated from the oils. The wax collected on the plates is called **slack wax,** and it contains 50% wax and 50 oil. This is chilled to free it from oil. The yellow wax is filtered to make a white, semitranslucent, refined wax, which is odorless and tasteless. For large-scale operations, solvent methods of wax extraction are used. Paraffin wax is soluble in ether, benzene, and essential oils. **Match wax** has a melting point of 105 to 112°F (40 to 44°C); white crude wax, 111 to 113°F (43 to 44°C); yellow crude, 117 to 119°F (46 to 48°C); and special white, 124 to 126°F (51 to 52°C). The refined waxes are in various melting-point ranges from 115 to 136°F (45 to 58°C) and are used for coating paper and for blending in coating and impregnating compounds. They are also used in candles and other products. The refined paraffin wax used for molded goods and for rubber compounding is a white solid having a melting point of 122°F (50°C) and a specific gravity 0.903.

Borneo wax has a very high melting point and a hard, crystalline structure which makes it valuable for coatings and for high-quality candles. By treatment of the waxes from U.S. petroleums to remove the low-melting constituents, a similar wax is obtained having branched-chain molecules and a fine crystalline structure. This is known as **microcrystalline wax. Aristowax** is a treated wax of this kind with melting points from 145 to 165°F (63 to 74°C). **Petrolite wax,** of Petrolite Corp., is a microcrystalline wax with melting point at 194°F (90°C). **Sunwax** is a microcrystalline wax in two grades, a brown with melting point of 175°F (79°C) and a yellow with melting point of 185°F (85°C). **Warcosine wax** is a white microcrystalline wax melting at about 153°F (67°C), while **Fortex wax** has a melting point at about 194°F (90°C). Microcrystalline wax does not emulsify as carnauba wax does, but when oxidized with a catalyst, it is emulsifiable and the melting point is raised so that it can be used in hard, self-polishing floor waxes. **Cardis wax** and **Polymekon wax** have melting points at 198 and 250°F (92 and 121°C). **Petronauba D** is an oxidized wax with a melting point of 192°F (89°C) used as a partial replacement for carnauba. Microcrystalline waxes are also compounded with polyethylene and other materials to increase strength, flexibility, and other properties. The paraffin waxes are sold under many trade names. **Arwax** may contain butyl rubber or polyethylene. **Advawax 2575,** for paper coatings, contains polyisobutylene. **Santowax,** of

Monsanto Chemical Co., is a high-melting-point microcrystalline wax. **Petrolatum** is a two-phase colloidal system or gel consisting of high-molecular-weight hydrocarbon oils dispersed in microcrystalline waxes. **Wax tailings** is a name for the distillate that comes from petroleum after the wax-bearing distillate is removed. It contains no wax, but at ordinary temperatures looks like beeswax. It is very adhesive and is employed in roofings and for waterproof coatings.

Synthetic paraffin wax, called **Ruhr wax,** is made in Germany from low-grade coals and other hydrocarbon sources. The waxes, with molecular weights of 900 to 1600, are white flakes, water-white when melted. They are odorless, free of sulfur and aromatics, with ash content below 50 ppm. The melting points are between 221 and 259°F (105 and 126°C). They are used as additives in paper coatings and printing inks and for mixing with refined paraffin.

Ozokerite, also known as **mineral wax,** and as **earth wax,** is natural paraffin found in Utah and in central Europe and used as a substitute or extender of beeswax, and in polishes, candles, printing inks, crayons, sealing waxes, phonograph records, and insulation. Ozokerite is a yellowish to black, greasy solid, melting at 131 to 230°F (55 to 110°C), and having a specific gravity of 0.85 to 0.95. It is soluble in alcohol, benzol, and naphtha, but not in water. The paraffin occurs in rocks, which are crushed, and the wax is melted out. The refined and treated ozokerite is called **ceresin** and is white to yellow and odorless. The melting point is up to 142°F (61°C). It is used for waxed paper, polishes, candles, and compounding.

A similar wax, called **montan wax,** or **lignite wax,** is produced in Germany from lignite. Montan wax is white to dark brown and has a melting point of 176 to 194°F (80 to 90°C), usually 181 to 185°F (83 to 85°C). The wax is obtained from the powdered lignite by solvent extraction with a mixture of benzene and ethyl alcohol and subsequent removal of the bitumen by oxidation and chromic acid. The brown coals of Oklahoma and Texas also contain as much as 13% montan wax. **IG wax S** is extracted and purified montan wax, but **IG wax V** is a synthetic substitute consisting of the octadecyl ether of vinyl alcohol, $C_{17}H_{34}CH_2 \cdot O \cdot CH \cdot CHOH$. Montan is valued for leather finishes, polishes, phonograph records, insulation compounds, and as a hard wax in candles. The mineral waxes are sold in white, waxy cakes or in flakes.

PARCHMENT. Originally, goatskin or sheepskin specially tanned and prepared with a smooth, hard finish for writing purposes. It was used for legal documents, maps, and fancy books, being more durable than the old papers. The extremely thin, high-quality parchment that was used for documents and handmade books was rubbed with pumice and flattened with lead. Parchment now is usually **vegetable parchment.** It is made from a base paper of cotton rags or alpha cellulose called **waterleaf** which contains no sizing or filling materials. The waterleaf is treated with sulfuric acid which converts a part of the cellulose to a gelatinlike amyloid. When the acid is washed off, the amyloid film hardens on the fibers and in the interstices of the paper. The strength of the paper is increased, and it will not disintegrate even when fully wet. The paper now has a wide usage in food packaging as well as for documents as a competitor of the resin-treated wet-strong papers. The wet strength and grease resistance are varied by differences in acid treatment and subsequent sizing. **Patapar** is a vegetable parchment marketed in many grades. Parchment papers are also waterproofed by dipping in the solution of copper hydroxide and ammonium hydroxide known as **Schweitzer's reagent,** and then hot-rolling. **Vellum** is a thick grade of writing paper made from high-grade rag pulp pebbled to imitate the original calfskin parchment called vellum.

PASTEBOARD. A class of thick paper used chiefly for making boxes and cartons, and for spacing and lining. It may be made by pasting together several single sheets, but more usually by macerating old paper and rolling into heavy sheets. It may also be made of straw, certain grasses, and other low-cellulose paper materials, and it is then known as **strawboard.** Colloquially, the term *pasteboard* applies to any paperstock board used for making boxes, including the hard and stiff boards made entirely from pulp, and the term *pasteboard* is not liked in the paper industry. The bulk of packaging boards are now pulp boards treated with resin and are called **carton boards.** **Cardboard** is usually a good-quality chemical pulp or rag pasteboard used for cards, signs, or printed material, or for the best-quality boxes. Ivory board, for art printing and menu cards, is a highly finished cardboard clay-coated on both sides. **Bristol board** is a high-class white cardboard, supercalendered with China clay, or it may be made by pasting together sheets of heavy ledger paper; but the name is also applied to any high-grade printing or drawing board over 0.06 in (0.15 cm) thick. **Index bristol** is always made solid on a Fourdrinier machine to prevent splitting in use or warping. The original board made in Bristol, England, was made in this way. **Jute board,** used for folding boxes, is a regular product of the paper mills, and is a strong, solid board made of kraft pulp. **Chipboard** is a cheap board made from mixed scrap paper, used for boxes and book covers. When made with a percentage of mechanical wood pulp, it is called **pulpboard.** A heavy, rope-pulp paper or board, usually reddish and used for large expansion filing envelopes, is called **paperoid.**

PEANUT OIL. Also known as **groundnut oil.** A pale-yellow oil with a distinctive nutty taste and odor, obtained from the pressing of the seed kernels of the peanut, a legume of the genus *Arachis,* of which there are many species. It was native to Brazil, brought to Africa in slave ships and then to the United States. It is now grown in many countries. The Spanish peanut, cultivated in temperate climates, has small seeds, while the common variety, *A. hypogaea,* known in the United States as the **Virginia peanut,** has pods up to 1.5 in (3.8 cm) long with seeds twice the length of the Spanish varieties. The **Spanish peanut,** however, is easily grown and gives a high yield per acre. The **Brazilian peanut,** *A. nambyquarae,* has pods up to 3 in (7.5 cm) in length. Vast quantities of peanuts are roasted and marketed as food nuts or for confections or are ground to make edible **peanut butter.** The best grades of cold-pressed oils are marketed as edible oils, but the oil is also used industrially for soaps, in diesel-engine fuels, and for blending in lubricating and varnish oils. The **arachidic acid,** $CH_3(CH_2)_{18}COOH$, contained in the oil to the extent of 4%, however, makes a hard soap. The oil also contains 52 to 65% oleic, 21 to 25 linoleic, besides palmitic, stearic, and lignoceric acids. The specific gravity is from 0.916 to 0.922, saponification value 189 to 196, and iodine value 83 to 101. The oil known as **arachis oil,** or as **Katchung oil** when imported from the Orient, is from the peanut *A. hypogaea.* It is used in lubricating, for varnishes, and for softening leather. **Peanut meal,** left after extraction of the oil, is sold as stock feed, or that from the final extraction of the inedible oil is used for fertilizer. **De-oiled peanuts** are marketed as low-calorie, nonfat food. Most of the oil is removed by solvent extraction, but the nuts retain the high-protein value, color, and flavor. The calorific value is reduced about 80%.

PEAT. An earthy mass formed by the rapid accumulation of quick-growing mosses and plants and valued as a fuel in countries where fuels are expensive. Large quantities are used for fuel in Finland, Switzerland, Ireland, and some other countries. In Russia large amounts are used to produce fuel gas and for processing the tar into chemicals. In the United States it is used for fertilizer, insulation, and packing. The dried **peat moss** is used for making insulating board. Peat bogs, or beds, are found mainly in moist districts in temperate climates. The top layers are only slightly decayed, are brown, and are of low specific gravity. But at greater depths peat is nearly black and is very compact. In the peat of southern New Jersey there are layers of trees as large as 5 ft (1.5 m) in diameter, buried for centuries.

The reserves of peat are very large in the states bordering on the Great Lakes, those in the state of Minnesota alone being estimated at

7×10^9 tons (6.35×10^9 metric tons). Fresh peat often contains as high as 80% moisture and must be dried before use. Wicklow dried peat contains about 71% volatile matter, 27% being fixed carbon, and 28% coke. The calorific value of peat is about 5,000 Btu (5.3×10^6 J). It is sometimes semicarbonized and made into fuel briquettes. **Charred peat** is peat that has been subjected to a temperature to cause partial decomposition. It is marketed as fertilizer. Peat is also distilled, yielding mainly gas and a high percentage of tar. **Peat wax,** extracted from peat in England, is a hard wax with characteristics similar to those of montan wax, and is a substitute for it.

PENTAERYTHRITOL. A tetrahydric alcohol of composition $HOH_2C \cdot C(CH_2OH)_3$ produced by the condensation of formaldehyde and acetaldehyde. It is a white, crystalline solid melting at about 504°F (262°C). The commercial grade is 85 to 90% pure. It is employed for the production of explosives, plastics, drying oils, and chemicals. Pentaerythritol will combine with the fatty acids of vegetable oils to form esters that are superior to linseed oil as drying oils. Combined with the fatty acids of linseed oil, it will give a drying oil that will dry completely in 6 h compared with 16 h for a bodied linseed oil; and as a varnish oil it gives higher gloss and greater water resistance. **Synthetic waxes** are also made by combining pentaerythritol with long-chain, saturated, fatty acids, and these waxes have higher melting points than beeswax or carnauba wax, but do not have the natural gloss of carnauba. **Pentawax 177 is a pentaerythritol stearate.** It is a light-brown wax melting at 127°F (53°C), used for coating paper, in printing inks, and in cosmetics. **Pentamull 126** is an ester of pentaerythritol and oleic acid. It is an amber-colored oil used as an emulsifying agent. **Pentex** is a technical grade of pentaerythritol containing 85% monopentaerythritol and the balance higher polymers. It produces fast-drying and glossy varnishes. **Pentaerythritol tetrastearate** is used for cosmetics. **Pentaerythritol tetranitrate** and **pentaerythritol triacrylate** are available commercially for similar applications. The **Pentalyn resins** of Hercules, Inc., which are employed to replace copals in varnishes, are pentaerythritol esters of rosin. **Pentalyn 802A** is a phenol-modified pentaerythritol rosin ester for gloss printing inks and traffic paints. It is pale in color and has high resistance to chemicals and wear.

PEPPER. One of the oldest and most important of the spices. **Black pepper,** the common household spice, is the ground, dried, unripe fruit of the evergreen shrub or vine, *Piper nigrum,* of India and southeast Asia. There are two grades of Indian pepper, **Alleppey** and **Tellicherry**; the latter is bolder and heavier and the more expensive grade. The fruits are small, berrylike drupes. They change in ripening from green to bright red to yellow. When dry, the unripe berries are reddish brown or black. The vine comes into full production in 3 years and lives for 20 to 30 years. A vine yields 5 to 10 lb (2.3 to 4.5 kg) of pepper. **White pepper,** preferred for the preparation of commercial foods, is from the nearly ripe berries. It has a yellow to gray color. Pepper is used as a condiment and stimulates the flow of gastric juices. White pepper is not as pungent as black pepper. Commercial pepper is often a blend of two kinds. **Pepper oil,** used for flavoring, is a yellowish essential oil of specific gravity 0.873 to 0.916 with a pepperlike odor and flavor but not pungent as pepper is. It is extracted from the common pepper berries. **Pepperoyal,** of Griffith Laboratories, is pepper flavor extracted from black pepper and converted to minute soluble globules that disperse easily in foodstuffs. Soluble pepper, of Fritzsche Bros., used for food processing, is a liquid solution of black-pepper oleoresin from which there is no precipitation of piperine crystals during processing or cold storage. The synthetic **piperidine,** $C_5H_{11}N$, is a colorless liquid with an odor resembling pepper. It yields crystalline salts, and it occurs in natural pepper in combination with **piperic acid** in the form of the alkaloid **piperine,** $C_{17}H_{19}NO_3$, which is the chief active constituent of pepper. **Piperazine,** made synthetically, is used in medicine as an anthelmintic and as an intermediate for pharmaceuticals. It is a six-membered heterocyclic ring compound with two nitrogen atoms in the para position. **B-cap,** of Evans Chemetics, is a cinnaylidene acetoyl piperide, a synthetic with a pepperlike flavor used in prepared foods.

Long pepper, esteemed in some countries for preserves and curries, is more aromatic and is sweeter than common pepper. It is from the tiny fruits of the climbing plant *P. retrofractum* of Malaya and *P. longum* of India and Indonesia. **Ashanti pepper,** of western Africa, is from the vine *P. guineense,* also known as **Guinea pepper,** although this name is applied to **grains of paradise,** the pungent, peppery seeds of the perennial herb *Aframomum melegueta* of West Africa, which are used for flavoring and in medicine. The seeds are also called **alligator pepper** and **melegueta pepper,** and are used as pepper in Europe, but the plant is of the ginger family. **Cubeb** is the dried, unripe fruit of the climbing vine *P. cubeba* of India, Indonesia, and the West Indies. The berries resemble those of black pepper, but have a strong, peculiar odor and a bitter, aromatic taste. Cubeb is used in medicine and cigarettes. **Cubeb oil** is from the berries, which yield 10 to 16% of the pale-green oil with a pepperlike odor. It is used in perfumery and in soaps. **Paprika** is the ground dried fruit of the *Capsicum annuum* of Europe and America. When full, red, ripe pods are used, and the seeds, cores, and stems are removed, a uniform, maximum red color is produced. Yellow pods give low red value. Paprika is used as a condiment. **Chili pepper** is from the smaller, podlike berries of species of *Capsicum* which grow as small trees or shrubs. It is a tropical plant. The ground fruits as a condiment are known as **red pepper,** or **cayenne pepper.** In medicine it is known as **capsicum** and is used as a carminative and as a source of vitamin P. It is also used in soft drinks in place of ginger. The Samoan beverage known as **kava** is made by steeping in water the ground root of a species of pepper plant, *P. methysticum.* It has a peppery flavor. But the Kava of Borden, Inc., is not a pepper but an instant coffee powder processed to neutralize the coffee acids.

PEPPERMINT. An oil distilled from the perennial herb *Mentha piperita,* which grows in the temperate climates of America, Europe, and Asia. The oil has a pleasant odor and a persistent, cooling taste, and it is valued as a flavor and used in soaps, toothpastes, perfumes, and pharmaceuticals. The oil contains **menthol,** $C_{10}H_{20}O$, which is extracted for use as an antiseptic, in perfumery, and in medicine for colds and as an antispasmodic and anodyne. **Japanese peppermint** is from *M. arvensis,* grown extensively in Japan, Brazil, and the United States. The oil is less fragrant and is used for the production of menthol, and it has a higher menthol content. The plant is propagated from roots and grows to a height of 2 to 3 ft (0.6 to 1 m). It is cut when it blooms and partly cured like hay. The crude oil is obtained by steam distillation; the menthol is obtained by freezing and recrystallization, with a yield of 50% menthol crystals to total crude oil. The residue oil is called **cornmint oil.** It retains the peppermint flavor and is used in perfumery and flavoring. **Spearmint** is from *M. viridis,* grown largely in Michigan. The oil is sharper in odor and taste and is used chiefly in chewing gums. **Pennyroyal oil** is distilled from the dried leaves and tops of the small, annual plant *Hedeoma pulegiodes* or *M. pulegium,* which grows in the eastern United States. The oil is a counterirritant and is used in liniments. It is also used in insect repellants and for the production of menthol. The plant yields 0.7% oil, and the oil will yield 65% menthol with a melting point of 91 to 95°F (33 to 35°C), or 40% of 108°F (42°C) menthol. **Horsemint oil** is from the plant *M. canadensis,* used for the production of thymol. Menthol and menthol substitutes are also synthesized from coal tar. **Cyclonol** is a derivative of cyclohexanol. It lacks one H and one CH of the structural formula of menthol, but has the characteristic odor and cooling effect. **Levomenthol,** produced synthetically by Glidden Co., is used as a replacement for natural menthol. A racemic variety of menthol, containing both the dextro-

and levorotatory isomers in equal amounts, is made by Morton Thiokol Inc.

PERFUME OILS.   Volatile oils obtained by distillation or by solvent extraction from the leaves, flowers, gums, or woods of plant life, although a few are of animal origin. Perfumes have been used since earliest times, not only for aesthetic value, but also for antiseptic value and for religious purposes. Simple perfumes usually take their names from the name of the plant, but the most esteemed perfumes are blends, and the blending is considered a high art. It is done by tones imparted by many ingredients. Some oils with repugnant odors have an attractive fragrance in extreme dilution and a persistence which is valued in blends and for stabilization. Some oils with heavy odors, such as **coumarin,** are used in dilution to give body. Since many of the odors come directly from esters, aldehydes, or ketones, they can be made synthetically from coal-tar hydrocarbons and alcohols. Synthetics are now most used in perfumes, although some natural odors have not yet been duplicated synthetically, and about 30,000 aromatics have been developed. A perfume may contain 50 components, sometimes as high as 300, and the average perfume manufacturer employs about 3,000 components. Some of the chemicals are not odors, but give lasting qualities or enhance odor. Some are used as fixatives or blending agents. **Hydroquinone dimethyl ether,** $C_8H_{10}O_2$, has an odor of sweet clover but is used as a fixative in other perfumes.

In general, the aldehyde odors are fugitive, and some become acid in the presence of light or oxygen. Ketones are more stable. Esters are usually stable, but some are saponified in hot solution and cannot be used for soaps. Some esters, made from complex high alcohols, are used to give a fresh top note to floral perfumes. **Linalyl acetate,** produced from citral, is an example. Acid perfumes neutralize free alkali and cannot be used in soaps. Phenol odors alter the color of soaps, and the odor may also become disagreeable.

Some odors are never extracted from the flowers, but are compounded. **Crab apple,** for example, which is a peculiarly sweet odor, is compounded of 16 oils, including bois de rose, ylang-ylang, nutmeg oil, jasmine, musk, heliotropin, coumarin, and others. *Wisteria* is the honeylike odor of the mauve and white flowers of the climbing plant *Wisteria sinensis.* The oil is never extracted but is compounded from geranium, Peru balsam, benzoin, bois de rose, and synthetics. Some oils such as **lavender,** from the flowers *Lavandula vera,* have no value when used alone but require skillful blending to develop the pleasant odor. Apple and peach odors are **allyl cinnamate.** Synthetic **rose** is the ester of phenyl ethyl alcohol made from benzene and ethylene oxide. Although the natural rose odor is readily extracted, it is more expensive.

Fixatives are used for the finer perfumes. They are essential oils that are less volatile and thus delay evaporation. The animal oils, such as musk and civet, are of this class, and also the balsam oils. Some evil-smelling distillates from the chemical manufacture may also be used as fixatives. Musk is from the male musk deer of Tibet. It is one of the most expensive materials. Synthetic musk is as powerful as the natural. The synthetic musk of Du Pont is called **Astrotone. Musk ambrette** is made from metacresol. **Ambrette oil** has a strong, musklike odor distilled from the **musk seed,** or **amber seed,** of the plant *Hibiscus abelmoschus* of Ecuador, India, and Egypt. **Civet** is an odorous, yellow fluid from the civet cat of tropical Asia and Ethiopia. *Civettone* is a liquid with a clean odor and easily soluble in alcohol, distilled from civet. **Patchouli oil** is one of the best fixatives for heavy perfumes. It is a powerfully odorous, viscous liquid obtained by distilling the fermented leaves of the shrub *Pogostemon patchouli* of India, China, and the Philippines. The odor resembles sandalwood. **Cassie** is a valuable oil with an odor similar to violet obtained by maceration in oil of the flowers of the shrub *Acacia farnesiana* of the Mediterranean countries and the West Indies. It is used to scent pomades and powders. **Versilide,** of Givaudan Corp., is a cyclic ketone synthetic musk that is very stable in soaps and cosmetics and does not discolor.

Attar of rose is one of the most ancient and popular of perfume oils. The name is derived from the Persian *attar,* and is sometimes incorrectly given as **ottar** but with the same French pronunciation. The finest attar of rose is from Bulgaria, where it is distilled from the flowers of the **damask rose,** *Rosa damascena.* The fresh oil is colorless, but turns yellowish green. About 20,000 lb (9,072 kg) of flowers is needed to make 1 lb (0.45 kg) of essence, and it is so valuable that it is usually adulterated with geraniol or synthetic rose. In France the oil is obtained from the *R. centifolia.* **Rose water** is the scented water left after distillation, or is made by dissolving attar in water. The **Otto of baronia** of Australia is a high-grade **rose oil.**

Geranium oil is obtained from the leaves or flowers of *Pelargonium graveolens* of the Mediterranean countries and other species of geranium. It is used as an adulterant or substitute for rose oils in perfumes and soaps. **Zdravetz oil** is a geranium oil from *P. macnorhijum* of Cyprus, used in rose bouquet and lavender perfumes. Many geranium and rose oils are derived from geraniol obtained from citronella and other oils. A synthetic rose-geranium is **diphenyl methane,** $(C_6H_5)_2CH_2$, a colorless solid melting at about 77°F (25°C). **Benzophenone** is also used for rose-geranium perfumes. It is a

diphenyl ketone, $C_6H_5(CO)C_6H_5$, melting at 117°F (47°C). This material is also used for making fine chemicals. **Geraniol,** from citral, is a colorless liquid with a sweet, delicate rose odor. **Vetiver,** a very sweet-scented oil used in high-grade perfumes and in medicine, is distilled from the roots of the **khuskhus** plant, *Vetiveria zizamoides,* native to India but produced chiefly in Java, Réunion, and Haiti. **Opopanox,** used in incense and in medicine, is an oleoresin from the roots of the *Pastinaca opopanox* of the Orient and British Somaliland. **Frankincense,** used in incense and perfumes, and in medicine under the name of **olibanum,** is a gum resin from the tree *Boswellia carterii* of the Sudan and Somaliland. It comes in hard, yellow grains. **Kiounouk,** used as a fixative, is a clear, yellowish semiliquid obtained from olibanum. **Mecca balsam,** used in oriental types of perfume, is a greenish oleoresin from the plant *Commiphora opobalsamum* of Arabia. It has the odor of rosemary. **Rosemary** is an oil distilled from the fresh flowering tops of the sweet-smelling evergreen shrub *Rosemarinus officinalis* of the Mediterranean countries. It is used in eau de cologne, soaps, and medicine. **Jasmine oil,** a highly valued perfume material, is from the fragrant flowers of the shrub *Jasminum grandiflorum,* a species of jasmine grown in southern France especially for perfume. The oil is extracted from the fresh flowers by enfleurage. A synthetic **jasmine-rose oil,** which also has a **peach-apricot flavor** and a sweet taste, is **benzyl propionate,** $CH_3CH_2COOCH_2C_6H_5$. It is a liquid boiling at 428°F (220°C) and is used in perfumes and as a flavor. **Ylang-ylang,** or **cananga oil,** is a valuable essential oil from the flowers of the tree *Canangium odorata,* cultivated in Indonesia, Madagascar, and the Philippines. No more than 150 lb (68 kg) of flowers is obtained from a tree, but about 400 lb (181 kg) is required to produce 1 lb (0.45 kg) of oil. It contains linalol and geraniol. Another oil that rivals ylang-ylang in fragrance is **champaca oil,** from the flowers of the large tree *Michelia champaca,* of southern Asia.

Lavender oil, used with rosemary in eau de cologne, and also as **lavender water** in a mixture of the oil in water and alcohol, is obtained from the flowers of the shrub *Lavandula officinalis* of southern Europe. The dried flowers are fragrant and are used in sachets. **Spike lavender** is an inferior oil from the plant *L. latifolia* of France and Spain. It is used in perfumes and sometimes as a food flavor. **Herbandin** is a synthetic ester made from petroleum and used as a replacement or extender for natural lavender oil. It has a pronounced lavender odor. Espantone is a synthetic ketone with a spike lavender odor. **Bay oil,** or **myrcia oil,** used in the toilet alcohol known as **bay rum,** and also in perfumes, is distilled from the leaves of the bay tree, *Pimenta acris,* of the West Indies, 60 lb (27 kg) of leaves yielding 1 lb

(0.45 kg) of oil, and 1 gal (3.8 L) of bay oil being used to 100 gal (379 L) of rum to make bay rum. It contains eugenol and has a spicy odor.

Carnation oil is obtained by solvent extraction or by enfleurage from the flowers of *Dianthus caryophyllus,* of which there are more than 2,000 varieties grown in the Mediterranean countries. The less highly cultivated plants give the richest perfumes. **Violet oil** is derived by solvents or maceration in hot oils from the flowers of the blue and purple varieties of *Viola odorata.* **Synthetic violet** is made from **ionone,** $C_{13}T_{20}O$, derived from lemongrass oil. The ionones are made synthetically by condensation of citral with acetone. They are monoenol and dienol butones. **Velvione** is such an ionone with a powerful violet odor. The true violet odor is **irone,** a complex seven-ring compound. It can be obtained from the iris root and is one of the most odoriferous materials obtained from plants. **Orris** is the dry root of *Iris florentian,* and the powdered root is used in violet powders and as a flavor. **Oak moss** was one of the perfumes of ancient Egypt. It is obtained from the lichen *Evernia prunastri* and *E. furfuracea* growing on oak and spruce trees of southern Europe. The resinous extract has the odor of musk and lavender. It is used as a fixative in perfumes of the poppy type. **Rue oil,** used for sweet pea perfume, is distilled from the plants *Ruta graveolans* of France and *R. montana* of Algeria.

As with jewelry, the manufacture of perfumes is normally classified as a luxury industry. But the vital test of essentiality comes in wartime, and these aesthetic materials are always considered essential to the public morale, and hence basic. Even under desperate wartime conditions France never stopped the manufacture of perfumes, and during the life struggle of England in the Second World War the restrictive regulations placed upon perfume manufacture had to be abandoned quickly because of public pressure. Wartime restrictions placed on the imports of perfume oils into the United States during the Second World War were immediately abandoned.

PERILLA OIL. A light-yellow oil obtained from the seeds of the plant *Perilla ocimoides* of China and Japan, and employed in varnishes, core oils, printing ink, and linoleum. It dries to a harder, tougher, glossier film than linseed oil. The specific gravity is about 0.935, iodine value 200, and saponification value 191. It contains 41 to 46% linolenic acid, 31 to 42 linoleic, and 3 to 10 oleic acid. In Japan it is called **egoma oil.** The raw oil tends to form globules, but this is overcome by boiling or blowing. The blown oil is rapid-drying and is more weather-resistant than linseed oil.

PERMEABILITY ALLOYS. A general name for a group of nickel-iron alloys with special magnetic properties. These soft magnetic materials possess a magnetic susceptibility much greater than that of iron. An early alloy was composed of 78.5% nickel and 21.5 iron. It also contained about 0.37% cobalt, 0.1 copper, 0.04 carbon, 0.03 silicon, and 0.22 manganese. It is produced sometimes with chromium or molybdenum, under the name of **Permalloy,** and is used in magnetic cores for apparatus that operates on feeble electric currents, and in the loading of submarine cables. It has very little magnetic hysteresis.

Supermalloy, developed by Bell Laboratories for transformers, contains 79% nickel, 15 iron, 5 molybdenum, and 0.5 manganese, with total carbon, silicon, and sulfur kept below 0.5. It is melted in vacuum and poured in an inert atmosphere. It can be rolled as thin as 0.00025 in (0.00064 cm). The alloy has an initial permeability 500 times that of iron. **Supermendur** contains 49% iron, 49 cobalt, and 2 vanadium. It is highly malleable and has very high permeability with low hysteresis loss at high flux density. **Duraperm** is a high-flux magnetic alloy containing 84.5% iron, 9.5 silicon, and 6 aluminum. **Perminvar** is an alloy containing 45% nickel and 25 cobalt, intended to give a constant magnetic permeability for variable magnetic fields. **A-metal** is a nickel-iron alloy containing 44% nickel and a small amount of copper. It is used in transformers and loudspeakers to give nondistortion characteristics when magnetized.

Permenorm, first produced in Germany as **Orthonel** and developed at the U.S. Naval Ordnance Laboratory, contains 50% nickel and 50 iron. It is subjected to drastic rolling in one direction followed by a heat treatment to obtain chemical combination of the nickel and iron. It has a grain orientation which, when subjected to a magnetic field, produces a square hysteresis loop, indicating instantaneous magnetization. It is used as a core material and in magnetic amplifiers. **Deltamax** is this material. **Alfenol,** developed by the Naval Ordnance Laboratory, contains no nickel, but has 16% aluminum and 84 iron. It is brittle and cannot be rolled cold, but can be rolled into thin sheets at a temperature of 1067°F (575°C). It is lighter than other permeability alloys and has superior characteristics for transformer cores and tape-recorder heads. A modification of this alloy, called **Thermenol,** contains 3.3% molybdenum without change in the single-phase solid solution of the binary alloy. The permeability and coercive force are varied by heat treatment. At 18% aluminum, the alloy is practically paramagnetic. The annealed alloy with 17.2% aluminum has constant permeability.

Iron-aluminum alloys with 13 to 17% aluminum are produced in sheet form for transformers and relays. They have magnetic properties equal to the 50–50 nickel-iron alloys and to the silicon-iron alloys, and they maintain their magnetic characteristics under changes in ambient temperature.

Conpernik contains equal amounts of nickel and iron with no copper. It is called *constant-permeability nickel* and has little permeability variation. It differs from **Hipernik** only in heat treatment. When heat-treated, Hipernik has higher permeability than silicon steels for transformers up to a flux density of about 10 kG (1 T). Both alloys are used in transformer cores. **Hipernik V** is grain-oriented, giving a square hysteresis loop, low coercive force, 0.15 Oe (12 A/m), and a high ratio of residual magnetism to saturation. It comes in thin strip for use in instruments and computer elements. **Hiperco** contains 35% cobalt, 64 iron, and 1 chromium. It is ductile and easy to roll to extremely thin strip for instrument parts. The heat-treated metal has high permeability and low hysteresis loss, but in this condition it has low tensile strength and is brittle. **Vicalloy,** developed by Bell Laboratories, is a high-permeability alloy containing 36 to 62% cobalt, 6 to 16 vanadium, and the balance iron. It is cast and hot-swaged, then drawn into wire or tape as fine as 0.002 in (0.005 cm). It retains high permanent magnetism, with coercive forces to 250 Oe (19,895 A/m) and residual flux of 8,000 G (0.8 T), and it retains its magnetism to 600°F (316°C).

Iron-nickel permeability alloys are used as loading in cable by wrapping layers around the full length of the cable. When nickel-copper alloys are used, they are employed as a core for the cable. **Magnetostrictive alloys** are iron-nickel alloys that will resonate when the frequency of the applied current corresponds to the natural frequency of the alloy. They are used in radios to control the frequency of the oscillating circuit. Magnetostriction is the stress that occurs in a magnetic material when the induction changes. In transducers it transforms electromagnetic energy to mechanical energy. The magnetostrictive alloy **Terfenol-D** is a near single-crystal alloy of terbium, dysprosium, and iron, made by Etrema Products Inc. in rods and disks of 0.04- to 2-in (1- to 51-mm) diameter. It was developed by the U.S. Navy for sonar applications and is also used for high-power transducers and high-fidelity hearing aids. **Temperature-compensator alloys** are iron-nickel alloys with about 30% nickel. They are fully magnetic at −20°F (−29°C) but lose their magnetic permeability in proportion to rise in temperature; until at about 130°F (54°C), they are nonmagnetic. Upon cooling they regain permeability at the same rate. They are used in shunts in electrical instruments to compensate for errors due to temperature changes in the magnets.

PERSIMMON WOOD. The wood of the common persimmon trees, *Diospyros virginiana,* of the southeastern United States and the **black persimmon,** *D. texana,* of western Texas. It is used for shuttles, golf club heads, and tools, and it takes a fine polish. The tree belongs to the ebony family, of which there are listed about 160 species. The persimmon is a small tree not over 50 ft (15 m) high and 12 in (30 cm) in diameter. The wood is very hard, strong, and compact and retains its smoothness under long rubbing. The sapwood is light brown, 2 to 5 in (5 to 13 cm) wide, and the heartwood is black but very small, the thickness of a pencil, for example, in a 6-in (15-cm) tree. The density is 49 lb/ft^3 (785 kg/m^3). The fruit is cultivated for food, and there are only limited quantities of the wood. A fine wood used for fancy articles is **olive wood,** from the olive fruit tree of California. It is yellow with beautifully streaked, dark lines.

PETITGRAIN. An essential oil obtained by distillation of the leaves and small twigs of the **bitter orange** tree, *Citrus aurantium,* native to tropical Asia, but now grown in other countries. In Spain, it is known as the **Seville orange.** Paraguay is the chief producer of high-grade petitgrain, which is one of the best fixatives for fine perfumes and is also used in flavoring extracts. One pound (0.45 kg) of petit-grain is obtained from 100 to 150 lb (45 to 68 kg) of leaves. The fruits of the tropical bitter orange are large and of the finest golden appear-ance, but the pulp is very acidic. The juice is used only for blending in orange drinks. It contains a dilactone, **limonin,** which gives it a bit-ter taste. The rind is used in **marmalade** and **candied orange peel.** An essential oil distilled from the rind is known as **curaçao,** and it is used in perfumery and in **curaçao liqueur.** The flowers are very fra-grant, and from them **neroli oil** is distilled. Neroli is used in per-fumery blends and for mixing with synthetic perfumes. **Neroli Portugal** is inferior, and comes from the sweet orange *C. sinensis* by extraction. **Orange oil,** obtained by expression from the ripe rind of the sweet orange, is a less valuable oil. **Bergamot oil** is from the rind of the fruit of the small spiny tree *C. bergamia* of Italy. It has a soft, sweet odor and is used in perfume blends and soaps. The golden-yellow, pear-shaped fruit has an acidic, inedible pulp.

The **bioflavonoids,** used in cold remedies, are obtained from the white pulp, or **albedo,** of citrus fruits. They are alkaline-soluble crys-talline compounds, variations of **chromone,** a **benzpyrone,** and **flavone,** the **phenyl benzpyrone.** The pressed product from the pulp is acidified, crystallized, and dried. **Citroid** is a product of this kind. About 150 distinct chemicals have been produced from citrus fruits. The bioflavonoids are six-membered, double-ring compounds. Some are isolated and used directly, such as **naringin** from grape-fruit peel, which is an effective substitute for quinine. Others are syn-thesized easily. Chromone and flavone are really the parent sub-stances of many natural vegetable dyes, drugs, and tannins and are readily convertible to these materials.

PETROLATUM. A jellylike substance obtained in the fractional distillation of petroleum. Its composition is between $C_{17}H_{36}$ and $C_{21}H_{44}$, and it distills off above 577°F (303°C). It is also called **petroleum jelly.** It is used for lubricating purposes and for compounding with rubber and resins. When highly refined for the pharmacy trade, it is used as an ointment. The specific gravity is from 0.820 to 0.865. It is insoluble in water but readily soluble in benzine and turpentine. For lubricating purposes it should be refined by filtration only and not with acids, and it should not be adulterated with paraffin. The melting point should be between 115 and 130°F (46 and 54°C). Petrolatum of Grade O, used as a softener in rubber, is a pale-yellow, odorless semisolid of specific gravity 0.84 and melting point 115 to 118°F (46 to 48°C). **Sherolatum** and **Vaseline** are petroleum jelly, but petroleum jellies for pharmaceutical uses may be compounded with other materials.

PETROLEUM. A heavy, liquid, flammable oil stored under the surface of the earth, and originally formed as the by-product of the action of bacteria on marine plants and animals. It consists chiefly of carbon and hydrogen in the form of hydrocarbons, including most of the liquids of the **paraffin series,** C_5H_{12} to $C_{16}H_{34}$, together with some of the gases, CH_4 to C_4H_{10}, and most of the solids of the series from $C_{17}H_{36}$ to $C_{27}H_{56}$. It also contains hydrocarbons of other series. While petroleum is used primarily for the production of fuels and lubricating oils, it is one of the most valuable raw materials for a very wide range of chemicals. The name **petrochemicals** is used in a general sense to mean chemicals derived from petroleum, but it does not mean any particular class of chemicals. Sulfur and helium are byproducts from petroleum working.

Petroleums from different localities differ in composition, but tests of oils from all parts of the world give the limits as 83 to 87% carbon, 11 to 14 hydrogen, with sulfur, nitrogen, and oxygen in amounts from traces to 3%. Mexican and Texan oils are high in sulfur. The crude oil is split by distillation into naphtha, gasoline, kerosene, lubricating oils, paraffin, and asphalt. It may also be split by cracking, that is, by subjecting it to violent heating in the absence of air. This process yields a higher proportion of volatile products because of the breaking down of the more complex molecules by the high heat. Liquefied petroleum gases, including **propane, butane, pentane,** or mixtures, are marketed under pressure in steel cylinders as **bottled gas.** Propane, $CH_3CH_2CH_3$, is used in cook stoves. Butane, which has an additional CH_2 group, is used to enrich illuminating and heating gas. Propane and butane gases have heating values from 2,800 to 3,000 Btu/ft³ (106×10^6 to 113×10^6 J/m³). **Liquid gas** is also used for internal combustion engines, as a solvent, and for making many chemicals. **Pyrogen** is the trade name of gas obtained during the process of recovering gasoline from natural gas. It is marketed in cylinders for use in flame cutting torches. **Road oil,** used on dirt roads, is a heavy-residue oil from the refineries.

Certain highly refined oils used in medicine as laxatives and used in other specialized applications are referred to as **mineral oils.** **White mineral oil** is petroleum highly refined to color. Russian and Rumanian oils with high naphthene content were particularly suitable for this refining, and the oil was called **Russian oil.** Pennsylvania paraffin-base oils and Texas asphaltic-base oils are difficult to refine to color. The American white Russian oil is refined from mid-Continent and Gulf Coast oils which contain high naphthenes. It is used as a laxative, as a carrier of many drugs, in textile spinning, as a plasticizer for synthetic resins, and for sodium dispersions where the alkali metal would normally react with any impurities. **Klearol** and **Blandol** are viscosity grades of white mineral oil.

Petroleum from Baku was used from ancient times for lighting purposes, and the Bolivian oil was used in the sixteenth century for burning. The first commercial wells in the United States were opened in 1859 at Titusville, Pennsylvania. The chief production of petroleum is in Mexico, the United States, Russia, Rumania, Asia Minor, Peru, and northern South America, but large reserves exist in many other places. About half the world production of crude petroleum is in the western hemisphere, but the largest single field is in the Persian Gulf area. Only about 1% of the production of crude oil is from the Pennsylvania field, but much of the motor lubricating oil is from this oil. Many undersea deposits of petroleum are known, but some are at great depth; for example, the water at the Rockall Plateau, northeast of Ireland, is 5,000 ft (1,524 m) deep with 10,000 ft (3,048 m) of sediment above the oil deposit.

PEWTER. A very old name for tin-lead alloys used for dishes and ornamental articles, but now referring to the use rather than to the composition of the alloy. Tin was the original base metal of the alloy, the ancient **Roman pewter** having about 70% tin and 30 lead, although iron and other elements were present as impurities. Pewter, or **latten ware,** of the sixteenth century contained as much as 90% tin, and a strong and hard **English pewter** contained 91% tin and 9 antimony. This alloy is easily cold-rolled and spun, and it can be hardened by long annealing at 437°F (225°C), quenching in cold water, and tempering at 230°F (110°C). Pewter is now likely to contain lead and antimony, and very much less tin; when the proportion of tin is less than about 65%, the alloys are unsuited for vessels to contain food products, because of the separation of the poisonous lead. Antimony is also undesirable in food containers because of its poisonous nature, but when the tin content is low, antimony is needed to make the alloy susceptible to polishing. Ordinary pewter, with 6% antimony, 1.5 copper, and the balance tin, has a Vickers hardness of 23. With the addition of 1.5% bismuth, the hardness is 29. **Best pewter,** used for high-class articles, contains 100 parts tin, 8 antimony, 2 bismuth, and 2 copper. **Triple** has 83 parts tin, 17 antimony, or some lead to replace part of the antimony. Pewter should have a peculiar bluish-white luster when polished. It can be spun easily. Pewters containing much lead are dark in color and must be plated.

Britannia metal is a type of latten ware that also usually contains copper. The color is silvery white with a bluish tinge, or with a yellowish tinge if the copper is high. The ordinary composition is quite similar to some babbitts, 89% tin, 7.5 antimony, and 3.5 copper. It takes a fine polish and does not tarnish easily. It is easily worked by stamping, rolling, or spinning, or may be cast. Some zinc may be used in the casting alloys. English Britannia metal has 94% tin, 5 antimony, and 1 copper. **Hanover white metal** contains 87% tin, 7.5 antimony, and 5.5 copper. **Dutch white metal** has 81.5% tin, 8.5 antimony, and 9.5 copper. **Queen's metal** is a Britannia metal with a small amount of zinc. A typical composition is 88.5% tin, 7 antimony, 3.5 copper, and 1 zinc. The zinc helps to strengthen the alloy. **Minoflor** is another grade containing up to 9% zinc and up to 1 iron. When zinc is used in these metals, the antimony is reduced because both metals tend to make the alloy brittle. **Ashberry metal,** for tableware, contains some nickel and aluminum. **Tutanic metal,** a German utensil alloy, had 88 to 92% tin, up to 3 copper, and 6 to 7.5 lead. **Ludensheid plate** had 72% tin, 24 antimony, and 4 copper, with a trace of lead to increase fluidity. **Wagner's alloy** and **Koeller's alloy** were names for utensil alloys, the latter containing some bismuth to improve the casting.

PHENOL. Also known as **carbolic acid** or **monohydroxybenzene.** A colorless to white, crystalline material of sweet odor, having composition C_6H_5OH, obtained from the distillation of coal tar and as a by-product of coke ovens. It is also made by alkylating benzene with propylene to form **cumene,** which is **isopropylbenzene,** $C_6H_5CH(CH_3)_2$, and then oxidizing to cymene hydroperoxide and finally splitting to form phenol and acetone. Or, it can be made by oxidation of toluene and then a catalytic conversion to phenol. It is also produced by hydrogenation of the lignin from sulfite waste liquor. It is a valuable chemical raw material for the production of plastics, dyes, pharmaceuticals, syntans, and other products.

Phenol melts at about 109°F (43°C) and boils at 361°F (183°C). The pure grades have melting points of 102.2, 103.1, and 104°F (39, 39.5, and 40°C). The technical grades contain 82 to 84% and 90 to 92 phenol. The crystallization point is given as 104°F (40°C). The specific gravity is 1.066. It dissolves in most organic solvents. By melting the crystals and adding water, liquid phenol is produced, which remains liquid at ordinary temperatures. Phenol has the unusual property of penetrating living tissues and forming a valuable antiseptic. It is also used industrially in cutting oils and compounds and in tanneries. The value of other disinfectants and antiseptics is usually measured by comparison with phenol. The **phenol coefficient** of a disinfectant is the ratio of the dilution required to kill the Hopkins strain of typhoid bacillus in a specified time compared with the dilution of phenol required for the same organism under the same conditions and time. However, phenol is poisonous and gives dangerous burns on skin, so that as a disinfectant, it must be used with caution.

Phenol is a very reactive and versatile compound. In coal tar it occurs with many homologs, and many of the complex chemicals occurring in vegetable life are homologs or complexes of phenol. It can be easily nitrated, sulfonated, or halogenated. The hydroxyl hydrogen is readily replaceable by strong bases to produce **phenolates,** and these are readily convertible to ethers, such as **diphenyl ether,** or **diphenyl oxide,** $(C_6H_5)_2O$, which has a boiling point of 498°F (259°C) and is used as a heat-transfer medium. Ortho and para compounds are obtained by direct substitution, and these are called **substituted phenols. Thiophenol,** C_6H_5SH, also called **mercaptobenzene** and **phenyl mercaptan,** is an evil-smelling liquid boiling at 342°F (172°C), used for making dyes and pharmaceuticals. One of the most important applications of phenol is its condensation with formaldehyde to produce synthetic resins, and these resins can be varied greatly by alterations in the phenol. The term **phenols** is a class name for a wide variety of materials, as distinct from the normal phenol.

Koppers Co. markets **aklylated phenols** for the production of tough and oil-soluble phenol-formaldehyde resins. Tertiary **butyl phenol,** $C_{10}H_{14}O$, comes in flakes melting at 208°F (98°C). Tertiary **amyl phenol,** $C_{11}H_{10}O$, is also in flakes melting at 194°F (90°C). **Nonyl phenol** is a mixture of monoalkyl phenols with side chains of random-bracketed alkyl radicals on the molecule. It is a liquid of specific gravity 0.94, boiling at 572°F (300°C), used for making oil-soluble phenolic resins and lubricating oil additives. Texaco Chemical Co. and Rohm & Haas Co. market such products. **Phenyl phenol** is used in cosmetics to control odors and prevent bacterial deterioration, and as a preservative in paper, paints, and leather. **Bisphenol** is a hydroxy phenol, $C_{15}H_{16}O$, with high reactivity, used for producing modified phenolic resins and epoxy resins, as an antioxidant for oils, and as a stabilizer for resins. **Bisphenol A,** made by Mitsui Petrochemical Industries and Mitsubishi Petrochemical Co. (both of Tokyo), is a raw material for optical-grade **polycarbonate.** The latter is fabricated into the plastic substrates of compact disks. Santobrite, of Monsanto Chemical Co., is chlorinated phenol, or **sodium pentachlorophenol,** used as a preservative. **Biolite** is a formulation of Santobrite used for slime and mildew control in laundries and in paper-pulp plants. Para **nitrophenol,** made by treating phenol with cold dilute nitric acid, is used as a preservative for leather goods to prevent mold growth. **Lorothiodol,** of Hilton-Davis Chemical Co., is a sulfurized chlorinated phenol. It is a more powerful antiseptic than phenol, is odorless and nontoxic, and is used in soaps. **Pentaphen** is a para tertiary amyl phenol, $C_5H_{11}C_6H_4OH$, used for making pale-colored, oil-soluble resins by condensation with aldehydes. In a Nalco Chemical process, a proprietary blend of oxyalkylated alkyl phenols, fatty acids, and surfactants serves as a chemical penetrant for the recovery of almost all residual hydrocarbons in crude-petroleum storage tanks.

A wide range of related materials are produced from **diphenyl,** also called **biphenyl** and **phenylbenzene.** It has composition $(C_6H_5)_2$, is a liquid below 160°C (71°C), and boils at 491°F (255°C). It is a stable compound and is used as a heat-transfer medium. But it is made easily by heating benzene to eliminate H_2 and, because of its low cost, is a valuable intermediate chemical. Like phenol, it can be modified easily to produce innumerable compounds. **Tetramethylbenzene,** or **durene,** for example, is a 10-carbon aromatic liquid used for making **pyromellitic acid** and its anhydride for the production of polyester resins. Many other useful acids are also produced.

PHENOL-FORMALDEHYDE RESIN. A synthetic resin, commonly known as **phenolic,** made by the reaction of phenol and formaldehyde and employed as a molding material for the making of mechanical and electrical parts. It was the earliest type of hard, thermoset synthetic resins, and its favorable combination of strength, chemical resistance, electrical properties, glossy finish, and nonstrategic abundance of low-cost raw materials has continued the resin, with its many modifications and variations, as one of the most widely employed groups of plastics for a variety of products. The resins are also used for laminating, coatings, and casting resins.

The reaction was known as early as 1872 but was not utilized commercially until much later. A condensation product of 50 parts phenol and 30 parts 40% formaldehyde made under an English patent of 1905 was called **Resinite,** and was originally offered as a substitute for Celluloid. Various modifications were made by other inventors. **Redmanol** was one of the first U.S. products by Bakelite Corp.

Juvelite was made in Germany by condensing the phenol and formaldehyde with the aid of mineral acids, and **Laccain** was made under an English patent by using organic acids as catalysts. A Russian phenol resin, under the name of **Karbolite,** employed an equal amount of naphthalenesulfonic acid, $C_{10}H_7SO_3H$, with the formaldehyde.

The hundreds of different phenolic molding compounds can be divided into six groups on the basis of major performance characteristics. General-purpose phenolics are low-cost compounds with fillers such as wood flour and flock, and they are formulated for noncritical functional requirements. They provide a balance of moderately good mechanical and electrical properties and are generally suitable in temperatures up to 300°F (149°C). Impact-resistant grades are higher in cost. They are designed for use in electrical and structural components subject to impact loads. The fillers are usually paper, chopped fabric, or glass fibers. Electrical grades, with mineral fillers, have high electrical resistivity plus good arc resistance, and they retain their resistivity under high-temperature and high-humidity conditions. Heat-resistant grades are usually mineral- or glass-filled compounds that retain their mechanical properties in the 375 to 500°F (190 to 260°C) temperature range. Some of these, such as **phenylsilanes,** provide long-term stability at temperatures up to 550°F (288°C). Special-purpose grades are formulated for service applications requiring exceptional resistance to chemicals or water, or combinations of conditions such as impact loading and a chemical environment. The chemical-resistant grades, for example, are inert to most common solvents and weak acids, and their alkali resistance is good. Nonbleeding grades are compounded specially for use in container closures and for cosmetic cases.

The resins are marketed usually in granular form, partly polymerized for molding under heat and pressure, which complete the polymerization, making the product infusible and relatively insoluble. They may also come as solutions, or compounded with reinforcing fillers and pigments. The tensile strength of a molded part made form a simple phenol-formaldehyde resin may be only about 6,000 lb/in² (41 MPa), with a specific gravity of 1.27 and dielectric strength of about 450 V/mil (17.7×10^6 V/m). Reinforcement is needed for higher strength, and with a wood-flour filler the tensile strength may be as high as 10,000 lb/in² (69 MPa). With a fabric filler the tensile strength may be 15,000 lb/in² (103 MPa), or 18,000 lb/in² (124 MPa) with a mineral filler. The specific gravity is also raised, and the mineral fillers usually increase the dielectric strength.

Proper balance of fillers is important, since too large a quantity may produce brittleness. Organic fillers absorb the resin and tend to brittleness and reduced flexural strength, although organic fibers and fabrics generally give high impact strength. Wood flour is the usual filler for general-service products, but prepared compounds may have mineral powders, mica, asbestos, organic fibers, or macerated fabrics, or mixtures of organic and mineral materials. **Bakelite** was the original name for phenol plastics, but trade names now usually cover a range of different plastics, and the types and grades are designated by numbers.

The specific gravity of filled phenol plastics may be as high as 1.70. The natural color is amber, and because the resin tends to discolor, it is usually pigmented with dark colors. Normal **phenol resin** cures to single-carbon methylene groups between the phenolic groups, and the molded part tends to be brittle. Thus, many of the innumerable variations of phenol are now used to produce the resins, and modern phenol resins may also be blended or cross-linked with other resins to give higher mechanical and electrical characteristics. Furfural is frequently blended with formaldehyde to give better flow, lower specific gravity, and reduced cost. The alkylated phenols give higher physical properties. **Phenol-phosphor resin** is a phenol resin modified with phosphonitrilic chloride. When cured, the resin contains 15% phosphorus, 6 nitrogen, and less than 1 chlorine. The tensile strength is 7,000 lb/in^2 (48 MPa), and it withstands continuous temperatures to 500°F (260°C). Phenol resins may also be cast and then hardened by heating. The cast resins usually have a higher percentage of formaldehyde and do not have fillers. They are poured in a syrupy state in lead molds and are hardened in a slow oven.

Instead of making phenolic resins by polymerizing phenol formaldehyde using an acid catalyst, Enzymol International Inc. uses a peroxidase enzyme obtained from soybeans to polymerize phenols in an aqueous solution of organic solvents at 122 to 140°F (50 to 60°C). Hydrogen peroxide is added to activate the enzyme.

PHONOLITE. Also known as **clinkstone.** An aluminum-potassium-silicate mineral used in the production of glass, and in Germany for the production of aluminum. Phonolite is a variety of feldspar. It varies greatly in composition, the best of the Eifel Mountains mineral containing 20 to 23.25% alumina, 7 to 9 K$_2$O, 6 to 8 Na$_2$O, and 50 to 52 silica. A variety of this mineral, **nepheline,** from the Kòla Peninsula, is used in Russia to produce aluminum, with soda and potash as by-products. **Nepheline syanite** from Peterborough Co., Ontario, Canada, is used in the ceramic industry in pottery, porcelain, and tile to increase translucency and reduce warpage and crazing. From 3 to 5% added to structural clay increases the mechanical strength. As a substitute for potash feldspar in wall tiles, it increases the fluxing action and lowers the fusing point. **Agalmatolite,** a name derived from the Greek words meaning **image stone,** is the massive form of phonolite from which the Chinese carve figures and bas-reliefs. It has a soft, greasy feel and varies in color.

PHOSGENE. The common name for **carbonyl chloride,** $COCl_2$, a colorless, poisonous gas made by the action of chlorine on carbon monoxide. It was used as a poison war gas, called **D-stoff** by the Germans and **collongite** by the French. But it is now used in the manufacture of metal chlorides and anhydrides, pharmaceuticals, perfumes, isocyanate resins, and for blending in synthetic rubbers. It liquefies at 45.7°F (7.6°C) and solidifies at −180°F (−118°C). It is decomposed by water. When chloroform is exposed to light and air, it decomposes into phosgene. One part in 10,000 parts of air is a toxic poison, causing pulmonary edema. For chemical warfare it is compressed into a liquid in shells. **Lacrimite,** also a poison war gas, is thiophosgene mixed with stannic chloride. **Diphosgene,** $ClCOOC \cdot Cl_3$, called **green cross, superpalite,** and **perstoff,** is an oily liquid boiling at 262°F (128°C). It is an intense lachrymator, has an asphyxiating odor, and is a lung irritant.

Because of its toxicity, most phosgene is produced and employed immediately in captive applications by Dow Chemical Co., Du Pont Co., and BASF AG. The biggest use of the material is for **toluene diisocyanate (TDI),** which is then reacted into polyurethane resins for foams, elastomers, and coatings. Approximately 1.4 tons (1.3 metric tons) of phosgene is needed to make 1.1 tons (1 metric ton) of TDI. About 0.99 ton (0.9 metric ton) of phosgene is consumed to make 1.1 tons (1 metric ton) of **polymethylene polyphenylisocyanate,** also used for making polyurethane resins for rigid foams. **Polycarbonate** manufacturers require 0.46 ton (0.42 metric ton) of phosgene per ton (0.91 metric ton) of product resin. Polycarbonate is used for making break-resistant housings, signs, glazings, and electrical tools. Phosgene also is a reactant for **methyl isocyanate, diphenylmethane-4,4-isocyanate, acyl chloride, chloroformate esters, diethyl carbonate,** and **dimethyl carbamoyl chloride.** The isocyanates are used in pesticides, and the di- and polyisocyanates are used in adhesives, coatings, and elastomers.

PHOSPHOR BRONZE. Copper-base alloys with low phosphorus content, originally called **steel bronze** when first produced at the Royal Arsenal in Vienna. It was 92–8 bronze deoxidized with phosphorus and cast in an iron mold. It is now any bronze deoxidized by the addition of phosphorus to the molten metal. It may or may not contain residual phosphorus in the final state. Ordinary bronze frequently contains cuprous oxide formed by the oxidation of the copper during fusion. By the addition of phosphorus, a powerful reducing agent, a complete reduction of the oxide takes place. Phosphor bronzes have excellent mechanical and cold-working properties and low coefficients of friction, making them suitable for springs, diaphragms, bearing plates, and fasteners. In some environments, such as salt water, they are superior to copper. Phosphor bronzes have been known by many trade names, including **Duraflex,** a hard-rolled strip and wire product for springs; **Carbobronze,** hard-drawn tubing and rod for bearings; **Corvic,** a spring grade with a tensile strength of 95,000 lb/in^2 (655 MPa) and an electrical conductivity of 42% that of copper; and **Telnic,** a 1% nickel, 0.5 tellurium, 0.2 phosphorus grade. So-called **white phosphor bronze** is not a bronze, but a 72% lead, 15 phosphor tin, 12 antimony, 1 copper alloy.

Standard wrought phosphor bronzes are designated **C50100 to C54800.** Tin, which ranges from as much as 0.8 to 11% depending on the alloy, is the principal alloying element, although leaded alloys may contain as much lead (4 to 6%, for example) as tin. Phosphorus content is typically on the order of 0.1 to 0.35%; zinc, 0 to 0.3 (1.5 to 4.5% in C54400); iron, 0 to 0.1; and lead, 0 to 0.05 (0.8 to 6 in leaded alloys). The principal alloys were formerly known by letter designations representing nominal tin content: **phosphor bronze A,** 5% tin (C51000); **phosphor bronze B,** 4.75 tin (C53200); **phosphor bronze C,** 8 tin (C52100); **phosphor bronze D,** 10 tin (C52400); and **phosphor bronze E,** 1.25 tin (C50500). Phosphor bronze E, being almost 99% copper, is one of the leanest of these bronzes in the way of alloying ingredients and is used for electrical contacts, pole-line hardware, and flexible tubing. Its electrical conductivity is about half that of copper, and it is readily formed, soldered, brazed, and flash-welded. Thin, flat products have tensile yield strengths ranging from about 12,000 lb/in^2 (83 MPa) in the annealed condition to 75,000 lb/in^2 (517 MPa) in the extra-spring temper. More highly alloyed C54400 (4% tin, 4 lead, and 3 zinc, nominally) is about one-fifth as electrically conductive as copper, has good forming characteristics, and has 80% the machinability of C36000, a free-machining brass. Its ultimate tensile strength ranges from about 48,000 lb/in^2 (331 MPa) in the annealed condition to 100,000 lb/in^2 (690 MPa) in the extra-spring temper. Uses include bearings, bushings, gear shafts, valve components, and screw-machine products. An **alloy C94400,** which has been called a *phosphor bronze,* is suitable for sand castings and centrifugal castings. The nominally 81% copper, 11 lead, 8 tin alloy is 10% as electrically conductive as copper and, as sand-cast, has a typical ultimate tensile strength of 32,000 lb/in^2 (221 MPa) and a tensile yield strength of 16,000 lb/in^2 (110 MPa). It matches the machinability of C54400 and is used mainly for bushings and bearings.

PHOSPHOR COPPER. An alloy of phosphorus and copper, used instead of pure phosphorus for deoxidizing brass and bronze, and for adding phosphorus in making phosphor bronze. It comes in 5, 10, and 15% grades and is added directly to the molten metal. It serves as a powerful deoxidizer, and the phosphorus also hardens the bronze. Even slight additions of phosphorus to copper or bronze increase fatigue strength. Phosphor copper is made by forcing cakes of phosphorus into molten copper and holding until the reaction ceases. Phosphorus is soluble in copper up to 8.27%, forming Cu_3P, which has a melting point of about 1305°F (707°C). A 10% phosphor copper melts at 1562°F (850°C) and a 15% at about 1872°F (1022°C). Alloys richer than 15% are unstable. Phosphor copper is marketed in notched slabs or in shot. In Germany **phosphor zinc** was used as a substitute to conserve copper. **Metallophos** is a name for German phosphor zinc containing 20 to 30% phosphorus. The name *phosphor copper* is also applied to commercial copper deoxidized with phosphorus and retaining up to 0.50% phosphorus. The electrical conductivity is reduced about 30%, but the copper is hardened and strengthened. **Phosphor tin** is a master alloy of tin and phosphorus used for adding to molten bronze in the making of phosphor bronze. It usually contains up to 5% phosphorus and should not contain lead. It looks like antimony, with large glittering crystals, and is marketed in slabs.

PHOSPHORIC ACID. Also known as **orthophosphoric acid.** A colorless, syrupy liquid of composition H_3PO_4 used for pickling and rust-proofing metals; for the manufacture of phosphates, pyrotechnics, and fertilizers; as a latex coagulant; as a textile mordant; as an acidulating agent in jellies and beverages; and as a clarifying agent in sugar syrup. The specific gravity is 1.65 and melting point 164°F (73.6°C), and it is soluble in water. The usual grades are 90, 85, 75%, technical 50%, and dilute 10%. As a cleanser for metals, phosphoric acid produces a light etch on steel, aluminum, or zinc, which aids paint adhesion. **Deoxidine** is a phosphoric acid cleanser for metals. **Nielite D** is phosphoric acid with a rust inhibitor, used as a nonfuming pickling acid for steel. **Albrite,** from Albright & Wilson Americas, is available in 75, 80, and 85% concentrations in food and electronic grades, both high-purity specifications. **DAB** and **Phosbrite,** also from the same company, are called **Bright Dip** grades, for cleaning applications. **Phosphoric anhydride,** or **phosphorus pentoxide,** P_2O_5, is a white, water-soluble powder used as a dehydrating agent and as an opalizer for glass. It is also used as a catalyst in asphalt coatings to prevent softening at elevated temperatures and brittleness at low temperatures. **Granusic** is this material in granular form for removing water from gas streams. Production of phosphoric acid is by digestion of apatite ore, a tricalcium phosphate, with sulfuric acid, and purification. For industrial grades, **white phosphorus** or **yellow phosphorus** is burned in excess air, and the resulting phosphorus pentoxide is hydrated.

PHOSPHORUS. A nonmetallic element, symbol P, widely diffused in nature and found in many rock materials, in ores, in the soil, and in parts of animal organisms. Commercial phosphorus is obtained from phosphate rock by reduction in the electric furnace with carbon, or from bones by burning and treating with sulfuric acid. **Phosphate rock** occurs in the form of land pebbles and as hard rock. It is plentiful in the Bone Valley area of Florida, and it also comes from Tennessee, Idaho, and South Carolina. Vast quantities are mined in Morocco and Tunisia. Large deposits are found on many of the Pacific Islands, the Christmas Island resources being estimated at 30×10^6 tons (27×10^6 metric tons) and those on Nauru at 100×10^6 tons (91×10^6 metric tons). It is a calcium phosphate high in P_2O_5. The mineral **apatite,** widely distributed in the Appalachian range, in Idaho, Brazil, and French Oceania, is also a source of phosphorous, containing up to 20% P_2O_5, with iron oxide and lime. The Egyptian rock contains 62 to 70% tricalcium phosphate. The aluminocalceous phosphate rock of Senegal is treated to obtain a very soluble fertilizer known as **phosphal.** Florida hard phosphate rock contains 80% phosphate of lime. A ton (0.9 metric ton) of phosphorus is obtained from 7.25 tons (6.58 metric tons) of rock, requiring 30 lb (14 kg) of electrodes and 11,850 kW of electricity. The Tennessee Valley Authority produces about 8 tons (7 metric tons) of **expanded slag** for each ton (0.9 metric ton) of phosphorus produced from the phosphate rock. The slag from the smelter is run onto a forehearth at about 2000°F (1093°C) and treated with water, high-pressure steam, and air. The expanded slag formed is crushed to 0.375-in (0.95-cm) size, bulking 50 lb/ft³ (801 kg/m³). It is called **TVA slag** and is used for making lightweight concrete blocks. The **superphosphate** used for fertilizers is made by treating phosphate minerals with concentrated sulfuric acid. It is not a simple compound, but may be a mixture of **calcium acid phosphate,** $CaHPO_4$, and calcium sulfate. **Nitrophosphate** for fertilizer is made by acidulating phosphate rock with a mixture of nitric and phosphoric acids, or with nitric acid and then ammoniation and addition of potassium or ammonium sulfate. Such products are made by Cargill, Inc., and International Minerals & Chemicals Corp. Natural rock phosphate in finely ground form is also used as a fertilizer for legume crops, but the untreated natural rock is not readily soluble and is thus not as quick-acting as a fertilizer.

There are two common forms of phosphorus, yellow and red. The former, also called **white phosphorus,** P_4, is a light-yellow, waxlike solid, phosphorescent in the dark and exceedingly poisonous. Its specific gravity is 1.83, and it melts at 111°F (44°C). It is used for smoke screens in warfare and for rat poisons and matches. **Yellow phosphorus** is produced directly from phosphate rock in the electric furnace. It is cast in cakes of 1 to 3 lb (0.45 to 1.36 kg) each. **Red phosphorus** is a reddish-brown, amorphous powder, having a specific gravity of 2.20 and a melting point of 1337°F (725°C). Red phosphorus is made by holding white phosphorus at its boiling point for several hours in a reaction vessel. Both forms ignite easily. **Amorphous phosphorus,** or **crystalline black phosphorus,** is made by heating white phosphorus for extended periods. It resembles graphite and is less reactive than the red or white forms, which can ignite spontaneously in air. **Black phosphorus** is made by this process by Atomergic Chemetals Corp. **Phosphorus sulfide,** P_4S_3, may be used instead of white phosphorus in making matches. **Phosphorus pentasulfide,** P_2S_5, is a canary-yellow powder of specific gravity 1.30, or solid of specific gravity 2.0, containing 27.8% phosphorus, used in making oil additives and insecticides. It is decomposed by water.

Phosphorus is an essential element in the human body, a normal person having more than a pound of it in the system, but it can be taken into the system only in certain compounds. **Nerve gases** used in chemical warfare contain phosphorus which combines with and inactivates the choline sterase enzyme of the brain. This enzyme controls the supply of the hormone which transmits nerve impulses, and when it is inactivated, the excess hormone causes paralysis of the nerves and cuts off breathing. **Organic phosphates** are widely used in the food, textile, and chemical industries. **Tributyl phosphate,** for example, is a colorless liquid, used as a plasticizer in plastics and as an antifoaming agent in paper coatings and textile sizings. **Briquest** is an organic phosphate from Albright & Wilson Inc. that is employed for scale and corrosion control, ore flotation, pigment dispersion, and detergents. **Diethylchlorothiophosphate (DECTP)** is a highly reactive intermediate chemical from Ethyl Corp. Flour and other foodstuffs are fortified with **ferric phosphate,** $FePO_4 \cdot 2H_2O$. **Iron phosphate** is used as an extender in paints. **Tricalcium phosphate,** $Ca_3(PO_4)_2$, is used as an anticaking agent in salt, sugar, and other food products and to provide a source of phosphorus. The tricalcium phosphate used in toothpastes as a polishing agent and to reduce the staining of chlorophyll has formula $(10CaO \cdot H_2O \cdot 3P_2O_5)3H_2O$ and is a fine, white powder. **Dicalcium phosphate,** used in animal feeds, is precipitated from the bones used for making gelatin, but is also made by treating lime with phosphoric acid made from phosphate rock. **Diammonium phosphate,** $(NH_4)_2HPO_4$, is a mildly alkaline, white, crystalline powder used in ammoniated denti-frices, for pH control in bakery products, in making phosphors and to prevent afterglow in matches, and for flameproofing paper.

For manufacturing operations, phosphorus is generally utilized in the form of intermediate chemicals, but the phosphorus used for doping semiconductors and in electroluminescent coatings is 99.9999% pure. **Phosphorus trichloride,** PCl_3, is an important chemical for making phosphites. It is a colorless liquid boiling at 169°F (76°C). It decomposes in water to form phosphorus and hydrochloric acid. **Phosphorus oxychloride,** $POCl_3$, is a very reactive liquid used as a chlorinating agent and for making organic chemicals. In water it decomposes to form phosphoric and hydrochloric acids. **Phosphorus thiochloride,** $PSCl_3$, is a yellowish liquid containing 18.5% phosphorus and 18.6 sulfur. It is used for making insecticides and oil additives. **Phosphine,** PH_3, is produced by hydrolysis of a metal phosphide, such as **calcium phosphide** or **aluminum phosphide.** A toxic gas, it is widely used as a ligand in catalysis.

PHTHALIC ANHYDRIDE. A white, crystalline material of composition $C_6H_4(CO)_2O$, with a melting point of 267°F (130.8°C), soluble in water and in alcohol. It is made by oxidizing naphthalene, or it is produced from orthoxylene derived from petroleum. BASF Corp. markets the product both as flake and in the molten state. It is used in the manu-facture of alkyd resins and for the production of dibutyl phthalate and other plasticizers, dyes, and many chemicals. **Chlorinated phthalic anhydride** is also used as a compounding medium in plastics. It is a white, odorless, nonhygroscopic, stable powder containing 50% chlorine. It gives higher temperature resistance and increased stability to plastics. **Niagathal** is a chlorinated phthalic anhydride of Niagara Alkali Co. **Tetrahydrophthalic anhydride** is a white, crystalline powder with a molecular weight of 152.1, melting at 212°F (100°C), used to replace phthalic anhydride where a lighter color is desired. It is produced by condensing butadiene with maleic anhydride. In synthetic resin coatings it gives higher adhesion. **Terephthalic acid** may be obtained as a by-product in the production of phthalic anhydride from petroleum. It has a long-chain alkyl group having an amide linkage on one end and a methyl ester on the other. It is used for producing **polyethylene terephthalate** and other polyesters. The esters can also be made from **dimethyl terephthalate,** a molten material that burns readily when ignited. The dust can form explosive mixtures with air. The terephthalates are useful as textile and tire-cord fibers, plastic tape, and food-packaging polymers. Their sodium salt is used as a gelling agent for high-temperature lubricating greases for uses to 600°F (316°C). It forms fine **crystallites** of soft, flexible fibers in the grease. **Oronite GA10** is this material. **Isophthalic acid,** made by oxidation of ethyl benzene and orthoxylene, produces alkyd paint resins of greater heat stability than phthalic anhydride. **Maleic anhydride,** $(CHCO)_2O$, is a white, crystalline solid used to replace phthalic anhydride in alkyd resins to increase the hardness for baking enamels and to resist yellowing. Its use in papermaking, as sizing, is growing, as are markets in lubricating-oil additives, acidulation, where it is a flavoring agent, and agricultural chemicals. Maleic anhydride is also a building block for **L-aspartame,** used for making **NutraSweet,** the aspartame synthetic sweetener. Maleic anhydride was traditionally made from benzene, but ***n*-butane** has become the feedstock of choice because of its lower cost and because benzene is a carcinogen. In the United States, the transition was completed in the late 1980s. Recent uses for maleic anhydride, and **maleic acid** recovered from the catalytic oxidation of butane to maleic anhydride, are the production of chemical intermediates **1,4-butanediol (BDO), tetrahydrofuran (THF),** and **gamma-butyrolactone (GBL).**

PIASSAVA. Also called **Pará grass** and **monkey grass.** A coarse, stiff, and elastic fiber obtained from a species of palm tree, *Leopoldinia piassaba,* of Brazil, used for making brushes and brooms. The plant has long beards of bristlelike fibers, which are combed out and cut off the young plants. These fibers sometimes reach a length of 4 ft (1.2 m). The soft, finer fibers are made into cordage, and the coarser ones are used for brushes. Piassava is very resistant to water. The fiber for brush manufacture is separated into three classes, the heavy fibers being known as **bass,** the medium as **bassine,** and the fine as **palmyra.** The bass is used for heavy floor sweeps. The fiber of the palm *Attalea funifera,* which grows in the state of Bahia, Brazil, and is also called piassava, is a harder and stronger material than the piassava of Amazonas. It is used for marine cordage and is resistant to salt water. A substitute for piassava is **acury,** from the leaves of the palm *A. phalerata* of Matto Grosso. It is used for cordage and brushes, and the coarser fibers are used for brooms.

PICKLING ACIDS. Acids used for pickling, or cleaning castings or metal articles. The common pickling bath for iron and steel is composed of a solution of sulfuric acid and water, 1 part acid to 5 to 10 parts water being used. This acid attacks the metal and cleans it of the oxides and sand by loosening them. For pickling scale from stainless steels a 25% cold solution of hydrochloric or sulfuric acid is used, or hydrofluoric acid with the addition of anhydrous ferric sulfate is used. Hydrofluoric acid solutions are sometimes used for pickling iron castings. This acid attacks and dissolves away the sand itself. For bright-cleaning brass, a mixture of sulfuric acid and nitric acid is used. For a matte finish the mixed acid is used with a small amount of zinc sulfate. Copper and copper alloys can be pickled with sulfuric acid to which anhydrous ferric sulfate is added to speed the action, or sodium bichromate is added to the sulfuric acid to remove red cuprous oxide stains. Brass forgings are pickled in nitric acid to bring out the color. Since all of these acids form salts rapidly by the chemical action with the metal, they must be renewed with frequent additions of fresh acid. The French pickling solution known as **framanol,** used for aluminum, is a mixture of chromium phosphate and triethanolamine. The latter emulsifies the grease and oil, and the aluminum oxide film is dissolved by the phosphoric acid, leaving the metal with a thin film of chromic oxide.

The temperature of most pickling is from 140 to 180°F (60 to 82°C). An increase of 20°F (11°C) will double the rate of pickling. **Acid brittleness** after pickling is due to the absorption of hydrogen when the acid acts on iron, and is reduced by shortening the pickling time. **Inhibitors** are chemicals added to reduce the time of pickling by permitting higher temperatures and stronger solutions without hydrogen absorption. **Hibitite,** of Monsanto, is a brown liquid of composition $C_{27}H_{45}NO_{10}S_2$, used as a metal pickling inhibitor. Addition of a small amount of 2% tincture of iodine to a 5% sulfuric acid solution gives a 95% retardation of acid attack on steel without decreasing the rate of dissolution of rust. In plating baths, **fluoboric acid,** HBF_4, has high throwing power and has a cleansing effect by dissolving sand and silicides from iron castings and steel surfaces. It is a colorless liquid with specific gravity of 1.33 and decomposes at 266°F (130°C). **Pennsalt FA-42,** of Atochem North America Inc., is this material. It is a 42% solution of fluoboric acid for pickling and for control of acidity of plating baths.

Phosphoric acid is employed in hot solution as a dip bath for steel parts to be finished to a rough or etched surface. It leaves a basic iron phosphate coating on the steel which is resistant to corrosion and gives a rough base for the finish. **Coslettized steel** is steel rustproofed by dipping in a hot solution of iron phosphate and phosphoric acid. **Parkerized steel** is rustproofed steel treated in a bath of iron and manganese phosphates. **Bonderized steel** is steel treated with phosphoric acid and a catalyst to give a rough, tough, rust-resistant base for paints. **Granodized steel** is produced with zinc phosphate. In general, the coatings left on steel by phosphate treatments are extremely thin, not over 0.0002 in (0.0005 cm). The iron-manganese coatings are black, and the iron-zinc-phosphate coatings are gray.

Metals can be treated with alkaline solutions, too. Rust can be removed by caustic soda baths in which is mixed a sequestrant, such as **sodium gluconate** or **ethylenediaminetetracetic acid (EDTA),** to complex the dissolved iron and keep it from precipitating. **Hampene EDTA** is a chelating agent from Hampshire Div. of W. R. Grace & Co. In general, the alkali treatments work more slowly than acid-based ones.

PIG IRON. The iron produced from the first smelting of the ore. The melt of the blast furnace is run off into rectangular molds, forming, when cold, ingots called **pigs.** Pig iron contains small percentages of silicon, sulfur, manganese, and phosphorus, besides carbon. It is useful only for resmelting to make cast iron or wrought iron. Pig iron is either sand-cast or machine-cast. When it is sand-cast, it has sand adhering and fused into the surface, giving more slag in the melting. Machine-cast pig iron is cast in steel forms and has a fine-grained, chilled structure, with lower melting point. Pig irons are classified as Bessemer or non-Bessemer, according to whether the phosphorus content is below or above 0.10%. There are six general grades of pig iron: **low-phosphorus pig iron,** with less than 0.03%, used for making steel for steel castings and for crucible steelmaking; **Bessemer pig iron,** with less than 0.10% phosphorus, used for Bessemer steel and for acid open-hearth steel; **malleable pig iron,** with less than 0.20%, used for making malleable iron; **foundry pig iron,** with 0.5 to 1%, for cast iron; **basic pig iron,** with less than 1%, and low-silicon, less than 1%, for basic open-hearth steel; and **basic Bessemer,** with 2 to 3%, used for making steel by the basic Bessemer process employed in England.

Since silicon is likely to dissolve the basic furnace lining, it is kept as low as possible, 0.70 to 0.90%, with sulfur not usually over 0.095%. Pig irons are also specified on the basis of other elements, especially sulfur. The sulfur may be from 0.04 to 0.10%, but high-sulfur pig iron cannot be used for the best castings. The manganese content is usually from 0.60 to 1%. Most of the iron for steelmaking is now not cast but is carried directly to the steel mill in car ladles. It is called **direct metal.** Foundry pig iron is graded by the silicon content, No. 1 having from 2.5 to 3% and No. 3 from 1.5 to 2 silicon. **Silvery iron** is a name for pig iron of high silicon content because of its silvery fracture. **Puddling iron** is a grade of pig iron used for making wrought or puddled iron in a puddling furnace. A requirement is that the silicon be low, with manganese 0.5 to 1%.

Chateaugay iron is a low-phosphorus pig iron produced from New York State **magnetite ore.** The original ore as mined contains about 28% iron. The standard analysis of the pig iron is total carbon, 4%;

silicon, 0.75 to 4.0; sulfur, 0.03 maximum; phosphorus, 0.03 maximum; manganese, 0.10 to 0.15. Chateaugay iron is used for casting rolls, gears, and machine parts. **Norskalloy** is a name for pig iron produced from Norwegian ores containing vanadium and titanium. The standard grade contains from 4 to 4.5% total carbon, 0.5 to 1.5 silicon, 0.20 manganese, 0.20 to 0.25 phosphorus, 0.30 to 0.40 vanadium, and 0.40 to 0.80 titanium. From 15 to 20% Norskalloy pig is added to mixtures where vanadium is required. **Mayari iron** is pig iron made from Cuban ores which contain vanadium and titanium, or is pig iron made to duplicate the Cuban iron. These irons are considered especially suitable for heavy rolls or high-grade castings. Mayari pig contains 1.60 to 2.50% chromium, 0.80 to 1.25 nickel, 0.25 to 2.25 silicon, 0.10 to 0.20 titanium, 0.05 to 0.08 vanadium, 3.8 to 4.5 total carbon, 0.60 to 2 manganese, less than 0.05 sulfur, and under 0.10 phosphorus. **Nikrofer** is a German pig iron from Greek ore that is similar in composition.

PIGMENT. A substance, usually earthy or clayey, which when mixed with oil or other adhesive carrier and a solvent, forms a paint. Pigments usually give body as well as color to the paint, and the paint **hiding power** is measured by comparison tests when in the form of a mixed paint. If the hiding power of lithopone is taken as 100, the hiding power of zinc sulfide is 240, and that of titanium dioxide is 400. **Color standards** are distinct from hiding power. Pure magnesium oxide is used as the standard for the measurement of whiteness. The chemical interaction must also be considered in pigments. For example, zinc oxide increases wear life and mildew resistance in paints, but may tend to react and cause blisters. The use of pigments is not confined to paints. In ceramics, their primary choice is by color, but they usually also add other physical properties to the ceramic. In plastics they add body and strength, as distinct from dyes which do not add body.

A pigment is distinct from a **filler** in that a pigment must retain its opacity when wet. White, powdered quartz, used sometimes as a filler, is not a good pigment as it becomes glassy when wet. Fillers that retain their opacity are called **extenders,** or **auxiliary pigments,** and the final mixed pigment is called a **reduced color.** But an extender pigment, such as silica, that does not have good hiding power in itself will increase the hiding power and depth of color of a pigment if the extender is of such fine particle size that it will be dispersed in the voids between the pigment particles. Extremely fine silica will also cement itself chemically to lead pigments and add wearing qualities. **Hi-Sil** of PPG Industries is a silica with a particle size of 0.984 μin (0.025 μm). As a pigment for rubber, it adds strength and wear resistance to the rubber. **Extender RX-2022,** of Pfizer Inc., is iron phosphate in the form of white powder for use in paints. It has low hiding power, with refractive index of 1.7, but is a corrosion inhibitor and is used in undercoats. **Kaolin,** a natural aluminosilicate clay occurring in Georgia and Texas, is both a pigment and an extender. Georgia Kaolin offers kaolin clays specifically for paper, ceramics, paint, ink, plastics, and rubber applications. The clay crystals are usually less than 39 μin (1 μm) in size. **Hydrite** is a kaolin extender pigment that is white, chemically inert, and low in absorption. **Kaopaque** grades are high-purity kaolins that have been delaminated or sheared to further decrease the particle size, and bleached to increase brightness. They are used as coatings in paper. **Glomax** is a calcined kaolin, the calcining process removing the hydration water, increasing surface area, and burning off impurities. The clays have increased brightness and opacity, and substitute for titanium dioxide, either partially or completely, in many paper, paint, and plastics applications. **Camel-Wite** is a pigment extender based on calcium carbonate. It is a white, wet-ground pigment that matches the hiding power of ultrafine, precipitated calcium carbonate. **Antimony trioxide** is an opaque, yellow powder used in plastics and coatings as a flame-retardant pigment and opacifier. **Firemaster HHP** is this material. **Thermoguard S,** of Atochem North America, is antimony trioxide as a 99.8% pure, white powder with average particle size less than 39 μin (1 μm).

Pigments are mostly of mineral origin, the vegetable pigments such as logwood and the animal pigments such as cochineal being ordinarily classified as dyestuffs. Bone black, however, is an example of an animal pigment, and vine black is a typical vegetable pigment. **Vine black,** a fine pigment for inks, was originally made by charring grapevine stems, and was known as **Frankfurt black, but similar pigments are now made from fruit pits, nut shells, or wood pulp and are called** vegetable black. Pigments are also produced by dyeing clays with aniline dyestuffs. These are called **lakes. Dutch yellow,** for example, is a **yellow lake** made by adsorption of a yellow dye such as quercitron by an inert material such as calcium carbonate. Various chemicals such as copper acetate and potassium acetate are used as pigments. **Potassium acetate, CH_3COOK,** is a white powder also used in making crystal glass. Pigments should be ground fine enough so that all the powder will pass through a 325-mesh screen.

Natural pigments include ochre, umber, ground shale, hematite, and sienna. **Sepia** is a dark-brown pigment originally made from the black, inky secretion found in an internal ink sac of the *Sepia* mollusk. It is used in India inks of sepia or dark-brown color and was formerly employed directly as a writing ink. The **red-ochre** pigments are the natural red iron oxides of high oxide content. The yellows, or siennas, are the oxides mixed with considerable clay. The browns, or umbers, have manganese present in the clays. **Terre verte,** or **Verona green,** of Cyprus, is a fine, blue-green earth valued highly in the Middle Ages as a pigment. It contains 53% silica, 26 ferric oxide, 16 potash, and some magnesia, manganese, and other impurities. **Manganese green,** or **Cassel green,** is **barium manganate,** $BaMNO_4$, a green, poisonous powder of specific gravity 4.85, insoluble in water. **Mineral green,** or **Scheele's green,** is **copper arsenite,** $CuHAsO_3$. It is a light-green, amorphous powder used in paints and in textile printing. It is also used in medicine. **Orange pigments,** from yellow to brilliant red, with high tinting strength and great fastness, are made with mixtures of **lead chromate,** $PbCrO_4$, **lead molybdate,** $PbMoO_4$, and **lead sulfate,** $PbSO_4$. Pink to maroon is obtained in ceramic enamels with **calcium stannate,** $CaSnO_3 \cdot 3H_2O$, a white, crystalline powder that loses its water at 662°F (350°C). **Gloss white,** used as a reduced-color white pigment, for surface-coating pulp papers, and in printing inks, is a coprecipitation product consisting of 75% blanc fixe and 25 aluminum hydrate.

The most important yellow is **chrome yellow,** but it fades easily. However, a pigment's light-fastness and tinting effect depend on the crystal structure as well as the chemical composition. Normal **lead chromate** has a monoclinic crystal form, and it gives the strongest and most light-fast of the chrome yellows. Coprecipitated lead chromate and lead sulfate are orthorhombic and greenish in hue, giving primrose and lemon yellows poorer in light-fastness and rust-inhibiting action. A specialty pigment in this family is **Oncor M50,** a basic lead silico chromate made by NL Chemicals. **Strontium chromate,** $SrCrO_4$, gives a lemon yellow of good light-fastness. **Yellow ochre** is inferior as a color but durable. **Cadmium yellow** is cadmium sulfide and is a brilliant, permanent pigment but expensive. **Cadmium selenide** produces a bright **cadmium red** which when mixed with the sulfide, produces **cadmium orange.** The yellow pigment called **mosaic gold,** or **artificial gold,** is **stannic sulfide,** SnS_2, used in gilding and bronzing paints. **Stannous sulfide,** SnS, used for incorporation in bearings, is a black, crystalline material melting at 1616°F (880°C). Cadmium red and cadmium orange are produced by calcining selenium with cadmium sulfide. These **cadmium sulfoselenide** pigments give brilliant colors. Cadmium pigments are used in **camouflage paints** to give greater reflection of infrared rays. A building painted with a green containing cadmium has the same reflection as grass or leaves, and is indistinguishable in aerial photographs. An ancient lemon-brown pigment is **bistre.** It was obtained from the collected chimney soot of wood fires and much used by the old masters. It is very durable in watercolors. The most important green is **chrome green,** which is chrome yellow mixed with Prussian blue; it gives a good color but no permanence.

Ultramarine is the most important blue. It is used in paints and inks and as **bluing** for whitening paper, textiles, and organic materials by neutralizing the yellow cast. It is an ancient pigment, formerly made by grinding lapis lazuli, an azure-blue gemstone which is a sodium silicate sulfide. Ultramarine is now made by calcining a mixture of aluminum silicate and sodium sulfide, and it has the empirical formula $Na_7Al_6Si_6O_{24}S_2$. It is a deep-blue, water-soluble powder of 325 mesh, often marketed as a linseed-oil paste. A hydrated form goes by the mineral name **sodalite. Egyptian blue,** a chemical-resistant pigment, is a double silicate of calcium and copper, $CaO \cdot CuO \cdot 4SiO_2$. It was used by the ancients, and paintings 1,900 years old still retain the color. It is now made by fusing powdered quartz, chalk, copper oxide, and sodium carbonate. **Cobalt blue** is a good color but is expensive. **Prussian blue,** or **Chinese blue,** is ferric ferrocyanide, $Fe_4[Fe(CN)_6]_3$, a blue, amorphous powder. It is made by combining iron chloride and potassium ferrocyanide. **Celestial blue** is the light-blue pigment made by extending Prussian blue with barytes. **Milori blue,** used for coloring matches, inks, lacquers, and soaps, is ferric ferrocyanide with gypsum or barium sulfate. **Vermilion pigment** is mercury sulfide, which gives a fine color and is permanent, but it is expensive. High-grade blacks are usually **lampblack, bone black,** and **ivory black,** but may be extended with graphite. **Spanish black** is a name used in old texts for the black pigment made by burning cork. It is light and of soft texture. **Mineral black,** or **slate black,** is made by grinding black slate. **Metronite** is a white mineral composed of magnesium and calcium carbonates and magnesium silicate, used as a paint filler and extender. The pigment known as **Plessy's green** is **chromium phosphate,** $CrPO_4$, a bluish-green powder insoluble in water.

The **chemical colors** known as **phthalocyanines** give high tinting strength and resistance to deterioration by high-temperature baking. They are used for paints, inks, and plastics and are available as dry colors, in oil- or water-dispersible pastes, and in resin chips for plastics. They are chelated metallic salts of **tetrabenzoporphyrazine,** which is made from phthalamide or ammonium phthalic anhydride in the presence of iron, nickel, or copper salts. **Monastral blue,** of Du Pont, is **copper phthalocyanine,** a salt in which the copper is held in a chelate ring complex with four nitrogen atoms. The green is made by chlorination and is available from BASF Corp. **Fastolux pigments,** of Sun Chemical Corp., are phthalocyanines in fine particle size with a complete range of blue and green colors. The red and bluish-red colors of Du Pont are linear **quinacridones** made from terephthalic acid. The alpha crystal has a bluish-red color, the beta crystal has an intense violet color, and the gamma material has a true red color. The crystal structure can be controlled, and combinations give a range of brilliant, nonbleeding red and violet shades. The **Mercadium colors,** made by Hercules Inc., are compounds of the sulfides of mercury and cadmium to give colors from light orange to dark maroon. All of these pigments give permanence, light and chemical resistance, and a very high tinting strength. The tinting strength of the blue is 15 times that of ultramarine blue, and the green is superior to conventional pigments in brightness and permanence. **Heliogen blue** and **Heliogen green** are phthalocyanine pigments, and **Ramapo blue** and **Ramapo green** are the pigments with a barium resinate extender.

A reactive **fungicidal pigment** used in ship paints and antifouling paints is **cupric oxide hydrate,** a fine, dark-brown powder of composition $4CuO \cdot H_2O$. **Copper quinolinolate** is also used in fungicidal paints. **Cunilate** is this material. A yellow pigment used in anticorrosive and blister-resistant metal primers is **zinc tetroxychromate,** $4Zn(OH)_2 \cdot ZnCrO_4$. **Metallic pigments** are most frequently bronze powder or aluminum powder. They are used to increase light reflectivity as well as for appearance. **Stainless-steel flake** for pigment is marketed by Charles Hardy, Inc., as a fine powder in a paste with stearic acid and a solvent. Added to a clear lacquer or varnish, it gives a hard, silvery coating resistant to corrosive fumes. **Aluminum flake** gives high heat reflectivity as well as light reflectivity, and is used in silicone-based paints for high heats. Aluminum powder gives iridescent effects when dispersed in vinyl compounds. **Vinylum,** of Argus Chemical Laboratory, is such a powder in a vinyl copolymer.

For environmental reasons the trend in **colorants** for plastics has been away from the use of pigments based on heavy metals such as cadmium, chromium, lead, and selenium. Reed Plastics has replaced its heavy-metal products with seven series of **Reedlite** nylon-based colorants. Also organic are Hoechst Celanese's **PV-Gelb F2G,** a yellow No. 194 pigment, and **PV-Echtorange H4GL,** an orange No. 72 pigment. The yellow pigment, intended mainly for polyethylene film, replaces diarylide- and lead-chromium-based pigments. Also free of heavy metals is a series of low-plate-out **fluorescent concentrates** from Ferro Corp. **Holcomax** color concentrates, of Holland Colours (Holland), are dispersed in a copolymer compatible with various plastics. **Clear-Tint** and **Advantage,** of Milliken Chemical, are transparent and opaque colorants, respectively. For a shiny, metallic effect, Silberline, of Scotland, offers aluminum-containing pigments **Silvex** pastes for thermosets and **Silvet** granules for thermoplastics. Recent titanium dioxide white pigments include **UV-Titan P580,** of Kemira Inc.; **Tronox CR-837,** of Kerr-McGee Chemical; and **Tiona RCL-162,** of SCM Chemicals. **Iriodin** 201, 211, 221, and 231—yellow, red, blue, and green, respectively—are **pearlescent pigments** of Merck of Germany. **Stonewear Mica,** of this company, is a coarse-grain-effect pigment for use with Iriodin pigments for thermoset polyester basins and tubs. **Magna Pearl,** of Mearl Corp., is also a line of pearlescent pigments based on titanium-dioxide-coated mica. Also mica-based are **Afflair** pearlescent pigments of EM Industries, which serve as receptors of laser light and promote high-contrast laser marking.

Proprietary powder surface coatings with nylon, polypropylene, or polystyrene carriers are used in the injected-paint technology developed by Evode Powder Coatings of England, injection-molding-machine builder Battenfeld of Germany, and British automaker Rover. The surface material is injected first, followed by a shot of the core material. The resulting surface is a high-quality finish 1,378 μin (35 μm) or more thick with metallic, pearlescent, or other effects on plastic parts.

PINE. The wood of coniferous trees of the genus *Pinus,* of which there are 37 species in the United States. The **white pine, or northern white pine,** *P. strobus,* grows widely in Canada and in the northeastern United States. The trees are 80 to 100 ft (24 to 30 m) high, with trunks 3 to 9 ft (1 to 3 m) in diameter, reaching full size in 80 years. The wood is soft, straight-grained, and free from rosin. The heartwood is light brown, and the sapwood white. It is the chief wood for pattern making and is also extensively used for cabinetwork and general carpentry. **Cork pine** is a name for the clear, soft white pieces used for patterns. The white pine is now scarce in New England, and red spruce is used in its place. **Yellow pine** is a name for the wood of the **longleaf** or **longstraw pine** tree, *P. palustris,* of the southeastern states, and **shortleaf pine,** *P. echinata,* of the southeast and midwestern states, also called **North Carolina pine** and **rosemary pine.** The leaves, called *needles* or *straws,* of the longleaf pine are up to 18 in (46 cm) in length. The longleaf pine tree furnishes the best grades of yellow pine and is the chief source of turpentine. It is also called **Georgia pine, southern pine, hard pine,** and **hill pine.**

Slash pine, also known as **Cuban pine** and **swamp pine,** from the tree *P. caribaea,* **Caribbean pine,** which grows along the southern coasts of the United States and the Caribbean countries, is a yellow pine. It is one of the most rapidly growing forest trees in the United States and produces one of the heaviest, hardest, and strongest of all the conifers or softwoods. In Central America it is called **ocoté.** Slash pine is next to longleaf pine as a source of turpentine and rosin. As heartwood does not develop until the tree is 20 or more years old, slash pine forms a valuable source of paper pulp. The term **Arkansas pine** in the lumber trade includes mixtures of shortleaf, longleaf, slash, loblolly, and pond pines. **Lodgepole pine,** *P. contorta,* is from a small, slow-growing tree of the western United States, Canada, and Alaska. It is also called **knotty pine, scrub pine,** and **jack pine.** The wood is moderately lightweight, yellow to brown, with a narrow white sapwood. It is straight-grained with resin ducts and has large shrinkage. It is used for poles, ties, mine timbers, and rough construction. Also known as **jack pine** is the medium-size *P. banksiana* of central Canada, which is used for creosoted telephone poles. **Spruce pine,** or **cedar pine,** is a large tree, *P. glabra,* growing in a narrow area from southern Louisiana to Florida. **Virginia pine,** *P. virginiana,* also called **Jersey pine** and **scrub pine,** is a plentiful tree of the Atlantic states. The wood is soft, very knotty, and not durable. It is used for firewood, but much is used in low-cost houses.

  Ponderosa pine, also called **western yellow pine, western white pine,** and **Oregon white pine,** is from the tree *P. ponderosa.* The tree grows to a height of 175 ft (53 m) and a diameter of 6 ft (2 m). It grows throughout the mountain states from Mexico to Canada and is a source of turpentine and rosin. A similar western pine, **Jeffrey pine,** *P. jeffreyi,* contains heptane instead of turpentine in the oleoresin and is a more economical source of this material than petroleum. The lumber is usually mixed with ponderosa pine and sold as such. It is a moderately soft and lightweight wood with the heartwood light reddish brown, and quite similar to yellow pine. **Loblolly pine,** *P. taeda,* is called **North Carolina pine, Oldfield pine,** and **sap pine.** It grows from Virginia to northern Florida and to Texas; it is adapted to extensive areas and is easily propagated, receiving the name **field pine.** It is a type of shortleaf pine distinguished by three leaves or straws in each cluster, rough bark, and small, prickly burrs. It grows to a diameter of 12 in (30 cm) in 12 years. **Pitch pine** is the **pond pine,** *P. rigida,* of the southern states, but all yellow pines are called **pitch pine** in the export trade. **Norway pine,** of the north central states, is *P. resinosa.* The yellow pines are harder and more difficult to work than white pine. They are resinous and more durable. They also take a better polish and show a more figured grain. They are valued for flooring and general construction. White pine has a specific gravity, kiln-dried, of 0.38 and compressive strength perpendicular to the grain of 780 lb/in^2 (5.4 MPa); **western white pine,** *P. monticola,* has a specific gravity of 0.42 and a compressive strength of 750 lb/in^2 (5 MPa). **Deal** is a European name for the wood of the tree *P. silvestris,* also known as **Danzig pine, Baltic pine, Scotch fir, Scotch pine,** and **northern pine.** It is popularly called **Scots pine** in England and is one of the most plentiful of the European conifers, especially in Norway, Sweden, and Finland. It gives a straight pole, up to 70 ft (21 m) long, valued for telegraph poles.

Paraná pine is a soft, yellowish-white wood with rose veins from the tree *Araucaria brasiliensis* of southern Brazil. In Argentina it competes with American softwoods and is called **Brazilian pine** and **araucarian pine.** The specific gravity is 0.865. The tree is very tall, with branches only at the top, and a notable feature of the wood is the absence of knots. In the United States it is used for telephone-pole crossbars, and to replace birch for such articles as paintbrush handles. Other species of araucarian pine, or **Antarctic pine,** grow in southern Chile and Argentina. **Araucaria oil** is distilled from the wood. It is a viscous, reddish oil of roselike odor containing a high percentage of eudeomol and some geraniol. It has a more durable scent than guaiac wood oil for soaps.

The **araucarian pines** of new Guinea, A. *khinkii* and A. *cunninghamia,* are called **hoop pine. New Caledonia pine** is from the tree A. *cooki,* growing to a height of 200 ft (61 m) with no lower branches. White pine of New Zealand is from the very large tree *Podocarpus dacrydioides,* called also **kahikatea.** The sapwood is white, and the small heartwood is yellow. The wood is straight-grained, inodorous, easily worked, but not durable. Density is 29 lb/ft^3 (465 kg/m^3). It is used for boxes, crates, and packing. Another species, from the tree *P. ferrugineus,* called **miro,** is brownish, fine-grained, and easily worked and has high strength. The trees average 65 ft (20 m) in height and 20 in (51 cm) in diameter. The New Zealand species known as **black pine,** or **matai,** is from *P. spicatus.* The wood is yellowish brown, is straight-grained, and has a density of 38 lb/ft^3 (609 kg/m^3). **Red pine,** or **rimu,** is the chief timber of New Zealand, used for furniture, millwork, and kraft pulp. The tree *Dacrydium cupressinum* averages 100 ft (30 m) in height and 30 in (76 cm) in diameter. The wood is reddish brown with streaks, straight-grained, easily worked, and has a density of 37 lb/ft^3 (593 kg/m^3). **Silver pine, pink pine,** and **yellow silver pine** of New Zealand are from several other species of *Dacrydium* obtained only in limited amounts. The name *silver* is applied to the shiny, white woods, and the darker and mottled woods are called *pink.* They are very durable, cypresslike woods. **Mercus pine** is the wood of the tropical pine tree *P. merkusii,* of the East Indies, India, and the Philippines. It is called **Tinyu pine** in India and **Mindoro pine** in the Philippines. The wood is used in general construction. The tree yields a superior turpentine.

PINE OIL. An oil obtained from the wood of *Pinus palustris,* or longleaf pine, in the steam extraction of wood turpentine. It is used as a cold solvent for varnish gums and for nitrocellulose lacquers, and as a frothing agent in the flotation of ores. In paints and varnishes it aids dispersion of metallic pigments and improves the flow. It is also used in metal polishes and in liquid and powder scrubbing soaps, as the oil is a powerful solvent of dirt and grease. When free from water, pine oil has a yellowish color, but it is water-white when it contains dissolved water. It has an aromatic characteristic odor and is distinct from the pine oils distilled from pine leaves and needles and used in medicine. The distillate of the gum of the Jeffrey and Digger pines of California, called **abietine** in medicine, contains 96% heptane and is used as a cleaning agent and insecticide, and as a constituent of standard gasolines for measuring detonation of engines. Pine oil is obtained mainly from old trunks and branches and is a product formed by hydrolysis. Pine-oil disinfectants are made with steam-distilled pine oil. **Pine Power** is an oil with disinfectant properties from the Texize Div. of Dow Chemical Co. A similar product from Crystal Chemical Corp. is **Pine Tree Disinfectant Cleaner #6. Yarmor** is a refined pine oil of Hercules Inc. which is used to increase the detergency of soaps, for dyes, and as a solvent for oils and greases. **Hercosol** is a solvent made from pine oil by the same company. Synthetic pine oil made from gum turpentine by this company has a mild pine odor, a specific gravity of 0.9186, and a flash point of 154°F (68°C). It is technically the same as the natural and has the same uses. **Pine-root oil** was produced in Japan on a large scale for the manufacture of fuel oils. The terpenes of pine oil are converted to aromatic and hydroaromatic compounds by catalytic reaction. The edible **pine kernels** of Europe are the seeds from the large cones of *P. pinea* of southern Europe and Cyprus. **Pine-needle oil** is distilled from the **Siberian fir** tree, *Abies sibirica,* of northeastern Russia. It is also known as **Siberian pine oil.** It contains a high percentage of **bornyl acetate** and is used in soaps and perfumes.

PLANE WOOD. The wood of the plane tree, *Platanus orientalis,* native to Europe, and *P. occidentalis,* of North America. The latter species is also called **buttonwood** and **buttonball.** It is a yellowish, compact wood with a fine, open grain. The density is about 40 lb/ft^3 (641 kg/m^3). It resembles maple and gives a beautiful grain when quartered. It is employed in cabinetwork.

PLASTIC ALLOYS AND BLENDS. Alloys consisting of two thermoplastics compounded into a single resin. The two polymers must be melt-compatible. Some polymers are naturally compatible; others require the use of compatibilizing agents. The purpose of alloying polymers is to achieve a combination of properties not available in any single resin. Acrylonitrile butadiene styrene (ABS) is a common alloy ingredient, alloys including **polycarbonate/ABS** (PC/ABS), **ABS/polyurethane** (ABS/PUR), **ABS/nylon, ABS/polysulfone,** and **ABS/polyvinyl chloride** (ABS/PVC). Besides PC/ABS there are **polycarbonate/polybutylene terephthalate** (PC/PBT), **polycarbonate/acrylonitrile styrene acrylate** (PC/ASA), and **polyetherimide/polycarbonate** (PEI/PC) alloys. And besides ABS/PVC, there are **polyvinyl chloride/acrylic, polyvinyl chloride/chlorinated polyethylene** (PVC/CPE), and **acrylic/polyvinyl chloride** alloys. There are also **polyethylene** alloys, **polyphenylene oxide/polystyrene,** and **polyphenylene oxide/polybutylene terephthalate** (PPO/PBT) alloys.

PC/ABS alloys extend the exceptionally high impact strength of carbonate plastics to section thicknesses over 0.0625 in (0.16 cm). Tradenames of PC/ABS alloys include Cycoloy of General Electric Plastics; Pulse of Dow Plastics; Bayblend of Miles Inc.; and Celstran of Polymer Composites Inc. **Cycoloy C2950** HF features a 30% greater flow rate than **Cycoloy C2950** and a UL94 V-O flame retardance at a thickness of 0.06 in (1.5 mm) and a UL94 5VB at 0.10 in (2.5 mm). **Cycoloy LG9000** is a low-gloss, or matte-finish, grade for unpainted, interior auto parts. It provides 60% less gloss than standard grades and has a tensile strength of 7,900 lb/in^2 (54 MPa), a flexural strength of 12,300 lb/in^2 (85 MPa), a flexural modulus of 335,000 lb/in^2 (2,310 MPa), a heat deflection temperature of 225°F (107°C) at 264 lb/in^2 (1.8 MPa), and a notched Izod impact strength of 10 ft · lb/in (534 J/m). The Dow grades include **Pulse 830, 1725, 1735,** and **1745.** Pulse 1725 is a brominated (flame-retardant), ignition-resistant grade; Pulse 1735 is a nonbrominated, ignition-resistant grade. Pulse 1745 is brominated, but it does not contain polybrominated diphenyl ether, which may generate environmentally harmful dibenzodioxin and dibenzofuran. This alloy and bromine-free, flame-retardant **Bayblend DP2-1472** and **DP2-1510** are intended for computer and business-machine housings. **Celstran** PC/ABS alloys are reinforced with 25 or 40% long, glass fibers. **Electrafil ABS-1200/SD,** of DSM Engineering Plastics, combines ABS with a static-dissipative polymer for protection against electrostatic discharge. At 15% relative humidity, the static decay rate is less than 2 s compared with 99 s or more for unmodified ABS. Although the alloy's flexural modulus is reduced considerably—270,000 lb/in^2 (1,862 MPa) versus 340,000 lb/in^2 (2,344 MPa)—tensile strength and heat-deflection temperature are only slightly less, and impact strength is unaffected. ABS/PUR alloys combine the excellent abrasion resistance and toughness of the urethanes with the lower cost and rigidity of ABS.

ABS-PVC alloys are available commercially in several grades. One of the established grades provides self-extinguishing properties, thus eliminating the need for intumescent (nonburning) coatings in ABS applications, such as power tool housings, where self-extinguishing materials are required. A second grade possesses an impact strength about 30% higher than that of general-purpose ABS. Also ABS-PVC alloys can be produced in sheet form. The sheet materials have improved hot strength, which allows deeper draws than are possible with standard rubber-modified PVC base sheet. They also are nonfogging when exposed to the heat of sunlight. Some properties of ABS-PVC alloys are lower than those of the base resins. Rigidity, in general, is somewhat lower, and tensile strength is more or less dependent on the type and amount of ABS in the alloy.

Another sheet material, an alloy of about 80% PVC and the rest acrylic plastic, combines the nonburning properties, chemical resistance, and toughness of vinyl plastics with the rigidity and deep-drawing merits of the acrylics. The **PVC-acrylic alloy** approaches some metals in its ability to withstand repeated blows. Because of its unusually high rigidity, sheets ranging in thickness from 0.60 to 0.187 in (1.5 to 0.5 cm) can be formed into thin-walled, deeply drawn parts. **Kydex 200,** of Kleerdex Co., is an acrylic/PVC alloy in the form of sheet 0.04 to 0.25 in (1 to 6.4 mm) thick in various surface textures and colors. The tensile strength is 6,000 lb/in^2 (41 MPa), flexural modulus is 350,000 lb/in^2 (2,413 MPa), and the notched Izod impact strength is 4 to 6 ft · lb/in (214 to 320 J/m).

PVC is also alloyed with chlorinated polyethylene (CPE) to gain materials with improved outdoor weathering or to obtain better low-temperature flexibility. The **PVC-CPE alloy** applications include wire and cable jacketing, extruded and molded shapes, and film sheeting. **Acrylic-base alloys** with a polybutadiene additive have also been developed, chiefly for blow-molded products. The acrylic content can range from 50 to 95%, depending on the application. Besides blow-molded bottles, the alloys are suitable for thermoformed products such as tubs, trays, and blister pods. The material is rigid and tough and has good heat-distortion resistance up to 180°F (82°C).

Another group of plastics, polyphenylene oxide (PPO), can be blended with polystyrene to produce a **PPO-polystyrene alloy** with improved processing traits and lower costs than nonalloyed PPO. The addition of polystyrene reduces tensile strength and heat-deflection temperature somewhat and increases thermal expansion. **Gemax** is a PPO/PBT alloy of General Electric Plastics. **Noryl GTX,** also of this company, is a PPO/nylon alloy having a heat-deflection temperature of 302°F (150°C) and good stiffness without gloss or mineral fillers. Variations of this alloy combine various levels of stiffness, impact strength, and heat resistance. **Xenoy** is a PC/PBT alloy of the same company, and **Ultem LTX** alloys are PC/ASA alloys. LTX100A and

100B are aimed at thick-wall connectors and extrusions; LTX300A and 300B provide greater heat resistance. **Hostalloy 731,** of Hoechst Celanese, is an alloy of several polyethylenes of different molecular weights. It has a density of 0.034 lb/in^3 (941 kg/m^3), a tensile yield strength of 5,000 lb/in^2 (35 MPa), and a flexural modulus of 170,000 lb/in^2 (1,170 MPa). The alloy approaches the abrasion resistance and impact strength of ultrahigh-molecular-weight polyethylene but is more readily processed. **Hivalloy,** of Himont Inc., is a line of propylene-based copolymer and terpolymer alloys that combine the strength of the semicrystalline monomer with the attributes of various amorphous monomers. They are intended to compete with amorphous plastics such as acrylonitrile butadiene styrene and polycarbonate as well as with semicrystalline acetal, nylon, and thermoplastic polyester.

PLASTIC BRONZE. A name once applied by makers of bearing bronzes to copper alloys that are sufficiently pliable to assume the shape of the shaft and make a good bearing by running in. These bronzes have a variety of compositions, but the plasticity is always obtained by the addition of lead, which in turn weakens the bearing. In some cases the lead content is so high, and the tin content so low, that the alloy is not a bronze. These copper-lead alloys have been referred to as **red metals.** The plastic bronze ingot marketed by one large foundry for journal bearings contains 65 to 75% copper, 5 to 7 tin, and the balance lead. **Semiplastic bronze** usually contains above 75% copper and not more than 15 lead. **ASTM alloy No. 7** has about 10% lead, 10 tin, 1 zinc, 1 antimony, and 78 copper. The compressive strength is 12,500 lb/in^2 (85 MPa).

PLASTIC LAMINATES. Resin-impregnated paper or fabric, produced under heat and high pressure. Also referred to as **high-pressure plastic laminates.** Two major categories are decorative thermosetting laminates and industrial thermosetting laminates. Most of the **decorative thermosetting laminates** are a paper base and are known generically as **papreg.** Decorative laminates are usually composed of a combination of phenolic- and melamine-impregnated sheets of paper. The final properties of the laminate are related directly to the properties of the paper from which the laminate is made.

Early laminates were designated by trade names, such as **Bakelite** of Union Carbide, **Textolite** of General Electric, **Micarta** of Westinghouse, **Phenolite, Condensite, Dilecto, Haveg, Spauldite,** and **Synthane. Formica,** of Formica Corp., designated various types of laminates with a decorative facing layer for such uses as tabletops. Trade names now usually include a number or symbol to describe the type and grade. Textolite, for example, embraces more than 70 categories of laminates subdivided into use-specification grades, all produced in many sizes and thicknesses. **Textolite 11711** is an electronic laminate for such uses as multilayer circuit boards. It is made with polyphenylene oxide resin and may have a copper or aluminum cladding. The tensile strength is up to 10,000 lb/in^2 (69 MPa), and the dielectric strength is 400 V/mil (16×10^6 V/m). **Phenolyte Y240** is a paper-base laminate bonded with a polyester-modified melamine which gives high dielectric strength and arc resistance together with good punching or blanking characteristics in thicknesses up to 0.125 in (0.32 cm). **Doryl H17511,** of Westinghouse, has glass fabric laminations bonded with a modified phenolic resin based on diphenyl oxide and polyphenyl ether. This laminate has a flexural strength of 27,400 lb/in^2 (184 MPa) at 480°F (249°C). **Luxwood,** of Formica Corp., for furniture, is a 0.0625-in (0.16-cm) laminate with photographic reproductions of wood grains on the face, while **Beautywood** is this material in thicker sizes for wall panels.

Industrial thermosetting laminates are available in the form of sheet, rod, and rolled or molded tubing. Impregnating resins commonly used are phenolic, polyester, melamine, epoxy, and silicone. The base material, or reinforcement, is usually one of the following: paper, woven cotton or linen, asbestos, glass cloth, or glass mat. NEMA (National Electrical Manufacturing Association) has published standards covering over 25 standard grades of these laminates. Each manufacturer, in addition to these, usually produces a range of special grades.

Laminating resins may be marketed under one trade name by the resin producer and other names by the molders of the laminate. **Paraplex P resins,** of Rohm & Haas, for example, comprise a series of polyester solutions in monomeric styrene which can be blended with other resins to give varied qualities. But **Panelyte,** of St. Regis Paper Co., refers to the laminates which are made with phenolic, melamine, silicone, or other resin, for a variety of applications.

PLASTIC POWDER COATINGS. Although many different plastic powders can be applied as coatings, vinyl, epoxy, and nylon are most often used. Vinyl and epoxy provide good corrosion and weather resistance as well as good electrical insulation. Nylon is used chiefly for its outstanding wear and abrasion resistance. Other plastics frequently used in powder coating include chlorinated polyethers, polycarbonates, acetals, cellulosics, acrylics, and fluorocarbons.

Several different methods have been developed to apply these coatings. In the most popular process, **fluidized bed,** parts are preheated and then immersed in a tank of finely divided plastic powders, which are held in a suspended state by a rising current of air. When the powder particles contact the heated part, they fuse and adhere to the surface, forming a continuous, uniform coating. Another process,

electrostatic spraying, works on the principle that oppositely charged materials attract each other. Powder is fed through a gun, which applies an electrostatic charge opposite to that applied to the part to be coated. When the charged particles leave the gun, they are attracted to the part, where they cling until fused together as a plastic coating. Other powder application methods include flock and flow coating, flame and plasma spraying, and a cloud-chamber technique.

PLASTICS. A major group of materials that are primarily noncrystalline hydrocarbon substances composed of large molecular chains whose major element is carbon. The three terms—**plastics, polymers,** and **resins**—are sometimes used interchangeably to identify these materials. However, the term **plastics** has now come to be the commonly used designation.

The first commercial plastic, **Celluloid,** was developed in 1868 to replace ivory for billiard balls. **Phenolic plastics,** developed by Baekeland and named Bakelite after him, were introduced around the turn of the twentieth century. A *plastic material,* as defined by the Society of the Plastics Industry, is "any one of a large group of materials consisting wholly or in part of combinations of carbon with oxygen, hydrogen, nitrogen, and other organic and inorganic elements which, while solid in the finished state, at some stage in its manufacture is made liquid, and thus capable of being formed into various shapes, most usually through the application, either singly or together, of heat and pressure."

There are two basic types of plastics based on intermolecular bonding. **Thermoplastics,** because of little or no cross-bonding between molecules, soften when heated and harden when cooled, no matter how often the process is repeated. **Thermosets,** on the other hand, have strong intermolecular bonding. Therefore, once the plastic is set into permanent shape under heat and pressure, reheating will not soften it.

Within these major classes, plastics are commonly classified on the basis of base **monomers.** There are over two dozen such monomer families or groups. Plastics are also sometimes classified roughly into three stiffness categories: rigid, flexible, and elastic. Another method of classification is by the "level" of performance or the general area of application, using such categories as engineering, general-purpose, and specialty plastics, or the two broad categories of engineering and commodity plastics.

In general, plastics are lightweight, are dielectric, and provide low thermal conductivity. They are resistant to many environments, have low softening temperatures, and can be formed into complex shapes. They are also viscoelastic—that is, after an applied load is removed, plastics tend to continue to exhibit strain or deformation with time.

Plastics can be built of one, two, or even three different monomers, and are termed **homopolymers, copolymers,** and **terpolymers,** respectively. Their geometric form can be linear or branched. Linear or unbranched polymers are composed of monomers linked end to end to form a molecular chain that is like a simple string of beads or a piece of spaghetti. Branched polymers have side chains of molecules attached to the main linear polymer. These branches can be composed either of the basic linear monomer or of a different one. If the side molecules are arranged randomly, the polymer is **atactic**; if they branch out on one side of the linear chain in the same plane, the polymer is **isotactic**; and if they alternate from one side to the other, the polymer is **syndiotactic.**

Plastics are produced in a variety of different forms. Most common are **plastic moldings,** which range in size from less than 1 in to several feet (2 cm to several meters). Thermoplastics, such as polyvinylchloride and polyethylene, are widely used in the form of **plastic film** and **plastic sheeting.** The term *film* is used for thicknesses up to and including 10 mils (0.25 cm), while *sheeting* refers to thicknesses over that. **Rigid-rod polymers** are high-performance thermoplastics characterized by long, stiff molecules and high strength and stiffness. Until the 1992 development by Maxdem Inc. of injection and compression moldable and extrudable **Poly-X** grades, they could only be spun into fibers or cast as film from acid solutions. The moldable grades, described as substituted chains of benezene rings called polyparaphenylene, have tensile moduli of 0.9×10^6 lb/in^2 to 2.6×10^6 lb/in^2 (6,205 to 17,927 MPa). Potential uses include printed-circuit boards, electronic connectors, scratch-resistant windows, lenses, bearings, bushings, and valves.

Both thermosetting and thermoplastic materials are used as **plastic coatings** on metal, wood, paper, fabric, leather, glass, concrete, ceramics, or other plastics. There are many coating processes, including knife or spread coating, spraying, roller coating, dipping, brushing, calendering, and the fluidized-bed process. Thermosetting plastics are used in **high-pressure laminates** to hold together the reinforcing materials that comprise the body of the finished product. The reinforcing materials may be cloth, paper, wood, or glass fibers. The end product may be plain, flat sheets, or decorative sheets as in countertops, rods, tubes, or formed shapes.

PLASTICS ADDITIVES. Almost all plastics contain one or more additive materials to improve their physical properties and processing characteristics or to reduce costs. There are a wide range of additives for use with plastics, including antimicrobials, antistatic agents, clarifiers, colorants, fillers, flame retardants, foaming agents, heat stabilizers, impact modifiers, light stabilizers, lubricants, mold-release agents, odorants, plasticizers, reinforcements, and smoke retardants.

Fillers are probably the most common of the additives. They are usually used to either provide bulk or modify certain properties. Generally, they are inert and thus do not react chemically with the resin during processing. The fillers are often cheap and serve to reduce costs by increasing bulk. For example, **wood flour,** a common low-cost filler, sometimes makes up 50% of a plastic compound. Other typical fillers are chopped fabrics, asbestos, talc, gypsum, and milled glass. Besides lowering costs, fillers can improve properties. For example, asbestos increases heat resistance, and cotton fibers improve toughness.

Plasticizers are added to plastics compounds either to improve flow during processing by reducing the glass transition temperature or to improve properties such as flexibility. Plasticizers are usually liquids that have high boiling points. Polyvinyl chloride accounts for about 80%, but phthalates and polyesters are also used. **Stabilizers** are added to plastics to help prevent breakdown or deterioration during molding or when the polymer is exposed to sunlight, heat, oxygen, ozone, or combinations of these. Thus there are a wide range of compounds, each designated for a specific function. Stabilizers can be metal compounds, based on tin, lead, cadmium, barium, and others. And phenols and amines are added antioxidants that protect the plastic by diverting the oxidation reactions to themselves. The trend, however, is away from toxic lead and cadmium for environmental reasons. For heat stabilizers, the lead can be diluted by adding calcium-zinc compounds. These compounds are intended mainly for use with polyvinyl chloride for window profiles and pipe. Some are also alternatives for barium-zinc compounds, which are alternatives for cadmium heat stabilizers.

Catalysts, by controlling the rate and extent of the polymerization process in the resin, allow the curing cycle to be tailored to the processing requirements of the application. Catalysts also affect the shelf life of the plastics. Both metallic and organic chemical compounds are used as catalysts. **Colorants,** added to plastics for decorative purposes, come in a wide variety of pigments and dyestuffs. The traditional colorants are metal-base pigments such as cadmium, lead, and selenium; but here, too, environmental concerns have led to a trend away from these metals. More recently, liquid colorants, composed of dispersions of pigments in a liquid, have been developed. **Clarifiers** are used to impart clarity to plastics for packaging and other applications. **Sorbitol,** for example, is used for polypropylene in food packaging and other uses. **Coupling agents** based on organometallic titanates, zirconates, and aluminates bond to the surface of glass, carbon, aramid, and other reinforcements for plastics, making them more dispersible. **Silane** coupling agents, for example, are used for both thermosets and thermoplastics. **Flame retardants** are added to plastic products that must meet fire-retardant requirements, because polymer resins are generally flammable, except for such notable exceptions as polyvinyl chloride. In general, the function of fire retardants is limited to the spread of fire. They do not normally increase heat resistance or prevent the plastic from charring or melting. Some fire-retardant additives include compounds containing chlorine or bromine, phosphate-ester compounds, antimony thrioxide, alumina trihydrate, magnesium hydroxide, ammonium phosphates and melamine amyl phosphate, and zinc borate. Halogen compounds have long dominated but have lost favor in recent years due to concern over generating toxic combustion products, such as bromides, hydrochloric acid, and brominated dioxins and furans, on burning. Thus, the use of halogens has been reduced (the most widely used having contained as much as 80%) or eliminated.

Reinforcement materials in plastics are not normally considered additives. Usually in fiber or mat form, they are used primarily to improve mechanical properties, particularly strength. Although asbestos and some other materials are used, glass fibers are the predominant reinforcement for plastics.

PLATINUM. A whitish-gray metal, symbol Pt. It is more ductile than silver, gold, or copper and is heavier than gold. The melting point is 3217°F (1769°C), and the specific gravity is 21.45. The Brinell hardness of the annealed metal is 45, and its tensile strength is 17,000 lb/in^2 (117 MPa); when hard-rolled, the Brinell hardness is 97 and tensile strength 34,000 lb/in^2 (234 MPa). Electrical conductivity is about 16% that of copper. The metal has a face-centered-cubic lattice structure, and it is very ductile and malleable. It is resistant to acids and alkalies, but dissolves in aqua regia. Platinum is widely used in jewelry, but because of its heat resistance and chemical resistance it is also valued for electrical contacts and resistance wire, thermocouples, standard weights, and laboratory dishes. Generally too soft for use alone, it is almost always alloyed with harder metals of the same group, such as osmium, rhodium, and iridium. An important use of the metal, in the form of gauze, is as a catalyst. **Platinum gauze** is of high purity in standard meshes of 45 to 80 per inch (18 to 31 per centimeter), with wire from 0.0078 to 0.003 in (0.020 to 0.008 cm) in diameter. **Dental foil** is 99.99% pure and of maximum softness. **Platinum foil** for other uses is made as thin as 0.0002 in (0.0005 cm). **Platinum powder** comes in fine submesh particle size. It is made by chemical reduction and is at least 99.9% pure, with amorphous particles 12 to 138 μin (0.3 to 3.5μm) in diameter. Atomized powder has spherical particles of 50 to 200 mesh, and is 99.9% pure and free-flowing. **Platinum flake** has the powder particles in the form of tiny laminar platelets which overlap in the coating film. The particles in **Platinum flake No. 22** have an average diameter of 118 μin (3 μm) and thickness of 3.9 μin (0.1 μm).

Because of the high resistance of the metal to atmospheric corrosion even in sulfur environments, **platinum coatings** are used on springs and other functioning parts of instruments and electronic devices where precise operation is essential. Electroplating may be done with an electrolyte bath of **platinum dichloride,** $PtCl_2$, or **platinum tetrachloride,** $PtCl_4$. Hard plates may be produced with an alkaline bath of **platinum diamine nitrite,** $Pt(NH_3)_2(NO_2)_2$. Coatings are also produced by vapor deposition of platinum compounds; thin coatings, 0.0002 in (0.0005 cm) or less, are made by painting the surface with a solution of platinum powder in an organic vehicle and then firing to drive off the organic material, leaving an adherent coating of platinum metal.

Platinum occurs in small, flat grains or in pebbles usually in alluvial sands, and the native metal generally contains other metals of the platinum group. The largest nugget ever found came from South America and weighed 2 lb (0.9 kg). The chief sources of the metal are Russia and Colombia, with smaller amounts from Alaska, Canada, and South Africa. Some platinum is obtained from the copper-nickel ores of Canada and South Africa. There are no commercial ores of the metal, but the rare mineral **sperrylite** is found in Wyoming and in Ontario. It is a **platinum arsenide,** $PtAs_2$, found in small grains of a tin-white metallic luster. The only other known natural compound is the rare mineral **cupperite,** which is a **platinum sulfide,** PtS. The Russian platinum is 99.8 to 99.9% pure, with some iridium. Platinum is sold by the troy ounce (0.03 kg), 1 in^3 (16 cm^3) of the metal weighing 11.28 troy oz (0.34 kg).

PLATINUM ALLOYS. **Platinum-iridium alloys** are employed for instruments, magneto contacts, and jewelry. The alloys are hard, tough, and noncorrosive. An alloy of 95% platinum and 5 iridium, when hard-worked, has a Brinell hardness of 170; an alloy with 30% iridium has a Brinell hardness of 400. The 5 and 10% alloys are used for jewelry manufacture; the 25 and 30% alloys are employed for making surgical instruments. An alloy of 80% platinum and 20 iridium is used for magneto contact points, and the 90–10 alloy is widely used for electrical contacts in industrial control devices. The addition of iridium does not alter the color of the platinum. The 5% alloy dissolves readily in aqua regia; the 30% alloy dissolves slowly.

Platinum-cobalt alloys, with about 23% cobalt, are used for permanent magnets. **Platinum-nickel alloys,** with as much as 20% nickel, are noted for high strength. With 5% nickel, for example, tensile strength of the annealed alloy is about 90,000 lb/in^2 (621 MPa), and with 15% it increases to 130,000 lb/in² (896 MPa). Strength almost doubles with appreciable cold work.

Platinum-rhodium alloys are used for thermocouples for temperatures above 1100°C (2012°F). The standard thermocouple is platinum versus platinum–10% rhodium. Other thermocouples for higher operating temperatures use platinum-rhodium alloys in both elements. The alloys of platinum-rhodium are widely used in the glass industry, particularly as glass-fiber extrusion bushings. Rhodium increases the high-temperature strength of platinum without reducing its resistance to oxidation. **Platinum-rhodium gauze** for use as a catalyst in producing nitric acid from ammonium contains 90% platinum and 10 rhodium. **Platinum-rhenium alloys** are efficient catalysts for re-forming operations on aromatic compounds. The platinum alloys have lower electrical conductivity than pure platinum, but are generally harder and more wear-resistant and have high melting points. A platinum-rhenium alloy with 10% rhenium has an electrical conductivity of only 5.5% that of copper compared with 16% for pure platinum. Its melting point is 3362°F (1850°C), and the Rockwell T hardness of the cold-rolled metal is 91 compared with 78 for cold-rolled platinum. **Platinum-ruthenium alloy,** with 10% ruthenium, has a melting point of 3272°F (1800°C) and an electrical conductivity 4% that of copper.

Platinum-tungsten alloys, with 2 to 8% tungsten, have been used for aircraft-engine spark plug electrodes, radar-tube grids, strain gages, glow wires, switches, and heating elements. The tungsten markedly increases electrical resistivity while decreasing the temperature coefficient of resistivity. It also substantially increases tensile strength—to about 130,000 lb/in² (896 MPa) for platinum (8% tungsten alloy in the annealed condition)—and tensile strength more than doubles with appreciable cold work.

PLYWOOD. A laminated wood made up of thin sheets of wood glued together with the grains of successive layers set at right angles to give strength in both directions. Plywood is an outgrowth of the laminated wood known as *veneer,* which consists of an outside sheet of hardwood glued to a base of lower-cost wood. The term **veneer** actually refers only to the facing layer of selected wood used for artistic effect, or for economy in the use of expensive woods. Veneers are generally marketed in strip form in thicknesses of less than 0.125 in (0.32 cm) in mahogany, oak, cedar, and other woods. The usual purpose of plywood now is not aesthetic but to obtain high strength with low weight. The term *laminated wood* generally means heavier laminates for special purposes, and such laminates usually contain a heavy impregnation of bonding resin which gives them more of the characteristics of the resin than of the wood.

Plywood usually comes in 4- by 8-ft (1.2- by 2.4-m) panels and is always built up with an odd number of layers. The cross-ply construction gives strength in both directions and also gives symmetric shrinkage stresses. A three-ply softwood panel of equal ply thicknesses may shrink about 0.080 in (0.203 cm) in width and 0.100 in (0.254 cm) in length, but increasing the thickness of the core ply can equalize the shrinkage; or equalization may be obtained by increasing the number of plies. The odd number of plies gives a symmetry of construction about the core ply. Low-cost plywoods may be bonded with starch pastes, animal glues, or casein, and they are not water-resistant, but are useful for boxes and for interior work. Waterproof plywood for paneling and general construction is now bonded with synthetic resins, but when the plies are heavily impregnated with the resin and the whole is cured into a solid sheet, the material is known as a *hardboard* or as a *laminated plastic* rather than a plywood.

For construction purposes, where plywood is employed because of its unit strength and nonwarping characteristics, the plies may be of a single type of wood and without a hardwood face. The Douglas Fir Plywood Association sets up four classes of construction plywood under general trade names. **Plywall** is plywood in wallboard grade; **Plypanel** is plywood in three standard grades for general uses; **Plyscord** is unsanded plywood with defects plugged and patched on one side; and **Plyform** is plywood in a grade for use in concrete forms.

The bulk of commercial plywood comes within these classes, the variations being in the type of wood used, the type of bonding adhesive, or the finish. **Etch wood,** for example, is a paneling plywood with the face wire-brushed to remove the soft fibers and leave the hard grain for two-tone finish. **Paneling plywoods** with faces of mahogany, walnut, or other expensive wood have cores of lower-cost woods, but woods with good physical qualities are usually chosen. The tensile strength of a white ash three-ply plywood parallel to the grain of the faces is about 6,200 lb/in^2 (43 MPa), that of a mahogany plywood is 6,400 lb/in^2 (44 MPa), and that of a walnut plywood is 8,200 lb/in^2 (57 MPa). **Stabilite,** of Georgia Pacific Corp., is a wood laminate with the veneers impregnated with phenolic resin and bonded together with the grain of the plies parallel. It has the density of hardwood and the workability of cherry wood.

The **K-Veneer** used during the Second World War as a substitute for plywood was a 0.1875-in (0.478-cm) fir or hemlock sheet bonded to a heavy kraft paper. **Welchboard** is 0.375-in (0.953-cm) plywood with a smooth, grainless surface produced by curing on one side under heat and pressure a mixture of wood pulp and synthetic resin. A great variety of trade-name plywoods are marketed for paneling, but they do not always have the typical characteristics of plywood and are often

paneling boards rather than plywoods. Some have metal faces, or they may have special-purpose cores or backings. **Fybr-Tech,** developed by Technical Plywoods for aircraft paneling, has a 0.0156-in (0.038-cm) walnut veneer on a 0.25-in (0.635-cm) balsa wood core, with a 0.005-in (0.013-cm) vulcanized fiber back. The weight is 0.2 lb/ft^2 (0.98 kg/m^2). One of the earliest aluminum-faced plywoods was the English **Plymax** of Venesta, Ltd. **Metal-faced plywoods** are strong and can be riveted. **Metalite,** developed by United Aircraft Corp., has thin sheets of strong aluminum alloy bonded to both sides of a balsa wood core, the grain of the wood being perpendicular to the metal faces. **Siding-panel 15,** of Weyerhauser Co., has a Douglas fir plywood core and a facing of 0.01-in (0.025-cm) embossed aluminum with a vinyl resin finish. **Flexwood** consists of very thin sheets of veneer glued under heat and pressure to cotton sheeting, used as an ornamental covering for walls. **Algonite** is Masonite faced with fancy veneers. **Protekwood,** designed for protection against vermin attack, has a sheet of hardwood between two sheets impregnated with asphalt and resin and bonded with a urea-formaldehyde resin to a total thickness of 0.156 in (0.396 cm). **Parkwood** is a flexible, **woven veneer** made with thin strips of mahogany or other fancy wood pressed between sheets of transparent cellulose acetate or other plastic. **Novoply,** for panels, furniture, and structural parts, has a core of resin-impregnated wood chips bonded between hardwood veneers. Sheets are up to 0.75 in (1.91 cm) thick and have high strength.

POISON GASES. Substances employed in chemical warfare for disabling people, and in some cases used industrially as fumigants. They are all popularly called gases, but many are liquids or solids. Normally, information on military gases is kept secret, but the tear gases used by police are also poisonous, often causing serious damage to eyes, throat, and lungs. **Anesthetic gases** have not been used so far in chemical warfare, but are used in medicine. One of the simplest of these, **nitrous oxide,** N_2O, called **laughing gas,** produces a deep sleep. **Fluorthane,** or **ethyl fluoride,** is a volatile liquid like ether, but is nonexplosive, and it is used to replace ether in surgery. **Cyclopropane** is a potent anesthetic. It is not less than 99% by volume of C_3H_6.

Poison gases are classified according to their main effect on the human system, but one gas may have several effects. They are grouped as follows: **lethal gases,** intended to kill, such as phosgene; **lachrymators,** or **tear gases,** which have a powerful irritating effect on the eyes, causing temporary blindness and swelling of the eyes with a copious flow of tears; **vesicants,** or skin blisterers, such as lewisite and mustard gas; **sternutatory gases,** which induce sneezing; and **camouflage gases,** which are harmless, but cause soldiers to suffer the inconvenience of wearing gas masks and thus reduce their morale. Some of the gases have a sour, irritating odor and are also classified as **harassing agents.** Gases are also sometimes designated as **casualty agents** and further subdivided into persistent and nonpersistent. A **systemic gas** is one that interferes with one phase of the system, such as carbon monoxide, which paralyzes the respiratory function of the blood. A **laryrinthic gas** is one that affects an organ of the body, such as **dichlormethyl ether,** which affects the ears.

Effects of persistent gases, such as mustard, remain over the ground for as long as 7 days, but phosgene is quickly decomposed by dampness. **Obscuring agents,** such as white phosphorus, and the **toxic smokes,** such as diphenylaminochloroarsine, are also classified as war gases. Dusts of materials having catalytic properties, but not poisonous themselves, may be used to penetrate gas masks and create poisons, such as carbon monoxide, within the mask. Carbon and oil smokes may be used to choke the filters of gas masks and cause their removal. Absorbents used in gas masks are usually activated charcoal and soda lime. These will absorb or disassociate most of the toxic gases, but will not stop carbon monoxide. A mixture of powdered oxides of copper, manganese, silver, and cobalt, called **Hopcalite,** is used as a catalyst to oxidize carbon monoxide.

Lethal gases are divided into four classes: actual poisons, such as hydrocyanic acid, which kill with little pain; **asphyxiating gases,** such as phosgene, diphosgene, and chloropicrin, which affect the membranes of lungs, destroying them and allowing blood to fill the air sacs; poisons which destroy the lining of air passages and block the passages to lung tissues, such as mustard gas and ethyldichloroarsine; and poisons which affect the nose and throat, causing great pain, headache, and vomiting, such as diphenylchloroarsine. **Mustard gas,** $(CH_2ClCH_2)_2S$, also known as **blister gas, yperite,** and **yellow cross,** is an oily liquid which boils at 410°F (210°C) and vaporizes easily in the air. It destroys the cornea of the eyes, blisters skin, affects the lungs, and causes discharge from the nose and vomiting. One part in 14 million parts air is toxic and cannot be detected in dilutions by smell. Another powerful vesicant is **Bromlost,** which is **dibromethyl sulfide,** $(CH_2BrCH_2)_2S$. It is a solid melting at 70°F (21°C). **Sulvanite** is **ethylsulfuryl chloride,** $ClSO_3C_2H_5$. It is a colorless liquid boiling at 275°F (135°C).

Lewisite, $CHCL:CH \cdot AsCL_2$, is a liquid boiling at 374°F (190°C). It is a powerful vesicant, causing painful blisters on skin, pain in the eyes, nose irritation, permanent impairment of eyesight, and arsenic poisoning. It forms a heavy mist and has been called **dew of death.**

Chloropicrin, called **aquinite, klop,** and **nitrochloroform,** is **nitrotrichloromethane,** CCl_3NO_2. It is a persistent lachrymatory and lethal poison. It is a colorless liquid boiling at 234°F (112°C), with a specific gravity of 1.692. It is used as a soil fumigant to control insects and fungi. **Tonite** is **chloroacetone,** CH_3COCH_2Cl, a clear liquid vaporizing at 246°F (119°C). It is a powerful lachrymator and skin blisterer. As it is very reactive, it is also used in the synthesis of pharmaceuticals, dyes, and organic chemicals. Also used in large quantities as chemical intermediaries are the phosphorus chemicals, **phosphorus oxychloride** and **phosphorus trichloride.** Among the toxic cyanides of industrial importance are **hydrogen cyanide** and **cyanogen chloride. Thiodiglycol,** which is structurally close to mustard gas, is a raw material used widely in industry.

Bromoacetone is a colorless liquid of composition $CH_2BrCOCH_3$, with a specific gravity of 1.631 and boiling point 259°F (126°C). It is thrown in bombs or shells and disseminated as a mist which attacks the eyes. **Bromobenzyl cyanide** is a solid of composition $BrC_6H_4CH_2CN$, with a melting point of 77°F (25°C); when impure, it is a liquid, but it is not purified, as it is easily decomposed. It is very persistent and because of its odor is classified as a harassing agent. It was called **camite** by the French. Another tear gas, **chloroacetophenone,** $C_6H_5COCH_2Cl$, is a white, crystalline solid, specific gravity 1.321, and melting point 138°F (59°C), which, when thrown as a vapor, has a sweet, locustlike odor but produces pains in the eyes and temporary blindness. The tear gas called **Mace,** used for riot control, contains this substance. In contact with the skin, it causes severe dermatitis and often permanent damage to the eyes. The French gas **fraissite** is **benzyl iodide,** $C_6H_5CH_2I$, a liquid boiling at 439°F (226°C). The gas **papite** is acrolein with stannic chloride. **Calderite** is **benzyl bromide** with stannic chloride. **Xylyl bromide,** $CH_3C_6H_4CH_2Br$, is a colorless liquid boiling at 421°F (216°C). When disseminated as a mist from explosive bombs, it causes a copious flow of tears. The tear gas known as **CS gas** is **chlorobenzol malononitrile** and is destructive to nasal and lung tissues. Benzyl chloride also has been used as a tear gas.

Martonite, a powerful lachrymator, is a mixture of 20% chloroacetone and 80 bromoacetone. **Bretonite** is **iodoacetone,** CH_3COCH_2I, a brownish liquid boiling at 216°F (102°C), mixed with stannic chloride as a lachrymator. **Manguinite** is **cyanogen chloride,** CNCl, which boils at 55°F (13°C) and is a lachrymator. Mixed with arsenic trichloride to make it more toxic, it was used under the name of **vitrite. Campillit** is **cyanogen bromide,** CNBr, a white solid melting at 126°F (52°C) and vaporizing at 142°F (61.3°C). The fumes are highly toxic, paralyzing the nerve centers. **Diphenylchloroarsine,**

or **blue cross,** $(C_6H_5)_2AsCl$, is a **sneezing gas** which penetrates gas masks, forcing their removal. It affects chiefly the nose and throat, but is used with other more violent gases. **Adamsite** is a greenish, granular solid of composition $(C_6H_4)_2NHAsCl$ which has a pleasant odor but burns the nose and throat. Many of the lachrymators have important industrial uses. **Phenyl isocyanate,** C_6H_5NCO, is a water-white liquid of specific gravity 1.101 and boiling point 324°F (162°C), used for the production of alkyd resins, ureas, urethanes, and other chemicals. **Decontaminants,** used for combating the effects of poison gases, are neutralizing chemicals. The decontaminant known as STB, or **supertropical bleach,** is a mixture of chlorinated lime and ground quicklime.

Another recent poison-gas category is for **supertoxic lethal weapons,** defined as chemicals with a median lethal dose less than or equal to 6.4×10^{-6} oz/lb (0.5 mg/kg) by subcutaneous administration, or 0.156 lb/(min · ft³) [2,000 mg/(min · m³)] by inhalation. Among these are **Tabrun GA,** or ethyl N,N-dimethylphosphoramidocyanidate; **Sarin GB,** isopropyl methylphosphonofluoridate; **Soman GD,** 1,2,2-trimethylpropyl methylphosphonofluoridate; **BZ,** bis 2-chloroethyl sulfide; **VX,** ethyl S-2-diisopropylamino-ethyl methylphosphonothiolate; **pinacolyl alcohol,** 3,3-dimethylbutanol-2; **Saxitoxin**; mustard gas; and methylphosphonylfluoride.

POLONIUM.   A rare metallic element, symbol Po, belonging to the group of radioactive metals, but emitting only alpha rays. The melting point of the metal is about 489°F (254°C). It is used in meteorological stations for measuring the electrical potential of air. **Polonium-plated metal** in strip and rod forms has been employed as a static dissipator in textile-coating machines. The alpha rays ionize the air near the strip, making it a conductor and drawing off static electric charges. **Polonium 210** is obtained by irradiating bismuth, 100 lb (45 kg) yielding 0.04 oz (1 g) of polonium 210. It is used as a heat source for emergency auxiliary power such as in spacecraft. The metal is expensive, but can be produced in quantity from bismuth.

POLYARYLATES.   These high-heat-resistant thermoplastics are derived from aromatic dicarboxylic acids and diphenols. When molded, they become amorphous, providing a combination of toughness, dimensional stability, high dielectric properties, and uv stability. Polyarylates have heat-deflection temperatures up to 345°F (653°C) at 264 lb/in² (1.8 MPa). The resins can be injection-molded, extruded, and blow-molded, and sheet can be thermoformed. These resins also are blended with other engineering thermoplastics and reinforcements.

POLYCARBONATE RESINS. Traditionally made by reacting bisphenol A (BPA) and phosgene or by reacting a polyphenol with methylene chloride and phosgene. The monomer may be $OC_6H_4C(CH_3)_2C_6H_4OC:O$. In a continuous, solventless, melt-phase process recently developed by General Electric Plastics Japan and Japan's Mitsui Petrochemical Industries, polycarbonate is made by reacting diphenylcarbonate with BPA at 480 to 570°F (250 to 300°C) in the presence of a proprietary catalyst. This approach eliminates the need for toxic phosgene gas with BPA dissolved in an excess of the chloride. The molecular structure is in double-linked zigzag chains that give high rigidity. The resin is thermoplastic. It is crystalline with rhombic crystals.

Polycarbonate is a linear, low-crystalline, transparent, high-molecular-weight plastic. It is generally considered to be the toughest of all plastics. In thin sections, up to about 0.1875 in (0.478 cm), its impact strength is as high as 16 ft · lb (22 J). In addition, polycarbonate is one of the hardest plastics. It has good strength and rigidity and, because of its high modulus of elasticity, is resistant to creep. These properties, along with its excellent electrical resistivity, are maintained over a temperature range of about −275 to 250°F (−170 to 121°C). It has negligible moisture absorption, but it also has poor solvent resistance and, in a stressed condition, will craze or crack when exposed to some chemicals. It is generally unaffected by greases, oils, and acids. Polycarbonate plastics are easily processed by extrusion; by injection, blow, and rotational molding; and by vacuum forming. They have very low and uniform mold shrinkage. With a white light transmission of almost 90% and high impact resistance, they are good glazing materials. They have more than 30 times the impact resistance of safety glass. Other typical applications are safety shields and lenses. Besides glazing, polycarbonate's high impact strength makes it useful for air-conditioner housings, filter bowls, portable tool housings, marine propellers, and housings for small appliances and food-dispensing machines.

Lexan, of General Electric Plastics, is the most common trade name. **Lexan F-8000** is a polycarbonate sheet that can be laminated with **polyvinyl fluoride film,** such as Du Pont's **Tedlar,** and thermoformed. Among other applications, it is intended for aircraft cabin windows. **Lexan Nu-View** is a glazing laminate intended to keep mass-transit windows free of graffiti. It comprises **Lexan MR5** sheet with a 0.10-in (2.54-mm)- thick abrasion-resistant film on either face or both faces. If the film is marked, it can be easily removed, leaving a clean surface. **Krystaflex,** of Morton International, is a laminate of transparent polyurethane between two polycarbonate sheets. It is intended as a glass replacement in banks, prisons, and other facilities. **Panlite LN-2250,** of Japan's Teijin Chemical Ltd., is a low-bromine, flame-retardant polycarbonate.

POLYESTER THERMOPLASTIC RESINS.   There are several types of melt-processible thermoplastics, including polybutylene terephthalate, polyethylene terephthalate, polyethylene naphthalate, and aromatic copolyesters.

Polybutylene terephthalate (PBT) is made by the transesterfication of dimethyl terephthalate with butanediol through a catalyzed melt polycondensation. These molding and extrusion resins have good resistance to chemicals, low moisture absorption, relatively high continuous-use temperature, and good electrical properties (track resistance and dielectric strength). PBT resin is sensitive to alkalies, oxidizing acids, aromatics, and strong bases. Various additives, fillers, and fiber reinforcements are used with PBT resins, in particular flame retardants, mineral fillers, and glass fibers. PBT resins and compounds are used extensively in automotive, electrical and electronic, appliance, military, communications, and consumer product applications. **Valox** is the trade name of General Electric Plastics' PBT polyester. There are many other producers, including BASF Corp., DSM Engineering Plastics, and Hoechst Celanese.

Polyethylene terephthalate (PET) is a widely used thermoplastic packaging material. Beverage bottles and food trays for microwave and convection oven use are the most prominent applications. PET resins are made from ethylene glycol and either terephthalic acid or the dimethyl ester of terephthalic acid. Most uses for PET require the material's molecular structure to be oriented. Orientation of PET significantly increases tensile strength and reduces gas permeability and water vapor transmission. For packaging uses, PET is processed by blow molding and sheet extrusion. Typical trade names are **Cleartuf** and **Traytuf** (Goodyear Tire & Rubber Co.), **Tenite** and **Kodapak** (Eastman Chemical Products), **Melinar** (ICI Americas), and **Selar** (Du Pont). **Crisper,** a synthetic paper of Toyobo of Sanyo Corp. of America, is pearlescent PET film having excellent printability and good wear and ultraviolet resistance. **Spectar PETG,** of Eastman Chemical Products, is an extrudable copolyester for thick [to 0.5-in (13-mm)] sheet. Its impact strength is superior to that of acrylic, approaching that of polycarbonate. Color quality is equivalent to that of polycarbonate but less than that of acrylic, which is also more scratch-resistant. Injection-molding grades of PET, with fillers and/or reinforcements, also are available for making industrial products. Producers include Allied Signal, DSM Engineering Plastics, Du Pont, Eastman Chemical Products, General Electric Plastics, and Hoechst Celanese. **Rynite CR** grades, of Du Pont, contain nonmetallic conductive ingredients for superior electrical and thermal conductivity, heat dissipation, and electromagnetic shielding. **Ertalyte,** of Polymer Corp., comes in rod form for machined parts.

Polyethylene naphthalate (PEN), of Amoco Chemicals Co., is based on the naphthalate monomer dimethyl 2,6-naphthalate. It is intended for copolymers and blends in packaging applications, such as stretch-blow-molded bottles and, as a homopolymer, in thin and thick film for electrical and electronic applications. Relative to PET, PEN bottles have demonstrated similar crystallinity, much better oxygen and moisture barrier resistance, higher hot-fill temperature, and better appearance and dimensional stability after caustic washing. Replacing glass bottles in applications too demanding for PET is a potential application. Microwaveable packaging is a potential use for thick film. Magnetic tape is a current thin-film application. Candidate electrical and electronic applications for thick film include printed-circuit boards, capacitors, and membrane switches.

Ektar FB, of Eastman Chemical Products, is a family of poly 1,4-cyclohexylenedimethylene terephthalate (PCT) polymers. They are high-temperature, semicrystalline, thermoplastic polyesters with a melting point of 545°F (285°C) and service temperatures up to 340°F (171°C). Tensile strengths range up to 22,000 lb/in^2 (152 MPa), with flexural moduli up to 19×10^5 lb/in^2 (13,100 MPa) and heat-deflection temperatures at 264 lb/in^2 (1.8 MPa) up to 523°F (273°C). They are intended for mechanical and electrical parts.

Polyester film is generally made by the condensation of terephthalic acid and ethylene glycol. The extremely thin film, 0.00025 to 0.0005 in (0.00063 to 0.0013 cm), used for capacitors and for insulation of motors and transformers, has a high dielectric strength, up to 6,000 V/mil (236×10^6 V/m). It has a tensile strength of 20,000 lb/in^2 (138 MPa) with elongation of 70%. It is highly resistant to chemicals and has low water absorption. The material is thermoplastic, with a melting point of about 490°F (254°C). **Polyester fibers** are widely used in clothing fabrics. The textile fiber produced from dimethyl terephthalate is known as **Dacron.**

For magnetic sound-recording tape, polyester tape has the molecules oriented by stretching to give high strength. The 0.005-in (0.013-cm) tape has a breaking strength of 120 oz (3.4 kg) per 0.25 in (0.64 cm) of width. Electronic tape may also have a magnetic-powder coating on the polyester. But where high temperatures may be encountered, as in spacecraft, the magnetic coating is applied to metal tapes.

Aromatic polyesters (liquid crystals) have high mechanical properties and heat resistance. Commercial grades are aromatic polyesters, which have a highly ordered or liquid-crystal structure in solution and molten states. A high degree of molecular orientation develops during processing and, hence, anistropy in properties. Typically, these melt-processible resins can be molded or extruded to form products capable of use at temperatures over 500°F (260°C). Tensile strengths up to 35,000 lb/in^2 (241 MPa) and flexural moduli up to 4.6 × 10^6 lb/in^2 (31,717 MPa) are reported for liquid-crystal polymers. Chemical resistance also is excellent. Trade names for liquid-crystal polymers include **Vectra,** of Hoechst Celanese, **Xydar,** of Amoco Performance Products, and **Granlar,** of Granmont. Applications are in chemical processing and electronic, medical, and auto components.

POLYESTER THERMOSETTING RESINS. A large group of synthetic resins produced by condensation of acids such as maleic, phthalic, or itaconic with an alcohol or glycol such as allyl alcohol or ethylene glycol to form an unsaturated polyester which, when polymerized, will give a cross-linked, three-dimensional molecular structure, which in turn will copolymerize with an unsaturated hydrocarbon, such as styrene or cyclopentadiene, to form a copolymer of complex structure of several monomers linked and cross-linked. At least one of the acids or alcohols of the first reaction must be unsaturated. The polyesters made with saturated acids and saturated hydroxy compounds are called **alkyd resins,** and these are largely limited to the production of protective coatings and are not copolymerized.

The resins undergo polymerization during cure without liberation of water and do not require high pressure for curing. Through the secondary stage of modification with hydrocarbons, a very wide range of characteristics can be obtained. The most important use of the polyesters is as laminating and molding materials, especially for glass-fiber-reinforced plastic products. The resins have high strength, good chemical resistance, high adhesion, and capacity to take bright colors. They are also used, without fillers, as casting resins, for filling and strengthening porous materials, such as ceramics and plaster of paris articles, and for sealing the pores in metal castings. Some of the resins have great toughness and are used to produce textile fibers and thin plastic sheet and film. Other resins are used with fillers to produce molding powders that cure at low pressure of 500 to 900 lb/in^2 (3 to 6 MPa) with fast operating cycles.

Polyester laminates are usually made with a high proportion of glass-fiber mat or glass fabric, and high-strength reinforced moldings may contain a high proportion of filler. A resin slurry may contain as high as 70% calcium carbonate or calcium sulfate, with only about 11% of glass fiber added, giving an impact strength of 24,000 lb/in^2 (165 MPa) in the cured material. Bars and structural shapes of glass-fiber-reinforced polyester resins of high tensile and flexural strengths are made by having the glass fibers parallel in the direction of the extrusion. Rods and tubes are made by having the glass-fiber rovings carded under tension, then passing through an impregnating tank, an extruding die, and a heat-curing die. The rods contain 65% glass fiber and 35 resin. They have a flexural strength of 64,000 lb/in^2 (441 MPa) and a Rockwell M hardness of 65.

Physical properties of polyester moldings vary with the type of raw materials used and the type of reinforcing agents. A standard glass-fiber-filled molding may have a specific gravity from 1.7 to 2.0, a tensile strength of 4,000 to 10,000 lb/in^2 (28 to 69 MPa) with elongation of 16 to 20%, a flexural strength to 30,000 lb/in^2 (207 MPa), a dielectric strength to about 400 V/mil (16 $\times$ 10^6 V/m), and a heat distortion temperature of 350 to 400°F (177 to 204°C). The moldings have good acid and alkali resistance. But since an almost unlimited number of fatty-type acids are available from natural fatty oils or by synthesis from petroleum, and the possibilities of variation by combination with alcohols, glycols, and other materials are also unlimited, the polyesters form an ever-expanding group of plastics.

Het acid, of Hooker Chemical, is a complex chlorinated phthalic acid produced by hydrolyzing the product of the condensation of maleic anhydride with hexachlorocyclopentadiene made from pentane. This acid is reacted with glycols and maleic anhydride to give a hard polyester resin which is then cross-linked with styrene to give the liquid **Hetron resin,** which will cure with heat and a catalyst to an insoluble solid. The resin contains 30% chlorine. It is used for making laminated or reinforced plastics. Another chlorinated polyester resin is **FR resin.** It is flame-resistant, cures at normal temperatures, and is used for such layup lamination work as boat building and tank construction.

Some of the polyester-type resins have rubberlike properties, with higher tensile strengths than the rubbers and superior resistance to oxidation. These resins have higher wear resistance and chemical resistance than GRS rubber. They are made by reacting adipic acid with ethylene glycol and propylene glycol and then adding diisocyanate to control the solidifying action. They can be processed as rubber, but solidify more rapidly.

An extensive group of polyesters and alkyd resins with good heat stability can be made from **pyromellitic dianhydride,** $C_{10}H_2O_6$, a benzene tetracarboxylic dianhydride. **PMDA** is a white powder with melting point of 549°F (287°C). It reacts with alcohols, benzene, and other hydrocarbons. It is produced from **mellitic acid,** $C_6(COOH)_6$, which has a melting point of 550°F (288°C). It occurs in brown coal, peat, and the mineral **mellite,** or **honeystone,** which is hydrous aluminum mellate. **Tetrahydrophthalic anhydride** is easier to combine with styrene than phthalic or maleic anhydride and gives coating resins that are flexible and have quicker cure with high gloss.

POLYETHYLENES. Polyethylene thermoplastic resins include low-density polyethylenes (LDPEs), linear low-density polyethylenes (LLDPEs), high-density polyethylenes (HDPEs), and ethylene co-polymers, such as ethylene-vinyl acetate (EVA) and ethylene-ethyl acrylate (EEA), and ultra-high-molecular-weight polyethylenes (UHMWPEs). In general, the advantages gained with polyethylenes are light weight, outstanding chemical resistance, good toughness, excellent dielectric properties, and relatively low cost compared to other plastics. The basic properties of polyethylenes can be modified with a broad range of fillers, reinforcements, and chemical modifiers, such as thermal stabilizers, colorants, flame retardants, and blowing agents. Further, polyethylenes are considered to be very easy to process by such means as injection molding, sheet extrusion, film extrusion, extrusion coating, wire and cable extrusion coating, blow molding, rotational molding, fiber extrusion, pipe and tubing extrusion, and powder coating. Major application areas for polyethylenes are packaging, industrial containers, automotive, materials handling, consumer products, medical, wire and cable insulation, furniture, housewares, toys, and novelties.

The basic building blocks for polyethylenes are hydrogen and carbon atoms. These atoms are combined to form the ethylene monomer, C_2H_4, that is, two carbon atoms and four hydrogen atoms. In the polymerization process, the double bond connecting the carbon atoms is broken. Under the right conditions, these bonds re-form with other ethylene molecules to form long molecular chains. Ethylene copolymers **ethylene-vinyl acetate (EVA)** and **ethylene-ethyl acrylate (EEA)** are made by the polymerization of ethylene units with randomly distributed comonomer groups, such as **vinyl acetate (VA)** and **ethyl acrylate (EA).**

Three basic molecular properties affect most polyethylene properties: crystallinity (density), average molecular weight, and molecular weight distribution.

Molecular chains in polyethylenes in crystalline areas are arranged somewhat parallel to one another. In amorphous areas, they are randomly arranged. **High-density polyethylene** resins have molecular chains with comparatively few side chain branches. Therefore, the chains are packed more closely together. The result is crystallinity up to 95%. **Low-density polyethylene (LDPE)** resins have crystallinity from 60 to 75%. **Linear low-density polyethylene (LLDPE)** resins have crystallinity from 60 to 85%. For polyethylenes, the higher the degree of crystallinity, the higher the resin density. Higher density, in turn, influences numerous properties. With increasing density, heat-softening point, resistance to gas and moisture vapor permeation, and stiffness increase. However, increased density results in a reduction of stress cracking resistance and low-temperature toughness. LLDPE resins have a specific gravity of 0.915 to 0.940; LDPE resins range from 0.910 to 0.930; and HDPE from 0.941 to 0.965. **Dowlex NG 3347A** LLDPE, an ethylene-octene copolymer of Dow Plastics for stretch-film applications, provides up to 225% stretch with flow and puncture resistance. **Dowlex NG 1P** grades are intended for injection-molded containers. High-strength film grades of LLDPE, based on Union Carbide's Unipol polymerization technology, are said to provide much higher dart-impact strength than previous **Unipol** copolymers. **NTX** grades, of Mobil Polymers, are hexene LLDPE copolymers. Sumitomo Bakelite of Japan combines axially oriented Nylon 6 with ethylene-vinyl alcohol for flexible, stand-up water and juice pouches. New catalyst technology, such as Dow Plastics' and Exxon's "single-site" metallocene, is making LLDPE increasingly competitive in LDPE markets.

Atoms of different elements, such as carbon, hydrogen, etc., have different atomic weights. For carbon, the atomic weight is 12, and for hydrogen it is 1. Thus, the molecular weight of the ethylene unit is the sum of its six atoms (two carbon plus four hydrogen), or 28. Every polyethylene resin consists of a mixture of large and small chains, i.e., chains of high and low molecular weights. The molecular weight of the polymer chain generally is in the thousands. The average of these is called the *average molecular weight*. As average molecular weight increases, resin toughness increases. The same holds true for tensile strength and environmental stress cracking resistance (cracking brought on when molded parts are subjected to stresses in the presence of liquids such as solvents, oils, detergents, etc.). Because of increasing melt viscosity, polyethylene resins become more difficult to process as the average molecular weight increases. Melt index is a measure of the polymer's melt viscosity under standard conditions of temperature and pressure. Melt index is inversely related to the resin's average molecular weight: As average molecular weight increases, melt index decreases. Generally, a polyethylene resin with high molecular weight has a low melt index, and vice versa. Melt viscosity is an extremely important property since it affects the flow of the molten polymer. The resin's flow when melted increases with increasing melt index.

Molecular weight distribution is the relative distribution of large, medium, and small molecular chains in the polyethylene resin. When the distribution is made up of chains close to the average length, the resin is said to have a *narrow molecular weight distribution. Broad molecular weight distribution* polyethylenes are those resins with a wider variety of chain lengths. In general, resins with narrow molecular weight distributions will have good low-temperature impact strength and low warpage. Resins with broad molecular weight (MW) distributions generally will have greater stress cracking resistance and greater ease of processing.

High-molecular-weight low-density polyethylene resins are used to extrude high-clarity, tough film used in shrink packaging and for making heavy-duty bags. **High-molecular-weight high-density polyethylene** resins (MW 200,000 to 500,000) have excellent environmental stress cracking resistance, toughness, high moisture barrier properties, and high strength and stiffness. The HMW-HDPE resins are used in film, pressure pipe, large blow moldings, and sheet for thermoforming. **Ultra-high-molecular-weight polyethylenes** generally are considered to be those resins with molecular weights greater than 2×10^6 (materials with MW up to 6×10^6 are available). These resins have excellent abrasion resistance, stress cracking resistance, and toughness. They are, however, much more difficult to process than standard polyethylenes and require special forming techniques. UHMWPE resins are used in applications requiring high wear resistance, chemical resistance, and low coefficient of friction. Nonwoven fabric of UHMWPE **Spectra fibers** of Allied Signal is used in a thermosetting matrix for soldiers' helmets. Made by CGS Gallet SA of France, they are more ballistic-resistant than conventional steel helmets.

Polyethylene rubbers are rubberlike materials made by crosslinking with chlorine and sulfur, or they are ethylene copolymers. **Chlorosulfonated polyethylene** is white, spongy material. It has chlorine atoms and sulfonyl chloride groups spaced along the molecule. It is used to blend with rubber to add stiffness, abrasion resistance, and resistance to ozone, and for wire covering. **Ethylene-propylene rubber,** produced by various companies, is a chemically resistant rubber of high tear strength. The **ethylene butadiene resin** can be vulcanized with sulfur to give high hardness and wide temperature range. For greater elongation a terpolymer with butene can be made.

Polyethylene of low molecular weight is used for extending and modifying waxes, and in coating compounds especially to add toughness, gloss, and heat-sealing properties. Such materials are called **polyethylene wax,** but they are not chemical waxes. They can be made emulsifiable by oxidation, and they can be given additional properties by copolymerization with other plastics. The **polymethylene waxes** are microcrystalline and have sharper melting points than the ethylene waxes. They are more costly, but have high luster and durability. **Polybutylene plastics** are rubberlike polyolefins with superior resistance to creep and stress cracking. Films of this resin have high tear resistance, toughness, and flexibility and are used widely for industrial refuse bags. Chemical and electrical properties are similar to those of polyethylene and polypropylene plastics. **Polymethyl pentene** is a moderately crystalline, polyolefin plastic resin that is transparent even in thick sections. Almost optically clear, it has a light transmission value of 90%. Parts molded of this plastic are hard and shiny with good impact strength down to $-20°F$ ($-29°C$). Specific gravity (0.83) is the lowest of any commercial solid plastic. A major use is for molded food containers for quick-frozen foods that are later heated by the consumer.

POLYMERIC TRANSFLECTIVE MATERIALS. These materials consist of hundreds or more of alternating coextruded layers of two dissimilar, optically clear thermoplastics, such as acrylic and polycarbonate, for light management and decorative effects. Developed by Dow Plastics, they are referred to as **transflective materials** because they simultaneously transmit and reflect light. The composite structure provides a reflective surface ranging from diffuse to spectral or mirror-like. Changes to the front and back layers create various effects, such as pearlescence.

POLYPHENYLENE OXIDE. A plastic that is notable for its high strength and broad temperature resistance. There are two major types: **polyphenylene oxide (PPO)** and **modified PPO** (General Electric Plastics' **Noryl**). These materials have a deflection temperature ranging from 212 to 345°F (100 to 174°C) at 264 lb/in^2 (1.8 MPa). Their coefficients of linear thermal expansion are among the lowest for thermoplastics. Room-temperature strength and modulus of elasticity are high, and creep is low. In addition, they have good electrical resistivity. Their ability to withstand steam sterilization and their hydrolytic stability make them suitable for medical instruments, electric dishwashers, and food dispensers. They are also used in the electrical and electronic fields and for business-machine housings.

Tensile strength and modulus of phenylene oxides rank high among thermoplastics. They are processed by injection-molding, extrusion, and thermoforming techniques. Structural-foam grades, with their high rigidity, are suitable for large structural parts. Because of good dimensional stability at high temperatures and under moisture conditions, these plastics are readily plated without blistering.

POLYPHTHALAMIDE. **PPA** is a family of semicrystalline thermoplastics of Amoco Performance Products, with the trade name **Amodel,** having advantages over conventional nylons in resistance to humidity, auto fuels, water/glycol mixtures, and heat. The melting temperature is 590°F (310°C), and unreinforced, the resin has a specific gravity of 1.17, 0.81% 24-h water absorption, a tensile yield strength of 15,100 lb/in^2 (104 MPa), a flexural modulus of 475,000 lb/in^2 (3,275 MPa), a heat-deflection temperature of 248°F (120°C) at 264 lb/in^2 (1.8 MPa), and Izod notched impact strength of 1 ft · lb/in (53 J/m). An extra-tough grade is slightly lighter in weight, less water-absorbent (0.65%), and considerably less strong and rigid, but has an impact strength of 20 ft · lb/in (1,068 J/m). With 45% glass reinforcement, the specific gravity is 1.56, tensile strength (at break) 38,000 lb/in^2 (262 MPa), flexural modulus 2.1 × 10^6 lb/in^2 (14,480 MPa), heat-deflection temperature 549°F (287°C), and impact strength 2.5 ft · lb/in (133 J/m). **Thermocomp UF-1006,** of LNP Engineering Plastics, is a 30% glass-reinforced grade based on Amodel. Its continuous-use temperature is 545°F (285°C) versus 480°F (249°C) for similarly reinforced Nylon 6/6. RTP Co. makes a line of lubricated compounds based on PPA for gears and bearings. Polytetrafluoroethylene and/or silicone serve as lubricants, and glass, carbon, and/or aramid fibers as reinforcements. An **RTP 4000** grade with 15% aramid fiber, 14 PTFE, and 2 silicone is the most wear-resistant.

POLYPROPYLENE PLASTICS.   An important group of synthetic plastics employed for molding resins, film, and texture fibers. Developed in 1957 in Italy and Germany, they are produced as **polypropylene** by catalytic polymerization of propylene, or they may be copolymers with ethylene or other material. **Propylene** is a **methyl ethylene,** $CH_3CH:CH_2$, produced in the cracking of petroleum. It belongs to the class of unsaturated hydrocarbons known as **olefins,** which are designated by the word ending *-ene.* Thus propylene is known as **propene** as distinct from propane, the corresponding saturated compound of the group of **alkanes** from petroleum and natural gas. These unsaturated hydrocarbons tend to polymerize and form gums, and are thus not used in fuels although they have antiknock properties.

In **isotactic polypropylene,** the most common, the carbon atoms linked in the molecular chain between the CH_2 units each have a CH_3 and an H attached as side links, with the bulky side groups spiraled regularly around the closely packed chain. The crystalline structure imparts rigidity, and the material has a melting point of 329°F (165°C). Strength and rigidity can be increased further by nucleation or by broadening the molecular weight distribution and increasing isotacticity, resulting in resins referred to as *high-crystallinity polypropylenes.* **Atactic polypropylene** has a random molecular structure and lacks the crystallinity and rigidity required for many applications. It is used as an adhesive or is mixed with other materials, such as asphalt, for roofing applications. **Syndiotactic polypropylene,** whose side groups are arranged in an alternating symmetric pattern, is a recent development based on the use of metallocene catalysts. The material is crystalline and has a lower melting point than that of the isotactic type but better impact strength and clarity. It has been produced by Mitsui Toatsu Chemicals of Japan and is suitable for film and fibers.

Isotactic polypropylene includes homopolymers, copolymers, and random copolymers. The homopolymers are used mainly for fibers and filaments, oriented and cast films, blow-molded bottles, and injection moldings. Fiber and filament uses include textiles and carpets, bags, netting, and decorative ribbons. In film applications they provide high tensile and tear strengths, stiffness, clarity, gloss, and heat-sealing characteristics. Unfilled and unreinforced injection moldings have a density of about 0.033 lb/in^3 (913 kg/m^3) and a dielectric strength of 600 V/mil (23.6 $\times$ 10^6 V/m). Tensile strength is 4,500 to 6,000 lb/in^2 (31 to 41 MPa), flexural strength is 6,000 to 8,000 lb/in^2 (41 to 55 MPa), flexural modulus is 170,000 to 250,000 lb/in^2 (1,170 to 1,720 MPa), Izod notched impact strength is 0.4 to 1.4 ft $\cdot$ lb/in (21 to 75 J/m), and the heat-deflection temperature at 264 lb/in^2 (1.8 MPa) is 120 to 140°F (49 to 60°C). Talc, a common filler, appreciably increases rigidity, and glass fibers, the common reinforcement, markedly increase strength and rigidity. Homopolymers also provide the "living hinge" effect, that is, the ability to withstand repeated flexing.

Reflex, a line of flexible polypropylenes of Rexene Corp., combines about 80% atactic material for flexibility and 20% isotactic material for strength and chemical resistance. There are several homopolymers and copolymers, with tensile modulus ranging from 5,000 to 55,000 lb/in^2 (34 to 379 MPa) and elongation to 1000% and more. They are intended to replace plasticized polyvinyl chloride in autoclavable health care products and other plastics in extruded profiles and injection-molded parts.

Heterophasic copolymers, formed by the polymerization of a rubber phase such as ethylene propylene, increase impact strength. Random copolymers, made by the random addition of ethylene, have greater clarity, a lower but broader melting range, and lower flexural modulus. Their impact strength is greater than that of the homopolymers but less than that of the heterophasic copolymers. The **Accutuf** line of Amoco Chemical provides impact strengths ranging to 13 ft $\cdot$ lb/in (694 J/m) with flexural modulus up to 205,000 lb/in^2 (1,410 MPa).

Biaxially oriented polypropylene (BOPP) film (film stretched in two ways) has greatly improved moisture resistance, clarity, and stiffness. It is used for packaging tobacco products, snack foods, baked goods, and pharmaceuticals. Metallized grades also are available for packaging designed for extended shelf life.

Foamed polypropylenes include **expandable polypropylene (EPP)** bead and injection-molded structural foam. EPP provides greater energy absorption and flexibility than expandable polystyrene. Structural PP foam moldings consist of a solid outer skin and foam core. They are used to achieve greater stiffness in larger, lightweight parts (strength-to-weight ratios are 3 to 4 times greater than those for solid parts).

Polypropylene fiber was originally produced in Italy under the name of **Merkalon.** Unless modified, it is more brittle at low temperatures and has less light stability than polyethylene, but it has about twice the strength of high-density linear polyethylene. Monofilament fibers are used for filter fabrics, and have high abrasion resistance and a melting point of 310°F (154°C). Multifilament yarns are used for textiles and rope. **Polypropylene rope** is used for marine hawsers, floats on water, and does not absorb water as Manila rope does. It has a permanent elongation, or set, of 20%, compared with 19% for nylon and 11% for Manila rope, but the working elasticity is 16%, compared with 25% for nylon and 8% for Manila. The tensile strength of the rope is 59,000 lb/in^2 (407 MPa). Fine-denier multifilament **polypropylene yarn** for weaving and knitting dyes easily and comes in many colors. **Chlorinated polypropylene** is used in coatings, paper sizing, and adhesives. It has good heat and light stability, high abrasion resistance, and high chemical resistance.

POLYSTYRENES. A thermoplastic resin used for molding, in lacquers, and for coatings, polystyrene is formed by the polymerization of monomeric styrene, which is a colorless liquid of composition $C_6H_5CH:CH_2$, specific gravity 0.906, and boiling point 293°F (145°C). It is made from ethylene, and is ethylene with one of the hydrogen atoms replaced by a phenyl group. It is also called **phenyl ethylene** and **vinyl benzene.** As it can be made by heating **cinnamic acid,** $C_6H_5CH: CHCO_2H$, an acid found in natural balsams and resins, it is also called **cinnamene.** In the form of **vinyl toluene,** which consists of mixed isomers of methyl styrene, the material is reacted with drying oils to form alkyd resins for paints and coatings.

The polymerized resin is a transparent solid very light in weight with a specific gravity of 1.054 to 1.070. The tensile strength is 4,000 to 10,000 lb/in^2 (28 to 69 MPa), compressive strength 12,000 to 17,000 lb/in^2 (83 to 117 MPa), and dielectric strength 450 to 600 V/mil (18 $\times$ 10^6 to 24 $\times$ 10^6 V/m). Polystyrene is notable for water resistance and high dimensional stability. It is also tougher and stronger at low temperatures than most other plastics. It is valued as an electrical insulating material, and the films are used for cable wrapping.

When produced from **methyl styrene,** parts have a Rockwell M hardness to 83, with tensile strengths to 8,900 lb/in^2 (61 MPa), and have a stiffness that makes them suitable for such products as cabinets and housings. Dielectric strength is also high, above 800 V/mil (32×10^6 V/m), and the resin is thus used for electronic parts. The heat distortion temperature is 215°F (101°C).

Styrenes are subject to creep. Therefore, the long-term bearing strength (over 2 weeks) is only about one-third the short-time tensile strength. Since their maximum useful service temperature is about 160°F (71°C), their use is restricted chiefly to room-temperature applications. Because of their low cost and ease of processing, polystyrenes are widely used for consumer products. The impact grades and glass-filled types are used quite widely for engineering parts and semistructural applications. Because of good processing characteristics, polystyrenes are produced in a wide range of forms. They can be extruded; injection-, compression-, and blow-molded; and thermoformed. They are also available as film sheet and foam.

Polystyrenes can be divided into several major types. General-purpose grades, the lowest in cost, are characterized by clarity, colorability, and rigidity. They are applicable where appearance and rigidity, but not toughness, are required. Impact grades of polystyrenes are produced by physically blending styrene and rubber. Grades are generally specified as medium, high, and extra-high. As impact strength increases, rigidity decreases. Medium-impact grades are used where a combination of moderate toughness and translucency is desired. High-impact polystyrenes have improved heat resistance and surface gloss. The extra-high-impact grades are quite low in stiffness, and their use is limited to parts subject to high-speed loading. High-impact grades include Chevron Chemicals' **Valtra** and Dow Plastics' **ARS** (advanced resin system). Highly crystalline **syndiotactic polystyrene,** based on metallocene catalysts, was developed by Idemitsu Petrochemical of Japan in a joint effort with Dow Plastics. It has a melting point of 518°F (270°C), which is similar to that of Nylon 6/6 and about 3 times that of conventional, or amorphous, polystyrene. Tensile and flexural strengths are 5,100 lb/in^2 (35 MPa) and 9,200 lb/in^2 (64 MPa), respectively, and, with 30% glass reinforcement, 17,000 lb/in^2 (117 MPa) and 26,800 lb/in^2 (185 MPa).

Styrene can be polymerized with butadiene, acrylonitrile, and other resins. The terpolymer, **acrylonitrile-butadiene-styrene** (abbreviated **ABS**), is one of the common combinations. **Styrene-acrylonitrile (SAN)** has excellent resistance to acids, bases, salts, and some solvents. It also is among the stiffest of the thermoplastics, with a tensile modulus of 400,000 to 550,000 lb/in^2 (2,758 to 3,792 MPa). **Acrylate-styrene-acrylonitrile (ASA)** has very good weathering resistance (nonyellowing), good toughness, and stress cracking resistance. **Styrene-butylene resins** are copolymers that mold easily and produce thermoplastic products of low water absorption and good electrical properties. They have strength equal to that of the vinyls with greater elongation.

Foamed polystyrene is used in many forms, including extruded sheet (which is then thermoformed, e.g., egg cartons, trays); expandable polystyrene (EPS) beads, which contain a blow agent (usually pentane) and which are processed into low-density [0.7 to 10 lb/ft^3 (11 to 160 kg/m^3)] foamed products, such as hot and cold drink cups and protective packaging; and block and heavy sheet, used for thermal insulation.

PONTIANAK.   A gum from the trees *Dyera costulata* and *D. laxifolia* of Borneo and Malaya. The commercial pontianak is a grayish-white mass like burned lime and contains 60% water, with only 10 to 25% rubberlike materials. It is a rubber, but has a high content of resin similar to balata and gutta percha, and is classified with the lower guttas. It is used in the friction compounds employed for coating transmission belting, in insulations and varnishes, for mixing with gutta percha, and also to adulterate or replace chicle for chewing gums. It is also called **jelutong. Pontianak copal** is from varieties of *Agathis* trees of Borneo. Its peculiar turpentinelike qualities come from the method of tapping. It is valued for varnishes.

POPLAR.   The wood of several species of the tree *Populus*. The **black poplar,** or **English poplar,** *P. Nigra,* of Europe, is a large tree with blackish bark. The wood is yellowish white with a fine, open grain. It is soft and easy to work. The density is about 25 lb/ft^3 (400 kg/m^3). It is used for paneling, inlaying, packing cases, carpentry, and paper pulp. **Lombardy poplar** is a hybrid variety of this species. It is a tall, columnar tree that is male only and can be propagated only from rootstocks. It is grown in the United States for shelter belts, but in some countries is grown in fruit districts as a wood for packing boxes. **White poplar,** *P. alba,* is a larger tree native to the United States. The wood is similar to that of the black poplar. **Cottonwood** is another species of poplar. **Gray poplar** is from the tree *P. canescens,* of Europe. The color of the wood is light yellow. It has a tough, close texture somewhat resembling that of maple. It is used for carpentry and flooring. The wood of the canary whitewood is called **Virginia poplar,** or simply *poplar,* but belongs to a different family of trees. Aspen is also called poplar.

PORCELAIN.   Porcelains and **stoneware** are highly vitrified ceramics that are widely used in chemical and electrical products. **Electrical**

porcelains, which are basically classical clay-type ceramics, are conventionally divided into low- and high-voltage types. The high-voltage grades are suitable for 500 V and higher and are capable of withstanding extremes of climatic conditions. **Chemical porcelains** and stoneware are produced from blends of clay, quartz, fledspar, kaolin, and certain other materials. **Vita porcelain,** a feldspathic type, is used for dental restorations. Porcelain is more vitrified than stoneware and is white. A hard glaze is generally applied. Stonewares can be classified into two types: a dense, vitrified body for use with corrosive liquids, and a less dense body for use in contact with corrosive fumes. **Chemical stoneware** may range from 30 to 70% clay, 5 to 25 feldspar, and 30 to 60 silica. The vitrified and glazed product will have a tensile strength up to 2,500 lb/in^2 (17 MPa) and a compressive strength up to 80,000 lb/in^2 (552 MPa). **Industrial stoneware** is made from specially selected or blended clays to give desired properties.

Both chemical porcelains and stoneware resist all acids except hydrofluoric. Strong, hot, caustic alkalies mildly attack the surface. These ceramics generally show low thermal-shock resistance and tensile strength. Their universal chemical resistance explains their wide use in the chemical and processing industries for tanks, reactor chambers, condensers, pipes, cooling coils, fittings, pumps, ducts, blenders, filters, and so on.

Ceratherm, of U.S. Stoneware Co., is an acid-resistant and heat-shock-resistant ceramic having a base of high-alumina clay. It is strong and nonporous and is used for pump and chemical-equipment linings. The chemical stoneware of General Ceramics Inc., made with a mullite-zircon body, is white, dense, strong, and thermal-shock-resistant. The ceramic marketed under the name of **Prestite** by Westinghouse Electric Corp. for insulators and molded electrical parts is blended of flint, feldspar, kaolin, and ball clay. It is nonporous and moistureproof without a glaze, has high dielectric strength, has a tensile strength of 5,000 lb/in^2 (34 MPa), and has a compressive strength of 48,000 lb/in^2 (331 MPa). The ceramic produced by Rostone Corp. under the name of **Rosite,** for molded electrical parts and panels, is a calcium-aluminum-silicate mixed with asbestos. It withstands temperatures to 900°F (482°C), is resistant to alkalies, and has a compressive strength up to 15,000 lb/in^2 (103 MPa).

POROUS METALS. Metals with uniformly distributed controlled pore sizes, in the form of sheets, tubes, and shapes, used for filtering liquids and gases. They are commonly made by powder metallurgy, and the pore size and density are controlled by the particle size and the pressure used. Stainless steel, nickel, bronze, silver, and other metal powders are used, depending on the corrosion resistance required of the filter. Pore sizes can be as small as 7.9 μin (0.2 μm), but the most generally used filters have pores of 157, 315, 472, and 984 μin (4, 8, 12, and 25 μm). Pore sizes have a uniformity within 10%. The density range is from 40 to 50% of the theoretical density of the metal. Standard **filter sheet** is 0.30 to 0.60 in (0.76 to 1.52 cm), but thinner sheets are available. Sheets as thin as 0.004 in (0.010 cm), and with void fractions as high as 90%, have been made for fuel cells and catalytic reactors. **Porous steel** is made from 18–8 stainless steel, with pore openings from 787 to 2,559 μin (20 to 65 μm). The fine-pore sheet has a minimum tensile strength of 10,000 lb/in^2 (69 MPa), and the coarse has a strength of 7,000 lb/in^2 (48 MPa). **Felted metal,** developed by the Armour Research Foundation, is porous sheet made by felting metal fibers, pressing, and sintering. It gives a high strength-to-porosity ratio, and the porosity can be controlled over a wide range. In this type of porous metal the pores may be from 0.001 to 0.015 in (0.003 to 0.038 cm) in diameter, and of any metal to suit the filtering conditions. A felted fiber filter of 430 stainless steel with 25% porosity has a tensile strength of 25,000 lb/in^2 (172 MPa).

Gasar materials are porous metals, ceramics, and glasses produced in the Ukraine by melting in a furnace filled with hydrogen or other gases. Controlling gas pressure and cooling time governs pore size, shape, orientation, and volume. Pore size can range from 197 μin (5 μm) to 0.4 in (10 mm) and volume from 5 to 75%. Developed in 1960 at the Ukraine's Dnepropetrovsk Metallurgical Institute, applications include rocket kerosene-fuel atomizers, water-purification oxygenators, bronze bearings, magnesium aerospace panels, nickel chemical filters, and high-temperature ceramic catalyst supports in rocket and jet engines.

POTASH.   Also called **pearl ash.** A white, alkaline, granular powder, which is a **potassium carbonate,** K_2CO_3 or $K_2CO_3{:}H_2O$, used in soft soaps, for wool washing, and in glass manufacture. It is produced from natural deposits in Russia and Germany and also produced from wood and plant ashes. The U.S. production is largely from potash salts of New Mexico, from the brines of Searles Lake, California, and from solar evaporation in Utah, and usually contains iron, clay, and salt as impurities. **Kalium** is a high-purity grade that results from an evaporation-crystallization step; it has about 62.4% K_2O and less than 1 salt. The material is marketed by Kalium Chemicals. The material as produced by Hooker Chemical Co. is a free-flowing, white powder of 91 to 94% K_2CO_3, or is the hydrate at 84%, or calcined at 99% purity. The specific gravity of potash is 2.33 and melting point 1668°F (909°C). **Hartsalz,** mined in the Carpathian Mountains and used for producing potash, is a mixture of sodium chloride, potassium chloride, and magnesium sulfate. It is also a source of magnesium. The extensive potassium mineral deposits at Strassfurt and Mülhausen contain **sylvite,** KCl; **carnallite,** KCl · $MgCl_2$ · $6H_2O$; **kainite,** K_2SO_4 · $MgSO_4$ · $MgCl_2$ · $6H_2O$; and **leonite,** K_2SO_4 · $MgSO_4$ · $4H_2O$. There are at least 1×10^9 tons (9×10^8 metric tons) of the potash mineral **Wyomingite** in the deposits near Green River, Wyoming. It is a complex mineral containing leucite, phlogopite, diopside, kataphorite, and apatite. It has 11.4% K_2O, with sodium oxide, magnesium oxide, phosphorus pentoxide, and other oxides. The sylvite ore mined at Carlsbad, New Mexico, contains KCl and NaCl. It is electrically refined to 99.95% KCl, and is used to produce caustic potash. Electrolysis of the chloride solution yields caustic potash.

POTASSIUM. An elementary metal, symbol K, atomic weight 39.1, also known as **kalium.** It is silvery white, but oxidizes rapidly in air and must be kept submerged in ether or kerosene. It has a low melting point, 145°F (63°C), and a boiling point of 1392°F (756°C). The specific gravity is 0.855 at 68°F (20°C). It is soluble in alcohol and in acids. It decomposes water with great violence. Potassium is obtained by the electrolysis of potassium chloride. Potassium metal is used in combination with sodium as a heat-exchange fluid in atomic reactors and high-temperature processing equipment. A **potassium-sodium alloy** contains 78% potassium and 22 sodium. It has a melting point of 12°F (-11°C) and a boiling point of 1393°F (756°C) and is a silvery, mobile liquid. **Cesium-potassium-sodium alloys** are called **BZ Alloys. Potassium hydride** is used for the photosensitive deposit on the cathode of some photoelectric cells. It is extremely sensitive and will emit electrons under a flash so weak and so rapid as to be imperceptible to the eye. **Potassium diphosphate,** KH_2PO_4, a colorless, crystalline, or white powder soluble in water, is used as a lubricant for wool fibers to replace olive oil in spinning wool. It has the advantages that it does not become rancid as oil does and can be removed without scouring. Potassium, like sodium, has a broad range of use in its compounds, giving strong bonds. Metallurgically it is listed as having a body-centered-cubic structure, but the atoms arrange themselves in pairs in the metal as K_2, and the structure is cryptocrystalline.

POTASSIUM CHLORATE. Also known as **chlorate of potash** and **potassium oxymuriate.** A white, crystalline powder, or lustrous, crystalline substance, of composition $KClO_3$, employed in explosives, chiefly as a source of oxygen. It is also used as an oxidizing agent in the chemical industry, as a cardiac stimulant in medicine, and in toothpaste. It melts at 675°F (357°C) and decomposes at 752°F (400°C) with the rapid evolution of oxygen. It is odorless but has a slightly bitter, saline taste. The specific gravity is 2.337. It is not hygroscopic, but is soluble in water. It imparts a violet color to the flame in pyrotechnic compositions.

Potassium chloride is a colorless or white, crystalline compound of composition KCl, used for molten salt baths for the heat treatment of steels. **Trona** is a fertilizer- and explosives-grade material from Kerr-McGee Chemical Corp. **Morton** is a pharmaceutical-quality potassium chloride that has no iodine contamination. It is sold by Morton Salt. The specific gravity is 1.987. A bath composed of three parts potassium chloride and two parts barium chloride is used for hardening carbon-steel drills and other tools. Steel tools heated in this bath and quenched in a 3% sulfuric acid solution have a very bright surface. A common bath is made up of potassium chloride and common salt and can be used for temperatures up to 1652°F (900°C).

POTASSIUM CYANIDE. A white, amorphous or crystalline solid of composition KCN, employed for carbonizing steel for case hardening and for electroplating. The specific gravity is 1.52, and it melts at about 1550°F (843°C). It is soluble in water and is extremely poisonous, giving off the deadly hydrocyanic acid gas. For cyaniding steel the latter is immersed in a bath of molten cyanide and then quenched in water, or the cyanide is rubbed on the red-hot steel. For this use, however, sodium cyanide is usually preferred, because of its lower cost and the higher content of CN in the latter. Commercial potassium cyanide is likely to contain a proportion of sodium cyanide. **Potassium ferrocyanide,** or **yellow prussiate of potash,** can also be used for case-hardening steel. It has composition $K_4Fe(CN)_6$ and comes in yellow crystals or powder. The nitrogen as well as the carbon enters the steel to form the hard case. **Potassium ferricyanide,** or **red prussiate of potash,** is a bright-red, granular powder of composition $K_3Fe(CN)_6$, used in photographic reducing solutions, in etching solutions, in blueprint paper, and in silvering mirrors. **Redsol crystals,** of American Cyanamid Co., is the name of this chemical for use as a reducer and mild oxidizing agent, or toner, for photography. **Potassium cyanate,** KCNO, is a white, crystalline solid used for the production of organic chemicals and drugs. It melts at 590°F (310°C). The **potassium silver cyanide** used for silver plating comes in white, water-soluble crystals of composition $KAg(CN)_2$. **Sel-Rex,** of Bart-Messing Corp., is this material. **Potassium gold cyanide** has a similar function in gold plating. **Platina,** of Platina Laboratories, comes as colorless tablets that are soluble in both water and alcohol.

POTASSIUM NITRATE. Also called **niter** and **saltpeter,** although these usually refer to the native mineral. A substance of composition KNO_3, it is used in explosives, for bluing steel, and in fertilizers. A mixture of potassium nitrate and sodium nitrate is used for steel-tempering baths. The mixture melts at 482°F (250°C). Potassium nitrate is made by the action of potassium chloride on sodium nitrate, or **Chile saltpeter.** It occurs in colorless, prismatic crystals, or as a crystalline, white powder. It has a sharp saline taste and is soluble in water. The specific gravity is 2.1, and the melting point is 639°F (337°C). It is found in nature in limited quantities in the alkali region of the western United States. Potassium nitrate contains a large percentage of oxygen which is readily given up and is well adapted for pyrotechnic compounds. It gives a beautiful, violet flame in burning. It is used in flares and signal rockets. **Potassium nitrite** is a solid of composition KNO_2 used as a rust inhibitor, for the regeneration of heat-transfer salts, and for the manufacture of dyes.

POTATO. The bulbous tubers of the roots of the annual plant *Solanum tuberosum,* native to Peru but now grown in many parts of the world. It is used chiefly as a direct food, but is also employed for making starch and alcohol. The potato was brought to Europe in 1580, and received the name of **Irish potato** when brought to New England in 1719 by Irish immigrants. The plant is hardy and has a short growing cycle, making it adaptable to many climates. The tuber contains about 78% water, 18 starch with some sugar, 2 proteins, 1 potash, and only about 0.1 fats. The average water loss in storage is about 11%. There are more than 500 varieties of potato cultivated. **Dehydrated potato,** produced as powder, flake, and porous granules, is widely used in restaurants and institutions and is marketed under various trade names for home use.

The **sweet potato** is the root bulb of the trailing perennial vine *Ipomoea batata,* native to tropical America, but now grown extensively in the southern United States and in warm climates. In South America it is known by the Carib name **batata.** Like the white potato, the sweet potato has a high water content, but is rich in sugars. There are many varieties and two general types: one with a dry, mealy, yellow flesh and the other with a soft, gelatinous flesh higher in sugars. The latter type is called **yam** in the United States, but the true yam is a larger tuber from the climbing plant *Diascorea alata,* grown widely in the West Indies. The sweet potato is used as a direct food, but large amounts are also employed for making preserves, starch, and flour for confectionery. **Alamalt** is cooked and toasted sweet potato ground to a powder for use in confectionery. It adds flavor as well as sugar to the confectionery. **Sweet-potato flake,** used in foodstuffs, is cooked and dehydrated sweet potato in orange-colored flakes with the flavor of candied sweet potato. It is reconstituted with milk or water.

The **taro** is the root tuber of the large, leafy plant *Colocasia esculenta,* which constitutes one of the chief foods of southeast Asia and Polynesia. There are more than 300 varieties grown. The tuber is high in starch, has more proteins than the potato, but has an acrid taste until cooked. The pasty starch food known as **poi** in the Pacific Islands is made from taro. In Micronesia the taro is called **jaua,** and the **mwang plant,** *Cyrtosperma chamissonis,* is called taro. This plant is larger, and the rootstock weighs as much as 50 lb (23 kg), while the taro does not exceed 5 lb (2 kg). Taro matures in 6 months, while the mwang requires 2 years. The **dasheen** is a variety of taro grown in the southern United States. The **yautia,** grown in the West Indies, resembles the taro, but is from the large plant *Xanthosma sagittifolium.* It is high in starch and has greater food value than the potato.

PRECIOUS METALS. The metals gold, silver, and platinum, which are used for coinage, jewelry, and ornaments, and also for industrial applications. Expense or rarity alone is not the determining factor; rather, a value is set by law, with the coinage having an intrinsic metal value as distinct from a copper coin, which is merely a token with little metal value. The term *noble metal* is not synonymous, although a metal may be both precious and noble, such as platinum. Although platinum was once used in Russia for coinage, only gold and silver fulfill the three requisites for coinage metals. Platinum does not have the necessary wide distribution of source. The **noble metals** are gold, platinum, iridium, rhodium, osmium, and ruthenium. Unalloyed, they are highly resistant to acids and corrosion. Radium and certain other metals are more expensive than platinum but are not classified as precious metals. Because of the expense of the platinum noble metals, they may be alloyed with gold for use in chemical crucibles. **Platino** is an alloy of 89% gold and 11 platinum. **Palau** is the name of an alloy of gold and palladium, and **rhotanium** is a rhodium-gold alloy.

PREFINISHED METALS. Sheet metals that are precoated or treated at the mill so as to eliminate or minimize final finishing by the user. The metals are made in a ready-to-use form with a decorative and/or functional finish already applied. A large number and variety of prefinished metals have been developed. **Prepainted metals** are produced using various organic coatings on many common ferrous and nonferrous metals. Extra durability or special decorative effects are provided by **plastic-metal laminates.** Polyvinyl chloride, polyvinyl fluoride, and polyester are the plastics commonly used.

Black-coated steel is used to give a high thermal emittance in electronic equipment. The base metal is aluminum-deoxidized steel containing 0.13% carbon, 0.45 manganese, 0.04 maximum phosphorus, and 0.05 maximum sulfur. The steel is coated with a 5% by weight layer of nickel oxide, which is reduced in a hydrogen furnace to form a spongy layer of nickel. This sponge is impregnated with a carbon slurry to form a black, carbonized surface.

Preplated metals consist of a thin, electrodeposited plate of one metal or alloy on a base, or substrate, usually sheet, of another metal. Steel is the most common base metal, and it is commonly plated with brass, chromium, copper, nickel, nickel-zinc, or zinc. Other common preplated metals are chromium-plated brass, copper, nickel, or zinc; and nickel-plated brass, copper, or zinc. The surface of the plate may be mirror-bright, satin, bright satin, embossed, antique, or black, or may have some other finish. The plate, usually 0.001 to 0.005 in (0.025 to 0.127 mm) thick, is sufficiently ductile to withstand shearing, bending, drawing, and stamping operations. Common joining methods include lock-seaming, stud welding, adhesive bonding, and spot welding. One of the earliest groups of preplated metals included **Brassoid, Nickeloid,** and **Chromaloid** of American Nickeloid, which were brass-, nickel-, and chromium-plated zinc sheet. Other producers of preplated metals have included Apollo Metals, National Sheet Metal, and Thomas Steel Strip.

Prefinished metals are now available with almost any metal plated or bonded to almost any other metal, or single metals may be had prefinished in colors and patterns. They come in bright or matte finishes and usually have a thin paper coating on the polished side, which is easily stripped off before or after forming. The metals are sold under a variety of trade names and are used for decorative articles, appliances, advertising displays, panels, and mechanical parts.

PRESERVATIVES. Chemicals used to prevent oxidation, fermentation, or other deterioration of foodstuffs. The antioxidants, inhibitors, and stabilizers used to retard deterioration of industrial chemicals are not usually called preservatives. The most usual function of a preservative is to kill bacteria, and this may be accomplished by an acid, an alcohol, an aldehyde, or a salt. A legal requirement under the Food and Drug Act is that a preservative be nontoxic in the quantities permitted. **Sugar** is the most commonly used preservative for fruit products. **Sodium chloride** is used for protein foods. **Sodium nitrate** is reduced to sodium nitrite in curing meats, and the nitrite has an inhibitory action on bacterial growth, the effect being greatest in acid flesh. **Potassium sorbate,** $KOCOCH:CH:CHCH_3$, a white, water-soluble powder, inhibits the growth of many molds, yeasts, and bacteria which cause food deterioration, and it is used in cheese, syrups, pickles, and other prepared foods.

The inhibitory effect of organic acids is due chiefly to the undissociated molecule. **Acetic acid** is normally more toxic to bacteria than lactic acid; but when sugar is present, the reverse is true, and citric acid then has little toxicity. The inhibitory action of inorganic acids is due mainly to the pH change which they produce. The antimicrobe effect of the **vanillic acid** esters generally increases with increasing molecular weight. Only small quantities of chemicals are usually needed for preservation. **Isobutyl vanillate,** an ester of vanillic acid, is effective as a preserving agent in milk and some other foods when only 0.10 to 0.15% is used. Preservatives are also marketed for external application to foodstuffs in storage, though these are more properly classified as fumigants. Beverages containing fruit juices or little carbonation are preserved with less than 0.05% **sodium benzoate.** Methyl, propyl, butyl, and ethyl **parabens; potassium sorbate; sodium dihydroacetate;** and **imidazolidinyl urea** are all industrial **microbials** used in the food, cosmetic, and pharmaceutical industries. **Abiol,** grade CFTA, is a water-soluble, stable, wide-pH-range-acting preservative of the imidazolidinyl urea family. Grade NF is **imidurea.** Both, available from Tri-K Industries, Inc., are particularly effective in conjunction with the parabens.

PRIMER. A surfacing material employed in painting or finishing to provide an anchorage or adhesion of the finishing material. A primer may be colorless, or it may have color. In the latter case, it is sometimes called an **undercoat.** A primer is distinct from the filler coat used on woods to fill the pores and thus economize on the more expensive finish. Primers for industrial or production finishing are of two types: air-dry and baking. The air-dry types have drying-oil vehicle bases and are usually called **paints.** They may be modified with resins. They are not used as extensively as the baking-type primers, which have resin or varnish vehicle bases. These dry chiefly by polymerization. Some primers, known as **flash primers,** are applied by spraying and dry by solvent evaporation within 10 min. In practically all primers, the pigments impart most of the anticorrosion properties to the primer and, along with the vehicle, determine its compatibility with and adherence to the base metal.

Primer coats of **red lead paint** were formerly much used on construction steel to give corrosion resistance, but chromate or phosphate primers are now more common. **Barium potassium chromate** gives a pale-yellow coating with good anticorrosion properties for steel, aluminum, and magnesium. **Zinc yellow paints** may also be used as primer coats on metal. **Zinc chromate** is used as a primer on steel. It has a tendency to dissolve when moisture penetrates the paint, and this dissolved chromate retards corrosion of the steel. **Zinc phosphate primers** applied to iron and steel give corrosion resistance and improve paint adhesion. **Manganese phosphate** forms a dense, crystalline coating on steel, which acts as a corrosion-resistant base for paint. A mixture of 95% zinc powder and 5% epoxy resin binder in a solvent gives a gray, metallic finish, and the zinc blocks corrosion by galvanic action. In addition to the pigment, various corrosion inhibitors may be used in primer paints. **Ammonium ferrous phosphate,** $NH_4FePO_4 \cdot H_2O$, has a platelike structure which gives impermeability to the film as well as adding corrosion resistance. It is greenish. A primer is especially required in the finishing of sheet-metal objects that are likely to receive dents or severe service, but it is not usually necessary for castings or roughened surfaces. For sheet-metal work, baked enamels were formerly much used for the primers for lacquer finishes, but synthetic resin primers give good adhesion and are less expensive.

TPO-Direct Primer (AP-202), of Akzo Coatings, is a nonisocyanide, two-component epoxy primer for thermoplastic olefin elastomeric parts, such as auto-bumper fascia. Its key advantage is that adhesion promoters are not required prior to painting.

PROTEIN. A nitrogen organic compound of high molecular weight, from 3,000 to many millions. Proteins are made up of complex combinations of simple amino acids, and they occur in all animal and vegetable matter, but are also made synthetically. They form a necessary constituent of foods and feeds, but are also used for many commercial products, but some proteins are highly poisonous. The poison of the cobra and that of the jellyfish are proteins.

Different types of plant and animal life have different types of proteins. At least 10 different proteins are known to be essential to human body growth and maintenance, but many others may have subsidiary functions since the amino acids are selective chelating agents, separating copper, iron, and other elements from the common sodium, calcium, and potassium compounds entering the system. The simple proteins are made up entirely of the acids, but the complex or conjugated proteins also contain carbohydrates and special groups, while the cystine of hair and wool also contains sulfur. The constituent amino acids of the protein molecule are linked together with a peptide bond, and the linkage forms the backbone of the molecule, but the arrangement is not similar to the high polymers usually associated with plastics with one type of polymer, or group, repeating itself. The linkage is formed by the loss of carbon dioxide rather than by the loss of water, as in plastics.

The simplest proteins are the **protamines** with molecular weights down to about 3,000. They are strongly basic, are water-soluble, and contain no sulfur. **Clupeine** in herring and **salmine** in salmon and trout are examples. The **histones** which occur in white blood corpuscles contain sulfur and are more complex. The albumins of eggs and milk are soluble in water and coagulate with heat. They also occur in plants, as in the **leucosin** of wheat. **Prolamines** are **vegetable proteins,** as the zein of corn and the **gliadin** of wheat. They are not an adequate human protein food without **animal proteins.**

Biologically, the edible proteins are classified as first-class and second-class, the first being from animal and the second from vegetable origin. Meat and fish proteins are both complete, or first-class, proteins, but the digestibility of fish protein is slightly higher than that of beef protein, while oyster protein is high in growth-promoting value. The synthesis of globulin and antibody formation for resistance to disease depends upon the utilization of various amino acids most readily obtained from first-class proteins. The term **protein isolates** used in the food industry refers to proteins from soybean or other sources and containing 85% or more of protein.

Lysine, essential for human nutrition, is found in soy proteins, but it is also made synthetically as a water-soluble white powder, and is added to bakery products to raise the protein value to nearly that of animal protein. **Isoleucine,** a bitter amino acid, occurs in casein. It is an **aminomethyl valeric acid** which is fermented by yeast to amyl alcohol. **Sclero proteins,** or **albuminoids,** contain much sulfur, are insoluble in water, resist hydrolysis, and are the most complex of the proteins. They occur in skin, ligaments, horn, wool, and silk. The complex, indigestible protein of poultry feathers, used for making brush fibers, is also broken down to produce digestible proteins used in feeds. **Glycine,** or **glycocoll,** is an **aminoacetic acid,** H_2NCH_2COOH, formed by the hydrolysis of complex proteins and also made synthetically. The **methylated glycine,** called **betaine,** occurs in plants and is obtained from sugar-beet molasses.

Hydrolyzed proteins are used in flavoring foodstuffs. The Japanese condiment **adjinimoto,** made from wheat gluten, is largely **sodium glutamate,** a salt of **glutamic acid,** $C_5H_9O_4$, which also occurs in seeds and beets. **Monosodium glutamate** is sold under trade names such as **Zest** of A. E. Staley Mfg. Co. and **Ac'cent.** It is a white, crystalline powder derived from soybeans and sugar beets. In small amounts it has no flavor, but intensifies the taste and flavor of foodstuffs. This effect may be greatly increased by adding a chemical compound such as disodium inosinate or disodium guanylate to replace part of the glutamate. **Mertaste,** of Merck & Co., is such a compound which enhances the flavor and acts as a synergist. In large amounts the glutamates have an offensive odor and flavor and are considered toxic. **Insulin,** used in medicine, contains **glutamine,** the half amide of glutamic acid, and also **cystine,** the disulfide of **cysteic acid,** an aminopropionic acid essential for nutrition. **Cystein** is aminomercaptopropionic acid. It is used with whey in bakery dough to react with yeast enzymes and give faster rising at lower temperatures. **Royal jelly,** used in face creams, is a protein complex high in vitamin B. It is a secretion of bees to nourish the egg of the queen bee, but has no apparent therapeutic value.

Wheat gluten, made from flour as a spray-dried powder, contains about 82% protein. It is used as an additive to improve the texture and shelf life of baked goods. **Soybean protein** is marketed as a highly refined, odorless, and tasteless powder for use in confectionery and other foods to retain freshness and add food value. **Proset-Flake** is a soybean protein in flake form to give firmer texture. **Lupin,** from the lupin bean, has more protein than soybean and more than twice the calcium of most nondairy foods. The bean can be grown in poor soils and fixes nitrogen in the soil, reducing the need for chemical fertilizer. Lupini Pasta is a lupin-based food product, and animal feed is a potential product. **Mushroom powder** is used to add proteins to foodstuffs where the flavor is also desirable, as in sauces. It is made by sheeting the pulped mushroom, *Agaricus campestris,* on a double-drum drier and then grinding the sheet to a fine powder. **Animal protein factor,** used in animal feeds, and marketed commercially in fish solubles, is an amino acid combination containing several vitamins. It is also used in feeds for single-stomach animals, such as the hog, which cannot synthesize within themselves all the amino acids necessary for health. It is also used in poultry feeds, as it contains the hatchability factor not adequately supplied by grains. **DL-methionine,** of Dow Chemical Co., is a synthetic amino acid of composition $CH_3SCH_2CH_2CH(NH_2)CO_2H$, used in feeds for fattening poultry. **MPF granules,** of General Mills, Inc., are granules of concentrated proteins, vitamins, and minerals for adding to bakery and meat products. The whey from milk in powder form is added to bread to increase the protein content, and it also refines the texture of bakery products. In porous granule form it is used as a **flow agent** to add butter, margarine, or oils to bakery products as well as protein. **Kraflow,** of Kraft General Foods, consists of **whey granules** which will absorb 8 times their volume of oils.

Protein plastics are produced by the isolation of precipitation of proteins from animal or vegetable products and hardening or condensing into stable compounds that can be molded into sheet or fiber. The oldest of the protein plastics is **casein plastic** which was used to replace bone for toothbrush handles and buttons. It is still an important molding plastic, as it is tough and can be made in pastel shades, but it is more costly than many other molding plastics. The proteins from soybean meal or other vegetable products are condensed with aldehydes or with various mineral salts or acids to form plastics. These plastics are distinct from those made from the fatty acids of soybean or other oils. **Peanut fiber,** under the name of **Ardil,** is made in England by precipitating the protein from peanut meal with an acid at low temperatures, so as not to denature the meal. It is then dissolved in caustic soda and formed through spinnerets into a hardening bath. The fiber is soft and resilient, moisture-absorbent as wool, moth-resistant, and it will dye easily. It is mixed with wool in weaving fabrics. **Vicara** is a protein fiber produced from the zein of corn. The fiber is light yellow, soft, tough, and strong. In fabrics it has the warmth of wool, is resistant to mildew, and withstands temperatures to 310°F (154°C). It can be blended with cotton, wool, or rayons. Proteins obtained by alkaline extraction from cottonseed are also used to produce woollike fibers. **Azlon** is a general name for protein fibers. **Chromated protein** finishes that provide a corrosion-resistant undercoat on iron or steel are produced by coating the metal with casein or albumin and then impregnating with a chromate solution which hardens the film.

Enzymes are complex proteins consisting of hundreds of amino acids and are nature's catalysts, generally in the presence of water. In industry, the Austrian firm Chemie Linz employs a **lipase** to produce optically active 2-chloropropionic acid, a raw material for phenoxypropionic herbicides. **Peroxidase,** an enzyme extracted from horseradish, works in organic solvents to convert suitable phenolic raw materials into plastics, such as **Novolac.** An active isomer of **ibuprofen,** the anti-inflammatory drug, has been generated by the action of a **protease** enzyme on an ibuprofen precursor, an ester. **Durazyme,** a protease of Novo Nordisk of Denmark, is widely used in detergents.

Several enzymes have been developed recently for the pulp paper and food-processing industries. **Polyzyme,** of Novo Nordisk, and **Xylanase,** of Gencore International, reduce the need for chlorine in pulp and paper bleaching. **Resinase,** a lipase of Novo Nordisk's Japanese subsidiary and Jujo Paper Co. of Tokyo, removes from pulp the triglyceride-heavy resin that can blemish paper products and stick to process equipment. A lipase enzyme derived from *Rhizopus delemar* is used by Ajinomoto Co. of Tokyo in a process that combines rapeseed and palm oils to produce vegetable cooking oil. The cooking oil resists oxidation as well as palm oil does, but remains liquid at lower temperature—5°F (−15°C) versus 104°F (40°C). Using a reactor packed with immobilized lipase, Japan's Fugi Oil Co. produces a vegetable oil similar to cocoa butter.

PUMICE. A porous, frothlike volcanic glass which did not crystallize due to rapid cooling, and frothed with the sudden release of dissolved gases. Powdered or ground pumice is used as an abrasive for fine polishing, in metal polishes, in scouring compounds and soaps, and in plaster and lightweight concrete and pozzolanic cement. In very fine powder it is called **pounce,** when used for preparing parchment and tracing cloth. **Pouncing paper** is paper coated with pumice used for pouncing, or polishing felt hats. Pumice is grayish white, and the fine powder will float on the surface of water. The natural lump pumice contains 65 to 75% silica, 12 to 15 alumina, and 4 to 5 each of soda and potash. It is produced chiefly in California and New Mexico. **Pumicite** is a **volcanic ash** similar in composition to pumice, found in large beds in Nebraska, Kansas, and Colorado. Its chief distinction is that it is fine-grained and has sharp edges suitable for abrasive purposes. The natural material is dried, pulverized, and screened so that 98.8% will pass a 325-mesh screen. **Seismotite** is a trade name for pumice used as an abrasive in scouring compounds. **Slag pumice,** or **artificial pumice,** is made in Germany by treating molten slag with less water than is required for granulation. It is used as an aggregate in lightweight concrete and as a heat insulator. **Obsidian** will change into pumice when melted, but *obsidian* is a general name for **volcanic glass** and varies in composition. It is an extrusive igneous rock that gets its glassy nature from its method of cooling, and some obsidian has a composition similar to that of granite. It is colored black from magnetite, or brown to red from hematite. Obsidian was used by the ancients for instruments and by the American Indians for arrowheads and knives. Semitransparent, smoky-colored obsidian nodules of Arizona ore are called **marekanite** and are cut for Indian silver jewelry.

Hawaiian obsidian, or **tachylite,** also known as **basalt glass,** is a volcanic glass from Oahu, Hawaii. It is jet-black, takes a fine polish, and is used for making ornamental articles. The type of obsidian found in Oregon and California, known as **perlite,** is flash-roasted to form a bubblelike, expanded material about 15 times the original size which is crushed to a white, fluffy powder. Perlite contains about 75% silica. In California it has been called **calite.** The powder has a density of only 4 to 12 lb/ft^3 (64 to 192 kg/m^3) and is used in lightweight wallboard, acoustical tile, and insulation and as a lightweight aggregate in concrete. **Perlalex** is the finely ground powder used for removing smears from drawing paper. **Rhyolite perlite** of California is expanded to ovaloid particles of 23,000 to 33,000 μin (590 to 840μm) containing complete vacuums and having a density of only 0.78 lb/ft^3 (12.4 kg/m^3). This material has about 70% silica, 15 alumina, 2 Fe_2O_3, 1.5 CaO, 2.75 Na_2O, 1 MgO, 4 K_2O, and 4 water. **Grellex,** of Great Lakes Carbon Corp., is expanded California perlite in fine particle size for use in plastics, adhesives, and insulation.

Other materials besides perlite may be expanded by heat. **Expanded clay** is made from common brick clays by grinding and screening to 48 to 80 mesh and feeding through a gas burner at about 2700°F (1482°C). The ferric oxide is changed to ferrous oxide, liberating oxygen and CO_2. Strong, light bubbles about 0.020 in (0.051 cm) in diameter are formed. **Kanamite** is this material. It has a density of 17 to 25 lb/ft^3 (272 to 401 kg/m^3) and is used as an aggregate for lightweight concrete.

PURPLEHEART.　　The wood of several species of trees, notably *Peltogyne paniculata* of the Guianas. The color of the wood is brown, the heartwood turning purple on exposure. The grain is open and fine. The wood has a density of about 53 lb/ft^3 (849 kg/m^3) and is very hard, strong, and durable. It is used for machine and implement parts, inlays, furniture, and turnery.

PUTTY.　　A mixture of calcium carbonate with linseed oil, with sometimes white lead added. It is used for cementing window glass in place and as a filler for patterns. Litharge is often added to putty for steel sash. Another putty for steel contains red lead, calcium carbonate, and linseed oil. The dry pigment for putty, **whiting putty,** according to ASTM specifications, contains 95% calcium carbonate and 5 tinting pigment. **White lead putty** contains 10% or more white lead mixed with the calcium carbonate. Federal specifications call for a minimum of 11% boiled or processed linseed oil with a maximum of 89% pigment as a white lead–whiting putty. **Putty powder** is a mixture of lead and tin oxides, or a mixture of tin oxide and oxalic acid, or it may be merely an impure form of tin oxide. It is used in enameling, for polishing stone and glass, and as a mild abrasive for dental polishes. **Caulking putty,** used for setting window and door frames, is made of asbestos fibers, pigments, and drying oils, or with rubber or resins. The older caulking compounds used in the building industry for sealing between window and door frames and masonry were made with drying oils and inert fillers, but they had poor adhesion and weathering qualities. **Caulking compounds** are now composed of synthetics with usually a polysulfide rubber and a lead peroxide curing agent. They are heavy pastes of 75 to 95% solids. **Koplac 1251-5,** of Koppers Co., Inc., for automobile body patching and general utility putties, is a low-viscosity polyester resin with 32% styrene. In putty formulations it is used with 50% talc and a catalyst. It hardens rapidly at room temperature.

PYRARGYRITE. An ore of silver, known also as dark **ruby silver.** It is a sulfantimonite of silver, Ag_3SbS_3, containing 22.3% antimony and 59.8 silver. It is found in various parts of Europe and in Mexico, Colorado, Nevada, and New Mexico. It occurs in crystals or massive, and in grains. Its Mohs hardness is 2.5 and specific gravity 5.85. The color is dark red to black, showing ruby red in thin splinters. **Proustite** is another ore of silver occurring in silver veins associated with other metals. It is found in the mines of Peru, Mexico, and Chile, and in Nevada and Colorado. It is also called *light ruby silver* and is a sulfarsenite of silver of composition Ag_3AsS_3, containing theoretically 65.4% silver. It occurs massive, compact, in disseminated grains. The hardness is 2 to 2.5, specific gravity 5.55, and the color is ruby red with an adamantine luster.

PYRETHRUM. The dried flowers of several species of chrysanthemum, of which *Chrysanthemum cinerariaefolium* of Yugoslavia and Japan is the best known. It is a slender perennial, about 15 in (38 cm) high, with daisylike flowers. The powder is used as an insecticide chiefly in sprays. The crude pyrethrum from Kenya contains 1.3% **pyrethrin** as compared with only 0.9% in the Japanese. **Persian powder** is pyrethrum from the species *C. coccineum* of southwestern Asia. **Lethane,** of Rohm & Haas Co., is a synthetic aliphatic diacyanate used as a substitute for pyrethrum. It is 30 to 40% more powerful than pyrethrum in insect sprays. Pyrethrin contains **pyrethronic acid** and **cyclopentane,** $(CH_2)_5$. The active principles of natural pyrethrum flowers have been designated as pyrethrin and **cinerin,** and the synthetic material marketed is a homolog of cinerin. In high concentration, it is more effective than natural pyrethrum. A substitute for pyrethrum for the control of corn worms is **styrene bromide,** or **bromostyrene,** an oily liquid of composition $C_6H_5CBr:CH_2$. **Ryanodine,** of composition $C_{25}H_{35}NO_9$, is 3 times more toxic than pyrethrin. It is extracted from the stem wood of the **ryania,** a shrub of Trinidad.

PYROPHORIC ALLOYS. Metals which produce sparks when struck by steel, used chiefly for gas and cigarette lighters. The original pyrophoric alloy, or **sparking metal,** was known as **Auer metal.** It was patented by Auer von Welsbach in 1903, and contained 35% iron and 65 mischmetal. The French **kunheim metal** contained 10% magnesium and 1 aluminum instead of iron. A very durable alloy for cigarette lighters is **zirconium-lead alloy** containing 50% of each metal. Titanium can replace part of the zirconium, and tin can replace part of the lead, but alloys with less than 25% zirconium are not pyrophoric. The 50–50 alloy has a crystalline structure. Some liquids are pyrophoric. **Trimethyl aluminum,** $Al(CH_3)_3$, a colorless liquid made by sodium reduction of methyl aluminum chloride, is used as a pyrophoric fuel.

PYROPHYLLITE   An aluminum silicate mineral found in North Carolina, used as a substitute for talc. It is similar to talc in structure and appearance, but its composition, $Al_2Si_4O_{10}(OH)_2$, is more nearly like that of kaolin. It is white, gray, or brown, with a pearly or greasy luster, specific gravity 2.8, and Mohs hardness 1 to 2. Compact varieties of the mineral are made into slate pencils and crayons. A fine-grained compact rock mined in South Africa, composed of about 90% pyrophyllite, with rutile and other minerals, is called **wonderstone** and is used for tabletops and switchboard panels. It is resistant to weathering, acids, and heat, and it can be planed, sawed, or turned in the lathe. It then becomes harder on exposure. Wonderstone is an ancient indurated clay resembling fireclay in which the colloidal matter has been destroyed by heat, pressure, and age. Unfired refractory bricks are made of dry-pressed pyrophyllite. They have high spalling resistance and do not shrink. **Pyrax,** of R. T. Vanderbilt Co., is pyrophyllite in fine, white powder of 100 mesh, with specific gravity of 2.6, used as a filler in rubber. **Silical** is pyrophyllite from Newfoundland in fine-powder form as a dusting talc for rubber. The Japanese employ great quantities of pyrophyllite in the making of firebrick, fireclay, and crucibles. The mineral used averages of 86.7% pyrophyllite, 12.8 kaolin, and 0.5 diaspore.

QUARTZ.   The most common variety of silica. It occurs mostly in grains or in masses of a white or gray color, but often colored by impurities. Pure crystalline quartz is colorless and is called **rock crystal.** Quartz usually crystallizes in hexagonal prisms or pyramids. Many crystals are obtained from nodules, called **geodes,** which are rounded, hollow rocks with the crystals grown on the inside surface of the cavity. These rocks range in size from very small to hundreds of pounds. The crystals are not always quartz, but may be grown from minor constituents of the rocks. A geode in limestone usually has a shell of silica, and the interior crystals are of quartz or calcite, but some geodes contain crystals of gem quality containing metal coloring constituents. The hard, rigid **beta quartz crystals** have a latticelike molecular structure in which each silicon atom is linked to four separate oxygen atoms, each oxygen atom being linked to two different silicon atoms. The formula of quartz crystal, therefore, is $(Si_2O_7)_x$, which is a pattern of the lattice, while the unit crystal of silica is SiO_2 and the silica grains are cryptocrystalline. The grains in sand are often less than 0.04 in (0.10 cm), but crystals up to 20 in (51 cm) have been found. The specific gravity is 2.65. Pure crystals have a dielectric strength of 1,500 V/mil (59×10^6 V/m) and a dielectric constant of 3.8, with good corona resistance.

Quartz crystals have the property of generating an electric force when placed under pressure and, conversely, of changing dimensions when an electric field is applied. This property is termed *piezoelectric*. A **piezoelectric crystal** is made up of molecules that lack both centers and planes of symmetry. Many materials other than quartz have this property, such as rochelle salt and ammonium dihydrogen phosphate, but most are water-soluble or lack hardness. The best quartz crystals are hexagonal prisms with three large and three small cap faces. For electric use the crystals must have no bubbles, cracks, or flaws, and they should be free from twinning, or change in the atomic plane. **Brazilian quartz** crystals are cut into plates of different sizes to initiate and receive various frequencies on multiple-message telephone wires, and to obtain selectivity in radio apparatus. Owing to its peculiar refractive powers, quartz crystal is also employed for the plates in polarization instruments and in lenses. Quartz crystals for radio-frequency control are marketed in three forms: rough-sawed blanks, cut to specified angles; semifinished blanks, machine-lapped to approximate size; and electrically finished blanks, finished by hand and electrically tested. **Synthetic quartz crystals** of large size and high purity and uniformity are grown from seed crystals suspended in an alkaline silica solution at high temperature and pressure. The synthetic crystals are purified by imposing an electric current across the crystal at 932°F (500°C), which sweeps out the sodium and lithium impurities by electrolysis. The addition of lithium nitride to the sodium hydroxide solution used in the hydrothermal growing process increases the Q value of the crystal to the range of natural quartz crystals. These synthetic crystals are used in precision oscillators and highly selective wave filters.

Barium titanate crystals are used to replace quartz for electronic use. **Ethylenediamine tartrate** crystals may be used to replace quartz for telephone and sonar work.

Quartz is harder than most minerals, being Mohs 7, and the crushed material is much used for abrasive purposes. Finely ground quartz is also used as a filler, and powdered quartz is employed as a flux in melting metals. When quartz is fused, it loses its crystalline structure and becomes a **silica glass** with a specific gravity of 2.2, compressive strength 210,000 lb/in^2 (1,448 MPa), tensile strength 4,000 lb/in^2 (28 MPa), Mohs hardness 5, and dielectric strength 410 V/mil (16×10^6 V/m). The chemical formula of this material is sometimes given as SiO_3, but is really SiO_2 repeated in a lattice structure but different from that of quartz crystal. **Fused quartz,** or **quartz**

glass, is used for bulbs, optical glass, crucibles, and tubes and rods in furnaces. Its softening and working temperature is about 3040°F (1671°C), and it fuses at 3193°F (1755°C). The translucent material, made from sand, has a specific gravity of 2.7, with much lower strength. It withstands rapid changes of temperature without breaking. Fused quartz made from rock crystal is transparent to visible light, while fused silica is normally translucent or opaque. **Vitreosil** is fused quartz, containing 99.8% silica. It comes opaque, translucent, and transparent. It transmits ultraviolet and short wavelengths, has high electrical resistance, and has a coefficient of expansion about one-seventeenth that of ordinary glass. **Quartz tubing** for electronic use comes in round, square, hexagonal, and other shapes. The softening point is 3033°F (1667°C). Tubing as small as 0.003 in (0.008 cm), produced by Monsanto Co., is flexible and as strong as steel.

Quartz fiber originally was made by extruding the molten quartz through a stream of high-pressure hot air which produced a fluffy mass of fine fibers of random lengths. Quartz fibers are now made with many differing compositions and methods of manufacture. Fibers used for wool or mat have a diameter of 39 to 591 μin (1 to 15 μm). Those used for continuous filament may be as small as 0.0035 in (0.009 cm). **Astroquartz,** of J. P. Stevens & Co., may be corded as a thread with outside diameters from 0.014 to 0.020 in (0.036 to 0.051 cm). The thread is used for sewing insulation blankets and separation curtains. **Quartz yarn** made from these filaments is used for weaving into tape and fabric. **Quartz paper,** or **ceramic paper,** developed by the Naval Surface Weapons Center and used to replace mica for electrical insulation, is made from quartz fiber by mixing with bentonite and sheeting on a papermaking machine. It has high dielectric strength and withstands temperatures to 3000°F (1649°C). **Micro quartz** is felted, fine quartz fibers for insulation. The felted material has a density of 3 lb/ft^3 (48 kg/m^3) and is capable of service temperatures to 2000°F (1093°C).

Since quartz crystallizes more slowly than many other minerals, the natural crystals may include other minerals which were crystallized previously. **Sagenite** is a form of crystalline quartz containing hairlike crystals crossing in a netlike manner. A variety of fibered quartz with a pale-amethyst color which shows deep red by transmitted light, found in Russia and Colorado, is called **onegite. Rutilated quartz** is clear quartz penetrated by rutile crystals. A smoky, dark quartz of this type is the **Venushair stone. Aventurine** is a form of quartz crystal containing the inclusion in the form of flakes or spangles. It comes from the Ural Mountains and from India and is prized for gems. For costume jewelry it is made synthetically in great quantities under the name of **goldstone** by melting the inclusions into quartz glass.

Amethyst, topaz, and many other gemstones are quartz. The golden-yellow topaz of Mexico and Brazil is a type of quartz called **citrine.** The yellow variety called **imperial topaz** in Brazil is rare, but yellow-brown stones are common. **Pink topaz** is also rare, but can be made by heating yellow-brown stones with a risk of breaking. Inferior-colored amethysts may also be made into yellow or orange citrine by heating. The **rose quartz** of South Dakota is prized in the beautiful rose color, but in the large deposits the shades may run from milky white through pale pink to deep rose-red. The best stones are used for gems, as are also the translucent pink crystals from Maine. Other grades are cut into vases, ornaments, and architectural facings. **Chalcedony** is a cryptocrystalline quartz with a waxy luster deposited in rock veins from colloidal solution, or in concentric rings on rocks. Its fibers are biaxial instead of the uniaxial of quartz. The chalcedony of South Dakota known as **beckite** fluoresces under ultraviolet light. Chalcedony was an ancient gemstone and was used for intaglios and seals and for figurines and vases. Some chalcedony from New Mexico and Arizona is stained and cut for costume jewelry. **Chrysoprase** is a translucent, apple-green variety of cryptocrystalline quartz colored with hydrated nickel silicate found in Silesia. It is highly valued for mural decorations and as a gemstone.

The so-called **massive topaz** used as a refractory material instead of kyanite is not true topaz or quartz. The massive topaz mined in North Carolina contains about 50% Al_2O_3 and 40 SiO_2, with iron oxide. When calcined for refractory use, it has the same composition as kyanite. The topaz from the wolframite mines of São Paulo, Brazil, used for refractories, has a high alumina content and a high fluorine content. The purer crystals have a melting point of 3416°F (1880°C). The quartz known as **cristobalite,** used as a refractory, differs from ordinary quartz only in crystal structure. It has a melting point of 3140°F (1725°C). **Jasper** is a variety of quartz colored red with iron oxide. It is cut and polished as an ornamental building stone. **Egyptian jasper** is brown with dark zones. In ancient times many of the gemstones were silica stones, and the **Athiaenon stone** from Cyprus was jasper of bright colors. The **jasper iron ore** of Michigan has an iron content of about 33% with less silica than taconite, making it easier to crush, but concentration must be done by flotation, which is more expensive than the magnetic separation of taconite.

Quartzite is a rock composed of quartz grains cemented together by silica. It is firm and compact and breaks with uneven, splintery fractures. Most of the quartzites used are made up of angular grains of quartz and are white or light in color with a glistening appearance. It often resembles marble, but is harder and does not effervesce in acid. Quartzite is employed for making silica brick, abrasives, and siliceous linings for tube mills. It is also rather widely used as a structural stone and as a broken stone for roads. It is found as a widely distributed common rock. **Medina quartzite,** from Pennsylvania, contains 97.8% silica. The melting point is about 3092°F (1700°C).

QUASSIA.   Also known as **bitterwood.** The wood of the Jamaica quassia tree, *Picroena excelsa,* and of the **Surinam quassia,** *Quassia amara,* of the West Indies and northern South America. The **Jamaica quassia** is a large tree, sometimes called **bitter ash** because the leaves resemble those of the common ash. The wood is yellow, light, dense, and tough. It is odorless, but has an intensively bitter taste. The wood is imported mostly as chips for the production of the extract which is used in medicine as a bitter tonic, and in insecticides. It is also used as an ingredient in stock-feed tonics for cattle. In tropical countries the wood is valued for furniture, as it is resistant to insects. The wood of the Surinam quassia is darker in color, heavier, and harder, but has similar properties. **Quassin,** extracted from quassia, is used to denature alcohol.

QUEBRACHO.   The wood of the **quebracho colorado,** or **red quebracho,** tree, *Aspidospera quebracho,* found only along the west bank of the Parana and Paraguay rivers in Argentina and Paraguay. It contains about 24% tannin. The wood is exceedingly hard and has a brownish-red color, often spotted and stained almost black. Quebracho is valued as a firewood in Argentina, and is used for crossties and posts, but is too brittle for structural work. It takes a fine polish and is very durable, carvings of this wood being in perfect condition after 300 years. The density is 78 lb/ft^3 (1,250 kg/m^3). **Quebracho extract,** from the wood, is a hard, resinous, brownish-black, and extremely bitter solid containing 62% soluble tannins. One and one-tenth ton (1 metric ton) of wood yields about 551 lb (250 kg) of solid extract. The liquid extract contains 25 to 35% tannin. It is employed in tanning leather and is rapid-acting, but is seldom used alone, as it makes a dark leather. It is mixed with alum and salt, or with chestnut extract. Some extract is used in boiler compounds, but one of the larger uses has been for the treatment of oil-well-drilling muds. **Aerosol Q,** of American Cyanamid Co., is powdered quebracho and an organic colloid for oil-well muds. **White quebracho,** *Schinopois lorentsii,* is a smaller tree than the red quebracho, growing over a wider area of Argentina, Brazil, and Paraguay. It produces a similar tannin. Some **urunday extract** is produced in Argentina for export instead of quebracho. The **urunday wood** is red and very hard, but not as brittle as quebracho, and is valued for cabinetwork. The tannin from the wood is similar to quebracho extract.

RADIOACTIVE METALS. Metallic elements which emit radiations that are capable of penetrating matter opaque to ordinary light. They give out light and appear luminous, also having an effect on photographic plates. The metal radium is the most radioactive of all the natural elements, and was much used for luminous paints for the hands of watches and instrument pointers. Because of the emission of dangerous gamma rays, however, it has been replaced for this purpose by radioactive isotopes of other metals. These isotopes, such as **cobalt 60,** used as a source of gamma rays, and **krypton 85,** for beta rays, are marketed selectively. Radioactive metals are used in medicine, for luminous paints, for ionization, for breaking particle bonds in powdering minerals, for polymerization and other chemical reactions, and for various electronic applications.

The metals which are naturally radioactive, such as uranium and thorium, all have high atomic weights. The radiating power is atomic and is unaffected in combinations. **Radium** and other radioactive metals are changing substances. Radium gives out three types of rays; some of the other elements give out only one or two. The measure of the rate of radioactivity is the curie, which is the equivalent of the radioactivity of 0.0022 lb (1 g).

Each radioactive metal has a definite breakdown period, measured in half-life. **Actinium,** which is **element 89,** has a half-life of 21.7 years. It emits alpha particles to decay to **actinium K,** which is the radioactive isotope of **francium,** and then emits beta particles. Radioactive metals break down successively into other elements. By comparison of changing atomic weights, it has been deduced that the metal lead is the ultimate product, and uranium the parent metal under existing stability conditions. But heavier metals, now no longer stable under present conditions, have been produced synthetically, notably plutonium. The heavy **element 103** was first produced in 1961 and named **lawrencium** in honor of the inventor of the cyclotron. Not all radiation produces radioactive materials, and by controlled radiation useful elements may be introduced into alloys in a manner not possible by metallurgy. The crystal lattice of an alloy can be expanded, or atoms displaced in the lattice, thus altering the properties of the alloy. In like manner, the molecules of plastics may be cross-linked or otherwise modified by the application of radiation. For example, ethylene bottles may be irradiated after blowing to give higher strength and stiffness. Radioactive isotopes are also used widely in chemistry and in medicine and as sources of electric power.

RADIUM. The best-known radioactive metal, symbol Ra, scattered in minute quantities throughout almost all classes of rocks, but commercially obtainable only from the uranium ores monazite, carnotite, and uraninite. It is a breakdown product, and it disintegrates with a half-life of 1,590 years. The metal is white, but it tarnishes rapidly in air. The melting point is about 1292°F (700°C). It was discovered in 1898 by Curie, and the original source was from the pitchblende of the Sudetenland area of Austria after extraction of thorium oxide, but most of the present supply comes from the carnotite of Zaire and from the pitchblende of western Canada. One gram of radium and 7,800 lb (3,538 kg) of uranium are obtained from 370 tons (336 metric tons) of pitchblende. The ratio of radium to uranium in any uranium ore is about 1:3,000,000. Radium is marketed in the form of bromides or sulfate in tubes and is extremely radioactive in these forms.

In a given interval of time, a definite proportion of the atoms break up with the expulsion of α, ß, and γ rays. When an alpha particle is emitted from radium, the atom from which it is emitted becomes a new substance, the inert gas **radon,** or **element 86,** with a half-life of 3.82 days. During its short life, it is a definite elemental gas, but it deposits as three isotopes in solid particles, decaying through polonium to lead. Radium is most widely known for its use in therapeutic medicine. It is also used for inspecting metal castings for flaws. **Radium-beryllium powder** is marketed for use as a neutron source.

RAMIE. A fiber used for cordage and for various kinds of coarse fabrics, obtained from the plant *Urtica nivea,* of temperate climates, and *U. tenacissima,* of tropical climates. The former plant has leaves white on the underside, and the latter has leaves all green. The name **rhea** is used in India to designate the latter species. It is also grown in China, Egypt, Brazil, and Florida. The plants grow in tall, slender stalks like hemp and belong to the nettle family. The bast fibers underneath the bark are used, but are more difficult to separate than hemp fiber owing to the insolubility of the adhesive gums. The fibers are 8 times stronger than cotton, 4 times stronger than flax, and nearly 3 times stronger than hemp. They are fine and white and are as silky as jute. They are not very flexible and are not in general suitable for weaving, but their high wet strength, absorbent qualities, and resistance to mildew make the fibers suitable for warp yarns in wool and rayon fabrics. The yarn is used also for strong, wear-resistant canvas for such products as fire hose. The fiber is valued for marine gland packings and for twine. The composition is almost pure cellulose, and the tow and waste are used for making cigarette paper. **China grass** is the hand-cleaned but not degummed fiber. It is stiff and greenish yellow. **Grass cloth** is woven fabric made in China from ramie. **Swatow grass cloth,** imported into the United States, is made of ramie fibers in parallel strands, not twisted into yarns.

RAPE OIL. Also known as **rapeseed oil, colza oil,** and recently **canbra oil.** An oil obtained from seeds of the mustard family, Cruciferae. The genus *Brassica,* a form of turnip, species of which are referred to as *B. campestris, B. rapa, B. napus,* and *B. hirta,* is grown in India, Pakistan, Europe, and Canada. **Rapeseed** is one of the principal oil seeds of the world. It is widely used as an edible oil, for making factice, and for mixing with lubricating and cutting oils and for quenching oils. The seeds are very small, with 1 oz (28 g) having as many as 40,000 seeds. The seeds contain 40% oil. The edible oil is cold-pressed and refined with caustic soda. The burning and lubricating oils are refined with sulfuric acid. The refined oil has a pungent, mustardlike odor that can be removed by deodorization. The iodine value is about 100, the specific gravity 0.915, and the flash point 455°F (235°C). The oil contains palmitic, oleic, linoleic, and stearic acids and 43 to 50% of the typical acid, **erucic acid,** also called **brassidic acid,** $C_{21}H_{41}COOH$. It has a melting point of 93°F (34°C). It occurs also in grape seed oil. For edible oils, the erucic acid is reduced, generally to less than 5%; the high-erucic-acid oils are used industrially as lubricant additives. Genetic variants with no erucic acid have also been made. Edible low-erucic-acid **canola oils** for food applications are made by Agro Ingredients, Inc. Colza oil is a rape oil extracted from French seed, used to mix with mineral oils to make cutting oils. The name *colza* now refers to any refined rape oil. **Chinese colza oil,** from *B. campestris chinoleifera,* contains the mustard volatile oil. The specific gravity is 0.91, saponification value 174, and iodine number 100.3. From 15 to 20% of blown rapeseed oil is mixed with mineral oil for lubricating marine engines. **Crambe seed oil,** from *Crambe abyssinica,* an Asiatic mustard, contains 55 to 60% erucic acid. The erucic can be broken down to perlargonic acid used as a substitute for dibasic acids such as azelaic and brassylic acids. **Cameline oil,** called also **dodder oil** and **German sesame oil,** has the same uses as rape oil. It is from the plant *Camelina sativa* grown in central Europe. The seeds contain 35% oil which contains oleic and palmitic acids and erucic acid. The seed itself is high in mineral and protein content and is used in birdseed mixtures.

RAPID PROTOTYPING MATERIALS. Materials used primarily to rapidly create precise models of prototype parts using computer-based systems and computer-aided-design data. They are also used to make tools and limited quantities of parts.

Several systems are laser-based. Stereolithography, of 3D Systems, Inc., involves sequential curing of a liquid **photopolymer** on a descending platform by an ultraviolet laser beam. Two such polymers are Ciba Geigy's **Cibatool 5170** epoxy for use with helium-cadmium lasers and **Cibatool 5180** for use with argon lasers. **Exactomer** resins, of Allied Signal, use **vinyl ethers** and cationic photoinitiators to start polymerization. DTM Corp. uses a low-power CO_2 laser beam to trace part outlines in a thin layer of powder and to sinter the powder particles. Again, layer upon layer of material is built up on a descending platform until the model is completed. Materials, called **Laserite,** include **investment-casting wax, nylon,** and **polycarbonate.** This system can also be used to form **iron-matrix-composite** tools. In this case, thermoplastic-coated carbon-steel powder particles are tacked in place and transferred to a furnace where the coating is burned off and the powder sintered to a porous shape, and the shape infiltrated with copper. A similar laser is used by Helisys Inc. in what is called *laminated object manufacturing.* Here the laser cuts thin paper or film, dispensed on a roll, to the part outline, and the cut pieces are deposited on a descending table until the model is completed. The materials are heat-activated, adhesive-coated, bleached **kraft paper** or polyester film in various colors. The paper results in a model resembling wood. The film creates water-resistant models.

In fused deposition modeling, by Stratasys, Inc., a thermoplastic or wax filament from a spool is heated, extruded, and deposited in thin layers onto a base by a robot-held dispensing head. The model is built from the base up. The materials, which include **machinable wax, investment-casting wax, Plastic P200** polyolefin, and nylonlike **Plastic P300** and **Plastic P301,** are heated just sufficiently to flow, and solidify instantly upon deposition. **Acrylonitrile-butadiene-styrene** parts, having a tensile strength of 3,000 lb/in^2 (21 MPa), and **polyester** parts also have been made. A robotic extrusion system developed by IBM accepts thermoplastic pellets, including an elastomer and a machinable nylonlike material. In solid ground modeling, by Cubital Ltd. (Israel) and Cubital America, successive layers of a liquid photopolymer are exposed and cured by an ultraviolet lamp through a glass mask generated ionographically using toner and representing part cross sections. The exposed resin is then removed, and the cavity filled with **water-soluble wax.** The layer is cooled and milled to accurate thickness, and the next layer is formed. When all the layers have been built up, the wax is dissolved, leaving the model shape. Direct shell production casting, developed at the Massachusetts Institute of Technology and marketed by Soligen Technology Inc., is based on three-dimensional printing. This system bypasses the modeling stage and is used directly to make metal-infiltrated, **ceramic-matrix-composite** cores and molds for casting parts. A jet, similar to an ink jet but using a **colloidal silica** binder, sprays successive patterns of the part shape onto **alumina** powder, which is compacted between powder charges. When the final layer is formed, the partially consolidated shape is removed, the binder burned off, and the shape sintered and infiltrated with metal. Aluminum and stainless-steel parts have been cast using molds and cores made in this way.

RARE-EARTH METALS. A group of trivalent metallic elements that occur together. They are also called **rare earths,** because of the difficulty of extracting them, not because of their rarity. They include elements 57 through 71, from lanthanum to luterium, and **yttrium, element 39,** and **thorium, element 90,** because these are also together in monazite, the chief ore. The **cerium metals** are a group of rare-earth metals consisting of elements with atomic numbers 57 through 63, including the metal **cerium.** This group is also referred to as the **light rare earths.** The metal **ytterbium** (atomic number 70) also may be included in this group because of its light weight.

Thorium is separated by a relatively easy process, and the others remain grouped as the cerium metals, to be extracted as metals or compounds for special purposes justifying high costs. The separate metals are regularly marketed in pellets and in 325-mesh powder of 99.9% purity for pyrophoric and electronic uses, and as oxides of 99.9% purity. Cerium metal has an iron-gray color, is only slightly harder than lead, and is malleable. It has a specific gravity of 6.77 and a melting point of 1480°F (804°C).

After extraction of the thorium oxide from monazite, the chief rare-earth ore, the residual matter is reduced by converting the oxides to chlorides and then removing the metals by electrolysis. The product obtained is an alloy containing about 50% cerium together with lanthanum, didymium, and the other rare-earth metals. It is usually called **mischmetal,** the German name for mixed metal, and its original use was for making pyrophoric alloys. **Cerium standard alloy** of Cerium Metals Corp. is a mischmetal containing 50 to 55% cerium, 22 to 25 lanthanum, 15 to 17 neodymium, and the balance a mixture of yttrium, terbium, illinium, praseodymium, and samarium, with 0.5 to 0.8 iron.

Mischmetal is used in making aluminum alloys and in some steels and irons. In cast iron it opposes graphitization and produces a malleable iron. It removes the sulfur and the oxides and completely degasifies steel. In stainless steel it is used as a precipitation-hardening agent. An important use of mischmetal is in magnesium alloys for castings. From 3 to 4% of mischmetal is used with 0.2 to 0.6 zirconium, both of which refine the grain and give sound castings of complex shapes. The cerium metals also add heat resistance to magnesium castings.

Ceria, cerium oxide, or **ceric oxide,** CeO_2, is a pale-yellow, heavy powder of specific gravity 7.65, used in coloring ceramics and glass for producing distortion-free optical glass. It is used also for decolorizing crystal glass, but when the glass contains titania, it produces a canary-yellow color. **Cerous oxide,** Ce_2O_3, is a greenish powder of specific gravity 7.0 and refractive index 2.19. About 3% of the oxide in glass makes the glass completely absorbent to ultraviolet rays. It is also an excellent opacifier for ceramics. **Cerium fluoride,** CeF_3, is used in arc carbons to increase brilliance. **Cerous nitrate,** $Ce(NO_3)_3 \cdot 6H_2O$, is a red, crystalline powder used in gas-mantle manufacture. Cerium salts are used for coloring glass. **Ceric titanate,** $Ce(TiO_3)_2$, gives a golden-yellow color, and **ceric molybdate** gives a blue color.

Neodymium has a specific gravity of 7.01 and a melting point of 1875°F (1024°C). It is used in magnesium alloys to increase strength at elevated temperatures and is used in some glasses to reduce glare. **Neodymium glass,** containing small amounts of neodymium oxide, is used for color television filter plates since it transmits 90% of the blue, green, and red light rays and no more than 10% of the yellow. It thus produces truer colors and sharper contrasts in the pictures and decreases the tendency toward gray tones. Neodymium is also a dopant for **yttrium-aluminum-garnet,** or **YAG,** lasers as well as for glass lasers. **Praseodymium** has a specific gravity of 6.77 and a melting point of 1715°F (935°C). **Lanthanum** is a white metal, malleable and ductile, with a specific gravity of 6.16, and melts at 1688°F (919°C). Like the other cerium metals, it oxidizes easily in air and is easily soluble in acids. **Lanthanum oxide,** La_2O_3, is a white powder used for absorbing gases in vacuum tubes. **Lanthanum boride,** LaB_6, is a crystalline powder used as an electron emitter for maintaining a constant, active cathode surface. It has high electrical conductivity.

Didymium is not an element, but is a mixture of rare earths without cerium. It averages 45% La_2O_3; 38 **neodymium oxide,** Nd_2O_3; 11 **praseodymium oxide,** Pr_6O_{11}; 4 **samarium oxide,** Sm_2O_3; and other oxides. It is really the basic material from which the rare metals are produced. In glass it gives a neutral gray color, and it is used in glass for welders' goggles, as it absorbs yellow light and reduces glare and eye fatigue. It is available as **didymium carbonate,** a pink powder soluble in acids; as **didymium oxide,** a brown acid-soluble powder; and as **didymium chloride** in pink lumps soluble in water and in acids.

Dysprosium has a specific gravity of 8.56 and a melting point of 2700°F (1482°C). Its corrosion resistance is higher than that of other cerium metals. It also has good neutron-absorption ability, with a neutron cross section of 1,100 barns. The metal is paramagnetic. It is used in nuclear reactor control rods, in magnetic alloys, and in ferrites for microwave use. It is also used in mercury-vapor lamps. With argon gas in the arc area, it balances the color spectrum and gives a higher light output. **Samarium** has a higher neutron cross section, 5,500 barns, and is used for neutron absorption in reactors. Samarium has a specific gravity of 7.54 and a melting point of 1925°F (1052°C).

Ytterbium metal is produced in lumps and ingots. It has a specific gravity of 6.96 and a melting point of 1515°F (824°C). **Yttrium** is more abundant in nature than lead, but is difficult to extract. It is found associated with elements 57 to 71, although its atomic number is 39. It has a silvery luster, a specific gravity of 4.47, and a melting point of 1550°F (843°C). It is the lightest of the cerium metals except scandium. The metal is corrosion-resistant to 752°F (400°C). It has a hexagonal, close-packed crystal structure. **Ytterbium oxide,** Yb_2O_3, and **yttrium oxide,** Y_2O_3, are the usual commercial forms of these metals. The two metals occur in the mineral **gadolinite, or ytterbite,** $4BeO \cdot FeO \cdot Y_2O_3 \cdot 2SiO_2$, which also contains **gadolinium, erbium, europium, holmium,** and **rhenium.** The mineral is found in Greenland, Sweden, Norway, and Colorado. Erbium has been obtained only in small quantities as a dark-gray powder. The metal has a specific gravity of 9.06 and a melting point of about 2700°F (1482°C). It forms the rose-red **erbium oxide,** or **erbia,** Er_2O_3, and other, highly colored, reddish salts. At high temperatures erbia glows with a greenish light.

Yttrium also occurs in the scarce mineral **nuevite** found in California. The mineral also contains titanium, tantalum, iron, and quartz, and is similar to the **keilhauite** found in Norway. **Fergusonite,** a brown mineral with a vitreous fracture, found sparsely in the Appalachian hills from New England to South Carolina, is a columbate and tantalate of yttrium with cerium, erbium, and uranium. In Europe it is known as **bragite** and **tyrite.** The mineral known as **bastnasite** in California is a fluorocarbonate of cerium and lanthanum, and the deposit at Mountain Pass, California, is sufficient to supply the commercial needs of all the cerium metals. About 0.01% of rare-earth metals remain in the waste after apatite ores are processed in the making of phosphoric acid fertilizer. These metals are extracted by solvent or ion-exchange methods. Yttrium oxide is available as a fine, white powder. **Yttralox** is a transparent ceramic made from yttrium oxide and has a melting point above 4000°F (2204°C). It is used for special high-temperature lenses, infrared windows, lasers, and high-intensity lamps. For the brilliant reds for television phosphors, small amounts of europium are added to **yttrium vanadate.**

Gadolinium oxide, or **gadolinia,** Gd_2O_3, has high neutron absorption and is used for shields in atomic power plants. As a molded ceramic, it has a specific gravity of 7.0, low thermal conductivity, and a melting point of 4262°F (2350°C). **Lutetium** is a heavy refractory metal with a specific gravity of 9.85 and a melting point of about 3100°F (1704°C). **Lutetium oxide,** Lu_2O_3, of 99% purity, is produced for atomic uses. **Holmium** has a specific gravity of 8.8 and a melting point of about 2700°F (1482°C). It is used in glass to transmit radiant energy for wavelength-calibration instruments. **Thulium** is a heavy metal with a specific gravity of 9.32 and a melting point of about 2800°F (1538°C). It is used for radiographic applications. **Thulium 170** is a soft gamma-ray emitter and is used as a radiation source. **Thulium oxide,** Tm_2O_3, is radioactive and is used as a power source for small thermoelectric devices. **Scandium** is a silvery-white metal found in the mineral **thortveitite,** $(ScYt)_2Si_2O_7$, of Norway, which contains 42% **scandium oxide,** Sc_2O_3. It also occurs in small amounts in the fergusonite of Montana, and in lepidolite, muscovite, beryl, and the amphiboles. The metal has a melting point of 2552°F (1400°C) and a specific gravity slightly higher than that of aluminum.

RARE GASES.   Also known as **inert gases** in the metallurgical industry, and as **noble gases.** A general name applied to the five elements helium, neon, argon, krypton, and xenon. They are rare in that they are highly rarified gases at ordinary temperatures and are found dissipated in minute quantities in the atmosphere and in some substances. All have zero valence and normally make no chemical combinations, but by special catalyzations, except in the case of **helium,** the outer proton bonding may be broken and compounds produced. The rare gases are colorless, odorless, and tasteless at ambient temperatures. However, they exhibit very different properties when cooled to extremely low temperatures. When saturated helium, or **helium 4,** is cooled to below 2.17 K, it becomes a **superfluid.** One unique property of superfluids is the ability to pass undetected through very small openings. **Helium 3,** the unsaturated counterpart, differs in that it is magnetic.

Neon is procured from the air by liquefaction. When it is energized, it emits light and is used for signs and in glow lamps. The specific gravity, compared with air, is 0.674. It liquefies at −414°F (−248°C). It is colorless, but gives a reddish-orange glow in lamps to which an electric current is applied. Neon is also used in voltage-regulating tubes for radio apparatus and responds to low voltages. In television the neon lamp gives fluctuations from full brilliancy to total darkness as many as 100,000 times per second. Colored electric advertising signs are often referred to as neon signs, but the colors other than orange are produced by different gases. **Argon** gives a purple light when an electric current is passed through it. It occurs free in the atmosphere to the extent of 0.935%. Its liquefying point is about $-305°F$ ($-187°C$). It is obtained by passing atmospheric nitrogen over red-hot magnesium, forming magnesium nitride and free argon. It is also obtained by separation from industrial gases. Argon is employed in incandescent lamps to give increased light and to prevent vaporization of the filament. It is also used as a shielding gas in welding, as an assist gas in laser cutting, as an inert blanket for nuclear fuels, and, at high pressure, in hybrid-type airbag inflators for cars.

Krypton, which occurs in the air to the extent of 1 part in 1 million, gives a pale-violet light. It is a heavy gas, with a specific gravity of 2.896. It is used as a filler for fluorescent lamps to decrease filament evaporation and heat loss and to permit higher temperatures in the lamp. The $3×10^9$ candlepower aircraft-approach lights first used on the Berlin airlift contained krypton. **Krypton 85,** obtained from atomic reactions, is a beta-ray emitter with a half-life of 9.4 years. It is used in luminous paints for activating phosphors and as a source of radiation. It comes combined in solid form with a hydroquinone to give higher concentration of energy and more convenience in use and in disks of 0.23 and 0.5 in (0.58 and 1.27 cm) in diameter encased in acrylic plastic for use with phosphors as luminous sources. The light is yellow-green. **Xenon,** another gas occurring in air to the extent of 1 part in 11 million, gives a sky-blue to green light. It is the heaviest of the rare gases, with a specific gravity of 4.561 compared with 1 for air. Its liquefying point is about $-162°F$ ($-108°C$). When atomic reactors are operated at high power, xenon tends to build up as a reaction product, poisoning the fuel and reducing the reactivity. **Xenon lamps** for military use give a clear, white light known as *sunlight* plus north-sky light. This color does not change with the voltage, and thus the lamps require no voltage regulators. An 800-W xenon lamp delivers 2,000 lm, 4 times as much as a 1,000-W incandescent lamp. The xenon lamp reflects each half cycle, so that shutterless projectors are possible. Krypton and xenon have lower thermal conductivity and lower electrical resistance than argon. A helical arc of xenon is used to activate ruby optical masers. Xenon is a mild anesthetic, the accumulation from air helping to induce natural sleep, but it cannot be used in surgery since the quantity needed produces asphyxiation.

RARE METALS. A term given to metals that are rare in the sense that they are difficult to extract and are rare and expensive commercially. They include the elements **astatine, technetium,** and **francium.** The silvery metal technetium, element 43, has been produced by bombardment of molybdenum with neutrons. It is available in the form of

technetium carbonyl, $Tc_2(CO)_{10}$, which is stable in air and soluble in most organic solvents but reacts with halogens. Although **radium** is a widely distributed metal, it is classified as a rare metal. All of the ultra-heavy metallic elements, such as plutonium, which are produced synthetically, are classified as rare metals. They are called **transuranic metals** because they are above the heavy-metal uranium in weight. They are all radioactive.

Element 99, called **einsteinium,** was originally named **ekaholmium** because it appears to have chemical properties similar to holmium. It is produced by bombarding uranium 238 with stripped nitrogen atoms. It decays rapidly to form the lighter **berkelium,** or element 97. **Neptunium,** element 93, **californium,** element 98, and **illinium,** element 61, are also made atomically. The last also has the names **florentium** and **promethium.**

Plutonium is made from uranium 238 by absorption of neutrons from recycled fuel. The metal, 99.8% pure, is obtained by reduction of **plutonium fluoride,** PuF_4, or **plutonium chloride,** $PuCl_3$. It has a melting point of 147°F (64°C). The surface reacts in air to form the nonadherent **plutonium oxide,** PuO_2, which becomes airborne and is pyrophoric and poisonous. **Plutonium 238, 239,** and **240** emit chiefly alpha rays. Plutonium 238 has a low radiation level and is used as a heat source for small water-circulating heat exchangers for naval undersea diving suits. Heat generated by radioactive decay and converted to electrical energy has been used to power spacecraft.

Plutonium 241 emits beta and gamma rays. Since all the allotropic forms are radioactive, it is a pure nuclear fuel in contrast to uranium, which is only 0.7% directly useful for fission. It is thus necessary to dilute plutonium for control. For fuel elements it may be dispersed in stainless steel and pressed into pellets at about 1600°F (871°C), or pellets may be made of **plutonium carbide. Plutonium-iron alloy,** with 9.5% iron, melts at 770°F (410°C). It is encased in a tantalum tube for use as a reactor fuel. **Plutonium-aluminum** alloy is also used. These alloys have hard compounds of PuFe and PuAl in the matrix, and the plutonium is insoluble. While plutonium 241 has a half-life of only 14 years, the beta emitters plutonium 239 and 240 have half-lives of 24,300 and 16,600 years, respectively. Element 102, called **nobelium,** has a half-life of only 0.2 h. Other transuranic metals produced synthetically are **americium,** element 95, and **curium,** element 96. Curium is used as a heat source in remote applications. **Curium 244** is obtained as **curium nitrate** in the reprocessing of spent reactor fuel. It is converted to **curium oxide.** The by-product americium is used as a component in neutron sources. Other transuranic metals that have been produced by nuclear reactions and synthesis include **fermium** (element 100), **mendelevium** (element

101), **lawrencium** (element 103), **rutherfordium** or **kurchatovium** (104), and **hahnium** or **nielsbohrium** (105).

RATANY. Also known by the original Inca name of **payta,** and in medicine as **krameria.** The root of the shrub *Krameria triandria,* which grows in Peru and is used for tanning leather and in medicine as an astringent. The root comes in diameters up to 1 in (2.5 cm) and in pieces up to 3 ft (0.9 m) in length. It contains about 40% tannin which is extracted by hot water. It gives the leather a deep-brown color and is usually mixed with other tannins.

RATTAN. The long, slender stem of the palm *Calamus rotang* and other species, of Sri Lanka, Malaya, and Laos. There are more than 40 varieties. The Malay word is *rottan,* meaning cane. It is tough, flexible, strong, and durable, and it is used for canes, umbrella handles, and furniture. When split, it is used for car seats, baskets, baby-carriage bodies, furniture, whips, and heavy cordage. Commercial rattans are in pieces 5 to 20 ft (1.5 to 6.1 m) long. A substitute for rattan is **jacitará,** from the plant *Desmoncus macroacanthus,* of Brazil. It is used for seating. Vinylidene chloride plastic is now widely used as a substitute for rattan for seating.

RAYON. A general name for artificial-silk textile fibers or yarns made from **cellulose nitrate, cellulose acetate,** or cellulose derivatives. In general, the name *rayon* is limited to the viscose, **cuprammonium,** and acetate fibers, or to fibers having a cellulose base. Other synthetic fiber groups have their own group names, such as **azlon** for the protein fibers and nylon for the polymeric amine fibers, in addition to individual trade names.

Viscose rayon is made by treating the cellulose with caustic soda and then with carbon disulfide to form cellulose xanthate, which is dissolved in a weak caustic solution to form the viscose. With the cuprammonium process, the cellulose is digested in an ammonia solution of copper sulfate, and the solution is forced through the spinnerets into dilute acid for hardening. Rayons manufactured by the different processes vary both chemically and physically. They are resistant to caustic solutions which would destroy natural silk. They are also mildewproof, durable, and easily cleaned. But they do not have the permeability and soft feel of silk. The **acetate rayons** are more resistant than the viscose or cuprammonium. The lack of permeability of the fibers is partly overcome by having superfine fibers so that the yarns are permeable. The one-denier viscose staple produced by FMC Corp. by stretching the fiber after it leaves the spinnerets is finer than Egyptian cotton and can thus be made into yarns that are permeable between the fibers. Fabrics made from the superfine yarns have the appearance of sheer silk. **Multicell rayon** of this company, for making nonwoven fabrics and lining and filter papers on regular papermaking machines without the addition of a binder, is a multicellular, short-staple fiber cut to uniform 0.25-in (0.64-cm) length. The fibers lock themselves firmly in place with contacting fibers, and the 1.5-denier fiber makes a soft, opaque sheet of paper. The density of the sheets and the strength per unit weight decrease with increase in fiber denier. **Fiber 40,** of this company, is a type of rayon called **Avril,** made by a special pulping process which decreases the tendency to shrink or felt when the fiber is wet. **Avlin** is a multicellular rayon fiber that gives a tight, firm bulk to fabrics. **Fabray,** of Stearnes Technical Textiles Co., is a thin, lightweight, porous, nonwoven fabric of rayon fiber used for throw-away garments.

The objectionable high gloss of synthetic fibers is reduced by pigmentation. **Glos** was an early name for rayon because of this gloss, but the name now has been abandoned. Mixtures of rayon and other fibers have some use for dress fabrics as well as underwear. The material is also used for automotive tire fabrics because of its strength. **Stable fiber** is fiber cut to length for the spinning system to be used.

High-tenacity rayon is produced by stretching the fibers so that the molecular chains run parallel to the filament axis, and a number of small crystalline regions act as anchors for the cellulose chains. **Tire cord** stretched in this way has greater tensile strength.

RECYCLATE PLASTICS. Resins made at least partially from recycled thermoplastics. Polyurethane-bonded foam made from flexible polyurethane scrap is used for carpet underlays, gymnasium mats, auto sound damping and mud guards, particleboards, and, mixed with rubber chips, adhesive pavements on athletic fields. Polystyrene foam with as much as 50% recycled polystyrene is used for extruded food trays and containers. Polyethylene film scrap and stretch and shrink wrap, bottles, and trash, dry-cleaning, and merchandizing bags are reprocessed into film products and used as fillers in molded products. **Petra 140,** of Allied Signal, is a 40% glass-reinforced polyethylene terephthalate made from recycled soda bottles. It has a tensile strength of 26,000 lb/in^2 (179 MPa) and a heat-deflection temperature of 437°F (225°C) at 264 lb/in^2 (1.8 MPa). Discarded plastic bottles are also reduced to green and white flakes, melted, and spun into yarn for Synchilla sweaters by Patagonia. Dow Plastics has a method for incorporating as much as 10% process scrap in reaction-injection-molded (RIM) auto parts. Miles Inc. has a process for compression-molding parts from 100% RIM scrap regrind. Polyvinyl chloride recycled from flexible PVC wire and cable is used to mold hidden auto parts that dampen noise and resist water.

Recyclates are widely used for **plastic lumber,** or **synthetic wood.** Advantages over natural wood include weight reduction, especially by foaming, and rot resistance. They also resist splintering and color fade. **Replex,** of RPE of Italy, is a recyclate of 60% by weight polyolefins and 40% polyethylene terephthalate, polyvinyl chloride, and polystyrene. It is used for tabletops and lighting units of desktop furniture. In England, decorative building panels are made from packaging waste, largely low- and high-density polyethylene (HDPE), polypropylene, and rigid polyvinyl chloride. Sheet, made by Stanley Smith & Co. from recycled HDPE, is similar in tensile strength and creep resistance to that made of virgin material. Also in England, Plexite Ltd. converts reclaimed polystyrene foam to synthetic wood that looks like rain forest hardwoods. **Tectan,** made by Tetra Pak subsidiary EVD of Germany, is a form of **chipboard** made from recycled, multilayer beverage cartons. This machinable and thermoformable sheet contains about 20% polyethylene and is used for desks, cupboards, and bar stools. Mobil Chemical uses all recycled material—50% polyethylene film (grocery sacks and stretch wrap) and 50% sawdust—to extrude profiles 12 ft (3.7 m) long. They are colorable and paintable, resist moisture and ultraviolet, feel like wood, and can be machined with woodworking tools.

Polycarbonate (PC)/acrylonitrile-butadiene-styrene blends, or alloys, using 25% recyclate are used for personal-computer housings. **PC23MS-200,** of MCR Polymers, contains at least 25% recyclate from PC compact disks and polyethylene terephthalate beverage bottles. **DMDA-1343 NT polyethylene,** of Union Carbide, contains 28% color-sorted recyclate and has physical properties similar to those of virgin stock. **NPP05,** of Newplast SA of Switzerland, is a thermoplastic amalgam of heterogeneous polymers in pellet form made from unsorted domestic plastic waste and is suitable for extrusion, compression-molding, and injection-molding machines. **Encore resins,** of Hoechst Celanese, are a family of plastics based on 100% reclaimed thermoplastics such as acetal, polyester, polyphenylene sulfide, Nylon 6/6, and liquid-crystal polymer. Unfilled **DR-1000 acetal,** an Encore resin, has a specific gravity of 1.41, a tensile strength at yield of 7,900 lb/in^2 (54 MPa), 18% elongation at break, a flexural modulus of 342,000 lb/in^2 (2,358 MPa), a notched Izod impact strength of 1.3 ft · lb/in (69 J/m), and a heat-deflection temperature of 230°F (110°C) at 264 lb/in^2 (1.8 MPa).

Glass fibers used to reinforce plastics are also recyclable. **CSX hybrid fibers,** of Phoenix Fiberglass, are reclaimed from sheet molding compound waste and suitable for mixing with virgin fibers. Deja Inc. makes footware from recycled tire rubber, wetsuit trimmings, old file folders, coffee filters, seat cushions, soda bottles, and other materials. Used Rubber USA makes wallets and belts from old inner tubes and scrap rubber. **EEKO recyclable synthetic leather** is used for handbags.

RED BRASS. A series of copper alloys including one wrought alloy (C23000) and several cast alloys (C83300 to C84800). **Red brass C23000** contains 84 to 86% copper and the balance zinc except for small amounts (0.05 maximum) of lead and iron. The cast alloys, which include **leaded red brasses,** have a zinc content that can range from 1.0 to 2.5% (C83500) to 13 to 17% (C84800), and all but C83400, which is limited to 0.20 tin, contain substantial amounts of this element: 1.0 to 6.5, depending on the alloy. Lead content of the cast alloys can be as little as 0.50% maximum (C83400) or as much as 8.0 (C84400). Some of these alloys also contain about 1% nickel, and most contain smaller amounts of other metals.

C23000 is available in most wrought mill forms and is used for condenser and heat-exchanger tubing, plumbing lines, electrical conduit, fasteners, and architectural and ornamental applications. It is quite ductile in the annealed condition, and can be appreciably hardened by cold work. Annealed, thin, flat products have tensile properties of 39,000 to 45,000 lb/in^2 (269 to 310 MPa) ultimate strength, 10,000 to 18,000 lb/in^2 (69 to 124 MPa) yield strength, and 42 to 48% elongation. When it is cold-worked to the spring temper, ultimate strength increases to 84,000 lb/in^2 (579 MPa) and yield strength to 62,000 lb/in^2 (427 MPa), while elongation decreases to 3%. The alloy has a density of 0.316 lb/in^3 (8,747 kg/m^3) and, annealed, has an electrical conductivity about 37% that of copper. It has good resistance to fresh and seawaters and is not prone to dezincification.

All of the cast alloys are suitable for sand casting, and some for centrifugal and/or investment casting. As cast, the ultimate tensile strength is typically 32,000 to 37,000 lb/in^2 (221 to 255 MPa), yield strength is 10,000 to 17,000 lb/in^2 (69 to 117 MPa), and elongation 25 to 35%. C83300 is used for terminal ends for electrical cables, and C83400 is used for rotating bands. The other cast alloys find broader applications, including pump parts and housings, valves, flanges, pipe fittings, general plumbing hardware, and fasteners.

Standard **gilding metal** (C21000), used for making cheap jewelry and small-arms ammunition, contains 94 to 96% copper. It has a golden-red color, is stronger and harder than copper, but has only about half the electrical conductivity. The English **cap copper,** used for cartridge caps, has 97% copper and 3 zinc. **BES gilding metal** comes in three grades, 80, 85, and 90% copper, but the 80–20 alloy has a definite golden-yellow color and is not in the class of red brasses. The

80–20 brass (C24000) is one of the standard alloys of the brass mills under the name of **low brass.** It was early used for jewelry under the name of **Dutch metal.** It is very ductile and is easily drawn, takes a high polish, and has high strength. It is still much used for cheap jewelry. **Chain bronze,** used for flat-link jack chains and for costume jewelry, contains 87 to 89% copper, 0.60 to 1.25 tin, and the balance zinc. The tin increases the hardness, and the alloy has good strength and corrosion resistance.

Many old names used for designating the golden or reddish high-copper brasses, especially for cheap jewelry making, are still in occasional use. **Pinchbeck metal,** originally made by C. Pinchbeck, an English jeweler, contained 88% copper and 12 zinc. This is the same as the **Guinea gold** used for traders' jewelry. **Manila gold,** or **traders' gold,** was about the same with some lead to heighten the color. **Ormolu gold,** a name still used by brass platers, was any composition that would give a golden color. **Rich gold metal** was the 90–10 alloy, and this alloy is still being used for decorative purposes under the name of **copper-rich brass. Manheim gold,** containing 83% copper, 10 zinc, and 7 tin, was a German alloy for cheap jewelry. It is a considerably harder alloy than ordinary red brasses. **Tournay metal** was a French alloy widely used for buttons when brass buttons were in vogue. It was essentially the same as the original **tombac metal** used by the Chinese for buttons, and was the 85–15 alloy containing considerable arsenic to give a brilliant grayish tone to the metal. The alloy is quite brittle. The name *tombac* is the Malay word *tombaga* used to designate gold-colored jewelry alloys. **Chrysochalk** is another old name for gilding metal containing enough lead to give a dull gold tone. Japanese low-priced jewelry alloys often contained silver or gold to balance the color and make them resistant to tarnishing. **Shadke** was a high-copper alloy with some gold, and **Shakudo** contained about 4% gold and 1 silver. The gold color of the red brasses is enhanced by pickling in nitric acid.

RED LEAD. A common lead pigment, also erroneously called **minium.** It is a **lead tetroxide,** Pb_3O_4, forming a bright-red or orange-red powder of specific gravity 0.096, insoluble in water. As a pigment, it has great covering power and brilliancy, but red lead that has not been completely oxidized and contains litharge must be applied immediately after mixing to avoid combining with the oil. It is used as a heavy protective paint for iron and steel. Red lead is also used in storage-battery plates and leaded glass. With linseed oil it is used as a lute in pipe fitting. **Orange mineral** is a pure form of red lead made from white lead and has an orange color. Chemically it is the same as red lead, but it has a different structure, giving it a more brilliant color. Red lead is made from lead metal by drossing and then heating in a furnace. **Fume red lead** is a fine grade made from **fume litharge,** which is made by oxidizing molten lead and is passed off as a yellow smoke or fume. Fume red lead is notable for the extreme fineness of its particles, and it bulks greater than ordinary red lead. Fume red lead is marketed for pigment as **superfine red lead.**

REDWOOD.　 Also called **sequoia, California redwood,** and **Humboldt redwood.** The wood of the tree *Sequoia sempervirens,* native to the west coast of the United States. The wood is light, soft, and spongy but has comparatively high strength and is resistant to decay and insect attack. The trees grow in a narrow coastal strip in California and are immense, reaching a diameter of 20 ft (6.1 m) and a height of 350 ft (107 m) in 2,000 years. The so-called Big tree, however, is the *S. gigantea* or *S. washingtoniana.* It grows in the mountains at elevations of 5,000 to 8,500 ft (1,524 to 2,591 m), but is not cut for lumber. Planks of redwood can be readily obtained 6 ft (1.8 m) in width. The specific gravity is 0.374 to 0.387. It has a tensile strength of 7,000 to 11,000 lb/in² (48 to 76 MPa). The heartwood varies from light-cherry to dark-mahogany in color, and the narrow sapwood is almost white. The wood is used in all kinds of common construction. **Redwood bark fiber** is the shredded fiber of the bark of the redwood. It has excellent felting properties, is water-resistant and fire-resistant, and is an excellent insulator for house walls and refrigerators. It is also used in wool mixtures for blankets and overcoatings. The fiber is short, but has a natural twist that facilitates spinning. The fiber contains a high content of lignin and makes a good filler for heat-curing plastics. **Palco wool** is redwood bark fiber, and **Palco board** is a lightweight insulation board for cold-storage chambers made from the shredded bark fibers. Palcotan is a sodium salt of **palcotannic acid,** a weak sulfonated tannic acid, used as a binder in ceramic clays, a stabilizer for asphalt emulsions, and in latex and paste adhesives. **Redwood tannin,** produced by soaking redwood chips in hot water and dissolving out the tannin with ethyl acetate, is used as an oxidation inhibitor in hydrocarbons.

REFRACTORIES.　 Materials, usually ceramics, are employed where resistance to very high temperature is required, as for furnace linings and metal-melting pots. Materials with a melting point above 2876°F (1580°C) are called *refractory,* and those with melting points above Seger cone 36, or 3254°F (1790°C), are called *highly refractory.* But, in addition to the ability to resist softening and deformation at the operating temperatures, other factors are considered in the choice of a refractory, especially load-bearing capacity and resistance to slag attack and spalling. Heat transfer and electrical resistivity are sometimes also important. Many of the refractories are derived directly from natural minerals, but synthetic materials are much used. To manufacture refractory products, powders of the raw materials are mixed and usually dry-pressed to form the desired shape.

Clay is the oldest and most common of the refractories. The natural refractories are kaolin, chromite, bauxite, zirconia, and magnesite, often marketed under trade names. Refractories may be acid, such as silica, or basic, such as magnesite or bauxite, for use in acid- or basic-process steel furnaces. Graphite and chromite are neutral refractories. Magnesia is insoluble in the slag of open-hearth furnaces and is used for linings. Magnesia fuses at 3929°F (2165°C), chromite at 3722°F (2050°C), and alumina at 3670°F (2021°C). The fusing point of the refractory, however, is usually dependent on the binder, as all binders or impurities lower the melting point. **Arco refractory brick,** of General Refractories Co., with 60 to 80% alumina, withstands temperatures from 3290 to 3335°F (1810 to 1835°C). The melting point of a 99.5% pure alumina is given as 3725°F (2052°C), and the decomposing point of a 98% pure silicon carbide as 4175°F (2301°C). **Chrome-magnesite bricks** are made usually with 75% chrome ore and 25 dead-burned magnesite.

The chief artificial refractories are silicon carbide and aluminum oxide. **Refrax,** of Carborundum Co., is silicon carbide held together by crystallization without a binder. It withstands temperatures to 4064°F (2240°C), at which point it decomposes. The crushing strength of the brick is 12,500 lb/in^2 (85 MPa). This type of material can be made only in simple shapes, but is also made into rolls for roller-type furnaces. **Refrax FS** is fused silica bonded with silicon nitride. It is used for formed parts up to 23 in (58 cm) in diameter for such applications as brazing fixtures, and has dimensional stability and thermal shock resistance to 2250°F (1232°C). **Silfrax** is in grades with 40 to 78% SiC, and porosities from 9 to 18%. **Monofrax,** of the same company, is a refractory block for lining glass furnaces. It is composed of 98% alpha and beta alumina crystals interlocked in a dense structure of 200 lb/ft^3 (3,204 kg/m^3), with only a small amount of bond and impurities. A grade with more open structure has a density of 175 lb/ft^3 (2,804 kg/m^3), and **Monofrax K** contains 80% alumina and the balance chromite crystals saturated with alumina. These materials are resistant to abrasion to temperatures above 3000°F (1649°C). They have porosities from 20 to 29%. **Aluminite** is furnished in blocks for temperatures to 2000°F (1093°C). **Korundal XD brick,** of Harbison-Walker Refractories Co., is corundum bonded with mullite. For open-hearth and electric-furnace roofs it will withstand a 25-lb/in^2 (0.17-MPa) load at a temperature of 3000°F (1649°C) without spalling. **Lo-Sil brick** of Kaiser Aluminum, for aluminum-melting furnaces, is made with a high alumina content with very low silica to avoid pickup of silicon in the molten aluminum. Phosphate-bonded refractories having at least 50% alumina content, relatively low density, and 3- to 5-in (76- to 127-mm) thickness are used to thermally insulate steel ducts and bypass stacks of incinerators.

The **silica brick** used in coke ovens has a high coefficient of expansion below about 155°F (68°C), so that ovens must be heated gradually over a period of 6 to 8 weeks to prevent cracking of the brick. **Fire sand** is a sand composed of 98% silica and is very refractory. The natural silica refractories used to replace fireclay for high temperatures should contain at least 97% silica and not yield too fine a powder on crushing. In order of merit, the materials used are chalcedony, old quartzites, and vein quartz. The refractory known in Europe as **klebsand,** used for steel furnace linings, has 87% SiO_2, 8.6 Al_2O_3, 0.3 TiO_2, and some ferric oxide. **Ganister** is a natural refractory mineral used for furnace linings. In compact form it was used for furnace hearths. It contains about 95% silica, about 1.5 Al_2O_3, and a small amount of lime as a binder. The chief deposits are in the quartzites of Wisconsin, Alabama, and Colorado. An artificial ganister is made with silica and clay. **Ganisand** is a ganister having a fusing point at 3250°F (1788°C). **Dinas silica** is an English ganister with about 97% silica, having a melting point of 1680°C. Silica brick made from quartzite containing 98% silica has lime added as bond, and the resulting brick usually contains about 96% silica, 2 lime, and 2 alumina and ferric oxide. The bricks are rigid under load and are resistant to attack by acid slags and to spalling under rapid temperature changes. **Vega silica brick,** of Harbison-Walker Refractories Co., has no more than 0.4% of oxides other than lime and withstands temperatures of 3090°F (1699°C) under load. **Foamed silica blocks** for lining tanks and for refractory insulation are made from pure fused silica. They have a density of 10 to 15 lb/ft³ (160 to 240 kg/m³) with an impermeable closed-cell structure, have a compressive strength of 130 to 210 lb/in² (0.9 to 1.4 MPa), and withstand temperatures to 2200°F (1204°C). **Foamsil,** of Pittsburgh Corning Corp., is this material.

Pinite, from Nevada, is a secondary material derived from the alteration of feldspar and other rocks, and it is used for kiln linings in cement plants. It is a hydrous silicate of alumina and potash, and the massive material resembles steatite. It will bond alone as clay does and has low shrinkage. At 2057°F (1125°C), the mineral inverts to mullite. **Agalmatolite** is a massive pinite and can be used in the same way. **Bull-dog** is an old name for a refractory which is a mixture of ferric oxide and silica made by roasting tap cinder (a basic silicate of iron) with free access of air.

REFRACTORY CEMENT. A large proportion of the commercial refractory cements used for furnace and oven linings and for fillers are fireclay-silica-ganister mixtures with a refractory range of 2600 to 2800°F (1427 to 1538°C). Cheaper varieties may be mixtures of fireclay and crushed brick, fireclay and sodium silicate, or fireclay and silica sand. An important class of refractory cements is made of silicon carbide grains or silicon carbide–fire sand with clay bonds or synthetic mineral bonds. The temperature range of these cements is 2700 to 3400°F (1482 to 1871°C). **Silicon carbide cements** are acid-resistant and have high thermal and electric conductivity. For crucible furnaces the silicon carbide cements are widely used except for molten iron. Alumina and **alumina-silica cements** are very refractory and have high thermal conductivity. Calcined kaolin, diaspore clay, mullite, sillimanite, and combinations of these make cements that are neutral to most slags and to metal attacks. They are electrical insulators. **Chrome-ore cements** are difficult to bond unless mixed with magnesite. **Plastic 695** is a **chrome-magnesite cement** made of treated magnesite and high-grade chrome ore. It sets quickly and forms a hard, dense structure. The melting point is above 3600°F (1982°C). It is used particularly for hot repairs in open-hearth furnaces. **Magnefer** is the name of a dead-burned dolomite refractory, while **Basifrit** is a magnesia refractory for resurfacing. **Zircon-magnesite cement** is made with 25% refined zircon sand, 10 milled zircon, 15 fused magnesia, and 50 low-iron dead-burned magnesite bonded with sodium silicate. A wide range of refractory cements of varying compositions and characteristics are sold under trade names, and these are usually selected by their rated temperature resistance. For example, **Hadesite** is composed of refractory clay and aggregates, mineral wool, and a binder, and is recommended for temperatures up to 1900°F (1038°C). **Carbofrax cement,** of Carborundum Co., is silicon carbide with a small amount of binder in various grades for temperatures from 1600 to 3200°F (871 to 1760°C), depending on the fineness and the bond. **Firefrax cement** of the same company is an aluminum silicate, sometimes used in mixtures with ganister for lining furnaces. It is for temperatures to 3000°F (1649°C). **Alfrax cement** is fused silica also in various grades for temperatures from 1650 to 3300°F (899 to 1816°C). **Mullfrax cement** has a base of electric-furnace mullite and is for temperatures from 2200 to 3200°F (1204 to 1760°C), while **Mullite S cement** is of converted kyanite for ferrous and nonferrous melting furnaces for temperatures to 3150°F (1732°C). **Ankorite,** of Harbison-Walker Refractories Co., is a high-alumina hot-setting refractory mortar for laying superduty firebrick. **Thermolith,** of this company, is chrome-based cold-setting mortar for laying magnesite, chrome, and forsterite brick.

REFRACTORY HARD METALS.   True chemical compounds of two or more metals in the form of crystals of very high melting point and high hardness. Because of their ceramiclike nature, they are often classified as ceramics. These materials were originally called **Hartstoffe** in Germany, and they do not include the hard, metallic carbides, some of which, with metal binders, have similar uses; nor do they include the hard cermets. The refractory hard metals may be single large crystals, or crystalline powder bonded to itself by recrystallization under heat and pressure. In general, parts made from them do not have binders or contain only a small percentage of stabilizing binder. The **intermetallic compounds,** or **intermetals,** are marketed regularly as powders of particle size from 150 to 325 mesh for pressing into mechanical parts or for plasma-arc deposition as **refractory coatings,** and the powders are referred to chemically, such as **borides, beryllides,** and **silicides.** The oxides and carbides of the metals are also used for sintering and for coatings, and to make cermets.

Zirconium boride is a microcrystalline, gray powder of composition ZrB_2. When compressed and sintered to a specific gravity of about 5.3, it has a Rockwell A hardness of 90, a melting point of 5396°F (2980°C), and a tensile strength of 35,000 to 40,000 lb/in^2 (241 to 276 MPa). It is resistant to nitric and hydrochloric acids, to molten aluminum and silicon, and to oxidation. At 2200°F (1204°C) it has a transverse rupture strength of 55,000 lb/in^2 (379 MPa). It is used for crucibles and rocket nozzles.

Chromium boride occurs as very hard crystalline powder in several phases: the orthorhombic crystal CrB melting at 2732°F (1500°C), the hexagonal crystal Cr_2B melting at 3362°F (1850°C), and the tetragonal crystal Cr_3B_2 melting at 3560°F (1960°C). Chromium boride parts produced by powder metallurgy have a specific gravity of 6.20 to 7.31, with a Rockwell A hardness of 77 to 88. They have good resistance to oxidation at high temperatures, are stable to strong acids, and have high heat-shock resistance up to 2400°F (1316°C). The transverse rupture strength is from 80,000 to 135,000 lb/in^2 (552 to 931 MPa). **Colmonoy,** of Wall Colmonoy Corp., is chromium boride, CrB, used for oil-well drilling. A sintered material, used for gas-turbine blades, contains 85% CrB with 15 nickel binder. It has a Rockwell A hardness of 87 and a transverse rupture strength of 123,000 lb/in^2 (848 MPa).

Molybdenum boride, Mo_2B, has a specific gravity of 9.3, a Knoop hardness of 1,660, and a melting point of about 3020°F (1660°C). **Tungsten boride,** W_2B, has a specific gravity of 16.7 and a melting point of 5018°F (2770°C). **Titanium boride,** TiB_2, is lightweight with a specific gravity of 4.5. It has a melting point of about 4700°F

(2593°C). Molded parts made from the powder have a Knoop hardness of 3,300 and a flexural strength of 35,000 lb/in^2 (241 MPa), and they are resistant to oxidation to 1800°F (982°C) with a very low oxidation rate above that point to about 2500°F (1371°C). They are inert to molten aluminum. Intermetal powders of **beryllium-tantalum, beryllium-zirconium,** and **beryllium-columbium** are also marketed, and they are lightweight and have high strength. Sintered parts resist oxidation to 3000°F (1649°C).

Molybdenum disilicide, MoSi$_2$, has a crystalline structure in tetragonal prisms, a Knoop hardness of 1,240, and excellent oxidation resistance up to at least 2912°F (1600°C). It is a potential candidate for high-temperature composites for aircraft-turbine parts. The decomposition point is above 3398°F (1870°C). It can be produced by sintering molybdenum and silicon powders or by growing single crystals from an arc melt. The specific gravity of the single crystal is 6.24. The tensile strength of sintered parts is 40,000 lb/in^2 (276 MPa), and compressive strength is 333,000 lb/in^2 (2,296 MPa). The resistivity is 11.3 $\mu\Omega \cdot$ in (29 $\mu\Omega \cdot$ cm). It is used in rod form for heating elements in furnaces. The material is brittle, but can be bent to shape at temperatures above 2000°F (1093°C). **Super Hot Rod,** of Norton Co., is molybdenum disilicide rod. **Kanthal Super,** of Kanthal Corp., is a similar material. In an inert atmosphere the operating temperature is 2912°F (1600°C). Furnace gases containing active oxygen raise the operating temperature to about 3092°F (1700°C), while gases containing active hydrogen lower it to about 2462°F (1350°C).

Tungsten disilicide, WSi$_2$, is not as hard and not as resistant to oxidation at high temperatures, but has a higher melting point, 3722°F (2050°C). **Titanium nitride,** TiN, is a light-brown powder with a cubic lattice crystal structure. Sintered bars are extremely hard and brittle, with a Mohs hardness above 9 and a melting point of 5342°F (2950°C). It is not attacked by nitric, sulfuric, or hydrochloric acids, and it is resistant to oxidation at high temperatures. In recent years, **titanium nitride coatings** have been used to markedly extend the life of tool-steel cutters and forming tools. The coatings, golden in color, are deposited by chemical or physical vapor deposition. **Titanium carbonitride (TiCN)** and **chromium nitride (CrN)** are also used to increase wear resistance. **Aluminum nitride (AlN),** when molded into shapes and sintered, forms a dense, nonporous structure with a Mohs hardness of 6. It resists the action of molten iron or silicon to 3100°F (1704°C), and molten aluminum to 2600°F (1427°C), but is attacked by oxygen and carbon dioxide at 1400°F (760°C). At least 1% oxygen causes AlN properties to deteriorate rapidly. However, if oxygen is removed, AlN is highly thermally conductive, lightweight, and a good insulator. A newly developed process has removed oxygen concentration from AlN down to 0.5%, and subsequent nitriding and a self-purification reaction have produced commercial-grade AlN. Doping **low-oxygen AlN** powder with 3% yttria, densifying to high density, and pressureless sintering for 80 h at 3362°F (1850°C) more than double thermal conductivity—to 98 to 110 Btu/(h · ft · °F) [170 to 190 W/(m · K)] versus 45 Btu/(h · ft · °F) [77 W/(m · K)]—relative to yttria-free **polycrystalline AlN.** However, **single-crystal AlN** of equivalent oxygen content has a thermal conductivity of 116 Btu/(h · ft · °F) [200 W/(m · K)].

Nickel aluminide, NiAl, is a chemical compound of the two metals, and when molded and sintered into shapes, it has good oxidation resistance and heat resistance at high temperatures. **Borolite 1505** is a nickel aluminide sintered material with a specific gravity of 5.9 and a transverse rupture strength of 150,000 lb/in² (1,034 MPa) at 2000°F (1093°C), which is twice that of cobalt-bonded titanium carbide at the same temperature. The melting point is 3000°F (1649°C). It resists oxidation at 2000°F (1093°C). It is used for highly stressed parts in high-temperature equipment. **Metco 405,** of Metco, Inc., is this compound in wire form for welding, flame coating, and hard surfacing. **Metco 404** is aluminum powder coated with nickel. This powder may be mixed with zirconia or alumina, which will increase the hardness and heat resistance of the nickel aluminide coating. **Columbium aluminide** is used as a refractory coating as it is highly resistant for long periods at 2600°F (1427°C). **Tin aluminide** is oxidation-resistant to 2000°F (1093°C), but a liquid phase forms at about this point.

Tribaloy intermetallic materials, developed by Du Pont, are composed of various combinations of nickel, cobalt, molybdenum, chromium, and silicon. For example, one composition is 50% nickel, 32 molybdenum, 15 chromium, and 3 silicon. Another is 52% cobalt, 28 molybdenum, 17 chromium, and 3 silicon. The structure consists of a hard intermetallic Laves phase in a softer matrix. Supplied as alloy-metal powder, welding rod, or casting stock, they can be cast, deposited as a hard-facing surface, plasma-sprayed, or consolidated by powder metallurgy. The materials have exceptional wear and corrosion resistance properties in corrosive media and in air up to 2000°F (1093°C). Typical applications are in pumps, valves, bearings, seals, and other parts for chemical process equipment. Also, the materials are suited for marine and saltwater applications and for parts subject to wear in atomic energy plants.

REFRIGERANTS. Gases, or very-low-boiling-point liquids, used for the heat-absorbent cycle in refrigerating machines. The ideal refrigerant, besides having a low boiling point, should be noncorrosive, nonflammable, and nontoxic. It must be free of water, since as little as 40 parts per million of water may cause freezing in the system. Ammonia is a common refrigerant. Ethyl chloride, methyl ether, carbon dioxide, and various chlorinated and fluorinated hydrocarbons marketed under trade names such as the **Freons** of Du Pont Co. and the **Genetrons** of Allied Chemical Corp. are also used. **Dichlorodifluoromethane,** a nonflammable, colorless, odorless gas of composition CCl_2F_2 is Freon 12. It liquefies at $-21.7°F$ ($-30°C$). **Freon E3** is a fluorated hydrocarbon which boils at $306°F$ ($152°C$) and is stable to $570°F$ ($299°C$). It is pourable at $175°F$ ($79°C$). It is used for dielectric insulating in electronic equipment. **Trichloro-monofluoromethane,** CCl_3F, is used as a refrigerant in industrial systems employing centrifugal compressors, and in indirect expansion-type air-conditioning systems. The boiling point is $74.7°F$ ($24°C$), and freezing point $-168°F$ ($-111°C$). The condensing pressure at $86°F$ ($30°C$) is 18.3 lb/in^2 (0.13 MPa), and the net refrigerating effect of the liquid is 67.5 Btu/lb (157 kJ/kg). **Genetron 11** is this material. **Genetron 101** has composition $CH_3{:}CClF_2$, with boiling point at $15.4°F$ ($-9.2°C$) and freezing point at $-203°F$ ($-131°C$). The chlorinated constituent of these chlorofluorocarbons has been implicated in the depletion of stratospheric ozone, and alternatives are now being sought that are less environmentally harmful.

Of the **chlorofluorocarbons,** CFC-11 (CCl_3) and CFC-12 have the highest ozone-depletion potential (ODP), with CFC-11 having an ODP of 1.0. **Hydrochlorofluorocarbons** (HCFCs) are also ozone-depleting but much less so and, thus, are viewed as temporary replacements. HCFC-22 ($CHClF_2$), for example, has an ODP of 0.055, and HCFC-123 ($CHCl_2CF_3$) has an ODP of 0.02. **Hydrofluorocarbons** (HFCs), being chlorine-free, have no ozone-depletion potential. A large variety of HFC mixtures are viewed as possible replacements for HCFCs and CFCs. Allied Signal's **Genetron AZ-50** is an azeotropic mixture of HFC-125 and HFC-143a, and Du Pont's **Suva HP 62** is a blend of HFC-125, HFC-134a (CF_3CH_2F), and HFC-143a. **Suva Blend-MP** is a blend of HFC-152a (CHF_2CH_3), HCFC-22, and HCFC-124.

RESINS. Historically, **resins** is the term applied to an important group of substances obtained as gums from trees or manufactured synthetically. It is also frequently used interchangeably with the term **plastics.** The common resin of the pine tree is called **rosin.** The natural resins are soluble in most organic solvents and are used in varnishes, adhesives, and various compounds. **Oleoresins** are natural resins containing essential oils of the plants. **Gum resins** are natural mixtures of true gums and resins and are not as soluble in alcohol.

They include rubber, gutta percha, gamboge, myrrh, and olibanum. Some of the more common natural resins are rosin, dammar, mastic, sandarac, lac, and animi. **Fossil resins,** such as amber and copal, are natural resins from ancient trees, which have been chemically altered by long exposure. The synthetic resins differ chemically from natural resins, and few of the natural resins have physical properties that make them suitable for mechanical parts.

Galbanum, used in medicine, is a gum resin from the perennial herb *Ferula galbaniflua* of western Asia. It comes in yellowish to brownish tears. **Myrrh** is a yellow-reddish, aromatic gum resin from *Commiphora malmol* and other species of small trees of India, Arabia, and northeast Africa. There are more than 80 species. The Hebrew word *mur* means bitter, and the gum has a bitter taste. It consists of a mixture of complex acids and alcohols but mainly the mucilaginous **arabin,** with from 3 to 8% of a volatile oil of formula $C_{10}H_{14}O$. It is used in incense and perfumes, and in medicine as a tonic. The gum is called **mulmul** and **ogo** in eastern Africa and **herabol** in India. **Sweet myrrh,** or **bisabol myrrh,** is a very ancient perfume and incense. It comes from the tree *C. erythraea* of Arabia. **Herabol myrrh** is from the small tree *C. myrrha* of India and is a brown to black solid. It is used in medicine as a stimulant and antiseptic and in perfumes and incense. **Asafetida** is a gum resin with a foul odor and acrid taste obtained from the roots of the perennial herb *F. assafoetida* and other species. It is used in Asia for flavoring foods, but in the United States it is employed in medicines and perfumes.

Creosote bush resin is amber-colored, soft, and sticky, from the leaves and small twigs of the **greasewood bush,** *Larrae tridentata,* or **creosote bush,** *L. divaricata,* of the desert regions of Mexico and the southwestern United States. It is used in adhesives, insecticides, core binders, insulating compounds, and pharmaceuticals. When distilled, the resin yields **nordihydroguaiaretic acid,** $(C_6H_3 \cdot Cl \cdot COOH)_2CH_2$, called **NDGA,** formerly used as a preservative but now prohibited by the FDA for use in foods. It is a white, crystalline solid melting at 363°F (184°C), soluble in fats and slightly soluble in hot water. **Okra gum** is from the pods of the plant *Hibiscus esculentus,* native to Africa but now grown in many countries. It is edible and is used as a thickening agent in foodstuffs and pharmaceuticals. It has antioxidant properties and also acts as a stabilizer. As a gelling agent, it forms a network molecular structure with branched chains, giving slipperiness valued for foodstuff spreads. It is a tan-colored, water-soluble powder. Okra gum is also used in plating baths as a brightener.

Ammoniac is a gum resin from the stems of *Dorema ammoniacum,* a desert perennial plant of Persia and India. It forms in hard, brittle, brownish-yellow tears, and has a peculiar fetid odor and an acrid taste. It is used in adhesives, in perfumery, and as a stimulant in medicine. In pharmacy it is called **ammoniacum. Oil ammoniac** is a yellow liquid distilled from the gum. The specific gravity is 0.890, boiling point is 527°F (275°C), and it is soluble in alcohol and benzol.

RESORCINOL.   A colorless crystalline material with composition $C_6H_4(OH)_2$ and melting point 230°F (110°C). It is very soluble in water and in alcohol. It is used in the production of plastics; in the manufacture of fluorescein; in the production of xanthene and azo dyes, particularly the fast **Alsace green**; in medicine as an antiseptic; and in making the explosive lead styphnate. Resorcinol polymerizes with formaldehyde to form the **resorcinol-formaldehyde plastics** that will cure at room temperature and with only slight pressure. They are used in strong adhesives for plywood and wood products, and do not deteriorate from acid action as some other plastics do. **Resorcinol adhesives** remain water-soluble during the working period for 2 to 4 h and are then insoluble and chemical-resistant. Derivatives of resorcinol are used in medicine as specific bactericides, such as **diverinol,** which is propyl resorcinol, and **olivertol,** which is amyl resorcinol.

RHENIUM.   An elementary metal, symbol Re, present in small quantities in many minerals. Rhenium has a specific gravity of 21.4, being almost twice as heavy as lead. The melting point is 5756°F (3180°C). It is a hard, silvery-white metal and takes a high polish. As a plating metal, the white color is darker than that of rhodium. It has good chemical resistance, but is soluble in nitric acid and oxidizes in air at temperatures of about 1112°F (600°C) and greater. The crystal structure is closely packed hexagonal, making it more difficult to work than the cubic-structured tungsten; but the crystal grains are tiny, and small amounts of rhenium added to tungsten improve ductility and increase high-temperature strength of tungsten used for lamp filaments and wire.

Rhenium is among the strongest and most rigid of metals, having an ultimate tensile strength of about 280,000 lb/in^2 (1.931 MPa) and a tensile modulus of 68 × 10^6 lb/in^2 (469,000 MPa) at room temperature, with good ductility (35% reduction of area). Also, it is quite heat-resistant, having an ultimate tensile strength of 134,000 lb/in^2 (924 MPa) and a tensile modulus of 54 × 10^6 lb/in^2 (372,000 MPa) at 1600°F (871°C). The metal is used mainly for electrical contacts, thermocouples, filaments in electronic components, and as an alloying element to increase the ductility of tungsten and molybdenum for electrical and electronic applications. At 70°F (21°C), rhenium has an electrical resistivity of 7.5$\mu\Omega \cdot$ in (19$\mu\Omega \cdot$ cm), a coefficient of thermal expansion of $3.7 \times 10^{-6}/°F$ (6.7/K), and a specific heat of 0.035 Btu/(lb · °F) [0.147 kJ/(kg · K)]. Rhenium can be obtained from molybdenite, which contains 0.0001 to 0.05% of the metal. But the usual source is from the flue dusts and from the sublimed **rhenium oxide,** Re_2O_7, of stack gases in the smelting of copper and other ores. It is precipitated from the flue dust of molybdenum-bearing copper ores in the form of **potassium perrhenate,** $KReO_4$, or of **ammonium perrhenate,** NH_4ReO_4. The Russians also obtain some rhenium from the Ural platinum ores. The stable **rhenia,** used in alloying, is **rhenium trioxide,** ReO_3, which comes as a red powder.

The metal is obtained as a dense, silvery powder which can be compacted, sintered,and cold-rolled with frequent annealing. It is marketed in the form of rod, strip, foil, and wire. **Rhenium-tungsten alloys,** with up to about 25% rhenium, are ductile and are used for high heat resistance. The melting point of the alloy, 5650°F (4023°C), is lower than that of pure tungsten, but it has higher recrystallization temperature. Rhenium has a higher electrical resistivity than tungsten, has high arc resistance, and does not become brittle after prolonged heating, as tungsten does. The alloys are used for electronic filaments and small structural parts. Rhenium metal is used as an undercoat for tungsten coatings on graphite nozzles to prevent the formation of tungsten carbide and thus give the full heat resistance of the tungsten. Rhenium or rhenium-tungsten electric contacts give very long service life. Rhenium-tungsten versus tungsten thermocouples are good for service to 5072°F (2800°C). The **tungsten-rhenium alloy** of Englehard Industries, Inc., for thermocouple use, contains 26% rhenium.

A **molybdenum (50%–rhenium (50%) alloy** is almost 3 times stronger than pure molybdenum and somewhat more rigid. At 70°F (21°C), it has an ultimate tensile strength of 240,000 lb/in² (1,655 MPa) and a tensile modulus of 53×10^6 lb/in² (365,000 MPa). Its electrical resistivity is about the same as that of rhenium but much greater than that of molybdenum. Because of their resistance to high temperatures and hydrogen embrittlement and their good thermal conductivity, Mo-Re alloys are candidates for the actively cooled engine combustor of the proposed national aerospace plane.

RHODIUM. A rare metal, symbol Rh, found in platinum ores such as the nickel-copper ores of Canada and **pyroxenite** of South Africa. It is very hard and is one of the most infusible of the metals. The melting point is 3565°F (1963°C). It is insoluble in most acids, including aqua regia, but is attacked by chlorine at elevated temperatures and by hot fuming sulfuric acid. Liquid rhodium dissolves oxygen, and ingots are made by argon-arc melting. At temperatures above 2192°F

(1200°C), rhodium reacts with oxygen to form **rhodium oxide,** Rh_2O_3. The specific gravity is 12.44. Rhodium is used to make the nibs of writing pens, to make resistance windings in high-temperature furnaces, for high-temperature thermocouples, as a catalyst, and for laboratory dishes. As a catalyst, it is used with platinum and palladium in auto catalytic converters to convert nitrogen oxide emission to nitrogen gas. It is the hardest of the platinum-group metals, the annealed metal having a Brinell hardness of 135. Rhodium also has considerable strength and rigidity, ultimate tensile strengths ranging from 138,000 to 300,000 lb/in² (952 to 2,068 MPa), and tensile modulus from 42×10^6 to 55×10^6 lb/in² (290,000 to 379,000 MPa), depending on condition or hardness. Rhodium is also valued for electroplating jewelry, electric contacts, hospital and surgical instruments, and especially reflectors. For electrodepositing rhodium, an electrolyte of **rhodium sulfate,** $Rh_2(SO_4)_3$, is used. The coatings are wear-resistant and tarnish-resistant. The plated metal has a pinkish-white luster of high corrosion resistance and a light reflectivity of 80%. Decorative finishes are seldom more than 0.0002 in (0.0005 cm) thick, but plates for electric contacts may be up to 0.005 in (0.013 cm). **Rhodium carbonyl** has the general formula $Rh(CO)$, but with several variations in the number of CO groups. They are used for depositing **rhodium coatings.**

The most important alloys of rhodium are **rhodium-platinum.** They form solid solutions in any proportion, but alloys of more than 40% rhodium are rare. Rhodium is not a potent hardener of platinum but increases its high-temperature strength. It is easily workable and does not tarnish or oxidize at high temperatures. These alloys are used for thermocouples and in the glass industry. A **rhodium-iron** alloy with equiatomic proportions of the metal has an ordered crystal structure and changes from antiferromagnetic to ferromagnetic in an electric field. Rhodium is sold by the troy ounce (31 g), 1 in³ (16 cm³) weighing 6.56 troy oz (204 g).

RICE. The white seed of the large annual grass *Oryza sativa,* growing to a height of 2 to 4 ft (0.6 to 1.2 m). It is a tropical plant native to Asia, but grows in hot, moist regions well into the temperate zones and is cultivated in many parts of the world. Rice fields are flooded after planting to control weeds. The rice seed grows in an inflorescence composed of a number of fine branches, each terminating in a single grain enclosed in a brown husk. Rice forms the staple food of more than half the populations of the world. **Wild rice** was used by the Indians of North America before the first Asiatic rice was brought to South Carolina in 1694. Rice is high in starch and low in proteins. It is used as a direct food, also as flour, as cereal, in puddings, and for the manufacture of starch and for alcoholic beverages. **Rice hulls** are used as stock feed, and **rice straw** is used for packing, hats, and other articles. **Maxflo** and **Profix,** of EnviroGuard, Inc., are filtration aids made from **rice hull ash,** a by-product of rice milling. The textured surface of the ash speeds solids separation from wastestreams. A low-ash version has some caloric value so that it can be used to sop up waste oils during filtering and produce a recyclable waste fuel. Rice in the husk before hulling is known by the Hindu name of **paddy. Brown rice** is rice that has been cleaned but not polished. Broken grains are sold in the India trade as **coodie** or **khood,** and about 20% of the rice produced from paddy is broken. **Patna rice** does not refer to the Patna district of India, but to a variety of rice with bold and hard grains especially suited for soups, as it holds its shape in boiling. **Malekized rice,** developed by General American Transportation Co., is produced by steaming unpolished rice to force the soluble part of the bran and the vitamins into the core of the grain, and then sealing the rice kernal by gelatinization, after which it is polished. The treated rice holds its shape and does not become gummy when cooked, and the nutritional value is improved. The beverage known in Japan as **saké** is **rice wine** containing 14% alcohol, made by fermenting rice with the mold **tané koji. Rice bran oil,** used as a salad and cooking oil and in lubricants, is produced from rice bran. By wet-milling whole rice in a rice oil solutions, a higher yield of oil and wax is obtained together with a yield of proteins. It is clear, odorless, and neutral, with a pleasant flavor, and is resistant to oxidation and rancidity. **Rice wax** is produced from rice bran by hot hexane extraction after cold extraction of the oil. It is a hard, brown, lustrous wax with a melting point of 174°F (79°C) used in polishes. **Synthetic rice,** used in Japan as a rice extender, is made from wheat flour, potato starch, and powdered rice.

ROSEWOOD.   The wood of several species of *Dalbergia* of northern South America, but chiefly from the **jacaranda tree,** *D. nigra.* It is used for fine cabinetwork, pianos, novelties, and expensive furniture. It should not be confused with the wood of the tree *Physocalymma floridum,* which also comes from Brazil and is there called **pao rosa** or rosewood. The color of rosewood is dark brown to purple, and it takes a beautiful polish. It has a characteristic fragrance. Very hard with a coarse, even grain, it has a density of 54 lb/ft^3 (865 kg/m^3). The tree grows to a height of 125 ft (38 m). **Brazilian rosewood** is the preferred wood for guitar bodies, but its use was banned in 1969 by the Convention on International Trade in Endangered Species. Another Brazilian wood, **caroba,** from the large tree *Jacaranda copaia,* is also called **jacaranda** and is sometimes confused with rosewood. The tree has purple flowers while the true rosewood has white flowers. Caroba wood is chocolate-colored and is used for fine furniture and knife handles. **Indian rosewood** is from the tree *D. sissoo* of India. It is also called **sissoo** and is a beautiful, brown hardwood employed for carvings. In Europe it is used for parquet floors. **Borneo rosewood,** also known as **ringas,** is the wood of several species of trees of genus *Melanorrhoea* of Borneo. The wood has a deep-red color with light and dark streaks. It has a close texture suitable for carving. **Satinee** is a type of rosewood from the tree *Ferolia guianensis* of the order Rosaceae, native to tropical America, particularly the Guianas. The wood is reddish brown, has a density of 54 lb/ft^3 (865 kg/m^3), is fairly hard, has a fine grain, and takes a lustrous polish. It is used for cabinetwork. **Bois de rose oil,** or **rosewood oil,** is not from rosewood, but is extracted from the heartwood of the tree *Aniba panurensis* of Brazil and the Guianas, though the wood of this tree is also used as a cabinet wood. The oil is also called **Cayenne linaloe,** or **Cayenne oil.** It has a delicate rose odor with a suggestion of orange and mignonette valued in perfumes. It contains a high percentage of **linalol,** a colorless alcohol with a soft, sweet odor, also found in the rose, lilac, lily, lavender, petitgrain, and other plants. It also contains geraniol. **Linaloe oil,** or **Mexican linaloe,** is distilled from the heartwood of the trees *Bursera delpechianum* and *B. aloeoxylon.* It contains less linalol and also terpineol and geraniol. Linalol is closely related to geraniol and nerol. Bois de rose is also made synthetically from geraniol. **Oriental linaloe** is distilled from selected, highly perfumed parts of the wood of the large tree *Aquilaria agollocha* of eastern India, Burma, and Java. The odor of the oil is like rose, ambergris, and sandalwood. Like the linaloes of the American continent, the oil is a pathological product and comes only from old trees. It is also called **aloe wood oil** and **agar attar,** and is a very ancient perfume. The beautifully figured and fragrant, reddish wood of this tree, called **aloes wood, eagle wood,** and **paradise wood,** is used for ornamental articles. True original rosewood oil known as **rhodium oil** was distilled from the wood of the plant *Convolvulus scoparius* of the Canary Islands.

ROSIN. The common resin of several varieties of pine tree, found widely distributed in North America and Europe. It is obtained by cutting a longitudinal slice in the tree and allowing the exudation to drip into containers. The liquid resin is then distilled to remove the turpentine, and the residue forms what is known as **gum rosin,** or **pine gum. Wood rosin** is obtained by distillation of old pine stumps. It is darker than gum rosin and is inferior for general use.

Rosin contains seven acids with very similar characteristics, but consists chiefly of **abietic acid,** $C_{19}H_{29}COOH$. Normally, when gum rosin is heated, the natural **pimaric acid** isomerizes to form abietic acid; but in the production of turpentine and rosin from pine sap, the turpentine is removed by steam distillation, and various acids are then extracted. Pimaric acid is closely related to abietic acid. It reacts with maleic and anhydride, and **maleopimaric acid** is used in printing inks and coatings. Rosin has a specific gravity of about 1.08 and a melting point of about 180°F (82°C), and it is soluble in alcohol, turpentine, and alkalis. It is used in varnishes, paint driers, soluble oils, paper sizing, and belt dressings; for compounding with rubber and other resins; and for producing many chemicals.

Rosin is generally graded commercially by letters according to color. The darkest grade is B, and the lightest is W. Extra grades are A, nearly black, and WW, water-white. Thirteen color grades are designated under the Naval Stores Act. The dark grades of wood rosin are considered inferior. They have a high melting point and low acid number and are used for making rosin oil; for battery wax, thermoplastics, and dark varnish; and for linoleum manufacture. The ruby-red wood rosin, obtained by extraction from fat pine wood, has high acid number, 155, and low melting point, 175°F (79°C). It is used for printing inks, paper size, and adhesives. Rosin is usually marketed in barrels of 280 lb (127 kg). **Naval stores** is an old name for rosin and turpentine. **Pexite** and **Pamite** are rosin grades from Hercules Inc. that cover the various color standards set by the U.S. Department of Agriculture. **Pelletized rosin** consists of free-flowing, dustless pellets produced by coating droplets of molten rosin with inert powder. **Colophony** is an old pharmacy name for rosin before distillation of rosin oil. Rosin was referred to by early writers as **Greek pitch,** but the ancient incendiary known as **Greek fire** was tow or pine sawdust impregnated with rosin, pitch, and sulfur. **Burgundy pitch** was originally the resin of the Norway spruce, *Picea abies,* used in medicine, but the name was later applied to a rosin, rubber, and mineral oil compound used for friction tape.

Hardened rosin is a weak resinate made by adding 6 to 8% high-calcium lime to melted rosin. It is used in some varnishes. **Fosfo rosin,** of Newport Industries, Inc., is a **lime-hardened rosin.** It is an FF rosin treated with 4.75% calcium hydrate, which raises the melting point, decreases the free rosin acids, and decreases the tendency to crystallize. It is used in paints, varnishes, and molded products. **Soda-treated rosins,** with about 1% Na_2O, but no free alkali, are used for soap, paper size, and disinfectants. **Rosin size** is alkali-treated rosin in dry powder or emulsion form for sizing paper. **Dresinite** is such a sodium or potassium salt of rosin. **Cyfor,** of American Cyanamid Co., is a rosin size fortified with a synthetic resin to give increased water and acid resistance to paper. **Rosin ester,** or **ester gum,** is prepared by heating rosin with glycerin. It is lighter in color than rosin, has a higher softening point, and has a much lower acid number, usually 7 to 9. It is used with tung oil in enamels and varnishes and in adhesives. **Resin V** is a rosin glycerin ester gum. Rosin esterified with glycerin has lower molecular weight and is not as stable as rosin esterified with pentaerythritol or other tetrahydric alcohol; but modified rosin ester gums develop hardness quickly in nitrocellulose and are used for such purposes as furniture lacquers. **Cellolyn 102,** of Hercules Inc., is a modified ester gum of this type, and **Lewisol 28** is a maleic alkyd modified rosin ester used for hard, glossy furniture lacquers. **Hydroabietyl alcohol,** used as a plasticizer and tackifier for rubber and for sizing textiles, is a colorless, tacky liquid made by reduction of rosin. The 85% alcohol has a specific gravity of 1.008 and a flash point of 369°F (187°C). It is also used for making rosin esters. **Abitol,** of Hercules, is this material. **Abalyn** is a methyl ester of abietic acid, **methyl abietate,** made by treating rosin with methyl alcohol. It is a liquid rosin used as a plasticizer.

Hydrogenated rosin has greater resistance to oxidation than common rosin, has less odor and taste, and has a pale color that is more stable to light. It is used in protective coatings, in paper size, in adhesives, in soaps, and as a tackifier and plasticizer in rubber. Because of its saturated nature, it cannot be used for rosin-modified plastics. The average acid number is 162, saponification value 167, and softening point 157°F (69°C). **Staybelite resin,** of Hercules, may be glycerol ester or ethylene glycol ester of hydrogenated rosin. Another modified rosin, a perhydrogenated variety, is **Foral Ax.** And a disproportionated rosin from the same company is **Rosin 731.** **Vinsol resin** is a hard, high-melting-point, dark resin produced from the distillation of wood, or is the black residue left after rosin is extracted with petroleum solvents. It is soluble in alcohols, has a melting point of 239°F (115°C), and is used for insulating varnishes where light color is not essential, and for compounding in thermoplastics. **Hercolyn** is Abalyn hydrogenated to saturate the double bonds with hydrogen. **Flexalyn,** of the same company, is a pale-colored, very tacky, semisolid resin produced by the esterification of rosin acids with diethylene glycol, and it has a complex chemical structure. It is used in adhesives to give added tack and strength. Although rosin should be considered a natural resin, it is used primarily in coatings as a synthetic ester. The principal materials are the glycerol ester of rosin, or *ester gum,* and the **pentaerythritol ester,** or **PE ester gum.** The **Pentalyn resins** of Hercules are pentaerythritol esters. **Pentalyn M** is a phenol-formaldehyde-modified pentaerythritol ester; it has a melting point of 329°F (165°C), and when used in linseed oil varnishes, it gives a tough coating. Maleic modified ester gums have higher melting points and show better color retention in varnishes than simple rosin esters. **Ester gum varnishes** are used primarily in low-cost, interior finishes that are not exposed to the weather.

Rosin is hardened by polymerization to form a dipolymer of abietic acid. The product is then pale in color, and has a lower acid number and a higher melting point than rosin. **Poly-pale resin,** of Hercules, is a polymerized rosin with melting point of 208 to 217°F (98 to 103°C), acid number 152 to 156, and saponification value 157 to 160. It can be substituted for natural copals in paints, and in gloss oil it gives water resistance and high viscosity. In the making of metallic resinates, it gives higher melting points, higher viscosity, and better solubility than natural rosin. Another modified rosin of this company is **Dymerex resin.** It consists chiefly of dimeric rosin acids, is highly soluble, is resistant to oxidation, and has a high softening point at 282°F (139°C). It is used in synthetic resins and protective coatings. **Rosin amine D,** of the same company, is a primary amine made from rosin. It is a yellow, viscous liquid that wets glass and siliceous materials. It is soluble in most organic solvents and emulsifies in water. It is used in cutback asphalts, in road asphalts, in asphalt cements, in ceramic inks, in foundry core binders, and in paper pulp to improve adhesion of resins. **Nuroz** is a polymerized rosin with a melting point of 169°F (76°C) and acid value 161. It has high resistance to oxidation and is used in varnishes and soaps.

Rosin oil is an oil produced by the dry distillation of rosin at a temperature of 392 to 680°F (200 to 360°C). There are two qualities of the oil: a light spirit, **pinolin,** which forms from 1 to 5% of the rosin, and a bluish, heavy oil, which forms 80 to 84%. It contains abietic acid and has an acid value of about 28. The commercial oil has a specific gravity of 1.020 with a flash point of 320 to 338°F (160 to 170°C). The refined oil is a yellow liquid with a pleasant odor and is used for blending with turpentine. It is also employed as a plasticizer in rubber, as a tack producer in rubber cements, and in synthetic molding resins. When treated with lime, it may be used to mix with lubricating oils. The light distillate is used sometimes in pharmacy under the name of **oil of amber.** Blended rosin oil is a mixture with mineral oils.

ROUGE. A hydrated iron oxide used for polishing metals and in break-in lubricants for aluminum bronze bearings. It has a Mohs hardness of 5.5 to 6.5 and is made by calcining ferrous sulfate and driving off the sulfur. The color is varying shades of red; the darker the color, the harder the rouge. The grains are rounded, unlike the grains of crocus. The pale-red rouge is used for finishing operations; the other grades are used for various polishing of metal surfaces. **Stick rouge** is made of finely crushed powder. Although the word *rouge* means red, materials of other colors are used for buffing and are called rouge. **Black rouge,** also called **Glassite,** is magnetic iron oxide made by precipitating ferrous sulfate with caustic soda. It is used for buffing but is not popular because it stains the skin. **Green chrome rouge** is **chromium oxide,** CrO, made by the strong heating of chromic hydroxide. It is used for buffing stainless steels. When used as a paint pigment, it is called **Guignet's green. Satin rouge** is a name applied to lampblack when used as a polishing medium in the form of brick for polishing silverware. **Crocus** is a name applied to mineral powders of a deep-yellow, brown, or red color made into cakes with grease for polishing. **Polishing crocus** is usually red ferric oxide used for a buffing glass and jewelry. **Crocus cloth** is a fabric coated with red iron oxide, marketed in sheets and used for polishing metals.

RUBBER. A gum resin exudation of a wide variety of trees and plants, but especially of the tree *Hevea brasiliensis* and several other species of *Hevea* growing in all tropical countries and cultivated on plantations in southern Asia, Indonesia, Sri Lanka, Zaire, and Liberia, from which **natural rubber elastomers** are made.

The gum resin was formerly referred to as **India rubber,** and the name given it by Charles Goodyear was **gum elastic.** The first highly compounded rubber for insulation, developed in 1867, was called **Kerite.** Brazilian rubber is sometimes called **Pará rubber. Caoutchouc** was an early name for the crude rubber then cured over a fire into a dark, solid mass for shipment. **Castilla rubber,** or **castilloa,** is from the large tree *Castilla elastica,* and was the original rubber of the Carib and Mayan Indians, but was cultivated only in Mexico and in Panama where it was called **Panama rubber.** The latex and rubber are identical with **heavea rubber** after purification. **Euphorbia rubber** is from vines of the genus *Euphorbia,* of which there are 120 species in tropical Africa. Much **mangabeira rubber** was formerly produced in the Amazon Valley. It is the latex of the mangabeira tree which comprises various species of the genus *Hancoria* and yields the edible fruit **mangaba.** The latex is coagulated with alum or sodium chloride, but the native Indians coagulated it with the latex of the **caxiguba** tree, *Ficus anthelmintica,* giving a better rubber. The rubber is softer than hevea rubber, but ages better. The low-grade **Assam rubber** is from a species of **ficus tree,** *F. elastica,* of India and Malaya. **Ceara rubber** comes from the small, rapid-growing tree, *Manihot glaziovii,* native to the semidesert regions of Brazil but now grown in India and Sri Lanka. The rubber is of good grade.

Rubber latex is a colloidal emulsion of the gathered sap, containing about 35% of rubber solids, blended from various sources to give average uniformity. The latex is coagulated with acid and milled into ribbed sheets called **crepe rubber,** or into sheets exposed to wood smoke to kill bacteria and called **smoked sheet rubber.** These sheets constitute the commercial **crude rubber,** although much rubber latex is used directly, especially for dipped goods such as gloves, toys, and balloons; for coatings; and for making foam rubber. Rubber has the property of being vulcanized with sulfur and heat, removing the tackiness and making it harder and more elastic in the low-sulfur compounds. All natural rubber except adhesive rubbers is thus vulcanized rubber. Ordinary **soft rubber** contains only 3 to 6% sulfur, but usually also contains softeners, fillers, antioxidants, or other compounding agents, giving varying degrees of elasticity, strength, and other qualities. When as much as 30% sulfur is added, the product is called **hard rubber. Vapor-cured rubber** is rubber vulcanized by sulfur chloride fumes and neutralized with magnesium carbonate. It is used for thin goods only. **Acid-cured rubber** is rubber cured in a bath of sulfur chloride in a solvent.

The tensile strength of rubber of low vulcanization is 800 to 1,200 lb/in^2 (6 to 8 MPa) of the original cross section. A good soft rubber can be stretched as much as 1,000% without rupture and will return to close to the original length with little permanent set. The specific gravity is about 1.05, but with fillers may be as high as 1.30. When the term **vulcanized rubber** is now used, it generally refers to **hard rubber** vulcanized to a rigid but resilient solid, used for electrical parts and tool handles. **Ace hard rubber** has a specific gravity of 1.27, a tensile strength of 8,700 lb/in^2 (60 MPa), dielectric strength of 485 V/mil (19×10^6V/m), heat distortion temperature of 172°F (78°C), and water absorption 0.04%. **Vulcanite** and **Ebonite** are old names for hard rubber. **Reclaimed rubber** is produced largely from old tires and factory scrap. It is usually lower in cost than new rubber, but it is easier to process and is employed in large quantities even when the price is higher. It is sold in sheets, slabs, pellets, and powder, but much of the **rubber powder,** or **granulated rubber,** used for adhesives and molding is not reclaimed rubber but is made by spray-drying latex. In a reclamation system developed by American Resource Recovery Corp., waste tires depolymerized in a high-temperature vacuum chamber are converted into oil, carbon black, steel, and ash.

Several types of modified natural rubber are used in the production of coatings, protective films, and adhesives. These types are **chlorinated rubber, rubber hydrochloride,** and **cyclized rubber** or

isomerized rubber. Chlorinated rubber, for example, modified with any one of a number of plastic resins, provides maximum protection against a wide range of chemicals, and the coatings are widely used in chemical plants, in gas works, and as tank-car linings.

Red rubber is now simply rubber colored red, but was originally rubber vulcanized with antimony pentasulfide which broke down with the heat of vulcanization, yielding sulfur to the rubber and coloring it red with the residual antimony trisulfide. Many trade-name accelerators, fillers, and stiffeners are marketed for rubber compounding. **Crumb rubber** is any rubber in the form of porous particles that can be dissolved easily without milling, cutting, or pelletizing. It is used in adhesives and plastics. **Magnetic rubbers,** produced in sheets and strips of various magnetic strengths, are made of synthetic rubbers compounded with magnetic metal powders.

RUBIDIUM. A rare metallic element, symbol Rb, atomic weight 85.45, belonging to the group of alkali metals. The chief occurrence of rubidium is in the mineral lepidolite. There is no real rubidium ore, but the element is widely disseminated over the earth in tiny quantities. It is a necessary element in plant and animal life and is found in tea, coffee, tobacco, and other plants. It is a silvery-white metal, with a specific gravity of 1.53, melting point, of 102°F (39°C), and boiling at 1270°F (688°C). It takes fire easily in air and decomposes water. Of all the alkali metals it is next to cesium in highest chemical activity. It can be obtained by electrolysis, but has few industrial applications owing to its rarity. Its chief use is in electronics. For photoelectric cells it is preferred to cesium, and a very thin film is effective. Like potassium, it has a weak radioactivity by the emission of beta particles, the beta emission being only about one-thousandth that of an equal weight of uranium.

RUBY. A red variety of the mineral **corundum** which ranks with the best grades of precious stones as a gemstone, while the off-color stones are used for watch and instrument bearings. Most of the best rubies come from upper Burma, Thailand, and Cambodia, but the center of natural-ruby cutting is near Bombay. Some deep-red rubies are found in east Africa, and they also occur in western North Carolina. The carmine-red, or pigeon's-blood, stones are the most highly prized. Before the advent of the synthetic ruby, the larger stones were more valuable than diamond. The pink to deep-red colors of the ruby are due to varying percentages of chromic oxide. **Star rubies** contain also a small amount of titania which precipitates along crystallographic planes of the hexagonal crystal and shows as a movable six-ray star when the gem is cut with the axis normal to the base of the stone. **Spinel ruby** is not corundum but magnesium aluminate and is a spinel, often occurring in the same deposits.

Synthetic rubies are equal in all technical qualities to the natural, and synthetic star rubies surpass the natural stones in perfection and quality. Most of the ruby used for instrument bearings is synthetic corundum colored with chromic oxide, since the instrument makers prefer the red color, but the name *ruby* is often applied regardless of color. The U.S. practice in bearing manufacture is to start with a cylindrical rod of the diameter of the desired bearing and to slice to the required thickness. The single-crystal rods are flame-polished. For industrial uses the name *ruby* is often applied to the synthetic material even when it is not red. Synthetic rubies with 0.05% chromium are also used for **lasers,** or light amplifiers, to produce high-intensity light pulses in a narrow beam for communications. Lasers have a wide range of uses, and various other materials are used for specific purposes. Crystals of **potassium dihydrogen phosphate** are used to control the direction of narrow beams of light. The crystal is mounted on the face of the cathode-ray tube. Platelets of zinc oxide with a phosphor, when activated at 15,000 to 20,000 V by a pulsated electron stream, give a beam of pure ultraviolet light at 14,764 nin (375 nm) for use in chemical synthesis.

Rubies are also used in **masers** to detect radio signals for space rockets at great distances. The word *maser* means *m*icrowave *a*mplification by *s*timulated *e*mission of *r*adiation. Ruby has the same physical and chemical properties as the sapphire and corundum. But the color inclusions do affect the electronic properties. The **ruby-sapphire crystals** produced by Linde for **optical masers** are grown with a core of ruby and an overlay sheath of sapphire. The ruby core containing 0.05% chromic oxide emits a beam of extremely high frequency, 4.2×10^{14} cycles, for sending messages. The sapphire has a high refractive index, 1.76, and when the sheath is surrounded by a helical xenon arc, the light from the arc excites the chromium atoms in the ruby to emit a concentrated beam of parallel rays from one end.

The cutting of rubies, sapphires, and other hard crystals into tiny shaped bearings was formerly a specialized hand industry, and large stocks of cut stones were kept in the National Stockpile for wartime emergencies. But the slicing and shaping of hard crystals are now widely dispersed on a production basis to meet the needs of the electronic industries, and the equipment is regularly manufactured. For example, the **Accu-Cut wheels** of Aremco Products, for slicing and notching crystals and hard electronic ceramics, are metal disks 0.010 to 0.030 in (0.025 to 0.076 cm) thick and 1 to 4 in (2.54 to 10.2 cm) in diameter, with diamond grit metallurgically bonded to the periphery.

RUTHENIUM. A hard, silvery-white metal, symbol Ru, having a specific gravity of 12.4, a melting point of about 4190°F (2310°C), and a Brinell hardness of 220 in the annealed state. The metal is obtained from the residue of platinum ores by heat reduction of **ruthenium oxide,** RuO_2, in hydrogen. Ruthenium is the most chemically resistant of the platinum metals and is not dissolved by aqua regia. It is used as a catalyst to combine nitrogen in chemicals. As **ruthenium tetroxide,** RuO_4, it is a powerful catalyst for organic synthesis, oxidizing alcohols to acids, ethers to esters, and amides to imides. Ruthenium has a close-packed hexagonal crystal structure. It has a hardening effect on platinum, 50% addition of ruthenium raising Brinell hardness from 30 to 130 and the electrical resistivity to double that of pure platinum. **Ruthenium-platinum alloys** are used for electric contacts, electronic wires, chemical equipment, and jewelry. The alloy with 5% ruthenium has a tensile strength, annealed, of 60,000 lb/in² (414 MPa) with elongation of 34% and Brinell hardness of 130. The hard metal has a Brinell hardness of 210. The alloy with 10% ruthenium as a tensile strength of 85,000 lb/in² (586 MPa), and a Brinell hardness of 190 in the soft condition and 280 when hard-drawn.

Several ruthenium intermetallic compounds hold promise for potential high-temperature, aircraft-turbine parts because of their high melting temperature and evidence of room-temperature ductility. Identified at General Electric's Research & Development Center are **aluminum ruthenium** with a specific gravity of 7.95 and a melting temperature of 4100°F (2060°C); **ruthenium scandium,** 7.40, 3992°F (2200°C); and **ruthenium tantalum,** 14.83, 3776°F (2080°C). AlRu and RuSc are the most promising because of their light weight and better ductility. $Al_{47}Ru_{53}$ is the most oxidation-resistant, and $Al_{48}RuY$ could be used at temperatures up to 2280°F (1250°C).

RYE. The seed of the plant *Secale cereale* used as a food grain, but in the United States and Great Britain it is valued chiefly for the production of whiskey and alcohol and for feeding animals. Only 4% of the world production is in the United States. The grain looks like wheat, and the stalks of the plant are slender and tough, growing to a height of up to 6 ft (1.8 m). But flour made from the grain produces bread that is dark in color, bitter, and soggy. When used for flour in the United States, it is mixed with wheat and other flours. The plant has the advantage that it will grow on poor soil, in arid regions, at high altitudes, and in regions of severe winter. It is thus a grain of poor agricultural countries, and it has been called the **grain of poverty. Rye straw** is the dried and sun-bleached stalks of the plant. It is very tough and resilient and is the most valued of all the commercial **straw** derived from grains. It is used for packing, bedding, and the manufacture of strawboard.

SAGO FLOUR. A starch extracted from the pith of the sago palm, *Metroxylon sagu,* of Indonesia and Malaya, and from the aren palm. Sago is valued industrially for sizing and filling textiles because, like tapioca, it holds mineral fillers better than other starches. It gives a tougher and more flexible feel than tapioca, but its tan color limits its use. From 600 to 800 lb (272 to 363 kg) of crude sago is obtained from a tree, which is destroyed in the process. **Pearl sago,** used for food, is the same material made into dough and forced through a sieve. **Bullet sago,** a lower grade, is used for sizing. The aren palm, or **sugar palm,** is the species which yields arenga fiber. It contains only 20% as much sago as the sago palm, but a juice called **taewak** is produced from the cut flower stems and is used to make **palm wine,** or **arak.** The juice is also boiled down to produce a brown **palm sugar** used for sweetening.

SALMON OIL. A pale-yellow oil obtained as a by-product in the salmon canning industry and employed as a drying oil for finishes and in soaps. There are different classes of the oil, depending upon the type of salmon. The oil contains an average of 23.5% **arachidonic acid,** $C_{19}H_{31}COOH$; 16.2 clupanodonic; 11.5 linoleic; 17.1 oleic; 15 palmitic; 10.6 **palmitoleic acid,** $C_{15}H_{29}COOH$; 4 myristic; and 2 stearic acids. The specific gravity is 0.926. It has a high iodine number, up to 160, but does not form an elastic skin on drying and is not a good varnish oil untreated. It is, however, a valuable source of fatty acids for paint-oil blends and for plastics. The salmon is a valuable food fish and is extensively canned. There are five commercial species of North Pacific salmon of the genus *Oncorhynchus*; the **steelhead trout,** or **salmon trout,** *Salmo gairdneri,* is of the Atlantic. The **Atlantic salmon,** caught off Newfoundland, is *S. salar.* The **red salmon,** or **sockeye salmon,** is *O. nerka*; the **pink salmon** is *O. gorbuscha*; and the **Chinook,** or **king salmon,** is *O. tschawytscha.* The catch is in the rivers on both sides of the Alaskan peninsula and in the Columbia River where the fish enter the rivers to spawn. **Australian salmon,** which is the chief fish canned in Australia, is of a different genus, *Arripis trutta.*

SALT. The common name for **sodium chloride,** known in mineralogy as **halite** but chemically a salt; it is any compound derived from an acid by replacing hydrogen atoms of the acid with the atoms of a metal. **Common salt,** or sodium chloride, is widely used as a preservative, for flavoring food, in freezing mixtures, for salt-brine quenching baths, and for the manufacture of soda ash and many chemicals. Common salt has such a variety of uses that its curve of consumption practically parallels the curve of industrial expansion. It is a stable compound of composition NaCl, containing theoretically 60.6% chlorine, but it usually contains impurities such as calcium sulfate and calcium and magnesium chlorides. The hardness of salt is 2.5, specific gravity 2.1 to 2.6, and melting point 1472°F (800°C). It is colorless to white, but when impure, it may have shades of yellow, red, or blue. It occurs in crystalline, granular masses with cubical cleavage, known as **rock salt,** or **mineral salt.** Vast deposits of salt are found underground in Louisiana, and a large area of Kansas is underlain with a salt deposit reaching 800 ft (244 m) in thickness. The rock salt occurring in immense quantities on the island of Hormuz in the Persian Gulf contains 97.4% NaCl, 1.83 $CaCO_3$, and only very small amounts of magnesium chloride, iron oxide, and silica. **Bay salt** is an old name for salt extracted from seawater, now known as **solar salt.** Seawater also contains more than 20% of magnesium chloride and magnesium, calcium, and potassium sulfates, which are extracted to give a purity of at least 99% sodium chloride. A short ton (907 kg) of **seawater** contains about 55 lb (25 kg) of common salt. But some **sea salt** containing all the original elements is marketed for corrosion tests. **Sea-Rite salt** of Lake Products Co. is a synthetic sea salt containing all the elements of natural sea salt except those of less than 0.0004%. From the salt wells of Michigan, magnesium, bromine, and other elements are extracted, and the salt brine is an important source of these elements.

Commercial salt is marketed in many grades, depending chiefly on the size of the grain. The term **industrial salt** refers rather to the method of packing and shipping than to a grade distinct from **domestic salt,** but most of the industrial salt is rock salt; the bulk of the domestic salt is evaporated salt. Producers of salt for the food processing industries usually guarantee a quality 99.95% pure since small amounts of calcium, magnesium, copper, and iron in the salt may give a bitter taste, discolor some foods, or cause oxidation rancidity in foods containing fats. **Crystal Flake salt,** of Diamond Salt, is 99.5% pure with less than 1.5 parts per million of copper or iron. **Micronized salt,** tradenamed **MicroSized salt,** of the same company, is 99.9% pure with thin flake particles of super fineness which have a solubility 5 times greater than that of granulated salt. **Improved Fine Flake salt** has a precisely sized grain for better flow properties as a builder, carrier, and diluent in dishwashing detergents, dry bleaches, floor care products, and cosmetics. The **CMF** salt is purified to have less than 20 parts per million of calcium and magnesium, to increase the sudsing properties of bar soaps, shampoos, and conditioners in personal-care products. **Flour salt** and **coarse-grain salt** grades are used as abrasives. **Dendritic salt,** of Morton Salt, is highly purified salt evaporated by a process that produces tiny dendritic crystals instead of the regular cubic from. It has a faster dissolving rate. This company produces **Calacid salt,** a formulated salt in tablet form for regulating acidity in canned foods. It contains 60.9% sodium chloride with citric acid, calcium sulfate dihydrate, and sodium carbonate for effervescence. **See 100,** of Diamond Salt, is a noncaking, free-flowing salt for use in automatic food processing equipment. It contains 5 ppm of prussiate of soda, which blocks the interchemical action of the salt crystals without decreasing solubility, and 0.04% glycerin to increase the free-flowing properties.

Domestic consumption of salt for direct human consumption is large in all countries. It is required in the bloodstream up to about 3.5% and is rapidly exhausted in hot weather. **Salt tablets,** used in hot weather or by workers in steel mills, are made with about 70% salt and 30 dextrose. To make domestic salt free-flowing in humid weather, 2% of calcium sulfate may be added. Salt obtained by simple evaporation of seawater contains salt-resistant bacteria which are capable of developing in salted hides or fish and injuring the material. Mineral salt is thus preferred for these purposes. The glasslike salt crystals, called **struvite,** that form in canned fish and some other products in storage are not common salt, but are crystals of magnesium ammonium phosphite hexahydrate, $Mg(NH_4)PO_4 \cdot 6H_2O$. They are harmless, but objectionable in appearance, and common salt is added to inhibit their growth. In making **salt brines** for steel treating and other industrial purposes, 100 parts of water at ordinary temperature will dissolve 36 parts of common salt, but no more than 15% is ordinarily used because of the corrosive effect.

Sodium hypochlorite, or **sodium oxychlorite,** $NaOCl$, is a stable, noncorrosive salt used in tanneries. **Merclor D** is a trade name of Monsanto Co. for this material in water solution. **Javel water** is a name given in the laundry industry to a water solution of $NaOCl$ used as a **bleach.** Household bleach contains 5% $NaOCl$. When used as a bleach in the textile industry, it is called **chemic.** In lower chlorine concentrations it is also used to disinfect potable water in municipalities. **Chloramone,** named after its manufacturer, is such a product. **Sodium chlorite,** $NaClO_2$, is a white to yellow, crystalline, water-soluble powder used as a bleaching agent for textiles and paper pulp. It is stable up to 302°F (150°C). It yields ClO_2 in solution, is an oxidizing agent, and attacks the coloring matter without injuring the fibers. It is also used for waterworks purification. **Textone** is a trade name for sodium chlorite as a bleach for textiles. **Sodium chlorate,** $NaClO_3$, is used in large quantities as a weed killer and for cotton defoliation and is also used for paper pulp and textile bleaching. It comes in water-soluble, colorless crystals melting at 482°F (250°C). It is used as the electrolyte in the chemical machining of metals.

SAND. An accumulation of grains of mineral matter derived from the disintegration of rocks. It is distinguished from gravel only by the size of the grains or particles, but is distinct from clays which contain organic materials. Sands that have been sorted out and separated from the organic material by the action of currents of water or by winds across arid lands are generally quite uniform in size of grains. Usually commercial sand is obtained from river beds or from sand dunes originally formed by the action of winds. Much of the earth's surface is sandy, and these sands are usually quartz and other siliceous materials. The most useful commercially are **silica sands,** often above 98% pure. Silica sands for making glass must be free from iron. The sand mined near Hot Springs, Arkansas, called **amosil,** is 99.5% pure silica and comes in transparent, rounded grains of 118-μin (3-μm) average size. **Beach sands** usually have smooth, spherical to ovaloid particles from the abrasive action of waves and tides and are free of organic matter. The white beach sands are largely silica but may also be of zircon, monazite, garnet, and other minerals, and are used for extracting various elements. **Monazite sand** is the chief source of thorium. The **black sands** of Oregon contain chromate, and those of Japan contain magnetite. Kyanite is found in the Florida sands.

Sand is used for making mortar and concrete and for polishing and sandblasting. Sands containing a little clay are used for making molds in foundries. Clear sands are employed for filtering water. Sand is sold by the cubic yard (0.76 m^3) or ton (0.91 metric ton) but is always shipped by weight. The weight varies from 2,600 to 3,100 lb/yd^3 (1,538 to 1,842 kg/m^3), depending on the composition and size of grain. Construction sand is not shipped great distances, and the quality of sands used for this purpose varies according to local supply. **Standard sand** is a silica sand used in making concrete and cement tests. The grains are free of organic matter and pass through a 20-mesh sieve, but are retained on a 30-mesh. **Engine sand,** or **traction sand,** is a high-silica sand of 20 to 80 mesh washed free of soft bond and fine particles, used to prevent the driving wheels of locomotives or cars from slipping on wet rails.

Molding sand, or **foundry sand,** is any sand employed for making molds for casting metals, but especially refers to sands that are refractory and also have binding qualities. Pure silica is ideal for heat resistance, but must contain enough alumina to make it bind together. Molding sands may contain from 80 to 92% silica, up to about 15 alumina, about 2 iron oxide, and not more than a trace of lime. Some molding sand contains enough clay or loam to bond it when it is tamped into place. The amount of bond in **Grant sand** and in **Tuscarawa sand** is 17 to 18%. About 33% of these natural sands pass through a No. 100 screen, and 20 through a No. 150 screen. The finer the grain, the smoother the casting, but fine-grained sand is not suitable for heavy work because of its impermeability to gases. Sands without natural bond are more refractory and are used for steel molding. Sands for steel casting must be silica sands containing 90% silica, or preferably 98, and are mixed with 2 to 10 fireclay. For precision casting, finely ground aluminum silicate is used in the silica sand mixes, and it requires less of a bonding agent. **Calamo,** of Harbison-Walker Refractories Co., is aluminum silicate for this purpose.

Zircon sand has high heat resistance and is used for alloy steel casting. **Zircon flour** is finely milled zirconite sand used as a mold wash. **Zirconite sand** for molding is 100 to 200 mesh in its natural state. It is 70% heavier than silica sand and has a higher heat conductivity that gives more rapid chilling of castings. The zirconite sands have melting points from 3650 to 3850°F (2010 to 2121°C). Common molding sands may contain from 5 to 18% of clay materials and may be mixtures of sand, silt, and clay, but they must have the qualities of refractoriness, cohesiveness, fineness of grain, and permeability. To have refractory quality, they must be free of calcium carbonate, iron oxide, and hydrocarbons. **Core sands** also have these qualities, but they are of coarser grain and always require a bond that will bake solidly but will break down easily at the temperature of pouring. About 25% of a medium molding sand will be retained on a 150-mesh sieve, and about 10% on a 200-mesh sieve. Sand with rounded grains is preferred, and the grains must be very uniform in size to prevent filling. When a molding sand is burned out, it is made suitable for reuse by adding bond; but when fireclay is used as a bond, it adheres to the sand grains and that makes it unsuitable for reuse. **Parting sand** is a round-grained sand without bond used on the joints of molds. **Foundry parting** is usually tripoli or bentonite. Cores are made with sand mixed with core oils. **Greensand cores** are unbaked cores made with molding sand.

Sandblast sand is sand employed in a blast of air for cleaning castings, removing paint, cleaning metal articles, giving a dull, rough finish to glass or metal goods, or renovating the walls of stone or brick buildings. Sandblast sand is not closely graded, and the grades vary with different producers. The U.S. Bureau of Mines gives the following usual range: No. 1 sand should pass through a 20-mesh and be retained on a 48-mesh screen; No. 2 should pass through a 10-mesh and be retained on a 28-mesh screen; No. 3 through a 6-mesh and be retained on a 14-mesh screen; No. 4 should pass through a 4-mesh and be retained on an 8-mesh screen. Number 1 sand is used for light work where a smooth finish is desired; No. 4 sand is employed for rough cast-iron and cast-steel work. Sharp grains cut faster, but rounded grains produce smoother surfaces. The sand is usually employed over and over, screening out the dust. The dust and fine used sand may be blasted wet. This is known as *mud blasting* and produces a dull finish.

SANDALWOOD.   The heartwood of the evergreen tree *Santalum album* and other species of southern Asia. The heartwood is usually equivalent to about one-third of the log. It is sweet-scented and is used for chests, boxes, and small carved work. The chips and sawdust are used for incense and for oil production. **Sandalwood oil** is a yellowish essential oil of specific gravity 0.953 to 0.985, distilled from the wood, which yields 5 to 7% of the oil. It is used in medicine, perfumery, and soaps. West Indian sandalwood oil is called **anyris oil.** **Australian sandalwood** oil is from the tree *S. spicatum.* It has a very strong and lasting sandalwood odor. **Sandela,** of Givaudan Corp., is a synthetic sandalwood oil. It is a polycyclic alcohol product with the odor and properties of the natural oil.

SANDARAC.   Known also as **white gum,** or **Australian pine gum.** A white, brittle resin obtained as an exudation from various species of the coniferous tree *Callitris,* known as **Cyprus pine.** The north African sandarac is from the tree *C. quadrivalis* of the Atlas Mountains, and it resembles the resin from the Australian tree *C. arenosa.* The trees in Morocco are tapped from May to June, and two months later the small tears of gum are gathered. Sandarac is used in varnishes and is soluble in turpentine and alcohol. It melts at 275 to 284°F (135 to 140°C). It gives a hard, white spirit varnish used for coating labels and for paper and leather finishes. The product is available from O.G. Innes Corp. Ground sandarac, under the name of **pounce,** was formerly used as a pouncing powder and for smoothing parchment and tracing cloth, but is now replaced by pumice.

SANDPAPER.   Originally a heavy paper coated with sand grains on one side, used as an abrasive, especially for finishing wood. Sharp grains obtained by crushing quartz later replaced sand, and the product was called *flint paper.* But most abrasive papers are now made with aluminum oxide or silicon carbide, although the term *sandpapering* is still employed in wood polishing. Quartz grains, however, are still much used on papers for the wood industries. For this purpose the quartz grains are in grades from 20 mesh, known as No. 3.5

through No. 3, 2.5, 2, 1.5, 0, 00, and 000. All the No. 000 grains pass through a 150-mesh sieve, with 25% retained on a 200-mesh sieve and 80% on a 325-mesh sieve. Good **sandpaper quartz** will contain at least 98.9% silica. The paper used is heavy, tough, and flexible, usually 70- or 80-lb (32- or 36-kg) paper, and the grains are bonded with a strong glue. A process is also employed to deposit the grains on end by electrostatic attraction so that the sharp edges of the grains are presented to the work.

SANDSTONE. A consolidated sand rock, consisting of sand grains united with a natural cementing material. The size of the particles and the strength of the cement vary greatly in different natural sandstones. The most common sand in sandstone is quartz, with considerable feldspar, lime, mica, and clayey matter. The cementing material is often fine chalcedony. **Silica sandstones** are hard and durable but difficult to work. **Calcareous sandstone,** in which the grains are cemented by calcium carbonate, is called **freestone** and is easily worked, but it disintegrates by weathering. Freestone is homogeneous and splits almost equally well in both directions. **Chert,** formerly used as an abrasive and, when employed in building and paving, known under local names as **hearthstone, firestone,** and **malmstone,** is a siliceous stone of sedimentary origin. It has a radiating structure and splintery fracture and is closely allied to flint. In color it is light gray to black or banded. The colors of sandstones are due to impurities, pure siliceous and calcareous stones being white or cream-colored. The yellow to red colors usually come from iron oxides, black from manganese dioxide, and green from glauconite. **Crab Orchard stone** of Tennessee is high in silica with practically no CaO and is often beautifully variegated with red and brown streaks. It splits in uniform slabs and is used for facing. The compressive strength is high, up to 24,000 lb/in^2 (165 MPa), and the density is 165 lb/ft^3 (2,643 kg/m^3). The water absorption is less than 2%.

About half of the commercial sandstone block in the United States comes from Ohio. It has a density of 140 lb/ft^3 (2,243 kg/m^3) and a compressive strength of 10,000 lb/in^2 (69 MPa), but the average of much other sandstone is 135 lb/ft^3 (2,163 kg/m^3) with a compressive strength of 12,000 lb/in^2 (83 MPa). Sandstones for building purposes are produced under innumerable names, usually referring to the locality. The **bluestone** of New York State is noted for its even grain and high crushing strength, up to 19,000 lb/in^2 (131 MPa). It contains about 70% silica sand with clay as the binder. **Amherst sandstone** from Ohio contains up to 95% silica with 4 aluminum oxide and is colored gray and buff with iron oxides. **Flexible sandstone,** which can be bent, comes from North Carolina. It is **itacolumite** and has symmetrically arranged quartz grains which interlock and rotate against one another in a binder of mica and talc.

Holystone is a block of close-grained sandstone, formerly used for rubbing down the decks of ships and still used for rubbing down furniture and concrete work. **Briar Hill stone** and **Macstone** are trade names for building blocks consisting of lightweight concrete faced with a slab of sandstone. **Kemrock** is a sandstone impregnated with a black furfural resin and baked to a hard finish. It is used for tabletops and chemical equipment to resist acids and alkalies. The term **reservoir rock** refers to friable, porous sandstone that contains oil or gas deposits. The porosity of such sandstone or compacted sand of Pennsylvania is from 15 to 20%, while that of California and the Gulf Coast is 25 to 40%. A sandstone of 20% porosity may contain as much as 75,000 bbl of oil per acre-foot (1,234 m^3).

SANDWICH MATERIALS. A type of **laminar composite** composed of a relatively thick, low-density core between faces of comparatively higher density. **Structural sandwiches** can be compared to I beams. The facings correspond to the flanges, the object being to place a high-density, high-strength material as far from the neutral axis as possible, thus increasing the section modulus. The bulk of a sandwich is the core. Therefore, it is usually lightweight for high strength-to-weight and stiffness-to-weight ratios. However, it must also be strong enough to withstand normal shear and compressive loadings, and it must be rigid enough to resist bending or flexure.

Core materials can be divided into three broad groups: cellular, solid, and foam. Paper, reinforced plastics, impregnated cotton fabrics, and metals are used in cellular form. Balsa wood, plywood, fiberboard, gypsum, cement-asbestos board, and calcium silicate are used as solid cores. **Plastic foam cores**—especially polystyrene, urethane, cellulose acetate, phenolic, epoxy, and silicone—are used for thermal insulating and architectural applications. **Foamed inorganics** such as glass, ceramics, and concrete also find some use. **Foam cores** are particularly useful where the special properties of foams are desired, such as in insulation. And the ability to foam in place is an added advantage in some applications, particularly in hard-to-get-at areas.

Of all the core types, however, the best for structural applications are the **rigid, cellular cores.** The primary advantages of the cellular core are that (1) it provides the highest possible strength-to-weight ratio and (2) nearly any material can be used, thereby satisfying virtually any service condition.

There are, essentially, three types of cellular cores: honeycomb, corrugated, and waffle. Other variations include small tubes or cones and mushroom shapes. All these configurations have certain advantages and limitations. **Honeycomb sandwich materials,** for example, can be isotropic, and they have a high strength-to-weight ratio, good thermal and acoustical properties, and excellent fatigue resistance. **Corrugated-core sandwich** is anisotropic and does not have as wide a range of application as honeycomb, but it is often more practical than honeycomb for high production and fabrication into panels.

Theoretically, any metal that can be made into a foil and then welded, brazed, or adhesive-bonded can be made into a cellular core. A number of materials are used, including aluminum, glass-reinforced plastics, and paper. In addition, stainless steel, titanium, ceramic, and some superalloy cores have been developed for special environments.

One of the advantages of sandwich construction is the wide choice of facings, as well as the opportunity to use thin sheet materials. The facings carry the major applied loads and therefore determine the stiffness, stability, and, to a large extent, strength of the sandwich. Theoretically, any thin, bondable material with a high tensile- or compressive-strength–weight ratio is a potential facing material. The materials most commonly used are aluminum, stainless steel, glass-reinforced plastics, wood, paper, and vinyl and acrylic plastics, although magnesium, titanium, beryllium, molybdenum, and ceramics have also been used.

SAPELE. The figured woods of various species of trees of tropical Africa which are mixed with khaya and exported from west Africa as **African mahogany.** Sapele woods are harder and heavier than red khaya, but shrink and swell more than khaya with changes in moisture. They are also darker with a purplish tinge. **Sapele mahogany,** also called **scented mahogany** and **West African cedar,** is from the *Entandrophragma cylindricum,* a very large tree growing on the Ivory Coast and in Ghana and Nigeria. On the Ivory Coast it is called **aboundikro.** Another species, *E. angolense,* is called **Tiama mahogany** on the Ivory Coast and in Nigeria is known as **brown mahogany** and **gedunohor.** A less heavy wood, from the tree *E. utile,* is known on the Ivory Coast as **Sipo mahogany** and in the Cameroons as **Assie mahogany.** It is one of the chief woods exported as mahogany from the Cameroons. The wood known on the Ivory Coast as **heavy mahogany** and **omu** in Nigeria is from the tree *E. candollei,* and it is much heavier than other sapeles. **Nigerian pearwood,** from species of *Guarea,* notably *G. cedrata* and *G. thompsonii,* is also exported as African mahogany. The woods are more properly called **guarea.** The color is pale pink to reddish. The weight is about the same as for sapele. The wood is of a finer texture than khaya, but is not figured like sapele of khaya. Another wood marketed as African mahogany is **lingue,** from the tree *Afzelia africana* of the west coast of Africa from Senegal to Nigeria. The wood is light brown, turning dark when seasoned, and is beautifully figured.

SAPONIN. Glycosides of the empirical formula $C_{32}H_{54}O_{18}$ which have the property of frothing with water. They are found in soap bark, soap nut, licorice, and other plants; when separated out, saponin is a white, amorphous powder of a disagreeable odor. Before the advent of the synthetic detergents saponin was important for replacing soaps in washing compounds where high sudsing was undesirable, and it was used in industrial scouring compounds, soapless shampoos, and tooth powders. It is still used in some detergents, in fire extinguishers, as an emulsifying agent, and for synthesizing other complex chemicals. It is found to some extent in sweets, such as **Turkish halva.** Saponin is not a single compound, but is a great group of **alicyclic compounds,** or five-member or more highly complex ring compounds having **aliphatic,** or fatty acid, properties. The saponins occur directly in plants where they have a triterpene structure and may be either converted to or derived from a great variety of acids, vitamins, and other products by photosynthesis or catalyzation. They are closely related to the styrols of animal life, and in both plant and animal life slight catalytic rearrangements with nitrogen produce the natural **venoms** and poisonous compounds. The saponins thus form one of the most useful of the basic chemical groups for biological and pharmaceutical work. Chemically, they are called **polymethylene compounds** and can be synthesized from petroleum. In the drug industry they are called **sapogenines.** Diosgenin and its derivative, hecogenin, are saponic compounds that are important raw materials for making steroids.

The saponins can be obtained from many plants. **Soap nut** is the fruit of the trees *Sapidus mukorossi* and *S. laurifolia* of northern India known locally as **ritha.** The soap nut has been used as a detergent in washing fabrics since ancient times. The ground nut is prized as an herbal shampoo, generically known as **Shikakai.** The nut has 56% of pericarp and 44 of seed, and saponin is found in the pericarp. It is extracted with solvents from the dried, powdered fruit. Saponin is soluble in water but insoluble in petroleum spirits. **Soapbark,** also called **morillo bark,** is the dried, inner bark of the tree *Quillaja saponaria* of the west coast of South America. It was used by the Incas, and the botanical name comes from the Inca word *quillean,* meaning to wash. The bark produces suds in water, but the powdered bark is highly sternutatory owing to fine crystals of calcium oxalate. It is marketed in brownish-white pieces and is used as a source of saponin. It has been used in beverages to produce froth, but is highly toxic, affecting the heart and respiration. In medicine it is called **quillaja** and is used as an irritant and expectorant. **Soapwort** consists of the leaves of the plant *Saponaria officinalis,* growing in North America. The leaves contain saponin which dissolves out in water to produce a lather useful for cleaning silk and fine woolens. **Soapberry** is the fruit of the tropical tree *Sapindus saponaria,* used in hair and toilet preparations. The soapberry of the American southwest consists of the fleshy berries of the small tree *S. drummondii.* **Soaproot** is the bulb root of the plant *Chlorogalum pomeridianum* of California. **Mexican soaproot** is the thick rootstock of *Yucca baccata* and of the **wild date,** *Y. glauca,* growing in the dry regions of Mexico and the southwestern United States. Both plants are called **yucca,** and in Mexico they are called **amole.** The Indians used the roots, which were called **vegetable soap** by the settlers, for washing. In the processing of yucca leaves to obtain fiber, about 20% of a powder is obtained which contains 3% saponin. **Yucca powder** is used in scouring compounds or for the extraction of saponin. The long, stout stems of the **soap plant,** *Chenopodium californicum,* also yield saponin.

SAPPHIRE.   A transparent variety of the mineral corundum. When it has the beautiful blue hue for which it is noted, it ranks with diamond, ruby, and emerald among precious gemstones. The off-color stones are cut for pointers and wearing points of instruments. The specific gravity of sapphire is 3.98, and the Mohs hardness is 9. The blue color is from iron and titanium oxides and is rarely uniform throughout the stone in the natural material. The green is produced with cobalt, and the yellow comes from nickel and magnesium. The **pink sapphire** contains a tiny proportion of chromic oxide, and larger amounts produce the dark-red ruby. The best gem sapphires come from India. A valuable **black sapphire** comes from Thailand. Industrial stones are found in Montana. Most natural sapphires are small, but a large one known as the **Star of Artaban** weighs 300 carats.

   Synthetic sapphire is produced by flame-fusing a pure alumina powder made from calcined ammonium aluminum sulfate. The fused material forms a boule as a single crystal. The average boule is 200 carats, but sometimes they are as large as 400 carats, or about 0.75 in (1.9 cm) in diameter and 2 in (5.1 cm) long. The rods are single crystals up to 0.23 in (0.58 cm) in diameter and 18 in (45.7 cm) long. **Sapphire balls,** for bearings and valves, are produced by Linde to great accuracy in diameters from 0.0625 to 0.625 in (0.16 to 1.59 cm). The balls are single crystals with a Knoop hardness from 1,525 to 2,000, a coefficient of friction of 0.140, and a compressive strength of

300,000 lb/in^2 (2,069 MPa). The material has a melting point of 3686°F (2030°C) and is resistant to acids and alkalies. It also has a very low coefficient of expansion and is used for ring and plug gages, and for such wear parts as the thread guides on textile machines. Stones free from strains, and as large as 0.75 in (1.9 cm) square, for use as lenses, prisms, and optical windows, are made by Bell Laboratories by recrystallization at high temperature and pressure. They transmit light better than quartz into the infrared and ultraviolet areas, and sapphire is used as an **infrared detector** in antiaircraft missiles. The dielectric constant of sapphire is also high, about 10.6.

Sapphire whiskers are **alumina fibers** 39 μin (1 μm) in diameter and 0.125 in (0.3 cm) long. They may be matted without bond into **ceramic papers** for electrical insulation and filters. As a reinforcement for light metals, **sapphire fibers** increase tensile and fatigue strengths. **Whiskerloy AA20** is an aluminum alloy with 20% sapphire fiber. It has a tensile strength of 48,000 lb/in^2 (331 MPa) and a modulus of 18×10^6 lb/in^2 (124,110 MPa).

SCHEELITE. An ore of the metal tungsten, occurring usually with quartz in crystalline rocks associated with wolframite, fluorite, cassiterite, and some other minerals. It is found in various parts of the United States, Brazil, Asia, and Europe. Scheelite is **calcium tungstate,** $CaWO_4$, containing theoretically 80.6% tungsten trioxide and 19.4 lime. It is called *powellite* when it contains some molybdenum to replace a part of the tungsten. It occurs massive granular or in crystals. The color is white, yellow, brown, or green, with a vitreous luster. Chinese scheelite from Kiangsi averages 65% WO_3 and can be used directly for adding tungsten to steel. **Tungstic acid** is a yellow powder of composition H_2WO_4 made from the ore by treating with hydrochloric acid. It is not soluble in water, but is soluble in alkalies and in hydrofluoric acid, and it is used as a mordant in dyeing, in plastics, and for making tungsten wire by reducing. Tungstic acid is also obtained as a by-product in the manufacture of alkalies from the brine of Owens Lake and is a source of tungsten, 1,000 lb (454 kg) of acid yielding about 800 lb (363 kg) of metallic tungsten. **Phosphotungstic acid,** $H_3[P(W_3O_{10})_4] \cdot 5H_2O$, called **heavy acid,** is used as a catalyst in difficult synthesis operations on complex ring-compound chemicals. The molecular weight is 2,879, of which three-fourths is tungsten. The three hydrogens produce the strong acid activity. The acid is soluble in water and in organic solvents.

Pure crystals of scheelite suitable for scintillation-counter phosphors for gamma-ray detection are found, but the natural crystal is rare. Calcium tungstate is grown synthetically as a clear, water-white crystal of tetragonal structure in rods and boules with the axis oriented perpendicular to the growth axis of the rod. It has a specific gravity of 6.12, a Mohs hardness of 4.5 to 5, a melting point of 2795°F (1535°C), and a refractive index of 1.9368. It has a blue luminescence under ultraviolet light. The crystals can be made in shiny, crystalline scales, and the material is also used in fluorescent pigments. **Cadmium tungstate,** $CdWO_4$, is similarly grown in clear, yellowish-green, monoclinic crystals with a refractive index of 2.25, and it is superior to calcium tungstate for scintillation counters. The crystals can also be grown with a cleavage much like mica, and it is used in fluorescent pigments.

SCOURING ABRASIVE. Natural sand grains or pulverized quartz employed in scouring compounds and soaps, buffing compounds, and metal polishes. Federal specifications require that the abrasive grains used in grit cake soap and scouring compounds all pass a No. 100 screen; the grains for scouring compounds for marble floors must all pass a No. 100, and 95% pass a No. 200 screen. For ceramic floors 90% must pass a No. 80, and 95% must pass a No. 60 screen. Very fine air-floated quartz is employed in metal polishes, and all grains pass a 325-mesh screen, but the extremely fine powders of metal oxides for polishes and fine finishes are generally called *soft abrasives* and are not classified as scouring materials.

SCREW STOCK. A common term for steels and nonferrous metals, notably brass, having superior machinability and used to make screws and small turned parts on automatic screw machines. In steels, free-cutting, or free-machining, quality is imparted by the addition of small amounts of alloying elements, such as sulfur, phosphorus, lead, tellurium, bismuth, calcium, or selenium, sometimes in various combinations. In brass, lead is the usual free-machining additive.

SEAL OIL. An oil resembling sperm oil obtained from the blubber of the **oil seal,** *Phoca vitulina,* a sea mammal native to the Atlantic Ocean. The oil has a saponification value as high as 195 and an iodine value up to 150, and it was once valued for lubricating and cutting oils but is now scarce. In the nineteenth century, as many as 400 ships at a time operated from Newfoundland in seal catching, but the unrestricted catch resulted in the destruction of the herds, and North Atlantic sealing was reduced to three ships by the middle of the twentieth century. The industry now centers on South Georgia in the south Atlantic as an adjunct to the whale industry, but considerable oil and **seal meal** come as by-products of the Alaskan fur seal industry.

Some seal oil is obtained from Steller's sea lion, a large-eared seal found from southern California to the Bering Sea. The adult male weighs up to 2,200 lb (998 kg). The blubber is about 75% oil, with an iodine value of 143 and saponification value of 190. From 40 to 50% of the carcass is a dense, dark-red, edible meat, but in the United States **seal meat** is used only in animal foods. **Seal leather,** from the skin, is used for fancy specialty articles, but it has too many defects for general use. The product known as **sealskin** is a valuable fur skin from the fur seal, about 80% of which are caught off the Pribilof Islands where they return in June to breed. No killing is now permitted at sea. Each bull seal has as many as 50 females, and the killing is usually restricted to the surplus males. About 30% of the skins are black fur which brings the highest price. Next in value is the **Matara fur,** or dark-brown, which is 60% of the catch. The **Safari fur** is light brown.

SEAWEED.   A plant growing in the sea, belonging to the extensive plant division known as **algae.** About 17,000 varieties of seaweed are listed, but only a few are exploited commercially. Algae are non-seed-bearing plants containing photosynthetic pigments. They have no vascular or food-conveying system and must remain submerged in the medium from which they acquire their food. They occur in both fresh and salt waters.

The **brown seaweeds,** which are the true kelps, grow in temperate and polar waters. They produce **algin, fucoidin,** and **laminarin.** The red seaweeds are the **carrageens,** which produce **carrageenin,** and the **agarophytes,** which yield **agar** and **agaroid.** They grow in warm waters. But color is an indefinite classification; the chlorophyll in the green Irish moss is often so masked by other pigments that the weed may be purplish black. All the seaweed colloids, or phycocolloids, are polysaccharides, having galactose units linked in long chains of molecular weights from 100,000 to 500,000, varying in their chemical structure. They are anionic polyelectrolytes, with negative radicals on each repeating polymer unit. **Irish moss,** also called **chondrus** (pronounced chone-droosh), **pearl moss,** and **carrageen,** is a dwarf variety of brown seaweed, *Chondrus crispus,* and *Gigantina mamillosa,* found off the west coast of Ireland and in New England. The weed used mostly for alginic acid is the brown kelp *Laminaria saccharium, L. ditata,* and other species, found off the Hebrides. It is a cold-water plant.

The seaweed *G. stellata,* of the north Atlantic, is also used to produce agar and algin. It is bleached and treated to produce gelatin used in foodstuffs, as a clarifying agent, and as a sizing for textiles. It is a better suspending and gelatinizing medium than agar for foodstuffs and cosmetic emulsions. At least 25 mineral salts are known to be present in seaweed as well as several vitamins. In the utilization of the seaweed as gelatin or alginate these are left in the **kelp meal** which is marketed as poultry and stock feed. In Asia the whole plant is cooked and eaten. **Seaweed flour,** made in Germany from **Iceland seaweed,** *Phaeophyceen,* is the ground, dry seaweed containing all the minerals and vitamins. It is mixed with wheat and rye flours to make **algenbrot,** a bread with higher food value and better keeping qualities than ordinary wheat bread. But more than 8% gives a peculiar flavor to the bread. The Irish name *dulse* is applied to the dried or cooked seaweed, *Rhodymenia palmata,* used in the Canadian Maritime Provinces for food. It is purple and rich in iodine and mineral salts. Other species, known as **laver** and **murlins,** are also used in Iceland, Ireland, and Scotland for food. When used for producing iodine in Scotland, the seaweed goes under the general name of **tangle.** Much of the 4,500 mi (7,607 km) of coastline of Scotland contains brown kelp. The kelp found along the Chilean and Peruvian coasts, when dried for making algin products, is called by the Quechua name **cochayuyo. Kombu,** used by the Japanese for food, is a brown seaweed from the coast of Hokkaido.

Dry seaweed contains up to 30% **alginic acid;** the water-soluble salts of this acid are called *algin.* It belongs to the group of complex, open-chain **uronic acids** which occur widely in plant and animal tissues and are related to the proteins and pectins. All the algins are edible, but they pass unchanged through the alimentary tract and add no food value. Carageenin is much used as a stabilizer for chocolate in milk. Laminarin is used as **laminarin sulfate** as a blood-clotting agent. **Sodium alginate** is used as a stabilizer and ice-crystal retarder in ice cream, as an emulsifier in medicines, and to replace gum arabic. It is a colorless, water-soluble gum made by dissolving algin in sodium carbonate solution and neutralizing with hydrochloric acid. **Protan jelly,** used for coating fish for freezing, is algin in a dilute edible acid. When frozen, the jelly is impervious to air and prevents oxidation. It can be washed off with water.

Kelgin, of the Kelco Div. of Merck & Co., Inc., is sodium alginate used as a foodstuff stabilizer, and **Keltrex** is the material in granular form for textile coating. **Kelset,** of the same company, is an algin used as a suspending agent for foodstuffs. It does not alter flavor or texture and is stable up to 200°F (93°C). **Dariloid** is sodium alginate to replace gelatin in ice cream; **Kelcosol** is an alginate to replace starch in foodstuffs. One part of algin can replace six parts of starch, and it does not smother flavor as starch does. **Protakyp K,** of Croda, Inc., used as a thickener for textile printing inks, is an alginate compatible with gums. **Viscobond** is a modified sodium alginate for finishing cellulosic textiles. **Kim-Ko gel** is a light-buff, scaly powder easily soluble in water, made from Irish moss, used as a collodial gelling agent. It has a pH of 6.4 to 7.2. **Carrigar** is a purified alginate for pharmaceuticals and foodstuffs. It contains the natural mineral salts and has food value. It has high capacity for water absorption, making rigid, sugar-free jellies with less than 2% in solution. **Algaloid** and **Agagel** are algins of this type.

Alginic fibers are silklike fibers made by forcing a sodium alginate solution through spinnerets into a calcium chloride bath and insolubilizing with beryllium acetate; but the fiber is soluble in sodium soaps, and the fabrics must be dry-cleaned. Soluble alginic yarns are used for making fancy fabrics where uneven spacing of threads is desired without change in the loom. The alginate yarn is washed out of the fabric after weaving, leaving the desired spacing.

SELENIUM. An elementary metal, symbol Se, found native in cavities in Vesuvian lavas and in some shales. The volcanic tuff of Wyoming contains 150 parts per million of selenium, and the black shale of Idaho has up to 1 lb (0.45 kg) of selenium per ton (0.9 metric ton). It also occurs in many minerals, chiefly in **cucairite,** $(AgCu)_2Se$, **naumannite,** Ag_2Se, **zorgite,** $(ZnCu)_2Se$, and in crooksite and clausthalite. Production in the United States and Canada is largely as a by-product of copper refining, the blister copper anodes containing 0.03 to 0.04%, and the refinery slimes having a content of 8 to 9%. The commercial recovery is 0.66 lb/ton (0.33 kg/metric ton) of copper. In England it is recovered from the residues of roasting iron sulfide ores in sulfuric acid production.

Like sulfur, selenium exists in various forms. Six allotropic forms are recognized, but four well-defined forms are usually listed. **Amorphous selenium,** produced by reducing selenous acid, is a finely divided, brick-red powder with a specific gravity of 4.26. It yields the vitreous form on heating. **Vitreous selenium** is a brownish-black, brittle, glassy mass with a specific gravity of 4.28. It is a dielectric and is electrified by friction. The monoclinic **crystalline selenium** is produced by crystallization from carbon disulfide, and it is a deep-red glass material with a specific gravity of 4.46 and a melting point of 347°F (175°C). The hexagonal crystalline selenium is produced by heating the monoclinic. It is a stable metal and is a good conductor of electricity. It has a specific gravity of 4.79 and melts at 422°F (217°C). All of the forms become gaseous at 1270°F (688°C). Selenium is marketed as a blackish powder, the high grade being 99.99% pure and the commercial grade 99.5% pure.

Selenium metal is odorless and tasteless, but the vapor has a putrid odor. The material is highly poisonous and is used in insecticides and ship-hull paints. Foods grown on soils containing selenium may have toxic effects, and some weeds growing in the western states have high concentrations of selenium and are poisonous to animals eating them. Selenium burns in air with a bright flame to form **selenium dioxide,** SeO_2, which is in white, four-sided, crystalline needles. The oxide dissolves in water to form **selenous acid,** H_2SeO_3, resembling sulfurous acid but very weak. Oxidation of this acid forms **selenic acid,** H_2SeO_4, a strong acid resembling sulfuric acid. By burning loco weed and converting to the acids, selenium has been extracted from the weeds.

The photoelectric properties of selenium make it useful for light-measuring instruments and for electric eyes. Amorphous or vitreous selenium is a poor conductor of electricity, but when heated, it takes the crystalline form and its electrical resistance is reduced, and it changes electrical resistance when exposed to light. The change of electrical conductivity is instantaneous, even the light of small lamps having a marked effect since the resistance varies directly as the square of the illumination. The pure amorphous powder is also used for coating nickel-plated steel or aluminum plates in rectifiers for changing alternating current to pulsating direct current. The coated plates are subjected to heat and pressure to change the selenium to the metallic form, and the selenium coating is covered with a layer of cadmium-bismuth alloy. Selenium rectifiers are smaller and more efficient than copper oxide rectifiers, but they require more space than silicon rectifiers and are limited to an ambient temperature of 185°F (85°C).

Selenium is also used in steels to make them free-machining, up to 0.35% being used. Up to 0.05% of selenium may also be used in forging steels. From 0.6 to 0.85 oz (0.017 to 0.024 kg) of selenium per ton (0.9 metric ton) of glass may be used in glass to neutralize the green tint of iron compounds. Large amounts produce pink and ruby glass. Selenium gives the only pure-red color for signal lenses. Pigment for glass may be in the form of the black powder, **barium selenite,** $BaSeO_3$, or as **sodium selenite,** Na_2SeO_3, and may be used with cadmium sulfide. Selenium is also used as an accelerator in rubber and to increase abrasion resistance. **Vandex,** of R. T. Vanderbilt Co., is a selenium powder used as a rubber vulcanizer. **Novac,** used for curing synthetic rubbers, is selenium dibutyl dithiocarbonate in the form of a liquid easily dispersed in the rubber. **Selsun,** of Abbott Laboratories, is **selenium sulfide** suspended in a detergent, used to control dandruff in hair. In copper alloys, selenium improves machinability without hot-shortness. **Selenium copper** is a free-cutting copper containing about 0.50% selenium. It machines easily, and the electrical conductivity is nearly equal that of pure copper. The tensile strength of annealed selenium copper is about 30,000 lb/in^2 (207

MPa). Small amounts of selenium salts are added to lubricating oils to prevent oxidation and gumming.

SEMICONDUCTORS. Materials that are capable of being partly conductors of electricity and partly insulators and are used in rectifiers for changing alternating current to pulsating direct current, and in transistors for amplifying currents. They can also be used for the conversion of heat energy to electric energy, as in the solar battery. In an electric conductor, the outer rings of electrons of the atoms are free to move and provide a means of conduction. In a semiconductor the outer electrons, or **valence electrons,** are normally stable, but when a **doping element** that serves to raise or lower energy is incorporated, the application of a weak electric current will cause displacement of valence electrons in the material. Silicon and germanium, each with a single stable valence of four outer electrons, are the most commonly used semiconductors. Elements such as boron, with a lower energy level but with electrons available for bonding and thus accepting electrons into the valence ring, are called **hypoelectronic elements.** Elements such as **arsenic,** which have more valence electrons than are needed for bonding and may give up an electron, are called **hyperelectronic elements.** Another class of elements, like **cobalt,** can either accept or donate an electron, and these are called **buffer atoms.** All of these types of elements constitute the doping elements for semiconductors.

In a nonconducting material, used as an **electrical insulator,** the energy required to break the valence bond is very high, but there is always a limit at which an insulator will break the bond and become a conductor with high current energy. The resistivity of a conductor rises with increasing temperature, but in a semiconductor the resistivity decreases with temperature rise, and the semiconductor becomes useless beyond its temperature limit. **Germanium** can be used as a semiconductor to about 200°F (93°C), **silicon** can be used to about 400°F (204°C), and **silicon carbide** can be used to about 650°F (343°C).

Metals for use as semiconductors must be of great purity, since even minute quantities of impurities would cause erratic action. The highly purified material is called an **intrinsic metal,** and the desired electron movement must come only from the doping element, or **extrinsic conductor,** that is introduced. The semiconductors are usually made in single crystals, and the positive and negative elements need be applied only to the surfaces of the crystal, but methods are also used to incorporate the doping element uniformly throughout the crystal.

The process of electron movement, although varying for different uses and in different intrinsic materials, can be stated in general terms. In the silicon semiconductor, the atoms of silicon with four outer valence electrons bind themselves together in pairs surrounded by eight electrons. When a doping element with three outer electrons, such as boron or indium, is added to the crystal, it tends to take an electron from one of the pairs, leaving a hole and setting up an imbalance. This forms the **p-type semiconductor**. When an element with five outer electrons, such as antimony or bismuth, is added to the crystal, it gives off electrons, setting up a **conductive band,** which is the **n-type semiconductor.** Fusing together the two types forms a *pn* junction, and a negative voltage applied to the *p* side attracts the electrons of the three valence atoms away from the junction so that the crystal resists electronic flow. If the voltage is applied to the *n* side, it pushes electrons across the junction and the electrons flow. This is a **diode,** or **rectifier,** for rectifying alternating current into pulsating direct current. When the crystal wafers are assembled in three layers, *pnp* or *npn,* a weak voltage applied to the middle wafer increases the flow of electrons across the whole unit. This is a **transistor.** Germanium and silicon are bipolar, but silicon carbide is unipolar and does not need a third voltage to accelerate the electrons.

Semiconductors can be used for rectifying or amplifying, or they can be used to modulate or limit the current. By the application of heat to ionize the atoms and cause movement they can also be used to generate electric current; or in reverse, by the application of a current they can be used to generate heat or remove heat for heating or cooling purposes in air conditioning, heating, and refrigeration. But for uses other than rectifying or altering electric current, the materials are usually designated by other names and are not called semiconductors. **Varistors** are materials, such as silicon carbide, whose resistance is a function of the applied voltage. They are used for such applications as frequency multiplication and voltage stabilization. **Thermistors** are thermally sensitive materials. Their resistance decreases as the temperature increases, which can be measured as close as $0.002°F$ ($0.001°C$), and they are used for controlling temperature or to control liquid level, flow, and other functions affected by rate of heat transfer. They are also used for the production or the removal of heat in air conditioning and may then be called **thermoelectric metals.**

Indium antimonide, InSb, has a cubic crystal structure, with three valence electrons for each indium atom and five for each antimony atom. Between each atom and its four nearest neighbors there are four electron-pair bonds, and there are an average of four electrons per atom in the compound. It is used for **infrared detectors** and for amplifiers in galvanomagnetic devices. **Indium arsenide,** InAs, also has a very high electron mobility and is used in thermistors for heat-current conversion, since the number of electrons free to constitute the electric current increases about 3% with each 1°F (0.56°C) rise in temperature. It can be used to 1500°F (816°C). Some materials can be used only for relatively low temperatures. Copper oxide and pure selenium have been much used in current rectifiers, but they are useful only at moderate temperatures, and they have the disadvantage of requiring much space. **Indium phosphide,** InP, has a mobility higher than that of germanium and can be used in transistors above 600°F (316°C). **Aluminum antimonide,** AlSb, can be used at temperatures to 1000°F (538°C). In **lead selenide,** PbSe, the mobility of the charge-carrying electrons decreases with rise in temperature, increasing resistivity. It is used in thermistors.

The thermoelectric generation, set up when the junction of two dissimilar thermoelectric metals is heated and which is used in thermostats for temperature measuring and control, is essentially the same as the energy conversion and heat pumping with p-type and n-type materials. The difference is in mechanical applications. When a semiconductor is operated thermoelectrically as a heat pump, the electric charge passing through the heat-absorbing junction is carried by electrons in the n-type material and holes in the p-type material, and the charge carriers both move away from the junction and carry away heat, thus reducing temperature at the junction. By reversing the current, heat is produced. Each material has a definite temperature difference, or gradient, and the efficiency is proportional to the temperature difference across the material, while the power rating is proportional to the square of the temperature difference. Thus, a material with low efficiency may have a high power rating if it can be operated at a high enough temperature, but some materials do not maintain chemical stability at high temperatures. Also, for many uses it is undesirable to operate at advanced temperatures.

Bismuth telluride, Bi_2Te_3, maintains its operating properties between −50 and 400°F (−46 and 204°C), which is the most useful range for both heating and refrigeration. When doped as a p-type conductor, it has a temperature difference of 1115°F (601°C) and an efficiency of 5.8%. When doped as an n-type conductor, its temperature difference is lower, 450°F (232°C), but the efficiency within this range is more than doubled. **Lead telluride,** PbTe, has a higher efficiency, 13.5%, and a temperature difference of 1080°F (582°C), but it is not usable below 350°F (177°C) and is employed for conversion of the waste heat from atomic reactors at about 700°F (371°C).

Gallium arsenide has high electron mobility and can be used as a semiconductor. When polycrystalline semiconductors are used in thin films against a metal barrier, the minimum grain size of the deposited film must equal the thickness of the film so that the carrier is not intercepted by a grain boundary. **Cadmium sulfide,** CdS, is thus deposited as a semiconductor film for **photovoltaic cells,** or **solar batteries,** with film thickness of about 79 μin (2 μm). Thin-film cells of **copper-indium diselenide** and gallium can convert 15.5% of sunlight into electricity, and tandem cells of **gallium-indium phosphide** and gallium arsenide can convert 29.5%. When radioactive isotopes, instead of solar rays, are added to provide the activating agent, the unit is called an **atomic battery,** and the large area of transparent backing for the semiconductor is not needed.

Manganese telluride, MnTe, with a temperature difference of 1800°F (982°C), has also been used as a semiconductor. Many other materials can be used, and semiconductors with temperature differences at different gradients can be joined in series electrically to obtain a wider gradient, but the materials must have no diffusion at the junction. If intermetal compounds are of such a nature as to have a *pn* balance, no doping is needed; but usually they are not in perfect balance, causing scattering, and balancing is necessary. Materials for thermoelectric use are usually doped higher than for semiconductors, but increased doping reduces resistivity, and for high emf and low power, only small amounts are used.

Cesium sulfide, CeS, has good stability and thermoelectric properties at temperatures to 2000°F (1093°C) and has a high temperature difference, 2030°F (1110°C). It can thus be used as a high-stage unit in conversion devices. High conversion efficiency is necessary for **transducers,** while a high dielectric constant is desirable for capacitors. Low thermal conductivity makes it easier to maintain the temperature gradient, but for some uses high thermal conductivity is desirable. **Silver-antimony-telluride,** $AgSbTe_2$, has a high energy-conversion efficiency for converting heat to electric current, and it has a very low thermal conductivity, about 1% that of germanium.

Mechanical stress, as well as heat stress, produces an electric charge in balanced semiconductors, and they can be used for controlling pressure. The semiconductor-type intermetals are also used in magnetic devices, since the ferroelectric phenomenon of heat conversion is the electrical analog of ferromagnetism. **Chromium-manganese-antimonide** is nonmagnetic below about 482°F (250°C) and magnetic above that temperature. Various compounds have different critical temperatures. Below the critical temperature, the distance between the atoms is less than that which determines the lineup of magnetic forces; but with increased temperature the atomic distance becomes greater, and the forces swing into a magnetic pattern.

Organic semiconductors fall into two major classes: well-defined substances, such as molecular crystals and crystalline complexes, isotatic and syndiotactic polymers; and disordered materials, such as atactic polymers and pyrolitic materials. Few of these materials have yet found commercial application. **Anthracene,** an example of the molecular crystal type, has perhaps been given the most study. Its transport properties are not unlike those of silicon and germanium. **Amorphous silicon** containing hydrogen is promising for use in solar cells because of its low cost and suitable electrical and optical properties.

SENNA. The dried, small leaves and the pods of the bushy plant *Cassia acutifolia,* the **Alexandrian senna,** and *C. angustifolia,* the **Tinnevelly senna,** of India, Arabia, and north Africa. The plants are cultivated in India, but the Sudan material comes mostly from wild plants. The sun-dried leaves and pods are shipped in bales. They are used directly as a laxative by steeping in water, or the extract is used in pharmaceuticals. It contains the yellowish, noncrystalline **cathartine,** a powerful purgative. Another species of the plant, *C. auricula,* yields **avarem bark,** which is an important tanning material in India. It is similar to algarobilla in action.

SERPENTINE. A mineral of theoretical formula $3MgO \cdot 2SiO_2 \cdot 2H_2O$, containing 43% magnesium oxide. It is used for building trim and for making ornaments and novelties. The chips are employed in terrazzo and for roofing granules. Actually, the stone rarely approaches the theoretical formula and usually contains 2 to 8% iron oxide with much silica and aluminum. It has an asbestoslike structure. The attractively colored and veined serpentine of Vermont is marketed under the name of **verde antique marble.** The massive verde antique of Pennsylvania is used with dolomite in refractories. **Antigorite** is a form of serpentine found in California which has a platy rather than a fibrous structure. The serpentine of Columbus County, Georgia, contains 36 to 38% MgO and 2 to 5 chrome ore. It is used as a source of magnesia.

SESAME OIL. A pale-yellow, odorless, bland oil obtained from the seeds of the tropical plant *Sesamum orientale* and other species, grown in India as *S. indicum,* and in China, Africa, and Latin America, and used for soaps, foodstuffs, and blending industrial oils. It is distinct from German sesame oil. The seeds from different species and localities vary greatly in size and color, from yellowish white to reddish brown to black. The oil from Nigeria is called **benne oil,** and the seed **benniseed.** In India it is known as **til oil,** and the seed as **til seed.** In Madras it is called **gingelli.** In Mexico it is called **ajonjoli.** The seeds contain 50% of oil of a specific gravity of 0.920 to 0.925, with saponification value 188 to 193. **Super Refined oil** is a high-purity, colorless, odorless oil for use as a pharmaceutical vehicle. Croda, Inc. is the marketer of this variety.

Sesame is one of the most ancient of food grains, but has been little grown in the United States because of the difficulty of harvesting and collecting the seed. **Sesame seed** is used as a flavor garnish for breads and bakery products. **Sesame protein,** extracted from the seed and marketed as a powder, contains the amino acids methionine and cystine which are low in other vegetable foods. It is low in lycine. It is valued for blending in foodstuffs. The oil contains the natural antioxidants **sesamol** and **sesamolin,** making it very stable, and it is also highly unsaturated. These are present in concentrations of less than 1%. The protein is marketed as a flour containing 60% protein and less than 1 fat, and in granules containing 55 protein and up to 18 fat. **SesaLac 86** is a spray-dried blend of sesame protein and whey for use in beverages, bakery products, and prepared foods, giving a more complete amino acid combination than milk powder.

SHALE. A rock formed by deposition of colloidal particles of clay and mud, and consolidated by pressure. It is fine-grained and has a laminated structure, usually containing much sand colored by metal oxides. Unlike sandstones, shales are not usually porous, most shales being hard, slatelike rocks. Slate is a form of shale that has been subjected to intense pressure. Some shales are calcareous or dolomitic and are used with limestone in making portland cement. These are called **marlstone. Oil shale** is a hard shale with veins of greasy solid known as **kerogen,** which is oil mixed with organic matter. Crude **shale oil** is a black, viscous liquid containing up to 2% nitrogen and a high sulfur content. But when oil shale is heated above 750°F (399°C), the kerogen is cracked into gases condensable to oils, gases, and coke. Some shales also yield resins and waxes, and the **Kvarmtorp shale** of Sweden contains small amounts of uranium, vanadium, and molybdenum. The regular commercial by-products of **Swedish shale oil** recovery are sulfur, fuel gas, ammonium sulfate, tar, and lime. Oil shales are widely distributed in many parts of the world and are regularly distilled in most of the countries in Europe, the yield varying from 15 to 100 gal (57 to 379 L) per short ton (907 kg). Scottish shales give an average yield of 24.5 gal (93 L) crude oil and 35.7 lb (16 kg) ammonium sulfate per short ton (907 kg). Shale occurs in strata and is mined as coal is. The oil shales of Colorado contain **dawsonite,** a hydrated carbonate of aluminum and sodium, and **nacolite,** a sodium bicarbonate mineral. Alumina, soda ash, and other by-products can be produced from the shales. **Bituminous shale** was originally called **boghead coal** in England and **torbane mineral** in Scotland. In the Green River Basin of northwestern

Colorado about 1,000 mi² (2,590 km²) is underlain by oil shale 500 ft (152 m) thick averaging 15 gal/ton (0.06 L/kg). The lower portion, about 100 ft (30 m) thick, averages 30 gal/ton (0.13 L/kg). Recovery in a continuous retort extracts 94% of the oil, which is then cracked by heat and treated with hydrogen to remove impurities and improve the quality before it is sent to the refinery. It is estimated that the shale of the Mahogany Ledge in Colorado, extending into Utah and Wyoming, has a content of 1.2 trillion bbl of extractable oil. No oil is visible in the shale, but it is present as the solid kerogen, which yields oil when heated. Other deposits occur in Nevada, Tennessee, Indiana, Ohio, and Kentucky. **Oil sands** of Alberta, Canada, Utah, and California are free-flowing sands impregnated with bituminous oil. A deposit on the Athabasca River covers 1,800 mi² (4,662 km²) and is 165 ft (50 m) thick. Vast quantities of oil are available from these sands.

SHARK LEATHER. A durable, nonscuffing leather used for bookbindings, handbags, and fancy shoes, made from the skin of sharks. The shark is the largest of the true fishes, but has a skin unlike fishskin. When tanned, the surface is hard, the epidermis thicker than cowhide, and the long fibers lie in a crossweave. The shark is split on the back instead of the belly, as in cowhides, and the skins measure from 3 to 20 ft² (0.3 to 2 m²), averaging 10 ft² (1 m²). The hard denticle, called the *shagreen,* is usually removed, after which the leather is pliable but firm, the exposed grain not pulling out. **Shagreen leather** is a hard, strong leather with the grain side covered with globular granules made to imitate the sharkskin. Eastern shark leather has a deep grain with beautiful markings. The eastern shark includes about a dozen species of shark caught off the Florida and Cuban coasts, except the **nurse shark** and the **sawfish** which are graded separately. The **whale shark** attains a length of 50 ft (15 m) and a weight of several tons. It is an offshore species, feeding on small organisms, and is harmless to people. The **basking shark** and the **white shark** grow to 40 ft (12 m). The nurse shark measures 6 to 10 ft (1.8 to 3 m). **Olcotrop leather,** from a species of shark, has a smooth, fine grain with regular markings. **Galuchat leather,** or **pearl sharkskin,** is from the Japanese ray. It is used for trim on pocketbooks. **Boroso sharkskin, rousette leather,** or **Morocco leather** is from a small shark of the Mediterranean, but the name is also applied to a vegetable-tanned Spanish goatskin on which a pebbly grain is worked up by hand boarding. It is now made from ordinary goatskin by embossing.

Most of the sharkskin is now a by-product of the catch for oil, which is used for medicinal purposes. The shark liver is about one-fourth the total weight of the animal, and **shark-liver oil** is 30 times higher in vitamin A than cod-liver oil. The oil is also used for soap, lubricant, and heat-treating oil, though normally it is too expensive for these purposes. The Mexican shark-oil industry centers at Mazatlán, and about 25 species are caught off the west coast. **Vitamin oil** from South Africa is from the liver of the **stockfish,** *Merluccius capensis,* which is an important food fish. The liver contains 30% oil.

SHEEPSKIN. The skin of numerous varieties of sheep, employed for fine leather for many uses. The best sheepskins come from the sheep yielding the poorest wool. When the hair is short, coarse, and sparse, the nourishment goes into the skin. The merino types having fine wool have the poorest pelts. Wild sheep and the low-wool crossbreeds of India, Brazil, and South Africa have close-fibered, firm pelts comparable in strength with some kidskin, and retain the softness of sheepskin. This type of sheepskin from the hair sheep is termed **cabretta** and is used almost entirely for making gloves and for shoe uppers. None is produced in the United States. The lambs grown in the mountains of Wales, Scotland, and the western United States also furnish good skins. The commercial difference between sheepskin and **lambskin** is one of weight only. Sheepskins usually run 3 to 3.5 lb (1.4 to 1.6 kg) per skin without wool, and lambskins are those below 3 lb (1.4 kg). Sheepskins are tanned with alum, chrome, or sumac. The large, heavy skins from Argentina and Australia are often split, and the grain side tanned in sumac for bookbinding and other goods; the flesh side is tanned in oil or formaldehyde and marketed as chamois. The fine-grained sheepskins from Egypt, when skived and specially treated, are known as **mocha leather. Uda skins** and **white fulani skins,** from Nigerian sheep, are used for good-quality grain and suede glove leather. **Sheepskin shearlings** are skins taken from heavy-wooled sheep a few weeks after shearing. The wool is about 1 in (2.54 cm) in length. They are tanned with the wool on, and the leather is used for aviation flying suits and for coats.

SHEET MOLDING COMPOUND. Typically **SMC** is an unsaturated, thermosetting polyester reinforced with glass fibers for strength and rigidity and modified with additives to ease handling, improve surface quality, control shrinkage, provide color and environmental resistance, and ease mold release. It is used in the form of soft wads, typically thickened with oxides to ease handling, and compression-molded in metal dies at about 300°F (149°C) at pressures up to 1,000 lb/in² (6.9 MPa). Depending on fiber content, tensile strength ranges from 5,000 to 30,000 lb/in² (34 to 207 MPa), flexural strength from 10,000 to 50,000 lb/in² (69 to 345 MPa), and notched Izod impact strength from 7 to 25 ft · lb/in (374 to 1,335 J/m). SMC is widely used for panels of autos and truck cabs, underbody auto components, and many other applications. **Hi-Flex SMC,** of Budd Co., contains 20% glass fiber, a thermoplastic-elastomer copolymer as a low-profile additive for superior surface quality, a polyurethane additive to enhance flexibility, and a calcium carbonate filler. The density is 0.054 lb/in^3 (1,495 kg/m^3), and the coefficient of thermal expansion is 5.6 $\times$ 10^{-6}/°F (10.1 $\times$ 10^{-6}/K). Tensile and flexural strengths are 8,500 and 15,800 lb/in^2 (59 and 109 MPa), respectively, and the flexural modulus is 650,000 lb/in^2 (4,482 MPa). The impact strength is 21 ft · lb/in (1,121 J/m).

The high pressure needed to mold traditional SMC is due to the oxide thickening additives used to ease room-temperature handling of the soft mass. **Crystal Impreg,** a crystalline unsaturated polyester resin of Scott Bader of England, can be substituted for part of the traditional polyester to ease handling equally and reduce molding pressure to as low as 150 lb/in^2 (1 MPa). This resin melts at 200°F (93°C), thus decreasing the viscosity of the compound at molding temperature. Such **low-pressure sheet molding compounds** also speed molding. Resulting properties are about the same as for traditional SMC. For example, with 27% glass fibers, density is 0.066 lb/in^3 (1,827 kg/m^3), tensile strength 11,900 lb/in^2 (82 MPa), flexural strength 24,500 lb/in^2 (169 MPa), and impact strength 20 ft · lb/in (1,084 J/m). **Aropreg 3230,** of Ashland Chemical, is a low-pressure SMC.

Arotech 2000 compounds, of the same company, are **vinyl ester SMCs.** The base resin is heat-resistant to 300°F (149°C), the 2001 is a zero-shrink grade, and the 2002 provides good surface quality and shrink control. Quantum Composites offers a urethane-toughened vinyl ester SMC as well as a **phenolic SMC,** both with a high content of glass fibers. **Cyglas 695,** of Cytec Industries, is a reinforced vinyl ester used for auto-engine valve covers.

SHELLAC. A product of *Tachardia lacca,* an insect that lives on various trees of southern Asia. The larvae of the lac insect settle on the branches, pierce the bark, and feed on the sap. The lac secretion produced by the insects forms a coating over their bodies and makes a thick incrustation over the twig. Eggs developed in the females are deposited in a space formed in the cell, and the hatched larvae emerge. This swarming continues for 3 weeks and is repeated twice a year. The incrustation formed on the twigs is scraped off, dried in the shade, and is the commercial **stick lac.** It contains woody matter, lac resin, lac dye, and bodies of insects. **Seed lac** is obtained by screening, grinding, and washing stick lac. The washing removes the lac dye. **Lac dye** was once an important dyestuff, giving about the same colors as cochineal but not as strong. It gives a fast, bright-red tint to silk and to wool, but is now replaced by synthetics. **Ari lac** is stick lac collected before the young insects have swarmed, and it contains living insects. Lac harvested after the swarming is called **phunki lac** and contains dead bodies of the insects. Average yield of stick lac from kusum trees is 12 lb (5.4 kg), from the ber tree 3 lb (1.4 kg), and from the palas tree 2 lb (0.9 kg). About 80% of lac production is in the state of Bihar in India, but it is also obtained from Bengal, the Central Provinces, and Assam. The stick lac from Burma and Thailand is brought to India for making shellac.

Shellac is prepared from seed lac by melting or by extraction with solvents. The molten material is spread over a hot cylinder and stretched, and the cooled sheet is broken into flakes of shellac. **Button lac** is made by dropping molten lac on a flat surface which spreads it into button-shaped cakes 3 to 4 in (7.6 to 10.2 cm) in diameter. **Kiri** is the refuse from the filtering bags. It is marketed in pressed cakes and contains 50 to 60% lac with resin and dirt. The yield of shellac from stick lac is about 57%. When pure, shellac varies from pale orange to lemon yellow, but the color of commercial shellac may be due to a high content of common resin. **White shellac** is made by bleaching with alkalies. **Garnet lac** is the material with lac dye left in. Color may also be balanced with pigments. **Orange shellac** contains up to 1% powdered orpiment, and the yellow may have smaller quantities. Shellac is composed of polyhydric acids which condense with loss of water to form long-chain esters, thus giving polyester resins in the final coating. It also contains resin and wax. **Aleuritic acid,** $OH \cdot CH_3(CH_2)_5(CH \cdot OH)_2(CH_2)_7COOH$, extracted from shellac, reacts with alcohols to produce odoriferous esters used in perfumes. It is a yellowish solid melting at 214°F (101°C). It can be made synthetically and is also used in cellulose lacquers.

Hard lac has the soft constituents removed by solvent extraction. For electrical use the wax content should be below 3.5%. By solvent extraction of the seed lac, the wax may be reduced to 1%. Shellac is graded by color and by its freedom from dirt. The first grade contains no resin, but other grades may contain up to 12%. Most Indian exports of seed lac to the United States are of the special grade which has a high bleach index. **Cut shellac** is shellac dissolved in alcohol, but usually mixed with a high percentage of resin. Shellac has good adhesive properties and high dielectric strength, and is used in adhesives, varnishes, floor waxes, insulating compounds, and some molding plastics. Hard-face wax polishes contain a high percentage of shellac, up to 80%, to conserve carnauba and other waxes.

SILICA. A mineral of general composition SiO_2, **silicon dioxide,** which is the most common of all materials, and in the combined and uncombined states is estimated to form 60% of the earth's crust.

Many sands, clays, and rocks are largely composed of small silica crystals. When pure, silica is colorless to white. The unit crystal, or molecule, of ordinary silica has the formula SiO_2, and the single-crystal grains are thus molecularly cryptocrystalline with no electron bonded lattice. But the chemical formula of fused silica and quartz is given as Si_2O_7, which is the pattern of a continuous lattice in which each silicon atom is surrounded by two oxygen atoms and each oxygen atom is surrounded by four silicon atoms. The varieties of natural silica are **crystalline silica,** such as quartz and **tridymite; cryptocrystalline silica** (minute crystals), such as flint, chert, chalcedony, and agate; and **amorphous silica,** such as opal. Silica is insoluble in water when anhydrous and is also insoluble in most acids except hydrofluoric. Crystallized silica in the form of quartz has a Mohs hardness of 7 and a specific gravity of 2.65. Amorphous **silica glass** has a specific gravity of 2.21. It is a transparent fused silica. **Vitreous silica** is a silica glass of high transparency. When impurities are no more than 1 ppm, it is the most transparent of the glasses and has high transmission of ultraviolet rays.

Pure **fused silica** has a melting point of 3182°F (1750°C), but softens slightly at 2552°F (14000°C). In chemical and heat ware, it is used up to 2012°F (1100°C). The coefficient of expansion is very low, 3×10^{-7}/°F (5.4×10^{-7}/K), and the dielectric strength is 500 V/mil (20×10^6V/m). The infrared transmission of a 96% pure silica glass extends to 157 μin (4 μm) while retaining the lower reach of the band to 15.7 μin (0.4 μm). Fused silica is used for chemical parts as it withstands severe thermal shock and is resistant to acid except hydrofluoric and hot phosphoric. **Amersil,** in the form of pipes and shapes, withstands continuous temperatures of 2700°F (1482°C). Fused silica parts may be made by pressing and sintering silica powder or by casting. Large cast parts for crucibles, molds, and furnace hearths are made by remelting a powder produced by melting 99.9% pure silica sand and then crushing and grinding the glass. Cast parts have a tensile strength of 1,500 lb/in^2 (10 MPa) and compressive strength of 20,000 lb/in^2 (138 MPa) and withstand repeated heating and cooling from 2000°F (1093°C). The material is white.

Fibrous silica, used for high-temperature insulation in jet aircraft, is produced from silica minerals in the same manner as rock wool and then extracting the nonsilica content of the fiber. **Refrasil,** of H. I. Thompson Co., is this material. The fibers have a diameter of 0.00023 in (0.00058 cm), fuse at 3100°F (1704°C), and withstand continuous temperatures to 2000°F (1093°C). It is produced as fibers, batts, cloth, and cordage. **Silica fiber** in diameters as small as 0.00003 in (0.000076 cm) comes in random matted form or in rovings. **Irish Refrasil** is 98% silica and has a green color. It is used for ablative protective coatings. It resists temperatures to 2800°F (1588°C). **Silica flour,** made by grinding sand, is used in paints, as a facing for sand molds, and for making flooring blocks. **Silver bond silica** is water-floated silica flour of 98.5% SiO_2, ground to 325 mesh. In zinc and lead paints it gives a hard surface. **Pulverized silica,** made from crushed quartz, is used to replace tripoli as an abrasive. **Ultrafine silica,** a white powder having spherical particles of 157 to 984 μin (4 to 25 μm), is made by burning silicon tetrachloride. It is used in rubber compounding, as a grease thickener, and as a flatting agent in paints. **Aerosil,** of Cabot Corp., is this material. **Silica powder,** of Praxair Inc., is a white, amorphous powder with maximum particle size of 1,969 nin (50 nm). Other natural amorphous silicas come in an average particle size of 59 μin (1.5 μm) with no particles larger than 394 μin (10 μm). **Quso,** of Philadelphia Quartz Co., is a soft, white powder with small particles, 394 to 787 nin (10 to 20 nm). It is used in cosmetics and paper coatings and as an anticaking agent in pharmaceuticals. As a filler in plastics, it gives a plasticizing action that aids extrusion. These fine silicas are also marketed as dust-free agglomerate particles which disperse easily in solution to the discrete hydrophyllic particle. **Arc silica,** of PPG Industries, used as a flatting agent in clear lacquers, is produced directly from silica sand in an arc furnace at 5432°F (3000°C). It has crystals of 0.59 μin (0.015 μm) agglomerated into translucent grains, 79 to 118 μin (2 to 3 μm). **Valron,** of Du Pont, originally called **Estersil,** is ester-coated silica powder of 0.3- to 0.4-in (8- to 10-mm) particle size, for use as a filler in silicone rubbers, printing inks, and plastics. **Ludex,** of the same company, is another **colloidal silica** with the fine particles negatively charged by the incorporation of a small amount of alkali. It forms a **sol,** or high-concentration solution, without gelling. **Min-U-Sil,** of Pennsylvania Glass Sand Corp., for making molded ceramics, has tiny crystalline particles. **Syton,** of Monsanto, is a water dispersion of colloidal silica for treating textiles. Translucent silica particles deposited on the fibers increase the coefficient of friction, giving uniformly high-strength yarns.

A **polymer-impregnated silica, Polysil,** produced by Westinghouse, has twice the dielectric strength of porcelain as well as better strength. It is also cheaper to make, and its composition can be tailored to meet specific environmental and operating conditions.

Silica aerogel is a fine, white, semitransparent silica powder, the grains of which have a honeycomb structure, giving extreme lightness. It has a density of 2.5 lb/ft^3 (40 kg/m^3) and is used as an insulating material in the walls of refrigerators, as a filler in molding plastics, as a flatting agent in paints, as a bodying agent in printing inks, and as a reinforcement for rubber. It is produced by treating sand with caustic soda to form sodium silicate and then treating with sulfuric acid to form a jellylike material called **silica gel,** which is washed and ground to a fine, dry powder. It is also called **synthetic silica. Syloid** is this material. It is a fluffy, white powder with a pH of 7.2. **Silica hydrogel** is a colorless, translucent, semisolid **hydrated silica** of composition $SiO_2 \cdot xH_2O$, bulking about 44 lb/ft^3 (705 kg/m^3). It contains 28% solids and 72 water. It becomes fluid by mixing with water and regels on standing. It is used for paper and textile coatings, ointments, and water suspensions of silica. **Hi-Sil,** of PPG Industries, and **Santocel,** of Monsanto, are silica gels. **Mertone WB-2,** of the same company, is silica gel used as a coating material for blueprint papers to deepen the blue and increase legibility. When silica gel is used as a pigment, the vehicle surrounds the irregular particle formation, producing greater rigidity and hardness of paint surface than when a smooth pigment is used. For insulation use, the thermal conductivity of silica gel powder is given as 0.1 Btu/(h · ft^2 · °F [0.57 W/(m^2 · K)] at -115°F (-81°C).

Silicon monoxide, SiO, does not occur naturally but is made by reducing silica with carbon in the electric furnace and condensing the vapor out of contact with air. It is lighter than silica, having a specific gravity of 2.24, and is less soluble in acid. It is brown powder valued as a pigment for oil painting, as it takes up a higher percentage of oil than ochres or red lead. It combines chemically with the oil. **Monox** is a trade name for silicon monoxide. **Fumed silica** is a fine, translucent powder of the simple amorphous silica formula made by calcining ethyl silicate. It is used instead of carbon black in rubber compounding to make light-colored products, and to coagulate oil slicks on water so that they can be burned off. It is often called **white carbon,** but the "white carbon black" of Cabot Corp. called **Cab-O-Sil,** used for rubber, is a silica powder made from silicon tetrachloride. **Cab-O-Sil EH5,** a fumed colloidal form, is used as a thickener in resin coatings. The thermal expansion of amorphous fused silica is only about one-eighth that of alumina. Refractory ceramic parts made from it can be heated to 2000°F (1093°C) and cooled rapidly to subzero temperatures without fracture. **QLF silicon oxide,** of Airco Coating Technology, is a vapor-deposited barrier coating for resistance to oxygen and moisture in paperboard/polyethylene laminate products.

SILICON. A metallic element, symbol Si, used chiefly in its combined forms. Pure silicon metal is used in transistors, rectifiers, and electronic devices. It is a semiconductor and is superior to germanium for transistors, as it withstands temperatures to 300°F (149°C) and will carry more power. Rectifiers made with silicon instead of selenium can be smaller and will withstand higher temperatures. Its melting point when pure is about 2615°F (1434°C), but it readily dissolves in molten metals. It is never found free in nature, but combined with oxygen, it forms silica, SiO_2, one of the most common substances in the earth. Silicon can be obtained in three modifications. **Amorphous silicon** is a brown powder with a specific gravity of 2.35. It is fusible and dissolves in molten metals. When heated in air, it burns to form silica. **Graphitoidal silicon** consists of black, glistening spangles and is not easily oxidized and is not attacked by the common acids, but is soluble in alkalies. **Crystalline silicon** is obtained in dark, steel-gray globules or crystals or six-sided pyramids of specific gravity 2.4. It is less reactive than the amorphous form, but is attacked by boiling water. All these forms are obtainable by chemical reduction. High-quality crystalline silicon is the most efficient material for photovoltaic cells used to generate electricity from sunlight. Amorphous silicon films are also used and are less costly, but the cells are less efficient because the atoms are random. Silicon is an important constituent of commercial metals. Molding sands are largely silica, and silicon carbides are used as abrasives. Commercial silicon is sold in the graphitoidal flake form, or as ferrosilicon, and silicon-copper. The latter forms are employed for adding silicon to iron and steels. Commercial refined silicon contains 97% pure silicon and less than 1 iron. It is used for adding silicon to aluminum alloys and for fluxing copper alloys. High-purity silicon metal, 99.95% pure, made in an arc furnace, is too expensive for common uses, but is employed for electronic devices and in making silicones. For electronic use, silicon must have extremely high purity, and the pure metal is a nonconductor with a resistivity of 118,000 $\Omega \cdot$ in (300,000 $\Omega \cdot$ cm). For semiconductor use it is "doped" with other atoms, yielding electron activity for conducting current. **Epitaxial silicon** is higher purified silicon doped with exact amounts of impurities added to the crystal to give desired electronic properties. Thus, silicon doped with boron has resistivities in grades from 394 to 3,940 $\Omega \cdot$ in (1,000 to 10,000 $\Omega \cdot$ cm). **Silicon ribbon** of Westinghouse, for semiconductors, consists of dendritic silicon crystals grown into thin continuous sheets 0.5 in (1.3 cm) wide, thus eliminating the need to saw slices from ingots. Pure **single-crystal silicon ribbon** of Dow Chemical is as thin as 49 μin (1.25 μm) and is made as a membrane formed by surface tension between two growing dendritic crystals. Float-zoned single-crystal silicon is 100 times purer than semiconductor-grade silicon. It is used in wafer form for laser and infrared detectors in guided bombs and missiles and for high-power switching devices, such as thyristors.

Silicon does not possess a metallic-type lattice structure and, like antimony, is a semimetal and lacks plasticity, but is more akin to the diamond in structure. Because of its feeble electronegative nature, it has a greater tendency to form compounds with nonmetals than with metals. Silicon forms **silicon hybrids** of general formula Si_xH_{2x+2}, similar to the paraffin hydrocarbons, but they are very unstable and ignite in air. But a mixture of ferrosilicon and sodium hydroxide, called **hydrogenite,** which yields hydrogen gas when water is added, is used for filling balloons. Silicon, like carbon, has a valence of 4 and links readily to carbon in SiC chain formations. The SiC bond acts as the C—C bond of organic chemistry, but silicon does not enter into animal or plant structures.

SILICON BRONZE. A family of wrought copper-base alloys (C64700 to C66100) and one **cast copper alloy** (C87200), the wrought alloys containing from 0.4 to 0.8% silicon (C64700) to 2.8 to 4.0 silicon (C65600), and the cast alloy 1.0 to 5.0, along with other elements, usually lead, iron, and zinc. Other alloying elements may include manganese, aluminum, tin, nickel, chromium, and phosphorus. The most well-known alloys are probably **silicon bronze C65100, or low-silicon bronze B,** and **silicon bronze C65500, or high-silicon bronze A,** as they were formerly called. As these names imply, they differ mainly in silicon content: 0.8 to 2.0% and 2.8 to 3.8, respectively, although the latter alloy also may contain as much as 0.6 nickel. C87200 contains at least 89% copper, 1.5 silicon, and as much as 5 zinc, 2.5 iron, 1.5 aluminum, 1.5 manganese, 1 tin, and 0.5 lead. Regardless of alloying ingredients, copper content is typically 90% or greater.

Both of the common wrought alloys are quite ductile in the annealed condition, C65500 being somewhat more ductile than C65100, and both can be appreciably strengthened by cold working. Annealed, tensile yield strengths are on the order of 15,000 to 25,000 lb/in^2 (103 to 172 MPa) depending on mill form, with ultimate tensile strengths to about 60,000 lb/in^2 (414 MPa) and elongations of 50 to 60%. Cold working can increase yield strength to as much as 70,000 lb/in^2 (483 MPa). Electrical conductivity is 12% for C65100 and 7 for C65500 relative to copper, and thermal conductivity is 33 Btu/(ft · h · °F) [57 W/(m · K)] and 21 Btu/(ft · h · °F) [(36 W/m · K)], respectively. The alloys are used for hydraulic-fluid lines in aircraft, heat-exchanger tubing, marine hardware, bearing plates, and various fasteners.

Silicon bronze C87200, which is suitable for centrifugal, investment, and sand-, plaster-, and permanent-mold casting, also has been known by the trade names **Everdur, Herculoy,** and **Navy Tombasil.** Typical as-sand-cast tensile properties are 55,000 lb/in^2 (379 MPa) ultimate strength, 25,000 lb/in^2 (172 MPa) yield strength, and 30% elongation. Brinell hardness is 85, electrical conductivity 6%, and, relative to free-cutting brass, machinability is 40%. Uses include pump and valve parts, marine fittings, and bearings.

SILICON CARBIDE. A bluish-black, crystalline, artificial mineral of composition SiC having a Knoop hardness of 2,500. It is used as an abrasive as loose powder, coated abrasive cloth and paper, wheels, and hones. It withstands temperatures to its decomposing point of 4175°F (2301°C) and is valued as a refractory. It retains its strength at high temperatures and has low thermal expansion, and its heat conductivity is 10 times that of fireclay. It is used for butterfly valves that control the flow of hot blasts through the tuyeres of blast furnaces, and for high-pressure, high-temperature mechanical seals in polymer-processing reactors. Silicon-carbide particulates are used as reinforcements in aluminum-alloy composites, and silicon-carbide fibers and whiskers serve as reinforcements in emerging metal-matrix and ceramic-matrix composites. The material is also a potential matrix material for composites. Fibers are used to strengthen and toughen glass ceramics. Thermal-insulation blankets of spacecraft, which can withstand repeated exposure to temperatures as high as 3632°F (2000°C), comprise layers of silicon-carbide and aluminoborosilicate fabrics, and silicon-carbide thread is used to stitch the fabrics. The material also holds promise for integrated circuits able to withstand higher temperatures than silicon-based ICs and for mirrors of superior mechanical, thermal, and optical properties in space systems, solar collectors, and astronomical telescopes. Silicon carbide is made by fusing sand and coke at a temperature above 4000°F (2204°C). It can also be made from polymer precursors and by vapor-phase reactions. One such precursor, developed at Rensselaer Polytechnic Institute, is **hydridopolycarbosilane.** When it is heated to 1832°F (1000°C), 90% of the polymer converts to the carbide.

Unlike aluminum oxide, the crystals of silicon carbide are large, and they are crushed to make the small grains used as abrasives. They are harder than aluminum oxide, and as they fracture less easily, they are more suited for grinding hard cast irons and ceramics. The standard grain sizes are usually from 100 to 1,000 mesh. The crystalline powder in grain sizes from 60 to 240 mesh is also used in lightning arrestors. **Carborundum,** of Standard Oil Engineered Materials Co., **Crystolon,** of Norton Co., and **Carbolon,** of Exolon Co. of Canada Ltd., are trade names for silicon carbide. Many other trade names are used, such as **Carborite, Carbolox, Carbolite, Carbobrant, Storalon, Sterbon,** and **Natalon. Ferrocarbo** is a silicon carbide of Carborundum Co. in briquettes for adding to the iron cupola charge. It breaks down in the cupola above 2000°F (1093°C) to form nascent carbon and silicon for adding to the iron and also for deoxidizing. It produces more-uniform iron castings. **Alsimag 539** is a fine-grained silicon carbide in the form of molded parts for brazing fixtures and furniture for kilns for high-temperature sintering. The **siliconized graphites** produced by Pure Carbon Co., named **Purebide,** are graphite materials with surfaces chemically converted to silicon carbide. They have the wear resistance of silicon carbide, but retain some of the lubricity of graphite. Cost savings are achieved by machining graphite into intricate shapes before conversion, and subsequently impregnating parts to control leakage or modify strength and/or wear properties.

When used as a refractory in the form of blocks or shapes, silicon carbide may be ceramic-bonded or self-bonded by recrystallization. A standard silicon carbide brick has about 90% SiC, with up to 8 silica. The specific gravity is about 3.2. It has very high resistance to spalling. The thermal conductivity is about the same as that of mullite, and the coefficient of expansion is about $2.6 \times 10^{-6}/°F$ ($4.7 \times 10^{-6}/K$). **Carbex** is a **silicon carbide firebrick** of General Refractories Co. **Refrax silicon carbide** of Carborundum Co. is bonded with silicon nitride. It is used for hot-spray nozzles, for heat-resistant parts, and for lining electrolytic cells for smelting aluminum. **Silicon carbide KT,** of the same company, is molded without a binder. It has 96.5% SiC with about 2.5 silica. The specific gravity is about 3.1, and it is impermeable to gases. Parts made by pressing or extruding and then sintering have a flexural strength of 24,000 lb/in² (165 MPa) and compressive strength of 150,000 lb/in² (1,034 MPa). The Knoop hardness is 2,740. It is made in rods, tubes, and molded shapes, and the rough crystal surface can be diamond-ground to a smooth, close tolerance. The operating temperature in inert atmospheres is to 4000°F (2204°C) and in oxidizing atmospheres to 3000°F (1649°C). For reactor parts, it has a low neutron-capture cross section and high radiation stability. The thermal conductivity is 2.5 times that of stainless steel. **Crystolon R** of Norton Co. is a stabilized silicon carbide bonded by recrystallization. It has a specific gravity of 2.5, a tensile strength of 5,500 lb/in² (38 MPa), compressive strength of 25,000 lb/in² (172 MPa), and Knoop hardness of 2,500. The porosity is 21%. It is for parts subject to temperatures to 4200°F (2316°C), and it withstands high thermal shock. **Crystolon C** is a self-bonding silicon carbide for coating molded graphite parts to give high wear and erosion resistance. The coatings, 0.003 to 0.020 in (0.008 to 0.051 cm) thick, produced by high-temperature chemical reaction, form an integral part of the graphite surface. **Vitropore** filter candles, of Pall Corp., are made from rigid silicon carbide and are used to recover particulates from hot gas streams. They are especially effective in recovering catalysts from fluid-catalytic-cracking processes. **Diasil,** of Crestel Pty of Australia, comprises low-value, crushed diamond dust in silicon carbide and is used for cutting, drilling, and grinding ceramics.

Silicon carbide foam is a lightweight material made of self-bonded silicon carbide foamed into shapes. The low-density foam has a density of 17 lb/ft^3 (272 kg/m^3), a porosity of 90%, and tensile and compressive strengths of 30 lb/in^2 (0.2 MPa). The high-density foam of 33 lb/ft^3 (529 kg/m^3) has a tensile strength of 85 lb/in^2 (0.6 MPa) and compressive strength of 750 lb/in^2 (5 MPa). Its porosity is 80%. It is inert to hot chemicals and can be machined.

Silicon carbide crystals are used for semiconductors at temperatures above 650°F (343°C). As the cathode of electronic tubes instead of a hot-wire cathode, the crystals take less power and need no warm-up. In the silicon carbide crystal, both the silicon and the crystalline carbon have the covalent bond in which each atom has four near neighbors and is bonded to each of these with two electrons symmetrically placed between the atoms; but since there is an electronegative difference between silicon and carbon, there is some ionic bonding which results in a lesser mobility for lattice scattering. The silicon carbide semiconductor crystals of Westinghouse have less than 1 part of impurities to 10 million, and the junction is made by diffusing aluminum atoms into the crystal at a temperature of 3900°F (2149°C), making a p-type junction.

Silicon carbide fiber is one of the most important fibers for high-temperature use. It has high strength and modulus and withstands temperatures even under oxidizing conditions up to 3272°F (1800°C), though the fibers show some deterioration in tensile strength and modulus properties at temperatures above 2192°F (1200°C). It has advantages over carbon fibers for some uses, having greater resistance to oxidation at high temperatures, superior compressive strength, and greater electrical resistance. **SCS silicon-carbide** fibers, of Textron Specialty Materials, maintain strength at temperatures over 2500°F (1371°C) and are useful as reinforcements for ceramic-matrix composites.

There are two commercial processes for making continuous silicon carbide fibers: (1) by coating silicon carbide on either a tungsten or a carbon filament by vapor deposition to produce a large filament [3,937 to 5,906 μin (100 to 150 μm) in diameter], or (2) by melt-spinning an organic polymer containing silicon atoms as a precursor fiber followed by heating at an elevated temperature to produce a small filament [394 to 1,181 μin (10 to 30 μm) in diameter]. Fibers from the two processes differ considerably, but both are used commercially.

Silicon carbide whiskers as small as 276 μin (7 μm) in diameter can be made by a number of different processes. Although these whiskers have the disadvantage in some applications of not being in continuous-filament form, they can be made with higher tensile strength and modulus values than continuous silicon carbide filament.

SILICON CAST IRON. An **acid-resistant cast iron** containing a high percentage of silicon. When the amount of silicon in cast iron is above 10%, there is a notable increase in corrosion and acid resistance. The acid resistance is obtained from the compound Fe_3Si, which contains 14.5% silicon. The usual amount of silicon in acid-resistant castings is from 12 to 15%. The alloy casts well but is hard and cannot be machined. These castings usually contain 0.75 to 0.85% carbon. Lesser amounts decrease acid resistance. Too much carbon also separates out as graphite in silicon irons, causing faulty castings. Increasing the content of silicon in iron reduces the melting point progressively from 2786°F (1530°C) for pure iron to 2282°F (1250°C) for iron containing 23% silicon. A 14 to 14.5% **silicon iron** has a silvery-white structure, a compressive strength of about 70,000 lb/in^2 (483 MPa), and Brinell hardness 299 to 350, and it is resistant to hot sulfuric acid, nitric acid, and organic acids. Silicon irons are very wear-resistant and are valued for pump parts and for parts for chemical machinery. They are marketed under many trade names. **Duriron,** of Duriron Co., contains 14.5% silicon and 1 carbon and manganese. The tensile strength is 16,000 lb/in^2 (110 MPa) and density 0.253 lb/in^3 (7,003 kg/m^3).

SILICON-COPPER. An alloy of silicon and copper used for adding silicon to copper, brass, or bronze, also employed as a deoxidizer of copper and for making hard copper. Silicon alloys in almost any proportion with copper, and is the best commercial hardener of copper. A 50–50 alloy of silicon and copper is hard, extremely brittle, and black. A 10% silicon, 90 copper alloy is as brittle as glass; in this proportion silicon copper is used for making the addition to molten copper to produce hard, sound copper-alloy castings of high strength. The resulting alloy is easy to cast in the foundry and does not dross. Silicon-copper grades in 5, 10, 15, and 20% silicon are also marketed, being usually sold in slabs notched for breaking into small sections for adding to the melt. A 10% silicon-copper melts at 1500°F (816°C); a 20% alloy melts at 1152°F (623°C).

SILICON-MANGANESE. An alloy employed for adding manganese to steel and as a deoxidizer and scavenger of steel. It usually contains 65 to 70% manganese and 12 to 25 silicon. It is graded according to the amount of carbon, generally 1, 2, and 2.5%. For making steels low in carbon and high in manganese, silicomanganese is more suitable than ferromanganese. A reverse alloy, called **manganese-silicon,** contains 73 to 78% silicon and 20 to 25 manganese, with 1.5 maximum iron and 0.25 maximum carbon. It is used for adding manganese and silicon to metals without the addition of iron. Still another alloy is called **ferromanganese-silicon,** containing 20 to 25% manganese, about 50 silicon, and 25 to 30 iron, with only about 0.50 or less carbon. This alloy has a low melting point, giving ready solubility in the metal.

Silicon-spiegel is an alloy of silicon and manganese with iron employed for making additions of silicon and manganese to open-hearth steels, and also for adding manganese to cast iron in the cupola. A typical analysis gives 25 to 30% manganese, 7 to 8 silicon, and 2 to 3 carbon. Both the silicon and manganese act as strong deoxidizers, forming a thin, fusible slag, making clean steel.

SILICON NITRIDE. Si_3N_4 is a hard, lightweight, heat- and creep-resistant polycrystalline ceramic having low coefficients of friction and thermal expansion and good resistance to corrosion and thermal shock. Powder, the starting stock for parts production, is commonly made by the nitridation of metallic silicon. Other methods include gas-phase ammonolysis of silicon tetrachloride, carbothermic reduction of silicon dioxide, and thermal decomposition of silicon diimide. Parts are usually made by reaction bonding without sintering or hot-pressing and liquid-phase sintering. Reaction bonding involves reacting a consolidated and shaped mass of pure silicon powder with nitrogen at high temperature. Resulting parts, commonly designated **reaction-bonded silicon nitride (RBSN)**, are 15 to 20% porous, thus only moderate in strength, but essentially shrink-free, thus quite accurate as formed. Hot-pressing powder, using powder with sintering additives, followed by sintering results in parts commonly designated **hot-pressed silicon nitride (HPSN)**. These are nearly full-density parts of more-robust mechanical performance. Density ranges from 0.111 to 0.122 lb/in^3 (3,072 to 3,377 kg/m^3), and the coefficient of thermal expansion is 1.67×10^{-6} to 2.17×10^{-6}/°F (3×10^{-6} to 3.9×10^{-6}/K). Tensile modulus ranges from 36×10^6 to 47×10^6 lb/in^2 (248,000 to 324,000 MPa), and the flexural strength is 65,200 to 159,500 lb/in^2 (450 to 1,100 MPa). At 2552°F (1400°C), the modulus ranges from 25.4×10^6 to 36.2×10^6 lb/in^2 (175,000 to 250,000 MPa), with the flexural strength as high as 87,000 lb/in^2 (600 MPa).

Noralide, of Norton Co., is an HPSN used for ball and roller bearings. Such bearings, used in machine-tool spindles and instruments, are noted for their light weight, low friction, and good wear and fatigue resistance. Other silicon nitride applications include valves, seals, and cutting tools. **Ceralloy 147,** of Ceradyne, Inc., is a cast silicon nitride material for check-valve balls and mechanical seals. Its key features are resistance to abrasion, oxidation, corrosion, and thermal shock. **Ceralloy 147-3 Needlelok** is toughened by interlocking, needlelike grains. It has a Vickers hardness of 1,600, a tensile strength of 113,000 lb/in^2 (780 MPa), and a fracture toughness of 5,600 (lb/in^2) · $\sqrt{\text{in}}$ (6.2 MPa · $\sqrt{\text{m}}$), and it is used for oil-drilling applications. **Roydazide,** of Materials Research Corp., is for coatings as well as parts production.

Silicon nitride fibers have been made by reacting silicon oxide and nitrogen in the presence of a reducing agent in an electrical-resistance furnace at 2552°F (1400°C). Discontinuous fibers are used as reinforcements in composites for specialty aircraft and electrical parts, and in radomes (microwave windows). Whiskers have been made by the reaction of nitrogen and a mixture of silicon and silica. For the emerging ceramic-matrix composites, silicon nitride is a potential matrix and reinforcement material.

SILICON STEEL. All grades of steel contain some silicon, and most contain from 0.10 to 0.35% as a residual of the silicon used as a deoxidizer. But from 3 to 5% silicon is sometimes added to increase the magnetic permeability, and larger amounts are added to obtain wear-resisting or acid-resisting properties. Silicon deoxidizes steel, and up to 1.75% increases the elastic limit and impact resistance without loss of ductility. Silicon steels within this range are used for structural purposes and for springs, giving a tensile strength of about 75,000 lb/in^2 (517 MPa) and 25% elongation. A common low-silicon structural steel contains up to 0.35% silicon and 0.20 to 0.40 carbon, but the **structural silicon steels** are ordinarily **silicon-manganese steel,** with the manganese above 0.50%. Low-carbon steels used as structural steels are made by careful control of carbon, manganese, and silicon and with special mill heat treatment. **LT-75,** of Lukens Steel, contains 0.2% carbon, up to 1.35 manganese, and 0.3 silicon. The tensile strength is 90,000 lb/in^2 (621 MPa), with elongation of 24%. European silicon structural steels contain 0.80% or more silicon, with manganese above 0.50, and very low carbon. The silicon alone is a graphitizer and, to be most effective, needs the assistance of manganese or other carbide-forming elements. It is useful in high-strength, low-alloy steels and has a wide range of utility when used in alloy steels. Considerable addition of silicon above 1.75% increases the hardness and the corrosion resistance, but reduces the ductility and makes the steel brittle. The lower grades can be rolled, however, and silicon-steel sheet is used for electric transformer laminations. Silicon forms a chemical combination with the metal, forming an iron silicide.

The value of silicon steel as a **transformer steel** was discovered by Hadfield in 1883. Silicon increases the electrical resistivity and decreases the hysteresis loss, making silicon steel valuable for magnetic circuits where alternating current is used. **Electrical steel,** or **electric sheet,** is sheet steel for armatures and transformers, in var-

ious grades from 1 to 4.5% silicon. **Hipersil** is a high-permeability silicon steel, and **Cubex** is a silicon steel containing 3% silicon which
has been processed so that each cubic crystal of the steel structure is
oriented with the faces symmetric, giving alignment in four directions
instead of the normal two. The steel is easily magnetized across as
well as along the sheet. In transformers it lowers energy losses, and
also gives greater flexibility in designing shapes. One silicon iron is
double-oriented, with the cubic crystals of the iron in exact alignment
in all directions with the sides of the cubes parallel to the sides and
ends of the sheets. It gives high permeability with low induction loss.
Relay steel, used for relays and magnets, contains 0.5 to 2.75% silicon. **Orthosil** is silicon steel sheet, 0.004 in (0.010 cm) thick, for electrical laminations. **NK Super E-core,** of NKK Corp. of Japan, is a
6.5% silicon electrical steel in which about half of the silicon is diffused into the sheet after cold rolling.

   Several **cold-work steels** and **shock-resisting tool steels** contain 1 to 2.25% silicon. Of the cold-work air-hardening type, A10 contains 1.25% silicon. Of the cold-work oil-hardening type, O6 contains
1% silicon. And of the shock-resisting type, S2 contains 1%; S4 and
S5, 2%; and S6, 2.25%.

SILICONES.   A group of resinlike materials in which silicon takes the
place of carbon of the organic synthetic resins. Silicon is quadrivalent,
like carbon. But while carbon also has a valence of 2, silicon has only
one valence of 4, and the angles of molecular formation are different.
The two elements also differ in electronegativity, and silicon is an
amphoteric element, having both acid and basic properties. The
molecular formation of the silicones varies from that of the common
plastics, and they are designated as **inorganic plastics** as distinct
from the **organic plastics** made with carbon.

   In the long-chain organic synthetic resins, the carbon atoms repeat
themselves, attaching on two sides to other carbon atoms, while in
the silicones the silicon atom alternates with an oxygen atom so that
the silicon atoms are not tied to each other. The simple **silane** formed
by silicon and hydrogen corresponding to methane, CH_4, is also a gas,
as is methane, and has the formula SiH_4. But, in general, the silicones do not have the SiH radicals, but contain CH radicals as in the
organic plastics. Basically, silicon is treated with methyl chloride and
a catalyst to produce a gas mixture of silanes, $(CH_3)_x(SiCl)_{4-x}$. After
condensing, three silanes are fractioned, methyl chlorosilane,
dimethyl dichlorosilane, and trimethyl trichlorosilane. These are the
common building blocks of the **siloxane** chains, and by hydrolyzing
them cyclic linear polymers can be produced with acid or alkali catalysts to give fluids, resins, and rubbers. **Silicone resins** have, in gen-

eral, more heat resistance than organic resins, have higher dielectric strength, and are highly water-resistant. Like organic plastics, they can be compounded with plasticizers, fillers, and pigments. They are usually cured by heat. Because of the quartzlike structure, molded parts have exceptional thermal stability. Their maximum continuous-use service temperature is about 500°F (260°C). Special grades exceed this and go as high as 700 to 900°F (371 to 482°C). Their heat-deflection temperature for 265 lb/in^2 (1.8 MPa) is 900°F (482°C). Their moisture absorption is low, and resistance to petroleum products and acids is good. Nonreinforced silicones have only moderate tensile and impact strength, but fillers and reinforcements provide substantial improvement. Because silicones are high in cost, they are premium plastics and are generally limited to critical or high-performance products such as high-temperature components in the aircraft, aerospace, and electronic fields.

A great variety of molecular combinations are available in the silicone polymers, giving resins of varying characteristics, and those having CH radicals with silicon bonds are termed **organosilicon polymers. Silicon tetramethyl,** $Si(CH_3)_4$, is a liquid boiling at 79°F (26°C). **Trichlorosilane,** $HSiCl_3$, is also called **silicochloroform,** and it corresponds in formation to chloroform. By replacing the hydrogen atom of this compound with an alkyl group, the **alkylchlorosilanes** are made which have high adhesion to metals and are used in enamels. **Methyl chlorosilane,** $(CH_3)_2SiCl_2$, is a liquid used for waterproofing ceramic electrical insulators. The material reacts with the moisture in the ceramic, forming a water-repellent coating of methyl silicone resin and leaving a residue of hydrochloric acid which is washed off.

Silicone insulating varnishes will withstand continuous operating temperatures at 350°F (177°C) or higher. **Silicone enamels** and paints are more resistant to chemicals than most organic plastics, and when pigmented with mineral pigments, they withstand temperatures up to 1000°F (538°C). For lubricants the liquid silicones are compounded with graphite or metallic soaps and operate between −50 and 500°F (−46 and 260°C). The silicone liquids are stable at their boiling points, between 750 and 800°F (399 and 427°C), and have low vapor pressures, so that they are also used for hydraulic fluids and heat-transfer media. **Silicone oils,** used for lubrication and as insulating and hydraulic fluids, are methyl silicone polymers. They retain a stable viscosity at both high and low temperatures. As hydraulic fluids, they permit smaller systems to operate at higher temperatures. In general, silicone oils are poor lubricants compared with petroleum oils, but they are used at high temperatures, 302 to 392°F (150 to 200°C), low speeds, and low loads.

Silicone resins are blended with alkyd resins for use in outside paints, usually modified with a drying oil. **Silicone-alkyd resins** are also used for baked finishes, combining the adhesiveness and flexibility of the alkyd with the heat resistance of the silicone. A **phenyl ethyl silicone** is used for impregnating glass-fiber cloth for electrical insulation, and it has about double the insulating value of ordinary varnished cloth.

Silicone rubber is usually a long-chain **dimethyl silicone** which will flow under heat and pressure, but can be vulcanized by cross-linking the linear chains. Basically, it consists of alternate silicon and oxygen atoms with two methyl groups attached to each silicon atom. The tensile strength is 300 lb/in^2 (2 MPa), but with fillers it is raised to 600 lb/in^2 (4 MPa). It is usually compounded with silica and pigments. It is odorless and tasteless, is resistant to most chemicals but not to strong acids and alkalies, resists heat to 500°F (260°C), and remains flexible to -70°F (-57°C). The dielectric strength is 500 V/mil (20 $\times$ 10^6 V/m). Silicone adhesive sealants have similar advantages and bond well to various metals and nonmetallics.

Ordinary silicone rubber has the molecular group H $\cdot$ CH$_2$ $\cdot$ Si $\cdot$ CH$_2$ $\cdot$ H in a repeating chain connected with oxygen linkages, but in the **nitrile-silicone rubber** one of the end hydrogens of every fourth group in the repeating chain is replaced by a C:N radical. These polar nitrile groups give a low affinity for oils, and the rubber does not swell with oils and solvents. It retains strength and flexibility at temperatures from -100°F (-73°C) to above 500°F (260°C) and is used for such products as gaskets and chemical hose. As lubricants, silicones retain a nearly constant viscosity at varying temperatures. **Fluorosilicones** have fluoroalkyd groups substituted for some of the methyl groups attached to the siloxane polymer of dimethyl silicone. They are fluids, greases, and rubbers, incompatible with petroleum oils and insoluble in most solvents. The greases are the fluids thickened with lithium soap, or with a mineral filler.

SILK. The fibrous material in which the silkworm, or larva of the moth *Bombyx mori,* envelops itself before passing into the chrysalis state. Silk is closely allied to cellulose and resembles wool in structure, but unlike wool, it contains no sulfur. The natural silk is covered with a wax or silk glue which is removed by scouring in manufacture, leaving the glossy **fibroin,** or raw-silk fiber. The fibroin consists largely of the amino acid **alanine,** CH$_3$CH(NH$_2$)CO$_2$H, which can be synthesized from pyruvic acid. **Silk fabrics** are used mostly for fine garments, but are also valued for military powder bags because they burn without a sooty residue.

The fiber is unwound from the cocoon and spun into threads. Each cocoon has from 2,000 to 3,000 yd (1,829 to 2,743 m) of thread. The chief silk-producing countries are China, Japan, India, Italy, and France. **Floss silk** is a soft silk yarn practically without twist, or is the loose waste silk produced by the worm when beginning to spin its cocoon. **Hard silk** is thrown silk from which the gum has not been discharged. **Soft silk** is thrown silk yarn, degummed, dyed or undyed. **Souple silk** is dyed skein silk from which little gum has been discharged. It is firmer but is less lustrous. **Organizine silk** is from the best grade of cocoons. **Marabout silk,** used for making imitation feathers, is a white silk, twisted and dyed without discharging the gum. **Silk waste** is silk other than that reeled from the cocoon. It includes cocoons not fit for reeling, partly unwound cocoons, broken filaments, mill waste, and discarded noils. It is used in the spun-silk yarn industry. **Noils** consist of the short, staple knotty combings.

In China the cultivation of the silkworm is claimed to date back to 2640 B.C. Silk was first woven in Rome about 50 B.C. The eggs of the silkworm were smuggled into Europe in the year 552. **Sericulture,** or silkworm culture, is a highly developed industry. The larvae, which have voracious appetites, are fed on mulberry leaves for 24 days, after which they complete their cocoons in 3 to 4 days. In 7 to 70 days these are heated to kill the chrysalis to prevent bursting of the shell. The reeling is done by hand and by machine. **Wild silk** is from a night peacock moth which does not feed on the mulberry. It is coarser and stronger, but darker in color and less lustrous. **Tussah silk** is a variety of wild silk from South China and India. **Charka silk** is raw silk produced in Bengal on native hand-reeling machines. **Byssus silk** is a long fiber from a mussel of Sardinia and Corsica which spins the thread to attach itself to rocks. The fiber is golden brown, soft, lustrous, and elastic, and not dissolved by acids or alkalies. It was formerly used for fine garments but is no longer obtained commercially. **Canton silk** is soft and fluffy, but is greenish and lacks firmness. It is from *B. textor* and is used for weft yarns and in crepes. The silk grown in India and known as **Indian silk** is the finest of all silks with fibers 0.0004 in (0.0016 cm) compared with 0.001 in (0.003 cm) for Japanese silk. Before World War II Japan produced most of the silk of the world from a cultivated moth of the tussah variety, *Antheria yama mai.* **Shantung silk** is from a tussah moth, *A. pernyi,* which feeds on oak leaves.

The fabric called **shantung** is a rough-textured, plain-woven silk with irregular fillings. It is heavier and more bumpy than pongee. **Grosgrain** is a heavy, close-woven, corded fabric of silk. It is used for tapestry and in narrow widths for ribbons. **China silk,** or **habutai,** is an unweighted, all-silk fabric of close, firm, but uneven texture woven of low-quality, unthrown **raw silk** in the gum, but it is also imitated with textiles with a silk warp and a rayon filling. The lightweight grades of 3, 3.5, and 4 momme, or 371, 317, and 273 ft^2/lb (76, 65, and 56 m^2/kg), are classified as sheer fabrics and are used for impregnated fabrics for umbrellas, raincoats, and hospital sheetings. Unimpregnated habutai is used for curtains, lampshades, handkerchiefs, and caps. Heavyweight habutai of 12 momme [53 ft^2/lb (18 m^2/kg)] is used for parachutes. **Pongee** is a rough-textured, plain-woven, silk fabric with irregular filling yarns. It is made in natural color or dyed and, like China silk, has a gummy feel. **Bolting cloth,** for screening flour, is a fine, strong, silk fabric. The yarn is a fine-thread, hard-twist tram thrown in the gum from high-quality raw silk. The fabric has a lino weave with two warp threads swiveled around the weft. It comes in various meshes, the finest having 166 to 200 threads/linear inch (65 to 79 threads/linear cm). It is produced on handlooms in Switzerland and France. **Cartridge cloth** is a thin, strong fabric for powder bags for large-caliber guns. It is made of silk waste and noils. The silk is consumed in the explosion without leaving residues that would cause premature explosion of the subsequent charge. It also does not deteriorate in storage in contact with the powder.

The **kente** cloth of Ghana is a silk fabric of fine weave in delicate colors, hand-woven in long, narrow strips which are sewn together to make a pattern. **Satin** is a heavy silk fabric with a close twill weave in which the fine warp threads appear on the surface and the weft threads are covered up by the peculiar twill. Common satin is of eight-leaf twill, the weft intersecting and binding down the warp at every eighth pick, but 16 to 20 twills are also made. In the best satins a fine quality of silk is used. It was originally called **zayton,** derived from the Arab name of the Chinese trading post where the fabric was produced. Varieties of imitation satin are made with a cotton weft. Satins are dyed to many colors and much used for linings and trimmings.

Qiana, of Du Pont, originally called Fiber Y, produces **synthetic resin fabrics** with the feel and drape of silk. They are resilient and take dyes readily. The fiber is a polyamide based on an alicyclic diamine. **A-Tell** is a Japanese textile fiber of great silkiness. It is a polyethylene oxybenzoate, the molecule having both ester and ether linkages. Another Japanese fiber is 50% polyvinyl chloride and 50 polyvinyl alcohol. Called **Cordelan,** it produces fabrics with the feel of wool.

SILVER. A white metal, symbol Ag, very malleable and ductile, and classified with the precious metals. It occurs in the native state, and also combined with sulfur and chlorine. Copper, lead, and zinc ores frequently contain silver; about 70% of the production of silver is a by-product of the refining of these metals. Mexico and the United

States produce more than half of the silver of the world. Canada, Peru, and Bolivia are also important producers. Although nearly 90% of the silver produced in Arizona comes from copper ores, most of that produced in California is a by-product of gold quartz mining. Silver is the whitest of all the metals and takes a high polish, but easily tarnishes in air because of the formation of a silver sulfide. It has the highest electrical and heat conductivity: 108% IACS relative to 100% for the copper standard and about 244 Btu/(ft · h · °F) [422 W/(m · K)], respectively. Cold work reduces conductivity slightly. The specific gravity is 10.7, and the melting point is 1764°F (962°C). When heated above the boiling point (3925°F, 2163°C), it passes off as a green vapor. It is soluble in nitric acid and in hot sulfuric acid. The tensile strength of cast silver is 41,000 lb/in^2 (283 MPa), with a Brinell hardness of 59. The metal is marketed on a troy-ounce value.

Since silver is a very soft metal, it is not normally used industrially in its pure state, but is alloyed with a hardener, usually copper. **Sterling silver** is the name given to a standard high-grade alloy containing a minimum of 925 parts in 1,000 of silver. It is used for the best tableware, jewelry, and electrical contacts. This alloy of 7.5% copper work-hardens and requires annealing between roll passes. Silver can also be hardened by alloying with other elements. The old alloy **silanca** contained small amounts of zinc and antimony, but the name sterling silver is applied only to the specific **silver-copper alloy.**

The standard types of commercial silver are fine silver, sterling silver, and coin silver. **Fine silver** is at least 99.9% pure and is used for plating, making chemicals, and for parts produced by powder metallurgy. **Coin silver** is usually an alloy of 90% silver and 10 copper, but when actually used for coins, the composition and weight of the coin are designated by law. Silver and gold are the only two metals which fulfill all the requirements for coinage. The so-called **coins** made from other metals are really official tokens, corresponding to paper money, and are not true coins. Coin silver has a Vickers hardness of 148 compared with a hardness of 76 for hard-rolled pure silver. It is also used for silverware, ornaments, plating; for alloying with gold; and for electric contacts. When about 2.5% of the copper in coin silver is replaced by aluminum, the alloys can be age-hardened to Vickers 190. Silver is not an industrial metal in the ordinary sense. It derives its coinage value from its intrinsic aesthetic value for jewelry and plate, and in all civilized countries silver is a controlled metal.

Silver powder, 99.9% purity, for use in coatings, integrated circuits, and other electrical and electronic applications, is produced in several forms. Amorphous powder is made by chemical reduction and comes in particle sizes of 35 to 591 μin (0.9 to 15 μm). Powder made electrolytically is in dendritic crystals with particle sizes from 394 to

7,874 μin (10 to 200 μm). Atomized powder has spherical particles and may be as fine as 400 mesh. **Silver-clad powder** for electric contacts is a copper powder coated with silver to economize on silver. **Silver flake** is in the form of laminar platelets and is particularly useful for conductive and reflective coatings and circuitry. The tiny, flat plates are deposited in overlapping layers permitting a metal weight saving of as much as 30% without reduction in electrical properties. **Nickel-coated silver powder,** for contacts and other parts made by powder metallurgy, comes in grades with 0.25, 0.5, 1, and 2% nickel by weight.

The **porous silver** of the Pall Corp. comes in sheets in standard porosity grades from 79 to 2,165 μin (2 to 55 μm). It is used for chemical filtering. **Doré metal** used for jewelry is silver containing some gold, but the material known as doré metal, obtained as a by-product in the production of selenium from copper slimes, is a mixture of silver, gold, and platinum. Silver plating is sometimes done with a **silver-tin alloy** containing 20 to 40 parts silver and the remainder tin. It gives a plate having the appearance of silver but with better wear resistance. Silver plates have good reflectivity at high wavelengths, but reflectivity falls off at about 13,780 nin (350 nm), and is zero at 118,110 nin (3,000 nm), so that it is not used for heat reflectors. **Silvar,** of Texas Instruments, is a silver and nickel-iron composite made by infiltrating silver into a porous preform of the alloy. It is intended for heat-sink and thermal management applications in electronics.

Silver-clad sheet, made of a cheaper nonferrous sheet with a coating of silver rolled on, is used for food processing equipment. It is resistant to organic acids but not to products containing sulfur. **Silver-clad steel,** used for machinery bearings, shims, and reflectors, is made with pure silver bonded to the billet of steel and then rolled. For bearings, the silver is 0.010 to 0.35 in (0.025 to 0.889 cm) thick, but for reflectors the silver is only 0.001 to 0.003 in (0.003 to 0.008 cm) thick. The **silver-clad stainless steel** of American Cladmetals Co. is stainless-steel sheet with a thin layer of silver rolled on one side for electrical conductivity.

Silver iodide is a pale-yellow powder of composition AgI, best known for its use as a nucleating agent and for seeding rain clouds. **Silver nitrate,** formerly known as **lunar caustic,** is a colorless, crystalline, poisonous, and corrosive material of composition $AgNO_3$. It is used for silvering mirrors, for silver plating, in indelible inks, in medicine, and for making other silver chemicals. The high-purity material is made by dissolving silver in nitric acid, evaporating the solution and crystallizing the nitrate, then redissolving the crystals in distilled water and recrystallizing. It is an active oxidizing agent. **Silver chloride,** AgCl, is a white, granular powder used in silver-plating solutions. This salt of silver and other halogen compounds of silver, especially **silver bromide,** AgBr, are used for photographic plates and films. The image cast on the plate by the lens breaks down the atomic structure of the compound in proportion to the intensity of light waves received and time of exposure. Electrons gather on the positive lower side of the bromide grains, causing the formation of black threads of silver when the film is placed in a developing solution of **ferrous oxalate,** FeC_2O_4, or other reducing chemical. The comparative values, or tones, in the picture come from the different color wavelengths in the white light and the different intensities of incoming waves. Measured in seconds, the action of violet light, the shortest wavelength, on the compound is more than 40 times greater than the action of the long wavelength of red light. To prevent further action by light, the film is transferred to a fixing bath of sodium thiosulfate which dissolves out the unreduced silver bromide. **VerdeFilm,** developed by Xerox Corp. for commercial printing, requires only electrostatic sensitizing prior to use and no silver halide or chemical developers. Its use avoids wastewater disposal problems associated with developers, fixers, and solubilized silver salts.

Silver chloride crystals in sizes up to 10 lb (4.5 kg) are grown synthetically. The crystals are cubic and can be heated and pressed into sheets. The specific gravity is 5.56, index of refraction 2.071, and melting point 851°F (455°C). They are slightly soluble in water and soluble in alkalies. The crystals transmit more than 80% of the wavelengths from 1,969 to 7,874 μin (50 to 200 μm). **Cerargyrite,** sometimes called **horn silver,** an ore of silver, found in the upper zone of silver veins in Nevada, Colorado, Idaho, Peru, Chile, and Mexico, is a silver chloride containing theoretically 75.3% silver, with sometimes some mercury. The Mohs hardness is 2.3 and specific gravity 5.8. It is massive, resembling wax, with a pearl-gray color.

Silver sulfide, Ag_2S, is a gray-black, heavy powder used for inlaying in metal work. It changes its crystal structure at about 355°F (179°C), with a drop in electrical resistivity, and is also used for self-resetting circuit breakers. **Silver potassium cyanide,** $KAg(CN)_2$, is a white, crystalline, poisonous solid used for silver-plating solutions. **Silver tungstate,** Ag_2WO_4, **silver manganate,** $AgMnO_4$, and other silver compounds are produced in high-purity grades for electronic and chemical uses.

SILVER SOLDER. High-melting-point solder employed for soldering joints where more than ordinary strength and, sometimes, electrical conductivity are required. Most silver solders are **copper-zinc brazing alloys** with the addition of silver. They may contain from 9 to 80% silver, and the color varies from brass yellow to silver white.

Cadmium may also be added to lower the melting point. Silver solders do not necessarily contain zinc, and may be alloys of silver and copper in proportions arranged to obtain the desired melting point and strength. A silver solder with a relatively low melting point contains 65% silver, 20 copper, and 15 zinc. It melts at 1280°F (693°C), has a tensile strength of 64,800 lb/in^2 (447 MPa), and elongation 34%. The electrical conductivity is 21% that of pure copper. A solder melting at 1400°F (760°C) contains 20% silver, 45 copper, and 35 zinc. **ASTM silver solder No. 3** is this solder with 5% cadmium replacing an equal amount of the zinc. It is a general-purpose solder. **ASTM silver solder No. 5** contains 50% silver, 34 copper, and 16 zinc. It melts at 1280°F (693°C) and is used for soldering electrical work and refrigeration equipment.

Any tin present in silver solders makes them brittle; lead and iron make the solders difficult to work. Silver solders are malleable and ductile and have high strength. They are also corrosion-resistant and are especially valuable for use in food machinery and apparatus where lead is objectionable. Small additions of lithium to silver solders increase fluidity and wetting properties, especially for brazing stainless steels or titanium. **Sil-Fos,** of Handy & Harmon, is a phosphor-silver brazing solder with a melting point of 1300°F (704°C). It contains 15% silver, 80 copper, and 5 phosphorus. Lap joints brazed with Sil-Fos have a tensile strength of 30,000 lb/in^2 (207 MPa). The phosphorus in the alloy acts as a deoxidizer, and the solder requires little or no flux. It is used for brazing brass, bronze, and nickel alloys. The grade made by this company under the name of **Easy solder** contains 65% silver, melts at 1325°F (718°C), and is a color match for sterling silver. **TL silver solder** of the same company has only 9% silver and melts at 1600°F (871°C). It is brass yellow in color and is used for brazing nonferrous metals. **Sterling silver solder,** for brazing sterling silver, contains 92.8% silver, 7 copper, and 0.2 lithium. Flow temperature is 1650°F (899°C).

A **lead-silver solder** recommended by Indium Corp. of America to replace tin solder contains 96% lead, 3 silver, and 1 indium. It melts at 590°F (310°C), spreads better than ordinary lead-silver solders, and gives a joint strength of 4,970 lb/in^2 (34 MPa). **Silver-palladium alloys** for high-temperature brazing contain from 5 to 30% palladium. With 30%, the melting point is about 2250°F (1232°C). These alloys have exceptional melting and flow qualities and are used in electronic and spacecraft applications.

SISAL. The hard, strong, light-yellow to reddish fibers from the large leaves of the sisal plant, *Agave sisalana,* and the henequen plant, *A. fourcroydes,* employed for making rope, cordage, and sacking. About

80% of all binder twine is normally made from sisal, but sisal ropes have only 75% of the strength of Manila rope and are not as resistant to moisture. Sisal is a tropical plant, and grows best in semiarid regions. The agave plant is native to Mexico, but most of the sisal comes from Haiti, east Africa, and Indonesia. The retting, separation, and washing of the fiber are done by machine, and less than 5% of the weight of the leaf results in good fiber. **Mexican sisal** is classified in seven grades from the Superior white fiber 41 in (105 cm) in length to Grade C-1, short-spotted fiber 24 in (60 cm) in length. **Yucatan sisal,** or **henequen,** is from the henequen plant and is reddish, stiffer, and coarser, and is used for binder twine. The Indian word *henequen* means knife, from the knifelike leaves. The plant is more drought-resistant than sisal. Henequen also comes from Indonesia as the spotted or reddish grades of sisal. **Maguey,** or **cantala,** is from the leaves of *A. cantala* of India, the Philippines, and Indonesia. It is used principally for binder twine. The fibers are white, brilliant, stiff, and lightweight. The fibers are not as strong as sisal, but have a better appearance and greater suppleness. **Zapupe fiber,** of Mexico, is from *A. zapupe.* The fiber is similar to sisal, finer and softer than henequen. **Salvador sisal,** of El Savador, is from *A. letonae.* The leaves are more slender than those of Mexican sisal, and the fiber is softer and finer. It is used for cordage and fabrics.

The fibers of sisal are not as long or as strong as those of Manila hemp, and they swell when wet, but they are soft and are preferred for binder twine either alone or mixed with Manila hemp. **Sisal fiber** is also used instead of hair in cement plasters for walls and in laminated plastics. **Corolite** is a molded plastic made with a mat of sisal fibers so as to give equal strength in all directions. **Agave fibers** from other varieties of the plant are used for various purposes, notably **tampico,** from *A. rigida,* which yields a stiff, hard, but pliant fiber employed for circular power brushes, and **istle,** a similar stiff brush fiber from several plants. Tampico is valued for polishing wheels, as the fibers hold the grease buffing compositions, and it is not brittle but abrades with flexibility. **Jaumave istle** is from *A. funkiana* of Mexico. It yields long, uniform fibers finer than tampico. **Lechuguilla** is a type of istle from *A. lechuguilla.*

There are at least 50 species of agave in Mexico and the southwestern United States which yield valuable by-products in addition to fiber. From some varieties saponin is obtained as a by-product. From a number of thick-leaved species the buds are cut off, leaving a cavity from which juice exudes. This juice is fermented to produce **pulque,** a liquor with a ciderlike taste containing about 7% alcohol. The juice contains a sugar, **agavose,** $C_{12}H_{22}O_{11}$, which is used in medicine as a laxative and diuretic. **Agava,** of Agava Products, Inc., is a dark-brown, viscous liquid extracted from the leaves of agave plants, used as a water conditioner for boiler-water treatment. It is a complex mixture of sapogenines, enzymes, chlorophyllin, and polysaccharides.

A fine strong fiber is obtained from the long leaves of the **pineapple,** *Ananus comosus,* native to tropical America. The plant is grown chiefly for its fruit, known in South America under its Carib name **ananá** and marketed widely as canned fruit and juice, preserves, and confections. **Pineapple concentrate** is also sold as a flavor enhancer, as much as 10% being added to apricot, cherry, or other fruit juices without altering the original flavor. For fiber production the plants are spaced widely for leaf development and are harvested before the leaves are fully mature. The retted fibers are long, white, and of fine texture and may be woven into water-resistant fabrics. The very delicate and expensive **piña cloth** of the Philippines is made from **pineapple fiber.** The fabrics of Taiwan are usually coarser and harder.

SLAG. The molten material that is drawn from the surface of iron in the blast furnace. Slag is formed from the earthy materials in the ore and from the flux. Slags are produced in the melting of other metals, but iron blast-furnace slag is usually meant by the term. Slag is used in cements and concrete, for roofing, and as a ballast for roads and railways. Finely crushed slag is used in agriculture for neutralizing acid soils. **Blast-furnace slag** is one of the lightest concrete aggregates available. It has a porous structure and, when crushed, is angular. It is also crushed and used for making pozzuolana and other cements. Slag contains about 32% silica, 14 alumina, 47 lime, 2 magnesia, and small amounts of other elements. It is crushed, screened, and graded for marketing. **Crushed slag** weighs 1,900 to 2,100 lb/yd^3 (1,127 to 1,245 kg/m^3), or is about 30% lighter than gravel. **Honeycomb slag** weighs only about 30 lb/ft^3 (481 kg/m^3). The finest grade of commercial slag is from 0.1875 in (0.48 cm) to dust; the run-of-crusher slag is from 4 in (10 cm) to dust. **Basic phosphate slag,** a by-product in the manufacture of steel from phosphatic ores, is finely ground and sold for fertilizer. It contains not less than 12% phosphoric oxide, P_2O_5, and is known in Europe as **Thomas slag. Foamed slag** is a name used in England for honeycomb slag used for making lightweight, heat-insulating blocks. A superphosphate cement is made in Belgium from a mixture of basic slag, slaked lime, and gypsum.

SLATE. A shale having a straight cleavage. Most shales are of sedimentary origin, and their cleavage was the result of heavy or long-continued pressure. In some cases slates have been formed by the consolidation of volcanic ashes. The slaty cleavage does not usually coincide with the original stratification. Slate is of various colors: black, gray, green, and reddish. It is used for electric panels, chalkboards, slate pencils, tabletops, roofing shingles, floor tiles, and treads. The terms **flagstone** and **cleftstone** are given to large, flat sections of slate used for paving, but the names are also applied to blue sandstones cut for this purpose. Slate is quarried in large blocks, and then slabbed and split. The chief slate-producing states are Pennsylvania, Vermont, Virginia, New York, and Maine. **Roofing slates** vary in size from 12 by 6 in (30 by 15 cm) to 24 by 14 in (61 by 36 cm), and from 0.125 to 0.75 in (0.32 to 1.91 cm) in thickness, and are usually of the harder varieties. The roofing slate from coal beds is black, fine-grained, and breaks into brittle thin sheets. It does not have the hardness or weather resistance of true slate. As late as 1915 more than 85% of all slate mined was used for roofing, but the tonnage now used for this purpose is small. **Ribbon slate,** with streaks of hard material, is inferior for all purposes. Lime impurities can be detected by the application of dilute hydrochloric acid to the edges and noting if rapid effervescence occurs. Iron is a detriment to slates for electric purposes. The average compressive strength of slate is 15,000 lb/in^2 (103 MPa) and the density 175 lb/ft^3 (2,804 kg/m^3). **Slate granules** are small, graded chips used for surfacing prepared roofing. **Slate flour** is ground slate, largely a by-product of granule production. It is used in linoleum, caulking compounds, and asphalt surfacing mixtures. **Slate lime** is an intimate mixture of finely divided, calcined slate and lime, about 60% by weight lime to 40 slate. It is employed for making porous concrete for insulating partition walls. The process consists in adding a mixture of slate lime and powdered aluminum, zinc, or magnesium to the cement. The gas generated on the addition of water makes the cement porous.

SMOKE AGENTS. Chemicals used in warfare to produce an obscuring cloud of fog to hide movements. Smokes may be harmless and are then called **screening smokes,** or **smoke screens,** or they may be toxic and called **blanketing clouds.** There are two types of smokes: those forming solid or liquid particles and those forming fogs or mists by chemical reaction. **White smokes,** which do not have light-absorbing particles, such as carbon, are formed by chemical reaction and have the best opacity or screening action. The first naval smoke screens were made by limiting the admission of air to the fuel in the boilers, and the first Army smoke pots contained mixtures of pitch, tallow, saltpeter, and gunpowder. The British **smoke candles** contained 40% potassium nitrate, 29 pitch, 14 sulfur, 8 borax, and 9 coal dust. They gave a brown smoke, but one that lifted too easily.

Fog or military screening may be made by spraying an oil mixture into the air at high velocity. The microscopic droplets produce an impenetrable fog which remains for a long period. **White phosphorous** gives a dense, white smoke, called **WP smoke,** by burning to the pentoxide and changing to phosphoric acid in the moisture of the air. Its vapor is toxic. Smoke from **red phosphorus** is known as **RP smoke. Sulfuric trioxide,** SO_3, is an effective smoke producer in humid air. It is a mobile, colorless liquid vaporizing at 113°F (45°C) to form dense, white clouds with an irritating effect. The French **opacite** is **tin tetrachloride,** or **stannic chloride,** $SnCl_4$, a liquid that fumes in air. When hydrated, it becomes the crystalline pentahydrate, $SnCl_4 \cdot 5H_2O$. The smoke is not dense, but it is corrosive and it penetrates gas masks. **Sulfuryl chloride,** SO_2Cl_2, is a liquid that decomposes on contact with the air into sulfuric and hydrochloric acids. **FS smoke** is made with a mixture of chlorosulfonic acid and sulfur trioxide. **Silicon tetrachloride,** $SiCl_4$, is a colorless liquid that boils at 140°F (60°C), and fumes in the air, forming a dense cloud. Mixed with ammonia vapor, it resembles a natural fog. The heavy mineral known as **amang,** separated from Malayan tin ore, containing ilmenite and zircon, is used in smoke screens. **Titanium tetrachloride,** $TiCl_4$, is a colorless to reddish liquid boiling at 277°F (136°C). It is used for smoke screens and for skywriting from airplanes. In most air it forms dense, white fumes of **titanic acid,** H_2TiO_3, and hydrogen chloride. The commercial liquid contains about 25% titanium by weight.

A common smoke for airplanes is **oleum.** It is a mixture of sulfur trioxide in sulfuric acid, which forms fuming sulfuric acid, or **pyrosulfuric acid,** $H_2S_2O_7$. The dense liquid is squirted in the exhaust manifold. **Zinc smoke** is made with mixtures of zinc dust or zinc oxide with various chemicals to form clouds. **HC smoke** is zinc chloride with an oxidizing agent to burn up residual carbon so that the smoke will be gray and not black. **Signal smoke** is colored smoke used for ship distress signals and for aviation marking signals. They are mixtures of a fuel, an oxidizing agent, a dye, and sometimes a cooling agent to regulate the rate of burning and to prevent decomposition of the dye. Unmistakable colors are used so that the signals may be distinguished from fires, and the dyes are mainly anthraquinone derivates, together with mixtures of azo, azine, and diphenyl-methane compounds.

SNAKESKINS. The snakeskins employed for fancy leathers are in general the skins of large, tropical snakes which are notable for the beauty or oddity of their markings. Snakeskins for shoe-upper leathers, belts, and handbags are glazed like kid and calfskin after tanning. Small cuttings are used for inlaying on novelties. The leather is very thin, but is remarkably durable and is vegetable-tanned and finished in natural colors, or is dyed. **Python skins** are used for ladies' shoes. **Regal python skins** from Borneo, the Philippines, and the Malay Peninsula sometimes measure 30 ft (9 m) in length and have characteristic checked markings. Diamond-backed rattlesnakes are raised on snake farms in the United States. The meat is canned as food, and the skins are tanned into leather. Only the back is used for leather, as the belly is colorless.

SOAP. A cleansing compound produced by saponifying oils, fats, or grease with an alkali. When caustic soda is added to fat, glycerin separates out, leaving **sodium oleate,** $Na(C_{17}H_{33}O_2)$, which is soap. But since oils and fats are mixtures of various acid glycerides, the soaps made directly from vegetable and animal oils may be mixtures of oleates, palmitates, linoleates, and laurates. **Soap oils** in general, however, are those oils which have greater proportions of nearly saturated fatty acids, since the unsaturated fractions tend to oxidize to form aldehydes, ketones, or other acids, and turn rancid. If an excess of alkali is used, the soap will contain free alkali; and the greater the proportion of the free alkali, the coarser is the action of the soap. ASTM standards for milled toilet soap permit only 0.17% free alkali. Soap makers now employ refined and bleached oils, which are then hydrolyzed into fatty acids and glycerol prior to saponification with caustic. This allows the fatty acids to be distilled, resulting in a more stable product. **Sodium soaps** are always harder than **potassium soaps** with the same fat or oil. Hard sodium soaps are used for chips, powders, and toilet soaps. Soft, caustic potash soaps are the liquid, soft, and semisoft pastes. Mixtures of the two are also used. Soaps are made by either the boiled process or the cold process. **Chip soap** is made by pouring the hot soap onto a cooled revolving cylinder from which the soap is scraped in the form of chips or ribbons which are then dried to reduce the moisture content from 30 to 10%. **Soap flakes** are made by passing chips through milling rollers to make thin, polished, easily soluble flakes.

    Powdered soap is made from chips by further reducing the moisture and grinding. **Milled soaps** are made from chips by adding color and perfumes to the dried chips and then passing through milling rollers and finally pressing in molds. **Toilet soaps** are made in this way. Soap is used widely in industrial processing, and much of the production has consisted of chips, flakes, powdered, granulated, and scouring powders. Soaps have definite limitations of use. They are unstable in acid solutions and may form insoluble salts. In hard waters they may form insoluble soaps of calcium or magnesium unless a phosphate is added. Many industrial cleansers, therefore, may be balanced combinations of soaps, synthetic detergents, phosphates, or alkalies, designed for particular purposes.

About half of all soap is made with tallow, 25% with coconut oil, and the remainder with palm oil, greases, fish oils, olive oil, soybean oil, or mixtures. A typical soap contains 80% mixed oils and 20 coconut oil, with not over 0.2 free alkali. Auxiliary ingredients are used in soap to improve the color, for perfuming, as an astringent, or for abrasive or harsh cleaning purposes. Phenol or cresylic acid compounds are used in **antiseptic soap.** The soft soaps and liquid soaps of USP grade have a therapeutic value and may be sold under trade names.

Solvents are added to industrial soaps for scouring textiles or when used in soluble oils in the metal industry. Zinc oxide, benzoic acid, and other materials are used in facial soaps with the idea of aiding complexion. Excessive alkalinity in soaps dries and irritates the skin, but **hand grit soap** usually has 2 to 5% alkaline salts such as borax or soda ash and 10 to 25% abrasive materials. Softer **hand soap** may contain marble flour. Silicate of soda, used as a filler, also irritates the skin. **Face soaps,** or **toilet soaps,** contain coloring agents, stabilizers, and perfuming agents. For special purposes, **cosmetic soaps** contain medications. **Deodorant soaps** contain antibacterial chemicals, such as **triclosan,** which inhibit the production of bacteria on the skin. Experts disagree on whether antibacterial ingredients are harmful to the skin. Some, such as **Dove,** are a blend of detergents and soap. **Castile soap** is a semitransparent soap made with olive oil. **Marseilles soap** and **Venetian soap** are names for castile soap with olive oil and soda. Ordinary soft soaps used as bases for toilet soap are made with mixtures of linseed oil and olive oil. Linseed oil, however, gives a disagreeable odor. Soybean oil, corn oil, and peanut oil are also used, although peanut oil, unless the arachidic acid is removed, makes a hard soap. **Tall oil soaps** are sodium soaps made from the fatty acids of tall oil. They are inferior to sodium oleate in detergency, but superior to sodium rosinate. Many toilet soaps contain excess unsaponified oil, fatty acid, or lanolin and are known as **superfatted.**

Saddle soap is any soap used for cleaning leather goods which has the property of filling and smoothing the leather as well as cleaning. The original saddle soaps were made of palm oil, rosin, and lye, with glycerin and beeswax added. Oils for the best soaps are of the nondrying type. High-grade **soft soap** for industrial use is made with coconut or palm kernel oil with caustic potash. But soft soap in paste form is generally made of low-titer oils with caustic soda, usually linseed, soybean, or corn oil. The lauric acid of coconut oil gives the coconut-oil soaps their characteristic of profuse lathering, but lauric acid affects some skins by causing itching, and soaps with high coconut-oil content and low titer are also likely to break down in hot water and wash ineffectively. Palm-kernel oil develops free acids, and upon aging the soap acquires the odor of the oil. Palm oil produces a crumbly soap. It does not lather freely, but is mild to the skin. Olive oil is slow-lathering, but has good cleansing powers. It is often used in textile soaps. Cottonseed oil is used in some laundry soaps, but develops yellow spots in the soap. Corn oil with potash makes a mild soft soap. Soybean oil also makes a soft soap. Rosin is used to make yellow laundry soaps. ASTM standards for **bar soap** permit up to 25% rosin. Sulfonated oils do not give as good cleansing action as straight oils, but are used in shampoos where it is desirable to have some oil or greasiness. Blending of various oils is necessary to obtain a balance of desired characteristics in a soap. Hand soaps may be made with trisodium phosphate or with disodium phosphate, or **sodium perborate,** $NaBO_3 \cdot H_2O$, known as **perborin,** all of which are crystalline substances which are dissolved in water solution. **Soap powder** is **granular soap** made in a vacuum chamber or by other special processes. It usually contains 15 to 20% soap and the balance sodium carbonate. **Scouring powder** is an intimate mixture of soap powder and an insoluble abrasive such as pumice. **Floating soaps** are made light by blowing air through them while in the vats. **Soapless shampoos** and tooth powders contain saponin or chemical detergents. **Liquid soaps** are made by saponification with potassium and ammonium hydroxide, or triethanolamine, to produce more-soluble products. The floating soaps, such as **Ivory** from Procter & Gamble Co., are made by injecting air into the molten soap.

SOAPSTONE. A massive variety of impure talc employed for electric panels, gas-jet trips, stove linings, tank linings, and as an abrasive. It can be cut easily and becomes very hard when heated because of the loss of its combined water. The waste product from the cutting of soapstone is ground and used for the same purposes as talc powder. **Steatite** is a massive stone rich in talc that can be cut readily, while soapstone may be low in talc. When free of iron oxide and other impurities, block steatite is used for making spacer insulators for electronic tubes and for special electrical insulators. **Block steatite** suitable for electrical insulation is mined in Montana, India, and Sardinia. Steatite is also ground and molded into insulators. It can be purified of iron and other metallic impurities by electrolytic osmosis. When fluxed with alkaline earths instead of feldspar, the molded steatite ceramics have a low loss factor at high frequencies, and have good electrical properties at high temperatures. The white-burning refractory steatite of the Red Sea coast of Egypt averages 60% silica and 30.5 magnesia, with 1 iron oxide and 1.5 CaO.

Alberene stone, quarried in Virginia, is blue-gray. The medium-hard varieties are used for building trim and for chemical laboratory tables and sinks, and the hard varieties are employed for stair treads and flooring. Alberene stone marketed by the Alberene Stone Corp. as a basic refractory substitute for chrome or magnesite for medium temperatures has a fusion point of 2400°F (1316°C). **Virginia greenstone** is a gray-green soapstone resistant to weathering, used as a building stone. **Talc crayons** for marking steel are sticks of soapstone.

SODA ASH.   The common name for anhydrous **sodium carbonate,** Na_2CO_3, which is the most important industrial alkali. It is a grayish-white, lumpy material which loses any water of crystallization when heated. For household use in hydrous crystallized form, $Na_2CO_3 \cdot 10H_2O$, it is called **washing soda, soda crystals,** or **sal soda,** as distinct from **baking soda,** which is **sodium hydrogen carbonate,** or **sodium bicarbonate,** $NaHCO_3$. Sal soda contains more than 60% water. Another grade, with one molecule of water, $Na_2CO_3 \cdot H_2O$, is the standard product for scouring solutions. Federal specifications call for this product to have a total alkalinity not less than 49.7% Na_2O. Commercial high-quality soda ash contains 99% minimum Na_2CO_3, or 58 minimum Na_2O. It varies in size of particle and in bulk density, being marketed as extra-light, light, and dense. The extra-light has a density of 23 lb/ft³ (368 kg/m³) and the dense has a density of 63 lb/ft³ (1,009 kg/m³). **Laundry soda** is soda ash mixed with sodium bicarbonate, with 39 to 43% Na_2O. **Modified sodas,** used for cleansing where a mild detergent is required, are mixtures of sodium carbonate and sodium bicarbonate. They are used in both industrial and household cleaners. **Tanners' alkali,** used in processing fine leathers, and **textile soda,** used in fine wool and cotton textiles, are modified sodas. **Flour bland,** used by the milling industry in making free-flowing, self-raising food flours, is a mixture of sodium bicarbonate and tricalcium phosphate.

Soda ash is made by the Solvay process, which consists of treating a solution of common salt with ammonia and with carbon dioxide and calcining the resulting filter cake of sodium bicarbonate to make **light soda ash. Dense soda ash** is then made by adding water and recalcining. Soda ash is less expensive than caustic soda and is used for cleansing, for softening water, in glass as a flux and to prevent fogging, in the wood-pulp industry, for refining oils, in soapmaking, and for the treating of ores. **Caustic ash,** a strong cleaner for metal scouring and for paint removal, is a mixture of about 70% caustic soda and 30 soda ash. **Flake alkali,** of PPG Industries, contains 71% caustic soda and 29 soda ash. Soda ash is also used as a flux in melting iron to increase the fluxing action of the limestone, as it will carry off 11% sulfur in the slag. **Soda briquettes,** used for desulfurizing iron, are made of soda ash formed into pellets with a hydrocarbon bond. **Hennig purifier** is soda ash combined with other steel-purifying agents made into pellets.

The natural hydrous sodium carbonate of Egypt and Libya is called **nitron.** Natural soda ash is obtained in Wyoming from beds 5 to 10 ft (1.5 to 3.0 m) thick located 1,200 ft (366 m) underground, which contain 47% Na_2CO_3 and 36 $NaHCO_3$, designated as **trona,** $Na_2CO_3 \cdot NaHCO_3 \cdot 2H_2O$. By calcination the excess CO_2 is driven off, yielding soda ash. The salt brine of Owens Lake, California, is an important source of soda ash. The brine, which contains 10.5% Na_2CO_3 and 2.5 **sodium borate decahydrate,** is concentrated and treated to precipitate the trona. The Salt Lake area of Utah is a source of trona. Soda ash and sodium carbonate may be sold under trade names. **Purite** is a sodium carbonate. **Tronacarb** is an industrial grade, and **Tronalight,** as the name suggests, is a light soda ash. Both products are made by Kerr-McGee Chemical Corp.

SODIUM. A metallic element, symbol Na and atomic weight 23, occurring naturally only in the form of its salts. The most important mineral containing sodium is the chloride, NaCl, which is common salt. It also occurs as the nitrate, Chile saltpeter, as a borate in borax, and as a fluoride and a sulfate. When pure, sodium is silvery white and ductile, and it melts at 208°F (97.8°C) and boils at 1620°F (882°C). The specific gravity is 0.97. It can be obtained in metallic form by the electrolysis of salt. When exposed to the air, it oxidizes rapidly, and it must therefore be kept in airtight containers. It has a high affinity for oxygen, and it decomposes water violently. It also combines directly with the halogens, and is a good reducing agent for the metal chlorides. Sodium is one of the best conductors of electricity and heat. The element has five isotopes, and **sodium 24,** made by neutron irradiation of ordinary sodium, is radioactive. It has a half-life of 15 h and decays to stable **magnesium 24** with the emission of one beta particle and two gamma rays per atom.

The metal is a powerful desulfurizer of iron and steel even in combination. For this purpose it may be used in the form of **soda-ash pellets** or in alloys. **Desulfurizing alloys** for brasses and bronzes are **sodium-tin,** with 95% tin and 5 sodium, or **sodium-copper. Sodium-lead,** used for adding sodium to alloys, contains 10% sodium and is marketed as small, spheroidal shot. It is also marketed as **sodium marbles,** which are spheres of pure sodium up to 1 in (2.54 cm) in diameter coated with oil to reduce handling hazards. **Sodium bricks** contain 50% sodium metal powder dispersed in a paraffin binder. They can be handled in air and are a source of active sodium. Sodium in combination with potassium is used as a heat-exchange fluid in reactors and high-temperature processing equipment. A **sodium-potassium alloy,** containing 56% sodium and 44 potassium, has a melting point of 66°F (19°C) and a boiling point of 1517°F (825°C). It is a silvery, mobile liquid. **High-surface sodium** is sodium metal absorbed on common salt, alumina, or activated carbon to give a large surface area for use in the reduction of metals or in hydrocarbon refining. Common salt will adsorb up to 10% of its weight of sodium in a thin film on its surface, and this sodium is 100% available for chemical reaction. It is used in reducing titanium tetrachloride to titanium metal. **Sodium vapor** is used in electric lamps. When the vapor is used with a fused alumina tube, it gives a golden-white color. A 400-W lamp produces 42,000 lm and retains 85% of its efficiency after 6,000 h.

Sodium compounds are widely used in industry, particularly sodium chloride, sodium hydroxide, and soda ash. **Sodium bichromate,** $Na_2Cr_2O_7 \cdot 2H_2O$, a red, crystalline powder, is used in leather tanning, textile dyeing, wood preservation, and pigments. When heated, it changes to the anhydrous form which melts at 673°F (356°C) and decomposes at about 752°F (400°C). In a hot water solution with sulfuric acid, sodium bichromate gives a golden-brown, brasslike finish to zinc parts. The sodium bichromate liquor from alkali production is used for making pigments. When combined with lead compounds, the bichromate precipitate is yellow. The addition of **iron blue,** or ferric cyanide, develops greens. **Sodium metavandate,** $NaVO_3$, is used as a corrosion inhibitor to protect some chemical-processing piping. It dissolves in hot water, and a small amount in the water forms a tough, impervious coating of magnetic iron oxide on the walls of the pipe. **Sodium iodide** crystals are used as scintillation probes for the detection and analysis of nuclear energies. **Sodium oxalate** is used as an antienzyme to retard tooth decay. In the drug industry, sodium is used to compound with pharmaceuticals to make them water-soluble salts. Gas-generating pellets or wafers of **sodium azide** are used as a propellant in auto airbag inflators. Upon ignition, the gas burns and liberates large amounts of nitrogen gas, inflating the bag. Sodium is a plentiful element, easily available, and is one of the most widely used.

SODIUM CYANIDE. A salt of hydrocyanic acid of composition NaCN, used for carbonizing steel for case hardening, for heat-treating baths, for electroplating, and for the extraction of gold and silver from their ores. For carburizing steel it is preferred to potassium cyanide because of its lower cost and its higher content of available carbon. It contains 53% CN, as compared with 40% in potassium cyanide. The nitrogen also aids in forming the hard case on the steel. The 30% grade of sodium cyanide, melting at 1156°F (679°C), is used for heat-treating baths instead of lead, but it forms a slight case on the steel. Sodium cyanide is very unstable, and on exposure to moist air it liberates the highly poisonous **hydrocyanic acid** gas, HCN. For gold and silver extraction it easily combines with the metals, forming soluble double salts, $NaAu(CN)_2$. Sodium cyanide is made by passing a stream of nitrogen gas over a hot mixture of sodium carbonate and carbon in the presence of a catalyst. It is a white, crystalline powder, soluble in water. The white **copper cyanide** used in electroplating has composition $Cu_2(CN)_2$, containing 70% copper. It melts at 887°F (475°C) and is insoluble in water, but is soluble in sodium cyanide solution. **Sodium ferrocyanide,** or yellow **prussiate of soda,** is a lemon-yellow, crystalline solid of composition $Na_4Fe(CN)_6 \cdot 10H_2O$, used for carbonizing steel for case hardening. It is also employed in paints, in printing inks, and for the purification of organic acids; in minute quantities, it is used in salt to make it free-flowing. It is soluble in water. **Calcium cyanide** in powder or granulated forms is used as an insecticide. It liberates 25% of hydrocyanic acid gas. **Cyanogas,** of American Cyanamid Co., is gaseous HCN from calcium cyanide.

SODIUM HYDROXIDE. Known commonly as **caustic soda,** and also as **sodium hydrate. Lye** is an old name used in some industries and in household uses. It is a white, massive, crystalline solid of composition NaOH used for scouring and cleaning baths, for etching aluminum, in quenching baths for heat-treating steel, in cutting and soluble oils, in making soaps, and in a wide variety of other applications. It is usually a by-product in the production of chlorine from salt. The specific gravity is 2.13 and melting point 604°F (318°C). It is soluble in water, alcohol, and glycerin. Sodium hydroxide is sold in liquid and in solid or powder forms on the basis of its Na_2O content. A high-grade commercial caustic soda contains 98% minimum NaOH equivalent to 76 minimum Na_2O. The liquid contains 50% minimum NaOH. **Pels,** of PPG Industries, is a caustic soda in bead form. It is less irritating to the skin when used in detergents. **Phosflake,** of PPG Industries, used in washing machines, is a mixture of caustic soda and trisodium phosphate. **Caustic potash** is **potassium hydroxide,** KOH, which has the same uses but is more expensive. Caustic potash is a white, lumpy solid. It is soluble in water and makes a powerful cleansing bath for scouring metals. It is marketed as solid, flake, granular, or broken, and also is 40 to 50% liquid solutions. It is used in soaps and for bleaching textiles. When used in steel-quenching baths, it gives a higher quenching rate than water alone and does not corrode the steel as a salt solution does.

SODIUM NITRATE. Also called **soda niter** and **Chile saltpeter.** A mineral found in large quantities in the arid regions of Chile, Argentina, and Bolivia, where the crude nitrate with iodine and other impurities is called **caliche.** It is used for making nitric and sulfuric acids, for explosives, as a flux in welding, and as a fertilizer. The composition is $NaNO_3$. It is usually of massive, granular, crystalline structure with a Mohs hardness of 1.5 to 2 and specific gravity of 2.29. It is colorless to white, but sometimes colored by impurities. It is readily soluble in water. In other parts of the world it occurs in beds with common salt, borax, and gypsum. Sodium nitrate is also made by nitrogen fixation and is marketed granulated, in crystals, or in sticks. It is colorless and odorless, and it has a specific gravity of 2.267 and a melting point of 601°F (316°C). It has a bitter, saline taste. **Sodan,** used for spraying on soils, is a clear liquid solution of sodium nitrate and ammonium nitrate containing 20% nitrogen. **Norway saltpeter,** used in fertilizers and explosives, is **calcium nitrate,** $Ca(NO_3)_2$, in colorless crystals soluble in water. Calcium nitrate of fine crystal size is used as a coagulant for rubber latex.

SODIUM SILICATE. A water-soluble salt commonly known as **water glass** or **soluble glass.** Chemically, it is **sodium metasilicate** of composition Na_2SiO_3 or $NaSiO_3 \cdot 9H_2O$. Two other forms of the silicate are also available, **sodium sequisilicate,** $3Na_2O \cdot 2SiO_2$, and **sodium orthosilicate,** $2Na_2O \cdot SiO_2$. All of these are noted for their powerful detergent and emulsifying properties and for their suspending power. The material has good adhesion, and large quantities are used in water solutions for industrial adhesives. It is also used to inhibit corrosion in potable- and industrial-water systems, forming an oxidation-resistant film on pipe walls. If corrosion has begun, pH-neutral reconditioning solutions can remove the rust or scale without pH adjustment of flush water. When solid, sodium silicate is glassy in appearance and dissolves in hot water. It melts at 1864°F(1018°C). It is obtained by melting sodium carbonate with silica, or by melting sand, charcoal, and soda. The fused product is ground and dissolved in water by long boiling. **Potassium silicate** is made in the same way, or a complex soluble glass is made by using both sodium and potassium carbonates. Potassium silicate is more soluble than sodium silicate. **Kasil,** of Philadelphia Quartz Co., is a potassium silicate in fine powder containing 71% SiO_2 and 28.4 K_2O. It is used in ceramic coatings and refractory cements. **Corlok** is potassium silicate free of fluorides and sodium compounds. It is resistant to strong oxidizing acids, has good bond strength, and is used as a cement for acid tanks. **Ammonium silicate** has an ammonium group instead of the sodium. **Quram 220,** of Philadelphia Quartz Co., is this material in the form of white powder or in opalescent solution. The intermediate silica grades act like sodium silicates and are used as binders for refractory ceramics.

Sodium silicate is marketed as a viscous liquid or in powder form. It is used as a detergent, as a protection for wood and porous stone, as a fixing agent for pigments, for cementing stoneware, for lute cements for such uses as sealing lightbulbs, for waterproofing walls, for greaseproofing paper containers, for coating welding rods, as a filler for soaps, and as a catalyst for high-octane gasoline. It increases the cleansing power of soaps but irritates the skin. However, it is used in cleansing compounds because it is a powerful detergent. **Brite Sil** is a spray-dried sodium silicate powder which dissolves more easily and more uniformly.

Sodium silicate is also used for insulating electric wire. It is applied in solution, and the coated wire is then heated, leaving a flexible coating. Mixed with whiting, it is used as a strong cement for grinding wheels. Sodium metasilicate marketed by Philadelphia Quartz Co. as a cleaner of metals is a crystalline powder. Hot solutions of this salt in water are caustic and will clean grease readily from metals. **Drymet** is the anhydrous sodium metasilicate. It is a fine, white powder with total alkalinity of 51% Na_2O. It is easily soluble in water and is used as a detergent in soap powders to give free-flowing, noncaking properties. The anhydrous material for a given detergent strength weighs little more than half the weight of the hydrous powder. **Dryorth** is the anhydrous material of 60% alkalinity. It is a powerful detergent and grease remover. **Crystamet** is the material with 42% water of crystallization. It is a free-flowing white powder. **Penchlor** is an **acidproof cement** made by mixing cement powder with a sodium silicate solution. It is used for lining chemical tanks and drains. **Aquagel,** of Silica Products Co., is a hydrous silicate of alumina, used in the same manner for waterproofing concrete.

SOLDER. An alloy of two or more metals used for joining other metals together by surface adhesion without melting the base metals as in welding and without requiring as high a temperature as in brazing. However, there is often no definite temperature line between **soldering alloys** and brazing alloys. A requirement for a true solder is that it have a lower melting point than the metals being joined and an affinity for, or be capable of uniting with, the metals to be joined.

A common solder is called **half-and-half,** or **plumbers' solder,** and is composed of equal parts of lead and tin. It melts at 360°F

(182°C). The density is 0.318 lb/in^3 (8,802 kg/m^3), the tensile strength is 5,500 lb/in^2 (38 MPa), and the electrical conductivity is 11% that of copper. **SAE solder No. 1** has 49.5 to 50.0% tin, 50 lead, 0.12 maximum antimony, and 0.08 maximum copper. It melts at 359°F (181°C). Much commercial half-and-half, however, usually contains larger proportions of lead and some antimony, with less tin. These mixtures have higher melting points, and solders with less than 50% tin have a wide melting range and do not solidify quickly. Sometimes a wide melting range is desired, in which case a **wiping solder** with 38 to 45% tin is used. A narrow-melting-range solder, melting at 362 to 365°F (183 to 185°C), contains 60% tin and 40 lead. A 42% tin and 58 lead solder has a melting range of 362 to 448°F (183 to 231°C). **Slicker solder** is the best quality of plumbers' solder, containing 63 to 66% tin and the balance lead. The earliest solders were the Roman solders called **argentarium,** containing equal parts of tin and lead, and **tertiarium,** containing 1 part tin and 2 lead. Both alloys are still in use, and throughout early industrial times tertiarium was known as **tinman's solder.**

Good-quality solders for electrical joints should have at least 40% tin, as the electrical conductivity of lead is only about half that of tin, but conductivity is frequently sacrificed for better wiping ability, and the wiping solders are usually employed for electrical work. **Soft solders** should not contain zinc because of poor adhesion from the formation of oxides. Various melting points to suit the work are obtained with solders by varying the proportions of the metals. The **low-melting solders** are those that melt at 446°F (230°C) or lower, and the **high-melting solders** melt at higher temperatures. The flow point, at which the solder is entirely liquid, is often considerably above the melting point. Tin added to lead lowers the melting point of the lead until, at 356°F (180°C), at 68% tin, the melting point rises as the tin content increases until the melting point of pure tin is reached. A standard solder with 48% tin and 52 lead melts at 360°F (182°C). A 45–55 solder melts at 440°F (227°C). Cheap solders may contain much less tin, but they have less adhesion. **SAE solder No. 4** contains 22.5 to 23.5% tin, 75 lead, and 2 maximum antimony. It melts at 370°F (188°C). A tin-silver-copper alloy, developed at the University of Iowa (Ames), is lead-free and intended to replace tin-lead solders. Made by blending spherical powder particles with a fluxing agent, the **lead-free solder** melts at 421°F (216°C) and wets similarly to tin-lead solders. The silver and copper form hard, intermetallic phases, reinforcing the tin and strengthening the solder.

Solders with low melting points are obtained from mixtures of lead, tin, cadmium, and bismuth. **Bismuth solder** is also more fluid, as bismuth lowers surface tension. Bismuth, however, hardens the alloy, although to a lesser extent than antimony. A bismuth solder containing equal parts of lead, tin, and bismuth melts at 284°F (140°C). **Cerrolow alloys,** of Cerro Metal Products Co., are bismuth solders containing sufficient indium to be designated as **indium solders. Cerrolow 147,** which melts at 142°F (61°C), contains 48% bismuth, 25.6 lead, 12.8 tin, 9.6 cadmium, and 4 indium. **Cerrolow 105,** melting at 100°F (38°C), contains 42.9% bismuth, 21.7 lead, 8 tin, 5 cadmium, 18.3 indium, and 4 mercury. Cadmium solders have low melting points, are hard, and are usually cheaper than tin solders; but they have the disadvantage of blackening and corroding, and the fumes are toxic. **Cadmium-zinc solders** were used in wartime because of the scarcity of tin. A solder containing 80% lead, 10 tin, and 10 cadmium has about the same strength as a 50–50 tin-lead solder and has greater ductility, but is darker in color. **Cadmium-tin solder,** with high cadmium, is used to solder magnesium alloys. Soft solders for soldering brass and copper, especially for electric connections, may be of tin hardened with antimony. **Solder wire** for this purpose contains 95% tin and 5 antimony. Thallium may be used in high-lead solders to increase strength and adhesion.

Hard solder may be any solder with a melting point above that of the tin-lead solders; more specifically, hard solders are the brazing solders, silver solders, or aluminum solders. **Aluminum solders** may contain up to 15% aluminum. A solder prepared by the National Bureau of Standards contains 87% tin, 8 zinc, and 5 aluminum. It has good strength and ductility. **Alcoa solder 805,** for joining aluminum to steel or other metals, has 95% zinc and 5 aluminum. The melting range is 715 to 725°F (379 to 385°C). For soldering aluminum to aluminum, an alloy of 91% tin and 9 zinc is used.

The solder known as **Richard's solder** is a yellow brass with 3% aluminum and 3 phosphor tin. Solders containing nickel are used for soldering nickel silver, and silver and gold solders are used for jewelry work. Silver solder in varying proportions is also used as a high-melting-point solder for general work, and small amounts of silver are sometimes used in lead-tin solders to conserve tin, but the melting point is high. **Lead-tin solders** with more than 90% lead and some silver, in use during war emergency, had high melting points and poor spreading qualities. Indium improves these solders, and a solder with 96% lead, 3 silver, and 1 indium has a melting point of 590°F (310°C) and a tensile strength of 4,970 lb/in^2 (34 MPa). **Cerroseal 35,** of Cerro Metal Products Co., contains 50% tin and 50 indium. It melts at 240°F (116°C), has low vapor pressure, and will adhere to ceramics. **Alkali-resistant solders** are **indium-lead alloys.** A solder with 50% lead and 50 indium melts at 360°F (182°C) and is very resistant to alkalies, but lead-tin solders with as little as 25% indium are resistant to alkaline solutions, have better wetting characteristics, and are strong. Indium solders are expensive. Adding 0.85% silver to a 40% tin soft solder gives equivalent wetting on copper alloys to a 63% tin solder, but the addition is not effective on low-tin solders. A **gold-copper solder** used for making high-vacuum seals and for brazing difficult metals such as iron-cobalt alloys contains 37.5% gold and 62.5 copper. The **silver-palladium solders** have high melting points, 2246°F (1232°C) for a 30% palladium alloy, and good flow, and corrosion resistance. A **palladium-nickel alloy** with 40% nickel has a melting point of about 2258°F (1237°C). The brazing alloys containing palladium are useful for a wide range of metals and metal-to-ceramic joints.

Cold solder, used for filling cracks in metals, may be a mixture of a metal powder in a pyroxylin cement with or without a mineral filler, but the strong cold solders are made with synthetic resins, usually epoxies, cured with catalysts, and with no solvents to cause shrinkage. The metal content may be as high as 80%. **Devcon F,** of Devcon Corp., for repairing holes in castings, has 80% aluminum powder and 20 epoxy resin. It is heat-cured at 150°F (66°C), giving high adhesion. **Epoxyn solder** is aluminum powder in an epoxy resin in the form of a putty for filling cracks or holes in sheet metal. It cures with a catalyst. The metal-epoxy mixtures give a shrinkage of less than 0.2%, and they can be machined and polished smooth.

SOLVENT. A material, usually a liquid, having the power of dissolving another material and forming a homogeneous mixture called a **solution.** The mixture is physical, and no chemical action takes place. A solid solution is such a mixture of two metals, but the actual mixing occurs during the liquid or gaseous state. Some materials are soluble in certain other materials in all proportions, while others are soluble only up to a definite percentage and the residue is precipitated out of solution. Homogeneous mixtures of gases may technically be called solutions, but are generally referred to only as mixtures.

The usual industrial applications of solvents are for putting solid materials into liquid solution for more convenient chemical processing, for thinning paints and coatings, and for dissolving away foreign matter as in dry-cleaning textiles. But they may have other uses, such as absorbing dust on roadways and killing weeds. They have an important use in separating materials, for example, in the extraction of oils from seeds. In such use, a **clathrate** is a solid compound added to the solution containing a difficult-to-extract material, but which is trapped selectively by the clathrate. The solid clathrate is then filtered out and processed by heat or chemicals to separate the desired compound. **Antifoamers** are chemicals, such as the silicones, added to solvents to reduce foam so that processing equipment can be used to capacity without spillover. **Antifoam 71,** of General Electric, is a silicone emulsion that can be used in foodstuffs in proportions up to 100 parts per million. **Solvent-solvents** are solvents used for second-stage extraction of difficult-to-extract metals such as gold, uranium, and thorium. **Tributyl phosphine oxide,** $(C_4H_9)_3PO$, a white, crystalline powder, is such a material used in benzene or kerosene solution for extracting metals from the acids employed in ore extraction.

The usual commercial solvents for organic substances are the alcohols, ether, benzene, and turpentine, the latter two being common solvents for paints and varnishes containing gums and resins. The so-called **coal-tar solvents** are light oils from coal tar, distilling off between 293 and 356°F (145 and 180°C), with specific gravity 0.850 to 0.890. **Solvent oils,** from coal tar, are amber to dark liquids with distillation ranges from about 302 to 644°F (150 to 340°C), with specific gravity 0.910 to 0.980. They are used as solvents for asphalt varnishes and bituminous paints. **Shingle stains** are amber to dark grades of solvent oils of specific gravity 0.910 to 0.930.

A valuable solvent for rubbers and many other products is **carbon bisulfide,** CS_2, also called **carbon disulfide,** made by heating together carbon and sulfur. It is flammable and toxic. When pure, it is nearly odorless. The specific gravity is 1.2927 and boiling point 116°F (46.5°C). **Ethyl acetate,** $CH_3COOC_2H_5$, made from ethyl alcohol and acetic acid, is an important solvent for nitrocellulose and lacquers. It is liquid, boiling at 171°F (77°C). Ashland Chemical Co. makes it in three grades, containing 85 to 88, 99, and 99.5% ethyl acetate. Some producers offer ethyl acetate solvent grades for urethanes. One of the best solvents for cellulose is **cuprammonium hydroxide.** Amyl and other alcohols, amyl acetate, and other volatile liquids are used for quick-drying lacquers, but many synthetic chemicals are available for such use. **Dioxan,** a water-white liquid of specific gravity 1.035 and composition $CH_2CH_2OCH_2CH_2O$, is a good solvent for cellulose compounds, resins, and varnishes, and is used also in **paint removers,** which owe their action to their solvent power. Dow Chemical Co. makes a similar material, called **diethylene ether.** Ferro Corp.'s brand is an uninhibited variety, **1,4-dioxane,** that has a purity of 99.97% and comes packed under nitrogen. **Ethyl lactate,** used as a solvent for cellulose nitrate, is a liquid with boiling point of 150°C and specific gravity of 1.03. **Octyl alcohol,** a liquid of composition $CH_3(CH_2)_6CH_2OH$, specific gravity 1.429, and boiling point of 383°F (195°C), has a high solvent power for nitrocellulose and resins. **Diafoam** is a secondary octyl alcohol used as a defoaming agent in plastics and lacquers. **Methyl hexyl ketone,** $CH_3(CH_2)_5COCH_3$, is a powerful, high-boiling-point solvent which also acts as a dispersing agent in inks, dyestuffs, and perfumes. It is a water-white liquid boiling at 343°F (173°C). It is made by Penta Manufacturing Co.

The chlorinated hydrocarbons have powerful solvent action on fats, waxes, and oils and are used in degreasing. Of major commercial significance are **perchloroethylene (PCE), trichloroethylene (TCE),** and **1,1,1-trichloroethane (1,1,1-TCE).** The biggest industrial use of PCE is as a dry-cleaning solvent because of its nonflammability, and high solvency, vapor pressure, and stability. The largest applications of TCE and 1,1,1-TCE have been in metal cleaning, which also consumes significant quantities of PCE. Because 1,1,1-TCE has been implicated in ozone depletion of the stratosphere, its use is being discontinued. **Actrel ED,** of Exxon Chemical, is a line of non-ozone-depleting solvents for cleaning electronics. These specially processed hydrocarbons are effective in removing resin residues from printed-circuit boards and other organic and ionic contaminants from electronic components. Water is a solvent for most acids and alkalies and for many organic and inorganic materials. Acids or alkalies that decompose the material are not solvents for the material. Solvents are used to produce a solution that can be applied, as in the case of paints, and the evaporation of the solvent then leaves the material chemically unchanged. They may also be employed to separate one substance from another, by the selection of a solvent that dissolves one substance but not the other. **Dichlorethyl ether,** a yellowish liquid with a chloroformlike odor, of composition $ClCH_2CH_2OCH_2CH_2Cl$, is a good solvent for fats and greases and is used in scouring solutions and in soaps. **Dichlorethylene** is a liquid of composition $C_2H_2Cl_2$, specific gravity 1.278, and boiling point about 126°F (52°C). It is used as a solvent for the extraction of fats and for rubber.

Dichloromethane, known also as **methylene chloride** and **carrene,** is a colorless, nonflammable liquid of composition CH_2Cl_2, boiling at 104°F (39.8°C). It is soluble in alcohol and is used in paint removers, as a dewaxing solvent for oils, for degreasing textiles, and as a refrigerant. A low-boiling solvent for oils and waxes is **butyl chloride,** $CH_3CH_2CH_2CH_2Cl$. It is a water-white liquid of specific gravity 0.8875, boiling at 173°F (78.6°C). **Isocrotyl chloride** is a liquid of composition CH_3:$C(CH_2)_2 \cdot CHCl$, with specific gravity 0.919 and boiling point 154°F (68°C), used for cleaning and degreasing. **Cyclohexane,** $(CH_2)_6$, made by the hydrogenation of benzene, is a good solvent for rubbers, resins, fats, and waxes. It is a water-white, highly flammable liquid of specific gravity 0.777, boiling point 177°F (80.8°C), and flash point 10°F (−12°C). This solvent is marketed in England as **Sextone. Nadene** of Allied-Signal Co. is **cyclohexanone,** $CH_2(CH_2)_4C \cdot O$. It is a powerful general solvent, and is used as a coupling agent for immiscible compounds. The **Sulfolanes** of Shell

Oil Co. are selective solvents for separating mixtures having different degrees of saturation, and they can be removed easily by water wash. **Dimethyl sulfolane** is produced from pentadiene by reacting with SO_2 and hydrogenation. **Cyclohexanol,** also called **hexalin** and **hexahydrophenol,** $C_6H_{11}OH$, is a solvent for oils, gums, waxes, rubber, and resins. It is made by the hydrogenation of phenol, and is a liquid, boiling at 316°F (158°C).

Dichlorethyl, $CH_2(OCH_2 \cdot CH_2 \cdot Cl)_2$, is a water-insoluble high-boiling solvent for cellulose, fats, oils, and resins. The boiling point is 424°F (218°C), and specific gravity 1.234. The **nitroparaffins** constitute a group of powerful solvents for oil, fats, waxes, gums, and resins. Blended with alcohols, they are solvents for cellulose acetate, producing good flow and hardening properties for nonblushing lacquers. **Nitromethane,** CH_3NO_2, is a water-white liquid, with specific gravity 1.139, boiling point 214°F (101°C), and freezing point −20°F (−29°C). It is also used as a rocket fuel. At 500°F (260°C) it explodes into a hot mixture of nitrogen, hydrogen, carbon monoxide, carbon dioxide, and water vapor, but with a catalyst the disintegration can be controlled into a smooth, continuous explosion. **Nitrofuel** is an industrial grade used in automobile racing and in model engines. It is also a raw material for chemical synthesis and a stabilizer for halogenated alkanes. Nitrofuel is made by Angus Chemical Co. **Nitroethane,** $CH_3CH_2NO_2$, has a specific gravity 1.052, boiling point 237°F (114°C), and freezing point −130°F (−90°C).

Because of the ozone depletion potential and health and environmental concerns regarding chlorine solvents, various compounds for cleaning with aqueous and semiaqueous solvents have come about. **NST,** a low-temperature **polyethylene glycol ether** with **sodium pyrophosphate,** of Oakite Products Inc., is used for ultrasonic cleaning of aluminum, copper, and brass parts. **Micro,** a high-temperature sulfonated compound with **ethylenediamine tetracetic acid,** of International Products Corp., is applied in ultrasonic cleaning of iron and steel parts. A blend of **Solvent 140,** a high-flash-point hydrocarbon mineral spirit, and 5% **dipropylene glycol methyl ether (DPM)** is effective in manually cleaning 304L stainless steel. Adding DPM makes the water-immiscible solvent miscible with water, enhancing its ability to displace waterborne machining coolants. Both the 140 and blend are flammable and slow to evaporate, however, necessitating special precautions. **Zestron** solvents, of Dr. O. K. Wack Chemie GmbH of Germany, use **propylene glycol ether,** have a boiling point of about 338°F (170°C), and are used to clean circuit boards in electronics production. A semiaqueous system of Dow Europe uses modified propylene glycol ethers for this application. In the United States, Dow offers semiaqueous solvent **Dowanol PX-165,** which contains polar and nonpolar molecules, is nontoxic and biodegradable, and has no ozone depletion potential. BASF AG of Germany offers semiaqueous cleaning solvents based on **n-methyl pyrrolidine,** a nonhalogenated solvent said to be a suitable alternative to **methyl chloroform** (TCA) for degreasing and to **methylene chloride** for paint stripping.

A **plasticizer** is a liquid or solid that dissolves in or is compatible with a resin, gum, or other material and renders it plastic, flexible, or easy to work. A sufficient quantity of plasticizer will result in a viscous mixture which consists of a suspension of solid grains of the resin or gum in the liquid plasticizer. The plasticizer is in that sense a solvent, but unlike an ordinary solvent the plasticizer remains with the cured resin to give added properties to the materials, such as flexibility. **Dibutyl phthalate,** a water-white, oily liquid of specific gravity 1.048, boiling point 644°F (340°C), and composition $C_{16}H_{22}O_4$, is a plasticizer for Buna N rubber and polyvinyl chloride plastics. **Monoplex DOA,** of Rohm & Haas Co., used to give flexibility to vinyl resins at low temperatures, is **diisooctyl adipate.** It has a flash point of 400°F (204°C) and freezing point of 131°F (55°C). **Diisononyl phthalate** and **diisodecyl phthalate** are high-molecular-weight plasticizers for flexible polyvinyl chloride. An **aprotic solvent** is a solvent that contains no hydrogen, such as **selenium oxychloride,** a liquid of composition $SeOCl_2$. Such solvents are used in electronic applications where the energy deflection of free protons would be undesirable. **Phosphorus oxychloride,** $POCl_3$, is an aprotic solvent used with neodymium in **liquid lasers** to give high light-beam efficiency.

SORBITOL. A **hexahydric alcohol,** $(CH_2OH)_5CHOH$, which occurs naturally in many fruits, but is now made on a large scale by the direct hydrogenation of corn sugar, or dextroglucose. It is a white, odorless, crystalline powder of faint sweet taste. It melts at 208°F (97.7°C) and is easily dissolved in water. It is used as a humectant, softener, and blending agent; for the production of synthetic resins, plasticizers, and drying oils; and as an emulsifier in cosmetics and pharmaceuticals. It is digestible and nutritive and is used in confectionery to improve texture and storage life by inhibiting crystal growth of the sugar, and in dietary foods as a substitute for sugar. **Sorbo** and **Arlex,** of ICI Americas, Inc., are water solutions of sorbitol. **Neosorb** is a granular form from Roquette Corp. that is used for tabletting drugs. **Mannitol** is an isomer form of the alcohol and is produced in granular form for pharmaceuticals and foodstuffs as a binder. In the form of a free-flowing powder, it is used as an **anticaking agent** in pharmaceuticals and foodstuffs where a silica or other mineral-based agent is undesirable. The **polysorbates** are esters of sorbitol. **Polysorbate 80,** of Hodag Chemical Corp., is such a material used as an emulsifier in prepared mixed food for improving texture and stability. **Hex,** a metal-cleaning and protective agent, is a phosphoric acid ester of sorbitol.

Sorbitol additives are also used to impart clarity to polypropylene for packaging food and other products. **EC-1,** of EC Chemical of Japan and marketed outside Asia by Milliken Chemicals as **Millad 3905,** is based on dibenzylide technology. Though it provides excellent odor and taste quality, limited clarity and significant plateout (premature volatilization in processing) has retarded its use. **Gell All MD,** of New Japan Chemical, and **Millad 3940,** of Milliken, are based on dimethylbenzylidine technology. They improve clarity and reduce plateout but pose odor and taste problems. Also based on this technology is Schering Polymer Additives' (United Kingdom) **Geniset MD.** It improves clarity, reduces plateout, and lessens odor and taste problems. **NC-4,** of Mitsui Toatsu of Japan and based on bis (*p*-ethylbenzylidene) technology, virtually eliminates plateout, significantly improves clarity, and, by removing alkyl impurities, improves odor and taste. Milliken's **Millad 3988,** a sorbitol-acetal additive, also excels in terms of clarity, plateout, and odor and taste performance.

SOUND AND VIBRATION INSULATORS. Materials used for reducing the transmission of noise. Insulators are used to impede the passage of sound waves, as distinct from isolators used under machines to absorb the vibrations that cause the sound. For factory use the walls, partitions, and ceilings offer the only media for the installation of sound insulators. All material substances offer resistance to the passage of sound waves, and even glass windows may be considered as insulators. But the term refers to the special materials placed in the walls for this specific purpose. Insulators may consist of mineral wool, hair felt, foamed plastics, fiber sheathing boards, or simple sheathing papers. Sound insulators are marketed under a variety of trade names, such as **Celotex,** made from bagasse, and **Fibrofelt,** made from flax or rye fiber. Wheat straw is also used for making insulating board. Sound insulators are often also heat insulators. **Linofelt** is a sound- and heat-insulating material used for walls. It consists of a quilt of flax fiber between tough waterproof paper. It comes in sheets 0.3125 to 0.75 in (0.80 to 1.91 cm) thick. **Torofoleum** is a German insulating material made from peat moss treated with a waterproofing agent. It withstands temperatures up to 230°F (110°C), is porous, and has a density of less than 1 lb/ft^3 (0.005 kg/m^3). **Fiber metal,** of Technetics Corp., comprises randomly interlocked similar metal fibers, with the fibers bonded by sintering at all contact points.

Similar to nonwoven textile felts, its trade name is **Feltmetal,** and it is available in sheet form in various fibers, thickness, and porosity. Stainless steel (316 and 347) and aluminum-alloy fibers are used mainly for noise reduction of aircraft turbines, turbine blowers, and high-speed fans. Noise reduction is by *resistive absorption,* by which the amplitude of sound waves is reduced by converting most of the acoustic energy into heat. Other applications include abradable seals, using nickel alloy (Hastelloy X) fibers, and high-temperature thermal insulation, using an iron, chromium, aluminum, and yttrium alloy.

Vibration insulators, or **isolators,** to reduce vibrations that produce noises, are usually felt or fiberboards placed between the machine base and the foundation, but for heavy pressures they may be metal wire helically wound or specially woven, deriving their effectiveness from the form rather than the material. **Keldur,** of Quimby & Co., is a fibrous insulating material made up in sheets 0.75 in (1.91 cm) thick, with a resilient binder. **Korfund isolator,** of Korfund Dynamics Co., is a resilient mat of cork treated with oil and bound in a steel frame. It will take loadings up to 4,000 lb/ft^2 (19,528 kg/m^2). **Vibro-Insulator,** of Karman Rubber Co., is an isolator of Ameripol synthetic rubber. Plastic foams in sheet or flexible tape form are also used as isolators for instruments. **Isoloss,** of E-A-R Specialty Composites, is a high-density urethane foam for shock absorption. **Viscolas,** of the same company, is sheet or molded viscoelastic polymers for shock absorption. **Isodamp,** also of E-A-R, is vinyl-based sheet and foam for vibration and shock control.

SOYBEAN OIL. Also known as **soya bean oil.** A pale-yellow oil obtained by expression from the seeds of the plant *Glycine soya,* native to Manchuria but grown in the United States. Soybean oil is a linolenic acid oil; in contrast, the other three major oilseed oils—cottonseed, peanut, and sunflower—are oleic-linolenic acid oils, because they contain more than 50% of these fatty acids. It is primarily a food oil but has an undesirable off-flavor unless highly purified. It is also used as a drying oil for linoleum, paints, and varnishes, or for mixing with linseed oil, although the untreated oil has only half the drying power of linseed oil. It is also used in core oils and in soaps. The bean contains up to 20% oil. The average yield factor is 15%, but by trichlorethylene extraction a bushel of beans will yield 11 lb (5 kg) of oil and 46 lb (21 kg) of high-protein meal containing less than 1% oil. The oil content decreases in warm climates. Southern-grown soybeans contain 2 to 5% less oil than those grown in Illinois. The usual conversion factor is 8.5 lb (4 kg) of oil and 48 lb (22 kg) of meal per bushel of beans. The oil is easy to bleach, has good consistency as a food oil, and does not become rancid easily, but has less flavor stability than many other oils. There are 280 varieties of the bean grown in the United States and 2,500 varieties listed. The pods contain two or three beans which range in color from light straw through gray and brown to nearly black. Most varieties are straw-colored or greenish yellow. The stalks and leaves of the plant contain much nitrogen, and about half of the crop is usually plowed under for fertilizer.

The specific gravity of the oil is about 0.925, iodine value 134, and it should have a maximum of not more than 1.5% free fatty acids and not more than 0.3 moisture and volatile matter. The fractionated oil yields 15% cut soybean oil of an iodine value of 70 to 90, used for soaps, lubricants, and rubber compounding; 72% selected-acid oil of an iodine value of 145 to 155, used for varnish and paint oils alone or in blends with other oils, or for glycerin making; 13% bottoms, used for soaps, lubricants, and giving a by-product pitch used in insulation and mastic flooring. **Snowflake oil,** of Archer-Daniels-Midland Co., is a heavy-bodied, oxidized soybean oil for paints. It has a specific gravity of 0.986 to 0.989 and iodine number from 64 to 95. Special kettle-bodied and blown grades for use in coatings, caulks, and putties are available from Werner G. Smith, Inc. **Soyalene,** of the same company, is an alkali-refined soybean oil for varnishes. The specific gravity is 0.924, and the iodine number is 130. **Epoxidized soybean oil** is used in vinyl and alkyd resins as a plasticizer and to increase heat resistance. A very large use of soybean oil is in the making of margarine.

Soybean meal is the product obtained by grinding the soybean chips from the expeller process, or the **soybean oil cake** from the hydraulic process. The meal is marketed as stock feed or fertilizer. It is chiefly used as a protein feed for dairy cattle, but it is inferior to fish meal for poultry, as it lacks the mineral salts and vitamins of fish meal. Soybean meal hardened with formaldehyde is used as a filler with wood flour in plastics to give better flow in molding. **Gelsoy** is a protein gel extracted from soybean meal. It is used in foodstuffs as a thickening agent, and is also used as a strong adhesive. **Genistein,** found in soybean curd, called **tofu,** and in soy milk, soy protein isolates, and most soy flours, may be an anticarcinogen. **Soy sauce** does not contain the substance, but its principal flavor component contains a substance called **HEMF,** also a possible anticarcinogen.

Soybean flour for bakery food products for the U.S. market is made from meal that has been treated by acidulated washing to remove the soluble enzymes and sugars that carry the taste. Meal produced by heat processing averages 40% protein and 20 fats, while meal from solvent extraction has 42 to 50 protein and a maximum of 2.5 fats. Further processing of the meal to remove sugars and other materials varies the final protein content of the flour, and meals from different types of beans vary in content. The protein content can be increased by removing the soy hulls before (front-end dehulling) or after (tail-end dehulling) solvent extraction. The **Promax** and **Isopro** soybean flours, of Griffith Laboratories, for high-protein additions to foodstuffs, contain 70% protein with all flavor removed, and are high in lysine. They have a pH of 5.5 and 7.0, respectively. **Soy protein,** of General Mills, used in canned soups and meat products, is toasted to eliminate all enzyme activity. It contains 50% protein with 2 lecithin and 3 lysine. **Promine,** of Central Soya Co., is a 93% concentrate of soybean proteins, used for thickening and enriching soup mixes. **Supro 610,** of Ralston Purina Co., is a spray-dried powder, 95% protein, with a light cream color and no bitter flavor.

SPECULUM METAL.   An alloy formerly used for mirrors and in optical instruments. It contains 65 to 67% copper, the balance tin. It takes a beautiful polish and is hard and tough. An old Roman mirror contained about 64% copper, 19 tin, and 17 lead; and an Egyptian mirror contained 85% copper, 14 tin, and 1 iron. The old Greek mirrors were carefully worked out with 32% tin and 68 copper. They had 70% of the reflecting power of silver, with a slight red excess of reflection that gave a warm glow, without the blue of nickel or antimony. This alloy is now plated on metals for reflectors. A modern telescope mirror contains 70% copper and 30 tin. **Chinese speculum** contains about 8% antimony and 10 tin. **Speculum plate,** which has been advocated by the Tin Research Institute for electroplating, to give a hard, white, corrosion-resistant surface for food processing equipment and optical reflectors, has 55% copper and 45 tin. It is harder than nickel and retains its reflectivity better than silver.

SPERM OIL.   The waxy oil extracted from the head cavity of the sperm whale, *Physeter breviceps* and *P. catadon,* and the **Bottlenose whale,** *P. macrocephalus.* Sperm whales have teeth and feed in deep water on squid and large animal life. The male sperm whale attains a length of 60 ft (18 m) and the female about 38 ft (12 m). The spermaceti is first separated out, leaving a clear, yellow oil. It is purified by being pressed at a low temperature. It is graded according to the temperature of pressing. A good grade of sperm oil has a specific gravity of 0.875 to 0.885 and a flash point above 440°F (227°C). Oils from other whale species, such as the humpback, fin, and sulfurbottom, have a specific gravity of 0.91 to 0.93. Inferior grades of sperm oil may be from sperm whale blubber. Commercial sperm oil is likely to be one-third head oil and two-thirds body oil. Sperm oil differs from fish oil and whale oil in consisting chiefly of liquid waxes of the higher fatty alcohol esters and not fats. Sperm oil absorbs very little oxygen from the atmosphere and resists decomposition even at temperatures above 400°F (204°C), and it will pour below its cloud point of 38 to 45°F (3 to 7°C). It wets metal surfaces easily. It is thus a valuable lubricating oil. It was formerly used as a lamp oil, burning with a white shining flame. It is also an excellent soap oil. **Sperm 42,** of Werner G. Smith, Inc., is a sperm oil with carbon chains of C_{10} to C_{22}, and it is emulsifiable in cold or warm water. **Sulfonated sperm oil** is used as a wetting agent for textiles, and it is also valued for cutting oils, crankcase oil, and high-pressure lubricants. **Smithol 25,** of the same company, is a synthetic fatty-acid oil resembling sperm oil and having the same uses. It is a light-colored, odorless oil with high viscosity, a low pour point of $-16°F$ ($-27°C$), and an iodine value of 105. **Maysperm** is a sulfurized sperm-oil replacement from Mayco Oil & Chemical Co. It is used as a lubricant additive.

Spermaceti is the white, crystalline flakes of fatty substance, or wax, that separate out from sperm oil on cooling after boiling. It is **cetyl palmitate,** a true wax, and does not yield glycerin when saponified. It is purified by pressing, and the triple-refined is snow white. It is also separated out from **dolphin-head oil.** Spermaceti is odorless and tasteless, has a melting point of 43°C, and is insoluble in water, but soluble in hot alcohol. It burns with a bright flame. It was formerly used for candles but now is employed chiefly as a fine wax for ointments and compounds. Sperm oil and spermaceti are inedible and indigestible. **Cetyl alcohol,** $C_{16}H_{33}OH$, originally obtained from spermaceti, is now made synthetically from **ethyl palmitate.**

A synthetic spermaceti wax from Sherex Chemical Co., called **Starfol Wax GG,** is used in cosmetic emollients, drawing compounds, finishing aids, lubricants, and leather treatment. **Synaceti 116** is a fine-chemical and pharmaceutical grade from Werner G. Smith, Inc. White flakes of a 90% cetyl palmitate grade, **Kessco 653,** from Stepan Chemical Co., are employed as a viscosity modifier. **Straplitz** is a similar material from Strahl & Pitsch, Inc. A substitute wax that is compositionally different from spermaceti is extracted from the jojoba plant. **Hydroba-70** is such a material, produced by Jojoba Growers & Processors, Inc.

SPICE. An aromatic vegetable substance, generally a solid used in powdered form, employed for flavoring foods. There is no sharp dividing line between **flavors** and spices, but in general a spice is a material that is used to stimulate the appetite and increase the flow of gastric juices. Spices are not classified as foods in themselves, having little food value, but as food accessories. Pepper is distinctly a spice, though not grouped with the spices. Some spices are also used widely as flavors and in perfumes, and in medicine either for antiseptic or other values or to disguise the unpleasant taste of drugs. A **condiment** is a strong spice, or a spice of sharp taste, although the word is often erroneously applied to any spice. A **savory** is a fragrant herb or seed used for flavor in cooking. Spices are obtained from the stalks, bark, fruits, flowers, seeds, or roots of plants. **Microground spices,** used to give uniform distribution in the quantity manufacture of foodstuffs, are spices ground to microscopic fineness in a roller mill. The most popular spices in the United States, in the order of quantity used, are: cinnamon, nutmeg, ginger, cloves, allspice, poppy seed, and caraway seed. Since ground spices lose flavor rapidly by the loss of volatile oils, the particles are sometimes coated with dextrose or a water-soluble gum. **Spisoseals** are ground spices with coated particles.

Allspice, also known as **pimento** and **Jamaica pepper,** is the dried, unripe fruit of the small evergreen tree *Pimenta officinalis* of the myrtle family growing in the West Indies and tropical America. The fruit is a small berry which when dried is wrinkled and reddish brown. It has a flavor much like a combination of clove, nutmeg, and cinnamon. **Pimento oil** is a fragrant essential oil distilled from the berries, which contain 4%. It contains eugenol and cineol and is used in flavors, in bay rum, and in carnation perfumes. **Coriander** is the dried fruit of the perennial plant *Coriandrum sativum* grown in the Mediterranean countries and India. It is one of the oldest spices and has a pleasant, aromatic taste. **Oil of coriander,** extracted from the dried seed, is used in medicine, beverages, and flavoring extracts. It has a higher aromatic flavor than the fruit. **Savory** is a fragrant herb of the mint family, *Satureia hortensis,* used in cooking, and in medicine as a carminative. It contains **carvacrol,** a complex phenol also occurring in caraway and camphor. The word *savory* also designates other herbs used directly in foods as flavors.

Celery seed, used as a savory, is from the plant *Apium graveolens.* The best-quality leafstalks, known as **celery,** are bleached white and eaten raw or cooked. The plant is widely grown for seed in France and Spain. **Celery-seed oil** is a pale-yellow oil extracted from the seeds and used as a flavor and in perfumery. **Fennel** is the dried, oval seed of the perennial plants *Foeniculum vulgare* and *F. dulce.* The stalks of the latter are blanched and eaten as a vegetable in Europe. Fennel is used as a flavoring in confectionery and liqueurs, and as a carminative in medicine. **Fennel oil** is a pale yellowish essential oil with specific gravity of 0.975, distilled from the seed. It has an aromatic odor and a camphorlike taste with a secondary sweetish, spicy taste. It contains **fenchone,** $C_{10}H_{16}O$, an isomer of camphor, with also pinene, camphene, and **anethole,** or **anise camphor,** $C_3H_5C_6H_4OCH_3$. The latter is used in dentifrices and pharmaceuticals. **Fenugreek** is the seed from the long pods of the annual legume *Trigonella foenum-grae-*

cum, native to southern Europe. It is used in curries, in medicine, and for making artificial maple flavor. **Oregano,** used as an ingredient in chili powder and as a spice in a variety of dishes, is the pungent herb *Coleus amboinica.*

Dill seed, from the herb *Anethum graveolens,* of the parsley family, is used as a condiment for pickles. **Dill leaves** are used as seasoning for soups, sauces, and pickles. **Dill oil,** extracted from the whole herb, is used as a flavor in the food industry. It resembles caraway oil and has a finer flavor than **dill-seed oil,** which is more plentiful, but **dill flavor** prepared from the whole seed is stronger. Dill is grown in the central United States and in central Europe. **Cardamon** is the highly aromatic and delicately flavored seeds of the large perennial herb *Elettaria cardamomum,* of India, Ceylon, and Central America. The seeds are used in pickles, curries, and cakes, and the oil is employed as a flavor. **Garlic** is the root bulb of the lily *Allium sativum,* used as a condiment, and also used in medicine as an expectorant under the name of **allium.** It contains **allyl sulfide,** a liquid of composition $(CH_2:CHCH_2)_2S$, which gives it a pungent odor and taste. **Allicine,** extracted from garlic, is used in medicine as an antibacterial. It is an oily liquid with a sharp garlic odor.

Cumin is the seed of *Cuminum cyminum,* the true cumin, and *Nigella sativa,* the **black cumin,** both of India. The seed is used in confectionery and in curries. A kind of black cumin known as **shiah zira,** from the plant *Carum indicum* of India, is superior in taste and fragrance to ordinary cumin. **Caraway** is the spicy seed of the biennial herb *C. carvi* of Europe and north Africa. The seeds are used on cookies. **Caraway oil,** distilled from the seeds, contains carvone and limonene, and in combination with cassia gives a pleasant odor. It is used in soap, perfumes, and mouthwashes. **Sage** is the grayish-green, hairy leaves of the shrublike plant *Salvia officinalis* used as a spice. It is cultivated extensively in the Mediterranean region. **Oil of sage** is used in perfumery. **Clary sage oil** is distilled from the flowers of *S. sclarea* of France, Italy, and north Africa. It has the odor of a mixture of ambergris, neroli, and lavender, and is used in flavoring vermouth liquor and muscatel wines, and in eau de cologne. **Sassafras** is sometimes classified as a spice but is a flavor. It is the aromatic spicy bark of the root of the tree *Sassafras albidum* which grows wild in the eastern United States. It is used mostly for making root beer, but also for flavoring tobacco, and in patent medicines. **Sassafras oil** is an oil extracted from the whole roots, which contain 2% of the yellow oil, and is used in medicine, perfumery, and soaps. It produces artificial heliotrope. The oil contains **safrol,** $C_{10}H_{10}O_2$, also produced from brown camphor oil. **Brazilian sassafras oil,** or **ocotea oil,** is distilled from the root of the tree *Ocotea cymbarum,* also of the laurel family. The root yields about 1% of an oil which contains 90% safrol, and has the odor and flavor of American sassafras oil. **Sarsaparilla** is an oil obtained from the long brown roots of the climbing vine *Smilax regellii* of Honduras, *S. aristolochiaefolia* of Mexico, and other species, all growing in tropical jungles. The roots are used in medicine. The oil is used as a flavor. It is odorless, but has an acrid sweet taste. It contains saponins.

Wintergreen oil is from the leaves of the small evergreen plant *Gaultheria procumbens* of the middle Atlantic states. The oil does not exist in the plant but is formed by the reaction between a glucoside and an enzyme when the chopped leaves are steeped in water. It is largely **methyl salicylate,** $C_8H_8O_3$. It is used in flavoring candies and soft drinks and in medicine. **Hop oil,** used to give hop flavor to cereal beverages, and also in perfumes, is obtained from **lupulin,** a glandular powder found in the female inflorescence of the hop plant, *Humulus lupulus.* **Hops** are used directly in making beer, and the oil is produced from the discard hops which contain 0.75% oil. **Anise seed** is from the annual plant *Pimpinella anisum* grown in the Mediterranean countries and in India. The best grades come from Spain. The seed is used in flavoring in the baking industry. The distilled oil, **anise oil,** is used in perfumes and in soaps, and in the liqueur known as **anisette.** The oil contains choline and is used in medicine as a carminative and expectorant.

SPINEL. A **magnesium aluminate,** $MgO \cdot Al_2O_3$, occurring as octahedral crystals of varying colors due to impurities of iron, manganese, or chromium. The best transparent stones are used as gems. Spinel is found as crystals or rolled pebbles in gem gravels with corundum stones; the **ruby spinel** often occurs with the true ruby. It has a deep-red color, but the variety **almandine** is violet.

Synthetic spinel was originally made in Germany to replace ruby and sapphire for instrument bearings because it is easier to cut and thus conserves diamond abrasive. Spinel is produced by Linde in the forms of drawing dies, gages, wearing parts, orifices, and balls. The composition is $MgO \cdot 3.5Al_2O_3$, and the crystal structure is cubic. The specific gravity is 3.61, the melting point is about 3704°F (2040°C), and the Mohs hardness is 8. Like corundum, it is not attacked by common acids or by sodium hydroxide. The spinel powder from which the crystals are flame-grown is made by calcining a mixture of pure ammonium sulfate and ammonium magnesium sulfate. Much synthetic spinel is used for synthetic gems, the colors being obtained with metal oxides. Small amounts of chromic oxide give the tinted crystals of sapphire, while up to 6% is used for the dark ruby colors. Blue is obtained with oxides of iron and titania, and green is from cobalt oxide. **Golden topaz** is colored with nickel and magnesium oxides. The **aquamarine spinel** is tinted with a complex mixture of nickel, cobalt, vanadium, and titanium oxides.

SPODUMENE.   A mineral of composition $Li_2O \cdot Al_2O_3 \cdot 4SiO_2$, with some potassium and sodium oxides. It is the chief ore of the metal lithium, but it requires a higher temperature for sintering than lepidolite, and the sinter is more difficult to leach. It is found in South Dakota and the Carolinas, and has an average content of 4% Li_2O, ranging from 2.9 to 7.6%. Crystals of spodumene in South Dakota are 8 to 10 ft (2 to 3 m) long and 1 ft (0.3 m) in thickness, appearing like logs of wood, with as high as 6.5% lithium. The specific gravity is 3.13 to 3.20, and the melting point is 2543 to 2597°F (1395 to 1425°C). Spodumene is three times more active than feldspar as a flux in ceramics, giving fluidity, increasing surface tension, and eliminating pinholes. A mixture of 25% spodumene with 75 feldspar is an active vitrifying agent in ceramics. The melting point of the mixture is 2030°F (1110°C), which is below the usual minimum temperature used for chinaware; it thus forms a glaze. **Lithospar** is a name for feldspar and spodumene from the pegmatites of King's Mountain, North Carolina. In Germany lithium is obtained from the lithium mica **zinnwaldite,** which is a mixture of potassium-aluminum orthosilicate and lithium orthosilicate with some iron, and contains less than 3% Li_2O. **Kryolithionite,** a mineral found in Greenland, has composition $Na_3Li_3(AlF_6)_2$ and contains up to 11.5% Li_2O. It has a crystal structure resembling garnet. A transparent, emerald-green spodumene in small crystals, known as **hiddenite,** is found in North Carolina and is cut into gemstones.

SPONGE.   The cellular skeleton of a marine animal of the genus *Spongia,* of which there are about 3,000 known species, only 13 of which are of commercial importance. It is employed chiefly for wiping and cleaning, as it will hold a great quantity of water in proportion to its weight, but it also has many industrial uses such as applying glaze to pottery. Sponges grow like plants, attached to rocks on the sea bottom. They are prepared for use by crushing to kill them, scraping off the rubbery skin, macerating in water to remove the gelatinous matter, and bleaching in the sun. Tarpon Springs, Florida, is the center of U.S. sponge fishing, but most of the best sponges have come from the Mediterranean and Red Seas.

The prepared sponge is an elastic, fibrous structure chemically allied to silk. It has sievelike membranes with small pores leading into pear-shaped chambers. The best sponges are spheroidal, regular, and soft. Commercial sponges for the U.S. market must have a diameter of 4.5 in (11.4 cm) or more. Most of the Florida sponges are the **sheepswool sponge,** *Euspongia lachne,* used for cleaning and industrial sponging. The **Rock Island sponge,** from Florida, and the **Key wool sponge** are superior in texture and durability to the **Bahama wool sponge,** which is coarser, more open, and less absorbent. The **Key yellow sponge** is the finest grade. The **grass sponge,** *E. graminea,* of the Caribbean, is inferior in shape and texture. The fine **honeycomb sponge,** *Hippiospongia equina,* of the Mediterranean Sea, is of superior grade and has been preferred as a bath sponge. About 80% of the north African catch consists of honeycomb sponges, with the remainder **Turkey cup sponge,** *E. officinalis,* and **zimocca sponge,** *E. zimocca.* The Turkey cup is rated as the finest, softest, and most elastic of the sponges, but the larger of the zimocca sponges are too hard for surgical use and are employed for industrial cleaning. Sponges for industrial and household uses have now been largely replaced by foamed rubbers and plastics.

SPONGE IRON.   Iron made from ferrous sand and pressed into briquettes, which can be charged directly into steel furnaces instead of pig iron. It was originally made on a large scale in Japan where only low-grade sandy ores were available. Sponge iron is made by charging the sand continuously into a rotary furnace to drive off the light volatile products and reduce the iron oxide to metallic iron, which is passed through magnetic separators, and the finely divided iron briquetted. Unbriquetted sponge iron, with a specific gravity of 2, is difficult to melt because of the oxidation, but briquetted material, with a specific gravity of 6, can be melted in electric furnaces. Sponge iron, to replace scrap in steelmaking, is also made from low-grade ores by reducing the ore with coke-oven gas or natural gas. It is not melted, but the oxygen is driven off, leaving a spongy, granular product. As it is very low in carbon, it is also valuable for making high-grade alloy steels.

A form of sponge iron employed as a substitute for lead for coupling packings was made in Germany under the name of **sinterit.** The reduction is carried out in a reducing atmosphere at a temperature of 2192 to 2462°F (1200 to 1350°C), instead of heating the iron oxide with carbon. Since the porous iron corrodes easily, it is coated with asphalt for packing use. **Iron sponge,** employed as a purifier for removing sulfur and carbonic acid from illuminating gas, is a sesquioxide of iron obtained by heating together iron ore and carbon. It has a spongy texture and is filled with small cells.

SPRENGLE EXPLOSIVES.   Chlorate compounds that have been rendered reasonably safe from violent explosion by separating the chlorate from the combustible matter. The potassium chlorate, made into porous cartridges and dipped, just before use, in a liquid combustible such as nitrobenzene or dead oil, was called **rack-a-rock.** It is a mixture of 79% chlorate and 21 nitrobenzene. **Rack-a-rock special** contains, in addition, 12 to 16% picric acid. Sprengle explosives were formerly used as military explosives, are very sensitive to friction and heat, and are now valued only for mining or when it is desired to economize on nitrates. **Cheddite** is a French explosive consisting of a chlorate with an oily material, such as castor oil, thickened by a nitrated hydrocarbon dissolved in it. A typical cheddite has 80% potassium chlorate, 8 castor oil, and 12 mononitronaphthalene. With sodium chlorate it is less sensitive to detonation and more powerful but is hygroscopic. Potassium chlorate cheddite is a soft, yellowish, fine-grained material, and is a slow, mild explosive which will split rocks rather than shatter them. **Minelite** is a chlorate with paraffin wax. **Steelite** is a chlorate explosive with rosin. **Prométhée,** or **Explosive O_3,** is another French chlorate explosive. In this explosive the oxygen carrier consists of 95% potassium chlorate and 5 manganese dioxide, and the combustible contains 50% nitrobenzene, with turpentine and naphtha. It is extremely sensitive and will explode by friction. **Silesia** is a German high explosive used for blasting. It is potassium chlorate with rosin, with some sodium chlorate to make it less sensitive. **Hellhoffite** is a mixture of nitric acid and **dinitrobenzene,** which are combined in a shell on impact. **Panclasites** are a class similar to the Sprengle type, in which carbon disulfide, nitrobenzene, and petroleum oil are combined with liquid nitrogen peroxide.

SPRING STEEL. A term applied to any steel used for springs. The majority of springs are made of steel, but brass, bronze, nickel silver, and phosphor bronze are used where their corrosion resistance or electrical conductivity is desired. Carbon steels, with from 0.50 to 1.0% carbon, are much used, but vanadium and chromium-vanadium steels are also employed, especially for heavy car and locomotive springs. Special requirements for springs are that the steel be low in sulfur and phosphorus, and that the analysis be kept uniform. For flat or spiral springs that are not heat-treated after manufacture, hard-drawn or rolled steels are used. These may be tempered in the mill shape. Music wire is widely employed for making small spiral springs. A much-used straight-carbon spring steel has 1% carbon and 0.30 to 0.40 manganese, but becomes brittle when overstressed. **ASTM carbon steel** for flat springs has 0.70 to 0.80% carbon and 0.50 to 0.8 manganese, with 0.04 maximum each of sulfur and phosphorus. Motor springs are made of this steel rolled hard to a tensile strength of 250,000 lb/in^2 (1,724 MPa). **Watch spring steel,** for mainsprings, has

1.15% carbon, 0.15 to 0.25 manganese, and in the hard-rolled condition, has an elastic limit above 300,000 lb/in^2 (2,069 MPa).

Silicon steels used for springs have high strength. These steels average about 0.40% carbon, 0.75 silicon, and 0.95 manganese, with or without copper, but the silicon may be as high as 2%. **Flexo steel,** used for automobile leaf springs and recoil springs, contains 2% silicon, 0.75 manganese, and 0.60 carbon. The elastic limit is 100,000 to 300,000 lb/in^2 (690 to 2,069 MPa), depending on drawing temperature, with Brinell hardness 250 to 600.

Manganese steels for automotive springs contain about 1.25% manganese and 0.40 carbon, or about 2 manganese and 0.45 carbon. When heat-treated, the latter has a tensile strength of 200,000 lb/in^2 (1,379 MPa) and 10% elongation. Part of the manganese may be replaced by silicon, and the **silicon-manganese steels** have tensile strengths as high as 270,000 lb/in^2 (1,862 MPa). The addition of chromium or other elements increases ductility and improves physical properties. **Uma spring steel** is a chromium-manganese steel with 1 to 1.2% chromium, 0.80 to 1 manganese, and about 0.50 carbon. In the rolled condition it has an ultimate strength of 135,000 lb/in^2 (931 MPa) and Brinell hardness up to 332. Manganese steels are deep-hardening but are sensitive to overheating. The addition of chromium, vanadium, or molybdenum widens the hardening range.

Wire for coil springs ranges in carbon from 0.50 to 1.20%, and in sulfur from 0.028 to 0.029. Bessemer wire contains too much sulfur for spring use. Cold working is the method for hardening the wire and for raising the tensile strength. A 0.85% carbon rod, with an ultimate strength of 140,000 lb/in^2 (965 MPa), when drawn with four or five passes through dies, will have a strength of 235,000 lb/in^2 (1,620 MPa). Wire drawn down to a diameter of 0.015 in (0.038 cm) may have an ultimate strength of 400,000 lb/in^2 (2,758 MPa). The highest grades of wire are referred to as *music wire*. The second grade is called **hard-drawn spring wire.** The latter is a less expensive, basic open-hearth steel with manganese content of 0.80 to 1.10%, and an ultimate strength up to 300,000 lb/in^2 (2,069 MPa). Specially treated carbon steels for springs are sold under trade names such as **Enduria** and **Resilla,** the latter being a silicon-manganese spring steel.

For jet-engine springs and other applications where resistance to high temperatures is required, stainless steel and high-alloy steels are used. But while these may have the names and approximate compositions of standard stainless steels, for spring-wire use their manufacture is usually closely controlled. For example, when the carbon content is raised in high-chromium steels to obtain the needed spring qualities, the carbide tends to collect in the grain boundaries and cause intergranular corrosion unless small quantities of titanium, columbium, or other elements are added to immobilize the carbon. **Blue Label stainless** is Type 302 stainless steel of highly controlled analysis for coil springs. **Alloy NS-355** is a stainless steel having a typical analysis of 15.64% chromium, 4.38 nickel, 2.68 molybdenum, 1 manganese, 0.32 silicon, 0.12 copper, with the carbon at 0.14. The modulus of elasticity is 29.3×10^6 lb/in^2 (202,000 MPa) at 80°F (27°C) and 24×10^6 lb/in^2 (165,000 MPa) at 800°F (427°C). **17-7 PH stainless steel** has 17% chromium, 7 nickel, 1 aluminum, and 0.07 carbon. Wire has a tensile strength up to 345,000 lb/in^2 (2,379 MPa). **Spring wire** for high-temperature coil springs may contain little or no iron. **Alloy NS-25,** for springs operating at 1400°F (760°C), contains about 50% cobalt, 20 chromium, 15 tungsten, and 10 nickel, with not more than 0.15 carbon. Annealed wire, drawn to a 30% reduction, has a tensile strength of 240,000 lb/in^2 (1,655 MPa) with 8% elongation. **Matreloy,** for high-temperature springs, contains 39% chromium, 4 molybdenum, 2 titanium, 1 aluminum, and the balance nickel. The rolled metal has a yield strength of 275,000 lb/in^2 (1,896 MPa) and, at 1400°F (760°C), retains a strength of 120,000 lb/in^2 (827 MPa).

SPRUCE. The wood of various coniferous trees of northern Europe and North America. Spruce is a leading commercial wood of northern Europe and is exported from the Baltic region as **white fir** and **white deal.** It is also called **Norway spruce** and **spruce fir.** The wood is white and has a straight, even grain. It is tough and elastic and is more difficult to work than pine. The density is 36 lb/ft^3 (577 kg/m^3). Norway spruce is *Picea abies,* and this tree yields the **Jura turpentine** of Europe. Spruce is used for making paper pulp, for packing boxes, and as a general-utility lumber. White spruce is from the tree *P. canadensis,* of the United States and Canada. It has quite similar characteristics. **Red spruce,** *P. rubra,* is the chief lumber spruce in the eastern United States. It is also called **yellow spruce, West Virginia spruce,** and **Canadian spruce. Black spruce,** *P. mariana,* of New England, eastern Canada, and Newfoundland, is used for making paper pulp. It is also called **blue spruce, bog spruce,** and **spruce pine. White spruce,** or **shingle spruce,** is from *P. glauca.* It is also called **skunk spruce** because of the peculiar odor of the foliage. All of these three species are called **eastern spruce,** and they grow from Nova Scotia to Tennessee and westward to Wisconsin except that red spruce does not grow in the lake states. All are mountain trees and are slow-growing. **Silver spruce, yellow spruce, Sitka spruce,** or **western spruce** is from the enormous tree *P. sitchensis,* of the west coast of the United States and Canada. It is soft and lightweight, but strong, close-grained, and very free from knots. The wide sapwood is creamy white, and the heartwood pinkish to brownish. The weight is less than that of eastern spruce, but it has high strength in proportion to weight. The trees reach a height of 280 ft (85 m) and a diameter of 10 ft (3 m) in 600 years, but growth is rapid in early life. The wood is used for boxes, crates, millwork, and paper pulp. It is particularly adapted for groundwood pulp, giving higher strength in paper than most groundwood. The various species of commercial spruce have an average specific gravity, when kiln-dried, of 0.40, a compressive strength of 840 lb/in^2 (6 MPa) perpendicular to the grain, and a shearing strength of 750 lb/in^2 (5 MPa) parallel to the grain. It combines stiffness and strength per unit weight and has a uniform texture free from pitch. **Japanese spruce** is from *Abies mariesii,* and **Himalayan spruce** is from *P. morinda.* The latter resembles Norway spruce.

Spruce gum is the gum exudation of *P. rubra, P. mariana,* and *P. canadensis* of the northeastern United States and Canada. It exudes from cuts in the trees as a transparent, viscous liquid which hardens when it loses the volatile oil. It occurs on all parts of the tree, and the nodules of gum are sometimes as large as an egg. **Spruce oil** is extracted from the needles. In colonial days the young twigs were boiled, and the liquid, mixed with molasses, was used as a beverage. The gum is brown or reddish black and has a turpentinelike odor and a bitter, pungent taste. It is used in cough medicines and chewing gum.

SQUILL. Also known as **red squill** and **sea onion.** A reddish powder used chiefly for the control of rats in warehouses and docks. It is classified as a cardiac glycoside, a category that includes **digitalis.** It is obtained from the onionlike bulb of the perennial plant *Urginea maritima,* which grows on the beaches of Italy and other Mediterranean countries. The bulb is pear-shaped, from 1 to 6 lb (0.5 to 15 kg) in weight and 6 to 12 in (15 to 30 cm) in diameter. The outer scales are dry, brittle, and reddish brown, and the inner scales are cream color to deep purple. **Red squill powder** is a powerful emetic to humans or animals other than rats or mice. As rats and mice do not vomit, they are poisoned by it, while it is harmless to poultry and domestic animals. It is also used in medicine. It contains calcium oxalate, and in contact with skin it gives a sensation like nettle poisoning. **White squill** is another variety used in medicine as an emetic, heart tonic, and expectorant. The substitute for red squill known as **Antu,** of Du Pont, is **naphthyl thiourea,** a gray powder of little odor or taste about 100 times more poisonous to rats than squill and not normally injurious to domestic animals. Another poison more toxic to rats than squill is **Ratbane 1080,** which is **sodium fluoracetic acid** made synthetically. The poison occurs as natural fluoracetic acid in the **gifblaar** plant, *Dichapetalum cymosum,* of South Africa, which has been used locally for killing rodents. The poison, however, also kills domestic animals and can thus be employed only in restricted places. The rodent poison known as **Warfarin** is a complex dicoumarol made under license of the Wisconsin Alumni Research Foundation. **Pival,** of Atlantic Research Corp., is 2-pivalyl-1,3-indandione. It kills rats but is not toxic to other animals. **Pivalyn** is the same material in a water-soluble sodium salt form.

STAINLESS STEEL. A large and widely used family of **iron-chromium alloys** known for their corrosion resistance—notably their "nonrusting" quality. This ability to resist corrosion is attributable to a chromium-oxide surface film that forms in the presence of oxygen. The film is essentially insoluble, self-healing, and nonporous. A minimum chromium content of 12% is required for the film's formation, and 18 is sufficient to resist even severe atmospheric corrosion. Chromium content, however, may range to about 30% and several other alloying elements, such as manganese, silicon, nickel, or molybdenum, are usually present. Most stainless steels are also resistant to marine atmospheres, freshwater, oxidation at elevated temperatures, and mild and oxidizing chemicals. Some are also resistant to saltwater and reducing media. They are also quite heat-resistant, some retaining useful strength to 1800°F (981°C). And some retain sufficient toughness at cryogenic temperatures. Thus, stainless steels are used in a wide range of applications requiring some degree of corrosion and/or heat resistance, including auto and truck trim, chemical and food processing equipment, petroleum-refining equipment, furnace parts and heat-treating hardware, marine components, architectural applications, cookware and housewares, pumps and valves, aircraft and aircraft-engine components, springs, instruments, and fasteners.

Stainless steels were first made in the United States in 1914 under English and German patents. The original composition had 13.5% chromium and 0.35 carbon. The original Krupp austenitic, or **KA steel,** or simply **austenitic steel,** had 20% chromium and 7 nickel, which was later balanced at 18–8. The **eighteen-eight chromium-nickel steels** were called **super stainless steels** in England to distinguish them from the plain chromium steels. Today, wrought stainless steels alone include some 70 standard compositions and many special compositions. They are categorized as austenitic, ferritic, martensitic, or precipitation-hardening (PH) stainless steels, depending on their microstructure or, in the case of the PH, their hardening and strengthening mechanism. There are also many cast stainless steels having these metallurgical structures. They are known as cast corrosion-resistant steels, cast heat-resistant steels, and cast corrosion- and heat-resistant steels. Several compositions are available in powder form for the manufacture of stainless-steel powder-metal parts.

Except for the PH stainless steels, wrought stainless steels are commonly designated by a three-digit numbering system of the American Iron and Steel Institute. Wrought **austenitic stainless steels** constitute the 2XX and 3XX series, and the wrought **ferritic stainless steels** are part of the 4XX series. Wrought **martensitic stainless steels** belong to either the 4XX or 5XX series. Suffix letters, such as L for low carbon content or Se for selenium, are used to denote special compositional modifications. Cast stainless steels are commonly known by the designations of the Alloy Casting Institute of the Steel Founders Society of America, which began with letters CA through CN and are followed by numbers or numbers and letters. Powder compositions are usually identified by the designations of the Metal Powder Industries Federation.

Of the austenitic, ferritic, and martensitic families of wrought stainless steels, each has a general-purpose alloy. All of the others in the family are derivatives of the basic alloy, with compositions tailored for special properties. The **stainless steel 3XX series** has the largest number of alloys, and **stainless steel 302, a stainless "18–8" alloy,** is the general-purpose one. Besides its 17 to 19% chromium and 8 to 10 nickel, it contains a maximum of 0.15 carbon, 2 manganese, 1 silicon, 0.4 phosphorus, and 0.03 sulfur. 302B is similar except for greater silicon (2 to 3%) to increase resistance to scaling. **Stainless steels 303** and **303Se** are also similar except for greater sulfur (0.15% minimum) and, optionally, 0.6% molybdenum in 303, and 0.06 maximum sulfur and 0.15 minimum selenium in 303Se. Both are more readily machinable than 302. **Stainless steels 304** and **304L** are low-carbon (0.08 and 0.03% maximum, respectively) alternatives, intended to restrict carbide precipitation during welding and, thus, are preferred to 302 for applications requiring welding. They may also contain slightly more chromium and nickel. **Stainless steel 304N** is similar to 304 except for 0.10 to 0.16% nitrogen. The nitrogen provides greater strength than 302 at just a small sacrifice in ductility and a minimal effect on corrosion resistance. **Stainless steel 304 SCQ,** of Carpenter Technology, is electroslag remelted and has the same mechanical properties as conventional air-melted 304 but greater cleanliness, that is, greater freedom from nonmetallic inclusions. It provides better resistance to fluid leakage under vacuum or high-pressure conditions in thin-wall vessels. **Stainless steel 305** has 0.12% maximum carbon but greater nickel (10.5 to 13) to reduce the rate of work hardening for applications requiring severe forming operations. **Stainless steel 308** contains more chromium (19 to 21%) and nickel (10 to 12) and, thus, is somewhat more corrosion-

and heat-resistant. Though it is used for furnace parts and oil refinery equipment, its principal use is for welding rods because its higher alloy content compensates for alloy content that may be reduced during welding.

Stainless steels 309, 309S, 310, 310S, and **314** have still greater chromium and nickel contents. 309S and 310S are low-carbon (0.08% maximum) versions of 309 and 310 for applications requiring welding. They are also noted for high creep strength. 314, which like 309 and 310 contains 0.25% maximum carbon, also has greater silicon (1.5 to 3), thus providing greater oxidation resistance. Because of the high silicon content, however, it is prone to embrittlement during prolonged exposure at temperatures of 1200 to 1500°F (649 to 816°C). This embrittlement, however, is only evident at room temperature and is not considered harmful unless the alloy is subject to shock loads. These alloys are widely used for heaters and heat exchangers, radiant tubes, and chemical and oil refinery equipment.

Stainless steels 316, 316L, 316F, 316N, 317, 317L, 321, and **329** are characterized by the addition of molybdenum, molybdenum and nitrogen (316N), or titanium (321). 316, with 16 to 18% chromium, 10 to 14 nickel, and 2 to 3 molybdenum, is more corrosion- and creep-resistant than 302- or 304-type alloys. 316L is the low-carbon version for welding applications; 316F, because of its greater phosphorus and sulfur, is the "free-machining" version; and 316N contains a small amount of nitrogen for greater strength. 317 and 317L are slightly richer in chromium, nickel, and molybdenum and, thus, somewhat more corrosion- and heat-resistant. Like 316, they are used for processing equipment in the oil, chemical, food, paper, and pharmaceutical industries. 321 is titanium-stabilized to inhibit carbide precipitation and provide greater resistance to intergranular corrosion in welds. 329, a high-chromium (25 to 30%), low-nickel (3 to 6) alloy with 1 to 2 molybdenum, is similar to 316 in general corrosion resistance but more resistant to stress corrosion. **Stainless steel 330,** a high-nickel (34 to 37%), normal-chromium (17 to 20), 0.75 to 1.5 silicon, molybdenum-free alloy, combines good resistance to carburization, heat, and thermal shock.

Stainless steels 347 and **348** are similar to 321 except for the use of columbium and tantalum instead of titanium for stabilization. Also, 348 contains a small amount (0.2%) of copper. Both have greater creep strength than 321, and they are used for welded components, radiant tubes, aircraft-engine exhaust manifolds, pressure vessels, and oil refinery equipment. 384, with nominally 16% chromium and 18 nickel, is another low-work-hardening alloy used for severe cold-heading applications.

The **stainless steel 2XX series** of austenitics comprises 201, 202, and 205. They are normal in chromium content (16 to 19%), but low in nickel (1 to 6), high in manganese (5.5 to 15.5), and with 0.12 to 0.25 carbon and some nitrogen. 201 and 202 have been called the low-nickel equivalents of 301 and 302, respectively. 202, having 17 to 19% chromium, 7.5 to 10 manganese, 4 to 6 nickel, and a maximum of 1 silicon, 0.25 nitrogen, 0.15 carbon, 0.06 phosphorus, and 0.03 sulfur, is the general-purpose alloy. 201, which contains less nickel (3.5 to 5.5%) and manganese (5.5 to 7.5), was prominent during the Korean war due to a nickel shortage. 205 has the least nickel (1 to 1.75%) and the most manganese (14 to 15.5), carbon (0.12 to 0.25), and nitrogen (0.32 to 0.40) contents. It is said to be the low-nickel equivalent of 305 and has a low rate of work hardening that is useful for parts requiring severe forming operations.

Like stainless steels in general, austenitic stainless steels have a density of 0.28 to 0.29 lb/in^3 (7,750 to 8,027 kg/m^3). Unlike some other stainless steels, they are essentially nonmagnetic, although most alloys will become slightly magnetic with cold work. Their melting point range is 2500 to 2650°F (1371 to 1454°C), specific heat at 32 to 212°F (0 to 100°C) is about 0.12 Btu/(lb · °F) [502 J/(kg · K)], and electrical resistivity at room temperature ranges from 27 to 31 μΩ · in (69 to 78 μΩ · cm). Types 309 and 310 have the highest resistivity, and 201 and 202 the lowest.

Most are available in many mill forms and are quite ductile in the annealed condition, tensile elongations ranging from 35 to 70%, depending on the alloy. Although most cannot be strengthened by heat treatment, they can be strengthened appreciably by cold work. In the annealed condition, the tensile yield strength of all the austenitics falls in the range of 30,000 to 80,000 lb/in^2 (207 to 552 MPa), with ultimate strengths in the range of 75,000 to 120,000 lb/in^2 (517 to 827 MPa). But cold-working 201 or 301 sheet just to the half-hard temper increases yield strength to 110,000 lb/in^2 (758 MPa) and ultimate strength to at least 150,000 lb/in^2 (1,034 MPa). Tensile modulus is typically 28 × 10^6 to 29 × 10^6 (193 × 10^3 to 199 × 10^3 MPa) and decreases slightly with severe cold work. As to high-temperature strength, even in the annealed condition most alloys have tensile yield strengths of at least 12,000 lb/in^2 (83 MPa) at 1500°F (815°C), and some (308, 310) about 20,000 lb/in^2 (138 MPa). Types 310 and 347 have the highest creep strength, or stress-rupture strength, at 1000 to 1200°F (538 to 649°C). Annealing temperatures range from 1750 to 2100°F (954 to 1149°C), initial forging temperatures range from 2000 to 2300°F (1093 to 1260°C), and their machinability index is typically 50 to 55, 65 for 303 and 303Se, relative to 100 for 1112 steel.

Among the many **specialty wrought austenitic stainless steels** are a number of nitrogen-strengthened stainless steels: **Nitronic 20,**

32, 33, 40, 50, and **60** from Armco; **18-18 Plus** and **Marinaloy HN** and **22** from Carpenter Technology; and **SAF 2205** and **253MA** from Sandvik. Nitrogen, unlike carbon, has the advantage of increasing strength without markedly reducing ductility. Some of these alloys are twice as strong as the standard austenitics and provide better resistance to certain environments. All are normal or higher than normal in chromium content. Some are also normal or higher than normal in nickel content, while others are low in nickel and, in the case of 18-18 Plus, nickel-free. Nitronic 20, a 23% chromium, 8 nickel, 2.5 manganese alloy, combines high resistance to oxidation and sulfidation and was developed for engine exhaust valves. Unlike austenitics in general, it is hardenable by heat treatment. Solution treating (2150°F, 1177°C), water quenching, and aging (1400°F, 760°C) provide tensile strengths of 84,000 lb/in² (579 MPa) yield and 142,000 lb/in² (979 MPa) ultimate. **Stainless steel SAF 2205,** an extra-low-carbon (0.03%), 22 chromium, 5.5 nickel, 3 molybdenum alloy, is a **ferritic-austenitic stainless steel** alloy with high resistance to chloride- and hydrogen-sulfide-induced stress corrosion, pitting in chloride environments, and intergranular corrosion in welded applications. Because of their superior corrosion resistance, such **duplex stainless steels** are also called **super duplex stainless steels.** Besides SAF 2205, others include **stainless steels 2304, 3RE60, 255,** and **2507.** As a class, they contain 18 to 25% chromium, 5 to 7 nickel, 0.2 to 4 molybdenum, and 0 to 0.3 nitrogen. The 2304 also contains as much as 0.6% copper, and 255 has 2% copper. All are low in carbon: 0.03 or 0.04%.

The specialty austenitic **stainless steel AL-6X,** of Allegheny Ludlum, is a low-carbon, 20% chromium, 24 nickel, 6 molybdenum alloy developed specifically for resistance to seawater pitting and crevice corrosion, and has found wide use for seawater condenser tubing. **Stainless steel AL-6XN,** with 0.22% nitrogen, is one of about a dozen alloys called **super austenitic stainless steels,** which are more highly alloyed than ordinary austenitic stainless steels. There are two major families: the 6% molybdenum group and the high-nickel group. Besides AL-6XN, the 6% Mo group includes **stainless steels 25-6 Mo, 1925 hMo, UR SB8, 254 SMO,** and **654 SMO.** They contain 20 to 25% chromium, 18 to 25 nickel, 5.5 to 7.5 molybdenum, 0.5 to 1.5 copper, 0.2 to 0.5 nitrogen, and 0.02 to 0.03 carbon. They are considerably stronger than the ordinary austenitics without sacrificing ductility and toughness, have good resistance to general corrosion, high resistance to pitting and crevice corrosion in wet chloride environments, and adequate resistance to stress-corrosion cracking in chloride environments. The high-Ni group includes **stainless steels Alloy 20, Alloy 28, Alloy 31, Alloy 825, 20 Mo-4,** and **20 Mo-6.**

They contain 31 to 42% nickel, 20 to 27 chromium, 2.5 to 6.5 molybdenum, 1 to 3.5 copper, and 0.015 to 0.07 carbon. Alloy 20 and 20 Mo-4 also have 0.7 and 0.3% columbium plus tantalum, respectively. Alloy 31 has 0.02% nitrogen, and Alloy 825 also has 0.9% titanium. Alloy 20 and Alloy 825 have excellent resistance to aggressive mineral acids but are not especially resistant to wet chloride environments. Alloy 31 and 20 Mo-6, the richest in molybdenum (6.5 and 5.9%, respectively), are appreciably more resistant to chloride corrosion.

Crucible Steel's **stainless steel 303CC** is a low-sulfur, low-carbon, more machinable version of the standard alloy for screw-machine products used in beverage dispensers. The lower sulfur eliminates taste and odor problems associated with the standard alloy in such products, and the lower carbon compensates for the loss in machinability associated with the reduction in sulfur. **Stainless steel 317LM** contains slightly more molybdenum than standard 317L and is used for flue-gas scrubbers and in the paper and textile industries. **Stainless steel JS 777,** a 21% chromium, 25 nickel, 4.5 molybdenum alloy from Jessop Steel, is a high-copper (2%) version of the company's **stainless steel JS 700** for greater resistance to sulfuric acid in scrubbers and coal gasification and desulfurization systems. Sandvik's **stainless steel SANICRO 28** is an extra-low-carbon (0.02%) high-alloy (27 chromium, 31 nickel, 3.5 molybdenum, 1 copper) austenitic stainless for resistance to chlorides and free fluorides. It is said to be especially resistant to chloride pitting, crevice corrosion, and intergranular corrosion in welds. **Stainless steel Cryotech 302,** a wire alloy from A1-Tech Specialty Steel, is similar to standard 302 in composition, but cooling to $-320°F$ ($-196°C$) prior to drawing in the 0 to $50°F$ (-18 to $10°C$) range induces greater tensile strength [to 295,000 lb/in^2 (2,034 MPa)] than conventional wire drawing.

Stainless steel 20Cb-3, of Carpenter Technology, contains 20% chromium, 34 nickel, 3 to 4 copper, 2 to 3 molybdenum, and a small amount of columbium for stabilization against loss of corrosion resistance due to intergranular attack in welding. It resists sulfuric acid, especially in high concentrations at high temperatures, as well as acetate solvents, boric acid, cadmium and ferrous sulfates, and zinc chloride. Two other austenitics of this company are noted mainly for superior galling resistance in self-mated applications, general metal-to-metal wear resistance, and scale resistance at temperatures up to $1800°F$ ($982°C$). **Stainless steel Gall-Tough** contains 15 to 18% chromium, 4 to 6 each of nickel and manganese, 3 to 4 silicon, 0.08 to 0.2 nitrogen, and maximum amounts of 0.15 carbon, 0.04 phosphorus, and 0.04 sulfur. It is stronger and more oxidation-resistant than 304 stainless and, depending on the environment, comparable in corrosion resistance. Annealed bar has an ultimate tensile strength of about

161,000 lb/in^2 (1,110 MPa), a tensile yield strength of 60,000 lb/in^2 (414 MPa), and a Charpy V-notch impact strength of 240 ft · lb (325 J). At 800°F (427°C), the yield strength is 29,000 lb/in^2 (200 MPa). **Stainless steel Gall-Tough Plus** has 16.5 to 21% chromium, 6 to 10 nickel, 4 to 8 manganese, 2.5 to 4.5 silicon, 0.5 to 2.5 molybdenum, 0.05 to 0.25 nitrogen, and maximum amounts of 0.15 carbon and 0.04 each of phosphorus and sulfur. It is stronger than 316 stainless, with equivalent oxidation resistance and equivalent or superior corrosion resistance in chloride environments. Annealed bar has an ultimate tensile strength of 113,000 lb/in^2 (783 MPa), a tensile yield strength of 61,000 lb/in^2 (423 MPa), and a Charpy V-notch impact strength of 298 ft · lb (404 J). At 800°F (427°C), the yield strength is 30,000 lb/in^2 (208 MPa). Cold-drawn bar of Gall-Tough and Gall-Tough Plus is much stronger and less tough.

The wrought ferritic stainless steels are magnetic and less ductile than the austenitics. Although some can be hardened slightly by heat treatment, they are generally not hardenable by heat treatment. All contain at least 10.5% chromium, and although the standard alloys are nickel-free, small amounts of nickel are common in the nonstandard ones. Among the standard alloys, **stainless steel 430** is the general-purpose alloy. It contains 16 to 18% chromium and a maximum of 0.12 carbon, 1 manganese, 1 silicon, 0.04 phosphorus, and 0.03 sulfur. **Stainless steels 430F** and **430FSe,** the "free-machining" versions, contain more phosphorus (0.06% maximum) and sulfur (0.15 minimum in 430F, 0.06 maximum in 430FSe). Also, 430FSe contains 0.15% minimum selenium, and 0.6 molybdenum is an option for 430F. The other standard ferritics are **stainless steels 405, 409, 429, 434, 436, 442,** and **446.** Types 405 and 409 are the lowest in carbon (0.08% maximum) and chromium (11.5 to 14.5 and 10.5 to 11.75, respectively), the former containing 0.10 to 0.30 aluminum to prevent hardening on cooling from elevated temperatures, and the latter containing 0.75 maximum titanium. 429 is identical to 430 except for less chromium (14 to 16%) for better weldability. 434 and 436 are identical to 430 except for 0.75 to 1.25 molybdenum in the former and this amount of molybdenum plus 0.70 maximum columbium and tantalum in the latter, these additives improving corrosion resistance in specific environments. 442 to 446 are the highest in chromium (18 to 23% and 23 to 27, respectively) for superior corrosion and oxidation resistance, and in carbon (0.20 maximum). Also, 446 contains more silicon (1.50% maximum).

These standard alloys melt in the range of 2600 to 2790°F (1427 to 1532°C), have specific heats of 0.11 to 0.12 Btu/(lb · °F) [460 to 502 J/(kg · K)] thermal conductivities of 12 to 15.6 Btu/(ft · h · °F) [21 to 27 W/(m · K)] at 212°F (100°C), and electrical resistivities of 23 to 26

$\mu\Omega \cdot$ in (59 to 67 $\mu\Omega \cdot$ cm) at 70°F (21°C). In the annealed condition, tensile yield strengths range from 35,000 to 40,000 lb/in² (241 to 276 MPa) for 405 to as high as 60,000 lb/in² (414 MPa) for 434, with ultimate strengths of 65,000 to 85,000 lb/in² (448 to 586 MPa) and elongations of 20 to 33%. For 1% creep in 10,000 h at 1000°F (538°C), 430 has a stress-rupture strength of 8,500 lb/in² (59 MPa). Typical applications include automotive trim and exhaust components, chemical processing equipment, furnace hardware and heat-treating fixtures, turbine blades, and molds for glass.

There are many **specialty wrought ferritic stainless steels.** Crucible Stainless Steel's **Sea-Cure** is a low-carbon titanium-stabilized alloy containing 26% chromium, 3 molybdenum, and 2.5 nickel developed for seawater condenser tubing. A still-higher-chromium (29%) alloy for such applications is Allegheny Ludlum's vacuum-induction-melted **stainless steel 29Cr-4Mo,** which also finds use in chemical and petroleum refining equipment. Both alloys are quite resistant to chlorides. In **stainless steel 29Cr-4Mo-2Ni** alloy, another VIM product from Allegheny Ludlum, the nickel adds resistance to sulfuric acid. The nickel-bearing alloy is also stronger, 90,000 lb/in² (621 MPa) versus 64,000 lb/in² (441 MPa) in terms of tensile yield strength, but somewhat less ductile and slightly less resistant to stress corrosion. Still another VIM ferritic alloy from this firm is the 26Cr-1Mo alloy known as **E-Brite.** It is produced to carbon contents of only 0.001 to 0.002% and low nitrogen levels (0.010%) and is noted for high resistance to pitting and virtual immunity to stress corrosion in chloride and caustic media. **Stainless steel 3CR12,** a specialty ferritic of Cromweld Steels Ltd., contains 10.5 to 12% chromium and maximum amounts of 1.5 nickel, 1.5 manganese, 1 silicon, 0.6 titanium, 0.04 phosphorus, 0.3 sulfur, 0.03 nitrogen, and 0.03 carbon. Tensile strengths are 67,000 lb/in² (462 MPa) ultimate and 41,000 lb/in² (283 MPa) yield, with 20% elongation and good formability and weldability.

The specialty ferritics also include two kinds of virtually nickel-free **stainless steel 18Cr-2Mo** alloys: (1) low-interstitial, titanium- and/or columbium-stabilized sheet alloys and (2) resulfurized free-machining bar alloys. The low-interstitial type, also designated **stainless steel 444,** is insensitive to intergranular corrosion after welding or exposure to high temperatures, more resistant than the austenitic 304 to chloride pitting and crevice corrosion, and virtually immune to chloride-induced stress-corrosion cracking. Producers include Allegheny Ludlum, Crucible, and LTV Steel, and the alloy has been used for catalytic converters on light trucks, solar-panel collector plates, beverage storage and processing tanks, heat-treating equipment, and processing vessels for molten nonferrous metals. The free-machining bar alloys, such as **stainless steel 182-FM** from Carpenter Technology and **Uniloy 18-2FM** from Universal Cyclops, are similar in machinability to standard 416 stainless steel (a martensitic alloy having improved machinability) but superior to 303 in resistance to pitting and stress corrosion. They are used for screw-machine products.

Wrought martensitic stainless steels are also magnetic and, being hardenable by heat treatment, provide high strength. Of those in the **stainless steel 4XX series,** 410, which contains 11.5 to 13.0% chromium, is the general-purpose alloy. The others—403, 414, 416, 416Se, 420, 420F, 422, 431, 440A, and 440C—have similar (403, 414) or more chromium, 16 to 18% in the 440s. Most are nickel-free or, as in the case of 414, 422, and 431, low in nickel. Most of the alloys also contain molybdenum, usually less than 1%, plus the usual 1% or so maximum of manganese and silicon. Carbon content ranges from 0.15% maximum in 403 through 416 and 416Se, to 0.60 to 0.75 in 440A, and as much as 1.20 in 440C. 403 is the low-silicon (0.50% maximum) version of 410; 414 is a nickel (1.25 to 2.50)-modified version for better corrosion resistance. 416 and 416Se, which contain 12 to 14% chromium, also contain more than the usual sulfur or sulfur, phosphorus, and selenium to enhance machinability. 420 is richer in carbon for greater strength, and 420F has more sulfur and phosphorus for better machinability. 422, which contains the greatest variety of alloying elements, has 0.20 to 0.25% carbon, 11 to 13 chromium, low silicon (0.75 maximum), low phosphorus and sulfur (0.025 maximum), 0.5 to 1.0 nickel, 0.75 to 1.25 of both molybdenum and tungsten, and 0.15 to 0.3 vanadium. This composition is intended to maximize toughness and strength at temperatures to 1200°F (649°C). 431 is a higher-chromium (15 to 17%) nickel (1.25 to 2.50) alloy for better corrosion resistance. The high-carbon, high-chromium 440 alloys combine considerable corrosion resistance with maximum hardness. The **stainless steel 5XX series** of wrought martensitic alloys—501, 501A, 501B, 502, 503, and 504—contain less chromium, ranging from 4 to 6% in 501 and 502, to 8 to 10 in 501B and 504. All contain some molybdenum, usually less than 1%, and are nickel-free.

These steels are hardenable by heat treatment in a manner similar to that for alloy steels in general. Hardening temperatures range from 1600 to 1950°F (871 to 1066°C), and subsequent tempering is performed at temperatures of 300 to 1400°F (149 to 760°C). Annealing temperatures range from 1500 to 1650°F (816 to 899°C). For the 4XX alloys, Brinell hardness ranges from about 180 to 250 in the annealed condition and to about 250 to 600 in the hardened and tempered condition. In general, 403, 410, and 416 are the least hard and the 440s the hardest in either condition. In general, their physical properties, such as melting range, specific heat, thermal conductivity, and electrical resistivity, are not markedly different than those of the ferritic stainless steels. Tensile strengths, however, after hardening and tempering, are far greater. Most of the 4XX alloys can provide yield strengths greater than 150,000 lb/in^2 (1,034 MPa) and some, such as the 440s, more than 250,000 lb/in^2 (1,724 MPa). The martensitic stainless steels, however, are less machinable than the austenitic and ferritic alloys, and they are also less weldable. Forging temperatures range from 1900 to 2250°F (1038 to 1232°C). Most of the alloys are available in a wide range of mill forms, and typical applications include turbine blades, springs, knife blades and cutlery, instruments, ball bearings, valves and pump parts, and heat exchangers.

Stainless steel Pyrowear 675, of Carpenter Technology, is an alternative to 440C and carburizing alloy steels for gears and bearings. A double-vacuum-melted product, it contains 13% chromium, 5.4 cobalt, 2.6 nickel, 1.8 molybdenum, 0.65 manganese, 0.6 vanadium, 0.4 silicon, and 0.07 carbon. It is as corrosion-resistant as 440C and can be carburized to a case hardness of Rockwell C 60 while providing a tough core of Rockwell C 40. At this hardness the core has an ultimate tensile strength of 185,000 lb/in^2 (1,276 MPa), a tensile yield strength of 143,000 lb/in^2 (986 MPa), and 19% elongation. Its fracture toughness—140,000 to 150,000 lb/in$^2 \cdot \sqrt{\text{in}}$ (154 to 165 MPa $\cdot \sqrt{\text{m}}$)—is said to be better than that of any martensitic stainless steel. **Stainless steel DD400,** of Minebea Co. of Japan, is used in the United States for bearing balls and races. This high-carbon (0.61%) steel contains 12.9% chromium, 0.67 manganese, 0.32 silicon, 0.24 copper, 0.08 nickel, 0.008 molybdenum, and 0.007 aluminum. It is said to provide superior performance to 440C due to the absence of primary carbides after quenching and tempering.

The wrought **PH stainless steels,** also called **age-hardenable stainless steels,** date back to the 1940s and the development of **Stainless W** by United States Steel Corp. Three basic types are now available: austenitic, semiaustenitic, and martensitic. Regardless of the type, the final hardening mechanism is precipitation hardening, brought about by small amounts of one or more alloying elements, such as aluminum, titanium, copper, and, sometimes, molybdenum. Their principal advantages are high strength, toughness, corrosion resistance, and relatively simple heat treatment.

Of the **austenitic PH stainless steels, A-286** is the principal alloy. Developed by Allegheny Ludlum and also referred to as an iron-base superalloy, it contains about 15% chromium, 25 nickel, 2 titanium, 1.5 manganese, 1.3 molybdenum, 0.3 vanadium, 0.15 aluminum, 0.05 carbon, and 0.005 boron. It is widely used for aircraft turbine parts and high-strength fasteners. Heat treatment [solution treating at 1800°F (981°C), water or oil quenching, aging at 1325 to 1350°F (718 to 732°C) for 16 to 18 h and air cooling] provides an ultimate tensile strength of about 150,000 lb/in^2 (1,034 MPa) and a tensile yield strength of about 100,000 lb/in^2 (690 MPa), with 25% elongation and a Charpy impact strength of 64 ft · lb (87 J). The alloy retains considerable strength at high temperatures. At 1200°F (649°C), for example, tensile yield strength is 88,000 lb/in^2 (607 MPa). The alloy also has good weldability, and its corrosion resistance in most environments is similar to that of 3XX stainless steels.

The **semiaustenitic PH stainless steels** are austenitic in the annealed or solution-treated condition and can be transformed to a martensitic structure by relatively simple thermal or thermomechanical treatments. They are available in all mill forms, although sheet and strip are the most common. True semiaustenitic PH stainless steels include Armco's **PH 14-8Mo, PH 15-7Mo,** and **17-7PH.** Allegheny Ludlum's **AM-350** and **AM-355** are also so classified, although they are said not to truly have a precipitation-hardening reaction. The Armco steels are lowest in carbon content (0.04% nominally in PH 14-8Mo, 0.07 in the others). PH 14-8Mo also nominally contains 15.1% chromium, 8.2 nickel, 2.2 molybdenum, 1.2 aluminum, 0.02 manganese, 0.02 silicon, and 0.005 nitrogen. PH 15-7Mo contains 15.2% chromium, 7.1 nickel, 2.2 molybdenum, 1.2 aluminum, 0.50 manganese, 0.30 silicon, and 0.04 nitrogen. 17-7PH is similar to PH 15-7Mo except for 17% chromium and being molybdenum-free. AM-350 contains 16.5% chromium, 4.25 nickel, 2.75 molybdenum, 0.75 manganese, 0.35 silicon, 0.10 nitrogen, and 0.10 carbon. AM-355 has 15.5% chromium, 4.25 nickel, 2.75 molybdenum, 0.85 manganese, 0.35 silicon, 0.12 nitrogen, and 0.13 carbon. In the solution-heat-treated condition in which these steels are supplied, they are readily formable. They then can be strengthened to various strength levels by conditioning the austenite, transformation to martensite, and precipitation hardening. One such procedure, for 17-7 PH, involves heating at 1400°F (760°C), air cooling to 60°F (16°C), then heating to 1050°F (565°C) and air cooling to room temperature. In their heat-treated conditions, these steels encompass tensile yield strengths ranging from about 180,000 lb/in^2 (1,241 MPa) for AM-355 to 260,000 lb/in^2 (1,793 MPa) for PH 15-7Mo.

After solution treatment, the **martensitic PH stainless steels** always have a martensitic structure at room temperature. These steels include the progenitor of the PH stainless steels, **Stainless W,** Armco's **PH 13-8Mo, 15-5 PH,** and **17-4 PH,** and Carpenter Technology's **Custom 455.** Of these, **PH 13-8Mo** and **Custom 455,** which contain 11 to 13% chromium and about 8 nickel plus small amounts of other alloying elements, are the higher-strength alloys, providing tensile yield strengths of 210,000 lb/in^2 (1,448 MPa) and 235,000 lb/in^2 (1,620 MPa), respectively, in bar form after heat treatment. The other alloys range from 15 to 17% in chromium and 4 to 6 in nickel, and typically have tensile yield strengths of 175,000 to 185,000 lb/in^2 (1,207 to 1,276 MPa) in heat-treated bar form. They are used mainly in bar form and forgings, and only to a small extent in sheet. Age hardening, following high-temperature solution treating, is performed at 800 to 1250°F (427 to 677°C).

Cast stainless steels are divided into two classes: those intended primarily for uses requiring corrosion resistance and those intended mainly for uses requiring heat resistance. Both types are commonly known by the designations of the Alloy Casting Institute of the Steel Founders Society of America, and these designations generally begin with the letter C for those used mainly for corrosion resistance and with the letter H for those used primarily for heat resistance. All are basically iron-chromium or iron-chromium-nickel alloys, although they may also contain several other alloying ingredients, notably molybdenum in the heat-resistant type, and molybdenum, copper, and/or other elements in the corrosion-resistant type. The **corrosion-resistant cast stainless steel** type follows the general metallurgical classifications of the wrought stainless steels, that is, austenitic, ferritic, austenitic-ferritic, martensitic, and precipitation hardening. Specific alloys within each of these classifications are austenitic (CH–20, CK–20, CN–7M), ferritic (CB–30 and CC–50), austenitic-ferritic (CE–30, CF–3, CF–3A, CF–8, CF–8A, CF–20, CF–3M, CF–3MA, CF–8M, CF–8C, CF–16F, and CG–8M), martensitic (CA–15, CA–40, CA–15M, and CA–6NM), and precipitation hardening (CB–7Cu and CD–4MCu). The chromium content of these alloys may be as little as 11% or as much as 30, depending on the alloy. The **heat-resistant cast stainless steel** types may contain as little as 9% chromium (Alloy HA), although most contain much greater amounts, as much as 32 in HL. Although nickel content rarely exceeds chromium content in the corrosion-resistant type, it does in several heat-resistant types (HN, HP, HT, HU, HW, and HX). In fact, nickel is the major ingredient in HU, HW, and HX. Several of the heat-resistant types can be used at temperatures as high as 2100°F (1149°C).

Diffused stainless steel is a sheet steel with a low-carbon ductile steel core and a diffused chromium-iron alloy surface. It is produced from low-carbon steel by heating the sheets in a retort containing a chromium compound, which diffuses into the metal at a temperature of about 2000°F (1093°C). The chromium alloys with the steel, the alloy on the surface containing as much as 40% chromium, which tapers off to leave a ductile nonchromium core in the sheet. **Black stainless steel,** for electronic applications, is produced by immersing sheet steel in a bath of molten potassium dichromate and sodium dichromate. The steel has a shiny, black finish.

Metal fibers are used for weaving into fabrics for arctic heating clothing, heated draperies, chemical-resistant fabrics, and reinforcement in plastics and metals. **Stainless steel yarn** made from the fibers is woven into **stainless steel fabric** that has good crease resistance and retains its physical properties to 800°F (427°C). The fiber may be blended with cotton or wool for static control, particularly for carpeting. Fibers of 316 or 347 stainless steels are used in **fiber metal,** also known as **Feltmetal,** for noise reduction.

STARCH. A large group of natural carbohydrate compounds of the empirical formula $(C_6H_{10}O_5)_x$, occurring in grains, tubers, and fruits. The common cereal grains contain from 55 to 75% starch, and potatoes contain about 18%. Starches have a wide usage for foodstuffs, adhesives, textile and paper sizing, gelling agents, and fillers; in making explosives and many chemicals; and for making biodegradable detergents such as sodium tripolyphosphate. Starch is a basic need of all peoples and all industries. Much of it is employed in its natural form, but it is also easily converted to other forms, and more than 1,000 different varieties of starch are usually on the U.S. market at any one time.

Most of the commercial starch comes from corn, potatoes, and mandioca. Starches from different plants have similar chemical reactions, but all have different granular structure, and the differences in size and shape of the grain have much to do with the physical properties. **Cornstarch** has a polygonal grain of simple structure. It is the chief food starch in the western world, although sweet-potato starch is used where high gelatinization is desired, and tapioca starch is used to give quick tack and high adhesion in glues. **Tapioca starch** has rounded grains truncated on one side and is of lamellar structure. It produces gels of clarity and flexibility, and because it has no cereal flavor, it can be used directly for thickening foodstuffs. **Rice starch** is polygonal and lamellar, and has very small particles. It makes an opaque stiff gel and is also valued as a **dusting starch** for bakery products, although it is expensive for this purpose. **White-potato starch** has conchoidal or ellipsoidal grains of lamellar structure. When cooked, it forms clear solutions easily controlled in viscosity, and gives tough, resilient films for coating paper and fabrics. Prolonged grinding of grain starches reduces the molecular chain, and the lower weight then gives greater solubility in cold water. Green fruits, especially bananas, often contain much starch, but the ripening process changes the starch to sugars.

In general, starch is a white, amorphous powder having a specific gravity from 0.499 to 0.513. It is insoluble in cold water but can be converted to **soluble starch** by treating with a dilute acid. When cooked in water, starch produces an adhesive paste. Starch is easily distinguished from dextrins as it gives a blue color with iodine while dextrins give violet and red. The starch molecule is often described as a chain of glucose units, with the adhesive **waxy starches** as those with coiled chains. But starch is a complex member of the great group of natural plant compounds consisting of starches, sugars, and cellulose, and originally named carbohydrates because the molecular formula could be written as $C_n(H_2O)_x$; but not all now-known carbohydrates can be classified in this form, and many now-known acids and aldehydes can be indicated by this formula.

Starch can be fractionated into two polymers of high molecular weight. **Amylose** is a straight-chain fraction having high adhesive properties for coatings and sizings, and **amylopectin** is a branched-chain fraction best known as a suspending agent for foodstuffs. Amylose is chemically identical with cellulose, but the chain units of the molecule have an alpha linkage and are coiled, while the cellulose molecule is rigid. It has a molecular weight of 150,000, while amylopectin has a molecular weight above 1 million. The 1–4 alpha linkage of amylopectin with random branches at the 6-carbon position makes the material easily dispersible in cold water but resistant to gelling. Amylopectin is thus best suited for thickening, but because it can be combined and cross-linked with synthetic resins and is highly resistant to deterioration, it is used with resins for water-resistant coatings for paper and textiles.

Tapioca is the starch from the root of the large tuber *Manihot utilissima,* now grown in most tropical countries. It is called **cassava** in southern Asia, **manioc** in Brazil, **mandioca** in Paraguay, and **yuca** in Cuba. This perennial vegetatively propagated shrub was cultivated as far back as 2,500 years ago, and there is some indirect evidence that it has been grown for 4,000 years in the Americas. Its fresh roots contain 30 to 40% dry matter and have a starch content of approximately 85% of the dry matter. It is used in enormous quantities for food in some countries, and in some areas much is used for the production of alcohol. In the United States it is valued for adhesives and coatings, and only a small proportion in globules and flakes, known as **pearl tapioca,** is used in foodstuffs. **Gaplek,** used for cattle feed in Asia, is not the starch, but is dried and sliced cassava root. Tapioca starch may be sold under trade names. **Kreamgel,** used as a thickener for canned soups, sauces, and pastries, is refined tapioca that gives clear solutions without imparting odor or flavor.

Potato starch, produced from the common **white potato,** *Solanum tuberosum,* has been the most important starch in Europe, but in the United States it is usually more expensive than cornstarch.

It forms heavier hot pastes than tapioca. It is also free of flavor and is used as a thickener in foods. It does not crystallize easily. **Arogum,** of Morningstar-Paisley, Inc., is potato starch used to give tough, resilient coatings on paper and textiles, and **Arojel P** is pregelatinized potato starch used as a beater additive to improve the strength and scuff resistance of kraft paper. **Sweet-potato starch** is from the tuber *Opomoea batata.* An average of 10 lb (4.5 kg) of starch is produced per bushel. The root has poor shipping qualities, and the starch is expensive, but it has excellent colloidal qualities and gelatinizes completely at 165°F (74°C). It is used in some foodstuffs. It has a pleasant, sweetish flavor, and in Latin countries great quantities are marketed in the form of a stiff gel as a dessert sweet known as **dulce de batata.**

Arrowroot starch is from the tubers of the *Maranta arundinacea* of the West Indies. It is easily digested and is used in cookies and other food products, especially baby foods. **Florida arrowroot** is from *Zamia floridana.* **East Indian arrowroot** is from the plant *Curcuma angustifolia,* which belongs to the ginger family. Arrowroot from St. Vincent, used in instant-pudding mixes and icings, is marketed as a precooked powder of about 200 mesh. It swells in cold water and does not add flavor.

The starches do not crystallize as sugar does, and they may be added to some confections to minimize crystallization. They are also used as binders in candies and in tablet sugar, but any considerable quantity in such products is considered as an adulterant. Metabolism of starch in the human system requires conversion to sugars, and the taking in of excessive quantities of uncooked starch is undesirable. **Modified starches** are starches with the molecule altered by chemical treatment to give characteristics suitable for particular industrial requirements. The modified starches and especially prepared starches are usually sold under trade names. **Superlose** is amylose from cornstarch, and **Auperlose** is amylose from potato starch. **Ramalin** is amylopectin. **Amylon,** of National Starch & Chemicals Corp., is cornstarch containing 57% amylose, and **Kosul** is cornstarch high in amylopectin. **Textaid,** of the same company, is a modified starch which reacts with water to form a grainy structure. It is used in comminuted meat products to give a firm texture. The **ColFlo** thickening agents, stable and soluble in frozen foods, are modified, waxy cornstarches, high in amylopectin. **Pregelatinized starches** are preheat-treated starches that require no cooking for use in dry food mixes or adhesives. **Snow Flake starch** is a cornstarch of this type.

Wheat starch is a fine, white starch made by separating out the gluten of wheat flour by wash flotation. It is used in prepared mixes for foam-type cakes and pie crusts to improve texture, add volume, and reduce the amount of shortening needed. It replaces up to 30% of the wheat-flour content of the mix. **Starbake starch,** of Hercules, is wheat starch. **Paygel,** of General Mills, is also wheat starch, but **alant starch,** or **inulin,** $(C_6H_{10}O_5)_6 \cdot H_2O$, is not a starch in the ordinary sense, but is an insoluble sugar which occurs as the reserve polysaccharide in many plants. It is obtained from the roots of the **artichoke,** *Helianthus tuberosis,* native to America but now grown widely in Europe. Unlike starch, the molecule has fructose units held in glucoside linkage, and hydrolysis converts it to fructose.

Starch acetate, or **acetylated starch,** is used for textile sizing, in adhesives, and for greaseproofing paper. The insertion of acetate radicals reduces the tendency of the molecular chains to cling together. The acetylated starches are gums which gelatinize at lower temperatures than starch, and produce stable, nonlumping pastes which give strong, flexible films. **Miralloid** and **Mira-Film,** of A. E. Staley Mfg. Co., are acetylated cornstarches. **Morgum** is a hydroxyethyl **etherized starch** which gives high film strength in coatings. The **Kofilms** of General Mills are acetylated cornstarches which give greaseproof, craze-resistant coatings on paper and textiles.

Laundry starches are usually ordinary starches, but silicone resin emulsions may be added to starches to permit higher ironing temperatures, improve slipperiness, and improve the hand of the starched fabric. The so-called **permanent starches,** for household use, that are not removed by washing, are not starch, but are emulsions of polyvinyl acetate. **Oxidized starch,** a resistant starch for coatings, is made by the chloro-oxidation of a starch solution. **Sumstar 190** is a **diallyl starch** made by acid oxidation of cornstarch. Small amounts of the powder added to kraft, tissue, or toweling pulp increase the wet and dry strengths and the folding endurance of the papers. An **ammoniated starch** called **Q-Tac starch** is cornstarch reacted with quaternary ammonium groups. A less than 1% solution improves paper strength. **Sulfonated starches** are used as dirt-suspending agents with detergents for cleaning textiles. **Nu-Film** is a starch of this type. **Clear Flo starch** is a modified starch containing a carboxyl group and a sulfonic acid group in the molecule. It has high hydrating capacity and gelatinizes sharply at low temperatures. It is used in adhesives and water paints. **Cato starch** is a carboxymethyl starch used in paper sizing to add strength. **Dry Flo starch** is modified to contain a **hydrophobic radical,** such as CH_2, which makes the material insoluble in water but soluble in oils. It is used in paints.

Many enzymes hydrolyze starch to maltose, but some enzymes convert the starch to the hard, tough glucosides known as **mannans,** such as the **mannose** of the ivory nut. **Phospho mannan,** produced by the fermentation of starch, is such a material used in adhesives.

Granular starch, used in enzyme conversion processing, is in dense, granular particles produced by flash drying. **Easy-Enz starch** is such a starch. **Cationic starch** is a starch with the molecules of stable negative polarity to give higher adhesion on the cellulose fibers of paper or textiles. **Molding starch,** for adding to sugar candies to give sharp molding characteristics, is starch containing an edible oil.

The **phosphate starch** of American Maize Products Co. is an orthophosphate ester of cornstarch, marketed in sodium salt form as a light-tan, dry powder. It has high thickening power and makes a clearer paste than cornstarch. It has superior water-binding properties at low temperatures. Frozen foods made with it do not curdle or separate when thawed, and canned foods thickened with the starch can be stored for long periods without clouding. It is also used as a briquetting binder for charcoal.

Starch sponge is an edible starch in the form of a coarse-textured, porous, crispy, spongelike material, used for confections by impregnating with chocolate or sweets. In crushed form it is added to candy or cookies. It is produced by freezing, thawing, and pressing starch paste. The freezing insolubilizes the starch so that no soluble starch goes off when the water is pressed out. **Lycasin** and **Polysorb** are hydrogenated starch hydrolysates produced by Roquette Corp. for food and feed applications. **Nitrostarch,** or **starch nitrate,** $C_{12}H_{12}O_{10}(NO_2)_3$, is a fine, white powder made by treating starch with mixed acid. It is highly explosive and is used for blasting, as a military explosive, and in signal lights. **Grenite** is nitrostarch mixed with an oil binder for use in grenades. **Trojan explosive** is a mixture of 40% nitrostarch with ammonium and sodium nitrates and some inert material to reduce the sensitiveness. **Sepol** starch coagulants, of Grace Dearborn, are used to break emulsions in waste treatment of lubricating oils, soluble oils, and synthetic coolants, including oils containing dissolved solids and water.

STATUARY BRONZE. Copper alloys used for casting statues, plaques, and ornamental objects that require fine detail and a smooth, reddish surface. Most of the famous large bronze statues of Europe contain from 87 to 90% copper, with varying amounts of tin, zinc, and lead. Early Greek statues contained from 9 to 11% tin with as much as 5% lead added apparently to give greater fluidity for crisp details. A general average bronze will contain 90% copper, 6 tin, 3 zinc, and 1 lead. Statuary bronze for cast plaques used in building construction contains 86% copper, 2 tin, 2 lead, 8 zinc, and 2 nickel. The nickel improves fluidity and hardens and strengthens the alloy, and the lead promotes an oxidized finish on exposure. The statuary bronze used for hardware has 83.5% copper, 4 lead, 2 tin, and 10 zinc.

STEARIC ACID. A hard, white, waxlike solid of composition $CH_3(CH_2)_{16}COOH$, obtained from animal and vegetable fats and oils by splitting and distilling. The hard cattle fats are high in stearic acid, but other fats and oils contain varying amounts. It is also called **octodecanoic acid,** and it can be made by hydrogenation of oleic acid. Stearic acid has a specific gravity of 0.922 to 0.935 and a melting point of about 130°F (54°C), and it is soluble in alcohol but insoluble in water. It is marketed in cakes, powder, and flakes. **Emory 3101-D** is **isostearic acid** which has the solubility and physical properties of oleic acid while retaining the heat and oxidation stability of stearic acid. **Pearl stearic acid** is the material in free-flowing bead powder. The acid is used for making soaps, candles, paint driers, lubricating greases, and buffing compositions, and for compounding in rubbers, cosmetics, and coatings.

Successive pressings remove liquid oils, thus raising the melting point and giving a whiter, harder product of lower iodine value. **Oleo oil** is a yellow oil obtained by cold-pressing the first-run cattle tallow. **Tallow oil** is the oil following the first two grades of oleo oil. **Industrene 4518** is the single-pressed grade, available as a molten liquid or in flakes, from Humko Chemical Div. of Witco Corp. **Industrene 5016** is the double-pressed variety. **Oleostearin,** used for treating leather, is the stearin remaining after extraction of the oils.

Stearin is the glyceride of stearic acid. **Acetostearins** are the monoglycerides acetylated with acetic anhydride. They are closely related to fats, but are nongreasy and are plastic even at low temperatures. The highly acetylated stearins melt below body temperature and are edible. Acetostearins are used as plasticizers for waxes and synthetic resins to improve low-temperature characteristics. **Stearite** is a trade name for **synthetic stearic acid** made by the hydrogenation of unsaturated animal and fish oils. It is used in rubber compounding, as it is more uniform than ordinary stearic acid. **Hystrene,** of Humko Chemical, is purified and hardened stearic acid in grades of 70, 80, and 97% stearic acid, with the remainder palmitic acid, used for candles, cosmetics, and stearates. However, **Hystrene 5016** is a triple-pressed oil. **Intarvin** is a synthetic edible fat made from stearic acid by converting it to **margaric acid,** or **dauric acid,** $C_{16}H_{33}COOH$, and then esterifying with glycerin. It is used as a fat for diabetics as it does not undergo the beta oxidation to lose two carbon atoms at a time and produce acetoacetic acid in the system as do the even-carbon food acids.

Wilmar 272 is refined stearic acid in flake form for use in candles and coatings. **Hydrofol** is a double- and triple-pressed rubber grade that is also used in coatings and candles. It is produced by Sherex

Chemical Co. **Flexchem B** is **sodium stearate,** $NaC_{18}H_{35}O_2$, in the form of a water-soluble, white powder which is insoluble in oils. It is used as a bodying agent in cosmetic creams. **Myvacet** is an acetostearin used as an edible plastic coating for poultry, cheese, and frozen fish and meats to prevent loss of the natural color and flavor. It is a white, waxy solid with melting points from 99 to 109°F (37 to 43°C), but it also comes as an oil with congealing point of 45°F (7°C) for use as a release agent on bakery equipment. **Alfol** is a high-purity, synthetic, linear primary alcohol from Vista Chemical Co. A similar stearyl alcohol, **Adol,** is from Sherex Chemical and is used as a chemical intermediate. **Cachalot** is a food-grade product of M. Michel and Co.

Stearin pitch is a brown-to-black by-product residue obtained in the splitting and distillation of fats and oils in the manufacture of soaps, candles, and fatty acids. While the word *stearin* implies that it contains only stearic acid, it usually comes from a variety of oils and has mixed acids, and it may take the name of the oil, such as **linseed pitch** or **palm pitch.** It is used in varnishes and cold-molding compositions.

STEEL. Iron alloyed with small amounts of carbon, 2.5% maximum, but usually much less. The two broad categories are **carbon steels** and **alloy steels,** but they are further classified in terms of composition, deoxidation method, mill-finishing practice, product form, and/or principal characteristics. Carbon is the principal influencing element in carbon steels, although manganese, phosphorus, and sulfur are also present in small amounts, and these steels are further classified as **low-carbon steels** (up to 0.30% carbon), **medium-carbon steels** (0.30 to 0.60), and **high-carbon steels** (more than 0.60). The greater the amount of carbon, the greater the strength and hardness, and the less the ductility. Alloy steels are further classified as **low-alloy steels, alloy steels,** and **high-alloy steels,** those having as much as 5% alloy content being the most widely used. The most common designation systems for carbon and alloy steels are those of the American Iron and Steel Institute and the SAE, which follow a four- or five-digit numbering system based on the key element or elements, with the last two digits indicating carbon content in hundredths of a percent.

Plain carbon steels (with 1% maximum manganese) are designated 10XX; resulfurized carbon steels, 11XX; resulfurized and rephosphorized carbon steels, 12XX; and plain carbon steels with 1 to 1.65% manganese, 15XX. Alloy steels include manganese steels (13XX), nickel steels (23XX and 25XX), nickel-chromium steels (31XX to 34XX), molybdenum steels (40XX and 44XX), chromium-molybdenum steels (41XX), nickel-chromium-molybdenum steels (43XX,

47XX, and 81XX to 98XX), nickel-molybdenum steels (46XX and 48XX), chromium steels (50XX to 52XX), chromium-vanadium steels (61XX), tungsten-chromium steels (72XX), and silicon-manganese steels (92XX). The letter B following the first two digits designates boron steels, and the letter L leaded steels. The suffix H is used to indicate steels produced to specific hardenability requirements. **High-strength, low-alloy** steels are commonly identified by a 9XX designation of the SAE, where the last two digits indicate minimum tensile yield strength in 1,000 lb/in^2 (6.9 MPa).

In contrast to **rimmed steels,** which are not deoxidized, **killed steels** are deoxidized by the addition of deoxidizing elements, such as aluminum or silicon, in the ladle prior to ingot casting. Thus we have such terms as **aluminum-killed steel.** Deoxidation markedly improves the uniformity of the chemical composition and resulting mechanical properties of mill products. **Semikilled steels** are only partially deoxidized, thus intermediate in uniformity to rimmed and killed steels. **Capped steels** have a low-carbon steel rim characteristic of rimmed-steel ingot and central uniformity more characteristic of killed-steel ingot, and are well suited for cold-forming operations.

Steels are also classified as air-melted, vacuum-melted, or vacuum-degassed. **Air-melted steels** are produced by conventional melting methods, such as open hearth, basic oxygen, and electric furnace. **Vacuum-melted steels** are produced by induction vacuum melting and consumable electrode vacuum melting. **Vacuum-degassed steels** are air-melted steels that are vacuum processed before solidification. Vacuum processing reduces gas content, nonmetallic inclusions, and center porosity and segregation. Such steels are more costly, but have better ductility and impact and fatigue strengths.

Steel-mill products are reduced from ingot into such forms as blooms, billets, and slabs, which are then reduced to finished or semifinished shape by hot-working operations. If the final product is produced by hot working, the steel is known as **hot-rolled steel.** If the final product is shaped cold, the steel is known as **cold-finished steel** or, more specifically, **cold-rolled steel,** or **cold-drawn steel.** Hot-rolled mill products are usually limited to low- and medium-nonheat-treated carbon steels. They are the most economical steels, have good formability and weldability, and are widely used. Cold-finished steels, compared with hot-rolled products, have greater strength and hardness, better surface finish, and less ductility. Wrought steels are also classified in terms of mill-product form, such as **bar steels, sheet steels,** and **plate steels. Cast steels** refer to those used for castings, and **PM (powder metal) steels** refer to powder compositions used for PM parts. Steels are also known by their key characteristic from the standpoint of application, such as **electrical steels, corrosion-resis-**

tant stainless steels, low-temperature steels, high-temperature steels, boiler steels, pressure-vessel steels, etc.

STEEL POWDER. Powder used mainly for the production of steel PM parts made by consolidating the powder under pressure and then sintering, and, to a limited extent, for steel-mill products, principally tool-steel bar products. For PM parts, the powder may be admixed for the desired composition or prealloyed; that is, each powder particle is of the desired composition. For mill products, prealloyed powder is used primarily. Steel powder is widely used to make small to moderate-size PM parts, having compositions closely matching those of wrought steels. Among the more common are carbon steels, copper steels, nickel steels, nickel-molybdenum steels, and stainless steels.

Ancorsteel 41 AB, of Hoeganaes Corp., is a premixed, highly compressible, low-alloy steel powder containing 0.5% carbon, 0.9 manganese, 0.85 molybdenum, and 0.75 chromium. Formed parts are intended for surface hardening by carburizing, nitriding, carbonitriding, or nitrocarburizing. **Stainless Steel Plus,** of the Specialty Metals Division of Ametek, are powders of 303L, 304L, or 316L stainless steels blended with 10% powder of 15 nickel, 8 tin and copper. They are said to provide greater corrosion resistance than conventional stainless steel powders. Powders for injection-molding PM parts are often of iron-nickel or stainless steel but of very fine particle size.

STEEL WOOL. Long, fine fibers of steel used for abrading, chiefly for cleaning utensils and for polishing. It is made from low-carbon wire that has high tensile strength, usually having 0.10 to 0.20% carbon and 0.50 to 1 manganese. The wire is drawn over a track and shaved by a stationary knife bearing down on it, and it may be made in a continuous piece as long as 100,000 ft (30,480 m). Steel wool usually has three edges but may have four or five, and strands of various types are mixed. There are nine standard grades of steel wool, the finest of which has no fibers greater than 0.005 in (0.0027 cm) thick, the most commonly used grade having fibers that vary between 0.002 and 0.004 in (0.006 and 0.010 cm). Steel wool comes in batts, or in flat ribbon form on spools usually 4 in (10 cm) wide. **Stainless steel wool** is also made, and **copper wool** is marketed for some cleaning operations.

STILLINGIA OIL. A drying oil obtained from the kernels of the seeds of the tree *Stillingia sebifera,* cultivated in China and the southern United States. The seeds contain about 23% of a light-yellow oil resembling linseed oil but of somewhat inferior drying power. The oil has a specific gravity of 0.943 to 0.946 and iodine value of 160. It has the peculiar property of expanding with great force at the congealing point. Stillingia oil is edible, but deteriorates rapidly, becoming bitter in taste and disagreeable in odor. **Stillingia tallow,** also known as **Chinese vegetable tallow,** is obtained by pressing from the coating, or mesocarp, of the seeds, yielding about 25 to 35% fat. Sometimes the whole seed is crushed, producing a softer fat than the true tallow. The tallow contains palmitic and oleic acids and is used in soaps and for mixing with other waxes. Some stillingia trees are grown in Texas.

STRIPPABLE COATINGS.   Coatings that are applied for temporary protection and can be readily removed. They are composed of such resins as cellulosics, vinyl, acrylic, and polyethylene; they can be water-base, solvent-base, or hot-melt. The choice of base depends on the surface to be protected. Water-base grades are neutral to plastic and painted surfaces, whereas solvent-base types affect those surfaces. Clear vinyl **strippable coatings,** perhaps the most widely used, are usually applied by spraying in thicknesses of 0.03 to 0.04 in (0.08 to 0.10 cm). **Acrylic strippable coatings** impart a clear, high-gloss, high-strength, temporary film to metal parts. **Polyethylene strippable coatings** are relatively low-cost and can be used on almost all surfaces except glass. **Cellulosic strippable coatings** are designed for hot-dip application. Film thicknesses range widely and can go as high as 0.2 in (0.51 cm). The mineral oil often present in these coatings exudes and coats the metal surface to protect it from corrosion over long periods.

STRONTIUM.   A metallic element of the alkaline group. It occurs in the minerals **strontianite,** $SrCO_3$, and **celestite,** $SrSO_4$, and resembles barium in its properties and combinations, but is slightly harder and less reactive and is not as white in color. It has a specific gravity of 2.54 and a melting point of about 1418°F (770°C), and it decomposes in water. The metal is obtained by electrolysis of the fused chloride, and small amounts are used for doping semiconductors. Its compounds have been used for deoxidizing nonferrous alloys, and were used in Germany for desulfurizing steel. But the chief uses have been in signal flares to give a red light, and in hard, heat-resistant greases. **Strontium 90,** produced atomically, is used in ship-deck signs as it emits no dangerous gamma rays. It gives a bright sign, and the color can be varied with the content of zinc, but it is short-lived. Strontium is very reactive and used only in compounds.

   Strontium nitrate is a yellowish-white, crystalline powder, $Sr(NO_3)_2$, produced by roasting and leaching celestite and treating with nitric acid. The specific gravity is 2.96, the melting point is 1193°F (645°C), and it is soluble in water. It gives a bright, crimson flame and is used in railway signal lights and in military flares. It is also used as a source of oxygen. The **strontium sulfate** used as a brightening agent in paints is powdered celestite. The ore of Nova Scotia contains 75% strontium sulfate. **Strontium sulfide,** SrS, used in luminous paint, gives a blue-green glow, but it deteriorates rapidly unless sealed in. **Strontium carbonate,** $SrCO_3$, is used in pyrotechnics, ceramics, and ceramic permanent magnets for small motors. **Strontium hydrate,** $Sr(OH)_2 \cdot 8H_2O$, loses its water of crystallization at 212°F (100°C) and melts at 707°F (375°C). It is used in making lubricating greases and as a stabilizer in plastics. **Strontium fluoride** is produced in single crystals for use as a laser material. When doped with samarium, it gives an output wavelength around 25,600 nin (650 nm).

STYRAX. A grayish-brown, viscous, sticky, aromatic balsam obtained from the small tree *Liquidambar orientalis* of Asia Minor. It is also called **Levant styrax.** It is used in cough medicines and for skin diseases, as a fixative for heavy perfumes, and for flavoring tobacco and soaps. American styrax is obtained by tapping the sweet gum, *L. styraciflua,* of Alabama, a tree producing 8 oz (0.2 kg) of gum per year. It is a brownish semisolid and has the same uses as Levant styrax. It is shipped from Central America under the name **liquidambar,** and in the southern United States is called **sweet gum** and **storax.** The gum is not present in large amounts in the wood, but its formation is induced by cuts. **Benzoin** is another balsam obtained from several species of *Styrax* trees. It is a highly aromatic solid with an odor like vanilla, and is used in medicine and in perfumes and incense. **Sumatra benzoin** is from the tree *S. benzoin* and comes in reddish-brown lumps or tears. In medicine it was originally called **gum Benjamin. Siam benzoin,** from southern Asia, is from the trees *S. tonkinense* and *S. benzoides.* It is in yellowish or brownish tears. The Sumatra benzoin contains cinnamic acid, while the Asiatic gum contains benzoic acid. **Benzoic acid,** or **phenylformic acid,** C_6H_5COOH, formerly produced from benzoin, is now made synthetically from benzol and called **carboxybenzene.** It is a white, crystalline solid melting at 252°F (122°C), soluble in water and in alcohol. It is used as a food preservative, as an antiseptic, for flavoring tobacco, as a weak acid mordant in printing textiles, and in the manufacture of dystuffs, pharmaceuticals, and cosmetics. Because it is poisonous, not more than 0.1% is used in food preserving in the form of its salt, **benzoate of soda,** or **sodium benzoate,** C_6H_5COONa, which is a white, crystalline powder. A potassium salt is also available from Mallinckrodt, Inc. **Sorbic acid,** $CH_3CH:(CH)_2:CHCOOH$, a solid melting at 273°F (134°C), occurs in unripe apples, but is made synthetically. As a preservative and antimold agent it is more effective than benzoic acid, is nontoxic, and is readily absorbed in the human system. It is used in cheese and other foods. **Sorbistat** is a food grade with antimicrobial activity against yeast, mold, and bacteria from Ashland Chemical Co. For food preservation it is used in the form of the water-soluble salt **potassium sorbate.** In a concentration of 0.2% it does not affect taste or aroma. **Preservastat,** a sorbic acid from Tri-K Industries, Inc., can be used at approximately 25% lower levels than potassium sorbate to achieve the same results. It is available as a powder, as granules, or in crystal form. **Anisic acid,** $CH_3OC_6H_4COOH$, used for pharmaceuticals, is the methyl ether of hydroxybenzoic acid. It is produced synthetically from carbon tetrachloride and phenol, and is a solid melting at 363°F (184°C). It is also called **methoxybenzoic acid, umbellic acid,** and **dragonic acid.**

SUEDE.   Also called **napped leather.** A soft-finished, chrome-tanned leather made from calf, kid, or cowhide splits, or from sheepskin. It is worked on a staking machine until it is soft and supple, and then buffed or polished on an abrasive wheel. It has a soft nap on the polished side and may be dyed any color. Suede is used for shoe uppers, coats, hats, and pocketbooks, but is now largely imitated with synthetic fabrics. **Artificial suede,** or **Izarine,** of Atlas Powder Co., has a base of rubber fabric. Fine cotton fibers dyed in colors are cemented to one side, and the underside of the sheet is beaten to make the fibers stand out until the cement hardens. The fabric looks and feels like fine suede. Some suede is also made by chemical treatment of sheepskins without staking. It has a delicate softness, but is not as wear-resistant as calfskin.

SUGAR.   A colorless to white or brownish, crystalline, sweet material produced by evaporating and crystallizing the extracted juice of the sugarcane or the sugar beet. Refined sugar is practically pure **sucrose,** $C_{12}H_{22}O_{11}$, and in addition to being a sweetening agent for many foods it is a valuable carbohydrate food and a food preservative. When used with cooked fruits to make jams and jellies, it is both a preservative and an added food. Lack of sugar in the diet develops ketosis, the disease of diabetics, and results in the wasting away of muscles, using up of reserve fats, and the production of poisonous ketones. When the blood-sugar level is low, a feeling of hunger is induced which may not be satisfied even by overeating. A small amount of sugar curbs the appetite and obviates surplus eating of proteins and fats that create obesity. Natural **brown sugar** contains about 2% of the minerals found in the plant, calcium, iron, phosphorus, magnesium, and potassium, and although these are valuable as foods, they are lost in the refining process.

Sugar is at present most valued as a food and for the production of by-product alcohol from the residue molasses, but the sucrose molecule is a convenient starting point for the production of many chemicals. However, the production and distribution of sugar have been hemmed in by restrictive laws based on its use for food. The sucrose molecule has two complex rings, a glucosido and a fructose. It can be regarded as a type of **fructosido-glucose,** but the fructose in sucrose has a different structure, a **furanose,** or five-ring form, instead of the **pyranose,** or six-ring structure, of ordinary fructose. Hydrolysis of sucrose with acid gives dextrorotatory glucose and fructose, and the mixture is called **invert sugar. Numoline** is a noncrystallizing invert sugar made by hydrolyzing sucrose to split the molecule into levulose and dextrose. It is used in confectionery and bakery products. Oxidation of sucrose produces oxalic acid and **saccharic acid,** $(HCOH)_4(COOH)_2$, which can be reduced to adipic acid. Glycerin can be made from sugar by hydrogenation to sorbitol and then splitting. Thus, because of the great versatility of the sucrose molecule, and the ease with which the sugar can be grown, sugar is one of the most valuable chemical raw materials. **Sucrose benzoate** is a benzoic acid derivative of sucrose used as a plasticizer and modifier for synthetic resins for lacquers and inks.

Sugarcane, *Saccharum officinarum,* is a tropical plant, originating in Asia and first brought to the Canary Islands in 1503 and thence to the West Indies. The plant will not withstand frost, but can be grown in a few favored regions outside of the tropics such as Louisiana. It is now grown on plantations in Cuba, Hawaii, Brazil, the Philippines, Indonesia, Puerto Rico, Peru, and many other countries. The cane or stalks of the plant are crushed to extract the juice, which is then concentrated by boiling, crystallized, and clarified with activated carbon or other material. The yield of sugar in Hawaii is about 14 tons (12,698 kg) of raw sugar per acre (4,047 m²). Analysis of sugarcane gives an average of 13.4% sucrose by weight of cane. The average yield by milling is 91% of the contained sucrose, but yields as high as 98.8% are obtained by diffusion extraction of the cut-cane chips.

The **sugar beet** is a white-rooted variety of the common beet, *Beta vulgaris,* and grows in temperate climates. It is cut up and boiled to extract the juice, and the production and refining of the sugar are essentially the same as for cane sugar. There is no difference in the final product, although **raffinose,** or **melitriose,** $C_{18}H_{32}O_{16}$, a tasteless trisaccharide, occurs in the sugar beet, and may not be completely changed to sucrose by hydrolysis, so that a greater quantity may sometimes be needed to obtain equal sweetening effect. The pectins and starches of the sugar beet are not extracted by the use of the slicing and diffusion method.

Refined commercial sugar contains 99.98% sucrose and is graded by screening to crystal size. The best qualities are the larger crystals from the first and second runs. The soft sugars are from further crystallizing, until the noncrystallizing brown sugars are reached. Raw sugar testing 96° by the polariscope is the grade used as a basis for raw-sugar quotations. Commercial sugar may have starch added. The ultrafine 6X **confectioner's sugar** usually contains 4% cornstarch as a noncaking agent, and block sugars may contain starch as a binding agent, but starch reduces the sweetening powder.

Cane sugar is the high-grade syrup or liquid sugar, while **molasses** is the heavy residual syrup left after the crystallization. **Edible molasses** is the yellow to brownish, light, purified residue syrup. **Blackstrap molasses** is the final, inedible, unpurified residue heavy syrup, used for the production of ethyl alcohol. It contains 50 to 60% sugar by weight, mostly sucrose but some glucose. A purified grade which retains the minerals is marketed as an edible blackstrap molasses.

Molasses powder, used for bakery products, is made by spray drying. It is a free-flowing, noncaking powder. **Liquid sugar,** much used in food manufacturing because it saves handling costs, comes in various liquid densities and in various degrees of invert. The liquid sugars are usually not pure sucrose, and are called **multisugars.** For food manufacturing the calcium and other minerals may be left in, and they then have a yellow color. Multisugars with 90% sucrose and 10% levulose and dextrose crystallize in hard, aggregate clusters, desirable in some confections. **Flo-Sweet** is liquid sugar. **Sucrodex** is liquid sugar containing one-third dextrose and two-thirds sucrose, with a solubility of 72% compared with only 45 for dextrose and 67 for sucrose. **Inverdex,** for canning and for fountain syrups, is about 85% invert sugar and 15 dextrose. **Amberdex,** used for cakes and cookies, is an amber-colored 50–50 mixture of sucrose and dextrose with the edible materials left in. **Caramel,** used for flavoring and coloring foodstuffs and liquors, has a deep-brown color and a characteristic taste. It is burnt sugar marketed as a liquid or powder.

The **papelon** of South America is solidified edible molasses. **Gur** is unrefined brown sugar of India, and the **pilancillo** of Mexico is unrefined brown sugar. **Treacle** is an English name for edible molasses. The refuse from sugar cane, called **bagasse,** is used as fuel and for making paper and insulating board. **Beet pulp,** after extraction of the juice, is marketed as cattle feed. Despite restrictive controls over the world supply of sugar, much sucrose is being used in the production of chemical products. **Nonionic detergents,** which are odorless, biodegradable powders with low toxicity, are made by reacting sucrose with fatty acid esters of volatile alcohols. Allyl sucrose is used as a shellac substitute. **Sucrose acetate isobutyrate** is available in three grades: semisolid, in ethyl alcohol, in toluene. The pure chemical is a clear, viscous liquid boiling at 550°F (288°C), used as a plasticizer in synthetic resins to improve extrusion and to give flexibility and adhesiveness in coatings. As much as 70% is used in nitrocellulose to give tough, flexible melt coatings. **Nitto ester** is **sucrose ester** made with sugar and stearic acid. It is used as a food additive.

A type of edible sugar syrup is also obtained from the juice of a variety of **sorghum grass,** *Sorghum vulgare,* native to South Africa, but now grown in the southern United States. The juice or syrup, called **sorghum syrup,** or **sorgo syrup,** is light in color, has a characteristic delicate flavor, and contains gums and starch, which prevent crystallization. It also contains other sugars besides sucrose, and considerable mineral salts of value as foods. The total sugar in the juice is from 9 to 17%, varying with the age of the plant. It is used in some sections to replace sugar and is employed in some confectionery to give a distinctive flavor.

Apple syrup, or **apple honey,** used as a sweetening agent in the food industry, for curing hams, and as a substitute moistening agent for tobacco, is made from cull apples. The reduced syrup is treated to remove the bitter calcium malate. It contains 75% solids of which 65% consists of the sugars levulose, dextrose, and sucrose. **Palm sugar,** or **jaggary,** is the evaporated sap of several varieties of palm, including the coconut and the palms from which kittool, gomuti, and palmyra fibers are obtained. The sap contains about 14% sugar. It is much used in India and the Pacific Islands. The **palm wine** known as **arrack** is made by fermenting the juice, called **taewak,** of the flower stems of the **aren palm** of Java. A liter (1.06 qt) of taewak yields about 0.2 lb (0.09 kg) of brown palm sugar. **Wood molasses** is made by concentrating and neutralizing the dilute sugar solution produced by pressure hydrolysis of wood chips using dilute sulfuric acid at high temperature. The molasses has a slightly bitter taste, but is used for stock feed and for industrial purposes. **Wood sugars** contain **xylose,** $CHO(HCOH)_3CH_2OH$, which belongs to the great group of **pentosans** occurring in plant life. They have the same general formula with different numbers of the HCOH group. Oxidation converts them to the respective acid, as **xylonic acid** from xylose, or arabinic acid from the arabinose of gum arabic. They can also be converted to the lactones, and are related to the furanes, so that the wood sugars have a wide utility for the production of chemicals.

Other plants yield **sweetening agents,** but few are of commercial importance. The leaves of the **caá heé,** a small plant of Paraguay, are used locally for sweetening Paraguayan tea. The name, pronounced kah-áh aye-áye, means sweet herb, and it has a more intense sweetening effect than sugar. **Miracle Fruit powder,** of International Minerals and Chemical Corp., is a complex protein-based chemical derived from the fruit pulp of the **Agbayun** shrub, *Synsepalum dulcificum,* of west Africa. It has a strong sweetening effect and a pleasant natural flavor. The 6-carbon sugar derivative known as **glucoronic lactone,** used as an antiarthritic drug, is derived from dextrose. **Amino sugar,** or **glucosamine,** has an NH_2 group in the molecule in place of the alpha hydroxyl group of glucose. This sugar occurs in marine animals.

Synthetic sweetening agents of no food value are used in diabetic foods and in dietetic foods for the treatment of obesity. Many of these synthetic sweeteners are toxic in excess and are cumulative in the human system. Thus, **dietary foods** that depend on the substitution of chemicals in place of sugar should be taken only with caution and under medical direction. **Saccharin,** produced from coal tar, is **benzoic sulfinide,** $C_6H_4SO_2NHCO$. It is 450 times sweeter than sugar and has no food value, but it has a disagreeable aftertaste. It is a water-insoluble white powder, but its salts, **sodium saccharin,** and **calcium saccharin,** are soluble in water and are 300 times sweeter than sugar. Saccharin is also used as a pH indicator, and as a brightener in nickel-plating baths.

The **cyclamates** were widely used in beverages and diet foods, but are now recognized as toxic drugs and are restricted. **Sodium cyclohexylsulfamate,** or **sodium cyclamate,** $Na(C_6H_{12}NO_3S)_2 \cdot 2H_2O$, is used in dietetic foods and in some soft drinks as it has no food value. It is 30 times sweeter than sugar, but at the 25% sweetening level of sugar it has an undesirable aftertaste, and at the sugar-sweetness level the off-taste predominates. For both sugar-free and salt-free diets, the calcium salt **calcium cyclamate** is used. **Sucaryl,** of Abbott Laboratories, is sodium cyclamate, and **Cyclan,** of Du Pont, is calcium cyclamate. **Hexamic acid,** a white, crystalline powder which is cycle hexylsulfamic acid, is used as a supplement sweetener and intensifier with the cyclamates and saccharin. **Aspartame,** also known as **Nutrasweet,** is a low-calorie sweetener used alone or in combination with sugar or saccharin in some breakfast cereals, diet soft drinks, and other ready-mixed beverages. **Peryllartine** is the sweetest known substance, being 2,000 times sweeter than sucrose. It is a complex aldehyde derived from **terpenylic acid,** which occurs in combined forms in turpentine and many essential oils.

A number of other artificial sweeteners are also being developed. **Acesulfame-potassium,** known as **Sunette** in the United States, is available in a table-top formula, **Sweet-One,** and as an ingredient in chewing gum and dry beverage mixes. Produced by Hoechst Celanese Corp., it is 200 times sweeter than sugar and slightly more so than aspartame. But since acesulfame-potassium suffers from a slight after-taste, it is usually mixed with other sweeteners; the combination works synergistically, with the mixture being sweeter than either component. **Xylitol** is a naturally derived sugar alcohol made from birch bark. It offers few benefits over sugar, since it is about as sweet and has the same number of calories. Produced in Finland, it is marketed in the United States by American Xyrofin primarily for specialty diet foods, such as for diabetics and infants, in oral hygiene, and pharmaceutical products. **Alitame,** formed from the amino acids **L-aspartic acid** and **D-alanine** by Pfizer, Inc., is 2,000 times sweeter than sucrose, just as heat-stable, and has a shelf life up to 4 times that of aspartame. **Sucralose,** being developed by McNeil Specialty Products Co., is 600 times sweeter than sugar, from which it is derived. **Natural thaumatin,** a protein that is 5,000 times sweeter than sugar, is used mainly as a flavor enhancer. **Isomalt,** a modified sugar, is used in chocolates and confectioneries in Europe and Asia, and is produced by West Germany's Subungsmittel GmbH. It is also used as a bulking agent with the highly sweet products. **Lev-O-Cal** is a left-handed L-sugar that is less sweet and has fewer calories than the right-handed, or normal, sugar. **Polydextrose** is another low-calorie bulking agent.

SUGAR PINE. The common name of the wood of the *Pinus lambertiana,* a coniferous tree growing in California and Oregon. The tree grows ordinarily to a height of 150 to 175 ft (46 to 53 m) with a diameter of 4 to 5 ft (1.2 to 1.5 m). Occasional trees are more than 200 ft (61 m) in height and 12 ft (3.7 m) in diameter, and are often free of limbs up to 75 ft (23 m) from the ground. It is the largest of the pines. Sugar pine is durable, has moderate strength and fairly even grain, and is not subject to excessive shrinkage or warping. Because of the latter quality it has come into use to replace the scarcer eastern pines for patterns. It does not darken on exposure as western pine does. It is widely employed for construction work and for **factory lumber** for doors, frames, boxes, and wooden articles. Sugar pine is classified into three standard classes of grades according to freedom from knots and faults as select, commons, and factory, or shop. The selects are designated as Nos. 1 and 2 clear, C select, and D select. The commons are graded as Nos. 1, 2, 3, and 4; the factory as No. 3 clear, No. 1 shop, No. 2 shop, and No. 3 shop. The shops are judged with the idea that they will be cut up into small pieces, and are consequently classified by the area of clear cuttings that can be obtained.

SULFAMIC ACID. A white, crystalline, odorless solid of composition $HSO_3 \cdot NH_2$, very soluble in water, but only slightly soluble in alcohol. The melting point is 352°F (178°C). The acid is stronger than other solid acids, approaching the strength of hydrochloric. It is used in bating and tanning leather, giving a silky, tight grain in the leather. An important use is for cleaning boiler and heat-exchanger tubes. It converts the calcium carbonate scale to the water-soluble calcium sulfamate, which can then be flushed off and combined with sodium chloride; it also converts the rust to ferric chloride and then to the water-soluble iron sulfamate. Pettibone Chemicals offers high-purity grades for bleaching paper pulp and textiles, organic synthesis, gasliberating compositions, and as a catalyst for urea-formaldehyde resins. **Ammonium sulfamate** is the ammonia salt of the acid, used as a cleanser and anodizer of metals, as a weed killer, and for flameproofing paper and textiles. **Lead ammonium sulfamate,** $Pb(SO_3NH_2)_2$, used in lead plating, is very soluble in water and has high throwing power. **Aminoethylsulfamic acid,** $NH_2CH_2CH_2OSO_3H$, is used for treating paper and textile fibers to increase wet strength and water repellency. **Tobias acid,** used in making azo dyes, is **naphthylamine sulfonic acid,** $NH_2C_{10}H_6SO_3H$, in white needles decomposing at 446°F (230°C).

SULFONATED OIL. A fatty oil that has been treated with sulfuric acid, the excess acid being washed out and only the chemically combined acid remaining. The oil is then neutralized with an alkali. Sulfonated oils are water-soluble and are used in cutting oils and in fat liquors for leather finishing. **Sulfonated castor oil** is called **Turkey red oil. Leatherlubric** is the trade name of E. F. Houghton & Co. for sulfonated sperm oil used for leather. **Solcod** is the sulfonated cod oil of the same company. **Sulfonated stearin** and **sulfonated tallow** are also used in leather dressing. They are cream-colored pastes readily soluble in hot water. **Mahogany soap** is a name for oil-soluble petroleum sulfonates used as dispersing and wetting agents, corrosion inhibitors, emulsifiers, and to increase the oil absorption of mineral pigments in paints. **Petronate** is a petroleum sulfonate containing 62% sulfonates, 35 mineral oil, and 3 water. **Phosphorated oils,** or their sulfonates, may be used instead of the sulfonates as emulsifying agents or in treating textiles and leathers. They are more stable to alkalies. **Phosoils** are phosphorated vegetable oils. **Aquasol,** of American Cyanamid Co., is a sulfonated castor oil used as an emulsifying agent. **Cream softener** is a name used in the textile industry for sulfonated tallow.

SULFUR. One of the most useful of the elements, symbol S. Its occurrence in nature is little more than 1% that of aluminum, but it is easy to extract and is relatively plentiful. In economics, it belongs to the group of "S" materials—salt, sulfur, steel, sugars, starches—whose consumption is a measure of the industrialization and the rate of industrial growth of a nation. Sulfur is obtained from volcanic deposits in Sicily, Mexico, Chile, and Argentina, and along the Gulf Coast in Louisiana, Texas, and Mexico it is obtained from great underground deposits in the cap rock above salt domes. Offshore deposits worked in the Gulf of Mexico are 2,000 ft (610 m) under the bottom. Strict environmental laws are driving the production of sulfur recovered as a by-product of various industrial operations. It is also obtained by the distillation of iron pyrites, as a by-product of copper and other metal smelting, and from natural gas. The **sterri** exported from Sicily for making sulfuric acid is broken rock rich in sulfur. **Brimstone** is a very ancient name still in popular use for solid sulfur, but the District Court of Texas has ruled that sulfur obtained from gas is not subject to tax as brimstone.

Sulfur forms a crystalline mass of a pale-yellow color, with a Mohs hardness of 1.5 to 2.5, a specific gravity of 2.05 to 2.09, and melting point of 232°F (111°C). It forms a ruby vapor at about 780°F (416°C). When melted and cast, it forms **amorphous sulfur** with a specific gravity of 1.955. The tensile strength is 160 lb/in^2 (1 MPa), and compressive strength is 3,300 lb/in^2 (23 MPa). Since ancient times it has been used as a lute for setting metals into stone. Sulfur also condenses into light flakes known as **flowers of sulfur,** and the **hydrogen sulfide gas,** H_2S, separated from sour natural gas, yields a sulfur powder. **Flotation sulfur** is a fine, free-flowing sulfur dust with particle sizes less than 157 μin (4 μm), recovered in gas production from coal. Commercial crude Sicilian sulfur contains from 2 to 11% of impurities and is sold in three grades. Refined sulfur is marketed in crystals, roll, or various grades of powder, and the Sicilian superior grade is 99.5% pure. This is the grade used in rubber manufacture. **Crystex** is a sulfur, 85% insoluble in carbon bisulfide, used in rubber compounding. The **sulfur powder** of Electronic Space Products, Inc., used for semiconductors, is 99.9999% pure.

Sulfur has twice the atomic weight of oxygen but has many similar properties and has great affinity for most metals. It has six valence electrons, but also has valences of 2 and 4. The **crystalline sulfur** is orthorhombic, which converts to monoclinic crystals if cooled slowly from 248°F (120°C). This form remains stable below 248°F. When molten sulfur is cooled suddenly, it forms the amorphous sulfur which has a ring molecular structure and is plastic, but converts gradually to the rhombic form. Sulfur has a wide variety of uses in all industries. The biggest outlet is for sulfuric acid, mainly for producing phosphate fertilizers. **Agri Sul,** from Eagle-Picher Industries, Inc., is available in prilled form or as a water-degradable grade, as a source of elemental sulfur for correcting sulfur deficiencies in crops and grasses. **Newsulfur,** from Tri-K Industries, Inc., is a nonirritating variety that can be mixed by hand with any kind of ingredient. It is used for making gunpowder and for vulcanizing rubber, but for most uses it is employed in compounds, especially as sulfuric acid or sulfur dioxide. A vast number of so-called thio compounds have been produced. The **thio alcohols,** or **mercaptans,** have an SH group instead of the OH of true alcohols, and they do not react as alcohols, but the **thio esters** are made directly from the mercaptans. **Thionyl chloride,** $SOCl_2$, a yellow liquid, is a typical compound used as a source of sulfur in synthesis. Most of the thio compounds have an offensive odor. **Vegetable sulfur** does not contain sulfur, but is **lycopodium,** a fine, yellow powder from the spores of the **club moss,** a fernlike plant, *Lycopodium clavatum,* which grows in North America and Europe. It belongs to the group of **lipochromes,** or coloring matter of plants related to lycopene and carotene.

   Sulfur dioxide, or **sulfurous acid anhydride,** is a colorless gas of composition SO_2, used as a refrigerant, as a preservative, in bleaching, and for making other chemicals. It liquefies at about $14°F$ ($-10°C$). As a refrigerant it has a condensing pressure of 51.7 lb (23.5 kg) at $86°F$ ($30°C$). The gas is toxic and has a pungent, suffocating odor, so that leaks are detected easily. It is corrosive to organic materials but does not attack copper or brass. The gas is soluble in water, forming **sulfurous acid,** H_2SO_3, a colorless liquid with suffocating fumes. The acid form is the usual method of use of the gas for bleaching.

SULFURIC ACID.   An oily, highly corrosive liquid of composition H_2SO_4, having a specific gravity of 1.841 and a boiling point of $626°F$ ($330°C$). It is miscible in water in all proportions, and the color is yellowish to brown according to the purity. It may be made by burning sulfur to the dioxide, oxidizing to the trioxide, and reacting with steam to form the acid. It is a strong acid, oxidizing organic materials and most metals. Sulfuric acid is used for pickling and cleaning metals, in electric batteries and plating baths, for making explosives and fertilizers, and for many other purposes. In the metal industries it is called **dipping acid,** and in the automotive trade it is called **battery acid. Fuming sulfuric acid,** or **oleum,** of 100% purity, was called **Nordhausen acid.** The grade of sulfuric acid known as **oil of vitriol,** or **vitriol,** is $66°Bé$, or 93.2% acid. A 96% grade, called **Particulo,** is available from General Chemical Co. **Sulfur trioxide,** or **sulfuric anhydride,** SO_3, is the acid minus water. It is a colorless liquid boiling at $115°F$ ($46°C$) and forms sulfuric acid when mixed with water. It is used for sulfonation. **Sulfan,** of Allied-Signal Corp., is sulfuric anhydride. **Sulfodox** is a liquid sulfur dioxide of high purity from Specialty Products Corp. It is used for acidifying and bleaching. **Chlorosulfonic acid,** $HClSO_3$, has equal amounts of sulfur trioxide and hydrochloric acid, and is a vigorous dehydrating agent, used also in chlorosulfonating organic compounds. It has a specific gravity of 1.752 and boils at 311°F (155°C). Mixed with sulfur trioxide, it has been called **FS smoke** for military smoke screens.

Niter cake, which is **sodium acid sulfate,** $NaHSO_4$, or **sodium bisulfate,** contains 30 to 35% available sulfuric acid and is used in hot solutions for pickling and cleaning metals. It comes in colorless crystals or white lumps, with a specific gravity of 2.435 and melting point 572°F (300°C). **Sodium sulfate,** or **Glauber's salt,** is a white, crystalline material of composition $Na_2SO_4 \cdot 10H_2O$, used in making kraft paper, rayon, and glass. It was first produced from Hungarian spring water by Johann Glauber, and when obtained from mineral springs, it is called **crazy water crystals.** The **burkeite,** sodium sulfate–sodium bicarbonate double salt, which separates out of Searles Lake brine, is used to produce sodium sulfate and other chemicals as by-products. **Salt cake,** Na_2SO_4, is impure sodium sulfate used in the cooking liquor in making paper pulp from wood. It is also used in freezing mixtures. **Synthetic salt cake,** used for making kraft pulp, is produced by sintering soda ash and sulfur. **Chrome cake** is a greenish by-product salt cake which contains some chromium as an impurity. It is used in papermaking. **Kaiseroda** is a German name for salt cake of high purity obtained as a by-product from the production of magnesium chloride from potash minerals. **Sodium sulfite,** Na_2SO_3 or $Na_2SO_3 \cdot 7H_2O$, is a white to tan, crystalline powder very soluble in water but nonhygroscopic. **Santosite,** of Monsanto, is a grade of sodium sulfite containing 93% sodium sulfite with the balance chiefly sodium sulfate.

Sodium sulfide, Na_2S, is a pink, flaky solid, used in tanneries for dehairing and in the manufacture of dyes and pigments. The commercial product contains 60 to 62% Na_2S, 3.5 NaCl, and other salts, and the balance water of crystallization. **Sodium sulfhydrate,** NaSH, is in lemon-yellow flakes. It has much less alkalinity than sodium sulfide, and is used in tanneries in unhairing solutions and for making thiourea and other chemicals. It contains 62.6% by weight of sulfur and is an economical material for sulfonating. **Sodium dithionate,** $Na_2S_2O_6 \cdot 2H_2O$, is used in leather tanning, as an assist in textile dyeing and printing, and for making other chemicals. It comes in transparent, prismatic crystals of bitter taste. **Sodium thiosulfate,** $Na_2S_2O_3 \cdot 5H_2O$, known as **hypo,** is a white, crystalline compound having a specific gravity of 1.73 and a melting point of 113°F (45°C). It is used in photography to fix films, plates, and papers. **White vitriol** is zinc sulfate in colorless crystals soluble in water and melting at 102°F (39°C). It is used for making zinc salts, as a mordant, for zinc plating, and as a preservative in adhesives.

SUMAC. The dried, ground leaves of the bush *Rhus coriaria* of Sicily, or *R. typhina* of the eastern United States, used for tanning leather. The Sicilian leaves contain up to 30% tannin, and the U.S. leaves up to 38%. It contains gallotannin and ellagitannin and gives a rapid tan. Sumac provides a light, strong leather of fine, soft grain and has a bleaching action which can produce a white leather. It is used for book and hatband leathers. Sumac grows profusely in the eastern states, but the gathering of the leaves is not organized commercially.

SUNFLOWER OIL. A pale-yellow drying oil with a pleasant odor and taste obtained from the large seeds of the common sunflower plant, *Helianthus annuus,* of which there are many varieties. The plant is native to Peru but is now grown in many parts of the world, particularly in California, Canada, Argentina, Chile, Uruguay, and Russia. It requires boron in the soil. The specific gravity of the oil is 0.925. Sunflower oil is used in varnish and soap manufacture or as a food oil. Refined and unrefined grades, with trade name **Trisun,** are available for these applications from SVO Enterprises. The by-product cake is used chiefly for cattle feed, but **sunflower meal** is also blended with wheat flour or cornmeal in foods. It is higher in vitamin B than soybean flour. **Sunflower seeds** are also used as poultry feed. **Madia-seed oil** is quite similar to sunflower oil and has the same uses. It is obtained from the seeds of the plant *Madia sativa,* native to California. The seeds contain 35% oil, and the cold-pressed oil has a pleasant taste. **Watermelon-seed oil,** produced in Senegal as **bereff oil,** is an edible oil similar to sunflower. It contains about 43% linoleic, 27 oleic, 19.5 stearic, and 5 palmitic acids.

The leaves of selected varieties of some species of sunflower contain from 1 to 6% **sunflower rubber** and up to 8 resin. The *H. occidentalis, H. giganteus, H. maximiliani,* and *H. strumosus* are cultivated in Russia both for the oil seed and for the rubber in the leaves. These perennials yield leaves up to 10 years. Another similar rubber-bearing plant of southern Russia is *Asclepias cornuti,* known as **vatochnik.** It is a perennial, producing leaves for 10 to 15 years. The leaves yield 1 to 6.5% rubber and large percentages of resin.

SUN HEMP. The bast fiber of the plant *Crotalaria juncea.* It is used for cordage and rope in place of jute, but is lighter in color and is more flexible, stronger, and more durable than jute. It resembles true hemp, but is not as strong. It is more properly called **sann hemp** from the Hindu word *sann.* It is also known as **sunn fiber, Indian hemp,** and **Bombay hemp.** The plant, which is a shrub, is cultivated extensively in India. It grows to a height of about 8 ft (2.4 m), with slender branches yielding the fiber. The method of extraction is the same as for true hemp. The best fibers are retained locally for making into cloth. It is also used in the United States for making cigarette paper and for oakum. **Madras hemp** is from another species of the same plant.

SUPERALLOYS. Iron-based, nickel-based, or cobalt-based alloys noted primarily for high strength and oxidation and corrosion resistance at high temperatures. Because of their excellent high-temperature performance, they are also known as **high-temperature, high-strength alloys.** Their strength at high temperatures is usually measured in terms of stress-rupture strength or creep resistance. For high-stress applications, the iron-base alloys are generally limited to a maximum service temperature of about 1200°F (649°C), whereas the nickel-and cobalt-based alloys are used at temperatures to about 2000°F (1093°C) and higher. In general, the nickel alloys are stronger than the cobalt alloys at temperatures below 2000°F, and the reverse is true at temperatures above 2000°F. Superalloys are probably best known for aircraft turbine applications, although they are also used in steam and industrial turbines, nuclear power systems, and chemical and petroleum processing equipment. A great variety of cast and wrought alloys are available, and in recent years, considerable attention has been focused on the use of powder-metallurgy techniques as a means of attaining greater compositional uniformity and finer grain size.

The **iron-based superalloys** include solid-solution alloys and precipitation-hardening (PH), or precipitation-strengthened, alloys. Solid-solution types are alloyed primarily with nickel (20 to 36%) and chromium (16 to 21), although other elements are also present in lesser amounts. **Superalloy 16-25-6,** for example, the alloy designation indicating its chromium, nickel, and molybdenum contents, respectively, also contains small amounts of manganese (1.35%), silicon (0.7), nitrogen (0.15), and carbon (0.06). **Incoloy 800, 801,** and **802,** of Inco Alloys International, Inc., contain slightly less nickel and slightly more chromium with small amounts of titanium, aluminum, and carbon. **Incoloy 803,** of Inco, was developed for pyrolysis tubing in severe ethylene furnaces and other petrochemical applications. It comprises 32 to 37% nickel, 25 to 29 chromium, 0.15 to 0.6 aluminum, 0.15 to 0.6 titanium, 0.06 to 0.1 carbon, with maximum amounts of 1.5 manganese, 1 silicon, 0.75 copper, and 0.015 sulfur, balance iron. The alloy has high resistance to oxidation and carburization. Protective scales, developed by high-temperature exposure, provide the self-healing quality. The alloy has an ultimate tensile strength of 88,000 lb/in^2 (607 MPa) and 50% elongation. At 1800°F (980°C), the tensile strength is 15,000 lb/in^2 (103 MPa). **N-155,** or **Multimet,** an early sheet alloy, contains about equal amounts of chromium, nickel, and cobalt (20% each), plus 3 molybdenum, 2.5 tungsten, 1 columbium, and small amounts of carbon, nitrogen, lanthanum, and zirconium. At 1350°F (732°C), this alloy has a 1,000-h stress-rupture strength of about 24,000 lb/in^2 (165 MPa).

PH iron-based superalloys provide greater strengthening by precipitation of a nickel-aluminum-titanium phase. One such alloy, which may be the most well known of all iron-based superalloys, is **A-286.** It contains 26% nickel, 15 chromium, 2 titanium, 1.25 molybdenum, 0.3 vanadium, 0.2 aluminum, 0.04 carbon, and 0.005 boron. At room temperature, it has a tensile yield strength of about 100,000 lb/in^2 (690 MPa) and a tensile modulus of 21.1 × 10^6 lb/in^2 (145,000 MPa). At 1200°F (649°C), tensile yield strength declines only slightly, to 88,000 lb/in^2 (607 MPa), and its modulus is about the same or slightly greater. It has a 1,000-h stress-rupture strength of about 21,000 lb/in^2 (145 MPa) at 1350°F (732°C). Other PH iron-based superalloys are **Discoloy, Haynes 556** (whose chromium, nickel, cobalt, molybdenum, and tungsten contents are similar to those of N-155); **Incoloy 903** and **Pyromet CTX-1,** which are virtually chromium-free but high in nickel (37 to 38%) and cobalt (15 to 16); and **V-57** and **W-545,** which contain about 14 chromium, 26 to 27 nickel, about 3 titanium, 1 to 1.5 molybdenum, plus aluminum, carbon, and boron. V-57 has a 1,000-h stress-rupture strength of about 25,000 lb/in^2 (172 MPa) at 1350°F and greater tensile strength, but similar ductility, than A-286 at room and elevated temperatures.

Nickel-based superalloys are solid-solution, precipitation-hardened, or oxide-dispersion-strengthened. All contain substantial amounts of chromium, 9 to 25%, which, combined with the nickel, accounts for their excellent high-temperature oxidation resistance. Other common alloying elements include molybdenum, tungsten, cobalt, iron, columbium, aluminum, and titanium. Typical solid-solution alloys include **Hastelloy X** (22 to 23% chromium, 17 to 20 iron, 8 to 10 molybdenum, 0.5 to 2.5 cobalt, 2 aluminum, 0.2 to 1 tungsten, and 0.15 carbon); **Inconel 600** (15.5 chromium, 8 iron, 0.25 copper maximum, 0.08 carbon); and **Inconel 601, 604, 617,** and **615,** of Inco Alloys International, Inc., the latter containing 21.5 chromium, 9 molybdenum, 3.6 columbium, 2.5 iron, 0.2 titanium, 0.2 aluminum, and 0.05 carbon. At 1350°F (732°C), wrought Hastelloy X (it is also available for castings) has a 1,000-h stress-rupture strength of about 18,000 lb/in^2 (124 MPa) and has high oxidation resistance at temperatures to 2200°F (1204°C). **Inconel 625** has a low-cycle (10^5) fatigue strength of 110,000 to 120,000 lb/in^2 (760 to 830 MPa). **Inconel 625 LCF** has low carbon, silicon, and nitrogen contents to improve resistance to low-cycle fatigue. **Inconel 725** is an age-hardenable version of Inconel 625, providing comparable corrosion resistance but greater strength. **Inconel 783** is an oxidation-resistant, low-expansion, nickel-cobalt-iron superalloy for aircraft turbine parts.

The precipitation-strengthened alloys, which are the most numerous, contain aluminum and titanium for the precipitation of a second strengthening phase, the intermetallic $Ni_3(Al,Ti)$ known as **gamma prime** (γ') or the intermetallic Ni_3Cb known as **gamma double prime** (γ''), during heat treatment. One such alloy, **Inconel X-750** (15.5% chromium, 7 iron, 2.5 titanium, 1 columbium, 0.7 aluminum, 0.25 copper maximum, and 0.04 carbon), has more than twice the tensile yield strength of Inconel 600 at room temperature and nearly 3 times as much at 1400°F (760°C). Its 1,000-h stress-rupture strength at 1400°F is in the range of 20,000 to 30,000 lb/in² (138 to 207 MPa). Still great tensile yield strength at room and elevated temperatures and a 25,000 lb/in² (172 MPa) stress-rupture strength at 1400°F are provided by **Inconel 718** (19% chromium, 18.5 iron, 5.1 columbium, 3 molybdenum, 0.9 titanium, 0.5 aluminum, 0.15 copper maximum, 0.08 carbon maximum), a wrought alloy originally that also has been used for castings. **Inconel 718SPF** is for superplastic forming, as the designation implies. A vacuum-induction-melted and electroslag-remelted alloy, it is produced to a fine grain size (ASTM 12), and reduced carbon and columbium contents to minimize carbide precipitation during forming. Gas pressure of only 300 lb/in² (2 MPa) at 1750°F (954°C) and low strain rates are sufficient to form complex shapes. Because of the low strain rates, forming cycles are long: 1 to 3 h with mill-annealed sheet, which has an ultimate tensile strength of 162,000 lb/in² (1,117 MPa), a yield strength of 118,000 lb/in² (814 MPa), and 33% elongation. Aging increases the yield strength to 192,000 lb/in² (1,324 MPa). At 1200°F (649°C), the yield strength is 160,000 lb/in² (1,103 MPa). Among the strongest alloys in terms of stress-rupture strength is the wrought or cast **IN-100** (10% chromium, 15 cobalt, 5.5 aluminum, 4.7 titanium, 3 molybdenum, 1 vanadium, less than 0.6 iron, 0.15 carbon, 0.06 zirconium, 0.015 boron). Investment cast, it provides a 1,000-h stress-rupture strength of 75,000 lb/in² (517 MPa) at 1400°F (760°C), 37,000 lb/in² (255 MPa) at 1600°F (871°C), and 15,000 lb/in² (103 MPa) at 1800°F (982°C). Other precipitation-strengthened wrought alloys include **Astroloy**; **D-979**; **IN 102**; **Inconel 706** and **751**; **M252**; **Nimonic 80A, 90, 95, 100, 105, 115,** and **263**; **René 41, 95,** and **100**; **Udimet 500, 520, 630, 700,** and **710**; **Unitemp AF2-1DA**; and **Waspaloy.** Other cast alloys, mainly investment-cast, include **B-1900**; **IN-738**; **IN-792**; **Inconel 713C**; **M252**; **MAR-M 200, 246, 247,** and **421**; **NX-188**; **René 77, 80,** and **100**; **Udimet 500, 700,** and **710**; **Waspaloy**; and **WAZ-20.**

A few of the cast alloys, such as MAR-M 200, are used to produce **directionally solidified castings,** that is, investment castings in which the grain runs only unidirectionally, as along the length of turbine blades. Eliminating transverse grains improves stress-rupture properties and fatigue resistance. Grain-free alloys, or **single-crystal alloys,** also have been cast, further improving high-temperature creep resistance. **CMSX-4,** a single-crystal alloy developed by Cannon-Muskegon Corp., is used for aircraft turbine blades. At 1400°F (760°C), it has a tensile yield strength of 140,000 lb/in^2 (965 MPa), and it retains useful strength up to 2125°F (1163°C). Regarding powder-metallurgy techniques, emphasis has been on the use of prealloyed powder made by **rapid solidification techniques (RST)** and **mechanical alloying (MA),** a high-energy milling process using attrition mills or special ball mills. **Dispersion-strengthened nickel alloys** are alloys strengthened by a dispersed oxide phase, such as thoria, which markedly increases strength at very high temperatures but only moderately so at intermediate elevated temperatures, thus limiting applications. **TD-nickel, or thoria-dispersed nickel,** was the first of such superalloys, and it was subsequently modified with about 20% chromium, **TD-NiCr,** for greater oxidation resistance. The recent **MA 754** and **MA 6000E** alloys combine dispersion strengthening with yttria and gamma-prime strengthening. **MA 6000,** of Inco Alloys International, Inc., contains 69% nickel, 15 chromium, 4.5 aluminum, 4 tungsten, 2.5 titanium, 2 molybdenum, 2 tantalum, 1.1 yttria, 0.5 carbon, 0.15 zirconium, and 0.01 boron. It provides high creep strength up to 2100°F (1150°C) and, in England, is used for solid blades of industrial gas turbines.

    Cobalt-based superalloys are for the most part solid-solution alloys, which, when aged, are strengthened by precipitation of carbide or intermetallic phases. Most contain 20 to 25% chromium, substantial nickel and tungsten and/or molybdenum, and other elements, such as iron, columbium, aluminum, or titanium. One of the most well known, **L-605,** or **Haynes 25,** is mainly a wrought alloy, though also used for castings. In wrought form, it contains 20% chromium, 15 tungsten, 10 nickel, 3 iron, 1.5 manganese, and 0.1 carbon. At room temperature, it has a tensile yield strength of about 67,000 lb/in^2 (462 MPa), and at 1600°F (871°C) about 35,000 lb/in^2 (241 MPa). Its 1,000-h stress-rupture strength at 1500°F (815°C) is 18,000 lb/in^2 (124 MPa). The more recent **Haynes 188** (22% chromium, 22 nickel, 14.5 tungsten, 3 iron, 1.5 manganese, 0.9 lanthanum, 0.35 silicon, and 0.1 carbon), which was developed for aircraft-turbine sheet components, provides roughly similar strength and high oxidation resistance to about 2000°F (1093°C). **MP35N** (35% nickel, 35 cobalt, 20 chromium, 10 molybdenum) is a work-hardening alloy used mainly for high-temperature, corrosion-resistant fasteners. **MP159** (25% nickel, 19 chromium, 9 iron, 7 molybdenum, 3 titanium, 0.2 aluminum) is a work and precipitation-hardening alloy for such fasteners. It has an ultimate tensile strength of 260,000 lb/in^2 (1,793 MPa), 205,000 lb/in^2 (1,413 MPa) at 1100°F (593°C). Another alloy, **S-816,** contains equal amounts of chromium and nickel (20% each), equal amounts of molybdenum, tungsten, columbium, and iron (4 each), and 0.38 carbon. Primarily a wrought alloy, though also used for castings, it has a 1,000-h stress-rupture strength of 21,000 lb/in^2 (145 MPa) at 1500°F (815°C). Other casting alloys include **AiResist 13, 213,** and **215; Haynes 21** and **31,** the latter also known as **X-40; Haynes 151; J-1650; MAR-M 302, 322, 509,** and **918; V-36;** and **W1-52.** Their chromium content ranges from 19% (AiResist 215) to 27 (Haynes 21), and some are nickel-free or low in nickel. Most contain substantial amounts of tungsten or tantalum, and various other alloying elements. Among the strongest in terms of 1,000-h stress-rupture strength at 1500°F are Haynes 21 and 31: 42,000 lb/in^2 (290 MPa) and 51,000 lb/in^2 (352 MPa), respectively.

SUPERBRONZE. A name applied to brasses containing both aluminum and manganese. They are ordinarily high brasses with 2 to 3% manganese and 1 to 6 aluminum, with sometimes also some iron. They have greatly increased strength and hardness over the original brasses, but the ductility is reduced and they are difficult to work and machine. The early superbronze was known as **Heusler alloy.** Muntz metal is also frequently modified with manganese, iron, and aluminum. The alloys are used where high strength and corrosion resistance are required, and they are often marketed under trade names. The name *superbronze* is a shop term rather than a technical classification, and thus the name is often applied to any hard, high-strength, heat-treatable, copper-base alloy.

SUPERCONDUCTORS. Materials having no electrical resistivity, thus maximum electrical conductivity, at or below a specific temperature, typically well below zero degrees. As long as the material remains at or below its superconducting temperature, strong magnetic fields can be generated for use in many applications, including levitating trains. Until recently, temperatures approaching absolute zero [−459°F (−272°C)] were required. Some of the metals exhibiting superconductivity at such temperatures are **columbium, lead, iridium, mercury, tantalum, tin,** and **vanadium,** as well as many alloys and compounds. Alloys considered among the best commercially available include **columbium-tin, columbium-titanium,** and **lead-molybdenum-sulfur.** Columbium-titanium, in the form of flexible wire, is probably the most widely used. It has a superconducting temperature of −441°F (−263°C) and is generally limited to magnetic fields below

80,000 G (8 T), particle accelerators being one application. Magnets of greater strength—200,000 G (20 T)—have been made of columbium-tin. Columbium alloyed with tin and titanium is used for magnets in magnetic resonance imaging and magnetic-energy storage devices. Columbium-titanium wire coils, cooled by liquid helium to $-452°F$ $(-269°C)$, have been proposed for use in subway tunnels as a means of preventing voltage sags from increasing numbers of accelerating trains. **Gallium-arsenide,** grown under certain conditions, is super-conductive at $-440°F$ $(-262°C)$ and, being compatible with semiconductor chips, could find electronic applications.

In recent years various ceramic compounds have been found to exhibit superconductivity at higher temperatures. These include $YBa_2Cu_3O_{7-x}$, also designated YBCO-123, with a superconducting temperature of $-294°F$ $(-181°C)$; $Bi_2Sr_2CaCu_2O_{10-x}$, or BSCCO 2212, also $-294°F$; $Bi_2Sr_2Ca_2Cu_3O_{12-x}$, or BSCCO 2223, $-262°F$ $(-163°C)$; $Tl_2Ba_2Ca_2Cu_3O_{12-x}$, or TBCCO 2223, $-235°F$ $(-148°C)$; and $HgBa_2Ca_2Cu_3O_{10-x}$, or HBCCO 1223, $-221°F$ $(-140°C)$. The principal advantage of such "high-temperature" superconductors is that liquid nitrogen $[-321°F$ $(-196°C)]$ instead of liquid helium $[-452°F$ $(-269°C)]$ could be used as the cooling medium. Potential applications include sonar transducers in submarines, high-power motors for utilities and processing industries, power transmission cables, magnetic bearings for energy-storage flywheels, electric current limiters, and electric current leads. A 3.28-ft (1-m) long conductor made by American Superconductor Corp. of a BSCCO-type material can carry more than 2,300 A of direct current at $-321°F$, or more than twice that typical of conventional underground copper conductors.

Despite the great interest in such ceramic superconductors, metal ones have the advantage of being more ductile and, thus, more readily fabricated into flexible forms, such as wire. Here, too, recent developments indicate increasing superconducting temperatures. At AT&T's Bell Laboratories, a composition of nickel, boron, yttrium, and carbon has demonstrated superconductivity at 418°F (214°C). Also, the compound $(NH_3)_4Na_2CsC_{60}$, a **carbon fullerene** modified with alkali metals, has been found to change from insulator to superconductor at $-405°F$ $(-243°C)$.

SUPERCRITICAL FLUIDS. SCFs are fluids which, when compressed and heated above a critical pressure and temperature, have diffusion properties similar to those of gases and densities similar to those of liquids and, thus, are efficient solvents. Critical parameters (pressure, temperature, and density, respectively) for **carbon dioxide** are 1,072 lb/in² (7.4 MPa), 88°F (31°C), and 0.017 lb/in³ (471 kg/m³); for **water,** 3,209 lb/in² (22 MPa), 706°F (374°C), and 0.012 lb/in³ (332 kg/m^3); for **hydrogen,** 188 lb/in^2 (1.3 MPa), $-400°$F ($-240°$C), and 0.0012 lb/in^3 (33 kg/m^3); for **ammonia,** 1,636 lb/in^2 (11.3 MPa), 270°F (132°C), and 0.0085 lb/in^3 (235 kg/m^3); and for **methyl alcohol,** or **methanol,** 1,154 lb/in^2 (8 MPa), 464°F (240°C), and 0.01 lb/in^3 (277 kg/m^3).

Current and potential applications for SCFs, especially carbon dioxide and water, stem largely from regulatory pressures on ecology and safety, and health trends. They were first used in the 1970s for decaffeinating coffee and tea, replacing trichloroethylene and methylene chloride, have replaced ethylene chloride in spice extraction, and have been recently introduced for metal cleaning, deasphalting, and spray painting without volatile organic compounds. Emerging or potential uses for CO_2 include ethanol purification; extraction of acetone from antibiotics, fat and cholesterol from egg yolks, and vitamin E from soybean oil; and soil remediation. CO_2, being nonpolar, is effective in removing nonpolar contaminants from virtually any matrix and may prove economical for on-site removal of organic soil contaminants that do not easily volatize. In pilot tests at the Westinghouse Hanford nuclear site, CO_2 at 6,000 lb/in^2 (41 MPa) and 140°F (60°C) extracted more than 95% of the diesel oil, polychlorinated biphenyls (PCBs), and bis (2-ethyl hexyl) phthalate.

Water at 3,400 lb/in^2 (23 MPa) and 650 to 1100°F (343 to 593°C) has potential for treating various organic substances by supercritical oxidation. Organic liquids and gases mix with the water and are transformed into CO_2 and water while inorganics dissolve only slightly, allowing them to concentrate and be recovered. Virtually all of the organics are destroyed. Potential abounds for waste treatment at chemical, pulp and paper, and weapons plants. At a U.S. Navy facility, a unit operating at 3,400 lb/in^2 (23 MPa) and 650°F (343°C) is intended to destroy mixtures containing paint, oil, chlorinated solvents, PCBs, and other hazardous compounds. Water is also in contention for soil remediation, having proved effective in removing virtually all of the polycyclic aromatic hydrocarbons in soil. It is aimed at some 2,000 sites in the United States and Canada where town gas has been made from coal. Kerforschungszentrum Karlsruhe of Germany is combining supercritical CO_2 extraction with either supercritical water oxidation or indirect electro-oxidation to cleanse metal-laden sludge at a glass-grinding facility and for treatment of halogenated organic wastestreams.

SUPERPOLYMERS. Many plastics developed in recent years can maintain their mechanical, electrical, and chemical resistance properties at temperatures over 400°F (213°C) for extended periods. Among these materials are **polyimide, polysulfone, polyphenylene sul-**

fide, **polyethersulfone, polyarylsulfone, novalac epoxy, aromatic polyester,** and **polyamide-imide.** In addition to high-temperature resistance, they have in common high strength and modulus of elasticity, and excellent resistance to solvents, oils, and corrosive environments. They are also high in cost. Their major disadvantage is their processing difficulty. Molding temperatures and pressures are extremely high compared to conventional plastics. Some of them, including polyimide and aromatic polyester, are not molded conventionally. Because they do not melt, the molding process is more of a sintering operation. Thus, parts are often molded by resin producers, such as Du Pont for its line of **Vespel** polyimides, Furon Co. for its **Meldin 3000** polyimides, and Hoechst Celanese for its **Celazole** polybenzimidazole. One indication of the high-temperature resistance of the superpolymers is their glass transition temperature of well over 500°F (260°C), as compared to less than 350°F (177°C) for most conventional plastics. In the case of polyimides, the glass transition temperature is greater than 800°F (427°C), and the material decomposes rather than softens when heated excessively. Aromatic polyester, a homopolymer also known as **polyoxybenzoate,** does not melt, but at 800°F (427°C) can be made to flow in a nonviscous manner similar to metals. Thus, filled and unfilled forms and parts can be made by hot sintering, high-velocity forging, and plasma spraying. Notable properties are high thermal stability, good strength at 600°F (316°C), high thermal conductivity, good wear resistance, and extrahigh compressive strength. Aromatic polyesters have also been developed for injection and compression molding. They have long-term thermal stability and a strength of 3,000 lb/in^2 (21 MPa) at 550°F (288°C). At room temperature, polyimide is the stiffest of the group with a top modulus of elasticity of 7.5 × 10^6 lb/in^2 (51,713 MPa), followed by polyphenylene sulfide with a modulus of 4.8 × 10^6 lb/in^2 (33,096 MPa). Polyarylsulfone has the best impact resistance of the superpolymers with a notched impact strength of 5 ft · lb/in (267 J/m).

Polyetherimide, such as General Electric Plastics' **Ultem,** is an amorphous thermoplastic that can be processed with conventional thermoplastic processing equipment. Its continuous-use temperature is 340°F (171°C), and its deflection temperature is 400°F (204°C) at 264 lb/in^2 (1.8 MPa). The polymer also has inherent flame resistance without the use of additives. This feature, along with its resistance to food stains and cleaning agents, makes it suitable for aircraft panels and seat component parts. Tensile strength ranges from 15,000 to 24,000 lb/in^2 (103 to 165 MPa). Flexural modulus is 480,000 lb/in^2 (3,310 MPa). Polyamide-imide has a glass transition temperature of 527°F (275°C), a tensile strength of 22,000 lb/in^2 (152 MPa), a flexural strength of 34,900 lb/in^2 (241 MPa), and heat-deflection temperature of 532°F (278°C) at 264 lb/in² (1.8 MPa). **Torlon** is such a material of Amoco Performance Products Inc. **Radel A,** of this company, is polyethersulfone, **Radel R** is polyphenylsulfone, **Udel** is polysulfone, and **Mindel** is a modified polysulfone. **Ryton** is a polyphenylene sulfide of Phillips Chemical Co., and **Supec** is one made by General Electric Plastics.

Polyimide foam is a spongy, lightweight, flame-resistant material that resists ignition up to 800°F (427°C) and then only chars and decomposes. Some formulations result in harder materials that can be used as lightweight wallboard or floor panels while retaining fire resistance. **Kapton** is Du Pont's polyimide film.

Aromatic polyketones are high-performance thermoplastics which include **polyetheretherketone** (PEEK), glass transition temperature of 290°F (143°C), and melting point of 635°F (335°C); **polyetherketone** (PEK), glass transition temperature of 310°F (154°C); **polyetherketoneketone** (PEKK), glass transition temperature of 310°F and melting point of 635°F; **polyaryletherketone** (PAEK), glass transition temperature of 338°F (170°C) and melting point of 718°F (381°C). Glass fiber reinforcement improves the strength, stiffness, and dimensional stability of these materials. **Victrex** is PEEK of Victrex USA Inc. **XC-2,** of ICI Fiberite, is PEEK prepreg tape for filament winding. It has a maximum service temperature of 600°F (316°C) and can be made in sheet form for compression molding. In addition, there are various ketone-based copolymers.

SYCAMORE. The wood of the tree *Acer pseudo-platanus,* which is also classified as a kind of maple, especially in England. The species cut as sycamore in the United States is largely *Platanus occidentalis.* The wood has a close, firm, tough texture and is yellowish, with a reddish-brown heartwood. The light-colored sapwood is up to 3 in (8 cm) thick in commercial trees. When quartered, the wood resembles quartered oak. The density is about 38 lb/ft³ (609 kg/m³). The surface is lustrous and takes a fine polish. It is used for veneers, flooring, furniture, cooperage, and handles and rollers. Two other species grown in the southwest are **California sycamore,** *P. racemosa,* and **Arizona sycamore,** *P. wrightii.*

TALC. A soft, friable mineral of fine colloidal particles with a soapy feel. It is a **hydrated magnesium silicate,** $4SiO_2 \cdot 3MgO \cdot H_2O$, with a specific gravity of 2.8 and a Mohs hardness of 1. It is white when pure, but may be colored gray, green, brown, or red with impurities. The pure white talc of Italy has been valued since ancient times for cosmetics. Talc is now used for cosmetics, for paper coatings, as a filler for paints and plastics, and for molding into electrical insulators, heater parts, and chemicalware. The massive block material, called **steatite talc,** is cut into electrical insulators. It is also called **lava talc. Talc dust** is a superfine, 400-mesh powder from the milling of steatite talc. It has an oily feel and is used in cosmetics. The more impure block talcs are used for firebox linings and withstand temperatures to 1700°F (927°C). Gritty varieties contain carbonate minerals and are in the class of soapstones. Varieties containing lime are used for making porcelain.

Talc of specified purity and particle size is marketed under trade names. **Asbestine** is a talc powder of 325 mesh for use as a filler. **Ceramitalc** is a talc powder used as a source of magnesia and to prevent crazing in ceramics. **Sierra Fibrene** is a California talc milled to 400 mesh. It is white and has a platy structure, and as an extender in paints, it wets easily and promotes pigment dispersion. **French chalk** is a high-grade talc in massive block form used for marking. The mineral occurs in the United States in the Appalachian region from Vermont to Georgia. **Georgia talc** for making crayons is mined in blocks. **Attasorb** and **Permagel** are finely powdered, cream-colored, hydrated **magnesium-aluminum silicate** from the mineral **attapulgite,** used for emulsifying and as a flatting agent and extender in paints. The material is also used in starch adhesives to improve shear strength. **Attacote** is the material in superfine particle size for use as an anticaking agent for hygroscopic chemicals. **Veegum F,** of R. T. Vanderbilt Co., is a fine, white, colloidal magnesium-aluminum silicate used as a suspending agent for oils and waxes.

Cordierite is a talclike mineral with a high percentage of magnesia used for refractory electronic parts. It is found sparsely in Norway, Finland, and Connecticut, usually in granite and gneiss, or in vitrified sandstones. When heated to 2624°F (1440°C), it is converted to sillimanite and glass. Synthetic cordierite is made by Muscle Shoals Electrochemical Corp. by mixing pure silica, magnesia, and alumina in various proportions and stabilizing with calcia. It is marketed as a powder for producing refractory insulating parts. Extruded cordierite serves as a substrate for the active catalyst metals in auto catalytic converters.

Agalite is a mineral having the same composition as talc but with a less soapy feel. It is used as a filler in writing papers, but is more wearing on the paper rolls than talc. The talc of northern New York, known as **rensselaerite,** does not have the usual talc slip, and has a fibrous nature. The hydrous aluminum silicate **pyrophyllite,** found in California, is similar to talc but with the magnesium replaced by aluminum. In mixtures with talc for wall tile it eliminates crazing. It is also substituted for talc as a filler for paints and paper. **Thix,** used as a thickening agent in emulsion paints, in cosmetics, and in textile finishes, is a refined, hydrous magnesium silicate marketed as a 200-mesh powder. It contains 56% silica, 26 magnesia, 2.8 calcia, 2.5 Na_2O, and 1.1 lithia.

Magnesium silicate, used as a filler in rubber and plastics, and also as an alkaline bleaching agent for oils, waxes, and solvents, is a white, water-insoluble powder of composition $MgSiO_4$, having a pH of 7.5 to 8.5. In the cosmetic trade it is known as **talcum powder.** **Magnesol,** of Westvaco Chemical Co., is finely ground magnesium silicate. **Brite-Sorb 30** is a synthetic magnesium silicate with high adsorption and filtering power. The **magnesium trisilicate** used in pharmaceuticals as an antacid is of extreme fineness, the superbulking grade having 65% of the particles less than 197 μin (5 μm) in size. The material known as **killas** from the tin mines of Cornwall is a slaty schist. It is finely ground and used as talc.

TALL OIL. An oily, resinous liquid obtained as a by-product of the sulfite paper-pulp mills. The alkali saponifies the acids, and the resulting soap is skimmed off and treated with sulfuric acid to produce tall oil. The name comes from the Swedish **talloel,** meaning pine oil. The crude oil is brown, but the refined oil is reddish yellow and nearly odorless. It has a specific gravity of 0.98, flash point of 360°F (182°C), and acid number about 165. The oil from Florida paper mills contains 41 to 45% rosin, 10 to 15 pitch, and the balance chiefly fatty acids. The fatty acids can be obtained separately by fractionating the crude whole oil. The oil also contains up to 10% of the phytosterol **sitosterol,** used in making the drug cortisone.

Tall oil is used in scouring soaps, asphalt emulsions, cutting oils, insecticides, animal dips, in making factice, and in plastics and paint oils. It is marketed in processed and concentrated form. **Seecotol** is crude tall oil from Georgia-Pacific. **Xtol,** from the same firm, is a distilled grade for use in surfactants, soaps, asphalt, alkyds, and as a chemical intermediate. **Acintol,** from Arizona Chemical Co., contains 60 to 68% fatty acids, 30 to 38 rosin acids, and has an acid value of 185. **Flextal** is processed tall oil containing 60% rosin acids. Detergents are made by reacting tall oil with ethylene oxide. Saturated alcohols are produced by high-pressure hydrogenation of tall oil. The high linoleic acid content makes tall oil suitable for making drying oils. **Lumitol** is a German vinyl plastic produced by reacting tall oil with acetylene. It is used for coatings. **Smithco RT,** of Archer-Daniels-Midland Co., used for varnishes and paints, is refined tall oil esterified with glycerin. **Smithco PE** is tall oil esterified with pentaerythritol. **Ardex PE,** of the same company, is a varnish oil that dries quickly to a hard film, made by esterifying tall oil with pentaerythritol. **Sulfonated tall oil** is used to replace sulfonated castor oil in coating mixes for paper to increase folding strength. **Opoil** is a crude tall oil, and **Facoil** is the refined oil with 60% fatty acid content and low rosin acid content. **Acolin, Acosix,** and **Aconon** are grades of refined tall oil. **Pamac,** of Hercules Inc., consists of tall oil monobasic fatty acids, used in resin coatings. **Lytor 100** is a tall-oil rosin for paper sizing, and for use as an adhesive and ink resin. It is produced by Georgia-Pacific.

TALLOW.   A general name for the heavy fats obtained from all parts of the bodies of sheep and cattle. The best grades of internal fats, or **suet,** are used for edible purposes, but the external fats are employed for lubricants, for mixing with waxes and vegetable fats, for soaps and candles, and for producing chemicals. The tallows have the same general composition as lard, but are higher in the harder saturated acids, with about 51% of palmitic and stearic acids, and lower in oleic acid. The edible grades known as **premier jus,** prime, and edible are white to pale yellow, almost tasteless, and free from disagreeable odor; but the nonedible or industrial tallows are yellow to brown unless bleached. The best grade of industrial tallow is Packers No. 1. **Peacock** is an acidless grade for metalworking, lubricants and additives, soaps, mold release, animal feed supplements, inks, and pigments; other premium and custom grades of Peacock are also produced by George Pfau's Sons Co. **White grease, yellow grease,** and **brown grease** may be hog fat, or they may be tallows with a titer below 104°F (40°C), the titer being the only commercial distinction between tallow and fat. Tallow is thus all animal fat above 104°F titer. **Beef tallow** is used to produce stearic acid, for leather dressing, lubricating greases, and making soap. **Mutton tallow** contains less liquid fat and is harder, but it becomes rancid more easily. Tallow for industrial use is generally highly purified and chemically treated, and marketed under trade names. **Adogen 442,** of Archer-Daniels-Midland Co., used as a softener for textiles, is a dimethyl **hydrogenated tallow.** It comes as a nearly white, odorless paste in isopropanol and water, and is dispersible in water or in organic solvents. **Wax A,** a mixed hydrogenated tallow from Chemol Co., is a tallow glyceride.

TANNING AGENTS.   Materials, known as **tannins,** used for the treatment of skins and hides to preserve the hide substance and make it resistant to decay. The tanned leather is then treated with fats or greases to make it soft and pliable. Tannins may be natural or artificial. The natural tannins are chiefly vegetable, but some mineral tanning agents are used. The vegetable tannins are divided into two color classes: the **catechol** and the **pyrogallol.** The catechol tannins are cutch, quebracho, hemlock, larch, gambier, oak, and willow. The

pyrogallol tannins are gallnuts, sumac, myrobalans, chestnut, valonia, divi-divi, and algarobilla. Catechol tannin is distinguished by giving a greenish-black precipitate with ferric salts; the pyrogallol tannins give a bluish-black precipitate. The catechol tannins, in general,
produce leathers that are more resistant to heat and decay than the
pyrogallols. Some tannins contain considerable coloring or dye matter, but the color that a tannin imparts to leather may be lightened or
darkened by raising or lowering the acidity of the tannin bath. In the
ink industry the catechol tannins are known as *iron-greening,* and the
pyrogallol tannins as *iron-bluing,* and the latter are used for making
writing inks. Catechol is also produced synthetically from coal tar. It
is a water-soluble **dihydric phenol** in white, crystalline granules
known as ortho-**dihydroxybenzene,** $C_6H_4(OH)_2$. It is used in some
inks and for making dyestuffs, medicinals, and antioxidants.

Alum tanning is an ancient process but was introduced in Europe
only about the year 1100, and the alum- and salt-tanned leather was
called **Hungary leather. Formaldehyde** is also used as a tanning
agent. Formaldehyde was patented as a tanning agent in 1898. A
later patent covered a rapid process of tanning sheepskins with alcohol and formalin and then neutralizing in a solution of soda ash.
Unlike vegetable agents, formaldehyde does not add weight to the
skin. It is often used as a pretanning agent to lessen the astringency
of the vegetable tannin and increase its rate of diffusion. **Melamine**
resins are used for tanning to give a leather that is white throughout
and does not yellow with age. Leather may also be tanned with
chromic acid or chrome salts, which make the fibers insoluble and
produce a soft, strong leather. Chrome alum, sodium or potassium
dichromates, or products in which chromic acid has been used as an
oxidizing agent may be used. **Chrome tanning** is rapid and is used
chiefly for light leathers. **Tanolin** is a name for basic **chromium
chloride** marketed in crystal form for use in the chrome tanning of
leather. **Santotan KR** is a trade name of Monsanto Co. for basic
chromium sulfate, $Cr_2(SO_4)_2(OH)_2$, used as a one-bath chrome-tan-
ning agent. This material is also used for treating magnesium-alloy
parts to give a gray to black surface color. **Panchrome,** an English
tanning agent, is a sulfur dioxide dichromate. **Chromalin** is a glyc-
erin-reduced dichromate. Chrome-tanned leather is more resistant to
heat than vegetable-tanned leathers, withstanding temperatures to
200°F (93°C). Chrome tanning is used for shoe-upper leathers and for
gloves, beltings, and packings. **Iron-tanned leather** is produced by
pretanning with formaldehyde, then tanning with ferric salts and
trisodium phosphate, and neutralizing with a solution of phthalic
anhydride and sodium carbonate. The leather is soft, will absorb
much oil and grease, and is suitable for use where a pliable leather is

desired. Glutar aldehyde gives a soft, bulky leather suitable for garments. It may be blended with chrome or vegetable tanning agents.

In tanning processes various supplementary materials may be used to give special properties to the leathers. Glucose or starch may be used to make the leather more plump. Hydrochloric acid is used in two-bath chrome tanning to enhance the feel and appearance of the leather. Synthetic tannins, or **syntans,** are largely condensation products made by condensing sulfonated phenols with formaldehyde. **Neradol D** is such a syntan. **Tansyn** is the trade name of an English syntan of this kind. **Permanol,** of Monsanto, is a sulfonic acid condensation syntan in liquid form used to produce light-fast white leathers. The free sulfuric acid is completely neutralized. Syntans do not add weight to leather and are seldom used alone. They are marketed under trade names. **Leukanol,** of Rohm & Haas Co., has a bleaching action and is used in combination with vegetable tannins to speed up the tanning and to give a light-colored leather. **Orotan,** of the same company, is a sulfonated phenol formaldehyde which makes a good shoe leather when used alone. **Tanigan,** a German tannin, is a complex condensation product produced from water pulp–mill liquor and formaldehyde or diphenyl methane.

TANTALUM. A white, lustrous metal, symbol Ta, resembling platinum. It is one of the most acid-resistant metals and is classified as a noble metal. Its specific gravity is 16.6, or about twice that of steel, and because of its high melting temperature [5425°F (2996°C)], it is called a *refractory metal.* In sheet form, it has a tensile yield strength of 50,000 lb/in^2 (345 MPa) and is quite ductile. At very high temperatures, however, it absorbs oxygen, hydrogen, and nitrogen and becomes brittle. Its principal use is for electrolytic capacitors, but because of its resistance to many acids, including hydrochloric, nitric, and sulfuric, it is also widely used for chemical processing equipment. It is attacked, however, by hydrofluoric acid, halogen gases at elevated temperatures, fuming sulfuric acid, and strong alkalies. Because of its heat resistance, tantalum is also used for heat shields, heating elements, vacuum-furnace parts, and special aerospace and nuclear applications. It is also a common alloying element in superalloys. The metal is used for prosthetic applications, and **tantalum carbide** is used in cemented-carbide cutting tools. **Tantalum alloys,** including tungsten and tungsten-hafnium compositions, such as **Ta-10W, T-111** (8% tungsten, 2 hafnium), and **T-222** (9.6% tungsten, 2.4 hafnium, and 0.01 carbon), are used for rocket-engine parts and special aerospace applications. The tensile yield strength of Ta-10W is about 158,000 lb/in^2 (1,089 MPa) at room temperature and 90,000 lb/in^2 (621 MPa) at 1600°F (871°C).

TANTALUM ORES. The most important ore of the metal tantalum is **tantalite.** When pure, its composition is $FeO \cdot Ta_2O_5$, but the American ore may contain only from 10 to 40% **tantalic oxide,** Ta_2O_5, and the Australian ore may contain as high as 70%. The ore is marketed on the basis of 60% tantalic oxide content. Tantalite occurs usually as a black, crystalline mineral with a specific gravity up to 7.3. It often contains manganese, tin, titanium, and sometimes tungsten; the tantalum may be replaced by columbium, which is similar to it. When the columbium content in the ore predominates, the mineral is called **columbite.** Tantalite also contains small amounts of germanium. The tantalite of the Congo usually contains tin. The ore from the Lukushi Basin contains 58% Ta_2O_5, 16.5 Cb_2O_5, 12.5 MnO, 4.5 Fe_2O_3, and 1.6 SnO_2, with some zirconium and titanium oxides. **Thoreaulite** of that region contains 72 to 74% Ta_2O_5 and 20 to 22 SnO_2. Tantalum metal is produced from tantalite by dissolving in acid and separating the tantalum salts from the columbium by precipitation. The tantalum salts are reduced to powder metal, which is then compressed into rods and sintered and rolled. The tantalite ore of Manitoba is embedded in pegmatite, and the crude ore contains about 0.25% Ta_2O_5.

A tantalum ore that is abundant at Wodgina, western Australia, is **mangano tantalite,** which contains about 69% tantalic oxide, 15 **columbium pentoxide,** Cb_2O_5, and 14 manganese protoxide, with a little tin oxide. The specific gravity of the ore is 6.34. **Microlite,** an ore found at Wodgina and in the McPhee Range of western Australia, contains 76% Ta_2O_5 and 4 to 7 Cb_2O_5. **Tanteuxenite,** another western Australian ore, contains 24 to 47% Ta_2O_5 and 4 to 14 Cb_2O_5. **Tapiolite,** of Australia, contains 82% Ta_2O_5 and 2 Cb_2O_5. **Euxenite,** of Idaho, contains about 28% columbium-tantalum oxide. The mineral **pyrochlore,** of Canada, is composed of complex oxides of tantalum, columbium, sodium, and calcium, and the metal oxides are obtained by acid extraction.

TAR. A black, solid mass obtained in the destructive distillation of coal, peat, wood, petroleum, or other organic material. When coal is heated to redness in an enclosed oven, it yields volatile products and the residue coke. Upon cooling, the volatile matter, tar, and water are deposited, leaving the coal gases free. Various types of coal yield tars of different qualities and quantities, depending on the relative levels of more than 200 compounds, including **guaiacol, cresol, xylenol, crotonic acid, maltol,** and **ketones.** Anthracite gives little tar, and cannel coal yields large quantities of low-gravity tar. In the manufacture of gas, the tar produced from bituminous coal is a viscous, black liquid containing 20 to 30% free carbon, and is rich in benzene, toluene, naphthalene, and other aromatic compounds. In the dry state, this tar has a specific gravity of about 1.20. Tar is also produced as a by-product from coke ovens. Crowley Tar Products Co. sells both crude and refined coal tars under the trade name **Impervotar.**

Coal tars are usually distilled to remove the light aromatics which are used for making chemicals, and the residue tar, known as **treated tar,** or **pitch,** is employed for roofing, road making, and bituminous paints and waterproofing compounds. **Coal-tar pitch** is the most stable bituminous material for covering underground pipes, as a binder for electrodes used in aluminum smelting, and as an impregnant in refractories. **Tarvia** is the trade name of a refined coal tar, marketed by Barrett Co. in various grades. **Tarmac** is practically the same material. **Bituplastic,** used for coating pipes and structures, is a refined coal-tar pitch that is odorless and quick-drying. **Bituvia** is a road tar, produced in various grades by Reilly Tar & Chemical Corp. **Coal-tar carbon** amounts to about 32% of the original tar. It is marketed in lump form for chemical use. The fixed carbon content is 92.5 to 95.6%, sulfur about 0.30, and volatile matter 3 to 6. **Calcined carbon,** from coal tar, contains less than 0.5% sulfur and 0.5 volatile matter.

The lightest distillate of coal tar, benzene, is used as an automotive fuel. **Coal-tar oils** are used as solvents and plasticizers. They consist of various distillates or fractions up to semisolids. **Tar oil** from brown coal tars was used for diesel fuel oil by extracting the phenols with methyl alcohol. **Bardol B,** of Allied Signal, is a clear, yellow, coal-tar fraction of specific gravity 1.0 to 1.04, used as a plasticizer for synthetic rubber, while **Carbonex** is a solid, black tar hydrocarbon in flake form used as a rubber plasticizer. The softening point is between 205 and 220°F (96 and 104°C). **Xylol** is a water-white liquid of specific gravity 0.860 to 0.870, distilling between 275 and 365°F (135 and 185°C). It is a mixture of **xylenes,** which are dimethyl benzene, $C_6H_4Me_2$.

Naphthalene and anthracene are among the distillates. **Anthracene** is a colorless, crystalline product of composition $C_6H_4:(C_2H_2):C_6H_4$ and melting point 423°F (217°C) used for the production of dyes, resins, plasticizers, tanning agents, and inhibitors. Crystals of anthracene are used for scintillation counters for gamma-ray detection. **Naphthalene oils,** distilled from coke-oven tars, contain 65% naphthalene. They are used for making phthalic anhydride, ß-naphthol, and dye intermediates. **Quinoline,** called also **benzazine** and **chinoline,** is a liquid with a tar odor. It has a double-ring molecular structure of empirical formula C_9H_7N, and it boils at 459°F (237°C). It is used for making antiseptics, pharmaceuticals, insecticides, and rubber accelerators.

Pine tar is a by-product in the distillation of pinewood. It is a viscous, black mass and is much used for roofing. **Rosintene** is a light grade from Crowley Tar Products Co. It is also sometimes called pitch, but pitch is the tar with the pine-tar oil removed, known as **pine pitch. Tarene** is a dry, free-flowing powder made by absorbing pine tar into a synthetic hydrous calcium silicate which absorbs 4 times its own weight of liquid tar. It is used for formulating with rubbers. **Navy pitch** and **ship pitch** are names that refer to specification pine pitch for marine use. It is medium hard to solid, has a specific gravity of 1.08 to 1.10, has a melting point not less than 148°F (64°C), is completely soluble in benzol, and has uniform black color, or red-brown in thin layers. **Wood tar** from the destructive distillation of other woods is a dark-brown, viscous liquid used as a preservative, deriving this property from its content of creosote. **Stockholm tar,** a name now out of commercial use, was a term employed in shipbuilding for the tar obtained from the crude distillation of pine stumps and roots.

TEA. The dried leaves of the shrubs *Camellia sinensis* and *Thea sinensis,* grown chiefly in southern Asia, Japan, Sri Lanka, Russian Caucasia, and Indonesia but also in Peru and in Tanzania. The plant requires a warm, subtropical, humid climate. Tea leaves are valued for making the beverage tea which contains the alkaloid caffeine and is stimulating. The leaves contain more caffeine than coffee berries, but the flavor is different. Like coffee, it also contains tannin, which dissolves out when the tea leaves are steeped too long, and is an astringent. In well-prepared tea, the tannins have been oxidized to the brown-and-red tannin which is not easily soluble and does not enter the properly steeped beverage to any great extent, although it gives the beverage its color. The flavor and aroma of tea depend largely upon the age of the leaves when picked and the method of drying. **Green tea** is made by drying the fresh leaves in the sun or artificially, while **black tea** is made by first fermenting the leaves and then drying. Rolling is done to break the leaves and release the juices. **Epigallocatechin,** a chemical found in green tea, is a strong antioxidant and anticarcinogen. The **oolong tea** of Taiwan is partly fermented and is intermediate between green and black. **Pouchong tea** is graded by mixing oolong with aromatic flowers such as jasmine. Tea is also graded by the size and age of the leaf. Flowery orange pekoe is the smallest leaf, **orange pekoe tea** the second, then pekoe, pekoe souchong, and souchong. Tea also varies with varieties grown in different climates so that Japan tea, China tea, and Ceylon tea have different flavors.

Commercial tea is usually a blend of different varieties to give uniformity under one trade name. The blending of tea is considered an art. **Brick tea,** made in China, is produced from coarse leaves and twigs which have been fermented. They are mixed with tea dust, treated with rice water, and pressed into bricks. **Cake tea,** or **puerh tea,** is produced in Yunnan. The leaves are panned, sun-dried, steamed, and pressed into circular cakes. **Tablet tea** is selected tea dust pressed into small tablets. **Tea waste** is the final dust from the tea siftings, and it is used for the production of caffeine. **Teaseed oil,** or **sasanqua oil,** is from the seeds of another species of the tea plant, *Thea sasanqua,* of Asia. The seeds contain 58% of a pale-yellow oil with a specific gravity of 0.916 used for lubrication, hair oil, soap, and pharmaceutical preparations. **Paraguayan tea,** or **yerba maté,** used in immense quantities as a beverage in Argentina, Paraguay, Brazil, and some other South American countries, consists of the dried, smoked leaves of the small evergreen tree *Ilex paraguayensis,* native to Paraguay and southern Brazil. The growing region is about the upper Paraná River. It was an ancient beverage of the Indians, and cultivation began on a large scale under the early missionaries. It contains a higher percentage of caffeine than tea or coffee, 3.88%, but less tannin. The flavor of the steeped beverage is different from that of tea. **Cassine** is a tealike beverage obtained from the twigs and leaves of two species of holly, *Ilex cassine* and *I. vomitoria,* found in the southern United States from Virginia to Texas. It was called **Yaupon** by the Indians and used medicinally and in religious rites. During the Civil War it was used in the south as a tea substitute. The beverage has an odor similar to tea but has a dark color with a sharp, bitter taste. It contains caffeine, tannin, and essential oils.

TEAK.   The wood of the tree *Tectona grandis,* of southern Asia. It resembles oak in appearance, is strong and firm, and in England is called **Indian oak.** It contains an oil that gives it a pleasant odor and makes it immune to the attacks of insects. It is used for boxes, chests, home furnishings, and woodwork on ships. The color is golden yellow, the grain is coarse and open, and the surface is greasy to the touch. It is one of the most durable of woods, and also has small shrinkage. The density is 40 lb/ft^3 (641 kg/m^3). In Burma large plantations grow teak for export. Trees grow to a height of 100 ft (30 m) with a diameter of 3 ft (0.9 m). The growth is slow, a 2-ft (0.6-m) tree averaging 150 years of age. The wood marketed as **African teak,** known also as **iroko,** is from the tree *Chlorophora excelsa,* of west Africa, and is unlike true teak. It is a firm, strong wood with a brownish color and a coarse, open grain. The weight is somewhat less than that of teak, and it is harder to work, but it is resistant to decay and to termite attack and is used in ship construction. **Surinam teak** is the wood of the tree *Hymenea courbaril* of the Guianas and the West Indies. It is also called **West Indian locust.** The wood is dark brown, hard, heavy, and difficult to work. It is not very similar to teak and not as durable. **Seacoast teak,** or **bua bua,** is a hard, yellow, durable wood from species of the tree *Guettarda* of Malaya. **Australian teak,** from New South Wales, is from the tree *Flindersia australis.* It is yellowish-red, close-grained, and hard with an oily feel resembling teak, but more difficult to work. **In wood,** of Burma, also called **eng teak,** is from the tree *Dipterocarpus tuberculatus,* from which gurjun balsam is obtained. The wood is reddish-brown, and it is not as durable as teak. Two woods of Brazil are used for the same purposes as teak: the **itaúba,** *Silvia itauba,* a tree growing to a height of about 75 ft (23 m) in the upland forests of the lower Amazon, and **itaúba preta,** *Oreodaphne bookeriana,* a larger tree growing over a wider area. The first is a greenish-yellow wood with compact texture and rough fiber, formerly prized for shipbuilding. The second resembles teak more closely and is used for cabinetwork.

TELLURIUM. An elementary metal, symbol Te, obtained as a steel-gray powder of 99% purity by the reduction of **tellurium oxide,** or **tellurite,** TeO_2, recovered from the residues of lead and copper refineries. It is marketed in slabs and sticks and is sometimes known as **sylvanium.** It occurs also with gold in Washington and Colorado as **gold telluride,** $AuTe_2$. The specific gravity is about 6.2 and the melting point 842°F (450°C). The chief uses are in lead to harden and toughen the metal, and in rubber as an accelerator and toughener. Less than 0.1% tellurium in lead makes the metal more resistant to corrosion and acids, and gives a finer grain structure and higher endurance limit. Tellurium-lead pipe, with less than 0.1% tellurium, has a 75% greater resistance to hydraulic pressure than plain lead. A **tellurium lead,** patented in England, contains 0.05% tellurium and 6 antimony. **Tellurium copper** (C14500, C14510, and C14520) is a free-machining copper containing 0.3 to 0.7% tellurium. It machines 25% more easily than free-cutting brass. The tensile strength, annealed, is 30,000 lb/in^2 (207 MPa), and the electric conductivity is 98% that of copper. A **tellurium bronze** containing 1% tellurium and 1.5 tin has a tensile strength, annealed, of 40,000 lb/in^2 (276 MPa), and is free-machining. Tellurium is used in small amounts in some steels to make them free-machining without making the steel hot-short, as do increased amounts of sulfur. But tellurium is objectionable for this purpose because inhalation of dust or fumes by workers causes garlic breath for days after exposure, although the material is not toxic. As a secondary vulcanizing agent with sulfur in rubber, tellurium in very small proportions, 0.5 to 1%, increases the tensile strength and aging qualities of the rubber. It is not as strong an accelerator as selenium, but gives greater resistance to the rubber. **Telloy** is the trade name of R. T. Vanderbilt Co. for **tellurium powder** ground very fine for rubber compounding.

TERNEPLATE.   Sheet or plate steel coated by hot-dip processes with a thin layer of lead containing 3 to 15% tin. *Terne* means dull and refers to the color of the coating as compared with bright tinplate. The coating, intended to enhance corrosion resistance primarily, also improves formability and solderability. Because of the toxicity of lead, however, special precautions are required in fabrication operations. The terms **short terne** and **long terne** pertain to plate products and sheet products, respectively. Standard coating-thickness designations for sheet products range from LT10 (no minimum) to LT110 [1.10 oz/ft^2 (336 g/m^2)] total weight both sides based on a triple-spot test. Automobile gasoline-fuel tanks have been the major application, although terneplate also has been used for fuel tanks of lawn mowers and outboard marine motors as well as for roofing and building construction, caskets, and other applications.

TERRA COTTA.   A general English term applied to fired, unglazed, yellow, and red clay wares; in the United States it refers particularly to the red-and-brown, square and hexagonal tiles made from common brick clay, always containing iron. Some special terra cottas are nearly white, while for special architectural work other shades are obtained. The clays are washed, and only very fine sands are mixed with them in order to secure a fine, open texture and smooth surface. Terra cotta is used for roofing and for tile floors, for hollow building blocks, and in decorative construction work. Good, well-burned terra cotta is less than 1.5 in (3.8 cm) thick. Terra cotta is light, having a density of 120 lb/ft^3 (1,922 kg/m^3), and withstands fire and frost.

TETRACHLOROETHANE.   A colorless liquid of the chemical formula $CHCl_2 \cdot CHCl_2$ employed as a solvent for organic compounds such as oils, resins, and tarry substances. It is an excellent solvent for sulfur, phosphorus, iodine, and various other elements. It is used as a paint remover and bleacher, as an insecticide, and in the production of other chlorine compounds. It is also called **acetylene tetrachloride** and is made by the combination of chlorine with acetylene. Tetrachlorethane boils at 291°F (144°C), freezes at −33°F (−36°C), is nonflammable, and has a specific gravity of 1.601. It is narcotic and toxic, and the breathing of the vapors is injurious. Mixed with dilute alkalies, it forms explosive compounds. In the presence of moisture it is very corrosive to metals. Mixed with zinc dust and sawdust, it is employed as a smoke screen.

THALLIUM. A soft, bluish-white metal, symbol Tl, resembling lead but not as malleable. The specific gravity is 11.85, and melting point 578°F (302°C). At about 600°F (316°C) it ignites and burns with a green light. Electrical conductivity is low. It tarnishes in air, forming an oxide coating. It is attacked by nitric acid and by sulfuric acid. The metal has a tensile strength of 1,300 lb/in² (9 MPa) and a Brinell hardness of 2. **Thallium-mercury alloy,** with 8.5% thallium, is liquid with a lower freezing point than mercury alone, -76°F (-60°C), and is used in low-temperature switches. **Thallium-lead alloys** are corrosion-resistant and are used for plates on some chemical equipment parts.

The metal occurs in copper pyrites and zinc ores, and the chief source is the flue dust of smelters from sphalerite ores. Four rare minerals are ores of thallium: **vrbaite,** $Tl_2S \cdot 3(AsSb)_2S_3$, is found in Macedonia; **lorandite,** $Tl_2S \cdot 2As_2S_3$, is found in Macedonia and Wyoming; **hutchinsonite,** $PbS \cdot (TlAg)_2S \cdot 2As_2S_3$, occurs in Switzerland and Sweden; and **crooksite,** $(CuTlAg)_2Se$, is found in Sweden. The salts of thallium are highly poisonous, the sulfide being used as a rat poison. **Thallium oxysulfide** is used in light-sensitive cells. It is also sensitive to infrared rays and is used for dark signaling. **Thallium sulfate,** $Tl_2(SO_4)_3$, is a crystalline powder used as an insecticide. It is more toxic than lead compounds. Thallium also gives high refraction to optical glass. Thallium bromide iodide crystals, grown synthetically, are used for infrared spectrometers.

The so-called **alkali-halide crytals** used in the **discriminator circuits** of **scintillators** for gamma spectrometry contain thallium. They separate the slow-decaying pulses of protons produced as fast neutrons from the electron pulses produced by gamma absorption. A French crystal, called **Scintibloc,** is **sodium iodide thallide,** NaI(Tl). The **cesium iodide thallide** crystal, CsI(Tl), gives a very blue light under electron excitation.

THERMOPLASTIC ELASTOMERS. A group of polymeric materials having some characteristics of both plastics and elastomers. Also called **elastoplastics** and **TPEs.** Requiring no vulcanization or curing, they can be processed on standard plastics processing equipment. They are lightweight, resilient materials that perform well over a wide temperature range. There are a half-dozen different types of elastoplastics. The **olefinics,** or **TPOs,** range in Shore A hardness from 55 to 90. Specialty flame-retardant and semiconductive grades are also available. The TPOs are used in autos for paintable body filler panels and air deflectors, and as sound-deadening materials in diesel-powered vehicles. The **styrenics** are block copolymers, composed of polystyrene segments in a matrix of polybutadiene or polyisoprene.

Lowest in cost of the elastoplastics, they are available in crumb grades and molding grades, hardness 35 to 95.

Thermoplastic urethanes are of three types: **polyester-urethane, polyether-urethane,** and **caproester-urethane.** All three are linear polymeric materials, and therefore do not have the heat resistance and compression set of the cross-linked urethanes. Shore A hardness typically ranges from 55 to 90. A harder grade, Shore D 70, is used for low-deflection, load-bearing applications.

Copolyether-ester thermoplastic elastomers are high-performance, high-cost materials. The five basic grades by Shore D hardness are 40, 47, 55, 63, and 72. Applications include tubing and hose, V belts, couplings, oil-field parts, and jacketing for wire and cable. Their chief characteristic is toughness and impact resistance over a broad temperature range. A thermoplastic elastomer of polyvinyl chloride and nitrile provides excellent oil resistance. **Trans-1, 4 polyisoprene** thermoplastic elastomer was developed as a replacement for refined balata rubber in golf ball covers.

THORIUM. A soft, ductile, silvery-white metal, symbol Th, occurs in nature to about the same extent as lead but so widely disseminated in minute quantities difficult to extract that it is considered a rare metal. It was once valued for use in incandescent gas mantles in the form of **thorium nitrate,** $Th(NO_3)_4$, but is now used chiefly for nuclear and electronic applications. **Thorium powder** is produced by calcium reduction of thorium oxide. The impure powder burns in the air with great brilliance. Pure thorium metal in sheet form has a specific gravity of 11.7, a melting point of 3090°F (1699°C), and a tensile strength of about 35,000 lb/in^2 (241 MPa). Even small amounts of impurities affect the physical properties greatly, and cold working increases the strength. The metal is dissolved by aqua regia or by hydrochloric acid.

Natural thorium consists largely of the alpha-emitting isotope **thorium 232,** and is a powerful emitter of alpha rays. Thorium produces fissile material, uraniuim 233, only when triggered by another fission material. Under neutron bombardment it forms **protactinium** which is nonfissile but decays slowly into fissile uranium 233. But, in rapid burning, the buildup of protactinium may be converted to the nonfissile uranium 234.

Thorium 230 is found in minerals that contain uranium and radium and was originally considered as a separate metal under the name of **ionium.** It is radioactive, emitting alpha rays. It has a half-life of 76,000 years, slowly converting to radium. The original production was from the fractionation of uranium ores. It was used as an additive in spark plug wire, but is too expensive for this purpose.

The chief **thorium ore** is the mineral **monazite,** occurring as sand or in granular masses, usually as sea sand. It is the chief source of thorium oxide and of the rare-earth metals. Most of the monazite comes from Brazil, India, and the East Indies. The monazite sands of Brazil contain 8% **thorium oxide,** or **thoria,** ThO_2. The ore of India may have as high as 10%, but is marketed on the basis of 8% oxide and 60 rare-earth metals. Thoria has a high melting point, 5522°F (3050°C), but its use as a refractory ceramic is limited because of its high cost and radioactivity. Monazite contains about 0.008 lb (3.5 g) of **mesothorium** per 1,000 tons (907 metric tons) and usually has 30 to 35% of the oxides of lanthanum, yttrium, neodymium, and praseodymium, and a small amount of europium. Mesothorium was originally considered a separate element, but is an isotope of thorium, with an atomic weight of 228 and a half-life of 6.7 years. The radiations from mesothorium are the same as those from radium—alpha, beta, and gamma rays. As it decomposes, it forms **radiothorium,** which is identical in chemical properties to thorium but emits a powerful alpha radiation. It is used in luminous paints and is a safer activator for this purpose than radium, but is scarcer and more expensive, and has a shorter life.

The type of monazite called **uranothorite,** from the Bancroft area of Canada, contains from 0.04 to 0.27% thorium oxide. The thorium is recovered from the waste liquors of the uranium treatment plant. The rare mineral **thorite,** found in Norway, is a **thorium silicate,** $ThSiO_2$. It occurs in crystals or massive, orange to black in color, and has a resinous luster and a specific gravity of about 5. **Thoria-urania ceramics** are used for reactor-fuel elements. They are reinforced with columbium or zirconium fibers to increase thermal conductivity and shock resistance. **Thorium-tungsten alloys** have been used for very high-voltage electronic filaments. The **incandescent mantle,** invented by Welsbach in 1893 and widely used during the period of gas lighting, consisted of a mixture of 98 to 99% thorium nitrate and 1 to 2 cerium oxide. The nitrate is converted to thorium oxide on ignition, with an increase of 10 times its original volume, and glows in the gas flame with an intense, white light.

THUYA. The wood of the tree *Thuya plicata,* also known as **western red cedar, giant arbor vitae, shinglewood,** and **Pacific red cedar.** The tree grows in cool, humid coastal regions from Alaska to northern California, and the wood is widely used for shingles, poles, and tanks. It is lightweight, soft, and weak, with a straight, coarse grain, but is durable. The sapwood is white and the heartwood reddish. The tree grows to great size, reaching to 200 ft (61 m) in height and 16 ft (5 m) in diameter at the age of 1,000 years. **Northern**

white cedar is the wood of the tree *T. occidentalis,* of the northeastern United States. It is also called **white cedar, arbor vitae, swamp cedar,** or simply **cedar.** The wood is soft, knotty, brittle, and weak, but very durable. It is used for shingles, poles, posts, and lumber for small boats. The sapwood is white and the heartwood light brown. The trees have a diameter of 1 to 3 ft (0.3 to 0.9 m) and a height of 25 to 75 ft (8 to 23 m). **Thuya leaf oil,** used as a fixative in perfumery, is a colorless oil with a bornyl acetate odor, distilled from the leaves.

TIN. A silvery-white, lustrous metal, symbol Sn, with a bluish tinge. It is soft and malleable and can be rolled into foil as thin as 0.0002 in (0.0051 cm). Tin melts at 450°F (232°C). Its specific gravity is 7.298, close to that of steel. Its tensile strength is 4,000 lb/in^2 (28 MPa). Its hardness is slightly greater than that of lead, and its electrical conductivity is about one-seventh that of silver. It is resistant to atmospheric corrosion, but is dissolved in mineral acids. The cast metal has a crystalline structure, and the surface shows dendritic crystals when cast in a steel mold. **Tin pest** is the breaking up of the metal into a gray powder which occurs below 66°F (19°C), and the metal is not used for applications at very low temperatures.

Tin is used in brasses, bronzes, and babbitts, and in soft solders. Tin with 0.4% copper is used as foil and for collapsible tubes. One of the most important uses is for the making of tinplate, an electroplating material. **Electroplated tin** has a fine, white color, gives a durable protective finish, and also has a lubricating effect as a bearing surface. **Standard tin** of the London Metal Exchange must contain over 99.75% tin. The common grade is known as **Grade A tin. Straits tin** is 99.895% pure. Federal specifications for **pig tin** are 99.80% minimum. **Block tin** is virgin tin cast in stone molds. Even small traces of impurities have an influence on the physical properties of tin. Lead softens the metal; arsenic and zinc harden it. An addition of 0.3% nickel doubles the tensile strength; 2 copper increases the strength 150%. Pure tin melts sharply, but small amounts of impurities broaden the melting point. **Tin powder,** used for making sintered alloys, is 99.8% pure, in powder from 100 to 300 mesh. The **tin crystals** used in the chemical industry are **tin chloride,** or **stannous chloride,** $SnCl_2 \cdot 2H_2O$, coming as large, colorless crystals or white, water-soluble flakes, melting at 475°F (246°C). They are also used for immersion tinning of metals and for sensitizing glass and plastics before metallizing. The chief source of tin is the mineral cassiterite, but Nigerian columbite may contain up to 6% tin oxide. The principal tin-producing countries are Indonesia, Malaya, Bolivia, China, and Nigeria; but tin mines have been worked in Cornwall since ancient times, and tin is also found in Canada and in irregular quantities in some other areas.

Tin oxide, or **stannous oxide,** is a fine, black, crystalline powder of composition SnO, made by oxidizing tin powder. It is used as an opacifier in ceramic enamels, as a ceramic color, as an abrasive, and as a coating for conductive glass. As a color in ceramics it is light-stable and acid-resistant. With magnesium and cobalt oxides it gives a sky-blue color called **cerulean blue.** It is also used with copper oxide to produce ruby glass.

Stannic oxide, SnO_2, is a white powder used in ceramic glazes as an opacifier and for color. As little as 1 to 2% gives fluidity and high luster to glass. With chromates and lime, it gives pinks and maroons in enamels, and with vanadium compounds it gives yellows. With gold chloride it gives brilliant-red jewelry enamels. **Protectatin** is the name of the Tin Research Institute for a thin, invisible film of oxide on tinplate to protect against sulfur staining and to give a base for paint. It is produced by dipping the tinplate in a solution of trisodium phosphate, sodium dichromate, and sodium hydroxide. **Potassium stannate,** $K_2SnO_3 \cdot 3H_2O$, or **sodium stannate,** $Na_2SnO_3 \cdot 3H_2O$, may be used for immersion tinning of aluminum. Both come as white, water-soluble crystals. The term **organotin** usually refers to butyl compounds of the metal used as catalysts, or heat and light stabilizers in vinyl polymers.

Stan-Guard 100, of Pfizer Inc., is a liquid **butyl-tin** compound containing sulfur and used as a stabilizer in rigid PVC sheet. A **butyl-tin maleate** powder is effective as a light stabilizer. **Hollicide LT-125** is a water-soluble organo-**trialkyl-tin** used as a bacteriocide in paper and textile processing. It can be used over a wide pH range.

TINPLATE. Soft-steel plate containing a thin coating of pure tin on both sides. A large proportion of the tinplate used goes into the manufacture of food containers because of its resistance to the action of vegetable acids and its nonpoisonous character. It solders easily, and also is easier to work in dies than terneplate, so that it is preferred over terneplate for making toys and other cheap articles in spite of a higher cost. Commercial tinplate comes in boxes of 112 sheets, 14 by 20 in (0.36 by 0.51 m), and is designated by the net weight per box when below 100 lb (45 kg). Heavy tinplate above 100 lb (45 kg) goes by number, as steel does, or by letter symbols. The weight of tin may be as high as 1.7% of the total weight of the sheet. **Coke plates** carry as little tin as is necessary to protect and brighten the plate for temporary use. The tin of the coat forms compounds of $FeSn_2$, Fe_2Sn, and FeSn with the iron of the plate, and on a coke plate this compound is 0.00006 to 0.00015 in (0.00015 to 0.00038 cm) thick. Best cokes carry more tin than do the standard cokes. **Charcoal plates** have heavier coats of tin designated by the letter A. The AAAAAA, or 6A, has the heaviest coating. Tinplate is made by the hot-dip process using palm oil as a flux, or by a continuous electroplating process. A base box contains 31,360 in^2 (20 m^2) of tinplate, and standard-dip tinplate has 1.5 lb (0.7 kg) of tin per base box, while electrolytic plate has only 0.25 lb (0.1 kg) of tin per base box and much **electrolytic tinplate** for container use has only 0.10 lb (0.05 kg) of tin per base box. Electrotinning gives intimately adherent coatings of any desired thickness, and the plate may have a serviceable coat as thin as 0.00003 in (0.00008 cm), or about one-third that of the thinnest possible dipped plate. A slight cold rolling of electrolytic tinplate gives a bright, smooth finish.

Taggers was originally a name for tinplate that is undersized, or below the gage of the plate in the package, but the name **taggers tin** is also applied to light-gage plate. These sizes are No. 38 gage, 55 lb (25 kg); No. 37, 60 lb (27 kg); and No. 36, 65 lb (29 kg). **Ductilite,** of Wheeling-Pittsburgh Steel Corp., is a tinplate that is not made by hot rolling in packs, but is cold-rolled from single hot-rolled strip steel. It is of uniform gage and does not have the thin edges of pack-rolled plate. It also has a uniform grain structure. **Weirite,** of Weirton Steel Corp., is cold-reduced coke tinplate. **Black plate,** used for cans in place of tinplate where the tin protection is not necessary, is not black, but is any sheet steel other than tinplate or terneplate in tinplate sizes. It may be chemically treated to resist rust or corrosion.

Electroplated **tin-zinc coatings,** developed at the International Tin Research Institute (ITRI) in England in the late 1940s, are more corrosion-protective of steel than zinc alone. However, bath control is difficult, and the sodium stannate and cyanide systems of the original bath are toxic. In the early 1990s, ITRI developed the Stanzec process, which is cyanide-free, nontoxic, and more controllable. The electrolyte, or bath, is a mixture of sodium or potassium stannate, sodium or potassium zincate, sodium or potassium hydroxide, plus complexing and stabilizing agents, brighteners, and grain refiners. Any alloy composition can be deposited by barrel, brush, or rack plating, and deposits of 70 to 80% tin with the balance zinc combine solderability with good corrosion resistance. A 75Sn–25Zn plate has a Vickers hardness of 37, and the coatings are typically ductile and suitable for painting. They also can be plated to brasses and high-copper alloys and are considered alternatives to plating with toxic cadmium in many applications.

TITANATES. Compounds made by heating a mixture of an oxide or carbonate of a metal and titanium dioxide. High dielectric constants, high refractive indices, and ferroelectric properties contribute primarily to their commercial importance. **Barium titanate** crystals, $BaTiO_3$, are made by die-pressing titanium dioxide and barium carbonate and sintering at high temperature. This crystal belongs to the class of **perovskite** in which the closely packed lattice of barium and oxygen ions has a barium ion in each corner and an oxygen ion in the center of each face of a cube with the titanium ion in the center of the oxygen octahedron. For piezoelectric use the crystals are subjected to a high current, and they give a quick response to changes in pressure or electric current. They also store electric charges and are used for capacitors. **Glennite 103,** of Gulton Industries Inc., is a **piezoelectric ceramic** molded from barium titanate modified with temperature stabilizers. **Bismuth stannate,** $Bi_2(SnO_3) \cdot 5H_2O$, a crystalline powder that dehydrates at about 284°F (140°C), may be used with barium titanate in capacitors to increase stability at high temperatures. **Ceramelex** is molded polycrystalline barium titanate. **Lead zirconate–lead titanate** is a piezoelectric ceramic that can be used at higher temperatures than barium titanate. **Lead titanate,** $PbTiO_3$, is used as a less costly substitute for titanium oxide. It is yellowish and has only 60% of the hiding power, but is very durable and protects steel from rust. **Butyl titanate** is a yellow, viscous liquid used in anticorrosion varnishes and for flameproofing fabrics. It is a condensation product of the tetrabutyl ester of ortho-titanic acid, and contains about 36% titanium dioxide. **Calcium titanate,** $CaTiO_3$, occurs in nature as the mineral perovskite. As a ceramic, it has a room-temperature dielectric constant of about 160. It is frequently used as an addition to barium titanate or by itself as a temperature-compensating capacitor. **Magnesium titanate,** $MgTiO_3$, crystallizes as an ilmenite rather than a perovskite structure. It is not ferroelectric, and is used with titanium dioxide to form temperature-compensating capacitors. It has also been used as an addition agent to barium titanate. **Strontium titanate,** $SrTiO_3$, has a cubic perovskite structure at room temperature. It has a dielectric constant of about 230 as a ceramic, and it is commonly used as an additive to barium titanate to decrease the Curie temperature. By itself, it is used as temperature-compensating material because of its negative temperature characteristics. **Strontium titanate,** used as a brilliant diamondlike gemstone, is a strontium mesotrititanate. Stones are made up to 4 carats. The refractive index is 2.412. It has a cubic crystal similar to the diamond, but the crystal is opaque in the X-ray spectrum. Crystalline **silico-titanate,** developed at Texas A&M University and Sandia National Laboratories, has potential use in the cleanup of radioactive wastes. As an inorganic ion-exchange agent, it promotes exchange of resident sodium ions for ions of radioactive elements. It has proved effective in removing cesium from neutral and highly acidic waste solutions. **Titanate fibers** can be used as reinforcement in thermoplastic moldings. The fibers, called **Fybex,** produced by LNP Engineering Plastics, Inc., can also be used in plated plastics to reduce thermal expansion, warpage, and shrinkage. Titanate fibers in plastics also provide opacity.

TITANIUM AND TITANIUM ALLOYS.   A metallic element, symbol Ti, occurring in a great variety of minerals. It was first discovered as an element in 1791 in a black magnetic sand at Manachin, Cornwall, England, and called **menachite,** from the name of the sand, **menachinite.** Its chief commercial ores are rutile and ilmenite. In rutile it occurs as an oxide. It is an abundant element but is difficult to reduce from the oxide. **High-purity titanium** (99.9%) has a melting point of about 3034°F (1668°C), a density of 0.163 lb/in^3 (4,512 kg/m^3), and tensile properties at room temperature of about 34,000 lb/in^2 (234 MPa) ultimate strength, 20,000 lb/in^2 (138 MPa) yield strength, and 54% elongation. It is paramagnetic and has low electrical conductivity and thermal expansion.

The commercial metal is produced from **sponge titanium,** which is made by converting the oxide to titanium tetrachloride followed by reduction with molten magnesium. The metal can also be produced in dendritic crystals of 99.6% purity by electrolytic deposition from titanium carbide. Despite its high melting point, titanium reacts readily in copper and in other metals and is much used for alloying and for deoxidizing. It is a more powerful deoxidizer of steel than silicon or manganese. An early German deoxidizing alloy known as **Badin metal** contained about 9% aluminum, 19 silicon, 5 titanium, and the balance iron. **Titanium copper,** used for deoxidizing nonferrous metals, is made by adding titanium to molten copper. The congealed alloy is broken into lumps.

One of the chief uses of the metal has been in the form of titanium oxide as a white pigment. It is also valued as titanium carbide for hard facings and for cutting tools. Small percentages of titanium are added to steels and alloys to increase hardness and strength by the formation of carbides or oxides or, when nickel is present, by the formation of **nickel titanide.** The first **titanium alloys** in the United States were produced in 1945 by the Bureau of Mines.

Titanium is one of the few allotropic metals (steel is another); that is, it can exist in two different crystallographic forms. At room temperature, it has a close-packed hexagonal structure, designated as the *alpha phase.* At around 1625°F (884°C), the alpha phase transforms to a body-centered cubic structure, known as the *beta phase,* which is stable up to titanium's melting point of about 3050°F (1677°C).

Alloying elements promote formation of one or the other of the two phases. Aluminum, for example, stabilizes the alpha phase; that is, it raises the alpha to the beta transformation temperature. Other alpha stabilizers are carbon, oxygen, and nitrogen. Beta stabilizers, such as copper, chromium, iron, molybdenum, and vanadium, lower the transformation temperature, therefore allowing the beta phase to remain stable at lower temperatures, and even at room temperature. Titanium's mechanical properties are closely related to these allotropic phases. For example, the beta phase is much stronger, but more brittle, than the alpha phase. Titanium alloys therefore can be usefully classified into three groups on the basis of allotropic phases: alpha, beta, and alpha-beta alloys.

Titanium and its alloys have attractive engineering properties. They are about 40% lighter than steel and 60% heavier than aluminum. The combination of moderate weight and high strengths, up to 200,000 lb/in^2 (1,379 MPa), gives titanium alloys the highest strength-to-weight ratios of any structural metal. Furthermore, this exceptional strength-to-weight ratio is maintained from $-420°F$ ($-216°C$) up to $1000°F$ ($538°C$). A second outstanding property of titanium materials is their corrosion resistance. The presence of a thin, tough, oxide surface film provides excellent resistance to atmospheric and sea environments as well as a wide range of chemicals, including chlorine and organics containing chlorides. Being near the cathodic end of the galvanic series, titanium performs the function of a noble metal. Titanium and its alloys, however, can react pyrophorically in certain media. Explosive reactions can occur with fuming nitric acid containing less than 2% water or more than 6 nitrogen dioxide and, on impact, with liquid oxygen. Pyrophoric reactions also can occur in anhydrous liquid or gaseous chlorine, liquid bromine, hot gaseous fluorine, and oxygen-enriched atmospheres.

Fabrication is relatively difficult because of titanium's susceptibility to hydrogen, oxygen, and nitrogen impurities, which cause embrittlement. Therefore elevated-temperature processing, including welding, must be performed under special conditions that avoid diffusion of gases into the metal. Heat is usually required in most forming operations.

Commercially pure titanium and many of the titanium alloys are now available in most common wrought mill forms, such as plate, sheet, tubing, wire, extrusions, and forgings. Castings can also be produced in titanium and some of the alloys, investment casting and graphite-mold (rammed graphite) casting being the principal methods. Because of titanium's highly reactive nature in the presence of such gases as oxygen, the casting must be done in a vacuum furnace. Because of their high strength-to-weight ratio primarily, titanium and titanium alloys are widely used for aircraft structures requiring greater heat resistance than aluminum alloys. Because of their exceptional corrosion resistance, however, they (unalloyed titanium primarily) are also used for chemical processing, desalination, and power generation equipment; marine hardware; valve and pump parts; and prosthetic devices.

There are five grades of **commercially pure titanium,** also called **unalloyed titanium:** ASTM Grade 1, Grade 2, Grade 3, Grade 4, and Grade 7. They are distinguished by their impurity content, that is, the maximum amount of carbon, nitrogen, hydrogen, iron, and oxygen permitted. Regardless of grade, carbon and hydrogen contents are 0.10 and 0.015% maximum, respectively. Maximum nitrogen is 0.03%, except for 0.05 in Grades 3 and 4. Iron content ranges from as much as 0.20% in Grade 1, the most pure (99.5) grade, to as much as 0.05 in Grade 4, the least pure (98.9). Maximum oxygen ranges from 0.18% in Grade 1 to 0.40 in Grade 4. Grade 7, 99.1% pure based on maximum impurity content, is actually a series of alloys containing 0.12 to 0.25% palladium for improved corrosion resistance in hydrochloric, phosphoric, and sulfuric acid solutions. Palladium content has little effect on tensile properties, but impurity content, especially oxygen and iron, has an appreciable effect. Minimum tensile yield strengths range from 25,000 lb/in^2 (172 MPa) for Grade 1 to 70,000 lb/in^2 (483 MPa) for Grade 4.

There are three principal types of titanium alloys: alpha or near-alpha alloys, alpha-beta alloys, and beta alloys. All are available in wrought form and some of each type for castings as well. In recent years, some also have become available in powder compositions for processing by hot isostatic pressing and other powder-metallurgy techniques. **Titanium alpha alloys** typically contain aluminum and usually tin. Other alloying elements may include zirconium, molybdenum, and, less commonly, nitrogen, vanadium, columbium, tantalum, or silicon. Though they are generally not capable of being strengthened by heat treatment (some will respond slightly), they are more creep-resistant at elevated temperature than the other two types, are preferred for cryogenic applications, and are more weldable but less forgeable. **Ti-5Al-2 Sn,** which is available in regular and **ELI grades** (extra-low interstitial) in wrought and cast forms, is the most widely used. In wrought and cast form, minimum tensile yield strengths range from 90,000 lb/in^2 (621 MPa) to 115,000 lb/in^2 (793 MPa) and tensile modulus is on the order of 15.5×10^6 to 16×10^6 lb/in^2 (106,873 to 110,320 MPa). It has useful strength to about 900°F (482°C) and is used for aircraft parts and chemical processing equipment. The ELI grade is noted for its superior toughness and is preferred for containment of liquid gases at cryogenic temperatures.

Other alpha or near-alpha alloys and their performance benefits include **Ti-8Al-1Mo-1V** (high creep strength to 900°F), **Ti-6Al-2Sn-4Zr-2Mo** [creep resistance and stress stability to 1100°F (593°C)], **Ti-6Al-2Cb-1Ta-0.8Mo** (toughness, strength, weldability), and **Ti-2.25Al-11Sn-5Zr-1Mo** [high tensile strength—135,000 lb/in² (931 MPa) yield, superior resistance to stress corrosion in hot salt media at 900°F]. Another alpha alloy, **Ti-0.3Mo-0.8Ni,** also known as **TiCode 12,** is noted for its greater strength than commercially pure grades and equivalent or superior corrosion resistance, especially to crevice corrosion in hot salt solutions.

Titanium alpha-beta alloys, which can be strengthened by solution heat treatment and aging, afford the opportunity of parts fabrication in the more ductile annealed condition and then can be heat-treated for maximum strength. **Ti-6Al-4V,** which is available in regular and ELI grades, is the principal alloy, its production alone having accounted for about half of all titanium and titanium-alloy production. In the annealed condition, tensile yield strength is about 130,000 lb/in² (896 MPa) and 13% elongation. Solution treating and aging increase yield strength to about 150,000 lb/in² (1,034 MPa). Yield strength decreases steadily with increasing temperature, to about 70,000 lb/in² (483 MPa) at about 950°F (510°C) for the aged alloy. At 850°F (454°C), aged bar has a 1,000-h stress-rupture strength of about 50,000 lb/in² (345 MPa). Uses range from aircraft and aircraft turbine parts to chemical processing equipment, marine hardware, and prosthetic devices. The alloy is also the principal alloy used for superplastically formed, and superplastically formed and simultaneously diffusion-bonded, parts. At 1650 to 1700°F (899 to 927°C) and low strain rates, the alloy exhibits tensile elongations of 600 to 1,000%, a temperature range also amenable to diffusion-bonding the alloy.

Following are other alpha-beta alloys and their noteworthy characteristics. **Ti-6Al-6V-2Sn:** high strength to about 600°F (315°C) but low toughness and fatigue resistance. **Ti-8Mn:** limited use for flat mill products, not weldable. **Ti-7Al-4Mo:** a forging alloy mainly, but limited use; 150,000 lb/in² (1,034 MPa) yield strength in the aged condition. **Ti-6Al-2Sn-4Zr-6Mo:** high strength, 170,000 lb/in² (1,172 MPa) yield strength, decreasing to about 110,000 lb/in² (758 MPa) at 800°F (427°C); for structural applications at 750 to 1000°F (400 to 540°C). **Ti-5Al-2Sn-2Zr-4Mo-4Cr** and **Ti-6Al-2Sn-2Zr-2Mo-2Cr:** superior hardenability for thick-section forgings; high modulus—about 17×10^6 to 18×10^6 lb/in² (117,215 to 124,110 MPa), respectively; tensile yield strength of about 165,000 lb/in² (1,138 MPa). **Ti-10V-2Fe-3Al:** best of the alloys in toughness at a yield strength of 130,000 lb/in² (896 MPa); can also be aged to a yield strength of about

172,000 lb/in² (1,186 MPa); intended for use at temperatures to about 600°F (315°C). **Ti-3Al-2.5V:** a tubing and fastener alloy primarily, moderate strength and ductility, weldable.

Beta titanium alloys, fewest in number, are noted for their hardenability, good cold formability in the solution-treated condition, and high strength after aging. On the other hand, they are heavier than titanium and the other alloy types, their density ranging from about 0.174 to 0.175 lb/in³ (4.84 g/cm³) for **Ti-13V-11Cr-3Al, Ti-8Mo-8V-2Fe-3Al,** and **Ti-3Al-8V-6Cr-4Zr-4Mo** to 0.183 lb/in³ (5,065 kg/m³) for **Ti-11.5Mo-6Zr-4.5Sn,** which is also known as **Beta III.** They are also the least creep-resistant of the alloys. Ti-13V-11Cr-3Al, a weldable alloy, can be aged to tensile yield strengths as high as 195,000 lb/in² (1,345 MPa) and retains considerable strength at temperatures to 600°F, but has limited stability at prolonged exposure to higher temperatures.

Timetal 21S, of Titanium Metals Corp., is a metastable beta alloy of composition **Ti-15Mo-3Al-2.7Cb-0.3Fe-0.2Si-0.13O** with maximum amounts of 0.05% carbon, 0.05 nitrogen, 0.015 hydrogen, and 0.4 residual elements. It is unique among titanium and titanium alloys in its resistance to Skydrol, a widely used aircraft hydraulic fluid. Also, its oxidation resistance at 1200°F (649°C) is far superior to that of commercially pure titanium. Aging at 1000°F (538°C) results in tensile yield strengths of 179,000 to 187,000 lb/in² (1,234 to 1,289 MPa). The alloy can be rolled to thin foil, a form useful for metal-matrix composites. **Timetal LCB** (low-cost beta), a **Ti-6.8Mo-4.5Fe-1.5Al** alloy of the same company, reduces formulating cost because iron need not be removed from the ore. It is a candidate for replacing steel spring wire and requires processing temperatures of only 300 to 400°F (149 to 204°C). The alloy has a tensile strength of 150,000 lb/in² (1,034 MPa) and a tensile modulus of 16.5 × 10⁶ lb/in² (113,768 MPa). Teledyne Wah Chang's **Tiadyne 3515,** also known as **Titanium Alloy C** and **Ti-1270,** contains 50% titanium, 35 vanadium, and 15 chromium. It is noted for high-temperature strength and the ability to resist combustion in air at temperatures and pressures far greater than for Ti-6Al-4V alloy. The average tensile yield strength is 137,000 lb/in² (945 MPa), 98,000 lb/in² (676 MPa) at 1000°F (538°C). It is available in rod, various flat products, and powder, and is also castable. **Ti-45Cb,** of this company, features superior resistance to oxidizing environments and combustion due to its high columbium content. It has a yield strength of 72,500 lb/in² (500 MPa), 29,000 lb/in² (200 MPa) at 752°F (400°C), and is used for autoclave vent lines in processing gold ores.

In an effort to spur nonaerospace uses, manufacturers have introduced several low-cost titanium alloys which are roughly similar in strength to aerospace alloy Ti-6Al-4V but which may sacrifice certain performance features required in aerospace applications. These alloys include **titanium alloy Auto-grade,** of Oregon Metallurgical Corp., **titanium alloys RM** and **VM** of RMI Titanium Co., and **Timetal-62S** of Titanium Metals Corp. To reduce cost, Auto-grade is initially forged and rolled in the beta region, then rolled in the alpha-beta range. RM, made of recycled material, has a nominal Ti-6Al-4V composition. VM, for virgin metal, is **Ti-6.4Al-1.2Fe,** the iron substituting for more-costly vanadium. Timetal-62S, Ti-6Al-1.7Fe-0.1Si, also uses iron instead of vanadium and costs about 25% less than the aerospace alloy.

Titanium alloys are leading candidates for metal-matrix composites, primarily for aircraft and aircraft engine applications. Silicon carbide fiber is a leading reinforcement. One approach, developed by Howmet Corp. and General Electric Aircraft Engines, is called *bicasting.* Preforms of alloys reinforced with the fiber are cast within a matrix alloy.

TITANIUM CARBIDE.   A hard, crystalline powder of composition TiC made by reacting titanium dioxide and carbon black at temperatures above 3272°F (1800°C). It is compacted with cobalt or nickel for use in cutting tools and for heat-resistant parts. It is lighter in weight and less costly than tungsten carbide, but in cutting tools it is more brittle. When combined with tungsten carbide in sintered carbide tool materials, however, it reduces the tendency to cratering in the tool. A general-purpose cutting tool of this type contains about 82% tungsten carbide, 8 titanium carbide, and 10 cobalt binder. **Kentanium,** of Kennametal, Inc., is titanium carbide in various grades with up to 40% either cobalt or nickel as the binder, used for high-temperature, erosion-resistant parts. For highest oxidation resistance, only about 5% cobalt binder is used. **Kentanium 138,** with 20% cobalt, is used for parts where higher strength and shock resistance are needed, and where temperatures are below about 1800°F (982°C). This material has a tensile strength of 45,000 lb/in^2 (310 MPa), compressive strength of 550,000 lb/in^2 (3,792 MPa), and Rockwell A hardness 90. **Kentanium 151A,** for resistance to molten glass or aluminum, has a binder of 20% nickel. **Titanium-carbide alloy,** of Ford Motor Co., for tool bits, has 80% titanium carbide dispersed in a binder of 10 nickel and 10 molybdenum. The material has a Rockwell A hardness of 93 and a dense, fine-grained structure. **Ferro-Tic,** of Chromalloy Corp., has titanium carbide bonded with stainless steel. It has a Rockwell C hardness of 55. **Machinable carbide** is titanium carbide in a matrix of Ferro-Tic C tool steel. **Titanium carbide tubing** is produced in round or rectangular form 0.10 to 3 in (0.25 to 7.6 cm) in diameter, by TEEG Research, Inc. It is made by vapor deposition of the carbide without a binder. The tubing has a Knoop hardness above 2,000 and a melting point of 5880°F (3249°C).

Grown single crystals of titanium carbide of Linde Co. have composition $TiC_{0.94}$, with 19% carbon. The melting point is 5882°F (3250°C), specific gravity 4.93, and Vickers hardness 3,230.

TITANIUM ORES. The most common titanium ores are ilmenite and rutile. Ilmenite is an iron-black mineral having a specific gravity of about 4.5 and containing about 52% **titanic oxide,** or titania, TiO_2. The ore of India is sold on the basis of titanium dioxide content, and the high-grade ore averages about 60% TiO_2, 22.5 iron, and 0.4 silica. **Ilmenite** is a **ferrotitanate,** $FeO \cdot TiO_2$, but much of the material called ilmenite is **arizonite,** $Fe_2O_3 \cdot 3TiO_2$. Titanium ores are widely distributed and plentiful. Ilmenite is found in northern New York, Florida, North Carolina, and in Arkansas, but the most extensive, accessible resources are found in Canada. The Quebec ilmenite contains 30% iron. The concentrated ore has about 36% TiO_2, and 41 iron, and is smelted to produce pig iron and a slag containing 70 TiO_2 which is used to produce titanium oxide. The beach sands of Senegal are mixed ores, the ilmenite containing 55 to 58% TiO_2, and the **zirconiferous quartz** containing 70 to 90 zirconia. The beach sands of Brazil are washed to yield a product averaging 71.6% ilmenite, 13 zircon, and 6 monazite. The Indian ilmenite also comes from beach sands. The ore of New York State averages 19% TiO_2.

Rutile is a **titanium dioxide,** TiO_2, containing theoretically 60% titanium. Its usual occurrence is crystalline or compact massive, with a specific gravity of 4.18 and 4.25 and Mohs hardness 6 to 6.5. The color is red to brown, occasionally black. Rutile is found in granite, gneiss, limestone, or dolomite. It is obtained from beach sand of northern Florida and Espirito Santo, Brazil, and is also produced in Virginia, and in Australia and India. Rutile and **brookite** and **Octahedrite,** or **anatase,** are produced in Arkansas and Massachusetts. The best Virginia concentrates are 92.5 to 98% TiO_2, but some are 42% from rock originally showing 18.5 TiO_2 in a body of feldspar. Rutile is marketed in the form of concentrates on the basis of 79 to 98.5% titanium oxide. It is used as an opacifier in ceramic glazes and to produce tan-colored glass. It is also employed for welding-rod coatings. On welding rods it aids stabilization of the arc and frees the metal of slag. **Tanarc,** used on welding rods as a replacement for rutile, is made from slag from Canadian titaniferous hematite, and it contains 70% TiO_2.

TITANIUM OXIDE. The white **titanium dioxide, or titania,** of composition TiO_2, is an important paint pigment. The best quality is produced from ilmenite, and is higher in price than many white pigments but has great hiding power and durability. Off-color pigments, with a light buff tone, are made by grinding rutile ore. The pigments have fine physical qualities and may be used wherever the color is not important. Titania is also substituted for zinc oxide and lithopone in the manufacture of white rubber goods, and for paper filler. The specific gravity is about 4. Mixed with blanc fixe, it is also marketed under the name of **Titanox. Zopaque** is a pure titanium oxide for rubber compounding. **Ti-Pure** of Du Pont is commercially pure titanium dioxide for pigment use. **Duolith,** of this company, is titanated lithopone pigment containing 15% titanium dioxide, 25 zinc sulfide, and 60 barium sulfate. A titanium dioxide photocatalytic film, developed by Toto Ltd. of Japan, has silver and copper compounds immobilized on the surface to kill bacteria when exposed to fluorescent light. It is used to coat ceramic tiles in hospitals, food and chemical processing plants, and other sanitary applications.

Titania crystals are produced in the form of pale-yellow, single-crystal boules for making optical prisms and lenses for applications where the high refractive index is needed. The crystals are also used as electric semiconductors, and for gemstones. They have a higher refractive index than the diamond, and the cut stones are more brilliant but are much softer. Knoop hardness is about 925, and the melting point is 3317°F (1825°C). The refractive index of the rutile form is 2.7, and that of the anatase is 2.5, while the synthetic crystals have a refractive index of 2.616 vertically and 2.903 horizontally.

Titanium oxide is a good refractory and electrical insulator. The finely ground material gives good plasticity without binders and is molded to make resistors for electronic use. **Micro sheet** is titanium oxide in sheets as thin as 0.003 in (0.008 cm) for use as a substitute for mica for electrical insulation where brittleness is not important. **Titania-magnesia ceramics** were made in Germany in the form of extruded rods and plates and pressed parts.

TOBACCO. The leaf of an unbranched annual plant of the genus *Nicotiana,* of which there are about 50 species and many varieties. It is used for smoking, chewing, snuff, insecticides, and production of the alkaloid nicotine. Commercial crops are grown in about 60 countries, but about one-third of world production is in the United States. Only two species are grown commercially, *N. tabacum,* a tropical plant native to the West Indies and South America, and *N. rustica,* grown by the Indians of Mexico and North America before 1492. About 85% of world production is now from *N. tabacum,* and there are more than 100 varieties of this plant.

Tobacco was not known in Europe until it was brought from the

West Indies by Columbus. Plants for cultivation were brought to Spain in 1558, and by 1586 smoking had become a general practice in western Europe. The first commercial shipments were made from Virginia in 1618, the growing of cultured varieties having begun in 1612. Smoking of tobacco was practiced by the Indians from Canada to Patagonia, and the natives of Haiti used powdered tobacco leaf as **snuff** under the name of **cohoba.** Like Indian corn, the tobacco plant had been domesticated for centuries, and the original wild ancestor of the plant is not known. Some Indian tribes, such as the Tobacco nation of southwest Ontario, specialized in the growing of tobacco types.

The quality of the tobacco leaf varies greatly with the soil and climate, the care of the plant, and the curing of the leaf; the nicotine content develops in the curing process. The narcotic effects are due to the alkaloid **nicotine,** $C_{10}H_{14}N_2$, a complex pyrrolidine, which is a heavy, water-white oil. The nicotine is absorbed through the mucous membranes of the nose and throat. The aroma and flavor come from the essential oils in the leaf developed during fermentation and curing. The more harmful effects to the eyes and respiratory system come from the **pyridine** C_5H_5N, a toxic aromatic compound that also occurs in coal tar, and from other elements of the smoke and not from the alkaloid. The burning of the tars may also produce **carcinogen** compounds which are complex, condensed, benzene-ring nuclei injurious to tissues.

Although *N. tabacum* is a less hardy plant than *N. rustica,* it adapts itself to a wide variety of climates and soils, and the types generated in given areas do not normally reproduce the same type in another area. The variety developed in the Near East, known as **Turkish tobacco** and valued as an aromatic blend for cigarettes, is a small plant with numerous leaves only about 3 in (7.6 cm) long, while the American tobaccos grown from the same species have leaves up to 3 ft (0.9 m) long. The nicotine content of Turkish tobacco is from 1 to 2%, while that of flue-cured Virginia tobacco is 2.5 to 3%, and that of burley and fire-cured American types is up to 4.5%. **Perique,** a strong, black tobacco much used in French and British pipe mixtures, is cultivated only in a small area of southern Louisiana. Other tobaccos brought into the area become perique in the second year, but when transplanted back, they do not thrive. *N. rustica* was the first tobacco grown in Virginia, but the tobacco now grown in the area and known as **Virginia tobacco** is *N. tabacum* brought from the West Indies, but now differing in type from West Indian tobacco. **Makhorka tobacco,** a black, air-cured type grown in Russia and Poland and very high in nicotine, is from *N. rustica.* Strong, black, highly fermented tobaccos high in nicotine, and considered as inferior in the United States, are preferred in France and some other countries.

Types of tobacco are based on color, flavor, strength, and methods of curing and fermentation, while grades are based on size, aroma, and texture, but the geographical growing area also determines characteristics. Commercial purchasing is done by the area and the Department of Agriculture type classification: fire-cured, dark air-cured, flue-cured, cigar wrapper, cigar binder, cigar filler, burley, Maryland, and perique, all of which are from *N. tabacum.* Grading is done by specialists, and a single area crop may produce more than 50 grades. In the manufacture of cigarettes, blending is done to attain uniformity, and some of the flavor and aroma may be from added ingredients. **Air-cured tobaccos** are alkaline, while **flue-cured tobaccos** are acid and the nicotine is less readily given off. *N. rustica* may contain as high as 10% nicotine and is thus more desirable for insecticide use or for the extraction of nicotine, but some strains of *N. tabacum* have been developed for smoking with as little as 0.3% nicotine.

Tobacco seed oil has an iodine value of 140 to 146 and is a valuable drying oil, but the production is low because the seed heads are topped in cultivation and seeds are developed only on the sucker growths. **Tobacco sauce,** used for flavoring chewing and smoking tobaccos, contains up to 10% nicotine, but since the nicotine is not desired in the flavoring, it is usually extracted for industrial use. Nicotine can be oxidized easily to **nicotinic acid** and to **nicotinonitrile,** both of which are important as antipellagra vitamins. Most of the nicotine used for insecticide is marketed as **nicotine sulfate** in water solution containing 40% nicotine. It is used as a sheep dip and as a contact insecticide. **Tobacco dust** is used for the control of plant lice. **Anabasine,** obtained in Russia from the Asiatic shrub *Anabasis aphylla,* has the same chemical composition as nicotine and is an isomer of nicotine. It is marketed in the form of a solution of the sulfate as an insecticide. It can also be obtained from *N. glauca,* a wild tree tobacco native to Mexico and the southeastern United States, or is made synthetically under the name of **neonicotine.**

TOLU BALSAM. A yellowish-brown, semisolid gum with a pleasant aromatic odor and taste, obtained from the tree *Myroxylon balsamum,* or *Toluifera balsamum,* of Venezuela, Colombia, and Peru. It is used in medicine, chiefly in cough syrups, and as a fixative in perfumes. A soft, tenacious, resinous substance that hardens on keeping, it is a mixture of free cinnamic and benzoic acids. **Balsam of Peru,** or **black balsam,** is a reddish-brown, viscous, aromatic liquid from bark of the tall tree *M. pereirae* of El Salvador. It is used in cough medicines and skin ointments, as an extender for vanilla, and as a fixative in perfumes. Some white-colored balsam is also obtained from the fruit of the tree. **Peru balsam** contains benzyl benzoate, benzyl cinnamate, and some vanillin. **Styrax** is an aromatic balsam from

Liquidambar orientalis, and **Zanthorrhoea balsams** are acaroid resins from the *X. australis* tree. It is used as a perfumery substitute for Peru balsam and Styrax.

TOLUOL.   Also called **toluene, methyl benzene,** and **methyl benzol.** A liquid of composition $C_6H_5CH_3$, resembling benzene but with a distinctive odor. It is obtained as a by-product from coke ovens and from coal tar. It occurs also in petroleum, with from 0.20 to 0.70% in Texas crude oil, which is not sufficient to extract. But toluol may be produced by dehydrogenation of petroleum fractions. It is used as a solvent, and for making explosives, dyestuffs, and many chemicals, and in aviation gasoline to improve the octane rating. Industrially pure toluol from coal tar distills off between 227 and 235°F (108.6 and 112.6°C), and is a water-white liquid with a specific gravity of 0.864 to 0.874, flash point 35 to 40°F (2 to 4°C), and freezing point about −139°F (−95°C). The fumes are poisonous. Nitration grades are at least 99.9% pure, and are used in synthesizing adhesives, agricultural chemicals, coatings, and in textiles. Such a material is Unocal Chemicals' **Amsco. B&J Brand,** from Burdick & Jackson, is ultrahigh purity and contains very low-luminescence impurities for liquid scintillation counters. It is also used for dissolving polymers for gel permeation chromatography, and has low residues in pesticide residue analyses. **Monochlorotoluene,** used as a solvent for rubber and synthetic resins, is a colorless liquid of composition $CH_3C_6H_4Cl$, boiling at about 320°F (160°C) and freezing at −49°F (−45°C). **T oil** is a sulfur toluene condensation product made under a British patent and used as a plasticizer for chlorinated rubber. **Notol No. 1,** of Neville Chemical Co., is a coal tar hydrocarbon high in aromatics used as a substitute for toluol as a lacquer solvent. The specific gravity is 0.825 and the boiling point between 177 and 280°F (81 and 138°C). **Tollac,** of the same company, is another hydrocarbon substitute for toluol. **Methyl cyclohexane,** $C_6H_{11}CH_3$, is a water-white liquid with a distilling range of 212 to 217°F (100 to 103°C), produced by hydrogenating toluol. It is used as a solvent for oils, fats, waxes, and rubbers. **Methyl cyclohexanol,** $C_6H_{10}CH_3OH$, another toluol derivative, is used as a cellulose ester solvent and as an antioxidant in lubricants. It is a straw-colored, viscous liquid distilling between 311 and 356°F (155 and 180°C). **Polyvinyl toluene** is a methyl form of styrene. It is polymerized with terphenyl stilbene to form plastic scintillators to count radiation isotopes.

TONKA BEAN.   Called in northeastern Brazil **cumarú bean.** The kernel of the pit of the fruit of the **sarrapia tree,** *Dipteryx odorata* or *Coumarouna odorata,* of northern South America, used for the production of **coumarin** for flavoring and scenting. It has an aroma resembling vanilla. The trees often reach a height of 100 ft (30 m) and begin to bear in 3 years. The fruit is like a mahogany-colored plum, but with a fibrous pulp. The pits, or nuts, contain a single shiny, black seed 1 in (2.5 cm) or longer. The chief production is in Venezuela, Brazil, Colombia, Trinidad, and the Guianas. The tonka bean from the tree *D. oleifera* of Central America has an unpleasant odor. Before shipping, the beans are soaked in rum or alcohol to crystallize the coumarin. The ground beans are again soaked in rum, and the aromatic liquid is used to spray on cigarette tobacco. The coumarin extract is also used as a perfume or flavor in soaps, liqueurs, and confectionery. The essential oil produced from the seed is called **cumarú oil.** A substitute for tonka bean is **deer's tongue leaf,** which is the long leaf of the herb *Trilisa odoratissima,* growing wild on the edges of swamps from Carolina to Florida. The leaf has a strong odor of coumarin when dry, and contains coumarin. It is used in cigarette manufacture, in flavoring, and to produce synthetic vanilla. It is also a normal constituent of **lavender oil.**

TOOL STEELS. Steels used mainly for cutters in machining, shearing, sawing, punching, and trimming operations, and for dies, punches, and molds in cold- and hot-forming operations. Some are also occasionally used for nontool applications. Tool steels are primarily ingot-cast wrought products, although some are now also powder-metal (PM) products. Regarding PM products, there are two kinds: (1) mill products, mainly bar, produced by consolidating powder into "ingot" and reducing the ingot by conventional thermomechanical wrought techniques, and (2) end-product tools, produced directly from powder by pressing and sintering techniques. There are seven major families of tool steels as classified by the American Iron and Steel Institute: (1) high-speed tool steels, (2) hot-work tool steels, (3) cold-work tool steels, (4) shock-resisting tool steels, (5) mold steels, (6) special-purpose tool steels, and (7) water-hardening tool steels.

   High-speed tool steels are subdivided into three principal groups or types: the molybdenum type, designated M1 to M46; the tungsten type (T1 to T15); and the intermediate molybdenum type (M50 to M52). Virtually all M types, which contain 3.75 to 9.5% molybdenum, also contain 1.5 to 6.75 tungsten, 3.75 to 4.25 chromium, 1 to 3.2 vanadium, and 0.85 to 1.3 carbon. M33 to M46 also contain 5 to 8.25% cobalt; and M6, 12% cobalt. The T types, which are molybdenum-free, contain 12 to 18% tungsten, 4 to 4.5 chromium, 1 to 5 vanadium, and 0.75 to 1.5 carbon. Except for T1, which is cobalt-free, they also contain 5 to 12% cobalt. Both M50 and M52 contain 4% molybdenum and 4 chromium; the former also contains

0.85% carbon and 1 vanadium, the latter 0.9 carbon, 1.25 tungsten, and 2 vanadium.

Although the tungsten types were developed first, around the turn of the century, the molybdenum types, developed in the 1930s when tungsten was scarce, are now by far the most widely used, and many of the T types have M-type counterparts. All of the high-speed tool steels are similar in many respects. They all can be hardened to at least Rockwell C 63, have fine grain size, and have deep-hardening characteristics. Their most important feature is hot hardness: They all can retain a Rockwell C hardness of 52 or more at 1000°F (538°C). The M types, as a group, are somewhat tougher than the T type at equivalent hardness; but otherwise, mechanical properties of the two types are similar. Cobalt improves hot hardness, but at the expense of toughness. Wear resistance increases with increasing carbon and vanadium contents. The M types have a greater tendency to decarburization and, thus, are more sensitive to heat treatment, especially austenitizing. Many of the T types, however, are also sensitive in this respect, and they are hardened at somewhat higher temperatures. The single T type that stands out today is T-15, which is rated as the best of all high-speed tool steels from the standpoint of hot hardness and wear resistance. Typical applications for both the M type and T type include lathe tools, end mills, broaches, chasers, hobs, milling cutters, planar tools, punches, drills, reamers, routers, taps, and saws. The intermediate M types are used for somewhat similar cutting tools but, because of their lower alloy content, are limited to less-severe operating conditions.

Hot-work tool steels are subdivided into three principal groups: (1) the chromium type (H10 to H19), (2) the tungsten type (H21 to H26), and (3) the molybdenum type (H42). All are medium-carbon (0.35 to 0.60%) grades. The chromium types contain 3.25 to 5.00% chromium and other carbide-forming elements, some of which, such as tungsten and molybdenum, also impart hot strength, and vanadium, which increases high-temperature wear resistance. The tungsten types, with 9 to 18% tungsten, also contain chromium, usually 2 to 4, although H23 contains 12% of each element. The one molybdenum type, H42, contains slightly more tungsten (6%) than molybdenum (5), and 4 chromium and 2 vanadium. Typical applications include dies for forging, die casting, extrusion, heading, trim, piercing and punching, and hot-shear blades.

There are also three major groups of **cold-work tool steels:** (1) high-carbon (1.5 to 2.35%); high-chromium (12), which are designated D2 to D7; (2) medium-alloy air-hardening (A2 to A10), which may contain 0.5 to 2.25% carbon, 0 to 5.25 chromium, 1 to 1.5 molybdenum, 0 to 4.75 vanadium, 0 to 1.25 tungsten, and, in some cases, nickel, manganese or silicon, or nickel and manganese; and (3) oil-hardening types (O1 to O7). They are used mainly for cold-working operations, such as stamping dies, draw dies, and other forming tools as well as for shear blades, burnishing tools, and coining tools. **Shock-resistant tool steels** (S1 to S7) are, as a class, the toughest, although some chromium-type hot-work grades, such as H10 to H13, are somewhat better in this respect. The S types are medium-carbon (0.45 to 0.55%) steels containing only 2.50 tungsten and 1.50 chromium (S1), only 3.25 chromium and 1.40 molybdenum (S7), or other combinations of elements, such as molybdenum and silicon, manganese and silicon, or molybdenum, manganese, and silicon. Typical uses include chisels, knockout pins, screw driver blades, shear blades, punches, and riveting tools.

There are three principal **mold steels:** (1) **P6,** containing 0.10% carbon, 3.5 nickel, and 1.5 chromium; **P20,** 0.35 carbon, 1.7 chromium, and 0.40 molybdenum; and **P21,** 0.20 carbon, 4 nickel, and 1.2 aluminum. P6 is basically a carburizing steel produced to tool-steel quality. It is intended for hubbing—producing die cavities by pressing with a male plug—then carburizing, hardening, and tempering. P20 and P21 are deep-hardening steels and may be supplied in hardened condition. P21 may be carburized and hardened after machining. These steels are tough but low in wear resistance and moderate in hot hardness, P21 being best in this respect. All three are oil-hardening steels, and they are used mainly for injection and compression molds for forming plastics, but they also have been used for die-casting dies.

Special-purpose tool steels include L2, containing 0.50 to 1.10% carbon, 1.00 chromium, and 0.20 vanadium; and L6, having 0.70 carbon, 1.5 nickel, 0.75 chromium, and sometimes 0.25 molybdenum. L2 is usually hardened by water quenching and L6, which is deeper-hardening, by quenching in oil. They are relatively tough and easy to machine and are used for brake-forming dies, arbors, punches, taps, wrenches, and drills. The **water-hardening tool steels** include W1, which contains 0.60 to 1.40% carbon and no alloying elements; W2, with the same carbon range and 0.25 vanadium; and W5, having 1.10 carbon and 0.50 chromium. All are shallow-hardening and the least qualified of tool steels in terms of hot hardness. However, they can be surface-hardened to high hardness and thus can provide high resistance to surface wear. They are the most readily machined tool steels. Applications include blanking dies, cold-striking dies, files, drills, countersinks, taps, reamers, and jewelry dies.

Domestic producers include Al Tech, Braeburn, Carpenter Technology, Columbia, Crucible, Cyclops, Guterl, Ingersoll Johnson, Jessop, Latrobe, Lukens, and Teledyne Vasco. Foreign producers with

U.S. outlets include Atlas, Böhler Edelstahl, Fagersta, Kobe, and Uddeholm. Besides the standard tool steels, several proprietary grades are available. Lukens, for example, makes a series of prehardened plate steels, designated MTD 1 to MTD 4, which are somewhat similar in composition to 41XX chromium-molybdenum steels. The 18% nickel **maraging steels,** although developed for structural applications, are also used as die-casting dies and metal- and plastic-forming dies. And Teledyne Vasco makes **matrix steels,** which are said to be matrix compositions of M2 and M42 high-speed tool steels with less carbon and alloy content. They are used for extrusion, compacting, and thread-rolling dies, and punches and saw blades. Although most die steels are made from wrought steels, cast steels are used.

To prolong tool life, tool-steel end products, such as mills, hobs, drills, reamers, punches, and dies, can be nitrided or coated in several ways. **Oxide coatings,** imparted by heating to about 1050°F (566°C) in a steam atmosphere or by immersion in aqueous solutions of sodium hydroxide and sodium nitrite at 285°F (140°C), are not as effective as traditional nitriding, but do reduce friction and adhesion between the workpiece and tool. The thickness of the coating developed in the salt bath is typically less than 0.0002 in (0.005 mm), and its nongalling tendency is especially useful for operations in which failure occurs in this way. **Hard-chromium plating** to a thickness of 0.0001 to 0.0005 in (0.0025 to 0.0127 mm) provides a hardness of DPH 950 to 1,050 and is more effective than oxide coating, but the plate is brittle and, thus, not advisable for tools subject to shock loads. Its toughness may be improved somewhat without substantially reducing wear resistance by tempering at temperatures below 500°F (260°C), but higher tempering temperatures impair hardness, thus wear resistance, appreciably. An antiseize **iron sulfide coating** can be applied electrolytically at 375°F (191°C) using a bath of sodium and potassium thiocyanate. Because of the low temperature, the tools can be coated in the fully hardened and tempered condition without affecting hardness. **Tungsten carbide** is another effective coating. One technique, developed by Rocklin Manufacturing and called **Rocklinizing,** deposits 0.0001 to 0.0008 in (0.0025 to 0.0203 mm) of the carbide using a vibrating arcing electrode of the material in a handheld gun. **Titanium carbide** and **titanium nitride** are the latest coatings. The nitride, typically 0.0003 in (0.008 mm) thick, has stirred the greatest interest, although the carbide may have advantages for press tools subject to high pressure. In just the past few years, all sorts of tools, primarily cutters but also dies, have been titanium-nitride-coated, which imparts a gold- or brasslike look. The coating can be applied by **chemical vapor deposition (CVD)** at

1750 to 1950°F (954 to 1066°C) or by **physical vapor deposition (PVD)** at 900°F (482°C) or less. Thus, the PVD process has an advantage in that the temperature involved may be within or below the tempering temperature of the tool steels so that the coating can be applied to fully hardened and tempered tools. Also, the risk of distortion during coating is less. Titanium nitride coaters include Aerobraze, Multi-Arc Vacuum Systems, Scientific Coatings, Star Cutter, Sylvester, and Ti-Coating. Another method being used to prolong tool life is to subject the tools to a temperature of −320°F (−196°C) for about 30 h. The **cryogenic treatment,** which has been called **Perm-O-Bond** and **Cryo-Tech** by Materials Improvement, is said to rid the steel of any retained austenite—thus the improved tool life. Others in this business include Amcry, Endure, and 3X Kryogenics.

TRAGACANTH GUM. An exudation of the shrub *Astragalus gummifer* of Asia Minor and Iran, used in adhesives or for mucilage, for leather dressing, for textile printing, and as an emulsifying agent. To obtain the gum, a small incision is made at the base of the shrub, from which the juice exudes and solidifies into an alteration product, not merely the dried juice. The gum derived from the first day's incision, known as **fiori,** is the best quality and is in clear, fine ribbons or white flakes. The second incision produces a yellow gum known as **biondo.** The third incision produces the poorest quality, a dark gum known as **sari.** Rainy weather during the incision period may cause a still inferior product. Tragacanth is insoluble in alcohol but is soluble in alkalies and swells in water. **Karaya gum** from southern Asia is from various species of *Sterculia* trees, especially *S. urens,* of India. It is also known as **Indian gum, Indian hog gum,** and **hog tragacanth.** The sticky gum is dried, and the chunks are broken and the pieces sorted by color. A single chunk may have colors varying from clear white to dark amber and black. The color is caused by tannin or other impurities. The No. 3 grade, the lowest, has up to 3% insoluble impurities. The gum is marketed in flakes and as a white, odorless, 150-mesh powder. The chief constituent is galactan. In general, the gum is more acid than tragacanth and is likely to form lumpy gels unless finely ground. It is widely used as a thickening and suspending agent for foodstuffs, drugs, cosmetics, adhesives, and textile finishes. Gum tragacynth thoroughly mixed with glycerin and water produces a thick paste, **tragacynth glycerite,** a useful excipient to bind tablet masses. For oral ingestion, suspensions of gum tragacynth have been formulated. One such product is a suspension of **procaine penicillin.** In hair lotion in which there is 10% isopropanol, gum tragacynth is able to withstand the alcohol without precipitation. It is also a thickener of the aqueous phase in oil-in-water systems, resulting in shelf-stable emulsions. Regular as well as low-calorie salad dressings, such as **Thousand Island, French,** and **Roquefort,** are such oil-in-water emulsions. The low-calorie versions have higher amounts of gum tragacynth, to provide the body that oil traditionally gives the regular dressing. **Tragacynth gums,** Type A.10, Type W, and Type L, are pharmaceutical grades in powder form, produced by Meer Corp.

The granules of **water-soluble gums,** such as karaya, tragacanth, and acacia, are swelled by water and dispersed in water in microscopic particles to form cells or filamentlike structures which hold the water like a sponge and will not settle out. This type of colloidal dispersion is called a **hydrasol,** and when thick and viscous is called a **gel.** From 2 to 3% of karaya or other gum will form a gel in water. These gums will gel in cold water, while gelatin requires hot water for dissolving. In a gel there is continuous structure with molecules forming a network, while in a **sol** the particles are in separate suspension and a sol is merely a dispersion. Some dispersions, such as albumen, cross-link with heat; others, like guar gum, cross-link with alkalies; still others, like pectin, link with sugar and an acid. Gums with weak surface forces form weak gels which are **pastes** or mucilage, and a high concentration is needed to produce a solid. Karaya has great swelling power, and is used in medicine as a bulk laxative. **Ghatti gum,** from the abundant tree *Anogeissus latifolia* of India, is entirely soluble in water to form a viscous mucilage. It is twice as effective as gum arabic as an emulsifier, but is less adhesive. It comes in colorless to pale-yellow tears of vitreous fracture, called also Indian gum, and is used in India for textile finishing. **Aqualized gum,** of Glyco Products Co., Inc., is tragacanth or karaya chemically treated to give more rapid solubility. Water-soluble gums are also produced synthetically. **Polyox gum,** of Union Carbide Chemicals Co., is a polymer of polyethylene oxide containing carboxylic groups giving water solubility when the pH is above 4.0. In paper coating with ammonia, the ammonia evaporates to leave a water-insoluble, grease-resistant film that is heat-sealing. It is also used in latex paints and in cosmetics.

Another water-soluble gum which forms a true gel with a continuous branched-chain molecular network is **okra gum,** produced as a 200-mesh tan powder. It is edible and is used for thickening and stabilizing foods and pharmaceuticals. It is also used in plating baths for brightening nickel, silver, and cadmium plates. It is extracted from the pods of the **okra,** *Hibiscus esculentus,* a plant of the cotton family. In the southern states the pods, called **gumbo,** are used in soups. The refined gum, after extraction of the oils and sugars, contains 40.4% carbon, 6.1 hydrogen, and 2.1 nitrogen, with the balance insoluble cellulose.

TRIPOLI. A name given to finely granulated, white, porous, siliceous rock, used as an abrasive and as a filler. True tripoli is an infusorial, diatomaceous earth known as **tripolite,** and is a variety of opal, or **opaline silica.** In the abrasive industry it is called **soft silica.** It is quarried in Missouri, Illinois, eastern Tennessee, and Georgia. Pennsylvania rottenstone is not tripoli, although it is often classified with it. The material marketed for oil-well drilling mud under the name of **Opalite,** is an amorphous silica. The Missouri tripoli ranges in color from white to reddish, and the crude rock has a porosity of 45% and contains 30% or more of moisture. It is air-dried and then crushed and furnace-dried. Tripoli is used in massive form for the manufacture of filter stones for filtering small supplies of water. Missouri tripoli is also used for the manufacture of foundry parting. Finely ground tripoli, free from iron oxide, is used as a paint filler and in rubber. The grade of tripoli known as O.G. (once ground) is used for buffing composition, D.G. (double ground) for foundry partings, and the air-float product for metal polishes. Tripoli grains are soft, porous, and free from sharp cutting faces, and they give a fine polishing effect. It is the most commonly used polishing agent. The word **silex,** which is an old name for silica and is also used to designate the pulverized flint from Belgium, is sometimes applied to finely ground white tripoli employed as an inert filler for paints. Much Illinois fine-grained tripoli is used for paint, and for this purpose it should be free from iron oxide. **Rottenstone** is a soft, friable, earthy stone of light gray to olive color, used as an abrasive for metal and wood finishing. It resembles Missouri tripoli and is derived from the weathering of siliceous-argillaceous limestone, with generally from 80 to 85% alumina, 4 to 15 silica, and 5 to 10 iron oxides. Rottenstone was largely imported from England, but one variety is found in Pennsylvania. It is finely ground and is marketed either as a powder or molded into bricks. The latter form is used with oil on rag-wheel polishing. A 250-mesh powder is used as a filler in molding compounds.

TRISODIUM PHOSPHATE. A white, crystalline substance of composition $Na_3PO_4 \cdot 12H_2O$, also known as **phosphate cleaner,** used in soaps, cleaning compounds, plating, textile processing, and boiler compounds. The commercial grade is not less than 97% pure, with total alkalinity of 16 to 19% calculated as Na_2O. The anhydrous trisodium phosphate is 2.3 times as effective as the crystalline form, but requires a longer time to dissolve. **Disodium phosphate** is a white, crystalline product of composition $Na_2HPO_4 \cdot 12H_2O$ used for weighting silk, boiler treatment, cheese making, and cattle feeds. The medicinal, or USP, grade has only seven molecules of water and has a different crystal structure. The commercial grade is 99.4% pure and is readily soluble in water. **Trisodium phosphate hemihydrate** is a granular, crystalline grade from FMC Corp. for degreasing and water conditioning. **Monosodium phosphate** is made by reacting soda ash with phosphoric acid in molecular proportions; it is used in similar applications to the disodium variety. **Sodium tetraphosphate,** $Na_6P_4O_{13}$, contains 39.6% Na_2O and 60.4 P_2O_5. It is the sodium salt of **tetraphosphoric acid** and is marketed in beads that are mildly alkaline and highly soluble in water. The specific gravity is 2.55 and it melts at 1,112°F (600°C). It is used in the textile industry as a water softener and to accelerate cleansing operations. It removes lime precipitation and sludge and saves soap. **Quandrafos,** of American Cyanamid Co., used to replace quebracho for reducing the viscosity of oil-well drilling mud, is sodium tetraphosphate, containing 63.5% P_2O_5. It makes the calcium and magnesium compounds inactive, and 0.06% of the material controls 16.1% of water in reducing viscosity. It also gives smooth flow with minimum water in paper coating and textile printing. **Metafos,** of the same company, has a higher percentage of P_2O_5—67%—and a lower pH, for use in textile printing where low alkalinity is needed. **Sodium pyrophosphate,** $Na_4P_2O_7$, is added to soap powders to increase the detergent effect and the lathering. It is also used in oil-drilling mud. The crystalline form, $Na_4P_2O_7 \cdot 10H_2O$, is very soluble in water and is noncaking, and it is used in household cleaning compounds. **Sodium tripolyphosphate,** $Na_5P_3O_{10}$, is a water-soluble, white powder used as a detergent, a water softener, and a deflocculating agent in portland cement to govern the viscosity of the shale slurry without excessive use of water. Large quantities of these phosphates are used in the processing of chemicals, textiles, and paper; and since they are toxic contaminants of ground and surface waters, mill wastes must be deactivated before they are discharged. The use of phosphates in detergents and soap powders has been banned in many areas since they lead to rapid algal growth in surface waters.

TULIPWOOD. Also called **yellow poplar, whitewood,** and **canary whitewood,** the wood of the tree *Liriodendron tulipifera* of Canada and the eastern United States. The tree grows to a height of 250 ft (76 m) and to diameters of more than 10 ft (3 m). It is used for furniture, veneer, millwork, toys, woodenware, boxes, crates, and pulpwood. Owing to its close texture and even coefficient of expansion, it has been used for expansion blocks in humidity regulators. It is yellowish, soft, and durable. The density is about 30 lb/ft³ (481 kg/m³). The lumber may be mixed with **cucumber magnolia,** *Magnolia acuminata,* and **evergreen magnolia,** *M. grandifolia,* but **magnolia woods** are lighter in color.

TUNG OIL. A drying oil which has almost double the rapidity of linseed oil. It is used for enamels and varnishes; in brake linings, plastic compounds, and linoleum; and for making pigment for India ink. Tung oil is pressed from the seeds of *Aleurites montana* and *A. fordii.* The names **wood oil** and **China wood oil** are loosely and erroneously used to designate tung oils, but true wood oil is an oleoresin from the **Keruing tree** of Malaya used for waterproofing and caulking boats, while tung oil is never from the wood. The oil has a powerful purgative action, and the Chinese word means stomach. The **Chinese tung oil** is from the nuts of the tree *A. montana,* the China wood oil tree, and *A. fordii.* The latter tree is hardier than *A. montana,* which requires a hot climate. The American tung oil is from the nuts of the tree *A. fordii* of the Gulf states, which gives an annual production of about 30 lb (14 kg) of oil per tree. The tree grows to a height of 25 ft (8 m) and bears for 5 years. The seeds, or nuts, contain 50 to 55% oil. This tree is also grown in South Africa and Argentina.

The color of tung oil varies from golden yellow to dark brown according to the degree of heat used in extraction. It has a pungent odor resembling that of bacon fat. A good grade of raw tung oil should have a specific gravity between 0.934 and 0.940, a saponification value of 190, and an iodine value of 163. The oil contains about 72% **eleostearic acid,** which has a very high iodine value, 274, and gives to the oil a greater drying power than is indicated by the iodine value of the oil itself. The oil has the property of drying throughout at a uniform rate, instead of forming a skin as linseed oil does; but it dries flat instead of glossy, like linseed oil, and is inclined to produce a wrinkled surface. It is mixed with rosin, since rosin has great affinity for it, and the two together are suitable for gloss varnishes. In combination with other drying oils, it improves water and alkali resistance, and is used mainly in quick-drying enamels and varnishes. The oil from *A. montana,* or **mu oil,** has a higher percentage of eleostearic acid than that from *A. fordii.* The **Japanese tung oil** is from the nuts of the larger tree *A. cordata.* The oil is superior to Chinese tung oil and is seldom exported. It does not gelatinize as Chinese tung oil does, when heated. It is used in Japan for varnishes, waterproofing paper, and soaps. The saponification value is 193 to 195, iodine value of 149 to 159, and specific gravity 0.934 to 0.940. The kernels of the nuts yield about 40% oil. The tree is grown also in Brazil and thrives in hot climates. **Candlenut oil** is from the seed nuts of *A. moluccana* of Oceania and southern Asia. It received its name from the fact that the Polynesians used the nuts as candles to light their houses. The oil is variously known as **kukui, kekune,** and **lumbang,** and as an artist's paint, oil is called **walnut oil** or **artist's oil.** The nut resembles the walnut but has a thicker shell. The oil has a specific gravity of 0.923, iodine value 165, and is between linseed and soybean oil in properties. It is high in linoleic and linolenic acids. The variety known as soft lumbang oil, or **bagilumbang oil,** from the tree *A. trisperma* of the Philippines, resembles tung oil and is high in eleostearic acid. The chief production of lumbang oil is in the Fiji Islands.

The **Safflower,** *Carthamus tinctorius,* is grown in California, France, and India, and in the latter country it is grown on a large scale for seeds, which yield up to 35% of the clear, yellowish **safflower oil** used in paints, leather dressings, in the manufacture of nonyellowing alkyd resins, and for foods. The oil has a high content, 73%, of linoleic acid, the highest of essential polyunsaturated acids of any vegetable food oil. It is odorless, with a bland taste, has a specific gravity of 0.915, and an iodine value of 150. **Safflower 22,** of Pacific Vegetable Oil Corp., is a conjugated paint oil made by isomerizing safflower oil. It has a rapid drying rate, color retention, and an ability to produce wrinkled finishes by adjustment of the amount of drier. It can thus replace tung oil. It takes up maleic anhydride readily, and is used for making modified alkyd finishes. **Wecoline SF,** of Drew Chemical, is a concentrate of safflower fatty acids with 67.3% linoleic acid and only 0.2 linolenic acid, for compounding in coatings. **Saff,** of Abbott Laboratories, is an emulsion of safflower oil used as a drug to lower blood cholesterol. Refined and deodorized oil shows 2.8×10^{-6} lb (1.3 mg) of "cholesterol-equivalent" sterols per 0.22 lb (100 g) of oil. The heads of the plant are dried and used as food colors, for dyeing textiles, and for cosmetic rouge.

TUNGSTEN.   A heavy, white metal, symbol W, with a specific gravity of 19.6, a density of 0.697 lb/in^3 (19,290 kg/m^3), and having the highest melting point, 6170°F (3410°C), of all the metals. **Wolframite** is the chief ore of the metal tungsten. Its composition is $(FeMn)WO_3$. When the manganese tungstate is low, the ore is called **ferberite**; when the iron tungstate is low, it is called **hübnerite.** The ore is concentrated by gravity methods to a concentrate containing 60 to 65% **tungstic oxide,** WO_3. To extract pure WO_3 from the concentrate, it is fused with sodium carbonate, Na_2CO_3, to form **sodium tungstate,** Na_2WO_3, which is dissolved in water. When an acid is added to the solution, the WO_3 precipitates out as a yellow powder. The metallic tungsten is obtained by reduction and is then pressed into bars and sintered. Wolframite occurs usually bladed or columnar in form. It has a specific gravity of 7.2 to 7.5, a Mohs hardness of 5, a black color, and a submetallic luster. It is found in the mountain states, Alaska, China, and Argentina, but it also widely distributed in various parts of the world in small quantities. Chinese **wolfram concentrates** contain 65% tungstic oxide; the Arizona concentrates contain an average of 67%. California and Nevada concentrates are scheelite containing from 60 to 67% tungstic oxide. The **sanmartinite** of Argentina is a variety containing zinc.

Tungsten has a wide usage in alloy steels, magnets, heavy metals, electric contacts, rocket nozzles, and electronic applications. Tungsten resists oxidation at very high temperatures and is not attacked by nitric, hydrofluoric, or sulfuric acid solutions. Flame-sprayed coatings are used for nozzles and other parts subject to heat erosion. **Tungsten alloys** are used for weights and counterbalances, radiation shielding, grinding tools, tooling, and high-temperature applications. **Copper-tungsten composites** and **silver-tungsten composites** serve as resistance-welding die inserts, electrode facings, electrical contacts, heat sinks, wear surfaces, and electrodes for electrical-discharge machining (EDM) and electrochemical machining.

Tungsten is usually added to iron and steel in the form of **ferrotungsten,** made by electric-furnace reduction of the oxide with iron or by reducing tungsten ores with carbon and silicon. Standard grades with 75 to 85% tungsten have melting points from 3200 to 3450°F (1760 to 1899°C). **Tungsten powder** is usually in sizes from 200 to 325 mesh, and may be had in a purity of 99.9%. Parts, rods, and sheet are made by powder metallurgy, and rolling and forging are done at high temperature. The rolled metal may have a tensile strength as high as 500,000 lb/in^2 (3,448 MPa) and a Brinell hardness of 290, whereas drawn wire may have a tensile strength to 590,000 lb/in^2 (4,068 MPa). The tungsten powder is used for spray coatings for radiation shielding and for powder-metal parts. **Tungsten wire** is used for spark plugs and electronic devices, and **tungsten filaments** are used in lamps. Tungsten wire as fine as 0.00018 in (0.00046 cm) is used in electronic hardware, and as thin as 0.0004 in (0.001 cm) for wire EDM. **Tungsten whiskers,** which are extremely fine fibers, are used in copper alloys to add strength. Copper wire, which normally has a tensile strength of 30,000 lb/in^2 (207 MPa), will have a strength of 120,000 lb/in^2 (827 MPa) when 35% of the wire is tungsten whiskers. **Tungsten yarns** are made up of fine fibers of the metal. The yarns are flexible and can be woven into fabrics. Continuous tungsten filaments, usually 394 to 591 μin (10 to 15 μm) in diameter, are used for reinforcement in metal, ceramic, and plastic composites. Finer filaments are used as cores, or substrates, for **boron filaments.**

The metal is also produced as arc-fused grown crystals, usually no larger than 0.375 in (0.952 cm) in diameter and 10 in (25.4 cm) long, and worked into rod, sheet, strip, and wire. **Tungsten crystals,** 99.9975% pure, are ductile even at very low temperatures, and wire as fine as 0.003 in (0.008 cm) and strip as thin as 0.005 in (0.013 cm) can be cold-drawn and cold-rolled from the crystal. The crystal metal has nearly zero porosity, and its electrical and heat conductivities are higher than those of ordinary tungsten. The normal electrical conductivity is about 33% that of copper, but that of the crystal is 15% higher. The molecules of tungsten appear as body-centered cubes, but in the pure metal the atoms normally bond uniformly in six directions, forming a double lattice so that each grain forms a true single crystal. At elevated temperatures, tungsten forms many compounds in chemicals and alloys. One **tungsten-aluminum alloy** is a chemical compound made by reducing tungsten hexachloride with molten aluminum.

Tungsten retains a tensile strength of about 50,000 lb/in^2 (345 MPa) at 2500°F (1371°C), but because of its heavy weight is normally used in aircraft or missile parts only as coatings, usually sprayed on. It is also used for X-ray and gamma-ray shielding. Electroplates of tungsten or tungsten alloys give surface hardnesses to Vickers 700 or above. **Cobalt-tungsten alloy,** with 50% tungsten, gives a plate that retains a high hardness at red heat. **Tungsten RhC** is a **tungsten-rhenium carbide alloy** containing 4% rhenium carbide. It is used for parts requiring high strength and hardness at high temperatures. The alloy retains a tensile strength of 75,000 lb/in^2 (517 MPa) at 3500°F (1927°C). **Ammonium metatungstate,** used for electroplating, is a white powder of composition $(NH_4)_6H_2W_{12}O_{40}$. It is readily soluble in water and gives solutions of 50% tungsten content. **Tungsten hexafluoride** is used for producing tungsten coatings by vapor deposition. At a temperature of 900°F (482°C) the gas mixed with hydrogen deposits a tungsten plate. **Tungsten hexachloride,** WCl_6, is also used for depositing **tungsten coatings** at that temperature in a hydrogen atmosphere. Smooth, dense **tungsten plates** can be deposited from **tungsten carbonyl,** $W(CO)_6$, at a temperature of 302°F (150°C). The carbonyl is made by reacting the hexachloride with carbon monoxide.

TUNGSTEN CARBIDE. An iron-gray powder of minute cubical crystals with a Mohs hardness above 9.5 and a melting point of about 5400°F (2982°C). It is produced by reacting a hydrocarbon vapor with tungsten at high temperature. The composition is WC, but at high heat it may decompose into W_2C and carbon, and the carbide may be a mixture of the two forms. Other forms may also be produced, such as W_3C and W_3C_4. Tungsten carbide is used chiefly for cutting tool bits and for heat- and erosion-resistant parts and coatings.

Briquetting of tungsten carbide into usable form was first patented in Germany and produced by Krupp Works under the name of **Widia metal.** It is made by diffusing powdered cobalt through the finely divided carbide under hydraulic pressure, and then sintering in an inert atmosphere at about 2732°F (1500°C). The briquetted material is ground to shape, and the pieces are brazed into tools. They withstand cutting speeds from 3 to 10 times those of high-speed steel, and will turn manganese steel with a Brinell hardness of 550, but are not shock-resistant. Pressed and sintered parts usually contain 3 to 20% cobalt binder, but nickel may also be used as a binder. The compressive strengths may be as high as 700,000 lb/in^2 (4,827 MPa) with rupture strengths to 200,000 lb/in^2 (1,379 MPa) or higher.

One of the earliest of the U.S. bonded tungsten carbides was **Carboloy,** of General Electric Co., used for cutting tools, gages, drawing dies, and wear parts. The sintered materials have been sold under many trade names such as **Dimondite, Firthite,** and **Firthaloy; Armide,** of Armstrong Bros. Tool Co.; **Wilcoloy**; and **Borium** and **Borod,** of Stoody Co. But the carbides are now often mixed carbides. **Carboloy 608** contains 83% chromium carbide, 2 tungsten carbide, and 15 nickel binder. It is lighter in weight than tungsten carbide, is nonmagnetic, and has a Rockwell A hardness to 93. It is used for wear-resistant parts and resists oxidation to 2000°F (1092°C). Titanium carbide is more fragile, but may be mixed with tungsten carbide to add hardness for dies. **Cutanit** is such a mixture. **Kennametal K601,** of Kennametal, Inc., for seal rings and wear parts, is a mixture of tantalum and tungsten carbides without a binder. It has a compressive strength of 675,000 lb/in^2 (4,654 MPa), rupture strength of 100,000 lb/in^2 (690 MPa), and Rockwell A hardness of 94. **Kennametal K501** is tungsten carbide with a platinum binder for parts subject to severe heat erosion. **Strauss metal,** of Allegheny Ludlum Steel Co., is tungsten carbide. **Tungsten carbide LW-1** is tungsten carbide with about 6% cobalt binder used for flame-coating metal parts to give high-temperature wear resistance. Deposited coatings have a Vickers hardness to 1,450 and resist oxidation at 1000°F (538°C). **Tungsten carbide LW-1N,** with 15% cobalt binder, has a much higher rupture strength, but the Vickers hardness is reduced to 1,150. **Metco 35C** is a fine powder of tungsten carbide and cobalt for flame spraying to produce a wear-resistant coating of carbide in a matrix of cobalt. **GPX 9660,** of Securamax International, is a tungsten carbide and cobalt coating applied by flame spraying to increase the wear resistance and, to some extent, the corrosion resistance of steel parts. A tungsten carbide and nickel formulation, **GPX 9657,** also increases wear resistance and provides better corrosion resistance.

TUNGSTEN STEEL.   Any steel containing tungsten as the alloying element imparting the chief characteristics to the steel. It is one of the oldest of the alloying elements in steel, the celebrated ancient Eastern sword steels having contained tungsten. Tungsten increases the hardness of steel, and gives it the property of red hardness, stabilizing the hard carbides at high temperatures. It also widens the hardening range of steel and gives deep hardening. Very small quantities serve to produce a fine grain and raise the yield point. The tungsten forms a very hard carbide and an iron tungstite, and the strength of the steel is also increased, but it is brittle when the tungsten content is high. When large percentages of tungsten are used in steel, they must be supplemented by other carbide-forming elements. Tungsten steels, except the low-tungsten chromium-tungsten steels, are not suitable for construction; but they are widely used for cutting tools, because the tungsten forms hard, abrasion-resistant particles in high-carbon steels. Tungsten also increases the acid resistance and corrosion resistance of steels. The steels are difficult to forge and cannot be readily welded when tungsten exceeds 2%. Standard **tungsten-chromium alloy steels 72XX** contain 1.5 to 2% tungsten and 0.50 to 1 chromium. Many tool steels rely on tungsten as an alloying element, and it may range from 0.50 to 2.50% in cold-work and shock-resisting types to 9 to 18 in the hot-work type, and 12 to 20 in high-speed steels.

TURPENTINE.   Also called in the paint industry **oil of turpentine.** An oil obtained by steam distillation of the oleoresin which exudes when various conifer trees are cut. Longleaf pine and slash pine are the main sources. It also includes oils obtained by distillation and solvent extraction from stumpwood and waste wood. Longleaf sapwood contains about 2% oleoresin, heartwood 7 to 10%, and stumpwood 25%. Most oleoresin is obtained from the sapwood of living trees, but it is not the sap of the tree. Heartwood resin is obtained only when the cut wood is treated with solvents. The oleoresin yields about 20% oil of turpentine and 80 rosin; both together are known as **naval stores.**

Wood turpentine, called in the paint industry **spirits of turpentine,** is obtained from waste wood, chips, or sawdust by steam extraction or by destructive distillation. Wood turpentine forms more than 10% of all American commercial turpentines. Wood turpentine has a peculiar characteristic sawmill odor, and the residue of distillation has a camphorlike odor different from that of gum turpentine. It differs very little in composition, however, from the true turpentine. Steam-distilled wood turpentine contains about 90% terpenes, of which 80% is alpha pinene and 10% is a mixture of beta pinene and camphene. Some wood turpentine is produced as a by-product in the manufacture of cellulose. **Sulfate turpentine** is a by-product in the making of wood pulp. It varies in composition as the less stable beta pinene is affected by the pulping process, and it is used largely in chemical manufacture. By hydrogenation it produces cymene from which **dimethyl styrene** is made. This material can be copolymerized to produce vinyl resins.

Turpentine varies in composition according to the species of pine from which it is obtained. It is produced chiefly in the United States, France, and Spain. The turpentine of India comes from the **chir pine,** *Pinus longifolia,* of the southern slopes of the Himalayas, also valued for lumber, and the **khasia pine,** *P. khasya.* The gum of the chir pine is different from U.S. gum, and the turpentine, unless carefully distilled, is slower-drying and greasy. French and Spanish turpentine, or **Bordeaux turpentine,** is from the **maritime pine,** *P. pinaster,* which is the chief source, and from **Aleppo pine,** *P. halepensis,* and **Corsican pine,** *P. lavicia.* In Portugal, the **stone pine,** *P. pinea,* is the source. The French maritime pine is also grown on plantations in Australia. Aleppo pine of Greece was the source of the naval stores of the ancients. **Venetian turpentine,** or **Venice turpentine,** is from the Corsican pine or European larch. It produces a harder film than U.S. turpentine. Artificial Venice turpentine is made by mixing rosin with turpentine. European pines do not give as high a yield as U.S. longleaf and slash pines.

American turpentine oil boils at 309°F (154°C), and the specific gravity is 0.860. It is a valuable drying oil for paints and varnishes, owing to its property of rapidly absorbing oxygen from the atmosphere and transferring it to the linseed or other drying oil, which leaves a tough and durable film of paint. Turpentine is also used in the manufacture of artificial camphor and rubber, and in linoleum, soap, and ink. **Gum thus,** used in artists' oil paints, is thickened turpentine, although gum thus was originally made from olibanum. Turpentine is often adulterated with other oils of the pine or with petroleum products, and the various states have laws regulating its adulteration for paint use.

The **pinene** in European turpentine is levorotatory while that in the U.S. is dextrorotatory. **Pinonic acid** is acetyl dimethyl cyclobutane acetic acid. It is produced by oxidation of the pinene and is a white powder used as a cross-linking agent for making heat-stable plastics.

Terpene alcohol, or **methylol pinene,** $C_{11}H_{17}OH$, is produced by condensing the beta pinene of gum turpentine with formaldehyde. **Nopol,** of Glidden Co., is terpene alcohol. It has the chemical reactions of both a primary alcohol and pinene, and it is used in making many chemicals. It is a water-insoluble liquid of specific gravity 0.963, boiling at 455°F (235°C). **Terpineol** is a name for refined terpene alcohols used largely for producing essential oils and perfumes. **Piccolyte resin** is a terpene thermoplastic varnish resin made from turpentine. The grades have melting points from 50 to 257°F (10 to 125°C). **Myrcene** is a polyolefin with three double bonds, which can be used as a substitute for butadiene in the manufacture of synthetic rubbers, or can be reacted with maleic anhydride or dibasic acids to form synthetic resins. It is made by isomerizing the beta pinene of gum turpentine. **Camphene** is produced by isomerizing the alpha pinene of turpentine. Camphor is then produced by oxidation of camphene in acid. Camphene was also the name of a lamp oil of the early nineteenth century made from distilled turpentine and alcohol. It gave a bright white light, but was explosive. The insecticide known as **Toxaphene,** of Hercules Inc., is made by chlorinating camphene to 68% chlorine, or to the empirical formula $C_{10}H_{20}Cl_8$. It is a yellow, waxy powder with a piney odor, melting at 149 to 194°F (65 to 90°C). It is soluble in petroleum solvents.

TURQUOISE.   An opaque-blue gemstone with a waxy luster. It is a hydrous phosphate of aluminum and copper oxides. It is found in the western United States in streaks in volcanic rocks, but most of the turquoise has come from the Kuh-i-Firouzeh, or turquoise mountain, of Iran, which is a vast deposit of brecciated porphyry, or feldspar igneous rock. The valuable stones are the deep blue. The pale blue and green stones were called **Mecca stones** because they were sent to Mecca for sale to pilgrims. **Bone turquoise,** or **odontolite,** used for jewelry, is fossil bone or tooth, colored by a phosphate of iron.

TYPE METAL.   Any metal used for making printing type, but the name generally refers to **lead-antimony-tin** alloys. Antimony has the property of expanding on cooling and thus fills the mold and produces sharp, accurate type. The properties required in a type metal are ability to make sharp, uniform castings; strength and hardness; fairly low melting point; narrow freezing range to facilitate rapid manufacture in type-making machines; and resistance to drossing. A common type metal is composed of 9 parts lead to 1 antimony, but many varieties of other mixtures are also used. The antimony content may be as high as 30%, 15 to 20% being frequent. A common **monotype metal** has 72% lead, 18 antimony, and 10 tin. Larger and softer types are made of other alloys, sometimes containing bismuth; the hardest small type contains 3 parts lead to 1 antimony. A low-melting-point, soft-type metal contains 22% bismuth, 50 lead, and 28 antimony. It will melt at about 310°F (154°C). Copper, up to 2%, is sometimes added to type metal to increase the hardness, but is not ordinarily used in metals employed in rapid-acting type machines. Some monotype metal has about 18% antimony, 8 tin, and 0.1 copper, but standard **linotype metal** for pressure casting has 79% lead, 16 antimony, and 5 tin. **Stereotype metal,** for sharp casting and hard-wearing qualities, is given as 80.0% lead, 13.5 antimony, 6 tin, and 0.5 copper. **Intertype metal** has 11 to 14% antimony and 3 to 5 tin. A typical formula for **electrotype metal** is 94% lead, 3 tin, and 3 antimony. The Brinell hardness of machine-molded type ranges from 17 to 23, and that of stereotype metal is up to 30. As constant remelting causes the separation of the tin and lead, and the loss of tin, or impoverishment of the metal, new metal must be constantly added to prevent deterioration of a standard metal into an inferior alloy. For many years lead-antimony-tin alloys have been used as a weld seam filler in auto and truck bodies. In this application they are commonly referred to as **body solder.** Because of advances in printing technology and auto manufacturing, use of these lead alloys is steadily declining.

ULTRAHIGH-STRENGTH STEELS. The highest-strength steels available. Arbitrarily, steels with tensile strengths of around 200,000 lb/in^2 (1,379 MPa) or higher are included in this category, and more than 100 alloy steels can be thus classified. They differ rather widely among themselves in composition and/or the way in which the ultrahigh strengths are achieved.

Medium-carbon, low-alloy steels were the initial ultrahigh-strength steels, and within this group, a **chromium-molybdenum steel** (4130) grade and a **chromium-nickel-molybdenum steel** (4340) grade were the first developed. These steels have yield strengths as high as 240,000 lb/in^2 (1,655 MPa) and tensile strengths approaching 300,000 lb/in^2 (2,069 MPa). They are particularly useful for thick sections because they are moderately priced and have deep hardenability. Several types of **stainless steels** are capable of strengths above 200,000 lb/in^2 (1,379 MPa), including a number of martensitic, cold-rolled austenitic, and semiaustenitic grades. The typical martensitic grades are types 410, 420, and 431, as well as certain **age-hardenable alloys.** The **cold-rolled austenitic stainless steels** work-harden rapidly and can achieve 180,000 lb/in^2 (1,241 MPa) tensile yield strength and 200,000 lb/in^2 (1,379 MPa) ultimate strength. **Semiaustenitic stainless steels** can be heat-treated for use at yield strengths as high as 220,000 lb/in^2 (1,517 MPa) and ultimate strengths of 235,000 lb/in^2 (1,620 MPa).

Maraging steels contain 18 to 25% nickel plus substantial amounts of cobalt and molybdenum. Some newer grades contain somewhat less than 10% nickel and between 10 and 14 chromium.

Because of the low carbon (0.03% maximum) and nickel content, maraging steels are martensitic in the annealed condition, but are still readily formed, machined, and welded. By a simple aging treatment at about 900°F (482°C), yield strengths as high as 300,000 and 350,000 lb/in^2 (2,069 and 2,413 MPa) are attainable, depending on specific composition. In this condition, although ductility is fairly low, the material is still far from being brittle.

Among the strongest of plain carbon sheet steels are the low- and medium-carbon sheet grades of Inland Steel, called **MarTinsite.** Made by rapid water quenching after cold rolling, they provide tensile yield strengths to 220,000 lb/in^2 (1,517 MPa) but are quite limited in ductility.

There are two types of ultrahigh-strength, low-carbon, hardenable steels. One, a **chromium-nickel-molybdenum steel,** named **Astralloy,** with 0.24% carbon is air-hardened to a yield strength of 180,000 lb/in^2 (1,241 MPa) in heavy sections when it is normalized and tempered at 500°F (260°C). The other type is an **iron-chromium-molybdenum-cobalt steel** and is strengthened by a precipitation-hardening and aging process to levels up to 245,000 lb/in^2 (1,689 MPa) in yield strength. High-alloy **quenched-and-tempered steels** are another group that have extrahigh strengths. They contain 9% nickel, 4 cobalt, and from 0.20 to 0.30 carbon, and develop yield strengths close to 300,000 lb/in^2 (2,069 MPa) and ultimate strengths of 350,000 lb/in^2 (2,413 MPa). Another group in this high-alloy category resembles **high-speed tool steels,** but is modified to eliminate excess carbide, thus considerably improving ductility. These so-called **matrix steels** contain tungsten, molybdenum, chromium, vanadium, cobalt, and about 0.5% carbon. They can be heat-treated to ultimate strengths of over 400,000 lb/in^2 (2,758 MPa)—the highest strength presently available in steels, except for heavily **cold-worked high-carbon steel** strips used for razor blades and drawn wire for musical instruments, both of which have tensile strengths as high as 600,000 lb/in^2 (4,137 MPa).

Aermet 100, of Carpenter Technology, is a nickel and cobalt steel strengthened by carbon, columbium, and molybdenum. Originally developed for aerospace applications, it combines high tensile yield strength [(250,000 lb/in^2 (1,724 MPa)] and fracture toughness [(115,000 lb/in^2 $\sqrt{\text{in}}$ (126 MPa $\sqrt{\text{m}}$)]. Uses include aircraft landing gears, racing-car shafts, racing-bicycle frames, mandrel-support shafts, punch-base supports, and special bolting systems.

URANIUM. An elementary metal, symbol U. It never occurs free in nature but is found chiefly as an oxide in the minerals pitchblende and carnotite where it is associated with radium. The metal has a specific gravity of 18.68 and atomic weight 238.2. The melting point is about 2071°F (1133°C). It is hard but malleable, resembling nickel in color, but related to chromium, tungsten, and molybdenum. It is soluble in mineral acids.

Uranium has three forms. The alpha phase, or orthorhombic crystal, is stable to 1220°F (660°C); the beta, or tetragonal, exists from 1220 to 1400°F (660 to 760°C); and the gamma, or body-centered cubic, is from 1400°F to the melting point. The cast metal has a Rockwell B hardness of 80 to 100, work-hardening easily. The metal is alloyed with iron to make **ferrouranium,** used to impart special properties to steel. It increases the elastic limit and the tensile strength of steels, and is also a more powerful deoxidizer than vanadium. It will denitrogenize steel and has carbide-forming qualities. It has been used in high-speed steels in amounts of 0.05 to 5% to increase the strength and toughness, but because of its importance for atomic applications its use in steel is now limited to the by-product nonradioactive isotope uranium 238. The **green salt** used in atomic work is **uranium tetrafluoride,** UF_4. **Uranium hexafluoride,** UF_6, is a gas used to separate uranium isotopes.

Metallic uranium is used as a cathode in photoelectric tubes responsive to ultraviolet radiation. Uranium compounds, especially the uranium oxides, were used for making glazes in the ceramic industry and also for paint pigments. It produces a yellowish-green, fluorescent glass, and a beautiful red with yellowish tinge is produced on pottery glazes. **Uranium dioxide,** UO_2, is used in sintered forms as fuel for power reactors. It is chemically stable and has a high melting point at about 5000°F (2760°C), but a low thermal conductivity. For fuel use the particles may be coated with about 0.001 in (0.003 cm) of aluminum oxide. This coating is impervious to xenon and other radioactive isotopes so that only the useful power-providing rays can escape. These are not dangerous at a distance of about 6 in (15 cm), and thus less shielding is needed. For temperatures above 2300°F (1260°C) a coating of pyrolitic graphite is used.

Uranium has isotopes from 234 to 239, and **uranium 235,** with 92 protons and 143 neutrons, is the one valued for atomic work. The purified natural metal contains only about one part U^{235} to about 140 parts of U^{238}, and about 100,000 lb (45,360 kg) of uranium fluoride, UF_6, must be processed to obtain 1 lb (0.45 kg) of $U^{235}F_6$. **Uranium 238,** after the loss of three alpha particles of total mass 12, changes to **radium 226.** The lead of old uranium minerals came from Ra^{226} by the loss of five alpha particles, and is **lead 206,** while the lead in thorium metals is **lead 208. Lead 207** comes from the decay of actinum and exists only in small quantities.

Natural uranium does not normally undergo fission because of the high probability of the neutron being captured by the U^{238} which then merely ejects a gamma ray and becomes U^{239}. When natural uranium is not in concentrated form, but is embodied in a matrix of graphite or heavy water, it will sustain a slow chain reaction sufficient to produce heat. In the fission of U^{235}, neutrons are created which maintain the chain reaction and convert U^{238} to plutonium. About 40 elements of the central portion of the periodic table are also produced by the fission, and eventually these products build up to a point where the reaction is no longer self-sustaining. The slow, nonexplosive disintegration of the plutonium yields neptunium. **Uranium 233** is made by neutron bombardment of thorium. This isotope is fissionable and is used in thermonuclear reactors.

Uranium yellow, also called **yellow oxide,** is a **sodium diuranate** of composition $Na_2U_2O_7 \cdot 6H_2O$, obtained by reduction and treatment of the mineral pitchblende. It is used for yellow and greenish glazing enamels and for imparting an opalescent yellow to glass, which is green in reflected light. **Uranium oxide** is an olive-green powder of composition U_3O_8, used as a pigment. **Uranium trioxide,** UO_3, is an orange-yellow powder also used for ceramics and pigments. It is also called **uranic oxide.** As a pigment in glass, it produces a beautiful greenish-yellow **uranium glass. Uranium pentoxide,** U_2O_5, is a black powder, and **uranous oxide,** UO_2, is used in glass to give a fine black color. **Sodium uranate,** Na_2UO_4, is a yellow to orange powder used to produce ivory to yellow shades in pottery glazes. The uranium oxide colors give luster and iridescence, but because of the application of the metal-to-atom work, the uses in pigments and ceramics are now limited.

URANIUM ORES. The chief source of radium and uranium is **uraninite,** or **pitchblende,** a black, massive or granular mineral with pitchlike luster. The mineral is a combination of the oxides of uranium, UO_2, UO_3, and U_3O_8, together with small amounts of lead, thorium, yttrium, cerium, helium, argon, and radium. The process of separation of radium is chemically complicated. Uraninite is found with the ores of silver and lead in central Europe. In the United States it occurs in pegmatite veins, in the mica mines of North Carolina, and in the carnotite of Utah and Colorado. The richest ores come from the Congo and from near Great Bear Lake, Canada. About 370 tons (336 metric tons) of Great Bear Lake ore produces 0.002 lb (1 g) of radium and 7,800 lb (3,538 kg) of uranium, and small amounts of polonium, ionium, silver, and radioactive lead. Numerous minor uranium ores occur in many areas. A low-grade ore of 0.1% U_3O_8 can be upgraded to as high as 5% by ion exchange. Black mud from the fjords of Norway contains up to 2 oz (0.06 kg) of uranium per long ton (1 metric ton).

Tyuyamunite, found in Turkman, averages 1.3% U_3O_8, with radium, vanadium, and copper. **Autunite,** or **uranite,** is a secondary mineral from the decomposition of pitchblende. The composition is approximately $P_2O_5 \cdot 2UO_3 \cdot CaO \cdot 8H_2O$. It is produced in Utah, Portugal, and south Australia. **Torbernite,** or **copper uranite,** $Cu(UO_2)_2P_2O_8 \cdot 12H_2O$, is a radioactive mineral of specific gravity 3.22 to 3.6 and Mohs hardness 2 to 2.5. **Sengierite** is a copper-uranium mineral found in the Congo. It occurs in small green crystals. **Casolite** is a yellow, earthy **lead uranium silicate,** $3(PbO \cdot UO_3 \cdot SiO_2)4H_2O$. **Pilbarite** is a **thorium lead uranate. Umohoite,** found in Utah, contains 48% uranium, with molybdenum, hydrogen, and oxygen. The name of the ore is a combination of the symbols of the contained elements. Uranium is also recovered chemically from phosphate rock. The phosphate waste rock of Florida contains from 0.1 to 0.4% U_3O_8. Most uranium ores contain less than 0.3% U_3O_8. Solvent methods of extraction are used.

UREA. Also called **carbamide.** A colorless to white, crystalline powder, $NH_2 \cdot CO \cdot NH_2$, best known for its use in plastics and fertilizers. The chemistry of urea and the carbamates is very complex, and a very great variety of related products are produced. Urea is produced by combining ammonia and carbon dioxide, or from **cyanamide,** $NH_2 \cdot C \cdot N$. It is a normal waste product of animal protein metabolism and is the chief nitrogen constituent of urine. It was the first organic chemical ever synthesized commercially. It has a specific gravity of 1.323 and a melting point of 275°F (135°C). Industrial grades are available either as prills or as a 50% solution from Columbia Nitrogen Corp. An ultrapure enzyme grade is produced in small quantities by Bethesda Laboratories, and material for electrophoresis by Bio-Rad Laboratories.

The formula for urea may be considered as $O \cdot C(NH_2)_2$, and thus as an amide substitution in **carbonic acid,** $O \cdot C(OH)_2$, an acid which really exists only in its compounds. The urea-type plastics are called **amino resins.** The **carbamates** can also be considered as deriving from **carbamic acid,** NH_2COOH, an **aminoformic acid** that likewise appears only in its compounds. The carbamates have the same structural formula as the bicarbonates, so that **sodium carbamate** has an NH_2 group substituted for each OH group of the sodium bicarbonate. The **urethanes,** used for plastics and rubber, are **alkyl carbamates** made by reacting urea with an alcohol, or by reacting isocyanates with alcohols or carboxyl compounds. They are white powders of composition $NH_2COOC_2H_5$, melting at 122°F (50°C).

Isocyanates are esters of **isocyanic acid,** $H \cdot N \cdot C \cdot O$, which does not appear independently. The **dibasic diisocyanate** of General

Mills, Inc., is made from a 36-carbon fatty acid. It reacts with compounds containing active hydrogen. With modified polyamines it forms **polyurea resins,** and with other diisocyanates it forms a wide range of urethanes. **Tosyl isocyanate,** of Upjohn Co., for producing urethane resins without a catalyst, is toluene sulphonyl isocyanate. The sulphonyl group increases the reactivity.

Methyl isocyanate, CH_3NCO, known as **MIC,** is a colorless liquid with a specific gravity of 0.9599. It reacts with water. With a flash point of less than 20°F (-6.6°C), it is flammable and a fire risk. It is a strong irritant and highly toxic. One of its principal uses is as an intermediate in the production of pesticides.

Urea is used with acid phosphates in fertilizers. It contains about 45% nitrogen and is one of the most efficient sources of nitrogen. Urea reacted with malonic esters produces **malonyl urea,** which is the **barbituric acid** that forms the basis for the many soporific compounds such as **luminal, phenobarbital,** and **amytal.** The malonic esters are made from acetic acid, and **malonic acid** derived from the esters is a solid of composition $CH_2(COOH)_2$ which decomposes at about 320°F (160°C) to yield acetic acid and carbon dioxide.

For plastics manufacture, substitution on the sulfur atom in thiourea is easier than on the oxygen in urea. **Thiourea,** $NH_2 \cdot CS \cdot NH_2$, also called **thiocarbamide, sulfourea,** and **sulfocarbamide,** is a white, crystalline, water-soluble material of bitter taste, with a specific gravity of 1.405. It is used for making plastics and chemicals. On prolonged heating below its melting point of 360°F (182°C), it changes to **ammonium thiocyanate,** or **ammonium sulfocyanide,** a white, crystalline, water-soluble powder of composition NH_4SCN, melting at 302°F (150°C). This material is also used in making plastics, as a mordant in dyeing, to produce black nickel coatings, and as a weedkiller. **Permafresh,** of Warwick Chemical Co., used to control shrinkage and give wash-and-wear properties to fabrics, is **dimethylol urea,** $CO(NHCH_2OH)_2$, which gives clear solutions in warm water.

Urea-formaldehyde resins are made by condensing urea or thiourea with formaldehyde. They belong to the group known as **aminoaldehyde resins** made by the interaction of an amine and an aldehyde. An initial condensation product is obtained which is soluble in water and is used in coatings and adhesives. The final condensation product is insoluble in water and is highly chemical-resistant. Molding is done with heat and pressure. The urea resins are noted for their transparency and ability to take translucent colors. Molded parts with cellulose filler have a specific gravity of about 1.50, tensile strength from 6,000 to 13,000 lb/in^2 (41 to 90 MPa), elongation 15%, compressive strength to 45,000 lb/in^2 (310 MPa), dielectric strength to

400 V/mil (16 × 10^6 V/m), and heat distortion temperature to 280°F (138°C). Rockwell M hardness is about 118. **Urea resins** are marketed under a wide variety of trade names. The **Uformite resins** of Rohm & Haas are water-soluble thermosetting resins for adhesives and sizing. The **Urac resins,** of American Cyanamid, and the **Casco resins** and **Cascamite,** of Borden Co., are urea-formaldehyde. Borden's products are available as liquids, 55 to 66% solids, and spray-dried powder grades. They are used as adhesives for plasterboard and plywood and in wet-strength paper. **Resi-mat,** from Georgia-Pacific, is a liquid resin binder for glass-mat roofing and insulation materials.

URETHANES. Also termed **polyurethanes.** A group of plastic materials based on polyether or polyester resin. The chemistry involved is the reaction of a diisocyanate with a hydroxyl-terminated polyester or polyether to form a higher-molecular-weight prepolymer, which in turn is chain-extended by adding difunctional compounds containing active hydrogens, such as water, glycols, diamines, or amino alcohols. The urethanes are block polymers capable of being formed by a literally indeterminate number of combinations of these compounds. The urethanes have excellent tensile strength and elongation, good ozone resistance, and good abrasion resistance. Combinations of hardness and elasticity unobtainable with other systems are possible in urethanes, ranging from Shore A hardness of 15 to 30 (printing rolls, potting compounds) through A 60 to 90 for most industrial or mechanical goods applications, to Shore D 70 to 85. Urethanes are fairly resistant to many chemicals such as aliphatic solvents, alcohols, ether, certain fuels, and oils. They are attacked by hot water, polar solvents, and concentrated acids and bases.

Urethane foams are made by adding a compound that produces carbon dioxide or by reaction of a diisocyanate with a compound containing active hydrogen. Foams can be classified somewhat according to modulus as flexible, semiflexible or semirigid, and rigid. No sharp lines of demarcation have been set on these different classes as the gradation from the flexible to the rigid is continuous. Density of flexible foams ranges from about 1.0 lb/ft^3 (16 kg/m^3) to 4 to 5 lb/ft^3 (64 to 80 kg/m^3), depending on the end use. Applications of flexible foams range from comfort cushioning of all types, e.g., mattresses, pillows, sofa seats, backs and arms, automobile topper pads, and rug underlay, to clothing interliners for warmth at light weight. Density of rigid urethane foams ranges from about 1.5 to 50 lb/ft^3 (24 to 800 kg/m^3).

Confor, of E-A-R Specialty Composites, is a line of temperature-sensitive urethane foams for cushioning and padding. Surfaces in contact with body heat, for example, soften and conform to body shape while other regions remain stiff and supportive. Unlike fast-recovery foams, recovery is slow. They come in several stiffness grades and colors. **Isoloss LS** foams, of E-A-R, have a fine-cell, high-density structure, featuring high-strength dimensional stability and low compression set. Uses include seals, isolaters, and energy-absorbing mounts. **Conathane UC-38** and **UC-39,** of Conap, Inc., are liquid, two-component, room-temperature-curing resins for prototype parts. UC-38 has a tensile strength of 5,000 lb/in² (34 MPa), 35% elongation, a Shore D hardness of 75, and a low shrinkage to yield precision parts. UC-39 features a demold time of only 1 h.

Thermoplastic polyurethanes (TPUs) include two basic types: esters and ethers. Esters are tougher, but hydrolyze and degrade when soaked in water. There also are TPUs based on polycaprolactone, which while technically being esters, have better resistance to hydrolysis. TPUs are used when a combination of toughness, flex resistance, weatherability, and low-temperature properties is needed. These materials can be injection-molded, blow-molded, and extruded as profiles, sheet, and film. Further, TPUs are blended with other plastic resins, including PVC, ABS, acetal, SAN, and polycarbonate.

Polyol and isocyanate, two highly reactive components of urethanes, are used to form flexible bumper fascia for cars by reaction injection molding (RIM). There are low- and high-modulus unreinforced grades, as well as glass-reinforced grades for greater rigidity. There are several trade names, including **Bayflex,** of Miles Inc. There are also grades for what is called *structural reaction injection molding* (SRIM). **Polydicyclopentadiene** resins can be tailored for RIM or SRIM. Besides milled glass and flaked glass, wollastonite and mica fillers can be used for reinforcement and to improve surface finish. **Rimlite,** of Miles, refers to the use of lightweight microspheres as fillers.

Urethane elastomers are made with various isocyanates, the principal ones being **TDI (tolylene diisocyanate)** and **MDI (4,4´-diphenylmethane diisocyanate)** reacting with linear polyols of the polyester and polyether families. Various chain extenders, such as glycols, water, diamines, or aminoalcohols, are used in either a prepolymer or a one-shot type of system to form the long-chain polymer. Recent formulations are more environmentally friendly, containing less solvent, more water, and less aromatic diisocyanates.

Textile fibers of urethane were first made in Germany under the name of **Igamide.** Flexible **urethane fibers,** used for flexible garments, are more durable than ordinary rubber fibers or filaments and are 30% lighter in weight. They are resistant to oils and to washing chemicals, and also have the advantage that they are white. **Spandex fibers** are stretchable fibers produced from a fiber-forming substance in which a long chain of synthetic molecules is composed of a segmented polyurethane. Stretch before break of these fibers is from 520 to 610%, compared to 760% for rubber. Recovery is not as good as in rubber. Spandex is white and dyeable. Resistance to chemicals is good, but it is degraded by hypochlorides.

There are six basic types of **polyurethane coatings,** or **urethane coatings,** as defined by the American Society for Testing and Materials Specification D16. Types 1, 2, 3, and 6 have long storage life and are formulated to cure by oxidation, by reaction with atmospheric moisture, or by heat. Types 4 and 5 are catalyst-cured and are used as coatings on leather and rubber and as fast-curing industrial product finishes. Urethane coatings have good weathering characteristics as well as high resistance to stains, water, and abrasion. To limit emission of volatile organic compounds, there has been a trend to waterborne and high-solids coatings.

A biocompatible, polyurethane-based, shape-memory polymer, developed by Japan's Mitsubishi Heavy Industries and available in the United States from Memry Corp., undergoes dramatic changes in hardness, flexibility, elastic modulus, and vapor permeability with small temperature changes. It has lower glass transition temperatures—standardized at 77°F (25°C), 95°F (35°C), 113°F (45°C), or 131°F (55°C)—than former **memory polymers** and far lower temperatures than conventional thermoplastics. Spoon handles, which can be heated and re-formed to suit the deformed hands of handicapped persons, is an early use. Potential uses include medical catheters (stiff for insertion, flexible once implanted), custom-fitting orthopedic braces and splints, actuator mechanisms, and textile coatings whose permeability varies with temperature change.

VALVE STEEL. Used for automobile intake and exhaust valves. The intake valve steel contains about 2% chromium, 3.75 silicon, 0.3 manganese, and 0.40 carbon, and the steel for exhaust valves has up to 24% chromium, 5 nickel, 3 molybdenum, 1 each of manganese and silicon, and 0.40 carbon. These steels are hot-sprayed with aluminum and heated to 1450°F (788°C) to diffuse the aluminum into the surface, to give an iron-aluminum compound about 0.001 in (0.0025 cm) thick which is resistant to hot corrosive gases.

VANADIUM. An elementary metal, symbol V, widely distributed, but found in commercial quantities in only a few places, chiefly Peru, Zimbabwe, southwest Africa, and the United States. The common ores of vanadium are carnotite, patronite, roscoelite, and vanadinite. Much of the commercial vanadium comes from Peruvian patronite and shales. Some Russian vanadium comes from the mineral **tyuya-**

munite, the calcium analog of carnotite. This analog also occurs in American carnotite as a greenish-yellow powder. Titaniferous ores of South Africa also furnish vanadium. But more than 60% of the known resources are in the United States. Carnotite occurs in Utah and Colorado, and the Arizona ore is vanadinite. The most important ore in the United States is **roscoelite.** It is a muscovite mica in which part of the aluminum has been replaced by vanadium. It occurs in micalike scales varying in color from green to brown. It has a specific gravity of 2.9. The ore mined in Colorado contains about 1.5% vanadium oxide, V_2O_5, and this oxide is extracted and marketed for making ferrovanadium and vanadium compounds. The slag from Idaho phosphorus workings contains up to 5% vanadium, which is concentrated to 13% and extracted as vanadium pentoxide. It is recovered from petroleum. Venezuelan crude oil, containing 130 ppm vanadium, yields 2,000 lb (907 kg) of vanadium pentoxide per 1×10^6 gal (3.79 $\times$ 10^6 L) of oil.

Vanadium is a pale-gray metal with a silvery luster. Its specific gravity is 6.02, and it melts at 3236°F (1780°C). It does not oxidize in air and is not attacked by hydrochloric or dilute sulfuric acid. It dissolves with a blue color in solutions of nitric acid. It is marketed by Vanadium Corp., 99.5% pure, in cast ingots, machined ingots, and buttons. The as-cast metal has a tensile strength of 54,000 lb/in^2 (372 MPa), yield strength of 45,000 lb/in^2 (310 MPa), and elongation 12%. Annealed sheet has a tensile strength of 78,000 lb/in^2 (538 MPa), yield strength 66,000 lb/in^2 (455 MPa), and elongation 20%, while the cold-rolled sheet has a tensile strength of 120,000 lb/in^2 (827 MPa) with elongation of 2%. Vanadium metal is expensive, but is used for special purposes such as for springs of high flexural strength and corrosion resistance. The greatest use of vanadium is for alloying. **Ferrovanadium,** for use in adding to steels, usually contains 30 to 40% vanadium, 3 to 6 carbon, and 8 to 15 silicon, with the balance iron, but may also be had with very low carbon and silicon. **Vanadium-boron,** for alloying steels, is marketed as a master alloy containing 40 to 45% vanadium, 8 boron, 5 titanium, 2.5 aluminum, and the balance iron; but the alloy may also be had with no titanium. **Van-Ad alloy,** for adding vanadium to titanium alloys, contains 75% vanadium and the balance titanium. It comes as fine crystals. The **vanadium-columbium alloys** developed by Union Carbide, containing 20 to 50% columbium, have a tensile strength above 100,000 lb/in^2 (690 MPa) at 1292°F (700°C), 70,000 lb/in^2 (483 MPa) at 1832°F (1000°C), and 40,000 lb/in^2 (276 MPa) at 2192°F (1200°C).

Vanadium salts are used to color pottery and glass and as mordants in dyeing. **Red cake,** or crystalline **vanadium oxide,** is a reddish-brown material, containing about 85% **vanadium pentoxide,** V_2O_5, and 9 Na_2O, used as a catalyst and for making vanadium compounds. Vanadium oxide is also used to produce yellow glass; the pigment known as **vanadium-tin yellow** is a mixture of vanadium pentoxide and tin oxide.

VANADIUM STEEL. Vanadium was originally used in steel as a cleanser, but is now employed in small amounts, 0.15 to 0.25%, especially with a small quantity of chromium, as an alloying element to make strong, tough, and hard low-alloy steels. It increases the tensile strength without lowering the ductility, reduces grain growth, and increases the fatigue-resisting qualities of steels. Larger amounts are used in high-speed steels and in special steels. Vanadium is a powerful deoxidizer in steels, but is too expensive for this purpose alone. Steels with 0.45 to 0.55% carbon and small amounts of vanadium are used for forgings, and cast steels for aircraft parts usually contain vanadium. In tool steels vanadium widens the hardening range, and by the formation of double carbides with chromium makes hard and keen-edged die and cutter steels. All of these steels are classified as **chromium-vanadium steels.** The carbon-vanadium steels for forgings and castings, without chromium, have slightly higher manganese.

Vanadium steels require higher quenching temperatures than ordinary steels or nickel steels. **SAE 6145 steel,** with 0.18% vanadium and 1 chromium, has a fine grain structure and is used for gears. It has a tensile strength of 116,000 to 292,000 lb/in^2 (800 to 2,013 MPa) when heat-treated, with a Brinell hardness of 248 to 566, depending on the temperature of drawing, and an elongation of 7 to 26%. In cast vanadium steels it is usual to have from 0.18 to 0.25% vanadium with 0.35 to 0.45 carbon. Such castings have a tensile strength of about 80,000 lb/in^2 (552 MPa) and an elongation of 22%. A nickel-vanadium cast steel has much higher strength, but high-alloy steels with only small amounts of vanadium are not usually classified as vanadium steels.

VANILLA BEANS. The seed pods of a climbing plant of the orchid family of which there are more than 50 known species. It is native to Mexico, but now also is grown commercially in Madagascar, Seychelles, Tahiti, Réunion, Mauritius, and tropical America. It is used for the production of the flavor **vanilla.** The species grown for commercial vanilla is *Vanilla planifolia,* a tall climbing herb with yellow flowers. It grows in humid, tropical climates. The flowers are pollinated by hand to produce 30 to 40 beans per plant. The green beans are cured immediately in ovens to prevent spoilage after a sweating process. During the curing the glucoside is changed by enzyme action into **vanillin,** which crystallizes on the surface and possesses the characteristic odor and flavor. The dark-brown cured pods are put up in small packs in tin containers. Vanillin also occurs naturally in **potato parings** and **Siam benzoin. Vanilla extract** is made by percolating the chopped bean pods in ethyl alcohol, and then concentrating the mixture by evaporating the alcohol at a low temperature to avoid impairing the flavor.

The species *V. pompana* is more widely distributed, but is not as fragrant. The vanilla grown in Tahiti has an odor of heliotrope which must be removed. At least 15 species of vanilla grow in the Amazon and Orinoco valleys. Vanilla was used by the Aztecs for flavoring chocolate. It is now used for the same purpose, and as a flavor for ice cream, puddings, cakes, and other foodstuffs.

Vanillin is also produced synthetically from **eugenol** derived from clove oil, and from guaiacol obtained by the alkylation of catechol or by the destructive distillation of wood. It is also made from **coniferin,** $C_{16}H_{22}O_8 \cdot 2H_2O$, a white, crystalline material of melting point 365°F (185°C) obtained from the sapwood of the northern pine. It is produced in Wisconsin from pulp-mill waste liquors by hydrating into sugars and oxidizing to vanillin. But the synthetic vanillin does not give the full, true flavor of vanilla, as a blend of other flavors is present in the natural product. The demand for vanilla as a flavor is always greater than the supply, so that even the grades rated as pure vanilla extract may be so adulterated or diluted as to lose the full, rich flavor. Vanillin is used as a chemical intermediate in the production of pharmaceuticals, such as **L-dopa, Trimethaprim,** and **Aldomet.**

Ethyl vanillate, $C_6H_3(OH)(OCH)_3(COOC_2H_5)$, is made from Wisconsin sulfite liquor. It is used in cheese to prevent mold, and as a preservative in tomato and apple juice. **Lioxin,** of Ontario Paper Co., is an impure 97% vanillin made from sulfite lignin. It is not suitable for use as a flavor, but is used as an odor-masking agent, as a brightener in zinc-plating baths, as an antifoam agent in lubricating oils, and for making syntans. **Veratraldehyde** is obtained by methylating vanillin and is used for brightening metals in the plating industry. **Vanitrope,** of Shulton, Inc., is a synthetic aromatic with a flavor 15 times more powerful than vanillin but with a resinous note resembling that of coumarin. It differs from vanillin chemically by having no aldehyde group, and is a **propenyl guaethol** related to eugenol. It is used as a vanilla extender. A blend of Vanitrope and vanilla, called **Nuvan,** is used as a low-cost vanilla flavor. **Vanatone** and **Vanarine,** of Fritzsche Bros., Inc., are blends of vanillin with aldehydes and esters to increase the flavor tone.

VAPOR-DEPOSITED COATINGS. Thin, single- or multilayer coatings applied to base surfaces by deposition of the coating metal from its vapor phase. Most metals and even some nonmetals, such as silicon oxide, can be vapor-deposited. **Vacuum-evaporated films,** or **vacuum-metallized films,** of aluminum are most common. They are applied by vaporizing aluminum in a high vacuum and then allowing it to condense on the object to be coated. Vacuum-metallized films are extremely thin, ranging from 0.002 to 0.1 mil (0.00005 to 0.003 mm). In addition to vacuum evaporation, vapor-deposited films can be produced by **ion sputtering, chemical-vapor plating,** and a **glow-discharge** process. In ion sputtering, a high voltage applied to a target of the coating material in an ionized gas medium causes target atoms (ions) to be dislodged and then to condense as a **sputtered** coating on the base material. In chemical-vapor plating, a film is deposited when a metal-bearing gas thermally decomposes on contact with the heated surface of the base material. And in the glow-discharge process, applicable only to polymer films, a gas discharge deposits and polymerizes the plastic film on the base material. In recent years, **titanium nitride,** deposited by chemical vapor deposition or physical vapor deposition, has been used to markedly increase the wear resistance of cutting tools and forming tools made of tool steels.

VARNISH. A solution of a resin in drying oil, which when spread out in a thin film dries and hardens by evaporation of the volatile solvent, or by the oxidation of the oil, or by both. A smooth, glossy coating is left on the surface. Varnishes do not contain pigments; when mixed with pigments, they become enamels. The most commonly used resin is ordinary rosin, and the most common drying oils are linseed and tung oils. **Spirit varnishes** are those in which a volatile liquid, such as alcohol or ether, is used as a solvent for the resin or oil. They dry by the evaporation of the solvent. **Oleoresinous varnishes** are those in which the resin is compounded with an oxidizable oil, such as linseed oil. The gums used in varnish, such as copal, dammar, and kauri, produce hardness and gloss to the film, and the **fossil resins,** such as kauri, give greater hardness and luster to varnishes than do the natural resins. The oils, such as tung and linseed, make it elastic and durable.

Other important ingredients of varnishes are driers, such as manganese oxide, to hasten the action of the drying oil, and thinning agents, or reducers, such as turpentine, naphtha, and benzol. Hydrated lime is added to varnishes to neutralize the acid in the resin, and to clarify and harden the varnish to prevent it from becoming sticky in warm weather. **Spar varnishes** are those made to withstand weather conditions. **Gloss oil** is a solution of hardened rosin in benzene or in turpentine with sometimes a small amount of tung oil to give a tougher film. It gives a high gloss but is not durable. **Long**

varnishes are those containing 20 to 100 gal (76 to 379 L) of oil to 100 lb (45 kg) of resin; a **short varnish** is one with less oil. The short varnishes are hard and glossier, but not as flexible or durable. Ordinarily, quick-drying varnish made with a natural resin is less durable than slow-drying; hardness and gloss are not guarantees of good varnish.

Varnish was originally only a colorless or nearly colorless coating material for furniture and fancy wood products to give a smooth, glossy surface for protection and to bring out the texture of the wood, and **marine varnish** was a high grade of spar varnish. Any color used was merely to accent the original color of the wood or to imitate the color of another wood of similar grain. **Phenol-formaldehyde varnish** is noted for all-round resistance to weathering under marine conditions. The finish has a tendency to yellow with age and lose gloss, but the varnish rarely cracks, peels, or chips. Sometimes called **bakelite varnish** because of its resistance to caustic and acid materials, the varnish can be washed with soap without impairing the finish. **Modified phenolic varnishes** are hard and abrasion-resistant, but are more susceptible to weathering. They are suitable for floors, furniture, and toys. Short-oil types may be used for **rubbing varnish. Insulating varnishes** were colorless varnishes for protecting drawings, paintings, and other products from moisture, or for electrical insulating. But the term *varnish* has come to mean any light-bodied, quick-drying, glossy finish as distinct from heavily pigmented, glossy enamels. **Synthetic varnishes** may now contain synthetic resins in oils, or they may be made entirely with synthetic resins in solvents. These include bonding and impregnating varnishes, based mainly on alkyd, acrylic, vinyl, epoxy, urethane, amino, and polyester resins. Soybean oil is used in the manufacture of oil-modified alkyd resin varnish, and safflower and sunflower oils in color-retentive urethane varnishes. **Electrical varnishes** are likely to be silicone, epoxy, or polyester resins that give good dielectric strength and adhesion.

VEGETABLE FATS. When specifically used, the term refers particularly to semisolid vegetable oils that are used chiefly for food. Vegetable oils and fats usually contain only small quantities of the fat-soluble vitamins A, D, and E, and after refining, they are usually devoid of vitamins. Thus, they are a better food in the producing countries. Climate in which the plant is grown has an effect on the nature of the oils. Warm climates favor the development of oleic acid while colder climates favor the less palatable linolenic acid. The low-melting-point oils are more easily assimilated in the body, but when these are hydrogenated to a melting point above 113°F (45°C), they become difficult to assimilate. Most of the more edible vegetable fats, as distinct from the more liquid food oils, are tropical products. **Suari fat** is a hard, white fat with a pleasant taste obtained from the kernels of the seeds of *Caryocar brasiliense* and other species of tropical America. The kernels yield 60 to 70% fat of a specific gravity of 0.989, melting point 86 to 99°F (30 to 37°C), and iodine value 41 to 50. **Ucuhuba tallow,** used in soaps and for candles, is a fat from the seeds of the trees *Virola surinamensis* and *V. sebifera* of Brazil. The seeds yield about 65% fat, but the extraction with petroleum ether also removes resinous material. The tallow has an iodine value of 10 to 15 and a melting point between 109 and 122°F (43 and 50°C). **Mahuba fat** is a hard, edible fat from the fruit of the tree *Acrodicilidium mahuba* of Brazil. **Gamboge butter,** known locally as **gurgi** and **murga,** is from the seeds of the fruit of the trees *Garcinia morella, G. hanburii,* and other species of Sri Lanka and India. The melting point is 93 to 99°F (34 to 37°C), specific gravity 0.90 to 0.913, and saponification value 196. It is used as a soap and food oil, and locally as an illuminant and an ointment. From these trees also comes the gum resin **gamboge** used in medicine as a cathartic, and also used as a brown dyestuff. It is alcohol-soluble.

Sierra Leone butter, also called **kanga** or **lamy,** is a pale-yellow fat from the seeds of the fruit of the tree *Pentadeama butyracea* of west Africa. The melting point is about 104°F (40°C), specific gravity 0.857 to 0.860, saponification value 186 to 198, and iodine number around 60. It is a soap oil. **Mafura tallow** is a bitter-tasting, heavy fat from the nuts of the tree *Trichilia emitica* of east Africa. It is used for soap, candles, and ointments. The specific gravity is 0.902, melting point about 104°F, and saponification value 201. **Shea nut oil,** also known as **shea butter, Bambuk butter, Galam butter,** and by various local names as **karité, kade,** and **kedempó,** is a fat obtained from the kernels of the fruit of the large tree *Bassis butyrospermum* of tropical west Africa. The kernels contain 45 to 55% fat, which when refined is white, stiffer than lard, with little odor or taste. The melting point is 91 to 108°F (33 to 42°C). It contains oleic and stearic acids and 3 to 4% lauric acid. It is used in Europe in butter substitutes, as a substitute for cocoa butter, and in candles. **Malabar tallow,** also called **dhupa fat** and **piney tallow,** used in Europe for soap and candles and in India for food, is from the kernels of the seed of the evergreen **pinne tree,** *Vateria indica* of south India. The tree also yields white dammar or Indian copal. The seeds give about 25% of a greenish-yellow, odorless, and tasteless fat of specific gravity 0.890, melting point 104°F, and saponification value 190. The fat is extracted by grinding the roasted seed, boiling in water, and skimming off. It is bleached by exposure.

Vegetable tallow, also called **bayberry tallow, capeberry wax,**

laurel wax, and **myrtle wax,** used extensively in Europe for soap-making and in the United States for blending in candles and with waxes, and for polishing leathers, is a waxy fat obtained from the outside coating of the berries of species of *Myrica* bushes of America, Europe, and Africa by boiling the berries in water. The berries yield 15 to 25% tallow. The species *M. cerifera* and *M. carolinensis* grow in the eastern coastal states, and *M. mexicana* grows in Central America. The melting point of the tallow is 104 to 115°F (40 to 46°C), specific gravity 0.995, and saponification value 205 to 212. The Central American product contains about 58% myristic acid, 36 palmitic acid, and 1.3 oleic acid. **Ocuba wax** is a waxy fat, but not chemically a wax, obtained from the seeds of the fruit of the shrub *Myristica ocuba* of Brazil. The seeds yield about 20% fat with a specific gravity of 0.920 and melting point of 104°F, which is used in candles. The fruit nut is surrounded by a thick skin which yields a water-soluble pink dye known as **ocuba red.**

Mahubarana fat is a pale-yellow, solid oil of melting point 104 to 111°F (40 to 44°C) obtained from the kernels of the fruit of trees of the genus *Boldoa.* The kernels contain 65% oil, which is used for soaps and candles. **Mocaya butter** is a fat from the kernels of the nuts of the **Paraguayan palm,** *Acrocomia sclerocarpa,* of tropical South America. The tree resembles a coconut palm, but the nuts grow in bunches. The pulp of the fruit contains 60% of a yellow oil similar to palm oil. The kernels yield 53 to 65% of the mocaya fat, which is softer than palm kernel oil. The specific gravity is 0.865, and saponification value 240. It has the same uses as palm kernel oil.

VEGETABLE OILS. An important class of oils obtained from plants, used industrially as drying oils, for lubricants, in cutting oils, for dressing leather, and for many other purposes. Many of the oils find wide usage in food products. Large tracts of land are under cultivation in all parts of the world for the production of the seeds and fruits from which the oils are obtained. **Linseed, cottonseed, palm, olive,** and **castor beans** are examples of these, and the oils are obtained by crushing. In some cases the oil-bearing material, copra or soybean, may be dehydrated before crushing, making it simpler to extract the oil, and giving a better residue meal for animal feed. The chief distinction between vegetable oils and fats is a physical one, oils being fluid at ordinary temperatures. **Aceituno oil** from the seeds of the Central American plant *Simarouba glauca* is a borderline vegetable oil due to its consistency, and it can be easily converted into a vegetable fat by slight hydrogenation. Vegetable oils can be thickened for various uses by oxidation, by blowing air through them, or they can be solidified by hydrogenation.

In the making of plastics and chemicals from fatty acids derived from vegetable oils, the cost of the oil may be as much as 50% of the final cost of the product, and price plays an important part in the choice of raw material. Oils produced in countries subject to political and economic disturbances may have sudden and great price fluctuations. Thus, domestic **soybean oil** may be substituted for castor oil in making nylon even though more chemical operations are needed, or the acids may be synthesized from petroleum hydrocarbons.

Food oils are chosen by their content of essential fatty acids, but taste is an important factor. Linseed oil is not used for food in the United States, although it has high food value and contains both linoleic and linolenic acids. **Safflower oil,** high in linoleic acid, ranks high as a food oil, only 0.003 lb (1.35 g) of oil being required to provide 0.002 lb (1 g) of essential fatty acids. Olive oil, high in oleic acid with only one double bond, requires 0.031 lb (14.2 g) of oil for 0.002 lb of essential acid. But olive oil requires less linoleic acid to counteract its effect than an equivalent amount of a saturated acid with no double bond. Butter requires the consumption of 0.04 lb (20 g) to obtain 0.002 lb of essential acids, while soybean, **corn,** and cottonseed oils, used in margarine, rank high as food oils.

Considerable oil is extracted from the kernels of the stones or pits of cherries, apricots, and other fruits as a by-product of the canning and drying of fruits. **Cherry kernel oil** is from cherry pits which contain 30 to 38% oil. The cold-pressed oil is yellow and has a pleasant flavor. It is used in salad oils and in cosmetics. The hot-pressed oil is used in soaps. The oil contains 47% oleic acid, 40 linoleic, 4 palmitic, 3 stearic, and some arachidic and myristic acids. **Apricot kernel oil** and **almond oil** have similar properties and applications. **Grapeseed oil** is obtained by pressing the by-product grape seeds from the wine industry. The seeds contain 10 to 15% oil, valued in Europe as an edible oil, but used in the United States mostly for paints and soaps. The oil contains about 52% linoleic acid, 32 oleic acid, and palmitic, stearic, and arachidic acids. The hot-pressed oil is dark green and not sweet, but the cold-pressed, refined oil is colorless and has a nutlike taste. Another name for grapeseed oil is **raisin seed** oil. **Tomato seed oil** is from the seeds of the **tomato,** *Lycopersicon esculentum,* the seeds being by-products of the manufacture of **tomato juice** and **tomato puree,** vast quantities of which are produced in the United States from pulp. The tomato plant is a perennial native to Central and South America, and was grown by the Aztecs under the name of **tomatl.** There are many varieties, and the fruits are true berries. The common red varieties are 2 to 3 in (5 to 8 cm) in diameter and contain a large number of seeds in the pulp. The seeds yield 17% oil by cold pressing, or 33% by solvent extraction. The cold-pressed oil is a clear liquid of 0.920 specific gravity, with an agreeable odor and bland taste. The iodine number is 113, and saponification value 192. It is used in salad oils, margarine, soaps, and as a semidrying oil for paints.

VELVET. A closely woven silk fabric with a short pile on one side formed by carrying the warp threads over wires and then cutting open the loops. Velvet is made in a great variety of qualities and weights, and it may have a cotton back in the cheaper grades or be made in wool. True velvet is all silk; but because of the number of imitations in other materials, this variety is usually designated as **silk velvet.** Velvet is dyed in various colors, the depth of color shown by the pile, giving it an air of richness. Its largest use is in dress goods and hangings, but it is used industrially for upholstery, fancy linings, and trim.

Plush is a name for fabrics woven of cotton, silk, linen, or wool, having a pile deeper than that of velvet. It is used for upholstery. Originally the pile of plush consisted of mohair or worsted yarns, but there is now no distinction except in the length of the pile. **Upholstery plush** is made in brocade designs by burning the pile with rollers to form a lower background. Plush is also dyed and curled to imitate furs.

Velveteen is an imitation velvet, woven of cotton. In the best grades, the pile is of mercerized yarns. Velveteen is woven into two systems of filling yarns and one system of warp yarns, the pile being made with the cut filling yarns instead of the warp yarns, as in velvet. It belongs to the class of **fustians** which also includes **moleskin** and **corduroy.** The latter is a sturdy pile fabric with heavy warp rib, dyed in the piece. It is also made in wool. The ribs run lengthwise, while **whipcord,** a hard-woven worsted fabric, has fine ribs running diagonally on the face. Velveteen is used for apparel, linings for jewelry and silverware boxes, shoe uppers, artificial flowers, and covering material. It is made either plain back or twill back, the plain back having a tendency to loosen and drop the pile.

VERMICULITE. A foliated mineral employed in making plasters and board for heat, cold, and sound insulation, as a filler in caulking compounds, and for plastic mortars and refractory concrete. The mineral is an alteration product of **biotite** and other micas, and is found in Colorado, Wyoming, Montana, and the Transvaal. It occurs in crystalline plates, specific gravity 2.3, and Mohs hardness 1.5, measuring sometimes as much as 9 in (23 cm) across and 6 in (15 cm) in thickness. The color is yellowish to brown. Upon calcination at 1750°F (954°C), vermiculite expands at right angles to the cleavage into threads with a vermicular motion like a mass of small worms; hence its name. The volume increases as much as 16 times, and the color changes to a silvery or golden hue. It is ground into pellet form. Plaster made with 60% vermiculite, 30 plaster of paris, and 10 asbestos will withstand red heat without disintegrating. **Therm-O-Flake brick** is made of granules of **exfoliated vermiculite** bonded with a chemical. It is lightweight, tough, and withstands temperatures to 2000°F (1093°C). The corklike pellets of vermiculite used for insulating fill in house walls are called **mica pellets. Zonolite,** of W. R. Grace & Co., is an exfoliated vermiculite. **Zonolite Verxite,** of the same company, is a spongy, granular powder form of **Verxite,** a thermally expanded vermiculite. It is used as a blending agent in animal feeds. A sound-absorbing building tile, called **Rockoustile,** is made of exfoliated mica. An expanded vermiculite of extremely fine mesh, under the name of **Mikolite,** is used as an extender in aluminum paint and in lubricating oils. **Exfoliated mica** is a name for expanded vermiculite. **Terra-Lite** is fluffy, powdered vermiculite for conditioning soils. It holds water, prevents soil crusting, and helps to maintain soil temperature below the critical 80°F (27°C).

VINYL RESINS. A group of products varying from liquids to hard solids, made by the polymerization of ethylene derivatives, employed for finishes, coatings, and molding resins; or it can be made directly by reacting acetic acid with ethylene and oxygen. In general, the term *vinyl* designates plastics made by polymerizing vinyl chloride, vinyl acetate, or vinylidene chloride, but may include plastics made from styrene and other chemicals. The term is generic for compounds of the basic formula RCH:CR′CR″. The simplest are the polyesters of vinyl alcohol, such as vinyl acetate. This resin is lightweight, with a specific gravity of 1.18, and is transparent, but it has poor molding qualities and its strength is no more than 5,000 lb/in² (34 MPa). But the vinyl halides, CH_2:CHX, also polymerize readily to form **vinylite resins,** which mold well, have tensile strengths to 9,000 lb/in² (62 MPa), high dielectric strength, and high chemical resistance; and a widely useful range of resins is produced by copolymers of vinyl acetate and vinyl chloride.

Vinyl alcohol, CH_2:CHOH, is a liquid boiling at 95.9°F (35.5°C). **Polyvinyl alcohol** is a white, odorless, tasteless powder which on drying from solutions forms a colorless, tough film. The material is used as a thickener for latex, in chewing gum, and for sizes and adhesives. It can be compounded with plasticizers and molded or extruded into tough and elastic products. **Hydrolyzed polyvinyl alcohol** has greater water resistance and higher adhesion, and its lower residual acetate gives lower foaming. **Soluble film,** for packaging detergents and other water-dispersible materials to eliminate the need to open the package, is a clear polyvinyl alcohol film. Textile fibers are also made from polyvinyl alcohol, either water-soluble or insolubilized with formaldehyde or another agent. Polyvinyl alcohol textile fiber is hot-drawn by a semimelt process and insolubilized after drawing. The fiber has a high degree of orientation and crystallinity, giving good strength and hot-water resistance.

Vinyl alcohol reacted with an aldehyde and an acid catalyst produces a group of polymers known as **vinyl acetal resins,** and separately designated by type names, as polyvinyl butyral and polyvinyl formal. The polyvinyl alcohols are called **Solvars,** and the polyvinyl acetates are called **Gelvas.** The vinyl ethers range from **vinyl methyl ether,** $CH_2{:}CHOCH_3$, to **vinyl ethylhexyl ether,** from soft compounds to hard resins. **Vinyl ether** is a liquid which polymerizes, or can be reacted with hydroxyl groups to form acetal resins. **Alkyl vinyl ethers** are made by reacting acetylene with an alcohol under pressure, producing **methyl vinyl ether, ethyl vinyl ether,** or **butyl vinyl ether.** They have reactive double bonds which can be used to copolymerize with other vinyls to give a variety of physical properties. The **polyvinyl formals, Formvars,** are used in molding compounds, wire coatings, and impregnating compounds. It is one of the toughest thermoplastics.

A **plastisol** is a vinyl resin dissolved in a plasticizer to make a pourable liquid without a volatile solvent for casting. The poured liquid is solidified by heating. **Plastigels** are plastisols to which a gelling agent has been added to increase viscosity. The **polyvinyl acetals, Alvars,** are used in lacquers, adhesives, and phonograph records. The transparent **polyvinyl butyrals, Butvars,** are used as interlayers in laminated glass. They are made by reacting polyvinyl alcohol with **butyraldehyde,** C_3H_7CHO. **Vinal** is a general name for vinyl butyral resin used for laminated glass.

Vinyl acetate is a water-white, mobile liquid with boiling point 158°F (70°C), usually shipped with a copper salt to prevent polymerization in transit. The composition is $CH_3{:}COO{:}CH{:}CH_2$. It may be polymerized in benzene and marketed in solution, or in water solution for use as an extender for rubber, and for adhesives and coatings. The higher the polymerization of the resin, the higher the softening point of the resin. The formula for **polyvinyl acetate resin** is given as $(CH_2{:}CHOOCCH_3)_x$. It is a colorless, odorless thermoplastic with a specific gravity of 1.189, unaffected by water, gasoline, or oils but soluble in the lower alcohols, benzene, and chlorinated hydrocarbons. Polyvinyl acetate resins are stable to light, transparent to ultraviolet light, and valued for lacquers and coatings because of their high adhesion, durability, and ease of compounding with gums and resins.

Resins of low molecular weight are used for coatings, and those of high molecular weight for molding. Vinyl acetate will copolymerize with maleic acrylonitrile, or acrylic esters. With ethylene it produces a copolymer latex of superior toughness and abrasion resistance for coatings.

Vinyl chloride, CH_2CHCl, also called **ethenyl chloride** and **chloroethylene,** produced by reacting ethylene with oxygen from air and ethylene dichloride, is the basic material for the **polyvinyl chloride** resins. It is a gas. The plastic was produced originally in Germany under the name of **Igelite** for cable insulation and as **Vinnol** for tire tubes. The tensile strength of the plastic may vary from the flexible resins with about 3,000 lb/in² (21 MPa) to the rigid resin with a tensile strength to 9,000 lb/in² (62 MPa) and Shore hardness of 90. The dielectric strength is high, up to 1,300 V/mil (52×10^6 V/m). It is resistant to acids and alkalies. Unplasticized polyvinyl chloride is used for rigid chemical-resistant pipe. Polyvinyl chloride sheet, unmodified, may have a tensile strength of 8,200 lb/in² (57 MPa), flexural strength of 12,600 lb/in² (87 MPa), and a light transmission of 78%. Sheet, thermoformed over vinyl foam, is commonly used for auto dashboard tops. Extrusions are widely used for house siding and window profiles.

Vinylidene chloride plastics are derived from ethylene and chlorine polymerized to produce a thermoplastic with softening point of 240 to 280°F (116 to 138°C). The resins are noted for their toughness and resistance to water and chemicals. The molded resins have a specific gravity of 1.68 to 1.75, tensile strength 4,000 to 7,000 lb/in² (28 to 48 MPa), and flexural strength 15,000 to 17,000 lb/in² (103 to 117 MPa). **Saran** is the name of a vinylidene chloride plastic of Dow Chemical Co., extruded in the form of tubes for handling chemicals, brines, and solvents to temperatures as high as 275°F (135°C). It is also extruded into strands and woven into a box-weave material as a substitute for rattan for seating. **Saran latex,** a water dispersion of the plastic, is used for coating and impregnating fabrics. For coating food-packaging papers, it is waterproof and greaseproof, is odorless and tasteless, and gives the papers a high gloss. Saran is also produced as a strong, transparent film for packaging. **Saran bristles** for brushes are made in diameters from 0.010 to 0.020 in (0.025 to 0.051 cm). Vinylidene chloride is also used to line metal pipe for chemical processing equipment.

Vinyl benzoate is an oily liquid of composition $CH_2{:}CHOOCC_6H_5$, which can be polymerized to form resins with higher softening points than those of polyvinyl acetate, but are more brittle at low temperatures. These resins, copolymerized with vinyl acetate, are used for water-repellent coatings. **Vinyl crotonate,** $CH_2{:}CHOOCCH{:}CHCH_3$, is a liquid of specific gravity 0.9434. Its copolymers are brittle resins, but it is used as a cross-linking agent for other resins to raise the softening point and to increase abrasion resistance. **Vinyl formate,** $CH_2{:}CHOOCH$, is a colorless liquid which polymerizes to form clear **polyvinyl formate resins** that are harder and more resistant to solvents than polyvinyl acetate. The monomer is also copolymerized with ethylene monomers to form resins for mixing in specialty rubbers. **Methyl vinyl pyridine,** $(CH_3)(CHCH_2)C_5H_3N$, is produced by Phillips Chemical Co. for use in making resins, fibers, and oil-resistant rubbers. It is a colorless liquid boiling at 148°F (64.4°C). The active methyl groups give condensation reactions, and it will copolymerize with butadiene, styrene, or acrylonitrile. **Polyvinyl carbazole,** under the name of **Lucivan,** was used in Germany as a mica substitute for high-frequency insulation. It is a brown resin, softening at 302°F (150°C).

The possibility of variation in the vinyl resins by change of the monomer, copolymerization, and difference in compounding is so great that the term *vinyl resin* is almost meaningless when used alone. The resins are marketed under a continuously increasing number of trade names. In general, each resin is designed for specific uses, but not limited to those uses.

Vinylidene fluoride, $CH_2{:}CF_2$, has a high molecular weight, about 500,000. It is a hard, white thermoplastic resin with a slippery surface, and it has a high resistance to chemicals. It resists temperatures to 650°F (343°C) and does not become brittle at low temperatures. It extrudes easily and has been used for wire insulation, gaskets, seals, molded parts, and piping.

VITAMINS. Organic chemical compounds which are vital building units, enzymes, or catalyzing agents in the growth and maintenance of animal bodies. They are produced by extraction from vegetable or animal products or are made synthetically and are marketed in solid or extract form for use in foodstuffs and pharmaceuticals. **Vitamin A,** called **carotene** because of its abundance in carrots, is an orange-yellow, needle-shaped crystalline substance with a complex molecular structure having the empirical formula $C_{40}H_{56}$. It is soluble in fats, but poorly soluble in water. Yellow and leafy, green vegetables are rich sources of carotene-bearing pigments, and carotene accompanies the green chlorophyll coloring of all plants. The more intense the green or yellow coloring, the greater the carotene content. **Lycopene,** the red coloring agent of tomatoes, has the same empirical formula as carotene, and both contain eight isoprene units, but it has a different structure. The color is due to large numbers of conjugated double bonds, and different colors are from different arrangements.

Cryptoxanthin, one of the four yellow carotene-carrying pigments, occurs in yellow corn, egg yolk, and green grasses. Animals convert carotene of green plants to vitamin A which is then obtained commercially from the tissues, especially from the liver. Deficiency of vitamin A in the human body causes night blindness, muscular weakness, and defective tooth structure, but an excess can cause body deformities and stillbirth.

Vitamin B is a complex of several vitamins, including **vitamin B_1** and **vitamin G.** The former cannot be formed in the normal processes of the human body and must be supplied in the diet. Plants manufacture and store it in seed. Lack of vitamin B_1 causes beriberi, fatigue, stiffness, headache, nervousness, and loss of appetite and, when chronic, causes enlargement of the heart, polyneuritis, and loss of coordination of muscular movements. The crystalline vitamin B_1 is called **thiamine chloride,** and in Europe is called **aneurin.** It is water-soluble, insoluble in most fats, and is destroyed by heat in the presence of moisture. In alkaline solutions the destruction is rapid. It is essential to the health of every living cell. Greater amounts are needed when lots of starch or sugar foods are eaten in order to prevent the formation of pyruvic acid, which produces noxious breath. **Pyruvic acid,** CH_3COCO_2H, a liquid boiling at 329°F (165°C), also called **glucic acid, pyroacemic acid,** and **propanone acid,** is **onion flavor. Onions** are root plants of the genus *Allium,* of which there are more than 200 species. They constitute a valuable food product but contain varying amounts of pyruvic acid, from 5.3 μmol/mil in the yellow **Spanish onion** to 18.6 in the strong **Ebenezer onion.**

Riboflavin is the accepted name for **vitamin B_2,** or **vitamin G.** The orange-yellow, needle-shaped crystals have a green fluorescence. Riboflavin, $C_{27}H_{20}N_4O_6$, is water-soluble. It is gradually destroyed by exposure to light, and is destroyed by many chemicals, or by high temperatures in the presence of alkalies. It is present in meats, eggs, barley malt, yeast, milk, green leafy vegetables, and grasses. Deficiency of riboflavin results in ill health, loss of hair, and dermatosis. **Nicotinic acid,** or **niacin,** is the pellagra-preventing member of the vitamin B complex. It can be made from the nicotine of the variety of tobacco *Nicotiana rustica.* Coffee contains some niacin, and meat extracts are rich in both niacin and riboflavin. **Biotin,** originally named **vitamin H,** is also a member of the B group, and has an enzyme action on starches and sugars. It occurs widely in nature as a **phytohormone** for the growth of organisms and plants. It is extracted from yeast, egg yolk, and liver by adsorbing on carbon. **Vitamin B_6,** $C_8H_{11}NO_3HCl$, called **pyridoxine,** is required to enable the human system to assimilate proteins. A deficiency causes nausea, muscular weakness, and anemia. **Vitamin B$_{12}$,** or **cobalamin,** is a high-molecular-weight complex containing five-membered nitrogen nuclei. It can replace protein as a growth factor.

Vitamin C, $C_6H_8O_6$, known also as **ascorbic acid,** or **ceritamic acid,** is unstable and is easily oxidized, especially in the presence of alkalies or in iron or copper vessels, so that in foods that have been long exposed to the air or overcooked it loses its value. It is thus probably the only vitamin likely to be deficient in the American diet, but the need is easily satisfied with fresh fruits, tomatoes, and green vegetables, and it is now added to frozen and canned foods as it also preserves the natural color of the products. **Isoascorbic acid,** or **erythorbic acid,** has the same composition as ascorbic acid but with the OH and H reversed on one carbon. It has the same antioxidant value, and is a lower-cost chemical, but has no vitamin C activity. It is used in meats to preserve the red color and in canned foods to prevent discoloring. **Mercate 5,** of Merck & Co., Inc., is isoascorbic acid for these purposes, and **Mercate 20** is **sodium isoascorbate. Cebicure,** for curing meats, is ascorbic acid.

The synthetic ascorbic acid is not claimed to be a complete cure for scurvy. The natural vitamin from lemons and limes also contains bioflavins which counteract the skin hemorrhage of scurvy. The juice from the **acerola plant** of Puerto Rico, used for scurvy, is 80 times richer in vitamin C than orange juice.

Vitamin D regulates the metabolism of calcium and phosphorus in the human body. Without it the body is subject to rickets, soft bones, or ill-formed bones and teeth. It is also used to counteract the germ of tuberculosis. It is found in fish and fish-liver oils and in some fruits. The vitamin D concentrate of General Mills, Inc., is made by the activation of crystalline ergosterol with low-velocity electrons, in vegetable oil. **Calciferol,** or **vitamin D$_2$,** is a synthetic antiarchitic marketed in crystalline form or in solution in corn oil. Its melting point is 241°F (116°C). Vitamin D$_2$ is formed in the body from cholesterol by the action of sunlight on the skin. **Vitamin E** is so widely distributed in foods that the effect is not well known. It is also called **tocopherol** as it is a tocopherol acetate. **Tofanin,** of Winthrop-Stearns, Inc., is vitamin E. **Vitamin K$_5$** is very stable. It is used in the foodstuffs industry instead of sulfur dioxide to control fermentation without affecting flavor, and in medicine to coagulate blood.

Vitamin P is found in capsicum and in lemon peel, and is used as a preventive of rheumatic fever. Although proper quantities of vitamins are necessary in the human body, overdoses are often toxic and poisonous. An excess of vitamin C, for example, causes irritability, vertigo, and vomiting. An excess of vitamin D causes metastatic calcification, or deposition of calcium in the arteries and kidneys, and concentrated vitamin D is classified as a toxic drug. Since metabolism may vary with each person and is also affected by physical condition, vitamins should not be taken as supplementary drugs without medical advice.

VULCANIZED FIBER. A wood, paper, or other cellulose fiberboard impregnated with a gelatinizing medium. It is not vulcanized in the same sense as rubber is vulcanized. It is made by various processes, and the medium may be sulfuric acid, zinc chloride solution, or cuproammonium solution. It may also be made by impregnating the cellulose fiber with a phenol-furfural resin dissolved in alcohol or other solvent. After dipping in the solution, the fiber is washed to remove excess alcohol, and then dipped in a zinc chloride solution which hydrolyzes it; and it is washed free of the chloride, dried, and rolled. The original vulcanized fiber, patented in 1899 and called **Cellulith,** was sulfite wood pulp molded into sheets or formed parts. The modern fiber in the hard grades is a tough, resilient, hornlike material in standard gray, red, and black colors. Soft, flexible grades are made for washers and gaskets. The four major NEMA grades are: electrical insulation, commercial, bone (high-density), and trunk. The commercial grade is in thicknesses from 0.005 to 1 in (0.013 to 2.54 cm), with lengthwise tensile strength of 7,500 lb/in^2 (52 MPa), flexural strength of 14,000 lb/in^2 (97 MPa), compressive strength of 20,000 lb/in^2 (138 MPa), and dielectric strength of 250 V/mil (10×10^6 V/m). Unless impregnated with a synthetic resin, it is not resistant to alkalies. The bone quality is a dense material with a specific gravity of 1.4, capable of being machined. The **hard vulcanized fiber** of Spaulding Fibre Co. was made from cotton rags gelatinized in a zinc chloride solution and built up in layers. The shear strength is to 15,000 lb/in^2 (103 MPa), and compressive strength 30,000 lb/in^2 (207 MPa). **Bone fiber** is characterized as dense and hard, while trunk fiber is tough and abrasion-resistant. Because of the moderate cost, vulcanized fiber still has many uses, but practically all the material for electrical use is now of the insoluble type made with synthetic resin impregnation and having higher dielectric strength. **Fish paper,** for electrical use, was originally vulcanized fiber in thicknesses down to 0.004 in (0.010 cm), but is now likely to be a resin impregnate. **Shoe fiber** is vulcanized fiber in leather color used for reinforcement in shoes. It is very resilient, and can be die-cut and nailed. Much of the fiber generally called vulcanized fiber is now impregnated with synthetic resins to meet conditions of chemical resistance, strength, and electrical properties. Vulcanized fiber is produced in the form of sheets, coils, tubes, and rods. Sheets are made in thicknesses of 0.0025 to 2 in (0.0064 to 5 cm) and approximately 48 by 80 in (122 by 203 cm) in size. Outside diameters of tubes range from 0.1875 to

4.375 in (0.478 to 11 cm). Rods are produced 0.09 to 2 in (0.239 to 5 cm) in diameter.

VULCANIZED OILS. Vegetable oils vulcanized with sulfur and used for compounding with rubber for rubber goods, or as a rubber substitute. Castor oil, corn oil, rapeseed oil, and soybean oil are used. Vulcanized oil is a white to brown, spongy, odorless cake, or a sticky plastic, with specific gravity of 1.04. The material was invented in France in 1847 and was known as **factice. Factice cake** is solidified, vulcanized oils, cut in slab form. It is an oil modifier of rubber, to add softness and plasticity. It also has some elasticity. **Brown factice** is made by treating the oil with sulfur chloride at 320 to 392°F (160 to 200°C). The softer grades are made with blown oils and low sulfur. The harder grades contain up to 20% sulfur. **White factice** is made from rapeseed oil, which is high in a characteristic acid, crucic acid, by slow addition of sulfur chloride up to 25% sulfur content. **Erasing rubbers** are rubber compounded with white factice or the factice alone. **Black factice** has mineral bitumen added to brown factice. **Neophax** is the trade name of Stamford Rubber Supply Co. for brown factice, and **Amberex** is the name for light-tan-colored factice. **Factex** of the same company is partly vulcanized oil dispersed in water. It produces a nontacky elastic film. When mixed with rubber latex to the extent of 30%, it gives a velvety feel to the vulcanized product and does not decrease the strength greatly. **Factice sheet** is specially processed factice made by treating warm oil with sulfur and then with sulfur chloride. The strength and elasticity are higher. **Mineral rubber** was a name applied to vulcanized oils mixed with bitumens, especially gilsonite.

WALNUT. A hardwood from the tree *Juglans regia,* native to Europe and Asia Minor, but now growing in many other places. The wood is firm, with a fine to coarse, open grain, and a lustrous surface. The density is about 45 lb/ft^3 (721 kg/m^3). The color is dark brown to black, and it takes a beautiful polish. Walnut has great strength, toughness, and elasticity. It also has great uniformity of texture and does not split easily. It is particularly adapted for carving. Walnut is valued as a cabinet wood, for fine furniture, and for gunstocks. The wood from *J. regia* is called **English walnut,** and the beautifully figured wood from Iran is known as **Circassian walnut. Black walnut,** or **American walnut,** is from the tree *J. nigra,* of the eastern United States. The color is darker, and it has a more uniform color than European walnut. It is handsomely grained and has the same general characteristics and uses as European walnut. It has a specific gravity, kiln-dried, of 0.56, a shear strength parallel to the grain of 1,000 lb/in^2 (6.9 MPa), and a compressive strength perpendicular to the grain of 1,730 lb/in^2 (12 MPa). **Butternut,** from the tree *J. cinerea,* resembles closely the wood of the black walnut except for its color, which is yellowish gray. The supply of this wood is limited.

East India walnut is the wood of the tree *Albizzia lebbek* of tropical Asia and Africa, used for furniture, paneling, and interior decorative work. It is hard, dense, and close-grained, with a density of about 50 lb/ft^3 (801 kg/m^3). The color is dark brown with gray streaks. The logs come as large as 30 in (76 cm) square and 20 ft (6 cm) long. The shipments may be mixed with the wood of *A. procera,* the **white siris wood** of India. This wood has a brown walnut color, is lustrous, and resembles true walnut more than does the East India walnut. **Mahoe,** also called **blue mahoe** and **majagua,** is the wood of the tree *Hibiscus elatus,* of tropical America. It has been used to replace true walnut for gunstocks and in cabinetwork. The wood has a gray-blue color, an aromatic odor, and is hard with a coarse, open grain. **Brazilian walnut,** or **frejó,** from the tree *Cordia goeldiana,* is a strong, tough, straight-grained wood used for cooperage and cabinetwork.

African walnut, or **amonilla,** is from the tree *Lovoa Klaineana* of Nigeria. The wood is brown, and has a fine texture and an interlocking grain that shows a striped figure when quartersawed. The density is about 40 lb/ft^3 (641 kg/m^3). It is used for flooring, paneling, veneers, and cabinetwork. The Brazilian wood known as **imbuia,** from the tree *Nectandra villosa,* is a close match to true walnut and is valued for cabinetwork, flooring, and furniture. The heartwood has an olive to brown color and takes a high polish. There are as many as 50 species of *Nectandra* trees in Brazil, varying widely in characteristics. The **canela preta,** from the tree *N. mollis,* is a wood with large, satiny stains on a dark-yellow background. It has a silvery luster when polished, has a spicy scent, and is very durable. It resembles bleached walnut.

Walnut oil is yellowish oil obtained by pressing the nut kernels of the common walnut. It is a good drying oil and is used especially for artists' paints. The specific gravity is 0.919 to 0.929 and iodine value 148. It is soluble in alcohol. The oil from the candlenut is also called walnut oil. **Walnut-shell flour,** made from the refuse shells of the walnut industry of California, is used as a filler in molded plastics and in synthetic adhesives to increase bonding strength. It contains cellulose with about 28% lignin, 5 furfural, 9 pentosans, 6 methyl hydroxylamine, and 2.5 sugars and starch. In colonial times walnut bark was used as a cathartic. It contains a **juglone,** or **nucin,** a complex **naphthoquinone,** $C_{10}H_6O_2$, a reddish crystalline compound also called **lapachol** as it occurs also in **lapacho,** an important hardwood of Argentina and Paraguay.

WALRUS HIDE. The skin of the walrus, a marine mammal, *Odontobaenus rosmarus* and *O. abesus,* native to the north Atlantic and Pacific oceans. The animals sometimes attain a length of 16 ft (5 m) and weigh up to 2,000 lb (907 kg), and the hide is obtainable in large pieces. They congregate in herds on the icebergs of the north. The skin is tanned and makes a leather with a beautiful natural grain. It is also very tough and was formerly much used for coach traces. It is now employed where a tough and ornamental leather is needed, but the animals are now scarce and the killing of them is controlled by law. Imitation **walrus leather** is made from specially tanned and heavy, embossed sheepskins, and is used mostly for bags and ornamental articles.

WATER. A nearly transparent liquid of composition H_2O. The specific gravity of pure water is taken as 1.0 at 39°F (4°C), and water is used as the standard for measuring the specific gravity of other liquid and solid materials. The boiling point is 212°F (100°C), and the freezing point is 32°F (0°C). The first essential use of water is for drinking and for the watering of plants to sustain life; but the largest consumptive use is in industrial processing, and a large supply of water is essential for manufacturing.

The per capita intake of water for human drinking is taken as 1 gal (3.8 L) per day, but, because of waste, the amount is larger. The consumption of water from the municipal systems of large American cities exceeds 150 gal (568 L) per capita per day, and that figure includes some industrial use. The employment of water for hydroelectric power is not considered as a consumptive use. About 80% of the supply in the United States is from surface water, and about 20% from groundwater.

Quality of water is important in many industrial operations. Factories may obtain water from municipal systems, from groundwater pumped from wells, or from surface water from streams. As industries concentrate, it becomes more important to protect water supply by dams and watersheds, and by preventing the pollution of streams by the return of unclean water. Typical municipal waters contain from 30 to 1,000 parts per million (ppm) of dissolved minerals, chiefly silica, iron, calcia, magnesia, potassium, sulfates, chlorides, and nitrates. Organic matter is also present in the water in varying amounts. So-called pure lake water averages above 150 ppm. Thus, pure water for chemical processing may require ion-exchange purification after filtering. Water for atomic reactors is thus purified to not more than 0.08 ppm.

The water molecule is one of the simplest of the chemical combinations of the natural elements, but it embraces such a vast complexity that it can be taken as an example to illustrate the basic principles of the combining habits of all the elements. The unit molecule H · O · H appears to be ovaloid over a wide range of temperatures. The bonding is very strong. With active metallic inclusions in water, the disintegration of the molecule may begin at an energy equivalent of about 1832°F (1000°C), but complete dissociation into its elements, hydrogen and oxygen, requires a temperature above 3632°F (2000°C). Water is a very stable **oxygen hydride.**

The molecule is too minute for measurement of hardness by any known methods, but it is deduced to be extremely hard. When a stream of water is projected through a tiny orifice at high pressure, it will cut through a tree as a power saw would. At the energy levels of the liquid stage, the molecules are very mobile and roll on each other; but when water drips from a faucet or falls as rain, it forms into rounded droplets. The cause is usually given as **surface tension,** but may be the electromagnetic attraction of the exposed proton pairs on the oxygen atoms. When the molecules are lubricated with a chemical, the water flows more easily. This property is used in fire-fighting hose. **Polyox FRA,** of Union Carbide, is a polymer derivative of ethylene oxide used for this purpose. When a 1% solution of the resin is injected into the line, the stream of water is projected more than twice the normal distance. Water can also be coated to form **dry water** of semisolid consistency in the form of droplets which are dry to the touch. This property is utilized in water-based cosmetic creams by adding silicic acid. **Waterglass** is a form of soluble sodium silicate. When the cosmetic is applied to the face, silicic acid dehydrolyzes to leave an adherent coating of extremely fine amorphous silica on the skin, and the thin film of released water evaporates quickly to give a cool, fresh feel. When heated and pressurized above its critical temperature and pressure, water is a supercritical fluid of high dissolving power and chemical reactivity, allowing fast oxidation and breakdown of various hazardous wastes.

Lumping and caking of flours and other powders are usually caused by absorption of water moisture from the air, but the addition of as little as 0.5% of an **anticaking agent** such as silicic acid will coat and dry the absorbed moisture and prevent caking. The material known as **anomalous water** has properties similar to those of pure water treated with silicic acid. It forms in dry globules of low mobility, and has a wider temperature range in the liquid state. It is produced by condensing distilled water in vacuum in a small quartz (latticed silica) tube. The accumulation is extremely slow, requiring 18 h to fill a tube 0.010 in (0.025 cm) in diameter. It is also called **polywater,** though the water molecule does not polymerize in the ordinary meaning of the term. The water molecule can also be absorbed within the molecules of many chemicals, both solid and liquid, and such materials are called **hygroscopic.** For example, phosphophenyl methyl phosphinic acid has a unit molecular group in the shape of an irregular toroid. With only enough water to fill the spaces in the toroid, the unit is rounded and mobile, and the material becomes a viscous liquid which is dissimilar to a water solution of a material. Large proportions of water may be encased in the cells and lattices of proteins or other organic materials, making the mass into a solid or semisolid in which as much as 97% of the volume is water. This process is used in the making of puddings such as **Jello** in which a small amount of edible gelatin encases the water. **Water-filled plastics** are usually thermoplastic resins in which 80% or more water is used as a filler. The water is stirred into the resin, and the casting sets to a hard, firm solid which can be machined, nailed, or screwed. The castings are used for models, ornaments, and such products as lamp stands.

There is no room within the water molecule for the containment of any other atom or molecule, but the liquid mass may be likened to the arrangement of a great pile of uniformly sized, ovaloid Danish pebbles, and the mass has comparatively large spaces between and among the units. All natural waters contain much foreign matter including oxygen and other gases from the air, organic material from plant and animal life, minerals picked up from contact with the earth, and often high proportions of large inorganic molecules such as **astrakanite,** $Na_2SO_4 \cdot MgSO_4 \cdot 4H_2O$, and carnallite. **Groundwater,** pumped from deep wells for industrial and municipal uses, was once considered the purest of waters, containing chiefly mineral salts; but in populated areas underground waters are now contaminated with seepage from fertilizers, insecticides, industrial chemicals, and sewage which includes froth-forming detergents. Water in streams, in addition to the natural minerals and plant-decay matter, now usually contains high contents of industrial and agricultural chemicals and sewage, plus biological compounds from the decayed proteins and albumins of human and animal wastes. Many of these can only be removed by costly, special ion-exchange methods. Bacteria may be inactivated by the addition of chlorine, but the skatolelike odors may remain. The content of air in natural water, given in terms of O_2, is a minimum of 9 ppm. This contained air is a biological necessity in water for human consumption and for the maintenance of fish and other marine life. The chemical process of decay of sewage in water depletes this oxygen and thus tends to destroy marine life. **Seawater,** and water in inland lakes such as the Dead Sea where evaporation greatly exceeds runoff, contains high amounts of mineral salts. The variety and percentage content vary in different areas and at different temperatures. The density of seawater at 68°F (20°C) may be 69 lb/ft^3 (1,105 kg/m^3) compared with about 62 lb/ft^3 (993 kg/m^3) for natural freshwater. About 80 elements have been found in seawater, and it is probable that all of the natural elements occur to some extent. About 30% of all commercial halite, or common salt, is now produced from seawater. By ordinary solar evaporation at about 80°F (27°C), the proportion of common salt, NaCl, precipitating after the calcite and gypsum may be about 12% of the total salts. As much as 65% of commercial magnesium metal is produced from seawater by precipitation and reduction of contained magnesium compounds such as epsomite, kainite, kieserite, **bischofite,** MgCl$_2 \cdot$ 6H$_2$O, and the magnesium sulfate known as **hexahydrite,** MgSO$_4 \cdot$ 6H$_2$O. Seawater is a source of pure water via multiflash evaporation or desalination plants, in areas devoid of **freshwater. Rain** is a source of freshwater, and freshwater is considered **potable water** if it contains less than 500 ppm of dissolved solids. **Heavy water** is a form of water in which the hydrogen atoms are replaced by the heavy stable isotope of hydrogen **deuterium.** Heavy water has a molecular weight of 20.028, melting point of 38.9°F (3.81°C), boiling point of 214.6°F (101.4°C), and a viscosity of 1.107 MPa $\cdot$ s. At the energy level of solidification the water molecules arrange themselves in precise order close together, and the frozen water, or **ice,** can be split in straight cleavage from a line scratched on the surface. In the solid assembly of molecules, there is no available space, and the contained impurities of water are thrown out in freezing, except in the dendritic snow molecule. Ice has load-bearing capacity, and thus the term **structural ice.** Ice bridges across frozen lakes and rivers in the Arctic have been made of ice reinforced with ferry cable and wood. Ice reinforced with random distribution of glass fibers, wood, sawdust, and other materials is called **icecrete.** Continuous strands of glass yarn add rigidity and, if prestressed, load-carrying capacity.

WATER REPELLENTS. Chemicals used for treating textiles, leather, and paper such as washable wallpaper, to make them resistant to wetting by water. They are different from waterproofing materials in that they are used where it is not desirable to make the material completely waterproof, but to permit the leather or fabric to "breathe." Water repellents must not form acids that would destroy the material, and they must set the dyes rather than cause them to bleed on washing. There are two basic types: a durable type that resists cleaning and a renewable type that must be replaced after the fabric is dry-cleaned. **Zelan,** a pyridinium-resin compound of Du Pont, is representative of the first type. **Quilon,** of the same company, is used for paper, textiles, and glass fabric and forms a strong chemical bond to the surface of the material by an attachment of the chromium end of the molecule through the covalent bond to the negatively charged surface. It is a stearotochromic chloride. The second type is usually an emulsion of a mineral salt over which a wax emulsion is placed; the treatment may be a one-bath process, or it may be by two separate treatments. **Aluminum acetate** is one of the most common materials for this purpose. Basic aluminum acetate is a white, amorphous powder of composition $Al(OH)(OOC \cdot CH_3)$. It is only slightly soluble in water but is soluble in mineral acids. **Niaproof,** of Niacet Corp., is a concentrated aluminum acetate for waterproofing textiles, and **Ramasit** and **Migasol** are similar materials. **Zirconium acetate,** a white, crystalline material of composition $ZrOH(C_2H_3O_2)_3$, and its sodium salt are used as water repellents. **Zirconyl acetate,** $ZrO(C_2H_3O_2)_2$, a light-yellow solution containing 13% ZrO_2, is used for both water repellancy and flame resistance of textile fibers. **Intumescent agents** are repellent coatings that swell and snuff out fire when they become hot. **Latex 744B,** of Dow Chemical Co., is a repellent of this type. It is a vinyl water emulsion compounded with pentaerythritol, dicyandiamide, and monosodium phosphate, and is used on textiles, wallboard, and fiber tile.

Silicones have established their value as water-repellent finishes for a range of natural and synthetic textiles. The silicone polymers may be added as a solution, an emulsion, or by spraying a fine mist; alternatively, intermediates may be added that either polymerize in situ or attach themselves to the fibers. These techniques result in the pickup of 1 to 3% of silicone resin on the cloth. Commercially, **dichloromethylsilane** polymer is added as a solution or emulsion to a fabric; this is heated in the presence of a catalyst, such as a zinc salt of an organic acid or an organotin compound, to condense the polymer and form a water-repellent sheath around each fiber. Soluol Chemical Co.'s **Aquagard 170** is a nonionic organopolysiloxane in the form of a white emulsion; it is cured at low temperatures by its catalyst **Curade 170,** a cationic metallic compound, producing a finish that is highly durable to dry cleaning. Similar techniques are employed for imparting water repellency to leather. Silicones containing Si—H groups are used for paper treatment. The treated paper has a measure of water repellency and, in addition, some antiadhesive properties. Fluorine-based polymers are also employed for treating fabrics. **Gore-Tex,** produced by W. L. Gore, is a polytetrafluoroethylene coating on nylon fabric; garments fashioned from this treated nylon are weatherproof and breathable. **Scotchgard,** from 3M, is a polymer containing fluoroalkyl groups that is effective for repelling both water and oil. **Scotchban,** from the same company, provides water, oil, and grease repellency to paper. **Zepel B,** from Du Pont Co., is a fluoropolymer dispersion in water that does not promote yellowing or discoloration of coated outerware. The **Quillon** series, also from Du Pont, consists of greenish solutions of chrome complexes in isopropanol that are water-repellent agents for packaging materials, nonwoven fabrics, and adhesive tapes. **Vinsol MM** from Hercules Inc. is a dark brown, free-flowing powder that is a sodium soap of a blend of Vinsol resin and a fatty acid. It was specially developed for use in masonry cements.

WATER SOFTENERS. Chemical compounds used for converting soluble, scale-forming solids in water into insoluble forms. In the latter condition they are then removed by setting or filtration. The hardness of water is due chiefly to the presence of carbonates, bicarbonates, and sulfates of calcium and magnesium; but many natural waters also contain other metal complexes which need special treatment for removal. Temporary **hard waters** are those that can be softened by boiling; permanent hard waters are those that require chemicals to change their condition. Sodium hydroxide is used to precipitate magnesium sulfate. Caustic lime is employed to precipitate bicarbonate of magnesium, and sodium aluminate is used as an accelerator. Barium carbonate may also be used. Prepared water softeners may consist of mixtures of lime, soda ash, and sodium aluminate, the three acting together. **Sodium aluminate,** $Na_2Al_2O_4$, is a water-soluble, white powder melting at 3002°F (1650°C), which is also used as a textile mordant, for sizing paper, and in making milky glass. Reynolds Metals Co. produces this material in flake form with iron content below 0.0056% for paints, water softeners, and paper coatings. Alum is used in settling tanks to precipitate mud, and zeolite is used extensively for filtering water. The liquids added to the washing water to produce fluffier textiles are **fabric softeners** and not water softeners. They are usually basic quaternary ammonium compounds such as distearyl dimethyl ammonium chloride with 16 and 18 carbon atoms, which are cationic, or positively charged. A thin coating is deposited on the negatively charged fabric, giving a lubricated cloth with a fluffy feel.

Water is also softened and purified with **ion-exchange agents,** which may be specially prepared synthetic resins. **Cation-exchange agents** substitute sodium for calcium and magnesium ions and produce soft waters. When the water is treated with a hydrogen derivative of a resin, the metal cations form acids from the salts. The carbonates are converted to carbonic acid which goes off in the air. When it is treated again with a basic resin derivative, or **anion-exchange agent,** the acids are removed. Water receiving this double treatment is equal to distilled water. Salt-cycle anion exchange substitutes chloride ions for other anions in the water, and when combined with cation exchange, it produces sodium chloride in the water in place of other ions.

In electrolytic ion exchangers for converting seawater to freshwater, the basic cell is divided into three compartments by two membranes, one permeable only to cations and the other only to anions. The sodium ions migrate toward the cathode, and the chlorine ions go toward the anode, leaving freshwater in the center compartment. **Ion-exchange membranes** for electrodialysis (salt splitting or separation), and also used in fuel cells, are theoretically the same as powdered exchange resins but with an inorganic binder. Such a membrane resin of the Armour Research Foundation is made by the reaction of zirconyl chloride and phosphoric acid, giving a chain molecule of zirconium-oxygen with side chains of dihydrogen phosphate. **Zeo-Karb,** a sulfonated coat, and **Zeo-Rex,** a sulfonated phenolformaldehyde resin, are cation exchangers of Permutit Co., while **De-Acidite** and **Permutit A** of the same company are anion exchangers. **Amberlite IRA-400,** of Rohm & Haas, is a strongly basic alkyl amine which will split neutral salts in the water and also remove silica. The German **Wofatit P** exchanger is a sodium salt of a phenol-formaldehyde resin. Ion-exchange agents are also used for refining sugar, glycerin, and other products, and for the purification of acids and the separation of metals. An **eluting agent** is a solvent used to elutriate the resin beds in the separation of metals, that is, to separate the heavier from the lighter particles, causing a metal ion on the resin to change place with hydrogen or with an ammonium group in the elutriant. **Zeolites** are crystalline aluminosilicates that display cation-exchange properties. The most common zeolite for softening uses is **zeolite 4A,** a sodium aluminosilicate made by Union Carbide Corp. **Zeolex** is a similar product from J. M. Huber Corp. **EZA Zeolite A** is a white powder from Ethyl Corp. that is employed as a replacement for sodium phosphates in laundry detergents.

WAX. A general name for a variety of substances of animal and vegetable origin, which are fatty acids in combination with higher alcohols instead of with glycerin, as in fats and oils. They are usually harder than fats, less greasy, and more brittle, but when used alone, they do not mold as well. Chemically, the waxes differ from fats and oils in being composed of high-molecular-weight fatty acids with high-molecular-weight alcohols. The most familiar wax is beeswax from the honeybee, but commercial beeswax is usually greatly mixed or adulterated. Another **animal wax** is spermaceti from the sperm whale. **Vegetable waxes** include Japan wax, jojoba oil, candelilla, and carnauba wax. These are sold under the trade name **Stralpitz** by Strahl & Pitsch, Inc. **Mineral waxes** include paraffin wax from petroleum, ozokerite, ceresin, and montan wax. The mineral waxes differ from the true waxes and are mixtures of saturated hydrocarbons.

The animal and vegetable waxes are not plentiful materials, and are often blended with or replaced by hydrocarbon waxes or waxy synthetic resins. But waxes can be made from common oils and fats by splitting off the glycerin and reesterifying selected mixtures of the fatty acids with higher alcohols. **Hywax 122** is a self-emulsifiable wax composed of cetyl, myristyl, and stearyl esters derived from animal and vegetable oils. **Mazawax** and **Macol,** from Mazer Chemicals, are fatty alcohol blends with various emulsifiers; they are all-purpose waxes for creams, lotions, hair relaxants, and hair dipilatory formulations. **Opalwax,** of Du Pont, is a **synthetic wax** produced by the hydrogenation of castor oil. It has about the same hardness as carnauba, specific gravity of 0.98, and melting point of 187 to 190°F (86 to 88°C), but it lacks the luster of carnauba. It is odorless and has a pearl-white color. It is very resistant to most solvents and is used for insulation, coatings, candles, and carbon paper. **Acrawax,** of Lonza, Inc., is a somewhat similar substitute for carnauba with higher melting point. **Stroba wax,** of the same company, is a synthetic wax with a base of stearic acid and lime. The melting point is 217 to 223°F (103 to 106°C). It is used in polishes, insulation, and as a flatting agent. Synthetic wax under the name of **Pentawax 286** is a true wax in that it is a combination of fatty acids with an alcohol. It is made from the long-chain acids of vegetable oils with pentaerythritol. It has a higher melting point than carnauba, 110°F (43°C), but does not form a self-polishing liquid wax as carnauba does. **Wax R21** is a metal-containing synthetic wax used in liquid floor waxes, temporary corrosion protection, release agents, and as a melting point booster. Other brands from the same manufacturer, Hoechst Celanese Corp., are **Hostalub** and **Ceridust,** which are specialty waxes based on polyolefins, paraffins, chemically modified montan, and micropowders. **Sheerwax** is made by catalytic hydrogenation of vegetable oils. It has the hardness and high melting point of carnauba wax and can be had in white color. Waxes are employed in polishes, coatings, leather dressings, sizings, waterproofing for paper, candles, and varnishes. They are softer and have lower melting points than resins, are soluble in mineral spirits and in alcohol, and are insoluble in water.

Some plastics have wax characteristics and may be used in polishes and coatings or for blending with waxes. Polyethylene waxes are light-colored, odorless solids of low molecular weight, up to about 6,000. Mixed in solid waxes to the extent of 50%, and in liquid waxes up to 20%, they add gloss and durability and increase toughness. In emulsions they add stability. **Acumist** is a micronized polyethylene wax that is a processing and performance additive for adhesives, coatings, color concentrates, cosmetics, inks, lubricants, paints, plastics, and rubber. It can also be constituted from low-molecular-weight homopolymer, oxidized homopolymer, or as a copolymer. Acumist is from Allied-Signal, Inc. **Marlex 20,** of Phillips Petroleum Co., is a methylene polymer used to blend with vegetable or paraffin waxes to increase the melting point, strength, and hardness. **Santowax R,** of Monsanto, is a mixture of terphenyls. It is a light-buff, waxy solid, highly soluble in benzene, and with good resistance to heat, acids, and alkalies. It is used to blend with natural waxes in candles, coatings, and insulation. **Epolene wax,** of Eastman Chemical Products, Inc., is a polyethylene. Waxes are not digestible, and the so-called **edible waxes** used as water-resistant coatings for cheese, meats, and dried fruits are not waxes, but are modified glycerides. **Monocet** is such a material. It is a white, odorless, tasteless, waxy solid melting at 104°F (40°C) and is an acetylated monoglyceride of fatty acids.

WEAR-RESISTANT STEEL. Many types of steel have wear-resistant properties, but the term usually refers to **high-carbon, high-alloy steels** used for dies, tooling, and parts subject to abrasion and for wear-resistant castings. They are generally cast and ground to shape. They are mostly sold under trade names for specific purposes. The excess carbon of the steels is in spheroidal form rather than as graphite. One of the earlier materials of this kind for drawing and forming dies, **Adamite,** was a **chromium-nickel-iron alloy** with up to 1.5% chromium, nickel equal to half that of the chromium, and from 1.5 to 3.5 carbon with silicon from 0.5 to 2. The Brinell hardness ranges from 185 to 475 as cast, with tensile strengths to 125,000 lb/in^2 (862 MPa). The softer grades can be machined and then hardened, but the hard grades are finished by grinding.

   Kinite has about 13% chromium, 1.5 carbon, 1.1 molybdenum, 0.70 cobalt, 0.55 silicon, 0.50 manganese, and 0.40 nickel. It is used for blanking dies, forming dies, and cams. **Martin steel** has 13% chromium, about 1 molybdenum, 0.80 cobalt, 0.35 vanadium, and 1.5 carbon. **T15 tool steel,** for extreme abrasion resistance in cutting tools, is classified as a super-high-speed steel. It has 13.5% tungsten, 4.5 chromium, 5 cobalt, 4.75 vanadium, 0.50 molybdenum, and 1.5 carbon. Its great hardness comes from the hard vanadium carbide and the complex tungsten-chromium carbides, and it has full red-hardness. The property of abrasion or wear resistance in steels generally comes from the hard carbides, and is thus inherent with proper heat treatment in many types of steel.

WELDING ALLOYS. Usually in the form of rod, wire, or powder used for either electric or gas welding, for building up surfaces, or for hard-facing surfaces. In the small sizes in continuous lengths, welding alloys are called **welding wire.** Nonferrous rods used for welding bronzes are usually referred to as **brazing rods,** as the metal to be welded is not fused when they are used. Welding rods may be standard metals or special alloys, coated with a fluxing material or uncoated, and are normally in diameters from 0.094 to 0.25 in (0.239 to 0.635 cm). Compositions of standard welding rods follow the specifications of the American Welding Society. Molded carbon, in sizes from 0.125 to 1 in (0.318 to 2.54 cm) in diameter, is also used for arc welding. **Intensarc** is a trade name for carbon rods. Low-carbon steel rods for welding cast iron and steel contain less than 0.18% carbon. High-carbon rods produce a hard deposit that requires annealing, but these are also used for producing a hard filler. High-carbon rods, with 0.85 to 1.10% carbon, will give deposits with an initial Brinell hardness of 575, whereas high-manganese rod deposits will be below Brinell 200 but will work-harden to above Brinell 500. For high-production automatic welding operations, carbon-steel wire may have a thin coating of copper to ease operation and prevent spattering. Stainless-steel rods are marketed in various compositions. **Flexarc welding rods** comprise a range of stainless steels with either titania-lime or straight-lime coatings. **Stainless C** is an 18–8 type of stainless steel having also 3.5% molybdenum. **Aluminum weld** is a 5% silicon aluminum rod for welding silicon-aluminum alloys, and the **Tungweld rods,** for hard surfacing, are steel tubes containing fine particles of tungsten carbide. **Kennametal KT-200** has a core of tungsten carbide and a sheathing of steel. It gives coatings with a Rockwell C hardness of 63. **Chromang,** for welding high-alloy steels, is an 18–8 stainless steel modified with 2.5 to 4% manganese. An iron-based alloy containing 25% nickel, 21 chromium, 7 manganese, 5 molybdenum, 1.6 copper, 0.19 nitrogen, 0.015 phosphorus and sulfur, and 0.01 oxygen is designed to produce austenitic welds that stay tough at $-450°F$ ($-270°C$). Developed at the National Institute of Standards and Technology, it provides about twice the fracture toughness of 308 and 316 stainless steels and is intended for welding superconducting magnets.

Amsco welding rods are grades of high-manganese steel giving Brinell harnesses from 500 to 700, and **Toolface** is a high-speed steel rod for facing worn cutting tools. **Superloy,** used for facing surfaces requiring extreme hardness, has the alloy granules in a soft steel tube. The welded deposit has a composition of 30% chromium, 8 cobalt, 8 molybdenum, 5 tungsten, 0.05 boron, and 0.20 carbon. **Tung-Alloy, Resisto-Loy,** and **Isorod** are other **hard-facing rods.** Resisto-Loy has a nonferrous content. **Croloy welding rods,** for welding alloy steels without high preheating, are low-alloy chromium-molybdenum steels. **Stoodite** is a high-manganese steel in the form of a rod used for hard facing. **Chromend 9M** is a rod for arc-welding hard deposits. It contains 8 to 10% chromium and 1.5 molybdenum, giving a weld with a Brinell hardness of 400. **Elkonite** is the name of a group of welding alloys made especially for welding machines. They are, in general, sintered tungsten or molybdenum carbides, combined with copper or silver, and are electrodes for spot welding rather than welding rods. **Tungsten electrodes** may be pure tungsten, thoriated tungsten, or zirconium tungsten, the latter two used for direct-current welding. **Thoriated tungsten** gives high arc stability, and the thoria also increases the machinability of the tungsten. **Zirconium tungsten** provides adhesion between the solid electrode and the molten metal to give uniformity in the weld.

Thermit is a mixture of aluminum powder and iron oxide used for welding large sections of iron or steel or for filling large cavities. The process, originally developed by Goldschmidt Thermit Co., consists of burning the aluminum to react with the oxide, which sets free the iron in molten form. To ignite the aluminum and start the reaction, a temperature of about 2800°F (1538°C) is required, which is reached with the aid of a gas torch or ignition powder, and the exothermic temperature is about 4600°F (2538°C). **Cast-iron thermit,** used for welding cast iron, is thermit with the addition of about 3% ferrosilicon and 20 steel punchings. **Red thermit** is made with red oxide, and **black thermit** with black oxide. **Railroad thermit** is thermit with additions of nickel, manganese, and steel.

The **Stellite hard-facing rods** are cobalt-based alloys that retain hardness at red heat and are very corrosion-resistant. The grades have tensile strengths to 105,000 lb/in² (724 MPa) and Rockwell C hardness to 52. **Inco-Weld A** is welding wire for stainless steels and for overlays. It contains 70% nickel, 16 chromium, 8 iron, 2 manganese, 3 titanium, and not more than 0.07 carbon. The annealed weld has a tensile strength of 80,000 lb/in² (552 MPa) with elongation of 12%. **Nickel welding rod** is much used for cast iron, and the operation is brazing, with the base metal not melted. **Nickel silver** for brazing cast iron contains 46.5% copper, 43.4 zinc, 10 nickel, 0.10 silicon, and 0.02 phosphorus. The deposit matches the color of the iron. **Colmonoy 23A** is a nickel alloy **welding powder** for welding cast iron and filling blow holes in iron casting by torch application. It has a composition of 2.3% silicon, 1.25 boron, 0.10 carbon, not over 1.5 iron, and the balance nickel, with a melting point of 1950°F (1066°C). For welding on large structures where no heat treatment of the weldment is possible, the welding rods must have balanced compositions with no elements that form brittle compounds. **Rockide rods** are metal oxides for hard surfacing.

WETTING AGENTS. Chemicals used in making solutions, emulsions, or compounded mixtures, such as paints, inks, cosmetics, starch pastes, oil emulsions, dentifrices, and detergents, to reduce the surface tension and give greater ease of mixing and stability to the solution. In the food industries, chemical wetting agents are added to the solutions for washing fruits and vegetables to produce a cleaner, bacteria-free product. Wetting agents are described in general as chemicals having a large hydrophilic group associated with a smaller hydrophilic group. Some liquids naturally wet pigments, oils, or waxes, but others require a proportion of a wetting agent to give mordant or wetting properties. Pine oil is a common wetting agent, but many are complex chemicals. They should be powerful enough not to be precipitated out of solutions in the form of salts, and they should be free of odor or any characteristic that would affect the solution. **Aerosol wetting agents,** of American Cyanamid Co., are in the form of liquids, waxy pellets, or free-flowing powders. **Aerosol OS** is a sodium salt of an alkyl naphthalene sulfonic acid. It is a yellowish-brown powder soluble in most organic solvents. This salt was called **Nekal** in Germany. The **Cyanamers** are also free-flowing powders from the same company; basically modified polyacrylates, they are soluble in water and less so in alcohol. The **Dresinols** of Hercules, Inc., are sodium or ammonium dispersions of modified rosin, with 90% of the particles below 39 μin (1 μm) in size. **Polyfon** is a sodium lignosulfonate produced from lignin waste liquor. It is used for dye and pigment dispersion, oil-well drilling mud, ore flotation, and boiler feedwater treatment.

WHALE OIL. An oil extracted by boiling and steaming the blubber of several species of whale that are found chiefly in the cold waters of the extreme north and south. Whales are mammals and are predaceous, living on animal food. The blubber blanket of fat protects the body, and the tissues and organs also contain deposits of fat. Most whale oil is true fat, namely, the glycerides of fatty acids, but the head contains a waxy fat. In the larger animals the meat and bones yield more fat than the blubber. Both the whalebone whales and the toothed whales produce whale oil. The **bluehead whales** of the south, *Silbaldus musculus,* are the largest and yield the most oil per weight. The whaling industry is under international control, and allocations are made on the basis of blue whale units averaging 20 tons (18 metric tons) of oil each. The blue whale is about 25 ft (8 m) long at birth and reaches 70 ft (21 m) in 2 years. This species often reaches 100 ft (30 m) with a weight of about 150 tons (136 metric tons) and will yield about 27 tons (24 metric tons) of oil. The **gray whale,** or **California whale,** of the northern Pacific, is a small 50-ft (15-m)

species. The **Greenland whale** of the north, *Balaena mysticetus,* and
the **finback whale** of the south, *Balaenoptera physalus,* produce
much oil. The **beluga,** or **white whale,** *Delphinapterus leucas,* and
the **narwhal,** *Monodon monoceros,* of the north polar seas, produce
porpoise oil. Both species of porpoise measure up to 20 ft (6 m) in
length.

Whale oil is sold according to grade, which depends upon its color
and keeping qualities. The latter in turn depends largely upon proper
cooking at extraction. Grades 0 and 1 are fine, pale-yellow oils, grade
2 is amber, grade 3 is pale brown, and grade 4 is the darkest oil.
Grade 1 has less than 1% free fatty acids, while grade 4 has from 15
to 60% with a strong, fishy odor. The specific gravity is 0.920 to 0.927,
saponification value 180 to 197, and iodine value 105 to 135. Whale
oil contains oleic, stearic, palmitic, and other acids in varying
amounts. But whales are now so scarce that the former uses of the
oils and meat are restricted, particularly in the United States.

Whale oils of the lower grades were used for quenching baths for
heat-treating steels, and in lubricating oils. The best oils are used in
soaps and candles, or for preparing textile fibers for spinning, or for
treating leather. In Europe whale oil is favored for making margarine
because it requires less hydrogen than other oils for hardening, and
the grouping of 16 to 22 carbon atom acids gives the hardened product greater plasticity over a wider temperature range. **Sod oil** is oil
recovered from the treatment of leather in which whale or other
marine mammal oil was used. It contains some of the tannins and
nitrogenous matter which make it more emulsifiable and more penetrant than the original oil.

Whale meat was used for food in Japan and in dog food in the
United States. When it is cured in air, the outside is hard and black,
but the inside is soft. In young animals the flesh is pale; in older animals it is dark red. It has a slight fishy flavor, but when cooked with
vegetables is almost indistinguishable from beef. It contains 15 to
18% proteins. **Whale-meat extract** is used in bouillon cubes and
dehydrated soups. It is 25% weaker than beef extract. **Whale liver
oil** is used in medicine for its high vitamin A content. It also contains
kitol, which has properties similar to vitamin A but is not absorbed
in all animal metabolism. **Whalebones** are the elastic, hornlike
strips in the upper jaw of the Greenland whale and some other
species. The strips are generally from 8 to 10 ft (2 to 3 m) long and
number up to 600. Those from the **bowhead whale** of the Arctic
Ocean are the longest slabs, measuring up to 13 ft (4 m) in length to
10 to 12 in (25 to 30 cm) wide at the bottom. **Finback whalebone** is
less than 4 ft (1 m) in length. The **humpback whale,** *Megaptera
longimana,* of the northern Pacific, is a **baleen whale** with no teeth

and with plates of baleen in the mouth to act as a sieve. It grows to a length of 50 ft (15 m). Whalebone is lightweight, very flexible, elastic, tough, and durable. It consists of a conglomeration of hairy fibers covered with an enamel-like fibrous tissue. It is easily split and when softened in hot water is easily carved. Whalebone has a variety of uses in making whips, helmet frames, ribs, and brush fibers. **Baleen** is a trade name for strips of whalebone used for whips, and for products where great flexibility and elasticity are required.

WHEAT. The edible seed grains of an annual grass of the genus *Triticum,* of which there are many species and thousands of varieties. Wheat was the basic food grain of the early civilizations of the Near East, and has remained the chief grain of the white races except in cold climates where rye grows better. The plains of the United States, Canada, Argentina, Australia, southern Russia, the Danube Valley, and northern India are the great wheat areas.

The types grown commercially are chiefly common wheat and durum wheat. **Common wheat,** *T. vulgare,* is the chief source of **wheat flour.** It has a stout head from which the grains can be separated easily. The hundreds of varieties are divided roughly into hard wheats and soft wheats, and red wheats and white wheats. The hard wheats usually have smaller grains, but are richer in proteins.

Spring wheat is wheat that is sown in the spring and harvested in late summer. **Winter wheat** is sown in the fall to develop a root system before winter and is then harvested in early summer. It is more resistant and gives a higher yield. **Durum wheat,** *T. durum,* has a thick head with long beards, and large, hard grains rich in gluten. The plant is hardy and drought-resistant, but the flour is too glutenous for U.S. bread and is much used for macaroni and in mixtures. Seven classes of wheat are designated in the official grain standards of the U.S. Department of Agriculture: hard red spring wheat; durum wheat; red durum wheat; hard red winter wheat; soft red winter wheat; white winter wheat; and mixed wheat. Each class permits mixtures of varieties. The minimum test weight of wheat is required to be 60 lb/bu (778 kg/m^3).

Most of the wheat production is ground for edible flour. Since wheat varies with the variety, climate, and soil, uniformity in the flour could formerly be obtained only by blending wheats from different areas to obtain an average; but uniformity is now obtained by an air-spinning process which separates the milled flour into fractions according to protein-starch ratios and then combining for the flour of uniform ratio. These are called **turbo-flours.** Wheat flour is not normally a uniform product even from one area, as it is made up of starch granules, fractured endosperm cells, and protein fragments. **Pregela-**

tinized flour is used for canned goods to reduce the time needed for dextrinizing. **Wheat-flour paste,** for textile coatings, is **hydroxyethylated flour** made by treating wheat flour with ethylene oxide. It requires little cooking to form a starchy product.

Wheat is also used for making beer, and at times is employed for producing starch and alcohol. Some wheat is used for stock feed, but most of the wheat for this purpose is of lower and condemned grades. **Buckwheat** consists of the seed grains of *Fagopyrum esculentum,* a plant of the same family as the rhubarb and dock. It is native to Asia and is one of the chief foods in Russia, but is used only in mixed flours in the United States. The flour is more starchy and has less protein than wheat. It is also darker in color and has a different flavor.

WHETSTONE. Stones of regular fine grains composed largely of chalcedony silica, often with minute garnet and rutile crystals. They are used as fine abrasive stones for the final sharpening of edge tools. Whetstones are also sometimes selected, fine sandstones from the grindstone quarries. The **chocolate whetstone** from New Hampshire is mica schist. The finest whetstones are called **oilstones.** A fine-grained **honestone,** known as **coticule,** comes from Belgium and is used for sharpening fine-edged tools. It is compact, yellow in color, and contains minute crystals of yellow manganese garnet, with also potash mica and tourmaline. Coticule is often cut double with blue-gray **phyllite** rock adhering to and supporting it. **Scythestones** are made from Ohio and Indiana sandstones and from the schist of Vermont. **Rubbing stones** are fine-grained Indiana sandstones.

WHISKERS. Very fine, single-crystal fibers that range from 118 to 394 μin (3 to 10 μm) in diameter and have length-to-diameter ratios of 50 to 10,000. Since they are single crystals, their strengths approach the calculated theoretical strengths of the materials. **Alumina whiskers,** which have received the most attention, have tensile strengths up to 3×10^6 lb/in^2 (20,700 MPa) and a modulus of elasticity of 62×10^6 lb/in^2 (427,000 MPa). Other whisker materials are **silicon carbide, silicon nitride, magnesia, boron carbide,** and **beryllia.**

WHITE BRASS. A bearing metal which is actually outside of the range of the brasses, bronzes, or babbitt metals. It is used in various grades, the specification adopted by SAE being tin, 65%; zinc, 28 to 30; and copper, 3 to 6. It is used for automobile bearings and is close-grained, hard, and tough. It also casts well. A different alloy is known under the name of white brass in the cheap jewelry and novelty trade. It has no tin, small proportions of copper, and the remainder zinc. It is a high-zinc brass and varies in color from silvery white to yellow, depending upon the copper content. An old alloy formerly used for casting buttons, known as **Birmingham platina, or platina,** contained 75% zinc and 25 copper. It had a white color but was very brittle. A yellowish metal known as **bath metal,** once widely used for casting buttons, candlesticks, and other articles, was a brass containing 55% copper and 45 zinc. **White nickel brass** is a grade of nickel silver. The white brass used for castings where a white color is desired may contain up to 30% nickel. The 60:20:20 alloy is used for white plaque castings for buildings. The high-nickel brasses do not cast well unless they also contain lead. Those with 15 to 20% nickel and 2 lead are used for casting hardware and valves. **White nickel alloy** is a copper-nickel alloy containing some aluminum. **White copper** is a name sometimes used for copper-nickel alloy or nickel brass. Nickel brasses known as **German silver** are copper-nickel-zinc white alloys used as a base metal for plated silverware, for springs and contacts in electrical equipment, and for corrosion-resistant parts. The alloys are graded according to the nickel content. Extra-white metal, the highest grade, contains 50% copper, 30 nickel, and 20 zinc. The lower grade, called *fifths,* for plated goods, has a yellowish color. It contains 57% copper, 7 nickel, and 36 zinc. All of the early German silvers contained up to 2% iron, which increased the strength, hardness, and whiteness, but is not desirable in the alloys used for electrical work. Some of the early English alloys also contained up to 2% tin, but tin embrittles alloys. The Federal Trade Commission prohibits the use of the term *German silver* in the marketing of silver-plated ware, but the name still persists in other industries.

WHITE GOLD.　　The name of a class of jewelers' white alloys used as substitutes for platinum. The name gives no idea of the relative value of the different grades, which vary widely. Gold and platinum may be alloyed together to make a white gold, but the usual alloys consist of 20 to 50% nickel, with the balance gold. Nickel and zinc with gold may also be used for white golds. The best commercial grades of white gold are made by melting the gold with a white alloy prepared for this purpose. This alloy contains nickel, silver, palladium, and zinc. The 14-karat white gold contains 14 parts pure gold and 10 white alloy. A superior class of white gold is made of 90% gold and 10 palladium. High-strength white gold contains copper, nickel, and zinc with the gold. Such an alloy, containing 37.5% gold, 28 copper, 17.5 nickel, and 17 zinc, when aged by heat treatment, has a tensile strength of about 100,000 lb/in^2 (690 MPa) and an elongation of 35%. It is used for making jewelry; has a fine, white color; and is easily worked into intricate shapes. Two nickel-free white gold alloys, developed by Handy and

Harmon, contain about 51% silver, 5 palladium, and either 2 zinc or 2 germanium. They were developed because some people are allergic to nickel leaching from 10-karat white gold alloys. **White-gold solder** is made in many grades containing up to 12% nickel, up to 15 zinc, with usually also copper and silver, and from 30 to 80 gold. The melting points of eight grades range from 1283 to 1553°F (695 to 845°C).

WHITE METALS. Although a great variety of combinations can be made with numerous metals to produce white or silvery alloys, the name usually refers to the lead-antimony-tin alloys employed for machine bearings, packings, and linings; to the low-melting-point alloys used for toys, ornaments, and fusible metals; and to the type metals. **Slush castings,** for ornamental articles and hollow parts, are made in a wide variety of soft white alloys, usually varying proportions of lead, tin, zinc, and antimony, depending on cost and the accuracy and finish desired. These castings are made by pouring the molten metal into a metal mold without a core, and immediately pouring the metal out, so that a thin shell of the alloy solidifies against the metal of the mold and forms a hollow product. A number of white metals are specified by the ASTM for bearing use. These vary in a wide range from 2 to 91% tin, 4.5 to 15 antimony, up to 90 lead, and up to 8 copper. The alloy containing 75% tin, 12 antimony, 10 lead, and 3 copper melts at 363°F (184°C), is poured at about 707°F (375°C), and has an ultimate compressive strength of 16,150 lb/in^2 (111 MPa) and a Brinell hardness of 24. The alloy containing 10% tin, 15 antimony, and 75 lead melts at 464°F (240°C) and has a compressive strength of 15,650 lb/in^2 (108 MPa) and a Brinell hardness of 22. The first of these two alloys contains copper-tin crystals; the second contains tin-antimony crystals. A white bearing metal produced under the name of **Asarcoloy** is composed of cadmium with 1.3% nickel. It contains $NiCd_7$ crystals, is harder and has higher compressive strength than babbitt, and has a low coefficient of friction. It has a melting point of 604°F (317°C). **SAE Alloy 18** is such a **cadmium-nickel alloy** with also small amounts of silver, copper, tin, and zinc. A **bismuth-lead alloy** containing 58% bismuth and 42 lead melts at 254°F (123.5°C). It casts to exact size without shrinkage or expansion and is used for master patterns and for sealing.

Various high-tin or reverse bronzes have been used as corrosion-resistant metals, especially before the advent of the chromium, nickel, and aluminum alloys for this purpose. **Trabuk** was a corrosion-resistant, high-tin bronze with about 5% nickel. **Fahry's alloy** was a **reverse bronze** containing 90% tin and 10 copper, used as a bearing metal, and the **Jacoby metal** used for machine parts had 85% tin, 10 antimony, and 5 copper. The scarcity and high cost of tin have made these alloys obsolete. The bearing alloy known in England as **motor bronze** is a babbitt with about twice the copper of a standard babbitt. One analysis gives tin, 84%; antimony, 7.5; copper, 7.5; and bismuth 1. An old alloy, used in India for utensils and known as **bidery metal,** contained 31 parts zinc, 1 lead, and 2 copper, fluxed with resins. It was finished with a velvety-black color by treating with a solution of copper sulfate. A **white metal sheet** now much used for making stamped and formed parts for costume jewelry and electronic parts is zinc with up to 1.5% copper and up to 0.5 titanium. The titanium with the copper prevents coarse-grain formation, raising the recrystallization temperature. The alloy weighs 2% less than copper, and it plates and solders easily. **Zilloy-20** is pure zinc with no more than 1% of other elements. In rolled strip it has a tensile strength up to 27,000 lb/in^2 (186 MPa) and elongation of 35%.

WILLOW. The wood of the trees *Salix coerulea* and *S. alba,* native to Europe, but grown in many other places. It is best known as a material for cricket bats made in England. The American willows are known as **black willow,** from the tree *S. nigra,* and **western black willow,** from the tree *S. lasiandra.* The wood is also employed for making artificial limbs and for articles where toughness and nonshrinking qualities are valued. The wood is brownish yellow; has a fine, open grain; and has a density of about 30 lb/ft^3 (481 kg/m^3). It is of the approximate hardness of cherry and birch. **Japanese willow** is from the tree *S. urbaniana.* It has a closer and finer texture and a browner color. Black willow has a maximum crushing strength parallel to the grain of about 1,500 lb/in^2 (10 MPa). **Salicin,** also called **salicoside** and **saligenin,** is a glucoside extracted from several species of willow bark of England and also from the American aspen. It is a colorless, crystalline material of composition $(OH)_4C_6H_7 \cdot OO \cdot C_6H_4CH_2OH$, decomposing at 394°F (201°C) and soluble in water and in alcohol. It is used in medicine as an antipyretic and tonic, and as a reagent for nitric acid. It hydrolyzes to glucose and salicyl alcohol, and the latter is oxidized to **salicylic acid,** $C_6H_4(OH)COOH$. **Aspirin, acetyl salicylic acid,** is used as an antipyretic and analgesic.

WIRE CLOTH. Stiff fabrics made of fine wire woven with plain, loose weave, used for screens to protect windows, for guards, and for sieves and filters. Steel and iron wire may be used—plain, painted, galvanized, or rustproofed—or various nonferrous metal wires are employed. It is usually put up in rolls in widths from 18 to 48 in (46 to 122 cm). **Screen cloth** is usually 12, 14, 16, and 18 mesh, but wire cloth in copper, brass, or Monel metal is made regularly in meshes from 4 to 100. The size of wire is usually from 0.009 to 0.065 in (0.023

to 0.165 cm) in diameter. Wire cloth for fine filtering is made in very fine meshes. **Mesh** indicates the number of openings per inch and has no reference to the diameter of wire. A 200-mesh cloth has 200 openings each way on a square inch, or 40,000 openings per square inch (6.4 cm^2). Wire cloth as fine as 400 mesh, or having 160,000 openings per square inch (6.4 cm^2), is made by wedge-shaped weaving, although 250 wires of the size of 0.004 in (0.010 cm) when placed parallel and in contact will fill the space of 1 in (2.5 cm). Very fine-mesh wire cloth must be woven at an angle since the globular nature of most liquids will not permit passage of the liquid through microscopic square openings. One wire screen cloth, for filtering and screening, has elongated openings. One way the 0.0055-in (0.0140-cm) wire count is 200 per inch (2.5 cm), while the other way the 0.007-in (0.018-cm) warp wire count is 40 per inch (2.5 cm).

Wire fabrics for reentry parachutes are made of heat-resistant nickel-chromium alloys, and the wire is not larger than 0.005 in (0.013 cm) in diameter to give flexibility to the cloth. Wire fabrics for ion engines to operate in cesium vapor at temperatures to 2400°F (1316°C) are made with tantalum, molybdenum, or tungsten wire, 0.003 to 0.006 in (0.008 to 0.015 cm) in diameter, with a twill weave. Meshes to a fineness of 350 by 2,300 can be obtained. Porosity uniformity is controlled by pressure calendaring of the woven cloth, but for extremely fine meshes in wire cloth it is difficult to obtain the uniformity that can be obtained with porous sintered metals.

Where accuracy of sizing is not important, as in gravel or ore screening, wire fabric is made with oblong or rectangular openings instead of squares to give faster screening. High-manganese steel wire is used for rock screens. For window screening in tropical climates or in corrosive atmospheres, plastic filaments are sometimes substituted for the standard copper or steel wire. **Lumite screen cloth** is woven of vinylidene chloride monofilament 0.015 in (0.038 cm) in diameter in 18 and 20 mesh. The impact strength of the plastic cloth is higher than that of metal wire cloth, but it cannot be used for screening very hot materials. **Lektromesh** is copper or nickel screen cloth of 40 to 200 mesh made in one piece by electrodeposition. It can be drawn or formed more readily than wire screen, and circular or other shapes can be made with an integral selvage edge.

WIRE GLASS. A sheet glass used in building construction for windows, doors, floors, and skylights, having woven wire mesh embedded in the center of the plate. It does not splinter or fly apart as common glass when subjected to fire or shock, and it has higher strength than common glass. It is made in standard thicknesses from 0.125 to 0.375 in (0.318 to 0.953 cm) and in plates 60 by 110 in (1.5 by 2.8 m) and 61 by

140 in (1.5 by 3.6 m). Underwriters' specifications call for a minimum thickness of 0.25 in (0.635 cm). Wire glass is made with plain, rough, or polished surfaces, or with ribbed or cobweb surface on one side for diffusing the light and for decorative purposes. It is also obtainable in corrugated sheets, usually 27.75 in (70.5 cm) wide. Wire glass 0.25 in (0.635 cm) thick weighs 2.25 lb/ft^2 (11 kg/m^2). Plastic-coated wire mesh may be used to replace wire glass for hothouses or skylights where less weight and fuller penetration of light rays are desired. **Cel-O-Glass,** of Du Pont, is a plastic-coated wire mesh in sheet form.

WOLLASTON WIRE. Any wire made by the Wollaston process of fine-wire drawing. It consists of inserting a length of bare drawn wire into a close-fitting tube of another metal, the tube and core then being treated as a single rod and drawn through dies down to the required size. The outside jacket of metal is then dissolved away by an acid that does not affect the core metal. **Platinum wire** as fine as 0.00005 in (0.00013 cm) in diameter is made commercially by this method, and gold wire as fine as 0.00001 in (0.00002 cm) in diameter is also drawn. Wires of this fineness are employed only in instruments. They are marketed as composite wires, the user dissolving off the jacket. **Taylor process wire** is a very fine wire made by the process of drawing in a glass tube. The process is used chiefly for obtaining fine wire from a material lacking ductility, such as antimony, or extremely fine wire from a ductile metal. The procedure is to melt the metal or alloy into a glass or quartz tube, and then draw down this tube with its contained material. Wire as fine as 0.00004 in (0.00012 cm) in diameter is made, but only in short lengths.

WOOD. A general name applied to the cut material derived from trees. A **tree,** as distinguished from a **bush,** is designated by the U.S. Forest Service as a woody plant with a single erect stem 3 in (7.6 cm) or more in diameter at 4.5 ft (1.4 m) above the ground, and at least 12 ft (3.7 m) high. But this definition is merely empirical since in the cold climate of northern Canada, perfect, full-grown trees 10 to 15 years old may be only 6 in (15 cm) high. **Timber,** in general, refers to standing trees, while **lumber** is the sawed wood used for construction purposes. In construction work the word *timber* is often applied to large pieces of lumber used as beams.

Wood is an organic chemical compound composed of approximately 49% carbon, 44 oxygen, 6 hydrogen, and 1 ash. It is largely cellulose and lignin. The wood of white pine is about 50% cellulose, 25 lignin, and the remainder sugars, resin, acetic acid, and other materials. Wood is produced in most trees by a progressive growth from the outside. In the spring, when sap flows rapidly, a rapid formation of large cells takes place, followed by a slower growth of hard and close cells in summer. In some woods, such as oak, there is a considerable difference in quality and appearance between the spring and summer woods. In some long-lived trees, such as Douglas fir, there is a decrease in strength between the outside wood with narrow rings and the wide-ringed wood of the interior. **Heartwood** is the dark center of the tree which has become set, and through which the sap has ceased to flow. **Sapwood** is the outer, live wood of the tree; unless treated, it has low decay resistance. The grain of sawed lumber results from sawing across the annual growth rings, varied to produce different grains.

Wood is seasoned either by exposing it to the air to dry or by kiln drying. The former method is considered to give superior quality, but it requires more time, is expensive, and is indefinite. Numerous tests made at the U.S. Forest Products Laboratory did not reveal any superiority in air-dried wood when kiln drying was well done. Solvent seasoning is a rapid process consisting of circulating a hot solvent through the wood in a closed chamber. California redwood, when seasoned with acetone at 130°F (54°C), yields tannin and some other chemicals as by-products. Seasoned wood, when dry, is always stronger than unseasoned wood. **Tank woods** are selected for resistance to the liquids to be contained. Tanks for vinegar and foodstuffs containing vinegar, such as pickles, are of white oak, cypress, or western red cedar. Beer tanks are of white oak or cypress. Tanks for aging wine are of redwood, oak, or fir. The traditional **violin woods** are spruce and curly maple, although sugar maple is also used.

The term **log** designates the tree trunk with the branches removed. **Balk** is a roughly squared log; **plank** is a piece cut to rectangular section 11 in (28 cm) wide; **deal** is a piece 9 in (23 cm) wide; and **batten** is a piece 7 in (18 cm) wide. **Board** is a thin piece of any width less than 2 in (5 cm). **Flitch** is half a balk, cut in two lengthwise. **Scantling** is a piece sawed on all sides. **Shakes** are longitudinal splits or cracks in the wood due to shrinkage or decay.

All woods are divided into two major classes on the basis of the type of tree from which they are cut. **Hardwoods** are from broad-leaved, deciduous trees. **Softwoods** are from conifers, which have needle- or scalelike leaves and are, with few exceptions, evergreens. These terms do not refer to the relative hardnesses of the woods in these two classes. Hardwood lumber is available in three basic categories: **factory lumber; dimension lumber,** or **dimension parts**; and **finished market products.** The important difference between factory lumber and dimension parts is that factory lumber grades reflect the proportion of the pieces that can be cut into useful smaller pieces, while the dimension grades are based on use of the entire piece. Finished market products are graded for their end use with little or no remanufacturing. Examples of finished market products are flooring, siding, ties, timbers, trim, molding, stair treads, and risers. The rules adopted by the National Hardwood Lumber Association are considered standard in grading factory lumber. The grades from the highest to the lowest quality are as follows: firsts, the top quality, and seconds, both of which are usually marketed as one grade called firsts and seconds (FAS); selects; and common grades No. 1, No. 2, No. 3A, and No. 3B. Sometimes a grade is further specified, such as FAS one face, which means that only one face is of the FAS quality. Another designation, WHND, sometimes used, means that wormholes are not considered defects in determining the grade. Dimension lumber, generally graded under the rules of the Hardware Dimension Manufacturers Association, are of three classes: solid dimension flat stock, kiln-dried dimension flat stock, and solid dimension squares. Each class may be rough, semifabricated, or fabricated. Rough dimension blanks are usually kiln-dried and are supplied sawn and ripped to size. Surfaced or semifabricated stock has been further processed by gluing, surfacing, etc. Fabricated stock has been completely processed for the end use. Solid dimension flat stock has five grades: clear—two faces, clear—one face, paint, core, and sound. Squares have three grades if rough (clear, select, sound) and four if surfaced (clear, select, paint, sound).

There are two major categories of softwood lumber: construction and remanufacture. **Construction lumber** is of three general types: stress-graded; non-stress-graded, also referred to as **yard lumber**; and **appearance lumber. Stress-graded lumber** is structural lumber never less than 2 in (5 cm) thick, intended for use where definite strength requirements are specified. The allowable stresses specified for stress-graded lumber depend on the size, number, and placement of defects. Because the location of defects is important, the piece must be used in its entirety for the specified strength to be realized. Stress-graded products include timbers, posts, stringers, beams, decking, and some boards.

Typical **non-stress-graded lumber** items include boards, lath, battens, cross-arms, planks, and foundation stock. **Boards,** sometimes referred to as **commons,** are one of the more important non-stress-graded products. They are separated into three to five different grades, depending upon the species and lumber manufacturing association involved. Grades may be described by number (No. 1, No. 2) or by descriptive terms (construction, standard). First-grade boards are usually graded primarily for serviceability, but appearance is also considered. Second- and third-grade boards are often used together for such purposes as subfloors and sheeting. Fourth-grade boards are selected not for appearance but for adequate strength. The appearance category of construction lumber includes trim, siding, flooring, ceiling, paneling, casing, and finish boards. Most appearance lumber grades are designated by letters and combinations of letters, and are also often known as select grades. Typical grades of **lumber remanufacture** are the factory grades and industrial clears. Factory select and select shop are typical high grades of factory lumber, followed by No. 1, No. 2, and No. 3 shop. Industrial clears are used for cabinet stock, door stock, and other products where excellent appearance, mechanical and physical properties, and finishing characteristics are important. The principal grades are B&BTR, C, and D.

Metallized wood is wood treated with molten metal so that the cells of the wood are filled with the metal. Fusible alloys, with melting points below the scorching point of the wood, are used. The wood is immersed in molten metal in a closed container under pressure. The hardness, compressive strength, and flexural strength of the wood are increased, and the wood becomes an electric conductor lengthwise of the grain. Woods are also metallized with a surface coating of metal by vacuum deposition.

Sugar pine is one of the most widely used **pattern woods** for foundry patterns. It replaces eastern white pine, which is scarcer and now usually more costly. Poplar is used for patterns where a firmer wood is desired; cherry or maple is employed where the pattern is to be used frequently or will be subject to severe treatment. Densified wood is also used for patterns required to be very wear-resistant. Mahogany is used for small and intricate patterns where a firm texture and freedom from warpage are needed. However, for small castings made in quantities on gates, aluminum or brass is more frequently used.

Excelsior is an old trade name, still used, for continuous, curly, fine wood shavings employed as a packing material for breakable articles. It is light and elastic, and it is also used as a cushioning and stuffing material. It is usually made from poplar, aspen, basswood, or cottonwood. A cord of wood produces about 1,500 lb (680 kg), but it may be made as a by-product from other woodworking. It is also called **wood fiber** and **wood wool,** but these terms more properly refer to fibers of controlled size and length used with a resin binder for molding into handles, knobs, and other **imitation wood** parts. Several plastics are suitable for imitation wood, also called **synthetic wood** and **plastic lumber.** And environmental concerns regarding global deforestation have increased prices of certain woods, making plastics more cost-competitive.

Some wood for special purposes comes from roots or from bushes. The **briar** used for tobacco pipes is from the roots of the **white heath,** *Erica arborea,* of north Africa. Substitutes for briar are the burls of the laurel and rhododendron. **Yareta,** used for fuel in the copper region of Chile, is a mosslike, woody plant which grows on the sunny northern mountain slopes at altitudes above 12,000 ft (3,658 m) and requires several hundred years to reach a useful size.

WOOD FLOUR. Finely ground dried wood employed as a filler and as reinforcing material in molding plastics and in linoleum, and as an absorbent for nitroglycerin. It is made largely from light-colored softwoods, chiefly pine and spruce, but maple and ash flours are preferred where no resin content is desired. Woods containing essential oils, such as cedar, are not suitable. Wood flour is produced from sawdust and shavings by grinding in burr mills. It has the appearance of wheat flour. The sizes commonly used are 40, 60, and 80 mesh; the finest is 140 mesh. Grade 1, used as a filler in rubber and plastics, has a particle size of 60 mesh and a specific gravity of 1.25, but 80 and 100 mesh are also used for plastic filler. Since wood flour absorbs the resin or gums when mixed in molding plastics and sets hard, it is sometimes mixed with mineral powders to vary the hardness and toughness of the molded product.

Vast quantities of **sawdust** are obtained in the sawmill areas. Besides being used as a fuel, it is employed for packing, for finishing metal parts in tumbling machines, for making particleboard, and for distilling to obtain resins, alcohols, sugars, and other chemicals. Some sawdust is pulped, and as much as 20% of such pulp can be used in kraft paper without loss of strength. Hickory, walnut, and oak sawdusts are used for meat smoking, or for the making of **liquid smoke,** which is produced by burning the sawdust and absorbing the smoke into water. For the rapid production of bacon and other meats, immersion in liquid smoke imitates the flavor of smoked meat. Some sawdust is used for agricultural mulch and fertilizer by chemical treatment to accelerate decay. **Bark fuel** is shredded bark, flash-dried and pelletized with powdered coal. **Particleboard,** made by compressing sawdust or wood particles with a resin binder into sheets, has uniform strength in all directions, and a smooth, grainless surface. When used as a core for veneer panels, it requires no cross-laminating. Mechanical pulp for newsprint can be made from sawdust but the quantity available is usually not sufficient. The material known as **ground wood,** of fine-mesh fibers, is made from cord wood, about 1 ton of fibers being produced from one cord of pulpwood. **Plastic wood,** usually marketed as a paste in tubes for filling cavities or seams in wood products, is wood flour or wood cellulose compounded with a synthetic resin of high molecular weight that will give good adhesion but not penetrate the wood particles to destroy their nature. The solvent is kept low to reduce shrinkage. When cured in place, the material can be machined, polished, and painted.

WOOD PRESERVATIVES. These fall into two general classes: oils, such as **creosote** and petroleum solutions of **pentachlorophenol**; and **waterborne salts** that are applied as water solutions. **Coal tar creosote,** a black or brownish oil made by distilling coal tar, is the oldest and still one of the more important and useful wood preservatives. Because it has recently been classified as a carcinogen, its use is expected to decrease. Its advantages are high toxicity to wood-destroying organisms; relative insolubility in water and low volatility, which impart to it a great degree of permanence under the most varied use conditions; ease of application; ease with which its depth of penetration can be determined; general availability and relatively low cost; and long record of satisfactory use.

Creosotes distilled from tars other than coal tar are used to some extent for wood preservation. For many years, either **cold tar** or **petroleum oil** has been mixed with cold tar creosote in various proportions to lower preservative costs.

Water-repellent solutions containing chlorinate phenols, principally pentachlorophenol, in solvents of the mineral spirit type have been used in commercial treatment of wood by the millwork industry since about 1931. Pentachlorophenol solutions for wood preservation generally contain 5% (by weight) of this chemical, although solutions with volatile solvents may contain lower or higher concentrations. Preservative systems containing water-repellent components are sold under various trade names, principally for the dip or equivalent treatment of window sash and other millwork. According to federal specifications the preservative chemicals may not contain less than 5% pentachlorophenol.

Standard wood preservatives used in water solution include **acid copper chromate, ammoniacal copper arsenite, chromated copper arsenate, zinc naphthenate, chromated zinc chloride,** and **fluor chrome arsenate phenol.** These preservatives are often employed when cleanliness and paintability of the treated wood are required. The chromated zinc chloride and fluor chrome arsenate phenol formulations resist leaching less than preservative oils, and are seldom used where a high degree of protection is required for wood in ground contact or for other wet installations. Several formulations involving combinations of copper, chromium, and arsenic have shown high resistance to leaching and very good performance in service. The ammoniacal copper arsenite and chromated copper arsenate are included in specifications for such items as building foundations, building poles, utility poles, marine piling, and piling for land and freshwater use. **Organic sulfones** are another class of wood preservatives offering high degrees of protection. One such product is **diiodomethyl p-tolyl sulfone,** with trade name **Amical,** from Angus Chemical Co.

WOOL. The fine, soft, curly hair or fleece of the sheep, alpaca, vicuña, certain goats, and a few other animals. The specific designation *wool* always means the wool of sheep. **Sheep's wool** is one of the most important commercial fibers because of its good physical qualities and its insulating value, especially for clothing, but it now constitutes only about 10% of the textile fiber market. It is best known for its use in clothing fabrics, called **woolens.** These are designated under a variety of very old general trade names such as a loosely woven fabric called **flannel,** or the fine, smooth fabric known as **broadcloth. Cheviot** is a close-napped, twill-woven fabric, and **tweed** is a woolen fabric with a coarse surface, usually with a herringbone-twill weave. **Serge** is a twill-woven worsted fabric. **Worsteds** are wool fabrics made from combed-wool yarn, usually from long, smooth wool. Wool is also employed for packings and for insulation, either loose or felted, and for making felts. The average amount of wool shorn from sheep in the United States is 8.1 lb (3.7 kg) per animal.

Wool differs from hair in fineness and its felting and spinning properties. The latter are due to the fine scales of the wool fibers. The finest short-staple wool has as many as 4,000 scales to the inch (2.5 cm), and the average long-staple wool has about 2,000 scales per inch (2.5 cm). These scales give wool its cohesive qualities. Some animals have both wool and hair, while others have wool only when young. There is no sharp dividing line between wool and hair.

Wool quality is by fineness, softness, length, and scaliness. Fiber diameters vary from 0.0025 to 0.005 in (0.0064 to 0.013 cm). Long wools are generally heavy. Fibers below 3 in (7.6 cm) in length are known as **clothing wool,** and those from 3 to 7 in (7.6 to 17.8 cm) are called **combing wools.** Long wools are fibers longer than 7 in (17.8 cm). The term **apparel wool** generally means clothing wool of fine weaving quality from known sources. **Fleece wool** is the unscoured fiber. It may contain as much as 65% grease and dirt, but this is the form in which wool is normally shipped because it then has the protection of the wool fat until it is manufactured. Wool is very absorbent to moisture and will take up about 33% of its weight of water, and in some areas moisture and dirty grease are added to fleece wool to increase weight. **Carpet wools** are usually long, nonresilient fibers from sheep bred in severe climates, such as the **Mongolian wool.** The only breed of sheep developed for wool alone is the merino. In Australia the **corriedale** and the **polworth sheep** are dual-purpose animals for wool and meat.

The finest of sheep wools come from the **merino sheep,** but these vary according to the age of breeding of the animal. The **Lincoln sheep** produces the longest fiber. It is lustrous but very coarse. Luster of wool depends upon the size and smoothness of the scales, but the chemical composition is important. The molecular chains are linked with sulfur, and when sulfur is fed to the sheep, in some deficient areas the quality of the wool is improved. Crimpiness in wool is due to the open formation of the scales. A fine merino will have 24 crimps per inch (2.5 cm), whereas a coarse crossbreed will have only 6 per inch (2.5 cm). Strength of wool fibers often depends upon the health of the animal and the feeding.

One-quarter of the world production of wool is in Australia. Argentina ranks second in production, with the United States third. But the United States is a lamb-eating nation, and a large proportion of the animals are slaughtered when 4 to 8 months old, and most of the others are kept only one season for one crop of wool. New Zealand, Uruguay, Russia, and England are also important producers. England is the center of wool-sheep breeding, with more varieties than any other country. In general, warm climates produce fine wools, and hot climates produce thin, wiry wools, but the fundamental differences come from the type of animal and the feeding. The **reused wool** from old cloth was originally called **shoddy,** but the name has an opprobrious signification in the United States, and is not used by manufacturers to designate the fabrics made from reclaimed wool. Shoddy is used in mixtures with new wool for clothing and other fabrics. **Extract wool** is shoddy that is recovered by dissolving out the cotton fibers of the old cloth with sulfuric acid. Short fibers of shoddy, less than 0.5 in (1.27 cm), are known as **mungo fibers.** They are used in woolen blends to obtain a napped effect. **Reprocessed wool** is fiber obtained from waste fabric which has not been used. **Noils** are short fibers produced in the combining of wool tops for making worsteds. They are used for woolen goods and felt. **Zeset,** of Du Pont, a shrinkproofing agent for wools, is a variant of **Surlyn T,** a terpolymer of 70% ethylene, 6 methacryloyl chloride, and 24 vinyl acetate. It prevents shrinkage and pilling under ordinary laundry methods, does not affect color, and increases the tensile strength of the fiber. But all resinous additives tend to harden the fiber and lessen the drape and feel. Conversely, each dry cleaning of wool fabric decreases the natural oil content and hardens the fiber.

WOOL GREASE.   A brownish, waxy fat of a faint, disagreeable odor, obtained as a by-product in the scouring of wool. The purified grease was formerly known as **degras** and was used for leather dressing, in lubricating and slushing oils, and in soaps and ointments; but it is now largely employed for the production of lanolin and its derivatives, chiefly for cosmetics. Wool grease contains **lanoceric acid**; **lanopalmic acid,** $C_{15}H_{30}O_3$; and **lanosterol,** a high alcohol related to cholesterol. All of these can be broken down into derivatives.

Lanolin is a purified and hydrated grease, also known as **lanain,** and in pharmacy as **lanum** and **adeps lanae.** It has a melting point of about 104°F (40°C) and is soluble in alcohol. Lanolin is basically a wax consisting of esters of sterol alcohols combined with straight-chain fatty acids, and with only a small proportion of free alcohols. It contains about 95% of fatty acid esters, but its direct use as an emollient depends on the 5% of free alcohols and acids. However, more than 30 derivatives are obtained from lanolin, and these are used in blends to give specific properties to cosmetics. They are often marketed under trade names, and some of the ingredients may be synthesized from raw materials other than wool grease, or chemically altered from wool-grease derivatives.

A variety of products used in cosmetics and pharmaceuticals are made by fractionation or chemical alteration of lanolin. They are also useful in compounding plastics and industrial coating, but are generally too scarce and expensive for these purposes. **Ethoxylated lanolin** and ethoxylated **lanolin alcohols** are used in water-soluble emulsions and conditioners. **Solulan** is a general trade name for these materials. **Lanolin oil** and **lanolin wax** are made by solvent fractionation of lanolin. **Viscolan** and **Waxolan** are these products. **Isopropyl lanolates,** with trade name **Amerlate,** are soft, hydrophylic solids which liquefy easily and are used in cosmetics as emollients, emulsifiers, and pigment dispersants. **Amerlate LFA** is derived from lanolin hydroxy acids containing iso-acids. The high hydroxyl content produces the emollient and emulsifying qualities. **Barium lanolate,** made by saponifcation, is used as an anticorrosion agent. It is antiphobic and is also used as an anticaking agent. In a 25% barium concentration it is used for hard lubricating grease.

Ethoxylan is an ethylene oxide derivative of lanolin, soluble in water and in alcohol, and used in shampoos. **Ceralan** is a waxy solid melting at 131°F (55°C) to an amber-colored, viscous liquid. It is a mixture of monohydroxyl alcohols, obtained by splitting lanolin, and contains 30% sterol, and free cholesterol. It forms water-in-oil emulsions and is used in cosmetics as a dispersing and stiffening agent and as an emollient. **Acetylated lanolin** is made by reacting lanolin with polyoxyethylenes. They are clear, nongreasy liquids soluble in water, oils, and alcohol. The acetylated lanolin is hydrophobic and oil-soluble, and is used as an odorless, nontacky emollient in cosmetics. **Acylan,** from Croda Chemicals, is a soft solid with a bland odor that is employed in baby products, hair grooms, creams, and pharmaceuticals. Oil-based solutions of Acylan are clear, forming soft, waxy, hydrophobic films. **Satulan** of the same firm is a **hydrogenated lanolin** useful in products for skin protection.

Veriderm, of Upjohn Co., is a substitute for lanolin as an emollient. It contains about the same percentage of triglycerol esters of fatty acids, free cholesterol, and saturated and unsaturated hydrocarbons as occurs in the natural human skin oils. Cholesterol is one of the most important of the complex **sterols,** or **zoosterols,** from animal sources. It is produced from lanolin, but also from other sources, and used in drugs and cosmetics. **Amerchol L-101** is a liquid nonionic cholesterol containing other sterols. Wool grease from the scouring of wool was originally called **Yorkshire grease. Moellon degras** is not wool grease, but is a by-product of chamois leather making. The sheepskins are impregnated with fish oil, and when the tanning is complete, they are soaked in warm water and the excess oil is pressed out to form the moellon degras.

WROUGHT IRON. Commercially pure iron made by melting white cast iron and passing an oxidizing flame over it, leaving the iron in a porous condition which is then rolled to unite it into one mass. As thus made, it has a fibrous structure, with fibers of slag through the iron in the direction of rolling. It is also made by the Aston process of shooting Bessemer iron into a ladle of molten slag. Modern wrought iron has a fine dispersion of silicate inclusions which interrupt the granular pattern and give it a fibrous nature.

The value of wrought iron is in its corrosion resistance and ductility. It is used chiefly for rivets, staybolts, water pipes, tank plates, and forged work. Minimum specifications for **ASTM wrought iron** call for a tensile strength of 40,000 lb/in^2 (276 MPa), yield strength of 24,000 lb/in^2 (165 MPa), and elongation of 12%, with carbon not over 0.08%, but the physical properties are usually higher. **Wrought iron 4D** has only 0.02% carbon with 0.12 phosphorus, and the fine fibers are of a controlled composition of silicon, manganese, and phosphorus. This iron has a tensile strength of 48,000 lb/in^2 (331 MPa), elongation 14%, and Brinell hardness 105. **Mn wrought iron** has 1% manganese for higher impact strength.

Ordinary wrought iron with slag may contain frequent slag cracks, and the quality grades are now made by controlled additions of silicate, and with controlled working to obtain uniformity. But for tanks and plate work, ingot iron is now usually substituted. **Merchant bar iron** is an old name for wrought-iron bars and rods made by faggoting and forging. **Iron-fibered steel** is soft steel with fine iron wire worked into it. **Staybolt iron** may be wrought iron, but was originally puddled charcoal iron. **Lewis iron,** for staybolts, is highly refined, puddled iron with a tensile strength of 52,000 lb/in^2 (359 MPa) and elongation of 30%.

The **Norway iron** formerly much used for bolts and rivets was a **Swedish charcoal iron** brought to America in Norwegian ships.

This iron, with as low as 0.02% carbon, and extremely low silicon, sulfur, and phosphorus, was valued for its great ductility and toughness and for its permeability qualities for transformer cores. Commercial wrought iron is now usually ingot iron or fibered low-carbon steel.

YARNS. Assemblages or bundles of fibers twisted or laid together to form continuous strands. They are produced with either filaments or staple fibers. Single strands of yarns can be twisted together to form ply or plied yarns, and ply yarns in turn can be twisted together to form cabled yarn or cord. Important yarn characteristics related to behavior are fineness (diameter or linear density) and number of twists per unit length. The measuring of fineness is commonly referred to as *yarn number.* Yarn numbering systems are somewhat complex, and they are different for different types of fibers. Essentially, they provide a measure of fineness in terms of weight per unit or length per unit weight.

Cotton yarns are designated by numbers, or **counts.** The standard count of cotton is 840 yd/lb (1,690 m/kg). Number 10 yarn is therefore 8,400 yd/lb (16,900 m/kg). A No. 80 sewing cotton is 80×840, or 67,200 yd/lb (135,500 m/kg).

Linen yarns are designated by the **lea** of 300 yd (274 m). A 10-count linen yarn is 10 × 300, or 3,000 yd/lb (6,048 m/kg).

The size or count of spun **rayon yarns** is on the same basis as cotton yarn. The size or count of rayon filament yarn is on the basis of the **denier,** the rayon denier being 492 yd (450 m), weighing 0.00011 lb (5 cg). If 492 yd of yarn weighs 0.00011 lb, it has a count of 1 denier. If it weighs 0.0011 lb (10 cg), it is No. 2 denier. Rayon yarns run from 15 denier, the finest, to 1,200 denier, the coarsest.

Reeled **silk yarn** counts are designated in deniers. The **international denier** for reeled silk is 547 yd (500 m) of yarn weighing 0.00011 lb. If 547 yd weighs 0.0022 lb (1 g), the denier is No. 20. Spun silk count under the English system is the same as the cotton count. Under the French system the count is designated by the number of **skeins** weighing 2.205 lb (1 kg). The skein of silk is 1,094 yd (1,000 m).

A **ply yarn** is one that has two or more yarns twisted together. A two-ply yarn has two separate yarns twisted together. The separate yarns may be of different materials, such as cotton and rayon. A six-ply yarn has six separate yarns. A ply yarn may have the different plies of different twists to give different effects. Ply yarns are stronger than single yarns of the same diameter. Tightly twisted yarns make strong, hard fabrics. Linen yarns are not twisted as tightly as cotton because the flux fiber is longer, stronger, and not as fuzzy as the cotton. **Filament rayon yarn** is made from long, continuous rayon fibers, and it requires only slight twist. Fabrics made from filament yarn are called **twalle. Monofilament** is fiber heavy enough to be used alone as yarn, usually more than 15 denier. **Tow** consists of multifilament reject strands suitable for cutting into staple lengths for spinning. Spun rayon yarn is yarn made from **staple fiber,** which is rayon filament cut into standard short lengths.

YUCCA FIBER. The fiber obtained from the leaves of a number of desert plants of the genus *Yucca* of the lily family native to the southwestern United States and northern Mexico. The fiber is similar to fibers from agave plants and is often confused with them and with istle. The heavier fibers are used for brushes, and the lighter fibers are employed for cordage and burlap fabrics. In Mexico the word **palma** designates yucca fibers and grades of istle as well as palm-leaf fibers. **Palma samandoca** is fiber from the plant *Samuela carnerosana,* the **date yucca.** It is also called **palma istle. Palmilla fiber** is from *Y. elata.* **Palma pita** is a fiber from *Y. treculeana.* **Pita fiber** used for coffee bags in Colombia and Central America is from a different plant. Other yucca fibers come from the plants *Y. glauca, Y. baccata,* and *Y. gloriosa.* Some varieties of *Y. baccata* also yield edible fruits. The roots of species of yucca yield saponin which is also obtained as a by-product in extracting the yucca fiber.

ZINC. A bluish-white, crystalline metal, symbol Zn, with a specific gravity of 7.13, melting at 788°F (420°C) and boiling at 1662°F (906°C). The commercially pure metal has a tensile strength, cast, of about 9,000 lb/in^2 (62 MPa) with elongation of 1%, and the rolled metal has a strength of 24,000 lb/in^2 (165 MPa) with elongation of 35%. But small amounts of alloying elements harden and strengthen the metal, and it is seldom used alone. Zinc is used for galvanizing and plating; for making brass, bronze, and nickel silver; for electric batteries; for die castings; and in alloyed sheets for flashings, gutters, and stamped and formed parts. The metal is harder than tin, and an electrodeposited plate has a Vickers hardness of about 45. Zinc is also used for many chemicals.

The old name **spelter,** often applied to slab zinc, came from the name *spailter* used by Dutch traders for the zinc brought from China. The first zinc produced in the United Stats in 1838 came from New Jersey ore. **Sterling spelter** was 99.5% pure. Special high-grade zinc is distilled, with a purity of 99.99%, containing no more than 0.006% lead and 0.004 cadmium. High-grade zinc, used in alloys for die casting, is 99.9% pure, with 0.07 maximum lead. **Brass special zinc** is 99.10% pure, with 0.06 maximum lead and 0.5 maximum cadmium. **Prime western zinc,** used for galvanizing, contains 1.60% maximum lead and 0.08 maximum iron. **Zinc crystals** produced for electronic uses are 99.999% pure metal.

On exposure to the air, zinc becomes coated with a film of carbonate and is then very corrosion-resistant. **Zinc foil** comes in thicknesses from 0.001 to 0.006 in (0.003 to 0.015 cm). It is produced by electrodeposition on an aluminum drum cathode and stripping off on a collecting reel. But most of the **zinc sheet** contains a small amount of alloying elements to increase the physical properties. Slight amounts of copper and titanium reduce grain size in sheet zinc. In cast zinc the hexagonal columnar grain extends from the mold face to the surface or to other grains growing from another mold face, and even very slight additions of iron can control this grain growth. Aluminum is also much used in alloying zinc. In zinc used for galvanizing, a small addition of aluminum prevents formation of brittle alloy layer, increases ductility of the coating, and gives a smoother surface. Small additions of tin give bright, spangled coatings.

Zinc has 12 isotopes, but the natural material consists of 5 stable isotopes, of which nearly half is **zinc 64.** The stable isotope **zinc 67,** occurring to the extent of about 4% in natural zinc, is sensitive to tiny variations in transmitted energy, giving off electromagnetic radiations which permit high accuracy in measuring instruments. It measures gamma-ray vibrations with great sensitivity and is used in the nuclear clock.

Zinc powder, or **zinc dust,** is a fine, gray powder of 97% minimum purity usually in 325-mesh particle size. It is used in pyrotechnics, in paints, as a reducing agent and catalyst, in rubbers as a secondary dispersing agent and to increase flexing, and to produce **Sherardized steel.** Sherardizing consists in hot-tumbling steel parts in a closed drum with the zinc powder. It is a form of galvanizing, and controlled zinc coatings of 0.1 to 0.4 oz/ft^2 (0.4 to 1.8 g/cm^2) of surface give good corrosion protection. In paints, zinc powder is easily wetted by oils. It keeps the zinc oxide in suspension and hardens the film. **Mossy zinc,** used to obtain color effects on face brick, is a spangly zinc powder made by pouring the molten metal into water. **Feathered zinc** is a fine grade of mossy zinc. **Photoengraving zinc** for printing plates is made from pure zinc with only a small amount of iron to reduce grain size and alloyed with not more than 0.2% each of cadmium, manganese, and magnesium. **Cathodic zinc,** used in the form of small bars or plates fastened to the hulls of ships or to underground pipelines to reduce electrolytic corrosion, is zinc of 99.99% purity with iron less than 0.0014 to prevent polarization. **Merrillite** is high-purity zinc dust.

ZINC ALLOYS. Alloys of zinc are mostly used for die castings for decorative parts and for functional parts where the load-bearing and shock requirements are relatively low. Since the zinc alloys can be cast easily in high-speed machines, producing parts that weigh less than brass and have high accuracy and smooth surfaces that require minimum machining and finishing, they are widely used for such parts as handles, and for gears, levers, pawls, and other small parts. Zinc alloys for sheet contain only small amounts of alloying elements, with 92 to 98% zinc, and the sheet is generally referred to simply as zinc or by a trade name. The **modified zinc sheet** is used for stamped, drawn, or spun parts for costume jewelry and electronics, and it contains up to 1.5% copper and 0.5 titanium. The titanium raises the recrystallization temperature, permitting heat treatment without coarse-grain formation.

Hartzink had 5% iron and 2 to 3 lead, but iron forms various chemical compounds with zinc and the alloy is hard and brittle. Copper reduces the brittleness. **Germania bearing bronze** contained 1% iron, 10 tin, about 5 each of copper and lead, and the balance zinc. **Fenton's alloy** had 14% tin, 6 copper, and 80 zinc; and **Ehrhard's bearing metal** contained 2.5% aluminum, 10 copper, 1 lead, and a small amount of tin to form copper-tin crystals. **Binding metal,** for wire-rope slings, has about 2.8% tin, 3.7 antimony, and the balance zinc. **Pattern metal,** for casting gates of small patterns, was almost any brass with more zinc and some lead added, but is now standard die-casting metal.

Zinc alloys are commonly used for die castings, and the zinc used is **high-purity zinc** known as **special high-grade zinc. ASTM AG40A (SAE 903)** is the most widely used; others include **AC41A (SAE 925), Alloy 7,** and **ILZRO 16.** All typically contain about 4% aluminum, small amounts of copper and very small amounts of magnesium. AG40A has a density of 0.24 lb/ft^3 (6,643 kg/m^3), an electrical conductivity 27% that of copper, a thermal conductivity of 65 Btu/(ft · h · °F) [113 W/(m · K)], an ultimate tensile strength of 41,000 lb/in^2 (283 MPa), and a Brinell hardness of 82. AC41A is stronger [48,000 lb/in^2 (331 MPa)] and harder (Brinell 91), a trifle less electrically and heat-conductive, and similar in density. The alloys have much greater unnotched Charpy impact strength than either die-cast aluminum or magnesium alloys, but are not especially heat-resistant, losing about one-third of their strength at temperatures above about 200°F (93°C). Both alloys have found wide use for auto and appliance parts, especially chromium-plated parts, as well as for office equipment parts, hardware, locks, toys, and novelties. Alloy 7 is noted primarily for its better castability and the smoother surface finish it provides. It is as strong as AG40A, though slightly less hard, and more ductile. ILZRO 16 is not nearly as strong [33,000 lb/in^2 (228 MPa)], but more creep-resistant at room and elevated temperatures.

The most recent casting alloys are three **high-aluminum zinc casting alloys** for sand and permanent-mold casting: **ZA-8, ZA-12,** and **ZA-27,** the numerals in the designations indicating approximate aluminum content. They also contain more copper than AG40A and AC41A, from 0.5 to 1.2% in ZA-12 to 2 to 2.5 in ZA-27, and a bit less magnesium. As sand-cast, ultimate tensile strengths range from 36,000 to 40,000 lb/in^2 (248 to 276 MPa) for ZA-8 and 58,000 to 64,000 lb/in^2 (400 to 441 MPa) for ZA-27. Unlike the common die-casting alloys, the ZA alloys also exhibit clearly defined tensile yield strengths: from 28,000 lb/in^2 (193 MPa) minimum for sand-cast ZA-8 to 53,000 lb/in^2 (365 MPa) for sand-cast ZA-27. Tensile modulus is roughly 12×10^6 lb/in^2 (83,000 MPa). Also, because of their greater aluminum content, they are lighter in weight than the die casting alloys. **Zinc-copper-aluminum alloys** developed at General Motors and designated **ACuZinc alloys,** are noted for high tensile strength and superior creep resistance. ACuZinc 5, with 5% copper and 3 aluminum, has a tensile strength of 59,000 lb/in^2 (407 MPa). ACuZinc 10, with 10% copper and 3.5 aluminum, has a creep strength of 8,000 lb/in^2 (55 MPa) at 120°F (49°C) for 0.2% creep in 10,000 h.

Manganese-zinc alloys, with up to 25% manganese, for high-strength extrusions and forgings, are really 60–40 brass with part of the copper replaced by an equal amount of manganese, and are classified with manganese bronze. They have a bright white color and are corrosion-resistant. **Zam metal,** for zinc-plating anodes, is zinc with small percentages of aluminum and mercury to stabilize against acid attack. **CorroBan,** of Pure Coatings Inc., is an electrolytically deposited coating of 82 to 89% zinc, balance nickel, which resists corrosion as well as cadmium plating. **Zinc solders** are used for joining aluminum. The **tin-zinc solders** have 70 to 80% tin, about 1.5 aluminum, and the balance zinc. The working range is 500 to 590°F (260 to 310°C). **Zinc-cadmium solder** has about 60% zinc and 40 cadmium. The pasty range is between 510 and 599°F (266 and 315°C).

A group of wrought alloys, called **superplastic zinc alloys,** have elongations of up to 2,500% in the annealed condition. These alloys contain about 22% aluminum. One grade can be annealed and air-cooled to a strength of 71,000 lb/in^2 (490 MPa). Parts made of these alloys have been produced by vacuum forming and by a compression molding technique similar to forging but requiring lower pressures.

ZINC CHEMICALS. With the exception of the oxide, the quantities of zinc compounds consumed are not large compared with many other metals, but zinc chemicals have a very wide range of use, being essential in almost all industries and for the maintenance of animal and vegetable life. Zinc is a complex element and can provide some unusual conditions in alloys and chemicals.

Zinc oxide, ZnO, is a white, water-insoluble, refractory powder melting at about 3587°F (1975°C), having a specific gravity of 5.66. It is much used as a pigment and accelerator in paints and rubbers. Its high refractive index, about 2.01, absorption of ultraviolet light, and fine particle size give high hiding power in paints, and make it also useful in such products as cosmetic creams to protect against sunburn. Commercial zinc oxide is always white, and in the paint industry is also called **zinc white** and **Chinese white.** But with a small excess of zinc atoms in the crystals, obtained by heat treatment, the color is brown to red.

In paints, zinc oxide is not as whitening as lithopone, but it resists the action of ultraviolet rays and is not affected by sulfur atmospheres, and is thus valued in outside paints. **Leaded zinc oxide,** consisting of zinc oxide and basic lead sulfate, is used in paints, but for use in rubber the oxide must be free of lead. The lead-free variety is also called **French process zinc oxide. Canfelzo** is one such product, from Pigment & Chemical Corp. In insulating compounds zinc oxide improves electrical resistance. In paper coatings it gives opacity and improves the finish. **Zinc-white paste** for paint mixing usually has 90% oxide and 10 oil. **Zinc oxide stabilizers,** composed of zinc oxides and other chemicals, can be added to plastic molding compounds to reduce the deteriorating effects of sunlight and other types of degrading atmospheres.

Zinc oxide crystals are used for transducers and other piezolectric devices. The crystals are hexagonal and are effective at elevated temperatures, as the crystal has no phase change up to its disassociation point. The resistivity range is 0.2 to 3.9 $\Omega \cdot$ in (0.5 to 10 $\Omega \cdot$ cm). Zinc oxide has luminescent and light-sensitive properties which are utilized in phosphors and ferrites. But the oxygen-dominated zinc phosphors used for radar and television are modifications of zinc sulfide phosphors. The **zinc sulfide phosphors** which produce luminescence by exposure to light are made with zinc sulfide mixed with about 2% sodium chloride and 0.005 copper, manganese, or other activator, and fired in a nonoxidizing atmosphere. The cubic crystal structure of zinc sulfide changes to a stable hexagonal structure at 1868°F (1020°C), but both forms have the phosphor properties. Thin films and crystals of **zinc selenide** with purities of 99.999% are used for photo- or electroluminescent devices. Zinc selenide is also used for optical lenses in CO_2 laser systems. **Zinc sulfide** is a white powder of composition $ZnS \cdot H_2O$, and is also used as a paint pigment, for whitening rubber, and for paper coating. **Cryptone** is zinc sulfide for pigment use in various grades, some grades containing barium sulfate, calcium sulfide, or titanium dioxide. Multilayer coatings of zinc sulfide and yttria protect zinc sulfide infrared sensor windows of missiles and military aircraft from harsh flight environments.

Zinc is an **amphoteric element,** having both acid and basic properties, and it combines with fatty acids to form metallic soaps, or with the alkali metals or with ammonia to form **zincates. Sodium zincate** is used for waterproofing asbestos-cement shingles. **Zinc stearate,** $ZN(C_{18}H_{35}O_2)_2$, is a **zinc soap** in the form of a fine, white powder used in paints and in rubber. A USP grade of 325 mesh is used in cosmetics. **Aquazinc** and **Liquizinc,** of Rubba, Inc., are zinc stearate dispersions in water used as an antitack agent in milling rubber. **Zinc acetate,** $Zn(C_2H_3O_2)_2$, is a white solid partly soluble in water, used as a mordant, as a wood preservative, in porcelain glazes, and as a mild antiseptic in pharmaceuticals.

Zinc sulfate, $ZnSO_4 \cdot 7H_2O$, is the chief material for supplying zinc in fertilizers, agricultural sprays, and animal feeds. For these purposes it is used in the form of white vitriol containing 22% zinc, or as the monohydrate, $ZnSO_4 \cdot H_2O$, containing 37% zinc. **Zink Gro** is a water-soluble grade for dry-blended fertilizers for correction of zinc deficiencies. It is from Eagle-Picher Industries, Inc. **Zinc chloride,** a white, crystalline, water-soluble powder, $ZNCl_2$, was formerly an important preservative for wood, and railway crossties treated with the material were called **Burnettized wood.** But it is highly soluble and leaches out of the wood, and is now chromated and copperized with sodium bichromate and cupric chloride. **Copperized CZC,** of Koppers Co., Inc., for treating wood against rot and termites, is copperized chromated zinc chloride. zinc chloride is also used for vulcanizing fiber, as a mordant, in mercerizing cotton, in dry batteries, in disinfecting, and in making many chemicals. **Spirits of salts** and **butter of zinc** are old names for the material.

Zinc chromate, used chiefly as a pigment and called **zinc yellow** and **buttercup yellow,** is stable to light and in sulfur atmospheres, but has a lower tinting strength than chrome yellow, although it is less subject to staining and discoloration. It is a crystalline powder of specific gravity 3.40. It is only slightly soluble in water, but will absorb 24 lb (11 kg) of linseed oil per 100 lb (45 kg). Zinc chromates are made by reacting zinc oxide with chromate solutions, and they may vary; but the usual composition is $4ZnO \cdot 4CrO_3 \cdot K_2O \cdot 3H_2O$. **Zinc bichromate,** $ZnCr_2O_7$, is an orange-yellow pigment. The **zinc peroxide** used in dental pastes and cosmetics as a mild antiseptic is a white powder, ZnO_2, containing 8.5% active oxygen. Organic salts of zinc that have achieved commercial prominence are **zinc naphthenate** and **zinc pyrithione.** The former is available in 6 and 8% grades for prevention of wood rot and decay, in solvent- and water-dispersible formulations. **Nap-All** and **M-Gards** are from Mooney Chemicals, Inc., and **Zinclear** is from Standard Tar Products Co. Olin Corp.'s **Zinc Omadine,** a zinc pyrithione, is employed as an antidandruff agent, for preserving cosmetics, in metalworking fluids, and as an antimicrobial on textiles.

Fluidized zinc titanate (FZT) can serve as a sorbent to remove 99% of the sulfur dioxide in power plants using sulfur-containing coal. In a process developed at Research Triangle Institute with the U.S. Department of Energy, the sorbent can be continuously recirculated and the sulfur absorbed recovered from the regenerator off-gas. Use of the sorbent is an alternative to cooling the coal gas to remove sulfur, then having to reheat it to produce electricity.

ZINC ORES. The metal zinc is obtained from a large number of ores, but the average zinc content of the ores in the United States is only about 3%, so that they are concentrated to contain 35 to 65% before treatment. The sulfide ores are marketed on the basis of 60% zinc content, and the oxide ores on the basis of 40% zinc content. **Sphalerite,** or **zinc blende,** is the most important ore and is found in quantities in Missouri and surrounding states and in Europe. Sphalerite is a zinc sulfide, ZnS, containing theoretically 67% zinc. It has a massive crystalline or granular structure and a Mohs hardness of about 4. When pure, its color is white; it colors yellow, brown, green, to black with impurities. The ores from New York State are round and concentrated by flotation to an average of 58% zinc and 32 sulfur, which is then concentrated by roasting to 68 zinc and 1 sulfur. It is then sintered to remove lead and cadmium and finally smelted with coke, and the zinc vapor condensed. The Silesian zinc blende, known as **wurtzite,** contains 15% zinc, 2 lead, and some cadmium.

Calamine is found in New Jersey, Pennsylvania, Missouri, and Europe. It is the ore that was formerly mixed directly with copper for making brass. The ore usually contains only about 3% zinc, and is concentrated to 35 to 45%, and then roasted and distilled. Calamine is **zinc silicate,** $2ZnO \cdot SiO_2 \cdot H_2O$. It is a mineral occurring in crystal groups of a vitreous luster, and it may be white, greenish, yellow, or brown. The specific gravity is 3.4, and Mohs hardness 4.5 to 5. It occurs in Arkansas with **smithsonite,** a **zinc carbonate** ore, $ZnCO_3$. **Franklinite** is an ore of both the metals zinc and manganese. Its approximate composition is $(FeZnMo)O \cdot (FeMn)_2O_3$, but it shows wide variation in the proportions of the different elements. It is found in the zinc deposits of New Jersey. The zinc is converted into zinc white, and the residue is smelted to form spiegeleisen. The mineral franklinite occurs in massive granular structure with a metallic luster and an iron-black color.

The ore **zincite** is used chiefly for the production of the zinc oxide known as zinc white employed as a pigment. Zincite has the composition ZnO, containing theoretically 80.3% zinc. The mineral has usually a massive granular structure with a deep-red to orange streaked color. It may be translucent or almost opaque. Deep-red specimens from the workings at Franklin, New Jersey, are cut into gemstones for costume jewelry. **Willemite** is an anhydrous silicate, Zn_2SiO_4, containing theoretically 58.5% zinc. When manganese replaces part of the zinc, the ore is called **troostite.** It is in hexagonal prisms of white, yellow, green, or blue; manganese makes it apple-green, brown, or red. The specific gravity is about 4 and Mohs hardness 5.5. The crushed ore is used in making fluorescent glass. The ore is widely dispersed in the United States.

ZIRCONIA. A white, crystalline powder which is **zirconium oxide,** ZrO_2, with a specific gravity of 5.7, Mohs hardness 6.5, and refractive index 2.2. When pure, its melting point is about 5000°F (2760°C), and it is one of the most refractory of the ceramics. It is produced by reacting zircon sand and dolomite at 2500°F (1371°C) and leaching out the silicates. The material is used as fused or sintered ceramics and for crucibles and furnace bricks. From 4.5 to 6% of CaO or other oxide is added to convert the unstable monoclinic crystal to the stable cubic form with a lowered melting point.

Fused zirconia, used as a refractory ceramic, has a melting point of 4620°F (2549°C) and a usable temperature to 4450°F (2454°C). The **Zinnorite** fused zirconia of Norton Co. is a powder that contains less than 0.8% silica and has a melting point of 4900°F (2704°C). A **sintered zirconia** can have a specific gravity of 5.4, a tensile strength of 12,000 lb/in² (83 MPa), compressive strength of 200,000 lb/in² (1,379 MPa), and Knoop hardness of 1,100. **Zircoa B** is stabilized cubic zirconia used for making ceramics. **Zircoa A** is the pure monoclinic zirconia used as a pigment, as a catalyst, in glass, and as an opacifier in ceramic coatings.

Zirconia brick for lining electric furnaces has no more than 94% zirconia, with up to 5 calcium oxide as a stabilizer, and some silica. It melts at about 4300°F (2371°C), but softens at about 3600°F (1982°C). The **IBC 4200** brick of Ipsen Industries, Inc., is zirconia with calcium and hafnium oxides for stabilizing. It withstands temperatures to 4200°F (2316°C) in oxidizing atmospheres and to 3000°F (1849°C) in reducing atmospheres. **Zirconia foam** is marketed in bricks and shapes for thermal insulation. With a porosity of 75% it has a flexural strength above 500 lb/in² (3 MPa) and a compressive strength above 100 lb/in² (0.7 MPa). For use in crucibles, zirconia is insoluble in most metals except the alkali metals and titanium. It is resistant to most oxides, but with silica it forms $ZrSiO_4$, and with titania it forms $ZrTiO_4$. Since structural disintegration of zirconia refractories comes from crystal alteration, the phase changes are important considerations. The monoclinic material, with a specific gravity of 5.7, is stable to 1850°F (1010°C) and then inverts to the tetragonal crystal with a specific gravity of 6.1 and volume change of 7%. It reverts when the temperature again drops below 1850°F (1010°C). The cubic material, with a specific gravity of 5.55, is stable at all temperatures to the melting point, which is not above 4800°F (2649°C) because of the contained stabilizers. A lime-stabilized zirconia refractory with a tensile strength of 20,000 lb/in^2 (138 MPa) has a tensile strength of 10,000 lb/in^2 (69 MPa) at 2370°F (1299°C). **Stabilized zirconia** has a very low coefficient of expansion, and white-hot parts can be plunged into cold water without breaking. The thermal conductivity is only about one-third that of magnesia. It is also resistant to acids and alkalies and is a good electrical insulator. **Diamond Z** refers to a line of "unbreakable" buttons made of zirconia, fired at 3200°F (1760°C), polished and coated to look like ivory. Developed by Adolph Coors Co., they are sold by ACX Technologies for high-priced shirts.

Toughening mechanisms, by which a crack in a ceramic can be arrested, complement processing techniques that seek to eliminate crack-initiating imperfections. **Transformation toughening** relies on a change in crystal structure (from tetragonal to monoclinic) that zirconia or zirconium dioxide (ZrO_2) grains undergo when they are subjected to stresses at a crack tip. Because the monoclinic grains have a slightly larger volume, they can "squeeze" a crack shut as they expand in the course of transformation. Due to ZrO_2's transformation-toughening abilities, which impart higher fracture toughness, research interest in engine applications has been high. In order for ZrO_2 to be used in high-temperature, structural applications, it must be stabilized or partially stabilized to prevent a monoclinic-tetragonal phase change. Stabilization involves the addition of calcia, magnesia, or yttria followed by some form of heat treatment. **PSZ ceramic,** the toughest known ceramic, is being investigated for diesel-engine applications.

A new zirconia ceramic being developed is **tetragonal zirconia polycrystal (TZP)** doped with Y_2O_3. Designated **Y-TZP**, it has the most impressive room-temperature mechanical properties of any zirconia ceramic. The commercial applications of TZP zirconia include scissors having TZP blades suitable for industrial use for cutting tough fiber fabrics, e.g., Kevlar, cables, and ceramic scalpels for surgical applications. One unique application is fish knives. The knife blades are Y-TZP and can be used when the delicate taste of raw fish would be tainted by slicing with metal-blade knives. Tungsten-carbide-reinforced Y-TZP, developed by Toray Industries and Nippon Tungsten Co. of Japan, has five times the thermal conductivity of Y-TZP and high hardness, strength, toughness, and heat resistance.

Magnesia-stabilized PSZ, Mg-PSZ, is fired at a higher temperature than Y-TZP and, thus, develops a larger grain size: 1,970 to 3,940 μin (50 to 100 μm) versus 11.8 to 31.5 μin (0.3 to 0.8 μm). Consequently, Mg-PSZ is slightly porous while Y-TZP is virtually free of porosity. However, this porosity does not affect its sealing behavior in valve applications. Mg-PSZ is not as strong as Y-TZP, but it is slightly tougher and, thus, more resistant to erosion by particle impingement. Also, Mg-PSZ has not exhibited susceptibility to low-temperature degradation in warm, moist environments even with just trace amounts of water vapor, which has limited Y-TZP to moisture-free valve applications.

Another zirconia ceramic–developed material is **zirconia-toughened alumina (ZTA). ZTA zirconia** is a composite polycrystalline ceramic containing ZrO_2 as a dispersed phase (typically about 15 volume %). Close control of initial starting-powder sizes and sintering schedules is thus necessary in order to attain the desired ZrO_2 particle dimensions in the finished ceramic. Hence the mechanical properties of the composite ZTA ceramics limit current commercial applications to cutting tools and ceramic scissors.

PSZ is also finding application in the transformation toughening of metals used in the glass industry as orifices for glass fiber drawing. This material is being termed **zirconia grain-stabilized (ZGS) platinum.**

Zirconia is produced from the zirconium ores known as **zircon** and **baddeleyite.** The latter is a natural zirconium oxide, but is obtainable commercially only from Minas Gerais, Brazil. It is also called **zirkite** and **Brazilite.** Zircon is **zirconium silicate,** $ZrO_2 \cdot SiO_2$, and comes chiefly from beach sands. The commercial sand is found in Florida, Brazil, India, Sri Lanka, Australia, and western Africa. The sands are also called **zirkelite** and **zirconite,** or merely **zircon sand.** The white zircon sand from India has a zirconia content of 62% and contains less than 1% iron. Beach sands of New South Wales are naturally concentrated to an average of 74% zircon, but Australian zircon is shipped on a basis of 65% zirconia. Zircon sand may be used directly for making firebricks, as an opacifier in ceramics, and for mold facings. Clear **zircon crystals** are valued as gemstones since the high refractive index gives great brilliance. The colorless natural crystals are called **Matura diamonds,** and the yellow-red are known as **jacinth.**

Zirconia fiber, used for high-temperature textiles, is produced from zirconia with about 5% lime for stabilization. The fiber is polycrystalline, has a melting point of 4700°F (2593°C), and withstands continuous temperatures above 3000°F (1649°C). These fibers are produced by Union Carbide as small as 118 to 394 μin (3 to 10 μm) and are made into fabrics for filter and fuel cell use. **Zirconia fab-**

rics are woven, knitted, or felted of short-length fibers and are flexible. **Ultratemp adhesive,** of Aremco Products, for high-heat applications, is **zirconia powder** in solution. At 1100°F (593°C) it adheres strongly to metals and withstands temperatures to 4400°F (2427°C). **Zircar,** of Union Carbide, is zirconia fiber compressed into sheets to a density of 20 lb/ft^3 (320 kg/m^3). It withstands temperatures up to 4500°F (2482°C) and has low thermal conductivity. It is used for insulation and for high-temperature filtering.

ZIRCONIUM.   A silvery-white metal, symbol Zr, having a specific gravity of 6.5 and melting at about 3362°F (1850°C). It is more abundant than nickel, but is difficult to reduce to metallic form as it combines easily with oxygen, nitrogen, carbon, and silicon. The metal is obtained from zircon sand by reacting with carbon and then converting to the tetrachloride, which is reduced to a sponge metal for further production of shapes. The ordinary **sponge zirconium** contains about 2.5% hafnium, which is closely related and difficult to separate. The commercial metal usually contains hafnium, but reactor-grade zirconium, for use in atomic work, is hafnium-free.

Commercially pure zirconium is not a high-strength metal, having a tensile strength of about 32,000 lb/in^2 (221 MPa), elongation 40%, and Brinell hardness 30, or about the same physical properties as pure iron. Because of its low neutron-capture cross section, thermal stability, and corrosion resistance, it is the standard metal for fuel-rod cladding and core components in nuclear reactors. It is employed mostly in the form of alloys but may be had in 99.99% pure single-crystal rods, sheets, foil, and wire for superconductors, surgical implants, and vacuum-tube parts. The neutron cross section of zirconium is 0.18 barn, compared with 2.4 for iron and 4.5 for nickel. The cold-worked metal, with 50% reduction, has a tensile strength of about 82,000 lb/in^2 (565 MPa), with elongation of 18% and Brinell hardness of 95. The unalloyed metal is difficult to roll and is usually worked at temperatures to 900°F (482°C). Though nontoxic, the metal is pyrophoric because of its heat-generating reaction with oxygen, necessitating special precautions in handling powder and fine chips resulting from machining operations.

The metal has a close-packed hexagonal crystal structure, which changes at 1583°F (862°C) to a body-centered cubic structure which is stable to the melting point. At 572 to 752°F (300 to 400°C) the metal absorbs hydrogen rapidly, and above 392°F (200°C) it picks up oxygen. At about 752°F it picks up nitrogen, and at 1472°F (800°C) the absorption is rapid, increasing the volume and embrittling the metal. The metal is not attacked by nitric (except red fuming nitric), sulfuric, or hydrochloric acids, but is dissolved by hydrofluoric acid.

Zirconium powder is very reactive, and for making sintered metals it is usually marketed as **zirconium hydride,** ZrH_2, containing about 2% hydrogen which is driven off when the powder is heated to 300°C. For making sintered parts, alloyed powders are also used. **Zirconium copper,** containing 35% zirconium, **zirconium nickel,** with 35 to 50% zirconium, and **zirconium cobalt,** with 50% zirconium, are marketed as powders of 200 to 300 mesh.

Small amounts of zirconium are used in many steels. It is a powerful deoxidizer, removes the nitrogen, and combines with the sulfur, reducing hot-shortness and giving ductility. **Zirconium steels** with small amounts of residual zirconium have a fine grain and are shock-resistant and fatigue-resistant. In amounts above 0.15% the zirconium forms zirconium sulfide and improves the cutting quality of the steel. **Zirconium alloys** generally have only small amounts of alloying elements to add strength and resist hydrogen pickup. **Zircoloy 2,** for reactor structural parts, has 1.5% tin, 0.12 iron, 0.10 chromium, 0.05 nickel, and the balance zirconium. Tensile strength is 68,000 lb/in² (469 MPa), elongation 37%, and Rockwell B hardness 89; at 600°F (316°C) it retains a strength of 30,000 lb/in² (207 MPa).

Small amounts of zirconium in copper give age-hardening and increase the tensile strength. Copper alloys containing even small amounts of zirconium are called **zirconium bronze.** They pour more easily than bronzes with titanium, and they have good electrical conductivity. Zirconium-copper master alloy for adding zirconium to brasses and bronzes is marketed in grades with 12.5 and 35% zirconium. A **nickel-zirconium** master alloy has 40 to 50% nickel, 25 to 30 zirconium, 10 aluminum, and up to 10 silicon and 5 iron. **Zirconium-ferrosilicon,** for alloying with steel, contains 9 to 12% zirconium, 40 to 47 silicon, 40 to 45 iron, and 0.20 maximum carbon, but other compositions are available for special uses. **SMZ alloy,** for making high-strength cast irons without leaving residual zirconium in the iron, has about 75% silicon, 7 manganese, 7 zirconium, and the balance iron. A typical zirconium copper for electrical use is **Amzirc.** It is oxygen-free copper with only 0.15% zirconium added. At 752°F (400°C) it has an electrical conductivity of 37% of electrolytic-tough-pitch copper (C11100), tensile strength of 52,000 lb/in² (359 MPa), and elongation of 9%. The softening temperature is 1076°F (580°C).

Zirconium carbide, ZrC_2, is produced by heating zirconia with carbon at about 3632°F (2000°C). The cubic crystalline powder has a hardness of Knoop 2,090 and melting point of 6404°F (3540°C). The powder is used as an abrasive and for hot-pressing into heat-resistant and abrasion-resistant parts. **Zirconium oxychloride,** $ZrOCl_2 \cdot 8H_2O$, is a cream-colored powder soluble in water that is used as a catalyst, in the manufacture of color lakes, and in textile coatings.

Zirconium-fused salt, used to refine aluminum and magnesium, is **zirconium tetrachloride,** a hygroscopic solid with 86% $ZrCl_4$. **Zirconium sulfate,** $Zr(SO_4)_2 \cdot 4H_2O$, comes in fine, white, water-soluble crystals. It is used in high-temperature lubricants, as a protein precipitant, and for tanning to produce white leathers. **Soluble zirconium** is **sodium zirconium sulfate,** used for the precipitation of proteins, as a stabilizer for pigments, and as an opacifier in paper. **Zirconium carbonate** is used in ointments for poison ivy, as the zirconium combines with the hydroxy groups of the urushiol poison and neutralizes it. **Zirconium hydride** has been used as a neutron moderator, although the energy moderation may be chiefly from the hydrogen.

    Zirconium alloys with high zirconium content have few uses except for atomic applications. **Zircoloy tubing** is used to contain the uranium oxide fuel pellets in reactors since the zirconium does not have grain growth and deterioration from radiation. **Zirconia ceramics** are valued for electrical and high-temperature parts and refractory coatings. **Zirconium-oxide powder,** for flame-sprayed coatings, comes in either hexagonal or cubic crystal form. **Zirconium silicate,** $ZrSi_2$, comes as a tetragonal crystal powder. Its melting point is about 3000°F (1649°C) and Knoop hardness is about 1,000.

Part 2

Structure and Properties of Materials

The Structure of Matter

Elements, or **atoms,** are the basic building blocks of all tangible materials in the universe. There are 92 natural elements, or material atoms, almost all of which are stable, from hydrogen, atomic number 1, or element 1, to uranium, or element 92. Elements of higher atomic weight than uranium are made, but they are unstable, their time decay being measured progressively as half-life.

The atom gets its name from the Greek word *atomos,* meaning indivisible, and it is not divisible by ordinary chemical means. The elements are used either alone or in combination for making useful products. They combine either as mechanical mixtures or as **chemical compounds.** In a **mixture** each element retains its original nature and energy, and the constituents of the mixture can be separated by mechanical means. In chemical compounds of two or more elements, the original elements lose their separate identities; the new substance formed has entirely different properties, and the atomic energy stored within the compound is not equal to the sum of the elemental energies. The atoms in chemical compounds are bonded by electrons. An **alloy** is usually a combination of chemical compounds and mixtures, the metal mixtures in the matrix being gaged by their maximum fused or liquid solubility, known as the **eutectic point.** With the elements the number of different compounds, or useful substances, that can be made by varying the combinations of elements and the proportions is infinite.

The known atoms are arranged progressively in a **periodic table** by **atomic number,** based on the **atomic weight** of the element with hydrogen as the unit of mass, though oxygen may be taken as the point of calculation. The atom is not a solid, but a region of energy particles in motion. At various energy levels the geometric shape of the electron orbit changes, and the apparent ring, or **electron shell** structure, is the energy-level extension of the orbital pattern. The distances and space covered are so vast in relation to the size of the particle, and the speeds are so great, that the interior of the atom might be considered mostly empty space. As a single atom is a billion or more times the size of an electron, it is estimated that if the space within the atom could be removed, a thimbleful of atoms would weigh millions of tons. If the copper atom were magnified 10 billion times, the electrons that the chemist employs to connect it with another atom of a molecule would still be too tiny to be seen. Thus, a solid metal used for construction is a region of relatively vast space populated by energy particles in perpetual motion.

The term **space chemistry** was first used at the beginning of the twentieth century by the Dutch physicist van't Hoff, the founder of modern physical chemistry, but the subject was not new. It may be said that modern atomic science, equipped with advanced experimental methods and testing instruments, has taken up where the Greeks, working only with geometry and the theoretical deductions of metaphysics, left off at their School of Numbers about 450 B.C. The Greeks reasoned that all matter came from one source, made from a qualitatively indeterminable primordial unit, the **monad,** now known as **energy.** It was stated to be incorporeal, but vital and always in motion. This idea of a nonmaterial basis of tangible materials, now necessary for modern scientific analysis of materials, is intrinsic in human logic. It came to the Greeks from the Ionians, survivors of the Cretan civilization antedating 3000 B.C., and appears in the Hebraic Genesis, in the Sanskrit Vedas, and in the Taoism of ancient China. Energy is in harmonic motion, in waves or rays, and may be said to become a particle of **mass** when the frequency is 1, that is, a closed unit cycle. All materials give off light when activated, and **light rays** have the fastest known speed, 186,000 mi/s (300,000 m/s).

More than 70 new elements, to element 168, have been projected, though not all have been synthesized. These are higher elements made by additions to natural elements. Atoms may also be broken down by the application of high energy. The process known as **fission** is usually by electric energy built up to extremely high voltage by resonant pulsation in a magnetic field in a manner akin to that of the generation of lightning in the clouds. More than 30 subatomic particles have been isolated. **Fissionable elements** are normally considered to be only those of high atomic weight and radioactivity, and relative unstability, but all elements are fissionable.

A subatomic unit may be considered as both a wave and a **particle.** The **nucleus** of the atom is a relative term. The **proton** is identical to the nucleus of the hydrogen atom, and is one unit of positive electricity. The nuclei of all other elements consist of combinations of protons and neutrons. The electrons of the various atoms appear to orbit around the nucleus, but the electron, though considered a negatively charged particle, is also a **beta ray,** and the axis of its vortex motion is in calculable relativity to the respective positron. A detached positron has only a momentary existence. In conjunction with an electron, it forms an atomlike structure known as positronium. A spheron may contain one or more neutrons, and atoms having different numbers of neutrons are called **isotopes** and are of different atomic weights and different physical properties.

The **helium atom** of mass 4, positive charge 2, and zero valence has two protons and two neutrons, with the protons apparently in opposite polarity. This combination is called an **alpha particle.** Alpha particles are emitted at high velocity from radioactive elements, expelling the detached electrons, and when captured are deposited as helium. These usually come from outer-ring spherons and not necessarily from the inner nucleus. The expelled electrons are beta rays. When these collide with a nucleus, high-frequency **X-rays** break off. **Gamma rays** are emitted from some radioactive elements. The difference between X-rays and gamma rays is their origin and wavelength. Gamma rays come from the nucleus; X-rays come from electrons striking matter. Few of the high-energy X-rays coming from the sun penetrate the atmosphere.

Gamma rays from the sun come only in infrequent bursts, and the **cosmic rays** from space are also entirely protons, or stripped ions of hydrogen. Cosmic rays appear to travel at about the speed of light. **Mesons** from cosmic rays appear to carry unit charges as beta rays do, but they have more energy and greater range. While beta rays are stopped in human skin, mesons can cause damage throughout the body. High-energy cosmic rays are stopped by the atmosphere, and only a small proportion penetrate to the earth's surface.

The **neutron** is a particle of neutral charge with a mass approximately that of a proton. A neutron has a mass 1,838 times that of an electron, while a proton has a mass 1,836 times that of an electron. High-energy bombardment of nuclei or an individual nucleus yields electron positrons, mesons, and neutrinos. In recent work, these seemingly fundamental particles have been subdivided into quarks and gluons.

In the technology of producing and processing materials, the atom is not subdivided, although in some operations of electrochemistry and electronics the electron is detached, and particles and rays are also employed, especially for activation. With respect to combining elements, **metallurgy** is high-energy chemistry. In a solid metal, as in other materials, the atom does not appear alone, and the physical properties of a metal or alloy derive chiefly from the molecular structure.

Elements having one, two, or three outside valence electrons are **metals.** In chemical reactions they can release these electrons and form positive metal ions. The elements having five, six, or seven outer electrons are **nonmetals.** An element with four outer electrons is a **semimetal** and can react as either a metal or a nonmetal. An element with eight outer electrons is said to have **zero valence** and is normally inactive, but by special energy application, or catalyzation, the linkage of the spherons can be broken and the electrons freed for chemical reactions.

The elements that make up all the planets and the stellar systems of the universe appear to be the same as those of the earth. There are many theories for the original formation of the material elements, but the subject pertains to astronomy rather than to materials technology, and involves the mathematics of progressive assembly of energy waves into monoquantic vortices which constitute mass. While elements do not have life in the same sense as the term is used for animals and plants, they do have intrinsic habits that can be controlled and altered by changing the environmental conditions. Elements are gregarious, and atoms separate only when activated by extremes of energy, as with high heat, and they tend to congregate even when dissipated in water or air.

Elements have orderly, calculable habits of combining into **molecules,** or geometrically shaped units bonded to their own kind or to atoms of other elements. Compounding the elements into useful materials is done by the addition or subtraction of energy with consideration of time and space. Even the automatic reactions of two elements in proximity, known as **chemical affinity,** and the seeming holding action of **stabilizing agents** depend upon a transfer of energy.

The term **crystal** is usually applied only to molecular structures which at normal temperatures are hard solids that form into pronounced geometric shapes or are capable of being split on precise planes. Solids without apparent planes are termed **amorphous.** But the crystal shapes tabulated for metals usually represent merely the typical position pattern of the atoms. **Single crystals** may be cut from natural crystals, grown by flame melting, or grown chemically by application of heat and pressure. **Seed crystals** used to initiate growth are grains or particles made up of many molecules, while a **unit crystal** is the unit molecule or, in some cases, the unit pattern of the lattice, and these determine the shape and nature of the structure. In microscopy the structures of aluminum and silver appear optically as similar cubes, but the unit crystal of aluminum in the solid state forms both a cube and a lattice, while the unit crystal of silver forms no cube and does not lattice, and the metal grains are cryptocrystalline. Usually, the smaller the grain size, the nearer the approach to the physical properties of the single crystal so that large single crystals are sometimes made by compacting extremely fine powders.

Quasicrystalline solids are a category of matter intermediate between crystals and amorphous materials, such as glasses. Termed **quasicrystals,** they consist of atoms in ordered arrays, but the patterns they form do not recur at precisely regular intervals.

All elements convert progressively from solid to gaseous form by the application of energy, usually by heat application, and vice versa by the extraction of heat. The terms **solid, liquid,** and **gas** are **phase changes** depending on the mobility of the molecule caused by changes in its three-dimensional shape. A **gaseous element** is one that is a gas at ordinary temperatures and pressures, such as hydrogen. At extremely low temperatures a **hydrogen crystal** should be a hard, white metal of cryptocrystalline structure with straight planes of cleavage.

Liquid hydrogen for rocket fuel normally has a molecule of conical shape in spin. When catalyzed by hot platinum, it changes to an ovaloid shape which can pass through a smaller molecular sieve, and it also requires 20% less storage space per unit of fuel. These forms are called **ortho hydrogen** and **meta hydrogen,** but are both H_2.

Phase changes often occur within the solid stage, and the change in dimensions of the material, called **creep,** is the effect from change in volume of the molecules. With some materials the liquid stage is so short as to be undetectable, appearing to pass directly from solid to vapor, and this transition is called **sublimination.** All molecules have energy transition points at which they break down to free the original elements or to interact and combine with other available elements to form new compounds. For example, iron molecules, having free electrons, disintegrate easily in the presence of air or moisture to form iron oxides. This process is called *corrosion*; in organic materials it is called *decay*. The molecule of gold has no free electrons and, because of its high energy, is not broken down easily by the influence of other elements. Thus it is said to be noncorrosive. In the case of aluminum, oxygen from air cross-links the free electrons on the surface of the grains and protects the metal from further corrosion.

In metallurgy and the metalworking industries, the elements are normally not used alone in a pure state, and as solids and liquids only in molecular forms. In casting metals and alloys from a melt, the time of solidification is short, and without the application of high energy, as in the form of high pressure, there is no growth into large single crystals. Growth is usually into particles, or grains, which may be single crystals or irregular conglomerates of unit crystals. In the contraction of cooling, however, grain boundaries may be so close as to be undetectable even at a magnification of 2 million to 1. Thus the impurities are likely to be in the unmatched open spaces among the crystals and not interstitial. But with some latticing molecules, such as copper, there is room within the lattice for smaller atoms or molecules, such as those of beryllium, without interference with the paths of bonding electrons. In the aluminum lattice there appears to be no such room.

Organic and other chemicals are usually produced from the elements by **synthesis,** that is, built up by progressive steps logically deduced from known data and theories concerning the natural habits and characteristics of the atoms and their elementary groups. A compound may thus be written as a **chemical formula** which expresses graphically the specific number and locations of the atomic elements in the compound. In some degree this system is also used in the production of **ceramics,** i.e., compounds or compound mixtures based on metallic oxides, where the resultant material is expressed in percentage proportions of the crystal formulas. Alloys are usually made by

The Natural Elements

Name	Atomic number	Symbol	Atomic weight $O = 16.000$	Melting point, °C
Actinium	89	Ac	......	1800
Aluminum	13	Al	26.97	660.0
Antimony	51	Sb	121.76	630.5
Argon	18	A	39.944	−189.3
Arsenic	33	As	74.91	814
Astatine	85	At	......	470
Barium	56	Ba	137.36	704
Beryllium	4	Be	9.02	1280
Bismuth	83	Bi	209.00	271.3
Boron	5	B	10.82	2300
Bromine	35	Br	79.916	−7.2
Cadmium	48	Cd	112.41	320.9
Calcium	20	Ca	40.08	850
Carbon	6	C	12.00	3700
Cerium	58	Ce	140.13	600
Cesium	55	Cs	132.91	28
Chlorine	17	Cl	35.457	−101
Chromium	24	Cr	52.01	1800
Cobalt	27	Co	58.94	1490
Columbium (niobium)	41	Cb	92.91	2000
Copper	29	Cu	63.57	1083.0
Dysprosium	66	Dy	162.46	
Erbium	68	Er	167.64	
Europium	63	Eu	152.0	
Fluorine	9	Fl	19.00	−223
Francium	87	Vi		
Gadolinium	64	Gd	157.3	
Gallium	31	Ga	69.72	29.78
Germanium	32	Ge	72.60	958
Gold	79	Au	197.2	1063.0
Hafnium	72	Hf	178.6	1700
Helium	2	He	4.002	−271.4
Holmium	67	Ho	163.5	
Hydrogen	1	H	1.0078	−259.2
Illinium	61	Il	140.0	
Indium	49	In	114.76	156.4
Iodine	53	I	126.92	114
Iridium	77	Ir	193.1	2454
Iron	26	Fe	55.84	1535
Krypton	36	Kr	83.7	−157
Lanthanum	57	La	138.92	826
Lead	82	Pb	207.22	327.4
Lithium	3	Li	6.940	186
Lutecium	71	Lu	175.0	

batch-mixing the elements, and the resultant material is expressed in weight percentages of the contained elements, not in terms of the molecular structure on which the physical properties of the alloy depend.

The Natural Elements (*Continued*)

Name	Atomic number	Symbol	Atomic weight $O = 16.000$	Melting point, °C
Magnesium	12	Mg	24.32	650
Manganese	25	Mn	54.93	1260
Mercury	80	Hg	200.61	−38.87
Molybdenum	42	Mo	96.0	2625
Neodymium	60	Nd	144.27	840
Neon	10	Ne	20.183	−248.6
Nickel	28	Ni	58.69	1455
Nitrogen	7	N	14.008	−210.0
Osmium	76	Os	191.5	2700
Oxygen	8	O	16.0000	−218.8
Palladium	46	Pd	106.7	1554
Phosphorous	15	P	31.02	44.1
Platinum	78	Pt	195.23	1773.5
Polonium	84	Po	......	1800
Potassium	19	K	39.096	63
Praseodymium	59	Pr	140.92	940
Protoactinium	91	Pa	231	
Radium	88	Ra	226.05	700
Radon	86	Rn	222	−71
Rhenium	75	Re	186.31	3000
Rhodium	45	Rh	102.91	1966
Rubidium	37	Rb	84.44	39
Ruthenium	44	Ru	101.7	2450
Samarium	62	Sm	105.43	1300
Scandium	21	Sc	45.10	1200
Selenium	34	Se	78.96	220
Silicon	14	Si	28.06	1430
Silver	47	Ag	107.880	960.5
Sodium	11	Na	22.997	97.7
Strontium	38	Sr	87.63	770
Sulfur	16	S	32.06	119.2
Tantalum	73	Ta	180.88	3000
Technetium	43	Ma	97.8	2300
Tellurium	52	Te	127.61	450
Terbium	65	Tb	159.2	327
Thallium	81	Tl	204.39	300
Thorium	90	Th	232.12	1700
Thulium	69	Tm	169.4	
Tin	50	Sn	118.70	231.9
Titanium	22	Ti	47.90	1820
Tungsten	74	W	184.0	3410
Uranium	92	U	238.14	1850
Vanadium	23	V	50.95	1735
Xenon	54	Xe	131.3	−112
Ytterbium	70	Yb	173.04	1500
Yttrium	39	Y	88.92	1490
Zinc	30	Zn	65.38	419.5
Zirconium	40	Zr	91.22	1700

Waves and Colors as Material Elements

Electromagnetic radiations

Tangible materials and **radiations** have a common energy origin, and thus bear a cosmic relation, but radiation is not matter in the ordinary sense of the term. Radiation is caused by vibrations, and is characterized by wavelengths rather than mass as is ordinary matter. Waves of high frequency and short wavelength result from the vibration of extremely small particles, such as electrons of the material atom, while those of low frequency and long wavelength arise from slow vibrations, such as those from a coil in a magnetic field.

Radiations are produced when materials are broken down or changed to another form, and there is then an actual loss of mass equal to the amount of energy emitted. In reverse, matter is produced when energy in the form of radiation is directed upon matter, and an actual increase in the mass of the matter results. All materials in nature are being constantly bombarded with various radiations, but it requires such an extremely large amount of energy to produce the most minute quantities of matter that the continuous changes in most materials are not noticeable in any historic period of time.

The spectrum of electromagnetic radiations extends from wavelengths of many hundred-millionths of a centimeter, or infinitely small, to wavelengths of many kilometers, or infinitely large. The velocity of these waves is the same for all lengths of wave, 186,000 mi/s (300,000 km/s). In the spectrum, the light waves which make objects visible to the human eye form only a small part. The human eye can see through only such materials as these light waves will penetrate. But electrical eyes can be made to operate in other wavelengths and record vision not seen by the human eye. Not all animals see with the same wavelengths, and some animals do not have normal eyes but receive vibrations through special receiving parts of the body. Different materials transmit, absorb, or reflect radiations differently. Quartz and glass, normally called transparent, transmit only a small band of light and heat waves, but will not pass very short radiations. By changing the composition of the glass the heat waves can be blocked, or some of the very short waves can be passed through. Some materials, like lead, will block the very short waves, and can be used for X-ray shields. Other materials, like beryllium, will pass only very short waves, and can be used for selective windows.

Silver will reflect 90% of visible light, while tin reflects only 70%, but silver loses reflectivity in sulfur atmospheres. Gold reflects only 61% of visible light, but has high reflectivity of infrared rays, useful for electronic purposes. All materials are sensitive to particular light waves and emit electrons when struck by those waves. Zinc is sensi-

Spectrum of Electromagnetic Radiations

Uses:

A-Radiology, radiotherapy, x-ray
   diagnosis

B-Physical and chemical analysis
   of matter

C-Therapeutic applications

D-Light

E-Heating, industrial baking

F-FM, radar, television, sonar

G-Short wave radio

H-Wireless

I-Long wave radio

J-Electric power, alternating
   current

Explanation:

The chart is logarithmic, the distance x from 10^9 to 10^{10} being a
billion times the distance y from 1 to 10

$10^1 = 10$	$10^3 = 1000$	$10^{-1} = 0.1$	$10^{-3} = 0.001$
$10^2 = 100$	$10^4 = 10,000$	$10^{-2} = 0.01$	$10^{-4} = 0.0001$

Wavelengths and frequencies. (*From J. F. Moulton, Jr.*)

tive to very short ultraviolet light; cesium is sensitive to green light;
potassium is sensitive to blue light. This property is the basis of electronic color selectors. It is also the basis for the operation of photoelectric cells, in which the liberated electrons constitute an electric
current. Such cells are widely used as automatic switches and for
electronic conversion of light intensities to sound waves.

Element colors at incandescence

Flame colorations caused by heating materials to incandescence indicate the presence of certain elements, as the light from each element in burning has a predominance of rays or wavelengths that are characteristic of that particular element. Some elements, such as sodium, show a distinct bright color because of a predominance of wavelengths within that color range in the visible spectrum, while others show pale or intermediate colors difficult to distinguish, usually because the rays have no predominating wavelength within the visi-

Predominant Flame Colors of Materials

Element	Color	Element	Color
Lithium	Deep red	Antimony	Blue-green
Strontium	Crimson	Copper	Green-blue
Calcium	Yellow-red	Arsenic	Light blue
Sodium	Bright yellow	Lead	Light blue
Barium	Yellow-green	Selenium	Blue
Molybdenum	Green-yellow	Indium	Deep blue
Zinc	Light green	Potassium	Purple-red
Boron	Green	Rubidium	Violet
Tellurium	Deep green	Cesium	Bluish purple
Thallium	Greenish blue		

Reflecting Powers of Various Metal Surfaces

	White light directly reflected, percent		Color Silver = 0
Silver	90		0
Chromium	61	Blue-green	12 units
Nickel	50	Red	16 units
Stainless steel	49	Blue-green	3 units
White bronze speculum	70	Red	1 unit

Reflecting Power of Various Colors in Paints

Color	Light reflection, percent	Color	Light reflection, percent
Flat white	85–89	Sky blue	58
Bone white	69–70	Light orchid	57
Canary yellow	68–72	Buff	47
Light ivory	70	Pea green	40
Aluminum	70	Tan	34
Cream	65–69	Peacock blue	34
Light green	66	Steel gray	30
Ivory	61–63	Brown	9
Peach	58–59		

ble spectrum but are mixtures of many wavelengths. Other elements, such as iron, have a predominance of rays that are not in the visible band, with wavelengths shorter or longer than those visible to the eye. Flame coloration is used in metallurgical laboratories to determine the content of alloys by burning small pieces and studying the light with a refractive prism. This property of the elements is also utilized in making carbon electrodes for electric-arc lights to give the full white light of sunshine, or short waves for therapy or industrial use, or long wavelengths for heat. For example, carbon alone gives a predominance of short wavelengths with the visible rays predominantly on the red side of the spectrum. When cerium metals are blended with the carbon, the visible light is balanced with the blue-violet to give a more even, white light. When the carbon is blended with iron, nickel, and aluminum, which are all on the low-wavelength side of the spectrum, lower-zone ultraviolet rays are obtained.

Terms used in material color designation

Hue is the predominant light wavelength reflected by the coloring material, and it determines the **color** designation.

Brightness, or **value,** is the percentage of light reflected. A brilliant white approaches 100% and a jet black approaches 0%. Black is the absence of light waves; white is a combination of all the various wavelengths. White light is broken down by refraction into separate wave bands, or hues, as in the natural rainbow. **Chroma** refers to the **intensity** of a color. **Tint** refers to color modified toward white, **shade** to one toward black.

The color circle is composed of 12 colors spaced at equal intervals: yellow, orange, red, violet, blue, green, etc., with intermediates between each. **Pigment colors** are obtained usually by **subtractive** mixing; for example, when blue and yellow are mixed, the blue absorbs the red, orange, and yellow rays, and the yellow absorbs the blue and violet rays, and so the resulting color is green.

Under proper illumination it is possible to detect with the eye exceedingly slight color differences, the number of distinguishable colors being estimated, by the U.S. Bureau of Standards, at 10,000,000.

Colors or hues vary slightly with different batches of paints, dyes, etc. For this reason products that must be matched exactly in hue are usually finished from the same batch or lot. Color matching of metals is also often important. For example, for installation of kitchens or other building equipment the stainless steel should preferably be from one lot since the color shades vary with the proportions of chromium, nickel, or manganese. These are "white" metals, but chromium has a blue tone, nickel has a yellow tone, and manganese

Angstrom units Logarithmic scale

1000 to 4000 — Ultraviolet
4000 to 7800 — Visible light
7800 to 120000 — Infrared

2000
1000

3000
2200 to 3200 — Therapeutic ultraviolet

4000
3200 to 6500 — Longer wave ultraviolet and visible

Violet
Blue
Green
Yellow
Orange
Red

5000
6000
7000
8000
9000
10000

6500 to 14000 — Penetrating visible and infrared

20000
30000
40000
50000
60000
70000
80000
90000
100000

14000 to 120000 — Nonpenetrating heat

Angstrom units

Zone of therapeutic ultraviolet

2200
2300
2400
2500
2600
2700
2800
2900
3000
3100
3200

Debated antirachitic rays — 2200 to 2700

Antirachitic (Steenbock) — 2500 to 3020

Preferred antirachitic (Heilbron-Morton-Kamm) — 2700 to 3000

Tan-producing (Rungling) — 3000 to 3200

Vital rays (Goodman) — 2900 to 3100

Erythema-producing rays (Haussen and Vahle) — 2800 to 3100

Heat and light rays from the carbon arc. Controlled rays are obtained with arc carbons by varying the core content of the carbon. With cerium metals in the core, a light approximating sunlight is obtained. Iron in the core gives only one-quarter the visible light of the plain carbon with the same current and voltage, but it gives strong ultraviolet rays. A carbon containing iron, nickel, and aluminum gives powerful ultraviolet rays between 250 and 302 nm. A carbon with strontium in the core gives penetrating infrared heat rays above 650 nm. (*Chart from National Carbon Co.*)

Metal melting range and color scale. (*Chart from the Linde Air Products Co.*)

has a purple tone. Welding alloys and solders are also matched to the color of the base metal by varying the proportions of metals with different tints.

Visibility at a distance varies with different colors. Red can be seen and recognized at long distances while blue can be seen at only short distances. The order of visibility of colors at a distance is red, green, white, yellow, blue. The legibility of a color, however, also varies with the background. Black on yellow is more legible than black on white, whereas green, red, or blue on white is more legible than black on white. Visibility and legibility are important in signs, packages, or products that must be distinguished easily.

Harmony of color or **tone design** is a complicated art. It comprises the color relationship to convey pleasing emotional reaction, and includes various terms. A **rich color** is a hue at its fullest intensity. A **warm color** is one in which the red-orange predominates. A **cool color** is one in which the blue-green predominates. In general, warm colors are pleasing or exciting, while cool colors are not so pleasing or are restful to the senses. A **receding color** is one giving the illusion of withdrawing into distance by a gradation toward another tone or hue. Color, from the standpoint of harmony and design, is a sensation effect. It is not inherent in the pigments, dyes, or other materials, but is the sensation effect from those light rays reflected to the eye by the material.

Property of Flavor in Materials

The quality of many materials is judged by the flavor. Flavor is the resultant of three senses: taste, smell, and feeling. Some materials, such as salt and quinine, may be detected by taste alone. Some, such as coffee and butter, depend largely upon smell. Flower perfumes are detected by smell alone. The flavor of fruits depends upon both taste and smell, and without smell it would be only sour or sweet. Pepper has little or no taste, but is detected by the aroma and by the sense of feeling.

The four standard components of taste are: sweet, sour, salt, and bitter. Taste buds are located in the tongue. The tip, back, and edges of the tongue can detect all four sensations, but the center of the tongue can detect only a sour taste, and the surrounding area can detect only salty and sour. **Sour taste** is caused by hydrogen ions, and **salty taste** is due to cations from the alkali metals, accented when anions from the halogens are present. **Sweetness** and **bitterness** may or may not be from ions, and the stimuli are more complex. But all taste is electrochemical, translated to the nerves as sensations. Some materials when injected into the blood can be tasted when the blood reaches the tongue.

Odor detection is electrochemical but not entirely so, since molecules which have the same shape may have the same odor though be unrelated chemically. The sense of smell is due to oscillations of the valence electrons in the molecules of the substance. The molecules of substances inhaled stimulate the tiny olfactory hairs high in the nasal cavity, and the effect is translated to the nerves as impressions or sensations. The sense of taste usually requires considerable material to register the sensation, but only the most minute molecular qualities are required to register smell. A normal person can detect a vast variety of odors, but for convenience the four fundamental odors have been designated as fragrant, acid, burnt, and caprylic. A material is then designated with a four-digit number to indicate the degree of each odor, each odor thus having 10 degrees or variations, thus given 9,999 variants. Various materials are taken as standards for the numbers, or 40 standards. The sense of smell is so discriminating that it can detect separate odors in highly complicated mixtures. Most odors are mixtures, and the art involved in the perfumery industry is to form harmonies that give a resultant pleasant sensation.

Flavors that affect the touch sensation are described as pungent, sharp, acrid, and cool. These are caused by actual pain as the biting of an acid or the cooling effect of deoxidation. **Greasiness** and **oiliness** are sensations of feel that affect taste but are not a part of it. **Texture** also is a feeling sense and not a part of flavor. The sensation of puckery of the tannin of some fruits is a definite constriction of membrane and is not taste. All of these have an effect upon the desirability of the material as a food, but in a manner apart from flavor. Too much sweetness, sourness, or saltiness will clog the taste buds, and they must be then rested before a true flavor can be detected, but the recovery is rapid. Temperature also has an effect, and true flavors are detected only at about the temperature of the body. The judgment and grading of coffee, tea, butter, etc., are done solely by the senses of experts in comparison with standards.

Fundamentals of Biotic Materials

The **biotics** constitute an extensive group of organic materials that are actual living microorganisms and cannot be expressed as chemical formulas. All of these minute bits of living matter are of plant and animal origin and may be considered as chemical factories. It is the chemicals that certain species secrete under certain conditions that make them industrially and medicinally useful. The chemical secretions are enzymatic and catalytic in character and thus enable various chemicals to react on contact. In industry they are used as chemical activating agents, as ferments, as leavening agents, and in various processing. In medicine they are known as **antibiotics** and are used to destroy the biotics, or bacteria, of diseases which locate themselves in the human body.

Biotics are found everywhere in myriad quantities. A cubic centimeter of raw earth may contain as many as 50,000 **fungi,** or **microphytes,** 500 million bacteria, and 250 million **actinomycetes,** the latter being living organisms that may be ascribed, with reservations, to either the plant or the animal kingdom, and are distinguished by their mass of long, silky filaments. All these organisms lead a jungle-like existence, attacking everything, decomposing plant debris to make humus, liberating nitrogen from proteins, liberating oxygen from rocks, liberating carbon dioxide and water from organic acids in plants and soil, and also preying on one another as jungle animals do. Without these organisms the life cycles of all living things could not be maintained. This same teeming population also furnishes the individual types of organisms that aid humanity in medicine and industry.

The number of species of these microorganisms is innumerable. Each species has at least one enemy species that it destroys on contact or by which it is itself destroyed. Each apparently has a certain definite range of activity, beyond the bounds of which it is useless. Where one biotic is used in the manufacture of a certain product, a similar but different species used under the same conditions produces an entirely different product, or may affect the quantitative yield of the product desired. The formation of ethyl alcohol by the fermentation of starch or sugar is caused by a biotic which is then itself killed when the alcohol produced has arrived at a certain concentration. Other biotics may be killed by the heat that their own work produces, as at the "crisis" point of certain fevers. In general, biotics can withstand excessive cold but are usually killed by relatively low heats.

Besides the use of biotics for the medicinal and industrial applications that are now known, there are believed to be enormous possibilities for their use in large-scale chemical processing in the future. But the isolation of microorganisms is tedious laboratory work involving the extraction of a pure strain from cultures containing many species, and once separated, the proper conditions to promote rapid multiplication must be discovered. Even at this stage, a biotic found to be useful in the manufacture of a certain product may simultaneously manufacture another unwanted product, difficult and costly to separate from the desired one. But from each biotic some kind of chemical is secreted, and that chemical is the ultimate end of biotic research.

Units of Measure

Useful conversion factors

1 acre = 43,560 square feet = 0.40469 hectare

1 nanometer = 0.001 micrometer = 0.03937 millionths of an inch

1 ardeb (Egypt) = 5.44 bushels

1 arshin (Russia) = 28 inches

1 barrel (U.S.), cement = 376 pounds

1 barrel, oils = 42 gallons

1 berkovets (Russia) = 361.13 pounds

1 board foot = 144 cubic inches

1 bolt, cloth = 40 yards = 36.576 meters

1 buncal (Indonesia) = 1.49 troy ounces

1 bushel = 2,150.4 cubic inches

1 bushel, imperial (British) = 1.0315 U.S. bushels

1 candy (India) = 784 pounds

1 carat, metric = 0.200 gram

1 carat (U.S.) = 0.2056 gram

1 chittak (India) = 900 grains

1 circular mill = 0.0000007845 square inch

1 cuarteron (Spain), oil = 0.133 liquid quart

1 cuartillo (Mexico), liquid = 0.482 liquid quart

1 cubic foot = 1,728 cubic inches

1 cuffisco (Sicily), oil = 5.6 gallons

1 dram = 1.7718 grams

1 dinero (Spain) = 18.5 grains

1 drachma (Turkey) = 49.5 grains

1 ell = 48 inches

1 feddan (Egypt) = 1.038 acres

1 firkin = 9 U.S. gallons = 34.068 liters

1 flask, mercury = 75 pounds = 34.02 kilograms

1 foot = 12 inches = 0.3048 meter

1 gallon = 231 cubic inches

1 gallon, imperial (British) = 1.20094 U.S. gallons

1 gallon, proof (British) = 1.37 U.S. proof gallons

1 gill = 0.25 pint = 0.118292 liter

1 grain = 0.06480 gram

1 gram = 15.43 grains = 0.03527 avoirdupois ounce

1 gram (Libya) = 165.3 pounds

1 hamlah (Egypt) = 165.1 pounds

1 hectare = 2.471044 acres

1 hogshead = 63 U.S. gallons

1 hundred weight (British) = 112 pounds

1 inch = 0.0833 foot = 2.54005 centimeters

1 iron, leather thickness measure = $\frac{1}{48}$ inch

1 kantar (Egypt) = 99.034 pounds

1 keel (British), coal = 21.2 long tons

1 kilogram = 2.205 pounds

1 koku (Japan) = 47.65 gallons = 5.119 bushels

1 kun (Korea) = 1.323 pounds

1 kwan (Japan) = 1,000 momme = 8.267 pounds

1 ligne, metal button measure = $\frac{1}{40}$ inch

1 liter = 1.057 liquid quarts

1 lug (Bahamas) = 30 pounds avoirdupois

1 meter, square = 1.196 square yards

1 mil, square = 0.000001 square inch

1 micrometer = 0.001 millimeter = 0.00003937 inch

1 mil = 0.001 inch = 0.0254 millimeter

1 mile = 5,280 feet = 1.69035 kilometers

1 mile, square = 640 acres

1 millimeter = 0.03937 inch

1 ounce, avoirdupois = 28.35 grams = 0.0625 pound

1 ounce, troy = 31.1 grams

1 peck = 0.25 bushel = 8.8096 liters

1 perch = 1 square rod = 30.25 square yards

1 picul (China) = 100 catties = 133.33 pounds

1 picul (Indonesia) = 136.2 pounds

1 picul (Japan) = 132.3 pounds

1 pint, dry measure = 33.6 cubic inches

1 pint, liquid measure = 28.875 cubic inches

1 pood (Russia) = 36.11 pounds

1 pound, avoirdupois = 16 ounces = 7,000 grains = 0.4536 kilogram

1 pound, troy = 12 ounces = 0.37324 kilogram

1 pound, Venetian = 1.058 avoirdupois pounds

1 quart, dry measure = 2 pints, dry = 1.1012 liters

1 quart, liquid = 57.749 cubic inches = 0.9463 liter

1 quintal (British) = 112 pounds

1 quintal, metric = 110 kilograms = 220.5 pounds

1 ream, paper measure = 500 sheets = 20 quires

1 rod = 5.5 yards

1 scruple, apothecary weight = 20 gains = 1.296 grams

1 shih tan (China) = 50 kilograms = 110.231 pounds

1 standard (British), timber = 1,980 board feet

1 standard (U.S.), timber = 16.67 cubic feet

1 tank (India), gemstones and pearls = 24 rati = 0.145 ounce

1 tierce, thin-staved cask = 42 gallons = 310 to 370 pounds

1 ton, long = 2,240 pounds = 1,016.047 kilograms

1 ton, metric = 1,000 kilograms = 0.9842 long ton = 1.102 short tons

1 ton, short = 2,000 pounds = 0.8929 long ton

1 vedro (Russia) = 3,249 gallons

1 yard = 3 feet = 0.9144 meter

Metric Length Measurements

Unit	Inches	Feet	Millimeters	Centimeters	Meters
One inch	1	0.0833	25.4	2.54	0.0254
One foot	12	1	304.8	30.48	0.3048
One millimeter	0.03937	0.00328	1	0.1	0.001
One centimeter	0.3937	0.0328	10	1	0.01
One meter	39.37	3.2809	1,000	100	1
One yard	36	3	914.4	91.44	0.9144

Standard Paper Sizes

Folio note	5.5 by 8.5 in (14.0 by 21.5 cm)
Pocket note	6 by 9.5 in (15.2 by 24.1 cm)
U.S. government writing	8 by 10.5 in (20.3 by 26.7 cm)
Commercial writing	8.5 by 11 in (21.6 by 27.9 cm)
Legal cap	8.5 by 14 in (21.6 by 35.6 cm)
Foolscap	12 by 16 in (30.5 by 40.6 cm)
Denny	16 by 21 in (40.6 by 53.3 cm)
Folio	17 by 22 in (43.2 by 55.9 cm)
Royal	19 by 24 in (48.3 by 70.0 cm)
Super royal	20 by 28 in (50.8 by 71.2 cm)
Elephant	23 by 28 in (58.4 by 71.2 cm)
Imperial	23 by 31 in (58.4 by 78.7 cm)

Temperature Conversion Scale

To change a temperature in degrees Celsius to degrees Fahrenheit, multiply by ⅝ and add 32, thus F = ⅝ (C) + 32. To change degrees Fahrenheit to degrees Celsius, subtract 32 and multiply by ⅝, thus, C = ⅝ (F − 32).

°C	°F	°C	°F	°C	°F	°C	°F	°C	°F
0	32	230	446	460	860	690	1274	920	1688
5	41	235	455	465	869	695	1283	925	1697
10	50	240	464	470	878	700	1292	930	1706
15	59	245	473	475	887	705	1301	935	1715
20	68	250	482	480	896	710	1310	940	1724
25	77	255	491	485	905	715	1319	945	1733
30	86	260	500	490	914	720	1328	950	1742
35	95	265	509	495	923	725	1337	955	1751
40	104	270	518	500	932	730	1346	960	1760
45	113	275	527	505	941	735	1355	965	1769
50	122	280	536	510	950	740	1364	970	1778
55	131	285	545	515	959	745	1373	975	1787
60	140	290	554	520	968	750	1382	980	1796
65	149	295	563	525	977	755	1391	985	1805
70	158	300	572	530	986	760	1400	990	1814
75	167	305	581	535	995	765	1409	995	1823
80	176	310	590	540	1004	770	1418	1000	1832
85	185	315	599	545	1013	775	1427	1005	1841
90	194	320	608	550	1022	780	1436	1010	1850
95	203	325	617	555	1031	785	1445	1015	1859
100	212	330	626	560	1040	790	1454	1020	1868
105	221	335	635	565	1049	795	1463	1025	1877
110	230	340	644	570	1058	800	1472	1030	1886
115	239	345	653	575	1067	805	1481	1035	1895
120	248	350	662	580	1076	810	1490	1040	1904
125	257	355	671	585	1085	815	1499	1045	1913
130	266	360	680	590	1094	820	1508	1050	1922
135	275	365	689	595	1103	825	1517	1055	1931
140	284	370	698	600	1112	830	1526	1060	1940
145	293	375	707	605	1121	835	1535	1065	1949
150	302	380	716	610	1130	840	1544	1070	1958
155	311	385	725	615	1139	845	1553	1075	1967
160	320	390	734	620	1148	850	1562	1080	1976
165	329	395	743	625	1157	855	1571	1085	1985
170	338	400	752	630	1166	860	1580	1090	1994
175	347	405	761	635	1175	865	1589	1095	2003
180	356	410	770	640	1184	870	1598	1100	2012
185	365	415	779	645	1193	875	1607	1105	2021
190	374	420	788	650	1202	880	1616	1110	2030
195	383	425	797	655	1211	885	1625	1115	2039
200	392	430	806	660	1220	890	1634	1120	2048
205	401	435	815	665	1229	895	1643	1125	2057
210	410	440	824	670	1238	900	1652	1130	2066
215	419	445	833	675	1247	905	1661	1135	2075
220	428	450	842	680	1256	910	1670	1140	2084
225	437	455	851	685	1265	915	1679	1145	2093

Hardness numbers

The Brinell method of determining hardness is by the indentation effect of a hard ball pressed into the surface of the metal to be tested. Tables of hardness numbers corresponding to the various indentation measurements are furnished by the makers.

The Scleroscope, or "Shore," method measures hardness by a comparison of the effect of the drop and rebound of a diamond-tipped hammer dropped from a fixed height. The resulting rebound is then read on a graduated scale.

The Rockwell hardness tester measures hardness by determining the depth of penetration under load of a steel ball or diamond cone in the material being tested. Rockwell hardness is expressed as a number, which is read on a graduated gage.

The Mohs hardness scale for abrasives and minerals is measured by scratch comparison, the mineral talc being taken as 1 and the diamond as 10 on the scale. This method is only an approximation for mineral comparison, and the Knoop indentor is used for measuring comparative hardness of hard materials.

The Vickers method is similar to the Brinell and Rockwell methods except that a diamond in the form of a pyramid is used as the penetrator. It is thus suitable for measuring metals of high hardness.

The Bierbaum microcharacter, or Bierbaum number, is used to determine the hardness by scratch. The width of a scratch made by drawing the point of a cube-shaped diamond across the surface under a 3-g load is measured with a microscope and determines the degree of hardness.

Index of refraction

Index of refraction indicates the relative amount of light transmitted by a material. As the index of refraction increases, the transmitted light decreases. The amount of light reflected may be considered as in inverse proportion to the amount transmitted, though much of the light may be dissipated. A vacuum transmits 100% of the light and reflects 0%, and has a refractive index of 1.00. A polished diamond with parallel sides will transmit only 83% of the light, and the reflected light makes the diamond shine. The sparkle of angle-cut diamonds and highly refractive cut glass is caused by the dissipated or deflected light emerging from the angles. (See the table on page 1023.)

Acidity and alkalinity scale

The degree of acidity or alkalinity of solutions is expressed by the pH value. Water is considered neutral and is given a pH value of 7.

Approximate Relationship of Vickers, Shore (Scleroscope), Rockwell, and Brinell Hardness Numbers

Vickers or Firth	Shore or Scleroscope	Rockwell C	Rockwell B	Brinell	Approximate tensile strength of steels, lb/in^2 MPa	Hardness class of steel
1,220	96	68	...	780		
1,114	94	67	...	745	329,000 to 380,000	
1,021	92	65	...	712	(2,268 to 2,620)	Hard to file
940	89	63	...	682		
867	86	62	...	653		
803	84	60	...	627		
746	81	58	...	601		
694	78	56	...	578		
649	75	55	...	555		
608	73	53	...	534		
587	71	51	...	514		
551	68	50	...	495		
534	66	48	...	477	165,000 to 317,000	Machining opera-
502	64	47	...	461	(1,138 to 2,186)	tions difficult
474	62	46	...	444		
460	60	44	...	429		
435	58	43	...	415		
423	56	42	...	401		
401	54	41	...	388		
390	52	39	...	375		
380	51	38	...	363		
361	49	37	...	352		
344	48	36	...	341		
335	46	35	...	331		
320	45	34	...	321		
312	43	32	...	311		
305	42	31	...	302		
291	41	30	...	293		
285	40	29	...	285		
278	38	28	...	277		
272	37	27	...	269		
261	36	26	...	262		
255	35	25	...	255		
250	34	24	100	248	78,000 to 159,000	Commercial
240	33	23	99	241	(538 to 1,096)	machining range
235	32	22	99	235		
226	32	21	98	229		
221	31	20	97	223		
217	30	18	96	217		
213	30	17	95	212		
209	29	16	95	207		
197	28	14	93	197		
186	27	12	91	187		
177	25	10	89	179		
171	24	8	87	170		
154	23	4	83	156		
144	21	0	79	143		

Index of Refraction

Material	Index	Material	Index
Diamond	2.42	Nylon	1.53
Ruby	1.80	Polyethylene	1.52
Sapphire (synthetic)	1.77	Pyrex (borosilicate glass)	1.52
Iceland spar	1.66	Window glass (soda-lime)	1.52
Flint glass (dense leaded)	1.66	Acrylic	1.50
Flint glass (dense barium)	1.62	Cellulose acetate	1.48
Vinylidene chloride	1.61	Cellulose acetate butyrate	1.47
Polystyrene	1.59	Ethyl cellulose	1.47
Flint glass (light leaded)	1.58	Fused quartz	1.46
Flint glass (light barium)	1.57	Water	1.33
Amber	1.55	Ice	1.30
Quartz crystal	1.54	Air	1.0003
Urea-formaldehyde	1.54	Vacuum	1.00

Acidity-Alkalinity Indicators

	pH range	
Meta cresol purple	1.2–2.8	Red to yellow
Thymol blue	1.2–2.8	Red to yellow
Bromophenol blue	3.0–4.6	Yellow to blue
Bromocresol green	4.0–5.6	Yellow to blue
Chlorophenol red	5.2–6.8	Yellow to red
Bromothymol blue	6.0–7.6	Yellow to blue
Phenol red	6.8–8.4	Yellow to red
Cresol red	7.2–8.8	Yellow to red
Thymol blue	8.0–9.6	Red to blue

Values below 7 are acid, each declining value being 10 times more acid than the previous value. A pH of 6 is 10 times more acid than a pH of 7, and a pH of 3 is 10,000 times more acid than a pH of 7. Solutions having values from 7 to 14 are alkaline by the same multiples of 10. A pH of 14 is 10 million times more alkaline than a pH of 7. Chemical indicators used to indicate the acidity or alkalinity of solutions are shown in the acidity-alkalinity table.

Viscosity of liquids

The **viscosity** of a liquid is its resistance to change in its form, or flow, caused by the internal friction of its particle components. Thus, the higher the viscosity, the less fluid it is. When a liquid is hot, there is less internal friction owing to the greater mobility and distance between the molecules, and a liquid will flow more readily than when it is cold. Thus, all the comparisons of viscosity should be at the same temperature. **Kinematic viscosity** is the ratio of viscosity to density. **Specific viscosity** is the ratio of the viscosity of any liquid to that of water at the same temperature. The reciprocal of viscosity is called **fluidity.** Viscosity is usually expressed in poises or centipoises, a poise being equal to 1 g/(cm · s).

The **specific gravity** of a liquid is the relative weight per unit volume of the liquid compared with the weight per unit volume of pure water. Water is arbitrarily assigned the value of 1.000 g/cm³. All liquids heavier than water thus have specific gravities greater than 1.000; liquids lighter than water have values less than 1.000. Usually, the specific gravity of a liquid is measured at 15°C or at room temperature. In practice, measurements are taken with a series of weighted and graduated glass cylinders called **hydrometers.** These float vertically, and the markings are usually in degrees Baumé.

Liquid	Viscosity, cp	Liquid	Viscosity, cp
Benzene (0°C)	0.906	Linseed oil (30°C)	33.1
Carbon tetrachloride (0°C)	1.35	Soybean oil (30°C)	40.6
Mercury (0°C)	1.68	Sperm oil (15°C)	42.0
Ethyl alcohol (0°C)	1.71	Sulfuric acid (0°C)	48.4
Water (0°C)	1.79	Castor oil (10°C)	2,420
Phenol (18°C)	12.7	Rape oil (0°C)	2,530

Color determination of lubricating oils

Color determination of lubricating oils and petroleum is made by comparison with standard colored disks. Light is dispersed through a 4-oz (0.1-kg) sample bottle of the oil to be tested, and the color is compared visually with the gelatin colors on the glass. The colors on the standard glass disks of the National Petroleum Association are as follows:

1	Lily white	4	Orange pale
1.5	Cream white	4.5	Pale
2	Extra pale	5	Light red
2.5	Extra lemon pale	6	Dark red
3	Lemon pale	7	Claret red
3.5	Extra orange pale		

Gasoline and fuel oil rating

The **cetane number** of a diesel fuel is numerically equal to the percentage by volume of cetane in a mixture of cetane and a-methylnaphthalene which will match the fuel in ignition quality. Cetane has the composition $CH_3(CH_2)_{14}CH_3$, and a-methylnaphthalene $CH_3 \cdot C_{10}H_7$. The cetane number of a fuel is given as the nearest whole number. Thus, if it required 49.8% of cetane in the mixture to match, the number would be 50.

The octane number of a fuel is the whole number nearest to the percentage by volume of isooctane, $(CH_3)_3C:CH_2CH(CH_3)_2$, in a blend of isooctane and normal heptane, $CH_3(CH_2)_5CH_3$, that the fuel matches in knock characteristics.

Phenol coefficient

Phenol is used as the standard for measuring the bacteria-killing power of all other disinfectants, and the relative bacteria-killing power is expressed as the **phenol coefficient.**

The phenol coefficient is the ratio of the dilution required to kill the Hopkins strain of typhoid bacillus in a specified time compared with the dilution of phenol required for the same organism in the same time. Usually, 2.5- and 15-minute time limits are used, and the coefficient is calculated from the average of the two. For example, if 1:80 and 1:110 dilutions of phenol kill in 2.5 and 15 minutes, respectively, as the necessary dilutions of the disinfectant under test are 1:375 and 1:650, then the phenol coefficient of the disinfectant is 5.3.

Physical and Mechanical Properties

Definitions of physical and chemical properties

Acid number. The weight in milligrams of potassium hydroxide required to neutralize the fatty acid in 1 g of fat or fatty oil.

Aliphatic. Having a straight, chainlike molecular structure.

Anhydrous. Having no water of crystallization in the molecule. A hydrated compound contains water of crystallization which can be driven off by heating.

Aromatic. Having a ringlike molecular structure.

Brittleness. The property of breaking without perceptible warning or without visible deformation.

Bursting strength. The measure of the ability of a material, usually in sheet form, to withstand hydrostatic pressure without rupture.

Compressibility. The extent to which a material, such as for gaskets, is compressed by a specified load. **Permanent set** is the unit amount, in percent, that the material fails to return to the original thickness when the load is removed. **Recovery** is the amount, in percent, of the return to original thickness in a given time, and is usually less under a prolonged load.

Conductivity. The rate at which a material conducts heat or electricity. Silver is the standard of reference, as it is the best of the known conductors.

Creep rate. The rate at which strain, or deformation, occurs in a material under stress or load. **Creep strength** is the maximum tensile or compressive strength that can be sustained by a material for a specified strain and time at a specified temperature. **Creep recovery** is a measure, in percent, of the decrease in strain, or deformation, when the load is removed.

Ductility.   The ability of a material to be permanently deformed by tension without rupture.

Elasticity.   The ability of a material to resume its original form after removal of the load which has produced a change in form. A substance is highly elastic if it is easily deformed and quickly recovers.

Elastic limit.   The greatest unit stress that a material is capable of withstanding without permanent deformation.

Elongation.   The increase in length of a bar or section under load, expressed as a percentage difference between the original length and the length at the moment of rupture or at a specific strain.

Factor of safety.   The ratio of the ultimate strength of a material to its working stress.

Fatigue strength.   The measure in pounds per square inch (megapascals) of the load-carrying ability without failure of a material subjected to a loading repeated a definite number of times. Fatigue strength is usually higher than the prolonged service tensile strength. **Fatigue life** is a measure of the useful life, or the number of cycles of loading, of a specified magnitude that can be withstood by a material without failure.

Flash point.   The minimum temperature at which a material or its vapor will ignite or explode.

Flow, or creep.   The gradual continuous distortion of a material under continued load, usually at high temperatures.

Fusibility.   The ease with which a material is melted.

Hardness.   A property applied to solids and very viscous liquids to indicate solidity and firmness in substance or outline. A hard substance does not readily receive an indentation.

Hygroscopic.   Readily absorbing and retaining moisture.

Impact strength.   The force in foot · pounds (joules) required to break a material when struck with a sudden blow.

Iodine value.   The number of grams of iodine absorbed by 100 g of fat or fatty oil. It gives a measure of the chemical unsaturation of an oil or fat. High iodine value, 117 to 206, in vegetable oils indicates suitability of the oil for use in paints. Low iodine value, not subject to oxidation, indicates nondrying quality suitable for soaps.

Malleability.   The property of being permanently deformed by compression without rupture, that is, the ability to be rolled or hammered into thin sheets.

Modulus of elasticity.   The ratio of the unit stress to unit strain in tension or compression within the elastic limit without fracture.

Modulus of rigidity.   The ratio of the unit stress to unit strain in shear or torsion within the elastic limit without fracture.

Plasticity.   The ability of a material to be permanently deformed at low load.

Porosity.   The ratio of the volume of the interstices of a material to the volume of its mass.

Reduction of area.   The percentage difference between the area of a bar before being subjected to stress and the area of the bar after rupture.

Resilience.   The energy of elasticity—the energy stored in a material under strain within its elastic limit which will cause it to resume its original shape when the stress is removed. The modulus of resilience is the capacity of a unit volume to store energy up to the elastic limit.

Saponification value.   The number of milligrams of potassium hydroxide required to saponify 1 g of fatty oil or grease.

Shrinkage.   The diminution in dimensions and mass of a material.

Softening point.   The Vicat softening point for thermoplastic materials is the temperature at which a flat-ended needle of 1-mm^2 area will penetrate a specimen to a depth of 1 mm under a load of 1,000 g when the temperature of the specimen is raised at a constant rate of 50°C/h.

Solubility.   Capacity for being dissolved in a liquid so that it will not separate out on standing, except the excess over the percentage which the liquid (solvent) will dissolve. A **suspension** is a physical dispersion of particles sufficiently large that physical forces control their dissolution in the liquid. A **colloidal solution** is a dispersion of particles so finely divided that surface phenomena and kinetic energy control their behavior in the liquid. A colloidal solution is close to a molecular combination.

Specific gravity.   The ratio of the weight of a given volume of a material to the weight of an equal volume of pure water at 4°C.

Specific heat.   The number of calories required to raise 1 g of a material 1°C in temperature.

Stiffness.   The ability of a material to resist deflection, as determined by its modulus.

Strain.   The distortion in a material by the action of an applied load.

Strength.   The ability of a material to resist applied loads.

Stress.   Force, or load, per unit area.

Tensile strength.   The maximum tensile load per square unit of original cross section that a material is able to withstand.

Thermal conductivity.   The number of calories transmitted per second between opposite faces of a cube, 1 cm by 1 cm by 1 cm, when the temperature difference between the opposite faces of the cube is 1°C.

Thermal expansion.   The coefficient of linear thermal expansion is the increase in unit length with each change of 1° in temperature.

Thermoplastic.   Capable of being molded and remolded without rupture by heat and pressure. When a material sets under heat and pressure into a hard solid not capable of being remolded, it is called **thermosetting**.

Toughness.   The ability of a material to resist impact, or absorb energy, without fracturing.

Ultimate strength.   The stress, calculated on the maximum applied load and the original area of cross section, which causes fracture of the material.

Yield point.   The minimum tensile stress required to produce continuous deformation in a solid material.

Specific Gravity of Materials

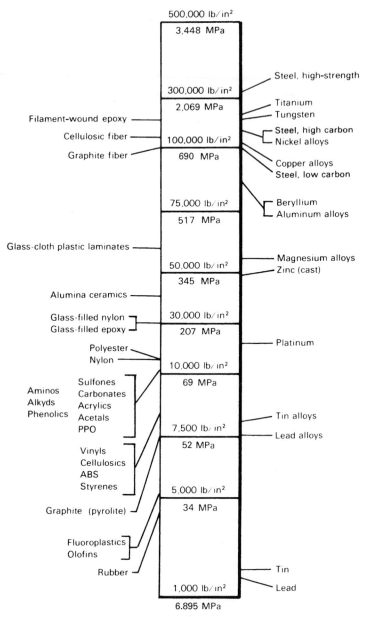

Ultimate Tensile Strength of Materials

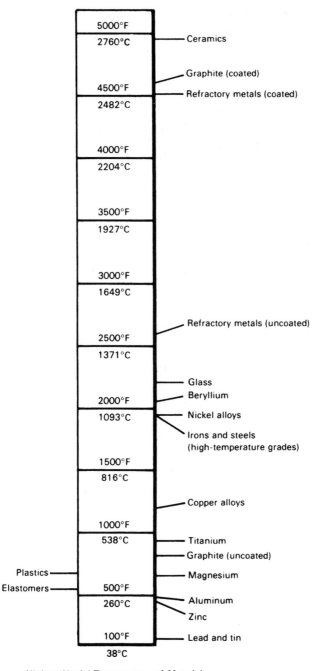

Highest Useful Temperature of Materials

Modulus of Elasticity in Tension of Materials

	lb/in²	MPa
Lead (cast)	700,000	4,827
Lead (hard-drawn)	1,000,000	6,895
Phenolic (fabric laminated)	1,000,000	6,895
Pine (static bending)	1,200,000	8,274
Ash (static bending)	1,300,000	8,964
Phenolic (paper base)	2,100,000	14,480
Tin (cast)	4,000,000	27,580
Tin (rolled)	5,700,000	39,300
Glass	8,000,000	55,160
Brass	9,000,000	62,100
Aluminum (cast)	10,000,000	68,950
Copper (cast)	11,000,000	75,850
Zinc (cast)	11,000,000	75,850
Zinc (rolled)	12,000,000	82,740
Cast iron (soft gray iron)	12,000,000	82,740
Brass (cast)	13,000,000	89,640
Bronze (average)	13,000,000	89,640
Phosphor bronze	13,000,000	89,640
Manganese bronze (cast)	14,000,000	96,530
Slate	14,000,000	96,530
Copper (soft, wrought)	15,000,000	103,400
Cast iron (average, with steel scrap)	16,000,000	110,300
Clock brass	16,600,000	114,500
Copper (hard-drawn)	18,000,000	124,100
Cast iron (hard, white iron)	20,000,000	137,900
Malleable iron	23,000,000	158,600
Wrought iron	27,000,000	186,200
Carbon steel	30,000,000	206,850
Alloy steel (nickel-chromium)	30,000,000	206,850
Nickel	30,000,000	206,850
Tungsten	60,000,000	413,700

Order of Ductility of Metals

1. Gold	6. Aluminum
2. Platinum	7. Nickel
3. Silver	8. Zinc
4. Iron	9. Tin
5. Copper	10. Lead

Melting and Welding Temperatures

	°F	°C
Direct electric arc	7232	4000°
Oxygen-acetylene torch	6332	3500°
Electric furnace	5432	3000°
Aluminum-iron oxide powder	5072	2800°
Combustion furnace	3092	1700°
Oxygen-hydrogen torch	2642	1450°
Plasma arc welding	59,432	33,000°
Electron-beam welding	>18,032	>10,000°
Laser-beam welding	>18,032	>10,000°

Mohs Hardness of Minerals

\multicolumn{2}{Original Mohs scale}		\multicolumn{2}{Modified Mohs scale}		
Hardness number	Mineral	Hardness number	Material	Bierbaum number
1	Talc	1	Talc	1
2	Gypsum	2	Gypsum	
3	Calcite	3	Calcite	15
4	Fluorite	4	Fluorite	
5	Apatite	5	Apatite	
6	Orthoclase	6	Orthoclase	
		7	Vitreous silica	
7	Quartz	8	Quartz or stellite	
8	Topaz	9	Topaz	
		10	Garnet	
		11	Fused zirconia	
9	Corundum	12	Fused alumina	
		13	Silicon carbide	
		14	Boron carbide	
10	Diamond	15	Diamond	10,000 (1-μm scratch)

Hardness Grades of Woods

1. Exceedingly hard	Lignum-vitae, ebony
2. Extremely hard	Boxwood, lilac, jarrah, karri
3. Very hard	Whitethorn, blackthorn, persimmon
4. Hard	Hornbeam, elder, yew
5. Rather hard	Ash, holly, plum, elm
6. Firm	Teak, chestnut, beech, walnut, apple, oak
7. Soft	Willow, deal, alder, Australian red cedar, birch, hazel
8. Very soft	White pine, poplar, redwood

Knoop Hardness of Hard Materials

Diamond	6,000 to 6,500
Boron carbide	2,250 to 2,260
Silicon carbide	2,130 to 2,140
Sapphire	1,600 to 2,100
Aluminum oxide (corundum)	1,635 to 1,680
Tungsten carbide (cobalt binder)	1,000 to 1,500
Spinel	1,200 to 1,400
Topaz	1,250
Quartz	710 to 790
Hardened steel	400 to 800
Glass	300 to 600

Comparative Hardness of Hard Abrasives

(Scale: Diamond 10, corundum 9)

South American brown bort	10.00
South American Ballas	9.99
Congo yellow (cubic crystals)	9.96
Congo clear white (cubic crystals)	9.95
Congo gray opaque (cubic crystals)	9.89
South American carbonadoes	9.82
Boron carbide	9.32
Black silicon carbide	9.15
Green silicon carbide	9.13
Tungsten carbide (13% cobalt)	9.09
Fused alumina (3.14% TiO_2)	9.06
Fused alumina	9.03
African crystal corundum	9.00
Rock-crystal quartz	8.94

Thermal Conductivity of Materials*

Conductivity measured in British thermal units transmitted per hour per square foot of material 1 in thick, per degree Fahrenheit difference in temperature of the two faces.

Silver	2,920.0	Diatomite block	0.58
Copper	2,588.0	Magnesia, 85%	0.51
Steel, 1.0 carbon	328.0	Wood pulp board	0.39
Building stone	12.50	Bagasse board	0.35
Slate, shingles	10.37	Cork, ground	0.31
Concrete, 1:2:4	6.10	Flax fiber	0.31
Glass, plate	5.53	Diatomite powder	0.308
Brickwork, mortar bond	4.00	Mineral wool	0.296
Gypsum plaster	2.32	Asbestos sheet	0.29
Brick, dry	1.21	Vermiculite	0.263
Airspace, 3.5 in	1.10	Wool	0.261
Pine wood	0.958	Hair felt	0.26
Clay tile	0.60	Cotton, compressed	0.206

*From Paul M. Tyler, U.S. Bureau of Mines.

Linear Expansion of Metals

Unit length increase per degree Celsius rise in temperature.

Cast iron	0.000010
Steel	0.000011
Cobalt	0.000012
Bismuth	0.000013
Gold	0.000014
Nickel	0.000014
Copper	0.000017
Brass	0.000019
Silver	0.000019
Tobin bronze	0.000021
Aluminum	0.000024
Zinc	0.000026
Tin	0.000027
Lead	0.000028
Cadmium	0.000029
Magnesium	0.000029

Melting Point of Materials Commonly Used for Heat-Treating Baths

Material	Melting point	
	°F	°C
35% lead / 65% tin	358	181
50% sodium nitrate / 50% potassium nitrate	424	218
Tin	450	232
Sodium nitrate	586	308
Lead	620	327
Potassium nitrate	642	339
45% sodium chloride / 55% sodium sulfate	1154	623
Sodium chloride (common salt)	1474	801
Sodium sulfate	1618	881
Barium chloride	1760	960

Forging Temperature of Steel

	Maximum forging temperature		Burning temperature	
	°F	°C	°F	°C
1.5% carbon steel	1920	1049	2080	1138
1.1% carbon steel	1980	1082	2140	1171
0.9% carbon steel	2050	1121	2230	1221
0.7% carbon steel	2140	1171	2340	1282
0.5% carbon steel	2280	1249	2460	1349
0.2% carbon steel	2410	1321	2680	1471
0.1% carbon steel	2460	1349	2710	1488
Silicomanganese spring steel	2280	1249	2460	1349
3% nickel steel	2280	1249	2500	1371
3% nickel-chromium steel	2280	1249	2500	1371
Air-hardening Ni-Cr steel	2280	1249	2500	1371
5% nickel (case-hardening) steel	2320	1271	2640	1449
Chromium-vanadium steel	2280	1249	2460	1349
High-speed steel	2370	1299	2520	1382
Stainless steel	2340	1282	2520	1382
Austenitic chromium-nickel steel	2370	1299	2590	1421

Relative values of electrical insulating materials

The usual comparisons of insulating values of materials are made on the basis of their dielectric strength. The **dielectric strength** of a material is the voltage that a material of a given thickness will resist. It is usually given in volts per mil (1 mil equals 0.001 in) or volts per meter. In any higher voltage the dielectric strength will permit a spark to pass through the material. The quoted dielectric strengths, however, are generally the minimum for the materials.

Material	Dielectric strength × 10^6 V/m	V/mil	Material	Dielectric strength × 10^6 V/m	V/mil
Mica, muscovite	39.4	1,000	Buna rubbers	20.3	515
Glass	35.5	900	Vinylidene chloride	19.7	500
Mica, phlogopite	31.5	800	Fish paper	19.7	500
Electrical porcelain	31.5	800	Methyl methacrylate	18.9	480
Steatite	29.6	750	Cellulose acetate	15.8	400
Hard rubber	27.6	700	Casein plastic	15.8	400
Silicone rubber	23.6	600	Shellac	15.8	400
Polystyrene	23.6	600	Varnished cambric	15.8	400
Pyroxylin	23.6	600			

Electrical Conductivity of Elements

Silver	100.00	Iron	14.57
Copper	97.61	Platinum	14.43
Gold	76.61	Tin	14.39
Aluminum	63.00	Tungsten	14.00
Tantalum	54.63	Osmium	13.98
Magnesium	39.44	Titanium	13.73
Sodium	31.98	Iridium	13.52
Beryllium	31.13	Ruthenium	13.22
Barium	30.61	Nickel	12.89
Zinc	29.57	Rhodium	12.60
Indium	26.98	Palladium	12.00
Cadmium	24.38	Steel	12.00
Calcium	21.77	Thallium	9.13
Rubidium	20.46	Lead	8.42
Cesium	20.00	Columbium	5.13
Lithium	18.68	Vanadium	4.95
Molybdenum	17.60	Arsenic	4.90
Cobalt	16.93	Antimony	3.59
Uranium	16.47	Mercury	1.75
Chromium	16.00	Bismuth	1.40
Manganese	15.75	Tellurium	0.001

The Electrochemical Series of Elements

In this table, the elements are electropositive to the ones which follow them, and will displace them from solutions of their salts.

1. Cesium	23. Nickel	45. Silicon
2. Rubidium	24. Cobalt	46. Titanium
3. Potassium	25. Thallium	47. Columbium
4. Sodium	26. Cadmium	48. Tantalum
5. Lithium	27. Lead	49. Tellurium
6. Barium	28. Germanium	50. Antimony
7. Strontium	29. Indium	51. Carbon
8. Calcium	30. Gallium	52. Boron
9. Magnesium	31. Bismuth	53. Tungsten
10. Beryllium	32. Uranium	54. Molybdenum
11. Ytterbium	33. Copper	55. Vanadium
12. Erbium	34. Silver	56. Chromium
13. Scandium	35. Mercury	57. Arsenic
14. Aluminum	36. Palladium	58. Phosphorous
15. Zirconium	37. Ruthenium	59. Selenium
16. Thorium	38. Rhodium	60. Idodine
17. Cerium	39. Platinum	61. Bromine
18. Didymium	40. Iridium	62. Chlorine
19. Lanthanum	41. Osmium	63. Fluorine
20. Manganese	42. Gold	64. Nitrogen
21. Zinc	43. Hydrogen	65. Sulfur
22. Iron	44. Tin	66. Oxygen

Flammability Characteristics of Common Gases and Liquids[1]

Material	Boiling point[2]		Flash point[3]		Autoignition temperature[4]		Flammability limit in air, volume %[5]		Maximum oxygen content, volume %[6]
	°F	°C	°F	°C	°F	°C	Lean limit	Rich limit	
Acetaldehyde	70	21.2	36	2.2	365	185	4.1	55	12
Acetone	133	56.2	0	−17.8	1000	538	2.6	12.8	11.6
Acetylene	−119	−84	—	—	571	300	2.5	81	—
Allyl chloride	113	45	−25	−31.7	737	392	3.3	11.1	12.6
Ammonia	−28	−33.3	—	—	1204	652	16	25	15
Benzene	176	80	12	−11.1	1044	563	1.4[7]	7.1	11.2
1, 3-Butadiene	24	−4.4	—	—	804	429	2	11.5	10.4
Butane	32	0	—	—	761	405	1.9	8.5	12.1
Butyl acetate	259	126	79	26.1	789	421	1.7	7.6	11.5
1-Butene	20	−6.7	—	—	723	384	1.6	9.3	11.4
2-Butene	34	1.1	—	—	615	324	1.8	9.7	11.7
N-butyl formate	225	107	64	17.8	612	322	1.7	8	12.4
Carbon disulfide	115	46.1	−22	−30	212	100	1.3	44	5.4
Carbon monoxide	−310	−190	—	—	1128	609	12.5	74	5.6
Cyclopropane	−27.4	−33	—	—	928	498	2.4	10.4	11.7
1, 1-Dichloroethylene	99	37.3	−5	−20.6	856	458	5.6	11.4	10
2, 2-Dimethylbutane	121	49.5	−54	−47.8	797	425	1.2	7	12.1
Ethane	−127	−88.4	—	—	959	515	3	12.5	11
Ethanol	173	78.4	−55	−48.4	793	423	4.3	19	10.6
Ethyl acetate	171	77.3	28	−2.2	798	426	2.2	11.5	11.2
Ethyl bromide	101	38.4	—	—	952	512	6.7	11.3	14
Ethyl chloride	54	12.1	−58	−50	966	519	3.8	15.4	13
Ethylene	−152	−102	—	—	842	450	3.1	32	10
Ethylene oxide	56.4	13.6	<0	<−17.8	804	429	3	100	—
Ethyl ether	94	34.4	−49	−45	356	180	1.9	48	—
Ethyl glycol	273.2	134.1	100.4	38	460	238	1.8	14	10.7
Ethyl formate	130	54.4	−4	−20	851	455	2.7	13.5	10.4
Ethylglycol acetate	312.8	156.1	134.6	57	716	380	1.7	—	11
Gasoline	—	—	−49	−45	996	536	1.4	7	—
Gasoline (60 octane)	—	—	−45	−42.8	536	280	1.4	7.6	11.6
Gasoline (92 octane)	—	—	—	—	734	390	1.5	7.6	11.6
Gasoline (100 octane)	—	—	−36	−37.8	853	457	1.4	7.4	11.6
Heptane	209	98.4	25	−3.89	433	223	1.2	6.7	11.6
Hexane	156.2	69.1	−7	−21.7	453	234	1.2	7.5	11.9
Hydrogen	−422	−252	—	—	1085	586	4	75	5
Isobutane	11	−11.7	—	—	864	463	1.8	8.4	12
Isopropanol	179.6	82.1	70	21.1	860	460	2	12	12
Isopropyl ether	154.4	68.1	−18	−27.8	830	443	1.4	21	10
Methane	−263	−164	—	—	999	538	5.3	14	12.1
Methanol	151	66.2	52	11.1	867	464	7.3	35	9.7
Methylethylketone	176	80	26	−3.3	960	516	1.8	12	11.4
Methylisobutylketone	242.6	117.1	73.4	23	858	459	1.4	7.5	12
Methyl acetate	135	57.3	14	−10	935	502	3.1	16	10.9
Methylamine	19.4	−7	—	—	806	430	4.9	20.7	10.7
Methyl butene	87.4	31.4	<20	<−6.7	—	—	—	—	11.4
Methyl chloride	−11	−23.9	—	—	1170	632	10.7	17.4	15
Methyl formate	89.6	32	−2	−18.9	853	456	5.9	20	10.1

Flammability Characteristics of Common Gases and Liquids[1] (*Continued*)

Material	Boiling point[2]		Flash point[3]		Autoignition temperature[4]		Flammability limit in air, volume %[5]		Maximum oxygen content, volume %[6]
	°F	°C	°F	°C	°F	°C	Lean limit	Rich limit	
N-butanol	244.4	118.1	95	35	649	343	1.4	11.2	11.3
Pentane	97	36.1	<−40	<−40	588	309	1.5	7.8	12.1
Propane	−44	−42.2	—	—	871	466	2.2	10	11.4
Propylene	−54.4	−48	—	—	770	410	2.4	10.3	11.5
Propylene oxide	93	33.9	−35	−37.3	869	465	2–2.1	21.5–22	~10
Toluene	230	110	39.2	4	996	536	1.4	7	9.1
tert-Butylamine	112	44.4	16	−8.9	716	380	1.7	8.9	<11
Vinyl chloride	9	−12.8	—	—	882	473	4	22	9
Xylene	284	140	85	29.5	867	464	1.1	7	8

[1]From Philippe B. Laut and David W. Johnstone, Air Liquide America, *Chemical Engineering,* June 1994.
[2]At standard atmospheric pressure.
[3]Minimum temperature at which vapors of a combustible liquid will be ignited by a flame under certain experimental conditions.
[4]Minimum temperature at which a material will spontaneously oxidize in air.
[5]Volume percent of combustible gas in air such that below the lean limit or above the rich limit the mixture is considered nonflammable.
[6]Maximum oxygen content in a combustible gas mixture below which the mixture is nonflammable.
[7]At 212°F (100°C).

Specific Gravity and Density of Materials

For rough estimating of common construction materials.

	Specific gravity	Density, lb/ft^3	Density, kg/m^3
Aluminum	2.7	165	2,643
Bronze	8.0	509	8,154
Cast iron	7.2	450	7,209
Copper	8.9	556	8,907
Glass	2.5	160	2,563
Lead	11.38	710	11,374
Magnesium	1.74	109	1,746
Nickel	8.9	556	8,907
Steel	7.8	490	7,850
Zinc	7.5	440	7,049
Ash, dry	0.63	40	641
Cedar, dry	0.36	22	352
Fir, dry	0.56	32	513
Maple, dry	0.65	43	689
Redwood, dry	0.42	26	417
White pine, dry	0.41	26	417
Granite	2.6	165	2,643
Limestone	2.5	165	2,643
Sandstone	1.8	110	1,762
Pressed brick	2.2	140	2,243
Common brick	1.9	120	1,922
Terra cotta	1.9	120	1,922
Concrete	2.3	144	2,307
Portland cement	3.0	183	2,932
Mortar	1.7	103	1,650
Earth, dry, loose		76	1,218
Earth, dry, packed		95	1,522
Sand and gravel		60	961
Asbestos		153	2,451
Marble	2.7	170	2,723
Shale		92	1,474
Tar	1.2	75	1,202
Bluestone	2.5	159	2,547

Index

ABOUT THE AUTHORS

GEORGE S. BRADY was a legendary engineer who worked on the Panama Canal. He is deceased. HENRY R. CLAUSER is a consultant and the editor of *Research Management*. JOHN A. VACCARI is a mechanical engineer, a consultant, and a former editor of *Materials Engineering*.